Page Numbers of Some Important Tables and Charts

W9-BCO-943

Names, Formulas, and Charges of Some Common Ions

POSITIVE IONS (Cations)		NEGATIVE IONS (Anions)	
aluminum	Al^{3+}	acetate	CH_3COO^-
ammonium	NH_4^+	bromide	Br^-
barium	Ba^{2+}	carbonate	CO_3^{2-}
cadmium	Cd^{2+}	hydrogen carbonate, bicarbonate	HCO_3^-
calcium	Ca^{2+}	chlorate	ClO_3^-
chromium(II)	Cr^{2+}	chloride	Cl^-
chromium(III)	Cr^{3+}	chlorite	ClO_2^-
cobalt	Co^{2+}	chromate	CrO_4^{2-}
copper(I)	Cu^+	dichromate	$Cr_2O_7^{2-}$
copper(II)	Cu^{2+}	fluoride	F^-
hydrogen, hydronium	H^+, H_3O^+	hydride	H^-
iron(II)	Fe^{2+}	hydroxide	OH^-
iron(III)	Fe^{3+}	hypochlorite	ClO^-
lead(II)	Pb^{2+}	iodide	I^-
lithium	Li^+	nitrate	NO_3^-
magnesium	Mg^{2+}	nitrite	NO_2^-
manganese(II)	Mn^{2+}	oxalate	$C_2O_4^{2-}$
mercury(I)	Hg_2^{2+}	oxide	O^{2-}
mercury(II)	Hg^{2+}	perchlorate	ClO_4^-
nickel	Ni^{2+}	permanganate	MnO_4^-
potassium	K^+	phosphate	PO_4^{3-}
scandium	Sc^{3+}	monohydrogen phosphate	HPO_4^{2-}
silver	Ag^+	dihydrogen phosphate	$H_2PO_4^-$
sodium	Na^+	sulfate	SO_4^{2-}
strontium	Sr^{2+}	hydrogen sulfate, bisulfate	HSO_4^-
tin(II)	Sn^{2+}	sulfide	S^{2-}
tin(IV)	Sn^{4+}	hydrogen sulfide, bisulfide	HS^-
zinc	Zn^{2+}	sulfite	SO_3^{2-}
		hydrogen sulfite, bisulfite	HSO_3^-

HEATH
CHEMISTRY

Teacher's Annotated Edition

J. Dudley Herron
Professor of Chemistry and Education
Purdue University
West Lafayette, IN

David A. Kukla
Chemistry Teacher and Chairman
Science Department
North Hollywood High School
North Hollywood, CA

Michael A. DiSpezio
Science Teacher and Chairman
Science Department
Cape Cod Academy
Osterville, MA

Clifford L. Schrader
Chemistry Teacher
Dover High School
Dover, OH

Julia Lee Erickson
Chemistry Teacher
Arundel High School
Gambrills, MD

D.C. HEATH AND COMPANY
Lexington, Massachusetts   Toronto, Ontario

Overview of Heath Chemistry

Use the following guide to examine the variety of flexible features included in the **Heath Chemistry** student edition.

page 25
Chapter lessons begin at the top of a page to help students organize their reading. **Concepts** at the start of every lesson aid students in focusing on the ideas to be studied.

pages 26–34
Numbered sections provide fully developed discussions of chapter topics in a logical sequence.

pages 99–101 and 192–193
Example problems carefully develop strategies for problem solving. Detailed explanations and alternative solutions teach students how to understand a problem and think through to a solution.

page 207
Review and Practice with every lesson provide immediate opportunities to work on problems and questions relating to the lesson.

pages 199, 226, and 607
Extensions and **Applications** expand basic concepts and teach students how chemistry is part of their world.

pages 217–219
The **Chapter Review** is a comprehensive and flexible resource. The **Chapter Summary** and **Chemically Speaking** recap major concepts and vocabulary introduced in the chapter. **Review** and **Interpret and Apply** include questions that evaluate learning at the recall, interpretative, and applied levels of comprehension.

Problems provide exercises like those developed in the chapter, while **Challenge** offers students more difficult problems to solve. **Synthesis** poses questions and problems that relate concepts from previous chapters. **Projects** provide an opportunity for individual exploration.

pages 396 and 612
Innovative chapters about **High-Tech Chemistry** and **Analytical Problem Solving** are a springboard for students to do independent library research and develop strategies for lab work.

pages 512–515
Full-color **illustrations** that complement and augment the text material promote better understanding of concepts.

page 595
Vocabulary words are printed in boldfaced type and defined in context to facilitate learning.

page 652
Careers in Chemistry featuring profiles of people at work in a variety of fields inform students about future opportunities.

page 676
The **Appendices** supplement selected topics within the text.

page 678
A **Glossary** of terms used in the text includes important words and a page reference to the place that each is first introduced.

Copyright © 1987 by D.C. Heath and Company
All rights reserved. No part of this publication may be reproduced or transmitted in any form by any means, electronic or mechanical, including photocopy, recording, or any information storage or retrieval system, without permission in writing from the publisher.

Published simultaneously in Canada
Printed in the United States of America
International Standard Book Number: 0-669-09854-X
2 3 4 5 6 7 8 9 0

TABLE OF CONTENTS

A Guide to Individual Chapters
(includes Chapter Organizer, Performance Objectives, Section Notes,
Answers to Questions, and Solutions to Problems)

Philosophy of the Program

All the world's a stage,
And all the men and women merely players,
They have their exits and their entrances,
And one man in his time plays many parts.

—*As You Like It*, Act II, Scene VII

Shakespeare's famous lines remind us that we all go through stages. One teacher describes three stages of his teaching career like this:

The first five years that I taught, I tried to impress my students with how much I knew, until I began to realize how little I really knew. With that realization, I spent the next five years really thinking things through and making connections among ideas. I really thought about how I knew various things that I had always believed were surely true. During this process I learned a great deal about how one comes to know, and during the rest of my teaching career, I have tried to teach my students something of what I learned. I have worried much less about how many facts they accumulate, but I have worried a great deal about how they learn and how I can help them to develop the intellectual skills and habits that separate true scholars from robots.

This book is written for third-phase teachers, teachers who are ready to help students become true scholars. It is a challenging occupation, but it is filled with rewards.

An effective chemistry course depends on the chemistry teacher. Often maligned, always overworked, grossly underpaid, and sometimes unloved, the classroom teacher remains the most important factor in the success or failure of education. A textbook is only a tool; it is never a teacher. We have developed a tool that can help you accomplish your goals for high school chemistry.

Goals for High School Chemistry

What are your goals? If you are like chemistry teachers in a recent class at Purdue, your goals are to

1. Prepare students to interpret everyday events in terms of chemical concepts and principles.
2. Teach students intellectual skills needed to address the problems that they will encounter as adults, particularly problems related to the prudent management of natural resources, protection of environmental quality, and providing inexpensive sources of energy.
3. Prepare students for college work in science and related fields.

Although last on the list, preparation for college seems to preoccupy teachers of the "college prep" course, chemistry. Just what is it that prepares students for college?

Why Students Fail in College

During the past thirty years, there has been little change in the reasons that college students fail.

1. College students fail because they are unprepared to assume responsibility for their own learning. They have not learned to manage time. They do not discipline themselves to study. Students have difficulty deciding whether they understand what is expected. Then they are not sure where to find the information that they need, or how to separate misleading or irrelevant information from that which is pertinent. They have difficulty synthesizing information from several sources and bringing it to bear on a problem. Assignments may simply go undone.
2. Many college students have poor communication skills. They are unable to interpret tables, diagrams, graphs, mathematical expressions, and specialized languages such as chemical equations. They express their own ideas ambiguously. Their writing often is poorly organized, grammatically incorrect, and riddled with contradictions.
3. Students lack originality. Although skilled at memorizing, applying specified algorithms in a routine manner, and repeating what they are told, they are stumped by novel tasks for which they have been given no algorithm. In addition, few students show an ability to evaluate facts, directions, or other information.
4. Students lack flexibility. Anyone can learn from a gifted teacher in a supportive environment, but colleges have many intelligent professors who are poor teachers and who provide weak learning environments. Success depends on the ability to use unlikable people in poor instructional environments to acquire useful knowledge.

When these causes of college failure are equated with the goals of chemistry teaching, there is cause for jubilation as well as despair; jubilation because the competencies that are needed for success in college are those needed for success in everyday life; despair because so little teaching is directly related to these competencies.

In writing this book, our goal has been to provide

you with a tool that you can use to help students develop those competencies required for success, either in continued education or in the everyday world. Then how is this book different from others that you have used?

Most chemistry texts require students to memorize words used by scientists to describe the natural world, but the meaning of those words is often unclear. Students are taught algorithms to solve routine problems encountered in chemistry, but little attention is given to why a rule or algorithm will produce a right answer. It often is difficult to see how the algorithms taught in school apply to the problems we encounter in everyday life. Because the logic behind the algorithm or rule is not understood any novel problem or exercise is incomprehensible, and the student fails.

In this text, students are guided to understand why a rule or algorithm can be used to solve a problem. Proportional reasoning, which is at the root of most chemistry concepts and computations, is emphasized throughout the text. Students who learn how to use proportional reasoning are far better prepared to understand chemical concepts or other complex ideas than are students who only memorize definitions and arbitrary rules.

The development of proportional reasoning or other reasoning skills requires time and practice in a variety of contexts. Consequently, fewer topics can be treated in a course designed to develop such intellectual skills. There is more material in this text than the average student will be able to understand in the time available. You will need to select those topics that lend themselves to the development of important reasoning skills. We have listed topics and chapters that we consider to be optional, but you must make the final decisions after considering the ability of your students and local educational needs.

Many teachers worry about omitting topics that they know students will encounter in later courses or that relate to issues of general interest. Experience has shown that focusing on thorough comprehension of a few basic concepts facilitates understanding other concepts in later courses or in application to everyday experience. Conversely, the cursory treatment of many topics hinders understanding of the logic inherent in mathematical and chemical language, of chemical computations, and of mental constructs such as the atomic model that chemists use to think about chemical changes.

Understanding versus Memorization

No teacher will maintain that rote memorization is their goal of instruction, but it is a common result. What will prevent it from being the result using this text for your course? One important factor will be your willingness to cover less material. Most students want to understand, but when information is presented faster than they can make sense of it, they resort to memorization in order to cope.

Most students view school as a place where students are given tasks to perform by teachers. It is the student's workplace. As in any workplace, when the workers do their job, they are paid. In school, pay is in the form of grades and credits that can be exchanged for diplomas, the needed tickets for entry to "the good life." Students do not readily see ideas and skills as the purpose of schooling, although they may concede that new ideas and skills may be acquired while completing the required work. Their purpose is to complete tasks and earn grades. Therefore, students often respond in the manner illustrated by this teacher's lament:

I try to emphasize understanding in my course, but my students don't want to understand. When I try to discuss the whys and wherefores, my students become impatient and tell me, 'I don't want to know all of that. Just tell me how to get the right answer.'

We will never accomplish the important, broader goals of education until we change instructional practices to overcome the causes of this prevalent attitude. As long as we fail to examine the thought processes that students use to arrive at their answers, we will fail to restructure poor intellectual skills or to reinforce good skills. As long as our texts focus only on correct answers, routine exercises, and prescribed laboratory activities, we will not accomplish the goals of instruction outlined at the beginning of this section.

The Teacher's Role

Beginning teachers often view education as the process by which accumulated information is transmitted from one generation to another. Although one purpose of education is to assist students in the construction of new ideas, the assumption that information can be transmitted intact from one individual to another is counterproductive. Each person must construct his or her own knowledge, and that process is influenced by many factors that are not under the direct control of a teacher. Experienced teachers know that the role of mediator or mentor is a way to facilitate learning. A teacher can provide the learner with feedback concerning the validity of the knowledge constructed by the learner, but a teacher can never transmit knowledge.

This view of teaching has important implications for the role of the teacher.

1. The teacher must know how the learner is thinking about ideas. An example will illustrate how important this is in helping students to construct knowledge in a rational manner.

A student was having trouble finding the mass of NaOH in 50 mL of a 1.5 M solution. First the student had obtained an answer using an algorithm based on the definition of molarity, the number of moles per unit volume. To verify the answer, the student had used the definition of density, mass per unit volume, to calculate mass of NaOH in the 50 mL of solution, and got inconsistent results. Since moles can be converted to mass via the mole mass of NaOH, the student believed there were two independent means to find the mass of NaOH in the solution. The student failed to realize that the volumes in the denominators of the two expressions have different referents. In molarity, it is the moles of NaOH (or mass, after conversion) that are compared to the volume occupied by the NaOH and the water with which it is mixed. In density, it is the mass of NaOH that is compared to the volume occupied by the NaOH alone. To help this student construct knowledge of these chemistry concepts so as to personally use them successfully to solve problems, the key is recognizing the student's current understanding of these ideas.

2. Once the teacher understands how the student has constructed knowledge, the teacher must focus attention on the discrepancies between the student's knowledge and accepted constructions. In the example given, the focus must be on the exact meaning of "volume" in the expressions of molarity and density.

   The discrepancies can be analyzed in various ways. Questions that focus attention on facts that the student presumably knows but has ignored, counterexamples, and even telling students what is different about their view and the accepted one can be effective.

3. Transcendent messages are important aspects of mediated learning. Transcendent messages are ones that deal with generalized knowledge that can be applied to tasks other than the one at hand. In working with the student in this example, the student was praised for looking for a way to verify a result by solving the problem by two independent means. This is a proven strategy for developing confidence in results of novel problems. The student also was reminded of the importance of attending carefully to the exact meaning or terms contained in formulas. When molarity is expressed as moles/liter and density is expressed as mass/cubic centimeter, the volumes expressed may appear to be comparable, but they are not. We must habitually ask questions like, "moles of what?", "liters of what?", and "mass of what?."

4. Mentors are guides who see themselves as older and wiser co-learners rather than purveyors of truth. Their teaching style often is indirect. They are not afraid to direct learning in areas that they do not totally understand, and they readily admit the limitations of their knowledge as they learn with their students.

Knowledge is Constructed

When we attempt to convey our ideas to students, we often are frustrated when we later learn that the idea they have acquired is far different from the idea we intended to convey. How many times have you had students say, "But you told us _____", followed by ideas totally foreign to your own thinking? When you stop to think about how learning must take place, this is not surprising.

There is no conduit from one brain to another. All we can do is disturb the environment. I have disturbed your environment by placing symbols on the page before you, but I have placed nothing in your brain. You have used your eyes to perceive the symbols on the page. Impulses have passed from the retina to your brain, where you interpret those signals under the direction of mental schemes that you constructed years ago. Just how you interpret these symbols and the resulting ideas is probably different from the interpretation made by anyone else.

Knowledge is constructed. What is constructed depends on what we have previously learned as well as on the stimuli we presently encounter. Effective instruction depends on our ability to understand how our students make sense out of the stimuli we present more than it depends on how we ourselves make sense out of those stimuli. Let us see how this applies to some of the things you will use in this text.

Proportional Reasoning and Unitary Rates

Proportional reasoning is an important intellectual skill that you will help students to develop in this course. A great many of the relationships that we use to make sense of the natural world are proportional ones. Research in intellectual development reveals that proportional reasoning normally develops around ages 12–16, but many adults (estimates are as high as 50%) have not developed this reasoning pattern. Proportional reasoning is involved in solving this problem:

"Mr. Short (a figure shown in the task as it is normally presented) was measured with buttons and found to be 4 buttons tall. Mr. Tall (not shown) was measured with the same buttons and found to be 6 buttons tall. If paper clips are used to measure Mr. Short (students normally do this measurement as part of the task), it is found that he is 6 clips tall. How tall would Mr. Tall be if measured with the same clips? How do you know?"

When adolescents do this task, many reply "8 clips" and explain that since Mr. Tall is two buttons taller than Mr. Short, he would be two clips taller as well. These students are unable to see the proportional relationship between the two heights and arrive at the correct answer of nine. Such students, many of whom may be in your class, are not prepared to construct meaningful interpretations of many ideas that they will encounter in this course. How can you assist them in developing proportional reasoning?

Simply telling students how we think about such problems or having them practice a rule that will produce a correct answer is unlikely to help. Students must build on what they already understand to construct more powerful thinking tools.

Even young children have learned to solve some problems that involve proportional relationships. For example, they know that they can find the cost of five candy bars when each bar costs 35 cents. What they are not able to do is generalize from this limited experience to solve more complicated problems, such as finding the number of candy bars they can buy with five dollars. Such a problem requires a more formal understanding of proportions and a language that can be used to represent these problems. The necessary language and understanding of proportional relationships can be developed from the limited capability that students have acquired as a result of informal learning.

In this text we have adopted the term, unitary rate, to describe ratios with a denominator of one. Thus, 35 cents/candy bar is an example of a familiar unitary rate. Students already know that a unitary rate can be used to calculate the value for any number of units in the numerator of a unitary rate that corresponds to the units named in the denominator: multiply the unitary rate by the appropriate number. For example, the cost of 5 candy bars:

$$5 \text{ bars} \times \$.35/1 \text{ bar} = \$1.75.$$

The logic of this solution is transparent because it has become familiar through repeated use in contexts that enabled students to check the validity of the result. The language used to describe the logic, in particular the inclusion of units to describe quantities, is likely to be unfamiliar. However, if the language is used in the context of familiar problems, the student learns it easily. Then the language can be used to describe less familiar problems, and the logic of unitary rates can be extended to solve more complex problems involving proportions.

Throughout the text we begin to develop the language and reasoning used by scientists. We ask students to recall ideas that they have used to make sense of everyday events, to apply these ideas to new tasks, and then develop those ideas into more powerful tools that can be used in increasingly complex situations. It is a slow and tedious process, but it is the way all knowledge must be constructed. When students develop understanding in this way, they can transfer their skills to new tasks as well as get correct answers to questions in chemistry class.

Teaching Problem Solving

The kind of ability that we have been talking about is frequently described as "problem solving." Recent research in cognitive psychology and artificial intelligence has provided new insights into important factors in problem solving. We have used that research to develop our approach to problems in this text. However, the success of our efforts depends on your understanding of what the text is designed to do and why.

To begin, we must distinguish between a problem and an exercise. Most tasks in this book are not problems for you. They are exercises. They are problems for your students. The difference is that you know a routine procedure to get the correct answer. You have no difficulty understanding what is required. This is not true for your students. When they see the task, they do not know what to do. They may have difficulty interpreting the task, and there are gaps between their understanding of the task and its goal. The existence of such gaps is what distinguishes a problem from an exercise.

The traditional approach to teaching students to solve problems is to demonstrate the known solution for the problem, have students practice the solution until it is familiar, and then present test questions like the problems students have solved. In other words, our approach has been to convert problems into exercises. It is effective as long as students never encounter different problems. If they do, they will be stuck once more until another teacher demonstrates a method to solve the new problem.

Teaching problem solving differs from teaching students an efficient way to get correct answers to routine problems. It involves teaching intelligent things to do when faced by unfamiliar tasks. As Grayson Wheatley, at Purdue puts it, "Problem solving is what you do when you don't know what to do."

A great deal of research has been done on problem solving. Thus we know a great deal about the differences between good and poor problem solvers. Much of this research is summarized by the following generalizations adapted from Whimbey, A., and Lockhead, J., *Problem Solving and Comprehension*. The Franklin Institute Press, Philadelphia, 1983.

GOOD PROBLEM SOLVERS

1. Good problem solvers believe that they can solve just about any problem if they work at it long enough.
2. Good problem solvers are persistent. They work for a long time on a problem before giving up.
3. Good problem solvers read carefully. They often read a problem several times before beginning to

work on it. They are certain that they know what is said before they begin.

4. Good problem solvers break complex problems into small steps and solve them one step at a time. If they don't see how to solve a step immediately, they try to find new relationships that will lead to a solution.

5. Good problem solvers organize their work so that they can come back at any point and follow steps they have taken.

6. Good problem solvers always check what they have done, both at the end of a problem and at various steps along the way.

7. Good problem solvers use mental representations. They sometimes draw pictures or try to visualize a concrete example. They jot down notes that translate what they have read into simpler words or symbols. If they can't solve a problem, they may try a simpler, related problem. They may
   a. substitute small numbers for large ones
   b. substitute whole numbers for variables
   c. restrict conditions given in the problem.

They guess and test. They may try out several approaches as they solve a problem.

POOR PROBLEM SOLVERS

1. Poor problem solvers don't believe that they can solve problems. They seem to feel that you either know it or you don't.

2. Poor problem solvers are impatient. If they don't see the answer quickly, they give up.

3. Poor problem solvers are careless readers. They often misread what is written. They may begin working the problem before they know what the problem asks.

4. Poor problem solvers jump to conclusions and guess. They expect to go immediately from what is given to the answer. If they can't, they give up.

5. Poor problem solvers organize their work carelessly or not at all. They are unable to retrace their thinking at the end of a problem.

6. Poor problem solvers seldom check their work.

7. Poor problem solvers have only one approach to problems. Usually they try to recall a formula. If they can't, they don't know what to do and give up.

You are undoubtedly aware of many of these differences. What you really want to know is how to make good problem solvers out of poor ones. Although much remains to be learned about teaching problem solving, there are some things that we can confidently suggest.

A. Establishing the Right Attitude

The attitude of both the teacher and the student will affect your success at teaching problem solving. The following suggestions deal with attitudes of both.

1. Determine from the beginning that you will attempt to teach problem solving rather than how to do exercises. This means that you will **not** begin instruction by demonstrating the efficient solution that you know. Rather, you will engage students in thinking about ways that they might approach a problem.

2. You must believe that students—all students—can solve problems. Some teachers believe that people are born bright or dull, and those who did not choose their parents wisely are unable to solve problems. There is considerable evidence to the contrary.

   In early research on wait time, Mary Budd Rowe observed that teachers waited up to five times as long for "bright" students as they waited for "slow" students to answer a question. When these teachers were trained to systematically lengthen the waiting time for a response, all students gave longer, more complex, and more appropriate answers. (Rowe, M. B., "Wait Time and Rewards as Instructional Variables: Their Influence on Language, Logic, and Fate Control", paper presented at the National Association for Research in Science Teaching, Chicago, April 1972.)

   A chemistry teacher in Indianapolis reported that groups of slower students came up with successful approaches to solving a problem in the lab with approximately the same frequency as groups of brighter students. Some recent research on grouping suggests that groups composed of individuals of comparable ability levels make more progress than heterogeneous groups.

3. Students must believe that the time spent on problem solving is worthwhile. If students perceive that their job is simply to complete tasks, they are unlikely to see the value in demanding mental activity such as problem solving. You must emphasize that the purpose of the activity is to develop intellectual skills that they can use to attack real problems.

The more realistic the problems that are presented, the more likely this message will come through. This text contains many examples of problems from the world outside of the science classroom. Teachers can look for additional opportunities to present students with real problems. For example, you can use a news report about the level of toxic wastes entering the local water treatment plant as a basis for questions on how to measure concentrations in parts per million, the allowable uncertainty in measurements used for calculations, and the reliability of the results. Local professionals might speak to the class about actual community problems they are trying to solve.

B. Understanding the Problem

Some of the suggestions that follow will be easier to understand in the context of an example. The following problem was used in research on problem solving at Purdue. It was correctly solved by only one third of the undergraduates and graduate students in the study.

A 1.00 g mixture of CuO and Cu_2O is quantitatively reduced to 0.839 g of Cu when hydrogen gas is passed over the hot mixture. What is the mass of CuO in the mixture?

Most students who attempted this problem immediately wrote the following equation and applied a familiar algorithm to generate an incorrect answer.

$$1CuO + 1Cu_2O + 2H_2O \longrightarrow 3Cu + 2H_2O$$

Poor problem solvers frequently jump into solving a problem before they understand it. This is probably fostered by the common classroom practice of presenting a problem and immediately demonstrating the "correct" solution. Here are steps to counteract this tendency.

1. Begin work on problems by having students work in small groups to decide what the problem is about. After groups have discussed the problem, let them share their understanding. Ask how they decided what the problem was about. Take time to focus on what students see in the problem statement that leads them to infer the goals of the task and conditions of the problem.
2. Ask students to identify specific information given in the problem that seems relevant to the task.
3. Ask students to identify information in the problem that seems irrelevant.
4. Ask students to identify needed information that is not given in the task statement. In the CuO problem, students need to know the mole mass of each oxide and the relationship between moles and mole mass.
5. Ask students how they can obtain needed information that is not given.
6. Do not insist on a total understanding of the problem before students begin doing something with it. We often fail to see all of the implications of a problem in the beginning, and a total understanding may come only after we have played with the ideas in some way. Encourage exploration. When working with the CuO problem, many students found it helpful to calculate the mole mass of each oxide early in the process. Other students drew diagrams to represent molecules of CuO and Cu_2O. Others wrote chemical and/or mathematical equations.

C. Problem Representation

In problem representation, the problem statement is translated into a mental representation of relationships defined by the problem. Ask the students to explore ways to think about the problem.

1. Encourage students to think of specific, concrete representations of the problem, such as what they would see in the chemical system described by the problem. In the CuO problem, some students found it helpful to think about actually pouring two powders together, placing that mixture in a glass tube, and passing hydrogen gas over the mixture as it was being heated by a bunsen burner.
2. Have students draw or use models. For example, several diagrams of CuO and Cu_2O might be drawn and the diagrams "decomposed" to see how the amount of copper changes as the number of CuO or Cu_2O molecules is varied.
3. Some students may think of mathematical models that can be used to represent the problem. For example, let x represent the mass of CuO in the mixture. Then the mass of Cu_2O in the mixture will be $1.00 g - x$.
4. Help students to reformulate the problem. For example, in the CuO problem, the mixture can be separated into two components in separate containers. Then the problem becomes one of finding the mass of copper produced in each of the containers when the copper oxides are reduced.

D. Problem Solution

After the students have thought of a useful way to represent the problem, have them explore various ways to solve it.

1. Encourage students to "guess and test." Assure students that scientists do a lot of this, based on enough understanding to make an educated guess. In the CuO problem, this might be a guess that the original mixture contained 0.5 g of each compound. These data can be used to calculate the mass of copper produced and then compare it to the value given in the problem. Other guesses can be tested in a similar manner.
2. Encourage students to look for a pattern. For example, after a series of guesses, ask what happens to the mass of copper produced when the estimated mass of CuO is systematically varied from 1.00 g to 0.5 g to 0.0 g.
3. Solve a simpler problem. For example, a mixture of 3 unicycles and bicycles are taken apart to produce 4 wheels. How many bicycles were there? How many bicycles would be in a mixture of 12 vehicles that produce 16 wheels when taken apart? Does the solution of this problem suggest a way to solve the original one?

E. Avoiding Silly Errors

Everybody makes silly errors. During real problem solving, errors in transcription, reading, computation, or making inferences are common. Good problem solvers as well as poor ones wrote down 0.893 g as the mass of the copper produced when 0.839 g was stated in the problem. Teachers do this, too. The difference is that good problem solvers detect these errors and correct them before they finish the problem.

Teachers frequently remind students to check their work, read carefully, and organize what they do. These exhortations may have little effect because students do not understand how to put them into practice. Here are some strategies to help students improve in this area.

1. After students have discussed a new kind of problem and have a general idea of how to proceed, you will want them to practice similar problems. (There is a place for algorithms in what we do.) At this point, have students pair up to work problems. Have one student act as a problem solver and the other act as a listener and checker. The checker should listen to what the solver says and work the problem along with the solver. If the solver makes an error in reading or transcription, the checker should call attention to the error by saying something like, "You did not read the problem correctly." The checker should not say, "You read 0.893 but the problem says 0.839." It is important that the problem solver find his or her own mistake. A person may repeat the same mistake three or four times before reading it correctly.

   After one problem has been solved, students should change roles and work another problem.

2. People often are unaware of why it is important to organize work carefully when solving problems. The reason is that there is a definite limit to the amount of information that we can hold in our conscious memory. The range is from five to seven bits, but what constitutes a bit is open to question.

Most real problems require attention to more information than anyone can hold in the working memory. Consequently, it is common to begin working a problem and then forget where you are going as you concentrate on one of the subgoals of the problem.

Students should be told about the limitations of working memory. They should know that everybody needs an external memory that can provide needed information later on. A common example of external memory is a grocery shopping list. Good problem solvers supplement working memory with an external memory, i.e., they make a written record of such things as the overall plan, procedures used to achieve subgoals, and tables of information that may be needed in the problem.

An external memory is only as good as the information it conveys. Teachers know that units must be attached to numbers because the units convey information about what the numbers mean. Our research suggests that one reason many students omit units is that they do not know what units like cubic centimeters, square feet, g/cm^3, or N/m^2 mean. They

do not realize that a number with such units attached is a measure of space occupied, surface area, mass of a cube, or force exerted on a surface. Until units are understood in this way, students ignorantly place no importance on including units in a record.

A good external record enables a person to retrace the logic of his or her problem-solving process by reviewing what was written. Once the student has learned to read the record produced by the factor-label (unit factor, unit analysis) algorithm, it is easy to retrace the logical steps and verify that the logic is sound. Using dimensional analysis to check the units for the answer provides an independent check of the logic. Unfortunately, many teachers teach factor-label as a foolproof algorithm (which it is not), telling the students to, "Set up the factors so the units cancel, and you will get the right answer." Throughout this text students are encouraged to use the factor-label algorithm and unit analysis to check their logic and verify a result. Factor-label is not presented as a foolproof algorithm.

F. Verification Strategies

Because it is so easy to make silly errors when tackling complex problems, good problem solvers customarily develop procedures to verify that their calculations are reasonable and correct. Research in problem solving reveals that poor problem solvers have few verification strategies and the ones that they have are often weak. Encourage your students to use these strategies.

1. Compare your result to that of others. Students know this strategy and use it unless they are discouraged by teachers. The more people who reach the same result independently, the more confidence we have in our result.

   If post laboratory discussions are conducted and students' results are recorded on the board, discrepant values are easily identified, and the student who produced them need not be told that something is wrong. Results of homework problems can be compared in a similar manner.

2. Work the problem by independent methods and compare the result. You may point out that important physical constants such as the value for Avogadro's number, the velocity of light, the mass of the electron, and the gravitational constant are measured by very different procedures in order to obtain confidence in the value. If students are encouraged to develop their own procedures for solving novel problems, several different procedures result. It is useful to point out that one procedure may be awkward and inconvenient, but an answer obtained using that awkward approach is a valuable contribution when it verifies the answer obtained by others.

3. Compare the result to other facts that you know.

When asked to find the mass of one oxygen atom, some students may answer 16 g. When asked about the size of an atom and the size of something having a mass of 16 g, all students can see that this is an unreasonable answer. Encourage such comparisons as a verification step.

4. Numerical values often can be estimated by rounding values used in the calculation to orders of magnitude and doing mental arithmetic. Point out the value of this verification strategy, particularly in the age of pocket calculators that give totally wrong answers when numbers are entered incorrectly.

5. Repeating a calculation is a weak verification strategy because one is likely to repeat logical errors made in the first place. However, it is better than nothing, and it frequently will reveal purely mathematical errors.

6. Frequent rereading of a problem statement to verify data used is another weak, but important strategy. Our research reveals that poor problem solvers may read the problem statement only once. Good problem solvers reread the problem or portions of it many times, often subconsciously. These students may report that they read the problem only once, but transcripts of their verbalization during problem solving reveals that they actually reread the problem several times.

7. Refuse to share your answer or the textbook answer until the students have used other means to verify that a result is reasonable and/or correct. Emphasize that there are no absolute standards of correctness. Consulting experts such as textbook authors and teachers is useful, but not foolproof. It is essential that students develop their own verification strategies.

The strategies suggested here have been shown to be useful in true problem solving, and they should be systematically taught. However, students should be aware that a particular strategy may help when solving one problem but not when solving another. Students should be encouraged to develop a variety of heuristics, to talk about the heuristics that they use and describe how they decided to try a particular strategy (even ones that were not helpful). Teachers occasionally can attempt to solve real problems in front of the class and talk aloud as the problem solving takes place. Students can record the teachers' heuristics and later discuss how these strategies helped or failed to help in the problem solving.

The suggestions given here are not exhaustive. Develop your own list as you work with students, and look for opportunities to teach problem solving strategies throughout the year.

The important point is that there is seldom a single, correct procedure for solving problems. When attention is focused on fixed procedures and practice, students do not develop intellectual skills needed to attack novel problems. There is a place for teaching rules and algorithms, but that is not enough. Students also must engage in real problem solving so they can learn how to apply rules and algorithms in the context of more complex tasks.

The Importance of Laboratory Work

Chemistry is an experimental science. Students should recognize that the principles and laws of nature that we accept as true are the result of extensive observations and speculative analyses refined over many investigations. Observations have been made and repeated under controlled conditions, and the results have been interpreted in generalized terms that hold under a particular set of conditions.

Most teachers recognize the importance of laboratory work in teaching this general truth about science. But many laboratory exercises used in beginning courses obscure rather than reveal the uncertainty in experimental work.

Many experiments that accompany this text are true experiments. Students are encouraged to deal with uncertainty and make judgments about the validity and general applicability of the results. This is an important part of teaching problem-solving skills.

Too little laboratory work is done in most science courses. Laboratory work is time consuming, lab facilities are expensive, and we are concerned about the safety of students. Furthermore, research on the value of laboratory work is discouraging. Generally it has shown that students who have laboratory work do no better on science exams than students who only have the text and class discussions.

This research is misleading. It is more a reflection on how we conduct laboratory activities and construct course exams than an indication of the value of laboratory work. Understanding uncertainty, its origin in experimental measurement, and its implications for interpreting information derived from scientific studies are important goals of science instruction. We know of no way to develop an understanding of uncertainty without having students measure and consider the uncertainty in what they observe. For this reason alone, laboratory work is important. Science is not exact, and students need to experience this.

Another reason that laboratory work is important has to do with the way knowledge is constructed. All knowledge grows out of experience. When science concepts such as reaction rates, heat of reaction, density, or pressure are divorced from concrete experience, the concepts are difficult to understand. Students memorize definitions and repeat them on exams. They memorize rules, such as $PV = nRT$, and they solve exercises using the rules. But the relationships described by the rules are poorly understood, and students are unable to apply the rules to novel tasks.

Similarly, the language that we use in chemistry, the formulas and equations, can be manipulated as nothing but symbols. However, the connections between the symbols, the macroscopic events represented in a problem, and the microscopic model of atoms and molecules on which the language is based must be understood. These connections are not made automatically. The laboratory helps students to connect real chemicals to formulas and real reactions to equations. The tests and laboratory activities in *Heath Chemistry* have been written to help students make these connections.

Teaching Communication Skills

Several suggestions are made in the laboratory guide that can help your students to improve their communication skills. Here are additional suggestions for this important goal.

1. Make each student a resident expert. Early in the course have students select their favorite element and become the resident expert on that element. Have them read about the element in the library and prepare a summary report on its properties, abundance, uses, cost, means of preparation in a pure form, and any other information that class members decide should be available. Compile the reports into a reference book that students can use throughout the year. When questions about an element arise, refer those questions to the resident expert for an answer.

   If you wish, you may evaluate reports individually, or you may have students read and evaluate each others' reports.
2. Science students should read books other than their text. Have students read a science-related book each semester and prepare a written or oral report.
3. Have students prepare a term paper on an aspect of chemistry that interests them. Chapter 14 of *Heath Chemistry* introduces several possible topics. Sources of information are suggested and questions are raised to be answered in the reports. Make assignments early in the year so students can prepare their papers long before the class reaches Chapter 14. Then use the student papers as the basis for class work in that chapter. You may have students submit their reports in writing and select the more interesting reports for presentation in class. Alternatively, all students may be asked to make oral reports.

   If students must prepare term papers for several of their classes, the work may be excessive. Students are likely to learn more from concentrating on one carefully prepared term paper than from doing several shoddy ones. Explore the possibility of cooperative projects with other teachers, such as

the English or the social studies teacher. In one school, all students prepared a term paper on a chemistry topic for a grade in both chemistry and English, one grade for content and the other for mechanics.

Papers on environmental pollution, the effect of scientific developments on the rising cost of health care as a result of new medical technology, or economic competition between industrialized nations might be excellent topics for chemistry/social studies papers.

4. Promote a school-wide writing program. Many adults, including highly educated scientists, have poor writing skills. Excellent progress has been made in developing writing skills in schools where there is a school-wide writing program. Conversely, research on writing indicates little progress is made when the task of developing good writing skills is left to the English department. Unless students perceive that all teachers expect good writing, they quickly revert to poor writing habits when they leave English class.

In most school-wide programs, a style manual prepared by the English department is provided to all students and staff. The manual is followed by students when they write papers in all classes, and teachers use the manual as a guide when they correct student papers. This ensures consistent rules throughout the school.

In most school-wide programs, students are expected to write some kind of paper in every class, and every teacher is expected to monitor writing. In this way, students receive extensive practice in writing, and no single teacher is burdened with providing feedback to students.

In some schools, a writing clinic for all students at all grade levels in all courses is established as part of the English department. Students may go to the clinic for advice when they are preparing papers. A teacher may recognize that some students are having difficulty with spelling, sentence construction, paragraph organization, punctuation, or other mechanics of writing and refer them to the clinic. Individualized modules for the various mechanical problems of writing have been developed and are used as the basic instructional materials in a writing clinic. Paraprofessionals who have expertise in writing often can be employed part time to staff a writing clinic.

Teaching Math

Science teachers complain that their job would be easier if the math department did theirs better! This may be true, but we must recognize that there are real differences between math as used in science and math that good mathematicians teach.

1. Mathematics deals with numbers while science deals with quantities. Math focuses on the properties of numbers and develops operations that can be applied to numbers. Science uses numbers to describe quantities. They are different, and this difference should be clarified in the science classroom. See section 1-7 of *Heath Chemistry*.

2. Mathematicians normally work with exact numbers, while scientists work with inexact measurements. In addition to teaching about uncertainty, you need to teach students that uncertainty in the numbers used in science has other consequences. For example, when graphs are prepared in math class, points representing exact numbers are plotted, and the graph is drawn through those points. In science, experimental data are plotted, and a best fit line is drawn in an effort to eliminate experimental uncertainty. Students must recognize that we have not changed what they were taught in math class, but the procedure must be modified to accommodate the uncertainty of science work.

3. Science deals with unusually large or small quantities. In math, properties of numbers and operations are normally taught using familiar whole numbers. Certainly the small whole numbers used in math class are more familiar to students than the exponential numbers used in science. Earlier we argued that students learn best when instruction is based on familiar ideas and then is expanded to the unfamiliar. We must teach students to extend what they have learned in math classes to the unusually large or small quantities in science. To begin, we must help students to visualize large and small numbers. Then we must show how the operations learned in math class will work on the unusual numbers in science.

4. Logarithms were once taught in math classes as a convenient way to do arithmetic with large and small numbers. Now that calculators are used to solve most problems, logarithms may not be taught in math classes. If they are taught, students probably will not have enough practice using them to fully understand their meaning or unique properties. You should expect to teach your students what they need to know about logarithms before you expect them to use logarithms.

5. Variables in math are often abstractions with no physical interpretation. X has no specific meaning. In science, variables normally refer to specific physical entities and the physical meaning of the variable often is important. It is unrealistic to assume that students will have learned to read mathematics in a math class in the same way that you want them to read mathematics in science. You must take time to have students read mathematical expressions in terms of the physical reality expressed.

When students are taught to read mathematical expressions, it is easier to give physical meaning to some than to others. For example,

$$D = 1.00 \text{ g/cm}^3, \quad \text{and} \quad P = 2.5 \text{ N/m}^2$$

are easily interpreted as "each cubic centimeter of water has a mass of 1.00 gram" and "the pressure produces a force of 2.5 Newtons on each square meter of surface." However,

$$R = 0.082 \, \frac{\text{L} \cdot \text{atm}}{\text{mol} \cdot \text{K}}$$

is not readily given a specific physical meaning because of the many variables incorporated in the proportionality constant. Whenever possible, the limits need to be acknowledged even as we emphasize the need to give a physical meaning to such expressions.

Pacing Chart
Priority Level Recommendations

This text will be used by a variety of students across the United States. You are in the best position to judge what your students are capable of mastering and what content serves their educational needs. In the final analysis, pacing must be an individual decision. The outline that follows is primarily for inexperienced teachers (both new teachers and older ones with little experience in chemistry). It is based on the experience of the authors who have over 40 years of experience teaching chemistry at the secondary level and 20 years of experience teaching at the university level. It represents our estimate of the class time required to teach each chapter and the priority we would give to various sections in the text.

Priority. There is more material in this text than the average student will comprehend in a beginning course. We assume that you will omit some of the text material. To assist you in making selections, we have indicated three levels of priority for the material.

The sections of highest priority are indicated by printing the section numbers in **bold black** type. We believe that any student who receives credit in an introductory high school chemistry course should master the ideas in these sections. Two considerations were used in assigning this priority: first, ideas that provide a necessary foundation for college chemistry or for understanding ideas likely to be encountered later in this course; second, ideas related to important societal problems.

Important but nonessential sections are indicated by bold blue type. Although none of these sections are absolutely essential, we would expect many of them to be included in an introductory course. The choice of which to include and which to omit depends on local student and teacher interest.

Sections printed in grey tone are optional materials that may be included if you have a group of able students who comprehend quickly. However, these sections should not be included at the expense of inadequate learning of the more fundamental topics. You will notice that many of these low-priority sections are quantitative. This does not mean that we give quantitative reasoning a low priority. On the contrary, we give it a high priority. You will see that many of those sections that we have marked for highest priority are quantitative sections. However, the quantitative aspects of quantum mechanics, equilibrium, and re-action kinetics can get very involved, and they are poorly understood by students, who have only recently encountered the concepts at a qualitative level. If these ideas are thoroughly understood at a qualitative level, the quantitative calculations can be developed later.

Time. The time required to teach a topic depends on so many factors that it is impossible to provide firm guidelines. We provide a range of class periods that should be devoted to a group of closely related chapter sections. The times shown should not be confused with priorities. For some sections with the highest priority we recommend no class time. In those cases we are simply suggesting that students should be able to comprehend the material through independent reading. Although we would always expect students to be able to review what they have read during class, we believe that they should obtain part of their knowledge through reading alone. It is not necessary to discuss every point in class, and doing so can be counterproductive. Why should a bright student read a homework assignment if he or she knows that everything will be explained in class?

In addition to the range of time suggested for each group of related sections, a Chapter Total is suggested for each chapter. You will notice that this total is greater than the sum of its parts. The times indicated for related sections are recommended for discussion of the text and related Review and Practice questions. The times indicated for the totals allow for related laboratory exercises and other ancillary materials. However, they do not allow for testing, pep rallies, field trips, library work, or like activities. Since the school year in most states is 180 days and since you will omit some sections, the total number of days suggested for all 24 chapters (117–174) should allow time for such interruptions.

While teaching the first chapter or two, you can learn what pace seems best for you and your students. By comparing the time you actually need with the estimates we made for those chapters, you can make more refined estimates for the rest of the school year.

We welcome feedback from you concerning pacing and priorities. We anticipate favorable reception of this book and later editions. Your suggestions will enable us to provide more explicit suggestions in future editions.

SECTION	CLASS DAYS	COMMENTS
CHAPTER 1		
1-1 1-2 1-3 1-4 1-5 1-6	 0–1	The material in these sections is important, but it is assigned a lower priority because it is review for many.
1-7 **1-8**	1–2	
1-9	0	
1-10 **1-11**	1–2	
1-12	0–1	
Total 7–10		
CHAPTER 2		
2-1 **2-2** **2-3**	0–1	
2-4 **2-5** **2-6**	2–3	
2-7 2-8 2-9	0–1	This material is important but may be review for most students.
2-10 **2-11** **2-12** 2-13 2-14 2-15	2–3	
2-16 2-17 2-18 2-19	1–2	
Total 9–11		
CHAPTER 3		
3-1	1–2	
3-2 3-3 3-4	0–1	Important material but may be review for most students.
3-5 **3-6**	0–1	

SECTION	CLASS DAYS	COMMENTS
3-7 **3-8**	0–1	
3-9 **3-10** **3-11** **3-12**	2–3	
3-13	0	
3-14 **3-15** **3-16** **3-17**	1–2	
Total 7–9		
CHAPTER 4		
4-1 **4-2** **4-3**	1–2	
4-4 **4-5** **4-6**	0–1	
4-7 4-8 4-9 4-10	1–2	
4-11	1–2	
Total 6–8		
CHAPTER 5		
5-1 **5-2** **5-3**	0	
5-4 **5-5** **5-6**	1–2	
5-7 5-8	1–2	
5-9 **5-10** **5-11**	0	
Total 3–6		
CHAPTER 6		
6-1 **6-2**		

SECTION	CLASS DAYS	COMMENTS
6-3 **6-4**	1–2	
6-5 **6-6** **6-7**	2–4	
6-8 6-9	1–2	
Total 6–10		

CHAPTER 7

SECTION	CLASS DAYS	COMMENTS
7-1 **7-2** 7-3 7-4	0–1	
7-5 7-6 **7-7** 7-8	1–2	
7-9 **7-10** 7-11	1–3	
7-12 7-13 **7-14**	1–2	
Total 5–7		

CHAPTER 8

SECTION	CLASS DAYS	COMMENTS
8-1 8-2 8-3 8-4	0–1	
8-5 **8-6** **8-7** 8-8	0–1	
8-9 **8-10** 8-11 8-12	0–1	
Total 3–5		

CHAPTER 9

SECTION	CLASS DAYS	COMMENTS
9-1 **9-2** 9-3 9-4	1–2	

SECTION	CLASS DAYS	COMMENTS
9-5 **9-6** **9-7** **9-8**	1–2	
9-9 **9-10** **9-11** **9-12** **9-13**	2–5	
9-14 9-15	0	
Total 4–8		

CHAPTER 10

SECTION	CLASS DAYS	COMMENTS
10-1 **10-2** **10-3** 10-4	1	
10-5 **10-6** **10-7** **10-8** 10-9	1–2	
10-10 **10-11**	2–3	
Total 4–6		

CHAPTER 11

SECTION	CLASS DAYS	COMMENTS
11-1 **11-2**	1–2	
11-3 **11-4** 11-5 11-6	2–4	
Total 4–7		

CHAPTER 12

SECTION	CLASS DAYS	COMMENTS
12-1 **12-2** **12-3** **12-4**	1–2	
12-5 **12-6** **12-7**	1–2	
12-8 **12-9**	0–1	

(continued)

SECTION	CLASS DAYS	COMMENTS
12-10 **12-11** **12-12**	1–2	
12-13 **12-14** 12-15	1–2	
Total 4–7		

CHAPTER 13

SECTION	CLASS DAYS	COMMENTS
13-1 **13-2** **13-3** **13-4**	0–2	
13-5 **13-6** **13-7** **13-8**	0–1	
13-9 **13-10**	0–1	
13-11 **13-12** **13-13** **13-14**	0–1	
Total 2–4		

CHAPTER 14

SECTION	CLASS DAYS	COMMENTS
14-1 **14-2** **14-3**	0	
14-4 **14-5**	0	
14-6 **14-7** Projects	1 5–10	
Total 5–10		

CHAPTER 15

SECTION	CLASS DAYS	COMMENTS
15-1 **15-2** **15-3** **15-4** **15-5** **15-6**	1–2	
15-7 **15-8** 15-9 **15-10**		

SECTION	CLASS DAYS	COMMENTS
15-11 **15-12**	1–2	
Total 2–5		

CHAPTER 16

SECTION	CLASS DAYS	COMMENTS
16-1 **16-2** 16-3 16-4 16-5	1–2	
16-6 **16-7** **16-8**	0–1	
16-9 **16-10** **16-11** **16-12**	0–1	
16-13 **16-14** **16-15** **16-16**	1–2	
Total 5–7		

CHAPTER 17

SECTION	CLASS DAYS	COMMENTS
17-1 **17-2** **17-3** **17-4**	1–2	
17-5 **17-6**	1–2	
17-7 **17-8** **17-9** **17-10**	1–2	
17-11 **17-12** **17-13**	1–2	
Total 4–8		

CHAPTER 18

SECTION	CLASS DAYS	COMMENTS
18-1 **18-2** **18-3** **18-4** **18-5** **18-6**	1–2	
18-7		

SECTION	CLASS DAYS	COMMENTS
18-8	1–2	
Total	3–5	
CHAPTER 19		
19-1 **19-2** **19-3**	1	
19-4 **19-5** **19-6** **19-7**	2–3	
19-8 19-9 19-10 19-11	2–3	
Total	3–6	
CHAPTER 20		
20-1 **20-2** **20-3** 20-4 **20-5**	1–2	
20-6 **20-7** 20-8 20-9 20-10	1–2	
20-11 **20-12** **20-13** **20-14**	1–2	
20-15 **20-16** 20-17 **20-18** 20-19	0–1	
Total	5–7	
CHAPTER 21		
21-1 **21-2**	1	
21-3	0–2	
21-4 **21-5** **21-6** **21-7**		

SECTION	CLASS DAYS	COMMENTS
21-8 21-9 **21-10**	1–2	
Total	8–10	
CHAPTER 22		
22-1 **22-2** **22-3** **22-4** **22-5**	1–2	Ideally all of the discussion related to this chapter will be done in connection with laboratory work.
Total	10–15	
CHAPTER 23		
23-1 **23-2** **23-3** **23-4**	1	
23-5 23-6 23-7 23-8 23-9 23-10	2–3	
23-11 **23-12** **23-13** **23-14** **23-15** **23-16**	1–2	
Total	5–8	
CHAPTER 24		
24-1 **24-2** **24-3** **24-4** **24-5** **24-6** **24-7** **24-8** **24-9** **24-10**	1–2	
Total	3–5	
Year Total	117–174	

Demonstrations

General Information

The demonstrations found in this section are correlated in each of the Chapter Organizer tables found in the Teacher's Guide.

Safety. In performing any demonstration wear a **lab apron, safety goggles,** and **plastic gloves** to protect yourself and to set an appropriate example for your students. Always wash your hands thoroughly and use a fingernail brush after completing a demonstration. Additional cautions are included where appropriate. Many other demonstrations could be used, but some are not safe.

Before doing demonstrations or experiments of your own we strongly suggest that you obtain and read

> *School Science Laboratories,*
> *A Guide to Some Hazardous Substances.*
> U.S. Consumer Product Safety Commission,
> Washington, D.C. 20207, 1984.

Use this document in selecting safe activities for yourself and your students.

Materials. Reagents needed are highlighted in bold-faced type. Equipment, other than standard test tubes, beakers, and burners, is also highlighted. Always run a demonstration yourself at least once before doing it in class.

CHAPTER 2

2-A Place a mixture of about **1 g of sulfur powder** and **0.25 g of powdered zinc** on a ceramic pad in the fume hood. Directions for mixing dry chemicals: The unignited Zn-S mixture generally is not hazardous. However, when mixing dry chemicals together, use caution.

CAUTION: *Do this in the fume hood.* Be sure the fume hood is operating properly. Wear safety goggles and heat-resistant gloves. A face shield also is recommended. *Do not use an excess of zinc.* Measure 0.25 g exactly.

Make sure that the area around the pile of zinc and sulfur is protected from flying sparks. Carefully, light the mixture with a burner turned upside down. Bright flames and smoke are produced.

Discuss what would happen to the amount of product if instead of 1 g of sulfur, you had used 500 g of sulfur without changing the amount of powdered zinc.

CHAPTER 3

3-A Display various substances chosen for their different densities. Some examples: **copper, lead, mercury, glass, water, oil, ethanol, plastic, rock, sand,** and more exotic substances like **helium gas, hydrogen gas, cadmium,** and **jewelry.** As a project, have students design experiments that can help them rank order the substances from least dense to most dense.

CHAPTER 4

4-A Obtain a **dozen nails of three different sizes** and determine the mass of each dozen. Ask students to find a unitary rate that expresses the mass of one nail. Use the unitary rate to calculate the mass of a gross of nails (twelve dozen).

$$144 \text{ nails} \times \frac{x \text{ grams}}{1 \text{ nail}} = 144x \text{ grams}$$

Use the unitary rate to determine the number of nails in one kilogram of nails.

$$1 \text{ kg nails} \times \frac{1000 \text{ g}}{1 \text{ kg}} \times \frac{1 \text{ nail}}{x \text{ g}} = y \text{ nails}$$

Compare the mass of nails made of steel to nails of equal length made of aluminum. Suggest that a student research the naming of types of nails. For example, a $2d$ nail is 2.5 cm long, a $6d$ nail is 5.1 cm long, and a $10d$ nail is 7.6 cm long.

4-B Determine the mass of a clean, dry, Pyrex® test tube. Briskly heat about **1.0 g of CuO** in a Pyrex® $(20 \times 150 \text{ mm})$ test tube using **methane** as a reducing agent.

CAUTION: Use the usual precautions when heating glass. Test tubes should be free of cracks and must be Pyrex® or Kimex®.

Light the CH_4 gas inside the test tube and let it burn with about a 5-cm flame. Briskly heat the CuO until it glows dull red (about 4 minutes). Turn off the burner but allow the reducing flame to continue burning for 5 minutes.

DATA:
#1 Mass of empty test tube
#2 Mass of test tube + CuO (before heating)
#3 Mass of test tube + contents (after heating)

Calculate the mass of Cu (**#3 − #1**) in CuO.
Calculate the mass of O (**#2 − #3**) in CuO.
Find the ratio of mass of Cu to mass of O.

The ratio of the molar masses of Cu and O should be close to the accepted value of 3.97 to 1.

4-C Allow students to make concrete observations of one mole. Using ten different substances from those listed at the end of this demonstration, put one mole of each in separate bottles. Have the students write the formula for each substance, the number of particles in the bottle, and observations of properties such as color and state. Ask the students to estimate the volume of one mole of each substance to give them practice in making estimates of metric quantities. If you use water, it could be placed in a sealed graduated cylinder so students can easily estimate the volume as 18.0 mL.

For homework ask the students to calculate the molar mass of each substance.

After the formulas, masses, and estimated volumes of each substance have been compared, have students calculate the volume of one mole of each substance using the density (shown in parenthesis). For example, the volume of one mole of water is

$$\frac{18.0 \text{ g } H_2O}{1 \text{ mol}} \times \frac{1.00 \text{ mL } H_2O}{1.00 \text{ g } H_2O} = 18.0 \text{ mL } H_2O/\text{mol}$$

Students may round the calculated value to the correct number of significant digits.

The list of substances includes many elements and compounds that will be used in experiments or demonstrations. They are common, and show a variety of colors, physical properties, and formula masses. Have the label contain either the formula or the name to review naming and formula writing.

SUBSTANCES (**density**)

Copper (8.92)	Water (1.00)
Aluminum (2.70)	Sodium chloride (2.16)
Iron (7.86)	Sucrose (1.58)
Lead (11.34)	Calcium carbonate (2.93)
Zinc (7.14)	Potassium sulfate (2.66)
Tin (5.75)	Magnesium sulfate heptahydrate (1.68)

4-D Demonstrate preparation of **1.00M solutions of sucrose, C$_{12}$H$_{22}$O$_{11}$; methanol, CH$_3$OH;** and **sodium chloride, NaCl.**

CAUTION: Methanol is highly flammable. Be sure there are no open flames in the area.

If a volumetric flask is not available, mark a beaker, flask, or bottle using a 250-mL graduated cylinder. Add 1.00 mole of the substance to the flask or bottle. Use the 250-mL graduated cylinder to add water while stirring until dissolution is complete and the final volume of the solution is 1.00 liter. Record and compare the volume of water used in each solution. Emphasize that the amount of water is not the same for each so-lution, but the number of molecules of solute in one liter of each solution is the same.

Have students make drawings which represent the microscopic events that are occurring during the solution process.

4-E Make a solution of sodium chloride in which there are **28.5 grams of NaCl** per 100.0 mL of solution. Store it in a capped bottle. The solution will look like pure water.

Call students' attention to the homogeneous nature of the solution. Slowly evaporate 10.0 mL of the solution to form large cubical crystals of sodium chloride similar to those in table salt. Determine the mass of sodium chloride.

Allow the heat of an overhead projector to evaporate drops of the NaCl solution on a watch glass. The crystals of NaCl will form quickly as you watch. The process occurs so fast that many very small crystals stick together in random ways and may not be recognizable as cubic in shape.

CHAPTER 5

5-A Ask two students to measure the mass of a clean, dry 150-mL beaker. Add about **2 grams of NaCl** and have the students determine the mass of the beaker and NaCl. Add 50 mL of distilled water to the 150-mL beaker. Place the beaker on an overhead projector and tell the students to watch as the solid NaCl disappears. Ask them where the NaCl is and how it can be recovered. Then, transfer the beaker to a warming tray.

The next day the students can observe the cubical crystals of NaCl. This is a physical change. The NaCl recovered is identical to the original NaCl in all of its characteristic properties. The size of the crystals may be slightly different but the NaCl is otherwise unchanged. Ask the students to predict the mass of the NaCl in the beaker. Have two students determine the mass of the beaker and NaCl.

Refer to the formation of sodium chloride from sodium metal and chlorine gas shown in Figure 5-5 in the text. The students should discuss the difference between the process of solution and a chemical reaction.

5-B Add **2.8 g of Na$_2$SO$_4$** to **200 mL of distilled water.** Use this solution to fill a **Hoffman electrolysis apparatus.** Attach a 6-volt battery or a direct current transformer. Oxygen will be generated at the positive electrode and hydrogen at the negative electrode. The volume of hydrogen produced will be twice the volume of the oxygen produced, as shown in this equation:

$$\text{electricity} + 2H_2O(l) \longrightarrow 2H_2(g) + O_2(g).$$

When about 10 mL of hydrogen have accumulated at the negative electrode, place a 10 × 150 mm Pyrex® ignition test tube over the stopcock tip and release the hydrogen into it. When a burning splint is inserted, the rapid reaction of the hydrogen with oxygen from the air to reform water will occur with a "pop."

A glowing splint will ignite when inserted in a test tube of oxygen collected in a similar way.

5-C Use forceps or crucible tongs to hold a **piece of copper wire, foil,** or some **copper turnings** in a burner flame until black copper(II) oxide is formed. For best results, be sure the copper is at least a centimeter above the tip of the inner blue cone of the flame. Have students record the macroscopic changes that occur and write the equation for the reaction.

$$2Cu(s) + O_2(g) \longrightarrow 2CuO(s)$$

| orange solid | colorless odorless gas | black solid |

Repeat the process with **magnesium metal** in place of copper.

CAUTION: Place a **blue tinted glass** or **plastic shield** between the burning magnesium and viewers. The reaction takes place at very high temperatures and produces harmful light in the ultraviolet part of the spectrum.

Show that the magnesium metal is flexible before the reaction and that the magnesium oxide formed is white and powdery.

This experiment can be done quantitatively in a crucible to show that the formula for the compound produced is MgO.

$$2Mg(s) + O_2(g) \longrightarrow 2MgO(s)$$

| grey solid | colorless odorless gas | white powdery solid |

5-D Add a **piece of zinc** to a 20 × 150 mm Pyrex® test tube containing **15 mL of 1.00M HCl** (85.5 mL concentrated HCl per liter of solution).

CAUTION: Hydrochloric acid is corrosive to the skin, eyes, and clothing. Wash any spills or splashes immediately with plenty of water. A face shield should be used when handling the acid.

The bubbles indicate that a chemical reaction is occurring. Ask the students to name the reactants and write the formulas for each on the chalkboard.

$$Zn(s) + HCl(aq) \longrightarrow$$

Review the concept of molarity from Chapter 4. The 1.00M HCl means there is 1.0 mole of HCl per liter of solution. The HCl solution looks like water because it is mostly water. Demonstrate that a piece of zinc does not react with water.

Now ask the students what they think the products of the reaction are. One product is obviously a gas. Ask the students to name all of the gases they can think of. All of the suggested gases can be eliminated except H_2 and Cl_2 because they are not contained in the reactants, Zn and HCl. If the gas is collected and its properties tested, the students can see that it is H_2.

Now you can ask students to finish the reaction equation.

If you suggest that there is only one more product, most students will see that it is composed of zinc and chlorine. This is an opportunity to review the principles of formula writing to show that the formula is $ZnCl_2$, not Zn_2Cl or $ZnCl$.

$$Zn + HCl \longrightarrow H_2 + ZnCl_2$$

Now balance the equation and add the states of the substances:

$$Zn(s) + 2HCl(aq) \longrightarrow H_2(g) + ZnCl_2(aq)$$

Put the test tube aside to be examined the following day. The test tube will contain a colorless liquid that looks like water and it is, in fact, mostly water. To show that the $ZnCl_2$ is dissolved in the water, pour a small amount of the water into a clean 150-mL Pyrex® beaker and allow the water to evaporate on a heater.

The complete equation now can be tied to macroscopic observations and written.

$$Zn(s) + 2HCl(aq) \longrightarrow H_2(g) + ZnCl_2(aq)$$

| grey solid | colorless solution | odorless bubbles | colorless solution |

$$ZnCl_2(aq) \longrightarrow ZnCl_2(s) + H_2O(g)$$

white solid

5-E Demonstrate by using a **lighted candle** that the products of a combustion reaction are carbon dioxide and water. To show that the reaction produces water, invert a cool 250-mL beaker over the flame. Water condenses.

Show that carbon dioxide is produced by holding a 250-mL flask over the flame for 30 seconds and then stoppering it. Add 25 mL of limewater, $Ca(OH)_2$, to the flask and to a second flask used as a control. To prepare limewater, add 1 gram of calcium oxide (lime) to 200 mL of deionized water and shake until the solution is saturated. Decant the solution. Stopper and shake both flasks. Carbon dioxide turns the limewater milky by forming finely divided insoluble calcium carbonate, $CaCO_3$.

$$CaO(s) + H_2O(l) \longrightarrow Ca(OH)_2(aq)$$

$$Ca(OH)_2(aq) + CO_2(g) \longrightarrow CaCO_3(s) + H_2O(l)$$

Blow into the control flask with a straw to show that the carbon dioxide we exhale turns the limewater milky in a similar way.

Wax is a complex mixture of many substances. The formula for wax can be expressed simply as $C_{18}H_{38}$, just as the formula for gasoline often is represented as C_8H_{18} even though it also contains components with differing formulas.

An equation for the reaction of the candle wax could be written as

$$2C_{18}H_{38}(s) + 55O_2(g) \longrightarrow$$
$$36CO_2(g) + 38H_2O(g) + energy.$$

CHAPTER 6

6-A Example 6-6 discusses the reaction of copper with silver nitrate. The chemicals required may be too expensive for general experimental use but this reaction is quite beautiful and can be done as a demonstration.

CAUTION: Silver nitrate is poisonous and corrosive to the skin and eyes. Wash any spills or splashes with plenty of water.

Prepare 300 mL of **0.12M AgNO₃** by dissolving 6.1 g of AgNO₃ in enough deionized water to make 300 mL of solution in a 400-mL beaker.

Measure the mass of a clean, dry 150-mL beaker. Use a 100-mL graduated cylinder to add 60 to 100 mL of the **0.12M AgNO₃(aq)** to the beaker. If you have 5 sections of chemistry, use about 60 mL each time. If you have 3 sections use about 85 mL for each demonstration.

Cut a 20-cm length of **number 14 copper wire.** Brush it lightly with steel wool until it is lustrous. Determine the mass. Form the copper wire into a coil with a hook at the end to put over the side of the beaker. Place the copper wire into the AgNO₃(aq). You can place the beaker on the overhead projector so the reaction will be magnified. Discuss the macroscopic observations and relate them to the equation for the reaction and the microscopic events we believe are occurring.

$$Cu(s) + 2AgNO_3(aq) \longrightarrow Cu(NO_3)_2(aq) + 2Ag(s)$$

Have the students make macroscopic and microscopic drawings of the reaction.

Draw a road map and calculate the mass of copper that will react.

$$62.5 \text{ mL } 0.12M \text{ AgNO}_3(aq) \longrightarrow \text{ ? g Cu}$$

(amount will vary) $\text{mL AgNO}_3 \times \dfrac{0.12 \text{ mol AgNO}_3}{1000 \text{ mL AgNO}_3}$

$\times \dfrac{1 \text{ mol Cu}}{2 \text{ mol AgNO}_3} \times \dfrac{63.5 \text{ g}}{1 \text{ mol Ca}} = \underline{\qquad} \text{ g Cu}$

Calculate the theoretical yield of silver.

$$62.5 \text{ mL } 0.12M \text{ AgNO}_3(aq) \longrightarrow \text{ ? g Ag}$$

(amount will vary) $\text{mL AgNO}_3 \times \dfrac{0.12 \text{ mol AgNO}_3}{1000 \text{ mL AgNO}_3}$

$\times \dfrac{2 \text{ mol Ag}}{2 \text{ mol AgNO}_3} \times \dfrac{107.9 \text{ g Ag}}{1 \text{ mol Ag}} = \underline{\qquad} \text{ g Ag}$

The following day the reaction will be complete. Record the macroscopic changes. Remove the remaining copper wire from the beaker after shaking and washing off the silver crystals. When the wire is dry, determine the mass of copper reacted.

Decant the blue solution of copper(II) nitrate, leaving the silver crystals in the beaker. Wash the silver crystals with 20 mL of distilled water 10 times and determine after drying the mass of the beaker and silver, and of the silver alone.

Calculate the % yield. If this is not the same as 100% within experimental uncertainty, there must be something in the beaker that is not silver. Discuss what impurities are possible. (Some unreacted copper may flake off the wire or the silver may not be completely dry.)

CHAPTER 7

7-A Demonstrate the difference in the amount of space occupied by a solid versus the same amount of substance in the gaseous state. Put some finely crushed **dry ice** through a **funnel** into a **balloon**.

CAUTION: Do not allow the dry ice to come in contact with your skin.

Dry off the balloon and it will inflate. Call attention to the fact that you added a very small amount of solid CO_2 to the balloon, and now it has increased tremendously in size.

Use the leftover CO_2 for Demonstration 7-E.

7-B Charles's Law—A simple apparatus will demonstrate the relationship between temperature and volume. You will need a capillary tube with an oil plug. Close off one end of a **capillary tube** about 20 cm long. Heat the tube in an **oven** for at least one hour until the tube is 150°C. Using a heat-resistant glove or other protection, grasp the tube and immediately dip the open tip into some 30-wt, non-detergent motor oil.

Do not lift the tube out until about 1 cm of the oil has been forced into the tip of the tube. Let the tube cool overnight. It is a good idea to make several of these tubes at one time. Store the tubes in a 100-mL graduated cylinder with about 10 mL of CaCl₂ at the bottom as a drying agent. Seal the cylinder with plastic wrap and a tight rubber band. If you take care of the tubes, they can be used from year to year.

On the day of the demonstration, have ice water in a battery jar and hot water (about 85°C) in a 1500-mL beaker. In front of the class, use two rubber bands to attach the tube with the oil plug to the zero centimeter end of a grooved wooden ruler. Set the closed end of the tube exactly at the zero centimeter mark.

Tell students you are going to see what happens to the volume of air trapped inside the tube when the temperature of the system is changed. Gently tapping the ruler and holding it vertically, read the height of the air column in the tube (the distance from the zero centimeter mark to the bottom of the oil plug). In this case, length is proportional to volume. Students record the length of the air column and the temperature of the room. Then immerse the ruler into ice water so that the open end of the tube is out of the water. After 5 minutes hold the tube vertically and read the length of the air column while the temperature of the water is recorded. Try to leave the tube in the water as much as possible when reading the length. Repeat the procedure in hot water.

The students now have three coordinates for a graph. Set up a graph for them, or have them set up one themselves, on which the y-axis is the length of the air column in centimeters (origin is zero cm) and the x-axis is the temperature (origin is −300°C in 50° increments up to +100°C). Students then drop a best fit line down to the x-axis through the points. The line will not meet the x-axis at the origin. Ideally the length will be 0 cm as the temperature is −273°C.

Discuss the relationship between the temperature and volume of a gas. You also may wish to discuss what happens if the temperature of the gas reaches −200°C. The gas liquifies and Charles's law is no longer valid.

7-C You will need a **Boyle's law apparatus** as shown in the picture and five books that are the same.

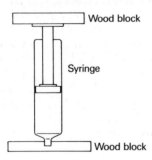

Wood block

Syringe

Wood block

The Boyle's law apparatus can be purchased from many scientific supply companies for under ten dollars. Plot the volume of the syringe versus the number of books (as an indication of increasing pressure). Have the students graph the results using the number of books on the y-axis, and volume in mL on the x-axis. The resultant line will be curved. Replot the

number of books versus the reciprocal of the volume on the x-axis. The result should be a straight line with formula $y = mx + b$ where $y =$ number of books (P_b) and $x = 1/\text{volume}$.

Rearrange the equation $P_b - b = m(1/V)$. $P_b - b$ is the total pressure on the system, which is the pressure due to the books and the atmospheric pressure. The atmospheric pressure is the distance from the origin to the y-intercept. Since the distance is in a negative direction, the expression $P_b - b$ gives an additive value for the total pressure on the system. Remember, even when no books are on the syringe, the pressure on the syringe is equal to the atmospheric pressure (plus the small amount due to the wooden block and the plastic plunger). The expression can be rearranged to yield

$$P_t = m/V (P_t = \text{total pressure})$$

Since m, the slope of the line, is a constant, the new expression is:

$$P = k/V \quad \text{or} \quad PV = k$$

7-D For a surprising demonstration dealing with Boyle's law, inflate a **marshmallow** by reducing the atmospheric pressure around it. You will need a **500-mL, thick-walled Erlenmeyer flask**, a **#6 one-hole rubber stopper**, a **90° bent glass tube**, **30 cm of thick-walled flexible tubing** and an **aspirator.** Connect the aspirator to a faucet. Connect the flexible tubing to the side arm of the aspirator, and connect the bent glass tube to the other end of the flexible tubing. Seal the glass tubing to the flexible tubing with some tightly twisted copper wire. Place the marshmallow in the flask and seal the flask with the rubber stopper attached to the glass tube. Turn on the faucet and watch the marshmallow expand due to the low pressure in the flask. After a little time the marshmallow begins to contract, perhaps due to the bursting of the enlarged air bubbles in it. To prevent water from splashing out of the sink, punch some holes in a one-gallon, plastic drinking-water bottle. Place the neck of the bottle over the bottom of the aspirator.

7-E A very convincing demonstration concerning the relationship between T and P of a gas can be performed using a **toilet-tank float** and a **pressure gauge.** The apparatus can be purchased for about $35 from many scientific supply houses. You will need the P-T apparatus, a 1500-mL beaker of hot water, and a Dewar jar (thermos) large enough to accommodate the tank float. The Styrofoam® shipping container for HCl can be used as a substitute for the Dewar jar. The jar is partially filled with a slurry of **acetone** and **small chunks of dry ice.** The temperature of this system is −78°C.

CAUTION: Acetone is highly flammable. Be sure there are no open flames around. Also, use of a fume hood is recommended. Do not allow the dry ice or the dry ice-acetone mixture to come in contact with your skin. Severe frostbite may result.

Tap the pressure gauge lightly, while the students record the room temperature and the pressure on the gauge. Carefully dip the entire float into the hot water. After five minutes, record the pressure and the temperature of the system. Repeat the procedure in the alcohol-dry ice slurry. The students then plot temperature on the x-axis (the origin is $-300°C$) and pressure on the y-axis. The data points should yield a straight line.

7-F Demonstrate some of the amazing effects of super low temperatures on some materials. Procure about **5 L of liquid nitrogen** ($-196°C$). You may have a cryogenics dealer in your area who delivers. A local university also may be a source for a small supply of material. Store the liquid nitrogen in a special Dewar jar. The loss of nitrogen will be minimal over a twenty-four hour period if the vented top is in place. DO NOT store the liquid nitrogen in a refrigerator or freezer.

CAUTION: Do not allow the liquid nitrogen to come in contact with your skin. It will cause severe frostbite.

While wearing protective gloves and using tongs, dip a **flower** into the liquid nitrogen for about 10 seconds. The flower will crumble like a crushed potato chip.

Skewer half of a **banana** with a **wooden dowel.** Dip the banana into the $N_2(l)$ for at least 20 seconds. Remove the banana from the liquid nitrogen and proceed to drive a nail into some soft wood using the half of the banana as the hammer.

CAUTION: Do not hold the banana in your unprotected hand.

Immerse a **hollow rubber ball** into the liquid nitrogen. Remove it with the tongs and hurl the ball against a wall. It will shatter into tiny pieces. Electrical resistance is greatly reduced at these temperatures. Immerse a **coil of wire** which is part of a battery circuit lighting up a pen light bulb. The bulb should glow brighter.

CAUTION: Do not come in direct contact with the liquid nitrogen. Be careful that shattered objects do not fly across the room and cause injury.

7-G For a change of pace, open a **bottle of perfume** in the room. Have students raise their hands as soon as they smell the scent. Time how long it takes the perfume to travel across the room.

7-H Demonstrate Graham's law of diffusion. On a ring stand, horizontally support a **glass tube** that is at least 1 cm in inside diameter and 0.75 meter long. Place **two drops of concentrated HCl** and **two drops of concentrated ammonium hydroxide** on

separate **balls of cotton.** Simultaneously place the balls of cotton in either end of the glass tube. After some time, a white ring of NH_4Cl will form inside the tube. Measure the distance from either end of the tube to the ring. Since both gases, HCl and NH_3, traveled different distances in the same amount of time, the distance each traveled can be used as the rate of diffusion.

Find the ratio of the distances traveled by the gases, and compare the square of this ratio to the ratio of the molar masses of the gases.

$$\left(\frac{\text{Distance } NH_3}{\text{Distance } HCl}\right)^2 = \frac{\text{molar mass } HCl}{\text{molar mass } NH_3}$$

You may not achieve an exact relationship between distance traveled and molar mass. There are factors working in this demonstration other than the kinetic energy of the two gases at the same temperature. The results should be good enough to use qualitatively as an example of gaseous diffusion.

CHAPTER 8

8-A If you have not done so already, now is a good time to demonstrate the nature of electrical charges. Use **pith balls** and a **glass rod** rubbed with **pure silk** and a **rubber rod** rubbed with **fur.** Touch two pith balls hanging together with the charged glass rod. The pith balls should separate since they both have acquired the same charge. Charge up the rubber rod and bring it close to one of the pith balls. The ball will be attracted to the rubber rod. However, after touching the rod the ball will be repelled. This demonstration works best in dry weather.

8-B The effect of a force on the path of two particles of different mass can be demonstrated rather easily. Place a **Ping-Pong ball** and a **golf ball** on a long flat surface, such as a table. Practice rolling the balls across the table so that they have the same speed. Turn on a **fan** so that the breeze blows perpendicularly across the path of the two balls. The Ping-Pong ball will be deflected much more than the golf ball. The greater the mass, the less the deflection. This principle also holds true for subatomic particles. Electrons are deflected to a much greater extent than protons or positive ions.

CHAPTER 10

10-A A large spectrum (24 cm) can be made from a **slide projector** and a **prism** in a darkroom. First, make a narrow slit through which the white light from the slide projector can pass. Carefully open a **2 × 2-inch slide.** Fit together **two pieces of exposed film**

so that they are separated by about 2 mm. Place the two pieces of film in the slide as illustrated. Carefully glue the slide back together.

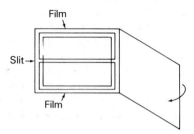

Place the slide in the projector so that the slit is vertical. Position a glass prism in the beam of light just in front of the projector. Project the spectrum on a screen.

Different **colored plastic filters** may be used to show students what happens to the spectrum when a filter is placed in the beam. A red filter will transmit mostly red light, a green filter will transmit green, blue, and some yellow.

10-B Give each student a **2 × 2-inch diffraction grating replica.** Thirty diffraction gratings can be purchased for about $7.00 from most scientific supply houses. The spectrum of various elements, such as H_2, He, Ar and Hg can be produced from a **spectrum power source** and **spectrum tubes.** (Ask a physics teacher for these items.)

CAUTION: Hg produces some ultraviolet light.

When the power source is turned on, the students can view the spectrum of the individual elements if they hold the diffraction grating about 2 cm from their eyes.

10-C A rough approximation of the probability of finding an electron in an *s* orbital can be demonstrated by making a **bull's-eye pattern on a piece of cardboard** about 40 cm on a side, as illustrated below:

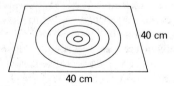

If **dried beans,** such as pinto or lentil, are dropped from about 10 cm directly above the center of the cardboard, the beans will distribute themselves around the cardboard so that the population distribution (beans per unit area) will gradually decrease with increasing distance from the center of the cardboard.

CHAPTER 12

12-A Demonstrate the conductivity of a molten substance such as KI. Heat about **10 g of KI** in a **crucible** over a **Fisher burner** in a fume hood. A **flashlight bulb** connected to a **circuit powered by two batteries** will light when the solution gets hot enough. The electrodes (bare copper wire inserted into glass tubing as a thermal insulator) are inserted into the molten KI. Allow the KI to completely melt around the wires.

CAUTION: This demonstration should not be done by students because of the extreme temperatures involved.

12-B Blow up **four similar balloons** so that they are the same size. Tie them off at the necks. Tie together two sets of two balloons each. Two balloons will make up an angle of about 180° and represent a linear molecule, for example, BeH_2.

When the two sets of balloons are put together so that the ties are touching each other, the balloons will be about 90° apart. You may want to throw the four balloons in the air and simultaneously hit two balloons that are at opposite ends of the four-balloon structure. With some practice the balloons will "explode" into a three-dimensional tetrahedral structure. CH_4, NH_3 and H_2O molecules can be demonstrated from this structure. You can make up several 3 × 5-file cards on which is written H or : to represent a hydrogen atom or a lone pair of electrons. Using double-stick tape, place the file cards on the large ends of the balloons to indicate bonded hydrogen atoms or lone pairs of electrons.

Five balloons, one set of two and one set of three, will yield a molecule whose central atom has five bonded atoms. Six balloons, three sets of two, can represent a molecule such as SF_6.

12-C The attraction of polar molecules in an electric field may be demonstrated easily. The demonstration works best when the humidity is low. Half fill a **buret** with water and another with ethanol (polar substances). Rub a **rubber rod** with **fur.** Allow a thin stream of water to flow by the rod, placed about 5 cm from the stream. The stream will be deflected toward the rod. The molecules will align themselves as they pass through the electric field and will be attracted to the rod. Repeat with the buret containing ethanol.

To repeat the procedure, use a nonpolar substance, such as **hexane** or **1,1,2-trichlorotrifluoroethane** (**TTE**). (DO NOT use toxic substances such as carbon tetrachloride or benzene.) The stream is not attracted to the rod because the molecules do not have dipoles with which they align themselves in the electric field. Some small attraction may be produced in the nonpolar substances because of induction.

CHAPTER 13

13-A Prepare a **saturated solution of Ca(OH)₂** (limewater) by adding 3.5 g of solid $Ca(OH)_2$ to 1 L of distilled water (3.5 g of CaO also may be used). Shake the bottle periodically over several hours to ensure thorough mixing and allow the excess $Ca(OH)_2$ to settle. Pour 200 mL of clear solution into each of two 250-mL beakers (A and B). Have a volunteer blow into beaker A and bubble CO_2 from a CO_2 gas generator into beaker B. A CO_2 generator can be made by placing a **metallic carbonate** (Na_2CO_3, K_2CO_3, $CaCO_3$, etc.) in a flask and adding **1M HCl.** A tube of compressed CO_2 or dry ice also can be used as sources of CO_2.

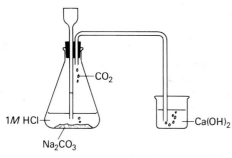

CO₂ gas generator

A milky precipitate, $CaCO_3$, will form in both containers:

$$Ca(OH)_2(aq) + CO_2(g) \longrightarrow CaCO_3(s) + H_2O(l)$$

If the student continues to blow into beaker A, the limewater solution will remain cloudy. In beaker B the solution will clear up as the addition of excess CO_2 dissolves in water to produce H_2CO_3 which reacts with $CaCO_3$ to form soluble $Ca(HCO_3)_2$:

$$Ca_2(g) + H_2O(l) \rightleftharpoons H_2CO_3(aq)$$
$$CaCO_3(s) + H_2CO_3(aq) \longrightarrow Ca(HCO_3)_2(aq)$$

The exhaled breath contains about 4% CO_2 which will not produce a concentration of H_2CO_3 high enough to dissolve the $CaCO_3$. The pure CO_2 from the generator is 25 times more concentrated (100%) and therefore dissolves the $CaCO_3$.

After removing 30 mL of the clear solution from beaker B, place the $Ca(HCO_3)_2(aq)$ on a magnetic hot plate/stirrer. Heat and stir the solution. As the solution approaches the boiling point, bubbles of CO_2 will come out of solution. This will be accompanied by a precipitate of $CaCO_3$.

$Ca(HCO_3)_2$ in water contributes to its hardness. Test both the $Ca(HCO_3)_2(aq)$ solution that was boiled (beaker B) and the one that was not boiled (beaker A) for hardness. Do this by adding a solution of soap and water a drop at a time and shaking the beakers. The number of drops of soap solution required to produce foam that lasts one minute is a measure of hardness.

The boiled solution will require fewer drops and will not be as hard. Hard water that contains primarily $Ca(HCO_3)_2(aq)$ is called temporary hard water because the hardness can be reduced by heating the water. Unfortunately, when the water is heated, the $CaCO_3$ is deposited as lime (boiler scale) and tends to clog pipes or ruin the heating elements in a water heater.

This demonstration also can be used to illustrate how rainwater containing CO_2 can dissolve limestone from the soil to form soluble $Ca(HCO_3)_2$. When water drips into an underground cave, the reduction of pressure promotes a loss of CO_2 from the solution and precipitates $CaCO_3(s)$, forming stalactites and stalagmites.

13-B To a 250-mL beaker on a hot plate/stirrer add **2.5 g of Mg(OH)₂** and 100 mL of deionized water. The milky appearance shows that the $Mg(OH)_2$ does not readily dissolve. The bulb of a conductivity apparatus will not light, showing that there are very few ions present in solution because the $Mg(OH)_2$ is only slightly soluble in water. The saturated solution is in equilibrium with the undissolved solid:

$$Mg(OH)_2(aq) \xrightarrow{H_2O} Mg^{2+}(aq) + 2OH^-(aq)$$

The addition of **10 drops of phenolphthalein** shows that there are enough hydroxide ions, OH^-, present to produce the characteristic purple color of the indicator. Add **10.0 mL of 1M HCl** and the phenolphthalein will become colorless because the hydroxide ions are neutralized by the acid.

$$Mg(OH)_2(aq) + 2HCl(aq) \longrightarrow MgCl_2(aq) + 2H_2O$$

CHAPTER 16

16-A To a clean, dry 250-mL graduate, add solid $CuSO_4 \cdot 5H_2O$ to reach the 50-mL mark. Boil 250 mL of **distilled water** to remove any dissolved CO_2. When the boiled water is cooled, pour it slowly down the side of the 250-mL graduate to minimize mixing. Stopper the graduate and seal the stopper with candle wax to prevent CO_2 from entering. Place the graduate in a well-lighted area in front of a white background where students can observe it daily. The volume of $CuSO_4(s)$ initially present is enough to saturate the solution at room temperature.

Ask students to make predictions of what they will observe tomorrow, next week, next month, and next year. They should predict that the color of the solution above the solid copper(II) sulfate will gradually become a darker blue. Few will predict the color gradient that is observed after several days. The darker color near the bottom indicates that more $CuSO_4$ has dissolved there. The resulting solution is more dense and gravity tends to retard upward migration of the dissolved $CuSO_4$. Have the students make a drawing of the microscopic events they believe are taking place.

A great teaching advantage in using this system is that concrete observations can be made that must be interpreted in terms of microscopic events. These can both be represented by an equation:

$$CuSO_4(s) \rightleftharpoons CuSO_4(aq)$$

When you want to show that $CuSO_4(aq)$ is an electrolyte, the microscopic events and the equation that represents them can be modified to

$$CuSO_4(s) \xrightarrow{H_2O} \rightleftharpoons Cu^{2+}(aq) + SO_4^{2-}(aq)$$

If you want to account for the blue color by using anhydrous $CuSO_4(s)$ the equation is

$$CuSO_4(s) \xrightarrow{H_2O} \rightleftharpoons Cu(H_2O)_4^{2+}(aq) + SO_4^{2-}(aq)$$
white   blue   colorless

It has been difficult to determine just how long it takes for this system to reach equilibrium. The last two times I set up this system, one student became so impatient after several months that he shook the graduate.

16-B Prepare about 50 mL of **1M solutions of NaCl, CaCl₂,** and **sucrose.** Assemble an **electrical conductivity apparatus** and test each solution for conductivity. To avoid an electric shock, always turn off the switch or pull the plug while rinsing the electrodes or changing the solutions. Explain the results in terms of the solvation of ionic and molecular substances.

16-C CAUTION: Lead compounds are very poisonous. Wear gloves, an apron, and safety goggles. Thoroughly wash off any spills or splashes. Lead waste must be stored in a solid waste container.

To show a dramatic change of solubility with temperature, prepare the following supersaturated solution. Mix **400 mL of 0.01M Pb(NO₃)₂** with **400 mL of 0.01M NaI.** (Pb(NO₃)₂ − 3.3 g/L; NaI − 1.5 g/L) Filter and save some of the precipitate. Heat the solution almost to boiling. Add precipitate until it just disappears as the solution reaches about 90°C.

As the solution cools, beautiful golden crystals of PbI_2 precipitate are produced. Swirling the flask produces a beautiful display of irridescent color. Leave the flask on the demonstration table for a few days. Save and use it again next year.

16-D To allow students to observe the rapid crystallization of a supersaturated solution, prepare the following solution. Dissolve **150 g of sodium acetate trihydrate crystals** in 100 mL of distilled water. Heat the solution to dissolve all the crystals. Cool the solution in a cold-water bath to reach room temperature. Keeping the solution free from dust or undissolved crystals, stopper the flask and save it for classroom presentation. While in full view of all students, add a seed crystal. The crystal will sink into the solution and then burst forth with sudden and rapid crystal growth. The flask will become warm.

16-E To allow students to observe the role of pressure on gas solubility, obtain a clear glass bottle of unopened, **chilled club soda.** Have the students examine the contents for any indication of gas content. Slowly twist open the bottle top, until a rush of bubbles can be observed. Quickly tighten the cap and the bubbles will stop. To demonstrate the effect of temperature on gas solubility, repeat the above experiment using a bottle of soda kept at room temperature. Although this is an excellent demonstration of gas solubility, it can get messy, so beware.

CHAPTER 17

17-A Pour equal volumes of **HCl** and **NaOH** that have the same concentration into separate beakers. Any amount from $1M$ to $6M$ is fine as long as they are equal.

CAUTION: These solutions are corrosive to the skin, eyes, and clothing. Wash spills and splashes immediately with water. A full face shield is suggested.

Measure the temperature of each solution and record it on the chalkboard. Pour the solutions together and measure the final temperature. Ask students why the temperature increased, where the heat came from, and so on.

Now repeat the experiment with double the volume of acid and base. Before pouring the two solutions together, ask students to predict the final temperature. Most will predict that the temperature increase will double because twice as much reaction takes place to produce twice as much heat. Mix the solutions, measure the temperature, and record it. (The temperature rise should be the same as in the first experiment.) Ask students to explain the result.

Continue the discussion until you feel that students appreciate the difference between heat and temperature; in particular, that they see why the temperature did not double even though the amount of heat produced did double in the second experiment.

17-B Some students will have seen a thermometer made in a previous science course, but others will not have seen it or will have forgotten. Construct a simple water thermometer (or better, have a student do the construction) by filling a test tube with colored water and inserting a **one-hole stopper** with a **long glass tube** through the hole. Use the thermometer to make a few temperature readings and calibrate it. Leave the thermometer in the room and use it for future readings of room temperature.

17-C Use **two or three large plastic foam cups** stacked together as a calorimeter. Prepare one calorimeter for each solid to be checked. Add 100 mL of water at room temperature to each of the calorimeters and place a **thermometer** in each to check the temperature. (Check the thermometers to be sure that they all read the same. If they do not, select thermometers that give the same reading when they are at the same temperature.)

Place pieces of **assorted metals** or other solids in a beaker of water and heat to boiling. The demonstration is more effective and easier for students to interpret if the solid samples have the same mass. (Samples with a mass of 10–25 g should give suitable results.) Allow the water containing the solid samples to boil for several minutes to allow them to come to thermal equilibrium. Remove the samples and place one in each calorimeter. Stir and take temperature readings until the temperature no longer rises. Compare the temperature rise in each calorimeter and relate it to the specific heat of each solid. (If you do not use solids of equal mass, it will be necessary to do calculations before comparisons can be made. That is why the demonstration is more effective if samples of equal mass are used.)

17-D Differences in the specific heat (heat capacity) of liquids can be demonstrated by slight modifications of the previous demonstration. Place **100 g of various liquids (water, mineral oil, saturated salt solutions, corn syrup)** in the calorimeter cups. Heat equal masses of an iron rod in boiling water. Transfer the pieces of iron to the various liquids and measure the temperature rise of the liquids. The specific heat of the liquid can be calculated if you wish, but simply observing the differences in the temperature rise in each liquid is sufficient to establish that the specific heat varies from one substance to another.

This demonstration could be used as a basis for student investigations. The apparatus is simple and the procedure is relatively safe, but many questions could be investigated. For example, how does specific heat of a solution vary with the concentration of solute? What should be kept constant in the investigation—mass of solvent, mass of solute, mass of solution? How does specific heat vary with the kind of solute? For example, do ionic and covalent solutes have the same effect? Is there any theory that explains the results?

17-E Endothermic reaction: Add about **5 g of NH_4Cl** to water at room temperature in a large test tube. Cover the tube with plastic wrap and shake it. The tube will be cold.

Exothermic reaction: Add about **2 mL of concentrated H_2SO_4** to water in a large test tube. Gently swirl the solution with a stirring rod. The tube will get hot.

CAUTION: Always add concentrated H_2SO_4 to water, never water to H_2SO_4. Concentrated H_2SO_4 is corrosive to the skin, eyes and clothing. Wash all spills and splashes immediately with plenty of water. A face shield should be used for this demonstration.

CHAPTER 18

18-A Blow a small amount of **lycopodium powder,** or **flour,** from a scoopula onto a **burner** flame.

CAUTION: Do not inhale the dust from lycopodium powder. It can be harmful to your lungs.

The powder will flash, burning up almost instantly. This demonstrates the relationship between surface area and rate of reaction.

18-B To demonstrate rate of reaction varying with nature of reactants: Add **1 g of $Fe(NO_3)_3$** to a large test tube half-filled with distilled water. Add **10 drops of 1M H_2SO_4.**

Add **1 g of $Na_2C_2O_4$** to a large test tube half-filled with distilled water. Add **10 drops of 1M H_2SO_4.**

Add a very weak solution of $KMnO_4$ a drop at a time to each test tube. The reaction is very fast in the Fe^{3+} solution but very slow in the $C_2O_4^{2-}$ solution (about 1 hour).

18-C Make up two large test tubes of $Na_2C_2O_4$ as in demonstration 18-B. Warm one tube to about 40°C before adding the MnO_4^-. Compare rate of reaction between room temperature and reaction at 40°C.

18-D Put a small piece of **charcoal** in a **burning spoon** and put it in the **bottle of O_2.** The charcoal is surrounded by oxygen but it does not react until activation energy is added. When you ignite the charcoal and put it in a bottle of pure oxygen, the same reaction takes place but it is five times faster.

$$C(s) + O_2(g) \longrightarrow CO_2(g) + energy$$

As the O_2 is consumed, the rate of the reaction decreases. If the bottle is closed, the reaction will eventually stop.

18-E Put a small candle stub on a burning spoon. Ignite the candle. Put the candle in the bottle of oxygen and compare the reaction rate with that of the candle in air. Eventually the reaction in the bottle will slow and stop. Compare the activation energy of carbon (from charcoal) and a candle.

18-F Take a **20–30-cm long glass tube** with an inside diameter of about 0.5 cm. Insert the glass tube in a bottle of **lycopodium powder** and rotate the tube to get the powder to a height of about 3 or 4 cm in the tube.

CAUTION: Do not inhale the dust from lycopodium powder. It can be harmful to your lungs.

Darken the room, light a burner and blow the lycopodium powder into the flame. This demonstration is called Dragon's Breath and can be compared to the reaction observed when a small pile of the powder on a ceramic pad is touched with a burner flame.

18-G   CAUTION: Hydrochloric acid is corrosive to the skin, eyes, and clothing. Wash any spills or splashes immediately with plenty of water. A face shield should be used when handling the acid.

Fill a 15 × 150-mm test tube with **3.0M HCl(aq).** Insert a **0.50 cm strip of magnesium** and measure the length of time it takes the magnesium to react.

Repeat with **1.0M, 0.50M, and 0.10M solutions of HCl.**

18-H   Add an Alka Seltzer® tablet to an inverted test tube of water at 0°C, 20°C, and 40°C, and observe the difference in reaction rates.

18-I   Bubbles of CO_2 gas are evidence of the oxidation of sodium potassium tartrate by hydrogen peroxide. When $KNaC_4H_4O_6$ and H_2O_2 solutions are mixed, no apparent reaction takes place. If systems are heated to increase the reaction rate, the rate remains imperceptible even at 60°C. When a catalyst is added to each system at a different temperature, a dramatic difference is observed.

Prepare **500 mL of 0.3M $KNaC_4H_4O_6$(aq)** by dissolving 42 g $KNaC_4H_4O_6 \cdot 4H_2O$ in 500 mL of distilled water. Prepare 200 mL of 6% H_2O_2 (available commercially as 20-volume hydrogen peroxide for use in bleaching hair). Prepare **50 mL of 0.3M $CoCl_2$** by dissolving 2.0 g of $CoCl_2 \cdot H_2O$ in 50 mL of distilled water.

Put 100 mL of the 0.3M $KNaC_4H_4O_6$(aq) in each of five 250-mL Pyrex® beakers. With the help of four student assistants, simultaneously add 40 mL of the 6% H_2O_2(aq) to each of the five beakers. The reaction is occurring at a rate so slow that it is imperceptible. Keep your solution at 25°C and have the students heat their solutions on a hot plate until the temperatures are 35°C, 45°C, 55°C, and 65°C respectively. Then remove them from the hot plate. The reaction rates are still imperceptible.

Simultaneously add 10 mL of 0.3M $CoCl_2$(aq) to each beaker. The solution at 65°C will turn from pink to green immediately, and in a few seconds will vigorously evolve bubbles of CO_2. The beaker at 55°C will turn from pink to green a little later and it, too, will bubble rapidly. Each of the three other beakers will follow in turn. The green color is identified as an activated complex of cobalt. It has been suggested that two cobalt atoms are linked to a peroxide radical forming a complex ion in which a 3+ oxidation state is assigned to cobalt. Have the students observe the temperature. The increase in temperature when the reaction rate is high will show that the reaction is exothermic.

When the reaction is complete, the evolution of CO_2 stops and the pink color of cobalt(II) returns, showing that the catalyst is not consumed. The beaker at 25°C is the last one to turn green and the last one to return to the pink color. Occasionally the pink color will not return completely until the following day.

The following day a sixth beaker could be prepared and used to show that the catalyst is regenerated. Heat the beaker to 50°C and add 20 mL of the solution from any one of the beakers in which the pink color has returned.

CHAPTER 20

20-A   Add 10 drops of 1M HCl solution to **50 mL of distilled water** in a 600-mL beaker. Place on an overhead projector. Add **20 mL of universal indicator.** Drop in one **Phillips® antacid tablet.** Stir slowly for 2 minutes as the pH increases and the color of the indicator changes from orange to blue.

20-B   Use a pH meter to demonstrate basicity of **sodium acetate solution,** acidity of **NH_4Cl solution.** Have students suggest other salts. Have them predict acid, base, or neutral. Then test solutions of these salts on the pH meter.

20-C   Use a pH meter to show the pH of **apple juice, 1.0M HCl, orange juice, distilled H_2O** (will not be 7.0), **ammonia cleaner, seawater** (if available), distilled water that is being aerated with breath from a **straw, soft drink** (carbonated), **liquid Plumr®,** and **baking soda solution.**

20-D   Put enough red cabbage into a beaker to fill it. Add **distilled water.** Bring it to a boil for about five minutes. Decant and save remaining liquid. Allow it to cool for five minutes. The liquid can be used as an indicator.

Put 40 mL of juice into a beaker on the overhead projector. Add **10 drops of 1.0M HCl** (85.5 mL acid/L), one drop at a time, swirling as acid is added. Then add **20 drops of 1.0M NaOH** (40 g/L), one drop at a time.

Notes: Neutral juice is blue. Decreasing the pH turns it purple, then red. A pH above 7 will yield a blue-green, then finally green color.

20-E   To 100 mL of 2M acetic acid solution (115 mL acid/L) add **2.0 g of sodium acetate.** Test for pH with the **pH meter.**

To 25 mL of this buffer slowly add some distilled water while the pH electrode is immersed in the beaker.

To 25 mL of this buffer slowly add **1M HCl** (85.5 mL acid/L) one drop at a time and note pH reading.

To 25 mL of this buffer, slowly add **1M NaOH** (40 g/L) one drop at a time and note the pH reading. Note: Dilution will not change the pH.

The addition of a strong acid (in limited quantities) will not change the pH.

The addition of a strong base will not change the pH. Discussion: This demonstration illustrates Le Chatelier's principle and the following equation:

$$CH_3COOH(aq) \longrightarrow H^+(aq) + CH_3COO^-(aq)$$

(large quantity from acetic acid)   (large quantity from sodium acetate)

Dilution with distilled water does not change the pH because both CH_3COOH and CH_3COO^- are equally diluted.

The addition of strong acid does not change the pH dramatically because the equation shifts to the left as new H^+ from the strong acid reacts with CH_3COO^- to form CH_3COOH.

The addition of a strong base does not change the pH because the equation shifts to the right. As OH^- is added from the strong base, H^+ from dissociating CH_3COOH neutralizes additional OH^-.

20-F Students have a difficult time understanding pH and ion concentration. Use a pulley and string to represent the idea of (pH + pOH) = 14. Then the idea of H_3O^+ increasing as OH^- decreases can be understood.

One drawback of this visual system is that it indicates the relationship is linear rather than logarithmic.

The demonstration can be used throughout acid-base concepts.

$$pH + pOH = 14$$
$$K_w = 1.00 \times 10^{-14} \text{ at } 25°C$$

At a pH of 7 $(H_3O^+) = (OH^-) = 1.0 \times 10^{-7}$ mol/L

$$pH = -\log(H_3O^+)$$

CHAPTER 21

21-A This is the reaction mentioned in Section 21-1 of the text. Make a **100 mL solution of CuCl$_2$ · 2H$_2$O** by adding about 10 g of the compound (1 teaspoon) to 100 mL of distilled water. (Do not use copper(II) sulfate. It is not acidic enough to cause the reaction.) Add a **10 x 10-cm square of aluminum foil** to the solution. The reaction takes about 5 minutes.

Discuss the reaction in terms of half-reactions.

$$Cu^{2+} + 2e^- \longrightarrow Cu$$
$$\underline{Al \longrightarrow Al^{3+} + 3e^-}$$
$$Cu^{2+} + Al \longrightarrow Cu + Al^{3+}$$

Discuss a method of balancing the two half-reactions by cancelling an equal number of electrons on both sides of the equation.

21-B CAUTION: Always add sulfuric acid to water, not water to sulfuric acid. Sulfuric acid is corrosive to the skin, eyes, and clothing. Wash all spills and splashes immediately with plenty of water. Use a face shield for this demonstration.

Prepare **400 mL of 2M H$_2$SO$_4$** by adding 45 mL of concentrated sulfuric acid slowly to 355 mL of distilled water in a 600 mL beaker. Hook **two clean Pb strips** over opposite sides of the beaker so that both strips are partly submerged in the acid solution. Attach **wire leads** to each strip and to positive and negative terminals of a **6-volt battery** or direct-current transformer. Charge up the cell by allowing a current to flow for about 5 minutes.

CAUTION: H_2 gas is produced. No source of fire or sparks should be brought close to the beaker while it is charging.

Disconnect the power source. Attach the wire leads to a **3.0 volt light bulb.** The bulb will light up and stay lit for a period of time. You may want to run a portable radio instead of the light bulb.

21-C CAUTION: Sulfuric acid is corrosive to the eyes, skin, and clothing. Wash any spills and splashes immediately with plenty of water. A face shield should be used in this demonstration.

Copper may be plated onto a **zinc strip** (or a steel key) by electrolysis. Make 250 mL of a copper plating solution by dissolving **50 g of CuSO$_4$ · 5H$_2$O** in 200 mL of distilled water. While stirring, carefully add **50 mL of concentrated sulfuric acid** to the solution.

Clean the object to be plated by dipping it in a **1.0M NaOH solution,** then in a **1M nitric acid solution.** Attach **wire leads** to the object on one end and to the negative terminal of a **6-volt battery** or direct current transformer on the other end. Place the object in the plating solution. Attach a wire lead to a **copper strip** and the positive terminal of the battery. Place the copper strip in the solution and make sure that the two pieces of metal do not touch. Turn on the power and gradually increase the voltage from 0 to 3 volts. Let the power run for about one minute.

For silver plating, prepare a silver plating solution by dissolving **42 g of Na$_2$S$_2$O$_3$ · 5H$_2$O** (thiosulfate), **5 g of Na$_2$S$_2$O$_5$** (sodium metabisulfite), **12 g of Na$_2$SO$_4$ · 10H$_2$O** (sodium sulfate), and **6 g of NaCH$_3$COO** (sodium acetate) in **250 mL of distilled water.** Add **10 g of pulverized AgCl** and stir until dissolved.

Clean a **strip of copper** with the cleaning solutions mentioned in the copper plating procedure. Attach the copper to the negative terminal of a **6-volt battery** or direct-current transformer. Use a **clean piece of** lead as the positive electrode. Place both pieces of metal in the plating solution so that they do not touch. Slowly increase the power from 0 to 0.5 volt for only one minute. The copper will be plated with silver.

CHAPTER 23

23-A CAUTION: The vapors associated with these reactants are hazardous. This demonstration must be performed in a fume hood.

Place **150 mL of 1,1,1-trichloroethane** and **3 mL of decanedioyl chloride** into a 600-mL beaker. Into a separate 600-mL beaker place **150 mL of 0.4***M* **NaOH** (16 g/L) and **70 drops of hexamethylene diamine** (premeasured). Add 4 drops of **phenolphthalein** to this beaker. Slowly pour the contents of the first beaker (containing trichloroethane solution) into the second beaker. Using tongs, lift up the skin that forms at the solution interface. You can wrap this fiber around a pencil and continue collecting the nylon polymer.

Resources

General References

Ebbing, Darrell D., *General Chemistry*, Boston: Houghton Mifflin Company, 1984.

Ege, Seyhan, *Organic Chemistry*, Lexington, MA: D. C. Heath and Company, 1984.

Herron, J. Dudley, *Understanding Chemistry: A Preparatory Course*, New York: Random House, 1986.

Holtzclaw, Henry F., Jr., et al., *General Chemistry*, Lexington, MA: D. C. Heath and Company, 1984.

McQuarrie, Donald A., and Rock, Peter A., *General Chemistry*, New York: W. H. Freeman and Company, 1984.

Zumdahl, Steven S., *Chemistry*, Lexington, MA: D. C. Heath and Company, 1986.

A list of film sources is included at the end of this section.

Teaching Physically Impaired Students

Journal of Chemical Education, March, 1981, has several articles on the education of physically impaired students.

Reese, Kenneth M., Ed., *Teaching Chemistry to Physically Handicapped Students*, December, 1985. Free booklet. American Chemical Society Committee on the Handicapped, Room 202, 1155 16th Street, NW, Washington, DC 20036.

Stearner, S. Phyllis, *Able Scientists—Disabled Persons*, 1984. $12.95, J. R. Associates, 2820 Oak Brook Road, Oak Brook, IL 60521.

CHAPTER 4

Herron, J. Dudley, "The Mole Concept," *Journal of Chemical Education*, November, 1975.

Slad, R., and Friedman, F., "Mole Concept Tips," *Journal of Chemical Education*, December, 1976.

Tyler, D. R., "Chemical Additives in Common Table Salt," *Journal of Chemical Education*, November, 1985.

Film

Gases and How They Combine. CHEM Study Film, color, 22 minutes.

CHAPTER 5

Film

Chemical Families. CHEM Study Film, color, 20 minutes.

CHAPTER 7

Davenport, Derek A., "How the Right Professor Charles Went Up in the Wrong Kind of Balloon," *CHEM Matters*, December, 1983.

Hudson, Reggie L., "Toy Flying Saucers and Molecular Speeds," *Journal of Chemical Education*, December, 1982.

Film

Gas Pressure and Molecular Collisions. CHEM Study Film, black and white, 21 minutes

Software

Molecular Velocity Distribution, Seraphim Computer Disks, disks 3 and 4

Gas Model. Seraphim Computer Disks

CHAPTER 8

Brescia, Frank, "The Rutherford Atom Revisited," *Journal of Chemical Education*, August, 1983.

Gribbin, John, *In Search of Schrodinger's Cat*, New York: Bantam Books, 1984.

Olenick, Richard, Apostol, Tom, and Goodstein, David, *The Mechanical Universe*, New York: Cambridge University Press, 1985.

Records, Roger M., "Developing Models: What Is the Atom Really Like?," *Journal of Chemical Education*," April, 1982.

"Viking Mission to Mars," NASA Facts, National Aeronautics and Space Administration, NF-76/6-75.

Film

Modeling. The Search for Solutions, color, 18 minutes

CHAPTER 9

Anderson, Earl V., "Troubled Nuclear Power Tries to Recover," *Chemical and Engineering News*, September 20, 1982.

Bartlett, Donald L., and Steele, James B., *Forevermore: Nuclear Waste in America*, New York: W. W. Norton and Company, 1985.

Bigelow, John E., "The Effect of Ionizing Radiation on Mammalian Cells," *Journal of Chemical Education*, February, 1981.

Black, Edwin F., and Libby, Leona M., "Commercial Food Irradiation," *Bulletin of the Atomic Scientists*, June–July, 1983.

Boslough, John, "Worlds within the Atom," *National Geographic*, May, 1985.

Dorfman, Leon M., "Principles and Techniques of Radiation Chemistry," *Journal of Chemical Education,* February, 1981.

Dworetzky, Tom, "Taking the Cover Off a Dangerous Pressure Cooker," *Discover,* October, 1984.

Schechter, Bruce, "In Quest of the Quark," *Discover,* July, 1981.

Sun, Marjorie, "Renewed Interest in Food Irradiation," *Science,* February 17, 1984.

Taub, Irwin A., "Radiation Chemistry and the Radiation Preservation of Food," *Journal of Chemical Education,* February, 1981.

Whitman, Mark, "Updating the Atomic Theory in General Chemistry," *Journal of Chemical Education,* November, 1984.

Film

Radiation, Naturally. Modern Talking Pictures

Transuranium Elements. CHEM Study Film, color, 23 minutes

CHAPTER 10

Bent, Henry A., "Old Wine in New Bottles: Quantum Theory in Historical Perspective," *Journal of Chemical Education,* December, 1984.

Hansch, T. W., Schawlow, A. L., and Series, G. W., "Spectrum of Atomic Hydrogen," *Scientific American,* March, 1979.

Pilar, Frank L., "4s Is Always above 3d," *Journal of Chemical Education,* January, 1978.

Shropshire, Richard, "Making Waves," *The Science Teacher,* November, 1985.

Wadlinger, Robert L., Lawler, James H., and Brent, Charles R., "An Elementary Pulsating-Sphere Hydrogen Atom Model," *Journal of Chemical Education,* April, 1985.

Film

The Hydrogen Atom as Viewed by Quantum Mechanics. CHEM Study Film, color, 13 minutes

CHAPTER 11

Film

Ionization Energy. CHEM Study Film, color, 22 minutes

Chemical Families. CHEM Study Film, color, 20 minutes

A Research Problem: Inert(?) Gas Compounds. CHEM Study Film, color, 19 minutes

CHAPTER 12

Carroll, James Allen, "Drawing Lewis Structures without Anticipating Octets," *Journal of Chemical Education,* January, 1986.

Sanderson, R. T., *Chemical Bonds and Chemical Energy,* New York: Academic Press, 1976.

Stranges, Anthony N., "Reflections on the Electron Theory of the Chemical Bond: 1900–1925," *Journal of Chemical Education,* March, 1984.

Film

Chemical Bonding, CHEM Study Film, color, 16 minutes

CHAPTER 13

Freeden, Earl, "New Perspectives on the Essential Trace Elements," *Journal of Chemical Education,* Vol. 62-11, 1985.

CHAPTER 14

Binnig, Gerd, and Rohrer, Heinrich, "The Scanning Tunneling Microscope," *Scientific American,* August, 1985.

Conwell, C. M., "The Difference between 1-dimensional and 3-dimensional Semiconductors," *Physics Today,* June, 1985.

Eiswith, Markus, and Schwanker, Robert J., "Phototropic Glass," *Journal of Chemical Education,* August, 1985.

"Fiber Optics: The Big Move in Communications and Beyond," *Business Week,* May 21, 1981.

Johnson, W. B., "The Coming Glut of Phone Lines," *Fortune,* January 7, 1985.

Klass, P. J., "New Microcircuits Challenge Silicon Use," *Aviation Week and Space Technology,* April 16, 1984.

Kogelnik, Herwig, "High-Speed Lightwave Transmission in Optical Fibers," *Science,* May 31, 1985.

Melliar-Smith, C. M., "Optical Fibers and Solid State Chemistry," *Journal of Chemical Education,* August, 1980.

Mickey, Charles D., "Solar Photovoltaic Cells," *Journal of Chemical Education,* May, 1981.

Oppenheimer, S., "Communicating by Light Beams," *Science Digest,* March, 1984.

Osgood, R. M., and Deutsch, T. F., "Laser-Induced Chemistry for Microelectronics," *Science,* February 15, 1985.

Peterson, I., "Chipping Away at Silicon Processing," *Science News,* March, 17, 1984.

Picraux, S. T., and Percy, P. S., "Ion Implantation of Surfaces," *Scientific American,* March, 1984.

Reid, T. R., "The Chip," *Science '85,* January–February, 1985.

Shuford, R. S., "An Introduction of Fiber Optics," *Byte,* December, 1984.

Space Shuttle, National Aeronautics and Space Administration, Washington, DC, 1976.

Thomsen, D. E., "Atom Detection Improves, On the Surface," *Science News,* August 18, 1984.

Thomsen, D. E., "Unhappy Wanderer," *Science News*, September 11, 1984.

Wallich, Paul, "Man and Machine: Making Friends at Last," *Sciquest*, March, 1982.

CHAPTER 15

"Experimenters' Notebook: Why Some Popcorn Kernels Fail to Pop," *CHEM Matters*, October, 1984.

CHAPTER 18

Last, Arthur, "Doing the Dishes: An Analogy for Use in Teaching Reaction Kinetics," *Journal of Chemical Education*, November, 1985.

Opportunities in Chemistry, "Control of Chemical Reactions," Washington, DC: National Academy Press, 1985.

Splittgerber, A. G., "The Catalytic Function of Enzymes," *Journal of Chemical Education*, November, 1985.

CHAPTER 19

Davenport, Derek A., "When Push Comes to Shove: Disturbing the Equilibrium," *CHEM Matters*, February, 1985.

CHAPTER 21

Alkire, Richard C., "Electrochemical Engineering," *Journal of Chemical Education*, April, 1983.

Burroughs, Tom, "Liberty under Repair," *Technology Review*, July, 1984.

Burroughs, Tom, "Statue of Liberty," *CHEM Matters*, April, 1985.

Chambers, James Q., "Electrochemistry in the General Chemistry Curriculum," *Journal of Chemical Education*, April, 1983.

Micky, Charles D., "Artifacts and the Electromotive Series," *Journal of Chemical Education*, April, 1980.

Film

Electrochemical Cells. CHEM Study Film, color, 22 minutes

Nitric Acid. CHEM Study Film, color, 18 minutes

CHAPTER 24

Basta, Nicholas, "Biopolymers Challenge Petrochemicals," *High Technology*, February, 1984.

Darnell, James E., Jr., "RNA," *Scientific American*, October, 1985.

Doolittle, Russell F., "Proteins," *Scientific American*, October, 1985.

Dusheck, Jennie, "Fish, Fatty Acids and Physiology," *Science News*, October 19, 1985.

Felsenfield, Gary, "DNA," *Scientific American*, October, 1985.

Monmaney, Terence, "Yeast at Work," *Science '85*, July/August, 1985.

Monmaney, Terence, and Morgan, Diana, "The Bug Catalog," *Science '85*, July/August, 1985.

Preuss, Paul, "Industry in Ferment," *Science '85*, July/August, 1985.

Snyder, Solomon H., "The Molecular Basis of Communication between Cells," *Scientific American*, October, 1985.

Tucher, Jonathan B., "Biochips: Can Molecules Compute?," *High Tech*, February, 1984.

Weinberg, Robert A., "The Molecules of Life," *Scientific American*, October, 1985.

Film Sources

CHEM Study—purchase or rental of super 8mm or 16mm films:

Ward's Natural Science Establishment, Inc.
Modern Learning Aids Division
P.O. Box 1712
Rochester, NY

Modern Learning Aids Division
c/o Arbor Scientific Ltd.
P.O. Box 113
Port Credit, Ontario, Canada

Modern Talking Pictures is a source of free-loan films. The borrower pays the cost of return postage and insurance.

Modern Talking Pictures
5000 Park St., North
St. Petersburg, FL 33709

The Search for Solutions is a series of nine films (18 minutes each) produced by Playback Associates and financed by Phillips Petroleum, and is loaned free except for return postage.

The Search for Solutions
Booking Center
708 Third Avenue
New York, NY 10017
Film confirmation: (800) 223-1928

Software

Seraphim Computer Disks
Eastern Michigan University
Ypsilanti, MI 48197

CHAPTER 1 Activities of Science

<table>
<tr><th colspan="6">CHAPTER ORGANIZER</th></tr>
<tr><th>USE WITH SECTION</th><th>DEMONSTRATION</th><th>WORKSHEET</th><th>LAB EXPERIMENT</th><th>SOFTWARE</th><th>TEACHER'S RESOURCE BINDER</th></tr>
<tr><td>1-1</td><td></td><td></td><td>1-A</td><td></td><td>Math Pretest</td></tr>
<tr><td>1-3</td><td>Appendix A</td><td></td><td>1-B</td><td></td><td></td></tr>
<tr><td>1-8</td><td></td><td>1-A</td><td></td><td>Disk 9</td><td></td></tr>
<tr><td>1-9</td><td></td><td></td><td></td><td>Disk 9</td><td></td></tr>
<tr><td>1-10</td><td></td><td>1-A, 1-B</td><td></td><td>Disk 9</td><td>Problems*</td></tr>
<tr><td>1-11</td><td></td><td></td><td>1-C</td><td>Disk 9</td><td></td></tr>
<tr><td>1-12</td><td></td><td>1-C</td><td></td><td></td><td>Chem Issue 1</td></tr>
<tr><td colspan="6">Chapter Test</td></tr>
</table>

*Use the *Problem-Solving Strategies* for students found in the Teacher's Resource Binder.

■ Objectives

When students have completed the chapter, they will be able to

1. Describe a given physical object or event.
2. Identify statements about a physical or natural event that represent observations and statements that represent inferences.
3. List at least five characteristics of a given object and use observed characteristics to classify objects or events.
4. Distinguish between properties that are characteristic of an object and properties that are not.
5. State whether something common is a form of matter, a form of energy, or neither, and name the properties that determine its classification.
6. Name two properties each of solids, liquids, and gases.
7. Distinguish between a number and a quantity.
8. Name the SI base units for length, mass, time, and temperature; the SI units for area and volume that are derived from the base unit of length; and the correct abbreviation for each unit.
9. Explain the meaning of commonly used metric prefixes such as kilo-, centi-, and milli-.
10. State the unitary rates derived from the equivalence between two units (e.g., 1 kg = 1000 g) and use unitary rates to convert any measurement from one unit to another.
11. Convert any metric measurement to its equivalent in an English unit when given an appropriate unitary rate.
12. Use unit analysis to verify that the units for the answer to a calculation are correct for the quantity being calculated.

■ Section Notes

1-1 The first day of class can set the tone for the entire year. Encourage students to believe that chemistry is something that they can do and want to do. Chemistry is serious, but it can also be fun.

I like to begin with my most spectacular demonstrations, prepared in advance. I drop dry ice into large beakers filled with colored water to produce clouds of "smoke". I tie hydrogen balloons to the lecture bench. I set up a "blue bottle" that turns blue when shaken and fades to colorless as it sets. I prepare my favorite pyrotechnics demonstrations.

When students arrive, my hair is tousled and I am wearing a lab coat that is dirty and full of holes. In a falsetto voice, I welcome the students to chemistry, the wonderful world of mad scientists causing beautiful color changes (as I do so) and loud explosions (as I do so).

After a few minutes of dramatic demonstrations without explanation, I straighten my hair and assume my normal voice. Then I explain that all of these things can and do take place in chemistry, but that the T.V. image of the mad scientist misrepresents what science is about. Science is, as Percy Bridgeman put it, "Doing one's darndest with one's mind, no holds barred." It is an attempt to understand why things happen, how they can be predicted and controlled, as well as enjoying the thrill of discovery and the sight of new events. It is partly plain hard work. It takes dedication and perseverance. In this course we try to reveal some of the thrill of discovery, but we also develop some of the intellectual skills used to make sense of the physical world.

Some teachers will feel uncomfortable with my introduction. If you do, use another, but make it exciting. Help your students to leave the first class with eager anticipation for the year ahead.

After your thrilling introduction, engage the class in observation activities. If you have done demonstrations, ask them to describe what they observed. Alternatively, have them describe the apparatus on your lecture bench, or open their book to the first photograph and describe it.

As the students report their observations, be alert for statements of inference rather than observation. Focus on the distinction between the two, and discuss the importance of reporting observations first. Science demands that observations be reported as objectively as possible, so others can know the basis of your inferences and can draw inferences of their own. If time permits, continue with an exercise on classification. If you have done several demonstrations, you might ask students to suggest different ways the materials and the events they saw could be classified. You might also invite students to formulate hypotheses to explain events. These activities will establish a useful introduction for the first reading assignment.

1-2 Here is a useful exercise in hypothesis formation and testing. Question 4 of the Review and Practice on page 3 refers to this exercise and could be used to introduce it.

On the first day of class have students fill out a sheet with personal information such as telephone number, address, favorite color, shoe size, etc. On the second day assign seats and announce to students that seats have been assigned in a systematic fashion. Their task is to determine the basis for the assignment. When students suggest that it is alphabetical order (usually the basis of the first assignment), insist that they be more specific. (By last name? From front to back? From left to right? What if two people have the same last name? etc.) Use the exercise to focus on the importance of exact description, to discuss how hypotheses are formed and tested, and eventually, what one does when some data are inconsistent with an hypothesis.

Periodically reassign seats on some new basis, and repeat the exercise. Make each successive assignment more difficult to figure out than the previous one.

1-7 The distinction between numbers and quantities is an important one. Be sure that students realize that the unit is just as important as the number.

1-10 See the section on Proportional Reasoning and Unitary Rates in the front matter for the rationale for using unitary rates. The term may be unfamiliar, but the concept is not new. Unitary rate is used extensively throughout the text when dealing with proportional relationships. It is probably the most important idea in the chapter.

Although worked examples are used throughout the text to suggest possible solutions to problems, students should be encouraged to devise their own solutions to problems whenever possible. Examples should not be presented as THE way to solve any problem. See suggestions for teaching problem solving skills in the front matter.

1-11 Many students learn to manipulate units without understanding their physical meaning. Here and elsewhere check your students' understanding by asking them to draw or represent the unit in some other way. For example, in the problem shown in the middle of p. 17, what is meant by 20 cm? (20 times this length: _____) By 510 cm²? 510 squares this size:

Also ask what must be true for the equality to stand. (25.5 cm and 20 cm must represent the length of sides for a rectangle. The equation describes a particular relationship between surface covered and the length of sides that enclose the surface.)

Review and Practice _____
page 3

1. See the appendix for the description of the burning match. Use this description as a sample in evaluating student responses.

2. A very different classification would be carbohydrate, fat, and protein. Sucrose, glucose, and starch are examples of carbohydrates because they are composed of carbon, hydrogen, and oxygen. The characteristic that all fats have in common is that they are esters of glycerol and large fatty acids such as palmitic or linoleic acid. Proteins are large polymers of amino acids such as alanine, cysteine, glycine, and aspartic acid.

   Many other classifications are possible. Students are likely to come up with classes that have an easy basis for classification, such as meats, vegetables, and dessert. Proteins, fats, and starch are classes whose properties students will be unable to describe. If this occurs, call attention to the fact that many names in common use specify classes of chemicals with properties related to molecular structure.

3. Answers will vary but students should note size differences, the static nature of the model, and the use of color in the model. A single atom would have no color.

   Models for atoms differ from the real thing in several ways. The models are much larger since atoms are microscopic. Models normally have rigid spherical boundaries while the boundaries of a real atom are probably not rigid. Models of atoms have the same size in molecules as when they are unbounded while real atoms do not. Models are normally colored while single atoms are colorless. Pictures are two dimensional; atoms are three dimensional. Pictures and models are static; atoms are dynamic. Some pictures show electrons moving in definite paths; the path of electrons is unknown.

4. First collect data on the people sitting in each seat. Then search for regularities in the data for the purpose of generating an hypothesis. For example height appears to decrease from left to right and from front to back. The hypothesis could be tested by determining if each seat assignment is consistent with the hypothesis. This process would produce a possible model, but not necessarily the model used. For example, it is possible that assignments were made on the basis of weight, but heights were also ordered because of the correlation between height and weight.

page 8

1. a, b, and f are not matter; they do not have mass nor do they occupy space.
2. The difficulties in this experiment are the slow rate of the change as well as the difficulty in monitoring oxygen intake and loss of CO_2 and H_2O through transpiration. Trace amounts of organic gases are also lost during plant growth.
3. a and i—solids
   b and d—liquids
   h and k—gases
   l—plasma, but students may not know this. The others are more difficult to classify. Items c and j have properties of solids but flow like liquids. Item e is a mixture of solid particles suspended in gases. Similarly, items f and g are liquid droplets suspended in gases.
      Even when we know properties of various classes, there are usually examples that are difficult to classify because some of the properties are not readily apparent. Students will encounter similar problems of classification when ionic and covalent bonds, acids and bases and other chemical concepts are discussed. They need to be aware that black and white classifications are often difficult to make.

page 12

1. a. years
   b. feet and inches
   c. pounds
2. Only like quantities can be added.
3. hecto
4. centi
5. $\frac{1}{1000}$

page 16

1. a. mg
   b. cm
   c. km
   d. kg
   e. cg
   f. dm
   g. μm
   h. Mg

2. a. millimeter
   b. centigram
   c. kilogram
   d. kilometer
   e. centimeter
   f. decigram
   g. microgram
   h. megameter

3. a. 100   0.01
   b. 0.001   1000
   c. 0.001   1000
   d. 100   0.01
   e. 10   0.1
   f. 1 000 000   0.000 001
   g. 0.000 001   1 000 000

4. a. ? dag $\longrightarrow$ 127 g $\times \dfrac{0.1\ \text{dag}}{1\ \text{g}}$ = 12.7 dag

   b. ? g $\longrightarrow$ 268 mg $\times \dfrac{0.001\ \text{g}}{1\ \text{mg}}$ = 0.268 g

   c. ? kg $\longrightarrow$ 354 g $\times \dfrac{0.001\ \text{kg}}{1\ \text{g}}$ = 0.354 kg

   d. ? m $\longrightarrow$ 247 km $\times \dfrac{1000\ \text{m}}{1\ \text{km}}$ = 2.47×10^5 m

   e. ? mm $\longrightarrow$ 3.4 m $\times \dfrac{1000\ \text{mm}}{1\ \text{m}}$ = 3.4×10^3 mm

   f. ? cm $\longrightarrow$ 0.025 m $\times \dfrac{100\ \text{cm}}{1\ \text{m}}$ = 2.5 cm

   g. ? Mg $\longrightarrow$ 5 245 g $\times \dfrac{0.000\ 001\ \text{Mg}}{1\ \text{g}}$
      = 0.005 245 Mg

   h. ? μg $\longrightarrow$ 0.000 000 15 g $\times \dfrac{1\ 000\ 000\ \text{g}}{1\ \mu\text{g}}$ = 0.15 μg

page 19

1. a. 10 mm = 1 cm
      100 mm² = 1 cm²
   b. 1000 m = 1 km
      1 000 000 m² = 1 km²
   c. 10 dm = 1 m
      100 dm² = 1 m²
   d. $10^4 \mu$m = 1 cm
      $10^8 \mu$m² = 1 cm²

2. The square millimeter, square centimeter, and square decimeter could be shown. A square meter is about half the area of a twin bed mattress. A square kilometer is slightly larger than the area covered by 16 city blocks (i.e., a square that is 4 city blocks long and 4 city blocks wide).

3. A cube has three dimensions—length, width, and height. A cubic decimeter is 10 cm × 10 cm × 10 cm. The number of cm³s that occupy the same space as a dm³ is 10 × 10 × 10 or 1000.

4. a. 100 cm = 1 m 10^6 cm³ = 1 m³

b. 1000 m = 1 km 10^9 m^3 = 1 km^3
c. 10 mm = 1 cm   1000 mm^3 = 1 cm^3
d. 10 cm = 1 dm   1000 cm^3 = 1 dm^3

5. 1000

Review, page 20

1. religion and philosophy, other answers are possible

2. Interpretations of events are influenced so much by what we know and believe that science insists that different individuals must make independent observations and reach similar conclusions before something is accepted as fact. Unless the first observer can communicate exactly what was done to make an observation, an independent observer will not be able to repeat an experiment or observation.

3. Whenever possible, they express regularities as mathematical equations. Graphs, tables, drawings, and photographs are also used. In any communication, they separate observations from inferences based on those observations.

4. What we observe and how we interpret it is influenced by what we already know and believe. Since no two people are likely to know and believe exactly the same things, the more people who agree about what happened, the more likely it actually happened.

5. It takes up space and has mass.

6. It does not take up space or have mass. Some experiments suggest that light does have mass, but it is extremely small.

7. Mass is a property of matter that gives it weight and inertia. Inertia is the resistance of matter to change, ie, being stopped or being moved. The heavier an object is, the more inertia it has; the more inertia it has, the more mass it has. To say matter has mass means that matter has inertia and requires a force to get it moving or stopped. Force is the energy needed to bring about a change.

8. classification

9. Plasmas have all the properties of gases except that they are composed of charged particles like electrons rather than uncharged atoms or molecules.

10. Potential energy was invented to preserve the belief that energy is never created or destroyed; it only is changed from one form to another. It is based on the observation of the potential for changing matter that exists in such things as coal or wood or an unused battery.

11. A quantity is composed of a number and an associated unit. Numbers describe amount, but not the nature of what is described. Quantities describe both amount and kind of what is observed.

12. Unlike quantities can not be added.

13. meter

14. kilogram

15. A unitary rate is a ratio of quantities. It always has a denominator of one plus the appropriate unit. Other ratios may only include numbers and the denominator may have any value.

16. 1

Interpret and Apply, page 21

1. Teachers and parents are normally older, and they have experienced different things in their lives than you have experienced in yours. Whether they are actually wiser as well as older is open to debate, but the differences in experience account for many disagreements!

2. People of the same age, people from similar home environments, and people from the same country are more likely to agree. The reasoning is that they are more likely to have similar experiences, and the experiences we have determine how we interpret events today.

3. The same point can be made by playing a game sometimes played by parents and their children or teachers and their pupils. The game is to do exactly what the person told us to do but not what we are pretty sure the person intended us to do! The point of both exercises is that communication is very difficult. We must work hard to express ideas so there is no doubt about what is intended.

4. A mass in space will still have inertia. A spring scale could be used to exert a force for a measured time. Measuring the change in velocity of the object would produce a measure of mass.

5. sublimation, the transition from solid to gas

6. The assumption that matter is neither created nor destroyed during any of its transformations would be violated.

7. Kinetic energy is the product of one half the mass of the object and its velocity squared. The truck has more mass; consequently, it has more kinetic energy.

8. The car could have more energy if it were moving much faster than the truck.

9. At the time this book was written, the exchange rate was approximately 0.75 Canadian dollars/1 U.S. dollar.

10. Place the 50-pound mass on the scale and see what it reads. If the scale is calibrated correctly, it should read 50 pounds. Now place the 20 kilogram mass on the scale and get the reading. It

should read 44 pounds. Then 20 kg = 44 lb. Dividing each side of the equation by 44, 0.45 kg = 1 lb. The unitary rate is 0.45 kg/lb.

Problems page 22

1. The new kinetic energy is $\frac{1}{4}$ the original.
2. # feet shoulder to wrist + # feet wrist to fingertip = # feet from shoulder to fingertip
3. # inches shoulder to wrist + # inches wrist to fingertip = # inches from shoulder to fingertip
4. a. millimeter   h. milligram
   b. megameter   i. decimeter
   c. kilometer   j. cubic decimeter
   d. kilogram   k. cubic meter
   e. centigram   l. cubic centimeter
   f. microgram   m. square millimeter
   g. megagram   n. square kilometer
5. a. example given
   b. 1 Mm = 1 000 000 m; 0.000 001 Mm = 1 m
   c. 1 km = 1000 m; 0.001 km = 1 m
   d. 1 kg = 1000 g; 0.001 kg = 1 g
   e. 1 cg = 0.01 g; 100 cg = 1 g
   f. 1 μg = 0.000 001 g; 1 000 000 μg = 1 g
   g. 1 Mg = 1 000 000 g; 0.000 001 Mg = 1 g
   h. 1 mg = 0.000 001 g; 1 000 000 mg = 1 g
   i. 1 dm = 0.1 m; 10 dm = 1 m
   j. 1 dm^3 = 0.001 m^3; 1000 dm^3 = 1 m^3
   k. 1 m^3 is the base unit.
   l. 1 cm^3 = 0.000 001 m^3; 1 000 000 cm^3 = 1 m^3
   m. 1 mm^2 = 0.000 001 m^2; 1 000 000 mm^2 = 1 m^2
   n. 1 km^2 = 1 000 000 m^2; 0.000 001 km^2 = 1 m^2
6. a. 570 g = 57 dag
   b. 37 000 mg = 37 g
   c. 4 700 g = 4.7 kg
   d. 0.138 km = 138 m
   e. 4.021 m = 4 021 mm
   f. 27 000 000 m = 27 Mm
   g. 4.62 g = 462 cg
   h. 0.014 g = 14 000 μg
   i. 3.7 × 10^{11} pm = 37 cm
   j. 3.2 km = 3 200 000 mm
7. a. 1 000 000 mm^2 = 1 m^2
   b. 0.000 001 km^2 = 1 m^2
   c. 0.01 cm^2 = 1 mm^2
   d. 1 000 000 μm^2 = 1 mm^2
   e. 0.000 001 Mm^2 = 1 km^2
   f. 0.001 cm^3 = 1 mm^3
   g. 0.000 001 m^3 = 1 cm^3
   h. 10^{-9} km^3 = 1 m^3
   i. 1 000 dm^3 = 1 m^3
   j. 10^{12} μm^3 = 1 cm^3
8. a. 0.005 280 m^2 = 5 280 mm^2
   b. 2.5 × 10^6 m^2 = 2.5 km^2
   c. 4.53 × 10^4 mm^2 = 453 cm^2
   d. 9.055 033 mm^2 = 9 055 033 μm^2
   e. 25 km^2 = 0.000 025 Mm^2
   f. 1.23 × 10^5 mm^3 = 123 cm^3
   g. 0.005 345 m^3 = 5 345 cm^3
   h. 4.6 × 10^7 m^3 = 0.046 km^3
   i. 1.000 m^3 = 1 000 dm^3
   j. 3.098 000 × 10^{-6} cm^3 = 3 098 000 μm^3
9. b and c

Challenge, page 22

1. Answers will vary. Such an experiment would need to include the measurement of the amount of food, water, and air ingested; measurement of solid, liquid, and gaseous wastes from digestive, urinary, and respiratory systems and also the skin; measurement of body growth and development, tissue repair and replacement, energy released from food; measurement of body synthesis of substances such as enzymes and hormones.
2. Measure any length in both inches and centimeters. Divide the length in centimeters by the length in inches. The result should be 2.54 cm/in within the uncertainty of the measurements done.
3. 16.4 cm^3 = 1 in^3
4. 10^{-18} Em = 1 m; 10^{18} am = 1 m
   Then 10^{-18} Em = 10^{18} am
              1 Em = 10^{36} am
5. 1 Em = 10^{18} m
   1 Em^3 = 10^{54} m^3
   The volume of a sphere is $\frac{4}{3}\pi$ r^3
   or $\frac{1}{6}\pi$ d^3.
   Given the diameter of Earth as 1.27 × 10^4 km^3 and assuming that Earth is a perfect sphere:
   V = $\frac{1}{6}\pi$(1.27 × 10^7 m^3)
   V = 1.07 × 10^{21} m^3
   One Em^3 is far greater than the volume of Earth.
6. 1 am = 10^{-18} m
   1 am^3 = 10^{-54} m^3
   The Bragg-Slater radius of the hydrogen atom is 25 pm (25 × 10^{-12} m). Then the volume of a hydrogen atom is:
   V = $\frac{4}{3}\pi$r^3
   V = $\frac{4}{3}\pi$(2.5 × 10^{-11} m^3)
   V = 6.5 × 10^{-32} m^3
   One am^3 is much smaller than the volume of the smallest atom.

CHAPTER 2 Finding Out About Matter

CHAPTER ORGANIZER

USE WITH SECTION	DEMONSTRATION	WORKSHEET	LAB EXPERIMENT	SOFTWARE	TEACHER'S RESOURCE BINDER
2-4		2-A	2-A		
2-6	2-A		2-B		
2-9		2-B, 2-C	2-C		
2-10		2-D			
2-11		2-E		Disk 1	
2-12				Disk 2, 3	
2-13				Disk 2, 3	
2-14		2-F		Disk 2, 3	Chem Issue 2
2-15		2-G		Disk 2, 3	
2-16		2-H		Disk 2, 3	
2-17		2-H		Disk 2, 3	
2-19		2-I		Disk 2, 3	
Chapter Test					

■ Objectives

When students have completed the chapter, they will be able to:

1. Name properties that are particularly useful in determining whether something is pure or impure, and describe observations that would convince them that the substance is pure.
2. Describe at least three ways in which a mixture can be separated, and name the property of the substance in the mixture that allows for the separation.
3. State evidence that determines whether a substance is a compound, an element, or a mixture.
4. Indicate whether a change in matter produced new chemical substances or rearranged substances that already existed, or whether it is possible to tell. State the additional macroscopic evidence needed in order to decide undetermined cases.
5. Indicate whether an observed change in matter is chemical or physical and state reasons for their conclusion.
6. Describe two ways that compounds can be separated into the elements that make up the compound.
7. Indicate whether a chemical change represents the combining of elements or compounds to form a new compound, or a compound decomposing to form elements, and state the macroscopic evidence needed to decide undetermined cases.
8. Explain why the properties differ in two compounds containing the same elements.
9. Describe the difference between the proportions of elements in a compound and in a mixture, and the microscopic difference between a mixture and a pure substance.
10. Draw diagrams to represent an element and a compound at the microscopic level.
11. State the differences between a molecule of an element and a molecule of a compound.
12. Using a microscopic model, describe how solids, liquids, and gases differ and how a substance changes from one physical state to another.
13. Use pictures of atoms and molecules to describe what happens when a specified chemical change occurs.
14. State how to determine whether a pure substance is an element or a compound.
15. Describe what is represented by the names and symbols on the periodic table, and use the names and symbols correctly.
16. Identify those elements on the periodic table that are metals and those that are nonmetals.

41T

17. Write the formula of a named compound or analyze a given formula for kind and number of atoms, and then construct or draw a model of the molecule.
18. Determine whether a model represents an element, a compound, or a mixture. Write the formula of a compound from the model.
19. Name a binary compound from its formula and write the formula when given the name.
20. Predict the ion charge of an element from its position on the periodic table.
21. Use ion charges to determine formulas.
22. Write the formula of the compound formed when two given elements combine.
23. Determine the ion charge on one element in a binary compound when the formula and ion charge of the other element are known.
24. Distinguish between symbols that represent a molecule and symbols that represent a polyatomic ion.
25. Name common compounds containing more than two elements, given their formulas, or write the formulas of common compounds, given their names.

■ Section Notes

2-1 The destructive distillation of wood is an excellent vehicle for discussing the concepts in this chapter. Wood splints are packed into a test tube fitted with a one-hole stopper and a short glass tube. A rubber hose connects the tube to a flask fitted with a two-hole stopper with glass tubing. The entrance tube should extend to the bottom of the flask and the exit tube should extend to just below the stopper. A rubber hose connected to the exit tube collects any gases produced by displacement of water in a second flask.

Begin heating the test tube containing the wood splints. Ask the students what is happening, if anything. Is some new substance forming? (yes) How do you know? (The properties of the wood are changed. The material collecting in the middle flask is liquid at room temperature; the wood is solid. The material collected over water remains gas at room temperature.) Is the wood decomposing? (probably) How do you know? (It isn't just boiling away since it does not return to solid when it cools. It isn't a mixture separating, since the wood looks dry and the volume of the gas produced is much greater than the initial volume of wood. Since the wood is enclosed in the test tube, there is little opportunity for it to react with air, and there is no evidence that the glass is deteriorating.)

Ask what the gas is in the collecting bottles. How might we find out? (Test the gas to see if it has the properties of air, carbon dioxide, etc.) Test the gas with a burning splint.

Save the brown liquid collected for further tests and discussion. In the meantime, fill a test tube with iron filings and repeat the experiment. Repeat again with a test tube filled with sodium bicarbonate. Repeat the questions asked about the wood.

This activity reinforces observation skills and focuses on the kinds of observations needed to decide whether a change is chemical or physical, and whether something new actually forms.

Now focus on the liquid collected from the wood. Is it pure? (no) How can we decide? (See if it can be separated into components.) Do a fractional distillation and record the changes in boiling point as the liquid boils away. Show the difference in color of the distillates. Return the distillates to the boiling flask and show that the original properties of the brown liquid are restored by mixing. When mixtures are separated, it is usually simple to mix them again; when compounds decompose, the reverse process is more difficult, although there are exceptions.

2-5 After demonstrating the electrolysis of water, you may want to invite students to investigate electrolysis as suggested in Project 2 on page 59. Electrolysis is a phenomenon that students can investigate with considerable success. The project can be started now and allowed to run for several days or weeks while other material is discussed. Teachers who want students to design their own experiments often find that the background needed to do so is extensive.

2-7 An important but difficult aspect of understanding chemistry is being able to visualize atoms and molecules. Beginning now and continuing throughout the course, encourage students to draw pictures and/or use models to represent matter. Periodically ask students to describe what something they are using would look like if atoms could be seen (a glass beaker, iron spatula, salt water solution, solid mixture).

This is NOT the time to describe atomic structure in all of its detail. Give students an opportunity to use the simple notion of spherical particles to describe and explain macroscopic events. They will be able to explain a surprising number of their observations. Presenting too much information about atomic structure this early makes it difficult for students to apply the atomic model to explain events.

2-8 It is very difficult to illustrate the microscopic difference between solids, liquids, and gases. Several companies sell apparatus to model the microscopic behavior of gases. Such models help students to visualize the dynamic nature of atoms. However, these models merely approximate what we believe happens at the microscopic level. After using a model and discussing solids, liquids, and gases, ask students to describe how the model fails and how it best simulates real atoms.

Some chemists define molecule as the smallest

piece of an element or compound. This definition is likely to cause confusion, for example, an atom of neon or helium could be a molecule. We recommend defining a molecule as a particle containing more than one atom.

Review and Practice
page 35

1. **a.** Heat is transferred from the water to the air.
   b. The water temperature will remain constant when it reaches room temperature.
   c. These are real data points. The curve shows a best-fit line.
   d. Paradichlorobenzene loses heat to the water; water loses heat to the room. A temperature gradient is established with the paradichlorobenzene at the higher temperature and the air at the lower temperature.
   e. A change of state occurs—melting or freezing.
   f. above 53°C—liquid; below 53°C—solid
2. Measuring a melting point and a boiling point could provide a clue.
3. Determine density—mass per unit volume.
4. Observe charateristic properties of the original sawdust and the final products. Different properties mean a different substance—chemical change. The brown liquid collected from the destructive distillation of wood can be further distilled to show its rise in temperature while distilling. This may be compared with the behavior of water or other pure liquids.
5. Boil the liquid and measure the temperature. A constant boiling temperature would indicate a pure substance.
6. Chemical
7. Make a muddy water mixture and see if it behaves like the river water.
8. Yes, mixtures scatter light.
9. The density of oxygen is 16 times the density of hydrogen.
10. 16
11. Since the amount of gold can vary from 10 to 22 carats, it must be a mixture.
12. A melting point determination could be used to test for pure gold.
13. If a pure substance is distilled, the distillate has the same properties as the original substance.

page 39

1. element
2. No, the atoms are not joined.
3. c and e; c represents molecules of an element; e represents molecules of a compound.

4. Gases; the molecules are far apart.
5. b and d; b is a mixture of atoms; d is a mixture of atoms, molecules of an element, and molecules of a compound.
6. **a.** b and c
   b. a
   c. e
7. The accuracy of the student's drawing is not particularly important. However, the drawings should show diatomic molecules coming into contact with diamond (all atoms alike) and a new molecule consisting of one atom of carbon and two atoms of oxygen connected. Students are unlikely to show the proper structure for diamond or for carbon dioxide. Point out that the way atoms are connected is important and that later chapters will present ways to predict the structure of molecules.

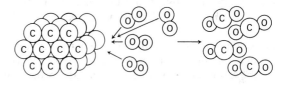

8. Atoms have definite masses. The fixed ratio of the masses implies a fixed ratio of atoms.

page 55

1. **a.** monosilicon monocarbide (silicon carbide)
   b. monocarbon disulfide (carbon disulfide)
   c. monosulfur hexafluoride (sulfur hexafluoride)
   d. monoxygen difluoride (oxygen difluoride) This is not an oxide as oxygen is the more metallic of the two elements.
   e. monosulfur dioxide (sulfur dioxide)
   f. monosulfur trioxide (sulfur trioxide). The sulfite ion would be written with a 2-charge when standing alone.
   g. dinitrogen tetraoxide
   h. mononitrogen monoxide (nitrogen monoxide)
2. **a.** ClO_2 **d.** NI_3
   b. Cl_2O **e.** P_2O_4
   c. IBr_3
3. **a.** KBr   **f.** NaI
   b. $MgCl_2$ **g.** $CaBr_2$
   c. CaO   **h.** K_2O
   d. AlI_3 **i.** $BeCl_2$
   e. BaF_2 **j.** Al_2O_3
4. **a.** 2+, tin(II) oxide
   b. 4+, tin(IV) bromide
   c. 2+, iron(II) sulfide
   d. 3+, iron(III) chloride
   e. 3+, chromium(III) oxide
   f. 2+, nickel(II) chloride
5. **a.** $SnCl_4$ **c.** SnO_2 **e.** FeF_2 **g.** Co_2O_3
   b. $SnCl_2$ **d.** Fe_2S_3 **f.** HgO

6. a. $Ca(C_2H_3O_2)_2$ **e.** $Al(OH)_3$
  b. Na_3PO_4 **f.** $Ca(HCO_3)_2$
  c. $CuSO_4$ **g.** $(NH_4)_3PO_4$
  d. $Fe_3(PO_4)_2$ **h.** $Li_2Cr_2O_7$

7. a. aluminum nitrate
  b. hydrogen carbonate
  c. tin(II) nitrate
  d. potassium dichromate
  e. copper(I) carbonate
  f. ammonium chloride
  g. nickel(I) oxalate

Review, page 57

1. Shine a strong light through a sample of the soft drink to see if light is scattered by particles in suspension. You also can allow the soft drink to stand for a period of time to see if matter settles out, or distill it to separate the mixture.

2. a. Distillation will work.
  b. Distillation will work to a point, but it will not produce a total separation. Students are unlikely to know any of the procedures that produce a total separation.
  c. If the pieces are large, simply pick them out by hand. If they are small, use a magnet to attract the iron.
  d. Add water. The sugar will dissolve; the glass will not. The sugar can be recovered by boiling away the water.
  e. It is possible to make a liquid that has a higher density than the aluminum but a lower density than the zinc. The aluminum would float on the liquid, but the zinc would sink. Students are unlikely to know about this procedure. The question is asked to make the point that it is often difficult to separate mixtures, even when the pieces are large.

SET I	SET II	SET III
3. potassium	**7.** bromine	**11.** Pb
4. Hg	**8.** Cu	**12.** gold
5. ZnI_2	**9.** PCl_5	**13.** NiB
6. Ag_2O	**10.** SO_3	**14.** LiI

Interpret and Apply, page 58

1. a. Chemical—the color change suggests that new compounds are formed.
  b. Physical—breaking things is considered to be a physical change because characteristic properties such as density, melting point, boiling point, ignition temperature, and color are unchanged.
  c. Physical—solution is normally considered to be a physical change, because it can be easily reversed by boiling away the solvent.

  d. Chemical—without knowing exactly how aspirin works, we cannot be sure, but it seems like a reasonable guess that some new substance is formed in the process.
  e. Physical—the water molecules remain intact. Phase changes such as this change from liquid to gas are normally considered to be physical.
  f. Chemical—the properties of cooked beans are clearly different from the properties of uncooked beans. For example, the taste is changed.
  g. Chemical—new cells are being produced in the grass, and water and carbon dioxide react in the process.
  h. Physical—like b, big pieces are made into smaller pieces.
  i. Chemical—you may have difficulty making this judgment, but the bleach reacts with stains to produce new compounds that dissolve in the water to be washed away.
  j. Both—the Drano reacts with fats to produce soaps. The aluminum flakes in the Drano react to produce hydrogen gas. These changes are chemical. The agitation caused by the hydrogen gas and the rise in temperature that melts the fats in the drain are physical changes that help unstop the sink.

2. a. The decrease in mass of the solid and the production of the gas lead me to believe that this is a compound that is decomposing.
  b. The increase in mass suggests that the metal has combined with something in the air, possibly an element.
  c. The decrease in mass suggests that some matter has left the log. Probably compounds in the log decomposed to form gases that escaped. Close inspection may show that some of the log was taken away in the digestive tracts of small animals, a physical change followed by a chemical change.
  d. This suggests that the black powder combined with the oxygen to form the colorless gas. The black powder probably is an element, and the product is a compound.
  e. If there were only one product, this could be an element that combines with oxygen gas to produce a compound. Since three products are formed, it seems more likely that the original solid was a compound that reacted with oxygen to produce several different products. This result also could come from an element reacting with oxygen, but since elements combine to form compounds with a definite composition, the three different products are more likely to result from reaction with a compound containing several elements than from a single element.

3. No. Compounds have a definite composition. The

two solids could be mixtures containing copper or they could be two compounds containing copper.

4. Not necessarily. Copper may form ions with a 1+ charge or ions with a 2+ charge. Copper can form two compounds with chlorine. You must know that the percent of copper in the two samples is the same before concluding that the samples are of the same compound.

5. No. Pure compounds have the characteristics described. It could be an element or a compound.

6. Heat it and see if you get a new substance. If you do, see if the new substance weighs more or less than the original substance. If nothing comes from heating it, try melting it and passing an electric current from a battery through the liquid. If new substances form at the two electrodes, it is probably a compound.

7. a. See Figure 2-22c.
  b. See Figure 2-22a.
  c. See Figure 2-23b.
  d. See Figure 2-17.
  e. Make a drawing like Figure 2-17, but show pairs of atoms connected rather than individual atoms. Like Figure 2-17, these pairs of atoms (molecules) would be close together but not in an ordered array.

8. a. (H)(S)(H)

  b. (N)(O)   (O)(N)(O)

  c. (H)(N)(H)
        (H)

9. a. (H)(H)   (H)(H)   (O)(H)   (H)(O)
       (O)   (O)(O)   (H)(O)(H)   (H)
      (O)

  b. (O)   (O)   (O)(O)

  c. (C)(O)   (O)(C)(O)

Problems, page 58

1. a. $AlCl_3$
  b. aluminum sulfide
  c. mercury(II) oxide
  d. $CoCl_2$
  e. carbon disulfide
  f. N_2O_4

2. a. dinitrogen trioxide
  b. $FeCl_2$
  c. Mg_3N_2
  d. chromium(III) oxide
  e. manganese dioxide; manganese(IV) oxide
  f. CoF_3

3. a. KBr
  b. silicon carbide; monosilicon monocarbide
  c. SCl_2
  d. tin(IV) oxide

  e. carbon tetrachloride
  f. AlN

4. a. Ba_3N_2
  b. calcium carbide
    Calcium dicarbide would be a better name, since no rule has been given to predict the number of carbon atoms that will combine with calcium.
  c. Cr_3Si_2
    Students will be unable to predict this formula from the rules given. The rules given enable us to predict the names of most common compounds, but additional rules are required before one can handle all nomenclature problems. The point made in this text is that systems of nomenclature do exist. Complete systems are far too complex to present in an introductory course.
  d. cobalt(II) oxide
  e. CuO
  f. gold(III) bromide
  g. $FeCl_3$
  h. tetraboron monocarbide
  i. NI_3
  j. pentanickel diphosphide
    Even though nickel is a metal, this name is better than nickel(II) phosphide because we have no rule that tells what the charge on the phosphide ion should be. This compound is, in fact, a covalent compound and should be named as shown, but students have no way of knowing that.

5. a. copper(II) phosphate
  b. hydrogen sulfite
    Sulfurous acid is the more common name, but it is not a systematic name.
  c. tin(II) chromate
  d. iron(II) sulfite
    Watch the spelling of sulfite.
  e. potassium hydrogen carbonate
    Potassium bicarbonate is the common name.
  f. magnesium carbonate
  g. calcium nitrate
  h. $Pb(Cr_2O_4)_2$ $CrO4$
  i. NaOH
  j. $(NH_4)_2CO_3$
  k. $MgSO_3$
    Compare this to sulfate and sulfide.
  l. MgS
  m. $MgSO_4$
  n. $Ca(C_2H_3O_2)_2$

6. a. This is the sulfite ion.
  b. sulfur trioxide, a molecule
  c. carbon disulfide, a molecule
  d. phosphorus pentoxide, a molecule
  e. phosphate ion
  f. carbon monoxide, a molecule
  g. carbonate ion

h. nitrogen dioxide, a molecule
i. nitrite ion
j. iron(III) sulfate, a molecule
More correctly, it is an ionic solid, but students are unlikely to know this.

Challenge, page 59

1. and 2. Various maps will be constructed. Here is an example. In evaluating students' maps, look to see that they have made sensible connections between terms.

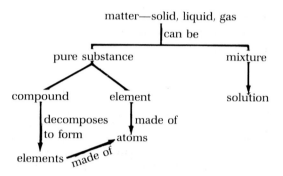

Projects, page 59

1. Answers will vary.
2. Changing the electrodes changes the products. For example, graphite electrodes produce some gas other than oxygen, probably carbon monoxide or carbon dioxide. Copper electrodes react and the solution turns blue. Some steel electrodes work well if the electrolyte is sodium carbonate, but not if it is acid or contains chloride ion.

Changing the amount of electrolyte affects the speed of the reaction. Changing the nature of the electrolyte changes the products formed. For example, chloride ion reacts to form chlorine gas or some compound, depending on the electrode used.

3. Some of the properties mentioned vary periodically, but others do not. In no case is the trend perfect.
4. Flotation, salting out, distillation, electrophoresis, paper and column chromatography are a few of the techniques that may be discovered.

CHAPTER 3 Numbers Large and Small

CHAPTER ORGANIZER					
USE WITH SECTION	**DEMONSTRATION**	**WORKSHEET**	**LAB EXPERIMENT**	**SOFTWARE**	**TEACHER'S RESOURCE BINDER**
3-2		3-A			Problems*
3-3		3-A			
3-4		3-A			
3-8		3-B			
3-12		3-A	3-A,B		
3-17	3-A	3-C	3-C	Disk 9	
Chapter Test					

*Use the *Problem-Solving Strategies* for students found in the Teacher's Resource Binder.

■ Objectives

When students have completed this chapter, they will be able to

1. State whether a relationship is directly proportional, inversely proportional, or neither from a table showing one variable as a function of another.

2. Convert any number from decimal to scientific (exponential) notation, and vice versa. Add, subtract, multiply, or divide numbers written in scientific notation.

3. Describe the uncertainty of a measurement that they have made and what is meant by uncertainty of measurement.

4. Identify the digits that are certain and the one

that is uncertain, given a number that represents a measurement.

5. Identify the more precise measurement, given two measured quantities.

6. Estimate the precision possible for any instrument they use in the laboratory.

7. Describe the difference between the precision and the accuracy of a measurement, using an example.

8. Determine the number of significant digits in a number resulting from a measurement, identify zero digits that are significant and those that are not, and identify numbers that have an unlimited number of digits and explain why they do.

9. Round the answer to addition, subtraction, multiplication, or division problems to the proper number of significant digits.

10. Order the data on a table of experimental measurements of one quantity as a function of another.

11. Prepare a graph from a table of data. The graph must contain all the information needed for a person to interpret it.

12. Calculate a unitary rate describing the change in one quantity that corresponds to a change of one unit of another quantity on a graph representing a proportional relationship.

13. Calculate the density of an object, given its mass and volume, or the volume of an object, given its mass and density, or the mass of an object, given its volume and density.

14. Calculate the density of a substance from a graph showing its mass as a function of its volume.

■ Section Notes

3-1 Many students have learned rules for manipulating exponential numbers (e.g., to multiply, add exponents) but do not know what they mean. Occasionally ask students to write small exponential numbers as $10 \times 10 \times 10$ or $\frac{1}{10} \times \frac{1}{10} \times \frac{1}{10}$ to be sure they know what is meant.

3-5 If you have a computer expert in class, you might ask the student to discuss the use of binary and hexadecimal numbers in computers.

3-15 Discuss the graph shown in Figure 3-11 and others drawn by students. Be sure that they understand how uncertainty in measurement results in uncertainty about where the point falls on the graph.

Review and Practice _____
page 64

1. $5 \text{ kg} \times \dfrac{1\,000 \text{ g}}{1 \text{ kg}} \times \dfrac{57.142\,857 \text{ gr}}{1 \text{ g}}$
   $= 285\,714.29 \text{ gr}$

2. **a.** Yes. # drops/volume is always 12.
   b. 12 drops/mL and 0.0833 mL/drop
   c. 12 drops/mL

page 67

1. **a.** 1.24×10^6
   b. 1.28×10^3
   c. 1×10^{18}
   d. 1.24×10^{-4}
   e. 2.1×10^{-10}

2. **a.** 132 000
   b. 2 060
   c. 602 000 000 000 000 000 000 000
   d. 0.001 32
   e. 0.000 003 45

page 69

1. Find the product of the decimal portion of the numbers. Find the product of the exponential part by adding exponents. Write the answer as the product of the decimal portion times ten to the appropriate power.

2. Find the quotient of the decimal part of the numbers. Find the quotient of the exponential part by subtracting the exponent in the denominator from the exponent in the numerator. Write the answer as the product of the decimal portion and ten to the appropriate power.

3. **a.** 10^8 **d.** 10^3
   b. 10^3 **e.** 10^8
   c. 10^{-7}

4. **a.** 10^{-2} **d.** 10^{-3}
   b. 10^{-13} **e.** 10^{20}
   c. 10

5. 3.906×10^1

6. 2.898×10^{12}

7. 6.3048×10^3

8. 1.6445×10^{-4}

9. 3.2591×10^{-22}

10. 3.9975×10^3

11. 8.5×10^1

12. 1.1290×10^5

13. 6.9876×10^{-20}

14. 1.9542×10^{42}

15. 1.8333×10^2

16. 3.6455×10^{-29}

17. 2.6422×10^{-3}

18. 8.5×10^{-9}

page 70

1. **a.** 6×10^3 **c.** 9×10^{-3}
   b. 5×10^2 **d.** 3.4×10^4

e. 2.31×10^{-3}
f. 5.06×10^3

2. a. -2×10^3 **d.** 1.9×10^4
   b. 1×10^2 **e.** 2.09×10^{-3}
   c. 1×10^{-3} **f.** -3.94×10^3

3. a. 1.0214×10^2
   b. 6.6171×10^{-4}
   c. 1.1970×10^{15}
   d. -1.3441×10^9
   e. 1.3459×10^9
   f. 5.985×10^{-8}
   g. 7.315×10^{-8}
   h. 4.4222×10^{-16}
   i. 1×10^{-1}
   j. 5×10^1

page 73

1. Since the purpose of the exercise is to have students practice doing conversions using unitary rates, you will want to look at each student's solution. The numerical answer obtained for each problem will depend on the value the student gets from a reference book, and these will vary from one reference to another.

page 77

1. Probably the five is uncertain; i.e., the estimate is to the nearest 10 000 000 kilometers.

2. 1.50×10^8 km

3. a. 4   **d.** 3
   b. 4   **e.** 3
   c. 7   **f.** 5

4. a. 2.135×10^1
   b. 8.705
   c. $1.212\,000 \times 10^2$
   d. 8.23×10^{-4}
   e. 9.10×10^{-2}
   f. $3.800\,2 \times 10^4$

5. a. 9.10×10^1 mm
   b. $3.800\,2 \times 10^{-1}$ km
   c. 2.5×10^3 g
   d. 1.1×10^{-3} cm
   e. 1.3×10^{-2} g

6. a. 1.1×10^{19} kg The original mass reading from which all of the calculations were done was 7.0g. There should be no more than two significant digits in any calculation based on that measurement.
   b. 5.9×10^{15} cars
   c. 1×10^6 cars/person

page 80

1. a. 16   **c.** 812 m²   **e.** 2.53 cm/in
   b. 6.3 cm   **d.** 29.1 m³

2. No

3. a. does not represent quantity
   b. a length 6.3 times as long as the standard centimeter
   c. an area that could be coverd by 812 squares measuring one meter on a side
   d. a volume that could be filled by 29.1 cubes measuring one meter on a side
   e. a unitary rate indicating 2.53 centimeters are spanned for each inch spanned

4. a. 97.9   **c.** 6.1 or 6.2
   b. 37   **d.** 3232 or 3231

5. Round to the digit representing the largest place value in which uncertainly was found in any measurement added or subtracted.

page 83

1. (5) 39   (12) 1.1×10^5
   (6) 2.9×10^{12} (13) 7.0×10^{-20}
   (7) 6.30×10^3 (14) 1.95×10^{42}
   (8) 1.6×10^{-4} (15) 2×10^2
   (9) 3.26×10^{-22} (16) 3.65×10^{-29}
   (10) 4.00×10^3 (17) 2.64×10^{-3}
   (11) 85   (18) 8.5×10^{-9}

2. a. 37.6
   b. 1.6×10^{-3}
   c. 36
   d. 2.8×10^2
   e. 8.2×10^9

3. Yes. You should get heads half of the time and tails half of the time. The rule is inconvenient to use, because you need a coin and must take time to flip it.

page 88

1.

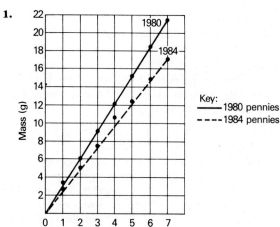

2. 4, 10.6

3. See graph for item 1.

page 89

1. The slope is 3.1 g/penny. It means that on the average the mass of 1980 pennies increases 3.1 g with each penny added.
2. The slope is 2.5 g/penny. It means that on the average the mass of 1984 pennies increases 2.5 g/penny with each penny added.
3. 1980 pennies have more mass than 1984 pennies. They do not have the same mass because they are made of different materials.

page 90

1. 2.17 g/cm^3; unitary rate
2. 45.6 g
3. 0.35 cm^3
4. the golf ball
5. The line would lie below the line for glass, i.e., its slope would be smaller because the density is smaller. The line should pass through the origin because plastic with no volume should have no mass.

Review, page 91

1. proportional
2. $A/B = k$
3. exponent; 4; base
4. 7
5. to make it easier to detect irregularities
6. g/cm^3; kg/m^3
7. precision, because they want to reduce the amount of uncertainty in the grades
8. accuracy, because there was a mistake in accepted values
9. a. 2.7×10^3
   b. 2.7×10^3; 27×10^2; 270×10^1
   c. 2700
10. 2.7×10^3 or 27×10^2
    The other expressions are unacceptable because they do not make clear that the measurement is uncertain in the second digit.
11. 0.014 is the relative uncertainty; absolute
12. 7 and 2; 5
13. A small slope means that the roof rises very little for its width; it is not steep. A large slope means that the roof rises a lot for its width; it is steep. To calculate the slope of the roof, measure the horizontal distance from the edge of the roof to a point directly beneath the ridge of the roof. Measure the distance from the ridge to this point. The distance from the point to the ridge divided by the horizontal distance to the edge is the slope of the roof.

14. A, c, e, and g can be considered to be extensive. The latter two can be debated. They do not depend on the amount of matter present, although they do depend on the amount of something else.
15. B, d, f, and h can be considered to be intensive. The latter two can be debated for the same reason given in question 14.
16. Neither I nor III lists data that are proportional. The data in II are directly proportional. The data in IV are inversely proportional.

Interpret and Apply, page 92

1–4. These questions require individual attention.
5. Probably not. The data represent measurements, and they contain uncertainty. The graph should be drawn to average out the uncertainty.
6. You could not simply look to see which looked steeper, because you may have used different scales. However, if you examined the scales carefully, you might be able to tell without actually doing a calculation.
7. There is uncertainty in both their measurements and their estimations.
8. There is uncertainty in measurement and fluctuations in conditions from one cooking session to another. For example, the relative humidity and the temperature of the room could affect some dishes.

Problems, page 923

1. 6; 10^{12}
2. 2.5 m in diameter; 2.5 km in diameter
3. 0.0012 mm; 0.0001 mm
4. 1.18×10^{-23} cm^3
5. 0.39 cm^3
6. 3.3×10^{22} atoms
7. a. 5.98×10^{-8} e. 3.4×10^2
   b. 3.26×10^{-5} f. 50
   c. 2.5×10^{17} g. 9×10^{-12}
   d. 5×10^{48} h. 3.9×10^5
8. a. infinite, defined
   b. 3
   c. infinite, defined
   d. 3
   e. infinite, defined
   f. 2
   g. 3
   h. infinite number, 22/7 is an infinite decimal
9. a. 5.4 g/cm^3 d. 5.3 kg/m^3
   b. 57 g   e. 1.1×10^2 cm^3
   c. 5.5×10^2 cm^3
10. a. 5.22
    b. 3

49T

c. 5.6   **f.** 62
d. 1.20×10^2 **g.** 3.1×10^3
e. 3.1   **h.** 5.000

11. All except 63, 119.56, 61.53, and 3081 are in scientific notation. $63 = 6.3 \times 10^1$
119.56 is 1.2×10^2.
61.53 is 6.2×10^1.
3.81 is 3.1×10^3.

1. 34.6   The average should reduce uncertainty.
2. 9 times; about 10^{13} km out in space
3. 4.2×10^8 cm³
4. 1.0×10^6 cm³

CHAPTER 4 The Mole

CHAPTER ORGANIZER

USE WITH SECTION	DEMONSTRATION	WORKSHEET	LAB EXPERIMENT	SOFTWARE	TEACHER'S RESOURCE BINDER
4-1	4-A, 4-B				Problems 1–2*
4-2			4-A		
4-3	4-C			Disk 4	Problems 3–4
4-4		4-A	4-B	Disk 4	Problems 5–6
4-5		4-A			
4-6		4-B		Disk 4	
4-7		4-C		Disk 5	
4-9			4-C	Disk 5	Problems 7–9
4-10				Disk 5	
4-11	4-D, 4-E	4-D		Disk 10	
4-12				Disk 10	Problems 10–12
Chapter Test					

*Use the *Problem-Solving Strategies* for students found in the Teacher's Resource Binder.

■ Objectives

When students have completed the chapter, they will be able to:

1. State the meaning of relative atomic mass and identify the standard for comparison.
2. Calculate the relative mass of several objects when given their actual mass and a relative mass for one of the objects.
3. State the number of particles in one mole.
4. Find the mass of one mole of a substance when given the formula for an element or compound.
5. Use the definition of a mole to obtain unit factors relating the mass of an element or compound to the number of particles present, or to the number of moles.
6. Convert the amount given to grams, moles, or molecules when given the formula and amount of a substance.
7. Differentiate between the number of atoms in one molecule of a diatomic element like fluorine (2) and the number of atoms in one mole of a diatomic element (1.2×10^{24}).
8. Find the empirical formula when given data for the composition of a compound.
9. Differentiate between a molecular formula and an empirical formula.
10. Calculate the percent composition of each element in a compound when given the formula of the compound and a table of atomic masses.
11. Use the definition of molarity as a unitary rate to obtain unit factors to calculate the mass or number of moles in a given volume of a solution of known concentration.

12. Describe how to prepare a given volume of a solution of a specified concentration.

■ Section Notes

4-1 The concept of relative mass is necessary for students to understand the meaning of the relative atomic masses of an atom, the mass of a mole, and the mass of one atom. The most meaningful knowledge is constructed by students themselves. Ask students to prepare a table of masses of equal numbers of objects and use these to determine the relative masses of each object like those given for fruit, atoms, and laboratory equipment. The number of each object need not be specified to determine the relative masses, but it must be known to find the mass of one object. Suggestions for groups:

A. kernels of rice, wheat, corn, peas
B. beads of plastic, glass, aluminum, copper, steel, lead
C. things typically bought by the dozen—eggs, doughnuts, oranges

To show how you find the mass of one atom or molecule, ask the students to include the numbers of each object that were used in their group. For example, in Group B possible data are:

OBJECT	MASS OF X OBJECTS (g)
plastic	1.31
glass	2.62
aluminum	2.70
steel	7.85
copper	8.96
lead	11.40

You can find the mass of one bead of any object. If the number of beads was twenty, each group may be called a score. If the number of beads was twelve, the group would be called a dozen.

4-2 The film "Gases and How They Combine" could be shown when studying this section. The film shows that the combining ratios of several gases is that of small whole numbers. The film uses the concepts of characteristic properties, uncertainty in measurements, and significant digits from Chapters 1, 2, and 3. It also addresses the question of how you can determine when a number is whole from Section 4-8.

4-3 The number of molecules in one mole was first estimated using the data on Brownian motion collected by Jean B. Perrin in 1908, and the equation Albert Einstein developed in 1905 to explain Brownian motion. Since then Avogadro's number has been determined by five completely independent methods and all give a similar value within experimental uncertainty. The most precise value for Avogadro's number, 6.022094×10^{23}, was determined by scientists at the National Bureau of Standards using the mass, density, and atomic size of silicon atoms. The size of the silicon atom was measured precisely using X-ray diffraction. Use this analogy to show students how Avogadro's number may be calculated.

Draw a box on the chalkboard and label the volume 2000 cm³. Draw several cubes in the box. Ask the students how many cubes are in the box if each cube has a volume of 5 cm³. Answer: 400 cubes. This shows that if you know the volume of the box and the volume of each cube, you can calculate the number of cubes in the box.

$$\text{Number of cubes} = \frac{\text{volume of box}}{\text{volume of each cube}}$$

This analogy can be used to calculate the number of atoms in one mole.

$$\frac{\text{Number of atoms}}{\text{in one mole}} = \frac{\text{volume of one mole}}{\text{volume of one atom}}$$

The volume of one mole of copper can be calculated from its density which is 8.96 g/cm³. For example:

$$\frac{63.5 \text{ g Cu}}{1 \text{ mol Cu}} \times \frac{1 \text{ cm}^3 \text{ Cu}}{8.96 \text{ g Cu}} = 7.09 \text{ cm}^3 \text{ Cu/mol Cu}$$

The copper atoms are 0.256 nanometers apart as shown by X-ray diffraction experiments. If the diameter of a copper atom is 0.256 nm, the volume of an atom can be calculated:

$$0.256 \text{ nm} \times \frac{1 \text{ cm}^3}{10^7 \text{ nm}} = 2.56 \times 10^{-8} \text{ cm}^3$$

If we assume the spheres occupy the volume of a cube whose edge is equal to the diameter of the sphere, the volume of the cube is:

$$V \text{ cube} = (2.56 \times 10^{-8} \text{ cm})^3$$
$$V \text{ cube} = 1.68 \times 10^{-23} \text{ cm}^3$$

The number of cubes in one mole

$$= \frac{7.09 \text{ cm}^3/\text{mol}}{1.68 \times 10^{-23} \text{ cm}^3/\text{atom}}$$
$$= 4.22 \times 10^{23} \text{ atoms/mol}$$

This is less than the accepted value for Avogadro's number because we have assumed the atoms are cubes.

If we assume the copper atoms are spheres of diameter 2.56×10^{-8} cm and radius of 1.28×10^{-8} cm, the volume of an atom is:

$$V \text{ atom} = \frac{4\pi r^3}{3} = \frac{4(3.14)(1.28 \times 10^{-8} \text{ cm})^3}{3}$$
$$= 8.78 \times 10^{-24} \text{ cm}^3$$

The number of atoms in one mole

$$= \frac{7.09 \text{ cm}^3/\text{mol}}{8.78 \times 10^{-24} \text{ cm}^3/\text{atom}}$$
$$= 8.08 \times 10^{23} \text{ atom/mol}$$

This value is slightly larger than 6.02×10^{23} because it does not consider the way in which the atoms are stacked in the solid crystal lattice. Packing arrangement A requires more space than B.

Have the students count the number of drops of water in 5.00 mL. The number of drops will be 100 ± 10. Calculate the number of drops in 1 mL and the number of mL in 1 drop. Use these unit factors to calculate the number of drops of water in:

a. 50 mL
b. 50 cm³
c. a tray of water 50 cm long, 30 cm wide, and 10 cm deep
d. a tank of water 15 m long, 10 m wide, and 1 m deep
e. a layer of water in a room 12 m long, 8 m wide, and 2 m deep
f. a layer of water 10 cm deep covering:
   1. the school parking lot
   2. a football field
   3. your town or city
   4. your county
   5. your state

Assign one problem each day and write the solution on the chalkboard each day. For example:

a. $50 \;\cancel{mL} \;\cancel{H_2O} \times \dfrac{20 \text{ drops } H_2O}{1 \;\cancel{mL} \;\cancel{H_2O}}$

$= 1.0 \times 10^3$ drops H_2O

b. $500 \;\cancel{cm^3} \;\cancel{H_2O} \times \dfrac{1 \;\cancel{mL}}{1 \;\cancel{cm^3}} \times \dfrac{20 \text{ drops } H_2O}{1 \;\cancel{mL} \;\cancel{H_2O}}$

$= 1.0 \times 10^3$ drops

c. Volume of water $= 50 \text{ cm} \times 30 \text{ cm} \times 10 \text{ cm}$

$= 1.5 \times 10^4 \text{ cm}^3$

$1.5 \times 10^4 \;\cancel{cm^3} \times \dfrac{20 \text{ drops } H_2O}{1 \;\cancel{cm^3} \;\cancel{H_2O}}$

$= 3.0 \times 10^5$ drops H_2O

This is equivalent to:

$50 \;\cancel{cm} \times 30 \;\cancel{cm} \times 10 \text{ cm} \times \dfrac{1 \;\cancel{mL}}{1 \;\cancel{cm^3}} \times \dfrac{20 \text{ drops}}{1 \;\cancel{mL}}$

$= 3.0 \times 10^5$ drops H_2O

d. Volume of water $= 15 \text{ m} \times 10 \text{ m} \times 1 \text{ m} = 150 \text{ m}^3$

$150 \;\cancel{m^3} \;\cancel{H_2O} \times \dfrac{10^6 \;\cancel{cm^3}}{1 \;\cancel{m^3}} \times \dfrac{20 \text{ drops } H_2O}{1 \;\cancel{cm^3} \;\cancel{H_2O}}$

$= 3.0 \times 10^9$ drops H_2O

Some students may understand the process better using this method:

length $15 \;\cancel{m} \times \dfrac{100 \text{ cm}}{1 \;\cancel{m}} = 1.5 \times 10^3 \text{ cm}$

width $10 \;\cancel{m} \times \dfrac{100 \text{ cm}}{1 \;\cancel{m}} = 1.0 \times 10^3 \text{ cm}$

height $1 \;\cancel{m} \times \dfrac{100 \text{ cm}}{1 \;\cancel{m}} = 1.0 \times 10^2 \text{ cm}$

Volume of water:

$1.5 \times 10^3 \text{ cm} \times 1.0 \times 10^3 \text{ cm} \times 1.0 \times 10^2 \text{ cm}$

$= 1.5 \times 10^8 \text{ cm}^3$

$1.5 \times 10^8 \;\cancel{cm^3} \;\cancel{H_2O} \times \dfrac{20 \text{ drops } H_2O}{1 \;\cancel{cm^3} \;\cancel{H_2O}}$

$= 3.0 \times 10^9$ drops H_2O

e. $12 \;\cancel{m} \times 8 \;\cancel{m} \times 2 \;\cancel{m} \; H_2O \times \dfrac{10^6 \;\cancel{cm^3}}{1 \;\cancel{m^3}}$

$\times \dfrac{20 \text{ drops } H_2O}{1 \;\cancel{cm^3} \;\cancel{H_2O}} = 1.9 \times 10^9$ drops H_2O

This is the number of drops in an average size chemistry classroom with the water 2 meters deep.

f. For the remaining problems have the students estimate the size of the parking lot, football field, their town, etc. The area of the town, county, or state can be obtained from a reference book for comparison. This is a good opportunity to discuss uncertainty in measurements, estimated uncertainty, and how you can determine when two measured or derived quantities are the same within experimental uncertainty.

These exercises can be used to develop the student's skill in:

1. making estimates in the metric system
2. using the factor-label method to solve problems
3. associating a number of particles with a way of calculating that number
4. estimating the number of significant digits in a calculated value.

4-4 Review the display of bottles containing one mole. Choose a different reagent and ask students to calculate the amount required for one mole. The number of molecules will be 6.02×10^{23} and the mass in grams will be the molar mass. The volume can be found if the density is known.

4-5 It is helpful to use the dozens, wheels, bicycles, cars' analogy in teaching the number of atoms of fluorine in one mole.

How many wheels are there in 4 dozen cars?

$4 \;\cancel{\text{dozen cars}} \times \dfrac{12 \text{ cars}}{1 \;\cancel{\text{dozen cars}}} = 48 \text{ cars}$

$48 \;\cancel{\text{cars}} \times \dfrac{4 \text{ wheels}}{1 \;\cancel{\text{car}}} = 192 \text{ wheels}$

or

$4 \;\cancel{\text{dozen cars}} \times \dfrac{12 \;\cancel{\text{cars}}}{1 \;\cancel{\text{dozen cars}}} \times \dfrac{4 \text{ wheels}}{1 \;\cancel{\text{car}}}$

$= 192 \text{ wheels}$

How many fluorine atoms are in 4 moles of fluorine?

$$4 \text{ moles of } F_2 \text{ molecules} \times \frac{6.02 \times 10^{23} \text{ molecules}}{1 \text{ mol } F_2}$$

$$= 2.408 \times 10^{24} \text{ molecules } F_2$$

$$\frac{2.408 \times 10^{24} \text{ molecules } F_2}{L} \times \frac{2 \text{ atoms } F}{1 \text{ molecule } F_2}$$

$$= 4.816 \times 10^{24} \text{ atoms } F$$

The mass of one atom is found by using Avogadro's number:

$$\frac{19.0 \text{ g } F}{1 \text{ mol } F \text{ atoms}} \times \frac{1 \text{ mol } F \text{ atoms}}{6.02 \times 10^{23} \text{ atoms}}$$

$$= \frac{3.16 \times 10^{23} \text{ g } F}{1 \text{ atom}}$$

4-6 Begin class by asking each student to give a unit factor for some substance. When you call the roll the students could answer with a unit factor.

TEACHER: "Eric Kinsey"

ERIC: $\dfrac{\text{"1 mol } C_{12}H_{22}O_{11}}{342 \text{ g } C_{12}H_{22}O_{11}\text{"}}$

TEACHER: "Jan Marlowe"

JAN: $\dfrac{\text{"6.02} \times 10^{23} \text{ molecules } K_2SO_4}{1 \text{ mol } K_2SO_4\text{"}}$

TEACHER: "Christine Marasky"

CHRISTINE: $\dfrac{\text{"98 g } H_3PO_4}{6.02 \times 10^{23} \text{ molecules } H_3PO_4\text{"}}$

You may find students will create responses.

KATE MOORE: $\dfrac{\text{"1 mol } C_{254}H_{377}N_{65}O_{75}S_6}{5.73 \times 10^3 \text{ g"}}$

This compound is one of the known enzymes.

4-9 Molecular units exist for substances in which the atoms in the molecules are joined by covalent bonds. The molecules are then bonded to each other by London forces or dipole forces that exist if the molecules are permanent dipoles. The London forces are very weak so these molecules tend to be gases at room temperature unless the molecules are quite large (molecular mass greater than 100). The differences between these types of molecules and ionic substances will be described in Chapter 15. Some substances do not exist as individual molecules. The formula $MgCl_2$ expresses the simplest ratio of the atoms and not a molecule.

4-10 The use of percent to express proportions is pervasive in daily life. Concentrations of solutions can be expressed as percent as well as by molarity. Molarity is more useful to chemists because it provides a knowledge of the number of molecules of solute in a given volume of solution.

Review and Practice

page 104

1. $\dfrac{12 \text{ amu}}{1 \text{ C atom}} \times \dfrac{2 \text{ C atoms}}{1 \text{ Mg atom}} = 24 \text{ amu/Mg atom}$

2. $\dfrac{20.00 \text{ g Ar}}{6.00 \text{ g C}} = 3.33 \text{ g Ar/g C}$

$\dfrac{3.33 \text{ g Ar}}{\text{g C}} \times 12.0 \text{ g C} = 40 \text{ g Ar}$

3. **MASSES OF LABORATORY EQUIPMENT**

OBJECT	ACTUAL MASS	RELATIVE MASS (divide each mass by 13.9)
stirring rod	13.9 g	1.00
100-mL beaker	50.2 g	3.61
250-mL beaker	74.8 g	5.38
evaporating dish	23.3 g	1.68
50-mL graduate	65.6 g	4.72

4. $1.25 \text{ mol } H_2SO_4 \times \dfrac{6.02 \times 10^{23} \text{ molecules } H_2SO_4}{1 \text{ mol } H_2SO_4}$

$= 7.52 \times 10^{23} \text{ molecules } H_2SO_4$

5. $1.81 \times 10^{25} \text{ molecules HCl}$

$\times \dfrac{1 \text{ mol HCl}}{6.02 \times 10^{23} \text{ molecules HCl}} = 30.1 \text{ mol HCl}$

6. $1.25 \text{ mol } SO_3 \times \dfrac{6.02 \times 10^{23} \text{ molecules } SO_3}{1 \text{ mol } SO_3}$

$\times \dfrac{3 \text{ atoms O}}{1 \text{ molecule } SO_3} = 2.26 \times 10^{24} \text{ O atoms}$

7. $1.2 \times 10^{25} \text{ } P_2O_5 \text{ molecules} \times \dfrac{5 \text{ O atoms}}{1 \text{ } P_2O_5 \text{ molecule}}$

$\times \dfrac{1 \text{ mol O}}{6.02 \times 10^{23} \text{ atoms O}} = 1.0 \times 10^2 \text{ mol O}$

8. $5.33 \text{ mol } FeCl_3 \times \dfrac{6.02 \times 10^{23} \text{ molecules } FeCl_3}{1 \text{ mol } FeCl_3}$

$\times \dfrac{1 \text{ atom Fe}}{1 \text{ molecule } FeCl_3} = 3.21 \times 10^{24} \text{ atom Fe}$

9. $5.33 \text{ mol } FeCl_3 \times \dfrac{6.02 \times 10^{23} \text{ molecules } FeCl_3}{1 \text{ mol } FeCl_3}$

$\times \dfrac{3 \text{ atoms Cl}}{1 \text{ molecule } FeCl_3} = 9.63 \times 10^{24} \text{ atoms Cl}$

page 109

1. Answers will vary. Some possible responses are:

$\dfrac{1 \text{ mol } H_2O}{6.02 \times 10^{23} \text{ molecules } H_2O}$, $\dfrac{1 \text{ mol } H_2O}{18.0 \text{ g } H_2O}$,

$\dfrac{18.0 \text{ g } H_2O}{6.02 \times 10^{23} \text{ molecules } H_2O}$

2. Answers will vary. Some possible responses are:

$\dfrac{1 \text{ mol } MnO_2}{86.9 \text{ g } MnO_2}$, $\dfrac{1 \text{ mol } MnO_2}{6.02 \times 10^{23} \text{ molecules } MnO_2}$,

$\dfrac{86.9 \text{ g } MnO_2}{6.02 \times 10^{23} \text{ molecules } MnO_2}$

3. Answers will vary. Some possible responses are:

$$\frac{142 \text{ g } Na_2SO_4}{1 \text{ mol } Na_2SO_4}, \quad \frac{6.02 \times 10^{23} \text{ molecules } Na_2SO_4}{1 \text{ mol } Na_2SO_4}$$

$$\frac{142 \text{ g } Na_2SO_4}{6.02 \times 10^{23} \text{ molecules } Na_2SO_4}$$

4. $3.15 \text{ mol } MnO_2 \times \dfrac{6.02 \times 10^{24} \text{ molecules } MnO_2}{1 \text{ mol } MnO_2}$

$= 1.90 \times 10^{24} \text{ molecules } MnO_2$

5. $3.15 \text{ mol } MnO_2 \times \dfrac{6.02 \times 10^{23} \text{ molecules } MnO_2}{1 \text{ mol } MnO_2}$

$\times \dfrac{2 \text{ O atoms}}{1 \text{ molecule } MnO_2} = 3.80 \times 10^{24} \text{ atoms O}$

6. Na_2SO_4

2 atoms Na $\times$ 23.0 g = 46.0 g
1 atom S $\times$ 32.1 g = 32.1 g
4 atoms O $\times$ 16.0 g = 64.0 g
_____
        Total   142.1 g = 1 mol Na_2SO_4

This can also be written: $2(23.0 \text{ g}) + 1(32.1 \text{ g})$
$+ 4(16.0 \text{ g}) = 142.1 \text{ g}$

7. $53 \text{ g } Na_2CO_3 \times \dfrac{1 \text{ mol } Na_2CO_3}{106 \text{ g } Na_2CO_3} = 0.50 \text{ mol } Na_2CO_3$

8. 1 mol Pb atoms = 207.2 g

$10 \text{ mol } H_2O \text{ molecules} \times \dfrac{18.0 \text{ g } H_2O}{1 \text{ mol } H_2O} = 180 \text{ g } H_2O$

Therefore, 1 mole of lead atoms has a greater mass.

9. $1.31 \text{ mol } AgNO_3 \times \dfrac{169.9 \text{ g } AgNO_3}{1 \text{ mol } AgNO_3} = 223 \text{ g } AgNO_3$

10. a. CO   12.0 g + 16.0 g = 28.0 g
b. C_8H_{18} 8(12.0 g) + 18(1.01 g) = 114.2 g
c. H_2SO_4 2(1.01 g) + 1(32.1 g) + 4(16.0 g) = 98.1 g
d. CS_2 1(12.0 g) + 2(32.1 g) = 76.2 g

11. a. $CaCO_3$ 1(40.1 g) + 1(12.0 g) + 3(16.0 g)
           = 100.1 g
b. $KMnO_4$ 1(39.1 g) + 1(54.9 g) + 4(16.0 g)
           = 158.0 g
c. MnO_2 1(54.9 g) + 2(16.0 g) = 86.9 g
d. $Ca_3(PO_4)_2$ 3(40.1 g) + 2(31.0 g) + 8(16.0 g)
           = 310.2 g

12. $124 \text{ g } C_8H_{18} \times \dfrac{1 \text{ mol } C_8H_{18}}{114.2 \text{ g } C_8H_{18}} = 1.08 \text{ mol } C_8H_{18}$

13. $56 \text{ g } CS_2 \times \dfrac{1 \text{ mol } CS_2}{76.1 \text{ g } CS_2} = 0.74 \text{ mol } CS_2$

14. $2.50 \text{ mol } H_2SO_4 \times \dfrac{98.1 \text{ g } H_2SO_4}{1 \text{ mol } H_2SO_4} = 245 \text{ g } H_2SO_4$

15. $2.50 \text{ mol } H_2SO_4 \times \dfrac{6.02 \times 10^{23} \text{ molecules } H_2SO_4}{1 \text{ mol } H_2SO_4}$

$= 1.50 \times 10^{24} \text{ molecules } H_2SO_4$

16. $245 \text{ g } H_2SO_4 \times \dfrac{6.02 \times 10^{23} \text{ molecules } H_2SO_4}{98.1 \text{ g } H_2SO_4}$

$= 1.50 \times 10^{24} \text{ molecules } H_2SO_4$

page 120

1. $4.20 \text{ g N} \times \dfrac{1 \text{ mol N}}{14.0 \text{ g N}} = 0.300 \text{ mol N}$

$12.0 \text{ g O} \times \dfrac{1 \text{ mol O}}{16.0 \text{ g O}} = 0.750 \text{ mol O}$

$\dfrac{0.750 \text{ mol O}}{0.300 \text{ mol N}} = \dfrac{2.50 \text{ mol O}}{1.00 \text{ mol N}} \times \dfrac{2}{2} = \dfrac{5.00 \text{ mol O}}{2.00 \text{ mol N}}$

Therefore, the formula is N_2O_5.

2. $4.80 \text{ g C} \times \dfrac{1 \text{ mol C}}{12.0 \text{ g C}} = 0.400 \text{ mol C}$

$0.40 \text{ g H} \times \dfrac{1 \text{ mol H}}{1.01 \text{ g H}} = 0.40 \text{ mol H}$

Therefore, the empirical formula is CH.

3. The molar mass for CH is 12 g + 1 g = 13 g

$(CH)_n = 78$
$(13)_n = 78$
$n = 6$

The molecular formula is C_6H_6 which is the formula for benzene.

4. $\% \text{ C} = \dfrac{\text{mass of carbon}}{\text{mass of compound}} = \dfrac{43.2 \text{ g}}{159.0 \text{ g}} = 27.2\%$

$\% \text{ O} = \dfrac{\text{mass of oxygen}}{\text{mass of compound}} = \dfrac{115.8 \text{ g}}{159.0 \text{ g}} = 72.8\%$

Example 4-18 gave 27.3% for carbon and 72.7% for oxygen. These are the same within experimental uncertainty.

5. Molar mass of $CS_2 = 1(12.0 \text{ g}) + 2(32.1 \text{ g}) = 76.2 \text{ g}$

$\% \text{ C} = \dfrac{12.0 \text{ g}}{76.2 \text{ g}} = 15.7\%$ $\% \text{ S} = \dfrac{64.2 \text{ g}}{76.2 \text{ g}} = 84.3\%$

6. Molar mass of $CaCO_3 = 1(40.1 \text{ g}) + 1(12.0 \text{ g})$
$+ 3(16.0 \text{ g}) = 100.1 \text{ g}$

$\% \text{ Ca} = \dfrac{40.1 \text{ g}}{100.1 \text{ g}} = 40.0\%$

$\% \text{ C} = \dfrac{12.0 \text{ g}}{100.1 \text{ g}} = 12.0\%$

$\% \text{ O} = \dfrac{48.0 \text{ g}}{100.1 \text{ g}} = 48.0\%$ 47.96

7. $1.00 \text{ L } H_2SO_4(aq) \times \dfrac{1.55 \text{ mol } H_2SO_4}{1 \text{ L } H_2SO_4(aq)}$

$= 1.55 \text{ mol } H_2SO_4$

8. $1.00 \text{ L } H_2SO_4(aq) \times \dfrac{1.55 \text{ mol } H_2SO_4}{1 \text{ L } H_2SO_4(aq)}$

$\times \dfrac{98.1 \text{ g } H_2SO_4}{1 \text{ mol } H_2SO_4} = 152 \text{ g } H_2SO_4$

9.

VOLUME OF SOLUTION (L)	MOLES OF NaCl	MOLES/VOLUME
0.13	0.63	4.8
0.34	1.67	4.9
0.46	2.25	4.9
0.58	2.84	4.9
0.82	4.00	4.9

All are the same within experimental uncertainty.

10. Solutions of equal molarity contain the same number of solute molecules per liter.

Review, page 121

1. 58.9 g

2. One atom of sulfur is 32.1 amu which is larger than one atom of aluminum that has a mass of 27.0 amu.

3. One mole of molybdenum = 95.9 g.

4. Since the volumes are equal at that temperature and pressure, both contain equal numbers of molecules.

5. One mole neon atoms is 6.02×10^{23} neon atoms.

6. One mole nickel atoms is 6.02×10^{23} nickel atoms.

7. One mole silver cyanide molecules is 6.02×10^{23} silver cyanide molecules.

8. Each diatomic element has two atoms in one molecule.

9. The diatomic elements are hydrogen, H_2; nitrogen, N_2; oxygen, O_2; fluorine, F_2; chlorine, Cl_2; bromine, Br_2; iodine, I_2; and astatine, At_2.

10. P_4, As_4

11. The empirical formula is CH. The molecular formula is C_6H_6.

Interpret and Apply, page 121

1. Equal volumes of gases at the same temperature and pressure contain equal numbers of molecules, so the number of nitrogen molecules is 3.2×10^{23}.

2. Hydrogen was selected as a standard because it had the smallest mass of any known element.

3. The relative atomic mass gives the ratio of the masses of single atoms. Hydrogen is diatomic so the flask that contains hydrogen contains the same number of molecules as the flask containing helium; but there are twice as many atoms.

4. Each rmu would be $\frac{3}{12}$ or $\frac{1}{4}$ of the previous value.

5. Hydrogen is 49% of the atoms. $\frac{22}{45} = 49\%$

6. Oxygen is 51% of the mass.

$$\% \text{ O} = 11(16 \text{ g}) = 176 \text{ g} \qquad \frac{176 \text{ g}}{342 \text{ g}} = 51\%$$

7. The MgO molar mass is $24.3 + 16.0 = 40.3$.

$$\% \text{ Mg} = \frac{24.3 \text{ g}}{40.3 \text{ g}} = 60.3\%$$

$$\% \text{O} = \frac{16.0 \text{ g}}{40.3 \text{ g}} = 39.7\%$$

The % of magnesium is larger.

8. The CH_4 molar mass is $12.0 + 4(1.01) = 16.0$.

$$\% \text{ C} = \frac{12.0 \text{ g}}{16.0 \text{ g}} = 75.0\%$$

$$\% \text{ H} = \frac{4.0 \text{ g}}{16.0 \text{ g}} = 25.0\%$$

Carbon is the largest % by mass.

Problems, page 122

1.

OBJECT	MASS IN GRAMS	RELATIVE MASS (divide all masses by 50.2)
stirring rod	13.9	0.277
100-mL beaker	50.2	1.00
250-mL flask	74.8	1.49
evaporating dish	23.3	0.464
50-mL graduate	65.6	1.31

2. $\dfrac{95.94 \text{ amu}}{1 \text{ molybdenum atom}} \times \dfrac{1 \text{ carbon atom}}{12.0 \text{ amu}}$
   $= 8.00$ carbon atoms

3.

GAS	MASS OF 300 L (g)	MASS OF GAS RELATIVE TO HYDROGEN (divide all masses by 0.24)
hydrogen	0.24	1.0
nitrogen	3.36	14.0
oxygen	3.93	16.4
fluorine	4.56	19.0
chlorine	8.40	35.0

4. $\dfrac{6.02 \times 10^{23} \text{ pennies}}{5 \times 10^9 \text{ people}} \times \dfrac{1 \text{ dollar}}{100 \text{ pennies}}$
   $= 1.2 \times 10^{12}$ dollars/person

This is more than a trillion dollars per person.

5. Area of Earth (sphere) $= 4\pi r^2$

$$= 4(3.14)(6.4 \times 10^6 \text{ m})^2$$
$$= 5.1 \times 10^{14} \text{ m}^2$$

$$5.1 \times 10^{14} \text{ m}^2 \times \frac{10^4 \text{ cm}^2}{1 \text{ m}^2} = 5.1 \times 10^{18} \text{ cm}^2$$

$$\frac{5.1 \times 10^{18} \text{ cm}^2}{6.02 \times 10^{23} \text{ people}} = 8.5 \times 10^{-6} \text{ cm}^2$$

This is an impossibly small area. A dot this size would not be visible to the unaided eye.

6. $(NH_4)_2CrO_4$

$2(14.0 \text{ g}) + 8(1.01 \text{ g}) + 1(52.0 \text{ g}) + 4(16.0 \text{ g}) = 152.0 \text{ g}$

7. $CoCl_2$

$1(58.9 \text{ g}) + 2(35.4 \text{ g}) = 129.7 \text{ g}$

8. **a.** K_2SO_4 $2(39.1 \text{ g}) + 1(32.1 \text{ g}) + 4(16.0 \text{ g})$
$= 174.3 \text{ g}$

b. H_3PO_4 $3(1.01 \text{ g}) + 1(31.0 \text{ g}) + 4(16.0 \text{ g}) = 98.0 \text{ g}$

c. NH_4Cl $1(14.0 \text{ g}) + 4(1.01 \text{ g})$
$+ 1(35.4 \text{ g}) = 53.4 \text{ g}$

d. Na_3PO_4 $3(23.0 \text{ g}) + 1(31.0 \text{ g}) + 4(16.0 \text{ g})$
$= 164.0 \text{ g}$

e. $Ni(CN)_2$ $1(58.7 \text{ g}) + 2(12.0 \text{ g}) + 2(14.0 \text{ g})$
$= 110.7 \text{ g}$

f. KH_2PO_4 $1(39.1 \text{ g}) + 2(1.01 \text{ g}) + 1(31.0 \text{ g})$
$+ 4(16.0 \text{ g}) = 136.1 \text{ g}$

g. $FeSO_4$ $1(55.8 \text{ g}) + 1(32.1 \text{ g}) + 4(16.0 \text{ g}) = 151.9 \text{ g}$

h. $Fe_2(SO_4)_3$ $2(55.8 \text{ g}) + 3(32.1 \text{ g})$
$+ 12(16.0 \text{ g}) = 399.9 \text{ g}$

i. Cu_2CO_3 $2(63.5 \text{ g}) + 1(12.0 \text{ g}) + 3(16.0 \text{ g})$
$= 187.0 \text{ g}$

j. $CuCO_3$ $1(63.5 \text{ g}) + 1(12.0 \text{ g}) + 3(16.0 \text{ g})$
$= 123.5 \text{ g}$

k. N_2O_3 $2(14.0 \text{ g}) + 3(16.0 \text{ g}) = 76.0 \text{ g}$

l. Al_2S_3 $2(27.0 \text{ g}) + 3(32.1 \text{ g}) = 150.3 \text{ g}$

9. a. $AgCl$ $1(107.9 \text{ g}) + 1(35.4 \text{ g}) = 143.3 \text{ g}$

b. $MgCrO_4$ $1(24.3 \text{ g}) + 1(52.0 \text{ g}) + 4(16.0 \text{ g})$
$= 140.3 \text{ g}$

c. K_2CrO_4 $2(39.1 \text{ g}) + 1(52.0 \text{ g}) + 4(16.0 \text{ g})$
$= 194.2 \text{ g}$

d. $Fe(CN)_2$ $1(55.8 \text{ g}) + 2(12.0 \text{ g}) = 2(14.0 \text{ g})$
$= 107.8 \text{ g}$

e. $CuSO_4$ $1(63.5 \text{ g}) + 1(32.1 \text{ g}) + 4(16.0 \text{ g})$
$= 159.6 \text{ g}$

f. CrO_3 $1(52.0 \text{ g}) + 3(16.0 \text{ g}) = 100.0 \text{ g}$

g. Na_2S $2(23.0 \text{ g}) + 1(32.1 \text{ g}) = 78.1 \text{ g}$

h. CCl_4 $1(12.0 \text{ g}) + 4(35.4 \text{ g}) = 153.6 \text{ g}$

10. F_2 $2(19.0 \text{ g}) = 38.0 \text{ g}$

11. $2.5 \text{ mol } H_2 \times \dfrac{2.02 \text{ g } H_2}{1 \text{ mol } H_2} = 5.0 \text{ g } H_2$

12. a. $\dfrac{80 \text{ g A}}{2 \text{ mol A}} = \dfrac{40 \text{ g A}}{1 \text{ mol A}}$

b. $\dfrac{9.0 \text{ g B}}{3.01 \times 10^{23} \text{ molecules B}}$
$\times \dfrac{6.02 \times 10^{23} \text{ molecules B}}{1 \text{ mol B}} = 18.0 \text{ g/mol B}$

c. $\dfrac{7.56 \text{ g C}}{1.57 \times 10^{23} \text{ molecules C}}$
$\times \dfrac{6.02 \times 10^{23} \text{ molecules C}}{1 \text{ mol C}} = 29.0 \text{ g/mol C}$

d. $\dfrac{1.35 \text{ g D}}{4.55 \times 10^{22} \text{ molecules D}}$
$\times \dfrac{6.02 \times 10^{23} \text{ molecules D}}{1 \text{ mol D}} = 17.9 \text{ g/mol D}$

13. a. $\dfrac{22 \text{ g Mn} \times 1 \text{ mol Mn}}{54.9 \text{ g Mn}} = 0.40 \text{ mol Mn}$

b. $14.0 \text{ g } N_2 \times \dfrac{6.02 \times 10^{23} \text{ molecules } N_2}{28.0 \text{ g } N_2}$
$= 3.01 \times 10^{23} \text{ molecules } N_2$

c. $17.1 \text{ g } Al_2(SO_4)_3$
$\times \dfrac{6.02 \times 10^{23} \text{ molecules } Al_2(SO_4)_3}{342.3 \text{ g } Al_2(SO_4)_3}$
$= 3.01 \times 10^{22} \text{ formula units}$

d. $16.0 \text{ g Ag} \times \dfrac{1 \text{ mol Ag}}{107.9 \text{ g Ag}} = 0.148 \text{ mol Ag}$

e. $17.5 \text{ g Cu} \times \dfrac{1 \text{ mol Cu}}{63.5 \text{ g Cu}} = 0.276 \text{ mol Cu}$

f. $\dfrac{27.0 \text{ g Al}}{6.02 \times 10^{23} \text{ atoms Al}} = 4.48 \times 10^{-23} \text{ g}$

g. $17.1 \text{ g } C_{12}H_{22}O_{11}$
$\times \dfrac{6.02 \times 10^{23} \text{ molecules } C_{12}H_{22}O_{11}}{342 \text{ g } C_{12}H_{22}O_{11}}$
$\times \dfrac{12 \text{ atoms C}}{1 \text{ molecule of } C_{12}H_{22}O_{11}}$
$= 3.61 \times 10^{23} \text{ atoms C}$

14. a. $28 \text{ g Na} \times \dfrac{6.02 \times 10^{23} \text{ atoms Na}}{23.0 \text{ g Na}}$
$= 7.3 \times 10^{23} \text{ atoms Na}$

$28 \text{ g Na} \times \dfrac{1 \text{ mol Na}}{23 \text{ g Na}} = 1.2 \text{ mol Na}$

b. $28 \text{ g Fe} \times \dfrac{6.02 \times 10^{23} \text{ atoms Fe}}{55.8 \text{ g Fe}}$
$= 3.0 \times 10^{23} \text{ atoms Fe}$

$28 \text{ g Fe} \times \dfrac{1 \text{ mol Fe}}{55.8 \text{ g Fe}} = 0.50 \text{ mol Fe}$

c. $150 \text{ g Zn} \times \dfrac{6.02 \times 10^{23} \text{ atoms Zn}}{65.4 \text{ g Zn}}$
$= 1.4 \times 10^{24} \text{ atoms Zn}$

$150 \text{ g Zn} \times \dfrac{1 \text{ mol Zn}}{65.4 \text{ g Zn}} = 2.3 \text{ mol Zn}$

d. $2.4 \text{ g Ca} \times \dfrac{6.02 \times 10^{23} \text{ atoms Ca}}{40.1 \text{ g Ca}}$
$= 3.6 \times 10^{22} \text{ atoms Ca}$

$2.4 \text{ g Ca} \times \dfrac{1 \text{ mol Ca}}{40.1 \text{ g Ca}} = 0.06 \text{ mol Ca}$

e. $150 \text{ g } Cl_2 \times \dfrac{1.204 \times 10^{24} \text{ atoms Cl}}{70.8 \text{ g } Cl_2}$
$= 2.55 \times 10^{24} \text{ atoms Cl}$

$2.5 \times 10^{24} \text{ atoms Cl} \times \dfrac{1 \text{ mol Cl}}{6.02 \times 10^{23} \text{ atoms Cl}}$
$= 4.2 \text{ mol Cl}$

f. $21 \text{ g } F_2 \times \dfrac{1.2 \times 10^{24} \text{ atoms F}}{38.0 \text{ g } F_2}$
$= 6.6 \times 10^{23} \text{ atoms F}$

$6.6 \times 10^{23} \text{ atoms F} \times \dfrac{1 \text{ mol F}}{6.02 \times 10^{23} \text{ atoms F}}$
$= 1.1 \text{ mol F}$

15. $1 \times 10^{-8} \text{ g Au} \times \dfrac{6.02 \times 10^{23} \text{ atoms Au}}{197 \text{ g Au}}$
$= 3 \times 10^{13} \text{ atoms Au or } 30\,000\,000\,000\,000$
This is thirty trillion atoms.

16. $2.5 \text{ g Cu} \times \dfrac{6.02 \times 10^{23} \text{ atoms Cu}}{63.5 \text{ g Cu}}$
$= 2.4 \times 10^{22} \text{ atoms Cu}$

17. $2.5 \text{ g H} \times \dfrac{1 \text{ mol H}}{1.01 \text{ g H}} = 2.5 \text{ mol H}$

$30 \text{ g C} \times \dfrac{1 \text{ mol C}}{12.0 \text{ g C}} = 2.5 \text{ mol C}$

The empirical formula is CH.

18. $32 \text{ g S} \times \dfrac{1 \text{ mol S}}{32.1 \text{ g S}} = 1.0 \text{ mol S}$

56T

$$32 \text{ g O} \times \frac{1 \text{ mol O}}{16.0 \text{ g O}} = 2.0 \text{ mol O}$$

The empirical formula is SO_2.

$$32 \text{ g S} \times \frac{1 \text{ mol S}}{32.1 \text{ g S}} = 1.0 \text{ mol S}$$

$$48 \text{ g O} \times \frac{1 \text{ mol O}}{16.0 \text{ g O}} = 3.0 \text{ mol O}$$

This empirical formula is SO_3.

19. g O = 0.142 g − 0.062 g = 0.080 g

$$0.062 \text{ g P} \times \frac{1 \text{ mol P}}{31.0 \text{ g P}} = 0.0020 \text{ mol P}$$

$$0.080 \text{ g O} \times \frac{1 \text{ mol O}}{16.0 \text{ g O}} = 0.0050 \text{ mol O}$$

Multiplying both by 1000, 2.0 mol P, 5.0 mol O
Therefore, the formula is P_2O_5.

20. $$1.82 \text{ g W} \times \frac{1 \text{ mol W}}{183.9 \text{ g W}} = 0.00990 \text{ mol W}$$

$$0.12 \text{ g C} \times \frac{1 \text{ mol C}}{12 \text{ g C}} = 0.010 \text{ mole C}$$

Multiplying both by 100, 1.0 mol W, 1.0 mol C
Therefore, the empirical formula is WC.

$$3.70 \text{ g W} \times \frac{1 \text{ mol W}}{183.9 \text{ g W}} = 0.0201 \text{ mol W}$$

$$0.12 \text{ g C} \times \frac{1 \text{ mol C}}{12 \text{ g C}} = 0.010 \text{ mol C}$$

Multiplying both by 100, 2.0 mol W, 1.0 mol C
Therefore, the empirical formula is W_2C.

21.

	DIVIDE ALL BY 1.32	MULTIPLY ALL BY 5
$19.01 \text{ g C} \times \dfrac{1 \text{ mol C}}{12.00 \text{ g C}}$		
= 1.584 mol C	1.20	6.00
$18.48 \text{ g N} \times \dfrac{1 \text{ mol N}}{14.00 \text{ g N}}$		
= 1.320 mol N	1.00	5.00
$25.34 \text{ g O} \times \dfrac{1 \text{ mol O}}{16.00 \text{ g O}}$		
= 1.584 mol O	1.20	6.00
$1.58 \text{ g H} \times \dfrac{1 \text{ mol H}}{1.008 \text{ g H}}$		
= 1.57 mol H	1.19	5.95

Therefore, the formula is $C_6N_5O_6H_6$.

22.

	DIVIDE ALL BY 0.600
$7.20 \text{ g C} \times \dfrac{1 \text{ mol C}}{12.0 \text{ g C}} = 0.600 \text{ mol}$	1.0
$1.20 \text{ g H} \times \dfrac{1 \text{ mol H}}{1.01 \text{ g H}} = 1.19 \text{ mol}$	1.98
$9.60 \text{ g O} \times \dfrac{1 \text{ mol O}}{16.0 \text{ g O}} = 0.600 \text{ mol}$	1.00

Therefore, the empirical formula is CH_2O.

$(CH_2O)_n = 180$
$(30)_n = 180$
$n = 6$

Therefore, the molecular formula is $C_6H_{12}O_6$.

23.

	DIVIDE BOTH BY 1.39	MULTIPLY BOTH BY 2
$16.66 \text{ g C} \times \dfrac{1 \text{ mol C}}{12.0 \text{ g C}}$		
= 1.39 mol	1.00	2.00
$3.49 \text{ g H} \times \dfrac{1 \text{ mol H}}{1.01 \text{ g H}}$		
= 3.46 mol	2.49	4.98

Therefore the empirical formula is C_2H_5.

$(C_2H_5)_n = 58$
$(29)_n = 58$
$n = 2$

The molecular formula is C_4H_{10}.

24. Total mass of compound 10.12 g × 17.93 g
= 28.05 g.

$$\% \text{ of Al} = \frac{10.12 \text{ g}}{28.05 \text{ g}} = 36.08\%$$

$$\% \text{ of S} = \frac{17.93 \text{ g}}{28.05 \text{ g}} = 63.92\%$$

25. $$63.6 \text{ g Fe} \times \frac{1 \text{ mol Fe}}{55.8 \text{ g Fe}} = 1.14 \text{ mol Fe}$$

$$36.4 \text{ g S} \times \frac{1 \text{ mol S}}{32.1 \text{ g S}} = 1.13 \text{ mol S}$$

Formula is FeS and the name is iron(II) sulfide.

26.

	DIVIDE BOTH BY 0.626
$79.9 \text{ g Cu} \times \dfrac{1 \text{ mol Cu}}{63.5 \text{ g Cu}} = 1.26 \text{ mol}$	2.01
$20.1 \text{ g S} \times \dfrac{1 \text{ mol S}}{32.1 \text{ g S}} = 0.626 \text{ mol}$	1.00

The formula is Cu_2S, and the name is copper(I) sulfide.

27.

	DIVIDE ALL BY 0.896
$52.6 \text{ g Ni} \times \dfrac{1 \text{ mol Ni}}{58.7 \text{ g Ni}} = 0.896 \text{ mol Ni}$	1.00
$21.9 \text{ g C} \times \dfrac{1 \text{ mol C}}{12.0 \text{ g C}} = 1.83 \text{ mol C}$	2.04
$25.5 \text{ g N} \times \dfrac{1 \text{ mol N}}{14.0 \text{ g N}} = 1.82 \text{ mol N}$	2.03

The empirical formula is NiC_2N_2, which can be written as $Ni(CN)_2$. The name is nickel(II) cyanide.

28. Since both contain 0.050 mole per liter, both contain 3.0×10^{22} molecules per liter.

29. $1.50 \text{ L Na}_2\text{SO}_4(\text{aq}) \times \dfrac{0.25 \text{ mol Na}_2\text{SO}_4}{1.00 \text{ L Na}_2\text{SO}_4(\text{aq})}$

$\times \dfrac{142.1 \text{ g Na}_2\text{SO}_4}{1 \text{ mol Na}_2\text{SO}_4} = 53.2 \text{ g Na}_2\text{SO}_4$

30. $1.50 \text{ L Na}_2\text{SO}_4(\text{aq}) \times \dfrac{0.25 \text{ mol Na}_2\text{SO}_4}{1 \text{ L Na}_2\text{SO}_4(\text{aq})}$

$\times \dfrac{142.1 \text{ g Na}_2\text{SO}_4}{1 \text{ mol Na}_2\text{SO}_4} = 53 \text{ g Na}_2\text{SO}_4$

Dissolve 53 grams of Na_2SO_4 in enough water to produce 1.5 liters of solution.

31. $1.00 \text{ L NaOH}(\text{aq}) \times \dfrac{3.00 \text{ mol NaOH}}{1.00 \text{ L NaOH}(\text{aq})}$

$\times \dfrac{40.0 \text{ g NaOH}}{1 \text{ mol NaOH}} = 120 \text{ g NaOH}$

Dissolve 120 g NaOH in enough water to produce 1.00 liter of solution.

32. $\dfrac{5.00 \text{ L KOH}(\text{aq})}{\text{L}} = \dfrac{6.00 \text{ mol KOH}}{1 \text{ L KOH}(\text{aq})} \times \dfrac{56.1 \text{ g KOH}}{1 \text{ mol KOH}}$

$= 1680 \text{ g KOH}$

Dissolve 1680 g of KOH in enough water to produce a final volume of 5.00 liters.

CAUTION: If the solution is actually prepared, a Pyrex flask should be used.

33. $3.50 \text{ L (NH}_4)_2\text{SO}_4(\text{aq}) \times \dfrac{1.55 \text{ mol (HN}_4)_2\text{SO}_4}{1 \text{ L (NH}_4)_2\text{SO}_4(\text{aq})}$

$\times \dfrac{132.2 \text{ g (NH}_4)_2\text{SO}_4}{1 \text{ mol (NH}_4)_2\text{SO}_4} = 717 \text{ g (NH}_4)_2\text{SO}_4$

34. $1.75 \text{ L Na}_2\text{CrO}_4(\text{aq}) \times \dfrac{2.00 \text{ mol Na}_2\text{CrO}_4}{1 \text{ L Na}_2\text{CrO}_4(\text{aq})}$

$= 3.50 \text{ mol Na}_2\text{CrO}_4$

Challenge, page 123

1. $FeCl_3$

$1.81 \times 10^{23} \text{ atoms Cl} \times \dfrac{1 \text{ molecule FeCl}_3}{3 \text{ atoms Cl}}$

$\times \dfrac{1 \text{ mol FeCl}_3}{6.02 \times 10^{23} \text{ molecules FeCl}_3} = 0.100 \text{ mol}$

2. $2.00 \text{ L HCl}(\text{aq}) \times \dfrac{3.00 \text{ mol HCl}}{1 \text{ L HCl}(\text{aq})} \times \dfrac{36.4 \text{ g HCl}}{1 \text{ mol HCl}}$

$\times \dfrac{1.00 \text{ mL}}{1.37 \text{ g}} \times \dfrac{100 \text{ g conc. HCl}}{37 \text{ g HCl}}$

$= 431 \text{ mL conc. HCl}$

Add 431 mL conc. HCl slowly to the 2.00-liter flask which contains about 1 liter of deionized water. Add enough deionized water while stirring to make the final volume 2.00 liters.

3. $3.00 \text{ L HCl}(\text{aq}) \times \dfrac{6.00 \text{ mol HCl}}{1 \text{ L HCl}(\text{aq})} = 18.00 \text{ mol HCl}$

The total volume is 3.00 L + 1.00 L or 4.00 L which contains 18.00 mol of HCl.

$\dfrac{18.00 \text{ mol HCl}}{4.00 \text{ L}} = \dfrac{4.50 \text{ mol HCl}}{1 \text{ L}} = 4.50\text{M HCl}$

Synthesis, page 123

1. $\dfrac{207.2 \text{ g Pb}}{1 \text{ mol Pb}} \times \dfrac{1 \text{ cm}^3 \text{ Pb}}{11.3 \text{ g Pb}} = 18.3 \text{ cm}^3/\text{mol Pb}$

2. $\dfrac{18.3 \text{ cm}^3 \text{ Pb}}{\text{mol Pb}} \times \dfrac{1 \text{ atom Pb}}{3.00 \times 10^{-23} \text{ cm}^3 \text{ Pb}}$

$= 6.10 \times 10^{23} \text{ atoms Pb/mol Pb}$

CHAPTER 5 Chemical Reactions

CHAPTER ORGANIZER

USE WITH SECTION	DEMONSTRATION	WORKSHEET	LAB EXPERIMENT	SOFTWARE	TEACHER'S RESOURCE BINDER
5-2	5-A, B			Disk 3	
5-3		5-A		Disk 3	
5-4	5-C, D			Disk 3	
5-5		5-B, C		Disk 3	
5-7	5-E	5-D	5-A, B		
5-9					
5-10		5-E			
5-11		5-F			
Chapter Test					

■ Objectives

When the students have completed the chapter, they will be able to

1. Describe how the products are different from the reactants in a chemical reaction.
2. Write and balance an equation for a reaction when given the names of the reactants and the products.
3. Recognize and distinguish among decomposition, combustion, water forming, synthesis, single replacement, and double replacement reactions.
4. Correctly complete and balance the equation when given the reactants of a known type of reaction.
5. Recognize an exothermic and an endothermic reaction and distinguish between them.
6. State two chemical concepts not specified by a balanced equation.

■ Section Notes

5-1 It is necessary for students to observe chemical reactions to enable them to differentiate between chemical reactions and physical changes.

If you ask students to describe the reaction of the penny with the nitric acid, many will say that the metal dissolved in the liquid. It is important for them to distinguish between the process of dissolving and a chemical reaction.

5-2 One of the most important tasks in teaching chemistry is to enable students to understand the macroscopic observations that indicate a chemical reaction has occurred, and to connect their observations to the equations for the chemical reaction. The equation represents what we infer is taking place on a molecular level. Many students have great difficulty connecting the balanced equation to the observations.

Start with the decomposition of water to show that it can form hydrogen and oxygen. Ask the students to compare the different properties of the reactant, water, and the products, hydrogen and oxygen. Write a word equation for the reaction:

$$\text{Water} \xrightarrow{\text{electricity}} \text{hydrogen} + \text{oxygen}$$

Substitute the formulas for the names and have the students recall that hydrogen and oxygen are diatomic elements:

$$H_2O \longrightarrow H_2 + O_2$$

Remind students that the subscripts in these formulas represent the ratio of the number of atoms of each element in each molecule (or formula unit), and that the subscripts are kept constant but the ratio of molecules is changed by changing the coefficients to balance the equation:

$$2H_2O \longrightarrow 2H_2 + O_2$$

5-3, 5-4, 5-5 Demonstrate the formation of copper(II) oxide. Write the word equation for the reaction on the chalkboard.

$$\text{copper} + \text{oxygen} \longrightarrow \text{copper(II) oxide}$$

Make a drawing of the atoms and molecules involved to represent the macroscopic changes. Show copper as a solid, oxygen as a diatomic gas, and copper(II) oxide as a solid:

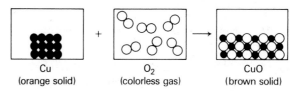

| Cu (orange solid) | | O_2 (colorless gas) | | CuO (brown solid) |

Change the word equation to a balanced chemical equation. Include the symbols that describe the physical state of each reactant and product:

$$2Cu(s) + O_2(g) \longrightarrow 2CuO(s)$$

Represent 12 copper atoms as shown in the first box, and ask how many molecules of oxygen should be shown in the second box to have the exact amount of oxygen required (six molecules are required). Show that this will produce twelve molecules of copper(II) oxide. Repeat this process with other reactions.

5-7 Students are more likely to understand classes of reactions if they begin by seeing at least five or six written examples of each. Begin with the synthesis reactions:

$$2Ca(s) + O_2(g) \longrightarrow 2CaO(s)$$
$$4Na(s) + O_2(g) \longrightarrow Na_2O(s)$$
$$4Fe(s) + 3O_2(g) \longrightarrow Fe_2O_3(s)$$

Ask a student to go to the chalkboard and write another equation that is similar. You may have to wait two or three minutes, but a student should be able to write

$$2Mg(s) + O_2(g) \longrightarrow 2MgO(s)$$
$$\text{or } 2Cu(s) + O_2(g) \longrightarrow 2CuO(s)$$

This activity can be creative and open-ended. Praise any efforts and gently correct any errors. For example, a student may write:

$$Al(s) + O_2(g) \longrightarrow AlO_2(s)$$

Ask if anyone in class sees any errors. If not, review formula writing and correct the product to Al_2O_3.

Eventually someone may write

$$Cr(s) + O_2(g) \longrightarrow CrO(s)$$
$$(CrO_2 \text{ and } CrO_3 \text{ both exist.})$$

The students may ask how one knows the correct formula of a compound when the metal involved can have more than one charge. Reply that the compound formed depends on several factors, including the reaction conditions, and that determining the correct formula depends on quantitative experiments like those described in Section 4-7.

Continue until every student can write and balance at least one equation for a synthesis reaction.

Ask the students to read aloud the equations on the board to review the naming of compounds and the meaning of the symbols. For example, the following equation may be read

$$Fe(s) + Cl_2(g) \longrightarrow FeCl_2(s) + energy$$

One mole of solid iron reacts with one mole of chlorine gas exothermically and yields one mole of solid iron(II) chloride and energy.

5-8 Write the equation for the reaction for magnesium and hydrochloric acid and ask the students to predict the products:

$$Mg(s) + HCl(aq) \longrightarrow$$

Ask them to predict what will be observed macroscopically. Add a 2-cm strip of magnesium ribbon to 5 mL of $1.00M$ HCl so students can determine if their predictions are correct.

Ask the students to make drawings of what is occurring on the microscopic level.

5-9 New substances with different properties are formed when a chemical reaction takes place. The wax, $C_{18}H_{38}$, in a burning candle reacts with oxygen from the air to form carbon dioxide and water:

$$C_{18}H_{38}(s) + O_2(g) \longrightarrow CO_2(g) + H_2O(g) + energy$$

Ask the students to balance the equation. Light a candle to show that this reaction also produces energy.

Review and Practice _____
page 133

1. a. carbon and oxygen
b. carbon—black, insoluble in water, solid at room temperature; oxygen—colorless gas at room temperature; carbon dioxide—colorless odorless gas, does not support combustion

c.

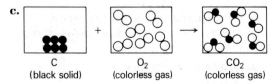

C
(black solid)
O_2
(colorless gas)
CO_2
(colorless gas)

d. $C(s) + O_2(g) \longrightarrow CO_2(g)$
e. It is balanced.

2. a. $2K + Cl_2 \longrightarrow 2KCl$
b. $2K + Br_2 \longrightarrow 2KBr$
c. $2K + F_2 \longrightarrow 2KF$
d. $4K + O_2 \longrightarrow 2K_2O$
e. $4Na + O_2 \longrightarrow 2Na_2O$
f. $6Na + N_2 \longrightarrow 2Na_3N$
g. $2Ca + O_2 \longrightarrow 2CaO$
h. $Ba + Br_2 \longrightarrow BaBr_2$
i. $3Sr + N_2 \longrightarrow Sr_3N_2$

3. a. $2K + 2H_2O \longrightarrow 2KOH + H_2$
b. $Ca + 2H_2O \longrightarrow Ca(OH)_2 + H_2$
c. $3Mg + N_2 \longrightarrow Mg_3N_2$
d. $Mg_3N_2 + 3H_2O \longrightarrow 3MgO + 2NH_3$
e. $2NH_4Cl + Ca(OH)_2 \longrightarrow 2NH_3 + 2H_2O + CaCl_2$
f. $(NH_4)_2SO_4 + 2NaOH \longrightarrow 2NH_3 + 2H_2O + Na_2SO_4$
g. $4FeS_2 + 11O_2 \longrightarrow 2Fe_2O_3 + 8SO_2$
h. $2MoS_2 + 7O_2 \longrightarrow 2MoO_3 + 4SO_2$
i. $MoO_3 + 3H_2 \longrightarrow Mo + 3H_2O$

4. a. $2Li + 2H_2O \longrightarrow 2LiOH + H_2$
b. $Ba + 2H_2O \longrightarrow Ba(OH)_2 + H_2$
c. $NH_4NO_3 + NaOH \longrightarrow NH_3 + H_2O + NaNO_3$
d. $2Cu_2S + 3O_2 \longrightarrow 2Cu_2O + 2SO_2$
e. $2Al + 3I_2 \longrightarrow 2AlI_3$

5. a. $2AgNO_3 + CuCl_2 \longrightarrow 2AgCl + Cu(NO_3)_2$
b. $BaCl_2 + (NH_4)_2CO_3 \longrightarrow BaCO_3 + 2NH_4Cl$
c. $2Mg(ClO_3)_2 \longrightarrow 2MgCl_2 + 6O_2$
d. $3FeCl_2 + 2Na_3PO_4 \longrightarrow Fe_3(PO_4)_2 + 6NaCl$
e. $ZnO + 2HCl \longrightarrow ZnCl_2 + H_2O$
f. $3CuSO_4 + 2Fe \longrightarrow Fe_2(SO_4)_3 + 3Cu$
g. $Br_2 + 2KI \longrightarrow 2KBr + I_2$
h. $2Al + 2NaOH + 6H_2O \longrightarrow 2NaAl(OH)_4 + 3H_2$

page 139

1. a. $2C_2H_2(g) + 5O_2(g) \longrightarrow 4CO_2(g) + 2H_2O(g)$
combustion
b. $Zn(s) + CuSO_4(aq) \longrightarrow ZnSO_4(aq) + Cu(s)$
single replacement
c. $Cl_2(aq) + 2KI(aq) \longrightarrow 2KCl(aq) + I_2(aq)$
single replacement
d. $2H_2O_2(l) \longrightarrow 2H_2O(l) + O_2(g)$
decomposition
e. $MgCl_2(s) \longrightarrow Mg(s) + Cl_2(g)$
decomposition
f. $Fe(s) + I_2(s) \longrightarrow FeI_2(s)$
synthesis
g. $16Cu(s) + S_8(l) \longrightarrow 8Cu_2S(s)$
synthesis
h. $C_6H_{12}O_6(aq) + 6O_2(g) \longrightarrow 6CO_2(aq) + 6H_2O(l)$
combustion
i. $FeCl_2(aq) + K_2S(aq) \longrightarrow FeS(s) + 2KCl(aq)$
double replacement
j. $H_2SO_4(aq) + 2NaOH(aq) \longrightarrow$
$Na_2SO_4(aq) + 2H_2O(l)$
double replacement, water forming
k. $Pb(NO_3)_2(aq) + K_2CrO_4(s) \longrightarrow$
$PbCrO_4(s) + 2KNO_3(aq)$
double replacement
l. $4Cr(s) + 3SnCl_4(aq) \longrightarrow 4CrCl_3(aq) + 3Sn(s)$
single replacement

m. $C_2H_5OH(l) + 3O_2(g) \longrightarrow 2CO_2(g) + 3H_2O(g)$
combustion

2. a. $Zn(s) + MgSO_4(aq) \longrightarrow$ N.R.
b. $Cd(s) + O_2(g) \longrightarrow CdO(s)$
c. $HgO(s) \longrightarrow Hg(l) + O_2(g)$
d. $HCl(g) + KOH(aq) \longrightarrow KCl(aq) + H_2O(l)$
e. $C_5H_{12}(l) + O_2(g) \longrightarrow CO_2(g) + H_2O(g)$
f. $Sr(s) + O_2(g) \longrightarrow SrO(s)$
g. $Br_2(aq) + CaCl_2(aq) \longrightarrow$ N.R.
h. $Zn(s) + Ni(NO_3)_2(aq) \longrightarrow$
$Zn(NO_3)_2(aq) + Ni(s)$
i. $ZnSO_4(aq) + SrCl_2(aq) \longrightarrow$
$SrSO_4(s) + ZnCl_2(aq)$
j. $AlCl_3(aq) + NaCO_3(aq) \longrightarrow$
$Al_2(CO_3)_3 + NaCl(aq)$
k. $Fe(s) + S_8(s) \longrightarrow FeS(s)$
l. $C_6H_6(l) + O_2(g) \longrightarrow CO_2(g) + H_2O(g)$
m. $Pb(s) + KNO_3(aq) \longrightarrow$ N.R.
n. $HNO_3(aq) + Sr(OH)_2(aq) \longrightarrow$
$Sr(NO_3)_2(aq) + H_2O(l)$

3. a. $Zn(s) + MgSO_4(aq) \longrightarrow$ N.R.
b. $2Cd(s) + O_2(g) \longrightarrow 2CdO(s)$
c. $2HgO(s) \longrightarrow 2Hg(l) + O_2(g)$
d. $HCl(aq) + KOH(aq) \longrightarrow KCl(aq) + H_2O(l)$
e. $C_5H_{12}(l) + 8O_2(g) \longrightarrow 5CO_2(g) + 6H_2O(g)$
f. $2Sr(s) + O_2(g) \longrightarrow 2SrO(s)$
g. $Br_2(aq) + CaCl_2(aq) \longrightarrow$ N.R.
h. $Zn(s) + Ni(NO_3)_2(aq) \longrightarrow$
$Zn(NO_3)_2(aq) + Ni(s)$
i. $ZnSO_4(aq) + SrCl_2(aq) \longrightarrow$
$SrSO_4(s) + ZnCl_2(aq)$
j. $2AlCl_3(aq) + 3Na_2CO_3(aq) \longrightarrow$
$Al_2(CO_3)_3(s) + 6NaCl(aq)$
k. $8Fe(s) + S_8(s) \longrightarrow 8FeS(s)$
l. $2C_6H_6(l) + 15O_2(g) \longrightarrow 12CO_2(g) + 6H_2O(g)$
m. $Pb(s) + KNO_3(aq) \longrightarrow$ N.R.
n. $2HNO_3(aq) + Sr(OH)_2(aq) \longrightarrow$
$Sr(NO_3)_2(aq) + 2H_2O(l)$

4. a. no reaction
b. synthesis, combustion
c. decomposition
d. water-forming
e. combustion
f. synthesis, combustion
g. no reaction
h. single replacement
i. double replacement
j. double replacement
k. synthesis
l. combustion
m. no reaction
n. water-forming

page 143

1. a. exothermic
b. exothermic
c. endothermic
d. exothermic

2. a. $Cl_2(aq) + 2KI(aq) \longrightarrow$
$2KCl(aq) + I_2(aq) +$ energy
b. $2Mg(s) + O_2(g) \longrightarrow 2MgO(s) +$ energy
c. energy $+ 2NaClO_3(s) \longrightarrow 2NaCl(s) + 3O_2(g)$
d. $2C_8H_{18}(l) + 25O_2(g) \longrightarrow$
$16CO_2(g) + 18H_2O(g) +$ energy

3. a. single replacement
b. synthesis, combustion
c. decomposition
d. combustion

4. $2H_2O_2 \longrightarrow 2H_2O + O_2$

5. $2Mg + O_2 \longrightarrow 2MgO$

6. $C_3H_8 + 5O_2 \longrightarrow 3CO_2 + 4H_2O +$ energy

7. Energy plus 2 moles of solid antimony plus 3 moles of solid iodine yields 2 moles of solid antimony(III) iodide.

8. endothermic

Review, page 145

1. products
2. reactants
3. the number of atoms of each element
4. **a.** 3
   b. 2 to 3
   c. 4
   d. solid
5. exothermic
6. synthesis, decomposition, combustion, single replacement, double replacement, water-forming
7. $\longrightarrow$, the arrow
8. $Fe(OH)_3$, Fe_2O_3
9. iron(III) hydroxide, iron(III) oxide
10. Energy is released; a white powder is formed.

Interpret and Apply, page 145

1. **a.** reactants—Al and O_2; product—Al_2O_3
   b. yes
   c. solid
2. exothermic
3. **a.** endothermic   **c.** endothermic
   b. exothermic   **d.** exothermic
4. a, c, d
5. **a.** decomposition   **c.** decomposition
   b. combustion   **d.** single replacement
6. b, in cars
7. **a.** synthesis, combustion
   b. synthesis
   c. synthesis
   d. combustion
   e. water-forming
   f. single replacement
   g. combustion

h. synthesis
i. decomposition
j. single replacement
k. decomposition
l. double replacement
m. water-forming
n. double replacement

8. c, i, k

9. d, e, g, j, k, l

10. e, f, j, l, m, n

11. d, g

Problems, page 146

1. a. $2Al(s) + 3Pb(NO_3)_2(aq) \longrightarrow$
   $\qquad 2Al(NO_3)_3(aq) + 3Pb(s)$
   b. $C_3H_8(g) + 5O_2(g) \longrightarrow 3CO_2(g) + 4H_2O(g)$
   c. $3Fe(s) + 4H_2O(g) \longrightarrow Fe_3O_4(s) + 4H_2(g)$
   d. $2Al(s) + 2NaOH(s) + 6H_2O(l) \longrightarrow$
   $\qquad 2NaAl(OH)_4(aq) + 3H_2(g)$
   ~~e. $NI_3(s) \longrightarrow N_2(g) + I_2(g)$~~
   f. $C(s) + H_2O(g) \longrightarrow CO(g) + H_2(g)$
   g. $2AlBr_3(aq) + 3Cl_2(g) \longrightarrow$
   $\qquad 2AlCl_3(g) + 3Br_2(l)$
   h. $2HNO_3(aq) + Ba(OH)_2(aq) \longrightarrow$
   $\qquad Ba(NO_3)_2(aq) + 2H_2O(l)$

2. a. $NH_4NO_2(s) + energy \longrightarrow N_2(g) + 2H_2O(l)$
   b. $3H_2(g) + N_2(g) \longrightarrow 2NH_3(g)$

3. a. $2C_6H_6(l) + 15O_2(g) \longrightarrow$
   $\qquad 12CO_2(g) + 6H_2O(g) + energy$
   b. $3Zn(s) + 2H_3PO_4(aq) \longrightarrow$
   $\qquad Zn_3(PO_4)_2(s) + 3H_2(g) + energy$
   c. $Pb(ClO_3)_2(aq) + 2KI(aq) \longrightarrow$
   $\qquad PbI_2(s) + 2KClO_3(aq)$
   d. $BaCO_3(s) \longrightarrow BaO(s) + CO_2(g)$
   e. $3Br_2(aq) + 2FeI_3(aq) \longrightarrow$
   $\qquad 2FeBr_3(aq) + 3I_2(aq)$
   f. $BaCl_2(aq) + H_2SO_4(aq) \longrightarrow$
   $\qquad BaSO_4(s) + 2HCl(aq)$
   g. $2C_4H_{10}(g) + 13O_2(g) \longrightarrow$
   $\qquad 8CO_2(g) + 10H_2O(g) + energy$
   h. $2Ca(s) + O_2(g) \longrightarrow 2CaO(s) + energy$
   i. $2HgO(s) \longrightarrow 2Hg(l) + O_2(g)$
   j. $Li_2SO_4(aq) + BaCl_2(aq) \longrightarrow$
   $\qquad BaSO_4(s) + 2LiCl(aq)$
   k. $F_2(g) + 2KCl(aq) \longrightarrow 2KF(aq) + Cl_2(aq)$
   l. $HC_2H_3O_2(aq) + LiOH(aq) \longrightarrow$
   $\qquad LiC_2H_3O_2(aq) + H_2O(l) + energy$

4. a. $Ni(s) + FeSO_4(aq) \longrightarrow N.R.$
   b. $Sr(s) + N_2(g) \longrightarrow Sr_3N_2(s)$
   c. $C_4H_8(l) + O_2(g) \longrightarrow CO_2(g) + H_2O(g)$
   d. $CoBr_2(s) \longrightarrow Co(s) + Br_2(l)$
   e. $H_3PO_4(aq) + Al(OH)_3(aq) \longrightarrow$
   $\qquad AlPO_4(s) + H_2O(g)$
   f. $CH_3OH(l) + O_2(g) \longrightarrow CO_2(g) + H_2O(g)$
   g. $Br_2(aq) + CuI_2(aq) \longrightarrow CuBr_2(aq) + I_2(aq)$

h. $Pb(NO_3)_2(aq) + NaCl(aq) \longrightarrow$
$\qquad PbCl_2(s) + 2NaNO_3(aq)$
i. $Zn(s) + CuCl_2(aq) \longrightarrow ZnCl_3(aq) + Cu(s)$
j. $AlCl_3(aq) + Pb(NO_3)_2(aq) \longrightarrow$
$\qquad PbCl_2(s) + Al(NO_3)_3(aq)$

5. a. $Ni(s) + FeSO_4(aq) \longrightarrow N.R.$
   b. $3Sr(s) + N_2(g) \longrightarrow Sr_3N_2(s)$
   c. $C_4H_8(l) + 6O_2(g) \longrightarrow 4CO_2(g) + 4H_2O(g)$
   d. $CoBr_2(s) \longrightarrow Co(s) + Br_2(l)$
   e. $H_3PO_4(aq) + Al(OH)_3(aq) \longrightarrow$
   $\qquad AlPO_4(s) + 3H_2O(l)$
   f. $2CH_3OH(l) + 3O_2(g) \longrightarrow 2CO_2(g) + 4H_2O(g)$
   g. $Br_2(aq) + CuI_2(aq) \longrightarrow CuBr_2(aq) + I_2(g)$
   h. $Pb(NO_3)_2(aq) + 2NaCl(aq) \rightleftharpoons$
   $\qquad PbCl_2(s) + 2NaNO_3(aq)$
   i. $Zn(s) + CuCl_2(aq) \longrightarrow ZnCl_2(aq) + Cu(s)$
   j. $2AlCl_3(aq) + 3Pb(NO_3)_2(aq) \rightleftharpoons$
   $\qquad 3PbCl_2(s) + 2Al(NO_3)_3(aq)$

6. a. single replacement
   b. synthesis
   c. combustion
   d. decomposition
   e. water-forming
   f. combustion
   g. single replacement
   h. double replacement
   i. single replacement
   j. double replacement

7. a. $Ba(ClO_3)_2(s) + energy \longrightarrow BaCl_2(s) + O_2(g)$
   b. $CH_4(g) + O_2(g) \longrightarrow CO_2(g) + H_2O(g) + energy$
   c. $Cl_2(aq) + CaI_2(aq) \longrightarrow CaCl_2(aq) + I_2(aq)$
   d. $Cr(s) + O_2(g) \longrightarrow CrO_3(s)$
   e. $HCl(aq) + Ba(OH)_2(aq) \longrightarrow BaCl_2(aq) + H_2O(l)$

8. a. $Ba(ClO_3)_2(s) + energy \longrightarrow BaCl_2(s) + 3O_2(g)$
   b. $CH_4(g) + 2O_2(g) \longrightarrow$
   $\qquad CO_2(g) + 2H_2O(g) + energy$
   c. $Cl_2(aq) + CaI_2(aq) \longrightarrow CaCl_2(aq) + I_2(aq)$
   d. $2Cr(s) + 3O_2(g) \longrightarrow 2CrO_3(s)$
   e. $2HCl(aq) + Ba(OH)_2(aq) \longrightarrow$
   $\qquad BaCl_2(aq) + 2H_2O(l)$

9. a. decomposition
   b. combustion
   c. single replacement
   d. synthesis, combustion
   e. water-forming

10. a. Solid magnesium chloride decomposes to form solid magnesium and chlorine gas.
    b. Aqueous lead(II) nitrate combines with aqueous potassium chromate to form solid lead(II) chromate and 2 moles of aqueous potassium nitrate.

Challenge, page 147

1. a. $Zn(s) + 2HCl(aq) \longrightarrow$
   $\qquad ZnCl_2(aq) + H_2(g) + energy$

b. $Zn(s) + H_2SO_4(aq) \longrightarrow$
$\qquad ZnSO_4(aq) + H_2(g) + energy$
c. $3Zn(s) + 2H_3PO_4(aq) \longrightarrow$
$\qquad Zn_3(PO_4)_2(s) + 3H_2(g) + energy$
d. $Mg(s) + 2HC_2H_3O_2(aq) \longrightarrow$
$\qquad Mg(C_2H_3O_2)_2(aq) + H_2(g) + energy$
2. single replacement
3. Examples:
$\qquad Fe(s) + 2HCl(aq) \longrightarrow$
$\qquad\qquad FeCl_2(aq) + H_2(g) + energy$
$\qquad 2Al(s) + 2H_3PO_4(aq) \longrightarrow$
$\qquad\qquad 2AlPO_4(s) + 3H_2(g) + energy$
4. a. $Na_2CO_3(s) + 2HCl(aq) \longrightarrow$
$\qquad\qquad 2NaCl(aq) + H_2O(l) + CO_2(g)$
b. $PbCO_3(s) + 2HNO_3(aq) \longrightarrow$
$\qquad\qquad Pb(NO_3)_2(aq) + H_2O(l) + CO_2(g)$
c. $CaCO_3(s) + 2HC_2H_3O_2(aq) \longrightarrow$
$\qquad\qquad Ca(C_2H_3O_2)_2(aq) + H_2O(l) + CO_2(g)$

d. $3K_2CO_3(s) + 2H_3PO_4(aq) \longrightarrow$
$\qquad\qquad 2K_3PO_4(aq) + 3H_2O(l) + 3CO_2(g)$
5. a. $2NH_4Cl + Pb(NO_3)_2 \longrightarrow PbCl_2 + 2NH_4NO_3$
b. $Zn + I_2 \longrightarrow ZnI_2$
c. $2Al + 3NiSO_4 \longrightarrow 3Ni + Al_2(SO_4)_3$
d. $PbCO_3 \longrightarrow PbO + CO_2$
e. $2AgNO_3 + Na_2CO_3 \longrightarrow Ag_2CO_3 + 2NaNO_3$

Synthesis, page 147

1. Any metal more active than hydrogen plus an acid will yield a metal compound and hydrogen gas.
2. $A + BC \longrightarrow AC + B$
3. Synthesis; this is the putting together of two elements.
   Combustion; this is the combination of a reactant with oxygen from the air to give off energy.

CHAPTER 6 Calculations Involving Reactions

CHAPTER ORGANIZER

USE WITH SECTION	DEMONSTRATION	WORKSHEET	LAB EXPERIMENT	SOFTWARE	TEACHER'S RESOURCE BINDER
6-2		6-A, B			Problems* 1–3
6-3					
6-5		6-C	6-A		Problems 4–5
6-6		6-D	6-B, C		Problems 6–8
6-7	6-A				
6-8		6-E	6-D, E		Problems 9–10
6-9		6-G			Problems 11–12
Chapter Test					

*Use the *Problem-Solving Strategies* for students found in the Teacher's Resource Binder.

■ Objectives

When students have completed this chapter, they will be able to

1. Use the balanced equation to determine the mole ratio of the reactants and products.
2. Use the mole ratio to calculate the correct stoichiometric number of moles of any other reactant or product when given the balanced equation and the number of moles of any reactant or product.
3. Calculate the correct stoichiometric amount of any other reactant or product in moles or grams when given the balanced equation and the amount of any reactant or product in moles or grams.
4. Use the principles of stoichiometry to determine the volume or molarity of solutions which are reacting when given the balanced equation.
5. Determine which reactant is in excess and calculate the amount of excess, and then use the limiting reactant to calculate the amount of each product when given the balanced equation and the amounts of each reactant.
6. Calculate the theoretical yield and the percent yield when given the balanced equation, the amounts of the reactants, and the actual yield.

■ Section Notes

6-1 Begin the chapter with a review of measurement, measurement techniques, and standards for measurement. In cooking, the standard units of measurement are cups, tablespoons, and teaspoons, the instruments found in every kitchen. But the cups and spoons must be close to the same size for recipes to succeed. It took some time before standardized measuring cups and spoons were developed.

The standards of measurement traditionally used in the United States, such as feet, miles, gallons, and pounds, came from various origins. The metric system is more consistent. The accuracy of the metric standards is greater, and conversion from one size unit to another is much simpler.

6-2, 6-3, 6-4 Introduce the principle of proportionality in chemical reactions. The students have made drawings with 12 copper atoms, 6 oxygen molecules, and 12 molecules of copper(II) oxide. Ask them to repeat the process for 6 copper atoms and ? oxygen molecules → ? copper(II) oxide molecules. Ask them to indicate the amount of oxygen and copper(II) oxide required for various amounts of copper atoms.

Emphasize the important principle that the balanced equation indicates the mole ratios of the reactions and products and is the basis for stoichiometry. The balanced equation for many reactions can be written based on a knowledge of formulas and types of reactions as discussed in Chapter 5. The balanced equation indicates the theoretical amounts of each reactant and product and is based on quantitative experiments like those in Section 4-7. Also, the percent yield must be determined experimentally. The equation does not usually specify the conditions required for the reaction to take place. It is best to avoid discussion of reaction conditions until the students have done enough calculations and experiments to make the process of stoichiometry routine. This topic will be re-examined in the discussion of reaction rates and equilibrium.

Once the ratio of moles is understood, review the method for measuring moles and the ratio of grams in a reaction.

$$1 \text{ mol Cu} + 0.5 \text{ mol O} \longrightarrow 1 \text{ mol Cu(II)O}$$
$$63.5 \text{ g Cu} + 16.0 \text{ g O} \longrightarrow 79.5 \text{ g Cu(II)O}$$

The mole ratio of copper to oxygen is 2 to 1. The gram ratio of copper to oxygen is 63.5 g to 16.0 g = 3.97 to 1.

Point out that there are 3 moles of reactants and 2 moles of products. The number of moles of reactants is not always equal to the number of moles of products. The mass of reactants is always equal to the mass of the products. You might use a bicycle analogy: 2 wheels (1 kg each) + 1 frame (3 kg) → 1 bicycle (5 kg). The number of parts is not conserved but the mass is conserved.

64T

6-5, 6-6, 6-7, 6-8, 6-9 If all of the suggested experiments for this chapter are not done by the students, do them as demonstrations. It is important for the students to be able to make macroscopic observations and quantitative measurements which will support the principles of stoichiometry.

Review and Practice
page 154

1. a. $4Na(s) + O_2(g) \longrightarrow 2Na_2O(s) + \text{energy}$
   b. 4 molecules Na + 1 molecule $O_2 \rightarrow$ 2 molecules Na_2O

   c.

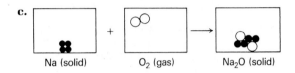

   Na (solid) O_2 (gas) Na_2O (solid)

   d. $4 \text{ mol Na} + 1 \text{ mol O}_2 \longrightarrow 2 \text{ mol Na}_2\text{O}$

   e. $12 \text{ mol Na} \times \dfrac{1 \text{ mol O}_2}{4 \text{ mol Na}} = 3 \text{ mol O}_2$

   f. $12 \text{ mol Na} \times \dfrac{2 \text{ mol Na}_2\text{O}}{4 \text{ mol Na}} = 6 \text{ mol Na}_2\text{O}$

2. a. $2Na(s) + F_2(g) \longrightarrow 2NaF(s)$

   b. $46.0 \text{ g Na} \times \dfrac{1 \text{ mol Na}}{23.0 \text{ g Na}} = 2.00 \text{ mol Na}$

   $38.0 \text{ g F}_2 \times \dfrac{1 \text{ mol F}_2}{38.0 \text{ g F}_2} = 1.00 \text{ mol F}_2$

   $84.0 \text{ g NaF} \times \dfrac{1 \text{ mol NaF}}{42.0 \text{ g NaF}} = 2.00 \text{ mol NaF}$

   $2.00 \text{ mol Na} + 1.00 \text{ mol F}_2 \longrightarrow 2.00 \text{ mol NaF}$

3. $4Na + O_2 \longrightarrow 2Na_2O$

   $5.00 \text{ mol Na} \times \dfrac{2 \text{ mol Na}_2\text{O}}{4 \text{ mol Na}} = 2.50 \text{ mol Na}_2\text{O}$

4. $CH_4 + 2O_2 \longrightarrow CO_2 + 2H_2O$

   $1.06 \text{ mol CH}_4 \times \dfrac{2 \text{ mol O}_2}{1 \text{ mol CH}_4} = 2.12 \text{ mol O}_2$

5. $2KClO_3 \longrightarrow 2KCl + 3O_2$

   $1.80 \text{ mol O}_2 \times \dfrac{2 \text{ mol KClO}_3}{3 \text{ mol O}_2} = 1.20 \text{ mol KClO}_3$

6. $4NH_3 + 7O_2 \longrightarrow 4NO_2 + 6H_2O$

   $1.22 \text{ mol NH}_3 \times \dfrac{7 \text{ mol O}_2}{4 \text{ mol NH}_3} = 2.14 \text{ mol O}_2$

   $1.22 \text{ mol NH}_3 \times \dfrac{4 \text{ mol NO}_2}{4 \text{ mol NH}_3} = 1.22 \text{ mol NO}_2$

   $1.22 \text{ mol NH}_3 \times \dfrac{6 \text{ mol H}_2\text{O}}{4 \text{ mol NH}_3} = 1.83 \text{ mol H}_2\text{O}$

page 164

1. a. $4 \text{ Na} + O_2 \longrightarrow 2 \text{ Na}_2\text{O}$

$92.0 \text{ g Na} \times \dfrac{1 \text{ mol Na}}{23.0 \text{ g Na}} \times \dfrac{1 \text{ mol O}_2}{4 \text{ mol Na}} \times \dfrac{32.0 \text{ g O}_2}{1 \text{ mol O}_2}$

$= 32.0 \text{ g O}_2$

b. $92.0 \text{ g Na} \times \dfrac{1 \text{ mol Na}}{23.0 \text{ g Na}} \times \dfrac{2 \text{ mol Na}_2\text{O}}{4 \text{ mol Na}}$

$\times \dfrac{62.0 \text{ g Na}_2\text{O}}{1 \text{ mol Na}_2\text{O}} = 124 \text{ g Na}_2\text{O}$

c. $9.20 \text{ g Na} \times \dfrac{1 \text{ mol Na}}{23.0 \text{ g Na}} \times \dfrac{1 \text{ mol O}_2}{4 \text{ mol Na}} \times \dfrac{32.0 \text{ g O}_2}{1 \text{ mol O}_2}$

$= 3.20 \text{ g O}_2$

d. $9.20 \text{ g Na} \times \dfrac{1 \text{ mol Na}}{23.0 \text{ g Na}} \times \dfrac{2 \text{ mol Na}_2\text{O}}{4 \text{ mol Na}}$

$\times \dfrac{62.0 \text{ g Na}_2\text{O}}{1 \text{ mol Na}_2\text{O}} = 12.4 \text{ g Na}_2\text{O}$

2. a. $2\text{Fe} + 3\text{Cl}_2 \longrightarrow 2\text{FeCl}_3$

b. $111.7 \text{ g Fe} \times \dfrac{1 \text{ mol Fe}}{55.85 \text{ g Fe}} = 2.000 \text{ mol Fe}$

$212.7 \text{ g Cl}_2 \dfrac{1 \text{ mol Cl}_2}{70.9 \text{ g Cl}_2} = 3.000 \text{ mol Cl}_2$

c. $2.00 \text{ mol Fe} \times \dfrac{2 \text{ mol FeCl}_3}{2 \text{ mol Fe}} \times \dfrac{162.2 \text{ g FeCl}_3}{1 \text{ mol}}$

$= 324.4 \text{ g FeCl}_3$

3. $4\text{Na} + \text{O}_2 \longrightarrow 2\text{Na}_2\text{O}$

$5.00 \text{ mol Na} \times \dfrac{2 \text{ mol Na}_2\text{O}}{4 \text{ mol Na}} \times \dfrac{62.0 \text{ g Na}_2\text{O}}{1 \text{ mol Na}_2\text{O}}$

$= 155 \text{ g Na}_2\text{O}$

4. $\text{CH}_4 + 2\text{O}_2 \longrightarrow \text{CO}_2 + \text{H}_2\text{O}$

$1.06 \text{ mol CH}_4 \times \dfrac{2 \text{ mol O}_2}{1 \text{ mol CH}_4} \times \dfrac{32.0 \text{ g O}_2}{1 \text{ mol O}_2}$

$= 67.8 \text{ g O}_2$

5. $2\text{KClO}_3 \longrightarrow 2\text{KCl} + 3\text{O}_2$

$1.80 \text{ mol O}_2 \times \dfrac{2 \text{ mol KClO}_3}{3 \text{ mol O}_2} \times \dfrac{122.5 \text{ g KClO}_3}{1 \text{ mol KClO}_3}$

$= 147 \text{ g KClO}_3$

6. $4\text{NH}_3 + 7\text{O}_2 \longrightarrow 4\text{NO}_2 + 6\text{H}_2\text{O}$

$1.22 \text{ mol NH}_3 \times \dfrac{4 \text{ mol NO}_2}{4 \text{ mol NH}_3} \times \dfrac{46.0 \text{ g NO}_2}{1 \text{ mol NO}_2}$

$= 56.1 \text{ g NO}_2$

7. $4\text{Na} + \text{O}_2 \longrightarrow 2\text{Na}_2\text{O}$

$115 \text{ g Na} \times \dfrac{1 \text{ mol Na}}{23.0 \text{ g Na}} \times \dfrac{2 \text{ mol Na}_2\text{O}}{4 \text{ mol Na}}$

$\times \dfrac{62.0 \text{ g Na}_2\text{O}}{1 \text{ mol Na}_2\text{O}} = 155 \text{ g Na}_2\text{O}$

8. $2\text{KClO}_3 \longrightarrow 2\text{KCl} + 3\text{O}_2$

$6.45 \text{ g KClO}_3 \times \dfrac{1 \text{ mol KClO}_3}{122.5 \text{ g KClO}_3} \times \dfrac{2 \text{ mol KCl}}{2 \text{ mol KClO}_3}$

$\times \dfrac{74.6 \text{ g KCl}}{1 \text{ mol KCl}} = 3.92 \text{ g KCl}$

9. $\text{Cu} + 2\text{AgNO}_3 \longrightarrow \text{Cu(NO}_3)_2 + 2\text{Ag}$

a. $9.35 \text{ g Ag} \times \dfrac{1 \text{ mol Ag}}{107.8 \text{ g Ag}} \times \dfrac{1 \text{ mol Cu}}{2 \text{ mol Ag}}$

$\times \dfrac{63.5 \text{ g Cu}}{1 \text{ mol Cu}} = 2.75 \text{ g Cu}$

b. $5.50 \text{ g Ag} \times \dfrac{1 \text{ mol Ag}}{107.8 \text{ g Ag}} \times \dfrac{1 \text{ mol Cu}}{2 \text{ mol Ag}}$

$= 0.0255 \text{ mol Cu}$

10. $\text{Zn} + 2\text{HCl} \longrightarrow \text{ZnCl}_2 + \text{H}_2$

a. $12.35 \text{ g Zn} \times \dfrac{1 \text{ mol Zn}}{65.38 \text{ g Zn}} \times \dfrac{2 \text{ mol HCl}}{1 \text{ mol Zn}}$

$= 0.3778 \text{ mol HCl}$

b. $0.3778 \text{ mol HCl} \times \dfrac{1000 \text{ mL HCl}}{3.00 \text{ mol HCl}}$

$= 125.9 \text{ mL HCl}$

11. $\text{Zn} + 2\text{HCl} \longrightarrow \text{ZnCl}_2 + \text{H}_2$

$12.35 \text{ g Zn} \times \dfrac{1 \text{ mol Zn}}{65.38 \text{ g Zn}} \times \dfrac{1 \text{ mol H}_2}{1 \text{ mol Zn}}$

$= 0.1889 \text{ mol H}_2$

12. $\text{CaC}_2 + 2\text{H}_2\text{O} \longrightarrow \text{C}_2\text{H}_2 + \text{Ca(OH)}_2$

$1.55 \text{ mol C}_2\text{H}_2 \times \dfrac{2 \text{ mol H}_2\text{O}}{1 \text{ mol C}_2\text{H}_2} \times \dfrac{18.0 \text{ g H}_2\text{O}}{1 \text{ mol H}_2\text{O}}$

$= 55.8 \text{ g H}_2\text{O}$

13. $\text{PbCO}_3 + 2\text{HNO}_3 \longrightarrow \text{Pb(NO}_3)_2 + \text{H}_2\text{O} + \text{CO}_2$

$7.52 \text{ g PbCO}_3 \times \dfrac{1 \text{ mol PbCO}_3}{267 \text{ g PbCO}_3} \times \dfrac{1 \text{ mol Pb(NO}_3)_2}{1 \text{ mol PbCO}_3}$

$\times \dfrac{331 \text{ g Pb(NO}_3)_2}{1 \text{ mol Pb(NO}_3)_2} = 9.32 \text{ g Pb(NO}_3)_2$

page 171

1. the behavior of the atoms, the temperature, the speed of the reaction, the amount of mixing, the pressure

2. a. oxygen (10 molecules in excess)
   b. 20 molecules H_2 + 10 molecules $\text{O}_2 \longrightarrow$
   20 molecules H_2O

3. ZnCl_2, H_2O, and either excess Zn or excess HCl; the H_2 gas produced will diffuse out of the beaker.

4. $2\text{Al} + 3\text{CuSO}_4 \longrightarrow 3\text{Cu} + \text{Al}_2(\text{SO}_4)_3$

a. $2\text{Al} + 3\text{CuSO}_4 \longrightarrow \text{Al}_2(\text{SO}_4)_3 + 3\text{Cu}$

$10.45 \text{ g Al} \times \dfrac{1 \text{ mol Al}}{27.0 \text{ g Al}} \times \dfrac{3 \text{ mol CuSO}_4}{2 \text{ mol Al}}$

$\times \dfrac{159.6 \text{ g CuSO}_4}{1 \text{ mol CuSO}_4} = 92.66 \text{ g CuSO}_4$ required

to react with 10.45 g Cu. There are 66.55 g CuSO_4 present, so Al is in excess.

b. $66.55 \text{ g CuSO}_4 \times \dfrac{1 \text{ mol CuSO}_4}{159.6 \text{ g CuSO}_4} \times \dfrac{2 \text{ mol Al}}{3 \text{ mol CuSO}_4}$

$\times \dfrac{27.0 \text{ g Al}}{1 \text{ mol Al}} = 7.51 \text{ g Al}$ required to react

with 66.55 g CuSO_4. The amount of excess is $10.45 - 7.50 = 2.95 \text{ g Al}$.

c. The limiting reactant, CuSO_4, is used to determine the mass of all products.

$$66.55\ g\ CuSO_4 \times \frac{1\ mol\ CuSO_4}{159.6\ g\ CuSO_4}$$

$$\times \frac{1\ mol\ Al_2(SO_4)_3}{3\ mol\ CuSO_4} \times \frac{342\ g\ Al_2(SO_4)_3}{1\ mol\ Al_2(SO_4)_3}$$

$$= 47.54\ g\ Al_2(SO_4)_3$$

$$66.55\ g\ CuSO_4 \times \frac{1\ mol\ CuSO_4}{159.6\ g\ CuSO_4}$$

$$\times \frac{3\ mol\ Cu}{3\ mol\ CuSO_4} \times \frac{63.5\ g\ Cu}{1\ mol\ Cu} = 26.48\ g\ Cu$$

5. $Pb(NO_3)_2 + 2NaCl \longrightarrow 2NaNO_3 + PbCl_2$

a. $15.50\ g\ Pb(NO_3)_2 \times \dfrac{1\ mol\ Pb(NO_3)_2}{331\ g\ Pb(NO_3)_2}$

$$\times \frac{2\ mol\ NaCl}{1\ mol\ Pb(NO_3)_2} \times \frac{58.44\ NaCl}{1\ mol\ NaCl}$$

$$= 5.47\ g\ NaCl\ required$$

Only 3.81 g NaCl are present. Therefore $Pb(NO_3)_2$ is in excess and the limiting reactant is NaCl.

b. $3.81\ g\ NaCl \times \dfrac{1\ mol\ NaCl}{58.44\ g\ NaCl} \times \dfrac{1\ mol\ Pb(NO_3)_2}{2\ mol\ NaCl}$

$$\times \frac{331.2\ g\ Pb(NO_3)_2}{1\ mol\ Pb(NO_3)_2} = 10.80\ g\ Pb(NO_3)_2$$

The amount of $Pb(NO_3)_2$ in excess is $15.50 - 10.80 = 4.70\ g\ Pb(NO_3)_2$.

c. The limiting reactant, NaCl, is used to calculate the mass of $PbCl_2$.

$$3.81\ g\ NaCl \times \frac{1\ mol\ NaCl}{58.44\ g\ NaCl} \times \frac{1\ mol\ PbCl_2}{2\ mol\ NaCl}$$

$$\times \frac{278.1\ g\ PbCl_2}{1\ mol\ PbCl_2} = 9.06\ g\ PbCl_2$$

6. $Cu + Cl_2 \longrightarrow CuCl_2$

$$12.5\ g\ Cu \times \frac{1\ mol\ Cu}{63.5\ g\ Cu} \times \frac{1\ mol\ CuCl_2}{1\ mol\ Cu}$$

$$\times \frac{134.4\ g\ CuCl_2}{1\ mol\ CuCl_2}$$

$$= 26.4\ g\ CuCl_2\ theoretical\ yield$$

$$\%\ yield = \frac{actual\ yield}{theoretical\ yield} = \frac{25.4\ g}{26.4\ g} = 96.2\%$$

7. $Fe + 2HCl \longrightarrow H_2 + FeCl_2$

$$6.57\ g\ Fe \times \frac{1\ mol\ Fe}{55.85\ g\ Fe} \times \frac{1\ mol\ FeCl_2}{1\ mol\ Fe}$$

$$\times \frac{126.8\ g\ FeCl_2}{1\ mol\ FeCl_2}$$

$$= 14.92\ g\ FeCl_2\ theoretical\ yield$$

$$\%\ yield = \frac{14.63\ g}{14.92\ g} = 98.1\%$$

Review, page 173

1. Three hydrogen molecules react with one nitrogen molecule.

2.

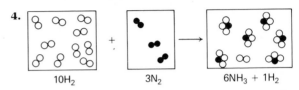

$3N_2 \quad\quad 9H_2 \quad\quad 6NH_3$

3. Two moles of O_2 react with one mole of CH_4.

4.

$10H_2 \quad\quad 3N_2 \quad\quad 6NH_3 + 1H_2$

5. Hydrogen is in excess. One molecule remains.

6. to make sure that all of the more expensive reactant is consumed; to increase the percent yield

7. If the chemicals are all 100% pure, all the chemicals react and no atoms are lost, the yield is 100%. If some atoms are lost or impurities are present, the yield will be less than 100%. If the apparent yield is more than 100%, there must be some impurity present in the product.

Interpret and Apply, page 173

1. $2H_2O_2 \longrightarrow 2H_2O + O_2$

a. $\dfrac{2\ mol\ H_2O_2}{2\ mol\ H_2O}$ or 1 to 1

b. $\dfrac{2\ mol\ H_2O_2}{1\ mol\ O_2}$ or 2 to 1

c. $1\ mol\ H_2O_2 \times \dfrac{1\ mol\ O_2}{2\ mol\ H_2O_2} = 0.50\ mol\ O_2$

2. b, e, f, h

3. a. before reaction   **b.** after reaction

$6N_2 + 12H_2 \quad\quad\quad 8NH_3 + 2N_2$

c. 8

d. N_2

e. 2 molecules N_2

4. a. 10 molecules NO_2

b. before reaction   after reaction

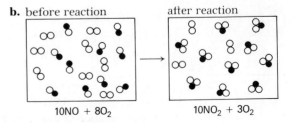

$10NO + 8O_2 \quad\quad\quad 10NO_2 + 3O_2$

c. oxygen—3 molecules

5. The gases in the reaction container will become red-brown.

6. a. Example 6-7, page 163, shows that the theoretical yield is 2.25 g Ag.

b. In order to react, the silver nitrate in solution must diffuse over to the copper wire. If not enough time has elapsed, some unreacted silver nitrate may remain in solution. Some silver crystals may be lost during the washing process.

Problems, page 174

1. $3Ag + 4HNO_3 \longrightarrow NO + 3AgNO_3 + 2H_2O$

a. $5.00 \text{ g Ag} \times \dfrac{1 \text{ mol Ag}}{107.8 \text{ g Ag}} \times \dfrac{4 \text{ mol HNO}_3}{3 \text{ mol Ag}}$

$\times \dfrac{63 \text{ g HNO}_3}{1 \text{ mol HNO}_3} = 3.90 \text{ g HNO}_3$

b. $5.00 \text{ g Ag} \times \dfrac{1 \text{ mol Ag}}{107.8 \text{ g Ag}} \times \dfrac{2 \text{ mol AgNO}_3}{2 \text{ mol Ag}}$

$\times \dfrac{169.8 \text{ g AgNO}_3}{1 \text{ mol AgNO}_3} = 7.88 \text{ g AgNO}_3$

c. $5.00 \text{ g Ag} \times \dfrac{1 \text{ mol Ag}}{107.8 \text{ g Ag}} \times \dfrac{1 \text{ mol NO}}{3 \text{ mol Ag}}$

$= 0.0155 \text{ mol NO}$

2. $N_2H_4 + 2H_2O_2 \longrightarrow N_2 + 4H_2O$

$100.0 \text{ g N}_2\text{H}_4 \times \dfrac{1 \text{ mol N}_2\text{H}_4}{32.03 \text{ g N}_2\text{H}_4} \times \dfrac{2 \text{ mol H}_2\text{O}_2}{1 \text{ mol N}_2\text{H}_4}$

$\times \dfrac{34.02 \text{ g H}_2\text{O}_2}{1 \text{ mol H}_2\text{O}_2} = 212.4 \text{ g H}_2\text{O}_2$

3. $3Zn + 2H_3PO_4 \longrightarrow ZN_3(PO_4)_2 + 3H_2$

a. $17.5 \text{ g Zn} \times \dfrac{1 \text{ mol Zn}}{65.4 \text{ g Zn}} \times \dfrac{2 \text{ mol H}_3\text{PO}_4}{3 \text{ mol Zn}}$

$= 0.178 \text{ mol H}_3\text{PO}_4$

b. $0.178 \text{ mol H}_3\text{PO}_4 \times \dfrac{1 \text{ L H}_3\text{PO}_4}{3.00 \text{ mol H}_3\text{PO}_4}$

$= 0.0593 \text{ L H}_3\text{PO}_4$

c. $17.5 \text{ g Zn} \times \dfrac{1 \text{ mol Zn}}{65.4 \text{ g Zn}} \times \dfrac{1 \text{ mol Zn}_3(\text{PO}_4)_2}{3 \text{ mol Zn}}$

$\times \dfrac{386.2 \text{ g Zn}_3(\text{PO}_4)_2}{1 \text{ mol Zn}_3(\text{PO}_4)_2} = 34.4 \text{ g Zn}_3(\text{PO}_4)_2$

d. $17.5 \text{ g Zn} + \dfrac{1 \text{ mol Zn}}{65.4 \text{ g Zn}} \times \dfrac{3 \text{ mol H}_2}{3 \text{ mol Zn}}$

$= 0.268 \text{ mol H}_2$

4. $2KClO_3 \longrightarrow 2KCl + 3O_2$

a. $5.45 \text{ g KClO}_3 \times \dfrac{1 \text{ mol KClO}_3}{122.5 \text{ g KClO}_3} \times \dfrac{3 \text{ mol O}_2}{2 \text{ mol KClO}_3}$

$\times \dfrac{32.0 \text{ g O}_2}{1 \text{ mol O}_2} = 2.14 \text{ g O}_2$

b. % yield $= \dfrac{\text{actual yield}}{\text{theoretical yield}} = \dfrac{1.95}{2.14} = 91.1\%$

c. Some oxygen may have escaped or dissolved in the water, or the $KClO_3$ may not have completely reacted.

5. $2Al + 3Cl_2 \longrightarrow 2AlCl_3$

a. $15.5 \text{ g Al} \times \dfrac{1 \text{ mol Al}}{27.0 \text{ g Al}} \times \dfrac{3 \text{ mol Cl}_2}{2 \text{ mol Al}}$

$\times \dfrac{70.9 \text{ g Cl}_2}{1 \text{ mol Cl}_2} = 61.0 \text{ g Cl}_2 \text{ required}$

This means that since there are 46.7 g Cl_2 present, Al is in excess.

b. $46.7 \text{ g Cl}_2 \times \dfrac{1 \text{ mol Cl}_2}{70.9 \text{ g Cl}_2} \times \dfrac{2 \text{ mol Al}}{3 \text{ mol Cl}_2}$

$\times \dfrac{27.0 \text{ g Al}}{1 \text{ mol Al}} = 11.9 \text{ g Al}$

$15.5 \text{ g} - 11.9 \text{ g} = 3.6 \text{ g Al in excess}$

c. $46.7 \text{ g Cl}_2 \times \dfrac{1 \text{ mol Cl}_2}{70.9 \text{ g Cl}_2} \times \dfrac{2 \text{ mol AlCl}_3}{3 \text{ mol Cl}_2}$

$\times \dfrac{133.3 \text{ g AlCl}_3}{1 \text{ mol AlCl}_3} = 58.5 \text{ g AlCl}_3$

6. a. $UF_4 + 2Mg \longrightarrow U + 2MgF_2$

b. $155 \text{ g UF}_4 \times \dfrac{1 \text{ mol UF}_4}{314 \text{ g UF}_4} \times \dfrac{2 \text{ mol Mg}}{1 \text{ mol UF}_4}$

$\times \dfrac{24.3 \text{ g Mg}}{1 \text{ mol Mg}} = 24.0 \text{ g Mg}$

c. $155 \text{ g UF}_4 \times \dfrac{1 \text{ mol UF}_4}{314 \text{ g UF}_4} \times \dfrac{1 \text{ mol U}}{1 \text{ mol UF}_4}$

$\times \dfrac{238 \text{ g U}}{1 \text{ mol U}} = 117 \text{ g U}$

7. a. $Pb + PbO_2 + 2H_2SO_4 \longrightarrow 2PbSO_4 + 2H_2O$

b. $10.45 \text{ g Pb} \times \dfrac{1 \text{ mol Pb}}{207.2 \text{ g Pb}} = 0.0504 \text{ mol Pb}$

$15.66 \text{ g PbO}_2 \times \dfrac{1 \text{ mol PbO}_2}{239.2 \text{ g PbO}_2}$

$= 0.0655 \text{ mol PbO}_2$

$25.55 \text{ g H}_2\text{SO}_4 \times \dfrac{1 \text{ mol H}_2\text{SO}_4}{98.1 \text{ g H}_2\text{SO}_4}$

$= 0.260 \text{ mol H}_2\text{SO}_4$

Dividing all the moles by the smallest amount (0.0504) gives 1 mol Pb, 1.30 mol PbO_2, and 5.16 mol H_2SO_4. The limiting reactant is Pb.

c. The limiting reactant is used to calculate the mass of a product.

$10.45 \text{ g Pb} \times \dfrac{1 \text{ mol Pb}}{207.2 \text{ g Pb}} \times \dfrac{2 \text{ mol PbSO}_4}{1 \text{ mol Pb}}$

$\times \dfrac{303.3 \text{ g PbSO}_4}{1 \text{ mol PbSO}_4} = 30.59 \text{ g PbSO}_4$

8. $12CO_2 + 11H_2O \longrightarrow C_{12}H_{22}O_{11} + 12O_2$

$455 \text{ g C}_{12}\text{H}_{22}\text{O}_{11} \times \dfrac{1 \text{ mol C}_{12}\text{H}_{22}\text{O}_{11}}{342 \text{ g C}_{12}\text{H}_{22}\text{O}_{11}}$

$\times \dfrac{12 \text{ mol CO}_2}{1 \text{ mol C}_{12}\text{H}_{22}\text{O}_{11}} = 16.0 \text{ mol CO}_2$

9. $C + O_2 \longrightarrow CO_2$

$99 \text{ g C} \times \dfrac{1 \text{ mol C}}{12.0 \text{ g C}} \times \dfrac{1 \text{ mol O}_2}{1 \text{ mol C}} \times \dfrac{32.0 \text{ g O}_2}{1 \text{ mol O}_2}$

$= 264 \text{ g O}_2$

10. $2CH_3OH + 3O_2 \longrightarrow 2CO_2 + 4H_2O$

$$34.2 \text{ g CH}_3\text{OH} \times \frac{1 \text{ mol CH}_3\text{OH}}{32 \text{ g CH}_3\text{OH}} \times \frac{3 \text{ mol O}_2}{2 \text{ mol CH}_3\text{OH}}$$

$$= 1.60 \text{ mol O}_2$$

Challenge, page 175

1. $Fe + Cl_2 \longrightarrow FeCl_x$

$Fe = 14.97$ g

$Cl = 43.47 \text{ g} - 14.97 \text{ g} = 28.50$ g

$$14.97 \text{ g Fe} \times \frac{1 \text{ mol}}{55.85 \text{ g Fe}} = 0.268 \text{ mol Fe}$$

$$28.45 \text{ g Cl} \times \frac{1 \text{ mol Cl}}{35.45 \text{ g Cl}} = 0.803 \text{ mol Cl}$$

Divide both by the smaller number of moles.

$$\frac{0.268}{0.268} = 1 \text{ mol Fe}, \quad \frac{0.803}{0.268} = 2.996 \text{ mol Cl}$$

Therefore the formula is $FeCl_3$

$2Fe + 3Cl_2 \longrightarrow 2FeCl_3$

2. $MnO_2 \longrightarrow O_2 + ?$

$10.68 - 9.37 = 1.31$ g oxygen

$$10.68 \text{ g MnO}_2 \times \frac{1 \text{ mol MnO}_2}{86.94 \text{ g MnO}_2} = 0.1228 \text{ mol MnO}_2$$

$$1.31 \text{ g O}_2 \times \frac{1 \text{ mol O}_2}{32.0 \text{ g O}_2} = 0.0409 \text{ mol O}_2$$

Divide both by the smaller number of moles.

$$\frac{0.1228}{0.0409} = 3.00 \text{ mol MnO}_2, \quad \frac{0.0409}{0.0409} = 1.00 \text{ mol O}_2$$

If the mole ratio is three MnO_2 to one O_2, the remaining solid must contain 3 moles of Mn for 4 moles of O_2. The equation is:

$3MnO_2 \longrightarrow O_2 + Mn_3O_4$

3. $Hg(NO_3)(aq) + 2NaI(aq) \longrightarrow$
$HgI_2(s) + 2NaNO_3(aq)$

$$75.0 \text{ mL Hg(NO}_3)_2 \times \frac{0.100 \text{ mol Hg(NO}_3)_2}{1000 \text{ mL Hg(NO}_3)_2}$$

$$= 0.007 \, 50 \text{ mol Hg(NO}_3)_2$$

$$150.0 \text{ mL NaI} \times \frac{0.100 \text{ mol NaI}}{1000 \text{ mL NaI}} = 0.0150 \text{ mol NaI}$$

Dividing both by the smaller number of moles:

$$\frac{0.007 \, 50}{0.007 \, 50} = 1 \text{ mol Hg(NO}_3)_2 \quad \frac{0.0150}{0.007 \, 50} = 2 \text{ mol NaI}$$

This is the correct mole ratio. Neither reactant is in excess and either can be used to calculate the product.

$$150.0 \text{ mL NaI} \times \frac{0.100 \text{ mol NaI}}{1000 \text{ mL NaI}} \times \frac{1 \text{ mol HgI}_2}{2 \text{ mol NaI}}$$

$$\times \frac{454.4 \text{ g HgI}_2}{1 \text{ mol HgI}_2} = 3.41 \text{ g HgI}_2$$

4. $Ni(NO_3)_2(aq) + K_2CO_3(aq) \longrightarrow$
$NiCO_3(s) + 2 KNO_3(aq)$

$$85 \text{ mL K}_2\text{CO}_3 \times \frac{0.25 \text{ mol K}_2\text{CO}_3}{1000 \text{ mL K}_2\text{CO}_3}$$

$$\times \frac{1 \text{ mol Ni(NO}_3)_2}{1 \text{ mol K}_2\text{CO}_3}$$

$$\times \frac{1000 \text{ mL Ni(NO}_3)_2}{0.55 \text{ mol Ni(NO}_3)_2} = 39 \text{ mL Ni(NO}_3)_2$$

5. $CuSO_4(aq) + 2NaOH(aq) \longrightarrow$
$Cu(OH)_2(s) + Na_2SO_4(aq)$

$$45 \text{ mL NaOH} \times \frac{1.50 \text{ mol NaOH}}{1000 \text{ mL NaOH}} \times \frac{1 \text{ mol CuSO}_4}{2 \text{ mol NaOH}}$$

$$\times \frac{1000 \text{ mL CuSO}_4}{0.60 \text{ mol CuSO}_4} = 56 \text{ mL CuSO}_4$$

6. $3Na_2CO_3(aq) + 2FeCl_3(aq) \longrightarrow$
$Fe_2(CO_3)_3(s) + 6NaCl(aq)$

$$82 \text{ mL FeCl}_3 \times \frac{0.25 \text{ mol FeCl}_3}{1000 \text{ mL FeCl}_3} \times \frac{3 \text{ mol Na}_2\text{CO}_3}{2 \text{ mol FeCl}_3}$$

$$\times \frac{1000 \text{ mL Na}_2\text{CO}_3}{0.45 \text{ mol Na}_2\text{CO}_3} = 68 \text{ mL Na}_2\text{CO}_3$$

7. $3Zn(s) + 2Cr(NO_3)_3(aq) \longrightarrow$
$3Zn(NO_3)_2(aq) + 2 Cr(s)$

$$425 \text{ cm}^3 \text{ Cr(NO}_3)_3 \times \frac{0.25 \text{ mol Cr(NO}_3)_3}{1000 \text{ cm}^3 \text{ Cr(NO}_3)_3}$$

$$\times \frac{3 \text{ mol Zn}}{2 \text{ mol Cr(NO}_3)_3} \times \frac{65.4 \text{ g Zn}}{1 \text{ mol Zn}} = 10.42 \text{ g Zn}$$

a. Zinc is in excess since 10.42 g are required to react with 425 cm³ of $Cr(NO_3)_3$ and 19.43 g are present. Mass of zinc in excess = 19.43 g − 10.42 g = 9.01 g.

b.

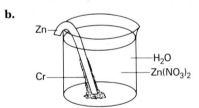

c. $425 \text{ cm}^3 \text{ Cr(NO}_3)_3 \times \dfrac{0.25 \text{ mol Cr(NO}_3)_3}{1000 \text{ cm}^3 \text{ Cr(NO}_3)_3}$

$$\times \frac{2 \text{ mol Cr}}{2 \text{ mol Cr(NO}_3)_3} \times \frac{52 \text{ g Cr}}{1 \text{ mol Cr}} = 5.5 \text{ g Cr}$$

Synthesis, page 175

1. a. Fe

b. The blue color of the copper(II) sulfate will decrease as the amount of iron metal decreases. Copper metal will be formed, attached to the iron at first and then, perhaps, falling to the bottom. If iron is in excess, the solution will become colorless. If copper(II) sulfate is in excess, all of the iron will be reacted and not be visible.

c. $Fe(s) + CuSO_4(aq) \longrightarrow FeSO_4(aq) + Cu(s)$

$$5.4 \text{ g Fe} \times \frac{1 \text{ mol Fe}}{55.8 \text{ g Fe}} = 0.0967 \text{ mol Fe}$$

$$135 \text{ mL CuSO}_4 \times \frac{0.65 \text{ mol CuSO}_4}{1000 \text{ mL CuSO}_4}$$
$$= 0.0878 \text{ mol CuSO}_4$$

Iron is in excess and $CuSO_4$ is the limiting reactant. The ratio of $CuSO_4$ to Cu is 1 to 1 so 0.0878 mol Cu is formed.

d. $135 \text{ mL CuSO}_4 \times \frac{0.65 \text{ mol CuSO}_4}{1000 \text{ mL CuSO}_4}$

$\times \frac{1 \text{ mol Cu}}{1 \text{ mol CuSO}_4} \times \frac{63.5 \text{ g Cu}}{1 \text{ mol Cu}} = 5.57 \text{ g Cu}$

e. $5.57 \text{ g Cu} \times \frac{6.02 \times 10^{23} \text{ atoms Cu}}{63.5 \text{ g Cu}}$

$= 5.28 \times 10^{22} \text{ atoms Cu}$

2. a. mass of Mo = 26.42 g

mass of O = 33.04 g − 26.42 g = 6.62 g

$26.42 \text{ g Mo} \times \frac{1 \text{ mol Mo}}{95.94 \text{ g Mo}} = 0.275 \text{ mol Mo}$

$6.62 \text{ g O} \times \frac{1 \text{ mol O}}{16.0 \text{ g O}} = 0.414 \text{ mol O}$

Dividing both by 0.275 $\longrightarrow$ 1 mol Mo and 1.50 mol O

Multiply both by 2 to keep the ratio the same.

2 mol Mo to 3 mol O

4 mol Mo to 3 mol O_2

$4 Mo + 3O_2 \longrightarrow 2Mo_2O_3$

b. molybdenum(v) oxide

3. a. $Cr + CoCl_2 \longrightarrow CrCl_2 + Co$

mass of Cr reacted = 21.55 g − 16.05 g = 5.50 g

$5.50 \text{ g Cr} \times \frac{1 \text{ mol Cr}}{52.0 \text{ g Cr}} = 0.106 \text{ mol Cu}$

$635 \text{ mL CoCl}_2 \times \frac{0.25 \text{ mol CoCl}_2}{1000 \text{ mL CoCl}_2}$

$= 0.159 \text{ mol CoCl}_2$

Dividing both by 0.106

$\frac{0.106}{0.106} = 1 \text{ mol Cr}, \frac{0.159}{0.106} = 1.50 \text{ mol CoCl}_2$

Multiplying both by 2 gives 2 mol Cr and 3 mol $CoCl_2$.

$$2Cr(s) + 3CoCl_2(aq) \longrightarrow 3Co(s) + 2CrCl_3(aq)$$

b. Pink color of the solution changes to blue-green. Crystals form on the chromium strip (students will not know this unless they have seen this reaction).

c.

4. a. $3AgNO_3(aq) + AlCl_3(aq) \longrightarrow$
$3AgCl(s) + Al(NO_3)_3(aq)$

b. $125 \text{ cm}^3 \text{ AgNO}_3 \times \frac{0.55 \text{ mol AgNO}_3}{1000 \text{ cm}^3 \text{ AgNO}_3}$

$= 0.069 \text{ mol AgNO}_3$

$85 \text{ cm}^3 \text{ AlCl}_3 \times \frac{0.25 \text{ mol AlCl}_3}{1000 \text{ cm}^3 \text{ AlCl}_3}$

$= 0.021 \text{ mol AlCl}_3$

c. Dividing both by 0.021

$\frac{0.069}{0.021} = 3.28 \text{ mol AgNO}_3, \frac{0.021}{0.021} = 1 \text{ mol AlCl}_3$

Therefore $AgNO_3$ is in excess and $AlCl_3$ is the limiting reactant.

d. $85 \text{ cm}^3 \text{ AlCl}_3 \times \frac{0.25 \text{ mol AlCl}_3}{1000 \text{ cm}^3 \text{ AlCl}_3} \times \frac{3 \text{ mol AgCl}}{1 \text{ mol AlCl}_3}$

$\times \frac{143.2 \text{ g AgCl}}{1 \text{ mol AgCl}} = 9.1 \text{ g AgCl}$

e.

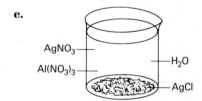

CHAPTER 7 Gases

CHAPTER ORGANIZER

USE WITH SECTION	DEMONSTRATION	WORKSHEET	LAB EXPERIMENT	SOFTWARE	TEACHER'S RESOURCE BINDER
7-1	7-A				
7-2				Disk 7	
7-3		7-A		Disk 7	
7-4		7-B			Problems 1–2
7-5	7-B			Disks 7, 9	Problems 3–4 Advanced Topic 1
7-6	7-C, 7-D	7-C, 7-D		Disk 7	Problem 5
7-7	7-E				
7-8				Disk 7	Problem 6
7-9	7-F				
7-10			7-A		Problems 7–8
7-11		7-E	7-B		Problems 9–10
7-12	7-G	7-F			
7-13	7-H	7-G			Problems 11–12
7-14		7-H			Problems 13–14
Chapter Test					

■ Objectives

When the students have completed the chapter, they will be able to

1. Distinguish between a solid, a liquid and a gas.
2. Describe the structure of a gas on the molecular level.
3. Calculate pressure in pascals, kilopascals, and atmospheres.
4. Calculate the partial pressure of a gas collected over water.
5. Describe the effect of a change in temperature on the volume of a gas.
6. Calculate the new volume of a gas when the pressure of the gas changes.
7. List the four postulates of the kinetic molecular theory.
8. Apply the ideal gas law to calculate the pressure or the volume of a gas.
9. Determine the molar volume of a gas under various conditions.
10. Describe some of the implications of the kinetic molecular theory, such as the velocity of molecules and Graham's law.
11. Calculate grams of product produced in reactions involving gases.

■ Section Notes

7-1 It is important that the students have a clear, mental picture (model) of a gas on the molecular level. The molecules are in constant and random motion. Oxygen molecules at room temperature move at about 1000 miles per hour. They collide with each other and the walls of their container from 2 to 5 billion times per second. Students also need to be aware of the tremendous amount of empty space between molecules.

By use of an analogy students can more fully understand the chaos in the movement of gas molecules. Increasing the scale by one billion changes the size of the oxygen atom from 10^{-8} cm^3 to the size of an average student and the speed from 1000 miles per hour to one trillion miles per hour. On this scale, only three gas molecules would occupy a space as large as a lab classroom and they would crash into the walls of the room and each other 2 to 3 billion times per second.

7-2, 7-3 The pascal is a unit of pressure equal to one newton per square meter. The pressure in pascals (N/m^2) of a column of mercury 1 m² × 760 mm high is calculated as follows. The volume of the column is 0.760 m³, or 7.60×10^5 cm³. The density of mercury is 13.596 g/cm³. Using the formula, density = mass/volume, the mass of the mercury column is 1.033×10^4 kg. Since force is $m \times a$, the force exerted by the column is $(1.033 \times 10^4$ kg) × (9.806 m/s²) or 1.013×10^5 kg m/s², which is 1.013×10^5 newtons. The pressure (force per unit area) of the column of mercury is

$$\frac{1.013 \times 10^5 \text{ N}}{1 \text{ m}^2}, \text{ or } 101\,300 \text{ pascals.}$$

The accepted (SI) unit for pressure, pascals, is difficult for students to visualize. Atmospheres also is an acceptable unit and can be used as an alternative. Millimeters of mercury or torr are units of pressure used in the past. Avoid them since they are not SI units.

7-5, 7-6 Charles's and Boyle's laws need careful development. Use the demonstrations provided for these sections so that the students have a clearer idea of what these laws describe. Charles's law is discussed first because it involves a direct relationship between temperature (in kelvins) and volume. Direct relationships are usually easier for students to grasp.

7-9, 7-10 The kinetic molecular theory is valid only for ideal gases. Most gases at room temperature and 1 atm of pressure behave as if they are ideal. They have a mean free path of about 600 times their own diameter. However, under conditions of very low temperature and very high pressure, gases deviate from the ideal, and are said to be real gases. Real gas molecules have volume that is part of the total volume of the gas. As a gas cools, attractive forces between molecules lower the pressure below what it would be ideally and also cause the molecules to liquify. The volume of the gas as it cools will become constant determined by the density of the liquid. Johannes van der Waals developed an equation (there are others) that compensates for these two factors:

$$(P + an^2/V^2)(V - nb) = nRT$$

The expression, an^2/V^2, compensates for the reduced pressure due to attractive forces and nb compensates for the increased volume due to the volume of the molecules themselves. Terms a and b are coefficients that vary from gas to gas.

7-12 The velocities of molecules determine the temperature of a gas. Temperature is a measure of kinetic energy. An increase in kinetic energy causes both temperature and velocity to increase. Kinetic energy is the cause, rise in temperature and velocity are effects. There is not a cause-and-effect relationship between increased temperature and increased velocity, but both are based on increases in kinetic energy.

This section relates average kinetic energy of molecules to temperature. A more rigorous treatment, which is derived from the kinetic molecular theory, yields the following equation:

$$\text{velocity} = \left(\frac{3RT}{\text{mol. mass}}\right)^{\frac{1}{2}}$$

For a given sample of gas, R and molar mass are constants. Temperature and velocity are related in that as the average velocity of gas molecules increases, so does the absolute temperature of the gas.

Review and Practice
page 186

1. Gas molecules are far apart from each other in comparison to the molecules in a liquid or a solid.
2. Weight is a force exerted by gravity on a mass. Pressure is a force per unit area.
3. On a mountaintop there is less atmosphere "overhead" than at sea level. The pressure is lower because there is less mass to contribute to the force per unit area.
4. The pressure of a gas is caused by the collision of gas molecules with the sides of the container. Pressure increases when the speed of the molecules increases.
5. $P_t = P_{N_2} + P_{H_2O}$
   99.4 kPa = P_{N_2} + 2.6 kPa
   96.8 kPa = P_{N_2}

page 198

1. The volume of a gas is directly proportional to its Kelvin temperature when pressure and moles are held constant.
2. $V = \dfrac{3.00 \text{ dm}^3}{325.3 \text{ K}} \times 246.0 \text{ K} = 2.27$ dm³
3. $V = \dfrac{1.1 \text{ kPa} \times 1.0 \text{ dm}^3}{315 \text{ kPa}} = 3.5 \times 10^{-3}$ dm³
4. $V = \dfrac{23.4 \text{ kPa} \times 5.71 \text{ L}}{3.40 \text{ L}} = 39.3$ kPa
5. $V = 100 \text{ cm}^3 \times \dfrac{150 \text{ K}}{200 \text{ K}} = 75$ cm³
6. $V = 150 \text{ kPa} \times \dfrac{100 \text{ mL}}{150 \text{ mL}} = 100$ kPa

page 200

1. $V = 3.5 \text{ dm}^3 \times \dfrac{2.0 \text{ atm}}{3.0 \text{ atm}} \times \dfrac{382 \text{ K}}{329 \text{ K}} = 2.71 \text{ dm}^3$

2. $V = 200 \text{ mL} \times \dfrac{92.1 \text{ kPa}}{98.5 \text{ kPa}} \times \dfrac{350 \text{ K}}{275 \text{ K}} = 238 \text{ mL}$

3. Temperature decreases with decreasing pressure and increasing volume just after the kernel pops.

page 207

1. a. Molecules are dimensionless points.
   b. Molecules are in constant motion.
   c. Collisions between molecules are perfectly elastic.
   d. Molecules exert no attractive force.

2. Absolute zero is the lowest possible temperature (the system has no kinetic energy). Absolute zero is 0 K and $-273°C$.

3. low pressure and relatively high temperature

4. Two values of R are
$8.31 \text{ L} \cdot \text{kPa/mol} \cdot \text{K}$ and $0.082 \text{ L} \cdot \text{atm/mol} \cdot \text{K}$.
The R values differ in units of atmospheric pressure.

5. $8.31 \dfrac{\text{L} \cdot \text{kPa}}{\text{mol} \cdot \text{K}} \times \dfrac{1000 \text{ mL}}{1 \text{ L}} = \dfrac{8.31 \times 10^3 \text{ mL} \cdot \text{KPa}}{\text{mol} \cdot \text{K}}$

6. $V = \dfrac{\dfrac{6.73 \text{ g}}{32.0 \text{ g/mol}} \times \dfrac{0.082 \text{ L} \cdot \text{atm}}{\text{mol} \cdot \text{K}}}{0.0367 \text{ atm}} \times 351 \text{ K} = 165 \text{ L}$

7. $n = \dfrac{(85.9 \text{ kPa})(27.3 \text{ L})}{(8.31 \text{ L} \cdot \text{kPa/mol} \cdot \text{K})(323.5 \text{ K})} = 0.872 \text{ mol}$

8. Molecules of a real gas move in straight-line motion unless they collide with something. Collisions between molecules are elastic.

page 208

1. Boyle's law, Charles's law

2. $R = PV/nT$. The units are $\text{L} \cdot \text{kPa/mol} \cdot \text{K}$ or $\text{L} \cdot \text{atm/mol} \cdot \text{K}$. (By convention, volume units are usually written before pressure units.)

3. 22.4 L; sometimes avoids need for more time-consuming calculations.

4. The unitary rate for this problem is $\dfrac{46.0 \text{ g}}{22.4 \text{ dm}^3}$ at STP. Therefore, 3.5 dm^3 would have a mass of
$3.5 \text{ dm}^3 \times \dfrac{46.0 \text{ g}}{22.4 \text{ dm}^3} = 7.2 \text{ g}$.

5. The mass of a molar volume of any gas is its molar mass. Therefore, the mass of a molar volume of CO_2 is 44.0 g.

6. temperature and the number of moles

page 216

1. Although the velocity of some molecules actually decreases, the average velocity increases as the gas is heated. Molecules bounce off each other and the walls of the container more frequently.

2. The relative rates of diffusion of two gases are proportional to the square root of the inverse ratio of their molar masses.

3. Let gas a = unknown and gas b = argon. Then

$\dfrac{\text{rate a}}{\text{rate b}} = \sqrt{\dfrac{\text{mol mass b}}{\text{mol mass a}}}$

$0.906 = \sqrt{\dfrac{39.9 \text{ g/mol}}{\text{mol mass a}}}$

$(0.906)^2 = \dfrac{39.9 \text{ g/mol}}{\text{mol mass a}}$

$0.821 = \dfrac{39.9 \text{ g/mol}}{\text{mol mass a}}$

$\text{mol mass a} = \dfrac{39.9 \text{ g/mol}}{0.821}$

$\text{mol mass a} = 48.6 \text{ g/mol}$

4. They are the same.

5. $PCl_3(g) + Cl_2(g) \longrightarrow PCl_5(g)$
Since the coefficients are all ones, the amount of PCl_5 formed is 7 L. There is an excess of 2 L of Cl_2 gas.   *7ℓ 9ℓ*

6. $PCl_5 \longrightarrow PCl_3 + Cl_2$
 23.2g The number of moles of Cl_2 produced is equal to the number of moles of PCl_5 that decompose.

$23.2 \text{ g} \times \dfrac{1 \text{ mol } PCl_5}{208.5 \text{ g } PCl_5} = 0.111 \text{ mol } PCl_5$

$0.111 \text{ mol } Cl_2 \text{ produced}$

22.4 because it is @STP $0.111 \text{ mol} \times \dfrac{22.4 \text{ dm}^3}{1 \text{ mol } Cl_2} = 2.49 \text{ dm}^3 \text{ of } Cl_2$

7. $2HCl(aq) + Zn(s) \longrightarrow H_2(g) + ZnCl_2(aq)$
Moles of Zn reacted are equal to moles of hydrogen gas produced.

$34.56 \text{ g} \times \dfrac{1 \text{ mol Zn}}{65.4 \text{ g}} = 0.528 \text{ mol Zn}$

$= 0.528 \text{ mol } H_2 \text{ produced}$

$0.528 \text{ mol} \times \dfrac{24.5 \text{ L}}{1 \text{ mol } H_2} = 12.9 \text{ L of } H_2 \text{ produced}$ *Because it is @ room temp*

Review, page 217

1. A gas is compressible, takes the shape of its container, and is much less dense than either liquid or solid states of matter.

2. The molecules of a gas are very far apart.

3. pascals, kilopascals, or atmospheres

4. the pressure exerted by an individual gas in a mixture of gases

5. the relationship between the temperature of a gas and the volume of the gas

6. The volume of a gas is inversely proportional to the pressure of the gas when temperature and moles of gas are held constant.

7. R acts to equalize the proportionality between PV and nT.

8. 22.4 dm^3 is the molar volume of any gas at STP. The value can be used to simplify some calculations in solving gas problems.

9. a. Molecules can be considered to be dimensionless points with no volume.
b. Molecules of a gas are in constant motion.
c. The collisions between molecules and the walls of the container are perfectly elastic.
d. Molecules exert no attractive force on each other.

10. Kinetic energy is the energy of motion. It is usually expressed as:

$$KE = \tfrac{1}{2}mv^2$$

The temperature of a gas is proportional to the average kinetic energy of a sample of gas.

Interpret and Apply, page 217

1. Pressure depends on the number of particles and the change in speed and direction of particles colliding with the walls of the container. The greater the number of collisions or the greater the velocity, the greater the pressure.

2. Gases are compressible because the volume of the molecules is very small in comparison to the volume of the gas. The molecules have a lot of room to get closer together as the gas is compressed.

3. At high pressure the volume of a gas is greater than predicted. The molecules themselves have volume that must be added to the volume predicted by the ideal gas law, which presumes the molecules are volumeless. The degree of deviation from the ideal depends on the number of moles of gas present.

4. The graph shows the distribution of the velocities of all molecules in a sample of gas. When heated, most gas molecules will increase in velocity and the graph will be skewed to the right. However, some molecules will decrease in velocity, as seen at the far left of the graph.

5. The helium atoms are moving faster. Since both gases have the same temperature, their average kinetic energies are equal:

$$\tfrac{1}{2}mv^2 \text{ (for He)} = \tfrac{1}{2}mv^2 \text{ (for O}_2)$$

Since the oxygen molecule is more massive than the helium atom, the velocity of the oxygen molecules must be less than that of the helium atoms.

6. Since the volume of an aerosol can is constant, the pressure of the gas is directly proportional to the absolute temperature of the gas inside the can. The pressure in an aerosol can in a fire would increase enough to cause the can to explode.

7. a. The number of moles are equal.
b. The number of molecules also are equal.
c. The CO_2 has a mass about 22 times the mass of H_2 gas.
d. When the temperature of the H_2 gas is increased, the kinetic energy increases and is greater than the kinetic energy of the CO_2. The number of moles and volume are still equal. The pressure of the H_2 gas is greater since pressure and temperature are proportional with volume constant.

8. As the balloon is inflated, the pressure, volume and number of moles increase. The temperature increases slightly. The pressure of the balloon eventually is greater than the balloon can contain and the balloon bursts. Air escapes into the atmosphere, cools, and pressure becomes uniform with the atmosphere.

Problems, page 218

1. a. $94\,000 \text{ Pa} = P_{H_2} + 1200 \text{ Pa}$
   $92\,800 \text{ Pa} = P_{H_2}$
b. $100.3 \text{ kPa} = P_{H_2} + 2.6 \text{ kPa}$
   $97.7 \text{ kPa} = P_{H_2}$

2. a. $V = \dfrac{85 \text{ cm}^3}{334 \, K} \times 308 \, K = 78 \text{ cm}^3$

b. $V = \dfrac{7.3 \text{ dm}^3}{501 \, K} \times 225 \, K = 3.3 \text{ dm}^3$

c. $V = \dfrac{(1.15 \times 10^3 \text{ cm}^3)(75.2 \text{ kPa})}{14.0 \text{ kPa}}$
   $= 6.18 \times 10^3 \text{ cm}^3$

d. $V = \dfrac{94.7 \text{ dm}^3 \times 1.00 \text{ kPa}}{101.3 \text{ kPa}} = 0.935 \text{ dm}^3$

e. $V = 4.03 \times 10^3 \text{ dm}^3 \times \dfrac{101.3 \text{ kPa}}{90.0 \text{ kPa}} \times \dfrac{200 \, K}{273 \, K}$
   $= 3.32 \times 10^3 \text{ dm}^3$

f. $V = 139 \text{ cm}^3 \times \dfrac{78.4 \text{ kPa}}{101.3 \text{ kPa}} \times \dfrac{273 \, K}{298 \, K} = 98.6 \text{ cm}^3$

3. $P = \dfrac{\dfrac{7.85 \text{ g}}{44.0 \text{ g/mol}} \times 8.31 \, \dfrac{\text{dm}^3 \cdot \text{kPa}}{\text{mol} \cdot K} \times 300 \, K}{19.6 \text{ dm}^3}$
   $= 22.7 \text{ kPa}$

4. $V = \dfrac{\dfrac{2.3 \text{ g}}{4.00 \text{ g/mol}} \times 8.31 \, \dfrac{\text{dm}^3 \cdot \text{kPa}}{\text{mol} \cdot K} \times 348 \, K}{101 \text{ kPa}}$
   $= 16 \text{ dm}^3$

5. unknown gas = a O_2 = b

$$1.5 = \sqrt{\dfrac{\text{mol mass a}}{32.0 \text{ g/mol}}}$$

73T

$$(1.5)^2 = \frac{\text{mol mass a}}{32.0 \text{ g/mol}}$$

$$2.25 = \frac{\text{mol mass a}}{32.0 \text{ g/mol}}$$

$$2.25 \times 32.0 \text{ g/mol} = \text{mol mass}$$
$$72 \text{ g/mol} = \text{mol mass}$$

6. $H_2 = a \qquad C_2H_6 = b$

$$\frac{\text{rate } H_2}{\text{rate } C_2H_6} = \sqrt{\frac{30.0 \text{ g/mol}}{2.0 \text{ g/mol}}}$$

$$\frac{\text{rate } H_2}{\text{rate } C_2H_6} = \sqrt{15} = 3.87$$

7. $V = 1.70 \times 10^3 \text{ dm}^3 \times \dfrac{99.3 \text{ kPa}}{20.6 \text{ kPa}} \times \dfrac{227 \text{ K}}{296 \text{ K}}$

$$= 6.28 \times 10^3 \text{ dm}^3$$

8. Since the coefficients of CO_2 to O_2 in the balanced equations are one to one, $2.5 \times 10^{10} \text{ dm}^3$ of CO_2 are produced.

9. a. $\frac{1}{4}$ atm   **c.** $\frac{3}{4}$ atm
   b. $\frac{1}{2}$ atm   **d.** $\frac{1}{2}$ atm

10. $P = \dfrac{9.3 \text{ kPa} \times 12.7 \text{ L}}{20.1 \text{ L}} = 5.9 \text{ kPa}$

11. $95.3 \text{ kPa} = P_{H_2} + 4.24 \text{ kPa}$
$91.1 \text{ kPa} = P_{H_2}$

12. $V = \dfrac{10.0 \text{ dm}^3 \times 100.2 \text{ kPa}}{325.5 \text{ kPa}} = 3.08 \text{ dm}^3$

13. $V = \dfrac{\dfrac{7.2 \text{ g}}{39.9 \text{ g/mol}} \times 0.082 \dfrac{\text{L} \cdot \text{atm}}{\text{mol} \cdot \text{K}} \times 351 \text{ K}}{0.53 \text{ atm}} = 9.8 \text{ L}$

14. The balanced equation for the reaction is:

$$Fe_2O_3 + 3CO \longrightarrow 2Fe + 3CO_2$$

Three moles of CO are needed for each mole of Fe_2O_3 that reacts. *gram → mol, mol → mol*

$$?L = 95.34 \text{ g} \times \frac{1 \text{ mol } Fe_2O_3}{159.694 \text{ g}} \times \frac{3 \text{ mol CO}}{1 \text{ mol } Fe_2O_3}$$

ordinary conditions $\times \dfrac{24.5 \text{ L CO}}{1 \text{ mol CO}} = 43.9 \text{ L of CO}$ *mol → g (ℓ)*

15. The balanced equation for the reaction is:

$$4NH_3 + 7O_2 \longrightarrow 4NO_2 + 6H_2O$$

For every 4 dm^3 of NH_3 that are burned, 7 dm^3 of O_2 are required.

$$?dm^3 = 20.3 \text{ dm}^3 NH_3 \times \frac{7 \text{ dm}^3 O_2}{4 \text{ dm}^3 NH_3}$$
$$= 35.5 \text{ dm}^3 \text{ of } O_2$$

16. $V = \dfrac{255 \text{ dm}^3 \times 67.4 \text{ kPa}}{145.1 \text{ kPa}} = 1.18 \text{ dm}^3$

Challenge, page 219

1. At STP one mole of O_2 gas occupies 22.4 cm^3.

$$\text{density} = \frac{32.0 \text{ g}}{22.4 \text{ dm}^3} = 1.43 \text{ g/dm}^3$$

2. At 25°C and standard pressure one mole of the gas will occupy 24.5 dm^3.

$$\text{density} = \frac{46.0 \text{ g}}{24.5 \text{ dm}^3} = 1.88 \text{ g/dm}^3$$

3. $1.25 \text{ g/dm}^3 = \dfrac{\text{molar mass}}{22.4 \text{ dm}^3}$

$$(1.25 \text{ g/dm}^3)(22.4 \text{ dm}^3) = \text{molar mass}$$
$$28.0 \text{ g/mol} = \text{molar mass}$$

4. One mole of Mg produces one mole of H_2.

$$1.31 \text{ g Mg} \times \frac{1 \text{ mol Mg}}{24.3 \text{ g}} \times \frac{1 \text{ mol } H_2}{1 \text{ mol Mg}} \times \frac{22.4 \text{ L}}{1 \text{ mol } H_2}$$
$$= 1.21 \text{ L of } H_2$$

5. $V = \dfrac{\dfrac{155 \text{ g} \times 0.15}{18.0 \text{ g/mol}} \times 8.31 \dfrac{\text{L} \cdot \text{kPa}}{\text{mol} \cdot \text{K}} \times 373 \text{ K}}{101 \text{ kPa}} = 39.6 \text{ L}$

6. a.

b. $\dfrac{\Delta V}{\Delta x} = \dfrac{5.79 \text{ L} - 4.28 \text{ L}}{410 \text{ K} - 303 \text{ K}} = \dfrac{1.51 \text{ L}}{107 \text{ K}} = 0.0141 \text{ L/K}$

c. $\dfrac{V}{T} = K \qquad \dfrac{4.28 \text{ L}}{303 \text{ K}} = 0.0141 \text{ L/K}$

d. $V/T = 0.0141 \text{ L/K}$

$$\frac{V}{1200 \text{ K}} = 0.0141 \text{ L/K}$$
$$V = 17 \text{ L}$$

7. $T = 85.9 \text{ cm}^3 \times \dfrac{308 \text{ K}}{70.0 \text{ cm}^3} = 378 \text{ K}$

8. $\text{density} = \dfrac{3.929 \text{ g}}{2.0000 \text{ dm}^3} = 1.964 \text{ g/dm}^3$

9. $\dfrac{43.9 \text{ cm}^3}{22\,400 \text{ cm}^3} \times 100 = 0.196\%$

Synthesis, page 219

1. $1000 \text{ g NaHCO}_3 \times \dfrac{1 \text{ mol NaHCO}_3}{84.0 \text{ g}}$

$$\times \frac{1 \text{ mol } NH_3}{1 \text{ mol NaHCO}_3} = 11.9 \text{ moles } NH_3$$

The mole ratios in the equation are all 1:1, so 11.9 moles each of NH_3 and CO_2 are needed.

$$V = \frac{11.9 \text{ mol} \times 8.31 \dfrac{\text{dm}^3 \cdot \text{kPa}}{\text{mol} \cdot \text{K}} \times 327 \text{ K}}{150.2 \text{ kPa}}$$
$$= 215 \text{ dm}^3 \text{ each of } NH_3 \text{ and } CO_2$$

74T

2. $N_2 + 2O_2 \longrightarrow 2NO_2$

The number of moles of N_2 reacted are

$$\frac{5.0 \times 10^8 \text{ dm}^3}{24.5 \text{ dm}^3/\text{mol}} = 2.04 \times 10^7 \text{ mol.}$$

The balanced equation states that 1 mol of N_2 is needed for every 2 mol of NO_2 produced:

$$2.04 \times 10^7 \text{ mol } N_2 \times \frac{2 \text{ mol } NO_2}{1 \text{ mol } N_2} \times \frac{46.0 \text{ g } NO_2}{1 \text{ mol } NO_2}$$

$$\times \frac{\text{metric ton}}{1 \times 10^6 \text{ g}} = 1900 \text{ metric tons } NO_2$$

3. As a rainstorm approaches, water vapor replaces more and more of the gas molecules in a given volume of air. Because the mass of a water molecule is less than the average mass of gas molecules, the air becomes less dense. Air pressure is reduced because the humid air is less dense.

CHAPTER 8 Composition of the Atom

CHAPTER ORGANIZER

USE WITH SECTION	DEMONSTRATION	WORKSHEET	LAB EXPERIMENT	SOFTWARE	TEACHER'S RESOURCE BINDER
8-2	8-A				
8-4	8-B	8-A			
8-6		8-B			
8-8		8-C			
8-9		8-D		Disk 1	
8-10		8-E, 8-F		Disk 1	
8-11		8-G			
8-12		8-H			
Chapter Test					

■ Objectives

When the students have completed the chapter, they will be able to

1. Describe the nature of scientific modeling.
2. Describe how J. J. Thomson investigated the charge-to-mass ratios of an electron and proton using a cathode-ray tube.
3. Explain how Thomson concluded that atoms contain protons and electrons.
4. Relate the nature of radioactivity to Ernest Rutherford's experiments on the nucleus of the atom.
5. Draw a model of the Thomson atom and the Rutherford atom.
6. Define atomic number, mass number, and isotopes.
7. Determine the atomic mass of an element, given the masses and percent abundance of the element's isotopes.
8. Describe how the mass spectrometer is used to determine the relative masses and abundances of isotopes.

■ Section Notes

8-1 The concept of scientific modeling is an important part of Chapters 8, 10, 11, and 12. Exposure to the historical development of atomic theory will help students to appreciate how many scientific ideas and models develop. A model is an attempt to explain reality. Each successive model improves the understanding of reality by integrating newly discovered phenomena. For example, Thomson incorporated the results of his experiments with electrons and protons into his model of the atom.

8-2 The electrical nature of the atom was suggested long before Thomson's experiments. Sir Humphrey

Davy and especially Michael Faraday investigated the relationship between the amount of a substance produced through electrolysis and the amount of electricity used to produce the substance. The same amount of electrical current is needed to produce a given amount of $H_2(g)$ from a solution of H_2SO_4 no matter what the size of the electrodes, the concentration of the acid, or the voltage applied. If a certain quantity of electricity produces a given number of atoms (hydrogen), then perhaps electricity is associated with particles just like with atoms themselves. Thomson's experiments confirmed the particulate nature of electricity, and also verified that atoms are composed of particles with an electric charge.

8-4 Charge-to-mass ratio does not vary for electrons of different atoms, since electrons are the same in charge and mass for all atoms. However, charge-to-mass ratios do vary for the positive ions of gaseous atoms stripped of at least one electron. The charges of these ions were found to be integers, usually $+1$ or $+2$. The heavier the mass of the ion, the less the deflection by the electric field. For example, an Ar^+ gaseous ion is deflected less than a Ne^+ ion. Thomson's experiments with the cathode ray tube were the basis for the mass spectrometer.

The positive beam produced from H_2 gas has the highest e/m ratio because H^+ is the ion with the smallest mass. Hydrogen gas produces only one positive beam, consisting of H^+ ions, or protons.

8-6 Radioactivity is mentioned here because it had a direct bearing on Rutherford's gold foil experiment. Chapter 9 deals with radioactivity in more detail. If you choose to skip Chapter 9, you may want to introduce nuclear decay at this time. Nuclear equations can be introduced with Sections 8-9 and 8-10, when students have a better understanding of the structure of the atom.

8-8 The existence of neutrons was predicted a decade before James Chadwick's discovery in 1932. Frederick Soddy and Francis Aston experimented with the cathode-ray tube and found that some gases formed positive ions with different masses. Neon, for example, formed two beams, one with mass 20 and the other with mass 22, leading to the discovery of the isotopes of neon. Soddy and Aston accounted for isotopes by postulating the existence of neutral particles in the nucleus of the atom.

8-9 The arrangement of elements in the periodic table is based on the atomic number of the elements. In 1912 Henry G. J. Moseley observed that the frequency of X rays emitted by excited atoms is directly related to their atomic number. When atoms are bombarded with a powerful electron beam, electrons in the lower energy levels of the atom are rejected. Outer energy electrons fill the spaces left by the ejected electrons. In the process X rays are given off. The frequency of X rays increases with an increase in atomic number. Therefore, the number of protons can be identified by the spectra of X rays given off by an element. The direct relationship between X ray spectra and atomic number made it possible to predict with a great deal of certainty: 1) the correct sequencing of elements by atomic number; 2) the fact that atomic numbers increase by integers; 3) a number of places in the periodic table left for unknown elements, such as technetium, promethium, and rhenium.

Moseley's scientific career ended abruptly and tragically in his twenties when he was killed in World War I.

Review and Practice
page 228

1. to explain phenomena and facilitate communication of knowledge; also, to picture reality
2. deflection of the beam by a magnet
3. The shadow cast by the anode showed the beam originated at the negative electrode.
4. the charge-to-mass ratio
5. e = charge in coulombs
   m = mass in grams
6. The proton has more mass and is positively charged; the electron is negatively charged.

page 233

1. The bulk of the atom is protons, with electrons scattered inside.
2. alpha particles, beta particles, and gamma rays
3. large angle deflections
4. by suggesting the existence of a small, dense nucleus
5. a subatomic particle with approximately the same mass as a proton and no charge

page 242

1. the proton
2. the number of protons plus the number of neutrons
3. $p = 30$; $e^- = 30$; $n = 35$
4. atoms of the same element with different numbers of neutrons
5. 10.8 amu
6. isotope identification and chemical analysis of the atmosphere on other planets; also identification of chemical composition, determination of isotope mass

7. analysis of the Martian atmosphere and soil for carbon-based compounds

Review, page 243

1. Dalton thought of the atom as a solid sphere. Atoms of different elements were spheres of different sizes.
2. the cathode
3. An electron is deflected in a magnetic field and in an electric field.
4. 9.11×10^{-28} gram
5. Rutherford was surprised that some alpha particles were deflected from the gold foil at very large angles.
6. A beam of particles was found to be produced by bombardment of beryllium with alpha particles. James Chadwick demonstrated that the particles were neutrally charged.
7. protons and neutrons
8. in the nucleus
9. There is a different number of neutrons in the nuclei of atoms of different isotopes. All isotopes of the same element contain the same number of protons and electrons.
10. to measure the number of isotopes, their precise masses, and their abundance in a sample
11. the charge and the mass
12. 2—helium; 20—calcium; 33—arsenic; 84—polonium
13. hydrogen

Interpret and Apply, page 243

1. Both are dependent on scientific models to represent something that cannot be seen. A model of a dinosaur is reconstructed from traces of evidence it has left behind. From these, information about how the animal lived may be surmised. A model of the atom can be developed from evidence of how it behaves under certain experimental conditions. The model is used to make predictions about the behavior of atoms under other conditions.
2. Thomson thought that the positive charge was evenly spread throughout the atom.
3. The charge on the electron would be larger.
4. $Ar(g) + energy \longrightarrow Ar^+(g) + e^-$
5. All other things being equal, the larger the nuclear charge, the larger the deflection of the alpha particle. The deflection of alpha particles by gold foil would be greater than the deflection caused by aluminum foil.
6. $^{26}_{12}Mg$, $^{57}_{26}Fe$, $^{123}_{51}Sb$

7. Na^+ would have a larger charge-to-mass ratio. Both ions have the same charge. However, Na^+ has a smaller mass than K^+. The ratio of charge to mass of K^+ would be smaller.
8. From Figure 8-17 students should be able to identify one or more neutron-proton ratios as stable. One possible answer is 50:40 (1.25). Other answers may include 51:40 (1.28), 52:40 (1.3), 54:40 (1.35), and 56:40 (1.4).
9. The right side of symbols is reserved for charge, written as superscripts, or for the number of atoms of an element in a formula, written as subscripts.

10.

11. Neutrons are not charged. Therefore, they are not affected by an electric or magnetic field.
12. He^{2+}

Problems, page 244

1. Since the mass of one electron is 9.11×10^{-28} g, one gram of electrons is the reciprocal of this number: 1.10×10^{27} electrons/gram.

   One mole of electrons would have a mass of
   $$\frac{9.11 \times 10^{-28} \text{ g}}{1 \text{ electron}} \times \frac{6.02 \times 10^{23} \text{ electrons}}{1 \text{ mol}}$$
   $$= 5.48 \times 10^{-4} \text{ gram/mol.}$$

2. Nine different water molecules can be made:

   1. $^1H^{16}O^1H$ 6. $^2H^{17}O^2H$
   2. $^1H^{16}O^2H$ 7. $^1H^{18}O^1H$
   3. $^2H^{16}O^2H$ 8. $^1H^{18}O^2H$
   4. $^1H^{17}O^1H$ 9. $^2H^{18}O^2H$
   5. $^1H^{17}O^2H$

3.

		HELIUM = 4	HELIUM = 3
a.	atomic number	2	2
b.	number of protons	2	2
c.	number of neutrons	2	1
d.	mass number	4	3
e.	nuclear charge	2+	2+

4. $$\frac{\text{Vol atom}}{\text{Vol nucleus}} = \frac{\frac{4}{3}\pi r^3 \text{ atom}}{\frac{4}{3}\pi r^3 \text{ nuc}} = \frac{r^3 \text{ atom}}{r^3 \text{ nuc}}$$
   $$= \frac{(10^{-8})^3}{(10^{-13})^3} = \frac{10^{-24}}{10^{-39}} = 10^{15}$$

5. The atomic mass of magnesium is
   $(23.98504 \times 0.7870) + (24.98584 \times 0.1013)$
   $+ (25.98259 \times 0.1117) = 24.31$ amu

6.

SYMBOL	Kr	Mn	U	Dy
atomic number	36	25	92	66
mass number	84	80	235	164
number of protons	36	25	92	66
number of neutrons	48	55	143	98
number of electrons	36	25	92	66

7. 1.0×10^{10} atoms $\times\ 0.0072 = 7.2 \times 10^7$ atoms

8. 53 protons, 78 neutrons, 54 electrons

Challenge, page 244

1. The enlargement is:

$$\frac{(1.5 \, \cancel{cm}^3 (\frac{4}{3}\pi))}{(0.77 \times 10^{-8} \, \cancel{cm})^3 (\frac{4}{3}\pi)} = 7.4 \times 10^{24}$$

2. Atomic mass does not increase between Ar and K, Co and Ni, Te and I, and in several places in the transuranium elements due to the abundance of isotopes for these pairs of elements. For example, the abundance of argon 40 is 99.60% while the abundance of less massive potassium 39 is 93.10%.

3. Several factors contribute to the greater instability of lighter radioisotopes. One factor depends on the neutron to proton ratio in the nucleus. A change of one or two protons affects the neutron to proton ratios less among heavy elements than among lighter elements. Therefore, most of the naturally occurring radioisotopes among lighter elements are highly unstable and disintegrate quickly. Many radioisotopes among heavier elements exist for a long time.

4. An alpha particle contains four subatomic particles, each having a mass of 1.67×10^{-24} g. The ratio of the mass of an alpha particle to the mass of an electron is

$$\frac{4 \times 1.67 \times 10^{-24} \, \cancel{g}}{9.11 \times 10^{-28} \, \cancel{g}} = 7330.$$

5. The spectrum has two groups of three lines each. One group will be the 1^+ ions of each of the three isotopes, and the second group will be the 2^+ ions. Within each group the line for the Si-28 isotope will curve the most and the line for the Si-30 will curve the least. (The heaviest lines will be those for Si-28, which has an abundance of 92%.)

6. The mass of the proton is

$$\frac{1.60 \times 10^{-19} \, \cancel{coulomb}}{9.58 \times 10^4 \, \cancel{coulomb}/gram} = 1.67 \times 10^{-24} \text{ gram}$$

7.

8. molar mass of the electron

$$= (9.11 \times 10^{-28} \text{ g/}e^-)(6.02 \times 10^{23} e^-/\text{mol})$$
$$= 5.48 \times 10^{-4} \text{g/mol}$$

molar mass of H atom = 1.0078 g/mol

mass of a mole of protons

$$= 1.0078 \text{ g/mol} - 5.48 \times 10^{-4} \text{ g/mol}$$
$$= 1.0073 \text{ g/mol}$$

$$\text{mass of one proton} = \frac{1.0073 \text{ g/}\cancel{mol}}{6.02 \times 10^{23} \text{ protons/}\cancel{mol}}$$

$$= 1.67 \times 10^{-24} \text{ g/proton}$$

Synthesis, page 245

1. At STP, one mole of gas occupies 22.4 dm³. Therefore, 1 cm³ contains 1/22 400 mol

$$= 4.5 \times 10^{-5} \text{ mol.}$$

This 1 cm³ contains $(4.5 \times 10^{-5})(6.0 \times 10^{23})$

$$= 2.7 \times 10^{19} \text{ molecules at STP.}$$

All other things being equal, the pressure in the tube is directly proportional to the number of particles present. When a glow starts at 0.01 atm, there will be 2.7×10^{17} molecules/cm³. When the glow disappears at 10^{-6} atm, there will be 2.7×10^{13} molecules/cm³.

2. He + energy $\longrightarrow$ He^{2+} + 2e$^-$

3. $\text{mass of one nucleus} = \dfrac{19.9984 \text{ g}}{6.02 \times 10^{23}}$

$$= 3.16 \times 10^{-23} \text{ g/nucleus}$$

vol. of one nucleus $= \frac{4}{3}\pi r^3$

$$= \frac{4}{3}\pi(5 \times 10^{-13} \text{ cm})^3 = 5 \times 10^{-37} \text{ cm}^3$$

$$\text{density of nucleus} = \frac{3.16 \times 10^{-23} \text{ g/}\cancel{nucleus}}{5 \times 10^{-37} \text{ cm}^3/\cancel{nucleus}}$$

$$= 6 \times 10^{13} \text{ g/cm}^3$$

4. $\text{charge density} = \dfrac{\text{charge}}{\frac{4}{3}\pi r^3} = \dfrac{\text{charge}}{4.19 \, r^3}$

$$\begin{array}{l}\text{charge density} \\ \text{(Thomson atom)}\end{array} = \frac{1.6 \times 10^{-19} \text{ coulomb}}{4.19 \, (10^{-8})^3 \text{ cm}^3}$$

$$= 3.8 \times 10^4 \text{ coulombs/cm}^3$$

$$\begin{array}{l}\text{charge density} \\ \text{(Rutherford atom)}\end{array} = \frac{1.6 \times 10^{-19} \text{ coulomb}}{4.19 \, (10^{-13})^3 \text{ cm}^3}$$

$$= 3.8 \times 10^{19} \text{ coulombs/cm}^3$$

The charge density of the Rutherford atom is fifteen magnitudes greater than the charge density of the Thomson atom.

CHAPTER 9 Nuclear Chemistry

CHAPTER ORGANIZER

USE WITH SECTION	DEMONSTRATION	WORKSHEET	LAB EXPERIMENT	SOFTWARE	TEACHER'S RESOURCE BINDER
9-2		9-A			
9-4		9-B, 9-C			
9-6					Advanced Topic 2
9-7		9-D			
9-10		9-E			
9-11		9-F			
9-13					Chem Issue 3
9-15		9-G			
Chapter Test					

■ Objectives

When the students have completed the chapter, they will be able to

1. Distinguish between a chemical change and a nuclear change.
2. List some of the sources of radiation.
3. Write a nuclear equation for an alpha or a beta decay.
4. Calculate the age of an object given its present radioactivity and the half-life of carbon-14.
5. Explain how radioactive isotopes are used as tracers.
6. Explain the significance of binding energy.
7. Explain how a nuclear power plant generates energy for electrical power.
8. Distinguish between nuclear fission and nuclear fusion.
9. Explain the problems involved with the disposal of nuclear wastes in the United States.
10. Describe the internal structure of protons and neutrons in terms of quarks.

■ Section Notes

9-1 All isotopes of an element, radioactive or not, have similar chemical properties. The present form of the actinide series in the periodic table of elements is based on the chemical properties of radioisotopes.

9-2 Students may be interested to know some of the effects of whole-body radiation exposure from a single dose in a short period of time. There is no visible effect from a dose of 0 to 25 rems. The white blood cell count drops slightly with a dose of 25 to 50 rems. Nausea, fatigue, vomiting, and loss of hair result from a dose of 100 to 300 rems. With a dose of 300 to 500 rems, the white blood cell count drops to zero, and hemorrhages and ulcers develop. A dose above 500 rems is fatal about 50 percent of the time.

9-3 Alpha emission occurs in some radioactive isotopes because their nuclei are unstable. Instability is predictable in isotopes with a mass number over 200 by a low neutron/proton ratio.

Beta emission generally occurs in nuclides that have a neutron/proton ratio that is too high to be stable. The daughter nuclides of both alpha and beta emissions have neutron/proton ratios that lie closer to the stability curve than the original isotope (see Figure 8-17).

Emission of gamma radiation during decay reduces the energy in the nucleus, resulting in a more stable nucleus. Gamma radiation usually accompanies alpha and beta radiation. However, beta emission sometimes does not result in a daughter nuclide in the nuclear ground state. The resulting excited-state daughter nuclide may decay by gamma emission. For example, an excited thorium-229 nuclide may decay to ground state thorium-229 by gamma emission.

9-4 The law of radioactive decay, first formulated by Rutherford and Soddy, states that 1) when a nucleus emits an alpha particle, the resulting isotope is two places to the left of the parent nuclide in the periodic table, and 2) when a nucleus emits a beta particle, the resulting isotope is one place to the right of the parent nuclide in the periodic table.

9-5 American scientists have suggested the names Rutherfordium for element 104 and Hahnium for 105. However, Soviet scientists have suggested Bohrium for 104 and Kurchatovium for 105. A compromise was reached by the IUPAC. Elements with atomic numbers greater than 103 are named after a system that uses labels for the digits of the atomic numbers.

1 = un	3 = tri		5 = pent	7 = sept	9 = enn
2 = bi	4 = quad		6 = hex	8 = oct	0 = nil

Therefore, element 1-0-4 is unnilquadium (Unq), and 1-0-8 is unniloctium (Uno).

9-10 The first successful fission reaction was accomplished by Otto Hahn and Fritz Strassman in Germany in 1938. Their experiments demonstrated that when slow-moving neutrons bombard uranium-235, the resulting nuclei were much lighter than the uranium. Scientists throughout the world were able to verify these findings. The release of energy from the fission of 1000 moles of uranium atoms in 0.0005 second was recognized as a potential source of destructive power. The race began to develop the atomic bomb.

9-13 In vitrification, radioactive waste is dissolved in molten glass, which is poured into graphite molds. When the glass cools, a solid, black, homogeneous cylinder results. The radioactive nuclides are trapped in the silicate structure of the glass. The cylinders are placed in containers and are ready for storage. Although the cylinders are still radioactive, the material they contain is not free to move through the environment via such mechanisms as wind or percolation through groundwater unless the glass structure is completely destroyed (pulverized), which is unlikely.

9-14 Protons and neutrons are particles that are categorized as baryons (subnuclear particles thought to be composed of three quarks). Baryons are a subgroup of hadrons. The other hadrons are mesons (quark antiquark pairs) and antibaryons (three antiquarks). No one has been able to isolate an individual quark.

Electrons are in the class of particles called leptons. Unlike quarks, leptons are not fractionally charged and they do not combine to form more complex particles. Particles that are classified as leptons are electrons, electron neutrinos, muons, taus, and others.

You may want to take time for class discussion of the many issues related to this chapter. Here are some suggestions for discussion.

If we *do not* use nuclear energy, how will we supply the energy that supports our present way of life?

If we *do* use nuclear energy, how will we handle the waste?

Is the knowledge that might be gained from a "super accelerator" worth the tremendous cost? Who should pay for it? What do we *not* buy with the money spent on such research?

Review and Practice _____
page 251

1. Nuclear reactions involve changes among protons and neutrons in the nucleus. Chemical reactions result from the interaction of electrons between atoms.
2. Ionizing radiation produces ion pairs in matter, nonionizing radiation does not.

IONIZING	NON-IONIZING
X rays	visible light
cosmic rays	infrared light
gamma rays	ultraviolet light

3. naturally occurring radiation in the environment
4. $^{218}_{84}Po \longrightarrow\ ^{214}Pb + ^{4}_{2}He$
5. $^{214}_{82}Pb \longrightarrow\ ^{214}_{83}Bi + ^{0}_{-1}e$
6. Bombard the nucleus with a particle of sufficient energy to produce instability.

page 257

1. changing one element into another by way of a nuclear reaction
2. Large nuclei are bombarded with smaller nuclei. The nuclei combine, forming an element with an atomic number greater than that of either original nucleus.
3. Tremendous amounts of energy are needed to cause atomic nuclei to combine.
4. Particle accelerators give particles enough kinetic energy to penetrate the target nuclei and cause transmutation.
5. the time it takes for half of the nuclei in a radioactive sample to decay
6. 5730 years

page 259

1. A radioactive nuclide used to follow chemical or physical changes in a material or process; A tracer behaves chemically like a nonradioactive form of the same element and can be detected at each stage in the chemical reaction.
2. Cancer cells are more vulnerable to the effects of radiation than some healthy cells.
3. Radioactive iodine accumulates in the gland and subsequently destroys tumor cells.

page 264

1. Matter is converted to energy.
2. mass and energy
3. the difference between the mass of the nucleus and the sum of the masses of its protons and its neutrons
4. In the reaction
$$^{211}_{84}Po \longrightarrow ^{207}_{82}Pb + ^4_2He$$
the loss of mass is
210.9405 g − (206.9309 g + 4.0015 g) = 0.0081 g
$E = (0.0081 \text{ g} \times 1 \text{ kg}/1000 \text{ g})(3.00 \times 10^8 \text{ m/s})^2$
$E = 7.29 \times 10^8 \text{ kJ}$
5. Binding energy is the amount of energy needed to break a nucleus into its individual protons and neutrons.
6. $E = (0.0990 \text{ g/mol} \times 1 \text{ kg}/1000 \text{ g})(3.00 \times 10^8 \text{ m/s})^2$
$E = 8.91 \times 10^{12} \text{ J/mol}$
$= 8.91 \times 10^{12} \text{ J/mol} \times 1 \text{ mol}/6.02 \times 10^{23} \text{ nuclei}$
$= 1.48 \times 10^{-11} \text{ J/nucleus}$

page 269

1. In fission, large nuclei are split; in fusion, small nuclei are fused together.
2. Neutrons bombard atomic nuclei causing them to split apart. More neutrons are released and bombard other nuclei, and so on.
3. Control rods placed between fuel rods absorb the neutrons and slow the reaction.
4. Water is a moderator. It slows the speed of the neutrons so that they can be absorbed by the nuclei. Water also transfers heat from the reactor to the steam turbine.
5. sustaining a sufficiently high temperature in a confined volume
6. There is so much; it is highly radioactive; it is dangerous for a long time.

page 272

1. a theoretical particle with a charge of either $\frac{1}{3}$ or $\frac{2}{3}$ that of an electron
2. proton = 2 up, 1 down
neutron = 2 down, 1 up
3. Leptons and quarks are both thought to be fundamental particles, that is, particles not composed of any smaller particles. An electron is a lepton.
4. antiparticles with the same characteristics as specific particles of matter, but with opposite charge

Review, page 273

1. In an alpha emission, an unstable nucleus emits a particle composed of two protons and two neu-

trons. The atomic number of the nucleus is reduced by 2. The mass number is reduced by 4.
2. If the mass of material is small, many neutrons will pass through without striking a nucleus to cause another disintegration. As the mass is increased, the chance of a neutron hitting a nucleus also increases.
3. The half-life of carbon-14 is 5730 years.
4. Green plants absorb carbon dioxide. A small amount of this carbon dioxide is composed of carbon-14 atoms. Plants will take up carbon-14 dioxide as readily as carbon-12 dioxide.
5. When a sample undergoes a nuclear change, a change in mass between reactants and products can be measured. Einstein's equation, $E = mc^2$, can be applied to calculate the amount of energy released.
6. The protons and neutrons in an atom or ion are not involved in a chemical reaction.
7. Answers will vary. Possible answers include

Np—planet Neptune	Pu—planet Pluto
Am—America	Cm—Curie
Bk—Berkeley, CA	Es—Einstein
Fm—Fermi	Md—Mendenleev
No—Nobel	Lr—Lawrence

8. The cyclotron operates on the principle of the attraction of a charged particle toward an opposite electric charge. Alternating current accelerates the charged particle each time it moves across the gap in the spiral path.
9. The average United States citizen is exposed to 100 millirems of radiation per year.
10. The total mass of the products is less than that of the starting material. The lost mass is converted to energy.
11. Nuclear waste may be buried in cannisters deep underground in beds of salt. Spent nuclear fuel rods may be reprocessed, using alpha emitters to make new rods, while remaining wastes are converted into a solid, usually glass, that can be stored underground.
12. In nuclear fusion, light nuclei are combined to make one heavier nucleus.

Interpret and Apply, page 273

1. In a nuclear reaction, a measurable amount of mass is changed into energy.
2. Yes. If the mole percent of carbon-14 increased for several years, then the rate of beta emissions in matter from plants that died during that time would be greater than usual. An object would be dated younger than it really is.
3. The mass defect is the difference in mass between the mass of individual particles that compose a

nucleus and the actual mass of the nucleus. The mass of the nucleus is less because energy is needed to bring protons and neutrons together in the nucleus.

4. Since the half-life of uranium-238 is approximately the age of Earth, about half of the original uranium-238 is still decaying.

5. Sulfur-35. The other isotope has such a short half-life that it would be difficult to work with it.

6. Low doses of radiation do not cause immediate measurable physiological change. Many factors influence the health of an individual, such as heredity, stress, food consumption, and environmental factors like air and water. It is difficult to filter out the effect of radiation from other factors.

7. A particle with three "up" quarks would probably have too large a positive charge to be stable.

8. Three "down" quarks produce a 1^- charge. (Two "antiup" quarks and one "antidown" quark also produce a 1^- charge.)

9. If Ca-47 and P-32 are substituted for nonradioactive calcium and phosphorus in $Ca_3(PO_4)_2$, they act as tracers that produce an image on a scanner. The image shows doctors where there are bone cells that may not be functioning properly.

Problems, page 274

1. a. $^{234}_{90}Th$ **d.** $^{4}_{2}He$
  b. $^{231}_{91}Po$ **e.** $^{0}_{-1}e$
  c. $^{211}_{82}Pb$

2. For strontium-90
  10 half-lives $\times$ 28 yrs/half-life = 280 yrs

  For iodine-131
  10 half-lives $\times$ 8.05 days/half-life = 80.5 days

3. 360 days/60 days/half-life = 6 half-lives
  In 6 half-lives, the radioactivity is reduced by $\frac{1}{2} \times \frac{1}{2} \times \frac{1}{2} \times \frac{1}{2} \times \frac{1}{2} \times \frac{1}{2} = 0.0156 \times 100 = 1.56\%$ of the original radiation.

4. The ratio of radioactive decay is
  15.3/0.96 = 15.9.

  The radioactivity has decreased by a factor of 15.9 or 16.

  $16^{1/2}$ = 4 half-lives
  The half-life of carbon-14 is 5730 years.
  5730 years/half-life $\times$ 4 half-lives = 23 000 years

5. $E = \left(1.68 \times 10^{-4} g \times \dfrac{1\ kg}{1000\ g}\right)\left(3.00 \times 10^8\ \dfrac{m}{s}\right)^2$
  $E = 1.5 \times 10^{10}\ J = 1.51 \times 10^7\ kJ$

6. The change in mass is
  221.9703 g $-$ (217.9628 g + 4.0015 g) = 0.0006 g.
  $E = \left(0.006\ g \times \dfrac{1\ kg}{1000\ g}\right)\left(3.00 \times 10^8 \dfrac{m}{s}\right)^2$
  $E = 5.40 \times 10^{11}\ J/mol = 5.40 \times 10^8\ kJ/mol$

7. $E = \left(0.003\ 40\ g \times \dfrac{1\ kg}{1000\ g}\right)\left(3.00 \times 10^8\ \dfrac{m}{s}\right)^2$
  $E = 3.06 \times 10^{11}\ J/mol = 3.06 \times 10^8\ kJ/mol$
  Then the binding energy per nucleus is
  $3.06 \times 10^8\ kJ/mol \times 1\ mol/6.02 \times 10^{23}$ nuclei
  $= 5.08 \times 10^{-16}\ kJ/nucleus.$

8. $E = \left(1.9353\ g \times \dfrac{1\ kg}{1000\ g}\right)\left(3.00 \times 10^8\ \dfrac{m}{s}\right)^2$
  $E = 0.0174 \times 10^{16}\ J/mol = 1.74 \times 10^{14}\ kJ/mol$

9. $^{235}_{92}U + ^{1}_{0}n \longrightarrow (^{236}_{92}U) \longrightarrow ^{137}_{52}Te + ^{97}_{40}Zr + 2\ ^{1}_{0}n$

Challenge, page 274

1. In one hour and 20 minutes the radioisotope goes through 80 min $\times$ 1 half-life/20 min = 4 half-lives. If 1.00 g is required at the end of four half-lives, then a minimum of 16.0 g of bismuth-214 is needed.

2. In 9540 years, the sample has gone through 9540 years $\times$ 1 half-life/1590 years = 6 half-lives. $(\frac{1}{2})^6 = 0.0156 =$ the fraction that remains.

3. Essentially there is none. In 60 minutes the decay has gone through 10 half-lives. After one half-life, $\frac{1}{2}$ is left. After ten half-lives $(\frac{1}{2})^{10} = 1/1024$ is left.

4. For the radioisotope with a half-life of one to eight minutes the experiment must be done quickly. The detection apparatus must be able to count many disintegrations per second. The experiment for the radioisotope with a half-life of 150–200 days requires a detection device sensitive enough to detect relatively few disintegrations per second. In the second experiment it is necessary to separate out background radiation from the disintegration of the isotope.

5. $(\frac{1}{2})^{10} = 1/1024$

Synthesis, page 275

1. The fusion reaction is many times more energetic than the burning of hydrogen gas. $1.69 \times 10^9\ kJ/241.8\ kJ = 6.99 \times 10^6$ times more than a mole of hydrogen gas. Therefore, the mass of hydrogen needed for burning is
  $\dfrac{2\ g}{1\ mol} \times 6.99 \times 10^6\ mol = 1.40 \times 10^7\ g\ H_2$

2. $\dfrac{4.6 \times 10^8\ kJ}{1\ mol} \times \dfrac{1\ mol}{226.0254\ g} = 2.0 \times 10^6\ kJ/g$

3. Using Graham's law the expression is
  $\dfrac{rate^{235}UF_6}{rate^{238}UF_6} = \left(\dfrac{352.041\ g}{349.035\ g}\right) = 1.0043$

CHAPTER 10 Electrons in Atoms

CHAPTER ORGANIZER

USE WITH SECTION	DEMONSTRATION	WORKSHEET	LAB EXPERIMENT	SOFTWARE	TEACHER'S RESOURCE BINDER
10-3	10-A, 10-B	10-A	10-A		Advanced Topic 3
10-6		10-B			
10-7	10-C				
10-10		10-C, 10-D			
10-11		10-E			
Chapter Test					

■ Objectives

When the students have completed the chapter, they will be able to

1. Characterize components of waves such as wavelength, frequency, and amplitude.
2. Determine that visible light is composed of waves of various frequencies.
3. Define a photon as a unit of light energy.
4. List the various forms of electromagnetic radiation.
5. Explain the significance of the unique bright-line spectra of different elements.
6. Relate the bright-line spectrum of hydrogen to energy levels of electrons in the hydrogen atom.
7. Describe the hydrogen atom using the quantum theory.
8. Describe the concept of orbitals.
9. Describe the orbitals of the hydrogen atom in order of increasing energy.
10. Explain the sequence of filling of orbitals in many-electron atoms.
11. Write electron configurations of many-electron atoms and monatomic atoms.

■ Section Notes

10-1 The Rutherford model of the atom introduced the nucleus but did not explain why atoms do not collapse. According to the laws of classical physics, forces of attraction should cause the electrons to plummet toward the nucleus and lose energy. No such energy change was observed in the atom and none of Rutherford's contemporaries had a reasonable explanation for its stability. Later, Niels Bohr proposed that electrons exist in certain energy states, or levels. His model of the atom successfully explained the bright-line spectrum of hydrogen.

10-2 In this section emphasize that an understanding of waves plays an essential role in the modern model of the atom. The "kitchen sink" experiment can be done by all students at home. Ask students for their findings the next day. In your discussion, make clear that the waves transmit energy to the cork, but the cork is not carried along with the wave in the direction of wave propagation.

10-3 Students have been exposed to the colors of the rainbow many times. The treatment here relates the concept of color to specific frequencies of light (see Table 10-1). A demonstration is particularly useful for helping students to see the refraction of white light into colors.

The discussion of Planck's work and photons of energy may cause some consternation among your students. The text deliberately avoids Albert Einstein's experiment on the photoelectric effect as being too complex for them. Yet, you may want to clarify the particulate nature of light in terms of what happens when light strikes a surface. The energy transmitted by the light is related to the length of exposure, intensity, and frequency. If the first two conditions remain constant, the light energy varies directly with frequency. The energy is in "packets" or photons. The energy carried by each photon is directly proportional to the frequency of light.

The spring analogy is limited, but helpful for an understanding of quantization and it also is helpful in explaining the nodes in the orbitals of atoms (see Section 10-7). The spring demonstration can be done in the classroom. Wavelengths with a denominator greater than 2 do not exist in the standing wave.

10-5, 10-6 Demonstration 10-A is dramatic and simple. It can serve as a good introduction to the section. Take time to develop the relationship between the bright-line spectrum of hydrogen and the energy levels of electrons in hydrogen atoms. Move the students on to the wave mechanical model of the atom and eventually to electron configurations and the periodic table. Avoid perpetuating the students' dependence on the Bohr planetary model of the hydrogen atom, which was replaced in the 1920's.

Bohr successfully related a simple mathematical expression to the potential energy of an electron in the hydrogen atom. $E = -k/n^2$ where $k = 1312 \text{ kJ/mol}$ for hydrogen, n is an energy level with an integer value of 1, 2, 3, 4. With this equation he was able to explain the bright-line spectrum of hydrogen in the visible, ultraviolet, and infrared regions. However, the equation is only valid for hydrogen or for any nucleus with only one electron. The use of this equation introduces about a 10% error for helium, and even greater deviations for other atoms.

10-7 The mathematical equations that describe the energy of electrons are modifications of equations used to describe other kinds of waves, such as standing waves and electromagnetic waves.

The probability of finding an electron in an orbital can be explained using the spring analogy of Section 10-3. The amplitude of the standing wave is analogous to the wave function in the Schrodinger equation. The probability of finding an electron in a region of space is proportional to the square of the wave function. Since the amplitude at a node of standing wave is zero, by analogy the probability of finding an electron at the node of an orbital (for example, $2s$ or $2p$) is zero. The probability of finding an electron in a region of space gradually increases with distance from the node, just as the amplitude gradually increases with distance from a node (up to a maximum amplitude, then the amplitude decreases again to another node).

10-8 You can help students understand the nature of quantum numbers by using a rectangular object such as an eraser. Three spatial relationships are needed to completely describe the eraser in comparison to some frame of reference, such as the walls, ceiling, and floor of the room. The size of an eraser is analogous to the principal quantum number, n. The shape of the eraser is analogous to the quantum number, l. The orientation of the eraser with reference to the floor is analogous to the quantum number, m_l. You may want to ask the students what "quantum number" changes when the eraser's orientation is changed with regard to the floor, or what "quantum number" changes if a larger eraser is used.

Solutions to the Schrodinger equation are given by three quantum numbers, (n, l, and m_l), under certain boundary conditions from which orbitals are derived. The n is an integer either 1, 2, or 3. The value of l is also an integer with a value of 0 to n-1. Therefore, when n is

3, l may be 0, 1, or 2. The m_l is an integer from $-l$ to $+l$. When l is 2, as in the d orbitals, m_l may be -2, -1, 0, $+1$, or $+2$. Thus five orbitals comprise the d orbitals. The value of l determines whether the orbital is:

$$s\ (l = 0),\ p\ (l = 1),\ d\,(l = 2),\ \text{or}\ f\,(l = 3).$$

10-10 The Pauli exclusion principle is an empirical statement based on the nature of the energy of electrons in orbitals and the periodic table. The fact that only two electrons may occupy an orbital is not necessarily derived from quantum mechanics. It is the arrangement of the elements in the periodic table in sets of two (two elements in the first period, eight in the next two periods, etc.) that gives weight to the Pauli exclusion principle. Two definitions of the Pauli exclusion principle may be used. Either one is compatible with the one in the text. However, if reference back to quantum numbers seems too confusing, you may want to use the following definition: A maximum of two electrons may occupy any single orbital.

The issue of $4s$ versus $3d$ can be confusing for students. Remind them that the orbitals of many-electron atoms are not hydrogen orbitals. Because of electron-electron repulsion in the many-electron atoms, the energies of the electrons in orbitals of the same principal quantum number, n, do not have the same energy as the single electron hydrogen system of orbitals. In the Aufbau process, $4s$ always fills before the $3d$, yet $4s$ is always above $3d$. This approach makes more sense to students when they study the common ions of the transition elements. It is the $4s$ electrons that are removed first in the transition elements. This system also makes more sense for the promotion and hybridization of bonding electrons in a sp^3d or sp^3d^2 system (see the appendix).

There are several exceptions to the Aufbau process in the transition elements. Cr, Cu, Mo, Ag, and Au contain either half-filled or completely filled d orbital systems. For example, the electron configuration of Cr is $1s^2\,2s^2\,2p^6\,3s^2\,3p^6\,4s^1\,3d^5$. The orbital system $4s^1\,3d^5$ has lower energy than $4s^2\,3d^4$.

10-11 The relationship between electron configurations and periodicity is treated more fully in Chapter 11. You may want to highlight this relationship now. The electron configurations of a group of the representative elements have similar outermost s and p orbital systems. Students can attempt to predict the valence electrons of more complex representative elements by finding the element of the periodic table by period and group number. For example, tin, Sn, is in period 5 and group 14. The valence electrons for Sn are $5s^2\,5p^2$ (with $4d$ electrons between).

A notable exception to the text's method of determining electron configurations of transition elements is Ti^{2+}. The electron configuration for this ion is $1s^2\,2s^2\,2p^6\,3s^2\,3p^6\,4s^2$.

Review and Practice
page 283

1. It suggested an unstable situation in which electrons would be pulled into the nucleus and the atom would collapse.

2. ocean or water waves, sound waves, light waves, radio waves

3. Frequency is the number of waves passing a point per unit of time. Wavelength and amplitude are not measured over time. Wavelength is the distance between corresponding points of a wave, and amplitude is the distance the wave deviates from an imaginary midline running in the direction of wave propagation.

4. The cork moves up and down in the same location but does not move in the direction of the wave.

5. A photon is a packet of energy of an electromagnetic wave. A photon of violet light carries more energy.

6. red, orange, yellow, green, blue, indigo, violet

7. Frequency times wavelength is equal to a constant, c (the speed of light). As frequency increases, wavelength decreases proportionally.

8. existing only with or at specific amounts of energy

9. Energy is dependent on (proportional to) frequency.

page 294

1. a series of colored lines, each with its own specific frequency

2. It is a way of describing the total energy of an electron in reference to the nucleus of the atom. Electrons can be found only in certain specific energy levels and no others.

3. When electrons of hydrogen drop from higher energy levels to lower ones, the energy lost is emitted as photons of specific energy and thus specific wavelength.

4. a region in space around the nucleus in which electrons are likely to be found

5. The bright-line spectrum of hydrogen; Bohr successfully predicted the line spectrum of hydrogen in the IR and UV regions.

6. Ionization energy is the energy needed to remove an electron from the influence of the nucleus. Spectral lines result from energy of electrons still associated with the nucleus.

7. The region of 0.99 probability of finding the electron in the $3s$ orbital is larger than that in the $2s$.

8. Energy of a specific wavelength is absorbed by the atom.

9. An orbital does not describe the path of an electron.

10. It is derived from the mathematics of particle waves. It also describes the behavior of electrons in atoms in terms of probability rather than specific paths.

11. They would be the same orbital.

12. An absorption spectrum is continuous with dark lines at the frequencies corresponding to the energies absorbed by the atoms. The dark lines occur in the location where the colored lines of an element's bright-line spectrum would be.

page 300

1. An electron configuration represents the ordered series of orbitals occupied by electrons in an atom.

2. If the $3d$ orbitals filled first, there would be more energy in the system due to electron-electron repulsion. The atom has lower energy as a whole if the $4s$ fills first.

3.

	$1s$	$2s$	$2p$			$3s$	$3p$		
Na	⊗	⊗	⊗	⊗	⊗	⊘	○	○	○
Cl	⊗	⊗	⊗	⊗	⊗	⊗	⊗	⊗	⊘
S	⊗	⊗	⊗	⊗	⊗	⊗	⊗	⊘	⊘
Mg	⊗	⊗	⊗	⊗	⊗	⊗	○	○	○

Na	$1s^2$	$2s^2$	$2p^6$	$3s^1$	
Cl	$1s^2$	$2s^2$	$2p^6$	$3s^2$	$3p^5$
S	$1s^2$	$2s^2$	$2p^6$	$3s^2$	$3p^4$
Mg	$1s^2$	$2s^2$	$2p^6$	$3s^2$	

4. Ca^{2+} $1s^2 2s^2 22p^6 3s^2 3p^6$
   Al^{3+} $1s^2 2s^2 2p^6$
   P^{3-} $1s^2 2s^2 2p^6 3s^2 3p^6$
   F^- $1s^2 2s^2 2p^6$
   Co^{2+} $1s^2 2s^2 2p^6 3s^2 3p^6 3d^7$

5. They have opposite spins ($+\frac{1}{2}$ and $-\frac{1}{2}$).

6. No two electrons can have the same four quantum numbers.

7. There is slightly less repulsion between electrons (and thus lower potential energy for the atom) if each one occupies a separate orbital until no empty d orbitals remain.

Review, page 301

1. Both kinds of waves have characteristic wavelengths. Standing waves have nodes whereas an ocean wave does not.

2. Answers will vary. Some possible answers are radio waves, microwaves, infrared, visible light, ultraviolet, X rays, and gamma rays.

3. Planck's constant is a proportionality constant that relates the energy in an electromagnetic wave to its frequency.

4. red, blue-green, blue and violet

5. Energy is given off as visible red light.

6. Each element has its own characteristic absorption spectrum. Astronomers analyze the discrete lines in an absorption spectrum of a star to determine the star's composition and other characteristics.

7. a region in space around the nucleus of an atom in which an electron is likely to be found

8. $3s$, three $3p$ and five $3d$ orbitals

9. $2p_y$

10. the energy needed to remove an electron from an isolated neutral atom; 1312 k J

11. Rutherford's model pictured a positively charged nucleus with electrons moving about it.

12. The boat does not move horizontally with the waves, but moves up and down in place as a wave passes.

13. n, l, and m

14. Depending on whether the ion is positive or negative, the total number of electrons represented would be more or less than the neutral atoms.

15. n describes the size of the orbital; l describes the shape, and m describes the orientation in space.

Interpret and Apply, page 301

1. A standing wave only can have specific wavelengths that are integer multiples of $\frac{1}{2}$.

2. ocean waves, electromagnetic waves such as radio waves, light waves, X rays, etc.

3. The equation expresses the relationship between the energy of an electromagnetic wave and its frequency.

4. Excited atoms lose energy. If a change occurs from the fifth energy level to the second, energy is given off as blue light.

5. The energy of an electron is quantized and the wavelength of a standing wave is also quantized.

6. Zero. The probability of finding an electron close to the nucleus in both a $1s$ orbital and a $3s$ orbital is great. However, the charge density in the $1s$ orbital is greater for a given region around the nucleus than for the $3s$ orbital.

7. **a.** The orbital is not equally dense throughout. The probability of finding an electron varies within the orbital.

   b. Orbitals do not describe the path of an electron around the nucleus. They describe a region in which the electrons are likely to be found.

8. c

9. A specific amount of energy must be added to the atom. Energy is given off in the form of a photon with specific energy corresponding to the red region of the visible light spectrum.

10. **a.** boron
    b. beryllium
    c. silicon

Problems, page 302

1. Energy must be added to decrease the wavelength.

2. $E_1 - E_5 = 1260$ k J/mol added

3. $E_{photons} = E_1 - E_4 = 1230$ k J/mol
   $= 1.230 \times 10^6$ J/mol

   $E = h\nu$

   $1.23 \times 10^6 \text{ J/mol} = \dfrac{6.6262 \times 10^{34} \text{ J} \cdot \text{s}}{\text{photon}} \times \dfrac{6.02 \times 10^{23} \text{ photons}}{\text{mol}} \times \nu$

   $\nu = \dfrac{1.23 \times 10^6 \text{ J/mol}}{(6.6262 \times 10^{-34} \text{ J} \cdot \text{s})(6.02 \times 10^{23} \text{ photons})}{\text{photon}} \quad \text{mol}$

   $= 3.08 \times 10^{15}$ cycles/s

4. As $\quad 1s^2\,2s^2\,2p^6\,3s^2\,3p^6\,4s^2\,3d^{10}\,4p^3$
   Kr $\quad 1s^2\,2s^2\,2p^6\,3s^2\,3p^6\,4s^2\,3d^{10}\,4p^6$
   Br $\quad 1s^2\,2s^2\,2p^6\,3s^2\,3p^6\,4s^2\,3d^{10}\,4p^5$
   p $\quad 1s^2\,2s^2\,2p^6\,3s^2\,3p^3$

5. $K^+ \quad 1s^2\,2s^2\,2p^6\,3s^2\,3p^6$
   $O^{2-} \quad 1s^2\,1s^2\,2p^6$
   $Br^- \quad 1s^2\,2s^2\,2p^6\,3s^2\,3p^6\,4s^2\,3d^{10}\,4p^6$
   $Ga^{3+} \quad 1s^2\,2s^2\,2p^6\,3s^2\,3p^6\,3d^{10}$

6.

	$1s^2$	$2s^2$	$2p^6$	$3s^2$	$3p^6$	$4s^1$
K	⊗	⊗	⊗⊗⊗	⊗	⊗⊗⊗	⊘

	$1s^2$	$2s^2$	$2p^6$	$3s^2$	$3p^2$
Si	⊗	⊗	⊗⊗⊗	⊗	⊘⊘⊘

	$1s^2$	$2s^2$	$2p^5$
F	⊗	⊗	⊗⊗⊘

	$1s^2$	$2s^2$	$2p^6$	$3s^2$	$3p^6$
Ar	⊗	⊗	⊗⊗⊗	⊗	⊗⊗⊗

7. Be $\quad 1s^2\,2s^2$
   Mg $\quad 1s^2\,2s^2\,2p^6\,3s^2$
   Ca $\quad 1s^2\,2s^2\,2p^6\,3s^2\,3p^6\,4s^2$
   Sr $\quad 1s^2\,2s^2\,2p^6\,3s^2\,3p^6\,4s^2\,3d^{10}\,4p^6\,5s^2$
   All the outermost s orbitals are occupied by two electrons.

8. Sc $\quad 1s^2\,2s^2\,2p^6\,3s^2\,3p^6\,4s^2\,3d^1$
   Ti $\quad 1s^2\,2s^2\,2p^6\,3s^2\,3p^6\,4s^2\,3d^2$
   Ni $\quad 1s^2\,2s^2\,2p^6\,3s^2\,3p^6\,4s^2\,3d^8$
   Zn $\quad 1s^2\,2s^2\,2p^6\,3s^2\,3p^6\,4s^2\,3d^{10}$
   The numbers of electrons in the d sublevel are increasing.

Challenge, page 302

1. $E = E_3 - E_2 = 182.8$ k J/mol

$$E = 182.8 \text{ k J/mol} \times \frac{1000 \text{ J}}{1 \text{ k J}} \times \frac{1 \text{ mol}}{6.02 \times 10^{23} \text{ atoms}}$$
$$= 3.03 \text{ J/atom}$$

2. Ne—ground; Na—excited; V—ground; Al—excited

3. N—ground; Na—excited; Ne—impossible; V—impossible

4. Rb $\quad 1s^2\, 2s^2\, 2p^6\, 3s^2\, 3p^6\, 4s^2\, 3d^{10}\, 4p^6\, 5s^1$
Sr $\quad 1s^2\, 2s^2\, 2p^6\, 3s^2\, 3p^6\, 4s^2\, 3d^{10}\, 4p^6\, 5s^2$
Y $\quad 1s^2\, 2s^2\, 2p^6\, 3s^2\, 3p^6\, 4s^2\, 3d^{10}\, 5s^2\, 4d^1$

5. $c = \lambda \times \nu$
105.4×10^6 cycles/s = frequency of radio station
$$\lambda = \frac{3.00 \times 10^{10} \text{ cm/s}}{105.4 \times 10^6 \text{ cycles/s}} = 2.85 \times 10^2 \text{ cm}$$

6. $c = \lambda \times \nu$
$$\nu = \frac{3.00 \times 10^{10} \text{ cm/s}}{590 \text{ nm}} \times \frac{10^7}{1 \text{ cm}}$$
$$= 5.08 \times 10^{14} \text{ cycles/s}$$

7. $E = E_2 - E_5 = 275.52 \text{ k J/mol}$
$E = h \times \nu$
$$\nu = \frac{275.52 \text{ k J/mol}}{3.98 \times 10^{-13} \text{ k J} \times \text{s/mol}} = 6.92 \times 10^{14} \text{ cycles/s}$$

Synthesis, page 303

1. O^{2-} $\quad 1s^2\, 2s^2\, 2p^6$
Mg^{2+} $\quad 1s^2\, 2s^2\, 2p^6$
Cl^- $\quad 1s^2\, 2s^2\, 2p^6\, 3s^2\, 3p^6$
Sc^{3+} $\quad 1s^2\, 2s^2\, 2p^6\, 3s^2\, 3p^6$
Cd^{2+} $\quad 1s^2\, 2s^2\, 2p^6\, 3s^2\, 3p^6\, 3d^{10}\, 4s^2\, 4p^6\, 4d^{10}$
As^{3-} $\quad 1s^2\, 2s^2\, 2p^6\, 3s^2\, 3p^6\, 3d^{10}\, 4s^2\, 4p^6$

2. When aluminum loses three electrons, the ion has the same electron configuration as Ne. Ne is a very unreactive element. Ga^{3+}

CHAPTER 11 The Periodic Table

CHAPTER ORGANIZER					
USE WITH SECTION	DEMONSTRATION	WORKSHEET	LAB EXPERIMENT	SOFTWARE	TEACHER'S RESOURCE BINDER
11-1					
11-2		11-A			Transparency Master 3
11-3		11-B			
11-4		11-C			Transparency Master 7
11-5		11-D			
11-6		11-D, 11-E	11-A	Disk 1	
Chapter Test					

■ Objectives

When the students have completed the chapter, they will be able to

1. Describe the periodic law.
2. Give a reasonable explanation for the shape of the periodic table.
3. Recognize the similarity in the chemical and physical properties of elements in the same family of elements.
4. Determine the location of families on the periodic table.
5. See the relationship between electron configuration and the position of an element on the periodic table.
6. Compare the outermost s and p orbital system of an element with its chemical properties.
7. Determine the trend in the periodic table of atomic and ionic radii.
8. Explain the forces that account for the trend in ionization energies.

■ Section Notes

11-1 You may want to begin this chapter with a discussion of the organization of the periodic table. Students may recall from Chapter 2 that the periodic table can be divided into metals and nonmetals. Ask them a series of questions like the ones below.

1. Is it sufficient to say that the table can be divided into two groups of elements: metals and non-metals? If yes, then why is the organization of the periodic table much more complex than simply showing two groups of elements?
2. What characteristic of the elements determines their sequence on the periodic table?
3. Are there any places on the periodic table where the atomic numbers or the molar masses are out of sequence?
4. Why are there only two elements in the first row (period) and eight in the second and third rows? Are there any similarities in the elements that are in each column of the table?
5. What would happen if a column of elements were to be eliminated in the periodic table? How would that affect the integrity of the periodic table?
6. What if one element were to be eliminated, for example, Li, and the rest of the elements then moved one place over to fill the hole left by lithium? How would moving the elements over one place affect the integrity of the periodic table? Would filling in for one element cause more damage to the integrity of the periodic table than eliminating a column of elements?

A discussion centered around the concepts implied in these or similar questions seems to stir curiosity in the students. Students begin to ask other questions about the periodic table. You may want to answer some of their questions now or tell them to look for some answers throughout their study of this chapter.

There are many fascinating stories about the development of the periodic table. One of the most intriguing figures in this history is John Newlands, an English chemist. In 1865, he developed the law of octaves, He placed the first seven elements then known in order of increasing molar masses. (The first tables, even Mendeleev's, were in order of increasing molar masses, not atomic numbers.) Then, Newlands placed the next seven elements in a row below the first and ended his table with the next seven elements. He immediately recognized the similarity in chemical properties of the first column of elements; H, F, and Cl, and a gain in the second column: Li, Na, and K. He wondered if he had discovered a fundamental law of nature, namely, the striking comparison between order among elements and the repetition of notes on the musical scale (A to G). Unfortunately, he tried to make too much of octaves in each row of his periodic table and the number of notes in a scale of music. He presented the rule of octaves to the Chemical Society of London in 1866. He was briskly ridiculed by members of the Society. One scientist caustically remarked whether Newlands had thought to compare the chemical properties of elements whose names begin with the letter "s." Three years later, however, Mendeleev successfully introduced the first modern periodic table.

The genius of Dmitri Mendeleev was that he proposed a periodic table that was based on the following premises:

1. He closely examined the current literature of molar masses. In several instances, namely chromium and indium, he discovered that errors in molar masses existed. He placed these elements in their correct location on the periodic table.
2. He introduced longer periods containing the transition elements.
3. He left a space in the periodic table if the next element did not have the properties of that column.

Mendeleev's bold proposals are good examples of the faith that scientists place in the order of the universe. The periodic table dramatically confirms this belief.

11-3   This section attempts to combine the electron configurations of multi-electron elements with the periodic table. Table 11-6 is important. It organizes the electron configurations of selected elements into their respective families. At this point, the students probably will not see the connection between electron configurations and the position of the elements on the periodic table. Help them to make this connection by pointing out that elements of a particular family have similar electron configurations in the outermost s and p orbital systems.

The Pauli exclusion principle (no two electrons have the same set of four quantum numbers) was derived empirically from the location of elements on the periodic table. Quantum mechanics does not necessarily lead to the Pauli exclusion principle. The principle has the form it does because of the arrangement of elements on the periodic table. The empirical basis of the Pauli exclusion principle can be demonstrated rather easily. The number of elements in each period of the periodic table is 2, 8, 8, 18, and 32. For the hydrogen atom, the number of orbitals possible in each energy level is:

$$1^2 = 1, \ 2^2 = 4, \ 3^2 = 9, \ 4^2 = 16$$

If the number of orbitals is doubled, they match the number of elements in the periods of the periodic table. Each orbital may contain a maximum of two electrons.

Students may be curious about the exceptions to the Aufbau process for the electron configurations of some transition metals, such as Cr and Cu. The electron configuration of Cr in the ground state is:

$$1s^2 \ 2s^2 \ 2p^6 \ 3s^2 \ 4s^1 \ 3d^5$$

A half-filled 4s and half-filled 3d system have lower energy that the expected $4s^2$ and $3d^4$. The electron configuration of Cu in the ground state is:

$$1s^2 \quad 2s^2 \quad 2p^6 \quad 3s^2 \quad 3p^6 \quad 4s^1 \quad 3d^{10}$$

For similar reasons, the electron configuration of Cu is not what is expected. Other exceptions exist, but they should receive only a mention. Students should be left with the idea that the periodic table was developed before electron configurations.

11-4 Much of chemistry depends on proper understanding of the relationship between electron configuration and chemical reactivity. There is nothing "magical" about the structure of the noble gases. A completely filled outermost s and p orbital system is predictive of nonreactivity both for the noble gases and other atoms that gain and lose electrons. It is misleading to students to personify the noble gases as "happy" or "content" atoms. It is just as misleading to suggest that in order to achieve a noble gas electronic structure, atoms of other elements *tend to* react by losing or gaining electrons. Atoms do not possess a psychological state of contentment, nor do they have mental powers of goal orientation. It is best to avoid such language because high school students take it literally.

Students are better served when you point out to them that energy effects dictate whether atoms gain or lose electrons in chemical reactions. These energy effects are explained in terms of ionization energy and electron affinity. In this section on chemical reactivity, merely point out to the students that reactivity is related mainly to the number of electrons present in an atom's outermost s and p orbital systems

11-5, 11-6 The periodic trends of atomic radii and ionization energy depend on two factors:

1. the force of attraction of electrons to the nucleus;
2. the force of repulsion among electrons.

It is the interplay between these two forces that determines the size of the atom and how much energy is required to remove an electron from an atom.

The measurement of ionization energies was of great importance in forming the modern theory of atomic structure. Ionization energies offer evidence for the relative energies of electrons in atoms.

There are several methods of determining ionization energies. One method involves the line spectrum of atoms. Electron energy levels gradually converge to a limit. Ionization energy is the energy needed to raise an electron to this limit. Beyond the limit, a continuous spectrum is seen. Once an electron has been removed from the atom, it is no longer bound by quantum energy states. Therefore, the effect is a continuous spectrum.

One of the exceptions to the trend that ionization energies increase across a period is the decrease in IE from nitrogen to oxygen. The decrease can be ex-

plained by the small but significant repulsion that occurs between the two electrons occupying the $2p_x$ orbital. This repulsion apparently is greater than the force of attraction of the oxygen nucleus. The repulsion means that one of the electrons of the $2p_x$ orbital is more easily removed than if the repulsion were not present, as in nitrogen.

Students often ask why the electrons with the highest energies are the electrons involved in ionization. A simple analogy is to view the atom as a bushel of apples. When you reach into the bushel to remove an apple, you usually remove an apple from the top of the bushel. It would take more energy to remove an apple from the bottom of a full bushel.

Review and Practice
page 311

1. The chemical and physical properties of elements are a cyclic function of atomic number.

2. Answers will vary. Possible responses are: Two groups of metals are the alkali metals and the Lanthanide series, or two groups of nonmetals are the noble gases and the halogens.

3. $Cs(s) + 2H_2O \longrightarrow Cs^+(aq) + 2OH^-(aq) + H_2(g)$

4. a. SrH_2 **d.** RbF
    b. SeH_2 **e.** GaF_3
    c. HI **f.** TeF_2

page 316

1. The electron configurations of all the noble gases have a completely filled outermost s and p orbital system. The exception is He which only has two electrons and has a filled $2s$ orbital.

2. Isoelectronic is the term used for the situations in which different species of atoms or ions have the same electron configuration.

3. O^{2-}: $1s^2 \quad 2s^2 \quad 2p^6$. The oxide ion is isoelectronic with Ne.

4.
Rb	alkali metal	1 electron
Te	oxygen group	6 electrons
Ca	alkali earth group	2 electrons
Xe	noble gas	8 electrons

5. Ni has eight electrons in its $3d$ system.

page 321

1. The size of the atoms decreases from left to right across the table.

2. With an increase in molar mass in any group, there is an increase in the energy levels of electrons. Electrons in higher energy levels will be found, on the average, farther from the nucleus than electrons in lower energy levels. The atoms, therefore, increase in size.

3. When another electron is added to a neutral non-metal atom, the force of repulsion of the other electrons in that energy level increases the distance that, on the average, the electrons are from the nucleus. The size of the atom, now a negative ion, increases.

4. Mg is 2^+. The ion is smaller than the atom.
   Cl is 1^-. The ion is larger than the atom.
   Al is 3^+. The ion is smaller than the atom.
   S is 2^-. The ion is larger than the atom.
   Cs is 1^+. The ion is smaller than the atom.
   I is 1^-. The ion is larger than the atom.
   O is 2^-. The ion is larger than the atom.

page 324

1. Ionization energy is the energy needed to remove an electron from a gaseous neutral atom.

2. The second electron removed is a $6s$ electron. This electron is weakly held. The third electron removed is a $5p$ electron and is part of a very stable, completely filled $5s$ and $5p$ system. Much energy is required to remove the third electron.

3. The effective nuclear charge in the noble gases is the greatest from any period. Therefore, the electrons are forcefully held by the nucleus.

Review, page 325

1. The answers will vary. Make sure that students understand that some elements have properties of both metals and nonmetals.

2. | HALOGEN FAMILY | | NOBLE GASES | |
   |---|---|---|---|
   | F | fluorine | He | helium |
   | Cl | chlorine | Ne | neon |
   | Br | bromine | AR | argon |
   | I | iodine | Kr | krypton |
   | At | astatine | Xe | xenon |
   | | | Rn | radon |

3. A period includes the elements across a horizontal row in the periodic table. A group consists of the elements that compose a vertical column.

4. Chemical and physical properties of elements are periodic. Electron configurations, ionization energies, radii, and electron affinity also can be said to be periodic.

5. Mendeleev knew to allow space for unknown elements because the next element in increasing molar mass did not have the same properties as the other elements in that column. He knew that some elements were not discovered because of periodicity.

6. Electrons in the outermost s and p orbitals are most likely to influence chemical behavior.

7. The size of atoms decreases.

8. $X(g) + \text{energy} \longrightarrow X^+(g) + e^-$

9. The most stable ion of aluminum is Al^{3+}.

10. Electron affinity is the energy given off or absorbed when a neutral gaseous atom takes on an electron.

Interpret and Apply, page 325

1. The stable oxides are $Sr\,O$, K_2O, Ga_2O_3.

2. The answers will vary. Some possible answers include periodicals, newspapers, the seasons, and class periods in school.

3. Rb is in the alkali metal group and is a metal. Nb is a transition metal. As is in group 15 or the nitrogen group and is a metalloid. Xe is a noble gas and is a nonmetal. Eu is in the lanthanide series and is a metal. Ag is a transition metal (This group is sometimes referred to as the coinage metals).

4. Elements with incomplete $4d$ orbital systems are found in the second row of the transition metals, such as Y, Zr, Nb, Mo, Tc, Ru, and Rh.

5. I^- contains 53 protons and 54 electrons. Ra^+ contains 88 protons and 86 electrons.

6. Noble gases are considered to be stable because they are nonreactive under most conditions. Their nonreactivity is due to their complete outermost s and p orbital systems.

7. Na^+ and Ne are isoelectronic. However, Na^+ has a greater effective nuclear charge than Ne, and therefore, Ne would be larger.

8. Element X is sulfur. Element Y is potassium. Element Z is nickel.

9. Be has a lower first ionization energy than Ca because the $6s$ electrons are shielded from the force of the nucleus more than the $4s$ electrons of Ca.

10. Ca would have the largest third ionization energy. Ca^{2+} has a larger effective nuclear charge than K^{2+}. Even though Ga has a larger nuclear charge than Ca, the third electron to be removed from Ga is shielded from the nuclear charge by inner electrons of Ga.

Problems, page 326

1. $I_2(s) + H_2O(1) \longrightarrow HOI(aq) + H^+(aq) + I^-(aq)$

2. | | x | y |
   |---|---|---|
   | a. | 1 | 3 |
   | b. | 1 | 1 |
   | c. | 1 | 2 |
   | d. | 3 | 2 |

3. Reasonable formulas are:
   a. SrS d. ClI
   b. GaF_3 e. $AsBr_3$
   c. $BeTe$

4. Exceptions to the increasing molar masses in the periodic table are Ar and K, Co and Ni, Te and I.

5. The size of atoms increases with increasing molar mass in a group. Therefore, a rough estimate of the radius of iodine from this given sequence is about 0.13 or 0.14 nm. The actual value is 0.133 nm.

6. $M(s) + X_2 \longrightarrow MX_2(s)$

7. The most common charge on the aluminum ion is 3^+. Aluminum has three electrons in the outermost s and p orbital systems. The tremendous increase in ionization energy between IE_3 and IE_4 indicates the number of electrons in the outermost orbital system.

Challenge, page 326

1. Element c should have the lowest IE_2. Element b is a noble gas that has a relatively high ionization energy. Element c is an alkali metal. The first ionization will require a small amount of energy. The second will be more difficult as the electrons in $2p$ orbitals are held more firmly than the outermost electron.

2. a. Cl has the highest IE_1.
b. Cl has the smallest radius.
c. Na has the most metallic character.

3. The smaller radius in each pair is:
a. Br   **c.** Ga^{3+}
b. Ne   **d.** S

4. a. 0.095 nm/0.181 nm = 0.52, hole is octahedral.
b. 0.060 nm/0.195 nm = 0.31, hole is tetrahedral.
c. 0.075 nm/0.184 nm = 0.41, hole is tetrahedral.

Synthesis, page 327

1. The equation for the reaction is
$$4Na(s) + O_2(g) \longrightarrow 2Na_2O(s)$$

$$25.0 \text{ g Na} \times \frac{1 \text{ mol Na}}{23.0 \text{ g Na}} \times \frac{2 \text{ mol Na}_2\text{O}}{4 \text{ mol Na}}$$
$$\times \frac{62.0 \text{ g Na}_2\text{O}}{1 \text{ mol Na}_2\text{O}} = 33.7 \text{ g}$$

2. a. B
b. C
c. B, A, E, D, C
d. B
e. A

CHAPTER 12 Chemical Bonding

CHAPTER ORGANIZER					
USE WITH SECTION	DEMONSTRATION	WORKSHEET	LAB EXPERIMENT	SOFTWARE	TEACHER'S RESOURCE BINDER
12-3		12-A, 12-B			
12-4	12-A		12-A		
12-6		12-C			
12-10					Advanced Topic 4
12-11	12-B		12-B		
12-12	12-C	12-D			
12-14		12-E			
12-15		12-F			
Chapter Test					

■ Objectives

When the students have completed the chapter, they will be able to

1. Determine what kinds of information are useful to know about chemical bonding.
2. Describe how a chemical bond is formed.
3. Distinguish between covalent, polar covalent, and ionic bonding using electronegativity differences of bonding atoms.
4. Write electron dot structures for atoms of representative elements.
5. Draw electron dot structures for selected molecules and polyatomic ions.
6. Apply the octet rule to describe molecular structures.
7. Define bond energies and explain how they can be used to compare bond strengths of different chemical bonds.
8. Determine the shapes of some molecules from the number of bonded pairs and lone pairs of electrons around the central atom.
9. Predict molecular polarity from the shapes of molecules.
10. Describe how knowledge of molecular polarity can be used to explain some physical and chemical properties of molecules.

■ Section Notes

12-1 Students often finish a chapter on chemical bonding with a dislike for Lewis dot structures and molecular shapes. They may not understand why they need to know so much of the material presented. Students may benefit from the long-range view that the chapter provides them with the tools to understand many of the other topics they will encounter in this chemistry course.

12-2 Impress upon students that there are not two distinct kinds of chemical bonding. Both ionic and covalent kinds of bonding can be explained from the same point of view. Electrons are attracted simultaneously to two nuclei. Chemical bonds occur when the potential energy of the bonded atoms (or ions) is lower than the isolated system.

There are two sources of repelling forces, electron-electron repulsion and nucleus-nucleus repulsion. Work must be done to bring two electrons or two nuclei closer together, leading to an increase in potential energy. If the attractive forces are greater than both repelling forces, bonding occurs. As electrons approach nuclei, energy is released, and the potential energy of the system decreases.

Ionic bonding occurs when the electrons of each ion are simultaneously attracted to both nuclei. Ionic crystals derive their strengths from ionic bonding in three dimensions. Ionic bond strengths depend on the size of the ions and their arrangement in the three-dimensional crystal lattice.

Covalent bonding involves the overlap of unfilled orbitals as they approach each other. In a covalent bond, the shared pair of electrons will be found between the two nuclei most of the time. The mutual attraction of the two nuclei for the shared electrons is greater than the electron-electron or nucleus-nucleus repulsions.

Two helium atoms will not bond because both atoms have filled orbitals. The helium atoms cannot get close enough to each other without a great deal of repulsion between electrons and nuclei.

For further information, you may want to consult a college-level chemistry book that explains the theory of molecular orbitals.

12-3 Electronegativities can help students understand the varying nature of chemical bonds. In covalent bonding the equal or almost equal sharing of electrons leads to a bond with no dipole. When the electronegativity difference is between 0.3 and 1.6, a charge separation occurs in the bond and a dipole is formed. An electronegativity difference of 1.7 denotes a bond with 50% ionic character, which is categorized as an ionic bond.

In a molecule containing more than one bond, the electronegativity difference of each bond can be considered separately. For example, NH_3 contains three polar covalent bonds.

12-5, 12-6, 12-7 The rules presented here for determining electron dot structures are different from most high school texts. With these rules, the students can learn the usefulness and limitations of the octet rule, and also learn to determine which atoms are bonded to each other in a molecule—no easy task for the beginning student. For example, students often are confused when asked to write the electron configuration for a molecule, such as CO_2. Two possible starting points are possible:

$$O—C—O \quad \text{or} \quad O—O—C$$

The rules state that the structure that looks most symmetrical is the correct one. This helps the student get started.

The octet rule has definite limitations. The dot structures of many molecules cannot be written using the octet rule. BF_3, NO, and PCl_5 are just a few examples. Also, some molecules, such as SO_2, which can be made to have an octet structure by the use of resonance structures, are better represented by not following the octet rule. In the case of SO_2 the following structure is thought to be more correct because experimental data about bond strength dictates the presence of two double bonds.

$$:\ddot{O} = \ddot{S} = \ddot{O}:$$

One would have to introduce the notion of formal charge in order to teach a more comprehensive picture of bonding. Formal charge is too complex for beginning students. James A. Carroll's article cited in the sources for this chapter emphasizes the hazards of trying to apply the octet rule too broadly. Yet, even with its limitations, the octet rule is a useful tool in understanding the bonding in many molecules, especially molecules involving C, N, and O atoms.

Another example of the limitations of the octet rule can be demonstrated with the O_2 molecule. The only Lewis structure that conforms to the rule has a double bond between the two oxygen atoms. Experimental measurement of bond lengths and bond strengths suggests the presence of a double bond. Two pairs of shared electrons would be expected. However, liquid oxygen can be pulled into a magnetic field (paramagnetism), a behavior characteristic of substances with *unpaired* electrons. Their presence is explained by molecular orbital theory, which describes the O_2 molecule as having two unpaired electrons, each occupying a separate πp^* antibonding molecular orbital.

Electron dot diagrams are useful in determining which atoms are bonded together to make molecules and in predicting single, double, and triple bonds in many molecules. However, the structures do not say anything about the geometry of molecules. For example, the dot structure for water may be written as

$$\begin{matrix} H \\ | \\ H-\ddot{O}: \end{matrix} \quad \text{or} \quad H-\ddot{O}-H$$

Either structure is correct. Students should not worry about the placement of bonded atoms around the central atom.

12-8 This section emphasizes the energetics of bonding. When a bond forms, energy is released as heat, which can be measured. The formation of bonds is always an exothermic reaction. When bonds are broken, the system gains potential energy. Equations representing the breaking of bonds are endothermic.

Bond energies are derived from heats of formation and energies associated with the decomposition of elements into isolated atoms. For example, the bond energy of a C—H bond should be one-fourth the energy needed to separate a mole of methane molecules into one mole of carbon atoms and four moles of hydrogen atoms. The heat of formation of CH_4 is

$$\text{C (graphite)} + 2H_2(g) \longrightarrow CH_4 \quad \Delta H^\circ = -75\,\text{kJ}$$

The energies involved in decomposing a mole of carbon (graphite) into a mole of isolated carbon atoms, and two moles of hydrogen gas into four moles of isolated hydrogen atoms are

$$\begin{aligned} \text{C (graphite)} &\longrightarrow \text{C}(g) & \Delta H^\circ &= 718\,\text{kJ} \\ 2H_2(g) &\longrightarrow 4H(g) & \Delta H^\circ &= 872\,\text{kJ} \end{aligned}$$

Then from the three equations, the heat of reaction from which the bond energy of C—H is calculated becomes

$$\begin{aligned} CH_4(g) &\longrightarrow \text{C (graphite)} + 2H_2(g) & \Delta H^\circ &= +75\,\text{kJ} \\ \text{C (graphite)} &\longrightarrow \text{C}(g) & \Delta H^\circ &= +719\,\text{kJ} \\ 2H_2(g) &\longrightarrow 4H(g) & \Delta H^\circ &= +872\,\text{kJ} \\ \hline CH_4(g) &\longrightarrow \text{C}(g) + 4H(g) & \Delta H^\circ &= +1666\,\text{kJ} \end{aligned}$$

The bond energy for one mole of C—H bonds is

$$\tfrac{1}{4} \times 1666\,\text{kJ} = 416\,\text{kJ}.$$

Bond energies in Table 12-2 represent averages of bond energies in a range of similar molecules rather than in one molecule.

12-10 Resonance structures seem very artificial to students. Yet several common molecules, such as SO_2 and O_3, and some common ions, such as CO_3^{2-} and NO_3^-, are best represented by resonance. Students may think that resonance presents an entirely new concept. Actually, resonance is an extension of the octet rule. Resonance helps to explain the bond strengths and bond lengths present in some molecules.

12-11 Three models are used to describe chemical bonding to students at an introductory level. Each approach has it strengths and weaknesses.

The valence bond model presumes that in bonding, atomic orbitals are kept intact. Bonding occurs when orbitals overlap. The bonding orbitals are occupied by oppositely spinning electrons that are attracted to two nuclei simultaneously. This approach involves the concepts of resonance structures and, at a more advanced level, hybridized orbitals (see the Appendix). The valence bond model offers the advantage of predicting molecule shape; for example, tetrahedral structures are based on sp^3 hybrid orbitals, trigonal planar structures are sp^2, and octahedral structures are sp^3d^2. One disadvantage of this model is the seemingly artificial ideas of resonance and hybridized orbitals.

The molecular orbital model of bonding (too complex for many high school students) presumes that when atoms bond, atomic orbitals lose their identity. New orbitals are formed that have the characteristics of the total molecule. The concept of molecular energy levels replaces the atomic orbital energy levels. The molecule is thought of as being built up from these molecular orbitals. The advantage to this model of bonding is its prediction of bond orders in simple molecules. The weakness of this model is that it does not adequately explain the geometry of molecules.

The valence shell electron pair repulsion theory presumes that the geometry of molecules is determined by the number of bonded pairs and lone pairs of electrons around the central atom.

These localized orbitals will take up positions around the central atom so that they will be as far apart from each other as possible. This arrangement maintains a minimum potential energy due to electron-electron repulsion. According to the theory, the greatest repulsion occurs between two lone pairs of electrons. A lone pair and a bonded pair of electrons will have less repulsion. The least repulsion is between two bonded pairs. When the central atom contains four bonded pairs of electrons as in CH_4, the bonded atoms take up positions at the corners of a tetrahedron. Their potential energy is the lowest possible when the bond angles are 109.5°. In the case of NH_3 (three bonded pairs and one lone pair), the tetrahedral arrangement is altered slightly by the greater repulsion of the lone pair of electrons. The potential energy of the system is lowest when the N—H bond angles are 107°. Even more tetrahedral distortion is present in H_2O. Potential energy is lowest when the H—O bond angle is 104.5°.

The strength of the VSEPR theory is its ability to predict the geometry of many molecules in a strikingly simple and straightforward fashion. The disadvantage is that the theory does not explain how atomic orbitals form the geometry of molecules. The students may want to explore the relationship between atomic orbitals and geometry by way of hybridized orbitals presented in the Appendix.

The geometry of many molecules can be determined from the number of bonded atoms and lone pairs of electrons surrounding the central atom. (See Fig. 12-26 in the student text.) However, it helps students to see the three-dimensional shapes of molecules in order to better think in three dimensions. Balloons offer a relatively simple and enjoyable alternative to commercial or teacher-made models. If you use the balloons in class, be sure that you emphasize two things:

1. Each balloon represents either a pair of bonded electrons or a lone pair of electrons.
2. The central atom is located at the position where the balloons are tied together. The central atom would take up a volume much larger than the tied necks of the balloons.

The VSEPR theory is an accurate predictor of geometric shape in molecules having a relatively small central atom. Molecules that have a large central atom from group 15 or 16 (4 units surrounding the central atom) do not have tetrahedral structures. A possible explanation is that because the central atom is rela-

tively large, there is less repulsion among the units of the central atom.

The names given to the shapes of molecules are determined from the positions of the bonded atoms around the central atoms. The lone pair(s) of electrons is (are) not represented in the name. For example, water is an angular molecule because the H—O—H bond is angular. The two unshared pairs of electrons are not considered in the name. Lone pairs of electrons also are usually not represented in space-filling models of molecules.

12-12 A principal objective of this section is the relationship between molecular geometry and molecular polarity. Uneven charge distribution in a molecule results in a molecular dipole moment. The dipole moment is determined by adding (in a vector sense) the individual dipoles of the polar bonds in the molecule. An analogy may help the students understand the role geometry plays in molecular polarity. If two tug-of-war teams pull with equal force on a rope tied to a string at an angle of 180°, the ring will not move. The resultant force is zero. The forces cancel each other.

$$\longleftarrow \quad O \quad \longrightarrow$$

This situation is comparable to the two dipoles in a linear molecule such as BeH_2. CH_4 presents a similar situation, but in three dimensions. In NH_3, however, there is an unequal distribution of charge in the molecule. The tug-of-war analogy is represented by the three forces pulling at an angle of 107°. They do not cancel each other, and the ring moves in the direction of the resultant force.

12-13, 12-14 London forces exist in nonpolar molecules because of instantaneous, short-lived, unequal distributions of electrons. London forces are weak in comparison to most dipole-dipole interactions between polar molecules, but they overcome the relatively small kinetic energies of noble gases at low temperature, explaining why the gases can liquefy.

Polar molecules are attracted to each other by dipole-dipole forces and London forces. The dipole-dipole forces are stronger than the London forces, allowing the molecules to attract each other even when they have relatively high kinetic energies. Generally, polar molecules have higher boiling points than nonpolar molecules of approximately equal molar masses.

In Chapter 16 the students also will use this information to understand why like dissolves like, that is, polar molecules dissolve polar molecules and nonpolar molecules dissolve nonpolar molecules.

1. the coming together of two atoms when a pair of electrons is simultaneously attracted to both of the nuclei

2. Potential energy decreases; attractive forces grow stronger as the particles approach each other, which lowers the energy of the system.

3. Bonds are categorized by the degree of sharing of a pair of electrons in the bond. The greater the difference in electronegativity of the two atoms, the more unequally the electrons are shared.

4. As two H atoms approach each other, their orbitals overlap. The electrons are drawn simultaneously to the two nuclei and are found between the two nuclei most of the time.

5. a measure of an atom's attraction for the shared pairs of electrons in a bond

6. HI = polar covalent; F_2 = covalent; CsCl = ionic; MgO = ionic; O_2 = covalent; KBr = ionic; AsH_3 = covalent; PbI_3 = polar covalent; PCl_3 = polar covalent

page 336

1. Electrons are shared more or less equally in covalent bonds and unequally in polar covalent bonds. In ionic bonds, the pair of electrons is said to be associated with the more electronegative atom most of the time.

2. Ionic bonding is likely to occur between elements in groups 1 and 17; covalent or polar bonding between groups 16 and 17.

3. Ionic bonds form when positive and negative ions attract each other.

4. Ions are arranged in characteristic patterns of tightly packed aggregates of alternating + and − ions. *Molecule* usually refers to a group of atoms bonded covalently.

page 343

1. simple structure using an atomic symbol and dots to represent an atom

2. It is the symbol of the element surrounded by the same number of dots as there are outer *s* and *p* electrons in atoms of that element.

3. K·   ·As: :Br·   ·Si·   ·Te: ·Al·

4.
:Cl: Cl: :F—Si—F: :O: H
with :F: above and below Si, and structure :F:

5.

H H H
H—C—C—C—H
H H H

H H   :O: 2−
C::C   :O—S—O: H:C:::N
H H   :O:

6. the sharing of two pairs of electrons between two atoms; CO_2 (Students may substitute dashed lines for shared pairs of electrons.)

page 348

1. when no single electron dot structure adequately represents the molecule

2. [resonance structures of carbonate ion shown]

3. the energy required to separate two bonded atoms

4. The bond energy of the double oxygen bond is much greater than that of the single bond. Two doubly bonded oxygen atoms would be more difficult to separate than two singly bonded atoms.

5. Quantum characteristics of molecules can be used to 1) classify bonds between atoms and 2) identify unknown chemicals by matching their energy absorption patterns against known patterns.

page 354

1. Bonded atoms and lone pairs of electrons will be arranged about a central atom as far apart as possible to minimize repulsion.

2. Their charges are the same, negative.

3. No. Electron dot structures are two-dimensional. The dot structure of CH_4 implies that the C—H bonds are 90° apart. The actual bond angle is 109.5°.

4. They have different numbers of bonded electrons and lone pairs of electrons around their central atoms.

5. GaH_3 = trigonal planar; GeH_4 = tetrahedral; PCl_3 = trigonal pyramidal; SO_3 = trigonal planar

6. linear

page 362

1. A hydrogen bond exists between the H atom in one polar molecule with an O, N, or F atom in it and another molecule of the same kind. Hydrogen

bonds exist between molecules rather than within molecules.

2. covalent, hydrogen bonds, dipole-dipole forces, London forces

3. Each of these arrangements allows for maximum distance and therefore, minimum repulsion between all the bonded pairs of electrons.

4. CO_2 and BeH_2; NH_3 and H_2O

5. Temporary unequal distribution of electrons in the molecules leads to momentary dipoles. Weak attractions between molecules result.

6. Biochemicals often function by interacting with other chemicals. Interactions are limited by whether or not the different molecules have compatible shapes.

Review, page 363

1. The potential energy of the system decreases.

2. a regularly repeating three-dimensional pattern of ions or molecules that form in the solid phase

3. It forms from the overlap of half-filled orbitals. A pair of electrons is shared, and the pair is found between the nuclei most of the time.

4. the degree to which a shared pair of electrons is attracted to one of the atoms in a chemical bond; used to classify bonds as covalent, polar covalent, or ionic

5. How many atoms are connected? Which atoms are bonded? How strong is the bond? What is the shape of the molecule?

6. Bonded nonmetallic atoms have eight electrons in their outermost energy levels.

7. A double bond exists when two atoms share four electrons. A triple bond exists when two atoms share six electrons.

8. one model of a molecule (or polyatomic ion) for which more than one electron dot structure is needed to describe bonding; CO_3^{2-} or NO_3^-

9. Potential energy increases.

10. Shared pairs or lone pairs of electrons take up positions around the central atom of a molecule so that they are as far apart from each other as possible.

11. The hydrogen atom bonded to nitrogen, oxygen, or fluorine in a molecule is attracted to that electronegative atom in another identical molecule.

12. In a polar bond, electrons are shared unequally, resulting in a dipole. In a nonpolar bond, electrons are shared equally, and there is no dipole.

13. If the dipoles are oriented around the central atom such that they cancel, the molecule is nonpolar.

14. Dipole-dipole forces result from the attraction of

the oppositely charged ends of two polar molecules. London forces are formed from the temporary unequal distribution of electrons in nonpolar molecules. (Hydrogen bonds are a special case of dipole-dipole forces.)

15. a) when the electronegativity difference between the two bonding atoms is 0.2 or less; b) when the electronegativity difference between the two bonding atoms is 1.7 or greater

Interpret and Apply, page 364

1. c, e, f

2. Hydrogen molecules are more stable. The hydrogen molecule has lower potential energy than two isolated hydrogen atoms.

3. Nonmetals on the right side of the periodic table form covalent bonds. The bonding between metals (from groups 1 and 2) and nonmetals (from groups 16 and 17) is often ionic.

4. When a bonded atom has an outermost s and p electron configuration that contains eight electrons, the atom is likely to be part of a stable molecule.

5. N_2

6. London forces are temporary dipoles in nonpolar molecules. Hydrogen bonds come about from permanent dipoles between certain polar molecules.

7. **a.** hydrogen bonds   **c.** dipole-dipole forces
   b. London forces   **d.** London forces

8. CH_3F is a polar molecule. The C—F bond has a dipole that is greater than the other three dipoles formed from the C—H bonds in the molecule.

9. CO_2 is a linear, nonpolar molecule. Carbon dioxide will form a solid (or liquid under certain conditions) by very weak London forces. The kinetic energy of CO_2 molecules at room temperature is too great to allow these London forces to be effective. Water is polar. Its molecules are held together by hydrogen bonds. The kinetic energies of most water molecules at room temperature are not great enough to overcome hydrogen bonding.

10. 1—polar covalent; 2—hydrogen bond; 3—covalent; 4—hydrogen bond; 5—covalent (double) bond; 6—polar covalent; 7—polar covalent

Problems, page 364

1. **a.** ionic
   b. ionic
   c. polar covalent
   d. ionic
   e. ionic
   f. polar covalent
   g. covalent

2.

Br₂ Br[Ar] ⊗ ⊗⊗⊗⊗⊗ ⊗⊗ ⊘
Br[Ar] ⊗ ⊗⊗⊗⊗⊗ ⊗⊗ ⊘

1s 2s 2p

N₂
N ⊗ ⊗ ⊘⊘⊘
N ⊗ ⊗ ⊘⊘⊘

1s 2s 2p 3s 3p

HCl
Cl ⊗ ⊗ ⊗⊗⊗ ⊗ ⊗⊗ ⊘
⊘
1s H

1s 2s 2p H 1s

H₂O₂
O ⊗ ⊗ ⊗ ⊘⊘
O ⊗ ⊗ ⊗ ⊘⊘
⊘
1s H

3.

H
|
H—C—Cl:
|
H

:Cl:
|
H—C—Cl:
|
H

:Cl:
|
:Cl—C—Cl:
|
H

:Cl:
|
:Cl—C—Cl:
|
:Cl:

H
|
:O—O: or   H—O—O—H
|
H

4.

:F:
|
:F—N—F:

:F—N=N—F:

H—C≡C—H

H—N—O—H
|
H

5.   :N≡N: [:C≡N:]⁻   [:N≡O:]⁺   :C≡O:

6.   [O=N—O]⁻ ⟷ [:O—N=O]⁻

O≡O—O: ⟷ :O—O=O

7. a. trigonal pyramidal   **d.** tetrahedral
b. angular   **e.** tetrahedral
c. octahedral

8. a, d, e

9. F₂ has the greatest bond energy; therefore, it is the most stable.

10.

[:O—H]⁻   [H—O—H]⁺ (with H above)   [S=C≡N]⁻

Challenge, page 365

1. a. NaF—ionic   **d.** no bond
b. CF₄—polar covalent   **e.** AsCl₃—polar covalent
c. F₂—covalent

2. Fe(CN)₆³⁻—octahedral; Al(H₂O)₆³⁺—octahedral; Zn(NH₃)₄²⁺—tetrahedral

3. a. NH₃ < NF₃   (NF₃ has a greater electronegativity difference.)
b. BF₃ < NH₃   (NH₃ is polar, BF₃ is nonpolar.)
c. CH₄ = CCl₄   (Both are nonpolar.)
d. HCl < HF   (HF has a greater electronegativity difference.)
e. SO₂ > BeH₂   (SO₂ is polar, BeH₂ is nonpolar.)

Synthesis, page 365

1. Both molecules have the same electron dot structure. NH₄⁺ is an ion and would be subject to solvation by water. CH₄ is nonpolar and would not be very soluble in water.

2. The table indicates a decrease, from top to bottom, in the melting points of these ionic substances. Since the radii of the anions increase from top to bottom, the distances between the atomic nuclei in the bonds increase correspondingly. As a result, there is a decrease in attractive force between the ions in each crystal, requiring less of an input of energy to separate them. Also, the attraction between ions is greatest at the top of the table (F is the most electronegative). Therefore, KF would have the lowest potential energy of the group and would have the highest melting point.

3. Ionization energies increase across the periodic table because of increasing nuclear charge and decreasing atomic size. A smaller atom is more likely to attract a pair of electrons in a bond because of the proximity of the shared pair of electrons to the nucleus. Ionization energy decreases from top to bottom of the table because of the shielding effect of the inner electrons and increased atomic size. The outer electrons are not as strongly held by the nucleus. Electronegativity also decreases from top to bottom in the table because of the shielding effect of the inner electrons. The shared pair of electrons is less likely to be attracted to the nucleus of a large atom than to a small one.

CHAPTER 13 Elements: A Closer Look

<table>
<tr><th colspan="6">CHAPTER ORGANIZER</th></tr>
<tr><th>USE WITH SECTION</th><th>DEMONSTRATION</th><th>WORKSHEET</th><th>LAB EXPERIMENT</th><th>SOFTWARE</th><th>TEACHER'S RESOURCE BINDER</th></tr>
<tr><td>13-1</td><td></td><td></td><td>13-A</td><td>Disk 1</td><td>Blackline Master 3</td></tr>
<tr><td>13-2</td><td></td><td>13-A</td><td></td><td></td><td></td></tr>
<tr><td>13-3</td><td>13A, 13B</td><td>13-B</td><td></td><td></td><td></td></tr>
<tr><td>13-4</td><td></td><td></td><td></td><td></td><td>Chem Issue 4</td></tr>
<tr><td>13-5</td><td></td><td>13-C</td><td>13-B</td><td></td><td>Advanced Topic 5</td></tr>
<tr><td>13-9</td><td></td><td></td><td></td><td>Disk 1</td><td></td></tr>
<tr><td>13-11</td><td></td><td>13-D</td><td></td><td>Disk 1</td><td></td></tr>
<tr><td>13-12</td><td></td><td>13-D</td><td></td><td>Disk 1</td><td></td></tr>
<tr><td>13-13</td><td></td><td>13-E</td><td></td><td>Disk 1</td><td></td></tr>
<tr><td>13-14</td><td></td><td>13-F</td><td></td><td>Disk 1</td><td></td></tr>
<tr><td colspan="6">Chapter Test</td></tr>
</table>

■ Objectives

When the students have completed this chapter, they will be able to

1. Describe the unique characteristics of a metallic bond.
2. Categorize the elements as metals, nonmetals or metalloids.
3. Identify a family of elements and list several characteristic properties for members of that family.
4. Name several alloys and describe the properties that are different from the pure elements.
5. Recognize allotropic forms of an element and state how allotropes are different from one another.
6. Name the properties of the transition metals that have led to the many applications of these elements and their compounds.

■ Section Notes

13-1 Begin by asking students to name characteristics of a metal such as being shiny, being a good conductor, and bending without breaking (malleable). They also are likely to suggest high density, high melting point, strong, and hard. They will learn that the latter properties vary greatly among metals. For example, mercury is a liquid at room temperature, sodium has a density that is less than water, and rubidium melts at less than 40°C and can be cut with a knife.

Metals have free electrons that account for the nature of metallic bonds and the ability of metals to conduct heat and electricity. Chemically defined, a metal is an element that loses electrons with ease when it reacts.

13-2 The alkali metals are all so reactive that they will react vigorously with water, liberating hydrogen and forming a basic solution. These elements lose electrons and form positive ions that are very soluble in water. If you have some sodium or lithium metal, show the students that they are stored under kerosene to prevent them from reacting with oxygen and water. Demonstrate the solubility of several compounds of sodium and potassium and show that these solutions are electrolytes.

13-3 These metals also are very reactive, although less reactive than other alkali metals. They also form many insoluble compounds including carbonates, sulfates, and phosphates. Look up the solubility of some of these compounds or demonstrate the low solubility of magnesium carbonate and calcium sulfate. The solubility of $Ca(HCO_3)_2$, important in cave formation, boiler scale, and several industrial processes, can be demonstrated.

13-4 The reactivity of aluminum can be demonstrated by putting a small piece of aluminum foil

(5 cm × 5 cm) in a beaker with $3M$ NaOH and a separate piece in a beaker with $3M$ HCl.

13-5, 13-6, 13-7, 13-8 The properties and uses of several of the transition metals are discussed here. Present research indicates that, except for scandium and titanium, all of the remaining fourth-row transition metals are essential trace elements, necessary for the proper functioning of the human body.

The chemical composition of gems could be used as a library research project.

Review and Practice

page 372

1. The outer electrons in metals are free to move into other nearby available energy levels because only a small amount of energy is required. These "free" electrons are bonded to all of the positive ions which remain. This "sea" of electrons bonds all the metal atoms and also is responsible for the strength, malleability, and conductivity that are characteristics of metals.
2. metal hydroxide, water, and energy
3. Magnesium forms a hard, stable coating of magnesium oxide which prevents further reaction.
4. a mixture of a metal with other metals or elements with a metal that alters the properties of the metal
5. The calcium carbonate in limestone is dissolved in the presence of CO_2 in rainwater. When CO_2 and H_2O are released into the air, the remaining $Ca(HCO_3)_2$ forms stalactites and stalagmites.

page 382

1. strength, high melting point, malleability, durability, colorful compounds
2. There are more electrons available for metallic bonding.
3. The molar mass increases regularly except for nickel.
4. photography, jewelry
5. good conductor, malleable, corrosion resistant, colorful compounds
6. Stainless steel is shiny, hard, corrosion resistant and is used in jewelry, pipes, sinks, tanks, and tools. Alnico is used to make strong, permanent magnets.
7. copper, silver, gold

page 385

1. silicon
2. boron
3. As—gray metal, metallic bonding, high melting point, malleable, strong As₄—yellow, molecular compound, low melting point
4. protect carpet, clothing, fabrics, furniture; heat-resistant paper and fabric; water-resistant fabric Allotropes of an element have the same physical state but different molecular structures.

page 391

1. the strong stable triple bond of shared electrons
2. Oxygen is an electron acceptor.
3. the halogens
4. Reactivity decreases.
5. They are stable and unreactive.
6. to increase the life of light bulbs and colored lights; to provide an inert atmosphere

Review, page 393

1. metals, nonmetals, metalloids
2. The metal ions are relatively fixed in positions (they can slide past one another) while electrons are free to move within the crystal.
3. to prevent them from reacting with water, nitrogen, and oxygen
4. sodium and potassium
5. soft, silvery-white, malleable, very reactive, low density, good conductors, 1^+ charge in bonding, produce hydroxides with water
6. sodium, potassium, lithium, magnesium, calcium, fluorine
7. reactive, 2^+ charge in bonding, good conductors, malleable, low density, produce weak hydroxides that are only slightly soluble, light weight, abundant
8. aluminum
9. $Mg(OH)_2$ is low in solubility and dissolves to form only a few hydroxide ions. If these hydroxide ions are neutralized by stomach acid (HCl), more magnesium hydroxide dissolves, thus regulating the neutralizing ability.
10. calcium carbonate, $CaCO_3$
11. A hard oxide coating forms which protects the remaining metal from further reaction. This coating makes these metals corrosion resistant and durable.
12. light weight, strong, corrosion resistant, malleable, good conductor
13. A metal is an element that loses electrons when bonding.
14. gold, silver, copper, platinum, iridium
15. limestone, $CaCO_3$, coke, C, and oxygen from the air
16. metalloids

17. silicon
18. They bond with the sulfur in protein molecules (enzymes) and deactivate the enzyme. This fatally disrupts the chemistry of the cell.
19. Silicon and oxygen are the primary elements in silicates, the most common and abundant part of the solid crust of Earth.
20. These are allotropes of phosphorous.
21. proteins
22. the halogens, F_2, Cl_2, Br_2, I_2, At
23. fluorine
24. The noble gases are very low in reactivity and do not combine with any other elements even at the very high temperatures used in welding.

Interpret and Apply, page 394

1. a. $3Na + AlCl_3 \longrightarrow Al + 3NaCl$
   b. $2Li + 2H_2O \longrightarrow 2LiOH + H_2$
   c. $2K + Br_2 \longrightarrow 2KBr$
   d. $SiO_2 + 2Ca \longrightarrow 2CaO + Si$
   e. $CaO + H_2O \longrightarrow Ca(OH)_2$
   f. $Al_2O_3 + 3H_2O \longrightarrow 2Al(OH)_3$

2. The transition metals are not as reactive and can easily be separated from their ores. The alkali metals and alkaline-earth metals are so active that they are difficult to isolate and keep pure.

3. $2CuS + 3O_2 \longrightarrow 2CuO + 2SO_2$
   $2CuO + C \longrightarrow 2Cu + CO_2$

4. Iron with 1% carbon is more brittle. The addition of carbon increases hardness but also increases brittleness.

5. The addition of pressure, the addition of energy in the form of an electric spark or heat, or the presence of seed crystals can cause a change in allotropic form.

6. $S_8(s) + 24F_2(g) \longrightarrow 8SF_6(g)$
   $S_8(s) + 4Br_2(l) \longrightarrow 4S_2Br_2(l)$
   The first equation gives the ratio of sulfur to fluorine as 1 to 24. Only 6 moles of fluorine are present, so sulfur is in excess. The second equation gives the ratio of sulfur to bromine as 1 to 4. If 6 moles of bromine are reacted, there will be no sulfur left over.

7. $:\!\ddot{I}\!:\!\ddot{I}\!:$

Problems, page 394

1. $8Ca + S_8 \longrightarrow 8CaS$
$$17.5\,g\,Ca \times \frac{1\,mol\,Ca}{40.1\,g\,Ca} \times \frac{1\,mol\,S_8}{8\,mol\,Ca} \times \frac{256.8\,g\,S_8}{1\,mol\,S_8}$$
$= 14.0\,g\,S_8$ required

Since there are 15.3 g S_8 present, S_8 is in excess by 1.3 g.

2. $Ca + 2H_2O \longrightarrow Ca(OH)_2 + H_2$
$$13.5\,g\,Ca \times \frac{1\,mol\,Ca}{40.1\,g\,Ca} \times \frac{1\,mol\,H_2}{1\,mol\,Ca} = 0.337\,mol\,H_2$$
$PV = nRT$
$$V = \frac{nRT}{P} \quad V = \frac{(0.337\,mol)(8.314\,L \cdot kPa)(316\,K)}{95.0\,kPa\,mol \cdot K}$$
$= 9.31\,L$

3. $2Mg + O_2 \longrightarrow 2MgO$
$$17.5\,g\,Mg \times \frac{1\,mol\,Mg}{24.3\,g\,Mg} \times \frac{2\,mol\,MgO}{2\,mol\,Mg}$$
$$\times \frac{40.3\,g\,MgO}{1\,mol\,MgO} = 29.0\,g\,MgO$$

4. $Mg(OH)_2 + 2HCl \longrightarrow MgCl_2 + 2H_2O$
$$6.75\,g\,Mg(OH)_2 \times \frac{1\,mol\,Mg(OH)_2}{58.3\,g\,Mg(OH)_2} \times \frac{2\,mol\,HCl}{1\,mol\,Mg(OH)_2}$$
$$\times \frac{36.5\,g\,HCl}{1\,mol\,HCl} = 8.45\,g\,HCl$$

5. $Ca + 2H_2O \longrightarrow Ca(OH)_2 + H_2$
$$13.5\,g\,Ca \times \frac{1\,mol\,Ca}{40.1\,g\,Ca} \times \frac{1\,mol\,H_2}{1\,mol\,Ca}$$
$$\times \frac{22.4\,L\,H_2\,at\,STP}{1\,mol\,H_2} = 7.45\,L\,H_2$$

6. $3K + AlCl_3 \longrightarrow 3KCl + Al$
$$145\,g\,Al \times \frac{1\,mol\,Al}{27.0\,g\,Al} \times \frac{3\,mol\,K}{1\,mol\,Al} \times \frac{39.1\,g\,K}{1\,mol\,K}$$
$= 6.30 \times 10^2\,g\,K$

7. $2C + O_2 \longrightarrow 2CO$
$$585\,mol\,CO \times \frac{2\,mol\,C}{2\,mol\,CO} \times \frac{12.0\,g\,C}{1\,mol\,C} = 7.02 \times 10^3\,g\,C$$

8. $FeO + CO \longrightarrow Fe + CO_2$
$$585\,mol\,CO \times \frac{1\,mol\,Fe}{1\,mol\,CO} \times \frac{55.85\,g\,Fe}{1\,mol\,Fe}$$
$$\times \frac{1\,kg\,Fe}{1000\,g\,Fe} = 32.7\,kg\,Fe$$

9. $2SO_2 + O_2 \longrightarrow 2SO_3$
$$25.8\,L\,SO_2 \times \frac{1\,mol\,SO_2}{22.4\,L\,SO_2} \times \frac{2\,mol\,SO_3}{2\,mol\,SO_2}$$
$= 1.15\,mol\,SO_3$

10. $2Na_2O_2 + 2H_2O \longrightarrow 4NaOH + O_2$
$$10.6\,g\,Na_2O_2 \times \frac{1\,mol\,Na_2O_2}{78\,g\,Na_2O_2} \times \frac{1\,mol\,O_2}{2\,mol\,Na_2O_2}$$
$= 0.0679\,mol\,O_2$
$$V = \frac{nRT}{P} = \frac{(.0679\,mol)(8.314\,L \cdot kPa)(301\,K)}{98.3\,kPa\,mol \cdot K}$$
$V = 1.73\,L\,O_2$

11. $Ti + 2F_2 \longrightarrow TiF_4$
$$35.6\,g\,Ti \times \frac{1\,mol\,Ti}{47.9\,g\,Ti} \times \frac{2\,mol\,F_2}{1\,mol\,Ti} = 1.49\,mol\,F_2$$
$$V = \frac{nRT}{P} = \frac{(1.49\,mol)(8.31\,L \cdot kPa)(295\,K)}{105\,kPa\,mol \cdot K}$$
$= 34.7\,L\,F_2$

12. $UO_2 + 4HF \longrightarrow UF_4 + 2H_2O$

$8.45 \text{ kg UF}_4 \times \dfrac{1000 \text{ g UF}_4}{1 \text{ kg UF}_4} \times \dfrac{1 \text{ mol UF}_4}{314 \text{ g UF}_4}$

$\times \dfrac{4 \text{ mol HF}}{1 \text{ mol UF}_4} \times \dfrac{20.0 \text{ g HF}}{1 \text{ mol HF}} = 2.15 \times 10^3 \text{ g HF}$

Challenge, page 395

1. $CaCO_3 + 2HC_2H_3O_2 \longrightarrow Ca(C_2H_3O_2)_2 + CO_2$

$34.6 \text{ g boiler scale} \times \dfrac{60 \text{ g CaCO}_3}{100 \text{ g boiler scale}}$

$\times \dfrac{1 \text{ mol CaCO}_3}{100 \text{ g CaCO}_3} \times \dfrac{2 \text{ mol HC}_2\text{H}_3\text{O}_2}{1 \text{ mol CaCO}_3}$

$\times \dfrac{60 \text{ g HC}_2\text{H}_3\text{O}_2}{1 \text{ mol HC}_2\text{H}_3\text{O}_2} \times \dfrac{100 \text{ g vinegar}}{4.5 \text{ g HC}_2\text{H}_3\text{O}_2}$

$\times \dfrac{100 \text{ mL vinegar}}{100 \text{ g vinegar}} = 554 \text{ mL vinegar}$

$MgCO_3 + 2HC_2H_3O_2 \rightarrow Mg(C_2H_3O_2)_2 + H_2O + CO_2$

$34.6 \text{ g boiler scale} \times \dfrac{40 \text{ g MgCO}_3}{100 \text{ g boiler scale}}$

$\times \dfrac{1 \text{ mol MgCO}_3}{84.3 \text{ g MgCO}_3} \times \dfrac{2 \text{ mol HC}_2\text{H}_3\text{O}_2}{1 \text{ mol MgCO}_3}$

$\dfrac{60 \text{ g HC}_2\text{H}_3\text{O}_2}{1 \text{ mol HC}_2\text{H}_3\text{O}_2} \times \dfrac{100 \text{ g vinegar}}{4.5 \text{ g vinegar}} \times \dfrac{100 \text{ mL vinegar}}{100 \text{ g vinegar}}$

$= 438 \text{ mL vinegar}$

Total volume of vinegar required = 554 mL + 438 mL = 992 mL

2. $2Mg(s) + O_2(g) \longrightarrow 2MgO(s)$

Assume that all the magnesium reacts with oxygen.

$2.16 \text{ g Mg} \times \dfrac{1 \text{ mol Mg}}{24.3 \text{ g Mg}} \times \dfrac{2 \text{ mol MgO}}{2 \text{ mol Mg}} \times \dfrac{40.3 \text{ g MgO}}{1 \text{ mol MgO}}$

$= 3.58 \text{ g MgO}$

Assume that all the magnesium reacts with nitrogen.

$3Mg(s) + N_2(g) \longrightarrow Mg_3N_2(s)$

$2.16 \text{ g Mg} \times \dfrac{1 \text{ mol Mg}}{24.3 \text{ g Mg}} \times \dfrac{1 \text{ mol Mg}_3\text{N}_2}{3 \text{ mol Mg}}$

$\times \dfrac{100.9 \text{ g Mg}_3\text{N}_2}{1 \text{ mol Mg}_3\text{N}_2} = 2.99 \text{ g Mg}_3\text{N}_2$

Let x = g of Mg reacting with nitrogen. Then $2.16 - x$ = g of Mg reacting with oxygen.

$\text{g Mg}_3\text{N}_2 = x\left(\dfrac{2.99}{2.16}\right)$ and $\text{g MgO} = (2.16 - x)\left(\dfrac{3.58}{2.16}\right)$

$(2.16 - x)\left(\dfrac{3.58}{2.16}\right) + x\left(\dfrac{2.99}{2.16}\right) = 3.43$

$(3.58 - 1.657\,x) + 1.384\,x = 3.43$

$0.15 = 0.273\,x$

$x = 0.55 \text{ g Mg reacting with nitrogen}$

% Mg reacting with nitrogen $= \dfrac{0.55}{2.16} = 25\%$

3. $2CuO + C \longrightarrow CO_2 + 2Cu$

$3500 \text{ kg ore} \times \dfrac{1000 \text{ g}}{1 \text{ kg}} \times \dfrac{6.0 \text{ g CuO}}{100.0 \text{ g ore}} \times \dfrac{1 \text{ mol CuO}}{79.5 \text{ g CuO}}$

$\times \dfrac{2 \text{ mol Cu}}{2 \text{ mol CuO}} \times \dfrac{63.5 \text{ g Cu}}{1 \text{ mol Cu}} = 1.7 \times 10^5 \text{ g Cu}$

4. $2H_2O_2(aq) \longrightarrow 2H_2O(l) + O_2(g)$

$15.0 \text{ cm}^3 \text{ solution} \times \dfrac{1.11 \text{ g}}{\text{cm}^3} \times \dfrac{20 \text{ g H}_2\text{O}_2}{100 \text{ g solution}}$

$\times \dfrac{1 \text{ mol H}_2\text{O}_2}{34.0 \text{ g H}_2\text{O}_2} \times \dfrac{1 \text{ mol O}_2}{2 \text{ mol H}_2\text{O}_2} = 0.049 \text{ mol O}_2$

$V = \dfrac{nRT}{P} = \dfrac{(0.049 \text{ mol})(8.31 \text{ L} \cdot \text{kPa})(308 \text{ K})}{95.0 \text{ kPa mol} \cdot \text{K}}$

$= 1.32 \text{ L O}_2$

5. $2H_2O_2(aq) \longrightarrow 2H_2O(l) + O_2(g)$

$15.0 \text{ cm}^3 \text{ solution} \times \dfrac{1.16 \text{ g}}{\text{cm}^3} \times \dfrac{30 \text{ g H}_2\text{O}_2}{100 \text{ g solution}}$

$\times \dfrac{1 \text{ mol H}_2\text{O}_2}{34.0 \text{ g H}_2\text{O}_2} \times \dfrac{1 \text{ mol O}_2}{2 \text{ mol H}_2\text{O}_2} = 0.077 \text{ mol O}_2$

$V = \dfrac{nRT}{P} = \dfrac{(0.077 \text{ mol})(8.314 \text{ L} \cdot \text{kPa})(300 \text{ K})}{95.5 \text{ kPa mol} \cdot \text{K}}$

$= 2.01 \text{ L O}_2$

6. $m_{He}v_{He}^2 = m_N v_N^2$

$4.00 \text{ g} \times v_{He}^2 = 28.0 \text{ g}(480 \text{ m/s})^2$

$v_{He}^2 = 7.00(480 \text{ m/s})^2$

$v_{He}^2 = 1.6 \times 10^6 \text{ m}^2/\text{s}^2$

$v_{He} = 1.3 \times 10^3 \text{ m/s}$

Since this exceeds 1200 m/s, the helium molecules will gradually "leak" off into space.

Synthesis, page 395

1. $UF_4 + 2Ca \longrightarrow U + 2CaF_2$

$UF_4 + 2Mg \longrightarrow U + 2MgF_2$

$1000 \text{ g U} \times \dfrac{1 \text{ mol U}}{238 \text{ g U}} \times \dfrac{2 \text{ mol Ca}}{1 \text{ mol U}} \times \dfrac{40.1 \text{ g Ca}}{1 \text{ mol Ca}}$

$\times \dfrac{1 \text{ kg}}{1000 \text{ g}} = 0.337 \text{ kg Ca}$

$1000 \text{ g U} \times \dfrac{1 \text{ mol U}}{238 \text{ g U}} \times \dfrac{2 \text{ mol Mg}}{1 \text{ mol U}} \times \dfrac{24.3 \text{ g Mg}}{1 \text{ mol Mg}}$

$\times \dfrac{1 \text{ kg}}{1000 \text{ g}} = 0.204 \text{ kg Mg}$

Since fewer kilograms of Mg are required, it would cost less to use Mg.

2. To produce 1 kg of uranium:

0.337 kg Ca × $11.00/kg Ca = $3.71 for Ca

0.204 kg Mg × $16.00/kg Mg = $3.26 for Mg

Magnesium would still cost less.

3. Na, K, Ba, and Li are all very active.

4. $: \overset{\cdot\cdot}{\underset{\cdot\cdot}{Br}} : \overset{\cdot\cdot}{\underset{\cdot\cdot}{Br}}$

nonpolar

5. Air is 80% N_2 and 20% O_2. A mole of air has a mass of 28.8 g.

28 g N_2/mol N_2 × 0.80 = 22.4 g N_2

32 g O_2/mol O_2 × 0.20 = 6.4 g O_2

22.4 g + 6.4 g = 28.8 g/mol air

28.8 g air/22.4 L air at STP = 1.29 g/L
A mole of H_2O has a mass of 18 g.
· 18 g H_2O(g)/22.4 L H_2O(g) at STP = 0.80 g/L
Since H_2O is less dense than air, adding water to air decreases the density. This is why the barometer reading decreases when a storm center arrives containing moist air.

CHAPTER 14 High-Tech Elements

CHAPTER ORGANIZER

USE WITH SECTION	DEMONSTRATION	WORKSHEET	LAB EXPERIMENT	SOFTWARE	TEACHER'S RESOURCE BINDER
14-1		14-A		Disk 1	
14-3			14-A		
14-5		14-B			
14-6		14-C			Chem Issue 5
14-7		14-C, 14-D			
Chapter Test					

■ Objectives

When students have completed this chapter, they will be able to

1. Describe the doping of silicon crystals to form *n*-type or *p*-type semiconductors.
2. Describe an integrated circuit.
3. Describe how a photovoltaic cell operates.
4. List high-tech uses of ceramics.
5. Conduct a literature search for a research project.

■ Section Notes

Introduction

This chapter cannot be a comprehensive survey of the application of chemistry to high-technology industries. There is simply too much to learn. New discoveries follow each other so quickly that no one is able to keep current in all fields. New technologies are replacing old ones at a staggering rate, as Alvin Toffler explained in his book *Future Shock*.

The material in this chapter is meant to give students the incentive to explore a particular field of chemistry more fully. If the students become "experts" in various areas, you will be relieved of the burden of providing information about widely divergent fields. The students become responsible for gathering and disseminating information. Instead of being the presenter of relevant information, use your skills as information evaluator. In this role you lead the students to decide how to find information, how to organize it, how to decide what is correct or incorrect, and

what is important or unimportant. The students should do research reports to sharpen their information evaluating abilities and to develop communication skills. As was suggested in the section on "Teaching Communication Skills" of the Philosophy of the Program, you could explore the possibility of making the reports a cooperative project with an English or social studies teacher. Evaluation of these reports should be based on effectiveness of communication rather than on content alone.

In order to accomplish these goals, you need to introduce students to research report writing long before the papers are due. Giving students ample time to prepare a report (several months) means that they will be more likely to do the reports well and that you will have time to help them as a resource person. If students work in small groups, for example two students per group, less class time is needed for the projects. You may want to build about 10 to 15 minutes into your lesson planning on consecutive Fridays for working on class presentations. You also may want to schedule deadlines for topic selections, outlines, and rough drafts so that you and the students are not overwhelmed at the last minute.

When you present this assignment to the students, there are several ideas you should consider:

1. What is an acceptable topic? You can refer the students to the suggested topics in the project section of this chapter or other chapters.
2. Who will do reports? Since gathering information and written and oral communications are important skills in science, all students should become

involved. Individual reports are time consuming. Groups of two or three may be considered practical. Larger groups are not recommended because often only one or two people in the group do most of the work.

3. What is the purpose of the project? You should refer the students to the lesson on "chemical research" at the end of this chapter. The students should not only present information, they should question scientific accuracy and may incorporate social, political, and economic concerns in the reports as well.

4. How will the reports be evaluated? Students should know on what criteria they are to be graded and what weight the grades will have in the context of your grading system. Fifty percent for the written report and fifty percent for the class presentation might give students incentive to do both parts well. If this report is going to be submitted for both a chemistry and an English or social studies grade, then students need to know how the report will be evaluated for each class.

5. How long should the report be? This is up to you and your students. The length of the written report and the oral presentation should be reflected in the weight of the grade.

6. How scholarly should the report be? The students should cite sources in the written report. Whether it will be a formal term paper or a more casual report should be decided before they begin.

Many students are reluctant to present an oral report. Yet, a class report helps the presenters summarize their findings for a specific audience. Some of the anxiety over an oral report can be diminished if you teach students some strategies that you practice in the classroom. In the oral report the students should introduce the topic and present relevant information so that the audience is aware of some pertinent facts. The use of visual aids enhances the presentation. The audience should be given a few minutes for questions.

14-1 In order to understand the conductivity differences in metal crystals (high conductivity), insulators (no conductivity), and semiconductors (some conductivity), the band theory is usually offered as an explanation. This theory is based on the molecular orbital theory. In a semiconductor there are molecular orbitals that are filled and empty having different energy levels. The empty orbitals form the conduction band and the filled orbitals form the valence band. The bands are separated by a moderately small energy gap, the forbidden zone. Some electrons in the valence band are able to cross the forbidden zone and enter the conduction band when the semiconductor material is heated. This is because the forbidden zone is sufficiently narrow to allow the electrons to be promoted. Semiconductors increase in conductivity when heated. The presence of impurities in the semiconductor material allows electrons to cross the forbidden zone more easily.

There are several techniques used today to dope a semiconductor crystal. In a process called diffusion, the semiconductor base material is heated to a temperature below its melting point. In an atmosphere high in dopant concentration, the dopant diffuses into the semiconductor to a depth of only several micrometers. A second process, called ion implantation, involves the bombarding of the semiconductor material with a high-energy beam of dopant ions. The advantage of this technique is that the semiconductor material need not be heated to achieve the proper dopant concentration.

The process of making integrated circuits would make an excellent research project.

14-5 There are advantages to using optical fibers in voice transmission for several reasons. Optical fiber transmission contains much less "noise" than copper wire transmission. Secondly, optical fibers carry more voice conversations simultaneously than copper wire. The difference lies in the mode of propagation. In the copper wire of a telephone system, analog signals have a frequency of about one million cycles per second. The human voice can produce vibrations whose frequency may reach 4000 cycles per second. Therefore, about 250 conversations can be carried by copper wire. If light pulses are used for transmission, many more conversations can be carried. The analog signals from the telephone are converted into digital pulses and transmitted as on/off pulses of light by a solid state laser. The maximum limit of conversations carried by one fiber is determined by the number of on/off pulses the laser can produce. Optical fibers used today carry about 1500 conversations simultaneously.

Review and Practice
page 404

1. arsenic
2. The circuit consists of electronic components on a tiny piece of semiconductor material.
3. The energy comes from a small voltage applied near the surface being studied.
4. An n-type semiconductor is in contact with a p-type semiconductor. When light of the proper frequency strikes the semiconductor material, an electric current is produced.

page 408

1. small crystals of aluminosilicate in a matrix of glassy cement

2. hard, porous, heat resistant, nonreactive, and inexpensive

3. SiO_2 with sodium or calcium oxides or carbonates

4. The light beam is continually reflected back and forth within the optical fiber.

5. light energy

Review, page 412

1. Si $\quad 1s^2 2s^2 2p^6 3s^2 3p^2$
  Ge $\quad 1s^2 2s^2 2p^6 3s^2 3p^6 4s^2 3d^{10} 4p^2$

2. The electrons in the outer energy levels of each silicon atom in a pure silicon crystal are involved in localized bonds with neighboring silicon atoms.

3. "Holes" are the locations of electron deficiencies in crystals. When electrons flow through the crystal, they fill the holes, producing other holes in the direction opposite to electron flow.

4. Integrated circuits are considered as the brains of electronic equipment.

5. photoengraving

6. Lasers allow manufacturers to deposit layers of materials only tens of atoms thick, producing highly miniaturized components.

7. Quartz is a true crystal, that is, it has a regular pattern throughout.

8. to power satellites and to produce electricity in homes

9. lead(II) oxide

10. The amount of information that can be transmitted at one time is much greater than in copper wire.

Interpret and Apply, page 412

1. outermost electron configurations
  Ga $\quad 4s^2 4p^1 \qquad$ B $\quad 2s^2 2p^1$
  As $\quad 4s^2 4p^3 \qquad$ Se $\quad 4s^2 4p^4$
  P $\quad 3s^2 3p^3$

As, P, and Se could be used to make *n*-type semiconductors. Ga and B would make *p*-type semiconductors.

2. Dopants can diffuse away from their specific locations. Radiation can affect the performance of a circuit. Electrons may not flow through the desired circuit. Components one atom thick would suffer most from all of the disadvantages above. Their reliability would be very short-lived.

3. Integrated circuits take up less space. They use less current. They require less maintenance.

4. The scanning tunneling microscope can detect details of atoms only on the surface of materials.

5. The scanning tunneling microscope allows a chemist to create certain reactions on the surface of materials and see the results of the reactions almost simultaneously.

6. The energy of photons is converted to electrical energy.

7. The binary code is transmitted through fiber optics by a pulsating monochromatic light from a solid state laser.

Problems, page 413

1. The left end of each unit will be separated by
$3 \times 10^{-6} + 5 \times 10^{-6}\,m = 8 \times 10^{-6}\,m$.

Since the total length available for the components is $4 \times 10^{-3}\,m$, the number of units that can be placed across the circuit is:
$$\frac{4 \times 10^{-3}\,m}{8 \times 10^{-6}\,m} = 500 \text{ components}$$

2. $WF_6 + 3H_2 \longrightarrow W + 6HF$

CHAPTER 15 The Condensed States of Matter

CHAPTER ORGANIZER

USE WITH SECTION	DEMONSTRATION	WORKSHEET	LAB EXPERIMENT	SOFTWARE	TEACHER'S RESOURCE BINDER
15-1					
15-2		15-A			
15-4		15-B			
15-5			15-A		
15-11		15-C			
15-12		15-D, 15-E		Disk 4	
Chapter Test					

■ Objectives

When the students have completed this chapter, they will be able to

1. Interpret the change in state of a substance in terms of the kinetic molecular theory.
2. Relate properties of a substance, such as melting point and vapor pressure, to the attractive forces between molecules.
3. Recognize that the vapor pressure of a given substance depends on temperature and that nonvolatile substances have low vapor pressures.
4. Explain the changes in energy that accompany a heating curve for a change in state.
5. Use a phase diagram to determine the state of a substance at a given set of conditions.
6. Describe the difference in bonding among molecular, network, and metallic solids.
7. Name three types of packing arrangements and either draw or construct models of them.
8. Define an alloy and describe some properties of an alloy that are different from the metals that compose it.
9. Name two factors that affect the shape of an ionic crystal.
10. Calculate the percent of water in a substance given the formula for the hydrate.
11. Explain why a chemical like $CaCl_2$ will fluctuate in mass from day to day because of the humidity.

■ Section Notes

15-2 If possible, borrow a vacuum pump and reduce the pressure above the water in a vacuum flask. Put a thermometer in the water so students can see that the temperature is unchanged even though the water is now boiling.

15-3 Someone who lives at a very high altitude may find it difficult to cook potatoes or vegetables in boiling water because the water is at a temperature that is not high enough to cook the food. Ask students how someone at a high altitude could cook food. Suggest using a pressure cooker. Discuss the need for a safety valve on the pressure cooker to release the pressure and avoid an explosion.

15-7 If possible, obtain and display several crystal shapes. Check with the earth science teacher for some crystals.

Have a student grow crystals as a project or construct models of crystals using Styrofoam balls, wood splints, and glue.

15-8 Invite a local dentist to discuss cavities, alloys, amalgams, tooth fillings and the role of fluorides in preventing tooth decay. Invite a local jeweler to discuss alloys, amalgams, and metalworking.

Review and Practice _____
page 425

1.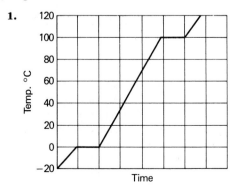

2. During the solid to liquid phase change, the temperature does not increase. During the liquid to gas phase change, the temperature does not change.

3. a. The added energy increases the kinetic energy of the molecules, and they rotate more rapidly.
    b. The energy is used to overcome the attractive forces between the molecules.

4. The boiling point increases as the atmospheric pressure increases.

5. The attractive forces between acetone molecules are weaker than those between ethanol molecules at 25°C.

6. a. 2582°C
    b. 1083°C
    c. 1499°C

7. a. The kinetic energy remains constant.
    b. The potential energy decreases as the molecules move closer together.

8.

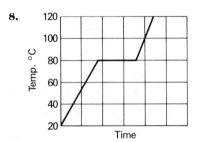

9. Some of the added heat is used to convert liquid to gas molecules—a process that does not increase the temperature.

10. Vapor molecules lose kinetic energy as the temperature decreases. At 78°C, substantial numbers of gas molecules are moving slowly enough to be attracted to each other and form a liquid. As the temperature continues to decrease, both the vapor and the liquid lose kinetic energy and additional condensation occurs.

page 433

1. covalent
2. $FeSO_4 \cdot 7H_2O$

Review, page 435

1. A phase is a state of matter: solid, liquid, or gas.
2. Evaporation occurs whenever a liquid molecule gains enough kinetic energy to overcome the attractive forces holding it in the liquid. It becomes a gas and escapes to the atmosphere above the liquid. Boiling occurs when the vapor pressure of the liquid equals atmospheric pressure. Then bubbles of gas vapor form within the liquid.

3. The vapor pressure of a substance is a characteristic property that depends only on the temperature of the substance. Raising the temperature increases the vapor pressure.

4. The greater the vapor pressure of a liquid, the more volatile it is. Highly volatile substances have high vapor pressure.

5. Kinetic energy is directly proportional to temperature.

6. The added energy is used to break the bonds between the solid molecules and not to increase the temperature (kinetic energy).

7. The triple point of water is the temperature and pressure at which all three phases of water exist simultaneously in a closed system. The triple point of water is used to define the temperature scale.

8. Wetting agents lower surface tension.

9. Each atom in a body-centered cubic arrangement has eight nearest neighbors. Atoms in hexagonal closest packing have twelve nearest neighbors.

10. polymorphous

11. copper and tin

12. alloys of other metals with mercury

13. tetrahedral

14. the relative numbers of positive and negative ions and the relative size of the ions

Interpret and Apply, page 436

1. a. gasoline   **c.** perfume
    b. acetone   **d.** water

2. The autoclave is sealed to increase the pressure. A pressure of a little more than 2 atmospheres is required to increase the boiling point to 121°C.

3. a. Temperature is increased.
    b. Some liquid water solidifies as the ice cube absorbs heat from the water, but the temperature stays constant.
    c. The warm water loses some kinetic energy to the ice cubes. Some ice melts, but the temperature stays constant if all of the ice does not melt.
    d. More ice forms when the −5°C ice cube is added. Some ice melts when the warm water is added.

4. a. A   **d.** C
    b. C   **e.** B and C
    c. B   **f.** C

5. Lower the external pressure to about 2.6 kPa.

6. They increase the cleaning ability of water by reducing its surface tension so cleansing agents can penetrate fibers better.

7. Atmospheric pressure is lower in Denver.

8. a. 7.33 kPa
 b. The vapor pressure of alcohol will not change.

9. Water molecules evaporate even at low temperatures.

10. Because the vapor pressure of carbon tetrachloride is high, all of it evaporates quickly and diffuses from the room. The vapor pressure of mercury is much lower. It may take months for mercury to evaporate and diffuse from the room. Meanwhile, anyone in the room breathes the fumes.

11. The water vapor condenses when it cools and makes the pressure inside the can much lower. The greater outside pressure collapses the can.

12. Pour water over your clothes. As the water evaporates, your skin is cooled.

13. Assuming equal heat transfer abilities (not true), steam will cause a more severe burn because it contains more energy.

14. cobalt

15. decrease because the atoms will pack closer together

16. harder

17. carbon, nitrogen, sulfur, phosphorus

Problems, page 437

1. 56°C

2. $MgSO_4 \cdot 7H_2O$ $24.3 + 32.1 + 64.0 + 7(18.0)$
$= 120.4 + 126.0 = 246.4$

$\% \ H_2O = \dfrac{126.0}{246.4} = 51.14\%$

3. $0.283 \ \text{mol CuSO}_4 \times \dfrac{5 \ \text{mol } H_2O}{1 \ \text{mol CuSO}_4} = 1.42 \ \text{mol } H_2O$

4. a. $MgSO_4 \cdot 7H_2O + \text{energy} \longrightarrow MgSO_4 + 7H_2O$

$23.6 \ \text{g MgSO}_4 \cdot 7H_2O \times \dfrac{1 \ \text{mol MgSO}_4 \cdot 7H_2O}{246.4 \ \text{g MgSO}_4 \cdot 7H_2O}$

$\times \dfrac{7 \ \text{mol } H_2O}{1 \ \text{mol MgSO}_4 \cdot 7H_2O} \times \dfrac{18.0 \ \text{g } H_2O}{1 \ \text{mol } H_2O}$

$= 12.1 \ \text{g } H_2O$

or $23.6 \ \text{g MgSO}_4 \cdot 7H_2O \times \dfrac{51.14 \ \text{g } H_2O}{100.0 \ \text{g MgSO}_4 \cdot 7H_2O}$

$= 12.1 \ \text{g } H_2O$

b. $23.6 \ \text{g} - 12.1 \ \text{g} = 11.5 \ \text{g MgSO}_4$

c. $12.1 \ \text{g } H_2O \times \dfrac{1 \ \text{mol } H_2O}{18.0 \ \text{g } H_2O} = 0.672 \ \text{mol } H_2O$

d. $11.5 \ \text{g MgSO}_4 \times \dfrac{1 \ \text{mol MgSO}_4}{120.4 \ \text{g MgSO}_4}$

$= 0.0955 \ \text{mol MgSO}_4$

or $0.672 \ \text{mol } H_2O \times \dfrac{1 \ \text{mol MgSO}_4}{7 \ \text{mol } H_2O}$

$= 0.096 \ \text{mol MgSO}_4$

e. Divide both by the smallest, 0.096.
H_2O $0.672/0.096 = 7$

$MgSO_4$ $0.096/0.096 = 1$ The ratio is 7 to 1.

5. $2.50 \ \text{L MgSO}_4(aq) \times \dfrac{1.25 \ \text{mol MgSO}_4(s)}{1 \ \text{L MgSO}_4(aq)}$

$\times \dfrac{1 \ \text{mol MgSO}_4 \cdot 7H_2O}{1 \ \text{mol MgSO}_4} \times \dfrac{246.4 \ \text{g MgSO}_4 \cdot 7H_2O}{1 \ \text{mol MgSO}_4 \cdot 7H_2O}$

$= 770 \ \text{g MgSO}_4 \cdot 7H_2O$

Add $770 \ \text{g MgSO}_4 \cdot 7H_2O$ to enough water to make the final volume of the solution 2.50 L.

6. a. $\dfrac{24.3 \ \text{g Mg}}{1 \ \text{mol Mg}} \times \dfrac{1 \ \text{cm}^3 \ \text{Mg}}{1.74 \ \text{g Mg}} = \dfrac{14.0 \ \text{cm}^3 \ \text{Mg}}{1 \ \text{mol Mg}}$

b. $\dfrac{63.5 \ \text{g Cu}}{1 \ \text{mol Cu}} \times \dfrac{1 \ \text{cm}^3 \ \text{Cu}}{8.96 \ \text{g Cu}} = \dfrac{7.09 \ \text{cm}^3 \ \text{Cu}}{1 \ \text{mol Cu}}$

c. $\dfrac{207.2 \ \text{g Pb}}{1 \ \text{mol Pb}} \times \dfrac{1 \ \text{cm}^3 \ \text{Pb}}{11.4 \ \text{g Pb}} = \dfrac{18.2 \ \text{cm}^3 \ \text{Pb}}{1 \ \text{mol Pb}}$

d. $\dfrac{23.0 \ \text{g Na}}{1 \ \text{mol Na}} \times \dfrac{1 \ \text{cm}^3 \ \text{Na}}{0.97 \ \text{g Na}} = \dfrac{24 \ \text{cm}^3 \ \text{Na}}{1 \ \text{mol Na}}$

e. $\dfrac{55.85 \ \text{g Fe}}{1 \ \text{mol Fe}} \times \dfrac{1 \ \text{cm}^3 \ \text{Fe}}{7.85 \ \text{g Fe}} = \dfrac{7.11 \ \text{cm}^3 \ \text{Fe}}{1 \ \text{mol Fe}}$

f. $\dfrac{183.85 \ \text{g W}}{1 \ \text{mol W}} \times \dfrac{1 \ \text{cm}^3 \ \text{W}}{19.3 \ \text{g W}} = \dfrac{9.53 \ \text{cm}^3 \ \text{W}}{1 \ \text{mol W}}$

g. $\dfrac{95.94 \ \text{g Mo}}{1 \ \text{mol Mo}} \times \dfrac{1 \ \text{cm}^3 \ \text{Mo}}{10.2 \ \text{g Mo}} = \dfrac{9.41 \ \text{cm}^3 \ \text{Mo}}{1 \ \text{mol Mo}}$

7. $BaCl_2 \cdot 2H_2O$ First find the formula mass.
$137.3 + 70.9 + 2(18.0) = 208.2 + 36.0 = 244.2$

$\% \ H_2O = \dfrac{36.0}{244.2} = 14.7$

Challenge, page 437

1.

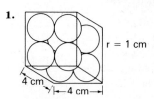

diameter of each sphere = 2.00 cm
edge of cube = 4.00 cm
volume of cube = $(4.00 \ \text{cm})^3 = 64.00 \ \text{cm}^3$
volume of 1 sphere = $\frac{4}{3}\pi(1.00 \ \text{cm})^3 = 4.19 \ \text{cm}^3$
volume of 8 spheres = $33.5 \ \text{cm}^3$

$\% \text{ volume of spheres} = \dfrac{33.5 \ \text{cm}^3}{64.0 \ \text{cm}^3} = 52.3\%$

2. When placed in a magnetic field (or near a magnet), the atoms in soft iron readily line up to become magnetic. They also become nonaligned easily because the bonds in iron are nondirectional. The carbon atoms in steel provide directional bonds that make it difficult to change the alignment of the atoms. Steel is more difficult to magnetize than pure iron, but once magnetized it retains the magnetism much longer.

3.
$$\frac{0.0585 \text{ g NaCl}}{3 \text{ days}} \times \frac{6.02 \times 10^{23} \text{ ions}}{58.45 \text{ g NaCl}} \times \frac{1 \text{ day}}{24 \text{ h}}$$
$$\times \frac{1 \text{ h}}{3600 \text{ s}} = 2.32 \times 10^{15} \text{ ions/s}$$

This is over 2 quadrillion ions per second.

Synthesis, page 437

1. Total pressure $= P_{\text{alcohol}} + P_{\text{acetone}}$
$= 0.1 \text{ atm} + 0.5 \text{ atm} = 0.6 \text{ atm}$

2. Pure iron is preferable because it is magnetized easily. However, when the current is shut off, it loses its magnetism rapidly.

3. $FeSO_4 + xH_2O + \text{energy} \longrightarrow FeSO_4 + xH_2O$
10.45 g hydrate $\longrightarrow$ 5.71 g anhydrous
$+ 4.74 \text{ g } H_2O$

$$5.71 \text{ g FeSO}_4 \times \frac{1 \text{ mol FeSO}_4}{152 \text{ g FeSO}_4} = 0.0376 \text{ mol FeSO}_4$$

$$4.74 \text{ g } H_2O \times \frac{1 \text{ mol } H_2O}{18.0 \text{ g } H_2O} = 0.263 \text{ mol } H_2O$$

Dividing by 0.0376, the ratio is 1 mol $FeSO_4$ to 6.99 mol H_2O. This ratio is 1 to 7 within experimental uncertainty, so the formula for the hydrate is $FeSO_4 \cdot 7H_2O$.

CHAPTER 16 Solutions

CHAPTER ORGANIZER

USE WITH SECTION	DEMONSTRATION	WORKSHEET	LAB EXPERIMENT	SOFTWARE	TEACHER'S RESOURCE BINDER
16-1					Advanced Topic 6
16-2	16-A			Disk 10	Chem Issue 4, 5, 6
16-4	16-B		16-A		
16-5		16-A		Disk 10	Problems 1–4
16-6	16-D			Disk 10	
16-12		16-B	16-B		Problems 5-7
16-13	16-C	16-C		Disk 10	
16-15		16-D			Problems 8-12
16-16		16-E	16-C	Disk 5	
Chapter Test					

▪ Objectives

When students have completed this chapter, they will be able to

1. Differentiate between a solution and a heterogenous mixture.
2. Give examples of gaseous, solid, and liquid solutions.
3. Understand the concept of solubility and how it applies to solution saturation.
4. Distinguish between the solvation of ionic solids and molecular substances.
5. Express solution concentration in molality.
6. Understand the effect of temperature on solubility and be able to interpret a solubility graph.
7. Distinguish among the characteristics of saturated, unsaturated, and supersaturated solutions.
8. Describe the relationship between pressure and gas solubility and how this may explain the behavior of an uncapped bottle of carbonated beverage.
9. Define heat of solution and apply this concept to the solvation of ammonium nitrate crystals.
10. Explain how solute particles may alter colligative properties.
11. Explain boiling point elevation and freezing point depression.

12. Differentiate between simple and fractional distillation.
13. Understand the relationship between reactant ions, spectator ions, precipitation, and solubility.
14. Write net ionic equations.
15. Understand the parameters of solubility equilibrium and use the solubility product constant in predicting chemical precipitation.

■ Section Notes

16-1   Stress the diversity and magnitude of the solution state. Have students examine various types of solutions such as brass, steel, carbonated beverages, vinegar, sea water and ethylene glycol solution. Have interested students research and report on the construction of the Brooklyn Bridge.

16-2   Students should be presented with mixtures containing miscible, partially miscible, and immiscible components.

16-3   To help students visualize the dissociation of an ionic solid into its component ions, examine the electrical conductivity of solutions.

16-7   Have any of the students who dive bring in their regulator and tank. Let them explain how divers make preplanned decompression stops to prevent the bends. Interested students should research and report on the usage of decompression chambers. The pressure/solubility relationship may be observed in a bottle of carbonated beverage.

16-12   Although many students drive to school, most do not understand the cooling system of an internal combustion engine. Hand out a diagram illustrating a cross section of such an engine. Identify the water jackets and explain the function of the water pump, radiator, and thermostat.

Review and Practice _____

page 446

1. **a.** $\dfrac{33\text{ g}}{23 + 35.5} = 0.56 \text{ mol}$ $\dfrac{0.56 \text{ mol}}{1 \text{ L}} = 0.56M$

b. $\dfrac{5.2\text{ g}}{14 + 3} = 0.31 \text{ mol}$ $\dfrac{0.31 \text{ mol}}{0.5 \text{ L}} = 0.62M$

c. $\dfrac{0.10\text{ g}}{72 + 12 + 96} = 0.00056 \text{ mol}$

$\dfrac{0.00056 \text{ mol}}{0.01 \text{ L}} = 0.056M$

d. $\dfrac{8.6\text{ g}}{35.5 + 1} = 0.24 \text{ mol}$ $\dfrac{0.24 \text{ mol}}{0.05 \text{ L}} = 4.8M$

e. $\dfrac{94\text{ g}}{2 + 32} = 2.8 \text{ mol}$ $\dfrac{2.8 \text{ mol}}{0.45 \text{ L}} = 6.2M$

2. **a.** $\dfrac{0.6\text{ g}}{12 + 142} = 0.004 \text{ mol}$ $\dfrac{0.004 \text{ mol}}{0.420 \text{ kg}} = 9 \times 10^{-3}m$

b. $\dfrac{0.45\text{ g}}{1 + 14 + 48} = 0.0071 \text{ mol}$

$\dfrac{0.0071 \text{ mol}}{0.905 \text{ kg}} = 7.8 \times 10^{-3}m$

c. $\dfrac{7.8\text{ g}}{24 + 71} = 0.082 \text{ mol}$ $\dfrac{0.082 \text{ mol}}{5.24 \text{ kg}} = 1.6 \times 10^{-2}m$

d. $\dfrac{52\text{ g}}{24 + 5 + 16 + 1} = 1.1 \text{ mol}$ $\dfrac{1.1 \text{ mol}}{0.160 \text{ kg}} = 7.1m$

e. $\dfrac{15.7\text{ g}}{72 + 6} = 0.201 \text{ mol}$ $\dfrac{0.201 \text{ mol}}{0.400 \text{ kg}} = 0.50m$

page 450

1. They contain more solute than a solution would normally hold at a particular temperature.
2. Dissolve as much NaCl as possible in boiling water, then cool the solution.
3. Atmospheric pressure is less than the pressure on the capped solution. Gas molecules have enough kinetic energy to escape the liquid.
4. pressure on the liquid, temperature of the liquid
5. Energy must be absorbed by the surroundings in order to overcome the energy that maintains the ammonium nitrate crystal structure.

page 453

1. One mole $MgCl_2$ dissociates to form 3 moles of ions as compared to one mole KCl which dissociates to form 2 moles of ions. Theoretically, $MgCl_2$ would have $1\frac{1}{2}$ times the effect on the freezing point of the solvent.
2. Volatile substances have high vapor pressures, nonvolatile substances have low vapor pressures.
3. Ethylene glycol lowers the freezing point of water and prevents water in the cooling system of a car from freezing at low temperatures.
4. fractional distillation

page 462

1. PbI_2
2. $AgNO_3(aq) + NaCl(aq) \rightleftharpoons AgCl(s) + NaNO_3(aq)$
3. **a.** $Ag^+(aq) + Cl^-(aq) \rightleftharpoons AgCl(s)$
   b. $Pb^{2+}(aq) + 2Br^-(aq) \rightleftharpoons PbBr_2(s)$
   c. $Ba^{2+}(aq) + SO_4^{2-}(aq) \rightleftharpoons BaSO_4(s)$
4. c and d are insoluble.
5. Net ionic equations show only those species reacting to form a precipitate, water, or gas.

6. a. $H^+(aq) + OH^-(aq) \rightleftharpoons H_2O(l)$
   b. $Bi^{3+}(aq) + 3OH^-(aq) \rightleftharpoons Bi(OH)_3(s)$
   c. $Pb^{2+}(aq) + SO_4^{2-}(aq) \rightleftharpoons PbSO_4(s)$
   d. no reaction
   e. $Fe^{2+}(aq) + S^{2-}(aq) \rightleftharpoons FeS(s)$
   f. $Sr^{2+}(aq) + CO_3^{2-}(aq) \rightleftharpoons SrCO_3(s)$
   g. no reaction
   h. $Ag^+(aq) + I^-(aq) \rightleftharpoons AgI(s)$
   i. $Ca^{2+}(aq) + SO_4^{2-}(aq) \rightleftharpoons CaSO_4(s)$
   j. $Ca^{2+}(aq) + SO_4^{2-}(aq) \rightleftharpoons CaSO_4(s)$
   k. $Ba^{2+}(aq) + SO_4^{2-}(aq) \rightleftharpoons BaSO_4(aq)$
   l. $Ca^{2+}(aq) + SO_4^{2-}(aq) \rightleftharpoons CaSO_4(s)$

7. $K_{sp} = [Ag^+][Cl^-]$

8. a. $K_{sp} = [Ag^+][Br^-]$
   b. $K_{sp} = [Ag^+]^2[SO_4^{2-}]$
   c. $K_{sp} = [Pb^{2+}][OH^-]^2$
   d. $K_{sp} = [Hg^+]^2[SO_4^{2-}]$
   e. $K_{sp} = [Al^{3+}]^2[SO_4^{2-}]^3$

9. $4 \times 10^{-8} > 4.8 \times 10^{-9}$, a ppt. forms

10. 9.2×10^{-6} moles per liter

11. AgCl—most soluble, AgI—least soluble

Review, page 463

1. Heterogenous materials are characterized by an observable segregation of component substances. Homogenous materials remain uniformly mixed.

2. In a 5% alcohol solution, water is the solvent. In a 70% alcohol solution, water is the solute.

3. No. On a microscopic level particles are constantly leaving the solution state, while solute initially restricted from the solution enters the solution.

4. In general, the amount and rate of dissolving will go up as the temperature increases.

5. The salt ions attract the oppositely charged ends of the water molecule. As the ions become isolated, the reduced effects of the dipole interionic forces allow them to diffuse into the solvent

6. Van der Waals and dipole forces

7. Polar substances will solvate polar substances. Nonpolar substances will solvate nonpolar substances.

8. Certain properties are dependent directly upon the number of particles of solute present within a given mass of solvent.

9. molality $= \dfrac{\text{moles of solute}}{\text{kilograms of solvent}}$

10. It will decrease the solubility.

11. It is the change in energy associated with solvation.

12. It is used as a method to separate the components of crude oil.

13. It will elevate the boiling point and depress the freezing point.

14. One of the products must be insoluble for a net ionic equation to be written.

15. It is an equation shorthand showing only reacting species.

16. It is an ion that remains unchanged during a reaction.

17. Additional information about the reaction is needed. Additional reactions not indicated by the net ionic equation may occur.

18. At equilibrium, the rate at which solute enters the solution is equal to the rate at which solute leaves the solution.

19. It allows the pressure to increase at the solution's surface. This will cause an increase in the solubility of the gas (CO_2) within the beverage.

Interpret and Apply, page 464

1. An unsaturated solution can support additional solute. A saturated solution is characterized by a state of dynamic equilibrium between dissolved and undissolved solute.

2. a. miscible   **e.** miscible
   b. immiscible   **f.** immiscible
   c. miscible   **g.** miscible
   d. immiscible

3. No. The nonpolar carbon tetrachloride does not demonstrate a significant attraction for ionic solids to cause solvation.

4. two phases—water/alcohol phase; oil/carbon tetrachloride phase

5. vapor pressure, boiling point and freezing point

6. Yes. Since salt would be restricted from sea ice, the solution left behind after the ice was removed would have a higher concentration of salt than untreated ocean water.

7. Since the lake does not contain a significant amount of salt, it will freeze sooner and thicker than an ocean-water bay.

8. The chemist would add a large mass of sodium thiosulfate to a given volume of water. The system would then be heated to increase its solubility. Once cooled, the system would support more solute than it normally would at that given temperature.

9. a) increase the solubility; b) no effect; c) no effect

10. Since the air that submariners breath would always be at constant atmospheric pressure, they would not undergo a change in blood gas solubility.

11. It is used to overcome the energy maintaining the solute's crystalline structure. In this reaction the tendency toward maximum disorder is greater than the tendency toward minimum energy.

12. It would increase the solution rate.

13. a. H^+—hydrogen ion; I^-—iodide ion
   b. Na^+—sodium ion; $CO_3{}^{2-}$—carbonate ion
   c. Ba^{2+}—barium ion; OH^-—hydroxide ion
   d. K^+—potassium ion; NO_3^-—nitrate ion
   e. $NH_4{}^+$—ammonium ion; Cl^-—chloride ion
   f. Ca^{2+}—calcium ion; $C_2H_3O_2^-$—acetate ion
   g. $NH_4{}^+$—ammonium ion; $PO_4{}^{3-}$—phosphate ion
   h. $NH_4{}^+$—ammonium ion; $SO_4{}^{2-}$—sulfate ion

14. A precipitate will form.

15. The concentration of B^- increases.

16. The product of the concentration of the Na^+ and Cl^- ions is now greater than their K_{sp} value, due to the additional Cl^- ion.

17. The additional Na^+ ion will cause the product of the concentration of Na^+ and Cl^- to exceed this K_{sp} value.

Problems, page 464

1. a. 98 g/mol × 0.5 mol/L = 49 g/L
   (49 g/L)(0.5 L) = 25 g
   b. 63 g/mol × 0.01 mol/L = 0.6 g/L
   (0.6 g/L)(0.5 L) = 0.3 g
   c. 36.5 g/mol × 6 mol/L = 220 g/L
   (220 g/L)(0.5 L) = 110 g
   d. 40 g/mol × 0.50 mol/L = 20 g/L
   (20 g/L)(0.5 L) = 10 g
   e. 163 g/mol × 0.1 mol/L = 16.3 g/L
   (16.3 g/L)(0.5 L) = 8 g
   f. 17 g/mol × 3 mol/L = 51 g/L
   (51 g/L)(0.5 L) = 26 g
   g. 56 g/mol × 5 mol/L = 280 g/L
   (280 g/L)(0.5 L) = 140 g

2. a. $\dfrac{0.05\ g}{44\ g/mol} = 0.001\ mol$ $\dfrac{0.001\ mol}{0.652\ kg} = 0.002m$
   b. $\dfrac{56\ g}{17\ g/mol} = 3.3\ mol$ $\dfrac{3.3\ mol}{0.050\ kg} = 66m$
   c. $\dfrac{3.21\ g}{84\ g/mol} = 0.0382\ mol$ $\dfrac{0.0382\ kg}{0.231\ kg} = 0.165m$
   d. $\dfrac{320\ g}{154\ g/mol} = 2.1\ mol$ $\dfrac{2.1\ mol}{3.5\ kg} = 0.60m$
   e. $\dfrac{157\ g}{18\ g/mol} = 8.72\ mol$ $\dfrac{8.72\ mol}{1.25\ kg} = 6.98m$
   f. $\dfrac{50.5\ g}{92\ g/mol} = 0.549\ mol$ $\dfrac{0.549}{0.742} = 0.740m$
   g. $\dfrac{3.0\ g}{254\ g/mol} = 0.012\ mol$ $\dfrac{0.012\ mol}{0.286\ kg} = 0.042m$

3. a. 0.238 mol/kg × 1 kg = 0.238 mol
   (0.238 mol)(163 g/mol) = 38.8 g
   b. 0.356 mol/kg × 0.5 kg = 0.178 mol
   (0.178 mol)(160 g/mol) = 28.5 g
   c. 0.550 mol/kg × 0.1 kg = 0.055 mol
   (0.055 mol)(78 g/mol) = 4.29 g

d. 2.25 mol/kg × 0.0300 kg = 0.0675 mol
   (0.0675 mol)(154 g/mol) = 10.4 g
e. 1.50 mol/kg × 0.75 kg = 1.13 mol
   (1.13 mol)(44 g/mol) = 49.7 g

4. 32 g/mol × 2 mol = 64 g
   $(6.02 \times 10^{23}$ particles/mol$)(4$ mol$)$
   $= 2.41 \times 10^{24}$ particles

5. 0.30 L × 0.40 mol/L = 0.12 mol
   54 g/mol × 0.12 mol = 6.5 g

6. $\dfrac{0.585\ g}{59\ g/mol} = 0.00992\ mol$ $\dfrac{0.00992\ mol}{0.025\ L} = 0.397M$

7. a. soluble **d.** insoluble **g.** insoluble
   b. soluble **e.** soluble
   c. soluble **f.** soluble

8. $K_{sp} = [Ag^+][Cl^-]$
   $1.8 \times 10^{-10} = [Ag^+][5.33 \times 10^{-1}]$
   $[Ag^+] = \dfrac{1.8 \times 10^{-10}}{5.3 \times 10^{-1}} = 3.4 \times 10^{-10}M$

9. $K_{sp} = [Al^{3+}][OH^-]^3$
   $K_{sp} = [Ca^{2+}]^3[PO_4{}^{3-}]^2$
   $K_{sp} = [Co^{2+}][SO_4{}^{2-}]$

Challenge, page 465

1. Since salt is restricted from the ice phase, the salinity of the remaining ocean water will increase. The relative increase in solute will further depress the freezing point temperature of the remaining water.

2. As the volatile components of a boiling solution escape into the gas phase, the relative concentration of the remaining nonvolatile solute increases. This increase in concentration causes a corresponding elevation in the solution's boiling point temperature.

3. Yes. The crystals would be solvated by water molecules, releasing three moles of ions per mole of salt. This added solute would cause a decrease in the freezing point temperature.

4. Distillation units (stills) can be used to obtain a high concentration alcohol distillate.

5. The crystals are formed by sugar molecules that have precipitated out of solution.

6. For each lead ion liberated in solution, two fluoride ions are liberated. Therefore if $[Pb^{2+}] = x$, then the $[F^-]$ must equal $2x$. Plug into the K_{sp} equation and evaluate the expression.
   $K_{sp} = [Pb^{2+}][F^-]^2$
   $3.7 \times 10^{-8} = x(2x)^2$
   $3.7 \times 10^{-8} = x(4x^2)$
   $3.7 \times 10^{-8} = 4x^3$
   $x^3 = 9.3 \times 10^{-9}$
   $x = \sqrt[3]{9.3 \times 10^{-9}}$
   $x = 2.1 \times 10^{-3}$
   $[Pb^{2+}] = 2.1 \times 10^{-3}$
   $[F^-] = 2x = 4.2 \times 10^{-3}$

7. $0.0540 \text{ mol/L} \times 24 \text{ g/mol} = 1.30 \text{ g/L}$

$$\frac{1 \text{ g}}{1.30 \text{ g/L}} = 0.770 \text{ L}$$

Synthesis, page 465

1. Molecular geometry often will determine the polarity of a particle. For example, although the C—Cl bond is polar, the CCl_4 molecule is nonpolar. It is a molecule's polarity or lack of it that will determine solvation.

2. The temperature of the solution must be lowered so that all compounds are in the liquid phase. The solution then may be distilled, separating its component substances by a difference in boiling temperatures.

CHAPTER 17 Thermodynamics

CHAPTER ORGANIZER					
USE WITH SECTION	**DEMONSTRATION**	**WORKSHEET**	**LAB EXPERIMENT**	**SOFTWARE**	**TEACHER'S RESOURCE BINDER**
17-1	17-A, 17-B		17-A		
17-2		17-A			Problems 1–3
17-3		17-B	17-B		Problems 4–6
17-4		17-B			
17-6	17-C, 17-D	17-C			Problems 17–8
17-8	17-E	17-D			Blackline Master 8
17-9		17-D			Problems 9–11
17-10					Blackline Master 9
17-13		17-E, 17-F			Problem 12
Chapter Test					

■ Objectives

When students have completed the chapter, they will be able to

1. Distinguish between the temperature of a substance and its heat content.
2. Explain how a substance at a lower temperature can have more heat than a substance at a higher temperature.
3. Measure the specific heat of a substance.
4. Calculate the heat content of a substance from its specific heat, temperature, and mass.
5. Identify the substance that would store the most energy when subjected to a given change in temperature using a table of specific heats.
6. Name the SI unit for energy and the unitary rate used to convert it to calories, or vice versa.
7. Name the energy associated with a change in state and calculate the energy required to change a given mass of substance from solid to liquid or liquid to gas.
8. Describe how the heat of fusion and the heat of vaporization affect the microscopic particles that make up matter.
9. List the required measurements and state how they would be used to calculate the heat of a reaction taking place in a calorimeter.
10. Describe what takes place at the microscopic level to account for the heat of a reaction.
11. List the specified conditions for a heat of reaction to be the standard heat of reaction.
12. Calculate the heat of reaction for a new equation resulting from a series of reactions with known heats of reaction.
13. Describe how heats of combustion can be used to estimate the energy available from foods.

112T

14. Indicate whether a change can occur spontaneously when the sign of Gibbs free energy for the change is known.
15. Describe the driving force for any spontaneous change and explain how an endothermic reaction can occur spontaneously.
16. Explain how a reaction can occur spontaneously even when the entropy of the system decreases.
17. Indicate whether the entropy of the system's surroundings for a reaction is increasing or decreasing, given the sign for ΔH of the reaction.
18. Explain how a chemical system can be used to do work, and determine which of two chemical reactions can do more work.

■ Section Notes

Introduction

A rigorous treatment of thermodynamics is beyond the scope of this book. Our intent is to make students aware of what might be learned from a study of thermodynamics rather than make them proficient at doing thermodynamic calculations. The basic concepts should be understood at qualitative levels, and students should realize that rigorous, quantitative descriptions of each concept do exist.

17-1 The distinction between heat and temperature is subtle and frequently misunderstood. Take time to discuss it. Have students prepare a simple water thermometer and calibrate it. Use it when you need to measure room temperature.

17-2 Ask a student (or a group of students) to prepare a demonstration to show the differences in specific heat of solids or liquids.

After specific heats are discussed, divide the class into groups of three to five and present the problem in Example 17-1 (or a similar one). Encourage students to come up with their own solutions. Discuss various solutions as a class, and focus attention on any novel problem-solving methods used. After the discussion, allow students to examine Example 17-1. Encourage students to devise similar problems (the more practical, the better) to present in class the next day. Allow students to work on the problems and argue about the best solutions.

17-3 If the problem-solving activity suggested for Section 17-2 goes well, encourage students to devise problems involving heat of fusion and/or heat of vaporization. The last two paragraphs of the section suggest some interesting problems. For example, how much heat would be released if a farmer sprays 1000 gallons of water into the air and it all freezes? What volume of air could be warmed 5 degrees by the heat released?

17-4 Use dynamic models or animated films to show changes at the molecular level when temperature changes.

17-5 If students have had a course in physics, you may wish to discuss the joule as a force times a distance and show how this definition of energy can be related to heat and/or electricity. If students have not had physics, mention that such relationships exist and that they will study them when they take a course in physics, but don't try to develop relationships.

If you have not demonstrated the calculation of specific heat, this would be a good place to do it. Do the demonstration qualitatively first. Then ask students how the same experiment could be done to measure the specific heat of a solid such as a rock or a piece of metal. Repeat the demonstration, collecting data suggested by the students as you go, and calculate the specific heat of a substance.

Have students review Example 17-2 after they have done calculations for the demonstration. At the conclusion of the demonstration, ask students to suggest reasons that the value calculated may differ from a value found in a handbook. Don't accept answers such as, "We may have made calculation errors," or "Our measurements may be off." Insist that students explain why their measurements might be off, in what direction they are likely to err, and what effect this would have on the calculated result. Some students will find this analysis difficult, but it is valuable in developing scientific reasoning.

17-7 Fill a balloon with hydrogen gas and fill a small test tube with water. Lower the test tube into a mixture of ice and salt so that the water in the tube freezes. Ignite the hydrogen balloon. Point out that in both cases bonds were formed and that energy was lost from the system as the new bonds were formed. Have students describe the bonds that formed in each case. Ask them how they know that bonds were formed. (Macroscopic evidence of the liquid turning to solid is probably sufficient evidence, but the burning hydrogen balloon provides no real clue. If burning hydrogen has been discussed previously, ask students to write the equation for the reaction and describe the change at the microscopic level. Molecular models will help.)

Ask students how they know that energy was released. Now the evidence in the case of the hydrogen balloon will be much easier to see than in the case of the freezing water. The fact that the temperature had to be lowered and heat always moves from an object at a higher temperature to an object at a lower temperature is perhaps the best evidence, but it certainly is indirect. Ask which released more energy, the formation of bonds between atoms or the formation of bonds between the water molecules. (Again, indirect arguments will be required.) Make sure that students understand that energy is always released when bonds form and always absorbed when bonds are broken. Help them to see that the energy required to break bonds between atoms within a molecule normally is greater than the energy required to break bonds between molecules in a solid. However, this is not always true. Sugar decomposes before it melts; so does baking soda.

17-8 Use the hydrogen balloon to introduce the idea of standard heat of formation. Point out that bonds between hydrogen atoms and bonds between oxygen atoms were broken and that bonds between hydrogen and oxygen atoms were formed to produce water. It is much easier to measure the energy change associated with the overall change than to measure it for each step. Introduce the idea of standard heat of formation by talking about the formation of water from hydrogen and oxygen in the balloon explosion. Contrast the conditions of your demonstration with the conditions that define the standard state.

17-9 If students are interested, calculate enthalpies of reaction for reactions that students have observed. Have students use the CRC Handbook or other reference to find standard heats of formation or heats of combustion for the compounds that appear in the equation. As a class, write equations for the formation or combustion of the compounds, and decide how the equations must be rearranged so they can be added to obtain the desired equation. Ask students what should be done to the ΔH value if an equation is reversed. If it is multiplied by two? Ask whether the value obtained would be observed under all experimental conditions or only under particular conditions. Review the conditions that pertain.

17-10 An appreciation of just how much energy a food calorie represents can be obtained by burning a sugar cube or a single peanut under a fruit juice can filled with water. (A sugar cube can be lit if ashes are rubbed on its surface before lighting.) Measure the temperature rise of the water and calculate the number of calories captured. Ask the students if this represents all of the energy released. (Much of it is lost to the surroundings but the point will be made.)

17-11 Set up the Cu/Zn cell described in both forms. Measurements of potential should probably be left until Chapter 21, but you may want to use the cell to light a bulb or run a small motor. Leave the cell for a few days and return to it when you discuss free energy. Show the "dead" cell, point out that the chemical reaction has reached equilibrium, and ask what the value of ΔG must be.

17-12 Although you will not want to discuss the Helmholtz function, you may wish to point out that the quantity discussed here pertains to chemical systems at constant pressure and that a comparable quantity can be calculated for systems at constant volume (bomb calorimeter). For most chemical reactions the values are similar, but they are not identical.

17-13 Entropy is a difficult concept, but it is an important one. A qualitative understanding of the concept is within the grasp of all students. An analogy of a teenager's room can be useful, and it does not distort the idea too much.

Most teenagers' parents complain about the "messy" room kept by their child. Ask students why this happens so much. After accepting such suggestions as, "My parents fuss at everything I do," or "It is impossible to keep it the way Mom wants it," shift attention to the fact that there are a limited number of ways the articles in their room can be arranged so the room will be called "neat" by parents, but there are millions of ways to arrange the same articles so the room is perceived as "messy". "Messy" is a much more likely state than "neat" (especially when things are arranged by chance) because there are so many more ways to achieve "messy" than "neat". Things move spontaneously toward those states that include the greatest number of possibilities.

After the analogy, discuss qualitatively the different ways energy can be distributed across a collection of atoms a) when the atoms are isolated by a gas, b) when they are condensed as a liquid, c) when they are condensed as a solid, d) when they are bonded together to form gaseous molecules. Point out that as the number of ways energy can be distributed increases, so does entropy.

Most books say that there are two factors that affect spontaneity, potential energy and entropy. Actually, there is only one condition, that the entropy of the universe increases. However, since a reduction in enthalpy of the system (exothermic reactions) will result in an increase in entropy of the surroundings, the two explanations lead to the same predictions.

Review and Practice _____
page 473

1. More energy is required to raise the temperature of 1 kg of iron one degree.

2. $?\,J = 1000 \text{ kg granite} \times \dfrac{800\,J}{\text{kg} \cdot \text{K}} \times 10\,K$

   $= 8.00 \times 10^6\,J$

   $?\,J = 1000 \text{ kg water} \times \dfrac{4180\,J}{\text{kg} \cdot \text{K}} \times 10\,K$

   $= 4.18 \times 10^7\,J$

   the one using water

3. No, the specific heats vary. The specific heat of water at 25°C is 4180 J/kg · K and the specific heat

of liquid ammonia is 4710 J/kg · K. The larger specific heat of ammonia will allow it to store more energy.

4. a. the iron nail
   b. The heat is more concentrated in the nail.
   c. temperature

5. It takes the form of kinetic energy as the particles move faster.

page 477

1. The unitary rates needed are 4.184 J/cal and 0.2390 cal/J.
   a. 4.184 J/cal
   b. 0.2390 cal/J (1 cal/4.184 J is not a unitary rate.)

2. $? J = 100 \text{ Calories} \times \dfrac{1000 \text{ calories}}{1 \text{ Calorie}} \times \dfrac{4.184 \text{ J}}{1 \text{ calorie}}$

$= 4.184 \times 10^5 \text{ J}$

3. $? J = 250 \text{ Calories} \times \dfrac{1000 \text{ calories}}{1 \text{ Calorie}} \times \dfrac{4.184 \text{ J}}{1 \text{ calorie}}$

$= 1.05 \times 10^6 \text{ kJ}$

4. Water is a very common liquid, and we know how to get it in a very pure form. Also, it is inexpensive.

5. 150 Calories × 0.3 = 45 Calories from carbohydrates.

6. a. true
   b. false
   c. false
   d. true (Students may not be able to answer this question, given only the material presented in this chapter.)

page 490

1. endothermic

2. Boiling away the water would take more energy. The heat of fusion for water is 334 000 J/kg and the heat of vaporization is 2 260 000 J/kg.

3. enthalpy

4.

$$C + O_2 \longrightarrow CO_2 \qquad \Delta H = -393.5 \text{ kJ/mol}$$
$$2H_2 + O_2 \longrightarrow 2H_2O \qquad \Delta H = -571.7 \text{ kJ/mol}$$
$$CO_2 + 2H_2O \longrightarrow CH_4 + 2O_2 \qquad \Delta H = +890.3 \text{ kJ/mol}$$
$$\overline{C + 2H_2 \longrightarrow CH_4 \qquad \Delta H = -74.9 \text{ kJ/mol}}$$

5. because parts of the foods that will burn in a calorimeter do not get digested and metabolized in the body

6. Some energy is always lost to the surroundings as heat. One hundred percent efficiency cannot be attained.

7. The reaction will not occur spontaneously.

8. Zero; neither the forward nor the reverse reaction is doing work.

9. a. It is spontaneous, since the ΔG is negative.
   b. It increases, since spontaneous change always results in greater entropy.

10. Negative; it is a spontaneous reaction.

11. a. $6 \text{ mol CuO} \times \dfrac{130.5 \text{ kJ}}{\text{mol CuO}} = 783.0 \text{ kJ}$

   b. $2 \text{ g CuO} \times \dfrac{1 \text{ mol CuO}}{79.5 \text{ g CuO}} \times \dfrac{130.5 \text{ kJ}}{1 \text{ mol CuO}} = 3.28 \text{ kJ}$

   c. $2 \text{ g Cu} \times \dfrac{1 \text{ mol Cu}}{63.5 \text{ g Cu}} \times \dfrac{130.5 \text{ kJ}}{1 \text{ mol Cu}} = 4.11 \text{ kJ}$

12. a. unburned
   b. an ear of corn
   c. unused

13. a. endothermic
   b. positive

Review, page 491

1. It breaks the bonds holding molecules together as a solid or liquid.

2. The mass of the water is far less in the cup of water. Total energy is a function of mass, temperature, and specific heat.

3. A Calorie (capital C) refers to 1000 calories (small c).

4. a. Endothermic; it requires energy.
   b. Positive; it is endothermic.

5. The energy your body gets is the difference between the energy content of the bar before digestion and metabolism and the energy content of the products of digestion and metabolism.

6. The reaction can occur spontaneously.

7. Zero; the reaction is doing no useful work.

8. The system must be capable of spontaneous reaction. The reaction must take place so that the energy change in the system is converted to useful work.

9. No. In fact, they must have an increase in entropy large enough to exceed the decrease in entropy of the surroundings. In all spontaneous changes, the total change in entropy (i.e., change in the system and surroundings) must be positive.

10. The energy comes from changes in the system and it goes to raise the temperature or produce some other change in the surroundings.

Interpret and Apply, page 491

1. When steam (gaseous water) strikes your hand, it will cool enough to change to liquid. The heat of vaporization is released and absorbed by your hand, burning it. The steam will then be liquid water at 100°C and lose as much additional energy as would water at 100°C.

2. Large amounts of energy are required to melt the solid without changing its temperature (heat of fusion). When the liquid freezes, that energy is released to warm the house.

3. The standard state of hydrogen and oxygen is a gas.

4. CO_2. Energy can be distributed in more ways, for example, by vibrations and rotations of the molecule as well as the translational motion of the particle.

5. The reaction is endothermic. ΔH is positive.

6. a. negative
   b. negative
   c. negative (b and c may be difficult for students to answer, given the presentation in the chapter.)

7. a. No; 298 K is not the freezing point of water.
   b. Negative; the reaction is exothermic.
   c. The ice must absorb energy from the surroundings.

8. a. Negative; the reaction is exothermic.
   b. Positive; the reverse reaction would be endothermic.
   c. Less; the reaction is exothermic.

9. a. Exothermic; ΔH is negative.
   b. +297 kJ The sign of ΔH is reversed.
   c. Less; the reaction is exothermic.

10. a. $H_2O(l) \longrightarrow H_2O(s) + 6.01$ kJ
   b. $H_2O(g) \longrightarrow H_2O(l) + 40.7$ kJ

11. a. $Mg(s) + \frac{1}{2} O_2(g) \longrightarrow MgO(s) + 607.1$ kJ
   b. $\frac{1}{2} N_2(g) + \frac{3}{2} H_2(g) \longrightarrow NH_3(g) + 46.0$ kJ
   c. $C_8H_{18}(l) + 12\frac{1}{2} O_2(g) \longrightarrow$
   $8\ CO_2(g) + 9\ H_2O(l) + 5470.6$ kJ

   (If students use whole number coefficients to balance the equations, the ΔH values also must be multiplied.)

Problems, page 492

1. For each degree change in temperature the water will absorb

$$4.78 \times 10^3\ \text{kg} \times \frac{4.180\ \text{kJ}}{\text{kg}} = 2.00 \times 10^4\ \text{kJ}$$

To find the mass of paraffin needed to absorb this energy for each degree change

$$2.00 \times 10^4\ \text{kJ} \times \frac{1\ \text{kg paraffin}}{2.900\ \text{kJ}}$$
$$= 6.90 \times 10^3\ \text{kg paraffin}$$

2. $1559.8\ \text{kJ} \times \dfrac{1\ \text{kg} \cdot \text{K}}{4.180\ \text{kJ}} \times \dfrac{1}{40\ \text{K}} = 9.3\ \text{kg } H_2O$

3. $400\ \text{kg oil} \times \dfrac{47.3\ \text{kJ}}{0.001\ \text{kg oil}} \times \dfrac{0.001\ \text{kg wood}}{18.8\ \text{kJ}}$
$= 1.01 \times 10^3\ \text{kg wood}$

4. a. $CO_2 \longrightarrow$ C(diamond) $+ O_2$ $\quad \Delta H = +395.4$ kJ
   $\text{C(graphite)} + O_2 \longrightarrow CO_2$ $\quad \Delta H = -393.5$ kJ
   $\overline{\text{C(graphite)} \longrightarrow \text{C(diamond)} \quad \Delta H = +1.9\ \text{kJ}}$
   b. absorbed

5. a. $NO \longrightarrow \frac{1}{2}N_2 + \frac{1}{2}O_2$ $\quad \Delta H = -90.4$ kJ

$\dfrac{\frac{1}{2}N_2 + O_2 \longrightarrow NO_2 \quad \Delta H = +33.8\ \text{kJ}}{NO + \frac{1}{2}O_2 \longrightarrow NO_2 \quad \Delta H = -56.6\ \text{kJ}}$

b. exothermic

6. a. $\dfrac{427\ \text{kJ}}{\text{mol NaOH}} \times 1.37\ \text{mol NaOH} = 585$ kJ

$\Delta H_f = -585$ kJ

b. $\dfrac{427\ \text{kJ}}{\text{mol NaOH}} \times \dfrac{1\ \text{mol NaOH}}{40\ \text{g NaOH}} \times 47.0\ \text{g NaOH}$
$= 502$ kJ
$\Delta H_f = -502$ kJ

7. a. $\dfrac{972\ \text{kJ}}{\text{mol } H_3PO_3} \times 16.7\ \text{mol } H_3PO_3 = 1.62 \times 10^4$ kJ

b. $\dfrac{972\ \text{kJ}}{\text{mol } H_3PO_3} \times \dfrac{1\ \text{mol } H_3PO_3}{82\ \text{g } H_3PO_3} \times 691\ \text{g } H_3PO_3$
$= 8.19 \times 10^3$ kJ
$\Delta H_f = -8.19 \times 10^3$ kJ

8. $1.0 \times 10^8\ \text{kJ} \times \dfrac{1\ \text{mol } N_2H_4}{627.6\ \text{kJ}} \times \dfrac{3.2 \times 10^{-2}\ \text{kg } N_2H_4}{1\ \text{mol } N_2H_4}$
$= 5.1 \times 10^3$ kg

9. $1.23\ \text{kg } H_2O \times \dfrac{4.180\ \text{kJ}}{\text{kg} \cdot \text{K}} \times 5.8\ \text{K} = 29.8$ kJ

10. a. $2Al + \frac{3}{2}O_2 \longrightarrow Al_2O_3$ $\quad \Delta H = -1670$ kJ
   $\text{Fe}_2O_3 \longrightarrow 2\text{Fe} + \frac{3}{2}O_2$ $\quad \Delta H = +822$ kJ
   $\overline{\text{Fe}_2O_3 + 2Al \longrightarrow Al_2O_3 + 2\text{Fe} \quad \Delta H = -848\ \text{kJ}}$

   b. $4\ \text{mol } \text{Fe}_2O_3 \times \dfrac{848\ \text{kJ}}{\text{mol } \text{Fe}_2O_3} = 3.39 \times 10^3$ kJ
   $\Delta H_f = -3.39 \times 10^3$ kJ

   c. $\dfrac{848\ \text{kJ}}{2\ \text{mol Fe}} \times \dfrac{1\ \text{mol Fe}}{5.58 \times 10^{-2}\ \text{kg Fe}} \times 1\ \text{kg Fe}$
   $= 7.60 \times 10^3$ kJ
   $\Delta H_f = -7.60 \times 10^3$ kJ

Challenge, page 493

1. With a mole mass of 161 g, the energy released per gram is

$\dfrac{1531\ \text{kcal}}{\text{mol fat}} \times \dfrac{1\ \text{mol fat}}{161\ \text{g fat}} = 9.51\ \text{kcal/g}$ (39.8 kJ/g)

9.51 kcal/g is 2.38 times the energy supplied by one gram of carbohydrate (4 kcal/g)?

2. $\dfrac{8.4 \times 10^6\ \text{kJ}}{\text{m}^2} \times \dfrac{1\ \text{mol C}}{393.7\ \text{kJ}} = \dfrac{2.13 \times 10^4\ \text{mol C}}{\text{m}^2}$

Brittanica lists the area of Arizona as 295 023 km² or 2.95×10^{11} m².

$2.95 \times 10^{11}\ \text{m}^2 \times \dfrac{2.13 \times 10^4\ \text{mol C}}{\text{m}^2} \times \dfrac{1.2 \times 10^{-2}\ \text{kg}}{1\ \text{mol C}}$
$= 7.5 \times 10^{13}$ kg C

Synthesis, page 493

1. a. Exothermic; heat is lost from the reactants.
   b. Negative; the reaction is exothermic.

c. Any exothermic reaction showing energy as a product would be appropriate.

2. a. $0.12 \text{ mol } H_2SO_4 \times \dfrac{908 \text{ kJ}}{1 \text{ mol } H_2SO_4} = 1.1 \times 10^2 \text{ kJ}$

$\Delta H_f = -1.1 \times 10^2 \text{ kJ}$

b. $\dfrac{908 \text{ kJ}}{\text{mol } H_2SO_4} \times \dfrac{1 \text{ mol } H_2SO_4}{98 \text{ g } H_2SO_4} \times 373 \text{ g } H_2SO_4$

$= 3.46 \times 10^3 \text{ kJ}$

$\Delta H_f = -3.46 \times 10^3 \text{ kJ}$

c. $0.5 \text{ L solution} \times \dfrac{3 \text{ mol } H_2SO_4}{L} \times \dfrac{98 \text{ g } H_2SO_4}{1 \text{ mol } H_2SO_4}$

$= 147 \text{ g } H_2SO_4$

$\dfrac{147 \text{ g}}{0.98} = 150 \text{ g concentrated acid}$

$\dfrac{908 \text{ kJ}}{\text{mol } H_2SO_4} \times \dfrac{1 \text{ mol } H_2SO_4}{98 \text{ g } H_2SO_4} \times 150 \text{ g } H_2SO_4$

$= 1.39 \times 10^3 \text{ kJ}$

$\Delta H_f = -1.39 \times 10^3 \text{ kJ}$

CHAPTER 18 Reaction Rates

CHAPTER ORGANIZER					
USE WITH SECTION	**DEMONSTRATION**	**WORKSHEET**	**LAB EXPERIMENT**	**SOFTWARE**	**TEACHER'S RESOURCE BINDER**
18-1			18-A		
18-5		18-A			
18-6		18-A,18-B			
18-7		18-C	18-B		
18-8		18-D			
Chapter Test					

■ Objectives

When the students have completed this chapter, they will be able to

1. Explain activation energy and the effects of concentration, temperature, and surface area on the reaction rate.
2. Draw energy diagrams that represent the activation energy and show the effect of a catalyst.
3. Explain the significance of the rate-determining step on the overall rate of a multistep reaction.
4. Describe the role of the rate constant in the theoretical determination of a reaction rate.

Review and Practice _____
Page 503

1. $2Na(s) + 2H_2O \longrightarrow$
   $2Na^+(aq) + 2OH^-(aq) + H_2(g) + \text{energy}$
2. potassium
3. $Mg(s) + 2HCl(aq) \longrightarrow$
   $MgCl_2(aq) + H_2(g) + \text{energy}$
4. Magnesium reacts faster with $3.00M$ HCl.

5. The powdered zinc has a larger surface area and reacts faster.
6. Increase the energy of collision by increasing the temperature. Improve the collision geometry with a contact catalyst.
7. Add energy by increasing the temperature.
8. activated complex
9. A catalyst provides a reaction pathway with a lower activation energy.
10. As the concentration increases the rate of the reaction increases.

page 505

1. **a.** no change
   b. The rate is increased.
   c. no change since washing is the rate-determining step

Review, page 507

1. Sodium reacts faster because of the nature of the reactants. Sodium has a lower activation energy.

2. The number of collisions of zinc with HCl would be increased, so the rate would be increased.

3. a. Rate is increased.
   b. Rate is increased.
   c. Rate is increased more than in b.
   d. Rate is decreased.

4. The temperature is higher and the chemical reactions that take place when milk spoils have a greater rate at a higher temperature.

5. An inhibitor decreases the rate of the reactions that cause the food to spoil.

6. Three hydrogen molecules would have to collide simultaneously with one nitrogen molecule. This is not very probable.

7. a. Assembling probably is the rate-determining step.
   b. The rate would increase. (Folding and inserting into envelopes may be the new rate-determining step.)
   c. Second step no change; third step no change; fourth step no change unless more people were added to step 1
   d. Step 4 would then be the rate-determining step.

Interpret and Apply, page 508

1. a. liters per hour
   b. metric tons per day
   c. milliliters per minute

2. Zinc reacts with $3M$ HCl faster because the concentration of HCl is higher.

3. Wood shavings react faster because there is more surface area.

4.

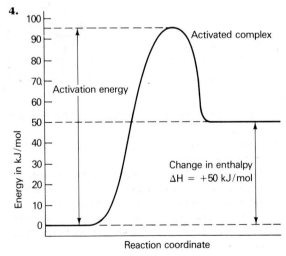

5. The activation energy is zero at room temperature.

6. The reaction of Cu with O_2 has a higher activation energy.

7. the messenger who goes by boat and on foot

8. a.
$$H^+ + H_2O_2 \longrightarrow H_3O_2^+$$
$$\underline{H_3O_2^+ + I^- \longrightarrow H_2O + HOI}$$
$$H^+ + I^- + H_2O_2 \longrightarrow H_2O + HOI$$
   b. The rate would be increased.
   c. No effect since reaction c is the rate-determining step.

9. The rate is slower during the second six-hour period because the concentration of H_2O_2 is lower.

Problems, page 508

1. a. 12 mL/20 s = 0.60 mL/1 s = 36 mL/1 min
   b. 24 mL if the rate is doubled for a 10° increase in temperature

2. Surface area = (1000 cm)^2 for each face = 10^6 cm^2. There are six faces so the area of the cube is 6×10^6 cm^2

Area of one 10-cm cube = $6 \times (10 \text{ cm})^2 = 6 \times 10^2$ cm^2. There are $(100)^3$ cubes or 10^6 cubes, so the total area will be $10^6 \times 6 \times 10^2$ cm$^2 = 6 \times 10^8$ cm^2

For one-cm cubes: The area of each cube is $6 \times (1 \text{ cm})^2 = 6$ cm^2. There are $(1000)^3$ cubes or 10^9 cubes so the total area is $10^9 \times 6$ cm^2 or 6×10^9 cm^2. The surface area is 10 times larger when the cubes are 10 times smaller.

3. 15 seconds

4. a. 15 mL/120 s = 7.5 mL/min
   b. 30 seconds; a 10°C increase in temperature may double the rate of the reaction, so 60 seconds are required to produce 15 mL. A second 10° increase to 40°C would again double the rate of the reaction, so it would take 30 seconds to produce 15 mL of oxygen gas.

5. 15 minutes

6. 2 molecules/min × 60 min/h × 24 h/day
       = 2880 molecules/day
   3.6×10^{-7} molecules/min × 60 min/h × 24 h/day
       = 5.184×10^{10} molecules/day

7. a. The reaction rate is doubled.
   b. The reaction rate is three times faster.
   c. The reaction rate is six times faster.

8. a. exothermic
   b. NH_3 = 17 g/mol
$$25 \text{ g NH}_3 \times \frac{1 \text{ mol}}{17 \text{ g NH}_3} = 1.47 \text{ mol}$$
$$\frac{1.47}{0.98} = 1.50, \text{ need 1.50 mol to get 25 g at}$$
       98% yield
$$1.5 \text{ mol NH}_3 \times \frac{3 \text{ mol H}_2}{2 \text{ mol NH}_3} = 2.25 \text{ mol H}_2$$
$$2.25 \text{ mol H}_2 \times 2.01 \text{ g/mol} = 4.52 \text{ g H}_2$$
   c. No, the overall equation implies that three molecules of H_2 must collide simultaneously with one molecule of N_2.

Challenge, page 509

1. $H_2(g) + I_2(g) \longrightarrow 2HI(g) + energy$
   a. Increase the number of collisions by increasing the concentration (decrease the volume or increase the pressure).
   b. Raise the temperature but increase the volume at the same time by an amount that will keep the number of collisions constant.

Synthesis, page 509

1. The pressure is lower at Pike's Peak so the water boils at a lower temperature. This means that the rate of denaturation of the protein (the reaction that occurs when an egg is cooked) is lower.

2. $2H_2O_2 \longrightarrow 2H_2O + O_2 \qquad 55 \text{ mL} \approx 55 \text{ g } H_2O_2$
   $55 \text{ g} \times 0.03 = 1.65 \text{ g } H_2O_2$ in 3% solution

   $\dfrac{1.65 \text{ g } H_2O_2}{34 \text{ g/mol}} = 0.048 \text{ mol } H_2O_2 \qquad 0.024 \text{ mol } O_2$

   $PV = nRT$

   $(95.5 \text{ kPa})(x) = (0.024)\left(8.31\dfrac{L \cdot kPa}{mol \cdot K}\right)(293 \text{ K})$

   $x = 0.61 \text{ L}$

3. $Mg + 2HCl \longrightarrow MgCl_2 + H_2$

   $\dfrac{0.22 \text{ g Mg}}{24 \text{ g/mol}} = 0.0092 \text{ mol Mg} \qquad 0.0092 \text{ mol } H_2$

   $(92 \text{ kPa})(x) = (0.0092)\left(8.31\dfrac{L \cdot kPa}{mol \cdot K}\right)(303 \text{ K})$

   $x = 0.25 \text{ L}$

   rate per second $= \dfrac{0.25 \text{ L}}{25 \text{ s}} = 0.01 \text{ L/s}$

CHAPTER 19 Chemical Equilibrium

CHAPTER ORGANIZER

USE WITH SECTION	DEMONSTRATION	WORKSHEET	LAB EXPERIMENT	SOFTWARE	TEACHER'S RESOURCE BINDER
19-6		19-A			
19-8		19-B, C, D			Problems 1–10
19-11		19-E	19-A		
Chapter Test					

■ Objectives

When the students have completed this chapter, they will be able to

1. Write both the forward and the reverse reactions and describe the macroscopic characteristics of each.
2. State the necessary conditions for equilibrium and the ways that equilibrium can be recognized.
3. Describe the microscopic events that occur when a chemical system is in equilibrium.
4. Propose microscopic events that account for observed macroscopic changes that take place during a shift in equilibrium.
5. State Le Chatelier's principle and be able to apply it to systems in equilibrium with changes in concentration, pressure, temperature, or the addition of a catalyst.
6. Write the equilibrium expression for a given chemical reaction.
7. Determine if the equilibrium constant will increase or decrease when the temperature is changed, given the equation for the reaction.
8. Calculate the equilibrium constant for a reaction, given the equilibrium concentrations of the reactants and products.
9. Calculate the concentration specified, given the equilibrium constant and appropriate information about the equilibrium concentrations.
10. Determine if the reactants or products are favored in a chemical reaction, given the equilibrium constant.

19-2 A closed system is an important condition of equilibrium. Evaluate several experiments that students have done to provide examples of closed systems.

19-3 Sealed tubes that contain NO_2 in equilibrium with N_2O_4 can be purchased commercially. When one tube is placed in an ice bath, the color becomes light yellow, showing a shift toward the N_2O_4. When another tube is placed in hot water, it becomes dark red-brown. When both are removed and cooled to room temperature, both become the same color.

19-4, 19-5, 19-6, 19-7 The effects of changing concentrations or pressure, and of a catalyst on a system in equilibrium, are not easy to demonstrate. Determining changes in equilibrium requires a system with readily observable properties. Use the Review and Practice questions to assess the students' understanding of Le Chatelier's principle.

The Haber process is important historically and the study of equilibrium may occur about the time that students are studying World War II in history. If a social studies teacher is interested, a unit on science and history could be a team-teaching topic.

Review and Practice

page 518

1. pressure, color, concentration of each reactant or product
2. No. All macroscopic properties of equilibrium, such as color, are constant.
3. As the color gets darker, the concentration of NO_2 is increasing and the concentration of N_2O_4 is decreasing.
4. The rates of opposing reactions are equal.

page 522

1. The concentration of I_2 is decreased as the H_2 molecules added react with I_2 to form HI.
2. The concentration of NO_2 decreases.
3. **a.** right
   b. no shift
   c. left

page 529

1. denominator

2. **a.** $K_{eq} = \dfrac{[H_2S]}{[H_2][S]}$ **b.** $K_{eq} = \dfrac{[PCl_3][Cl_2]}{[PCl_5]}$

c. $K_{eq} = \dfrac{[NO]^4[H_2O]^6}{[NH_3]^4[O_2]^5}$

d. $K_{eq} = \dfrac{[CO_2][H_2O]^2}{[CH_4][O_2]^2}$

3. $K_{eq} = \dfrac{[PCl_3][Cl_2]}{[PCl_5]} = \dfrac{[0.40][0.40]}{[0.32]}$

$K_{eq} = 0.50$ moles per liter

4. $K_{eq} = \dfrac{[PCl_3][Cl_2]}{[PCl_5]}$

$[PCl_5] = \dfrac{[PCl_3][Cl_2]}{K_{eq}} = \dfrac{[0.85][0.85]}{1.78}$

$[PCl_5] = 0.41$ moles per liter

5. $2NO_2 \rightleftharpoons N_2O_4$

Initial moles $NO_2 = 2$

Equilibrium moles $N_2O_4 = 1.7 \times 10^{-2}$

Moles NO_2 lost to form product
$2(1.7 \times 10^{-2}) = 3.4 \times 10^{-2}$

Moles at equilibrium
$(2 - 3.4 \times 10^{-2}) = 1.97$ moles in 1 liter flask

page 534

1. products
2. increased

3. **a.** $K_{eq} = \dfrac{[H_2S]}{[H_2]}$ **c.** $K_{eq} = \dfrac{[NO_2]^4[H_2O]^6}{[NH_3]^4[O_2]^7}$

   b. $K_{eq} = [CO_2]$

Review, page 535

1. $2HgO(s) \longrightarrow 2Hg(g) + O_2(g)$
2. For equilibrium to occur, the system must be closed.
3. pressure, concentration of reactants or products
4. Each time two atoms of mercury react with one molecule of oxygen to form two molecules of mercury(II) oxide, two molecules of mercury(II) oxide decompose to form two atoms of mercury and one molecule of oxygen.
5. a closed system, constant temperature, equal and opposing reactions
6. If a system in equilibrium is subjected to a change, processes occur that tend to counteract the imposed change.
7. decrease
8. A catalyst does not change the equilibrium concentrations.
9. right

10. **a.** $K_{eq} = \dfrac{[U]^u[V]^v}{[W]^w[X]^x}$ **c.** $K_{eq} = \dfrac{[CO][Cl_2]}{[COCl_2]}$

   b. $K_{eq} = \dfrac{[H_2][CO_2]}{[H_2O][CO]}$ **d.** $K_{eq} = \dfrac{[HCl]^2}{[H_2][Cl_2]}$

e. $K_{eq} = \dfrac{[CO_2][NO]}{[CO][NO_2]}$ **g.** $K_{eq} = \dfrac{[H_2][C_2H_4]}{[C_2H_6]}$

f. $K_{eq} = \dfrac{[Zn^{2+}]}{[Ag^+]^2}$

Interpret and Apply, page 536

1. a. [HI] increased
   b. [HI] unchanged
   c. [HI] increased
   d. [HI] unchanged
   e. $K_{eq} = \dfrac{[H_2][I_2]}{[HI]^2}$
   f. reactants

2. a. [CH_3OH] increased
   b. [CH_3OH] increased
   c. [CH_3OH] increased
   d. [CH_3OH] unchanged

3. a. right   **d.** left
   b. left   **e.** no change
   c. right

4. a. [NO_2] increased
   b. [NO_2] decreased
   c. [NO_2] decreased
   d. [NO_2] unchanged

5. a. [NO_2] decreased
   b. [NO_2] increased
   c. [NO_2] unchanged

6. Add SO_2, add O_2, lower the temperature, increase the pressure.

7. a. $2NO(g) + O_2(g) \rightleftharpoons 2NO_2(g) + 56.4$ kJ
   b. 1) [NO] and [O_2] increase, [NO_2] decreases.
       2) K_{eq} becomes smaller.
       3) increases
   c. 1) [NO_2] increases, [O_2] decreases.
       2) no change
       3) The rate of formation of NO_2 increases.

8. a. spectator ions:
       K^+, NO_3^-, SO_4^{2-}, Na^+
   b. ions causing color change:
       Fe^{3+}, darkens, causes shift to right; OH^-, lightens, causes shift to left.

9. a. lighter   **e.** no change
   b. darker   **f.** darker
   c. no change   **g.** no change
   d. lighter   **h.** darker

Problems, page 536

1. a. $K_{eq} = \dfrac{1}{1.20} = 0.833$

   b. $K_{eq} = \dfrac{1}{2.24} = 0.446$

c. $K_{eq} = \dfrac{1}{8.2 \times 10^{-2}} = 12.2$

2. $K_{eq} = \dfrac{[HBr]^2}{[H_2][Br_2]} = 1.02$

    [HBr] $= 0.50$ mol $\cdot$ L^{-1}. Let $x = [H_2] = [Br_2]$

    $\dfrac{[0.50]^2}{x^2} = 1.02$

    $x^2 = \dfrac{[0.50]^2}{1.02} = 0.24$ $x = 0.50$

    [H_2] $= 0.50M$

3. $K_{eq} = \dfrac{[NH_3]^2}{[N_2][H_2]^3}$

    [NH_3] $= 0.102$ [N_2] $= 1.03$ [H_2] $= 1.62$

    $K_{eq} = \dfrac{[0.102]^2}{[1.03][1.62]^3} = 2.38 \times 10^{-3}$

4. $K_{eq} = \dfrac{[NO_2]^2}{[N_2O_4]}$

    [NO_2] $= 1.75$ [N_2O_4] $= 3.6$

    $K_{eq} = \dfrac{[1.75]^2}{[3.6]} = 0.85M$

5. $K_{eq} = \dfrac{[CO_2][H_2]}{[CO][H_2O]}$

    $K_{eq} = \dfrac{[0.648][0.148]}{[0.352][0.352]} = 0.774$

    $K_{eq} = \dfrac{[0.234][0.234]}{[0.266][0.266]} = 0.774$

    $K_{eq} = \dfrac{[0.314][0.314]}{[0.186][0.686]} = 0.773$

6. $K_{eq} = \dfrac{[H_2][I_2]}{[HI]^2}$

    [H_2] $= 0.0075$ [I_2] $= 0.000\,043$ [HI] $= 0.0040$

    $K_{eq} = \dfrac{[0.0075][0.000\,043]}{[0.0040]^2} = 0.020$

7. $K_{eq} = \dfrac{[U][V]}{[W][X]}$

    At equilibrium, [W] $=$ [X] $= 0.80 - 0.60$
                       [W] $= 0.20$ [U] $=$ [V] $= 0.60$

    $K_{eq} = \dfrac{[0.60][0.60]}{[0.20][0.20]} = 9.0$

8. $K_{eq} = \dfrac{[NO]^2}{[N_2][O_2]}$

    [N_2] $=$ [O_2] $= 5.2$.   Let [NO] $= x$.

    $\dfrac{x^2}{[5.2][5.2]} = 6.2 \times 10^{-4}$

    $x^2 = 1.7 \times 10^{-2}$ $x = 0.13$ [NO] $= 0.13M$

9. $K_{eq} = \dfrac{[N_2][O_2]}{[NO]^2}$

    [NO] $= 0.13$ [N_2] $=$ [O_2] $= x$

    $\dfrac{x^2}{[0.13]^2} = 1.6 \times 10^3$

    $x^2 = 27$ $x = 9.0$ [N_2] $= 9.0M$

Challenge, page 537

1. $K_{eq} = \dfrac{[CO_2][H_2]}{[CO][H_2O]}$

$[CO] = 0.010M, \quad [H_2O] = 0.020M,$
$[CO_2] = 0.010M, \quad [H_2] = 0.010M$

$K_{eq} = \dfrac{[0.010][0.010]}{[0.010][0.020]} = 0.50$

The equilibrium constant is slightly less than one so the reactants are slightly favored.

2. $K_{eq} = \dfrac{[H_2]^2[S_2]}{[H_2S]^2}$

$[H_2S] = 7.06 \times 10^{-3}M, \quad [H_2] = 9.2 \times 10^{-3}$
$[H_2] = 0.002\,22$

$[S_2] = \dfrac{0.00222}{2} = 0.00111$

$K_{eq} = \dfrac{[0.00222]^2[0.00111]}{[7.06 \times 10^{-3}]^2} = 1.10 \times 10^{-4}\,mL^{-1}$

3. $PCl_5(g) \rightleftharpoons PCl_3(g) + Cl_2(g).$ At equilibrium
$[PCl_3] = [Cl_2] = x \quad [PCl_5] = 1.00 - x$

$K_{eq} = \dfrac{[PCl_3][Cl_2]}{[PCl_5]}$

$\dfrac{x^2}{1.00 - x} = 0.0211$

$x^2 = 0.0211 - 0.0211\,x$

$x^2 + 0.0211\,x - 0.0211 = 0$

Applying the quadratic equation

$x = \dfrac{-b \pm \sqrt{b^2 - 4ac}}{2a} \quad a = 1. \quad b = 0.0211.$
$\qquad\qquad c = -0.0211$

$x = \dfrac{-0.0211 \pm \sqrt{(0.0211)^2 - 4(1)(-0.02110)}}{2(1)}$

$x = \dfrac{-0.0211 \pm \sqrt{0.000445 + 0.0844}}{2}$

$\quad = \dfrac{-0.0211 \pm 0.291}{2}$

$x = 0.135 \quad [Cl_2] = 0.135M$

4. This is an endothermic reaction. If energy is added, the concentration of products will increase. This means that the equilibrium constant will increase when the temperature is increased, and decrease when the temperature is decreased for an endothermic reaction. The new equilibrium constant is lower so the new temperature is lower.

Synthesis, page 537

1. $K_{eq} = \dfrac{[CO_2][H_2]}{[CO][H_2O]}$

Initially

$[H_2O] = \dfrac{4.50\text{ g }H_2O}{L} \times \dfrac{1\text{ mol }H_2O}{18\text{ g }H_2O} = 0.25M$

$[CO] = \dfrac{7.00\text{ g CO}}{L} \times \dfrac{mol\text{ CO}}{28.0\text{ g CO}} = 0.25M$

Let $x = [CO_2] = [H_2]$
At equilibrium $[CO] = [H_2O] = 0.25 - x$

$\dfrac{x^2}{(0.25 - x)^2} = 3.59$

$x^2 = 3.59(x^2 - 0.50\,x + 0.625)$
$\quad = 3.59\,x^2 - 1.795\,x + 0.224$
$3.59\,x^2 - x^2 - 1.795\,x + 0.224 = 0$
$2.59\,x^2 - 1.795\,x + 0.224 = 0$
$a = 2.59, \ b = -1.795, \ c = 0.224$

$x = \dfrac{-b \pm \sqrt{b^2 - 4ac}}{2a}$

$\quad = \dfrac{1.795 \pm \sqrt{(1.795)^2 - 4(2.59)(0.224)}}{2(2.59)}$

$\quad = \dfrac{1.795\sqrt{3 \cdot 22 - 2.32}}{5.18} = \dfrac{1.795 \pm 0.949}{5.18}$

$x = \dfrac{2.74}{5.18} \ \text{ or } \ \dfrac{0.846}{5.18} \qquad x = 0.529 \ \text{ or } \ 0.163$

Since $[CO] = 0.25 - x$, x must be < 0.25.
Therefore, $x = 0.163$
$[CO_2] = 0.163M$

$0.163\text{ mol }CO_2 \times \dfrac{44\text{ g }CO_2}{1\text{ mol }CO_2} = 7.19\text{ g }CO_2$

CHAPTER 20 Acids and Bases

		CHAPTER ORGANIZER			
USE WITH SECTION	**DEMONSTRATION**	**WORKSHEET**	**LAB EXPERIMENT**	**SOFTWARE**	**TEACHER'S RESOURCE BINDER**
20-1	20-A				
20-3		20-A		Disk 8	
20-4				Disk 2	
20-5				Disk 2	
20-6				Disk 8	
20-7				Disk 8	
20-9					Problems 1–3 Advanced Topic 8
20-10					
20-11		20-B		Disk 8	
20-12	20-B			Disk 8	
20-13	20-C, D, F			Disk 8	Advanced Topic 7 Problems 4–7
20-14		20-C	20-A	Disk 8	Problems 8–10
20-15					CHEM Issue 6
20-18		20-D	20-B		CHEM Issue 4
20-19	20-E	20-E			
		Chapter Test			

■ Objectives

When the students have completed this chapter, they will be able to

1. Identify the properties common to the class of compounds known as acids and bases.
2. Use the Brönsted-Lowry definition of acids and bases to identify the proton donor, proton acceptor, conjugate acid, and conjugate base in a given equation.
3. Describe the industrial and practical uses of some acids and bases.
4. Determine the $[H_3O^+]$ and $[OH^-]$, given the mass of strong acid or base per liter of solution.
5. Calculate K_a for the system, given the equilibrium concentrations of a weak acid and the $[H_3O^+]$ in solution.
6. Calculate the $[H_3O^+]$, given the K_a and molar concentration of a weak acid.
7. Calculate the pH and pOH of a solution, given the $[H_3O^+]$ or $[OH^-]$.
8. Calculate the fourth value when given three of the four values—molarity of base, volume of base, molarity of acid, and volume of acid—used in a titration experiment, assuming a strong acid and strong base reaction.
9. Use hydrolysis to explain why the solution of a salt is not necessarily neutral.
10. Make a buffered solution and explain how such a solution maintains a constant pH, even with the addition of small amounts of strong acid or strong base.

■ Section Notes

20-4 The bottom line in most commercial processes is economics. However, this very fundamental consideration is rarely mentioned in the classroom.

Here we briefly introduce students to some real-world considerations. Although sulfuric acid is the cheapest to produce, other properties imparted by the presence of the acid anion often require other, more expensive acids.

20-6 Do not confuse acid or base strength with reactivity. Stainless steel, which is characteristically resistant to most chemical reactions, can be dissolved only by perchloric acid or hydrofluoric acid. Perchloric is the strongest acid known and is also one of the most reactive. Hydrofluoric acid, also reactive, is considered to be a weak acid.

Review and Practice

page 549

1. tastes sour; conducts an electric current; causes indicators to change color; produces H_2 gas when acid reacts with certain metals; neutralizes a base
2. Most acids are harmful.
3. An acid is a proton donor. A base is a proton acceptor.
4. Water can act as either an acid or a base.
5. $Ca(OH)_2$ is basic because it accepts protons from an acid.
   $$Ca(OH)_2 + 2HCl \longrightarrow CaCl_2 + 2H_2O$$
6. when moles of protons from the acid equal moles of protons accepted by the base
7. $KOH + HNO_3 \longrightarrow H_2O + KNO_3$
8. $HI + H_2O \longrightarrow H_3O^+ + I^-$

page 557

1. The presence of hydronium ion, H_3O^+, makes it acidic.
2. They differ in the completion of the reaction with water.
3. $HNO_3 + H_2O \longrightarrow NO_3^- + H_3O^+$
4. because not all of the molecules react
5. Cl^-
6. OH^- is stronger.
7. a proton acceptor
8. $Ba(OH)_2(s) \longrightarrow Ba^{2+}(aq) + 2OH^-(aq)$

page 558

1. $CH_3COOH + H_2O \rightleftharpoons H_3O^+ + CH_3COO^-(aq)$
2. The number of ions are fewer than the number of molecules.
3. $2H_2O \rightleftharpoons H_3O^+ + OH^-$
4. $[H_3O^+] = 1 \times 10^{-7}$;
   $[OH^-] = 1 \times 10^{-7}$

5. K_w describes the ion product constant. Numerical value $= 1 \times 10^{-14}$.
6. If $[OH^-]$ increases, $[H_3O^+]$ decreases since the two are inversely proportional.

page 567

1. pH is minus the log of the hydronium ion concentration
2. $[H_3O^+] = 0.0001M = 1 \times 10^{-4}M$
   $pH = -\log[H_3O^+] = -\log 1 \times 10^{-4}$
   $\log 1 + \log 10^{-4} = 0 + (-4)$
   $pH = 4$
3. $pH = -\log[H_3O^+] = 9$ $\log[H_3O^+] = -9$
   $[H_3O^+] = $ antilog of -9 $[H_3O^+] = 1 \times 10^{-9}$
4. When pH goes from 4 to 3, the $[H_3O^+]$ increases by a factor of 10. The $[H_3O^+]$ is ten times greater. At the same time, the $[OH^-]$ decreases from pOH of 10 to a pOH of 11. (pH + pOH = 14.) Therefore the solution is ten times less basic than it was originally.

page 571

1. $$30 \text{ mL} \times \frac{1 L}{1000 \text{ mL}} \times \frac{0.5 \text{ mol HCl}}{L} = 0.015 \text{ mol HCl}$$
   $$0.015 \text{ mol HCl} \times \frac{1 \text{ mol NaOH}}{1 \text{ mol HCl}} = 0.015 \text{ mol NaOH}$$
   $$\frac{0.015 \text{ mol NaOH}}{50 \text{ mL}} \times \frac{1000 \text{ mL}}{L} = 0.3M$$

2. $$32.5 \text{ mL} \times \frac{1 L}{1000 \text{ mL}} \times \frac{0.56 \text{ mol NaOH}}{L}$$
   $$= 0.0182 \text{ mol NaOH}$$
   $$HC_2H_3O_2 + NaOH \rightleftharpoons NaC_2H_3O_2 + H_2O$$
   $$0.0182 \text{ mol NaOH} \times \frac{1 \text{ mol } HC_2H_3O_2}{1 \text{ mol NaOH}}$$
   $$= 0.0182 \text{ mol } HC_2H_3O_2$$
   $$\frac{0.0182 \text{ mol } HC_2H_3O_2}{15 \text{ mL vinegar}} \times \frac{1000 \text{ mL}}{L} = 1.2M \text{ } HC_2H_3O_2$$

3. $$NH_3 + HCl \longrightarrow NH_4^+ + Cl^-$$
   $$48.25 \text{ mL} \times \frac{0.5246 \text{ mol HCl}}{L} \times \frac{1 L}{1000 \text{ mL}}$$
   $$= 0.02531 \text{ mol HCl}$$
   $$0.02531 \text{ mol HCl} \times \frac{1 \text{ mol } NH_3}{1 \text{ mol HCl}} = 0.02531 \text{ mol } NH_3$$
   $$\frac{0.02531 \text{ mol } NH_3}{22.0 \text{ mL}} \times \frac{1000 \text{ mL}}{L} = 1.15M \text{ } NH_3$$

4. $$H_2C_2O_4 + 2KOH \longrightarrow K_2C_2O_4 + 2H_2O$$
   $$6.25 \text{ g } H_2C_2O_4 \times \frac{1 \text{ mol}}{90 \text{ g}} = 0.0694 \text{ mol } H_2C_2O_4$$
   $$0.0694 \text{ mol } H_2C_2O_4 \times \frac{2 \text{ mol KOH}}{1 \text{ mol } H_2C_2O_4}$$
   $$= 0.139 \text{ mol KOH}$$
   $$\frac{0.139 \text{ mol KOH}}{32.2 \text{ mL}} \times \frac{1000 \text{ mL}}{L} = 4.31M \text{ KOH}$$

5. $5.0 \text{ g Mg(OH)}_2 \times \dfrac{1 \text{ mol}}{58.3 \text{ g}} = 0.0858 \text{ mol Mg(OH)}_2$

$$\text{Mg(OH)}_2 + 2\text{HCl} \longrightarrow \text{MgCl}_2 + 2\text{H}_2\text{O}$$

$0.0858 \text{ mol Mg(OH)}_2 \times \dfrac{2 \text{ mol HCl}}{1 \text{ mol Mg(OH)}_2}$

$= 0.172 \text{ mol HCl}$

$\dfrac{0.172 \text{ mol HCl}}{450 \text{ mL}} \times \dfrac{1000 \text{ mL}}{\text{L}} = 0.382M \text{ HCl}$

page 576

1. a. $\text{NH}_3 + \text{H}_2\text{O} \longrightarrow \text{NH}_4^+ + \text{OH}^-$
$2\text{NH}_4^+ + 2\text{OH}^- + \text{H}_2\text{SO}_4 \longrightarrow (\text{NH}_4)_2\text{SO}_4 + 2\text{H}_2\text{O}$
b. The salt can be extracted by evaporation.
c. $(\text{NH}_4)_2\text{SO}_4$ is acidic.

2. a. $[\text{Na}^+]$ increases, $[\text{CH}_3\text{COOH}]$ decreases, $[\text{H}_2\text{O}]$ increases, $[\text{CH}_3\text{COO}^-]$ increases.
b. $[\text{CH}_3\text{COO}^-]$ increases, $[\text{H}_3\text{O}^+]$ decreases, $[\text{CH}_3\text{COOH}]$ increases, $[\text{Na}^+]$ is unchanged.
c. $[\text{Na}^+]$ increases, no effect on other ions.

3. The addition of hydronium ions will shift a reaction in the direction that causes the formation of molecular acid. The addition of a base causes the acid molecules to react with water (shift in the opposite direction), thereby producing H_3O^+ to neutralize the OH^- from the base.

Review, page 577

1. Common properties: acids and bases neutralize each other and their solutions conduct electricity and cause indicators to change color. Different properties: acids taste sour and react with certain metals; bases taste bitter and do not react with metals.

2. $\text{HNO}_3 + \text{H}_2\text{O} \longrightarrow \text{NO}_3^- + \text{H}_3\text{O}^+$

3. $\text{CsOH} \longrightarrow \text{Cs}^+(aq) + \text{OH}^-(aq)$

4. $\text{HC}_2\text{H}_3\text{O}_2 + \text{H}_2\text{O} \rightleftharpoons \text{C}_2\text{H}_3\text{O}_2^- + \text{H}_3\text{O}^+$
    acid   base   conjugate conjugate
                         base   acid

5. $2\text{H}_2\text{O} \rightleftharpoons \text{H}_3\text{O}^+ + \text{OH}^-$

6. $\text{HC}_2\text{H}_3\text{O}_2 + \text{H}_2\text{O} \rightleftharpoons \text{H}_3\text{O}^+ + \text{C}_2\text{H}_3\text{O}_2^-$
$K_a = \dfrac{[\text{H}_3\text{O}^+][\text{C}_2\text{H}_3\text{O}_2^-]}{[\text{HC}_2\text{H}_3\text{O}_2]}$

7. Methyl red is suitable for a change at pH 5.0. Phenolphthalein is suitable for a change at pH 8.4.

8. a. $\text{KOH} + \text{HCl} \longrightarrow \text{KCl} + \text{H}_2\text{O}$
b. $2\text{K} + 2\text{HCl} \longrightarrow \text{H}_2 + 2\text{KCl}$
c. $2\text{K} + \text{Cl}_2 \longrightarrow 2\text{KCl}$

Interpret and Apply, page 577

1. $K_{eq} = \dfrac{[\text{H}_3\text{O}^+][\text{OH}^-]}{[\text{H}_2]^2}$ $K_w = [\text{H}_3\text{O}^+][\text{OH}^-]$
(H_2O)

2. The $[\text{H}_3\text{O}^+]$ due to the dissociation of water is negligible when an acid is added.

3. When pH = 10, pOH = 4; when pH = 8, pOH = 6.

4. K_a is less than one for weak acids because the concentration of ions is less than the concentration of molecules.

5. a. H_2PO_4^-
b. H_2SO_4
c. NH_4^+
d. H_2S

6. $\text{Mg}(s) + 2\text{HCl}(aq) \longrightarrow \text{MgCl}_2(aq) + \text{H}_2(g)$

7. $\text{H}_2\text{O} + \text{SO}_2 \longrightarrow \text{H}_2\text{SO}_3$

8. $\text{H}_2\text{S} + \text{H}_2\text{O} \longrightarrow \text{HS}^- + \text{H}_3\text{O}^+$
$\text{HS}^- + \text{H}_2\text{O} \longrightarrow \text{S}^{2-} + \text{H}_3\text{O}^+$

$\text{H}_2\text{CO}_3 + \text{H}_2\text{O} \longrightarrow \text{HCO}_3^- + \text{H}_3\text{O}^+$
$\text{HCO}_3^- + \text{H}_2\text{O} \longrightarrow \text{CO}_3^{2-} + \text{H}_3\text{O}^+$

$\text{H}_3\text{PO}_4 + \text{H}_2\text{O} \longrightarrow \text{H}_2\text{PO}_4^- + \text{H}_3\text{O}^+$
$\text{H}_2\text{PO}_4^- + \text{H}_2\text{O} \longrightarrow \text{HPO}_4^{2-} + \text{H}_3\text{O}^+$
$\text{HPO}_4^{2-} + \text{H}_2\text{O} \longrightarrow \text{PO}_4^{3-} + \text{H}_3\text{O}^+$

9. The end point and equivalence point should be the same. The indicator shows the end point during titration. If it is too far removed from the equivalence point, the two values will not be close enough to give accurate results.

10. If you overshoot the equivalence point, back titrate, which means adding a little bit more of the first solution until the color changes back to its original color.

11. a. For a strong acid, complete reaction occurs, so equilibrium favors products.
b. For a weak acid, few ions are formed, so equilibrium favors reactants.
c. If HA is stronger than HA_2, then $K_1 > K_2$.

12. NH_4^+, H_2S, HSO_4^-

13. $[\text{H}_3\text{O}^+]$ will be highest in KHSO_4 solution. It will be lowest in NH_4Cl solution.

14. a. $\text{NaCH}_3\text{COO}(aq) + \text{HF}(aq) \longrightarrow$
$\text{NaF} + \text{HCH}_3\text{COO}$
b. CH_3COO^-

15. a. $\text{H}_2\text{S} + \text{CO}_3^{2-} \longrightarrow \text{HS}^- + \text{HCO}_3^-$
b. CO_3^{2-} and HS^-

16. $\text{Zn} + \text{H}_2\text{SO}_4 \longrightarrow \text{ZnSO}_4 + \text{H}_2$
ZnSO_4 is the salt formed. This salt is extracted by evaporation of water.

17. $2\text{K} + 2\text{HI} \longrightarrow 2\text{KI} + \text{H}_2$
$2\text{K} + \text{I}_2 \longrightarrow 2\text{KI}$

18. Limestone is basic as the calcium hydroxide provides a source of OH^- ions. Gypsum and aluminum sulfate are salts composed of weak bases and strong acids so they hydrolyze to form acidic solutions. The addition of the base neutralizes the acidic soil. Conversely, the addition of the acidic salts neutralizes the basic soil.

Problems, page 578

1. $42.7 \text{ mL} \times \dfrac{0.498 \text{ mol HNO}_3}{\text{L}} \times \dfrac{1 \text{L}}{1000 \text{ mL}}$

$\qquad = 0.0213 \text{ mol HNO}_3$

$\qquad \text{KOH} + \text{HNO}_3 \longrightarrow \text{KNO}_3 + \text{H}_2\text{O}$

$0.0213 \text{ mol HNO}_3 \times \dfrac{1 \text{ mol KOH}}{1 \text{ mol HNO}_3} = 0.0213 \text{ mol KOH}$

$\dfrac{0.0213 \text{ mol KOH}}{30.0 \text{ mL}} \times \dfrac{1000 \text{ mL}}{\text{L}} = 0.71M \text{ KOH}$

2. $28.73 \text{ mL} \times \dfrac{0.15 \text{ mol HCl}}{\text{mL}} \times \dfrac{1 \text{L}}{1000 \text{ mL}}$

$\qquad = 4.31 \times 10^{-3} \text{ mol HCl}$

$4.31 \times 10^{-3} \text{ mol HCl} \times \dfrac{1 \text{ mol NaOH}}{1 \text{ mol HCl}}$

$\qquad = 4.31 \times 10^{-3} \text{ mol NaOH}$

$4.31 \times 10^{-3} \text{ mol NaOH} \times \dfrac{1000 \text{ mL}}{0.28 \text{ mol NaOH}}$

$\qquad = 15.4 \text{ mL NaOH}$

3. $0.01M \text{ HNO}_3 = 1 \times 10^{-2}M \text{ HNO}_3$

$\text{pH} = -\log[\text{H}_3\text{O}^+]$

$\qquad = -(\log 1 + \log 10^{-2}) = -(0 + -2) = 2$

4. $0.001M \text{ KOH} = 1 \times 10^{-3}M \text{ KOH} \quad \text{pOH} = 3$

$\text{pOH} + \text{pH} = 14 \quad \text{pH} = 14 - 3 = 11$

$[\text{H}_3\text{O}^+] = 1 \times 10^{-11}$

5. $2\text{H}_2\text{O} \rightleftharpoons \text{H}_3\text{O}^+ + \text{OH}^-$

$K_a = [\text{H}_3\text{O}^+][\text{OH}^-] \times [1 \times 10^{-7}][1 \times 10^{-7}]$

$\qquad = 1 \times 10^{-14}$

6. a. $\dfrac{200 \text{ g CH}_3\text{COOH}}{\text{L}} \times \dfrac{1 \text{ mol}}{60 \text{ g}} = 3.3M \text{ CH}_3\text{COOH}$

b. $K_a = \dfrac{[\text{H}_3\text{O}^+][\text{CH}_3\text{COO}^-]}{[\text{CH}_3\text{COOH}]} = 1.8 \times 10^{-5}$

Let $x = [\text{H}_3\text{O}^+]$ and $[\text{CH}_3\text{COO}^-]$

$\dfrac{[x]^2}{[3.3]} = 1.8 \times 10^{-5} \quad x^2 = 5.94 \times 10^{-5}$

$[\text{H}_3\text{O}^+] = 7.7 \times 10^{-3}$

7. $\text{mol HCl} = \left(\dfrac{50.0}{1000} \text{ liter}\right) \times \left(0.200 \dfrac{\text{mol}}{\text{liter}}\right)$

$\qquad = 1.00 \times 10^{-2}$

$\text{mol NaOH} = \left(\dfrac{49.0}{1000} \text{ liter}\right) \times \left(0.200 \dfrac{\text{mol}}{\text{liter}}\right)$

$\qquad = 0.980 \times 10^{-2}$

$\text{Excess mol HCl} = (1.00 \times 10^{-2}) - (0.98 \times 10^{-2})$

$\qquad\qquad = 0.02 \times 10^{-2}$

$[\text{H}_3\text{O}^+] = \dfrac{0.02 \times 10^{-2}}{0.099} = 0.002M$

$[\text{OH}^-] = \dfrac{1.00 \times 10^{-14}}{0.002} = 5 \times 10^{-12}M$

8. $\text{mol HCl} = \left(\dfrac{50.0}{1000} \text{ liter}\right) \times \left(0.200 \dfrac{\text{mol}}{\text{liter}}\right)$

$\qquad = 1.00 \times 10^{-2}$

$\text{mol NaOH} = \left(\dfrac{49.9}{1000} \text{ liter}\right) \times \left(0.200 \dfrac{\text{mol}}{\text{liter}}\right)$

$\qquad = 0.998 \times 10^{-2}$

$\text{Excess mol HCl} = (1.00 \times 10^{-2}) - (0.998 \times 10^{-2})$

$\qquad\qquad = 0.002 \times 10^{-2}$

$[\text{H}_3\text{O}^+] = \dfrac{2 \times 10^{-5}}{0.0999} = 2 \times 10^{-4}M$

$[\text{OH}^-] = \dfrac{1.00 \times 10^{-14}}{2 \times 10^{-4}} = 5 \times 10^{-11}M$

9. 1.0 mL, since mol HCl must equal mol NaOH

10. $[\text{H}_3\text{O}^+] = 1 \times 10^{-8}$

$K_w = 1 \times 10^{-14} = [\text{H}_3\text{O}^+][\text{OH}^-]$

$[\text{OH}^-] = \dfrac{1 \times 10^{-14}}{1 \times 10^{-8}} = 1 \times 10^{-6}$

The solution is basic.

11. $\text{HA(aq)} + \text{H}_2\text{O(l)} \rightleftharpoons \text{H}_3\text{O}^+ + \text{A}^-$

$[\text{H}_3\text{O}^+] = [\text{A}^-] = 4 \times 10^{-3}M$

$[\text{HA}] = 0.25M$

$K_a = \dfrac{[\text{H}_3\text{O}^+][\text{A}^-]}{[\text{HA}]} = \dfrac{[4 \times 10^{-3}]^2}{[0.25]} = \dfrac{1.6 \times 10^{-5}}{0.25}$

$\qquad = 6.4 \times 10^{-5}$

12. $\dfrac{0.056 \text{ g KOH}}{\text{L}} \times \dfrac{1 \text{ mol}}{56 \text{ g}} = \dfrac{0.001 \text{ mol KOH}}{\text{L}}$

$[\text{OH}^-] = 0.001 = 1 \times 10^{-3}M$

$K_w = [\text{OH}^-][\text{H}_3\text{O}^+] = 1 \times 10^{-14}$

$[\text{H}_3\text{O}^+] = \dfrac{1 \times 10^{-14}}{1 \times 10^{-3}} = 1 \times 10^{-11}M$

13. $500 \text{ mL} \times \dfrac{1 \text{ mol HBr}}{1000 \text{ mL}} = 0.5 \text{ mol HBr}$

$500 \text{ mL} \times \dfrac{1 \text{ mol LiOH}}{1000 \text{ mL}} = 0.5 \text{ mol LiOH}$

$\text{LiOH} + \text{HBr} \longrightarrow \text{LiBr} + \text{H}_2\text{O}$

$0.5 \text{ mol LiBr} \times 87 \text{ g/mol} = 43.5 \text{ g LiBr}$

Challenge, page 579

1. 1.34% ionized acid solution is weak.

2. $2.50 \text{ g NaHCO}_3 \times \dfrac{1 \text{ mol}}{84 \text{ g NaHCO}_3}$

$\qquad = 2.98 \times 10^{-2} \text{ mol NaHCO}_3$

$2.98 \times 10^{-2} \text{ mol NaHCO}_3 \times \dfrac{1 \text{ mol H}_2\text{SO}_4}{2 \text{ mol NaHCO}_3} =$

$\qquad 1.49 \times 10^{-2} \text{ mol H}_2\text{SO}_4$

$1.49 \times 10^{-2} \text{ mol} \times \dfrac{1000 \text{ mL}}{0.600 \text{ mol H}_2\text{SO}_4} = 24.8 \text{ mL}$

3. $\text{HF(aq)} \rightleftharpoons \text{H}^+\text{(aq)} + \text{F}^-\text{(aq)}$

Initial conc.	0.10	0	0 M
Equil conc.	0.092	0.008	0.008M

$K_a = \dfrac{[\text{H}^+][\text{F}^-]}{[\text{HF}]} = \dfrac{(0.008)^2}{(0.092)} = \dfrac{64 \times 10^{-6}}{9.2 \times 10^{-2}} = 6.9 \times 10^{-4}$

4. Amphoterism describes this dual behavior.

$\text{Al(OH)}_3\text{(s)} + \text{OH}^-\text{(aq)} \longrightarrow \text{acts as an acid}$

$\text{Al(OH)}_3 + 3\text{H}_3\text{O}^+\text{(aq)} \longrightarrow \text{acts as a base}$

Synthesis, page 579

1. The strong HCl secreted by the stomach would neutralize the caustic base of the Drano®. Vomiting in case of ingestion is not recommended because the esophagus would be damaged a second time by the caustic materials.

2. 5% by mass means $5\,g\,CH_3COOH/100\,g$ vinegar. Assume the density of vinegar is $1\,g/cm^3$.

$$\frac{5\,g}{100\,g\,vinegar} \times \frac{1\,g\,vinegar}{cm^3\,vinegar} \times \frac{1\,mol\,CH_3COOH}{60\,g\,CH_3COOH}$$

$$\times \frac{1000\,cm^3}{L} = 0.83M$$

CHAPTER 21 Electrochemistry

CHAPTER ORGANIZER

USE WITH SECTION	DEMONSTRATION	WORKSHEET	LAB EXPERIMENT	SOFTWARE	TEACHER'S RESOURCE BINDER
21-1	21-A	21-B			
21-2		21-A			Problems 1–2
21-3		21-C			Problems 3–6
21-4	21-B	21-D	21-A		
21-5					Blackline Master 12
21-6	21-C	21-E			Problems 7–10
21-8	21-C				
21-9					Advanced Topic 8
21-10		21-F	21-B		Advanced Topic 9
Chapter Test					

■ Objectives

When the students have completed this chapter, they will be able to

1. Write half-reactions for certain reactions.
2. Determine the reducing agent and oxidizing reagent for a redox reaction.
3. Determine the oxidation number for an atom or ion in an element, a molecule, or complex ion.
4. Define oxidation or reduction in terms of increase or decrease of oxidation number.
5. Balance redox reactions that take place in acid solutions.
6. Describe the components of an electrochemical cell.
7. Distinguish between electrical terms such as coulomb, ampere, and volt.
8. Define half-cell potential.
9. Calculate the cell potential for an electrochemical cell under standard conditions.

10. Describe how a dry cell supplies electricity.
11. Explain how a lead storage battery produces electricity.
12. Calculate the amount of substance reduced when a quantity of another substance oxidized in an electrochemical cell is given.
13. Describe the process of electrolysis.

■ Section Notes

21-1 The process of oxidation and reduction is described but not defined. It is easier for the students to think of oxidation and reduction in terms of oxidation number which is introduced in Section 21-2. Therefore, do not emphasize the traditional definition of oxidation as the loss of electrons and reduction as the gain of electrons. The usual connotation of reduction is to make smaller. To define chemical reduction as the gain of electrons does not make sense to the students.

21-2 Oxidation numbers are merely a tool of the electrochemist. The oxidation number of an atom or ion is a reference from which decisions can be made about the transfer of electrons in a redox reaction. The "pretend-that-all-bonds-are-ionic" game eliminates some of the arbitrariness of assigning oxidation numbers to covalently bonded atoms.

The meaning of reduction is more easily seen in oxidation numbers than in the traditional definition of reduction as the gain of electrons. Reduction is the reduction of the value of the oxidation number in a species, as when Cu^{2+} becomes Cu^0. Oxidation is the increase in the value of the oxidation number in a species, as when Cl^- becomes Cl^0. This relationship between oxidation and reduction is much easier to remember.

21-3 The method suggested here for balancing redox reactions seems quite involved for beginning students. They often get confused when adding up charges on each side of the equation. They need to gain confidence through practice.

Students need not rewrite the half-reactions many times over as implied in the text. To decrease repetition and copying errors, write out the half-reactions and leave spaces between species for the addition of electrons, hydrogen ions, water molecules and spectator ions. Example 21-5 is used here:

$$Zn \longrightarrow Zn^{2+} + 2e^-$$
$$2e^- + 8H^+ + 2VO_3^- \longrightarrow 2VO^{2+} + 4H_2O$$
$$\overline{8H^+ + Zn + 2VO_3^- \longrightarrow Zn^{2+} + 2VO^{2+} + 4H_2O}$$

Redox reactions also occur in alkali solutions with an excess of OH^-. This class of redox reactions is not presented in the text.

21-4 The main focus of the chapter is on the electrochemical cell. The cell is represented in terms of competition for electrons in much the same way acid-base reactions are presented in terms of competition for protons. Table 21-2 lists some metals in order of their tendency to lose electrons. Therefore, in the reaction

$$Zn + Cu^{2+} \longrightarrow Zn^{2+} + Cu,$$

zinc loses electrons more easily than copper. The reaction is spontaneous in the forward direction. In the competition for electrons, Zn^{2+} and Cu win out over Zn and Cu^{2+}.

The schematic representation of an electrochemical cell should be emphasized here. Students should understand that a voltaic cell requires reacting solutions, electrodes, a salt bridge, and electrical connections between the half-cells so that electrons will flow. At this point you also may introduce the idea that electrochemical cells do electrical work. Electrons will flow as long as the reacting solutions are not at equilibrium concentrations. When equilibrium is reached,

the electrons stop flowing in one direction, and the cell is no longer able to do electrical work.

21-5, 21-6 The concept of a reference point or benchmark is developed in Section 21-5. The choice of the hydrogen half-cell as the standard for half-cell potentials is arbitrary but reasonable. Half-cell potentials are not measurable in themselves. What is measured is the potential difference between two reacting species. You may want to point out to the students that the situation here is similar to the measurement of heat reactions. The heat content, H, of a substance is not measurable. However, the change in heat content from reactant to product is measurable as the heat of the reaction.

Table 21-4 lists E^0 values for some substances. These values are given for standard conditions of effective concentrations of $1M$ for solutions, and 1 atm for gases. If these concentrations change, the E for the cell changes. Walter Nernst developed the relationship between cell voltage and concentration:

$$E = E^0 - \frac{RT}{nF} \ln Q$$

where R is the gas constant, T is the temperature in kelvins, n is the number of electrons transferred in the balanced reaction equation, F is a conversion factor from volts to joules per mole, and $\ln$ indicates a natural logarithm. Q is the reaction quotient, which is expressed similar to the equilibrium law expression. At the standard temperature of 298 K, terms in the Nernst equation can be combined and expressed as follows:

$$E = E^0 - \frac{0.0591}{n} \log \frac{[\text{products}]}{[\text{reactants}]}$$

21-7 You may want to tell the students that many kinds of dry cells are available today. Some of the cells depend on the following reactions:

Mercury	$Zn + HgO \longrightarrow ZnO + Hg$
Nickel-cadmium	$Cd + 2NiOOH + 2H_2O \longrightarrow$ $2Ni(OH)_2 + Cd(OH)_2$
Silver-zinc	$Zn + 2AgO + H_2O \longrightarrow$ $Zn(OH)_2 + Ag_2O$

Review and Practice
page 583

1. **a.** reduction **b.** oxidation **c.** reduction **d.** oxidation

2. An oxidizing agent is the reactant which is reduced in a redox reaction.

3. A redox reaction is balanced when the number of atoms or ions of each element are equal on both sides of the equation. Electric charges must also balance.

4. a. $2Na(s) + Cl_2(g) \longrightarrow 2Na^+(s) + 2Cl^-(s)$
Cl_2 is reduced. Na is the reducing agent.
b. $Cu^{2+}(aq) + Mg(s) \longrightarrow Cu(s) + Mg^{2+}(aq)$
Cu^{2+} is reduced. Mg is the reducing agent.
c. $3Fe^{3+}(aq) + Al(s) \longrightarrow 3Fe^{2+}(aq) + Al^{3+}(aq)$
Fe^{3+} is reduced. Al is the reducing agent.
d. $2Au^{3+}(aq) + 3Cd(s) \longrightarrow 2Au(s) + 3Cd^{2+}(aq)$
Au^{3+} is reduced. Cd is the reducing agent.

page 588

1. Oxidation number is the real or apparent charge an atom or ion has when assigned a certain number of electrons.

2. Oxidation is an increase in the value of oxidation number of an element. Reduction is a decrease in the value of the oxidation number.

3. a. $C = 4^-$, $H = 1+$
b. $S = 4^+$, $O = 2^-$
c. $Mn = 4^+$, $O = 2^-$
d. $Cr = 3^+$
e. $P = 3^-$
f. $N = 5^+$, $O = 2^-$
g. $H = 1^+$, $O = 2^-$
h. $Na = 1^+$, $O = 1^-$
i. $Mg = 2^+$, $H = 1^-$
j. $Cr = 6^+$, $O = 2^-$
k. $K = 1^+$, $Mn = 7^+$, $O \doteq 2^-$

4. a. oxidation
b. reduction
c. neither (the oxidation number of Cr in each species is 6^+)
d. oxidation
e. reduction

page 591

1. a. $2I^- + Cl_2 \longrightarrow 2Cl^- + I_2$
b. $Co + 2Fe^{3+} \longrightarrow Co^{2+} + 2Fe^{2+}$

2. a. $6e^- + 14H^+ + Cr_2O_7^{2-} \longrightarrow 2Cr^{3+} + 7H_2O$
$3Hg \longrightarrow 3Hg^{2+} + 6e^-$
$$\overline{Cr_2O_7^{2-} + 3Hg + 14H^+ \longrightarrow}$$
$$2Cr^{3+} + 3Hg^{2+} + 7H_2O$$

b. $16H^+ + 2MnO_4^- + 10e^- \longrightarrow 2Mn^{2+} + 8H_2O$
$5H_2O_2 \longrightarrow 5O_2 + 10H^+ + 2e^-$
$$\overline{2MnO_4^- + 5H_2O_2 + 6H^+ \longrightarrow}$$
$$2Mn^{2+} + 5O_2 + 8H_2O$$

c. $4Zn \longrightarrow 4Zn^{2+} + 2e^-$
$8e^- + 10H^+ + NO_3^- \longrightarrow NH_4^+ + 3H_2O$
$$\overline{4Zn + NO_3^- + 10H^+ \longrightarrow 4Zn^{2+} + NH_4^+ + 3H_2O}$$

d. $2e^- + Cl_2 \longrightarrow 2Cl^-$
$2H_2O + Cl_2 \longrightarrow 2HClO + 2H^+ + 2e^-$
$$\overline{2Cl_2 + 2H_2O \longrightarrow 2Cl^- + 2HClO + 2H^+}$$

page 601

1. An anode is the site of oxidation. A cathode is the site of reduction.

2. The rate of electrical flow is a measure of the electrical charge (coulombs) flowing through a circuit per unit time (seconds). One coulomb per second

is an ampere. The tendency to flow is a measure of electrical pressure. It is the potential difference between the anode and cathode in an electrochemical cell. This tendency is usually measured in volts.

3. a. $Co \longrightarrow Co^{2+} + 2e^-$
$2Fe^{3+} + 2e^- \longrightarrow 2Fe^{2+}$
$$\overline{Co + 2Fe^{3+} \longrightarrow Co^{2+} + 2Fe^{2+}}$$
$E^0 = +0.28$ V
$E^0 = +0.77$ V
$\overline{E^0_{cell} = +1.05 \text{ V}}$

b. $H_2 \longrightarrow 2H^+ + 2e^-$ $\quad E^0 = +0.00$ V
$Cl_2 + 2e^- \longrightarrow 2Cl^-$ $\quad E^0 = +1.36$ V
$\overline{H_2 + Cl_2 \longrightarrow 2H^+ + 2Cl^-}$ $\quad E^0_{cell} = +1.36$ V

c. $10I^- \longrightarrow 5I_2 + 10e^-$
$2MnO_4^- + 16H^+ + 10e^- \longrightarrow 2Mn^{2+} + 8H_2O$
$$\overline{10I^- + 2MnO_4^- + 16H^+ \longrightarrow}$$
$$5I_2 + 2Mn^{2+} + 8H_2O$$
$E^0 = -0.53$ V
$E^0 = +1.52$ V
$\overline{E^0_{cell} = +0.99 \text{ V}}$

4. No. Both half-reactions are oxidations.

5. a. unitary rate or proportionality constant
b. The number of electrons is directly proportional to time.
c. It should be a straight line.

6. The aluminum spoon will be oxidized and the Fe^{2+} ions will be reduced to iron metal.

7. There will be no spontaneous reaction.
$E^0 = -1.22$ V for the reaction.

8. No, the container will dissolve.
$E^0 = +1.02$ V

9. $E^- = +0.30$ V

page 608

1. The alkaline battery is an electrochemical "dry" cell. It generates electricity from a redox reaction within the can whenever a complete circuit is available. Zn is the anode and MnO_2 is the cathode. The electrolyte is a solution of KOH in a gelling agent. The reaction is
$Zn + 2MnO_2 \longrightarrow ZnO + Mn_2O_3$

2. Sulfuric acid is the electrolyte in the lead storage battery. The acid solution also provides hydrogen ions for the reduction of lead(IV) oxide to lead (II) sulfate.

3. $2NaCl(l) + energy \longrightarrow 2Na(s) + Cl_2(g)$
$$2.0 \times 10^4 \text{ kg } Cl_2 \times \frac{1000 \text{ g } Cl_2}{1 \text{ kg } Cl_2} \times \frac{1 \text{ mol } Cl_2}{70.9 \text{ g } Cl_2}$$
$$\times \frac{2 \text{ mol } Na}{1 \text{ mol } Cl_2} \times \frac{23.0 \text{ g } Na}{1 \text{ mol } Na} \times \frac{1 \text{ kg } Na}{1000 \text{ g } Na}$$
$$= 1.3 \times 10^4 \text{ kg } Na$$

4. The reaction is
$$Mg + 2H^+ \longrightarrow Mg^{2+} + H_2.$$
$$5.00 \text{ g } Mg \times \frac{1 \text{ mol Mg}}{24.3 \text{ g Mg}} \times \frac{1 \text{ mol } H_2}{1 \text{ mol Mg}} \times \frac{2.02 \text{ g } H_2}{1 \text{ mol } H_2}$$
$$= 0.416 \text{ g } H_2$$

5. Iron rusts as Fe is oxidized to Fe^{2+}. Oxygen gas is reduced in the presence of H_2O to form OH^-. The Fe^{2+} and OH^- produce unstable $Fe(OH)_2$. This substance reacts with water and oxygen to form $Fe(OH)_3$.

Review, page 609

1. Answers will vary. Three oxidation reactions are
$$Fe \longrightarrow Fe^{2+} + 2e^-$$
$$2Br^- \longrightarrow Br_2 + 2e^-$$
$$Sn^{2+} \longrightarrow Sn^{4+} + 2e^-$$

2. a. Determine the charge on the molecule, NO_2^0.
   b. Determine the oxidation number of each atom of oxygen, $= -2$.
   c. Determine the oxidation number of nitrogen.
   $$\underline{\quad ? \quad} + -4 = 0$$
   The oxidation number of N is +4.

3. The number of electrons in the two half-reactions must be equal. The number of ions or atoms must be equal to each side of the equation.

4. Diagram of cell:

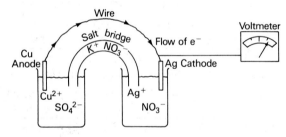

$$2Ag^+ + 2e^- \longrightarrow 2Ag \qquad E^0 = +0.80 \text{ V}$$
$$\underline{\quad Cu \longrightarrow Cu^{2+} + 2e^- \quad E^0 = -0.34 \text{ V}}$$
$$2Ag^+ + Cu \longrightarrow 2Ag + Cu^{2+} \quad E^0_{cell} = 0.46 \text{ V}$$

5. The salt bridge maintains the electric neutrality in the solutions of an electrochemical cell.

6. Voltage is the tendency of electrons to flow through an electrochemical cell. Other words for voltage are electric pressure or potential difference.

7. A half-cell potential (reduction potential) is the tendency of a half-cell reaction to gain electrons in comparison to the tendency of hydrogen ions to gain electrons. A positive E^0 value means that the half-cell has a greater tendency to be reduced than hydrogen ions. A negative value means that the half-cell has less tendency to gain electrons than hydrogen ions. E^0 is measured at a $1M$ effective concentration for solutions and 1 atm pressure for gases.

8. The reaction that occurs in an alkaline cell is
$$Zn + 2MnO_2 \longrightarrow ZnO + Mn_2O_3$$
Zn is the anode and MnO_2 is the cathode.

9. Electrolysis is the process by which an external source of electricity is applied to a chemical system to cause oxidation and reduction that ordinarily would not be spontaneous.

10. At the anode the following reaction takes place:
$$Fe \longrightarrow Fe^{2+} + 2e^-$$
At the cathode the reaction is
$$H_2O + \tfrac{1}{2}O_2 + 2e^- \longrightarrow 2OH^-$$

Interpret and Apply, page 609

1. a. $M \longrightarrow M^{1+} + e^-$ The substance has been oxidized.
   b. $X^{1-} + e^- \longrightarrow X^{2-}$ The ion has been reduced.

2. a. Mg—reduced state   **f.** S^{2-}—reduced state
   b. I^-—reduced state   **g.** Cr^{3+}—oxidized state
   c. O_2—oxidized state   **h.** Mn—reduced state
   d. Na^+—reduced state
   e. Al^{3+}—oxidized state

3. a. H = 1+   **e.** V = 5+
       S = 6+   O = 2−
       O = 2−   **f.** S = $\frac{10}{4}$+
   b. P = 0   O = 2−
   c. U = 6+   **g.** Cl = 3+
       O = 2−   O = 2−
   d. U = 5+   **h.** H = 1+
       O = 2−   S = 4+
                 O = 2−

4. a. reduction   **d.** reduction
   b. oxidation   **e.** neither
   c. reduction

5. $Cu^{2+} + 2e^- \longrightarrow Cu(s)$ reduction
   $Fe(s) \longrightarrow Fe^{2+} + 2e^-$ oxidation

6. To plate out any metal, positive metallic ions must be changed to neutral atoms:
$$M^{n+} + ne^- \longrightarrow M(s)$$
Such a reaction involves reduction. Since reduction takes place at the cathode, the spoon to be plated should be made the cathode. Pb may be a good choice for the other electrode.

7. a. Mg   **c.** Mg
   b. Co^{2+} **d.** Co^{2+}

8. a. oxidizing agent   **d.** reducing agent
   b. reducing agent   **e.** reducing agent
   c. oxidizing agent   **f.** reducing agent

9. The redox reaction that occurs in an electrochemical cell is a spontaneous process. The reaction that occurs in an electrolytic cell is a nonspontaneous process made possible by an external source of electricity.

10. No. Copper is less likely to be oxided than iron.

11. If the cells are connected in series, that is, anode to cathode, anode to cathode, etc., the voltage of the system is additive.

Problems, page 610

1.
$$2[Al \longrightarrow Al^{3+} + 3e^-]$$
$$3[2H^+ + 2e^- \longrightarrow H_2]$$
$$\overline{2Al + 6H^+ \longrightarrow 2Al^{3+} + 3H_2}$$

2.
$$5Sn^{2+} \longrightarrow 5Sn^{4+} + 10e^-$$
$$10e^- + 16H^+ + 2MnO_4^- \longrightarrow 2Mn^{2+} + 8H_2O$$
$$\overline{5Sn^{2+} + 2MnO_4^- + 16H^+ \longrightarrow 5Sn^{4+} + 2Mn^{2+}}$$
$$+ 8H_2O$$

3. a. $H_2O_2 + 2H^+ + 2e^- \longrightarrow 2H_2O$
$$2I^- \longrightarrow 2e^- + I_2$$
$$\overline{2I^- + H_2O_2 + 2H^+ \longrightarrow I_2 + 2H_2O}$$

b. $Cr_2O_7^{2-} + 14H^+ + 6e^- \longrightarrow 2Cr^{3+} + 7H_2O$
$$6Fe^{2+} \longrightarrow 6e^- + 6Fe^{3+}$$
$$\overline{Cr_2O_7^{2-} + 6Fe^{2+} + 14H^+ \longrightarrow 2Cr^{3+} + 6Fe^{3+}}$$
$$+ 7H_2O$$

c.
$$3Cu \longrightarrow 3Cu^{2+} + 6e^-$$
$$2NO_3^- + 8H^+ + 6e^- \longrightarrow 2NO + 4H_2O$$
$$\overline{3Cu + 2NO_3^- + 8H^+ \longrightarrow 3Cu^{2+} + 2NO + 4H_2O}$$

4. a.
$$2HBr \longrightarrow Br_2 + 2H^+ + 2e^-$$
$$2H^+ + H_2SO_4 + 2e^- \longrightarrow SO_2 + 2H_2O$$
$$\overline{2HBr + H_2SO_4 \longrightarrow Br_2 + SO_2 + 2H_2O}$$

b. $2NO_3^- + 8H^+ + 6e^- \longrightarrow 2NO + 4H_2O$
$$6Cl^- \longrightarrow 3Cl_2 + 6e^-$$
$$\overline{2NO_3^- + 8H^+ + 6Cl^- \longrightarrow 2NO + 3Cl_2 + 4H_2O}$$

c.
$$4Zn \longrightarrow 4Zn^{2+} + 8e^-$$
$$NO_3 + 10H^+ + 8e^- \longrightarrow NH_4^+ + 3H_2O$$
$$\overline{4Zn + NO_3^- + 10H^+ \longrightarrow 4Zn^{2+} + NH_4^+ + 3H_2O}$$

d. $2BrO^- + 4H^+ + 4e^- \longrightarrow 2Br^- + 2H_2O$
$$BrO^- + 2H_2O \longrightarrow BrO_3^- + 4H^+ + 4e^-$$
$$\overline{3BrO^- \longrightarrow 2Br^- + BrO_3^-}$$

5. All of the reactions in item 3 will occur spontaneously.

a. $E^0 = +1.77$ volts
$E^0 = -0.53$ volts
$\overline{E_{cell}^0 = +1.24 \text{ volts}}$

c. $E^0 = -0.34$ volts
$E^0 = +0.96$ volts
$\overline{E_{cell}^0 = +0.62 \text{ volts}}$

b. $E^0 = +1.33$ volts
$E^0 = -0.77$ volts
$\overline{E_{cell}^0 = +0.56 \text{ volts}}$

6. a. $3ClO_2^- + 3H_2O + 6e^- \longrightarrow 3ClO^- + 6OH^-$
$$2MnO_2 + 8OH^- \longrightarrow 2MnO_4^- + 4H_2O$$
$$+ 6e^-$$
$$\overline{3ClO_2^- + 2OH^- + 2MnO_2 \longrightarrow 3ClO^-}$$
$$+ 2MnO_4^- + H_2O$$

MnO_2 is oxidized and ClO_2^- is the oxidizing agent (is reduced).

b.
$$In \longrightarrow In^{3+} + 3e^-$$
$$BiO^+ + 2H^+ + 3e^- \longrightarrow Bi + H_2O$$
$$\overline{In + BiO^+ + 2H^+ \longrightarrow In^{3+} + Bi + H_2O}$$

In is oxidized and BiO^+ is the oxidizing agent (is reduced).

c.
$$4V^{2+} \longrightarrow 4V^{3+} + 4e^-$$
$$2H_2SO_3 + 2H^+ + 4e^- \longrightarrow S_2O_3^{2-} + 3H_2O$$
$$\overline{4V^{2+} + 2H_2SO_3 + 2H^+ \longrightarrow 4V^{3+} + S_2O_3^{2-}}$$
$$+ 3H_2O$$

V^{2+} is oxidized and H_2SO_3 is the oxidizing agent (is reduced).

d.
$$3Mn \longrightarrow 3Mn^{2+} + 6e^-$$
$$2IrCl_6^{3-} + 6e^- \longrightarrow 2Ir + 12Cl^-$$
$$\overline{3Mn + 2IrCl_6^{3-} \longrightarrow 3Mn^{2+} + 2Ir + 12Cl^-}$$

Mn is oxidized and $IrCl_6^{3-}$ is the oxidizing agent (is reduced).

7. The equation is
$$Ti + 2FeO \longrightarrow TiO_2 + 2Fe$$
$$57.3 \text{ g Ti} \times \frac{1 \text{ mol Ti}}{47.9 \text{ g Ti}} \times \frac{2 \text{ mol Fe}}{1 \text{ mol Ti}} \times \frac{55.8 \text{ g Fe}}{1 \text{ mol Fe}}$$
$$= 134 \text{ g Fe}$$

8. The equation is
$$2Cr + 3Pb^{2+} \longrightarrow 2Cr^{3+} + 3Pb$$
$$1.56 \text{ g Cr} \times \frac{1 \text{ mol Cr}}{52.00 \text{ g Cr}} \times \frac{3 \text{ mol Pb}}{2 \text{ mol Cr}} \times \frac{207.2 \text{ g Pb}}{1 \text{ mol Pb}}$$
$$= 9.32 \text{ g Pb}$$

9.
$$Cu \longrightarrow Cu^{2+} + 2e^-$$
$$2NO_3^- + 4H^+ + 2e^- \longrightarrow 2NO_2(g) + 2H_2O$$
$$\overline{Cu + 2NO_3^- + 4H^+ \longrightarrow Cu^{2+} + 2NO_2(g) + 2H_2O}$$

10. $Cd^{2+} + 2e^- \longrightarrow Cd \qquad E^0 = ?$
$Ni \longrightarrow Ni^{2+} + 2e^- \quad E^0 = -0.25 \text{ V}$
$\overline{Cd^{2+} + Ni \longrightarrow Cd + Ni^{2+} \quad E_{cell}^0 = +0.15 \text{ V}}$
The Cd^{2+} reduction potential is $+0.40$ volt.

11. a. $Mg \longrightarrow Mg^{2+} + 2e^- \quad E^0 = +2.37 \text{ V}$
$Sn^{2+} + 2e^- \longrightarrow Sn \qquad E^0 = -0.14 \text{ V}$
$\overline{Mg + Sn^{2+} \longrightarrow Mg^{2+} + Sn \quad E_{cell}^0 = +2.23 \text{ V}}$
Spontaneous as written.

b. $Mn \longrightarrow Mn^{2+} + 2e^- \quad E^0 = +1.18 \text{ V}$
$2Cs^+ + 2e^- \longrightarrow 2Cs \qquad E^0 = -2.92 \text{ V}$
$\overline{Mn + 2Cs^+ \longrightarrow Mn^{2+} + 2Cs \quad E_{cell}^0 = -1.74 \text{ V}}$
Nonspontaneous as written.

c. $Cu \longrightarrow Cu^{2+} + 2e^- \quad E^0 = -0.34 \text{ V}$
$Cl_2 + 2e^- \longrightarrow 2Cl^- \qquad E^0 = +1.36 \text{ V}$
$\overline{Cu + Cl_2 \longrightarrow Cu^{2+} + 2Cl^- \quad E_{cell}^0 = +1.02 \text{ V}}$
Spontaneous as written.

d. $Fe \longrightarrow Fe^{2+} + 2e^- \quad E^0 = +0.44 \text{ V}$
$2Fe^{3+} + 2e^- \longrightarrow 2Fe^{2+} \qquad E^0 = +0.77 \text{ V}$
$\overline{Fe + 2Fe^{3+} \longrightarrow 3Fe^{2+} \qquad E_{cell}^0 = +1.21 \text{ V}}$
Spontaneous as written.

12.
$$Zn + H_2SO_4 \longrightarrow Zn^{2+} + H_2S$$
$$4Zn \longrightarrow 4Zn^{2+} + 8e^-$$
$$H_2SO_4 + 8H^+ + 8e^- \longrightarrow H_2S + 4H_2O$$
$$\overline{4Zn + H_2SO_4 + 8H^+ \longrightarrow 4Zn^{2+} + H_2S + 4H_2O}$$

Challenge, page 611

1.

$$3Cu \longrightarrow 3Cu^{2+} + 6e^- \quad E^0 = -0.34\,V$$
$$\underline{2Cr^{3+} + 6e^- \longrightarrow 2Cr \quad\quad E^0 = -0.74\,V}$$
$$3Cu + 2Cr^{3+} \longrightarrow 3Cu^{2+} + 2Cr \quad E^0_{cell} = -1.08\,V$$

No visible spontaneous reaction occurs.

2. $Cu^{2+} + 2e^- \longrightarrow Cu \quad E^0 = +0.34$ volt

The other half-reaction must give up electrons and be positive or no more negative than -0.34 volt. Reaction (a) is the only one that satisfies the conditions.

3. The answer is reactions (a) and (d);

$$Fe^{2+} \longrightarrow Fe^{3+} + e^- \quad E^0 = -0.77 \text{ volt}$$

4. The balanced equation is

$$5HSO_3^- + 2IO_3^- \longrightarrow I_2(g) + 5SO_4^{2-} + 3H^+ + H_2O(l)$$

a. $1.00 \text{ mol KHSO}_3 \times \dfrac{2 \text{ mol NaIO}_3}{5 \text{ mol KHSO}_3}$

$\times \dfrac{198 \text{ g NaIO}_3}{1 \text{ mol NaIO}_3} = 79.2 \text{ g NaIO}_3$

b. $1.00 \text{ mol KHSO}_3 \times \dfrac{1 \text{ mol I}_2}{5 \text{ mol KHSO}_3} \times$

$\dfrac{254 \text{ g I}_2}{1 \text{ mol I}_2} = 50.8 \text{ g I}_2$

Synthesis, page 612

1.

$$H_2(g) \longrightarrow 2H^+ + 2e^- \quad E^0 = 0.00\,V$$
$$\underline{Pd^{2+} + 2e^- \longrightarrow Pd(s) \quad\quad E^0 = ?}$$
$$Pd^{2+} + H_2(g) \longrightarrow 2H^+ + Pd(s) \quad E^0_{cell} = +0.99\,V$$
$$Pd(s) \longrightarrow Pd^{2+} + 2e^- \quad E^0 = -0.99\,V$$

2. The reaction is

$$H_2SO_4 + Na_2CO_3 \longrightarrow CO_2 + H_2O + Na_2SO_4(aq)$$

a. $700.0 \text{ g H}_2SO_4 \times \dfrac{1 \text{ mol H}_2SO_4}{98.1 \text{ g H}_2SO_4}$

$\times \dfrac{1 \text{ mol Na}_2CO_3}{1 \text{ mol H}_2SO_4}$

$\times \dfrac{106 \text{ g Na}_2CO_3}{1 \text{ mol Na}_2CO_3} = 756 \text{ g Na}_2CO_3$

b. vol. $Na_2CO_3 = \left(\dfrac{7.14 \text{ mol}}{2.0 \text{ mol/L}}\right) = 3.57 \text{ L}$

3. The balanced reaction is

$$Cu + 4H^+ + 2NO_3^- \longrightarrow Cu^{2+} + 2NO_2 + 2H_2O$$

$0.100 \text{ kg Cu} \times \dfrac{1 \text{ mol Cu}}{0.0635 \text{ kg Cu}} \times \dfrac{4 \text{ mol HNO}_3}{1 \text{ mol Cu}}$

$\times \dfrac{1 \text{ L HNO}_3}{15.9 \text{ mol HNO}_3} = 0.40 \text{ L HNO}_3$

CHAPTER 22 Analytical Problem Solving

CHAPTER ORGANIZER

USE WITH SECTION	DEMONSTRATION	WORKSHEET	LAB EXPERIMENT	SOFTWARE	TEACHER'S RESOURCE BINDER
22-1		22-A	22-A		
22-2		22-B		Disk 6	
22-4			22-B		
22-5		22-C	22-C	Disk 6	
Chapter Test					

■ Objectives

When the students have completed this chapter, they will be able to

1. Examine data from a series of experiments and use it to identify the precipitate in an ionic reaction.

2. Use experimental data to predict the results of a reaction between two ionic solutions.

3. Relate the macroscopic observations in an ionic reaction to the microscopic events that are occurring and write a balanced equation that represents the reaction.

4. Identify an unknown by reasoning from experimental data.

5. Use a variety of techniques, including solubility and flame tests, to identify the ions present.

6. Apply concepts from the acid-base, electrochemistry, stoichiometry, and equilibrium chapters to determine the identity of an unknown solution.

■ Section Notes

22-1 Preparing all of the solutions required for this chapter on unknowns is a very time-consuming task. Take one class period to review solution preparation. Review the calculations necessary when preparing a solution from a solid reagent that contains water of hydration. Then assign each student a solution to prepare. Let the student examine the reagent bottle to determine if the compound is a hydrate or anhydrous. Have the student calculate the amount of solute required to prepare a specified volume of the solution for the following day. After checking and approving their calculations, have the students prepare the solution. Remind them to use distilled water.

22-2 More solutions are encountered to obtain more data like that in Table 23-2. Soon there is so much data that patterns must be noted to help organize the data. These regularities can be used to make predictions that extend beyond the range of the original data. These predictions can be confirmed by doing more experiments. If subsequent investigations (preferably in the lab but reference books also could be used) show a generalization is not correct, the pattern noted must be modified. The discovery can be included as an exception.

Review and Practice
page 615

1. $BaCl_2(aq) + K_2CO_3(aq) \longrightarrow BaCO_3(s) + 2KCl(aq)$
If KCl is a precipitate, it should be the same color each time. In equation 1, the precipitate is green; in equation 3, it is white. We can assume that the green precipitate is $NiCO_3$ in equation 1, and the white precipitate in equation 3 is $BaCO_3$.

2.

INSOLUBLE	SOLUBLE		?
$BaSO_4$	$NiCl_2$	$BaCl_2$	$CoCO_3$
$BaCO_3$	$CoCl_2$	K_2CO_3	K_2SO_4
$NiCO_3$	KCl		

3. Chlorides are usually soluble. Carbonates are often insoluble.

4. STEP 1. Add K_2CO_3.

green precipitate	$x = NiCl_2$
white precipitate	$x = BaCl_2$
red precipitate	$x = CoSO_4$
no reaction	$x = K_2CO_3$

STEP 2. Add $CoSO_4$.

no reaction	$x = NH_4I$
red precipitate	$x = K_2CO_3$

page 616

1 and 2.

$NiCl_2(aq) + K_2CO_3(aq) \longrightarrow$
$\qquad NiCO_3(s) + 2KCl(aq)$

$NiCl_2(aq) + 2NaOH(aq) \longrightarrow$
$\qquad Ni(OH)_2(s) + 2NaCl(aq)$

$NiCl_2(aq) + Pb(NO_3)_2 \longrightarrow$
$\qquad PbCl_2(s) + Ni(NO_3)_2(aq)$

$2NH_4I(aq) + Pb(NO_3)_2(aq) \longrightarrow$
$\qquad PbI_2(s) + 2NH_4NO_3(aq)$

$BaCl_2(aq) + CoSO_4(aq) \longrightarrow$
$\qquad BaSO_4(s) + CoCl_2(aq)$

$BaCl_2(aq) + K_2CO_3(aq) \longrightarrow$
$\qquad BaCO_3(s) + 2KCl(aq)$

$BaCl_2(aq) + Pb(NO_3)_2(aq) \longrightarrow$
$\qquad PbCl_2(s) + Ba(NO_3)_2(aq)$

$CoSO_4(aq) + BaCl_2(aq) \longrightarrow$
$\qquad BaSO_4(s) + CoCl_2(aq)$

$CoSO_4(aq) + K_2CO_3(aq) \longrightarrow$
$\qquad CoCO_3(s) + K_2SO_4(aq)$

$CoSO_4(aq) + 2NaOH(aq) \longrightarrow$
$\qquad Co(OH)_2(s) + Na_2SO_4(aq)$

$CoSO_4(aq) + Pb(NO_3)_2(aq) \longrightarrow$
$\qquad PbSO_4(s) + Co(NO_3)_2(aq)$

$K_2CO_3(aq) + NiCl_2(aq) \longrightarrow$
$\qquad NiCO_3(s) + 2KCl(aq)$

$K_2CO_3(aq) + BaCl_2(aq) \longrightarrow$
$\qquad BaCO_3(s) + 2KCl(aq)$

$K_2CO_3(aq) + CoSO_4(aq) \longrightarrow$
$\qquad CoCO_3(s) + K_2SO_4(aq)$

$K_2CO_3(aq) + Pb(NO_3)_2(aq) \longrightarrow$
$\qquad PbCO_3(s) + 2KNO_3(aq)$

$2NaOH(aq) + NiCl_2(aq) \longrightarrow$
$\qquad Ni(OH)_2(s) + 2NaCl(aq)$

$NaOH(aq) + CoSO_4(aq) \longrightarrow$
$\qquad Co(OH)_2(s) + Na_2SO_4(aq)$

$2HCl(aq) + Pb(NO_3)_2(aq) \longrightarrow$
$\qquad PbCl_2(s) + 2HNO_3(aq)$

$CuSO_4(aq) + BaCl_2(aq) \longrightarrow$
$\qquad BaSO_4(s) + CuCl_2(aq)$

$CuSO_4(aq) + K_2CO_3(aq) \longrightarrow$
$\qquad CuCO_3(s) + K_2SO_4(aq)$

$CuSO_4(aq) + 2NaOH(aq) \longrightarrow$
$\qquad Cu(OH)_2(s) + Na_2SO_4(aq)$

$CuSO_4(aq) + Pb(NO_3)_2(aq) \longrightarrow$
$\qquad PbSO_4(s) + Cu(NO_3)_2(aq)$

3.

INSOLUBLE		SOLUBLE	
$NiCO_3$	$Pb(OH)_2$	KCl	$Ba(NO_3)_2$
$BaCO_3$	$Cu(OH)_2$	$NaCl$	$Co(NO_3)_2$
$CoCO_3$	$PbCl_2$	$CoCl_2$	KNO_3
$PbCO_3$	PbI_2	$CuCl_2$	HNO_3
$CuCO_3$	$BaSO_4$	$NaNO_3$	K_2SO_4
$Ni(OH)_2$	$PbSO_4$	$Ni(NO_3)_2$	$Cu(NO_3)_2$
$Co(OH)_2$		NH_4NO_3	Na_2SO_4

4. Most carbonates are insoluble; chlorides are soluble except for $PbCl_2$; most hydroxides are insoluble; all nitrates are soluble; sulfates are soluble except $PbSO_4$ and $BaSO_4$.

5. STEP 1. Add $HCl(aq)$.
bubbles given off $\quad x = K_2CO_3$

133T

white precipitate $x = Pb(NO_3)_2(aq)$
hot reaction $x = NaOH(aq)$

STEP 2. Add $NaOH(aq)$.
hot reaction $x = HCl(aq)$
ammonia odor $x = NH_4I(aq)$
green precipitate $x = NiCl_2(aq)$
red precipitate $x = CoSO_4(aq)$
blue precipitate $x = CuSO_4(aq)$
white precipitate $x = Pb(NO_3)_2(aq)$

STEP 3. Add $CoSO_4(aq)$.
white precipitate $x = Pb(NO_3)_2(aq)$ or
$\qquad\qquad\qquad\qquad\quad BaCl_2(aq)$

STEP 4. Add $HCl(aq)$.
white precipitate $x = PbCl_2(aq)$
no reaction $x = BaCl_2(aq)$
If no reaction with
$HCl(aq)$, $NaOH(aq)$, or
$CoSO_4(aq)$ $x = KNO_3(aq)$

6. $CO_2(g)$

Review, page 622

1. **a.** $H_2SO_4(aq)$ **d.** $Sr(NO_3)_2(aq)$
   b. $KOH(aq)$ **e.** $(NH_4)_2SO_4(aq)$
   c. $Na_2CO_3(aq)$ **f.** $K_2SO_4(aq)$

2. **a.** H^+ (or H_3O^+)
   b. K^+
   c. Na^+

3. **d.** NO_3^{1-}
   e. SO_4^{2-}
   f. SO_4^{2-}

4. potassium

5. strontium

6. Ni^{2+}

7. $Hg_2(NO_3)_2(aq) + Li_2CO_3(aq) \longrightarrow$
   $\qquad HgCO_3(s) + 2LiNO_3(aq)$
   $Hg_2(NO_3)_2(aq) + (NH_4)_2CO_3(aq) \longrightarrow$
   $\qquad HgCO_3(s) + 2NH_4NO_3(aq)$

8. c, e

Interpret and Apply, page 622

1. $K^+(aq)$, $Cl^-(aq)$, $Mg^{2+}(aq)$, $SO_4^{2-}(aq)$

2. $Na^+(aq)$, $CO_3^{2-}(aq)$, $NH_4^{1+}(aq)$

3. $Na_2CO_3(aq)$ will form a white precipitate with $Sr(NO_3)_2(aq)$ which cannot be easily distinguished from $SrSO_4(s)$. To prevent this add an excess of $HNO_3(aq)$ to remove $CO_3^{2-}(aq)$.

4. $NaCl(aq)$

5. $H_2SO_4(aq)$
   no reaction $x = NaCl(aq)$
   hot reaction $x = NaOH(aq)$
   bubbles $x = Na_2CO_3(aq)$

6. $Ba(NO_3)_2(aq)$

7. $AgNO_3(aq)$
   no reaction $x = Ba(NO_3)_2(aq)$
   white precipitate $x = BaCl_2(aq)$
   black precipitate $x = Na_2S(aq)$

8. $Co(NO_3)_2(aq)$

9. $Na_2S(aq)$
   no reaction $x = ZnSO_4(aq)$
   black precipitate $x = Co(NO_3)_2(aq)$
   rotten egg odor $x = H_2SO_4(aq)$

10. $NaI(aq)$

11. $HCl(aq)$
    no reaction $x = NaI(aq)$
    hot reaction $x = NaOH(aq)$
    odorless bubbles $x = Na_2CO_3(aq)$

12. $Pb(NO_3)_2(aq)$

13. $HCl(aq)$
    no reaction $x = BaCl_2(aq)$
    odorless bubbles $x = Na_2CO_3(aq)$
    white precipitate $x = Pb(NO_3)_2(aq)$
    hot reaction $x = KOH(aq)$

Challenge, page 623

1. $2AgNO_3(aq) + CaCl_2(aq) \longrightarrow$
   $\qquad 2AgCl(s) + Ca(NO_3)_2(aq)$

$$20.0 \text{ mL AgNO}_3 \times \frac{0.50 \text{ mol AgNO}_3}{1000 \text{ mL AgNO}_3}$$
$$= 0.010 \text{ mol AgNO}_3$$

$$20.0 \text{ mL CaCl}_2 \times \frac{0.50 \text{ mol CaCl}_2}{1000 \text{ mL CaCl}_2}$$
$$= 0.010 \text{ mol CaCl}_2$$

Using the mole ratio from the equation, we can calculate the number of moles of $CaCl_2$ required.

$$0.010 \text{ mol AgNO}_3 \times \frac{1 \text{ mol CaCl}_2}{2 \text{ mol AgNO}_3}$$
$$= 0.0050 \text{ mol CaCl}_2 \text{ required}$$

Since 0.010 mol $CaCl_2$ is present, $CaCl_2$ is in excess and $AgNO_3$ is the limiting reactant. Use the limiting reactant to calculate the mass of the product.

$$0.010 \text{ mol AgNO}_3 \times \frac{2 \text{ mol AgCl}}{2 \text{ mol AgNO}_3} \times \frac{143.2 \text{ g AgCl}}{1 \text{ mol AgCl}}$$
$$= 1.4 \text{ g AgCl produced.}$$

The filtrate will contain the excess $CaCl_2$ and the $Ca(NO_3)_2$ produced.

Excess $CaCl_2$
$\quad$ = moles $CaCl_2$ present − moles $CaCl_2$ required
$\quad$ = 0.010 − 0.005 = 0.005 mol excess $CaCl_2$.

$$0.005 \text{ mol CaCl}_2 \times \frac{110.9 \text{ g CaCl}_2}{1 \text{ mol CaCl}_2} = 0.6 \text{ g CaCl}_2$$

Use the limiting reactant, $AgNO_3$, to calculate the products.

$$0.010 \text{ mol AgNO}_3 \times \frac{1 \text{ mol Ca(NO}_3)_2}{2 \text{ mol AgNO}_3}$$

$$\times \frac{164 \text{ g Ca(NO}_3)_2}{1 \text{ mol Ca(NO}_3)_2} = 0.8 \text{ g Ca(NO}_3)_2$$

The mass of the solid which remains in the beaker after the filtrate is evaporated is $0.6 \text{ g} + 0.8 \text{ g} = 1.4 \text{ g}$.

2. $\dfrac{12.4 \text{ g NaOH}}{500 \text{ mL}} = \dfrac{24.8 \text{ g NaOH}}{1000 \text{ mL}}$

$\dfrac{24.8 \text{ g NaOH}}{1000 \text{ mL}} \times \dfrac{1 \text{ mol NaOH}}{40.0 \text{ g NaOH}} \times \dfrac{1000 \text{ mL}}{1 \text{ L}}$

$= 0.62 \text{ mol/L}$

For water at 25°C

$[\text{H}^+][\text{OH}^-] = 1 \times 10^{-14}$

$[\text{OH}^-] = 0.62M$

$[\text{H}^+](0.62) = 1 \times 10^{-14}$

$[\text{H}^+] = 1.61 \times 10^{-13}$

pH = 13.8

3. $\text{NaOH(aq)} + \text{HCl(aq)} \longrightarrow \text{NaCl(aq)} + \text{H}_2\text{O(l)}$

40.0 mL of $1.00M$ HCl will neutralize 40.0 mL of $1.00m$ NaOH.

20.0 mL NaOH remain in 100 mL of solution.

$\dfrac{20.0 \text{ mL NaOH}}{1} \times \dfrac{1.00 \text{ mol NaOH}}{1000 \text{ mL}}$

$= 0.0200 \text{ mol NaOH}$

$\dfrac{0.0200 \text{ mol NaOH}}{100 \text{ mL}} = \dfrac{0.200 \text{ mol NaOH}}{1000 \text{ mL}}$

$= 0.200M \text{ NaOH}$

$[\text{H}^+][\text{OH}^-] = 1.0 \times 10^{-14}$

$[\text{H}^+](0.200) = 1.0 \times 10^{-14}$

$[\text{H}^+] = 5.00 \times 10^{-14}$

pH = 13.3

Synthesis, page 623

1. $\text{H}_2\text{SO}_4\text{(aq)} + 2\text{KOH(aq)} \longrightarrow \text{K}_2\text{SO}_4\text{(aq)} + 2\text{H}_2\text{O(l)}$

$10.0 \text{ mL KOH} \times \dfrac{0.50 \text{ mol KOH}}{1000 \text{ mL KOH}} \times \dfrac{1 \text{ mol H}_2\text{SO}_4}{2 \text{ mol KOH}}$

$\times \dfrac{1000 \text{ mL H}_2\text{SO}_4}{0.50 \text{ mol H}_2\text{SO}_4} = 5.0 \text{ mL H}_2\text{SO}_4$

2. $2\text{HNO}_3\text{(aq)} + \text{Ba(OH)}_2\text{(aq)} \longrightarrow \text{Ba(NO}_3)_2 + 2\text{H}_2\text{O(l)}$

$20.0 \text{ mL Ba(OH)}_2 \times \dfrac{0.05 \text{ mol Ba(OH)}_2}{1000 \text{ mL Ba(OH)}_2}$

$\times \dfrac{2 \text{ mol HNO}_3}{1 \text{ mol Ba(OH)}_2} \times \dfrac{1000 \text{ mL HNO}_3}{0.25 \text{ mol HNO}_3}$

$= 8 \text{ mL HNO}_3$

3. $\text{Na}_2\text{CO}_3\text{(aq)} + 2\text{HCl(aq)} \longrightarrow$
$\qquad 2\text{NaCl(aq)} + \text{H}_2\text{O(l)} + \text{CO}_2\text{(g)}$

$40.0 \text{ mL Na}_2\text{CO}_3 \times \dfrac{0.50 \text{ mol Na}_2\text{CO}_3}{1000 \text{ mL Na}_2\text{CO}_3}$

$\times \dfrac{1 \text{ mol CO}_2}{1 \text{ mol Na}_2\text{CO}_3} = 0.020 \text{ mol CO}_2\text{(g)}$

$0.20 \text{ mol CO}_2 \times \dfrac{22.4 \text{ L CO}_2}{1 \text{ mol CO}_2} = 0.45 \text{ L CO}_2 \text{ at STP}$

$PV = nRT$

$V = \dfrac{nRT}{P} = \left(\dfrac{0.020 \text{ mol}}{95 \text{ kPa}}\right)\left(\dfrac{8.31 \text{ L} \cdot \text{kPa}}{\text{mol} \cdot \text{K}}\right)298 \text{ K}$

$25°\text{C} = 298 \text{ K}$

$V = 0.52 \text{ L}$

4. $\text{CuSO}_4 \cdot 5\text{H}_2\text{O}$

$63.5 + 32.1 + 64.0 + 5(18.0) = 249.6$

$2.00 \text{ L CuSO}_4 \times \dfrac{0.25 \text{ mol CuSO}_4}{\text{L}}$

$\times \dfrac{249.6 \text{ g CuSO}_4 \cdot 5\text{H}_2\text{O}}{1 \text{ mol CuSO}_4}$

$= 124.8 \text{ g CuSO}_4 \cdot 5\text{H}_2\text{O}$

To 124.8 g $\text{CuSO}_4 \cdot 5\text{H}_2\text{O}$, add enough distilled water so the final volume is 2.00 L when the solution process is complete.

5. $\dfrac{4.9 \text{ g H}_2\text{SO}_4}{\text{L}} \times \dfrac{1 \text{ mol H}_2\text{SO}_4}{98 \text{ g H}_2\text{SO}_4} = 0.05M \text{ H}_2\text{SO}_4$

$\text{H}_2\text{SO}_4 \longrightarrow 2\text{H}^+\text{(aq)} + \text{SO}_4^{2-}\text{(aq)}$

$[\text{H}^+] = 2(0.05) = 0.10$

pH = 1.0

6. $\text{FeSO}_4 \cdot 7\text{H}_2\text{O}$

$55.8 + 32.1 + 64.0 + 7(18.0) = 277.9$

$1.00 \text{ L FeSO}_4 \times \dfrac{0.50 \text{ mol FeSO}_4}{1 \text{ L}}$

$\times \dfrac{277.9 \text{ g FeSO}_4 \cdot 7\text{H}_2\text{O}}{1 \text{ mol FeSO}_4}$

$= 139.0 \text{ g FeSO}_4 \cdot 7\text{H}_2\text{O}$

To 139.0 g $\text{FeSO}_4 \cdot 7\text{H}_2\text{O}$, add enough distilled water so the final volume is 1.00 L when the solution process is complete.

7. $\text{Pb(NO}_3)_2 + 2\text{KI(aq)} \longrightarrow \text{PbI}_2\text{(s)} + 2\text{KNO}_3\text{(aq)}$

$20.0 \text{ mL Pb(NO}_3)_2 \times \dfrac{0.50 \text{ mol Pb(NO}_3)_2}{1000 \text{ mL Pb(NO}_3)_2}$

$= 0.010 \text{ mol Pb(NO}_3)_2$

$20.0 \text{ mL KI} \times \dfrac{1.00 \text{ mol KI}}{1000 \text{ mL KI}} = 0.020 \text{ mol KI}$

The mole ratio is the correct stoichiometric ratio so there is no excess. Therefore, either reactant can be used to calculate the products.

$20.0 \text{ mL Pb(NO}_3)_2 \times \dfrac{0.50 \text{ mol Pb(NO}_3)_2}{1000 \text{ mL Pb(NO}_3)_2}$

$\times \dfrac{1 \text{ mol PbI}_2}{1 \text{ mol Pb(NO}_3)_2} \times \dfrac{461.2 \text{ g PbI}_2}{1 \text{ mol PbI}_2}$

$= 4.61 \text{ g PbI}_2$

8. $\dfrac{10.8 \text{ g NaNO}_3}{175 \text{ mL}} = \dfrac{61.7 \text{ g NaNO}_3}{1000 \text{ mL}}$

$\dfrac{61.7 \text{ g NaNO}_3}{1 \text{ L}} \times \dfrac{1 \text{ mol NaNO}_3}{85 \text{ g NaNO}_3} = 0.726M$

9. $\text{MgSO} \cdot 7\text{H}_2\text{O}$

$24.3 + 32.1 + 64.0 + 7(18.0) = 246.4$

$1.50 \text{ L MgSO}_4 \times \dfrac{0.50 \text{ mol MgSO}_4}{1 \text{ L}}$

$\times \dfrac{246.4 \text{ g MgSO}_4 \cdot 7\text{H}_2\text{O}}{1 \text{ mol MgSO}_4} = 185 \text{ g MgSO}_4 \cdot 7\text{H}_2\text{O}$

To 185 g $MgSO_4 \cdot 7H_2O$ add enough distilled water so the final volume is 1.50 L when the solution process is complete.

10. $2AgNO_3(aq) + Na_2SO_4(aq) \longrightarrow$
$\qquad Ag_2SO_4(s) + 2NaNO_3(aq)$
$K_{sp} = [Ag^+]^2[SO_4^{2-}] = 1.18 \times 10^{-5}$

$20.0 \text{ mL AgNO}_3 \times \dfrac{0.05 \text{ mol AgNO}_3}{1000 \text{ mL AgNO}_3}$

$= \dfrac{0.001 \text{ mol AgNO}_3}{50 \text{ mL}}$

$30.0 \text{ mL Na}_2SO_4 \times \dfrac{0.020 \text{ mol Na}_2SO_4}{1000 \text{ mL}}$

$= \dfrac{0.0006 \text{ mol Na}_2SO_4}{50 \text{ mL}}$

$\dfrac{0.001 \text{ mol AgNO}_3}{50 \text{ mL}} \times \dfrac{1000 \text{ mL}}{L} = 0.020M \text{ AgNO}_3$

$\dfrac{0.0006 \text{ mol Na}_2SO_4}{50 \text{ mL}} \times \dfrac{1000 \text{ mL}}{L} = 0.012M \text{ Na}_2SO_4$

$[Ag^+] = 0.020$
$[SO_4^{2-}] = 0.12$
$[Ag^+]^2[SO_4^{2-}] = 1.18 \times 10^{-5}$
$(0.020)^2(0.12) = 4.8 \times 10^{-5}$

Since 4.8×10^{-5} is greater than 1.18×10^{-5}, the solubility product is exceeded and a precipitate will form.

CHAPTER 23 Organic Chemistry

CHAPTER ORGANIZER

USE WITH SECTION	DEMONSTRATION	WORKSHEET	LAB EXPERIMENT	SOFTWARE	TEACHER'S RESOURCE BINDER
23-6			23-A		
23-10					
23-12		23-A, 23-B, 23-C	23-B		
23-15		23-D			
23-16	23-A	23-E			
Chapter Test					

■ Objectives

When students have completed this chapter, they will be able to

1. Identify some general characteristics of carbon compounds.
2. Describe how reserves of coal and petroleum were formed.
3. Explain why there is such a diversity and magnitude of organic compounds.
4. Identify representative members and describe the properties of the alkane, alkene, and alkyne series.
5. Name or identify organic compounds according to IUPAC rules.
6. Identify and differentiate structural isomers.
7. Describe what is meant by the term delocalized electrons in the context of the benzene ring.
8. Explain how petroleum is separated into its component substances.

9. Identify and list general characteristics of the following functional groups: alcohols, ethers, aldehydes, ketones, organic acids, esters, amines and amides.
10. Describe and differentiate between substitution and addition reactions.
11. Explain how oxidation and reduction may alter the structure of organic compounds.
12. Describe the chemical processes of addition and condensation polymerization.

■ Section Notes

23-2 Many planktonic organisms produce and store tiny droplets of oil. The low density of the oil allows the organism to float effortlessly in the upper sunlit surface waters. Many chemists believe that these oils comprise the raw material that is eventually transformed into petroleum deposits.

23-11 The oil found in shale rocks is locked in very large organic molecules called kerogen. Since kerogen is too viscous to flow out of the rocks, shale must be mined and crushed. The smaller pieces are then heated to 500°C, at which point kerogen decomposes, releasing its store of constituent oils. The process has not yet been made cost effective.

23-13, 23-14 The reactivity of the hydrocarbons will vary considerably depending upon the reagent and the type of reaction that is involved. All hydrocarbons readily undergo reaction with oxygen, especially when suitably initiated (combustion). The hydrocarbons react with a sufficient amount of oxygen to produce carbon dioxide and water.

Another example of a substitution reaction involves bromine. The reaction requires a catalyst or ultraviolet light:

$$H_3C—CH_3 + Br_2 \xrightarrow{\text{u.v. light}} H_3C—CH_2—Br + HBr$$

A comparison of addition reactions in alkenes and alkynes reveals that certain reactions occur faster among alkenes, whereas others are faster among alkynes. For example,

$$H_2C=CH_2 + Br \longrightarrow Br—CH_2—CH_2—Br$$

This addition reaction proceeds very readily and is classified as an electrophilic addition reaction. The reaction proceeds through a carbo-cation intermediate, which facilitates the reaction.

$$H_2C=CH_2 + Br_2 \longrightarrow Br—CH_2--C{\overset{H}{\underset{+\ H}{}}} + Br^-$$

stabilized intermediate
carbo-cation

Compare this reaction with an addition to an alkyne:

$$H—C\equiv C—H + Br_2 \longrightarrow {\overset{H}{\underset{Br}{}}}C=C{\overset{Br}{\underset{H}{}}}$$

The bromine addition reaction to an alkyne is slower than the addition to an alkene. With alkynes the intermediate formed is a "vinyl carbo-cation" which is not well stabilized.

$$H—C\equiv C—H + Br_2 \longrightarrow {\overset{H}{\underset{Br}{}}}C=C{\overset{H}{\underset{+}{}}} + Br^-$$

unstable intermediate
vinylic cation

Since the intermediate is difficult to form, we therefore interpret this as the reason for the slower rate of bromination of an alkyne compared to an alkene.

Other reactions, such as nucleophilic addition reactions, show a much faster rate of reaction with alkynes than with alkenes (e.g., addition of HCN to unsaturated hydrocarbons). Other examples of reactions are ozonolysis, epoxidation, reaction with hydrogen halides, addition of water with a catalyst, etc.

Review and Practice
page 628

1. It was the first time an organic compound was synthesized in the laboratory.
2. They are derived from decayed matter of both plants and animals that lived millions of years ago.
3. Bacteria decomposed the accumulated matter into new organic compounds, including methane.
4. high temperatures and pressures
5. a carbon compound containing only carbon and hydrogen atoms
6. Saturated compounds contain only single bonds. Unsaturated compounds contain at least one double or triple carbon-carbon bond.
7.

8. They tend to be either nonelectrolytes or weak electrolytes and to have low melting points.

page 633

1.

butane

2-methylpropane

2.

3. a = f; b = h; c = g; d = e
4. a = 2-methylpropane;
   b = 2,2-dimethylbutane;
   c = 3-ethyl-2-methylpentane
5. 32 hydrogen atoms; 102 hydrogen atoms

page 639

1. Major groups of hydrocarbons have different ranges of condensation temperatures.

2. C_nH_{2n}

3.

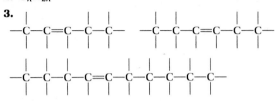

(The structures above are the *cis* isomers. Students may draw the *trans* isomers also.)

4. Cyclohexane has 12 atoms, cyclopentane has 10.

5.

Some possible answers include

Some isomers that the students draw may exist only on paper.

6. The bond distances and bond energies suggest that the carbon-carbon bonds in benzene are intermediate between the single and double bonds, whereas alkanes have all single bonds and alkenes contain bonds that are either single or double.

page 645

1. its number of atoms, structural arrangement, and number and type of functional groups

2. atoms, groups of atoms, or organization of bonds that determine a specific property

3. aldehyde; organic or carboxylic acid (carboxyl group); ketone (carbonyl group); ether

4. The central position of the oxygen atom leads to less of an opportunity for significant polarity. Intermolecular attractions are correspondingly low.

5. **a.** hydroxyl   **d.** oxygen
    b. amine   **e.** carbonyl
    c. carbonyl

6.

page 651

1. In substitution, an atom or group of atoms is replaced by another atom or group. In addition, new constituents are added across a double or triple bond.

2. $C_2H_4 + HCl \longrightarrow CH_3CH_2Cl$ chloroethane

3. a process by which extended chain structures are formed from smaller units of molecules

4. an amino acid

5. condensation

6. **a.** substitution
    b. addition
    c. esterification or condensation

Review, page 653

1. by the destructive distillation of coal

2. from the discarded belief that all the sources of carbon compounds originally came solely from living organisms

3. Carbon atoms can link together to form chains of varying length. The atoms of these compounds may be arranged in several different ways.

4. Due to the reactive nature of the double bond, alkenes generally are more reactive.

5. a compound whose molecules contain only single carbon-carbon bonds

6. $C_nH_{2n + 2}$

7. carbon dioxide, methane, water, nitrogen and hydrogen

8. They differ by the addition of the same structural unit.

9. carbon, hydrogen, oxygen, nitrogen, sulfur, phosphorus, and halogens

10. coal, petroleum, and natural gas

11. by the number of the carbon atom to which the branch is attached

12. 2 carbons in each of the ethyl groups, 1 carbon in the methyl group, and 3 carbons in the propyl group

13. a compound with the same chemical formula as another compound but a different arrangement of atoms

14. Ethyne; it is often used as a fuel in oxyacetylene torches.

15. a hexagon—⬡

16. Alkanes have only single carbon-carbon bonds, and alkenes contain carbon-carbon bonds that are either single or double. The bonds in benzene have characteristics somewhere between single and double bonds.

17. The polar nature of the hydroxyl group allows alcohols of low molecular masses to be soluble in water.

18. the carboxyl group, COOH

19. In aldehydes the carbonyl group is attached to a terminal carbon. In ketones the carbonyl group is attached to a carbon atom that is attached to two other carbon atoms.

20. fragrant foods such as fruit

21. protein molecules and nylon

22. the length and/or shape of the carbon chain and the presence of functional groups

23. large molecules composed of a repeating sequence of smaller molecular units (monomers)

24. aldehydes and ketones

Interpret and Apply, page 654

1. Answers will vary considerably. Two possibilities include

2. There is only one possible site for the double bond in the molecule.

3. by the substitution of chlorine atoms for each of methane's hydrogen atoms

4. Instead of liberating a wide array of carbon compounds, many of the organic compounds will be oxidized into undesirable substances.

5. Yes; an example is cyclohexene.

6. All of the carbon-carbon bonds are saturated.

7. Three. If there were less than three carbons, then the carbonyl functional group would have to be located on a terminal carbon, thus producing an aldehyde.

8. No. In ethers, the oxygen atom only can be bonded to two R-groups.

9. Oxidation refers to the addition of oxygen. Reduction refers to the removal of oxygen (or the addition of hydrogen).

10. an organic acid and an alcohol

11. because they are not associated with any one carbon atom

12. Most hydrocarbon molecules are nonpolar. Polar solvents will not dissolve nonpolar compounds.

13. No. Although they both contain five carbon atoms, they differ in the number of hydrogen atoms present.

14. Due to the stability of the aromatic ring, it is easier to substitute for hydrogen than to break any of the double bonds for an addition reaction.

Problems, page 654

1.

2.

3.

4.

5.

c. C—C—$\overset{\overset{\displaystyle NH_2}{|}}{C}$—C

d. $-\overset{|}{\underset{|}{C}}-\overset{|}{\underset{|}{C}}-\overset{\overset{\displaystyle O}{\|}}{C}-H$

6. H—$\overset{\overset{\displaystyle O}{\|}}{C}$—OH $-\overset{|}{\underset{|}{C}}-\overset{\overset{\displaystyle O}{\|}}{C}$—OH $-\overset{|}{\underset{|}{C}}-\overset{|}{\underset{|}{C}}-\overset{\overset{\displaystyle O}{\|}}{C}$—OH

7. a. $-\overset{|}{\underset{|}{C}}-\overset{\overset{\overset{\displaystyle -C-}{|}}{|}}{\underset{|}{C}}-\overset{|}{\underset{|}{C}}-\overset{|}{\underset{|}{C}}-$

b. $-\overset{|}{\underset{\overset{|}{Cl}}{C}}-\overset{|}{\underset{|}{C}}-\overset{\overset{\displaystyle Cl}{|}}{\underset{|}{C}}-\overset{|}{\underset{|}{C}}-\overset{|}{\underset{|}{C}}-\overset{|}{\underset{|}{C}}-$

c.

d. (ring structure)

8. a. $-\overset{|}{\underset{|}{C}}-\overset{\overset{\displaystyle OH}{|}}{\underset{|}{C}}-\overset{|}{\underset{|}{C}}-$

b. (chain structure with OH)

c. $-C-O-\overset{|}{\underset{|}{C}}-\overset{|}{\underset{|}{C}}-$

d. $-\overset{|}{\underset{|}{C}}-\overset{|}{\underset{|}{C}}-\overset{|}{\underset{|}{C}}-O-\overset{\overset{\displaystyle O}{\|}}{C}-\overset{|}{\underset{|}{C}}-$

9. a. 2,2,3-tribromo-3-methylhexane
b. 4-ethyl-2-methylheptane
c. 2,3-dichloropentane
d. 3,3,5-trimethylheptane

10. a. 2-methyl-2-butane   **e.** cyclohexane
b. propene   **f.** benzene
c. 2-pentyne   **g.** cyclohexene
d. ethyne

11. a. ethanoic   **e.** methoxybutane
   (acetic) acid   **f.** aminoethane
b. 2-butanone   (ethylamine)
c. methanal   **g.** 3-pentone
d. 2-propanol   **h.** 1-butanol

140T

i. 1,3-diaminopropane   **j.** butanoic acid
12. a. condensation polymerization
   b. addition polymerization
   c. substitution
   d. addition
   e. esterification
   f. oxidation

Challenge, page 656

1. one; two; two

2. (six chlorinated structures)

3. a. (cyclopentene ring)

b. (chain structure)

c. Br (benzene ring with Br)

d. C_2H_5, CH_3, C_3H_7 (substituted benzene)

e. OH, CH_3 (substituted benzene)

f. C_2H_5, CH_3 (cyclohexene)

g. $-\overset{|}{\underset{|}{C}}-C\equiv C-$

4. $-\overset{|}{\underset{|}{C}}-\overset{|}{\underset{|}{C}}-\overset{\overset{\displaystyle O}{\|}}{\underset{\displaystyle OH}{C}}$ + HO$-\overset{|}{\underset{|}{C}}-\overset{|}{\underset{|}{C}}- \longrightarrow$

$-\overset{|}{\underset{|}{C}}-\overset{|}{\underset{|}{C}}-\overset{\overset{\displaystyle O}{\|}}{\underset{\displaystyle O-\overset{|}{\underset{|}{C}}-\overset{|}{\underset{|}{C}}-}{C}}$ + H_2O

5. $\underset{H}{\overset{H}{>}}C=\overset{\overset{\displaystyle CH_3}{|}}{C}-\overset{\overset{\displaystyle O}{\|}}{C}-O-\overset{\overset{\displaystyle H}{|}}{\underset{\displaystyle H}{C}}-\overset{\overset{\displaystyle H}{|}}{\underset{\displaystyle H}{C}}-\overset{\overset{\displaystyle CH_3}{|}}{\underset{\displaystyle H}{C}}-\overset{\overset{\displaystyle O}{\|}}{C}-O-CH_3$

6. a. OH (biphenyl with OH)

b. $-\overset{|}{\underset{|}{C}}-O-\overset{\overset{\displaystyle O}{\|}}{C}-$

c. C_2H_5 C_2H_5
N
H

d. C_2H_5 CH_3
N
C_3H_7

e. NH_2
—C—C—C—C—
NH_2

f. Br Br
—C—C—
Cl Cl

g. Br—C—C=C
Br

7. a. methyl acetate   **c.** 1,4-cyclohexadiene
b. dimethylamine

8.

H H   H   H
H—C—C—OH   H—C—O—C—H
H H   H   H

The ether has a relatively low boiling point and a relatively low solubility in water compared to the alcohol. The central location of the oxygen atom in the ether provides less of an opportunity for a strong dipole, so the molecule is nonpolar. In the alcohol, the terminal location of the hydroxyl group leads to a polar molecule and the opportunity for hydrogen bonding, which increases the boiling point of alcohol and solubility in water.

Synthesis, page 657

1. Molecules with larger molar masses require greater amounts of energy to obtain velocities sufficient to exit the liquid phase.

2. Since the silicon and carbon atoms possess identical outer electron configurations, they demonstrate similar patterns of chemical behavior, including the ability to form long, intricate molecular chains and to react in similar proportions with other substances.

3. a. $S + O_2 \longrightarrow SO_2$; $2SO_2 + O_2 \longrightarrow 2SO_3$;
$SO_3 + H_2O \longrightarrow H_2SO_4$
  b. Emissions of sulfur dioxide can be reduced by limiting the burning of high-sulfur content fossil fuels or by removing the SO_2 components of the waste gases before they are released into the atmosphere.

4. Mixtures of related compounds with similar properties may be separated by the repeated vaporization and condensation (fractional distillation) of increasingly purer fractions.

5. Add additional methyl alcohol or acetic acid.

CHAPTER 24 Biochemistry

CHAPTER ORGANIZER					
USE WITH SECTION	**DEMONSTRATION**	**WORKSHEET**	**LAB EXPERIMENT**	**SOFTWARE**	**TEACHER'S RESOURCE BINDER**
24-1			24-A		
24-3		24-A			
24-4		24-B, 24-C			
24-7		24-D			
24-9					Chem Issue 7
24-10		24-E			
Chapter Test					

■ Objectives

When students have completed this chapter, they will be able to

1. Describe the general structure of a carbohydrate, a lipid, a protein, and a nucleic acid molecule.

2. State that carbohydrates and lipids are used as energy sources in living organisms.

3. Name the units that make up complex carbohydrate, protein, and nucleic acid polymers.

4. Describe how twenty amino acids can combine to form millions of different protein molecules.

5. State the nitrogen base pairs that can form in a double strand of DNA and in RNA.
6. Compare and contrast the molecular structures and actions of hormones and neurotransmitters.
7. Summarize the function of enzymes in regulating biochemical reactions.
8. State the difference between a vitamin and a mineral and give an example of each.
9. Describe the process of gene cloning.
10. Give three examples of how a product of gene cloning is being used in industry or medicine.

■ Section Notes

24-1, 24-2 These two sections will have more meaning for students if you tie them to weight loss and/or gain. Have students examine common weight loss recommendations in light of the information here. What does the body do with excess carbohydrates? What happens if energy requirements exceed carbohydrate intake? Students also may find it interesting to think about some of the nutritional practices suggested for athletes in different sports. For example, why do marathon runners often load up on carbohydrates the day before the race? Why not load up on lipids or proteins?

24-5 Recent research has begun to blur distinctions between hormones and neurotransmitters as more and more examples are discovered of molecules that act as both. The two enkephalins are examples. Researchers now know that a peptide neurotransmitter may have different effects, depending on the type of synapse at which it is acting. In 1977, scientists discovered that a nerve cell can release more than one kind of neurotransmitter. Another recent discovery is that different neurotransmitters can produce a variety of effects at the same synapse.

24-10 Currently there is a great deal of controversy about what constitutes prudent use of gene cloning techniques. As researchers become more adept at genetic manipulation, the chances become greater of creating organisms very different from their counterparts in nature. Researchers, government agencies and concerned citizens will be asked to make difficult and long-ranging decisions about how the products of genetic engineering will be treated.

Review and Practice _____
page 664

1. monosaccharides—glucose, fructose; disaccharide—sucrose; polysaccharides—cellulose, chitin, starch

2. A saturated fat molecule has no C—C double bonds and is fully hydrogenated. An unsaturated fat has one or more C—C double bonds.
3. Saturated fats have higher melting points than unsaturated fats.
4. High blood cholesterol levels have been implicated in heart disease.
5.

$$H_2N-\underset{\underset{R}{|}}{\overset{\overset{H}{|}}{C}}-C\overset{O}{\underset{OH}{\diagdown}}$$

6. The variety of sequences possible for twenty amino acid molecules makes it possible.
7. They are both nucleic acids. DNA is translated into RNA by enzyme action.

page 669

1. Neurotransmitters are quick-acting chemicals produced by nerve cells. Hormones are produced by the endocrine glands and act over longer periods of time. Some chemicals can be both.
2. **a.** peptides and steroids
   b. peptides
   c. norepinephrine, enkephalins
3. It is not used up in the reaction.
4. Minerals are inorganic, usually metal ions; vitamins are organic compounds.

page 673

1. the process by which many identical copies of a gene are made
2. It may be less susceptible to heat. It could catalyze novel reactions or work more quickly.
3. Proteins are easily destroyed by the common industrial conditions of high heat, pressure, mechanical stress, and organic solvents.
4. It has been used to change growth patterns, life cycles, tolerance to cold, and to correct defective genes.

Review, page 674

1. carbohydrates, lipids, proteins, nucleic acids
2. carbohydrates—primary energy source
   lipids—source of energy, component of the cell membrane
   proteins—catalysts for reactions in cells, structural component of cells
   nucleic acids—direction of protein synthesis, genetic information transfer
3. Heat disrupts the weak hydrogen bonds holding

the protein in a specific three-dimensional configuration and the protein becomes denatured.

4. a. amino acids
  b. monosaccharides (glucose)
  c. nucleotides

5. a. adenine, thymine, guanine, cytosine
  b. adenine, uracil, guanine, cytosine

6. adenine—thymine (A—T)
  guanine—cytosine (G—C)

7. adenine—uracil (A—U)
  guanine—cytosine (G—C)

8. The order of amino acids in a protein molecule determines its three-dimensional configuration which, in turn, determines its chemical and physical properties.

9. The lipids in butter are mostly saturated fatty acids as opposed to the unsaturated and polyunsaturated fatty acids found in vegetable oils.

10. Enzymes are protein catalysts that speed up biochemical reactions and regulate the rate at which they occur.

11. The differing size and structure of each purine and pyrimidine base allows it to pair with only one other base. The base sequence on one strand of DNA specifies the exact order of bases on a complementary strand of DNA.

Interpret and Apply, page 674

1.

$$CH_3—CH_2—CH_2OH$$
1-propanol

$$CH_3—\overset{\overset{\displaystyle OH}{|}}{CH}—CH_3$$
2-propanol

$$CH_3—O—CH_2—CH_3$$
an ether

2.

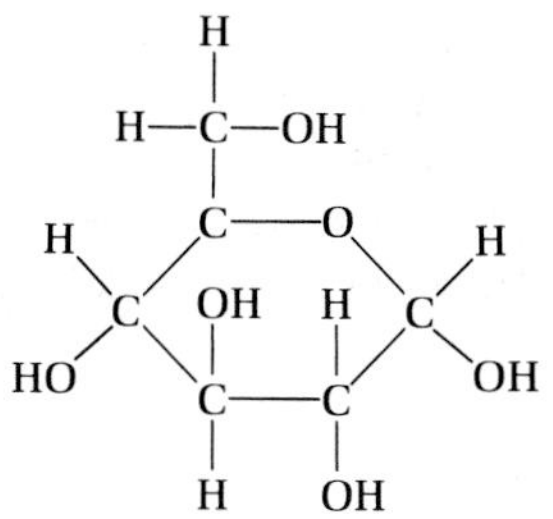

glucose

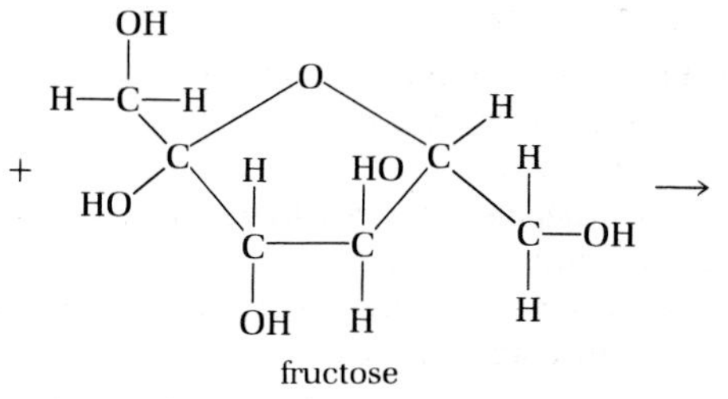

$+$

fructose

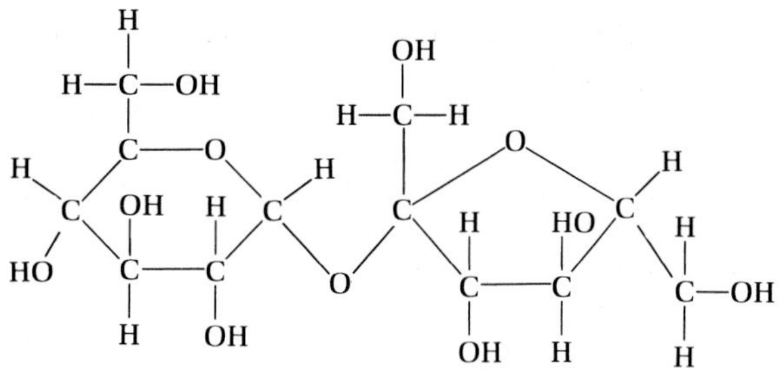

$\rightarrow$

sucrose

$+\ H_2O$

3. Usually the shape of the active site of an enzyme will provide the correct reaction conditions for only a specific type of molecule. This specificity may range to include all the molecules of a particular class sharing similar structures.

4. The hydrogen bonds holding the protein chain in a specific configuration are broken, and the molecule is said to be denatured.

5. All saturated fatty acids will have higher melting points, and unsaturated fats will have lower melting points.
  a. saturated; **b.** polyunsaturated; **c.** saturated

6.

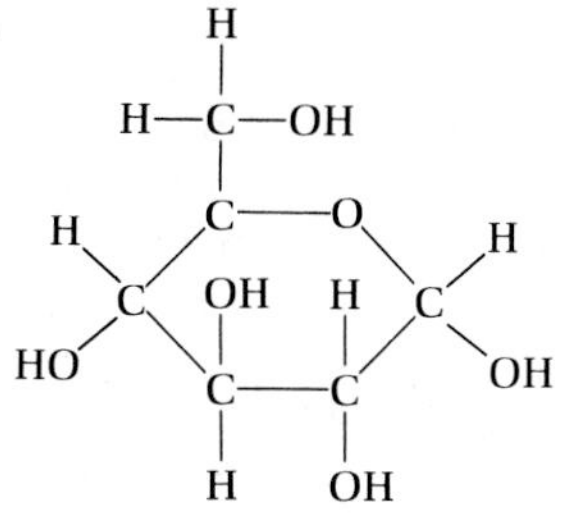

glucose

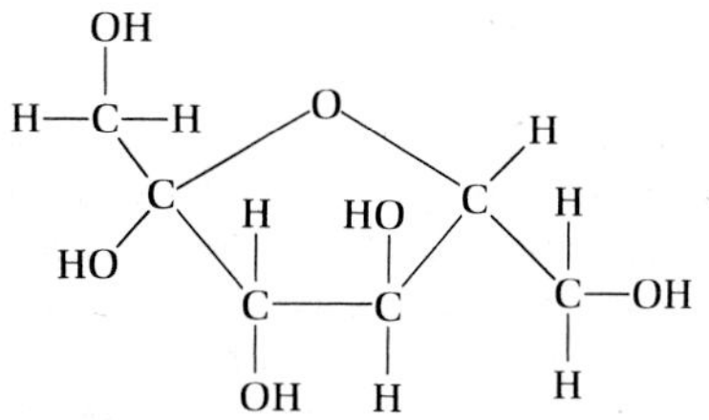

fructose

Students only will be able to compare the cyclic forms of glucose and fructose from the presentation in this chapter. Glucose has a six-member ring with one —CH₂OH side group. Fructose has a five-member ring with two —CH₂OH side groups.

7. b. Butter; butter has the highest lipid content of all the foods listed, and lipids provide approximately 2.3 times the amount of energy per gram that carbohydrates provide.

8. any change in the surrounding chemical environment of the enzyme that renders conditions less than optimal for the catalyst, such as heat (denaturization), change in pH, pressure, organic solvents

9. Nucleic acids direct the synthesis of proteins by a cell. DNA provides a template for the manufacture of messenger RNA which, in turn, determines the sequence of amino acids in a protein.

Problems, page 675

1. a. DNA base pairs A—T; G—C ATCTCGGAGTCC
  b. RNA base pairs A—U; G—C AUCUCGGAGUGG

2.

DNA nucleotide

ATP

  a. ATP has three phosphate groups and an additional —OH group attached to a ribose sugar molecule. A DNA nucleotide has one phosphate group attached to a deoxyribose sugar molecule.
  b. Each has a sugar molecule attached to a nitrogen base and one or more phosphate groups.

3. The energy required for cell processes primarily comes from the metabolism of glucose and also fatty acids. The energy is stored in the high energy bonds of ATP.

4. $2500 \text{ kcals} \times \dfrac{1 \text{ gram fat}}{39.6 \text{ kcals}} = 63.1 \text{ g fat}$

Challenge, page 675

1. The empirical molar mass of $C_6H_{10}O_5$ is 162.

  a. $? \text{ units in cellulose} = \dfrac{600\,000}{162}$

  $= 3700 \text{ monosaccharide units}$

$? \text{ units in starch} = \dfrac{4000}{162}$

$= 25 \text{ monosaccharide units}$

  b. length of cellulose molecule

$= 3700 \text{ units} \times \dfrac{0.50 \text{ nm}}{\text{unit}} = 1850 \text{ nm}$

length of starch molecule

$= 25 \text{ units} \times \dfrac{0.5 \text{ nm}}{\text{unit}} = 12.5 \text{ nm}$

Synthesis, page 675

1. $149 \text{ mg } CO_2 \times \dfrac{1 \text{ g}}{1000 \text{ mg}} \times \dfrac{1 \text{ mol } CO_2}{44.0 \text{ g } CO_2}$

$= 3.38 \times 10^{-3} \text{ mol } CO_2$

$3.39 \times 10^{-3} \text{ mol } CO_2 \times \dfrac{12 \text{ g C}}{1 \text{ mol } CO_2} \times \dfrac{1000 \text{ g}}{1 \text{ g}}$

$= 40.6 \text{ mg C}$

$45.4 \text{ mg } H_2O \times \dfrac{1 \text{ g}}{1000 \text{ mg}} \times \dfrac{1 \text{ mol } H_2O}{18.0 \text{ g } H_2O}$

$= 2.52 \times 10^{-3} \text{ mol } H_2O$

$2.52 \times 10^{-3} \text{ mol } H_2O \times \dfrac{2 \text{ g H}}{1 \text{ mol } H_2O} \times \dfrac{1000 \text{ mg}}{\text{g}}$

$= 5.0 \text{ mg H}$

$40.6 \text{ mg C} + 5.0 \text{ mg H} = 45.6 \text{ mg accounted for mg}$
oxygen $= 100 - 45.6 = 54.4 \text{ mg O}$

$54.4 \text{ mg O} \times \dfrac{1 \text{ g}}{1000 \text{ mg}} \times \dfrac{1 \text{ mol O}}{16 \text{ g O}}$

$= 3.40 \times 10^{-3} \text{ mol O}$

$40.6 \text{ mg C} \times \dfrac{1 \text{ g}}{1000 \text{ mg}} \times \dfrac{1 \text{ mol C}}{12 \text{ g C}}$

$= 3.38 \times 10^{-3} \text{ mol C}$

$5.0 \text{ mg H} \times \dfrac{1 \text{ g}}{1000 \text{ mg}} \times \dfrac{1 \text{ mol H}}{1 \text{ g H}}$

$= 5.0 \times 10^{-3} \text{ mol H}$

MOLE RATIOS	=	ACTUAL		RELATIVE	
C		3.38×10^{-3}		1	or 2
H		5.0×10^{-3}		1.47	or 3
O		3.4×10^{-3}		1	or 2

The empirical formula is $C_2H_3O_2$.

2. a. Negative; the reaction is an exothermic combustion reaction.
  b. exothermic

HEATH CHEMISTRY

J. Dudley Herron
Professor of Chemistry and
Education
Purdue University
West Lafayette, IN

David A. Kukla
Chemistry Teacher and Chairman
Science Department
North Hollywood High School
North Hollywood, CA

Clifford L. Schrader
Chemistry Teacher
Dover High School
Dover, OH

Michael A. DiSpezio
Science Teacher and Chairman
Science Department
Cape Cod Academy
Osterville, MA

Julia Lee Erickson
Chemistry Teacher
Arundel High School
Gambrills, MD

Content Reviewers

George Axelrad
Professor and Chairman
Chemistry Department
Queens College
Flushing, NY

Jerry A. Bell
Professor of Chemistry
Simmons College
Boston, MA

David W. Brooks
Professor of Chemistry
Education
University of Nebraska
Lincoln, NE

Joseph E. Davis
Curriculum Developer
Lawrence Hall of Science
Berkeley, CA

Sheldon L. Glashow
Professor of Physics
Harvard University
Cambridge, MA

W. Keith MacNab
Science Consultant
Middletown, CA

A. L. McClellan
Science Consultant
El Cerrito, CA

Paul R. O'Connor
Adjunct Professor
Chemistry Department
University of Minnesota
Minneapolis, MN

D. C. HEATH AND COMPANY
Lexington, Massachusetts / Toronto, Ontario

THE HEATH CHEMISTRY PROGRAM

Heath Chemistry,
  Pupil's Edition
Heath Chemistry,
  Teacher's Annotated Edition
Heath Chemistry Laboratory
  Experiments, Pupil's Edition
Heath Chemistry Laboratory
  Experiments, Teacher's Edition
Heath Chemistry Chapter Worksheets

Heath Chemistry Teacher's Resource
  Binder
Heath Chemistry Tests,
  Spirit Duplicating Masters
Heath Chemistry Computer Test Bank
Heath Chemistry Computer Test Bank,
  Teacher's Guide
Heath Chemistry Courseware
Heath Chemistry Lab Assistant

Executive Editor: Ellen M. Lappa
Project Editor: Toby Klang
Editorial Development: Mary M. Scofield, Marianne Knowles, Maureen Oates
Project and Cover Designer: Angela Sciaraffa
Photo Coordinator: Connie Komack
Production Coordinator: Donna Lee Porter
Readability Testing: J & F

Cover Photograph: In a heated mixture of mineral oil and water, the water (dyed red for visibility) boils first due to its lower boiling point. Photo by Richard Megna/Fundamental Photos.

Teacher Reviewers

Christina L. Borgford
Chemistry Teacher
Oregon Episcopal School
Portland, OR

Richard A. Brown, Ph.D.
Chemistry Teacher and Chairman
Science Department
Minnechaug Regional High School
Wilbraham, MA

Diane Wilson Burnett
Chemistry Teacher
Warren Central High School
Indianapolis, IN

Nancy E. Day
Chemistry Teacher
Belmont High School
Belmont, MA

Edmund J. Escudero
Chemistry Teacher
St. John's School
Houston, TX

Sister Agnes Joseph
Chemistry Teacher
Marian High School
Birmingham, MI

Jerrold Omundson
Chemistry Teacher and Chairman
Science Department
Memphis University School
Memphis, TN

Leonard B. Soloff
Chemistry Teacher
Kennedy High School
Granada Hills, CA

Copyright © 1987 by D. C. Heath and Company
All rights reserved. No part of this publication may be reproduced or transmitted in any form by any means, electronic or mechanical, including photocopy, recording, or any information storage or retrieval system, without permission in writing from the publisher.

Published simultaneously in Canada
Printed in the United States of America
International Standard Book Number: 0-669-09853-1
1 2 3 4 5 6 7 8 9 0

TO THE STUDENT

When students are asked why they enrolled in a high school chemistry course, they give a variety of responses. "A good friend said it was fun." "My parents said I should." "My counselor said I would need it for college." "I want to be an engineer, so I am taking all the science and math I can." "Chemistry is required for all students in the academic track." "Chemistry is required for pre-med, and I want to be a doctor."

These may be good reasons to study chemistry, but they ignore the real purpose. (You are matter and live in a world of matter.) The reason parents, counselors, college admissions officers, engineers, and doctors think you should study chemistry is that they realize the more you understand about the world around you, the more successful you can be regardless of your goal. A course in chemistry is a means to an end; it is not an end in itself.

When you finish this course, you should feel that you understand how matter behaves, that you can describe its behavior to others, and that you can learn more about matter through additional courses or independent research. If you pass the course without understanding, we have failed.

The suggestions that follow are meant to help you understand as you use this book.

Become an independent learner. At some point in your life, you will be on your own. When that time comes, you will obtain the majority of new information through reading. You will make sense of this information through your own reasoning. The better you learn to do that now, the more successful you will be later on.

Develop good study habits. Most people fail to achieve goals because they fail to do what they know is necessary for success. Poor students normally know what they must do to learn, but they fail to do it. Knowing what to do and doing it are two different things. The latter requires discipline as well as knowledge.

Schedule your study time. You must be disciplined about study. Make out a weekly schedule to include extra-class activities and study.

Read the assigned text before it is discussed in class. Students say it is a waste of time to read the assignment because teachers ✓ repeat everything in class. Smart students read the assignment in advance. As they read, they write down questions and summarize what it means to them. They then listen to the teacher's explanation and compare the two. When the interpretations differ, they search in the text for an explanation for these alternative interpretations.

Be an active reader. Good readers predict what a story, chapter, or section will be about by looking at the title, headings, or illustrations. As they read, they look for proof that their predictions about the reading are correct. If not, they make new predictions. At the end of a section, they attempt to summarize what they know about the subject, and they consciously review why they think so. Reading a chemistry book is not like reading a novel. It takes different skills. This book will help you understand symbolic representations like $Pb + PbO_2 + 2H_2SO_4 \rightarrow 2PbSO_4 + 2H_2O$, $PV = nRT$, in addition to pictures, graphs, and tables. These representations are just as important as words.

Review the ideas covered. If the class dealt ✓ with a new procedure or skill (e.g. writing formulas or balancing equations), practice the skill immediately. If you wait a day or two to begin, you will forget the details.

Answer the Review and Practice questions at the end of each section of text that you read. (Do this whether assigned or not.) These questions reinforce the ideas presented in the reading. If you can answer them without referring back to the reading, you probably understand what you read.

Do not study a lot of chemistry at one ✓ **sitting.** For most students, 30 minutes at a time is enough (though you may need slightly more or less). A reasonable time spent, five days a week, will be much more effective than two or three hours on a weekend.

Cram sessions before a major exam are ✓ **not very effective.** You cannot relax before an exam if you know you were not learning the material on a regular basis. On the other hand, if you keep up with your work every day, a short review just before the exam is all you need. Take the night off and relax. You will be alert for the exam, and you will make a higher score.

Overlearn basic skills through practice. Ideas that are used repeatedly and for example basic math skills, reading skills, units of measurement, the mole concept, formula writing, must be practiced to the point that you can almost use the ideas in your sleep. When basic skills are overlearned, all your conscious attention can be given to novel aspects of a problem, and you are more likely to solve it.

Be objective. Some students have an unrealistically high opinion of their abilities. When they get poor scores or do not understand what they read, they attribute their lack of success to poorly written test questions or a poorly written text. Other students have an unrealistically low opinion of their abilities. When they get poor scores, they decide they are stupid and cannot understand chemistry. Usually the problem stems from the use of unfamiliar terms in either the text or the class lecture. The key to success is having faith in yourself, organizing your work, and not giving up.

We honestly believe that every student who is reading this book can understand chemistry. If you share that faith, this book will provide an exciting beginning to your understanding of matter.

CONTENTS

Chapter 1 ACTIVITIES OF SCIENCE _____ 1

Chapter 2 FINDING OUT ABOUT MATTER _____ 26

Chapter 3 NUMBERS LARGE AND SMALL _____ 61

Chapter 23 ORGANIC CHEMISTRY _____ 625

Chapter 24 BIOCHEMISTRY _____ 659

Appendix A: DESCRIPTION OF A BURNING MATCH _____ 676

Appendix B: HYBRID ORBITALS _____ 676

Glossary _____ 678

Index _____ 684

Acknowledgments _____ 694

CHAPTER 1

You could call this picture Scientists at Work, even though it shows ordinary people doing ordinary things. Science is not so much what people do as how they go about doing it. In this chapter, you will look at the kinds of activities that fall into the realm of doing science. Along the way, you will have many opportunities to do science too. As a result, your reasoning and problem-solving skills will be improved. And you will have a better understanding of your world.

Activities of Science

Science Is a Way of Knowing

Science is a way of gaining knowledge, but it is only one way of knowing. Religion and philosophy are other ways of knowing—ways that contributed much to civilization before science was developed. These fields continue to contribute to today's body of knowledge. Science, religion, philosophy, literature, art, and other fields of study contribute different kinds of knowledge, and they do so by using different techniques. In today's society, you need to know the techniques that are used in science.

CONCEPTS

- observation as a basis for knowledge
- regularity through classification
- explanation through theories and models

1-1 Scientific Knowledge Is Based on Observation

Observation is the basis for all science. The art of careful observation takes practice to develop. For example, look at Figure 1-1. What do you see? At first you probably see nothing but black dots, but after a while you might see a Dalmatian dog sniffing the curb. If you did not see it before, you almost certainly do now. But what if you have never seen a dog or a street with curbs? You will interpret the picture differently than someone else, even now. Observing involves more than looking.

What you observe depends on what you already know. You construct knowledge for yourself, a little bit at a time. The knowledge that you get from observing depends on the messages that come through your senses—through sight, smell, sound, taste, and touch. However, the knowledge you get also depends on the messages that come from your brain. When you look at Figure 1-1, your eyes detect patterns of dots; your brain functions in matching that pattern with one you have seen before. That process is essential for learning from experience, but it can get in the way of learning too. If you are not careful, you observe what you expect to observe, even when it does not happen. Also, your expectations could differ greatly from someone else, so two people would not observe the same thing. For this reason, scientists insist that observations must be repeated several times by more than one person before they can be accepted.

It is important to distinguish between observations and inferences. In Figure 1-1, you *observe* a pattern of dots. That is what your sight detects. You also observe that some dots are larger than others, closer together than others, and so forth. However, you do *not* observe a dog sniffing at a street curb. Rather, you *infer* that the pattern of dots represents a dog sniffing at a curb. An **inference** is an interpretation of an observation.

Figure 1-1 When you first look at this picture, you probably see nothing but black dots. After a while, you should see an animal sniffing at the ground.

1

This point will not be clear without class discussion. Suggestions for class activities are found in the Teacher's Guide at the front of this book.

Making inferences is so much a part of observing that people often confuse them. People say they observed something when they actually inferred it from an observation.

Some people think that scientists (and science teachers) are too concerned about little details in science, but such an attitude is necessary if ideas are to be exchanged. If scientific knowledge is based on observation and no single observation can be trusted to be correct, the only way that scientific knowledge can grow is for scientists to communicate their experiments and observations in enough detail for others to repeat them.

As observations are repeated and confidence in them grows, the amount of knowledge that is accepted as fact quickly becomes unmanageable. A person can attend to a limited amount of new information at one time. You should keep this in mind as you study chemistry. Unless you adopt some of the strategies used in science to manage large amounts of information, you will find this course difficult.

1-2 Classification: Searching for Regularity

One way to cope with large amounts of information is to categorize and deal with the category rather than with the individual facts. This is the process you go through when you make a statement like "Birds lay eggs." You do not say that chickens lay eggs and ducks lay eggs and robins lay eggs and wrens lay eggs. You categorize all those animals as birds, so you can economize with words and say "Birds lay eggs." This process of placing similar things into categories is **classification**.

Classification would be impossible if we did not see regularity in diversity. Just think about birds and the regularity summarized by "Birds lay eggs." People noticed that wrens and chickens and turkeys and emus and many other animals lay eggs, but cows and dogs and people and goats do not. In other words, they noticed a regularity—something that was common to a large group of animals.

The regularities used to specify categories are a matter of choice and convenience. You could, for example, call anything that lays eggs a bird. If you did, you would call snakes, turtles, and most fishes, birds. Similarly, you could call anything that flies a bird. If you did, bats, kites, and airplanes would be called birds while emus and penguins would not.

Noticing regularities across observed events helps in communicating just as classification does. These regularities take many forms. Statements like "metals conduct electricity" or "foods that taste sour contain acids" describe regularities in observed events. These statements summarize and describe many separate observations.

Science proceeds with descriptions of regularity in observations, but great advances in science are usually made when people try to explain the regularity. A scientific **hypothesis** is a temporary explanation for an observed regularity. However, *explain* can be a devious word. By providing an explanation, we often think that we are talking about things as they really are, when in fact we are only talking about things as they might be. Science says we have explained something when the

CHEM THEME

"What is it?" is a question frequently asked in chemistry. The answer always results in having to classify. Solid, liquid, gas, or plasma are answers that tell you something about the physical state of matter. Acid, base, salt, alkali metal, transition metal, metalloid, halogen, and noble gas are answers that describe groups of elements and compounds that have similar properties. Other classifications result from asking, "How does matter change from one form to another?" Classifying matter and its changes will help you understand the physical world in which you live.

Suggestions for class activities on model building are found in the Teacher's Guide at the front of this book.

ACTIVITIES OF SCIENCE **3**

explanation or hypothesis accounts for past events and accurately predicts future ones. The hypothesis or explanation is called a **theory** in science.

Developing a good theory usually requires a great deal of experimentation and careful analysis of the results. Some important theories in science have developed over hundreds of years with the help of many scientists.

Theories often take the form of models. You probably think of a model as a physical object, such as a model train or model airplane. Although a model may be a physical object, **model** in science refers to any representation intended to convey information about another object or event. A model airplane is a physical representation that conveys information about a real airplane. Some models, however, are just mental pictures.

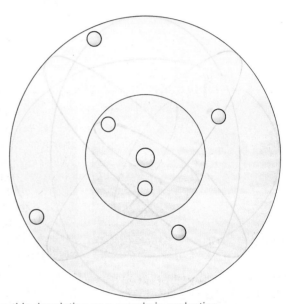

Figure 1-2 At left, water is being boiled in a paper container. What is your hypothesis as to why the paper does not burn? The illustration on the right shows an early model for an atom. This solar system representation was later altered as new information about the atom was discovered.

1. See the Appendix for the description of the burning match. Use this description as a sample in evaluating student responses.

2. One possible answer is vegetable, meat, and dessert. Potatoes, beans, and cabbage are classified as vegetables because they come from plants. Beef, pork, and chicken are meats because they come from red-blooded animals. Ice cream, cake, and pie are desserts because they are sweet and generally eaten at the end of a meal. Other responses appear in the Teacher's Guide at the front of this book.

Review and Practice_____

1. Strike a match. Write down all your observations. Compare your observations with those of a classmate. Identify statements that you think are inferences, and discuss them with your classmate to decide just what was observed and what was inferred.

2. Name three classes of foods and give three examples of each class. What are the properties of the examples that led you to classify them as you did?

3. You have probably seen pictures based on a model for atoms. How do you think these models differ from real atoms?

4. A teacher assigned seats on the first day of class but did not describe the basis for the assignment. How could you develop a model to explain the seat assignments?

3. Answers will vary, but students should pick up on the size differences, the static nature of the model, and the use of color on the model. A single atom would have no color. Additional responses appear in the Teacher's Guide.

4. First, collect data on the people sitting in each seat. Then search for regularities in the data for the purpose of generating a hypothesis. The hypothesis could be tested by determining if each seat arrangement is consistent with the hypothesis. This process would produce a possible model, but not necessarily the model used.

Chemists Study Matter

CONCEPTS

- matter has mass and occupies space
- matter is conserved
- energy changes matter
- energy is conserved

The activities of science should become clearer as you read descriptions of what chemists do and as you experiment with materials in the lab. Chemists are interested in matter. They are interested in dirt, air, and the matter that comprises your body. They want to know what matter is, what properties matter has that make it useful, and how matter can be changed from one form to another.

In question 1 on page 3, you were asked to strike a match and observe it. If you did not do it then, do it now and compare your observations to those in the Appendix.

There are a number of questions you can ask about these observations. Do you have the same kind of matter after you burned the match that you had before? How do you know? Do you have the same *amount* of matter after you burned the match that you had before? How do you know? If the matter is different, what made it change? How do you know? Can the matter that you have after you burned the match be changed back to a match again? If so, how? How do you know?

  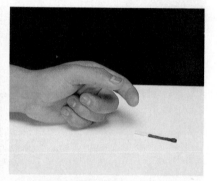

Figure 1-3 We learn about matter through observation. What observations can you make of a match before it is lit, as it is burning, and after it has gone out? How does the match change during this process?

These are the kinds of questions that chemists try to answer in gaining information about matter. They are the kinds of questions you will be expected to answer in this course, but before you can answer any questions about matter, you must formulate an idea of what matter is. You would agree that the match and the black stick that remain after burning are matter. You are accustomed to thinking of all solids and liquids as matter. But what about the flame? Is that matter? Is the smoke, the heat, the light, or the hiss matter? How do you decide?

1-3 Matter Has Mass and Takes Up Space

Matter is defined as anything that has mass and takes up space. You already know how to decide whether something takes up space, but how do you decide if it has mass.

Perhaps you know that mass is the term used to describe the amount of matter in a sample. If you stop there, you have a circular argument. Matter is anything that has mass; mass is a measure of matter. To break out of the circle, there must be an independent test for mass, but what is it?

Everyone has experienced objects falling, and everyone has observed that some objects are heavier than others. You also know that heavy objects are more difficult to move than light ones, and that once you get objects moving, the heavy ones are more difficult to stop. The resistance to change in the motion of an object—getting it moving, getting it stopped, or changing its direction—is called **inertia**.

You now have two ideas that describe regularities you have experienced. These ideas can be used to define mass. **Mass** is what gives an object the property of weight and inertia; the heavier an object, the more mass it has; the more inertia an object has, the more mass it has. Saying that matter is anything that has mass, means that matter is anything that has inertia and requires a force to get it moving or stopped.

At first it seems trivial to ask whether something takes up space. You can *see* whether something takes up space, or can you? Look at an empty drinking glass. Is it really empty? No, it is filled with air. The air is matter which has mass, and fills the space inside the glass. There is light in the room where you are reading. You might even say it fills the room but in what sense does light "fill" a room? Does light take up space as air does? Does it have mass? Is it matter? How can you know for sure?

Figure 1-4 Are the drinking glasses filled or empty? How could you test to see if the glasses contain any matter?

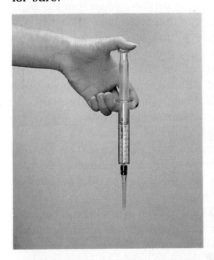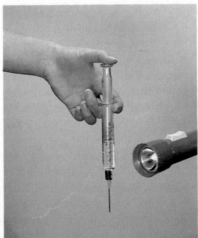

Figure 1-5 If you pushed on the plunger of the syringe on the left, the plunger would "push back" because it is filled with air. The light shining into the syringe on the right, offers no resistance to the plunger. If there is no air in the syringe, the plunger will be pushed in by the air on the outside of the syringe. There is no matter inside to occupy the space, so the plunger can fill the space.

If you put air into a syringe like the one pictured in Figure 1-5, close off the end, and push on the plunger, the air pushes back. You cannot close the space as long as the air is there. If, however, you take what appears to be an empty syringe, pull back the plunger, and allow light to shine in, the light offers no resistance when you push in the plunger. It appears that light does not occupy space in the same way that air does. Similarly, no balance is capable of detecting any difference in the mass of a syringe in the dark or one filled with light.

Because light fails the tests used to decide whether something is matter, we say that light is not a form of matter. However, if you study physics, you will find that some extraordinary tests do show light to have mass. Definitions in science are often changed when new knowledge makes it apparent that a different definition is more useful. Your definition of light may change as you learn more about it, but for now it is sufficient to say that light is not matter. We usually say that light is a form of energy.

1-4 Matter Is Conserved

The definition of matter can now be used to answer some questions about the burning match. One of those questions was whether you had the same amount of matter after the match burned as you did before.

With a balance, you could determine the mass of the match before and after burning and see that it lost mass. From this observation, it is tempting to say that matter is destroyed when it burns. Indeed, in everyday language we say things like "It burned up," meaning that something was "consumed" by fire. The implication is that the matter that existed before burning was destroyed. Your observations seem to support this inference, but scientists claim that matter is *not* destroyed when something burns. In fact, they argue that matter is neither created nor destroyed in any of its changes. They say that matter is conserved.

Some people have difficulty with this statement because they misunderstand what scientists mean when they say, "No matter is destroyed when it burns." (Even when you are careful with words, someone listening to you must interpret what you say in terms of *their* experience which differs from yours.) Scientists are not saying that the match is not destroyed. Clearly, the match is gone and it is replaced by a black stick with less mass. However, it can be shown that the black stick is only one product of the change that took place. The presence of smoke and an odor are observations indicating that something left the solid and mixed with the air in the room. Water, carbon dioxide, and other forms of matter that you neither saw nor smelled were also produced when the match burned. Although difficult, it is possible to collect these products along with the black stick. When this is done, it is found that *the mass of the matter collected after burning is the same as the mass of the matter before burning.*

Scientists often base what appears to be a confusing statement on observations that the general public never has an opportunity to make. Until the advent of atomic energy in the 1940s, no observation had detected a change in mass when matter changed from one form to another. It was generally accepted that matter is neither created nor destroyed in any change from one form to another, and this generalization was accepted as a general law of nature. The discovery of small mass changes in nuclear reactions, such as in an atomic explosion, meant that scientists had to modify their idea that matter is conserved to include the possibility that matter can be converted to energy in special cases. (Notice how statements that describe regularities in nature change as our understanding of nature changes.) Still, matter is conserved in the changes that you are likely to study.

One important aspect of science is that anyone can do it. If you think matter *disappears* when it burns without being replaced by some other kind of matter, think of a way to prove it. If you succeed, you will contribute to science by adding new knowledge about matter; if you fail, your work will strengthen the current beliefs about matter.

Figure 1-6 Water at its triple point. It looks strange to see water boiling and freezing at the same time, but that is what is happening. To achieve this, it is necessary to connect the container to a vacuum pump and reduce the pressure over the water so that it is very low (610 N/m²). As the water boils, its temperature will fall to its freezing point where it will stay until all the water freezes or boils away.

1-5 Matter Exists in Various States

The things called matter are the things that you know as solids, liquids, and gases, the three common **physical states** of matter. Most matter can exist in each of these states if the temperature and pressure are changed.

You consider iron a solid because it exists in that state at ordinary temperatures, but you know iron can be melted. It can even be boiled away as a gas if it gets hot enough. Similarly, water is a liquid, but you know it freezes to a solid in the freezer compartment of a refrigerator, and it changes to a gas when it boils on a stove or evaporates from your skin on a hot day.

A fourth state of matter, plasma, also should be included in a discussion of the states of matter. A **plasma** has all of the properties of a gas except that it is composed of charged particles like electrons rather than uncharged atoms or molecules. Plasmas exist on stars like the sun, in nuclear explosions, and in neon signs.

1-6 Energy Changes Matter and Is Conserved

You know that heat somehow causes the change that you observe in a match when it burns, and you probably know that heat is a form of energy. But what is energy?

One useful definition of **energy** is anything that is not matter and can cause a change in matter. Light from the sun certainly changes matter in causing green plants to grow and produce food. Sunlight causes your skin to redden and the pigment to darken. It can also cause the fabric of drapes to fade. Heat can change matter too. It can burn toast, cook eggs, boil water, warm your body, or ignite paper. Lightning can split a tree, kill a cow, burn out a TV, or cause your house to catch fire. We now know that lightning is just a gigantic electric spark. More usable forms of electric energy come in flashlight batteries or from electric generators in a power station.

Within limits, it is possible to change one form of energy into another. Coal can be burned to produce heat, which boils water to form steam, which turns a turbine to produce electricity, which can be used to light a bulb. In the process, chemical energy is converted to heat that is used to produce mechanical energy, which produces electric energy, which produces light. Through all of these changes, energy is conserved.

Saying that energy is conserved means that during an energy conversion, there is no loss or gain of energy. Although the form of the energy is changed, the amount remains the same.

Notice that energy is explained as something that *can* change matter. It may not be doing it. Matter is not being changed when an egg is held ten feet above a floor, but the potential is there. All you need to do is let go of the egg. The egg has potential energy because the potential for falling—and changing matter—is there. Similarly, you would think that a piece of wood or an unused battery has potential chemical energy. You cannot see any evidence of energy, but you know that the potential is there. If wood is ignited, heat is produced. If the battery is connected to a bulb, the bulb lights. **Potential energy** is present when no evidence of energy can be seen (matter is not changing), but the potential for changing matter is there. Potential energy was invented

Figure 1-7 Solar panels are used to convert radiant energy from the sun to kinetic energy.

CHEM THEME

Models take many forms. The most important model in chemistry is the *mental model* of atoms, the basic building blocks of all matter. The *idea* of individual particles bouncing around in a closed container provides a useful model to explain the behavior of gases. This idea, when described in careful detail, is called the kinetic theory, and it provides a useful description of how gases behave.

The behavior of matter is often described by mathematical equations. Such *equations* represent another kind of model, a kind that is particularly useful because it leads to quantitative predictions.

Whether models take the form of mental pictures of particles as in the case of atomic theory, the form of a series of assumptions as in the case of the kinetic theory, or the form of a mathematical equation, they provide powerful tools that are used to understand the world.

to preserve the belief that energy is never created or destroyed; it is only changed from one form to another.

Kinetic energy, the energy of motion, can be contrasted to potential energy. It is obvious that a moving object has energy—at least it is obvious when it hits you! The amount of energy that a moving object has (and the amount matter is changed by that energy) depends on the mass of the object and how fast it moves. Stated in the language of mathematics, kinetic energy is defined as

$$\text{kinetic energy} = \tfrac{1}{2}(\text{mass})(\text{velocity})^2$$

The equation shows that the kinetic energy of an object is equal to one half of its mass multiplied by the square of its velocity. The definition is not particularly important at this point, but it illustrates an important tool of science.

The most precise language is that of mathematics. The mathematical expression of kinetic energy shows not only that it depends on the mass and speed of an object, but exactly *how* it depends on mass and speed. In other words, it shows how much the kinetic energy will change when the mass is doubled or the speed of an object is reduced by one third.

Review and Practice

1. Decide which of the following are matter. If the item listed is not matter, explain how you could prove it to a classmate who disagrees with you. If you are not sure whether something listed is matter, explain what you need to know before deciding.
    - a. love
    - b. light
    - c. lips
    - d. eyelashes
    - e. a beach
    - f. a beep

2. Try designing an experiment to prove that matter is not created or destroyed when a plant seed germinates, sprouts, and grows into a mature plant.

3. What state of matter is represented by each of the following?
    - a. snow
    - b. vegetable oil
    - c. axle grease *s or l ?*
    - d. gasoline
    - e. smoke
    - f. clouds in the sky *colloid*
    - g. steam that you see over boiling water *colloid*
    - h. natural gas
    - i. talcum powder
    - j. toothpaste *s or l ?*
    - k. what you exhale
    - l. the beam of electrons in a TV

Black + white classification is often difficult.

1. *a, b,* and *f* are not matter; they do not have mass nor do they occupy space
2. The difficulties in this experiment are the slow rate of the change as well as the difficulty in monitoring oxygen intake and loss of CO_2 and H_2O through transpiration. Trace amounts of organic gases are also lost.
3. *a* and *i*—solids
b and *d*—liquids
h and *k*—gases; *l*—plasma, but students may not know this.
The others are more difficult to classify. Items *c* and *j* have properties of solids but flow like liquids. Item *e* is a mixture of solid particles suspended in gases. Similarly items *f* and *g* are liquid droplets suspended in gases.

Measurements in Science

Bill Cosby has a humorous monologue in which he plays a modern-day Noah building The Ark in suburbia. A booming voice commands Noah, "Build the Ark three hundred cubits long, fifty cubits wide, and thirty cubits high." A frightened Noah quickly answers, "Right!" and then thoughtfully asks, "What's a cubit?"

Noah's question calls attention to two important considerations in measurement. First, no measurement makes sense if you do not know what the unit means. Second, a unit is useless unless it means the same thing every time it is used.

A cubit is the distance from your elbow to the tip of your middle finger. This distance is not the same on your arm as it is on your teacher's arm. The cubit is not a useful unit in science because it is not standard.

CONCEPTS

- the difference between quantities and numbers
- measurements require standards
- advantages of SI units
- unitary rates and how to use them

1-7 Measurement Standards

All measurements are comparisons to some standard. If you are five feet, ten inches tall, your height is five times the length of a standard called a foot plus ten times the length of a standard called an inch. If you weigh 165 pounds, your weight is 165 times the weight of a standard called a pound. Standard measures are kept in the United States at the National Bureau of Standards. In practice, many secondary standards are used.

Measurements describe quantities such as length, mass, or temperature. A **quantity** is a property that can be measured and described by a number and a unit that names the standard used. 165 and 75 name *numbers;* 165 pounds and 75 kilograms name *quantities.*

This distinction between numbers and quantities is very important when you use mathematics in science. Mathematics deals with relationships among numbers; science deals with relationships among quantities.

Figure 1-8 Standard weights and measures are made of durable materials that resist changes. Why would these properties be important?

Nutritional information: Per serving
Serving size: 177 mL (6 fl. oz.)
Servings per container: 10.6
Calories: 110
Carbohydrate: 27 g   Protein: 0
Sodium: 5 mg   Fat: 0
% of U.S. Recommended Daily Allowance (U.S. RDA)
Vitamin C: 100%

Figure 1-9 How many measurement units appear on this fruit juice label?

Do You Know?

*One must keep in mind that "oz" on the cough syrup stands for "fluid ounces" and does not mean the same as the "oz" on the bottle of oregano in the spice rack. That bottle contains 1/2 **avoirdupois ounces**—one of three ounce units of mass.*

A number can refer to anything. When you say that two plus two is equal to four, it does not matter whether you are adding chickens, people, or peanuts. You could even be adding a mixture of all three; the mathematical statement would still be valid, and the idea can be expressed in the language of mathematics as:

$$2 + 2 = 4$$

Science deals with quantities. Therefore, the rules of mathematics must be adapted to handle quantities. New language is required, and care must be taken to ensure that the rules that apply to pure numbers still make sense when you apply them to quantities. Imagine you have two yardsticks and two standard rulers. These four measuring devices reach from one wall of a room to the other when they are laid end to end. How wide is the room? You can see that it will not make sense to simply add two plus two to get four. Two yards plus two feet are not equal to four of anything!

"First change the yards to feet," you say? Of course! In doing so, you are using one of the first rules of mathematics applied to quantities: *Only like quantities can be added or subtracted.*

Units are important because they name the quantity measured. There are many units in common use. Just wandering around the house, you can find many references to units. One cookbook having a recipe for chicken pilaf calls for a 4-*pound* chicken, 1 *teaspoon* of curry powder mixed with 1 *cup* of water, and 4 *tablespoons* of fat. The recipe says to cook the dish for $\frac{1}{2}$ *hour* in an oven set at 350 °*F* before draining the drippings. Rice is cooked·in the drippings for 20 *minutes*.

A bottle of diet cola boasts less than 1 *calorie* per bottle, a bottle of vitamin C might contain 250-*milligram* tablets, and a bottle of cough syrup might be an 8-*oz* bottle. Paper is purchased by the *ream*, pencils by the *gross*, eggs by the *dozen*, firewood by the *cord*, fruit by the *peck*, and diamonds—if you can afford them—by the *carat*. You see how units are an important part of everyday life.

1-8 Metric (SI) Units

Keeping units straight in science and knowing just what they mean can be a real chore. If scientists did not limit the kinds of units used, communication among scientists throughout the world would be seriously impaired. For that reason, scientists use *Le Systeme International d'unites*, more commonly referred to as SI. **SI** is a modification of the older metric system which was used in France. It is now the common system of measurement in virtually every country in the world.

The popularity of SI is due to two characteristics. First, SI has the same numerical base as the decimal number system. That is, every unit in the system is ten times the size of the next smaller unit, just as every place in a multi-digit number has ten times the value of any place to its right. Second, units for various quantities are defined in terms of units for simpler quantities whenever possible.

Now pick up any object—this book, for example—and consider what measurements you could make on that object. What quantities could you describe? When you answer that question, you will know what kinds of units you need.

You probably thought of length, width, and thickness. These are really all distances; they simply tell how far it is from one point to another, so one unit is enough. You could also measure how heavy this book is; its *mass*. You could describe the size of this page, its *area*, or you could describe the space occupied by this entire book, its *volume*. Perhaps you also thought of *temperature*. If you set the book on fire—something we trust you are not tempted to do—you could measure the heat produced. The authors are more aware of the *time* it took to write this book than anything else!

Every quantity that people presently know how to measure can be expressed in terms of only seven fundamental quantities, the **SI base units**. These are listed in Table 1-1 along with the quantity that the unit describes and the symbol for the unit.

Table 1-1

SI BASE UNITS		
QUANTITY	NAME	SI UNIT ABBREVIATION
length	meter	m
mass*	kilogram	kg
time	second	s
electric current	ampere	A
temperature	kelvin	K
amount of substance	mole	mol
luminous intensity	candela	cd

*Kilogram is the base unit in the system, but gram is the name modified by prefixes to obtain all other units of mass.

Those quantities in bold type will be used in this course. For now, first focus on length. As shown in Table 1-1, the basic unit of length is the meter. The meter is approximately the length of a baseball bat or a golf club. You will want to talk about lengths as small as the diameter of an atom (about 0.000 000 000 100 m)* and as large as the distance to the sun (about 150 000 000 000 m). It is convenient to define larger and smaller units so as not to use such large numbers. All SI units are derived from the base unit by using prefixes which stand for some multiple of ten. Table 1-2 lists the prefixes and their meaning.

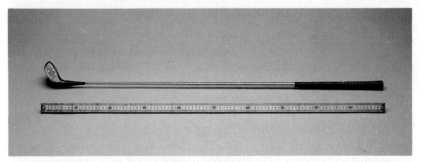

Figure 1-10 A golf club is about one meter long.

*Various styles are used to group digits in large and small numbers to make them easier to read. In this book, a space is used in place of a comma to group digits. Thus 10,000 appears as 10 000.

Those prefixes shown in bold type are frequently used in this text and must be memorized. Your teacher may ask you to memorize others.

Table 1-2

	SI PREFIXES				
FACTOR	PREFIX	ABBREVIATION	FACTOR	PREFIX	ABBREVIATION
10^{18}	exa	E	10^{-1}	**deci**	d
10^{15}	peta	P	10^{-2}	**centi**	c
10^{12}	tera	T	10^{-3}	**milli**	m
10^{9}	giga	G	10^{-6}	**micro**	μ
10^{6}	mega	M	10^{-9}	nano	n
10^{3}	**kilo**	k	10^{-12}	pico	p
10^{2}	hecto	h	10^{-15}	femto	f
10^{1}	deka	da	10^{-18}	atto	a

Review and Practice

1. a. years
   b. feet and inches
   c. pounds
2. only like quantities can be added
3. hecto
4. centi
5. 1/1000

1. In everyday speech, units are often left out because the quantity is clear from the context. What quantity is implied by each of the following?
   a. How old are you? Eighteen.
   b. How tall are you? Five, two and a half.
   c. How much do you weigh? A hundred and eighteen.

2. The length of a person's arm from the shoulder to the wrist is two feet. From the wrist to the tip of the finger is six inches. Why does it not make sense to find the distance from the shoulder to the tip of the finger by adding two and six?

3. What metric prefix means 100? *hecto*

4. What metric prefix means $\frac{1}{100}$? *centi*

5. What does *milli-* mean? $\frac{1}{1000}$

1-9 Application: Indirect Measurements

Measurements always involve comparison to a standard. Some measurements are straightforward; you measure the length of an object by comparing its length to the marks on a ruler that are some standard distance apart. However, many measurements cannot be made directly and require us to follow some standard procedure. Determination of water hardness is a good example.

"Hard" water is water with calcium and magnesium ions dissolved in it. These ions combine with soap to produce an insoluble scum. Hard water is "softened" by replacing these ions with sodium ions that do not combine with soap the way calcium and magnesium ions do. The more calcium and magnesium ions the water contains, the harder the water and the larger the softener unit needed.

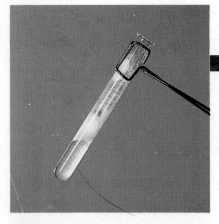

Figure 1-11 Soap combines with calcium and magnesium ions to produce an insoluble scum.

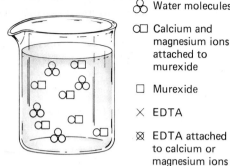

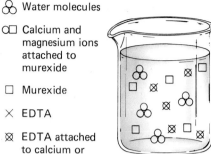

⚭ Water molecules

⊂⊐ Calcium and magnesium ions attached to murexide

☐ Murexide

✕ EDTA

⊠ EDTA attached to calcium or magnesium ions

If you buy a water softener for your home, it is best to test your water for hardness before making the purchase. There is no point in spending money for a larger softener unit than needed to treat the water. .The standard procedure for finding water hardness is to add a colored substance called murexide that attaches itself to calcium and magnesium ions in the water. Murexide has one color when it is attached to calcium or magnesium ions and another color when it is unattached. Another compound called EDTA has a stronger attraction for the calcium and magnesium ions than does murexide. When EDTA is added to the hard water containing murexide, it replaces the murexide molecules that are attached to the ions and causes the solution to change color. It is the amount of EDTA added to cause the color change that is actually measured, and from that measurement the amount of calcium and magnesium in the water is calculated.

Indirect measurements such as this one are commonly used to determine the amounts of various compounds in the blood, urine, and other body fluids, and are very important in medical diagnosis. The chemical changes on which the indirect measurement is based are frequently influenced by impurities in solution or the pH (acidity) of the solution. Therefore, it is important for the laboratory technician to follow a standard procedure in making the determination. Failure to follow the procedure exactly may result in serious errors in these types of measurements.

Figure 1-12 Hard water is being tested on the left to determine hardness. One such test involves using EDTA and murexide. EDTA molecules replace murexide molecules to cause a color change as shown in the model on the right.

1-10 Using Unitary Rates

Table 1-2 indicates that *centi-* is the prefix meaning "one one-hundredth." Then one centimeter is one one-hundredth of a meter. Applying mathematics to quantities and using the correct abbreviations, you can say that

$$1 \text{ cm} = 0.01 \text{ m}$$
$$\text{or}$$
$$100 \text{ cm} = 1 \text{ m}$$

The philosophy for using unitary rates in problem solving is described in the Teacher's Guide at the front of this book.

The first of these mathlike statements simply says that the length described as one centimeter is the same length that is described as one-hundredth of a meter. The second statement says that the length described as one hundred centimeters is the same length that is described as one meter. These equalities can be used to change from one unit to another.

Converting from one unit to another involves proportions, and the common language of proportions is difficult for many people to follow. For example:

A desk is 1.5 meters long. How can this length be expressed in centimeters? Using proportions, the problem could be described as:

$$x : 100 :: 1.5 : 1$$

This expression is read, "The unknown (x) is to one hundred as one and one half is to one." In equation form, the proportion would be written as

$$\frac{x}{100} = \frac{1.5}{1}$$

Solving the equation for x, you get 150 as the numerical answer. What are the units?

An easier way to do these conversions is by applying the rules of mathematics to quantities. This procedure goes by many names such as unit-factor, factor-label, and dimensional analysis. In this procedure, equations are written to describe equalities among quantities instead of equalities among numbers which means that *units are included in the equations*. In fact, you will see that the units become an important aid in checking the logic used in these problems.

Conversions from one unit to another can be solved using ratios. The logic is easier to follow when the ratio has a denominator of one. Such ratios will be called **unitary rates**. The following unitary rate describes the number of centimeters that *correspond* to one meter.

$$\frac{100 \text{ cm}}{1 \text{ m}}$$

This ratio is normally read as "one hundred centimeters per meter" or "there are one hundred centimeters for each meter." It simply tells how many of one quantity correspond to one of another.

To find the length of the desk in centimeters, multiply its length in meters by the number of centimeters in each meter.

$$1.5 \text{ m} \times \frac{100 \text{ cm}}{1 \text{ m}} = 150 \text{ cm}$$

You have used unitary rates for years. You say that the speed limit is fifty-five miles per hour (55 mi/1 h) or that candy costs two dollars per pound ($2/1 lb). You may work for four dollars per hour ($4/1 h) or take an exam that counts five points per question (5 points/1 question). All these ratios represent unitary rates that can be used in quantity equations to convert from one measurement to another.

The logic of calculations using unitary rates is so easy to follow that you normally do it subconsciously. If your teacher announced a quiz with four questions and told you that each question would count five points, you would calculate the total points on the quiz without ever writing it down. But if you *did* write it down, how would you describe the calculation in the language of mathematics?

Mathematics would simply indicate:

$$4 \times 5 = 20$$

You would then need to remember what the numbers represent. The mathematics of quantities includes the units in the equations so you are constantly reminded of what the numbers represent. The quiz problem would be represented like this:

$$4 \text{ questions} \times \frac{5 \text{ points}}{1 \text{ question}} = 20 \text{ points}$$

Notice that the same rules of mathematics that apply to numbers or unknowns in algebra also apply to the units in the quantity equation. "Question" appears as a unit in both the numerator and denominator of the expression on the left. Consequently, dividing the numerator and denominator by that unit results in cancellation. The answer of twenty points is the way you would respond if asked the total number of points on the quiz. The fact that the answer makes sense should increase your confidence that your calculation was correct.

If the calculations in science were always as simple as this one, there would be no need to introduce unitary rates and the mathematics of quantities. However, many of the problems encountered in science can become so complex that some systematic procedure is needed to check the logic to be sure that it makes sense. By learning to use unitary rates now and practicing the procedure with simple problems, you will be able to handle complex problems later on. To get you started, work through the following example.

Example 1-1

My mass is 73 kilograms. What is my mass expressed as grams?

▶ **Suggested solution**

First, you need to be clear about what is being asked. You are to find the number of grams that is equivalent to 73 kilograms. This equivalence is indicated by the following statement

$$? \text{ g} = 73 \text{ kg}$$

Knowing how many grams there are in one kilogram enables you to answer the question by multiplying the 73 kilograms by the number of grams in each kilogram. Table 1-2 shows that kilo- means 1000. Then

$$1 \text{ kg corresponds to } 1000 \text{ g}$$

Writing this information as a unitary rate,

$$\frac{1000 \text{ g}}{1 \text{ kg}}$$

You are now ready to multiply the 73 kilograms given in the statement of the problem by the number of grams in each kilogram.

$$? \text{ g} = 73 \text{ kg} \times \frac{1000 \text{ g}}{1 \text{ kg}} = 73\ 000 \text{ g}$$

Notice that the units in the numerator and denominator cancel after doing the indicated multiplication. The answer should be checked to see if it makes sense. The units are grams which is what you are asked to calculate. Furthermore, the number of grams is much larger than the number of kilograms and that too makes sense. Since a gram is smaller than a kilogram it should take far more gram masses to balance your mass than it would take kilogram masses to balance your mass.

Example 1-2

? mm = 2 m

▶ Suggested solution _____

The question is asking how many millimeters are equivalent to two meters. Table 1-2 indicates that milli- means one one-thousandth. Expressing this relationship as a unitary rate gives

$$\frac{1000\ mm}{1\ m}$$

You are now ready to multiply 2 meters by the number of millimeters in each meter.

$$2\ \cancel{m} \times \frac{1000\ mm}{1\ \cancel{m}} = 2000\ mm$$

Review and Practice _____

1. Using the abbreviations for the base units given in Table 1-1 and the abbreviations for prefixes in Table 1-2, write abbreviations for the following metric units:
  a. milligram
  b. centimeter
  c. kilometer
  d. kilogram
  e. centigram
  f. decimeter
  g. micrometer
  h. megagram

1. a. mg   e. cg
   b. cm   f. dm
   c. km   g. μm
   d. kg   h. Mg

2. Write the name of the metric unit abbreviated below.
  a. mm
  b. cg
  c. kg
  d. km
  e. cm
  f. dg
  g. μg
  h. Mm

2. a. millimeter   e. centimeter
   b. centigram   f. decigram
   c. kilogram   g. microgram
   d. kilometer   h. megameter

3. Calculate the equivalence between the following metric units as shown in Example 1-2.
  a. ? cg = 1 g   1 cg = ? g
  b. ? kg = 1 g   1 kg = ? g
  c. ? km = 1 m   1 km = ? m
  d. ? cm = 1 m   1 cm = ? m

3. a. 100   0.01
   b. 0.001   1000
   c. 0.001   1000
   d. 100   0.01
   e. 10   0.1
   f. 1 000 000   0.000 001
   g. 0.000 001   1 000 000

e. ? dg = 1 g   1 dg = ? g
f. ? μg = 1 g   1 μg = ? g
g. ? Mm = 1 m   1 Mm = ? m

4. Complete the indicated conversions using unitary rates as shown in Example 1-1.

a. ? dag = 127 g   e. ? mm = 3.4 m
b. ? g = 268 mg   f. ? cm = 0.025 m
c. ? kg = 354 g   g. ? Mg = 5 245 g
d. ? m = 247 km   h. ? μg = 0.000 000 15 g

4. a. 12.7 dag
b. 0.268 g
c. 0.354 kg
d. 2.47×10^5 m
e. 3.4×10^3 mm
f. 2.5 cm
g. 0.005 245 Mg
h. 0.15 μg

1-11 Derived Quantities: Area and Volume

If the base units shown in Table 1-1 were the only SI units, you would be limited in describing all the quantities you observe. Take, for example, the size of this page. How big is it? You could measure its length and width, but those measurements are useful only if you want to compare the size of this page with one from a different book. Which is larger, a page that is 18 centimeters wide and 22 centimeters long, or a page that is 4 centimeters wide and 1 meter long?

The answer to this question depends on what is meant by "size." In this case size refers to the surface of the page, its area. **Area** could be described in many ways, but it is customary to describe it as the number of squares of some given size needed to cover a surface.

This page is covered with squares that measure one centimeter on each side. Its area can be found in square centimeters by counting the squares. You do not usually determine area this way because you know the mathematical equation that describes area. You will get the same number by counting squares that you would get by multiplying the length in centimeters by the width in centimeters. Using the mathematics of quantities, you get

$$25.5 \text{ cm} \times 20 \text{ cm} = 510 \text{ cm}^2$$

Notice how the answer was obtained. First the numbers were multiplied to get 510. Then the units were multiplied to get centimeters times centimeters or centimeters raised to the power of two. The exponent of two simply indicates the number of times the unit was a factor. However, you also should see that cm² has a very sensible physical interpretation. The area of this page is covered by 510 *squares* that measure one centimeter on a side. The new unit, square centimeter, can be interpreted as the name for one of those squares.

We have derived a new unit which can be used to describe area from a unit of length, one of the quantities shown in Table 1-1. Since the base unit for length is the meter, the base SI unit for area is the square meter (m²). Other units, such as square centimeter and square kilometer, can then be derived just as centimeters and kilometers are derived from the base unit for length, the meter. However, it is not simply a matter of attaching the prefix, as the following Example illustrates.

Figure 1-13 The surface of a football field is described by an area. The gymnasium shown on the right can be described by measuring its volume.

CHEM THEME

Measurements are made to answer the question, "How much?" As you can see from the units described here, "how much" can be answered in terms of volume (the space occupied) or mass (the amount of matter). Throughout this course, you will learn how measurements are used to describe the amount of matter and energy involved when matter undergoes change.

Example 1-3

If you were to draw a square that measures one meter on each side and mark off the square in square centimeters, how many square centimeters are equal to one square meter?

▶ **Suggested solution**

If you had to answer the question by counting, it would be a tedious task. A simpler route would be to first write the equivalence between centimeters and meters.

$$100 \text{ cm} = 1 \text{ m}$$

Next, square both sides of the equation.

$$(100 \text{ cm})^2 = (1 \text{ m})^2$$
$$10\ 000 \text{ cm}^2 = 1 \text{ m}^2$$

Finally, verify the answer by counting the actual number of squares. (Actually, it is a good idea to test the procedure with several simpler problems using fewer squares to count until you are convinced that this procedure always works.)

SI units for all quantities that you may wish to measure can be derived from the base units in Table 1-1. Some derived quantities discussed in this course are shown in Table 1-3. Some of these have special names and symbols, such as joule (J), the unit for energy, but all derived quantities can be expressed in terms of the SI base units. The last column of Table 1-3 gives that expression.

As shown in Table 1-3, volume has units of cubic meters. Other units for volume can be derived in exactly the same way that units for area were derived. Follow the procedure of Example 1-3 and calculate SI units for volume.

Table 1-3

SI DERIVED QUANTITIES				
QUANTITY	NAME	SYMBOL	DERIVATION	EXPRESSED IN TERMS OF SI BASE UNITS
area	square meter	m^2	$length \times width$	m^2
volume	cubic meter	m^3	$length \times width \times height$	m^3
speed, velocity	meter per second	m/s*	$\dfrac{length}{time}$	m/s
wave number	1 per meter	m^{-1}*	$\dfrac{number}{time}$	m^{-1}
density	kilogram per cubic meter	kg/m^3	$\dfrac{mass}{length \times width \times height}$	kg/m^3
force	newton	N	$\dfrac{length \times mass}{time \times time}$	$m \cdot kg \cdot s^{-2}$
pressure	pascal	Pa	$\dfrac{force}{area}$	$m^{-1} \cdot kg \cdot s^{-2}$ or N/m^2
energy	joule	J	$force \times length$	$m^2 \cdot kg \cdot s^{-2}$ or $N \cdot m$

*These symbols illustrate two ways that denominators of ratios are represented. The units for speed are shown as m/s and are read as "meters per second" or meters divided by seconds. An alternative representation is $m \cdot s^{-1}$. Negative exponents indicate a reciprocal. Just as 10^{-1} means 1/10, s^{-1} means 1/s. Then $m \cdot s^{-1}$ means "meter multiplied by reciprocal seconds" or m × 1/s which is m/s.

In similar fashion, m^{-1} means the same as 1/m. It can be read in several ways. When written as m^{-1}, it is customary to say "reciprocal meters." If written as 1/m, it is common to say "one (wave) per meter." Both expressions mean the same thing.

Review and Practice

1. Use the procedure shown in Example 1-3 to complete the following equalities.
   a. $?\ mm^2 = 1\ cm^2$
   b. $?\ m^2 = 1\ km^2$
   c. $?\ dm^2 = 1\ m^2$
   d. $?\ \mu m^2 = 1\ cm^2$

2. Draw a square to represent each of the units described in item 1. If the square is too large or too small to draw, try to describe its size by referring to something familiar.

3. In calculating area from units of length, the length units are squared. To find volume, the units must be cubed. Why?

4. Calculate the following equalities among SI volume units.
   a. $?\ cm^3 = 1\ m^3$
   b. $?\ m^3 = 1\ km^3$
   c. $?\ mm^3 = 1\ cm^3$
   d. $?\ cm^3 = 1\ dm^3$

5. The last value you calculated is especially important because metric (but not SI) units for volume include the liter which is one cubic decimeter. How many cubic centimeters are there in one liter?

■1-12■ Your Role as a Chemist

Science is a way of answering questions, and you have already encountered a few of the many questions that chemists ask repeatedly while studying matter. Throughout this course you will encounter various attempts to answer questions such as the following:

What is this matter that I have? Is a clear liquid in a beaker water that I need to survive, or is it alcohol that can dull my senses so that I do

1. a. 10 mm = 1 cm
   $100\ mm^2 = 1\ cm^2$
   b. 1000 m = 1 km
   $1\,000\,000\ m^2 = 1\ km^2$
   c. 10 dm = 1 m
   $100\ dm^2 = 1\ m^2$
   d. $10^4\ \mu m = 1\ cm$
   $10^8\ \mu m^2 = 1\ cm^2$

2. The square millimeter, square centimeter; and square decimeter could be shown. A square meter is about half the area of a twin bed mattress. A square kilometer is slightly larger than the area covered by 16 city blocks.

3. A cube has three dimensions—length, width, and height. A cubic decimeter is 10 cm × 10 cm × 10 cm. The number of cm^3's that occupy the same space as a dm^3 is 10 × 10 × 10 or 1000.

4. a. 100 cm = 1 m
   $10^6\ cm^3 = 1\ m^3$
   b. 1000 m = 1 km
   $10^9\ m^3 = 1\ km^3$
   c. 10 mm = 1 cm
   $1000\ mm^3 = 1\ cm^3$
   d. 10 cm = 1 dm
   $1000\ cm^3 = 1\ dm^3$

5. 1000

not care to survive, or is it sulfuric acid that will ensure that I do not survive? Is there mercury or asbestos in the water I drink? Is there enough protein in my food? Is there enough phosphorus in the soil to grow corn? Is there too much pollution in the air that I breathe? These are just a few of the reasons chemists ask, "What is it?"

Identification is always a matching process. Is the paint on the victim's coat the same as the paint on the car of the hit-and-run suspect? From the police lineup to the analyst's test tubes, the detective always asks whether the characteristics of the culprit are identical to those of the suspect under investigation. (Do they match?) The more characteristics that match, the more positively the criminal is identified.

The chemist has learned to look for and trust a number of characteristics from simple things that can be measured such as density, melting and boiling points, solubility, pH, crystal structure, odor, and on occasion, even taste to more sophisticated measures such as the wavelengths of light absorbed by the substance, or how the substance reacts with other chemicals. If you are to identify matter, you must know about the properties of matter and how to measure them.

How much do I have? The chemist must often ask how much of a substance is present in addition to what kind. From studying stoichiometry, the gas laws, pH, titration, and a host of other procedures, you see how chemists have acquired tools for learning how much.

How can I change it? Never satisfied with what they have, people search for something new and different. They not only want to know what they have and how much, but perhaps how to turn lead into gold. Changing lead to gold was what the earliest chemists sought to do. The search continues for something new and different in the laboratories of duPont, Eastman Kodak, 3-M, Eli Lilly, Dow, Merck, and many other companies.

Most of the early efforts to make new materials were trial and error, and good fortune still plays a major role in discovery. However, it is preferable to work from an understanding of the nature of things rather than relying heavily on chance. Your study of atomic structure, kinetic theory, bonding theory, and equilibrium will all furnish a foundation from which predictions of change can be made with more confidence.

How much can I get and how fast? If you can get 12% yield for the same cost that a chemist gets 10%, you have a new line of work while the chemist may be looking for another job. If you know what affects the speed of a chemical change, you can speed up reactions giving you the products you want and slow down procedures that give products you do not want. The implications of this control are vast, from regulating body metabolism, to corrosion, to growing crops, to eliminating wastes, to killing bacteria, to curing cancer, to living for 500 years. The more chemists know about equilibrium and reaction kinetics, the closer they come to such answers.

Most of what you will study in chemistry easily can be related to just these four questions. Some of what you learn is simply for the purpose of communication or because it opens up possibilities for learning still more. Chemical formulas, equations, and much of the mathematics that you learn are aids for communication and thinking. Knowledge is useless if you cannot communicate it to others, and how can you answer questions using your knowledge if you do not think clearly?

Answers to all questions appear in the Teacher's Guide at the front of this book. Numbers in red indicate the appropriate chapter sections to aid you assigning these items.

Summary

■ Science is a way of knowing that is based on careful observation. Scientists look for regularities and summarize them through classification, testing hypotheses, and the development of theories.

■ Chemists study matter, anything that has mass and takes up space. Energy is used to change matter from one form to another. For example, a change of state involves energy.

■ The three common states of matter are solid, liquid, and gas. Plasma, the fourth state, is less common.

■ With the exception of some nuclear reactions, matter and energy are conserved in all changes from one form to another.

■ Measurements are made on matter by comparing an unknown quantity to a standard. Measurements result in a number and a unit. The unit is important because it identifies the standard used for comparison.

■ Many units are in common use, but scientists use SI units. SI units are defined so that all quantities can be expressed in terms of seven fundamental quantities or base units.

■ Prefixes are used with base units to name units that are larger or smaller than the base unit. Units for area and volume are derived from the base unit for length. Units for other quantities are derived by combining other base units.

■ The mathematics for quantities is similar to any mathematics, but units are included in calculations.

■ Unitary rates are used to convert from one unit to another and are often used in other calculations involving quantities. A unitary rate is a term used to describe a ratio between quantities where the quantity in the denominator has a numerical value of one.

Chemically Speaking

classification	physical state
energy	plasma
hypothesis	potential energy
inertia	quantity
inference	SI
kinetic energy	SI base unit
mass	theory
matter	unitary rate
model	

Review

1. Name two sources of knowledge other than science. (1-1)

2. Why is careful communication so important in science? (1-1)

3. Name two things that scientists frequently do to improve communication. (1-1, 1-2)

4. Why do scientists insist that observations be made by at least two independent observers before they accept the observation as valid? (1-2)

5. What evidence is there that air is matter? (1-3)

6. Why is light not described as matter? (1-3)

7. Tell how the following terms are related to each other: mass, matter, inertia, and force. (1-3)

8. What is the process called for categorizing air, water, heat, and light as matter or energy? (1-2, 1-3)

9. How do plasmas differ from gases? (1-5)

10. How did the idea of potential energy come about? (1-6)

11. What is the difference between a quantity and a number? (1-7)

12. What is wrong with adding 2 yd + 2 ft? (1-7)

13. What metric unit would you use to describe your height? (1-8)

14. If your bathroom scale were metric, what units would be used? (1-8)

15. How does a unitary rate differ from any other ratio? (1-10)

16. How many milliliters are there in one cubic centimeter? (1-11)

Interpret and Apply

1. Recall when you disagreed with a friend, parent, or teacher about something both of you observed. What can account for such disagreements? (1-1, 1-2)

2. Would you expect disagreements like those suggested by item 1 to occur more often between people of the same age or people of different ages? From similar or dissimilar home environments? From the same or different countries? Explain the reasoning you used in answering these questions. (1-1, 1-2)

3. Get a partner and do the following: One of you should build something with blocks or similar construction material and then write a description of what has been built so the other partner can reproduce it without looking at the original construction. Compare the two constructions and talk about what was said that helped describe the construction and what was said that seemed to hinder the construction process. (1-2)

4. In outer space, a balance is useless for measuring mass. How might mass be measured in an orbiting spaceship? (1-3)

5. What change of state is taking place when moth flakes disappear over the winter? (1-5)

6. What major assumption about matter would be violated if the moth flakes actually disappeared to nothing? (1-6)

7. A transport truck has far more kinetic energy than a car moving at the same speed. Why? (1-6)

8. Under what conditions could a car have more energy than the transport truck? (1-6)

9. Write a unitary rate to describe the exchange rate between U.S. and Canadian dollars. (Most daily newspapers report exchange rates in the business news.) (1-10)

10. If you have a bathroom scale, a 20-kilogram mass, and a 50-pound mass, describe how you could find a unitary rate to convert from pounds to kilograms. (1-10)

Problems

1. How would the kinetic energy of a moving object change if its velocity was reduced by one half? (1-6)

2. Expressing the quantities in feet, write a mathematical statement that says "the distance from my shoulder to my wrist plus the distance from my wrist to my fingertip is equal to the distance from my shoulder to the tip of my finger." (1-7)

3. Repeat item two, but this time express the quantities in inches. (1-7)

4. Write the name for the following units. (1-10, 1-11)

a. mm	f. μg	k. m^3
b. Mm	g. Mg	l. cm^3
c. km	h. mg	m. mm^2
d. kg	i. dm	n. km^2
e. cg	j. dm^3	

5. Write two equivalents between the units listed in item 4 and the base unit. For example,

1000 mm = 1 m and 1 mm = 0.001 m   (1-10, 1-11)

6. Complete the indicated conversions. (1-10, 1-11)

a. ? g = 57 dag	f. ? m = 27 Mm
b. ? mg = 37 g	g. ? g = 462 cg
c. ? g = 4.7 kg	h. ? g = 14 000 μg
d. ? km = 138 m	i. ? pm = 37 cm
e. ? m = 4 021 mm	j. ? km = 3 200 000 mm

7. Complete the indicated conversions. (1-10, 1-11)

a. ? mm^2 = 1 m^2	f. ? cm^3 = 1 mm^3
b. ? km^2 = 1 m^2	g. ? m^3 = 1 cm^3
c. ? cm^2 = 1 mm^2	h. ? km^3 = 1 m^3
d. ? μm^2 = 1 mm^2	i. ? dm^3 = 1 m^3
e. ? Mm^2 = 1 km^2	j. ? μm^3 = 1 cm^3

8. Complete the following conversions. (1-10, 1-11)

a. ? m^2 = 5 280 mm^2
b. ? m^2 = 2.5 km^2
c. ? mm^2 = 453 cm^2
d. ? mm^2 = 9 055 033 μm^2
e. ? km^2 = 0.000 025 Mm^2
f. ? mm^3 = 123 cm^3
g. ? m^3 = 5 345 cm^3
h. ? m^3 = 0.046 km^3
i. ? m^3 = 1 000 dm^3
j. ? cm^3 = 3 098 000 μm^3

9. Which of the following are acceptable ways to express a pascal in terms of SI base units? (1-11)

a. $\dfrac{m^{-1} \cdot kg}{s^2}$

b. $\dfrac{kg}{ms^2}$

c. $\dfrac{s^{-2} \cdot kg}{m}$

d. $kg \cdot m^{-1} \cdot s^{-2}$

e. $\dfrac{m^{-1} \cdot s^{-2}}{kg^{-1}}$

Challenge

1. The idea that matter is conserved in all ordinary processes is believed because no one has observed a case when it was not true and because this idea has been very helpful in explaining many puzzling observations. However, this idea has never been proven and never will be. Proof would require careful measurements of matter before and after every kind of change. To see why obtaining this proof is impractical, try designing an experiment to prove that matter is conserved when the food you eat is digested and used by your body.

2. Using a ruler calibrated in both centimeters and inches, calculate a unitary rate that you could use to convert inches to centimeters.

3. Find the number of cubic centimeters in one cubic inch.

4. *Atto* is the smallest prefix defined for SI units and *exa* is the largest. How many cubic attometers would there be in one cubic exameter?

5. Which is larger, a cubic exameter or the earth? (Where might you look to find the size of Earth?)

6. Which is smaller, a cubic attometer or a hydrogen atom? (The diameter of a hydrogen atom is 6 nanometers.)

3. Research the area of air pollution. What causes air pollution? How can it be prevented? What steps have auto manufacturers taken to reduce pollution? Find out what steps a local (smokestack) industry has taken to reduce air pollution.

4. Make arrangements for a tour of your municipal water treatment plant with a group of students. Find out how the quality control technicians make certain that treated water poses no threat to the local water supply. Then research various means of water treatment as practiced throughout the country.

Projects

1. Invent a new measuring system. To be a good one it should have these characteristics:
   a) All standards should be ones that can be kept for centuries without change.
   b) It should be easy to copy the standard to make practical measuring devices for everyday use.
   c) It should be possible to express all quantities in terms of a few standards.
   d) Conversion from one unit to another should be simple.

2. Although science can contribute to the solution of many social problems, it cannot solve those problems because the methods of science do not apply. Pick a social problem that interests you (pollution, nuclear war, drug abuse) and outline all aspects of the problem. After outlining the problem, decide if science might be applied to some aspect of the problem and suggest how. Identify those aspects of the problem that cannot be addressed by science and explain why.

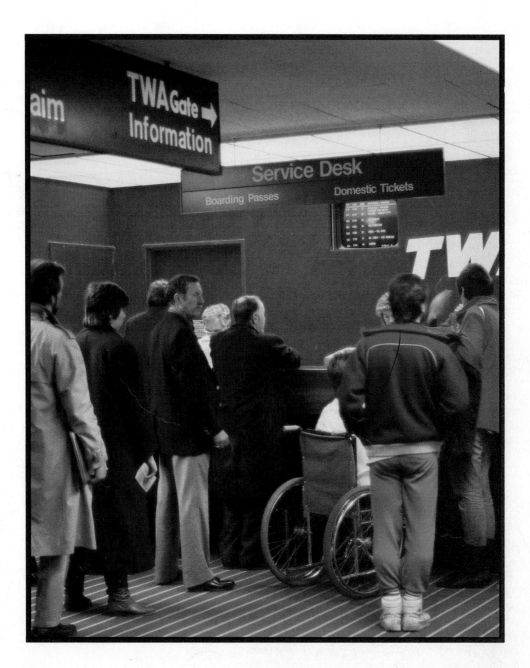

How do you get information when you need to know something? Some of the things you might do in trying to find out about something are similar to the steps scientists take in seeking information. You could ask people who are considered experts on a subject. You could use a literature search as another source of information. In this course, you will need to know about matter. This chapter is the start of your search for information to find out about matter.

Finding Out about Matter

Matter in the Macroscopic World

Civilization has made important advances when a few creative people have seen something where the majority claim nothing exists. Consider the picture in Figure 2-1. Perhaps your first impulse is to say, "What picture? There is nothing there!" But of course there is. There is paper and there is ink. Look closely with a magnifying glass, and you will see that what appears to be uniform color is actually a pattern of dots. Look at the dots under a microscope, and you will see that there is a pattern of red, blue, and yellow dots. What more might you see if you could look deeper into a yellow dot or a blue one? How is what you see under the microscope related to what you observe when you just look at the picture? How do you describe what you see so that others can understand results of careful observation? These are the questions that you will explore in taking a closer look at matter.

Unlike the previous picture, Figure 2-2 shows many things that you recognize immediately. You see food, plates, water, forks, people, candles, light, and a smile. In seeing these things, you have used the tools of science to observe and order observations through classification. You have even gone beyond what is seen to infer that the people are happy, constitute a family, and are preparing to eat.

Science applies the same skills you use to interpret the picture to make increasingly precise observations and detailed inferences. Detailed observation takes so much time and effort that it is necessary to

CONCEPTS
- macroscopic changes in matter
- mixtures and pure substances
- chemical and physical change
- compounds
- elements
- law of definite composition
- law of multiple proportions

Figure 2-1 Is there a picture?

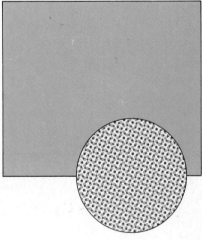

Figure 2-2 Many kinds of matter are represented in a familiar scene such as this one. What do you know about the matter pictured here?

specialize. Chemists specialize in matter; what it is, how one kind differs from another, what different kinds have in common, how one kind can be changed to another, and even how matter can be kept the same. Can a silk purse be made from a sow's ear? If so, how and is it worth the effort?

2-1 What You Know about Matter

By living in a world of matter like that shown in Figure 2-2, you have learned a tremendous amount about the subject. You know that water normally exists as a liquid that pours freely to fill any solid container, that water will change to solid when cooled, or to a gas when heated. You know that some solids such as water (ice) will shatter when struck by a hammer, but the metal in a fork or the wax in a candle will bend or flatten upon impact without shattering. If you heat a fork, it will eventually melt, change back to a solid when cooled, and, if heated in air, turn black; but the metal does not seem to burn. Not so with the wax in the candle. Burn a candle in air and you get fire that continues to burn until the wax is consumed. Such observations amazed you as a child. The fact that such a variety of changes can be understood using a few powerful ideas may be even more amazing to you now. To see how a few ideas can explain a lot of observations, you will learn to specialize; you will look carefully at a few familiar substances, classify them, and make generalizations about them based on the regularities you see.

Water is one of the most familiar kinds of matter. But what exactly is water, and is all water the same? If you collect water samples from a muddy stream and the ocean, you will claim the samples are different. They do not look the same, taste the same, or boil at the same temperature. A liter of one sample will not have the same mass as a liter of the other, because the samples have different densities.

In Chapter 1, you learned that properties such as color and taste are characteristic of matter; they describe matter specifically and can be used to identify matter. The temperature at which matter changes from a liquid to a gas, its **boiling point**, is still another property that you use to identify matter. Throughout this chapter, this course, and all other courses in science, you will add to this list of properties that can be used to identify matter and distinguish one kind from another. Part of the study of science is the accumulation of such information; another part is using the information you have to make sense of the world by describing it with increasingly greater precision.

2-2 Purifying Matter

You know that muddy water left standing in a glass for hours or days will separate into layers of dirt and clear water. This observation can be explained by saying that the water sample is not pure. You might say that it is a **mixture**—two (or more) kinds of matter that have separate identities. Matter that is easily separated into component parts is

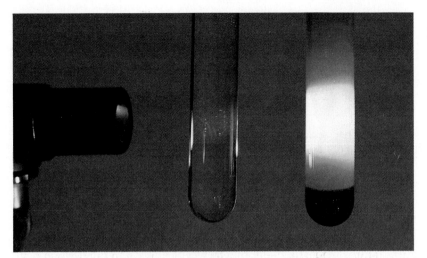

Figure 2-3 When a beam of light passes through pure water or a solution, the light is not scattered and is not visible when viewed from the side. When light passes through a liquid with minute particles mixed in it, the light is scattered and is visible when viewed from the side.

called a mixture or said to be impure. The tricky part is deciding which samples can be separated and which samples cannot.

Classifying matter into pure forms and mixtures is one step toward understanding matter and developing a language to describe that understanding. Like all understandings, it is based on experience. From the experience that some samples of matter are easily separated while other samples are not, chemists derive the idea of mixture and pure substance to describe matter.

If you compare the water from which dirt has settled with water from your kitchen tap, you can probably tell them apart by taste and possibly by appearance. If you shine a strong light through the samples as shown in Figure 2-3, the light is scattered by the settled muddy water, but it is not scattered by most tap water. (Some tap waters contain impurities that scatter light; pure water and solutions do not scatter light.) Is the scattered light due to some kind of matter mixed with the water? If so, you should be able to separate the mixture, but how?

If you have visited a water purification plant or purified water yourself, you know the answer. By adding alum and lime to the water, a gelatinous (jellylike) material is produced. As this gelatinous material settles out or is removed by filtering through beds of sand, small particles suspended in the water stick to the gelatinous material and are removed. These processes leave water that does not scatter light. The mixture has been separated into components but is the water now pure?

By adding things to pure water you can quickly find that there are mixtures that do not scatter light. For example, if table salt or sugar is added to water, it disappears. Taste suggests that the salt or sugar is still there, but it cannot be seen; neither will it scatter light. Thus light scattering cannot be used as a test of whether a material is pure. Mixtures like salt water or sugar water that look uniform throughout and do not scatter light are called **solutions**.

Mixtures are matter that can be separated into component parts that have a separate identity. If solutions such as salt water and sugar water are described as mixtures, it is implied that these solutions can be separated into component parts. If so, how?

Demonstrate the Tyndale effect using a projector or flashlight as a light source. Compare solutions, pure water, and water with a few drops of milk or soap added.

Figure 2-4 The addition of alum and lime to muddy water produces a gelatinous substance that can be used to trap impurities.

Figure 2-5 Ocean water or salt water made by mixing table salt and water can be separated by placing the mixture in the flask and boiling it. The water in the mixture will change to a gas, then move through the tube where it cools to become liquid once more. Because the boiling point of pure salt is well over 1000°C, it remains in the original flask after all the water has boiled off.

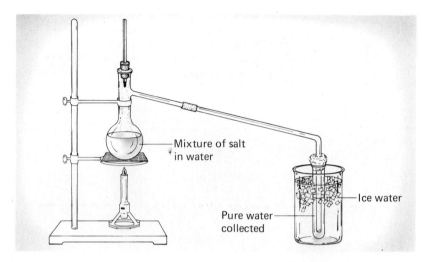

Mixture of salt in water

Ice water

Pure water collected

Do not slight this section. Interpreting macroscopic changes is difficult. You might do the destructive distillation of wood and ask students to decide whether a chemical change has occurred and explain how they know. Then examine the brown liquid to decide whether it is a mixture or a pure substance. Ask the same question about the charcoal left behind. The discussion could extend over two days to help students clarify concepts such as a pure substance, mixture, element, compound, chemical change, and physical change. The same apparatus can be used to heat iron filings, followed by the same questions. Baking soda and sulfur also can be heated, and students should be asked whether chemical changes occur and explain how they know. These activities may extend over 3–4 days of class.

If you ever have let a pan of salt water boil dry on the stove, you have a clue about how solutions can be separated. When the pan boils dry, the white solid that remains is the salt that was dissolved in the water. If this procedure is carried out so the water that boils away is recovered, solutions like salt water and sugar water can be separated into their component parts using a procedure called **distillation**.

Figure 2-5 shows a simple apparatus for separating mixtures by distillation. Although most mixtures containing water can be separated by distillation, some cannot. Household ammonia is a solution of ammonia and water that cannot be separated totally by distillation. Alcoholic beverages such as whisky are solutions containing ethanol (grain alcohol) that cannot be separated by any of the techniques described thus far.

Many useful ideas are simple in theory, but complicated in practice. The idea that matter can be separated into mixtures and pure substances is such an idea. By starting with examples such as muddy water and tap water, it is easy to see that the difference in the two can be explained by saying that one is a mixture of dirt and water but the other is not; perhaps it is pure. But when tap water is boiled, you find that it too can be separated into components, and you can conclude that it is a mixture of water and dissolved solids such as salt. Only when there is no way to separate the matter into components can you conclude that it is pure. Many mixtures are extremely difficult to separate. Such mixtures may be considered pure for many years until someone invents a new procedure or a better instrument for analysis and separation. Much of the recent awareness of air pollution, water pollution, carcinogens in foods, and similar environmental concerns have come about because new and better techniques have been found to detect those impurities.

Do You Know?

One of the gases produced when wood is heated in the absence of air can be condensed forming an alcohol. This alcohol, methanol, was once known as wood alcohol.

2-3 Characteristics of Pure Substances

If you measure the temperature of a boiling liquid while it is being distilled as shown in Figure 2-5, you will detect an important difference between mixtures and pure substances. As a solution of salt

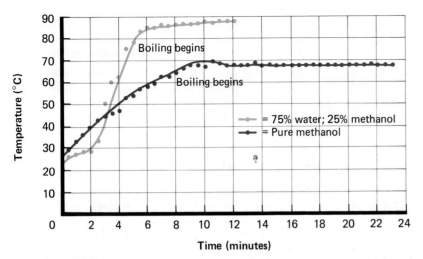

Figure 2-6 As pure methanol is heated, the temperature gradually rises to 65°C where it begins to boil. The temperature remains constant throughout the boiling. By contrast, a mixture containing 75% water and 25% methanol begins boiling at about 86°C, but the temperature continues to rise as the mixture boils.

water is boiled, the temperature gradually rises as the water boils away. However, if you boil the water collected in this way, you will find that the temperature will remain constant from the time the water first boils until it all disappears. Pure substances have a constant boiling point; mixtures ordinarily do not. Figure 2-6 contrasts the constant temperature at which pure methanol (wood alcohol) boils with the gradually increasing temperature observed when a mixture of methanol and water boils.

Unfortunately, nature seems to defy simple description. It would be easy to say that mixtures *never* have a constant boiling point, but a few do. The reason that water and grain alcohol cannot be separated by distillation is that a mixture containing 95.6% ethanol and 4.4% water has a constant boiling point of 78.2°C. Any other mixture of ethanol and water changes temperature as it boils.

Figure 2-7 shows the difference between mixtures and pure substances when they boil. Similar differences between mixtures and pure substances are observed when they freeze. The graphs in Figure

Figure 2-7 The graph on the left shows the temperature of paradichlorobenzene and water as the two substances cool using the equipment shown in Figure 2-8. As time passes, the temperature of the water drops as heat is radiated to the air, and the temperature of the paradichlorobenzene drops as it loses heat to the water. The paradichlorobenzene continues to lose heat to the water when it freezes, but the temperature does not drop any further until all of it has changed from a liquid to a solid. The graph on the right shows when a mixture of paradichlorobenzene and naphthalene is cooled in water, the observed temperature changes are similar to those seen in the graph on the left. However, when the mixture freezes, the temperature continues to fall.

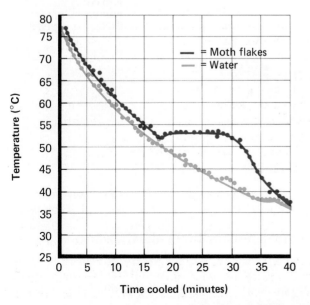

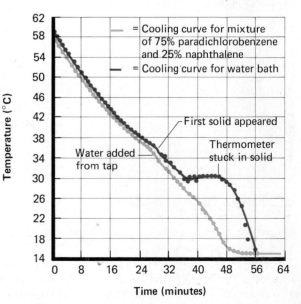

It is difficult for beginners to interpret such graphs. Class discussion is recommended. Better still, have students do the experiment and discuss their graphs in class.

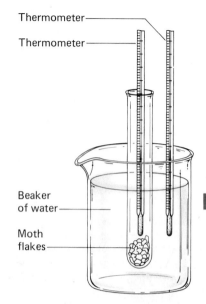

Thermometer

Thermometer

Beaker of water

Moth flakes

Figure 2-8 The data shown in Figure 2-7 were taken in this apparatus. The thermometer in the test tube is used to measure the temperature of the moth flakes. The thermometer in the beaker is used to measure the temperature of the water.

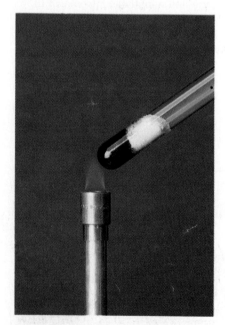

Figure 2-9 When sugar is heated, it decomposes to form a black solid and gaseous products. Some of the gaseous products condense to form a clear liquid. ·

2-7 contrast the behavior of pure paradichlorobenzene, a compound sold as a moth repellent, with a mixture of paradichlorobenzene and naphthalene, another moth repellent. The temperature at which a liquid changes to a solid is called its **freezing point**. The freezing point is the same as the **melting point**, the temperature at which a solid becomes a liquid. Study the two graphs until you can explain the difference between the freezing point of pure substances and the freezing point of mixtures.

2-4 Chemical and Physical Changes

Table 2-1 shows the melting point, boiling point, and density of some common substances that can be obtained in a pure form. Density is a property of matter that describes its mass per unit volume. The density of sucrose is shown as 1.58 g/cm^3. This means that one cubic centimeter of sucrose has a mass of 1.58 grams. You will study more about density as a property in Chapter 3. Notice that melting points and boiling points are not shown for sugar or baking soda. When sugar is heated, it bubbles and turns black, as shown in Figure 2-9. If the mouth of the test tube is cool, a colorless liquid forms on the walls of the test tube as sugar is heated. The sugar does not melt. When cooled, a black solid and colorless liquid remain. If the solid and liquid are mixed together, it becomes a wet mess that is not at all like sugar. There is no way to get the original sugar back. Instead of melting, the sugar changes into new substances with new properties. The black solid does not taste like sugar, look like sugar, or have the same density as sugar. The colorless liquid does not have the same properties as sugar either. Thus there is no longer any sugar in the test tube.

Table 2-1

REPRESENTATIVE PROPERTIES OF SOME PURE SUBSTANCES			
NAME	MELTING POINT (°C)	BOILING POINT (°C)	DENSITY (g/cm³)
sucrose (table sugar)	—	—	1.58
sodium bicarbonate (baking soda)	—	—	2.16
sodium chloride (table salt)	801	1413	2.16
water	0	100	1.00
iron	1535	3000	7.87
carbon (graphite)	3550	4827	1.9–2.3
copper	1083	2595	8.96

Similarly, when baking soda is heated, a colorless gas is driven away, leaving behind a white solid that looks like the original baking soda, but with different properties. Baking soda has a density of 2.159 g/cm^3; the white solid that remains when it is heated has a density of 2.532 g/cm^3. Baking soda can be added to bread dough to make the bread rise; the white solid that forms when it is heated will not make bread dough rise. Other properties of the two substances differ. Again, when baking soda is heated, some new kind of matter is formed. Changes that produce a new kind of matter with different properties are called **chemical changes**. This particular chemical change is called a **decomposition** because one kind of matter comes apart (decomposes) to form two or more kinds of matter.

Compare what happens when moth flakes are heated to what happens when sugar is heated. When moth flakes are heated, the solid material disappears and a liquid appears in its place. If the liquid is allowed to cool as shown in Figure 2-8, the liquid turns back to a solid that has the same properties as the original moth flakes. The new solid has the same smell as the original one, has the same density as the original solid, melts at the same temperature as the original solid, and so forth. Every observation indicates that the two solids are in fact the same kind of matter. Because of these observations you infer that the moth flakes melt when they are heated. Melting is the term used to describe the change of a solid to a liquid without the formation of any new kind of matter.

The chemical change taking place in sugar when it is heated differs from the melting of moth flakes in two ways: First, the change in the sugar is not easily reversed. Second, the black solid and the liquid formed when sugar is heated are nothing like the sugar used at the start. Melting and other changes that are easily reversed to get the original material back again are described as **physical changes**. They do not appear to produce new kinds of matter.

Boiling is a physical change. When water is boiled in the apparatus shown in Figure 2-5, the liquid collected has all the properties of the original water. The water has changed from a liquid to a gas and back again. However, it does not appear that any new kind of matter formed.

Even when ocean water is boiled to give a liquid and a solid with different properties from the original liquid, you might conclude that a mixture was separated rather than decomposed into new materials. You base this conclusion on two reasons: First, the change is easily reversed by mixing the solid and liquid to form a mixture just like the original ocean water. In most cases, physical changes are easily reversed. Second, ocean water has properties that seem to be a blend of the properties associated with the two substances formed when it is distilled. It has the same salty taste of the solid recovered, and it has most of the properties of the pure water that is obtained from it.

You may want to mention that the movement of charged particles completes the circuit and the bulb lights.

2-5 Compounds and Elements

Although many pure substances like sugar and baking soda decompose when they are heated, the majority do not. All other substances shown in Table 2-1 have melting points and boiling points that are constant, just like the melting point and boiling point of pure moth flakes. However, there are other things that can be done to these substances that may cause them to decompose.

Figure 2-10 shows a diagram of a simple apparatus that can be used to test electrical conductivity. When that apparatus is used to test the electrical conductivity of dry sodium chloride (table salt), the bulb

The discussion here implies that there is a clear distinction between physical and chemical changes which is not true. Read "Physical Versus Chemical Change" by Walter J. Gensler, *J. Chem. Educ.*, Vol. 47, No. 2 (February, 1970), pp. 154–155 for an excellent discussion of the issue.

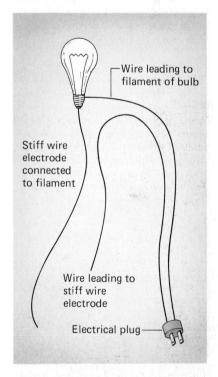

Figure 2-10 A simple electrical conductivity tester. If the two wires are connected by a piece of metal or any other matter that conducts electricity, the light bulb will glow. If they are connected by any matter that does not conduct electricity, the bulb will not light.

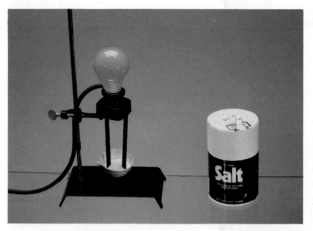

Figure 2-11 The dry salt on the left does not conduct electricity. If it did, the bulb would light when the salt connects the two wires on the conductivity tester. The melted salt on the right conducts electricity, causing the bulb in the conductivity tester to light. Salt dissolved in water will also conduct electricity.

Since atomic structure has not been discussed, do not go into detail about the difference between atoms and ions at this point. Students are aware that objects can acquire an electric charge, and some will know that the charge is due to an excess or deficiency of electrons, but it is not necessary to develop electron configuration at this point. Doing so is likely to interfere with the development of the other concepts presented.

does not light, as shown on the left in Figure 2-11. However, if the salt is melted before it is tested, it conducts electricity as shown on the right. Something else happens too.

Figure 2-12 shows a diagram of a container with two carbon rods immersed in melted salt and connected to a battery. Because melted salt conducts electricity, electricity will flow from the negative terminal of the battery to the carbon rod labeled (−), through the melted salt to the carbon rod labeled (+), and back to the positive terminal of the battery. If the battery is left connected, a silver liquid will form on the negative rod and drip to the bottom of the container, mixing with the melted salt. Meanwhile, a choking gas forms around the positive rod and bubbles out of the liquid salt. If the process is allowed to continue, all the melted salt will eventually disappear, leaving only the silver liquid and choking gas in its place. If these new substances are collected and their properties are checked, it is found that the silver liquid is sodium metal and the choking gas is chlorine.

The process just described is called electrolysis. **Electrolysis** involves passing an electric current through a substance, causing it to decompose into new kinds of matter.

It is not practical to do an electrolysis of sodium chloride in your school laboratory. The temperature required to melt salt is too high, and the chlorine gas produced is toxic. Sodium metal is very dangerous as well, because it reacts explosively with water. You can, however, do an electrolysis of water, the next substance in Table 2-1.

Figure 2-12 When salt is melted and an electric current is passed through the melt, the salt decomposes to form sodium metal at the negative electrode and chlorine gas at the positive electrode.

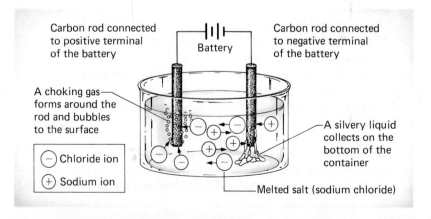

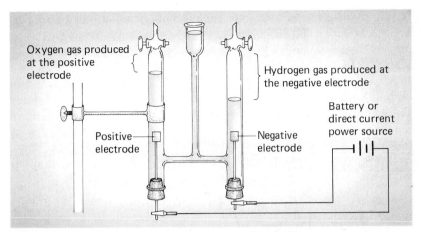

Oxygen gas produced at the positive electrode

Hydrogen gas produced at the negative electrode

Battery or direct current power source

Positive electrode

Negative electrode

Figure 2-13 Apparatus used for the electrolysis of water. The apparatus is designed to collect the two gases formed by the decomposition of water. The center tube of the apparatus is filled with water. The clamps on the outer tubes are open. When the apparatus is full, the clamps are closed and the two electrodes are connected to the battery. Oxygen gas is produced at the positive electrode and bubbles to the top of the tube. Hydrogen gas is produced at the negative electrode.

The apparatus you will need is shown in Figure 2-13. Since water is a very poor conductor of electricity, the reaction is too slow to see unless something that conducts electricity is dissolved in the water. Sulfuric acid is normally used, because it does not interfere with the decomposition of water. Other conducting substances can also be used, but many of them will react before the water does.

If the electrolysis of water is allowed to continue for a long time, the water will eventually change into hydrogen and oxygen gases, leaving behind nothing but the sulfuric acid added to make the water conduct.

Like the products formed when sugar and baking soda are decomposed by heat, the products of the electrolysis of sodium chloride and water have none of the properties of the original material, and they cannot be changed back to the original material simply by mixing them together. Consequently, electrolysis represents a chemical change in which matter decomposes to form new kinds of matter.

Decomposition of a pure substance and distillation of a mixture are both processes in which matter is separated into components, but in thinking about what happens you see that they are fundamentally different processes. In decomposition, a single, pure substance with constant, characteristic properties is somehow changed into new substances with different properties. Decomposition represents a chemical change. In distillation, the separated components exist in the original mixture as separate substances. The properties of the mixture are a blend of the properties of these components of the mixture. Distillation is a physical change that separates two or more things that already exist.

The last three substances listed in Table 2-1 cannot be decomposed by heat or electrolysis. In fact, no means have been found to separate these substances into new kinds of matter. This fact represents an important difference between substances in the table like iron, carbon, and copper, and the other substances like sugar, soda, salt, and water. Some words are needed to describe these different kinds of pure substances.

As you probably know, pure substances that can be decomposed into new kinds of matter are called **compounds**. They appear to be "compounded" or put together from simpler substances. The substances that they are composed of—sodium and chlorine in the case

It is recommended that you demonstrate the electrolysis of water. If you do not have platinum electrodes, monel metal or other iron alloys will work, providing sodium carbonate is used as the electrolyte.

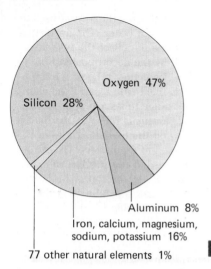

Oxygen 47%

Silicon 28%

Aluminum 8%

Iron, calcium, magnesium, sodium, potassium 16%

77 other natural elements 1%

Figure 2-14 Percent abundance of elements in the Earth's crust. What eight elements account for 99% of the Earth's crust?

This would be an excellent time to have students investigate the "pop" when small test tubes of hydrogen are lit under various conditions. See the *Heath Chemistry Laboratory Manual.*

CHEM THEME

Any sample of a given compound contains the same percentage by mass of each element. Samples of compounds which contain the same elements but in different percentages by mass are different.

Do You Know?

Freezing point and melting point are two different names for the same temperature. The term used depends on whether you start with a solid and melt it, or a liquid and freeze it. Having two technical terms for the same phenomenon is a common occurrence in chemistry.

of salt; hydrogen and oxygen in the case of water—are like iron, carbon, and copper. These pure substances cannot be decomposed. They are the elemental building blocks of all kinds of matter and are called **elements**.

There are now 109 known elements. These elements are listed in alphabetical order in Table 2-2 on page 40, and they are listed in a table on the inside back cover of this book. Several of these elements have been made in atomic reactors and do not exist in nature. Virtually all matter on Earth is made of only 85 elements, and just eight of these account for 99% of the Earth's crust as shown in Figure 2-14. All other elements in Earth's crust make up only 1% of its mass!

2-6 Compounds Have a Definite Composition

You now know that elements are a kind of matter that cannot be decomposed, and that compounds are matter put together by combining elements. However, not all combinations of elements are compounds.

The hydrogen and oxygen gases produced when water is decomposed can be mixed together in any proportions. One liter of hydrogen can be mixed with 100 liters of oxygen, with 50 liters of oxygen, or any other amount. The result will be a colorless, odorless gas like air. However, if the mixture is ignited it explodes. When it explodes, some of the hydrogen and oxygen combine to form water again. If that water (or any other water) is decomposed by electrolysis, the volume of hydrogen gas produced will always be twice the volume of oxygen gas produced as shown in Figure 2-13.

An important difference between mixtures of elements and compounds of elements is that the mixtures can have almost any composition that is desired, but the compounds will have a definite composition. This experimental fact is called the **law of definite composition**.

The definite composition of water was described in terms of volume: The volume of hydrogen gas obtained from water is always twice the volume of oxygen gas obtained. The definite composition also could be described in terms of mass. In the case of water, every nine grams of water contains one gram of hydrogen and eight grams of oxygen. Interestingly enough, the eight grams of oxygen occupy only half the volume that the one gram of hydrogen occupies when both elements are gases.

Although every compound has a definite composition, it is possible to make several compounds from the same elements. For example, a compound can be made from 100 g of copper and 25 g of oxygen. Under different conditions, a different compound containing 200 grams of copper and 25 grams of oxygen can be made. The fact that two or more compounds with different proportions of the same elements can be made, is known as the **law of multiple proportions**. Keep in mind that a *particular* compound does not have multiple proportions like a mixture; rather, the same elements can form *different* compounds, each with a definite composition, but each having a composition that differs from the others.

Review and Practice

1. Figure 2-7 (left) shows the temperature of paradichlorobenzene as the liquid is cooled in warm water. Explain the following from the graph.
   a. Why does the temperature of the water drop over time?
   b. If data were collected for a longer period of time, the temperature of the water would eventually become constant. Why? (Hint: If you leave a glass of water sitting in a room overnight, what will the water temperature be in the morning?)
   c. Several points on the graph do not fall on the smooth line drawn through the points. Why?
   d. The temperature of the paradichlorobenzene is always a little higher than the temperature of the water. Why?
   e. At 53°C the temperature of the paradichlorobenzene remains constant for almost ten minutes. What do you think is happening to the paradichlorobenzene during that time?
   f. In which region of the graph does the paradichlorobenzene exist as a liquid? As a solid? As a mixture of solid and liquid?

2. Sticking a thermometer into a glass of clear liquid will not reveal what the liquid is. Under what conditions could you measure the temperature of the material to get useful information about its identity?

3. Similarly, measuring the mass or volume of an object will give no clue to its identity. What could you do with mass and volume to produce a quantity that would describe a characteristic property?

4. If sawdust is heated in an apparatus like the one shown in Figure 2-5, the sawdust will turn black, and brown liquid will collect in the test tube. A gas with a bad odor will escape into the room. What would you do to decide whether this change is a chemical change or a physical change?

5. How could you test the brown liquid described in question 4 when sawdust is heated to see if the liquid is pure?

6. Would you describe the burning of charcoal as a physical change or a chemical change?

7. In Chapter 1, you learned that observations are often misleading and confidence is gained in ideas when several independent observations lead to the same conclusion. Suggest something you could do to increase your confidence that the muddy water from a river is a mixture of dirt in water.

8. Adding a drop of liquid soap or milk to pure water will cause it to scatter light. Does this observation increase your confidence that these liquids could be described as mixtures? Why?

9. In the discussion of the decomposition of water, you learned that 1 g of hydrogen gas occupies twice the volume of 8 g of oxygen gas. Which gas has the greater density? How do you know?

10. Using the data from question 9, complete the following statement to describe the facts given. "The density of (hydrogen, oxygen) is _____ times the density of (hydrogen, oxygen)."

11. Gold jewelry is not pure gold. It is made by combining less expensive metals with gold. The amount of gold can vary from 10 carats to 22 carats. Do you think that your gold jewelry is a compound or a mixture? Why?

12. How could you test your gold jewelry to determine whether it is a pure compound or a mixture?

13. What evidence suggests that distillation is a physical change rather than a chemical change?

1. a. Heat is transferred from the water to the air.
   b. The water temperature will remain constant when it reaches room temperature.
   c. These are real data points. The curve shows a best-fit line.
   d. Paradichlorobenzene loses heat to the water; water loses heat to the room. A temperature gradient is established with the paradichlorobenzene at the highest temperature and the air at the lowest temperature.
   e. A change of state occurs—melting or freezing.
   f. above 53°C—liquid; below 53°C—solid

2. Measuring a melting point and a boiling point could provide a clue.

3. Determine density—mass per unit volume.

4. Observe characteristic properties of the original sawdust and the final products. Different properties mean a different substance—chemical change.

The brown liquid collected from the destructive distillation of wood can be distilled further to show its rise in temperature while distilling. This may be compared with the behavior of water or other pure liquids.

5. Boil the liquid and measure the temperature. A constant boiling temperature would indicate a pure substance.

6. Chemical

7. Make a muddy water mixture and see if it behaves like the river water.

8. Yes, mixtures scatter light.

9. The density of oxygen is 16 times that of hydrogen.

10. 16

11. Since the amount of gold can vary from 10 to 22 carats, it must be a mixture.

12. A melting point determination could be used to test for pure gold.

13. If a pure substance is distilled, the distillate has the same properties as the original substance.

Matter Is Made of Atoms

CONCEPTS

- atoms
- microscopic nature of elements
- microscopic nature of compounds
- molecular models

Up to this point you have made observations by what you can see, feel, or smell. Such observations are **macroscopic observations** (*macro* means large; *scopic* means viewing or observing). Melting point, boiling point, heat of fusion, temperature, and mass are all properties of large chunks of matter and are called **macroscopic properties**. You can learn a lot about matter from such properties. Can you decide why matter behaves as it does from these properties? Usually not. If you could magnify matter in the same way that you magnified the picture that introduced this chapter to see deep inside a mixture, how would it differ from a pure substance? Is there something about the inner structure of matter that explains why a pure substance has a definite boiling point while a mixture does not? Is there some way to explain why sugar and salt can be broken down into simpler substances but iron and copper cannot?

"Explain" is a devious word. When you explain, you often think that you are talking about things as they really are, when in fact you are only talking about things as they might be. Some scientific explanations simply provide a way to think about why things happen. Sometimes they are accurate descriptions of what happens, but at other times they are not. You must realize that explanations in science may be intelligent guesses that are based on observations rather than established fact.

In this section, you will begin to develop a **microscopic model** (*micro* is a prefix meaning small; when used with SI units, it means one millionth) to explain the behavior of matter. Throughout this course the model will be expanded, enabling you to explain more things about the behavior of matter.

Exercises similar to the one in which students were asked to find a millionth of a sheet of paper may be useful to get students to realize just how tiny atoms are.

2-7 Atoms

You might think that matter is composed of some kind of small pieces. Is it? Most people seem to think so. Small particles like atoms and molecules are discussed in the popular literature as though they were as common as toads and warts.

People have not always believed that matter is made of atoms. It was not until the early 1800s that the idea became popular. The original idea of atoms was developed long before as a kind of armchair argument. If you take an iron nail and break it again and again, it seems reasonable that one will eventually obtain some "smallest possible piece" that can still be called iron. The word **atom** means this smallest possible piece of something.

Since you cannot see real atoms and it is easier to describe things that can be seen, atoms are normally represented with spheres of various sizes and colors. The size for the spheres is usually selected to suggest the relative size of the actual atom. It is difficult to keep everything to scale, and you should not assume that the spherical models are exact representations of real atoms. Chemists use spheres to help them imagine what is happening in the microscopic world.

Figure 2-15 The atoms in a compound can be visibly represented using spherical models. However as you study more about atoms you will realize how such models fall short of accurately representing real atoms.

2-8 Elements

Earlier, elements were defined as pure substances that cannot be broken down. Therefore, it seems reasonable to assume that an element contains only one kind of atom. If more than one kind were present, it should be possible to separate the two kinds of atoms, producing different substances with different properties.

If each element contains a different kind of atom, then there are as many kinds of atoms as there are elements, 109. Numbers can be assigned to each kind of atom, beginning with one and ending with 109; that number is called the atomic number. The table on the inside back cover of this book shows the atomic number for each element at the top of the rectangle representing the element.

An element can exist as a solid, liquid, or gas. How can the difference in these three states be explained using the idea of atoms?

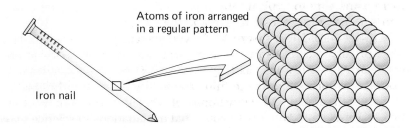

Atoms of iron arranged in a regular pattern

Iron nail

Figure 2-16 Iron is an element, and so we believe that pure iron is composed of only one kind of atom. In solid iron, such as the iron nail, we believe that these atoms are stuck together in a regular pattern similar to the one shown.

Figure 2-16 shows an iron nail and a representation of what a portion of the nail might look like if it could be magnified to show individual atoms. The atoms are stuck together so the solid holds its shape. However, atoms wiggle about or vibrate even in a solid, and they vibrate more as the temperature is increased.

When the temperature is increased enough, the atoms in a solid vibrate with such force that they overcome the forces that hold them together, allowing them to flow past one another. The force of gravity pulls the atoms down so that the liquid flows to take the shape of its container. Figure 2-17 illustrates what atoms might look like in a liquid. As before, chemists think that the atoms are vibrating, and since they can now move past one another, a particular atom may move from one side of the container to another.

When the temperature of a liquid is raised to its boiling point, atoms move with such energy that they totally escape from the liquid and move far apart as shown in Figure 2-18.

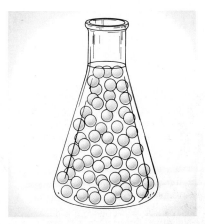

Figure 2-17 Atoms in a liquid are still close together, but they are free to move past one another. The atoms are no longer in a carefully ordered pattern, and they take up slightly more space in the liquid than in the solid.

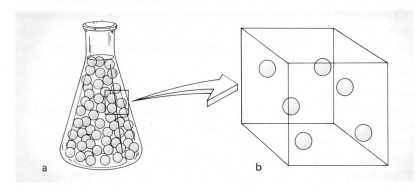

a   b

Figure 2-18 At high temperatures, atoms escape from a liquid and move very far apart. The gas phase of most substances is colorless. Atoms in a gas will move in a straight line until they collide with other gas atoms or the walls of the container.

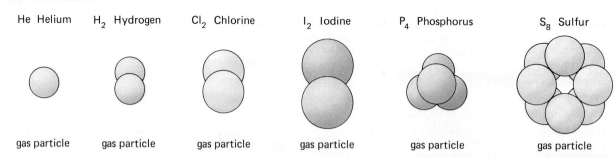

He Helium   H₂ Hydrogen   Cl₂ Chlorine   I₂ Iodine   P₄ Phosphorus   S₈ Sulfur

gas particle   gas particle   gas particle   gas particle   gas particle   gas particle

Figure 2-19 Some elements exist in various combinations of their atoms forming molecules. Notice that each molecule is composed of at least two atoms.

It is difficult to think in terms of atoms and molecules. Take time to discuss various changes that have been demonstrated (decomposition, distillation, melting, etc.) in terms of atoms. Use actual models whenever possible.

Students have no way of knowing what compounds are ionic and what compounds are molecular. Later students will calculate molar masses of these compounds since the term can apply to both types. Toward the end of the year when students have a basis for distinguishing between various kinds of compounds, you can clarify the difference between ionic and molecular compounds. At that time, point out that molar mass or formula mass is a general term that can be applied to any compound, whereas molecular mass is misleading when applied to ionic compounds.

Figure 2-20 (a) Solid water is composed of molecules bonded together in an ordered array. (b) When the solid melts, the molecules separate, but the atoms within each molecule do not. More energy is required to break the bonds holding atoms together in a molecule than to break the bonds holding the molecules together in the solid. (c) When water boils, the molecules move far apart to form gaseous water. The atoms making the water molecules are still joined together. Some compounds, such as sugar and baking soda, decompose before they melt or boil.

Most elements seem to exist as individual atoms as shown in Figure 2-16, but some elements exist as larger units. Figure 2-19 illustrates the gaseous particles of several elements. Particles made of more than one atom are called **molecules**. There is no simple way to tell which elements exist as individual atoms in a gas and which exist as larger clusters of atoms.

Elements have different melting points and boiling points. How can this fact be explained? Atoms vary in size and in mass. You know that it takes more energy to move a massive object than a light one, so one explanation for differences in melting points and boiling points of elements is that more energy is required to get large atoms moving with enough energy to break away from their neighbors. As you will learn later, other factors are involved, but it is generally true that the larger the particle, the higher the boiling point.

2-9 Compounds

You learned in Section 2-6 that compounds are made by combining elements in definite proportions. Then all compounds must be made of two or more kinds of atoms. Figure 2-19 shows individual molecules of several compounds. Molecules have definite shapes as well as definite composition, but you do not have to worry about why until later.

You can now begin to understand what happens when a compound decomposes into its elements. Since compounds are made of two or more kinds of atoms, the different atoms can be separated if enough energy is supplied to break the compound apart. Heat and electricity are forms of energy, and both heating and electrolysis can supply the energy needed to decompose the compound.

Like elements, compounds can exist as solids, liquids, and gases. Figure 2-20 shows water in these three states. Just keep in mind that the basic particle of water is a molecule rather than an atom. When ice melts or water boils, the molecules come unstuck, but the atoms do not. If the atoms within a water molecule came apart, there would be new substances (hydrogen and oxygen) with new properties.

physical

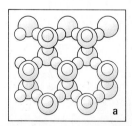

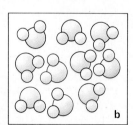

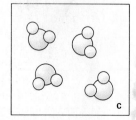

a   b   c

Not all compounds are made of molecules. Figure 2-21 is a diagram of a sodium chloride crystal. Sodium chloride melts to form positively charged sodium ions and negatively charged chloride ions. **Ions** are particles that have an electrical charge. The difference between ions and atoms is described in Chapter 12. For the present, it is enough to say that compounds that melt to form ions conduct an electric current, and compounds that do not melt to form ions do not conduct.

The only way to know which compounds are ionic (melt to form ions) and which compounds are molecular (melt as molecules) is to check them for conductivity.

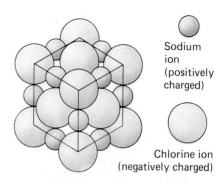

Sodium ion (positively charged)

Chlorine ion (negatively charged)

Figure 2-21 Solid table salt (sodium chloride) is made of crystals. A crystal contains equal numbers of sodium and chlorine ions. Like all compounds, salt has a definite composition. However, there is no unit that can be accurately described as a molecule. Each sodium atom is attracted to all the chlorine atoms around it and each chlorine atom is attracted to all the sodium atoms around it.

Review and Practice

1. Does Figure 2-22(a) represent an element or a compound?

2. Figure 2-22(b) shows two kinds of atoms. Does it represent a compound? Explain your answer.

3. Which diagrams in Figure 2-22 show only molecules? What is different about what is represented by these two diagrams?

4. All of the diagrams in Figure 2-22 represent the same physical state. Do they represent solids, liquids, or gases? How do you know?

5. Which diagrams in Figure 2-22 represent mixtures? What is different about the mixtures represented?

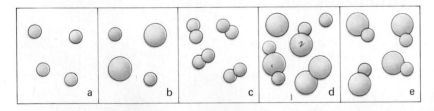

6. Which diagram(s) in Figure 2-23 could represent
   a. a solid compound?
   b. a gaseous element?
   c. neither a pure element nor a pure compound?

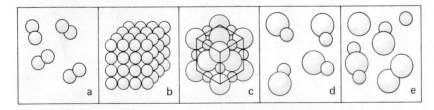

7. Diamond is pure carbon. At high temperatures, diamonds will burn. They combine with oxygen gas in the air to produce carbon dioxide gas. Draw diagrams to illustrate the changes that take place at the microscopic level when diamonds burn.

8. Water has a definite composition; there are eight grams of oxygen for each gram of hydrogen in water. Use the idea of atoms to explain why it has a definite composition.

1. element
2. No, the atoms are not joined.

Figure 2-22

3. c and e; c represents molecules of an element. e represents molecules of a compound.
4. Gases; the molecules are far apart.
5. b and d; b is a mixture of atoms. d is a mixture of atoms, molecules of an element, and molecules of a compound.
6. a. b and c
   b. a
   c. e

Figure 2-23

7. See the Teacher's Guide at the front of this book for drawings.
8. Atoms have definite masses. The fixed ratio of the masses implies a fixed ratio of atoms.

The Language of Chemistry

The diagrams of atoms and molecules used in the last section are useful, but they are awkward. It takes a long time to draw diagrams. Chemists have developed an easier and faster way to represent atoms and molecules. You will now embark on a study of the language of chemistry.

2-10 Chemical Symbols

The language of chemistry is based on abbreviations for the names of the elements. Such abbreviations are called **chemical symbols**. The simplest way to abbreviate is to use the first letter of the name of the element as its symbol. Since the names of several elements begin with the same letter, it is necessary to use two letters for many abbreviations. Table 2-2 lists the names and symbols for all known elements and gives some information about the origins of the symbols.

CONCEPTS

- chemical symbols
- periodic table
- writing chemical formulas
- ions and ionic charge
- periodicity of ionic charge
- polyatomic ions
- predicting ionic charge from formulas
- naming compounds containing ions
- naming compounds of two nonmetals

Do not burden students with unnecessary memorization. They will need to know the symbols for those elements you will be talking about, but there is little point in having them memorize all of them.

Table 2-2

NAMES AND SYMBOLS FOR ELEMENTS 1–103		
NAME	SYMBOL	ORIGIN OF SYMBOL
actinium	Ac	Gr. *Aktis, Aktinos*, beam, or ray
aluminum	Al	L. *alumen*, alum
americium	Am	the Americas
antimony	Sb	L. *Stibium*, mark
argon	Ar	Gr. *argos*, inactive
arsenic	As	L. *arsenicum*, Gr. *arsenikon*
astatine	At	Gr. *astatos*, unstable
barium	Ba	Gr. *barys*, heavy
berkelium	Bk	Berkeley, California
beryllium	Be	Gr. *berryllos, beryl*
bismuth	Bi	Ger. *Bisemutum*
boron	B	Ar. *Buraq*, Pers. *Burah*
bromine	Br	Gr. *bromos*, stench
cadmium	Cd	L. *cadmia*
calcium	Ca	L. *calx*, lime
californium	Cf	California
carbon	C	L. *carbo*, charcoal
cerium	Ce	the asteroid Ceres
cesium	Cs	L. *caesius*, sky blue
chlorine	Cl	Gr. *chloros*, greenish-yellow
chromium	Cr	Gr. *chroma*, color
cobalt	Co	Ger. *Kobold*, goblin
copper	Cu	L. *cuprium*, from Cyprus
curium	Cm	Marie and Pierre Curie
dysprosium	Dy	Gr. *dysprositos*, hard to get
einsteinium	Es	Albert Einstein
erbium	Er	Ytterby*, a town in Sweden
europium	Eu	Europe

Table 2-2 (con't)

NAMES AND SYMBOLS FOR ELEMENTS 1–103

NAME	SYMBOL	ORIGIN OF SYMBOL
fermium	Fm	Enrico Fermi
fluorine	F	L. and F. *fluerre*, flow or flux
francium	Fr	France
gadolinium	Gd	Gadolinite, a material named for a Finnish chemist
gallium	Ga	L. *Gallia*, France
germanium	Ge	L. *Germania*, Germany
gold	Au	L. *aurum*, shining dawn
hafnium	Hf	L. *hafnia*, Copenhagen
helium	He	Gr. *helios*, the sun
holmium	Ho	L. *Holmia*, Stockholm
hydrogen	H	Gr. *hydros*, water, and *genes*, forming
indium	In	Indigo
iodine	I	Gr. *iodes*, violet
iridium	Ir	L. *iris*, rainbow
iron	Fe	L. *ferrum*
krypton	Kr	Gr. *kryptos*, hidden
lanthanum	La	Gr. *lanthanein*, to be hidden
lawrencium	Lr	Ernest O. Lawrence
lead	Pb	L. *Plumbum*
lithium	Li	Gr. *lithos*, stone
lutetium	Lu	Lutetia, an ancient name for Paris
magnesium	Mg	Magnesia, a district in Greece
manganese	Mn	L. *magnes*, magnet
mendelevium	Md	Dmitri Mendeleev
mercury	Hg	hydragyrum, liquid silver
molybdenum	Mo	Gr. *molybdos*, lead
neodymium	Nd	Gr. *neos*, new, and *didymos*, twin
neon	Ne	Gr. *neos*, new
neptunium	Np	the planet Neptune
nickel	Ni	Ger. *nickel*, Satan
niobium	Nb	Niobe, daughter of Tantalus
nitrogen	N	L. *nitrum*, Gr. *nitron*
nobelium	No	Alfred Nobel
osmium	Os	Gr. *osme*, a small
oxygen	O	Gr. *oxys*, acid, and *genes*, forming
palladium	Pd	The asteroid Pallas
phosphorus	P	Gr. *phosphoros*, light-bearing
platinum	Pt	Sp. *platina*, silver
plutonium	Pu	the planet Pluto
polonium	Po	Poland
potassium	K	L. *kalium*
praseodymium	Pr	Gr. *prasios*, green and *didymos*, twin
promethium	Pm	Prometheus
protactinium	Pa	Gr. *protos*, first

Common elements whose symbol you should know are listed in bold type. Artificial elements, that have not been found in nature, are highlighted in blue.

Key to abbreviations: Gr., Greek; L., Latin; Ger., German; Ar., Arabic; Pers., Persian; F., French; Sp., Spanish

Table 2-2 (con't)

NAMES AND SYMBOLS FOR ELEMENTS 1–103

NAME	SYMBOL	ORIGIN OF SYMBOL
radium	Ra	L. *radius*, ray
radon	Rn	from radium
rhenium	Re	L. *Rhenus*, Rhine
rhodium	Rh	Gr. *rhodon*, rose
rubidium	Rb	L. *rubidius*, deepest red
ruthenium	Ru	L. *Ruthenia*, Russia
samarium	Sm	Samarskite, a mineral
scandium	Sc	L. *Scandia*, Scandinavia
selenium	Se	Gr. *Selene*, moon
silicon	Si	L. *ilex*, *silicis*, flint
silver	Ag	L. *argentum*
sodium	Na	L. *natrium*
strontium	Sr	Strontian, a town in Scotland
sulfur	S	L. *sulphurium*
tantalum	Ta	Gr. *Tantalos*
technetium	Tc	Gr. *technetos*, artificial
tellurium	Te	L. *tellus*, earth
terbium	Tb	Ytterby*, a town in Sweden
thallium	Tl	Gr. *thallos*, green shoot
thorium	Th	Thor
thulium	Tm	Thule, Scandinavia
tin	Sn	L. *stannum*
titanium	Ti	L. *Titans*
tungsten	W	Ger. *Wolfram*
uranium	U	the planet Uranus
vanadium	V	Vandis
xenon	Xe	Gr. *xenon*, stranger
ytterbium	Yb	Ytterby*, a town in Sweden
ytrium	Y	Ytterby*, a town in Sweden
zinc	Zn	Ger. *zink*
zirconium	Zr	Ar. *zargun*, gold color

*Ytterby is the site of a quarry that yielded many new elements.

Twelve elements each have just one letter for their symbol. All others use two letters. The first letter (and only the first letter) of the symbol is capitalized. Elements 104–109 will be referred to by number rather than using a letter symbol.

Some symbols are not letters taken from the English name of the element. That is because many elements were first named in Latin or Greek, and some elements were named by people who speak languages other than English.

Since symbols are used in place of the name of an element, you must learn to recognize an element by its symbol. You should know those marked in bold in Table 2-2 as they are ones you will see often. Your teacher may ask you to memorize others.

2-11 The Periodic Table

The table on the inside back cover of this book is known as the **periodic table**. It shows the symbols for each element, the atomic number, and the atomic mass. Atomic mass is the mass of an atom of the element. Since atoms are too small to be weighed directly, the atomic mass is obtained by indirect means that will be described in Chapter 4. There are many arrangements of the periodic table. All forms attempt to arrange the elements to show similarities and trends in their properties.

The periodic table summarizes a great deal of information about the elements. It is one of the most versatile tools for understanding chemistry. Throughout this book you will encounter trends that can be observed in the periodic table. One of those trends is metallic character.

The table in Figure 2-24 shows a heavy zig-zag line near the right that divides the elements into metals and nonmetals. As you can see, most elements are metals. **Metals** reflect light when they are polished, they can be bent or beaten flat, they have high melting points, and they are good conductors of heat and electricity. malleable, ductile

Metallic properties vary with position in the periodic table. In general, the more metallic elements are toward the left and bottom of the periodic table. The less metallic elements are toward the upper right of the table.

Figure 2-24 The periodic table. The stairstep line separates the metals from nonmetals by properties. The elements at the bottom of the table are classified as metals.

METALS — NONMETALS

1	2	3	4	5	6	7	8	9	10	11	12	13	14	15	16	17	18
1 H																	2 He
3 Li	4 Be											5 B	6 C	7 N	8 O	9 F	10 Ne
11 Na	12 Mg											13 Al	14 Si	15 P	16 S	17 Cl	18 Ar
19 K	20 Ca	21 Sc	22 Ti	23 V	24 Cr	25 Mn	26 Fe	27 Co	28 Ni	29 Cu	30 Zn	31 Ga	32 Ge	33 As	34 Se	35 Br	36 Kr
37 Rb	38 Sr	39 Y	40 Zr	41 Nb	42 Mo	43 Tc	44 Ru	45 Rh	46 Pd	47 Ag	48 Cd	49 In	50 Sn	51 Sb	52 Te	53 I	54 Xe
55 Cs	56 Ba	[57-71] *	72 Hf	73 Ta	74 W	75 Re	76 Os	77 Ir	78 Pt	79 Au	80 Hg	81 Tl	82 Pb	83 Bi	84 Po	85 At	86 Rn
87 Fr	88 Ra	[89-103] †	104	105	106	107	108	109									

*57 La	58 Ce	59 Pr	60 Nd	61 Pm	62 Sm	63 Eu	64 Gd	65 Tb	66 Dy	67 Ho	68 Er	69 Tm	70 Yb	71 Lu
†89 Ac	90 Th	91 Pa	92 U	93 Np	94 Pu	95 Am	96 Cm	97 Bk	98 Cf	99 Es	100 Em	101 Md	102 No	103 Lr

From its position in the periodic table, you would expect sodium to exhibit more properties of a metal than magnesium, and magnesium to be more metallic than aluminum. Judging by the trend from top to bottom of the table, you can predict that sodium has more metallic properties than lithium, which is above it, but fewer metallic properties than potassium, which is below it. This trend is particularly evident in columns crossed by the zig-zag line. For example, nitrogen is a gas at normal temperatures and has none of the properties associated with metals. Nitrogen is classified as a **nonmetal**. Phosphorus is a solid at normal temperatures, but it has no other properties associated with metals. Arsenic is a gray, brittle solid and a semiconductor. Antimony is a poor conductor of heat and electricity, is brittle, but has a metallic luster. Bismuth is a poor conductor of heat and electricity, but in earlier times was confused with tin and lead, two common metals.

2-12 Chemical Formulas

Comments on Nomenclature: Keep in mind that there is no need to learn a language unless you have something to say in that language. Few high school students will ever talk chemistry unless you provide the constant opportunities for them to do so. Teach enough of the language for students to understand the things covered during this course. Those who take chemistry in college will be confronted with nomenclature again. They will learn more there as they need it. The arguments to support these compromises are the following.
1. Teach the systematic naming discussed in this chapter, but do it several places in the text. One of the two systems could be introduced here and the other later. The alternative would be to name molecular compounds here, use molecular compounds in the discussions of Chapters 4–7, and introduce the other system in Chapter 8 after electrons in atoms are discussed. Defer the naming of complex (polyatomic) ions until still later. However, the best placement of the three discussions of nomenclature depends on just how material is developed.
2. Do not overdo nomenclature. Introduce only what is needed in order to communicate. The treatment of complex ions that is presented in this chapter is preferred over systematic naming. Nonsystematic names are used by almost every chemist for the common compounds that are discussed. Few students will ever need to know the details of systematic nomenclature for large molecules.

Symbols for elements are used in many ways. At times, they refer to any amount of the element. You might say, "The automobile is made of Fe." Here the symbol is used in place of the word iron, and it represents an indefinite amount of the element. At other times, the symbol is used to represent a single atom of the element which is the case when a symbol is used in writing chemical formulas.

It is common practice to represent a compound by a formula as well as a name. A **chemical formula**, such as H_2O, consists of the symbols of the elements in the compound combined with subscripts (small numbers) at the base and to the right of the symbols. The **subscript** indicates the number of atoms of that element in the compound. In Chapter 4, you will learn how experimental evidence is used to determine the number of atoms of each element in a molecule. Here you will see how formulas are written once you have that information.

The first two rules for writing formulas are:

Rule 1: *Represent each kind of element in a compound with the correct symbol for that element.*

Rule 2: *Use subscripts to indicate the number of atoms of each element in the compound. If there is only one atom of a particular element, no subscript is used.*

Applying these rules for a molecule composed of one atom of oxygen, O, and two atoms of hydrogen, H, the formula could be written OH_2. This formula says that a molecule of the compound is composed of O (oxygen) and H (hydrogen) and that there are two H atoms for each O atom.

You probably would not recognize OH_2 as water, but would recognize H_2O. To avoid confusion, the symbols are written in a particular order which is expressed as Rule 3.

Rule 3: *Write the symbol for the more metallic element first.*

Neither hydrogen nor oxygen is a metal. However, the location of hydrogen in the periodic table suggests that it should appear before oxygen in the formula. Similarly, for a compound containing oxygen and sulfur, the location of sulfur below oxygen in the periodic table suggests that it is more metallic than oxygen and should be written first in the formula. For example, SO_2 is the compound formed when sulfur burns. Figure 2-25 illustrates the three rules for writing formulas.

The formula represents one molecule of sulfur dioxide.

Sulfur is the more metallic element and its symbol is written first. No subscript is used since there is only one atom in the molecule.

Oxygen is less metallic than sulfur; its symbol is written last.

The subscript indicates that there are two oxygen atoms in each molecule.

Figure 2-25 The formula for sulfur dioxide illustrates the information conveyed by a chemical formula.

Using these three rules, you can write the correct formula for any compound if you know the elements it contains and the number of atoms of each element in one molecule (or formula unit) of that compound. However, the only way you can get that information is by experimental analysis. *empirical*

Many compounds have been analyzed experimentally to determine their formulas. Common patterns have been observed that allow us to predict the formulas for compounds of many elements. Those predictions are made using the periodic table as a guide.

2-13 Predicting Ion Charge from the Periodic Table

The periodic table is a powerful tool because it allows us to make many predictions without going into the laboratory to do an experiment. For example, you can predict the charge on many ions that are produced when ionic compounds form. The following rules describe those predictions.

Silver +1

Rule 1: *Ions formed from elements in Group 1 of the periodic table have a 1+ charge.*

Rule 2: *Ions formed from elements in Group 2 of the periodic table have a 2+ charge.*

Rule 3: *Aluminum ions have a 3+ charge. Other elements in Group 13 normally form ions with a 3+ charge, but there are exceptions.*

Rule 4: *Elements in all other groups form ions with various charges.* However, the following rules concerning a few of these elements will enable you to write formulas and assign names for most binary compounds. *Bi-* is a Greek prefix meaning two. **Binary compounds** are compounds containing two elements.
   a. *The nitride ion has a 3– charge.* (Nitrides are binary compounds of metals and nitrogen.)
   b. *The sulfide ion has a 2– charge.* (Sulfides are binary compounds of metals and sulfur.)

c. *The oxide ion has a 2− charge.* (Oxides are binary compounds of metals and oxygen.)
d. *All halogens (elements in Group 17) form halide ions with a 1− charge.* (Halides are binary compounds of a metal or hydrogen, and a halogen.)

Notice the *ide* ending for nitride, sulfide, oxide, and the halide ions. The *ide* ending is used in naming the second element of a binary compound. Recall that table salt is called sodium chlor*ide*.

Rule 5: *The algebraic sum of the charges in a compound is zero.*

You will now use these rules to predict the formulas for some binary compounds.

2-14 Using Ion Charge to Predict Formulas

Example 2-1

Predict the formula for a compound of calcium and chlorine.

▶ **Suggested solution**

Since calcium is in Group 2 of the periodic table, you know that the charge on the calcium ion will be 2+ (Rule 2). Chlorine is in Group 17 so the charge on the chloride ion will be 1− (Rule 4d). In order to balance the charge on the compound, there must be two chloride ions for each calcium ion (Rule 5). The formula and the procedure just described for predicting it are shown in Figure 2-26.

Figure 2-26

Calcium is in Group 2 of the periodic table. These ions have a charge of 2+ (Rule 2).

Chlorine is in Group 17 of the periodic table. Halogens form ions with a 1− charge (Rule 4d).

— **2+**   **1−** —

$$CaCl_2$$

Calcium chloride

There must be 2 chloride ions for each calcium ion in order for the algebraic sum of the charges to be zero (Rule 5).

Example 2-2

What is the formula of the compound formed from sodium and oxygen?

▶ Suggested solution _____

Sodium is in Group 1 of the periodic table so it always has a charge of 1+ (Rule 1). However, since you do not know how many sodium ions are in the formula, you cannot write the total positive charge. How do you go about writing the formula?

Well, you know that this compound is an oxide and the charge on oxygen will be 2− (Rule 4c). But you do not know the number of oxide ions in the compound. Now what?

Throughout your study of science and when you use science to solve real problems, you will often encounter problems that you have never seen before. You will need to figure out a way to solve them yourself. In order to do that, you need to learn some strategies that successful problem solvers use when they get stuck on a problem.

One useful strategy that can be applied to most any situation is to think of a similar problem that you already know how to solve. When I got stuck on this problem, it seemed to me that what I needed to do was similar to what I've done in finding the least common denominator when adding fractions. I remembered that the way to solve those problems is to multiply the denominators of the fractions to get the least common multiple. For example, in adding

$$\frac{1}{2} + \frac{1}{3} + \frac{3}{4} =$$

The least common denominator is the least common multiple of 2, 3, and 4. The product of 2, 3, and 4, which is 24, will certainly work, but 12 is the least common denominator for 2, 3, and 4. Thus, to add these fractions they must be converted to the equivalent values having a denominator of 12.

$$\frac{6}{12} + \frac{4}{12} + \frac{9}{12} =$$

Perhaps a similar method will apply to the problem stated in Example 2-2. From Rule 5, the total positive charge must be equal to the total negative charge.

Number of sodium ions		Charge on each ion		Total charge
?	×	1+	=	2+

Number of oxide ions		Charge on each ion		Total charge
?	×	2−	=	2−

Now you can see that two positive ions and one negative ion are needed to get totals that balance. The formula must be

$$Na_2O$$

Recalling how to find the least common denominator of fractions helped in figuring out a way to balance the charges on the ions in this compound. Let's hope it did not confuse you more than it helped! Part of problem solving involves recalling problems or situations that are familiar, and trying different methods for getting an answer. Do not feel that you have to use the same methods as your classmates or your teacher. Try strategies that make sense to you and discuss them with others. The important point is that when you are stuck DO NOT GIVE UP. All good problem solvers go through a lot of trial and error before they arrive at a sensible solution to a new problem.

Example 2-3

What is the formula for the compound formed from magnesium and oxygen?

▶ **Suggested solution**

Try the same strategy used in Example 2-2. According to Rule 2, Mg has a $2+$ charge. According to Rule 4c, the charge on oxygen is $2-$.

Number of positive ions		Charge on each ion		Total charge
?	$\times$	$2+$	$=$	$2+$

Number of negative ions		Charge on each ion		Total charge
?	$\times$	$2-$	$=$	$2-$

Using one ion of each will certainly work to give a formula of

MgO

empirical

In trying different strategies for writing formulas remember that a formula represents the simplest whole number ratio of atoms. Any time you write a formula in which the subscripts can be divided by the same number, you do not have the simplest formula. If you had obtained Mg_2O_2 as the formula in Example 2-3, you can see that both subscripts are divisible by 2. Always assume that the simplest formula is the correct one unless you have experimental data to the contrary.

2-15 Predicting Formulas Containing Polyatomic Ions

When NaCl, KI, or $CuBr_2$ come apart as ions, each atom in the compound forms a separate ion. For NaCl, it forms Na^+ and Cl^-; KI forms K^+ and I^-; $CuBr_2$ forms Cu^{2+} and $2Br^-$. At the other extreme, molecules such as sugar, $C_{12}H_{22}O_{11}$, do not ever come apart to form ions. The behavior of other compounds is in between these extremes.

Vinegar is a water solution of a compound with the formula $C_2H_4O_2$. Vinegar conducts an electric current so you know that it must contain ions. Experiments show that one of the ions is H^+. The other is $C_2H_3O_2^-$, which is composed of several atoms joined together with a $1-$ charge on the ion as a whole. Such ions are often called **polyatomic ions** (*poly* means many).

Sulfuric acid, the substance added to water to increase its conductivity in an electrolysis experiment, has the formula H_2SO_4. It breaks apart in water to give H^+ and HSO_4^-, another polyatomic ion with a $1-$ charge. Under certain conditions, this HSO_4^- ion breaks apart to form H^+ and SO_4^{2-}

It is difficult to predict what compounds will break apart into ions or what ions they will form until you have learned more about chemistry. In the meantime, Table 2-3 lists the names, formulas, and charges of several common ions. The table also lists a common compound containing each ion and the formula for that compound.

Table 2-3

COMMON POLYATOMIC IONS AND REPRESENTATIVE COMPOUNDS			
ION NAME	FORMULA	COMPOUND NAME	FORMULA
acetate	$C_2H_3O_2^-$	hydrogen acetate (acetic acid, vinegar)	$HC_2H_3O_2$
ammonium	NH_4^+	ammonium nitrate (in fertilizer)	NH_4NO_3
carbonate	CO_3^{2-}	calcium carbonate (limestone)	$CaCO_3$
hydrogen carbonate (bicarbonate)	HCO_3^-	sodium hydrogen carbonate (sodium bicarbonate, baking soda)	$NaHCO_3$
hypochlorite	ClO^-	sodium hypochlorite (bleach—Clorox®)	$NaClO$
chlorate	ClO_3^-	potassium chlorate	$KClO_3$
chromate	CrO_4^{2-}	potassium chromate	K_2CrO_4
dichromate	$Cr_2O_7^{2-}$	iron(III) dichromate	$Fe_2(Cr_2O_7)_3$
hydroxide	OH^-	sodium hydroxide (lye—Drano®)	$NaOH$
nitrate	NO_3^-	hydrogen nitrate (nitric acid)	HNO_3
nitrite	NO_2^-	calcium nitrite	$Ca(NO_2)_2$
oxalate	$C_2O_4^{2-}$	hydrogen oxalate (oxalic acid)	$H_2C_2O_4$
permanganate	MnO_4^-	potassium permanganate	$KMnO_4$
phosphate	PO_4^{3-}	sodium phosphate	Na_3PO_4
sulfate	SO_4^{2-}	calcium sulfate (in plaster of paris)	$CaSO_4$
sulfite	SO_3^{2-}	sodium sulfite	Na_2SO_3

The formulas for compounds containing polyatomic ions are predicted in exactly the same ways that you predicted formulas for other compounds containing ions. The positive (metallic) ion is written first, followed by the negative ion. Remember that the algebraic sum of the charges must be zero. The only thing new is that the charge on the polyatomic ion pertains to the ion as a whole, so parentheses must be placed around the ion before a subscript is written.

Example 2-4

What is the formula for a compound containing calcium ions and hydroxide ions?

▶ Suggested solution _____

The compound contains calcium ions and hydroxide ions. The metal ion is written first.

CaOH

This formula represents the ions, but it is incorrect because the charges are not balanced. Calcium is in Group 2 of the periodic table so its ion will have a 2+ charge. The hydroxide ion has a 1− charge (Table 2-3).

Number of calcium ions		Charge on each ion		Total Charge
?	×	2+	=	2+

Number of hydroxide ions		Charge on each ion		Total Charge
?	×	1−	=	2−

Using 1 calcium ion and 2 hydroxide ions gives a total charge that would algebraically equal zero.

Number of calcium ions		Charge on each ion		Total Charge
1	×	2+	=	2+

Number of hydroxide ions		Charge on each ion		Total Charge
2	×	1−	=	2−

In order to indicate two hydroxide ions, parentheses are used around the ion before writing the subscript.

$$Ca(OH)_2$$

How does writing the formula as above differ from writing it as $CaOH_2$? Why are the parentheses necessary?

2-16 Naming Compounds

Formulas are a convenient way to represent compounds, but compounds have names as well. You must learn how to connect the name with the formula and vice versa. There are millions of compounds, and it is impossible to memorize millions of names. It must be done systematically, or you will never succeed in making the connections.

Unfortunately, many common compounds were named before it became obvious that a systematic method would be needed. H_2O is called water by everyone (including chemists), even though its systematic name is dihydrogen monoxide. To add to the confusion the system for naming compounds has changed over the years. People who learned obsolete systems continue to use them through habit or stubbornness. Books printed years ago, but still in use, also help to perpetuate the use of old names.

Therefore, be advised that you will surely encounter chemical names other than those you learn to write here. Two current systems of nomenclature will be presented.

Students should realize that many rules in science are arbitrary conventions that have no inherent meaning. Communication is facilitated when everyone agrees to use the same conventions. Similar conventions are used in naming people. In Western cultures, the convention is to place the family name last; in Oriental cultures, the family name is given first.

To avoid undue confusion, the discussion of nomenclature is limited to current recommended practice. Although students may encounter names like ferrous and ferric in their reading, it is recommended that you do not teach this outmoded system of nomenclature. Although you will probably use names like sulfuric acid and nitric acid, point out that these are not systematic names.

2-17 Naming Ionic Compounds

Naming compounds like those whose formulas you have learned to predict is simple. You simply name the ions in the compound. The name of the negative ion in binary compounds always ends in *ide*. Use the periodic table and Table 2-3 as a guide in assigning names to compounds.

Example 2-5

What is the name of Na_2O?

▶ Solution

This compound contains sodium ions and oxide ions. The name is sodium oxide. Notice that the positive metal ion is named just like the neutral metal atom, but the name of the negative nonmetal ion is given a new ending. The *ide* suffix normally indicates a binary compound. However, there are a few polyatomic ions that end in *ide*, as indicated in the next example.

Example 2-6

What is the name of $Ca(OH)_2$?

▶ Solution

The compound contains calcium ions and hydroxide ions. The name is calcium hydroxide.

Example 2-7

What is the name of $(NH_4)_2SO_4$?

▶ Solution

If you recognized the ions, you know the name is ammonium sulfate. The ammonium ion is the only common polyatomic ion with a positive charge.

So far you have seen how to name compounds containing polyatomic ions and elements that have a charge that can be predicted. However, there are many elements that form more than one compound, and the charge on the metal ion varies from one compound to another. For example, iron and chlorine form two compounds, $FeCl_2$ and $FeCl_3$. If both compounds were called iron chloride, you would not know which compound is meant. The name must indicate which compound is described. The normal procedure is to indicate the charge on the metal ion by a Roman numeral. $FeCl_2$ is iron(II) chloride and $FeCl_3$ is iron(III) chloride. Notice that there is no space between the name of the metal and the parentheses indicating the charge on the iron ion.

2-18 Predicting Ion Charge from a Chemical Formula

In order to write the name for compounds containing metals that form more than one ion, you must be able to predict the charge on the ion from the formula. Remember that Rule 5 indicates that the algebraic sum of the charges must be zero. Since there are two chloride ions in $FeCl_2$, each with a charge of $1-$, the charge on the iron ion must be $2+$. Figure 2-27 summarizes this concept.

Figure 2-27 Iron(II) chloride

The charge on the iron ion must be $2+$ in order for the sum of the charges to be zero.

The charge on the chloride ion is $1-$ (Rule 4d).

$2+$ $1-$

One iron ion is assumed, because there is no subscript.

$$FeCl_2$$

Two chloride ions are indicated by the formula.

$$1 \times 2+ = 2+ \qquad 2 \times 1- = 2-$$

$$(2+) + (2-) = 0$$

There is only one iron ion. The charge must be $2+$ to give a total positive charge of $2+$

The total positive charge must be $2+$ in order for the algebraic sum of the charges to be zero. (Rule 5).

The total negative charge must be $2-$. (Two ions, each with a charge of $1-$.)

Example 2-8

Determine the charge on the iron ion in the compound $FeCl_3$ and then name the compound.

▶ Suggested solution

It is helpful to write the charge on the ions over the symbol in the formula to help keep everything straight.

$$\overset{?\ 1-}{FeCl_3}$$

From Rule 4d you know that the charge on the chloride ion is $1-$, but you do not know the charge on the iron. How can you find it?

Rule 5 tells you that the algebraic sum of the charges must be zero, so the charge on the iron must be positive and it must balance the negative charges on the three chloride ions.

Number of chloride ions		Charge on each ion		Total charge
3	$\times$	$1-$	$=$	$3-$

Number of iron ions		Charge on each ion		Total charge
1	$\times$	?	$=$	$3+$

The charge on the iron ion must be $3+$. Knowing the charge on iron, you can write the name iron(III) chloride.

Example   2-9

What is the charge on iron in Fe_2O_3?

▶ **Suggested solution** _____

This compound is an oxide, so the charge on oxygen is 2− (Rule 4c). From this information, you can calculate the total negative charge indicated by the formula.

Number of oxide ions		Charge on each ion		Total charge
3	×	2−	=	6−

The total positive charge must be 6+ and it is distributed across two iron ions.

Number of iron ions		Charge on each ion		Total charge
2	×	?	=	6+

The charge must again be 3+ because 2 × 3+ = 6+. The name of the compound is iron(III) oxide because the charge on iron is 3+ and the *ide* ending indicates a binary compound.

The system just described for naming compounds is normally used for ionic compounds. However, a different system is preferred for molecular compounds. Since even experienced chemists may be unable to predict which compounds are molecular and which are ionic, it is understandable that different names are often assigned to the same compound.

2-19 Naming Binary Compounds of Nonmetals

no metals present

The distinguishing feature of this second system of nomenclature is that Greek prefixes (Table 2-4) are used to indicate the number of atoms of each element in a molecule of the compound.

Table 2-4

GREEK PREFIXES			
PREFIX	**NUMBER**	**PREFIX**	**NUMBER**
mono-	1	penta-	5
di-	2	hexa-	6
tri-	3	hepta-	7
tetra-	4	octa-	8

The formula CO_2 will be used to illustrate the rules for naming binary compounds of nonmetals.

Rule 1: *Write the names of the elements in the same order that they appear in the formula. The more metallic element is named first.*

Figure 2-28

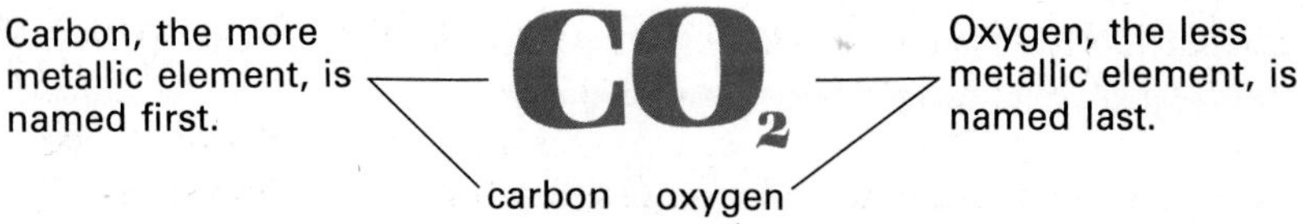

Carbon, the more metallic element, is named first.

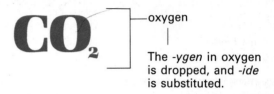

carbon   oxygen

Oxygen, the less metallic element, is named last.

Figure 2-28 shows that CO_2 contains carbon and oxygen. Carbon, the more metallic element, is written first.

Rule 2: *Drop the last syllable in the name of the final element and add -ide.*

If the elements were merely named, there would be confusion about whether we are referring to isolated elements or a compound containing those elements. The *-ide* ending is used for all binary compounds.

Figure 2-29

CO_2 ——oxygen

The *-ygen* in oxygen is dropped, and *-ide* is substituted.

Rule 3: *Add prefixes to the name of each element to indicate the number of atoms of that element in each molecule of the compound.*

Figure 2-30

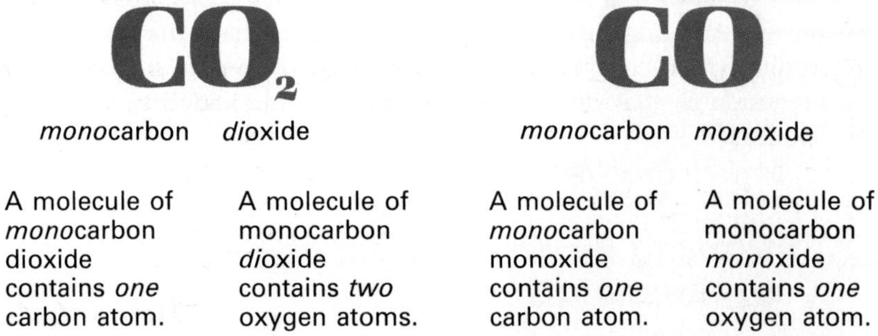

CO_2

*mono*carbon   *di*oxide

CO

*mono*carbon   *mono*xide

A molecule of *mono*carbon dioxide contains *one* carbon atom.

A molecule of monocarbon *di*oxide contains *two* oxygen atoms.

A molecule of *mono*carbon monoxide contains *one* carbon atom.

A molecule of monocarbon *mono*xide contains *one* oxygen atom.

Since there are two compounds containing carbon and oxygen, a name like carbon oxide could refer to either. By adding prefixes, a unique name is obtained for each compound. The two common compounds of carbon and oxygen are shown. The *mono-* prefix indicates one atom of the element; *di-* indicates two atoms of the element.

In practice, the *mono-* prefix is frequently omitted, particularly for the more metallic element in the compound. If no prefix is given, one atom of that element is understood. The two compounds shown in Figure 2-30 are called carbon dioxide and carbon monoxide.

Example   2-10

What is the name for N_2O_4?

▶ Solution _____

Since there is no metal in the compound N_2O_4 assume that the compound is not ionic. Use the system for naming compounds that is preferred for molecular compounds.

There are two nitrogen atoms in each molecule and four oxygen atoms in each molecule. The corresponding Greek prefixes are *di-* and *tetra-* so the name is dinitrogen tetraoxide. The *-ide* ending indicates that this is a binary compound.

Review and Practice

1. Write the name for each of the following compounds formed from two nonmetals.
   a. SiC
   b. CS_2
   c. SF_6
   d. OF_2 (Note that this is a binary compound containing oxygen, but it is *not* an oxide. Why?)
   e. SO_2
   f. SO_3 (What is the difference between this formula for a neutral compound and the formula for the sulfite ion?)
   g. N_2O_5
   h. NO

2. If you remember the meaning of the Greek prefixes, you should be able to write formulas for the following compounds. Try to do it without referring to Table 2-4.
   a. chlorine dioxide
   b. dichlorine monoxide
   c. iodine tribromide
   d. nitrogen triiodide
   e. diphosphorus tetraoxide

3. The following compounds contain ions whose charges can be predicted from the periodic table. Use the periodic table to predict the formula of each compound.
   a. potassium bromide
   b. magnesium chloride
   c. calcium oxide
   d. aluminum iodide
   e. barium fluoride
   f. sodium iodide
   g. calcium bromide
   h. potassium oxide
   i. beryllium chloride
   j. aluminum oxide

4. The metals in the following compounds form ions with different charges, so the charge cannot be predicted from the periodic table. Predict the charge on the metal ion and write the name for the compound.
   a. SnO
   b. $SnBr_4$
   c. FeS
   d. $FeCl_3$
   e. Cr_2O_3
   f. $NiCl_2$

5. Write the formula for the following compounds.
   a. tin(IV) chloride
   b. tin(II) chloride
   c. tin(IV) oxide
   d. iron(III) sulfide
   e. iron(II) fluoride
   f. mercury(II) oxide
   g. cobalt(III) oxide

6. Write formulas for the following compounds containing polyatomic ions.
   a. calcium acetate
   b. sodium phosphate
   c. copper(II) sulfate
   d. iron(II) phosphate
   e. aluminum hydroxide
   f. calcium hydrogen carbonate
   g. ammonium phosphate
   h. lithium dichromate

7. Name the following compounds containing polyatomic ions.
   a. $Al(NO_3)_3$
   b. H_2CO_3
   c. $Sn(NO_3)_2$
   d. $K_2Cr_2O_7$
   e. Cu_2CO_3
   f. NH_4Cl
   g. NiC_2O_4

1. a. monosilicon monocarbide (silicon carbide)
   b. monocarbon disulfide (carbon disulfide)
   c. monosulfur hexafluoride (sulfur hexafluoride)
   d. monoxygen difluoride (oxygen difluoride) This is not an oxide as oxygen is the more metallic of the two elements.
   e. monosulfur dioxide (sulfur dioxide)
   f. monosulfur trioxide (sulfur trioxide) The sulfite ion would be written with a 2− charge when standing alone.
   g. dinitrogen tetraoxide (pent)
   h. mononitrogen monoxide (nitrogen monoxide)
2. a. ClO_2 d. NI_3
   b. Cl_2O e. P_2O_4
   c. IBr_3
3. a. KBr   f. NaI
   b. $MgCl_2$ g. $CaBr_2$
   c. CaO   h. K_2O
   d. AlI_3 i. $BeCl_2$
   e. BaF_2 j. Al_2O_3
4. a. 2+, tin(II)oxide
   b. 4+, tin(IV)bromide
   c. 2+, iron(II)sulfide
   d. 3+, iron(III) chloride
   e. 3+, chromium(III)oxide
   f. 2+, nickel(II)chloride
5. a. $SnCl_4$ e. FeF_2
   b. $SnCl_2$ f. HgO
   c. SnO_2 g. Co_2O_3
   d. Fe_2S_3
6. a. $Ca(C_2H_3O_2)_2$ e. $Al(OH)_3$
   b. Na_3PO_4 f. $Ca(HCO_3)_2$
   c. $CuSO_4$ g. $(NH_4)_3PO_4$
   d. $Fe_3(PO_4)_2$ h. $Li_2Cr_2O_7$
7. a. aluminum nitrate
   b. hydrogen carbonate
   c. tin(II)nitrate
   d. potassium dichromate
   e. copper(I)carbonate
   f. ammonium chloride
   g. nickel(I)oxalate

Helen Read Steele

Title: Chemistry Teacher

Job Description: Teaches chemistry and biochemistry to high school students. Is responsible for helping students achieve a solid background in chemistry so that they can pursue further studies in science and make informed decisions about the nature of chemistry in their lives. Directs student learning in the classroom and laboratory.

Educational Qualifications: College degree with emphasis on courses in chemistry; mathematics, physics, languages, and the arts.

Future Employment Outlook: Excellent

Many different activities make up Helen Steele's day as a high school chemistry teacher. Besides teaching, she plans and supervises the training of teacher interns, hires and orients new teachers, develops curriculum, supervises science instruction, and writes the science budget. Ironically, Steele did not originally intend to enter teaching; she planned to attend medical school. After working in the chemical industry, she became attracted to teaching because of the challenge of working with students and the constant variety of experiences involved in teaching. Steele has a strong interest in the humanities and feels that scientists and students of science need to explore the legal and moral implications of technology.

Teaching is a profession that demands not only a firm knowledge of the subject but also experience in other areas. Steele feels that a science teacher should have a broad background that includes art, music, and languages. Most of all, teaching science requires an understanding of the scientific method. "Most students think chemistry is going to be hard, but I demonstrate that with organized, consistent effort, this is not so. Students don't enter the chemistry class or lab and leave just with facts and skills. They also leave with a

problem-solving method that applies to life. Without this method of thinking, we could not survive in the world today." Steele is quick to point out, however, that teaching also has difficult aspects. Along with a fascination with the work, a teacher must be able to deal with moments of failure.

Helen Steele says that teaching chemistry is demanding but satisfying. Long vacations provide time to spend with family and friends, to travel, to climb mountains, and to work in her community. She attends science conferences and conventions, has traveled with students to Italy and Greece, works with theater groups, serves on a rescue squad and in a preservation association, hikes, runs, reads, and farms, all while raising a family. Although Steele has multiple sclerosis, it has not interfered with her professional goal to be the best at what she does.

The job outlook for teachers of chemistry and other areas of science is excellent and will continue to improve. There is a shortage of science teachers, which is expected to become worse as the number of school children increases. For those who have a commitment to education and a desire to work with young people, teaching science may provide a secure and rewarding future.

Numbers in red indicate the appropriate chapter sections to aid you in assigning these items. Answers to all questions appear in the Teacher's Guide at the front of this book.

Summary

■ Our knowledge of matter is based on macroscopic observations. Important information about matter is gained by observing what happens to matter when it is heated, beaten, or an electric current is passed through it.

■ Matter can be pure or mixed with other kinds of matter. Mixtures often can be separated by filtration or distillation.

■ Pure substances have constant boiling points and constant melting points; mixtures do not. Pure substances that can be decomposed into simpler substances are called compounds. Pure substances that cannot be decomposed into simpler substances are called elements.

■ Compounds are made of elements in definite proportions; mixtures of elements can be made using any proportions.

■ Changes that produce new kinds of matter are chemical changes. Changes that do not produce new kinds of matter are physical changes.

■ 109 elements are known, but some of these are artificial elements that do not exist in nature. All matter on Earth is composed of about 85 elements; 99% of Earth is composed of only eight elements.

■ The same elements can be combined in different proportions to make many different compounds.

■ All matter is made of atoms. Elements contain only one kind of atom. Compounds contain two or more kinds of atoms joined together.

■ The periodic table is an arrangement of the elements that reveals trends in properties of the elements and allows us to make predictions about the behavior of the elements.

■ Chemical symbols are combined to write formulas that show the number of each kind of element in a compound.

■ Compounds are named systematically according to rules agreed upon by chemists. Several systems for naming compounds are still in use.

Chemically Speaking

atom
binary compound
boiling point
chemical change

law of multiple
 proportions
macroscopic observations
macroscopic properties

chemical formula
chemical symbol
compound
decomposition
distillation
electrolysis
element
freezing point
ion
law of definite
 composition

metal
melting point
microscopic model
mixture
molecule
nonmetal
periodic table
physical change
polyatomic ion
solution
subscript

pure substance
characteristic property

Review

1. A bottle of a soft drink looks pure, but it is not. How could you demonstrate that it is a mixture? (2-2, 2-4, 2-5)

2. Suggest a way to separate the following mixtures.
   a. salt dissolved in water   (2-2, 2-4, 2-5)
   b. alcohol dissolved in water
   c. pieces of iron and wood
   d. sugar and powdered glass
   e. large pieces of aluminum and zinc that look alike

Answer the items in each set *without referring to any section of the chapter*. If you miss *any* item in the first set, review the material related to the question, and then try the next set. (For example, if you did not recall the name for K, go back and review all the names and symbols you were asked to memorize before going on.)

Set I

3. What element is represented by K? (2-10)

4. What is the symbol for mercury? (2-10)

5. What is the formula of a molecule containing one atom of zinc and two atoms of iodine? (2-12)

6. What is the formula of a molecule containing one atom of oxygen and two atoms of silver? (2-12)

Set II

7. What element is represented by Br? (2-10)

8. What is the symbol for copper? (2-10)

9. What is the formula of a molecule containing one atom of phosphorus and five atoms of chlorine? (2-12)

10. What is the formula of a molecule containing three atoms of oxygen and one atom of sulfur? (2-12)

You might have students work with
molecular model kits in answering
items 7, 8, and 9.

Set III

11. What is the symbol for lead? (2-10)

12. What element is represented by Au? (2-10)

13. What is the formula of a molecule containing one atom of boron and one atom of nickel? (2-12)

14. What is the formula of a molecule containing one atom of iodine and one atom of lithium? (2-12)

Interpret and Apply

1. Indicate whether each of the following changes is chemical or physical, and explain why you answered as you did. (2-4)
   a. leaves changing color in autumn
   b. crushing a dry leaf in your hand
   c. an aspirin dissolving in your mouth or stomach
   d. an aspirin acting on the body to relieve a headache
   e. water boiling
   f. beans cooking
   g. grass growing
   h. cutting grass
   i. bleaching clothes
   j. using Drano® to unstop a sink

2. From the following list of observations, indicate whether the change is more likely to represent (1) an element combining to form a compound or (2) a compound decomposing into simpler com- (2-4, 2-5) pounds or elements. Give reasons for your decision.
   a. When a white solid is heated in a closed container, a gas appears to be produced, and the solid that remains after heating has less mass than the original solid.
   b. When a metal is heated in air, it changes color and gains mass.
   c. A log is massed and then left to rot. The rotted log has less mass than it did before rotting.
   d. A black powder is heated with oxygen. The only product is a colorless gas that has the same mass as the original powder and oxygen combined.
   e. A colorless solid is heated in oxygen. Three gaseous products are identified. The three products have the same mass as the original solid and the oxygen gas.

3. Two solids are analyzed for copper. One contains 32% copper by mass and the other contains 43% copper by mass. Could these solids be samples of the same compound? Cite evidence to support your answer. (2-6)

4. Two solids are analyzed and found to contain only copper and chlorine. Are the two solid samples the same compound? Cite evidence to support your answer. (2-6)

5. Some students obtain a new kind of solid as a result of an experiment. They find that the solid has a fixed melting point and boiling point and that it has a uniform density. They conclude that they have discovered a new element. Is the conclusion justified? (2-2, 2-5)

6. What tests would you perform on an unknown substance to decide if the substance is an element or a compound? (2-4, 2-5)

7. Draw a picture to represent the microscopic nature of each of the following elements as they exist at ordinary temperatures. Use the illustrations in this chapter as a guide. (2-7, 2-8, 2-9)
   a. nitrogen, the major component of air, is a diatomic gas
   b. neon, the element in neon lights, is a gas that exists as individual atoms
   c. tin is a solid metal
   d. mercury is a liquid metal
   e. bromine is a liquid consisting of diatomic molecules

8. Draw a picture to represent a single molecule of each of the following compounds (2-9)
   a. hydrogen sulfide, the gas that gives rotten eggs their odor
   b. nitrogen monoxide and nitrogen dioxide, two common air pollutants
   c. nitrogen trihydride (commonly known as ammonia), a common fertilizer, household cleaning fluid, and industrial chemical

9. Draw separate pictures (or make models) to represent each of the following. (2-2, 2-3, 2-7, 2-8, 2-9)
   a. a mixture of hydrogen and oxygen gas, and a compound formed from hydrogen and oxygen
   b. two oxygen atoms, and one oxygen molecule
   c. two different compounds that contain the same elements

Problems

(2-14 to 2-19)

1. Write the name or formula for each of the following.
   a. aluminum chloride   d. cobalt(II) chloride

b. Al_2S_3
c. HgO

e. CS_2
f. dinitrogen tetroxide

2. Write the name or formula for each of the following.
   a. N_2O_3
   b. iron(II) chloride
   c. magnesium nitride
   d. Cr_2O_3 (2-14 to 2-19)
   e. MnO_2
   f. cobalt(III) fluoride ✓

3. Write the name or formula for each of the following.
   a. potassium bromide
   b. SiC
   c. sulfur dichloride
   d. SnO_2 (2-14 to 2-19)
   e. CCl_4
   f. aluminum nitride

4. Write the formula or name for each of the following.
   a. barium nitride
   b. CaC_2
   c. chromium(III) silicide
   d. CoO
   e. copper(II) oxide
   f. $AuBr_3$ (2-14 to 2-19)
   g. iron(III) chloride
   h. B_4C
   i. nitrogen triiodide
   j. Ni_5P_2

5. Write the name or formula for the following compounds. (2-14 to 2-19)
   a. $Cu_3(PO_4)_2$
   b. H_2SO_3
   c. SnC_2O_4
   d. $FeSO_3$
   e. $KHCO_3$
   f. $MgCO_3$
   g. $Ca(NO_3)_2$
   h. lead(IV) chromate
   i. sodium hydroxide
   j. ammonium carbonate
   k. magnesium sulfite
   l. magnesium sulfide
   m. magnesium sulfate
   n. calcium acetate

6. Formulas are used to represent both molecules and polyatomic ions. What is different about the way each is represented? Write the name of each of the following formulas and indicate whether it represents a molecule or ion. (2-15)
   a. SO_3
   b. SO_3
   c. CS_2
   d. PO_5
   e. PO_4^{3-}
   f. CO
   g. CO_3^{2-}
   h. NO_2
   i. NO_2^-
   j. $Fe_2(SO_4)_3$

Challenge

1. A concept map is a diagram that summarizes relationships among ideas that we have in our head. For example, you have many ideas about school. They include ideas about teachers, students, principal, lessons, homework, and tests. One type of concept map for school is shown in Figure 2-31. Construct a similar map for matter. Include atom, molecule, element, compound, mixture, pure substance, solution, decompose, solid, liquid, and gas in your map. Draw lines that connect these words and write on the line

words that describe the relationships among the ideas connected.

Figure 2-31

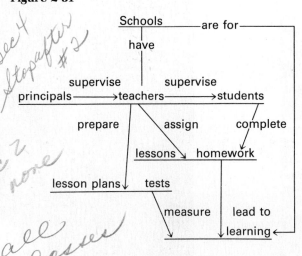

2. Construct a concept map for chemical language. Include the following terms in your map: chemical symbol, chemical formula, subscript, atom, molecule, element, compound, chemical name, and ionic charge. (You may include other terms as well.)

Projects

1. Pick an element and learn as much about it as you can. Write an advertising brochure for the element. Include information such as the following: Where is it found in nature? In what form is it found? How is it obtained in a pure form? What commercial value does it have? What dangers are associated with its use? What are some of its properties? Become the class authority on this element, and share your knowledge with your teacher and classmates.

2. Look up values for melting points, boiling points, densities, or other properties of the elements. Enter the values on a periodic table, and see if there are any obvious trends in the properties as you go across or down a column of the table.

3. Matter is normally found as mixtures. Find out how mixtures of chemicals are separated so that valuable materials can be separated in a pure form or toxic materials can be separated and disposed of safely. If there are industries in your community, you may be able to find out how the problem is solved in that industry. If not, you can read about it in the library.

CHAPTER 3

Chemists are interested in matter forming the stars in the distant universe to subatomic particles that make atoms. To study matter, chemists make measurements that range from the distance to the edge of the universe, about 60 000 000 000 000 000 000 000 000 meters, to the diameter of subatomic particles, less than 0.000 000 000 000 000 1 meter. Although such quantities may be written, it is difficult to imagine what they represent. Since most of your work in this course involves such large and small quantities, you need to try to imagine the relative size a large or small quantity represents.

Numbers Large and Small

Representing Large and Small Numbers

You have a good idea of what a hundred represents and you can probably visualize a thousand, but can you visualize a million of anything? Count a million of something. If you can, bring a million of something to class. If a million of what you count is too large or too expensive to bring, be prepared to explain how you counted it. For example, you may decide to "count" a million grains of sand, salt, sugar, or rice. If this million turns out to be a spoonful or a cup, you can bring it to class. If it turns out to be a ton, the cost might be prohibitive. In that case you can simply describe it: "There are a million grains of sugar in a five-pound bag," or "There are a million holes in the screen that covers the window in my bedroom."

To get you started, look at one method used to "count" a million grains of rice and find their mass. First, count 400 grains. Next, use a balance to find the mass of the 400 grains. From that, you can find the average mass of each grain and multiply by 1 000 000 to get the mass of a million grains of rice. What did you get?

Here are a few ideas that other students have tried:

1. determining the number of acoustical ceiling tiles that would contain one million holes.
2. determining the number of these books that would contain one million pages.
3. determining the mass of one million popcorn kernels.
4. programming a computer to print out one million A's. The printout was then brought to class so there would not be a shortage of A's at grading time!

Before you continue, count a million of something besides rice.

> **CONCEPTS**
>
> ■ large numbers and orders of magnitude
> ■ proportional relationships
> ■ using unitary rates in calculations
> ■ using scientific notation to represent numbers
> ■ calculations with scientific notation
> ■ the enormous size of 10^{23}

3-1 Unitary Rates Describe Proportions

In Chapter 1, calculations with quantities were introduced. At that time, a unitary rate was defined, and you learned how unitary rates can be used to convert from one metric unit to another. Now this idea will be extended to show how unitary rates can be used to keep track of other calculations.

A unitary rate is just a special way of describing a proportional relationship. By reading about the method used to count a million grains of rice, you get a better idea of what is meant by a proportional relationship.

Figure 3-1 The mass of 400 grains of rice is about 7 grams.

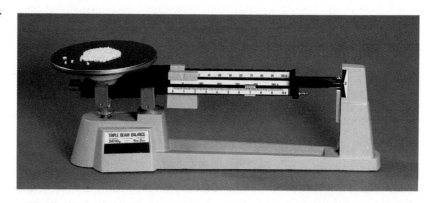

First, 400 grains of rice were placed on a balance. Their mass was 7.0 grams. As you know from experience, the mass of the rice depends on the number of grains. If you double the number of grains, you double the mass; if you triple the number of grains, you triple the mass; and so forth. Table 3-1 lists several counts of rice grains and their corresponding masses.

Table 3-1

NUMBER OF RICE GRAINS AND THEIR MASS	
NUMBER OF GRAINS	CORRESPONDING MASS
400	7.0 g
200	3.5 g
100	1.75 g
800	14.0 g
1600	28.0 g

Many students have not developed proportional reasoning, and it will take several weeks or months for this reasoning to be sensible to them. Refer to the Teacher Notes at the front of the book for suggestions on how to help students develop this important intellectual skill.

Two quantities that are related in this way are said to be **proportional**. Just what does this statement mean? From the table, it is clear that when the number of grains gets smaller, the corresponding mass gets smaller; when the number of grains gets larger, the corresponding mass gets larger; but that relationship is only half the story.

If any number of grains of rice listed in Table 3-1 is divided by the corresponding mass, the result is the same. For example:

$$\frac{400}{7.0} = 57.142857$$

Do the other divisions and you will see that you always get 57.142857. (Actually, you will get more digits if you do the division by hand. A calculator will give six, eight, or twelve digits depending on the brand that you use.) No matter how many grains of rice are placed on the balance, the number of grains of rice divided by their mass should be 57.142857. The result is constant.

Any two quantities that are proportional produce a constant number when one of the quantities is divided by the other. This is often expressed mathematically by

$$\frac{A}{B} = k$$

The letters A and B represent the two quantities, and k represents the constant number produced when one quantity is divided by the other. The expression is read as "A divided by B is equal to a constant." Just what the constant number turns out to be depends entirely on what quantities are divided.

Note that the units were intentionally left out in dividing to obtain the constant. However, in science, the units are included so that the constant has a sensible meaning.

$$\frac{400 \text{ grains}}{7.0 \text{ g}} = \frac{57.142857 \text{ grains}}{1 \text{ g}}$$

The constant can be read, "57.142857 grains per gram," and it means that there will be just over 57 grains of rice in each gram of rice.

The decision to divide the grains of rice by the mass was arbitrary. You could just as easily divide the mass by the number of grains. Dividing the first entries in Table 3-1, and including the units gives the following:

$$\frac{7.0 \text{ g}}{400 \text{ grains}} = \frac{0.0175 \text{ g}}{1 \text{ grain}}$$

If you do the other divisions, you will get the same number. In other words, this ratio is also constant. The constant is read as "0.0175 gram per grain," and it means that the mass reading on the balance should increase by 0.0175 gram each time a grain of rice is added.

Notice that the two proportionality constants that you calculated are unitary rates. The first describes the rate at which the number of grains increases as the mass of the rice is increased; the second describes the rate at which the mass increases as the number of grains is increased. The first rate is simply the inverse of the second. *Every unitary rate has an inverse that also describes a unitary rate.*

Now you will use the unitary rate just calculated to find the mass of 1 000 000 grains of rice.

Example 3-1

What is the mass of 1 000 000 grains of rice?

▶ **Suggested solution** _____

First, set up a relation that indicates what you are trying to find. You want the number of grams that correspond to one million grains.

> ? grams correspond to 1 000 000 grains

Notice that writing this relation involves some translation from the problem. Instead of writing "what mass," you can write "? grams." This quantity is then set equal to the quantity that you know, 1 000 000 grains. The problem, then, is to find the mass that corresponds to a million grains of rice. It is not necessary to write "grams of rice" and "grains of rice" because that gets awkward, but you must remember that what you are calculating is only true for the rice measured in this example.

Next, multiply by the unitary rate that tells you the mass of each grain of rice.

$$? \text{ grams} = 1\,000\,000 \text{ grains} \times \frac{0.0175 \text{ g}}{1 \text{ grain}} = 17\,500 \text{ g}$$

This expression contains two units grain(s) and gram(s). You treat the units algebraically. Since grain(s) appear in the numerator and denominator, these units cancel to leave units of grams for the answer. This is sensible, and it is what you wanted to find.

The mass can be expressed in some other units by using another unitary rate. For example, the mass in grams could be expressed in kilograms.

$$? \text{ kilograms} = 17\,500 \text{ g} \times \frac{0.001 \text{ kg}}{1 \text{ g}} = 17.5 \text{ kg}$$

Grams cancel to leave kilograms as units for the answer.

Review and Practice

1. $5 \text{ kg} \times \dfrac{1\,000 \text{ g}}{1 \text{ kg}} \times \dfrac{57.142\,857 \text{ gr}}{1 \text{ g}} =$
   $28\,714.29 \text{ gr}$
2. a. Yes. # drops/volume is always 12.
   b. 12 drops/mL and 0.0833 mL/drop
   c. 12 drops/mL

1. Use the unitary rate calculated in Section 3-1 to determine the number of grains of rice in a 5-kg sample.

2. A student collects the following data for the purpose of calibrating a dropper pipet in the lab. The calibration involves determining the volume of a drop of liquid (in this case water) delivered by the pipet. Review the data and answer the following questions.

NUMBER OF DROPS AND THEIR VOLUME

NUMBER OF DROPS	CORRESPONDING VOLUME (mL)
180	15
60	5
24	2
120	10
144	12
72	6

a. Are the two quantities in the table proportional? How do you know?
b. Calculate the two unitary rates that can be derived from these data.
c. Which unitary rate is most useful to the student in calibrating the pipet?

3-2 Scientific Notation

Most large numbers encountered in science have a long string of zeros at the end, and the small ones usually have a long string of zeros at the beginning. When you do calculations with these numbers, it is easy to make an error by losing some zeros. We need a way of writing these numbers so as not to lose track of the zeros. One technique is to write the number as a product of some number between one and ten multiplied by some power of ten. This form of the number is called **scientific notation**.

To understand scientific notation, you need to recall two things from mathematics; what is meant by exponents and how values between one and ten are used with exponents to describe a number of any size.

You probably recall that numbers like 3^2 represent exponential numbers. The 2 is called the **exponent**; the 3 is called the **base**. The exponent indicates the number of times the base would be multiplied by itself. Thus 3^3 is the same as $3 \times 3 \times 3$.

You will now use these two ideas to represent numbers in scientific notation.

Do You Know?

The logarithm of a number is an exponent. For common (base ten) logarithms it is the exponent to which ten must be raised to produce that number.

Example 3-2

Write 17 500 as a number between one and ten multiplied by some power of ten.

▶ **Suggested solution** _____

First, 17 500 is changed to a number greater than one but less than ten by moving the decimal point to the left four times. This process involves dividing by ten four times.

$$\frac{17\ 500}{10} = 1750$$

$$\frac{1750}{10} = 175$$

$$\frac{175}{10} = 17.5$$

$$\frac{17.5}{10} = 1.75$$

Dividing by ten four times gives 1.75, a number between one and ten. 1.75 is not the same as 17 500, so something else must be done to restore the original value. Multiplication is the inverse of division. An inverse operation cancels the effect of a previous operation. If a number is divided by ten and the result is multiplied by ten, the original number is restored.

$$\frac{17\ 500}{10} = 1750$$

$$1750 \times 10 = 17\ 500$$

Since 17 500 is divided by ten four times to get 1.75 you can get 17 500 back again if 1.75 is multiplied by 10 four times. In doing that, the product can be written in exponential form.

$$1.75 \times 10 \times 10 \times 10 \times 10 = 1.75 \times 10^4$$

The expression on the right is read as "1.75 multiplied by ten four times," or "1.75 times 10 to the fourth" which is equivalent to what is written on the left.

$$17\ 500 = 1.75 \times 10^4$$

You have written 17 500 as a number between one and ten multiplied by a power of ten. You have expressed it in scientific notation.

Scientific notation is only one form of exponential notation. 17 500 can be written as shown on the next page.

$$1750 \times 10^1$$
$$175 \times 10^2$$
$$17.5 \times 10^3$$
$$1.75 \times 10^4$$

All these represent the number 17 500 in exponential form. The last form shows a product of a number between one and ten multiplied by a base-10 exponential number. This form shows the number in scientific notation.

Small numbers can be expressed in scientific notation by reversing the order in which you multiply and divide by ten.

Example 3-3

Write 0.000 175 in scientific notation.

▶ **Suggested solution**

First, move the decimal point until you have a number that is greater than one but less than ten. The number will be 1.75, which you get by moving the decimal point to the right four times. This process is equivalent to multiplying by ten four times.

$$0.000\ 175 \times 10 \times 10 \times 10 \times 10 = 1.75$$

This multiplication could be shown using exponential notation:

$$0.000\ 175 \times 10^4 = 1.75$$

Multiplying 0.000 175 by ten four times clearly gives a different number. To "undo" the multiplication, you use the inverse operation of division and divide 1.75 by ten four times. Before doing that, recall how to say "divided by ten" using exponential notation. It is indicated by a negative exponent.

10^{-3} means $\frac{1}{10} \times \frac{1}{10} \times \frac{1}{10}$ or "divide by ten three times."

Reverse the multiplication by ten four times, but leave the result in exponential form.

$$1.75 \times \frac{1}{10} \times \frac{1}{10} \times \frac{1}{10} \times \frac{1}{10} = 1.75 \times 10^{-4}$$

The final result can be described in two ways: "Multiply 1.75 by one tenth four times," or "Divide 1.75 by ten four times." Performing either operation gives 0.000 175, the number started with.

$$1.75 \times 10^{-4} = 0.000\ 175$$

Practice converting from decimal notation to scientific notation with two or three classmates. Take turns writing a number for others to convert. When everyone agrees on the correct answer, someone else should take a turn writing the number to be converted.

Review and Practice

1. Express each of the following numbers in scientific notation.
   a. 1 240 000
   b. 1 280
   c. 1 000 000 000 000 000 000
   d. 0.000 124
   e. 0.000 000 000 210

2. The following numbers are written in scientific notation. Write them in the customary decimal form.
   a. 1.32×10^5 d. 1.32×10^{-3}
   b. 2.06×10^3 e. 3.45×10^{-6}
   c. 6.02×10^{23}

1. a. 1.24×10^6
   b. 1.28×10^3
   c. 1×10^{18}
   d. 1.24×10^{-4}
   e. 2.1×10^{-10}
2. a. 132 000
   b. 2 060
   c. 602 000 000 000 000 000 000 000
   d. 0.001 32
   e. 0.000 003 45

3-3 Multiplying and Dividing Exponential Numbers

Now that you know how to write numbers in exponential notation, you can do some mathematics with numbers in this form.

Recall what is meant by 10^4 and 10^3 and see if you can write the answer to the following multiplication.

$$10^4 \times 10^3 = ?$$

If you remembered that an exponent indicates how many times the base is taken as a factor, you can see that this equation could be written in the following manner. (Parentheses have been used to separate the part representing 10^4 and the part representing 10^3.)

$$(10 \times 10 \times 10 \times 10) \times (10 \times 10 \times 10) = ?$$

Written this way, it is easy to see that the answer should be 10^7.

$$10^4 \times 10^3 = 10^7$$

The product of two powers of ten is obtained by adding the exponents.

If you recall what a negative exponent represents, you will see that the answer to the following problem can be found in exactly the same way.

$$10^{-4} \times 10^3 = ?$$

If you write out the powers of ten as above, you get

$$\left(\tfrac{1}{10} \times \tfrac{1}{10} \times \tfrac{1}{10} \times \tfrac{1}{10}\right) \times (10 \times 10 \times 10) = \tfrac{1}{10}$$

Now it is easy to see the answer in exponential form.

$$10^{-4} \times 10^3 = 10^{-1}$$

If you remember how to add signed numbers, you can see that the exponents are added to obtain the product.

What do you think you will do to the exponents when you divide exponential numbers? Write out the tens indicated by the following division and see if you can write a rule for division of exponential numbers.

You should be sure that students derive a suitable rule. One acceptable statement is: The quotient of two powers of ten is obtained by subtracting the exponent of the denominator from the exponent of the numerator.

$$\frac{10^4}{10^3} = ?$$

The rule is the same whether the exponents are positive or negative.

$$\frac{10^{-4}}{10^3} = \frac{\frac{1}{10} \times \frac{1}{10} \times \frac{1}{10} \times \frac{1}{10}}{10 \times 10 \times 10} = 10^{-7}$$

So far all examples have involved only numbers that are powers of ten. Multiplying or dividing numbers like 3.14×10^{23} is no problem if you recall that changing the order in which a series of numbers is multiplied does not change the product. The following example illustrates this principle.

Example 3-4

What is the product of $(2 \times 10^3)(4 \times 10^2)$?

▶ **Suggested solution**

It is more important that students learn general strategies that can help them solve any unfamiliar problem than to learn how this particular problem is solved. Encourage students to share strategies that they use to solve problems and discuss their usefulness.

This problem could be solved by first converting the exponential numbers to decimal numbers.

$$(2 \times 10^3)(4 \times 10^2) =$$
$$2\,000 \times 400 = 800\,000 = 8 \times 10^5$$

That is easy when the exponents are small, but not when they are large like 10^{23}. Is there another way to solve the problem? The problem can be rearranged to change the order of multiplications.

$$(2 \times \mathbf{10^3})(4 \times \mathbf{10^2}) =$$
$$(2 \times 4)(\mathbf{10^3 \times 10^2}) = 8 \times \mathbf{10^5}$$

It looks like you can find the product of exponential numbers by finding the product of the decimal part and multiplying that by the product of the exponential part.

You can figure out how to divide exponential numbers in much the same way as you figured out how to do multiplication.

Example 3-5

What is the value of $\dfrac{2 \times 10^3}{4 \times 10^2}$?

▶ **Suggested solution**

The difficulty with this problem is knowing how to interpret what is written. If the expression is rearranged like before, which of the following is correct?

$$\frac{2 \times 4}{10^3 \times 10^2}$$

or

$$\left(\frac{2}{4}\right) \times \left(\frac{10^3}{10^2}\right)$$

The second interpretation is correct.

$$\frac{2 \times 10^3}{4 \times 10^2} = \left(\frac{2}{4}\right) \times \left(\frac{10^3}{10^2}\right) = 0.5 \times 10^1$$

To check, write the original problem as decimal numbers and then divide.

$$\frac{2 \times 10^3}{4 \times 10^2} = \frac{2000}{400} = 5$$

Five is the same as 0.5×10^1 so it seems to be correct.

Review and Practice

1. State the rule for multiplying exponential numbers.

2. State the rule for dividing exponential numbers.

3. Using the rule you stated in item 1, write the answer to the following as ten raised to the appropriate exponent.
   a. $10^3 \times 10^5 =$
   b. $10^{-5} \times 10^8 =$
   c. $10^{-3} \times 10^{-4} =$
   d. $10^{21} \times 10^{-18} =$
   e. $10^{14} \times 10^{-6} =$

4. Using the rule you stated in item 2, write the answer to the following as ten raised to the appropriate exponent.
   a. $10^3/10^5 =$
   b. $10^{-5}/10^8 =$
   c. $10^{-3}/10^{-4} =$
   d. $10^{-21}/10^{-18} =$
   e. $10^{14}/10^{-6} =$

5. $(2.1 \times 10^3)(1.86 \times 10^{-2}) =$

6. $(4.5 \times 10^{-4})(6.44 \times 10^{15}) =$

7. $(1.11 \times 10^{23})(5.68 \times 10^{-20}) =$

8. $(5.5 \times 10^{-8})(2.99 \times 10^3) =$

9. $(1.09 \times 10^{-25})(2.99 \times 10^3) =$

10. $(3.25)(1.23 \times 10^3) =$

11. $(8.5 \times 10^{-4})(10^5) =$

12. $\dfrac{2.1 \times 10^3}{1.86 \times 10^{-2}} =$

13. $\dfrac{4.5 \times 10^{-4}}{6.44 \times 10^{15}} =$

14. $\dfrac{1.11 \times 10^{23}}{5.68 \times 10^{-20}} =$

15. $\dfrac{5.5 \times 10^{-8}}{3 \times 10^{-10}} =$

16. $\dfrac{1.09 \times 10^{-25}}{2.99 \times 10^3} =$

17. $\dfrac{3.25}{1.23 \times 10^3} =$

18. $\dfrac{8.5 \times 10^{-4}}{10^5} =$

1. Find the product of the decimal portion of the numbers. Find the product of the exponential part by adding exponents. Write the answer as the product of the decimal portion times ten to the appropriate power.
2. Find the quotient of the decimal part of the numbers. Find the quotient of the exponential part by subtracting the exponent in the denominator from the exponent in the numerator. Write the answer as the product of the decimal portion and ten to the appropriate power.
3. a. 10^8
   b. 10^3
   c. 10^{-7}
   d. 10^3
   e. 10^8
4. a. 10^{-2}
   b. 10^{-13}
   c. 10
   d. 10^{-3}
   e. 10^{20}
5. 3.906×10^1
6. 2.898×10^{12}
7. 6.3048×10^3
8. 1.6445×10^{-4}
9. 3.2591×10^{-22}
10. 3.9975×10^3
11. 8.5×10^1
12. 1.1290×10^5
13. 6.9876×10^{-20}
14. 1.9542×10^{42}
15. 1.8333×10^2
16. 3.6455×10^{-29}
17. 2.6422×10^{-3}
18. 8.5×10^{-9}

3-4 Adding and Subtracting Exponential Numbers

Addition and subtraction of exponential numbers are just as simple as multiplication and division, so long as you remember not to mix apples and pears. *The exponents of the numbers added or subtracted must be the same.*

$$(2 \times 10^3) + (4 \times 10^3) = 6 \times 10^3$$

The following exponential numbers cannot be added without changing one of them so the exponents are the same.

$$(2 \times 10^3) + (4 \times 10^4) =$$

The solution is to rewrite one of the numbers. The following expressions are equivalent.

$$(0.2 \times 10^4) + (4 \times 10^4) = 4.2 \times 10^4 \quad \text{(or } 42 \times 10^3)$$
$$(2 \times 10^3) + (40 \times 10^3) = 42 \times 10^3 \quad \text{(or } 4.2 \times 10^4)$$

Subtraction of exponential numbers is done in exactly the same way. As in any subtraction, the answer may be a negative number.

$$(2 \times 10^3) - (4 \times 10^3) = -2 \times 10^3$$

Review and Practice

1. a. 6×10^3
   b. 5×10^2
   c. 9×10^{-3}
   d. 3.4×10^4
   e. 2.31×10^{-3}
   f. 5.06×10^3
2. a. -2×10^3 d. 1.9×10^4
   b. 1×10^2 e. 2.09×10^{-3}
   c. 1×10^{-3} f. -3.94×10^3
3. a. 1.0214×10^2
   b. 6.6171×10^{-4}
   c. 1.1970×10^{15}
   d. -1.3441×10^9
   e. 1.3459×10^9
   f. 5.985×10^{-8}
   g. 7.315×10^{-8}
   h. 4.4222×10^{-16}
   i. 1×10^{-1}
   j. 5×10^1

1. Do the indicated additions. Then check your answer by first converting the numbers to decimal notation and then doing the addition.
   a. $(2 \times 10^3) + (4 \times 10^3) =$
   b. $(3 \times 10^2) + (2 \times 10^2) =$
   c. $(5 \times 10^{-3}) + (4.0 \times 10^{-3}) =$
   d. $(6 \times 10^3) + (2.5 \times 10^4) =$
   e. $(2.2 \times 10^{-3}) + (1.1 \times 10^{-4}) =$
   f. $(5.6 \times 10^2) + (4.5 \times 10^3) =$

2. Do the indicated subtractions. Check your answers by first converting the numbers to decimal notation and then doing the subtraction.
   a. $(2 \times 10^3) - (4 \times 10^3) =$ d. $(6 \times 10^3) - (2.5 \times 10^4) =$
   b. $(3 \times 10^2) - (2 \times 10^2) =$ e. $(2.2 \times 10^{-3}) - (1.1 \times 10^{-4}) =$
   c. $(5 \times 10^{-3}) - (4.0 \times 10^{-3}) =$ f. $(5.6 \times 10^2) - (4.5 \times 10^3) =$

3. Solve the following problems without converting to decimal notation.
   a. $(2.24 \times 10^{-3})(4.56 \times 10^4) =$
   b. $\dfrac{8.9 \times 10^5}{1.345 \times 10^9} =$
   c. $(8.9 \times 10^5)(1.345 \times 10^9) =$
   d. $(8.9 \times 10^5) - (1.345 \times 10^9) =$
   e. $(8.9 \times 10^5) + (1.345 \times 10^9) =$
   f. $(6.65 \times 10^{-8}) - (6.65 \times 10^{-9}) =$
   g. $(6.65 \times 10^{-8}) + (6.65 \times 10^{-9}) =$
   h. $(6.65 \times 10^{-8})(6.65 \times 10^{-9}) =$
   i. $\dfrac{6.65 \times 10^{-8}}{6.65 \times 10^{-9}} =$
   j. $\dfrac{1.0 \times 10^{26}}{2.0 \times 10^{24}} =$

A	100 0001
B	100 0010
C	100 0011
D	100 0100
E	100 0101
1	011 0001
2	011 0010
3	011 0011
4	011 0100
5	011 0101
[	101 1011
\	101 1100
]	101 1101
↑	101 1110
←	101 1111

Figure 3-2 The American Standard Code for Information Interchange (ASCII) is the standard for coding information into the binary system. This standardization enables the computer industry to provide compatible equipment. Part of this code is shown.

3-5 APPLICATION: Number Systems

Number systems are developed for convenience. The base ten number system probably developed because we have ten fingers. People would count on their fingers and make a tally each time they used all their fingers. They then started again. The tallies represented groups of ten.

Computers, of course, do not have fingers. What they do have is switches that are either on or off. All computers count in twos and a binary number system consisting only of the digits 0 and 1 is best for describing numbers in a computer. Expressing numbers in a binary number system results in very long strings of ones and zeros to represent even small numbers. A compromise has been developed for expressing numbers in computers that takes into account the computer's method of doing arithmetic while allowing numbers to be expressed in shorter strings. The compromise is a hexadecimal system with a base of 16.

A base 16 number system requires more than the ten unique digits that you are accustomed to using (0 through 9). The first six letters of the alphabet are used to represent 10 through 15. Table 3-2 shows the first twenty numbers represented in binary (base 2), decimal (base 10), and hexadecimal (base 16) notation.

Table 3-2

BINARY, DECIMAL, AND HEXADECIMAL EQUIVALENTS		
BINARY	**DECIMAL**	**HEXADECIMAL**
0	0	0
1	1	1
10	2	2
11	3	3
100	4	4
101	5	5
110	6	6
111	7	7
1000	8	8
1001	9	9
1010	10	A
1011	11	B
1100	12	C
1101	13	D
1110	14	E
1111	15	F
10000	16	10
10001	17	11
10010	18	12
10011	19	13
10100	20	14

CHEM THEME

The development of the computer has provided an important new tool for representing data. It is now possible to study complex chemical changes using simulations. Graphics programs allow scientists to draw complex molecules and study their geometrical structures. Scientists are constantly looking for more sensible ways to represent data. So should you.

You undoubtedly find it difficult to read either the binary or hexadecimal numbers, because you are accustomed to thinking in the decimal number system. However, people who program computers in

Figure 3-3 This computer printout shows the hexadecimal system used in programming.

```
AA D5 AA D5 AA D5 AA D5
AA 85 D5 AA D5 AA D5 AA
D5 AA D5 AA D5 AA 85 D5
AA D5 AA D5 AA D5 AA D5
AA D5 AA 85 95 00 00 00
00 00 00 00 00 00 00 A0
85 D5 FF 7F 7F 7F 7F 7F
7F 7F 7F 7F AF 85 D5 81
80 80 80 80 80 80 80 80
80 AC 85 D5 81 80 80 80
80 80 80 80 80 80 AC 85
D5 81 80 80 80 80 80 80
80 80 80 AC 85 D5 81 80
80 80 80 80 80 80 80 00
AC 85 D5 81 80 80 80 80
80 80 80 80 00 AC 85 D5
81 80 80 80 80 80 80 80
```

machine language use hexadecimal numbers routinely and soon learn to do arithmetic in hexadecimals just as easily as you do arithmetic with decimal numbers.

3-6 How Big Is Avogadro's Number?

Powers of Ten by Philip and Phylis Morrison is an excellent book illustrating the effects of scale. If it is not available in your school library you may want to order it. Morrison, Philip and Phylis. *Powers of Ten*. Scientific American Books, distributed by W. H. Freeman, NY.

Although it is easier to work multistep problems by stringing together several unitary rates in a single equation, it usually confuses beginners to do so. It is recommended that each step be done independently in the beginning. As students gain confidence in the procedure, the better students often suggest the more efficient procedure of stringing factors together. Their suggestion can then be discussed in class and adopted by those students who follow the argument.

Do You Know?

The number of milliliters of water in the Pacific Ocean is nearly equal to Avogadro's number.

One very large number, 6.02×10^{23}, will appear many times in this course. This value is the number of hydrogen atoms in one gram of hydrogen, and it is called Avogadro's number after an Italian chemist. Why the number is important will become clear when you get to Chapter 4. Right now, we'll try to imagine how large that number is.

Earlier you calculated the mass of a million grains of rice to be 17.5 kg. What is the mass of 6.02×10^{23} grains of rice? To find out, you can use the mass of one million grains (10^6 grains) to calculate the mass of each grain. If you then multiply 6.02×10^{23} grains by the mass of each grain, you get the mass of that enormous amount of rice.

$$\frac{17.5 \text{ kg}}{10^6 \text{ grains}} = \frac{1.75 \times 10^{-5} \text{ kg}}{1 \text{ grain}}$$

$$? \text{ kg} = 6.02 \times 10^{23} \text{ grains} \times \frac{1.75 \times 10^{-5} \text{ kg}}{1 \text{ grain}} = 1.0535 \times 10^{19} \text{ kg}$$

This answer probably means little to you. Perhaps if you compared the mass of 6.02×10^{23} grains of rice to something familiar it would help. You already have a concept of how massive an automobile is. A typical car has a mass of about 1.8×10^3 kg. How many cars would have the same mass as all that rice?

$$? \text{ cars} = 1.0535 \times 10^{19} \text{ kg} \times \frac{1 \text{ car}}{1.8 \times 10^3 \text{ kg}} = 5.852778 \times 10^{15} \text{ cars}$$

Again, you have an answer that probably means little to you. If the cars are divided among all the people on Earth, would there be enough to go around? There are about 5 billion (5×10^9) people on

Figure 3-4 The vast number of cars on this lot represents a small fraction of 5.85×10^{15} cars. You would need 7.4×10^{12} lots of cars to equal that number.

Earth, so if the cars are distributed to everyone on Earth each person would get

$$\frac{5.852778 \times 10^{15} \text{ cars}}{5 \times 10^9 \text{ people}} = 1.170555 \times 10^6 \text{ cars/person}$$

That's over a million cars for every person on Earth! In other words, 6.02×10^{23} grains of rice would have the same mass as almost a million cars for every man, woman, and child on Earth! Now you must have an idea of the mass of Avogadro's number of rice grains.

In 1976, the annual rice production was 320 000 000 000 kg. At that rate, how many years would it take to grow 6.02×10^{23} grains of rice?

$$? \text{ years} = 1.05 \times 10^{19} \text{ kg} \times \frac{1 \text{ year}}{3.2 \times 10^{11} \text{ kg}} = 32\ 921\ 875 \text{ years}$$

It would take almost 33 million years to grow 6.02×10^{23} grains of rice. It seems safe to say that this is more rice than has ever been produced on Earth!

Do You Know?

The value of Avogadro's number, like all numbers based on measurement, is an estimate. In 1974, scientists at the National Bureau of Standards developed new techniques to calculate the number as $6.022\ 094\ 3 \times 10^{23}$. This value has an uncertainty of one part per million. The old value was $6.022\ 045\ 3 \times 10^{23}$ with an uncertainty of five parts per million.

Review and Practice

1. Use your imagination and see if you can come up with a better illustration of how big 6.02×10^{23} is. The following questions may give you some ideas. To answer them you will need to look up information in an encyclopedia or other reference books.
   a. If you made a chain of 6.02×10^{23} paper clips, how far would it reach? To the sun? To the edge of the Milky Way? To the edge of the universe?
   b. About twenty drops of water from a medicine dropper fill one cubic centimeter. Would 6.02×10^{23} drops fill the Pacific Ocean?
   c. Start counting and see how many numbers you say in a minute. At that rate, how long would it take you to count to 6.02×10^{23}?
   d. How many individual letters are there in an average book? (This book is probably average. You could use it to get an estimate.) How many average books would contain 6.02×10^{23} letters? Do you think there are that many books in print?

1. (Since the purpose of the exercise is to have students practice doing conversions using unitary rates, you will want to look at each student's solution. The numerical answer obtained for each answer will depend on the value the student gets from a reference book, and these will vary from one reference to another.)

Uncertainty in Measured Numbers

You know that the calculations you did in the last section were not exact. Several estimates were used. For example, the number of people on Earth was estimated at 5 billion. Nobody knows the exact number of people on Earth. Actually, few things are known exactly, and scientists want to indicate that fact when they record numbers. In this lesson, you will learn why it is necessary to use estimates, and you will learn a technique used to indicate the uncertainty in a measurement.

3-7 Some Reasons for Uncertainty

There are many reasons for uncertainty. The most common reason is that measuring tools are limited. Figure 3-5 shows a section of a ruler calibrated in centimeters and inches. If you use such a ruler to measure the length of something, the best you can do is estimate the length to the nearest hundredth of a centimeter. Even then you must imagine that the space between the closest marks is divided into equal parts and make a guess as to the length.

To illustrate this idea, look at Figure 3-5 and decide how far it is between the long arrows on the metric scale. Notice that the second long arrow does not fall on a line. The distance between the arrowheads is more than 10.5 cm, but it is less than 10.6 cm. It is common practice to imagine that the space between marks is divided into tenths and estimate a third digit.

Figure 3-5 When you read the scale on a measuring instrument such as an ordinary ruler, the final digit is an estimate of where the pointer falls between the smallest divisions on the scale.

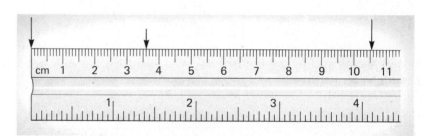

It appears that the arrow is about halfway between the marks or 0.5 of the total distance separating the marks. I estimate the distance between arrows as 10.55 cm. You may estimate 10.54 cm or 10.56 cm, and a classmate may make a slightly different estimate. *The last digit recorded in any measurement made in science is an estimate of this kind, and it is uncertain.*

Figure 3-6 shows an instrument known as a micrometer. (*Micro* means small and *meter* means measure.) A micrometer can be used to estimate a thickness to the nearest thousandth of a centimeter.

Using a micrometer and a ruler, the thickness of a book was measured. The best estimate with the ruler was 2.88 cm. Using the micrometer, the estimate was 2.878 cm. In both measurements, the final

Do You Know?

Micrometer also means 10^{-6} meter in SI.

digit is uncertain. However the micrometer allows you to estimate the length to a thousandth of a centimeter; the ruler allows you to estimate the length only to the nearest hundredth of a centimeter.

3-8 Accuracy and Precision

The measurement made with the micrometer is said to be more precise than the measurement made with the ruler. **Precision** refers to the uncertainty in a measurement; the less the uncertainty, the higher the precision.

Accuracy refers to how close one comes to an accepted value. Ordinarily the more precise measurement is the more accurate, but that is not necessarily so. A micrometer can be damaged so that it does not provide accurate readings. For example, the ends may wear with use or the "U" may be bent so that the scale does not read zero when there is nothing between the ends of the micrometer. In that case, the micrometer would continue to give *precise* readings, but the readings would not be *accurate*. Figure 3-7 illustrates the difference between accuracy and precision in another way.

Precision is actually used in two different ways. Because the micrometer measures to the nearest 0.001 cm and the ruler measures to the nearest 0.01 cm, we would normally say that the micrometer is more precise. It can provide a closer estimate. However, the relative uncertainty of a measurement made with the micrometer could be

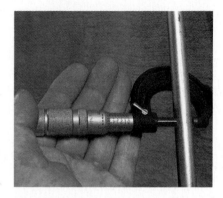

Figure 3-6 A micrometer can be used to measure small thicknesses. The object to be measured is placed between the jaws, and the movable jaw is rotated down until the jaws touch the object on both sides.

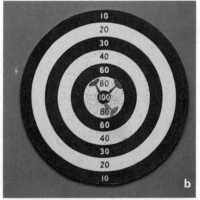

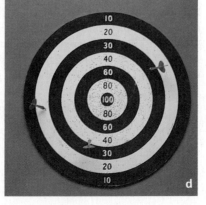

Figure 3-7 Accuracy and precision are illustrated by the holes formed in four targets. (**a**) The three shots fell close together. Thus, the shooting was very precise, but the accuracy was low because the shots missed the center of the target. (**b**) These shots show better accuracy because they are in the center of the target, but the precision is lower because there is more scattering of them. (**c**) These shots are both accurate and precise. (**d**) These shots are neither accurate nor precise.

greater than the relative uncertainty of a different measurement made with a ruler.

As the term suggests, **relative precision** or **relative uncertainty** describe the precision or uncertainty relative to something else. It is the amount of uncertainty compared to the total amount measured.

To understand relative uncertainty, consider the distance between the long arrows in Figure 3-5. It is 10.55 cm with an uncertainty of about plus or minus 0.01 cm. The relative uncertainty is the uncertainty divided by the total length.

$$\frac{0.01 \text{ cm}}{10.55 \text{ cm}} = 0.000\ 947\ 9$$

The thickness of this page was measured with a micrometer and a value of 0.016 cm was obtained with an uncertainty of plus or minus 0.001 cm. The *absolute* uncertainty of this measurement is only one tenth of the absolute uncertainty of the measurement made with the ruler, but the *relative* uncertainty is greater.

$$\frac{0.001 \text{ cm}}{0.016 \text{ cm}} = 0.0625$$

The relative uncertainty in 0.016 is 0.0625; it is larger than the relative uncertainty of 0.000 947 9 in 10.55. It is usually the relative uncertainty of a measurement that concerns us, and we need a simple way to describe it.

3-9 Significant Digits

Students tend to memorize rules for significant digits and apply them with little or no understanding of uncertainty. The important point is understanding that all numbers derived from measurement are uncertain. The importance of the issue becomes clear when students must make a decision that is impossible unless uncertainty is considered, e.g., looking at the results of an experiment in which each student determined the density of an object and deciding which students could have had the same material.

As you have seen, anytime a measurement is recorded, it includes all digits that are certain plus one uncertain digit. These certain digits plus the one uncertain digit are referred to as **significant digits**. The more significant digits you are able to record in a measurement, the less relative uncertainty there is in the measurement.

One confusing thing about significant digits is when to count zeros as significant digits. In Section 3-7, the distance between the long arrows in Figure 3-5 was recorded as 10.55 cm. As you might expect, the measurement has four significant digits. However, the thickness of the paper was recorded as 0.016 cm and that measurement has only two significant digits. The leading zeros are not counted. To see why, recall that the number of significant digits should indicate the relative uncertainty in a measurement. The measurement with the greater uncertainty should have fewer significant digits.

In calculating the relative uncertainty of the measurements, the uncertainty is divided by the total measurement. Look at this calculation again, but this time insert an intermediate step to show the relative uncertainty as a whole number fraction.

$$\frac{0.01 \text{ cm}}{10.55 \text{ cm}} = \frac{1}{1055} = 0.000\ 947\ 9$$

$$\frac{0.001 \text{ cm}}{0.016 \text{ cm}} = \frac{1}{16} = 0.0625$$

Notice that when the relative uncertainty is expressed as a whole number fraction, the measurement of 10.55 cm is still expressed as four digits; it has four significant digits. However, the measurement of 0.016 cm is expressed as two digits in the denominator of the whole number fraction; it has two significant digits. The number of significant digits actually indicates the number of digits that will appear in the denominator when you calculate a relative uncertainty.

Confusion about the number of significant digits can be avoided by using scientific notation. When this is done, the digits in the decimal part of the number represent the significant digits.

In scientific notation 0.016 is written as 1.6×10^{-2}. The decimal part of the number is 1.6; it has two digits with the last digit representing the uncertain digit in the measurement. At times the estimated digit in a measurement is zero. It should be written and counted as a significant digit.

Look at Figure 3-5 again. What is the distance from the end of the ruler to the short arrow? If you record an answer of 10 cm, you have the correct magnitude, but you have not indicated the uncertainty of the measurement. Someone looking at your measurement would interpret it to mean that the distance is closer to 10 cm than it is to 9 cm or 11 cm. In other words, one would assume that you only measured the length to the nearest centimeter. Obviously you can estimate the length closer than that. Writing the length as 10.0 cm would be better. One would then assume that you know the length is between 9.9 cm and 10.1 cm, but you actually know it more precisely than that. The best estimate is 10.00 cm, indicating that the length is estimated to the nearest hundredth of a centimeter.

Although there is no difference in the magnitude represented by 10 cm, 10.0 cm, and 10.00 cm, there is a difference in the uncertainty represented. *When quantities are written in science, they are written to express both magnitude and uncertainty. The last digit recorded is assumed to be uncertain.*

Review and Practice

1. An encyclopedia indicates that the distance to the sun is 150 000 000 kilometers. Which digit(s) do you think are uncertain? In other words, do you think that the distance was determined to the nearest kilometer, the nearest ten kilometers, hundred kilometers, or what?

2. If the distance to the sun is only known to the nearest million kilometers, write the distance in scientific notation so the last digit in the decimal part of the number is the uncertain digit.

3. The following measurements were properly recorded so that the final digit is the uncertain digit. How many significant digits are represented by each measurement?

   a. 21.35 cm
   b. 8.705 g
   c. 121.200 0 g
   d. 0.000 823 kg
   e. 0.091 0 m
   f. 38 002 cm

4. Rewrite the measurements in Question 3 in scientific notation. Be sure that you keep all significant digits.

1. Probably the five is uncertain; i.e., the estimate is to the nearest 10 000 000 kilometers.
2. 1.50×10^8 km
3. a. 4   d. 3
   b. 4   e. 3
   c. 7   f. 5
4. a. 2.135×10^1
   b. 8.705
   c. $1.212\ 000 \times 10^2$
   d. 8.23×10^{-4}
   e. 9.10×10^{-2}
   f. $3.800\ 2 \times 10^4$
5. a. 9.10×10^1 mm
   b. $3.800\ 2$ km $\times 10^{-1}$
   c. 2.5×10^3 g
   d. 1.1×10^{-3} cm
   e. 1.3×10^{-2} g
6. a. 1.1×10^{19} kg (The original mass reading from which all of the calculations were done was 7.0 g. There should be no more than two significant digits in any calculation based on that measurement.
   b. 5.9×10^{15} cars
   c. 1×10^6 cars/person

5. Perform the indicated conversions from one metric unit to another. Write the answer so that it shows the same uncertainty as the original measurement.

   a. 0.0910 m = ? mm   d. 0.011 mm = ? cm
   b. 38 002 cm = ? km   e. 13 mg = ? g
   c. 2.5 kg = ? g

6. Review all of the calculations in Section 3-6. Express the following using the appropriate number of significant digits.

   a. the mass of Avogadro's number of rice grains
   b. the number of cars with a mass equal to the mass of Avogadro's number of rice grains
   c. the number of cars per person with a mass equal to Avogadro's number of rice grains

3-10 Mathematical Operations with Uncertain Quantities

Significant digits are usually no problem, as long as you are recording a measurement that you make. You are unlikely to record more than one uncertain digit. Problems arise in keeping track of significant digits when you use those measurements to do a calculation. To illustrate the problem, recall some of the calculations from Section 3-6.

In an effort to understand the magnitude of 6.02×10^{23}, the mass of 400 grains of rice was measured. The 6.02×10^{23} grains of rice was multiplied by the mass of one grain derived from that measurement. The answer was recorded as 1.0535×10^{19} kg. This value implies a confidence in the mass to four digits and an estimate of the mass to five digits. Since the measurement allowed an estimate of the mass of each grain to three digits, the answer is misleading.

In the next calculation, the mass of a car was estimated to two digits. That estimate was used to calculate the number of cars with the same mass as 6.02×10^{23} grains of rice. The answer was recorded as 5.8527778×10^{15} cars. This answer implies that the number is known to eight digits, an unlikely event if the mass of one car is estimated to two digits.

Finally, this quantity is divided by the number of people on Earth to see how many cars could be given to each person. Since the Earth's population is estimated to the nearest billion people, it is unreasonable to report an answer as 1.170555×10^{6} people. Could you obtain an answer that is known to seven digits when a one-digit estimate is used to do the calculation? No, and the following example shows why.

Example 3-6

The length and width of a piece of paper is measured to the nearest tenth of a centimeter. The length is 21.3 cm and the width is 1.3 cm. Calculate the area and record an answer that has one and only one uncertain digit.

▶ Suggested solution _____

Area can be obtained by multiplying length times width. Using a calculator gives an answer of 27.69 cm^2. However, how do you know which digits are uncertain?

The following reasoning may help answer this question. If the 3 in 1.3 is uncertain, that means that the actual width could be 1.2 cm or 1.4 cm, and any digit that you get when you multiply by the three in 1.3 must be just as uncertain as the 3. The same would be true for the 3 in 21.3. If you multiply by hand and show all the uncertain digits in bold type, your result looks like the following:

$$\begin{array}{r} 21.\mathbf{3} \\ \times 1.\mathbf{3} \\ \hline \mathbf{639} \\ 21\mathbf{3} \\ \hline 2\mathbf{7.69} \end{array}$$

Because the 3 is uncertain, the digits that result from multiplication are uncertain.

The 3 is uncertain because the 3 in 21.3 is uncertain.

The 7, 6, and 9 in the total are all uncertain because uncertain digits were added.

The answer should be recorded as 28 cm^2. The measurements made only allow the area to be estimated to the nearest square centimeter and 28 cm^2 is closer to the calculated value of 27.69 cm^2 than is 27 cm^2.

This procedure gives the correct answer recorded to the proper number of significant digits, but it was certainly a lot of trouble. You cannot use your calculator to get the correct answer. Now you will see how the process can be streamlined.

The correct answer has two digits, and the smaller number used in the calculation also has two digits. If this relationship is always true, you could figure out how many digits to leave in the answer by simply counting the number of digits in the numbers used in the calculation.

If you try this procedure with several multiplications and divisions, you will find that it is true about 98% of the time, which is often enough to make the following rule worthwhile. *In any calculation involving multiplication or division, the answer should be rounded to the same number of digits as are found in the least precise number used in the calculation.*

If you try to apply the rule for multiplication and division to predict the answer to an addition or subtraction problem, you will not get the correct answer. The following example shows why.

Example 3-7

Find the perimeter of the sheet of paper described in Example 3-6.

▶ Suggested solution _____

The perimeter is the distance around the paper, or 1.3 cm + 21.3 cm + 1.3 cm + 21.3 cm. The calculator gives a sum of 45.2 cm. Should this value be rounded to 45 cm?

To check, use the same reasoning you used before. Assume that any digit added to an uncertain digit will be uncertain.

1.**3** cm	Since there are uncertain digits in the
21.**3** cm	tenths column, the answer is uncertain in
1.**3** cm	the tenths column as well.
21.**3** cm	
———	Since no other column contains an uncertain
45.**2** cm	digit, the answer is uncertain only in the tenths place.

The answer should be recorded as 45.2 cm. *The number of digits in the added numbers cannot be used to predict the number of digits to be retained in the answer.*

Review and Practice

1. The following problems have been solved using a calculator. Use the rule in Section 3-10 and your knowledge of units to express the answer with the proper units and only one uncertain digit.

   a. $\dfrac{21.3 \text{ cm}}{1.3 \text{ cm}} = 16.384\ 615$ *16*

   b. $\dfrac{6.34 \text{ cm}^2 \times 1.2 \text{ cm}}{1.217 \text{ cm}^2} = 6.251\ 437\ 9$ *6.2 cm*

   c. $13.21 \text{ m} \times 61.5 \text{ m} = 812.415$ *812 m²*

   d. $12.43 \text{ m} \times 2.34 \text{ m}^2 = 29.2105$

   e. $\dfrac{21.50 \text{ cm}}{8.50 \text{ in}} = 2.529\ 411\ 765$ *2.53 cm/in*

2. Repeat the calculations in item 1 in long hand and mark the uncertain digits. Did the rule always produce the desired result of one and only one uncertain digit retained in the answer?

3. All the answers to Question 1 except one represent quantities. Describe the quantities. (In Example 3-6, the answer is a quantity indicating that 28 squares measuring one centimeter on a side would cover the paper.)

4. Do the following additions and subtractions in long hand, and keep only one uncertain digit in the answer.

   a. $63.43 + 34.5 = 97.93$ c. $27.35 - 21.2 = 6.05$ ✓
   b. $124 - 87.2 = 36.8$ d. $3217 + 13.2 + 1.30 = 3231.50$ ✓

5. State a rule that you could use to round the result of addition or subtraction to the proper number of significant digits. Discuss the rule with several of your classmates to see if they agree it will always work.

Answers (margin)

1. a. 16
   b. 6.3 cm
   c. 812 m²
   d. 29.2 m³
   e. 2.53 cm/in
2. No
3. a. does not represent quantity
   b. a length 6.3 times as long as the standard centimeter
   c. an area that could be covered by 812 squares measuring one meter on a side
   d. a volume that could be filled by 29.2 cubes measuring one meter on a side
   e. a unitary rate indicating 2.53 centimeters are spanned for each inch spanned
4. a. 97.9   c. 6.1 or 6.0 *6.2*
   b. 37   d. 3232 or 3231
5. Round to the digit representing the largest place value in which uncertainty was found in any measurement added or subtracted.

Example 3-8

According to my measurement, a sheet of typing paper is 21.50 cm wide. What is the width in meters? In micrometers?

▶ Suggested solution

This problem is a conversion from one metric unit to another.

$$? \text{ m} = 21.50 \text{ cm} \times \frac{0.01 \text{ m}}{1 \text{ cm}} = 0.2150 \text{ m}$$

The only question is whether you are justified in recording the answer as 0.2150 m. If 0.01 m or 1 cm were actual measurements, you should assume that they are uncertain in the last digit. That would make the answer uncertain in the first digit.

But 0.01 and 1 are *not* the result of measurement. There is *no* uncertainty in them. One centimeter is *exactly* 0.01 meter. The uncertain quantity in the calculation is the measurement of 21.50 cm. It is uncertain in the hundredths place. It has four digits. The answer should also have four digits.

If you are confused by the zero in front of the decimal and think that 0.215 represents four significant digits, avoid the error by expressing the number in scientific notation. Written in that form, you should see that you must write 2.150×10^{-1} to show four significant digits. Leading zeros in a decimal number are there to locate the decimal point; they are not counted as significant digits.

3-11 Rounding Fives

An answer is always rounded to the nearest value containing only one uncertain digit. For example, in Example 3-6, the answer was rounded to 28 cm^2.

In the last Review and Practice, you were faced with the problem of rounding an answer that was just as close to one value as another. Both the 0 and the 5 in 6.05 were uncertain, so the answer should be rounded to the nearest tenth. But what is the nearest tenth? The value 6.05 is halfway between 6.0 and 6.1.

One of these choices is just as good as the other. However, there are circumstances that work out better if you round to the larger number as often as you round to the smaller number. Any rule that accomplishes that goal is fine. The following rule is frequently used because it is convenient. It should cause you to round to the larger number about half of the time, because half of the digits you use are even.

Round to the larger number if it is even; round to the smaller number if it is even. In other words, round to make the uncertain digit even.

3-12 Infinite Significant Digits

It is important to remember that significant digits have meaning only in relation to uncertain values. There are numbers that have no uncertainty. How many eggs are in a dozen? About 12? No, exactly 12. A dozen is defined as 12. Similarly, there is no uncertainty in the number of centimeters in one meter, the number of grams in a kilogram, or the number of shoes in a pair. Consequently, when such numbers are used in calculations, they do not affect the number of significant digits in the answer.

The conversion from meters to micrometers illustrates a case where it is essential to use scientific notation to avoid confusion.

$$? \ \mu m = 0.2150 \ m \times \frac{1\,000\,000 \ \mu m}{1 \ m} = 215\,000 \ \mu m$$

As before, the conversion factor is exact and does not affect the number of digits retained in the answer. There are four significant digits in 0.2150 and there should be four digits in the answer. Unfortunately, there is no way to write 215 thousand as a decimal number using fewer than six digits. The solution is to write the number in scientific notation.

In writing $2.150 \times 10^5 \ \mu m$, you know that the answer is known to four, and only four, digits.

3-13 APPLICATION: Determining the Triple Point of Mercury

In 1977, the National Bureau of Standards announced that they had redetermined the triple-point temperature of mercury. The triple point is the freezing point measured at a pressure at which solid, liquid, and gaseous material coexist. You can simply think of it as the freezing point. The value obtained was $-38.84168°C$ and represents a major improvement in the precision of the measurement.

To appreciate this recent accomplishment, you need to know a little history of temperature measurement. By the end of the 17th century, scientists (then known as natural philosophers) were using thermometers made by enclosing colored alcohol or mercury in long glass

Figure 3-8 Early thermometer used in measuring temperature.

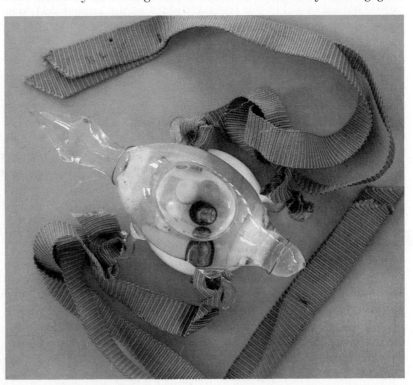

tubes similar to the thermometers still in use, but these thermometers did not have a standard scale scratched on them. Fahrenheit introduced his standard scale about 1714 and Celsius introduced his scale in 1742. The Celsius scale was fixed in reference to the freezing point and boiling point of water, but the method of measuring those points was not standardized. Thus one thermometer did not necessarily give the same reading as another.

At the suggestion of Henry Cavendish, the Royal Society of London appointed a committee of seven to study the problem. The committee made its report in 1777, just two hundred years before the National Bureau of Standards did its latest determination. This report defined standard atmospheric pressure as 29.8 inches of mercury and specified how boiling points measured at other pressures should be corrected. This work stimulated interest in the freezing point of mercury, the liquid used in most thermometers.

J. A. Braun had frozen mercury in a mixture of snow and nitric acid, but its freezing point could not be determined because there was no thermometer to measure it. At the request of the Royal Society, Governor Thomas Hutchins of Albany Fort at Hudson Bay repeated Braun's experiments with collaboration from Cavendish who provided thermometers that he had designed. The thermometers were made in 1776, but political events at that time made communication between England and North America somewhat difficult. The thermometers did not reach Hutchins until 1781 and the results of the experiments were first published in 1783, along with comments on the experiments made by Cavendish. In summary Cavendish said, "It follows, that all experiments agree in showing that the true point at which quicksilver freezes is $38\frac{2}{3}°$, or in whole numbers 39° below nothing." These values are in reference to the Fahrenheit scale. On the Celsius scale, the value would be $-39.26°C$, but the uncertainty in the experiment clearly does not justify four digits! The precision of Hutchins' work is a far cry from the seven-digit precision of the recent NBS determination.

Do You Know?

One of the major concerns of the National Bureau of Standards is reducing the uncertainty with which fundamental quantities are known.

Review and Practice

1. Repeat items 5–18 in the Review and Practice following Section 3-3 on page 69. Assume that the quantities given represent measurements and record the answers to the proper number of significant digits.

2. Do the indicated arithmetic and round the answers using the rule just given.
   a. $34.2 + 3.45 =$
   b. $1.5 \times 0.0011 =$
   c. $38 - 2.5 =$
   d. $\dfrac{1.375 \times 10^3}{5.0} =$
   e. $(1.1 \times 10^5)(7.5 \times 10^4) =$

3. A student suggested another rule for rounding fives: "Flip a coin. If it is heads, round to the higher value; if it is tails, round to the lower value." Would this rule cause you to round to the higher value as often as you round to the lower value? Explain. What disadvantage, if any, do you see in this rule?

1. (5) 39
   (6) 2.9×10^{12}
   (7) 6.30×10^3
   (8) 1.6×10^{-4}
   (9) 3.26×10^{-22}
   (10) 4.00×10^3
   (11) 85
   (12) 1.1×10^5
   (13) 7.0×10^{-20}
   (14) 1.95×10^{42}
   (15) 2×10^2
   (16) $3.6\cancel{4} \times 10^{-29}$
   (17) 2.64×10^{-3}
   (18) 8.5×10^{-9}

2. a. 37.6
   b. 1.6×10^{-3}
   c. 36
   d. 2.8×10^2
   e. 8.2×10^9

3. Yes. You should get heads half of the time and tails half of the time. The rule is inconvenient to use, because you need a coin and must take time to flip it.

Making Sense Out of Data

Science is little more than making observations and making sense out of what you observe. In Chapter 1, you learned several habits that scientists follow to make sense out of observations and to clearly communicate the results to others. In this chapter, the search continues for consistent interpretations of events. By considering uncertainty in measurements and the effect of uncertainty on quantities derived through calculation, we avoid the suggestion that data is more precise than it actually is.

Other techniques adopted by scientists help us make sense out of observations by revealing relationships that are not obvious. Several of these have to do with the way information is organized.

3-14 Making Tables

Table 3-1 on page 62 shows a number of rice grains and their corresponding mass, but it is not a good table. There is no order to the data. It is difficult to see that the mass increases as the number of grains increases. Table 3-3 shows the data ordered from smallest to the largest number of grains, and the corresponding order in the mass is more evident.

Table 3-3

NUMBER OF RICE GRAINS AND THEIR MASS	
NUMBER OF GRAINS	CORRESPONDING MASS
100	1.75 g
200	3.5 g
400	7.0 g
800	14.0 g
1600	28.0 g

It is not always clear how data should be recorded to make interpretation easy; several organizations may be tried before one is found that makes sense. As you make observations, you should try various ways of presenting the data and select the one that reveals relationships that seem important.

3-15 Graphing

Pictures, histograms, graphs, and physical models often reveal things that tables do not. Graphs are particularly useful when you suspect a proportional relationship like the one shown in Table 3-3.

Figure 3-9 represents the data from Table 3-3 in graphic form. A graph can reveal many things if it is carefully prepared and the reader knows how to interpret it. The following rules may help.

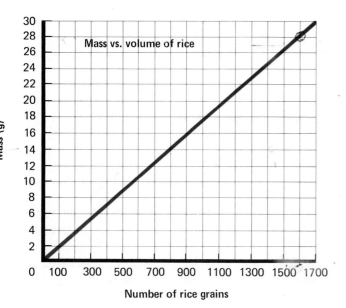

Figure 3-9 Graph showing the mass of rice as a function of the number of grains. The data used to locate the line appear to have no uncertainty since all points fall on the line.

1. The purpose of a graph is to communicate information in a concise manner. A graph conveys little information if it is not labeled properly. Rule 1 of good graphing is to *give your graph a descriptive title.*
2. *Indent the axes from the edge of the graph paper* and draw them with a straight edge.
3. *Label each axis and indicate the units used.* Numbers along each axis are useless if you do not show what they represent.
4. *Choose an appropriate scale* that allows you to get all data on the graph. Check the largest and smallest values to determine the range for each axis. For most graphs, there is no need to begin the scale at zero. Just be sure you begin with a number lower than your lowest data point.
5. *Choose a convenient scale.* A graph is easier to read (and plot) when each square represents a value of 1, 2, 5, or a multiple of 10 times these numbers: 10, 20, 50 or 0.2, 0.5, etc.
6. *Maintain the same scale for the length of the graph.*
7. *Locate points with an X or a dot with a small circle around it.* An X or a dot with a circle around it can be located after the graph line is drawn.
8. *Draw a smooth curve or straight line to represent the general tendency of the data points.* Graphs drawn in math classes represent absolute numbers, and each data point falls on the curve being plotted. This situation is seldom the case in science, where the data points represent experimental measurements. In science, the data used to plot the graph is uncertain.

We have had the most success in teaching students the importance of the rules for good graphing by passing out student graphs that omit titles, labels, marked points, etc., and asking students what would make it easier for them to interpret the graphs.

Graphs like Figure 3-9 are used in making predictions. A few measurements are recorded on the graph, and these points are used to predict what may be seen if more measurements were made. The line drawn in Figure 3-9 predicts where data points would be located if more rice were counted and its mass found. The more points plotted to predict the line, the better the prediction is likely to be. Once you are confident that enough points are plotted to locate the line, you can

The way data are represented influences our ability to see relationships and make new discoveries. The tables and graphs described here are two of the oldest and simplest means of representing data so that new relationships are more easily seen.

use the graph to predict what a measurement would be without bothering to measure it.

It is a good idea to look at a graph to see if it makes sense. Measurements can be incorrectly done. The popular adage about computers should be applied to graphs or any other transformation of data: garbage in, garbage out.

The upward slope of the line in Figure 3-9 indicates that the more grains you have, the greater the mass of the rice. This interpretation certainly makes sense. The line passes through the origin when extended downward, and that also makes sense. The (zero, zero) point predicts, "Zero grains of rice should have zero mass." That result is what you would expect from experience.

The one way that Figure 3-9 contradicts experience is that it is too good to be true. All the data points line up exactly. All points fall on the curve. This often happens when the data are artificial; it seldom happens with real data. Real data comes from measurements that are usually a little greater or less than the actual value.

The only data in Table 3-3 that is real is the data for 400 grains of rice, the number counted and weighed. The rest of the data was obtained by multiplying or dividing the measured values as discussed in Section 3-1. Had the rice been counted and weighed to get the data, the points would not line up as neatly as they do in Figure 3-9.

Figure 3-10 shows a graph of real data. The mass and volume of several pieces of glass were measured and used to plot the graph. A straight line was drawn to predict where data points should be, even though the line does not pass through all points representing actual measurements. The line was drawn as shown to represent what the data would look like if there were no uncertainty in the measurements.

Figure 3-10 Graph showing the mass of glass as a function of its volume. The line is not drawn through all data points because it is assumed that there was uncertainty in the measurements used to locate the points.

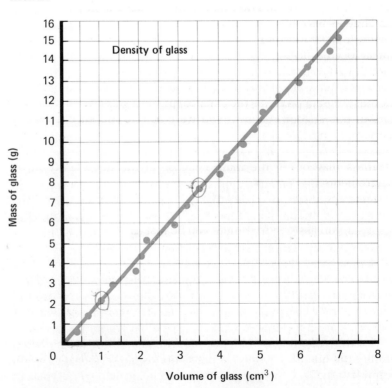

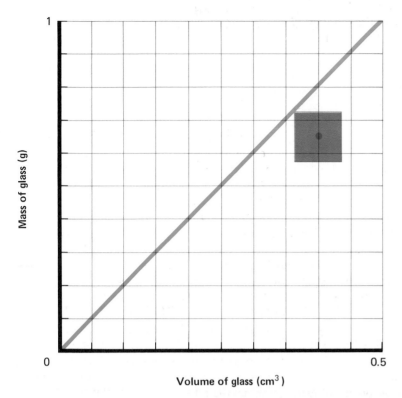

When a balance was used to measure the mass of the smallest piece of glass, the estimate was 0.73 g, but the balance used is only accurate to plus or minus 0.02 g. The true mass could be as small as 0.71 g or as large as 0.75 g. Similarly, the volume for this piece of glass was measured with a graduated cylinder as 0.4 cm^3. The graduate can be read to plus or minus 0.2 cm^3. The true volume could be as small as 0.2 cm^3 or as large as 0.6 cm^3.

Figure 3-11 shows the portion of Figure 3-10 where this first data point is located. In Figure 3-11, an area is shaded around the data point. The shaded area represents a mass ranging from 0.71 g to 0.75 g and a volume ranging from 0.2 cm^3 to 0.6 cm^3. You can be rather confident that the data point for the smallest piece of glass would be somewhere in this shaded region if you were able to measure it with no uncertainty. The uncertain measurements were used to make a "best guess" of where the point should be, but you should realize that this best guess may be off the mark because of the uncertainty of the measurements.

You have no way of knowing whether a measurement recorded is larger or smaller than the true value. However, in making many measurements, you assume that there will be just as many measurements that are too large as there are measurements that are too small. The chance of error in either direction is the same. Consequently, when the line is drawn to represent where points would be if there were no uncertainty in them, it is drawn so that there are as many points above the line as below it. The line is also drawn so the points that do not fall on the line are scattered along it.

The line described is called the *best fit* for the data. In careful scientific work, mathematical equations are used to locate this best-fit line. In this course you will do it by sight.

Review and Practice

1. See sample graph in the Teacher's Guide at the front of this book.
2. 4, 10.6
3. See sample graph in the Teacher's Guide at the front of this book.

1. The mass of several 1984 pennies was determined by finding the mass of a single penny, then two pennies, three pennies, and so forth. The data is recorded below. Plot a graph showing the number of pennies along the x-axis and the mass along the y-axis.

NUMBER OF PENNIES	MASS IN GRAMS
1	2.5
2	5.1
3	7.6
4	10.6
5	12.6
6	15.1
7	17.6

2. Which data points from item 1 appear to have the greatest uncertainty in the measurements?

3. The experiment was repeated with some 1980 pennies. The data for these pennies is shown below. Use the data to plot another line on the same sheet of graph paper. Be sure to label both lines so you will know which represents the 1984 pennies and which represents the 1980 pennies.

NUMBER OF PENNIES	MASS IN GRAMS
1	3.1
2	6.1
3	9.2
4	12.3
5	15.4
6	18.5
7	21.7

3-16 Using Graphs to Get Unitary Rates

If you plotted the graphs for the 1984 pennies and the 1980 pennies correctly, you should see that the points do not fall on the same line. Both lines pass through the origin, which makes sense. You would expect 0 pennies to have no mass, whether the pennies were made in 1980 or 1984.

The line for the 1980 pennies climbs faster. The term used to describe how fast the line goes up is **slope**. If you think about what you plotted, you realize that the line climbs because the mass of pennies increases as the number of pennies increases. Then the line for the 1980 pennies must climb faster because more mass is added when the number of 1980 pennies increases than when the number of 1984 pennies increases. In other words, 1980 pennies must have more mass than 1984 pennies!

Perhaps you recall how to calculate the slope of a line from your previous work in mathematics. To review the procedure, you will calculate the slope of the line in Figure 3-9.

Example 3-9

Find the slope of the line in Figure 3-9, and explain what it means.

▶ Suggested solution _____

The slope of a line shows how much the quantity plotted along the y-axis increases when the quantity plotted along the x-axis increases by one unit. To calculate the slope, you need to pick *two* points that fall on the line. (Some students use only one point. That appears to work when the line passes through the origin as this one does. However, it does not work in general.)

Two points that are easy to read from the graph are (1600 grains, 28 g) and (0 grains, 0 g). Notice that the data points include units. You have plotted quantities. Quantities have units and should be included.

Between these two points on the line, the value plotted along the y-axis increased 28 g. This value is often called *delta y* (Δy) and is the difference in the value of y at the two points selected (28 g − 0 g = 28 g). *Delta x* (Δx), the amount the quantity plotted along the x-axis increased between these same two points, is the difference between the two x values:

$$1600 \text{ grains} - 0 \text{ grains} = 1600 \text{ grains}$$

The slope can now be found by dividing delta y by delta x.

$$\frac{\Delta y}{\Delta x} = \frac{28 \text{ g}}{1600 \text{ grains}} = \frac{0.018 \text{ g}}{1 \text{ grain}}$$

As you can see, the slope of a line is the unitary rate which shows that the mass of rice increases by 0.018 g each time a grain of rice is added.

Review and Practice _____

1. Calculate the slope of the line that shows the mass of 1980 pennies as you increase the number of pennies and explain what it means.

2. Calculate the slope of the line that shows the mass of 1984 pennies as you increase the number of pennies and explain what it means.

3. What is the difference in 1980 pennies and 1984 pennies?

1. The slope is 3.1 g/penny. It means that on the average the mass of 1980 pennies increases 3.1 g with each penny added.
2. The slope is 2.5 g/penny. It means that on the average the mass of 1984 pennies increases 2.5 g/penny with each penny added.
3. 1980 pennies have more mass than 1984 pennies. They do not have the same mass because they are made of different materials.

3-17 Density: A Useful Unitary Rate

Pennies made in 1980 are the same size as those made in 1984, but they have more mass. Prior to 1981, pennies were made of copper; pennies made after that date were made of zinc with a thin copper coating. The newer pennies look like the older ones, but you can tell them apart because they have different masses. In this case, density can be used to describe this difference in mass. **Density** is defined as mass per unit volume (a unitary rate) or the ratio of mass to volume, but it is useful to think of density as the mass of a particular volume, say one cubic centimeter.

Density can be used to distinguish between many materials that look the same. For example, eyeglasses with plastic lenses look very much like eyeglasses with glass lenses. How could you tell them apart?

Try introducing density by passing around a plastic foam ball and lead sinker selected so that the ball has slightly more mass. Ask students which is heavier and then place the objects on a double pan balance to check. Ask students why they were fooled and how plastic foam and lead could be compared so the students would not be fooled.

Putting them into a fire to see if they burn would work, but it is hard on the glasses. Hitting them with a hammer or trying to scratch the lenses with a nail would also work, but these techniques also damage the eyeglasses. However, measuring the mass and volume of the lenses would not damage the eyeglasses, and the data could be used to calculate the density and identify the material in the lenses.

Notice that mass alone would not provide the information you need. Thick plastic lenses could have more mass than thin glass lenses. Mass is an extensive property. An **extensive property** depends on the amount of matter being measured. You need a measurement that would be the same regardless of how much material is measured. Properties that do not depend on the amount of material measured are called **intensive properties**. Density is an intensive property.

Density might be thought of as *mass intensity*. In SI units, density is expressed as the mass in kilograms of one cubic meter (1 m³) of material. A cubic meter is a very large piece of matter, so more commonly used metric units for density are grams per cubic centimeter (g/cm³).

If the copper in a 1980 penny has a density of 8.92 g/cm³, then each cubic centimeter of copper has a mass of 8.92 grams. Density is a unitary rate indicating how much the mass of some material increases each time the volume is increased by one cubic centimeter.

Like other unitary rates, the inverse of density can be expressed as another unitary rate.

$$\frac{1 \text{ cm}^3}{8.92 \text{ g}} = \frac{0.112 \text{ cm}^3}{1 \text{ g}}$$

This unitary rate indicates that each gram of copper will have a volume of 0.112 cm³. If you wish, you can give this unitary rate a name, just as its inverse is given the name density. In SI, this unitary rate is called *specific volume*.

1. 2.17 g/cm³; density
2. 45.6 g. 46g 3
3. 2.84 cm³ .35cm³
4. The golf ball
5. The line would lie below the line for glass, i.e., its slope would be smaller because the density is smaller. The line should pass through the origin because plastic with no volume should have no mass. It is nothing!

Review and Practice

1. Find the slope of the line in Figure 3-10. What is the name given to this slope?

2. Using the slope that you calculated in item 1, find the mass of a piece of glass that has a volume of 21 cm³.

3. In item 1 of the Review and Practice following Section 3-16 you found the average mass of a 1980 penny. 1980 pennies are made of copper, so the density of the 1980 penny is 8.92 g/cm³. What is the volume of a 1980 penny?

4. Figure 3-12 shows a plastic foam ball and a golf ball on each side of a double pan balance. The two objects have the same mass. Which is more dense?

5. Plastic eyeglass lenses are less dense than glass lenses. If you measured the mass and volume of several plastic lenses and plotted a graph like the one for glass in Figure 3-10, where would the best fit line for the data be located—above the line for glass or below it? Would it pass through the origin? How do you know?

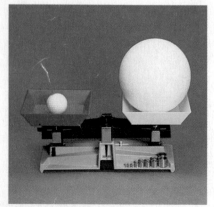

Figure 3-12 The plastic foam ball and golf ball have the same mass.

CHAPTER REVIEW

Numbers in red indicate the appropriate chapter sections to aid you in assigning these items. Answers to all questions appear in the Teacher's Guide at the front of this book.

Summary

■ Chemists study matter, be it large or small. Consequently, they generate numbers ranging from very large to very small, and one of their tasks is to visualize such numbers.

■ Many of the relationships that you study are proportional ones that can be described by unitary rates. These unitary rates are used to calculate values you do not know from values you do know.

■ Doing calculations with very large or very small numbers is difficult unless numbers are expressed in scientific notation.

■ Measurements produce estimates rather than exact values and the relative uncertainty of measurements is conveyed by the number of digits recorded. These digits are described as significant. When measurements are used in calculations, it is necessary to round answers to the correct number of digits so as not to imply that something is known more precisely than it actually is.

■ Much of what you learn from observations occurs when data are organized to reveal trends and relationships among two or more quantities. Carefully prepared tables and graphs often reveal important relationships. The slope of a straight-line graph is a unitary rate that describes how much one variable increases as the other variable increases by one unit.

■ Density is a unitary rate that indicates how much mass there will be in each cubic centimeter (or other unit volume) of matter. It is an intensive property; that is, it does not change with the amount of matter measured.

Chemically Speaking

accuracy
base
density
exponent
extensive property
intensive property
precision

proportional quantities
relative precision
relative uncertainty
scientific notation
significant digits
slope

Review

1. What kind of relationship is described by a unitary rate? (3-1)

2. Write a mathematical sentence that says "A is directly proportional to B." (3-1)

3. What term is used to describe the two in 3^2? What is the exponent in 6^4? What is the ten called in 10^{23}? (3-2)

4. If you measure the length of a desk and report that it is 187 cm, which digit is uncertain? (3-7)

5. Why are data in tables normally put in order from smallest to largest or vice versa? (3-14)

6. List at least two sets of units that could be used to describe the density of an object. (3-17)

7. A chemistry teacher graded a test by assigning scores in 5-point increments, giving grades of 60, 65, 70, and so on. Students complained that 1-point increments should be used to give grades of 60, 61, 62, and so on. Are the students concerned about the accuracy or the precision of the score? Explain your answer. (3-8)

8. A student complained that an error was made in summing the points earned for each question. The teacher had incorrectly added to get 65 rather than 70. Is the student concerned over the accuracy or precision of the score? Explain your answer. (3-8)

9. The number 2700 can be written as 2.7×10^3, 27×10^2, and 270×10^1. (3-2)
   a. Which expression(s) represents scientific notation?
   b. Which expression(s) represents exponential notation?
   c. Which expression(s) represents neither scientific nor exponential notation?

10. If 2700 represents a measurement estimated to two digits, which expression(s) in item 14 is/are acceptable representations of the measurement? Why are the others unacceptable? (3-9)

11. Two students estimate their mass as 72.5 kg each. They could say that the uncertainty in this measurement is 0.1 kg or that it is 0.1 kg/72.5 kg or (0.014). Does 0.014 represent the absolute uncertainty of the measurement or the relative uncertainty of the measurement? What kind of uncertainty is represented by 0.1 kg? (3-8)

12. Which digit(s) in the measurement of 72.5 kg represent significant digits? Which represent uncertain digits? (3-9)

13. Carpenters, roofers, and architects talk about the

91

slope of the roof on a house. What does it mean if they say that the roof has a small slope? A large slope? How could you calculate a numerical value for the slope of a roof? (3-16)

14. Which of the following represent extensive properties? How do you know? (3-17)
   a. A glass holds 250 mL of water.
   b. The density of water is 1.00 g/cm³.
   c. My mass is 72.5 kg.
   d. My body temperature is 37° C.
   e. My savings account earned $432 in interest last year.
   f. My savings account earns interest at the rate of 10% quarterly.
   g. I drove 235 miles yesterday.
   h. My speed was 55 miles per hour.

15. Which of the statements in item 14 describe intensive properties? (3-17)

16. The following tables show values for two variables *A* and *B*. Which table(s) contains data which are directly proportional? Which table(s) contains data which are inversely proportional? Which table(s) contains data which are neither directly nor inversely proportional? (3-1)

I		II		III		IV	
A	B	A	B	A	B	A	B
1	6	1	6	1	12	1	12
2	7	2	12	2	11	2	6
3	8	3	18	3	10	3	4
4	9	4	24	4	9	4	3

Interpret and Apply

1. Make the following measurements and organize your data in a table. (3-1, 3-14)
   a. Using a measuring cup and a watch with a second hand, collect data that show the volume of water that drips from a faucet over a period of time. Make at least ten readings of time and corresponding volume.
   b. If you have a kitchen scale, take several mass readings for one, two, three, etc., objects of uniform size. You could use spoons, forks, cups, candy, beans, or any convenient objects around the house.
   c. Make a chain of paper clips or other objects of uniform length and measure the length containing one, two, three, etc., of the objects.

2. Estimate the uncertainty in the measurements you made in item 1. (3-7)

3. Use the data collected in item 1 to plot a graph. (3-15)

4. Does the data you collected in item 1 describe a proportional relationship? How do you know? (3-1)

5. When you plotted a graph of your data, did all your data points fall on the line (or curve) that you drew to represent the points? Do you think they should? Why? (3-15, 3-16)

6. If you and a classmate collected the same kind of data and plotted two graphs, could you tell which graph had the greater slope by looking at them? Why? (3-16)

7. When carpenters buy lumber to build a deck or masons buy brick to build a wall, they usually have some material left over. (Occasionally they buy too little and run out!) Why does this situation occur? (3-7, 3-8)

8. A cook seldom gets exactly the same results when a dish is prepared several times, even when the same recipe is used. Why? (3-7, 3-8)

Problems

1. Magnification can be fun to experiment with. A low-power microscope lens magnifies ten times. This statement means that a line 1 mm long would appear to be 10 mm (1 cm) long when seen through the lens. If viewed through a ten-power (10X) eyepiece, the object would appear to be 10 cm long. Additional lenses would continue to magnify the object ten times. How many lenses would be needed to make a 1-millimeter square (1 mm²) appear to be 1 km square? How many 1-mm squares would fit on the 1-km square? (3-6, 3-8, 3-9, 3-10)

2. A red blood cell is about 0.0025 mm in diameter. How large would it look if magnified 10^6 times? If magnified 10^9 times? (3-6, 3-8, 3-9, 3-10)

3. By measuring a magnified image, you can often determine the size of an object with less uncertainty than by measuring the object directly. When magnified 100 times and measured with a ruler, a thread appeared to be 0.12 mm thick. What was its actual thickness? If the uncertainty in the measurement was 0.01 mm, what is the uncertainty in the estimate for the actual thickness? Do you think you could make a direct measurement of the thread with that precision? (3-8, 3-9, 3-10)

4. One cubic centimeter of copper contains about 8.46×10^{22} atoms. What is the volume of one copper atom? (3-8, 3-9, 3-10)

5. A 1980 penny has a mass of about 3.5 g. What is its volume? (3-17)

6. Use your answers to Questions 4 and 5 to find the number of atoms in a penny. (3-6, 3-17)

7. Do the indicated calculation and record your answer to the correct number of digits. (3-2, 3-3, 3-4, 3-9) ✓
   a. $(6.65 \times 10^{-8}) - (6.65 \times 10^{-9}) =$ 5.98×10^{-8} ✓
   b. $(3.30 \times 10^{-5})(98.74 \times 10^{-2}) =$ 3.26×10^{-5}
   c. $\dfrac{1.05 \times 10^{15}}{4.2 \times 10^{-3}} = 2.5 \times 10^{17}$ f. $\dfrac{1.0 \times 10^{26}}{2.0 \times 10^{24}} = 50.$ or 5.0×10^{1}
   d. $(5 \times 10^{25})(10^{23}) = 5 \times 10^{48}$ g. $(4.44 \times 10^{-8})(2 \times 10^{-4}) =$ 9×10^{-12}
   e. $\dfrac{3.4 \times 10^{22}}{10^{20}} = 3.4 \times 10^{2}$ h. $(3.4 \times 10^{5}) + (5 \times 10^{4}) =$ 4×10^{5}

8. How many significant digits are in the number underlined? If there is an infinite number, explain why. (3-9, 3-12)
   a. 1 gross = <u>144</u>
   b. my height = <u>69.7</u> in
   c. 1 km = <u>1000</u> m
   d. 1 yr = <u>365</u> days
   e. 1 yd = <u>36</u> in
   f. my body temperature is <u>37</u>° C
   g. pi = <u>3.14</u>
   h. <u>pi</u> = 3.14

22/7 = infinite decimal

9. The following table lists the mass, volume, and density of several objects. Supply the missing values. (3-17)

	Mass	Volume	Density
a.	45.2 g	8.4 cm³	— 5.4 g/cm³
b.	— 579	13.3 mL	4.3 g/cm³
c.	2 356 g	— 5.5×10² cm³	4.3 g/cm³
d.	1.5 kg	0.283 m³	— 5.3 kg/m³
e.	1.5 kg	— 1.1×10² cm³	13.6 g/cm³

10. Assume that the numbers in the following problems resulted from measurement and are recorded to reflect the uncertainty in the measurement. Round the answers to show the proper number of significant digits. (3-9, 3-10)
   a. $2.86 \times 1.824 = 5.21664$ 5.22
   b. $21/8 = 2.625$ 3
   c. $3.7 + 1.86 + 0.0024 = 5.5624$ 5.6
   d. $98.0 \times 1.22 = 119.56$ 120.
   e. $2.1 + 1.0 = 3.1$
   f. $65 - 3.47 = 62.53$ 63
   g. $4.6215/0.0015 = 3081$ 3100
   h. $10.00/2.000 = 5.000$

11. Convert all numbers in item 10 to scientific notation. (3-2)

Challenge

1. Three people measured the mass of a rock as 34.2 g, 35.0 g, and 34.6 g. What is the best estimate for the mass of the rock? Why?

2. The diameter of Earth is 1.271×10^{4} km at the poles. How many times would a marble measuring 1.271 cm need to be magnified by a power of ten to look as large as Earth? Where would the edge of Earth appear to be if it were magnified as much as the marble?

3. Pollutants in air and water are frequently measured in parts per million (ppm) or parts per billion (ppb). One part per million would mean that there is one gram of the pollutant in one million grams of air or water. At ordinary temperature and pressure, air has a density of 1.2×10^{-3} g/cm³. What volume of air would contain one gram of sulfur dioxide, a pollutant that causes acid rain, if the sulfur dioxide concentration is 2 ppm?

4. The density of water is 1.0 g/cm³. How much water would you need to drink to consume 1 mg of PCB if the PCB is present in a concentration of 1 ppb?

Synthesis

1. Cement has a density of about 3 g/cm³ once it has set. What is the mass of a cubic meter of cement? How many cubic centimeters of cement would you be able to lift?

2. A rock has a volume of 3.4×10^{-6} m³ and a mass of 1.09×10^{-2} kg. Could the rock be diamond? (What information will you need to look up in a handbook?)

Projects

1. Attention to toxic substances in food, water, and air has increased in recent years. For some toxic substance that is in the news, find out how it is detected and how much must be present to be detected.

2. Many laws controlling toxic substances are written to require elimination of the substance. Interview individuals who work in industries affected by these laws or interview environmentalists who have worked to enact the laws. Find out the implications and alternatives that are available to protect life without adversely affecting the economy?

Water is one of the simplest, most common, and most important chemical compounds. The ancient Greeks thought that water was one of four
elements (along with earth, air, and fire) that made up everything on Earth. You know
the smallest unit of water is a molecule composed of two elements that has the formula
H_2O. Molecules are very small, so small that the number of molecules in a glass of water
is larger than the total number of pages in every book on Earth. How do we know how
many molecules are in a glass of water? You cannot see water molecules so how would
you know how many water molecules are in a lake or ocean? A way is needed to
measure numbers of atoms and molecules and group them into a counting unit that is
useful. The unit for that measurement is the mole.

The Mole

Mass Relationships and Avogadro's Number

Paper is commonly sold in packages of 500 sheets called a ream. When the paper is packaged at the factory, the sheets are not counted. The mass of 500 sheets is determined and the paper is packaged according to mass.

"Counting" by mass is useful in dealing with large numbers of objects that are uniform in size. If you have ever collected aluminum cans for recycling, you know that the cans are not counted individually at the recycling center. Since payment is made on a per can basis, the cans are "counted" by weight. "Counting" in this way involves the selection of a standard, in this case an aluminum can, and determining its mass or weight. You now will see how this method is applied in dealing with large quantities of atoms. Even if atoms were large enough to count, imagine the difficulty in counting out Avogadro's number (6.02×10^{23}) of atoms.

CONCEPTS

■ "counting" indirectly through calculation
■ determination of relative mass
■ relative nature of atomic masses
■ Avogadro's hypothesis
■ the basis for atomic mass units
■ the mole as a counting unit

4-1 Relative Mass

If real atoms were as big as the plastic foam models used to represent them, there would be no problem in finding the mass of a single atom. You could put the model atom on a balance and determine its mass. Unfortunately the problem is not that simple! Yet, a way has been found to determine the mass of an atom like calcium with respect to an atom of carbon without knowing the mass of a single atom of either. This procedure involves comparing the masses of equal numbers of atoms.

Consider this analogy. A given number of oranges has a mass of 2160 grams (2.16 kg). An equal number of grapefruit has a mass of 3600 grams (3.60 kg). Assume that all of the oranges have the same mass and all of the grapefruit have the same mass, then the oranges have a mass that is 3/5 or 0.600 that of the grapefruit.

$$\frac{\text{oranges}}{\text{grapefruit}} \quad \frac{2160\,\text{g}}{3600\,\text{g}} = \frac{3}{5} = 0.600$$

Since there are an equal number of oranges and grapefruit, the mass of *one* orange is 3/5 or 0.600 that of the mass of one grapefruit. The mass of the orange is expressed in comparison to the mass of the grapefruit. The value 0.600 represents the relative mass of the orange. **Relative mass** of any object is expressed by comparing it mathematically to the mass of another object. The relative mass of an orange compared to a

Do You Know?

N$_2$O has the chemical name nitrous oxide but it is also known as laughing gas. This colorless, odorless gas was discovered by Joseph Priestley in 1772. Sir Humphry Davy found in 1799 that it could be used as an anesthetic. It was common for people who inhaled nitrous oxide to begin laughing uncontrollably and hence it became known as laughing gas. It became a fad for the high society in England to inhale laughing gas at small parties though usually not in mixed company.

grapefruit was found without knowing the mass of either a single orange or a single grapefruit.

The mass of an atom is called its **atomic mass**. One way of determining the atomic mass of an atom is to compare it to the mass of some other atom that is taken as a standard. At first the masses of the elements were compared to hydrogen because hydrogen is the lightest atom. The nitrogen atom was 14 times as heavy and so on. Later, oxygen was used as a standard for atomic mass and was assigned a value of exactly 16. Still later, a form of carbon with a mass of 12.0000 was chosen as the standard. These changes in the standard provided for more precise measurements. The masses of individual atoms are assigned a unit of relative measurement known as the **atomic mass unit** (amu). The atomic mass unit is defined simply as 1/12 the mass of a carbon-12* atom. The periodic table used by chemists today includes an atomic mass value for each element that is based on the carbon standard. In examining the periodic table, you may notice that many of the atomic masses are not whole numbers. The values you see are determined using sophisticated equipment. You will be studying more about atoms later in this course.

Since atoms are so small how were scientists able to compare the masses of equal numbers of atoms of different elements? The answer to this question comes from work done by scientists in the late 1700's and early 1800's.

4-2 Combining Volumes of Gases and Avogadro's Hypothesis

Joseph Louis Gay-Lussac was a French chemist who lived from 1778 to 1850. He conducted experiments to investigate how elements that are gases combine to form compounds that also are gases. Much of his work was done on a series of different gaseous compounds that were formed from the same two elements, nitrogen and oxygen. He noted that, for all compounds formed, the ratios of the volumes of gases used were simple, whole-number ratios. His analysis of data for various oxides of nitrogen are shown in Table 4-1. Notice that the volume ratios

Table 4-1

COMBINING VOLUMES OF GASES				
COMPOUND	VOLUME OF NITROGEN (in cm^3 or mL)	VOLUME OF OXYGEN (in cm^3 or mL)	APPROXIMATE VOLUME RATIO OF NITROGEN TO OXYGEN	ACCEPTED FORMULA
Compound 1	100	49.5	2 to 1	N$_2$O
Compound 2	100	108.9	1 to 1	NO
Compound 3	100	204.7	1 to 2	NO$_2$

*Carbon-12 denotes a particular type of carbon atom. This notation will be discussed further in Chapter 8.

are not precisely whole numbers. For compound 1, 100 to 49.5 is not exactly a 2 to 1 ratio. However, when more accurate measurements are made, the volume ratios are whole numbers within experimental uncertainty.

Look at the volumes of nitrogen and oxygen that react to form each compound. In Compound 1, the volume of nitrogen that combines is twice the volume of the oxygen. For Compound 2, equal volumes of nitrogen and oxygen react to form the compound. In Compound 3, the volume of nitrogen that combines is half the volume of the oxygen gas.

The volumes of gases that react to form each compound can be expressed as a simple ratio: in this case 2 to 1, 1 to 1, or 1 to 2. The importance of Gay-Lussac's results was recognized by Amadeo Avogadro, an Italian scientist and contemporary of Gay-Lussac.

In 1811, Avogadro wrote a paper in which he suggested that Gay-Lussac's data on the combining volumes of gases could be explained very simply, if one made a bold assumption:

Equal volumes of gases (at the same temperature and pressure) contain equal numbers of particles.

Avogadro's assumption was slow to gain acceptance in the scientific world. However, as time passed, other experiments involving combining volumes of gases supported his idea. In 1860, four years after Avogadro's death, another chemist named Stanislao Cannizzaro presented Avogadro's work at a large gathering of scientists. A few years later the idea that equal volumes of gases at the same temperature and pressure contain equal numbers of particles was finally accepted and eventually became known as **Avogadro's hypothesis**.

Avogadro's hypothesis can be applied to the question stated earlier—how to obtain the equal numbers of atoms of different substances needed in order to find their relative masses.

Do You Know?

Earth's atmosphere is about 78% nitrogen and 21% oxygen. At ordinary temperatures these two elements do not react. However, when air is taken inside an automobile engine to burn fuel, a mixture of compounds of oxygen and nitrogen given the general formula N_xO_y are formed. N_xO_y is a significant source of air pollution and cause of smog.

Table 4-2

MASS RELATIONSHIPS AMONG EQUAL VOLUMES OF SOME GASES			
GAS	MASS OF EVACUATED FLASK (in g)	MASS OF FLASK + GAS (in g)	MASS OF 1 L OF GAS (in g)	MASS OF GAS (approx.) RELATIVE TO HYDROGEN
hydrogen	157.35	157.43	0.08	1
nitrogen	157.35	158.47	1.12	14
oxygen	157.35	158.66	1.31	16
fluorine	157.35	158.87	1.52	19
chlorine	157.35	160.15	2.80	35

Table 4-2 contains data obtained from equal volumes of several gases. In this case, the volume chosen is one liter. All of the gases are at the same temperature and pressure. In looking at Table 4-2 you should notice that hydrogen has the least mass. Therefore, the mass of hydrogen is used here as the standard (given a value of 1) to find the relative masses of the other substances.

Recall the problem of oranges and grapefruit at the beginning of the chapter. The relative mass of one orange to one grapefruit is the same as the relative mass of a dozen oranges to a dozen grapefruit. The same reasoning is true for any individual gas atom listed in Table 4-2.

For example, the relative mass of one atom of nitrogen (14) compared to one atom of hydrogen is the same as the relative mass found when comparing all of the molecules in a liter of each gas.

4-3 How Many Is a Mole?

Do You Know?

Avogadro's number is so large that if you could count 100 particles every minute and counted twelve hours every day and had every person on Earth also counting, it would still take more than four million years to count a mole of anything!

A dozen is a convenient unit for expressing a frequently used quantity. However, one or two dozen atoms are too small to be seen with even the most powerful microscope. The term mole is used to talk about a number of atoms, molecules, ions, or electrons, just as dozen is used to talk about a number of eggs, oranges, or doughnuts. A **mole** (mol) is simply the amount of a substance that contains 6.02×10^{23} particles. The particle can be anything—atoms, molecules, green peas, or baseballs. When working with small particles such as atoms and molecules, 6.02×10^{23} is a convenient unit.

Like the oranges and grapefruit whose relative masses do not change whether you consider single fruits or dozens of fruits, relative masses of atoms do not change whether you consider individual atoms or moles of atoms. One mole of aluminum atoms has a mass in grams that equals the atomic mass, in amu's, of a single atom of aluminum. The number 6.02×10^{23} is an accepted standard for a mole and is called **Avogadro's number**, to honor the Italian scientist whose hypothesis led to its determination.

You would never want to count the number of particles in a mole individually. Recall how tedious it was to count a million of something in Chapter 3. If you have forgotten how large Avogadro's number is, look back at Section 3-5. Because 6.02×10^{23} atoms is too large to count, this number of atoms is determined by measuring the mass of the substance. In the case of atoms, the number of grams necessary to have one mole is the same as the atomic mass of that particular atom in amu's. For example, one mole of aluminum atoms has a mass of 26.98 grams.

Figure 4-1 In the photo on the left, you see the same number of grapes as grapefruit, one dozen of each. Even though there is a dozen of each, the masses are quite different. A mole of aluminum and a mole of lead, shown on the right, have the same numbers of atoms but different masses.

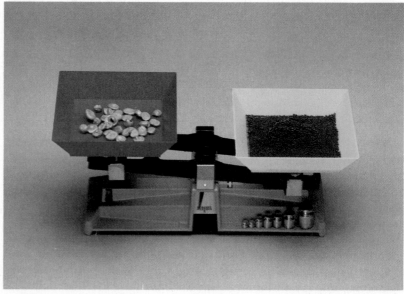

Figure 4-1 shows one dozen grapefruit and one dozen grapes. Remember, a dozen always represents the same number of particles, 12, but all dozens do not have the same mass. This same reasoning holds true for the mole too. Figure 4-1 also shows that one mole of lead has a greater mass than one mole of aluminum. A mole of lead has a mass of 207.2 grams. A mole of aluminum has a mass of 26.98 grams. The number of atoms in each mole is the same, but the mass differs. Note that since copper and aluminum are solids, the volume relationships stated in Avogadro's hypothesis do not hold. If Figure 4-1 showed mole quantities of nitrogen and oxygen, the gases would occupy equal volumes at the same temperature and pressure.

The following Examples will help you see the many relationships based on the mole. Notice that some of the Examples will involve simple problems that you can probably solve with mental calculations because they are familiar to you. These problems serve to show you that problems involving moles, molecules, and masses are solved in the same way.

CHEM THEME

The mole concept is a measurement that answers the question "How much do I have." It is particularly useful since it provides a basis of comparison among compounds in terms of mass, numbers of particles, and concentration.

Offer many concrete examples and allow students to observe quantities firsthand. You can use analogies to compare the mole to a score, gross, or football team (11 players).

Show many other examples of mole quantities in beakers or flasks. Choose a variety of substances exhibiting various colors and physical states. Some suggestions are water, ethanol, sulfuric acid, transition metal compounds, sugar, salt, hydrates, iodine, nickel, mercury, sulfur, tin, and bromine. Leave the substances in sealed flasks on display for several days. Label each flask with the substance, formula, and mass.

Example 4-1

How many dozen oranges do you have when you have 9 oranges?

▶ **Suggested solution**

You are asked to find:

? dozen oranges correspond to 9 oranges

This problem can be solved in a number of ways. One way would be to determine the unitary rate that expresses dozens per orange since a value for dozens is what you are asked to find. The expression that relates dozens and oranges is:

1 dozen corresponds to 12 oranges

Using this relationship, it is easy to see that

$$\frac{12 \text{ oranges}}{1 \text{ dozen}}$$

describes a unitary rate telling you how many oranges there are for each dozen. However, you need to know how many dozen are equivalent to one orange, the inverse of this relationship. The needed unit factor can be obtained by inverting the one shown and dividing by 12.

$$\frac{1 \text{ dozen}}{12 \text{ oranges}} = \frac{0.0833 \text{ dozen}}{1 \text{ orange}}$$

Notice that you simply need to divide the numerator and denominator of the fraction by 12 in order to get a denominator of one. Now multiply the number of oranges (given) by the unitary rate in dozens per orange to get the answer in dozens.

Invert unitary rates!

Many of the problems in this chapter are simple and straightforward. These problems require the application of mathematics to familiar situations. The purpose is to show students that the logic process they use in solving familiar problems can be applied to problems in chemistry. Try to use this technique in teaching problem solving.

Do You Know?

One can count the number of oranges in six dozen in a few minutes. However, six moles of oranges would have a mass as large as the entire Earth. The mole is a manageable unit for very small particles like atoms or molecules.

$$9 \text{ oranges} \times \frac{0.0833 \text{ dozen}}{\text{orange}} = 0.75 \text{ dozen}$$

You might think this is a long way to solve a simple problem that you could probably do in your head. You readily see that it was really unnecessary to convert

$$\frac{1 \text{ dozen}}{12 \text{ oranges}}$$

into a unitary rate. This factor could be used just as it is without first dividing the numerator and denominator by 12. The alternate set-up is

$$9 \text{ oranges} \times \frac{1 \text{ dozen oranges}}{12 \text{ oranges}} = 0.75 \text{ dozen oranges}$$

Notice that you still divide by 12, just as you did to find the unitary rate. You simply changed the order in which the math was done. The following shows that either procedure gives the same result.

Example 4-2

How many moles of oranges do you have if you have 3.01×10^{23} oranges?

▶ **Suggested solution**

The same logic that was used in Example 4-1 applies here. You are asked to find

? mol oranges that correspond to 3.01×10^{23} oranges

You need to determine a unitary rate for the number of moles per orange. The relationship between moles and numbers of oranges is:

1 mol oranges correspond to 6.02×10^{23} oranges

The unitary rate in moles per orange is:

$$\frac{1 \text{ mol}}{6.02 \times 10^{23} \text{ oranges}} = 1.66 \times 10^{-24} \text{ mol/orange}$$

Multiplying the number of oranges (given) by the number of moles per orange will give the number of moles.

$$3.01 \times 10^{23} \text{ oranges} \times \frac{1.66 \times 10^{-24} \text{ mol}}{\text{orange}} = 5.00 \times 10^{-1} \text{ mol}$$

An alternate setup using the unit factor equivalent of the unitary rate gives the following:

$$3.01 \times 10^{23} \text{ oranges} \times \frac{1 \text{ mol oranges}}{6.02 \times 10^{23} \text{ oranges}} = 0.500 \text{ mol}$$

There is a danger that students will focus on the mechanics of solving problems such as these and lose sight of the meaning. As problems are discussed in class, make a habit of asking students questions like: "What do we mean by a mole? What is a molecule? Could you see a CO_2 molecule? Could you see 1 mole of CO_2 molecules?"

Example 4-3

If there are 20 bicycles in the parking lot, how many dozen bicycle wheels are in the lot?

▶ **Suggested solution**

Confusion may arise when there are other factors to consider in solving the problem. You are asked to find:

? dozen bicycle wheels correspond to 20 bicycles

In solving this problem you have to consider two unit conversion factors.

1 bicycle has 2 wheels
1 dozen bicycle wheels correspond to 12 wheels

$$20 \text{ bicycles} \times \frac{2 \text{ wheels}}{1 \text{ bicycle}} \times \frac{1 \text{ dozen wheels}}{12 \text{ wheels}} = 3.3 \text{ dozen wheels}$$

Example 4-4

If a steel cylinder contains 3.01×10^{28} carbon dioxide molecules, how many moles of oxygen atoms are in the steel cylinder?

▶ **Suggested solution**

This problem is like the one for the bicycle. You are asked to find

? mol O atoms correspond to 3.01×10^{28} CO_2 molecules

There are 2 oxygen atoms in each CO_2 molecule.

$$3.01 \times 10^{23} \text{ } CO_2 \text{ molecules} \times \frac{2 \text{ O atoms}}{1 \text{ } CO_2 \text{ molecule}} \times \frac{1 \text{ mol O atoms}}{6.02 \times 10^{23} \text{ O atoms}}$$
$$= 1.00 \text{ mol O atoms}$$

If you got confused, take the time to determine the unitary rates for atoms per molecule and moles per atom. Or you can write out the necessary unit factors before attempting the math setup.

So far the Examples start with a given quantity which is used to calculate the number of moles, but you also can do the reverse. If the number of molecules can be converted to moles, then the number of moles can be converted to a number of molecules.

As you can see there are several ways to approach the problem. You should think about the problem and use the approach that makes sense to you. If you get an answer that is incorrect or an answer that does not make sense, you should review your reasoning and see if you can find where you went wrong.

Example 4-5

How many carbon dioxide molecules are in 0.75 mole of CO_2?

▶ **Suggested solution**

This time you are asked about the entire molecule rather than a part. You are asked to find:

$$? \ CO_2 \text{ molecules correspond to } 0.75 \text{ mol } CO_2$$

The unitary rate in molecules per mole for CO_2 is:

$$\frac{6.02 \times 10^{23} \text{ molecules}}{\text{mol}}$$

$$0.75 \ \cancel{\text{mol}} \ CO_2 \times \frac{6.02 \times 10^{23} \ CO_2 \text{ molecules}}{\cancel{\text{mol}}}$$

$$= 4.5 \times 10^{23} \ CO_2 \text{ molecules}$$

Example 4-6

If you have dealt with uncertainty, students will ask how many digits they should use for the atomic mass of carbon. You may have students work problems using several approximations (e.g., 10, 12, 12.0, 12.01, 12.011) to see that no error is introduced as long as you keep one more digit than you will round to in the final answer. Errors may be introduced if fewer digits are used in the calculation.

How many atoms are in 26.0 grams of carbon?

▶ **Suggested solution**

Now you are asked to look at the relationships among mass, moles, and number of atoms. The atomic mass of carbon is 12.0 amu, so 12.0 g of carbon is the mass of 6.02×10^{23} atoms. You are asked to find

$$? \text{ atoms correspond to } 26.0 \text{ grams of carbon}$$

The relationships are:

1 mole carbon corresponds to 12.0 grams
1 mole carbon corresponds to 6.02×10^{23} C atoms

These relationships can be used to answer the question. You may follow the logic better when each step is done independently. Using the first factor, 26.0 g of carbon can be converted to moles.

$$26.0 \ \cancel{g \ C} \times \frac{1 \text{ mol C}}{12.0 \ \cancel{g \ C}} = 2.17 \text{ mol C}$$

You now know that 26.0 g of carbon is the same as 2.17 mol carbon, and asking how many atoms are in 2.17 moles of carbon is equivalent to the original question. You can answer the question using the relationship between moles and number of particles.

$$2.17 \ \cancel{\text{mol C}} \times \frac{6.02 \times 10^{23} \text{ C atoms}}{\cancel{\text{mol C}}} = 1.31 \times 10^{24} \text{ C atoms}$$

You now know the number of carbon atoms in 2.17 mol carbon that has a mass of 26.0 g. By rereading the problem confirm that this is what you are asked to find.

Some students realize that the two steps used in solving this problem could have been expressed in one mathematical statement.

$$26.0 \text{ g } \cancel{C} \times \frac{1 \text{ mol } \cancel{C}}{12.0 \text{ g } \cancel{C}} \times \frac{6.02 \times 10^{23} \text{ C atoms}}{\text{mol } \cancel{C}} = 1.30 \times 10^{24} \text{ C atoms}$$

Other students use the relationships that are known to find a relationship between grams of carbon and number of atoms and use this new factor to solve the problem.

$$\frac{1 \text{ mol } \cancel{C}}{12 \text{ g C}} \times \frac{6.02 \times 10^{23} \text{ atoms C}}{1 \text{ mol } \cancel{C}} = \frac{6.02 \times 10^{23} \text{ atoms C}}{12 \text{ g C}}$$

$$26.0 \text{ g } \cancel{C} \times \frac{6.02 \times 10^{23} \text{ atoms C}}{12.0 \text{ g } \cancel{C}} = 1.30 \times 10^{24} \text{ C atoms}$$

Example   4-7

How many moles of carbon are in 26 grams of carbon?

▶ **Suggested solution**

? mol of carbon corresponds to 26 g of carbon

The mass of a mole of carbon atoms is the atomic mass in grams shown on the periodic table.

1 mol of carbon corresponds to 12.0 g of carbon

$$26 \text{ g } \cancel{C} \times \frac{1 \text{ mol C}}{12.0 \text{ g } \cancel{C}} = 2.2 \text{ mol C}$$

The logic and procedure are the same only the way the amount is expressed has been changed.

Example   4-8

How many oxygen atoms are in 0.75 mol CO_2?

▶ **Suggested solution**

? O atoms correspond to 0.75 mol CO_2

The formula, CO_2 shows that every CO_2 molecule contains two oxygen atoms.

1 CO_2 molecule has 2 oxygen atoms
1 mol O atoms correspond to 6.02×10^{23} O atoms

$$0.75 \cancel{\text{ mol } CO_2} \times \frac{6.02 \times 10^{23} \cancel{CO_2 \text{ molecules}}}{\cancel{\text{mol } CO_2}} \times \frac{2 \text{ O atoms}}{\cancel{CO_2 \text{ molecule}}}$$
$$= 9.0 \times 10^{23} \text{ oxygen atoms}$$

Example 4-9

What is the mass of one atom of aluminum?

▶ Suggested solution _____

? g correspond to an atom of aluminum

The mass of a mole of aluminum atoms is the atomic mass listed on the periodic table.

26.98 g correspond to 6.02×10^{23} atoms

Calculate the unitary rate for the number of grams per atom.

$$\frac{26.98 \text{ g}}{6.02 \times 10^{23} \text{ atoms}} = 4.48 \times 10^{-23} \text{ grams/atom}$$

This is a very small mass because atoms are very small. If the answer is written as a decimal equivalent, 4.48×10^{23} is

0.000 000 000 000 000 000 000 044 8 gram

It can be seen that the mass of an atom is very small indeed.

Review and Practice _____

1. 24
2. 40
3. Relative Mass
   rod   1.00
   beaker   3.61
   flask   5.38
   dish   1.68
   graduate   4.72
4. 7.52×10^{23} molecules
5. 30.1 mol HCl
6. 2.26×10^{24} O atoms
7. 1.0×10^2 mol O atoms
8. 3.21×10^{24} Fe atoms
9. 9.63×10^{24} Cl atoms

1. If magnesium atoms have twice as much mass as carbon-12 atoms, what is the mass of a magnesium atom?

2. Equal numbers of carbon-12 atoms and argon atoms have a mass of 6.00 grams and 20.00 grams respectively. What is the relative mass of an argon atom?

3. The actual masses of several objects are given. A stirring rod has been selected as a standard and given a relative mass value of 1.0. Complete the table with the relative mass of each object.

MASSES OF LABORATORY EQUIPMENT

OBJECT	ACTUAL MASS	RELATIVE MASS
stirring rod	13.9 g	1.0
100 mL beaker	50.2 g	
250 mL flask	74.8 g	
evaporating dish	23.3 g	
50 mL graduate	65.6 g	

4. How many H_2SO_4 molecules are in 1.25 moles of H_2SO_4?

5. How many moles of HCl do you have when you have 1.81×10^{25} HCl molecules?

6. How many oxygen atoms are in 1.25 moles of sulfur trioxide?

7. How many moles of oxygen atoms are in 1.2×10^{25} diphosphorus pentoxide molecules?

8. How many iron atoms are there in 5.33 moles of iron(III) chloride?

9. How many chlorine atoms are there in 5.33 moles of iron(III) chloride?

Mole Relationships

In Section 4-2 the relative masses for atoms were found without knowing the mass of a single atom. In the same way the relative masses of compounds can be found using what you know about atoms.

If you are given 36 oranges and asked how many dozen this amount represents, you would have little difficulty in answering three dozen. This problem is simple because you are very familiar with oranges and dozens. If you are asked to find the mass of 36 oranges if one dozen has a mass of 2160 grams, you can easily determine the answer using the unitary rate of grams per dozen.

$$3 \text{ doz} \times 2160 \text{ g/doz} = 6480 \text{ g}$$

If similar questions are asked about moles, molecules, and atoms, the answers do not occur to you as readily because you are not as familiar with these objects and concepts. It will be useful to have an effective algorithm for solving problems involving these concepts until they become commonplace. An algorithm is a method for solving a problem.

CONCEPTS

■ the distinction between molecules and formula units
■ determining molar mass
■ the relationship between molar mass and Avogadro's number
■ molar masses of diatomic elements
■ the mole as a basis of comparison

4-4 Molar Mass

The molar mass for a compound is found by adding the atomic masses in grams of all elements in the compound. For example, the formula for water is H_2O. It consists of two hydrogen atoms and one oxygen atom. Using the periodic table, you see the atomic mass of hydrogen is 1.0079 grams. For our purposes, this value is rounded to 1.01 grams. To determine the molar mass of water, you add 2.02 grams (2×1.01) to the mass of the oxygen atom which is shown in the table as 15.9994 grams. To the nearest hundredth digit, this value is 16.00 grams. The molar mass for H_2O is

$$2.02 \text{ g} + 16.0 \text{ g} = 18.02 \text{ g} \text{ (which may be rounded to 18.0 g)}$$

A large number of compounds differ from water in that they do not exist as molecules. Many compounds such as table salt, baking soda, alum, and lye consist of particles that do not behave in the same way as molecular substances. A model of a sodium chloride crystal was shown in Figure 2-21. Sodium chloride, NaCl, is made of alternating ions of sodium and chlorine in the ratio indicated by the formula NaCl. Recall that an ion is a charged atom having properties that differ from a neutral particle like a molecule. You saw in Chapter 2 that NaCl is not made of discrete NaCl units. Each sodium ion is surrounded by six chlorine ions. Each chlorine ion is surrounded by six sodium ions and this arrangement extends throughout the crystal.

NaCl is the simplest formula for salt in that it gives the ratio of sodium ions to chlorine ions as 1 to 1. The simplest formula that provides the smallest whole number ratio of ions in a compound is sometimes referred to as a **formula unit**. The mass in grams of one mole of atoms, ions, molecules, or formula units can best be described by the general term **molar mass**.

empirical formula

Example 4-10

What is the molar mass of sucrose, $C_{12}H_{22}O_{11}$?

▶ Suggested solution _____

The molar mass is the sum of all the atomic masses in the molecule. Sucrose contains 12 carbon atoms. Therefore,

the atomic mass for carbon		the number of × carbon atoms		the mass due = to carbon
12.0 g	×	12	=	144 g

For the hydrogen,

the atomic mass for hydrogen		the number of × hydrogen atoms		the mass due = to hydrogen
1.01 g	×	22	=	22.2 g

For the oxygen,

the atomic mass for oxygen		the number of × oxygen atoms		the mass due = to oxygen
16.0 g	×	11	=	176 g

Adding the masses of the three elements gives:

the mass of 12 carbon atoms	the mass of + 22 hydrogen atoms	the mass + of 11 oxygen atoms	the molar = mass of $C_{12}H_{22}O_{11}$
144 g	+ 22.2	+ 176 g	= 342 g

Example 4-11

What is the molar mass of aluminum sulfate, $Al_2(SO_4)_3$?

▶ Suggested solution _____

The molar mass of a compound is the sum of the molar masses of the elements in the compound. The molar masses for the elements can be found using the periodic table. Since you will normally work with experimental values that are uncertain in the second or third digit, it is generally acceptable to round atomic masses to the nearest tenth when calculating a molar mass. One mole of aluminum sulfate consists of the following:

2 moles of Al atoms each with a molar mass of 27.0 g
3 moles of S atoms each with a molar mass of 32.1 g
12 moles of O atoms each with a molar mass of 16.0 g

Add the masses of the Al atoms, the S atoms, and the O atoms.

$$2(27.0 \text{ g}) + 3(32.1 \text{ g}) + 12(16.0 \text{ g}) = 342.3 \text{ g}$$

4-5 Masses of Diatomic Molecules

What is the mass of 1.5 moles of hydrogen? This question, which seems simple, can have two intelligent answers. From the periodic table you can see that 1 mole of hydrogen has a mass of 1.0079 grams, which you round to 1.01 grams, so 1.5 moles of hydrogen should have a mass of about 1.5 grams. This answer is correct if you are talking about 1.5 moles of hydrogen *atoms*. Unfortunately, the word "hydrogen" can refer to hydrogen atoms or to ordinary hydrogen gas, which exists as H_2 molecules. In preferred usage, the name dihydrogen is used to refer to H_2. The formula H_2 represents a **diatomic molecule** which means it has two atoms in a molecule. Clearly 1.5 moles of H_2 contain 2×1.5 or 3 moles of hydrogen atoms. Therefore, the mass of 1.5 moles of H_2 gas is about 3 grams. Since very few chemists use preferred terminology it is important that you know which meaning for hydrogen is intended. The other elements that normally exist as diatomic molecules in the free state are listed in Table 4-3.

Table 4-3

ELEMENTS THAT EXIST AS DIATOMIC MOLECULES			
NAME	FORMULA	NAME	FORMULA
hydrogen	H_2	chlorine	Cl_2
oxygen	O_2	bromine	Br_2
nitrogen	N_2	iodine	I_2
fluorine	F_2	astatine	At_2

The following questions might be asked of you on a test.

1. What is the mass of one mole of oxygen gas?
2. What is the mass of one molecule of oxygen gas?
3. What is the mass of one mole of oxygen atoms?
4. What is the mass of one atom of oxygen?

The answers are:
1. 32.0 grams
2. 5.32×10^{-23} grams
3. 16.0 grams
4. 2.66×10^{-23} grams

You are confused about the mole concept if you give the same answer to the first two questions. To say that a molecule of oxygen gas has the same mass as a mole of oxygen gas is the same as saying that an orange has the same mass as a dozen oranges. Be sure that you can answer these four questions and that you understand how you got those answers.

4-6 A Mnemonic for Mole Relationships

It has been said that "the mole is at the heart of chemistry." Beginning students often have trouble remembering relationships based on the mole. The diagram shown in Figure 4-2 may help you remember the importance of the mole as well as the many relationships involving it.

In chemistry, you will have many occasions to express amounts of substances in these three ways:
1. in terms of the number of particles
2. in terms of the number of moles of particles
3. in terms of the mass of the particles

Figure 4-2 Use this diagram to help you remember all of the relationships based on the mole.

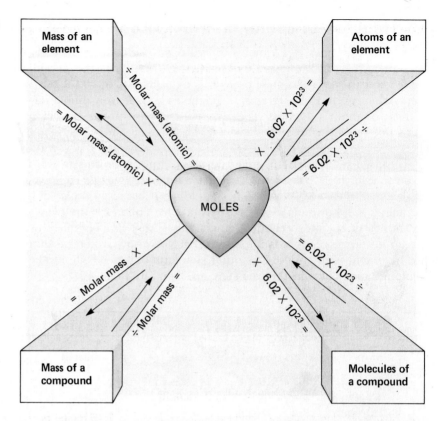

If you know the mass of an element in a sample, you can divide by the atomic mass to find the number of moles of that element in the sample. If you need to know the number of atoms in the sample, you can then multiply by 6.02×10^{23}. These operations are shown by the upper arrows in Figure 4-2.

If you know the number of atoms in a sample of an element and need to know the mass of the sample, you can follow the reverse procedure indicated by the upper arrows pointing toward the heart in Figure 4-2. The lower arrows represent similar computations for compounds.

If you find Figure 4-2 helpful, copy it on a card and use it as you work problems. Once you understand these relationships, you will not need this crutch. Try each problem without the card. Continue to practice until you can work through all of the items correctly without using the card.

For the problems you worked in the previous sections, unit factors were used to make the conversions between the given information and the desired answer. A mole conversion problem involving carbon dioxide could involve the following unit factors:

$$\frac{1 \text{ mol CO}_2}{6.02 \times 10^{23} \text{ molecules CO}_2}$$

$$\frac{1 \text{ mol CO}_2}{44 \text{ g CO}_2}$$

$$\frac{44 \text{ g CO}_2}{6.02 \times 10^{23} \text{ molecules CO}_2}$$

Unit factors also can be written for substances that are not molecular. For sodium carbonate, the unit factors are:

$$\frac{1 \text{ mol } Na_2CO_3}{6.02 \times 10^{23} \text{ formula units } Na_2CO_3}$$

$$\frac{1 \text{ mol } Na_2CO_3}{106 \text{ g } Na_2CO_3}$$

$$\frac{106 \text{ 44 g } Na_2CO_3}{6.02 \times 10^{23} \text{ formula units } Na_2CO_3}$$

Unit factors can be inverted so:

$$\frac{1 \text{ mol } Na_2CO_3}{106 \text{ g } Na_2CO_3} \quad \text{is the reciprocal} \quad \frac{106 \text{ g } Na_2CO_3}{1 \text{ mol } Na_2CO_3}$$
$$\text{of}$$

Review and Practice

1. Write three unit factors for water.

2. Write three unit factors for manganese(IV) oxide.

3. Write three unit factors for sodium sulfate. ✓

4. How many molecules of manganese(IV) oxide are in 3.15 moles of the compound?

5. How many oxygen atoms are there in 3.15 moles of manganese(IV) oxide? ✓

6. Use the periodic table to determine the mass of one mole of sodium sulfate?

7. How many moles of sodium carbonate of that substance are in 53 grams?

8. Which has a greater mass, one mole of lead atoms or ten moles of water molecules? ✓

9. What is the mass of 1.31 moles of silver nitrate? ✓

10. Using the atomic masses given in the periodic table, find the mass for each of the following molecules to the nearest tenth of a unit.
    a. carbon monoxide
    b. C_8H_{18}
    c. H_2SO_4
    d. carbon disulfide

11. Find the molar mass of the following:
    a. $CaCO_3$
    b. $KMnO_4$
    c. manganese(IV) oxide
    d. $Ca_3(PO_4)_2$

12. How many moles are in a 124-gram sample of C_8H_{18}?

13. How many moles are in a 56-gram sample of carbon disulfide?

14. What is the mass in grams of 2.50 moles of H_2SO_4?

15. How many molecules are in 2.50 moles of H_2SO_4? ✓

16. How many molecules are in 245 grams of H_2SO_4? ✓

Answers will vary for items 1, 2, and 3. Some responses are shown.

1. $\dfrac{1 \text{ mol } H_2O}{6.02 \times 10^{23} \text{ molecules } H_2O}$

$\dfrac{1 \text{ mol } H_2O}{18.0 \text{ g } H_2O}$

$\dfrac{18.0 \text{ g } H_2O}{6.02 \times 10^{23} \text{ molecules } H_2O}$

2. $\dfrac{1 \text{ mol } MnO_2}{86.9 \text{ g } MnO_2}$ 86.94

$\dfrac{1 \text{ mol } MnO_2}{6.02 \times 10^{23} \text{ molecules } MnO_2}$

$\dfrac{6.02 \times 10^{23} \text{ molecules } MnO_2}{86.9 \text{ g } MnO_2}$

3. $\dfrac{1 \text{ mol } Na_2SO_4}{6.02 \times 10^{23} \text{ molecules } Na_2SO_4}$

$\dfrac{142 \text{ g } Na_2SO_4}{6.02 \times 10^{23} \text{ molecules } Na_2SO_4}$ 142.02₄

$\dfrac{142 \text{ g } Na_2SO_4}{1 \text{ mol } Na_2SO_4}$

4. 1.90×10^{24} molecules 105.99

5. 3.80×10^{24} atoms

6. 142 g

7. 0.50 moles

8. 1 mol Pb

9. 223 grams

10. a. 28.0 g
    b. 114.2 g 66
    c. 98.1 g 08
    d. 76.2 g 13

11. a. 100.1 g
    b. 158 g
    c. 86.9 g
    d. 310.2 g

12. 1.08 mol

13. 0.74 mol

14. 245 g

15. 1.50×10^{24} molecules

16. 1.50×10^{24} molecules

Formula Calculations

CONCEPTS

■ determining empirical formulas from experimental data
■ determining smallest whole number ratios of atoms in a compound
■ moles in solution—molarity

Thus far in your study you have encountered a number of chemical formulas. In Chapter 3 you learned that the formula for water is symbolized by H_2O. Other formulas you have seen are carbon dioxide, CO_2; sugar, $C_{12}H_{22}O_{11}$; and lead(II) iodide, PbI_2. You know how to write formulas using tables showing ion charges. You should realize, however, that the actual formula for a compound is determined by laboratory analysis and not by making sure that the algebraic sum of the charges is zero. Experiments are done to measure the amount of each element in the compound. The interpretation of the results makes use of molar masses. The analysis of the data provides the simplest ratio of atoms in the compound. In the next sections, you will study this process in detail.

The analysis aspect of chemistry also involves dealing with many substances that are dissolved to make solutions. Nearly all of the chemical reactions that take place in your body occur in solution. For a dissolved substance, you need a method of relating the amount of substance in solution to the mole.

4-7 Finding An Empirical Formula

Empirical means based on experiment. Therefore, an **empirical formula** is one that is obtained from experimental data and represents the smallest whole number ratio of atoms in a compound. Recall once more what a formula represents. Carbon dioxide has the formula CO_2. The formula indicates that one molecule of CO_2 has one atom of carbon and two atoms of oxygen. It also is true that two molecules of CO_2 have two atoms of carbon and four atoms of oxygen. Increasing the number of molecules to 1000 would mean 1000 atoms of carbon and 2000 atoms of oxygen. One million molecules of CO_2 contain 1×10^6 (one million) atoms of carbon and 2×10^6 (two million) atoms of oxygen. A mole of CO_2 has 6.02×10^{23} molecules. There are 6.02×10^{23} carbon atoms in a mole of CO_2 and $2(6.02 \times 10^{23})$ atoms of oxygen. Taking these relationships further, it follows that 44 grams of CO_2 contain 12 grams of carbon and 32 grams of oxygen.

one molecule of carbon dioxide	one mole of carbon dioxide
mass = 44 amu or 7.31×10^{-23} g	mass = 44 g
CO_2	CO_2

1 atom C	2 atoms O	1 mole C	2 moles O
mass = 12 amu	mass = 32 amu	mass = 12 g	mass = 32 g

To determine the formula for a compound, it is not necessary to count the atoms in a single molecule. The same information is obtained by finding the number of moles of each element in a mole of the compound. The ratio of atoms, one carbon to two oxygen, for CO_2 is always the same as the ratio of moles. If you know the mass of each element, you can find the number of moles and determine the empirical formula. The following example shows one way an empirical formula can be determined for a compound.

Example 4-12

A typical charcoal briquette that is used in making a barbecue fire is composed of carbon and has a mass of 43.2 grams. When the charcoal lump is burned it combines with oxygen and the resulting compound has a mass of 159.0 grams. What is the empirical formula for the resulting compound?

▶ **Suggested solution**

The compound contains both carbon and oxygen. The mass of carbon in the compound is 43.2 grams. The mass of oxygen in the compound can be calculated by subtracting the mass of carbon from the mass of the compound.

159.0 grams − 43.2 g carbon = 115.8 g oxygen

Now find the number of moles of carbon and oxygen in the compound.

$$43.2 \text{ g C} \times \frac{1 \text{ mol C}}{12.0 \text{ g C}} = 3.60 \text{ mol C}$$

$$115.8 \text{ g O} \times \frac{1 \text{ mol O}}{16.0 \text{ g O}} = 7.24 \text{ mol O}$$

There are 2.01 moles of oxygen for every 1.0 mole of carbon. Do you see how these values were obtained? If not, the following reasoning may help.

You have 3.60 moles of carbon and 7.24 moles of oxygen. What can you divide by to get 1.00 mole of carbon? Divide by 3.60. You get 1.00 mole of carbon and 2.01 moles of oxygen. The general rule is to divide the number of moles of each element by the smaller of the numbers. Recall how you determined the relative mass of the lab equipment in item 3 on page 104. This method works just as well when analyzing more complex compounds that contain 3, 4, or more different elements.

The experiment shows that there are approximately 2 moles of oxygen for every mole of carbon in the compound. More precise measurements have shown that the ratio is 2.0000 to 1.0000 so it seems safe to assume that the formula is CO_2 for carbon dioxide.

Before presenting another example, here is a summary of what was done to find the formula for carbon dioxide.

1. The mass of each element in a sample of the compound is determined.
2. The mass of each element is divided by its molar mass to determine the number of moles of each element in the sample of the compound.
3. The number of moles of each element is divided by the smallest number of moles to give the ratio of atoms in the compound.

These steps can be used to find the formula for any compound for which at least one element is present as a single atom per molecule or formula unit. An additional step is required for compounds like N_2O_5. The following example illustrates this additional step.

Example 4-13

Charcoal is mixed with 15.53 grams of rust (an oxide of iron) and heated in a covered crucible to keep out air until all of the oxygen atoms in the rust combine with carbon. When this process is complete, a pellet of pure iron, with a mass of 10.87 grams remains. What is the empirical formula for rust?

▶ **Suggested solution**

To begin solving the problem, it is a good idea to organize the experimental data into a table.

mass of rust analyzed	15.53 g (stated in the problem)
mass of pure iron	10.87 g (stated in the problem)
mass of oxygen in rust	4.66 g (found by subtraction of given values)

As experimental data are obtained in the laboratory, it is a common practice to determine the mass of all elements but one and to find the mass of that final element by subtracting the mass of the known elements from the total mass of the compound. That is how the mass of oxygen was obtained.

The next step would be to calculate the number of moles of each element.

$$10.87 \text{ g Fe} \times \frac{1 \text{ mol Fe}}{55.8 \text{ g Fe}} = 0.195 \text{ mol Fe}$$

$$4.66 \text{ g O} \times \frac{1 \text{ mol O}}{16.0 \text{ g O}} = 0.291 \text{ mol O}$$

The sample of rust that was analyzed contained 0.195 mole of iron and 0.291 mole of oxygen. To continue, divide both mole quantities by the smaller of the two.

$$\frac{0.195 \text{ mol Fe}}{0.195 \text{ mol Fe}} = 1.00$$

$$\frac{0.291 \text{ mol O}}{0.195 \text{ mol Fe}} = 1.49 \text{ mol O/mol Fe}$$

There are 1.49 moles of oxygen atoms for every mole of iron atoms in the compound. The formula could be written as $Fe_1O_{1.49}$. However, because atoms are believed to combine in whole number ratios, multiply both numbers by some whole number. Multiplying by 2 gives 2.00 moles of iron and 2.98 moles of oxygen. Because of the uncertainty involved in the data, you can assume that 2.98 is the same as 3 within experimental uncertainty. You can use the techniques discussed in Section 3-9 to determine the uncertainty and find out whether this approximation is correct. The experiment stated in the problem can be done many times and with greater precision. If the experiment is repeated, the ratio obtained may be more like 2.9996 moles of oxygen to 2.0000 moles of iron and you can feel justified in asserting that the ratio can be expressed in whole numbers, giving a formula of Fe_2O_3 for rust.

4-8 When is a Number Whole?

In the last example 2.98 was rounded to 3; but 1.49 was not considered a whole number. Neither 1.49 nor 2.98 is exactly a whole number. When can you assume that the number should be rounded to a whole number?

The sources of uncertainty in any measurement were discussed in Chapter 3. Unless you know how much uncertainty there is in a number, you cannot know if you are justified in calling it a whole number. It is a common practice to use significant figures as a guide to determine what is a whole number. In Example 4-13, 2.98 was rounded to the whole number, 3, but 1.49 was not because the data were certain to the second digit. When 2.98 is rounded to two digits, you get 3.0. When 1.49 is rounded to two digits, you get 1.5. It makes sense to assume that uncertainty in the experiment could produce an answer of 2.98 when the true value is 3, but you would be stretching a point in assuming that the uncertainty would be great enough to produce an answer of 1.5 when the true value is either 1 or 2.

4-9 Empirical Versus Molecular Formulas

The **molecular formula** is always some multiple of the empirical formula. To find the multiple, you can divide the molar mass of the compound by the molar mass of the empirical formula. Example 4-14 serves to illustrate this relationship.

Example 4-14

A compound composed of hydrogen and oxygen is analyzed and a 10.00-gram sample of the compound yields 0.59 grams of hydrogen and 9.40 grams of oxygen. The molecular mass of this compound is 34 grams. Find the empirical formula and the molecular formula for the compound.

▶ Suggested solution

Determine the number of moles of each element.

$$0.59 \, g \, H \times \frac{1 \, mol \, H}{1.01 \, g \, H} = 0.58 \, mol \, H$$

$$9.40 \ g\cancel{O} \times \frac{1 \ \text{mol O}}{16.0 \ g \cancel{O}} = 0.59 \ \text{mol O}$$

The ratio of moles of hydrogen atoms to moles of oxygen atoms is 1 to 1. HO is the empirical formula for this compound. It shows the simplest ratio in which atoms combine. The empirical formula HO has a formula mass of $1 + 16$ or 17 grams. The molecular formula is some multiple of this ratio which is represented with the formula $(HO)_n$ where n is some whole number. You can write $(17)_n = 34$ and $n = 2$. The formula $(HO)_2$ would be commonly written as H_2O_2. The formula H_2O_2 has the same ratio of atoms and gives the correct molecular mass. H_2O_2 is the molecular formula for this compound. It is called hydrogen peroxide.

There is one additional point to be made about finding empirical formulas. Consider the iron compound discussed in Example 4-13. By multiplying 1 and 1.49 by 4 or 6, another whole number ratio within the uncertainty of the experiment could be obtained. Doing so would give formulas for rust of Fe_4O_6 or Fe_6O_9 as well as the answer you obtained in the example, Fe_2O_3. How do you know which of these formulas is correct? All of the formulas indicate 1 mole of iron combined with 1.5 moles of oxygen, except that the definition of an empirical formula is that it represents the *smallest* whole numbers that will give the ratio. Therefore you should choose the formula, Fe_2O_3, as the formula for rust.

If you also know the molar mass of the compound, the procedure outlined in Example 4-14 can be used to determine whether Fe_2O_3 is the true formula. Rust is an ionic compound that forms crystals of various size rather than molecules. In all ionic compounds, the empirical formula is taken as the true formula of the compound.

4-10 Percent Composition

When a compound is analyzed in the laboratory, the analysis involves finding the mass of each element in the compound. It is easier to compare the composition of compounds having the same elements if the composition is described by **percent composition** rather than to try to compare masses.

Assigning percents to elements in a compound is the same as assigning grades. If you score 25 points out of a possible 50 points on one test, and 45 out of a possible 60 points on another test, which score is better? You would not know unless you establish a basis for the comparison.

In a similar fashion, if a 36-gram sample of iron oxide produces 28 grams of iron and 8 grams of oxygen, while a 160-gram sample of iron oxide produces 112 grams of iron and 48 grams of oxygen, it is difficult to know whether the two samples are the same compound or different oxides of iron. You can see why a basis of comparison is needed. Now use the tests and the iron compounds in the following Examples to establish a basis of comparison.

Example 4-15

What are your grades in percent if your test scores were the following:

25 points out of a possible 50 points
45 points out of a possible 60 points

▶ Suggested solution

Percent means "per hundred parts." In this case percent indicates the score you would have gotten if the test had been graded on a 100-point basis. To find the percent, divide the portion by the total.

25/50 = 0.50 or 50/100
45/60 = 0.75 or 75/100

One way to interpret these results is to say that you got fifty hundredths of the items on the first test right and seventy-five hundredths of the items on the second test right. If each test had been worth 100 points your scores would be 50 and 75 respectively. Now look at the iron problem.

Example 4-16

The following data are obtained from analyzing two iron oxide samples. Are the samples the same compound?

a 36-g sample contains 28 g Fe and 8 g O
a 160-g sample contains 112 g Fe and 48 g O

▶ Suggested solution

You can begin by looking at the percentages of iron in both samples.

$$\frac{28 \text{ g Fe}}{36\text{-g sample}} = 0.78 = \frac{78}{100} = 78\%$$

$$\frac{112 \text{ g Fe}}{160\text{-g sample}} = 0.70 = \frac{70}{100} = 70\%$$

The percentages indicate that you have two different compounds.

Example 4-17

What are the formulas for two iron oxide compounds having the following compositions?

a 36 g sample which is 78% iron
a 160 g sample which is 70% iron

▶ **Suggested solution** _____

You need to know the percentage of oxygen in each compound. The easiest way to determine the percentage of oxygen is to subtract the percentage of iron from 100 in each case.

$$100 - 78 = 22$$
$$100 - 70 = 30$$

When finding the formula of a compound from percentage data, you assume that you have 100 grams of the compound. If the first sample is 78% iron, how many grams of iron are in 100 grams of the compound? You should guess 78 grams. Remember, percent means "per hundred parts." How many grams of oxygen are in 100 grams of the first sample?

Using 78 grams of iron and 22 grams of oxygen, the formula is found by calculating the moles of iron and oxygen.

$$78\ g\ Fe \times \frac{1\ mol}{55.8\ g} = 1.4\ mol\ Fe$$

$$22\ g\ O \times \frac{1\ mol}{16.0\ g} = 1.4\ mol\ O$$

The mole ratio is 1 to 1 which means the formula must be FeO for the first sample. The formula for the second sample is calculated as follows.

$$70\ g\ Fe \times \frac{1\ mol}{55.8\ g} = 1.3\ mol\ Fe$$

$$30\ g\ O \times \frac{1\ mol}{16.0\ g} = 1.9\ mol\ O$$

Dividing both numbers of moles by the smaller number produces a mole ratio of 1 to 1.5. In smallest whole numbers this mole ratio is 2 to 3. Thus the formula of the second sample is Fe_2O_3.

Example 4-18

Calculate the percent by mass of carbon and oxygen in carbon dioxide.

▶ **Suggested solution** _____

The percent of carbon in the compound CO_2 can be found by dividing the mass of the carbon by the mass of the compound. The mass of carbon is 12.0 g and the mass of the compound is 12.0 g + 32.0 g or 44.0 g. The percent carbon in CO_2 by mass is

$$\frac{12.0}{44.0} = 0.273\ \text{or}\ 273/100 = 27.3\%$$

Likewise the percent oxygen by mass can be found by dividing the mass of the oxygen by the mass of the compound.

$$\frac{32.0}{44.0} = 0.727 \text{ or } 727/100 = 72.7\%$$

To check the calculation, add the percents and see that they total 100.

4-11 Molarity

Many chemical compounds are stored, measured, and used as solutions. Medicines are commonly prepared by dissolving them in water so they may enter the blood stream more quickly. Household cleaners like bleach, ammonia, and vinegar are dissolved in water so they may be stored and measured more easily.

When salt, NaCl, is used in a reaction, it is commonly present as a salt solution. When a sodium chloride solution is poured into a silver nitrate solution, a reaction occurs as evidenced by the formation of a white solid. It is the NaCl in the solution that reacts and not the water. When a solution of NaCl is used, it is important to know how much of the NaCl in a solution reacts.

Concentration describes how much solute is in a given amount of solution. There are many ways to describe concentration. Vinegar is 5% acetic acid. Some fruit drinks are labeled as being 10% fruit juice. Rubbing alcohol is labeled as 70% isopropyl alcohol by volume. Percentage by mass or volume is a convenient way of expressing concentration for items purchased from the grocery. But chemists commonly describe the concentration of solutions by indicating the number of moles of solute dissolved in each liter of solution. The amount of solution containing the solute is commonly described in liters or cubic decimeters because the volume of a solution is measured in a graduated cylinder.

Now look at the mole and volume relationships for a solution of sodium chloride, NaCl. Table 4-4 gives the number of grams of sodium chloride that are contained in a given volume of solution. These masses were obtained by boiling the water off and determining the mass of the resulting residue (which is NaCl). The mass in each case was converted to moles using the procedure you learned in Section 4-3. Make the conversions yourself to be sure you understand how the values were obtained.

Table 4-4

MOLES OF SODIUM CHLORIDE IN A SOLUTION		
VOLUME OF SOLUTION (in L)	MASS NaCl (in g)	MOLES NaCl
0.13	37.1	0.634
0.34	97.5	1.67
0.46	131.4	2.25
0.58	166.4	2.84
0.82	234.3	4.01

Figure 4-3 The graph shows the moles of salt present at each volume for a salt solution. The best fit line passes through the origin and represents a direct proportion. The slope of the line is a unitary rate, moles per liter, which is molarity.

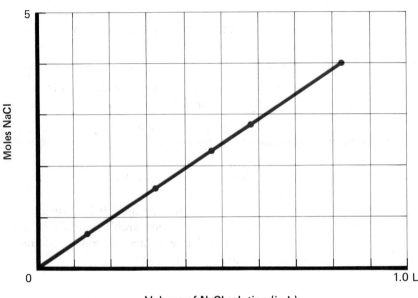

Graph of moles vs. volume of solution

A graph showing the relationship between moles and volume of solution is shown in Figure 4-3. Note that Figure 4-3 is a straight-line graph which passes through the origin. Therefore the relationship between moles NaCl and the volume of the solution is a direct proportion. If the moles of NaCl are divided by the volume of the solution for the data points listed in Table 4-4, you see that a constant is obtained. This constant is the same as the slope of the line which is a unitary rate. This unitary rate is given the name **molarity** and it describes the concentration of a solution in moles per liter.

Look at the significance of the data in Table 4-4 again. The moles per volume is a constant. Thus for a given solution, you can select any volume and still have the same concentration in moles per liter. The NaCl solution described here has a molarity of 4.90. The symbol for molarity is M. A 4.90M solution of NaCl means that there are 4.90 moles of NaCl in one liter of the solution.

Do not confuse the molarity of the solution with the number of moles of solute in a particular sample of the solution. For example, if you have 250 mL of a 4.90M NaCl solution, the molarity is 4.90 but the number of moles of NaCl in the solution can either be calculated or determined by using the graph in Figure 4-3. The calculation involves multiplying the molarity in moles per liter (a unitary rate) by the volume of the sample in liters.

$$4.90M = 4.90 \text{ mol/L}$$

$$\frac{4.90 \text{ mol}}{\cancel{L}} \times 0.250 \, \cancel{L} = 1.23 \text{ mol}$$

The following Example shows how the molarity expressed as a unitary rate can be used to determine the amount of the dissolved substance contained in a given volume of a solution.

Example   4-19

How many moles of HCl are contained in 1.45 L of a 2.25M solution?

▶ **Suggested solution** _____

> ? moles of HCl correspond to 1.45 liters of 2.25M soln
> 2.25M = 2.25 mol/L of soln

Multiplying the unitary rate in moles per liter by the number of liters will give an answer in moles.

$$1.45 \, \cancel{L} \times \frac{2.25 \, \text{mol}}{\cancel{L} \, \text{soln}} = 3.26 \, \text{mol HCl}$$

When smaller amounts of the solution are used it is common to measure the volume in milliliters. In solving problems using milliliters, you can either convert the volume to liters or convert the molarity to a unitary rate in moles per milliliter.

Demonstrate the preparation of 1.00M solutions of sugar, ethanol, and sodium chloride. Compare the volume of water needed to make each solution. Compare the types and numbers of particles present in each solution. Have students discuss their macroscopic observations during the dissolving process. Ask students how the dissolved solids might be recovered.

4-12 Preparing Molar Solutions

In the previous section, the molarity of a solution of sodium chloride was found by plotting experimental data. Because moles of solute was plotted along the x axis and liters of solution was plotted along the y axis, the slope of the graph is a unitary rate describing the number of moles of solute contained in each liter of solution.

It is certainly not necessary to plot a graph in order to know the molarity of a solution. All you need to know is the number of moles of solute contained in each liter of solution. That information is readily available when you are actually making up the solution.

Molarity is an accepted standard for concentration. The preparation of one liter of a 1.00M sugar solution involves the same process anywhere in the world. Molar solutions contain a known amount of the dissolved substance in a given volume of solution. This means that the amount of water required to prepare one liter of a 1.00M solution is not the same for every dissolved substance. For example, a 1.00M solution of sugar, $C_{12}H_{22}O_{11}$, is prepared by measuring 342 grams of sugar (one mole), and then adding enough water so the volume of the solution is 1.00 liter when the sugar completely dissolves. For a 1.00M solution of sugar, 790 milliliters of water are required. A 1.00M solution of methanol is prepared by measuring 32.0 grams of methanol and adding enough water so that the final volume is 1.00 liter when the solution process is complete. For the 1.00M methanol solution 962 milliliters are required. The volume of water used in the sugar solution is less than that for the methanol solution because of several factors. The primary factor is the size of the molecules being dissolved and the other factors include the molecular attraction and shape of the molecules. Therefore the directions for preparing a molar solution state that enough water is added to produce the final volume desired.

Figure 4-4 One liter of a 1M sugar solution requires 342 grams of sugar and a volumetric flask which is used in diluting the sugar with water to a volume of 1 liter. It will take about 790 mL of water to make the solution. It is not necessary to measure this volume exactly. It is only necessary that the final volume of the solution be one liter.

Example 4-20

Give directions for the preparation of 2.50 L of a 1.34M sodium chloride solution.

▶ **Suggested solution**

?g NaCl needed to make 2.50 liters of 1.34M NaCl

One way to solve the problem would be to calculate the number of moles of salt needed to make the solution. The number of moles is found by multiplying the molarity (moles per liter) by the volume of the solution (in liters).

$$2.50 \, \cancel{L} \times \frac{1.34 \text{ mol}}{\cancel{L} \text{ soln}} = 3.35 \text{ mol NaCl}$$

The number of moles is then converted to grams using the molar mass which is a unitary rate expressed in grams per mole. The molar mass of NaCl is 58.5 grams per mole.

$$3.35 \, \cancel{\text{mol}} \times \frac{58.5 \text{ g}}{\cancel{\text{mol}}} = 196 \text{ grams NaCl}$$

So the instructions for making the solution would be: Measure 196 grams of NaCl and add enough water so that the final volume is 2.50 liters when the solution process is complete.

Review and Practice

1. N_2O_5
2. CH
3. C_6H_6
4. Example 4-12 %C = 27.2
         %O = 72.8
   Example 4-18 %C = 273
         %O = 72.7
These are the same within experimental uncertainty.
5. 15.7% C, 84.3% S
6. 40% Ca, 12% C, 48% O
7. 1.55 mol
8. 152 g
9. All moles divided by volume equal 4.9.
10. All substances contain the same number of molecules per mole but different numbers of grams per mole.

1. A compound of nitrogen and oxygen is found to contain 4.20 grams of nitrogen and 12.0 grams of oxygen. Find the empirical formula for this compound.

2. A compound is composed of 4.80 grams of carbon and 0.40 gram of hydrogen. Calculate the empirical formula.

3. The molar mass of the compound in item 2 is 78 grams. What is the molecular formula?

4. Use the data in Example 4-12 to find the percent composition of carbon dioxide. How do the values obtained compare with those in Example 4-18?

5. Calculate the percent carbon and the percent sulfur in carbon disulfide.

6. Calculate the percent of each element in limestone, $CaCO_3$.

7. How many moles of H_2SO_4 are in 1.00 liter of a 1.55M H_2SO_4 solution?

8. How many grams of H_2SO_4 are in 1.00 liter of a 1.55M H_2SO_4 solution?

9. Use the data in Table 4-4 to prove to yourself that molarity is a constant regardless of the volume of the solution sample.

10. Why is moles per liter a better concentration unit than grams per liter when comparing substances in solution?

Numbers in red indicate the appropriate chapter sections to aid you in assigning these items. Answers to all questions appear in the Teacher's Guide at the front of this book.

Summary

■ While it is impossible to determine the mass of an atom directly, you can compare the mass of one atom to another to find the relative mass. For relative atomic masses carbon-12 is the standard.

■ Amadeo Avogadro assumed that equal volumes of gases at the same temperature and pressure contain equal numbers of particles. Avogadro's hypothesis allows us to find the mass of equal numbers of atoms by comparing the mass of equal volumes of gases.

■ The number of carbon-12 atoms in exactly 12.0000 grams of carbon is defined as one mole.

■ Avogadro's number represents the number of atoms, ions, molecules, or other particles in one mole. The mass of Avogadro's number of atoms, ions, molecules, or particles is called the molar mass.

■ The molar mass of a compound can be found by summing the masses of the atoms in that compound.

■ By experiment, one can find the mass of one element that will combine with another element to form a compound. Dividing these masses by the relative atomic mass of each element gives the number of moles of each element. The ratio of these numbers can be converted into a ratio leading to an empirical formula. The molecular formula may be some whole number multiple of the empirical formula. To find the molecular formula, one must know the molar mass of a molecule.

■ The formula for a compound and the relative atomic masses can be used to calculate the theoretical percent composition of that compound.

■ Percent composition data can be used to compare the composition of compounds having the same elements.

■ Concentration of solutions is best described by molarity. The molarity of a solution is a unitary rate representing the number of moles of solute per liter of solution.

Chemically Speaking

atomic mass	formula unit
atomic mass unit	molarity
Avogadro's hypothesis	molar mass
Avogadro's number	mole
concentration	molecular formula
diatomic molecule	percent composition
empirical formula	relative mass

Review

1. What is the molar mass of cobalt? (4-2, 4-4)

2. Which contains more mass, an atom of aluminum or an atom of sulfur? (4-2)

3. What is the molar mass of molybdenum? (4-2, 4-4)

4. What does Avogadro's hypothesis indicate about two 5.00-liter containers at the same temperature and pressure, one of which contains carbon dioxide gas and the other of which contains methane gas. (4-2)

5. How many atoms are in one mole of neon atoms? (4-3)

6. How many atoms are contained in one mole of nickel atoms? (4-3)

7. How many molecules are there in one mole of silver cyanide molecules? (4-4)

8. How many atoms are in one molecule for a diatomic element? (4-5)

9. Name and write the formulas for the diatomic elements. (4-5)

10. Phosphorus and arsenic are tetratomic elements. Write the formula for a molecule of each. (4-5)

11. The compound benzene has two formulas, CH and C_6H_6. Which is the empirical formula and which is the molecular formula? (4-9)

Interpret and Apply

1. If a 12.0-liter flask of oxygen contains 3.2×10^{23} molecules, how many molecules of nitrogen (at equal temperature and pressure) would be contained in the same flask? (4-2, 4-3)

2. Why was hydrogen selected as a standard in preparing the first relative atomic mass table? (4-2)

3. A 2.00-liter flask contains 1.00 gram of hydrogen gas and the same flask holds 2.00 grams of helium gas at the same pressure and temperature. The relative mass of helium to hydrogen is 2.00 to 1.00. Explain why these values do not agree with the mass ratio of 4.00 to 1.00 given in the periodic table. (4-2, 4-3, 4-5)

4. If the mass of a carbon-12 atom was redefined as exactly 3 rmu (redefined mass units), by what factor would the masses of the other atoms be changed when converted to rmu? (4-2)

5. For sucrose, $C_{12}H_{22}O_{11}$, which element constitutes the largest percent by number of atoms? (4-10)

6. For sucrose, which element contains the largest percent by mass? (4-10)

7. Which element in the compound magnesium oxide makes up the largest percent by mass? (4-10)

8. Which element in the compound methane, CH_4, makes up the largest percent by mass? (4-10)

Problems

1. Using the 100-mL beaker as a standard of 4.00, calculate the relative mass of the other pieces of equipment in question 3 of the Review and Practice on page 104. (4-1)

2. How many carbon atoms contain the same amount of mass as one molybdenum atom? (4-3)

3. The masses of 3 liters of each gas listed in Table 4-2 are shown as follows:

GAS	MASS OF 3 L (in g)	MASS OF GAS RELATIVE TO H_2
a. hydrogen	0.24	
b. nitrogen	3.36	
c. oxygen	3.93	
d. fluorine	4.56	
e. chlorine	8.40	

Calculate the relative mass of each gas compared to hydrogen as 1.00. (4-2)

4. If you had one mole of pennies to divide among all the people in the world, how many dollars would each person receive? Use 5 billion as the population of Earth. (4-3)

5. If there were a mole of people evenly distributed over the surface of Earth, including both land and water, how many square centimeters would be allotted to each person? (4-3)

6. What is the molar mass of ammonium chromate? (4-4)

7. What is the molar mass of cobalt(II) chloride? (4-4)

8. Calculate the molar mass of each of the following compounds. (4-4)
   a. K_2SO_4
   b. H_3PO_4
   c. NH_4Cl
   d. Na_3PO_4
   e. $Ni(CN)_2$
   f. KH_2PO_4
   g. iron(II) sulfate
   h. iron(III) sulfate
   i. copper(I) carbonate
   j. copper(II) carbonate
   k. dinitrogen trioxide
   l. aluminum sulfide

9. What is the mass of one mole of each of the following compounds? (4-4)
   a. AgCl
   b. $MgCrO_4$
   c. K_2CrO_4
   d. $Fe(CN)_2$
   e. $CuSO_4$
   f. chromium(VI) oxide
   g. sodium sulfide
   h. carbon tetrachloride

10. What is the molar mass of fluorine gas? (4-4, 4-5)

11. If 2.5 moles of hydrogen gas (H_2) react in an experiment, how many grams of hydrogen react? (4-4)

12. From the information given, calculate the molar mass of each compound. (4-4)
   a. 2 moles of compound A have a mass of 80 grams.
   b. 3.01×10^{23} molecules of compound B have a mass of 9 grams.
   c. 1.57×10^{23} molecules of compound C have a mass of 7.56 grams.
   d. 1.35 grams of compound D contain 4.55×10^{22} molecules.

13. a. How many moles are in a 22-gram sample of manganese? (4-3, 4-4)
   b. How many molecules are in 14.0 grams of nitrogen gas?
   c. How many molecules of aluminum sulfate are in 17.1 grams of the compound?
   d. How many moles of silver have a mass of 16.0 grams?
   e. If a piece of copper has a mass of 17.5 grams, how many moles of copper does it contain?
   f. What is the mass of one atom of aluminum?
   g. How many atoms of carbon are in 17.1 grams of sucrose, $C_{12}H_{22}O_{11}$?

14. For each of the following, find the number of atoms represented and the moles of atoms represented. (4-4)
   a. 28 grams of sodium
   b. 28 grams of iron
   c. 150 grams of zinc
   d. 2.4 grams of calcium
   e. 150 grams of chlorine gas
   f. 21 grams of fluorine gas

15. The most delicate balance can detect a mass of about 10^{-8} gram. How many gold atoms would be in a sample having this mass? (4-4)

16. How many atoms are in a copper penny if the penny has a mass of 2.5 grams? Assume the penny is pure copper. (4-4)

17. Acetylene gas is used for welding. Acetylene contains 30 grams of carbon for each 2.5 grams of hydrogen. What is the empirical formula for acetylene? (4-7)

18. There are two common oxides of sulfur. One con-

CHAPTER REVIEW

p121 1+6R   p121 1+3 I+A
p122 4+5 P

tains 32 grams of sulfur for each 32 grams of oxygen. The other oxide contains 32 grams of sulfur for each 48 grams of oxygen. What are the empirical formulas for the two oxides? (4-7)

19. A form of phosphorus called red phosphorus is used in match heads. When 0.062 gram of red phosphorus burns, 0.142 gram of phosphorus oxide is formed. What is the empirical formula of this oxide? (4-7)

20. There are two known compounds containing only tungsten and carbon. One is the very hard alloy tungsten carbide. Analysis of the two compounds gives 1.82 grams of tungsten per 0.12 gram of carbon for the first compound, and 3.70 grams of tungsten per 0.12 gram of carbon for the second compound. What are the empirical formulas for these compounds? (4-7)

21. A compound is composed of 19.01 grams of carbon, 18.48 grams of nitrogen, 25.34 grams of oxygen, and 1.58 grams of hydrogen. Find the empirical formula of this compound. (4-7)

22. A compound is composed of 7.20 grams of carbon, 1.20 grams of hydrogen, and 9.60 grams of oxygen. The molar mass of the compound is 180 grams. Find the empirical and molecular formulas for this compound. (4-7, 4-9)

23. A compound is composed of 16.66 grams of carbon and 3.49 grams of hydrogen. The molar mass of the compound is 58 grams. Find the empirical and molecular formulas for this compound. (4-7, 4-9)

24. Find the percent composition of a compound which is 10.12 grams of aluminum and 17.93 grams of sulfur. (4-10)

25. Find the simplest formula for a compound which is 63.6% iron and 36.4% sulfur. Write the name of the compound. (4-10)

26. Find the simplest formula for a compound which is 79.9% copper and 20.1% sulfur. Write the name of the compound. (4-10)

27. Find the simplest formula for a compound that is 52.6% nickel, 21.9% carbon, and 25.5% nitrogen. Write the name of the compound. (4-10)

28. A 0.050M solution of glycerine, $C_3H_8O_3$, and a 0.050M solution of lycine, $C_5H_{11}NO_2$, are prepared. Which solution contains the most dissolved molecules per liter? (4-11, 4-12)

29. How many grams of sodium sulfate, Na_2SO_4, are contained in 1.50 L of 0.25M solution? (4-12)

30. Describe how you would prepare 1.50 L of a 0.25M solution of sodium sulfate. (4-12)

31. Describe how you would prepare 1.00 liter of a 3.00M solution of sodium hydroxide. (4-12)

32. Describe how you would prepare 5.00 liters of a 6.00M solution of potassium hydroxide. (4-12)

33. How many grams of ammonium sulfate, $(NH_4)_2SO_4$, are required to prepare 3.50 L of a 1.55M solution? (4-12)

34. How many moles of sodium chromate, Na_2CrO_4, are contained in 1.75 L of a 2.00M solution? (4-12)

Challenge

1. How many moles of iron(III) chloride could be made with 1.81×10^{23} atoms of chlorine and the appropriate number of iron atoms?

2. Describe how you would prepare 2.00 liters of a 3.00M solution of HCl using a 100-mL graduated cylinder, a 2-L volumetric flask, and a stock solution of HCl which is 37% HCl (the rest is water) and has a density of 1.37 g/mL.

3. If 1.00 L of water is added to 3.00 L of a 6.00M solution of HCl, what is the new molarity of the acid solution?

Synthesis

1. If the density of lead is 11.3 g/cm^3, calculate the volume of one mole of lead.

2. If the volume of one lead atom is 3.00×10^{-23} cm^3, calculate the number of atoms in one mole of lead. Use the density given in the previous problem.

Projects

1. The story of the development of early atomic theory is a fascinating one, and it is told rather clearly in the words of those scientists who took part in the development. It is interesting to try to visualize atoms as Dalton, Gay-Lussac, and Avogadro imagined them. To understand their models, read articles on the history of science.

2. The kind of careful analysis needed in determining experimental uncertainty led to the discovery of argon and a Nobel prize for Baron Rayleigh and William Ramsey. Read an account of their discovery.

You do not have to be in a laboratory to observe the results of chemical reactions. Green plants use the sun's energy to combine water and carbon dioxide into sugar, and give off oxygen. In your cells, sugar and oxygen combine to produce energy, carbon dioxide, and water. Fermenting plant sugars can be used as a source of alternative fuels such as gasohol.

The more you observe nature the more apparent it is that you need a way to describe the chemical changes that you see. Food cooks, grass grows, cars corrode, teaspoons tarnish, and cells reproduce. Even the memories you may have after reading these words will be stored in your brain as a result of a chemical reaction.

Chemical Reactions

Writing and Balancing Chemical Equations

If you were given the task of describing a chemical reaction, where would you begin? Perhaps you might start by making macroscopic observations of the substances with which you chose to experiment. Then you might combine them and make more observations about the results. What would you look for? What could you measure? How would you find out what new substances you might have produced?

Chemists are constantly faced with the same problem of accurately communicating what they have observed and measured. Clear descriptions of experimental results lead to an understanding of the complex processes involved in chemical changes.

CONCEPTS

- reactants and products
- conservation of atoms and mass
- reading and balancing chemical equations

5-1 Describing Reactions

Ira Remsen Investigates Nitric Acid*

"While reading a textbook of chemistry, I came upon the statement, 'nitric acid acts upon copper.' I was getting tired of reading such absurd stuff and I determined to see what this meant. Copper was more or less familiar to me, for copper cents were then in use. I had seen a bottle marked 'nitric acid' on a table in the doctor's office where I was then 'doing time!' I did not know its peculiarities, but I was getting on and likely to learn. The spirit of adventure was upon me. Having nitric acid and copper, I had only to learn what the words 'act upon' meant. Then the statement, 'nitric acid acts upon copper,' would be something more than mere words. All was still. In the interest of knowledge I was even willing to sacrifice one of the few copper cents then in my possession. I put one of them on the table; opened the bottle marked 'nitric acid'; poured some of the liquid on the copper; and prepared to make an observation. But what was this wonderful thing which I beheld? The cent was already changed, and it was no small change either. A greenish blue liquid foamed and fumed over the cent and over the table. The air in the neighborhood of the performance became colored dark red. A great colored cloud arose. This was disagreeable and suffocating—how should I stop this? I tried to get rid of the objectionable mess by picking it up and throwing it out of the window, which I had meanwhile opened. I learned another fact—nitric acid not only acts upon copper but it acts upon fingers. The pain led to

*Frederick H. Getman, "The Life of Ira Remsen," *J. Chem. Educ.*, 1940.

Figure 5-1 What new substances are produced in the reaction of nitric acid and a copper penny?

another unpremeditated experiment. I drew my fingers across my trousers and another fact was discovered. Nitric acid acts upon trousers. Taking everything into consideration, that was the most impressive experiment, and, relatively, probably the most costly experiment I have ever performed. It resulted in a desire on my part to learn more about that remarkable kind of action. Plainly the only way to learn about it was to see its results, to experiment, to work in a laboratory."

The reaction of a copper penny with nitric acid made quite an impression on Ira Remsen—and on his trousers! Observing the formation of a cloud of red gas, a greenish-blue liquid, and a disagreeable odor is dramatic and interesting. Remsen's observations indicate that new substances were being formed during his experiment. While the observations of the chemical changes were simple to make (though sometimes painful), the chemical processes that occurred were quite complex.

Chemists seek an understanding of the processes that occur in a reaction as well as a concise way of communicating this information. This can be done by writing a chemical equation that describes the identity of the pure substances before and after the chemical change. The chemical equation also suggests what is taking place on an atomic or molecular level and can be used to express quantitative information. To learn how to write chemical equations, you will begin with a much simpler system.

5-2 Formation and Decomposition of Water

Water is produced when hydrogen gas combines with oxygen gas. The properties of the end product of the reaction, water, are very different from the properties of the starting elements, hydrogen and oxygen.

Try to imagine what is happening at the molecular level. Suppose that you could shrink in size so that you were about as big as a molecule of hydrogen or oxygen. You would then be smaller then by a factor of about 5 billion. You could see hydrogen and oxygen molecules whizzing past you, occasionally colliding and rebounding from each other. Every once in a while, collisions would break the bonds between atoms in hydrogen and oxygen molecules. New chemical bonds would form as the atoms rearrange to form water molecules.

Chemists have developed molecular models to help visualize these changes. Two hydrogen molecules and one oxygen molecule are represented in Figure 5-2. For these molecules to form water, the bonds

Demonstrate the Hoffman apparatus for the electrolysis of water. Have students predict which column contains hydrogen and which contains oxygen. Ask how they would test their predictions. (A glowing splint will ignite in oxygen and cause a "pop" with hydrogen.)

Figure 5-2 Two diatomic hydrogen molecules (four atoms) react with one diatomic oxygen molecule (two atoms) to produce two molecules of water.

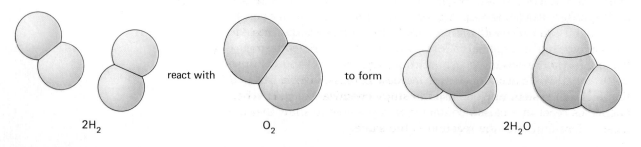

$2H_2$ react with O_2 to form $2H_2O$

between the atoms in the oxygen molecules and in the hydrogen molecules must be broken. As water molecules form, new chemical bonds form. A chemical change has taken place. Although the atoms regroup to form two water molecules, the total number of atoms does not change.

The water in a solution of water and sulfuric acid can be decomposed by passing an electric current through the solution. The way to do this is shown in Figure 5-3. Two pieces of metal called electrodes are placed in the solution. When the electrodes are connected to a battery, hydrogen gas appears at one electrode and oxygen gas at the other. If the apparatus is operated until two moles of water have decomposed, two moles of hydrogen gas and one mole of oxygen gas are produced.

Now compare the formation and decomposition of water. Figure 5-4 shows that the chemical change in the formation of water is exactly the reverse of the chemical change in water decomposition. In each drawing, there are two oxygen atoms on the left and two oxygen atoms on the right. You can see four hydrogen atoms on the left and four on the right. Atoms are neither gained nor lost. Atoms are conserved in chemical reactions.

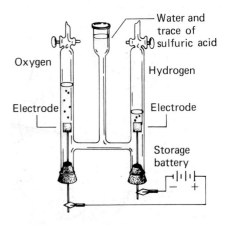

Figure 5-3 A direct electric current decomposes water into hydrogen and oxygen. The volume of hydrogen is twice the volume of oxygen.

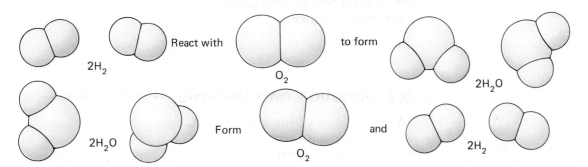

Figure 5-4 The formation of water from hydrogen and oxygen is exactly the reverse of the decomposition of water to form hydrogen and oxygen.

5-3 Conservation of Mass and Atoms

The French chemist, Antoine Laurent Lavosier (1743–1794), established an important principle based on observations of how some substances react. He stated that the total mass of the products of a reaction is equal to the total mass of the reactants. This relationship was not always readily apparent because many chemical changes were not understood. Experimental results were inconsistent. For example, when a candle burns, the mass of the candle decreases. Copper turns black when heated but shows an increase in mass. Both of these chemical changes involve oxygen. When a candle burns, the carbon and hydrogen that make up the wax are combined with oxygen to form carbon dioxide and water. These compounds escape into the atmosphere so that the mass of the candle decreases. When copper is heated, it combines with oxygen from the air to form solid black copper(II) oxide which accounts for the increase in mass. Lavosier showed that a closed system must be used to study chemical changes. When chemicals react in a closed container, it is possible to show that the mass before and after the reaction is the same.

The **law of conservation of mass** states that the total mass of all reactants before a chemical reaction must be the same as the total mass of all products after the chemical change has taken place. For this statement to be true, the total number of atoms must be constant. The substances present at the start of a reaction must contain the same number of atoms as the substances present at the end of the reaction. The total mass does not change because the atoms that take part in the reaction do not change mass. The bonds between the atoms are broken, and new bonds are formed but the same atoms are there.

This principle of conserving mass is an important one to keep in mind as you learn how to represent in writing the events that take place in a chemical reaction.

5-4 Writing Equations

CHEM THEME

A chemical is identified by its characteristic properties. When a chemical reaction takes place, new chemicals with different properties are formed.

Recall from Chapter 2 that macroscopic observations are those that can be made directly by an observer. Figure 5-5 shows the elements sodium and chlorine. Sodium is a soft, shiny metal that reacts quickly with air. It is kept under oil to keep it pure, and it can be transferred quickly into the bottle containing chlorine gas. The element chlorine is a pale yellow-green gas and is poisonous. When the sodium is put into the bottle of chlorine and warmed, the reaction shown in Figure 5-5 takes place. A new substance is formed that is a combination of sodium and chlorine. This substance, sodium chloride, has properties very different from those of the two elements that reacted. While sodium and chlorine are both toxic, the compound sodium chloride is not. It is a white crystalline solid and is soluble in water. You can swim in the ocean which contains 3 percent sodium chloride. It also tastes good on hamburgers! You buy it in the grocery store for pennies a pound under the name of table salt.

A word equation can be written to represent the macroscopic observations of what occurs in Figure 5-5:

Figure 5-5 The flask on the left contains the element chlorine, a pale yellow-green gas. The sodium metal in the middle is kept under oil. The sodium on the right is reacting vigorously with the chlorine to form white solid sodium chloride, NaCl.

$$\text{sodium} + \text{chlorine} \longrightarrow \text{sodium chloride}$$

This equation is read, "sodium reacts with chlorine to produce sodium chloride." The chemicals to the left of the arrow are called the **reactants**. The chemicals to the right of the arrow are called the **products**. The arrow is used to mean "produces" or "yields" and points in

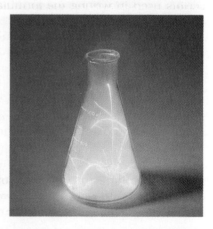

the direction of the change that occurs. Using the more convenient shorthand of symbols and formulas gives

$$Na + Cl_2 \longrightarrow NaCl$$

Note that this equation shows chlorine as one of the diatomic elements. The formula for sodium chloride, NaCl, is almost as familiar as the formula for water, H_2O.

The above equation for the formation of sodium chloride also represents microscopic events which cannot be observed directly. Therefore, it is also useful to have models that represent what you think is occurring. Figure 5-6 shows one atom of sodium reacting with a molecule of chlorine to produce sodium chloride.

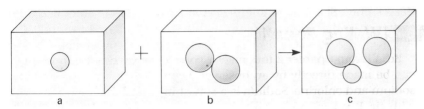

a   b   c

Figure 5-6 One atom of solid sodium in (a) reacts with the diatomic molecule of chlorine gas in (b). The result is one formula unit of sodium chloride and one unreacted atom of chlorine in (c).

5-5 Balancing Equations

When writing equations to represent chemical reactions, you must accurately represent what you believe is happening. After writing the symbols to represent reactants and products, the equation for the reaction must be balanced to show that mass is conserved. You will remember from Section 5-3 that the law of conservation of mass states that the total mass before a chemical reaction must be the same as the total mass after the chemical reaction has taken place. The same number of atoms must be present before and after the reaction. A balanced chemical equation shows equal numbers of each kind of atom present in the reactants and the products.

Balancing equations is in part a trial-and-error process, but it will be easier if you remember a few key points. First, you must write the correct formulas for all reactants and products. A review of the rules for formula writing in Section 2-13 might be helpful. Second, the subscripts used in writing the formulas for the reactants and products *must not* be changed during the process of balancing the equation.

The correct formulas for the reactants and products in the sodium chloride reaction are

$$Na + Cl_2 \longrightarrow NaCl$$

Numbers called **coefficients** are placed in front of the formulas to make the number of atoms on each side of the equation equal. If you write the coefficient 2 in front of the formula NaCl, as in

$$2NaCl$$

it implies two units of NaCl, each of which has one sodium atom and one chlorine atom, for a total of two sodium and two chlorine atoms. In the equation above, the number of chlorine atoms on the left is two,

Do You Know?

In 1915, the German army used poisonous chlorine gas as a weapon for the first time in Ypres, Belgium. Chlorine is denser than air. Once it was released, it stayed close to the ground. Protective gas masks containing activated charcoal were soon developed to remove chlorine from the air breathed into the mask.

as shown by the subscript in Cl_2. You can make the number of chlorine atoms on the right equal to two by using a coefficient of 2 in front of the sodium chloride:

$$Na + Cl_2 \longrightarrow 2NaCl$$

Now the number of chlorine atoms is equal on each side of the equation, but the number of sodium atoms is not. Using a coefficient of 2 for sodium completes the process:

$$2Na + Cl_2 \longrightarrow 2NaCl$$

The balanced equation shows 2 sodium atoms and 2 chlorine atoms on each side of the equation. Return to Figure 5-6, and you can see that the reaction of one atom of sodium with one molecule of chlorine leaves one uncombined chlorine atom. That chlorine atom can then combine with another sodium atom. The equation shows that the number of atoms is conserved and that mass is conserved.

The points to remember when balancing equations are:

1. Check to make sure all formulas are written correctly.
2. Do not change subscripts once the formula is written correctly.
3. Begin with an element that occurs only once on each side of the arrow.

Example 5-1

Write and balance the equation for the reaction of hydrogen and oxygen to produce water.

▶ Suggested solution _____

First, write the correct formulas for all reactants and products, recalling that hydrogen and oxygen are diatomic:

$$H_2 + O_2 \longrightarrow H_2O$$

The number of oxygen atoms is not equal on each side of the equation. Adding a coefficient of 2 in front of water would make the number of oxygen atoms equal:

$$H_2 + O_2 \longrightarrow 2H_2O$$

Now the number of hydrogen atoms is not equal on each side of the equation. Adding a coefficient of 2 in front of hydrogen completes the process of balancing the equation:

$$2H_2 + O_2 \longrightarrow 2H_2O$$

This balanced equation shows 4 hydrogen atoms and two oxygen atoms on each side of the equation.

Example 5-2

Write and balance the equation for the reaction of sodium and water to produce sodium hydroxide, NaOH, and hydrogen.

▶ Suggested solution _____

First, write the correct formulas for the reactants and products:

$$Na + H_2O \longrightarrow NaOH + H_2$$

The number of sodium atoms and oxygen atoms is equal on both sides of the equation. Hydrogen is the only element that does not have the same number of atoms on each side of the equation. Putting a coefficient of 2 in front of water and sodium hydroxide will give 4 atoms of hydrogen on each side:

$$Na + 2H_2O \longrightarrow 2NaOH + H_2$$

The number of oxygen atoms is equal on both sides of the equation. However, the sodium atoms are no longer equal. By placing a coefficient of 2 before sodium, the equation is balanced:

$$2Na + 2H_2O \longrightarrow 2NaOH + H_2$$

Example 5-3

Write and balance the equation for the reaction of ammonia, NH_3, and oxygen to produce nitrogen dioxide, NO_2, and water.

▶ Suggested solution _____

Write the correct formulas for all reactants and products:

$$NH_3 + O_2 \longrightarrow NO_2 + H_2O$$

To show the same number of hydrogen atoms on each side of the equation, a coefficient of 2 is put in front of ammonia and a coefficient of 3 is put in front of water:

$$2NH_3 + O_2 \longrightarrow NO_2 + 3H_2O$$

To make the number of nitrogen atoms equal, a coefficient of 2 is used for nitrogen dioxide:

$$2NH_3 + O_2 \longrightarrow 2NO_2 + 3H_2O$$

The number of oxygen atoms on the right is 7. Because the oxygen on the left side of this equation is in the form of diatomic molecules, there is no whole number coefficient that would make the number of oxygen atoms equal to 7. One way to solve this problem is to use the fractional coefficient $\frac{7}{2}$ for oxygen:

$$2NH_3 + \tfrac{7}{2}O_2 \longrightarrow 2NO_2 + 3H_2O$$

Now the equation shows 7 atoms of oxygen on both sides. Generally, equations are written using only whole number coefficients. Multiplying all the coefficients by 2 will eliminate the fractional coefficient of oxygen:

$$4NH_3 + 7O_2 \longrightarrow 4NO_2 + 6H_2O$$

The number of nitrogen, hydrogen, and oxygen atoms is still equal on both sides.

When students start doing calculations involving equations, they will find it easier if they have learned to express the mole ratios given by the equation as whole numbers.

5-6 Reading Equations

Equations provide much information about a chemical reaction:

$$2C(s) + O_2(g) \longrightarrow 2CO(g)$$

The above equation is usually read, "two moles of carbon combine with one mole of oxygen to produce two moles of carbon monoxide." Like many of the things discussed in chemistry, this equation indicates proportions. The proportional relationship can be emphasized by reading the equation as follows: "For every mole of oxygen, two times as many moles of carbon react to produce two times as many moles of carbon monoxide." The coefficients also represent the number of atoms and molecules of each substance involved in the reaction. Therefore, a third way of reading the equation would be, "two atoms of carbon combine with one molecule of oxygen (containing 2 atoms) to produce two molecules of carbon monoxide."

Figure 5-7 shows this proportional relationship. In the next chapter, you will see that a knowledge of equations will allow you to predict the amount of one chemical that will react with another.

Figure 5-7 This drawing represents four carbon atoms (a), combining with two diatomic molecules of oxygen (b), to produce four molecules of carbon monoxide (c). They combine in the ratio of one carbon atom to one oxygen atom. The atoms are drawn to represent their physical state as a solid or a gas.

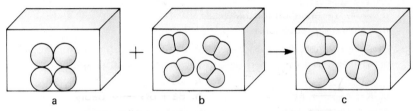

Table 5-1

SYMBOLS USED IN EQUATIONS	
SYMBOL USED	MEANING
+	Read *plus* or *and*. Used between two formulas to indicate reactants combined or products formed.
$\longrightarrow$	Read *yields* or *produces*. Used to separate reactants (on the left) from products (on the right). The arrow points in the direction of change.
(s)	Read *solid*. Written after a symbol or formula to indicate that the physical state of the substance is solid.
(l)	Read *liquid*. Written after a symbol or formula to indicate that the physical state of the substance is liquid.
(g)	Read *gas*. Written after a symbol or formula to indicate that the physical state of the substance is gaseous.
(aq)	Read *aqueous*. Written after a symbol or formula to indicate that the substance is dissolved in water.
CO_2	The subscripts (one for carbon which is understood, and 2 for oxygen) mean that each molecule of carbon dioxide contains 1 atom of carbon and 2 atoms of oxygen.
$3CO_2$	Read *3 molecules of carbon dioxide*. The coefficient 3 means 3 molecules of CO_2.
N.R.	Read *no reaction*. Indicates that the given reaction does not occur.

This equation also contains symbols which indicate the physical state of the chemicals. Many people might already know that carbon is a solid and oxygen and carbon monoxide are gases. However, many of the examples in the text will contain chemicals that are not familiar and for which the physical state is important in determining how the chemicals interact.

There are other symbols used in equations that provide additional information about the reaction. The meanings of some common symbols used in writing equations are summarized in Table 5-1.

Review and Practice

1. Answer the following for the reaction of solid carbon with oxygen gas to produce carbon dioxide gas:
  a. Name the reactants.
  b. List some macroscopic properties of the reactants and products.
  c. Draw models that represent the reaction of the atoms (use Figure 5-7 as a guide).
  d. Write the equation for the reaction.
  e. Balance the equation.

2. Balance the following equations:
  a. $K + Cl_2 \longrightarrow KCl$
  b. $K + Br_2 \longrightarrow KBr$
  c. $K + F_2 \longrightarrow KF$
  d. $K + O_2 \longrightarrow K_2O$
  e. $Na + O_2 \longrightarrow Na_2O$
  f. $Na + N_2 \longrightarrow Na_3N$
  g. $Ca + O_2 \longrightarrow CaO$
  h. $Ba + Br_2 \longrightarrow BaBr_2$
  i. $Sr + N_2 \longrightarrow Sr_3N_2$

3. Balance the following equations:
  a. $K + H_2O \longrightarrow KOH + H_2$
  b. $Ca + H_2O \longrightarrow Ca(OH)_2 + H_2$
  c. $Mg + N_2 \longrightarrow Mg_3N_2$
  d. $Mg_3N_2 + H_2O \longrightarrow MgO + NH_3$
  e. $NH_4Cl + Ca(OH)_2 \longrightarrow NH_3 + H_2O + CaCl_2$
  f. $(NH_4)_2SO_4 + NaOH \longrightarrow NH_3 + H_2O + Na_2SO_4$
  g. $FeS_2 + O_2 \longrightarrow Fe_2O_3 + SO_2$
  h. $MoS_2 + O_2 \longrightarrow MoO_3 + SO_2$
  i. $MoO_3 + H_2 \longrightarrow Mo + H_2O$

4. Write and balance the following word equations:
  a. lithium + water $\longrightarrow$ lithium hydroxide + hydrogen
  b. barium + water $\longrightarrow$ barium hydroxide + hydrogen
  c. ammonium nitrate + sodium hydroxide $\longrightarrow$ ammonia + water + sodium nitrate
  d. copper(I) sulfide + oxygen $\longrightarrow$ copper(I) oxide + sulfur dioxide
  e. aluminum + iodine $\longrightarrow$ aluminum iodide

5. Balance the following equations:
  a. $AgNO_3 + CuCl_2 \longrightarrow AgCl + Cu(NO_3)_2$
  b. $BaCl_2 + (NH_4)_2CO_3 \longrightarrow BaCO_3 + NH_4Cl$
  c. $Mg(ClO_3)_2 \longrightarrow MgCl_2 + O_2$
  d. $FeCl_2 + Na_3PO_4 \longrightarrow Fe_3(PO_4)_2 + NaCl$
  e. $ZnO + HCl \longrightarrow ZnCl_2 + H_2O$
  f. $CuSO_4 + Fe \longrightarrow Fe_2(SO_4)_3 + Cu$
  g. $Br_2 + KI \longrightarrow KBr + I_2$
  h. $Al + NaOH + H_2O \longrightarrow NaAl(OH)_4 + H_2$

1. a. carbon and oxygen
   b. carbon—black, insoluble in water, solid at room temperature; oxygen—colorless gas at room temperature
   c. See Teacher's Guide for diagram.
   d. $C(s) + O_2(g) \longrightarrow CO_2(g)$
   e. $C(s) + O_2(g) \longrightarrow CO_2(g)$
2. a. $2K + Cl_2 \longrightarrow 2KCl$
   b. $2K + Br_2 \longrightarrow 2KBr$
   c. $2K + F_2 \longrightarrow 2KF$
   d. $4K + O_2 \longrightarrow 2K_2O$
   e. $4Na + O_2 \longrightarrow 2Na_2O$
   f. $6Na + N_2 \longrightarrow 2Na_3N$
   g. $2Ca + O_2 \longrightarrow 2CaO$
   h. $Ba + Br_2 \longrightarrow BaBr_2$
   i. $3Sr + N_2 \longrightarrow Sr_3N_2$
3. a. $2K + 2H_2O \longrightarrow 2KOH + H_2$
   b. $Ca + 2H_2O \longrightarrow Ca(OH)_2 + H_2$
   c. $3Mg + N_2 \longrightarrow Mg_3N_2$
   d. $Mg_3N_2 + 3H_2O \longrightarrow 3MgO + 2NH_3$
   e. $2NH_4Cl + Ca(OH)_2 \longrightarrow 2NH_3 + 2H_2O + CaCl_2$
   f. $(NH_4)_2SO_4 + 2NaOH \longrightarrow 2NH_3 + 2H_2O + Na_2SO_4$
   g. $4FeS_2 + 11O_2 \longrightarrow 2Fe_2O_3 + 8SO_2$
   h. $2MoS_2 + 7O_2 \longrightarrow 2MoO_3 + 4SO_2$
   i. $MoO_3 + 3H_2 \longrightarrow Mo + 3H_2O$
4. a. $2Li + 2H_2O \longrightarrow 2LiOH + H_2$
   b. $Ba + 2H_2O \longrightarrow Ba(OH)_2 + H_2$
   c. $NH_4NO_3 + NaOH \longrightarrow NH_3 + H_2O + NaNO_3$
   d. $2Cu_2S + 3O_2 \longrightarrow 2Cu_2O + 2SO_2$
   e. $2Al + 3I_2 \longrightarrow 2AlI_3$
5. a. $2AgNO_3 + CuCl_2 \longrightarrow 2AgCl + Cu(NO_3)_2$
   b. $BaCl_2 + (NH_4)_2CO_3 \longrightarrow BaCO_3 + 2NH_4Cl$
   c. $Mg(ClO_3)_2 \longrightarrow MgCl_2 + 3O_2$
   d. $3FeCl_2 + 2Na_3PO_4 \longrightarrow Fe_3(PO_4)_2 + 6NaCl$
   e. $ZnO + 2HCl \longrightarrow ZnCl_2 + H_2O$
   f. $3CuSO_4 + 2Fe \longrightarrow Fe_2(SO_4)_3 + 3Cu$
   g. $Br_2 + 2KI \longrightarrow 2KBr + I_2$
   h. $2Al + 2NaOH + 6H_2O \longrightarrow 2NaAl(OH)_4 + 3H_2$

Observing Regularities in Reactions

CONCEPTS

- reaction types
- predicting reaction products
- activity of elements

The products of a chemical reaction are identified by doing experiments to test their characteristic properties. When that information is known, it becomes useful to chemists. Look back at the equation in Example 5-2. Then compare it with the first two equations in Review and Practice questions 3 and 4. There are similarities among these five equations. All are reactions between a metal and water, and all produce hydrogen and a hydroxide.

Rather than memorize each equation separately, it is useful to learn what regularities exist in how substances react. These regularities can be used to classify reactions into general categories. Recognizing types of reactions will enable you to complete an equation once you know the general pattern for each type. This process is similar to the memorization of the formulas and charges of a dozen polyatomic ions and thirty elements which allows you to write correct formulas for thousands of compounds.

This information on reaction types is presented as a method of classifying reactions into predictable and recognizable groups. Teachers should view this material as a reference aid and not as material to be memorized.

5-7 Types of Reactions

Synthesis is the combination of two or more substances to form a compound. It has the general form

$$A + B \longrightarrow AB$$

The formation of water is a synthesis reaction:

$$2H_2(g) + O_2(g) \longrightarrow 2H_2O(g)$$

This reaction was shown in Figures 5-2 and 5-4.

The formation of calcium hydroxide and iron(II) oxide are both synthesis reactions:

$$CaO(s) + H_2O(l) \longrightarrow Ca(OH)_2(s)$$
$$2Fe(s) + O_2(g) \longrightarrow 2FeO(s)$$

There are many types of reactions for which equations can be written that do not actually happen. Students will not have had enough experience to know which written equations represent actual reactions.

Decomposition is the opposite of synthesis. One substance breaks down to form two or more simpler substances. Decomposition reactions have the general form

$$AB \longrightarrow A + B$$

The decomposition of water produces hydrogen and oxygen:

$$2H_2O(g) \longrightarrow 2H_2(g) + O_2(g)$$

Upon heating, calcium hydroxide decomposes into calcium oxide and water:

$$Ca(OH)_2(s) \longrightarrow CaO(s) + H_2O(g)$$

CHEM THEME

An important part of the study of chemistry is the ability to predict the outcome of a chemical reaction before it occurs.

Students should have an opportunity to observe examples of each type of reaction. Observing several reactions of the same type allows them to see similar patterns of behavior in different chemical systems.

Combustion refers to a chemical reaction that usually gives off a large amount of energy in the form of heat and light and involves the reaction of a substance with oxygen from the atmosphere. This type of reaction is often referred to as burning. **Hydrocarbons**, compounds made of hydrogen and carbon, burn in oxygen to produce carbon dioxide and water. The general form for this type of reaction is

$$C_xH_y + O_2 \longrightarrow CO_2 + H_2O$$

The combustion or burning of a variety of hydrocarbons such as propane, gasoline, fuel oil, and natural gas, provide most of the energy for our homes, factories, and cars. The combustion of propane produces carbon dioxide and water:

$$C_3H_8(g) + 5O_2(g) \longrightarrow 3CO_2(g) + 4H_2O(g)$$

The combustion of heptane is similar:

$$C_7H_{16}(g) + 11O_2(g) \longrightarrow 7CO_2(g) + 8H_2O(g)$$

Substances other than hydrocarbons will burn in oxygen. For example, the combustion of magnesium to form magnesium oxide is usually conducted in the presence of oxygen in such a way that heat and a brilliant white light are produced:

$$2Mg + O_2 \longrightarrow 2MgO$$

You may notice that this equation is similar to the synthesis equation given for iron combining with oxygen to form iron(II) oxide:

$$2Fe + O_2 \longrightarrow 2FeO$$

In fact, both reactions can be described as either synthesis or combustion. Many of the reactions you will encounter can be classified in several ways. Describing the reaction of magnesium and oxygen as combustion calls attention to the large amount of visible energy it produces, although it is just as accurate to call it a synthesis reaction since two reactants are combining to form a product. It may seem less accurate to categorize as combustion the very slow reaction of iron with oxygen in the air to form brown iron(III) oxide (also called rust), however, it would be correct to do so. Even the reaction of hydrogen and oxygen to form water can be considered as either a combustion reaction or as a synthesis reaction. These categories of reactions can overlap, just as you can be classified both as a student and, at the same time, as a human being.

Energy changes take place in every chemical reaction as some chemical bonds are broken and others are formed. Energy changes during reactions and how they are written in equations are discussed in Sections 5-9 and 5-10.

Single replacement reactions occur when one element is replaced by another in a compound. These have the general formula

$$A + BC \longrightarrow AC + B$$

Bromine is replaced by chlorine in potassium bromide to form potassium chloride:

$$Cl_2(aq) + 2KBr(aq) \longrightarrow 2KCl(aq) + Br_2(aq)$$

Copper is replaced by iron in copper(II) sulfate:

$$Fe(s) + CuSO_4(aq) \longrightarrow FeSO_4(aq) + Cu(s)$$

Consider showing a film demonstrating reactions that are too dangerous or expensive to do in the school laboratory.

Do You Know?

When natural gas or methane, CH_4, is burned in insufficient air, not all of the carbon combines with oxygen. The unburned carbon particles form soot. You may have noticed these carbon deposits on glassware you have heated in the lab using natural gas as the fuel source.

Figure 5-8 Solid magnesium is reacting with oxygen gas to produce white solid magnesium oxide.

Do You Know?

Limestone buildings do not appear to dissolve, but stalagmites and stalagtites grow as the limestone above the cave dissolves over thousands of years. The slow rate of this change makes it seem that no reaction occurs.

Very few combinations of reactants give no reaction at all. There is usually some interaction though it may be at a rate that is not readily detectable. Students will find out more about this idea when they get to equilibrium and solubility products.

Table 5-2

ACTIVITY SERIES	
METALS	NONMETALS
decreasing activity	
lithium	fluorine
potassium	chlorine
calcium	bromine
sodium	iodine
magnesium	
aluminum	
zinc	
chromium	
iron	
nickel	
tin	
lead	
hydrogen	
copper	
silver	
mercury	
platinum	
gold	

Hydrogen is replaced by magnesium in hydrogen sulfate:

$$Mg(s) + H_2SO_4(aq) \longrightarrow MgSO_4(aq) + H_2(g)$$

Not all replacement reactions for which equations can be written will actually occur. The element that is doing the replacing must be able to displace the element occurring in the compound. The **activity** of an element is a measure of its ability to replace another element in a compound.

When nickel is mixed with magnesium chloride, no reaction takes place. Nickel is not as active an element as magnesium. To show that no reaction takes place, the equation is written as follows:

$$Ni(s) + MgCl_2(aq) \longrightarrow N.R.$$

The *N.R.* is used to mean that no reaction occurs. If, however, magnesium is mixed with aluminum chloride, a reaction does occur:

$$3Mg(s) + 2AlCl_3(aq) \longrightarrow 3MgCl_2(aq) + 2Al(s)$$

The results of many experiments are summarized with an activity series like that in Table 5-2. Any element will replace an element below it but will not replace an element above it.

Double replacement reactions occur when the elements in a solution of reacting compounds exchange places, or replace each other. The general form for these reactions is

$$AB + XY \longrightarrow AY + XB$$

In the most common double replacement reactions, the reactants are dissolved in water. There, the compounds come apart to form positive and negative ions which recombine to form a new compound. The new product quickly separates from the solution. For example, zinc bromide and silver nitrate react to form solid silver bromide.

$$ZnBr_2(aq) + 2AgNO_3(aq) \longrightarrow 2AgBr(s) + Zn(NO_3)_2(aq)$$

The zinc ions and nitrate ions remain dissolved in the solution as indicated by the (aq).

Iron(II) sulfate combines with potassium carbonate to form solid iron(II) carbonate and potassium sulfate:

$$FeSO_4(aq) + K_2CO_3(aq) \longrightarrow FeCO_3(s) + K_2SO_4(aq)$$

Lead(II) nitrate and potassium iodide form solid lead(II) iodide and potassium nitrate:

$$Pb(NO_3)_2(aq) + 2KI(aq) \longrightarrow PbI_2(s) + 2KNO_3(aq)$$

How can you know which ions form a precipitate and which ions remain in solution? You will learn more about predicting the products of this type of reaction in Chapter 16. For now, you can use the table in Appendix D which indicates whether a compound can stay dissolved in a solution or will form a solid precipitate.

Water-forming reactions make water as one of the products. The general form for this type of reaction is

$$HB + XOH \longrightarrow XB + HOH$$

Notice that HOH is another way to write the formula for water. The equation could also be written showing water as H_2O.

$$HB + XOH \longrightarrow XB + H_2O$$

Water-forming reactions are double replacement reactions. The reactants are dissolved in water where they come apart to form positive and negative ions which exchange places. In a water-forming reaction, one of the positive ions present is the hydrogen ion H^+ and one of the negative ions present is the hydroxide ion OH^-. These ions combine to form water.

The products of a water-forming reaction usually remain dissolved in the solution. Water is always one of the products, formed from the hydrogen ion and the hydroxide ion. The water molecules formed during the reaction mix with and cannot be distinguished from all other molecules in the reaction container.

Aqueous solutions of hydrogen chloride and sodium hydroxide react to form sodium chloride and water:

$$HCl(aq) + NaOH(aq) \longrightarrow NaCl(aq) + H_2O(l)$$

Aqueous solutions of hydrogen phosphate and barium hydroxide combine to form barium phosphate and water:

$$2H_3PO_4(aq) + 3Ba(OH)_2(aq) \longrightarrow Ba_3(PO_4)_2(s) + 6H_2O(l)$$

Water-forming reactions are some of the most important in chemistry and will be examined in detail in Chapter 20.

Do You Know?

The products of double replacement reactions are always written with the positive ion given first.

neutralization

5-8 Predicting the Products of a Reaction

It is very useful to be able to predict the products of a reaction when the reactants are given. Noticing regularities that can be classified allows you to make these predictions with greater ease.

Now look at some examples.

Example 5-4

Complete the equation for the synthesis reaction of magnesium and iodine.

▶ **Suggested solution** _____

First, write the formulas for the reactants, remembering that iodine is diatomic:

$$Mg + I_2 \longrightarrow$$

In a synthesis reaction, the reactants combine to form one or more new products. Refer to Section 2-13 and the rules for formula writing. Rule 2 indicates that ions from Group 2, like magnesium, will have a 2+ charge. Rule 4d states that iodine from Group 17 would form an ion with a 1− charge. The algebraic sum of the ionic charges in the formulas should be zero as indicated by Rule 5. Thus, the product of the synthesis reaction of magnesium and iodine must be magnesium iodide.

$$Mg + I_2 \longrightarrow MgI_2$$

The atoms on each side of the equation are counted to show that the equation is balanced as written.

Example 5-5

Write the equation for the combustion of liquid hexane, C_6H_{14}.

▶ Suggested solution _____

Combustion means the combination with oxygen, so the reactants are

$$C_6H_{14} + O_2 \longrightarrow$$

Hexane is a hydrocarbon, so the products of its combustion are water and carbon dioxide:

$$C_6H_{14} + O_2 \longrightarrow CO_2 + H_2O$$

The last step is to balance the equation:

$$2C_6H_{14} + 19O_2 \longrightarrow 12CO_2 + 14H_2O$$

Do You Know?

To burn 38 liters (10 gallons) of gasoline in an internal combustion engine requires 77 000 liters of oxygen or 390 000 liters of air. This is the amount of air breathed by 30 people in one day.

1. a. $2C_2H_2(g) + 5O_2 \longrightarrow$
$4CO_2(g) + 2H_2O(g)$
combustion
 b. $Zn(s) + CuSO_4(aq) \longrightarrow$
$ZnSO_4(aq) + Cu(s)$
single replacement
 c. $Cl_2(aq) + 2KI(aq) \longrightarrow$
$2KCl(aq) + I_2'(aq)$
single replacement
 d. $2H_2O_2(l) \longrightarrow 2H_2O(l) + O_2(g)$
decomposition
 e. $MgCl_2(s) \longrightarrow Mg(s) + Cl_2$
decomposition
 f. $Fe(s) + I_2(s) \longrightarrow FeI_2(s)$
synthesis
 g. $16Cu(s) + S_8(l) \longrightarrow 8Cu_2S(s)$
synthesis
 h. $C_6H_{12}O_6(aq) + 6O_2(aq) \longrightarrow$
$6CO_2(aq) + 6H_2O(l)$
combustion
 i. $FeCl_2(aq) + K_2S(aq) \longrightarrow$
$FeS(s) + 2KCl(aq)$
double replacement
 j. $H_2SO_4(aq) + 2NaOH(aq) \longrightarrow$
$Na_2SO_4(aq) + 2H_2O(l)$
double replacement, water forming
 k. $Pb(NO_3)_2(aq) + K_2CrO_4(aq) \longrightarrow$
$PbCrO_4(s) + 2KNO_3(aq)$
double replacement
 l. $4Cr(s) + 3SnCl_4(aq) \longrightarrow$
$4CrCi_3 + 3Sn(s)$
single replacement

Example 5-6

Write the equation for the replacement of aluminum by magnesium in aluminum chloride:

▶ Suggested solution _____

The reactants are magnesium and aluminum chloride:

$$Mg + AlCl_3 \longrightarrow$$

Table 5-2 shows that magnesium is a more active element than aluminum, so magnesium will replace the aluminum to form magnesium chloride and aluminum:

$$Mg + AlCl_3 \longrightarrow MgCl_2 + Al$$

The equation is then balanced:

$$3Mg + 2AlCl_3 \longrightarrow 3MgCl_2 + 2Al$$

Example 5-7

Complete and balance the equation for the formation of water from aqueous solutions of hydrogen chloride and strontium hydroxide.

▶ **Suggested solution**_____

First, write the correct formulas for the reactants, and indicate that they are dissolved in water:

$$HCl(aq) + Sr(OH)_2(aq) \longrightarrow$$

To form water, hydrogen replaces strontium to combine with the hydroxide. The table in Appendix D indicates that the strontium chloride remains dissolved in the water that was present in the reactants, and the water formed is in the liquid form:

$$HCl(aq) + Sr(OH)_2(aq) \longrightarrow SrCl_2(aq) + H_2O(l)$$

Finally, the equation is balanced by using a coefficient of 2 for HCl and H_2O:

$$2HCl(aq) + Sr(OH)_2(aq) \longrightarrow SrCl_2(aq) + 2H_2O(l)$$

Review and Practice_____

1. Balance the following equations and classify each of the reactions as synthesis, decomposition, combustion, single replacement, double replacement, or water-forming:
   a. $C_2H_2(g) + O_2(g) \longrightarrow CO_2(g) + H_2O(g)$
   b. $Zn(s) + CuSO_4(aq) \longrightarrow ZnSO_4(aq) + Cu(s)$
   c. $Cl_2(g) + KI(aq) \longrightarrow KCl(aq) + I_2(aq)$
   d. $H_2O_2(l) \longrightarrow H_2O(l) + O_2(g)$
   e. $MgCl_2(s) \longrightarrow Mg(s) + Cl_2(g)$
   f. $Fe(s) + I_2(s) \longrightarrow FeI_2(s)$
   g. $Cu(s) + S_8(l) \longrightarrow Cu_2S(s)$
   h. $C_6H_{12}O_6(aq) + O_2(g) \longrightarrow CO_2(aq) + H_2O(l)$
   i. $FeCl_2(aq) + K_2S(aq) \longrightarrow FeS(s) + KCl(aq)$
   j. $H_2SO_4(aq) + NaOH(aq) \longrightarrow Na_2SO_4(aq) + H_2O(l)$
   k. $Pb(NO_3)_2(aq) + K_2CrO_4(aq) \longrightarrow PbCrO_4(s) + KNO_3(aq)$
   l. $Cr(s) + SnCl_4(aq) \longrightarrow CrCl_3(aq) + Sn(s)$
   m. $C_2H_5OH(l) + O_2(g) \longrightarrow CO_2(g) + H_2O(g)$

2. Complete the equation given the reactants:
   a.! $Zn(s) + MgSO_4(aq) \longrightarrow$
   b.✓ $Cd(s) + O_2(g) \longrightarrow$
   c. $HgO(s) \longrightarrow$
   d. $HCl(aq) + KOH(aq) \longrightarrow$
   e.✓ $C_5H_{12}(l) + O_2(g) \longrightarrow$
   f.✓ $Sr(s) + O_2(g) \longrightarrow$
   g.✓ $Br_2(aq) + CaCl_2(aq) \longrightarrow$
   h.✓ $Zn(s) + Ni(NO_3)_2(aq) \longrightarrow$
   i. $ZnSO_4(aq) + SrCl_2(aq) \longrightarrow$
   j. $AlCl_3(aq) + Na_2CO_3(aq) \longrightarrow$
   k.✓ $Fe(s) + S_8(s) \longrightarrow$
   l. ✓ $C_6H_6(l) + O_2(g) \longrightarrow$
   m. $Pb(s) + KNO_3(aq) \longrightarrow$
   n. $HNO_3(aq) + Sr(OH)_2(aq) \longrightarrow$

3. Balance each equation in item 2.

4. Classify each equation in item 2 as synthesis, decomposition, combustion, single replacement, double replacement, or water-forming.

m. $C_2H_5OH(l) + 3O_2(g) \longrightarrow$
$$2CO_2(g) + 3H_2O(g)$$
combustion

2. a. $Zn(s) + MgSO_4(aq) \longrightarrow$ N.R.
   b. $Cd(s) + O_2(g) \longrightarrow CdO(s)$
   c. $HgO(s) \longrightarrow Hg(l) + O_2(g)$
   d. $HCl(aq) + KOH(aq) \longrightarrow$
      $$KCl(aq) + H_2O(l)$$
   e. $C_5H_{12}(l) + O_2(g) \longrightarrow$
      $$CO_2(g) + H_2O(g)$$
   f. $Sr(s) + O_2(g) \longrightarrow SrO(s)$
   g. $Br_2(aq) + CaCl_2(aq) \longrightarrow$ N.R.
   h. $Zn(s) + Ni(NO_3)_2(aq) \longrightarrow$
      $$Zn(NO_3)_2(aq) + Ni(s)$$
   i. $ZnSO_4(aq) + SrCl_2(aq) \longrightarrow$
      $$SrSO_4(s) + ZnCl_2(aq)$$
   j. $AlCl_3(aq) + NaCO_3(aq) \longrightarrow$
      $$Al_2(CO_3)_3(s) + NaCl(aq)$$
   k. $Fe(s) + S_8(s) \longrightarrow FeS(s)$
   l. $C_6H_6(l) + O_2(g) \longrightarrow$
      $$CO_2(g) + H_2O(g)$$
   m. $Na(s) + KNO_3(aq) \longrightarrow$ N.R.
   n. $HNO_3(aq) + Sr(OH)_2(aq) \longrightarrow$
      $$Sr(NO_3)_2(aq) + H_2O(l)$$

3. a. $Zn(s) + MgSO_4(aq) \longrightarrow$ N.R.
   b. $2Cd(s) + O_2(g) \longrightarrow 2CdO(s)$
   c. $2HgO(s) \longrightarrow 2Hg(l) + O_2(g)$
   d. $HCl(aq) + KOH(aq) \longrightarrow$
      $$KCl(aq) + H_2O(l)$$
   e. $C_5H_{12}(l) + 8O_2(g) \longrightarrow$
      $$5CO_2(g) + 6H_2O(g)$$
   f. $2Sr(s) + O_2(g) \longrightarrow 2SrO(s)$
   g. $Br_2(aq) + CaCl_2(aq) \longrightarrow$ N.R.
   h. $Zn(s) + Ni(NO_3)_2(aq) \longrightarrow$
      $$Zn(NO_3)_2(aq) + Ni(s)$$
   i. $ZnSO_4(aq) + SrCl_2(aq) \longrightarrow$
      $$SrSO_4(s) + ZnCl_2(aq)$$
   j. $2AlCl_3(aq) + 3Na_2CO_3(aq) \longrightarrow$
      $$Al_2(CO_3)_3(s) + 6NaCl(aq)$$
   k. $8Fe(s) + S_8(s) \longrightarrow 8FeS(s)$
   l. $2C_6H_6(l) + 15O_2(g) \longrightarrow$
      $$12CO_2(g) + 6H_2O(g)$$
   m. $Pb(s) + KNO_3(aq) \longrightarrow$ N.R.
   n. $2HNO_3(aq) + Sr(OH)_2(aq) \longrightarrow$
      $$Sr(NO_3)_2(aq) + 2H_2O(l)$$

4. a. single replacement
   b. synthesis, combustion
   c. decomposition
   d. water-forming
   e. combustion
   f. synthesis, combustion
   g. single replacement
   h. single replacement
   i. double replacement
   j. double replacement
   k. synthesis
   l. combustion
   m. single replacement
   n. water-forming

Energy Changes in Common Chemical Reactions

CONCEPTS
■ **exothermic reactions**
■ **endothermic reactions**
■ **common chemical reactions**

When a chemical reaction occurs, bonds between atoms are broken, and new and different bonds are formed. Since energy is required to break any chemical bond, no reaction will take place unless there is sufficient energy available to break some of the bonds holding atoms together in the reactants. For many reactions, the ordinary bumping of molecules at room temperature is sufficient. For other reactions, additional energy must be supplied in the form of heat, light, or electricity. Once the reaction begins, it may continue to require an energy input. The container in which an energy-requiring reaction is occurring may actually feel cold. For example, you may have been at a basketball game or a tennis match and observed a player place a cold pack on an injury. A cold pack contains chemicals that absorb heat energy when mixed.

Just as energy is required to break any chemical bond, energy is released when a bond forms. If the energy released when new bonds form is greater than the energy required to break the bonds in the reactants, there is a net release of energy during the reaction. The container holding an energy-releasing reaction may feel warm. If you go to an outdoor football game on a very cold day, you can keep warm using a "hot seat" or a hand warmer. Both release heat as a result of a chemical reaction. Other reactions release energy as light or, under the right conditions, as an electric current.

5-9 Exothermic Reactions

When a chemical reaction releases energy to the surroundings, the reaction is said to be **exothermic**. *Exo* is a prefix meaning "outside," and *therm* refers to heat. In Chapter 17, you will examine in detail the energy changes associated with chemical reactions. At this time, it is useful to recognize that energy changes occur because these changes are one of the observable signs that a chemical reaction is taking place. In some cases, such as in the combustion of magnesium or gasoline, the release of energy is obvious. The slow decomposition of hydrogen peroxide to form water and oxygen also is exothermic, but this fact might go unnoticed unless a thermometer is used to measure any temperature changes.

Even though temperature can be used to detect energy changes, you know that heat energy and temperature are not the same. Energy changes in matter are measured using the SI standard, the joule. A **joule** (J) is the amount of energy produced when a force of one newton acts over a distance of one meter. It is a small amount of energy. An ordinary match, burned completely, liberates about 1050 joules. A joule is also the amount of energy used by a 100-watt light bulb in 0.01

Figure 5-9 Products of an exothermic reaction are at a lower energy state than the reactants that lose some energy to the surroundings.

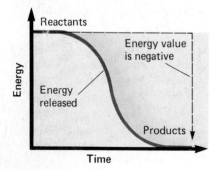

second. Throughout this course, energy values will be expressed in either joules or kilojoules (1000 J).

An exothermic reaction sometimes is written with energy as a product to show that the reaction produces energy. For example, the combustion of octane, a hydrocarbon, releases energy along with water and carbon dioxide:

$$2C_8H_{18}(l) + 25O_2(g) \longrightarrow 16CO_2(g) + 18H_2O(g) + energy$$

A certain amount of energy will be required to break the bonds in octane and oxygen. The energy released when the new bonds form is greater than that required to break the bonds. The net gain in energy is given off as heat.

Since the reaction releases energy, the products are at a lower energy than the reactants. Figure 5-9 shows the relative energy states of the reactants and products of the octane combustion. This reaction produces 5445.3 kilojoules per mole of octane. Exothermic reactions are assigned negative energy values because the energy produced by the reaction is lost to the surroundings.

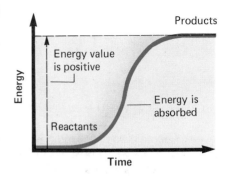

Figure 5-10 Energy is absorbed during an endothermic reaction so products are at a higher energy state.

5-10 Endothermic Reactions

When a chemical reaction requires more energy to break the bonds in the reactants than it gives off by forming the new bonds, the reaction is **endothermic**. The prefix *endo* means "inside" and is used with *therm* for heat. The energy is absorbed from the surroundings or from other chemicals, and a decrease in temperature is noted. As before, a thermometer is more sensitive in detecting changes than your sense of touch.

Equations for endothermic reactions often are written with energy as a reactant to indicate that energy is being absorbed. When barium hydroxide octahydrate and ammonium thiocyanate react, so much energy is absorbed that the temperature drops well below the freezing point of water. The equation for this reaction is

energy + $Ba(OH)_2 \cdot 8H_2O(s)$ + $2NH_4SCN(s) \longrightarrow$
$$Ba(SCN)_2(aq) + 2NH_4OH(aq) + 10H_2O(l)$$

As you might expect, endothermic reactions are assigned positive energy values. Note how the relative energies of the reactants and products for an endothermic reaction (Figure 5-10) differ from those of the exothermic reaction (Figure 5-9).

The cold pack used in first aid for athletes makes use of an endothermic reaction. The pack contains ammonium chloride and water separated by a barrier. When ammonium chloride is mixed with water, energy is absorbed from the water:

$$NH_4Cl(s) + H_2O + energy \longrightarrow NH_4Cl(aq)$$

When the inside bag (barrier) is broken, the chemical reaction absorbs enough energy to reduce the temperature of the cold pack. When the cold pack is placed on an injured area, swelling, pain, and tissue damage are minimized.

Figure 5-11 To activate this cold pack, the inner pouch that contains water is broken. The water then mixes with a chemical inside and energy is absorbed, lowering the temperature of the cold pack.

5-11 Everyday Chemical Reactions

Most of the chemical reactions discussed in this chapter occur only in a laboratory. They were used to explain the basic principles of chemical reactions because they represent fairly simple types of changes. The chemical reactions occurring around you daily, in your body, kitchen, garden, and car, are often quite complex.

The combustion of gasoline is an example of a complex chemical reaction that goes on around you every day. The process can be represented by the equation for the combustion of octane, C_8H_{18}, to produce carbon dioxide and water:

$$2C_8H_{18}(g) + 25O_2(g) \longrightarrow 16CO_2(g) + 18H_2O(g) + energy$$

Though the written equation is a fairly simple one, it actually describes a complicated set of chemical events. The reaction takes place only if there is enough oxygen to burn the octane that is present and if there is enough time for all of the molecules to come into contact. In a car going 60 kilometers per hour, this reaction has to take place in one thirtieth of one second. That may not be enough time for all of the molecules to react. Or, if the engine is not properly tuned, there may not be enough oxygen available, and some of the carbon atoms will combine with only a single oxygen atom to form carbon monoxide. If the engine is greatly out of tune, some carbon atoms will not combine with oxygen at all and may accumulate inside the engine.

Gasoline is not pure octane and usually contains many other chemicals such as heptane, C_7H_{16}, benzene, C_6H_6, and lead additives, $(C_2H_5)_4Pb$. Their presence affects the principal reaction of octane and oxygen though the results are not shown in the written equation.

The combustion of gasoline becomes even more complex because the oxygen in these reactions comes from the air. Air is about 20 percent oxygen and 80 percent nitrogen. At the high temperatures caused by the exothermic reactions, some of the nitrogen molecules also react.

Automobiles provide a second example of a common but complex chemical reaction. Rusting is a series of reactions involving iron, water, and oxygen. The steel that is used to make cars is mostly iron, and the reaction that turns the iron to rust primarily is a combination of iron with oxygen gas to produce iron oxide. Iron rusts much faster when water is present. The first step in the reaction is

$$2Fe(s) + 2H_2O(l) + O_2(aq) \longrightarrow 2Fe(OH)_2(s)$$

After the iron(II) hydroxide forms, it reacts further with water and oxygen to produce iron(III) hydroxide:

$$4Fe(OH)_2(s) + 2H_2O(l) + O_2(g) \longrightarrow 4Fe(OH)_3(s)$$

When iron(III) hydroxide is warmed by the sun, it decomposes to form iron(III) oxide and water:

$$2Fe(OH)_3(s) \longrightarrow Fe_2O_3(s) + 3H_2O(g)$$

Both iron(III) hydroxide and iron(III) oxide have the red-brown color that is characteristic of rust. What you see when a car rusts is probably a mixture of both of these substances.

Do You Know?

Exothermic reactions power rockets. The reaction of hydrazine, N_2H_4, and hydrogen peroxide, H_2O_2, is highly exothermic. The reactants are liquids, and their flow rates can be controlled easily. The reaction products, N_2 and H_2O, are gases. Gaseous products provide the maximum thrust for the rocket because the molecules are moving at very high speeds.

Figure 5-12 The iron in this car has reacted with oxygen and water and has been converted to iron(III) oxide, commonly called rust.

As you can see, the rusting of a car is actually a complex process. The equations written above give some information about it but do not indicate the amount of energy that is produced or required, which bonds are broken or formed, or other conditions necessary for the reaction to occur. Understanding some basic principles of chemistry will help make sense of the chemical reactions happening around you every day.

Review and Practice

1. Classify these reactions as exothermic or endothermic:
   a. $Cl_2(aq) + KI(aq) \longrightarrow KCl(aq) + I_2(aq) + energy$
   b. $Mg(s) + O_2(g) \longrightarrow MgO(s) + energy$
   c. $energy + NaClO_3(s) \longrightarrow NaCl(s) + O_2(g)$
   d. $C_8H_{18}(l) + O_2(g) \longrightarrow CO_2(g) + H_2O(g) + energy$

2. Balance each equation in item 1.

3. Classify each equation in item 1 as synthesis, decomposition, combustion, single replacement, double replacement, or water-forming.

4. Write and balance the equation for the decomposition of hydrogen peroxide to form water and oxygen. This reaction occurs when hydrogen peroxide is used as a disinfectant on a cut.

5. Write and balance the equation for the combination of a magnesium tennis racquet with oxygen to form a layer of magnesium oxide on the surface.

6. Write and balance the equation for the burning of propane, C_3H_8, to form carbon dioxide and water.

7. Write the words represented by each symbol in this equation:

$$energy + 2Sb(s) + 3I_2(s) \longrightarrow 2SbI_3(s)$$

8. Is the equation in item 7 exothermic or endothermic?

1. a. exothermic
   b. exothermic
   c. endothermic
   d. exothermic
2. a. $Cl_2 + 2KI \longrightarrow$
   $\qquad\qquad 2KCl + I_2 + energy$
   b. $2Mg + O_2 \longrightarrow 2MgO + energy$
   c. $energy + 2NaClO_3 \longrightarrow$
   $\qquad\qquad 2NaCl + 3O_2$
   d. $2C_8H_{18} + 25O_2 \longrightarrow$
   $\qquad\qquad 16CO_2 + 18H_2O + energy$
3. a. single replacement
   b. synthesis, combustion
   c. decomposition
   d. combustion
4. $2H_2O_2 \longrightarrow 2H_2O + O_2$
5. $2Mg + O_2 \longrightarrow 2MgO$
6. $C_3H_8 + 5O_2 \longrightarrow$
   $\qquad\qquad 3CO_2 + 4H_2O + energy$
7. energy plus 2 moles of solid antimony plus 3 moles of solid iodine yields 2 moles of solid ~~antimony(III) iodide~~
8. endothermic

Careers in Chemistry

Shericca Williams
Title: Biochemist
Job Description: Performs experiments with recombinant DNA technology (genetic engineering), using such methods as restriction enzyme digest, gel electrophoresis, and DNA sequencing. Previous experience involved using laboratory animals to study how the skin absorbs pesticides.
Educational Qualifications: College degrees in chemistry and biology; courses in advanced genetics and microbiology recommended.
Future Employment Outlook: Excellent

Shericca Williams became interested in science when she was in high school, but it was not until she started to work for a chemical research firm that her interest in chemistry surfaced. Williams returned to college to earn a second degree, this time in chemistry. Currently she is involved in studies dealing with genetic engineering in plants. The results of these experiments will be important as better quality crops that can be grown more efficiently are needed throughout the world.

Williams finds her job as a biochemist rewarding. "Keeping up with basic research and working in a lab, in contrast to working at a desk, just adds interest. There is a lot of team effort. For instance, when an experiment works it can be passed on to other departments. Our plant genetics research will directly help the farmer." Working conditions for scientists in research firms are excellent. Williams points out that her firm paid for 75 percent of the costs for her chemistry education. Being a scientist in a research firm is not necessarily a 9 to 5 job, however. "Sometimes I work quite late and occasionally carry on with an experiment on the weekends, but we have liberal vacations and holidays," says Williams.

The future for a biochemist is excellent. Because of important new areas, such as genetic engineering and environmental needs, there will be a great demand for persons with this training. Williams concludes: "If you have a sincere interest in science, a sense of thoroughness, patience, and a willingness to study to gain knowledge and improve your capabilities, then you should investigate this challenging field of chemistry."

Numbers in red indicate the appropriate chapter sections to aid you in assigning these items. Answers to all questions appear in the Teacher's Guide at the front of this book.

Summary

■ Chemical equations are written using the symbols and formulas for elements and compounds. These symbols represent the substances present and summarize the microscopic events that explain the macroscopic changes you observe.

■ The law of conservation of mass states that the total mass of reactants equals the total mass of products when a chemical reaction takes place. This law is expressed by using coefficients to show equal numbers of atoms of each element before and after the reaction takes place.

■ The physical state of the reactants and products is indicated by a symbol in parentheses after each formula. The common symbols are (s) for solid, (l) for liquid, (g) for gas, and (aq) to mean dissolved in water.

■ Similar reactions can be classified using a generalized equation. Five representative types of reactions were mentioned and their general equations are:

Synthesis	$A + B \longrightarrow AB$
Decomposition	$AB \longrightarrow A + B$
Combustion	$C_xH_y + O_2 \longrightarrow CO_2 + H_2O$
Single replacement	$A + BC \longrightarrow AC + B$
Double replacement	$AB + XY \longrightarrow AY + XB$
Water-forming	$HB + XOH \longrightarrow XB + H_2O$

Recognizing the type of reaction will help you to predict the products that will occur if you are given the reactants.

■ An equation may be written for a replacement reaction that does not take place when the reactants are mixed. An element will only be replaced by a more active element. If the intended replacing element is less active, no reaction will occur, and this is represented by *N.R.*

■ When a chemical reaction takes place, chemical bonds are broken and other bonds are formed. When this change results in a net release of energy, the reaction is exothermic, and this is indicated by writing energy as a product. When more energy must be added than is given off, the reaction is endothermic, and this is indicated by writing energy as a reactant.

■ Chemical changes occur constantly in everyday life. Most of these changes do not take place with pure substances or in one-step reactions. The written equations for these reactions are simple representations of much more complex processes.

Chemically Speaking

activity
coefficient
combustion
decomposition
double replacement
endothermic
exothermic
hydrocarbon

law of conservation of
  mass
products
reactants
single replacement
synthesis
water-forming

Review

1. What are the substances on the right side of the arrow called? (5-4)

2. What are the substances on the left side of the arrow called? (5-4)

3. For an equation to be balanced, what must be the same on both sides? (5-5)

4. Given the symbols: $2Ga_2O_3(s)$ (5-6)
   a. What is the subscript of oxygen?
   b. What is the ratio of gallium atoms to oxygen atoms?
   c. How many atoms of gallium are represented?
   d. What is the physical state of the compound?

5. When hydrogen reacts with oxygen, energy is given off. Is this reaction endothermic or exothermic? (5-10)

6. Name six types of reactions. (5-7)

7. Which symbol is read *yields* or *produces?* (5-6)

8. What are the two formulas for rust? (5-11)

9. Give the name for each formula for rust. (5-11)

10. Give two macroscopic observations of the reaction in Figure 5-8. (5-7)

Interpret and Apply

1. Given the equation:

$$4Al(s) + 3O_2(g) \longrightarrow 2Al_2O_3(s)$$

   a. Name the reactants and products. (5-4)
   b. Is the equation balanced? (5-5)
   c. What is the physical state of Al_2O_3? (5-6)

2. Is the reaction in Figure 5-8 endothermic or exothermic? (5-9)

3. Classify these reactions as exothermic or endothermic: (5-9, 5-10)
a. energy + $SO_2(g) \longrightarrow S(g) + O_2(g)$
b. $2C_8H_{18}(g) + 2O_2(g) \longrightarrow 16CO_2(g) + 18H_2O(g)$ + energy
c. energy + $P_4O_{10}(s) \longrightarrow P_4(s) + 5O_2(g)$
d. $Mg(s) + H_2SO_4(aq) \longrightarrow$
$MgSO_4(aq) + H_2(g)$ + energy

4. Which of the equations in item 3 are balanced? (5-5)

5. Give the type of each reaction in item 3. (5-7)

6. Which reaction in item 3 occurs in our everyday experience? (5-7)

7. Classify each of these reactions as synthesis, decomposition, combustion, single replacement, double replacement, or water-forming: (5-7)
a. $Ba(s) + O_2(g) \longrightarrow BaO_2(s)$ + energy
b. $PCl_3(s) + Cl_2(g) \longrightarrow PCl_5(s)$ + energy
c. $2Sb(s) + 3I_2(g)$ + energy $\longrightarrow 2SbI_3(s)$
d. $C_3H_8(g) + 5O_2(g) \longrightarrow 3CO_2(g) + 4H_2O(g)$
e. $H_3PO_4(aq) + 3LiOH(aq) \longrightarrow$
$Li_3PO_4(aq) + 3H_2O(l)$ + energy
f. $Fe(s) + CuSO_4(aq) \longrightarrow$
$FeSO_4(aq) + Cu(s)$ + energy
g. $CS_2(g) + 3O_2(g) \longrightarrow CO_2(g) + 2SO_2(g)$ + energy
h. $NH_3(g) + HCl(g) \longrightarrow NH_4Cl(s)$ + energy
i. $CaCO_3(s)$ + energy $\longrightarrow CaO(s) + CO_2(g)$
j. $3Mg(s) + 2CrCl_3(aq) \longrightarrow 3MgCl_2(aq) + 2Cr(s)$ + energy
k. $KNO_3(s)$ + energy $\longrightarrow 2KNO_2(s) + O_2(g)$
l. $Pb(NO_3)_2(aq) + Na_2SO_4(aq) \longrightarrow$
$PbSO_4(s) + 2NaNO_3(aq)$ + energy
m. $HNO_3(aq) + LiOH(aq) \longrightarrow$
$LiNO_3(aq) + H_2O(l)$ + energy
n. $KBr(aq) + AgNO_3(aq) \longrightarrow$
$AgBr(s) + KNO_3(aq)$ + energy

8. Which reactions in item 7 are endothermic? (5-9)

9. Which of the equations in item 7 are not balanced? (5-5)

10. Which of the reactions in item 7 take place in a solution? (5-6)

11. In which reactions in item 7 are all the reactants and products in the gaseous state? (5-6)

Problems

1. Balance these equations: (5-5)

a. $Al(s) + Pb(NO_3)_2(aq) \longrightarrow Al(NO_3)_3(aq) + Pb(s)$
b. $C_3H_8(g) + 5O_2(g) \longrightarrow 3CO_2(g) + 4H_2O(g)$
c. $Fe(s) + H_2O(g) \longrightarrow Fe_3O_4(s) + H_2(g)$
d. $Al(s) + NaOH(s) + H_2O(l) \longrightarrow$
$NaAl(OH)_4(aq) + H_2(g)$
e. $Ni(s) + I_2(s) \longrightarrow NiI(s)$
f. $C(s) + H_2O(g) \longrightarrow CO(g) + H_2(g)$
g. $AlBr_3(aq) + Cl_2(g) \longrightarrow AlCl_3(g) + Br_2(l)$
h. $HNO_3(aq) + Ba(OH)_2(aq) \longrightarrow$
$Ba(NO_3)_2(aq) + H_2O(l)$

2. Write and balance these equations: (5-4, 5-5)
a. Solid ammonium nitrite when heated produces nitrogen gas and water.
b. Hydrogen reacts with nitrogen gas to produce ammonia.

3. Complete and balance these equations: (5-5, 5-8)
a. $C_6H_6(l) + O_2(g) \longrightarrow$
b. $Zn(s) + H_3PO_4(aq) \longrightarrow$
c. $Pb(ClO_3)_2(aq) + KI(aq) \longrightarrow$
d. $BaCO_3(s) \longrightarrow$
e. $Br_2(aq) + FeI_3(aq) \longrightarrow$
f. $BaCl_2(aq) + H_2SO_4(aq) \longrightarrow$
g. $C_4H_{10}(g) + O_2(g) \longrightarrow$
h. $Ca(s) + O_2(g) \longrightarrow$
i. $HgO(s) \longrightarrow$
j. $Li_2SO_4(aq) + BaCl_2(aq) \longrightarrow$
k. $F_2(g) + KCl(aq) \longrightarrow$
l. $HC_2H_3O_2(aq) + LiOH(aq) \longrightarrow$

4. Complete each equation: (5-4)
a. $Ni(s) + FeSO_4(aq) \longrightarrow$
b. $Sr(s) + N_2(g) \longrightarrow$
c. $C_4H_8(l) + O_2(g) \longrightarrow$
d. $CoBr_2(s) \longrightarrow$
e. $H_3PO_4(aq) + Al(OH)_3(aq) \longrightarrow$
f. $CH_3OH(l) + O_2(g) \longrightarrow$
g. $Br_2(aq) + CuI_2(aq) \longrightarrow$
h. $Pb(NO_3)_2(aq) + NaCl(aq) \longrightarrow$
i. $Zn(s) + CuCl_2(aq) \longrightarrow$
j. $AlCl_3(aq) + Pb(NO_3)_2(aq) \longrightarrow$

5. Balance each equation in item 4. (5-5)

6. Classify each reaction in item 4 as synthesis, decomposition, combustion, single replacement, double replacement, or water-forming. (5-7)

7. Change these word equations to symbols: (5-4)
a. Barium chlorate, when heated, produces barium chloride and oxygen.
b. Methane combines with oxygen to produce carbon dioxide and water.

c. Chlorine replaces iodine in calcium iodide.

d. Chromium reacts with oxygen to produce chromium(VI) oxide.

e. Aqueous solutions of hydrogen chloride and barium hydroxide react to form barium chloride and water.

8. Balance each equation in item 7. (5-5)

9. Classify each reaction in item 7 as synthesis, decomposition, combustion, single replacement, double replacement, or water-forming. (5-7)

10. Change these symbols to word equations:
   a. $MgCl_2(s) \longrightarrow Mg(s) + Cl_2(g)$
   b. $Pb(NO_3)_2(aq) + K_2CrO_4(aq) \longrightarrow PbCrO_4(s) + 2KNO_3(aq)$

Challenge

1. Complete and balance these equations:
   a. $Zn(s) + HCl(aq) \longrightarrow$
   b. $Zn(s) + H_2SO_4(aq) \longrightarrow$
   c. $Zn(s) + H_3PO_4(aq) \longrightarrow$
   d. $Mg(s) + HC_2H_3O_2(aq) \longrightarrow$

2. What type of reaction is represented by the equations in item 1?

3. Write two more balanced equations for reactions of the same type as those in item 1 but using different substances.

4. Balance these equations:
   a. $Na_2CO_3(s) + HCl(aq) \longrightarrow$
   $NaCl(aq) + H_2O(l) + CO_2(g)$
   b. $PbCO_3(s) + HNO_3(aq) \longrightarrow$
   $Pb(NO_3)_2(aq) + H_2O(l) + CO_2(g)$
   c. $CaCO_3(s) + HC_2H_3O_2(aq) \longrightarrow$
   $Ca(C_2H_3O_2)_2(aq) + H_2O(l) + CO_2(g)$
   d. $K_2CO_3(s) + H_3PO_4(aq) \longrightarrow$
   $K_3PO_4(aq) + H_2O(l) + CO_2(g)$

5. Complete and balance these equations:
   a. ammonium chloride + lead(II) nitrate $\longrightarrow$
   b. zinc + iodine $\longrightarrow$
   c. aluminum + nickel(II) sulfate $\longrightarrow$
   d. lead(II) carbonate $\longrightarrow$
   e. silver nitrate + sodium carbonate $\longrightarrow$

Synthesis

1. All four equations in item 4 of the Challenge section are of the same type. Write and balance another equation of this type using different substances.

2. Write a general equation that represents all the equations in item 4 of the Challenge section.

3. Note the reaction:

$$S(g) + O_2(g) \longrightarrow SO_2(g)$$

It can be classified as either a synthesis or a combustion reaction. Choose one and give your reasons.

Projects

1. Prepare $1.00M$ solutions of HCl, H_2SO_4, H_3PO_4, and $HC_2H_3O_2$. React a small amount, about the size of half a pea, of nickel, zinc, magnesium, aluminum, and iron with about 15 mL of each acid. Write equations for all of the reactions. Make a table, and fill in your observations of each reaction. Compare the results, including how the reactions are similar and how they are different.

2. Put a small scoop of sodium carbonate, about the size of half a pea, in four test tubes. Add 15 mL of $1.00M$ HCl to one test tube. Add 15 mL of $1.0M$ H_2SO_4 to another test tube, 15 mL of $1.0M$ H_3PO_4 to the third, and 15 mL of $1.0M$ $HC_2H_3O_2$ to the last test tube of sodium carbonate. Repeat the process using lead(II) carbonate, calcium carbonate, and copper(II) carbonate. Write equations for each reaction. Design a data table that will clearly show your observations of each reaction. Compare the results, and make inferences about how the reactions are alike and how they are different.

CHAPTER 6

"**W**hat are you doing in the kitchen, son?"

"Oh, just practicing a little chemistry, Mom."

Chemistry in the kitchen? That thought might bring to mind cartoon images of foaming blue liquids overflowing their containers and explosions that send up clouds of thick smoke. However, most chemical reactions that go on in the kitchen do not produce such dramatic results. Cooking is similar to chemistry. It involves mixing the ingredients (reactants) given in a recipe (equation) to make a certain kind of food (product). The recipe is necessary to know how much of each ingredient to use.

In this chapter, you will learn how to use balanced equations to calculate the amounts of reactants needed to produce a specific amount of product. These calculations might be easier to do if you keep in mind that cooking and chemistry are very similar.

Calculations Involving Reactions

Quantitative Meaning of Equations

Remember the first time you tried to make something to eat using a recipe? You probably were a bit uncertain of what you were doing. You had to locate the right ingredients, figure out how to measure them, and then decide what to do with them. The next cooking attempt, though, probably was easier as you learned where things were and what some of the procedures were like. Perhaps you then were able to try a recipe that was more difficult.

Studying chemistry does get easier with practice. Each new idea builds on the ideas that you have already learned. You must remember an idea after you have studied it so you can understand what is coming up next. This chapter illustrates how related everything is in chemistry.

Do you remember how to write formulas? Do you remember how to write equations and balance them? Do you remember how to find the molar mass of a compound? You will need these skills in this chapter.

Do you remember all you learned about moles? Probably not. Then you will need to go back and review!

In this chapter, you will see how all of these skills are used to answer questions such as "How much oxygen does it take to burn a liter of gasoline?" and "How much rust can be made from a kilogram of iron?"

CONCEPTS

■ coefficients in a balanced equation represent the mole ratios
■ the mass ratio can be derived from the mole ratio
■ mole ratios in balanced equations are proportional in nature
■ stoichiometry is measuring amounts of chemicals involved in a reaction

6-1 The Cook as Chemist

Baking brownies is applied chemistry. The recipe is like a balanced equation that tells what the proper proportions are for combining the reactants, or ingredients. The recipe below gives the specific amount of each ingredient that is necessary to produce 24 brownies.

Brownies

$\frac{1}{2}$ cup butter
2 squares unsweetened chocolate (2 oz)
1 cup sugar
2 eggs
1 teaspoon vanilla

$\frac{2}{3}$ cup flour

$\frac{1}{2}$ teaspoon baking powder

$\frac{1}{4}$ teaspoon salt

Bake at 350°F for 30 minutes. Makes 24 brownies.

You could write this recipe to make it look more like a chemical equation.

$$\left.\begin{array}{l} \frac{1}{2} \text{ cup butter} + 2 \text{ squares chocolate} + \\ 1 \text{ cup sugar} + 2 \text{ eggs} + 1 \text{ teaspoon vanilla} + \\ \frac{2}{3} \text{ cup flour} + \frac{1}{2} \text{ teaspoon baking powder} + \\ \frac{1}{4} \text{ teaspoon salt} \end{array}\right\} \longrightarrow 24 \text{ brownies}$$

This recipe shows the ratios among the various ingredients used to make a specific number of brownies. Like a chemical equation, the information in the recipe is proportional. You use twice as much of each ingredient if you want 48 brownies. You can cut the recipe by $\frac{1}{2}$ if you only want 12 brownies. Cutting a recipe by a fractional amount may present some problems. You have to know how to take $\frac{1}{2}$ of each amount given. What is $\frac{1}{2}$ of $\frac{2}{3}$ cup flour or $\frac{1}{2}$ of $\frac{1}{4}$ teaspoon of salt? You may be faced with this very problem if you decide to try this recipe and there is only one egg in the house!

Recipes are based on observations and trial and error that have occurred over a long period of time. You can write any recipe you like, but it does not mean that the desired food will occur. If the proportions used to make cookies are not correct, the product can be more like rocks than cookies!

The recipe above does not provide all of the information you need. Should you mix all the ingredients at once? How thorough should the mixing process be? How is the dough divided into 24 portions?

Cooking is very similar to chemistry. Creativity is desirable but attention to detail also is important. Neither activity allows exact measurements, but precision and accuracy are necessary. Both use experimental results to develop recipes that can be used to predict the products of reactions.

6-2 Proportional Relationships

The recipe for brownies in Section 6-1 provides the specific amount of each ingredient needed to make a certain number of brownies. The recipe also provides information about the ratios among the ingredients so that you could double the recipe or cut it by half. An equation is like a recipe in that it also shows the ratios among the reactants and products as you first saw in Section 5-6.

The equation for the combustion of magnesium in oxygen to form magnesium oxide can be represented as shown in Figure 6-1.

MgO is an ionic solid, therefore the units should be formula units instead of molecules. To promote understanding and decrease confusion for the student, that distinction is not made here.

$$2 \text{ atoms Mg} + 1 \text{ molecule O}_2 \longrightarrow 2 \text{ molecules MgO}$$

Suppose you double the amounts of magnesium and oxygen used. You should expect to get twice as much magnesium oxide because mass is conserved.

$$4 \text{ atoms Mg} + 2 \text{ molecules O}_2 \longrightarrow 4 \text{ molecules MgO}$$

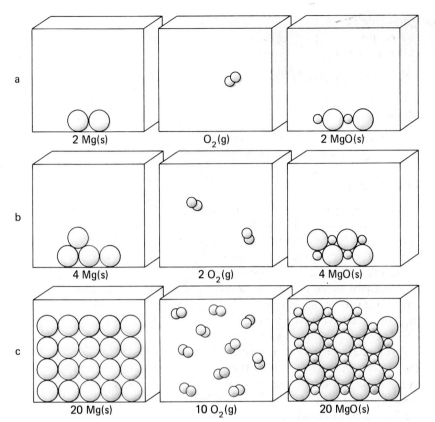

Figure 6-1 (a) Two atoms of Mg(s) combine with one molecule of O_2(g) to produce two molecules of MgO(s). (b) Four atoms of Mg(s) combine with two molecules of O_2(g) to produce four molecules of MgO(s). (c) Twenty atoms of Mg combine with ten molecules of O_2(g) to produce twenty molecules of MgO(s).

If you used ten times the original amount,

$$20 \text{ atoms Mg} + 10 \text{ molecules } O_2 \longrightarrow 20 \text{ molecules MgO}$$

You can continue to increase the amounts of magnesium and oxygen as long as the proportions are kept the same.

$$2\,000 \text{ atoms Mg} + 1\,000 \text{ molecules } O_2 \longrightarrow 2\,000 \text{ molecules MgO}$$
$$2 \times 10^{12} \text{ atoms Mg} + 1 \times 10^{12} \text{ molecules } O_2 \longrightarrow$$
$$2 \times 10^{12} \text{ molecules MgO}$$
$$2(6 \times 10^{23}) \text{ atoms Mg} + 1(6 \times 10^{23}) \text{ molecules } O_2 \longrightarrow$$
$$2(6 \times 10^{23}) \text{ molecules MgO}$$
$$2 \text{ (moles) atoms Mg} + 1 \text{ (mole) molecules } O_2 \longrightarrow$$
$$2 \text{ (moles) molecules MgO}$$

The balanced equation describes the ratio in which the substances combine. The ratio is obtained from the coefficients of the balanced equation. That combining ratio is the **mole ratio** of the combining substances. This mole ratio can be used to predict what would happen if the amount of one substance is changed.

Example 6-1

How many moles of oxygen are required to react with magnesium to produce 1 mole of magnesium oxide?

Do You Know?

When magnesium rapidly combines with oxygen, a great deal of energy is produced in the form of light. Camera flashbulbs contain very fine magnesium wires that are surrounded by oxygen. They produce a flash of light when sparked by the camera battery. To prevent damage to the eyes, harmful ultraviolet rays are filtered out by a coating on the bulb.

▶ Suggested solution _____

This question asks about a relationship between the reactants and product in a chemical reaction. What kind of statement describes such a relationship? It is the equation for the reaction. The first thing you need in order to answer questions like this is a balanced equation. Write and balance the equation for the reaction between magnesium and oxygen to form magnesium oxide.

$$2Mg + O_2 \longrightarrow 2MgO$$

Now translate the equation. Here are two translations:

1. Two *atoms* of magnesium combine with one *molecule* of oxygen to give two *molecules* of magnesium oxide.
2. Two *moles* of magnesium combine with one *mole* of oxygen (molecules) to give two *moles* of magnesium oxide.

The translation in terms of moles is more useful when practical calculations are to be made. The equation indicates that in this reaction, 1 mole of oxygen produces 2 moles of magnesium oxide. Then how many moles of oxygen are required to produce just 1 mole of magnesium oxide?

$$0.5 \text{ mol } O_2$$

You probably got the correct answer without doing any written calculations. Congratulations! You may not even be aware of the mathematics needed to get the answer. However, when the numbers are not so easy, you need to know what to do mathematically. The key to the mathematical solution is the relationship between moles of oxygen and moles of magnesium oxide given by the coefficients in the balanced equation. What is the relationship?

One mole of O_2 for every *two moles* of MgO

This represents a proportional relationship and can be used to construct unit factors. What are the two reciprocal unit factors that show this relationship?

$$\frac{1 \text{ mol } O_2}{2 \text{ mol MgO}} \quad \text{and} \quad \frac{2 \text{ mol MgO}}{1 \text{ mol } O_2}$$

When each of these unit factors is translated, the exact wording will vary, but the idea is expressed as "one mole of oxygen for every two moles of magnesium oxide" and "two moles of magnesium oxide for every one mole of oxygen."

Use the appropriate unit factor to finish setting up the following problem.

$$1 \text{ mol MgO} \times \underline{\hspace{2cm}} = ? \text{ mol } O_2$$

$$1 \text{ mol MgO} \times \frac{1 \text{ mol } O_2}{2 \text{ mol MgO}} = 0.5 \text{ mol } O_2$$

The balanced equation for the combustion of magnesium tells you that 2 moles of magnesium react with 1 mole of oxygen. This does not mean that 2 moles of magnesium and 1 mole of oxygen must be used to get a reaction. Smaller or larger amounts of reactants can be used to produce smaller or larger amounts of the product. Only the *ratio* of moles of magnesium to oxygen must always be 2 to 1.

6-3 From Grams to Moles and Back to Atoms

The coefficients in a balanced equation give the ratio in which the moles of each substance combine. They also can be used to predict the mass of one compound needed to react with another. The combining masses in grams can be calculated from the mole ratios by using unit factors. For example, in the combustion of magnesium to form magnesium oxide, the mole ratios are

$$2 \text{ mol Mg} + 1 \text{ mol O}_2 \longrightarrow 2 \text{ mol MgO}$$

Molar masses expressed as unit factors can be used to convert moles to grams of each substance, a process you learned in Chapter 4.

$$2 \text{ mol Mg} \times \frac{24.3 \text{ g Mg}}{1 \text{ mol Mg}} = 48.6 \text{ g Mg}$$

$$1 \text{ mol O}_2 \times \frac{32.0 \text{ g O}_2}{1 \text{ mol O}_2} = 32.0 \text{ g O}_2$$

$$2 \text{ mol MgO} \times \frac{40.3 \text{ g MgO}}{1 \text{ mol MgO}} = 80.6 \text{ g MgO}$$

The mass of each substance determined from the mole ratio can be used to rewrite the equation for the synthesis of magnesium oxide. When 48.6 grams of magnesium reacts with 32.0 grams of oxygen gas, 80.6 grams of magnesium oxide is produced.

$$48.6 \text{ g Mg} + 32.0 \text{ g O}_2 \longrightarrow 80.6 \text{ g MgO}$$

In the first equation above, the ratio of *moles* of magnesium to *moles* of oxygen is 2 to 1. However, the ratio of *grams* of magnesium to *grams* of oxygen is not 2 to 1. The reason is that magnesium and oxygen atoms are not equal in mass. This is similar to having a cube of wood and a cube of lead 2 centimeters on a side. The wood cube has less mass than the lead even though they are the same size. A mole of magnesium atoms has a mass of 24.3 grams, and a mole of oxygen atoms has a mass of 16.0 grams. The mass ratio of magnesium to oxygen is 1.52 to 1. In the problem above, the gram ratio of magnesium (48.6 g) to oxygen (32.0 g) also is 1.52 to 1.

6-4 Stoichiometry

The study of chemical calculations such as those you have been doing is called stoichiometry. The word is formed from a Greek word, *stoicheion*, meaning "element," and the *metry* suffix, which means "to measure." **Stoichiometry** involves measuring or calculating the amounts of elements or compounds involved in a chemical change.

A chemical equation is written in terms of *moles* of reactants and products. Predicting the moles of each substance can be done by using the mole ratio given in the balanced equation to construct unit factors as in Example 6-1. However, in the lab, substances are not measured in moles but in units of mass, like grams. You will need a way to translate information about proportions, given in moles, to the actual masses necessary for a reaction to occur.

The calculations you have done so far have been relatively simple. More complex problems await you, but they involve no new skills. They may, however, require several steps before you find the answer you want. Furthermore, a problem may not give all the information that you need. In order to solve the problem, you must realize what necessary information is missing and either recall it from memory or look it up.

The calculations in stoichiometry problems are not difficult once you recognize what you need to know in order to do the calculation and then recall the relationships that will get you from what is given to what you want to find. Patient analysis and a little trial and error usually provide some relationship that will yield the answer.

Test the students' skill in recognizing and balancing equations. Write two equations of the same type and ask each student to write a balanced equation like both but different from either. Look for and help the students correct errors in formula writing.

Review and Practice

1. Write the equation for the combination of sodium and oxygen to form sodium oxide, Na_2O.
   a. Balance the equation.
   b. Write the balanced equation in terms of molecules.
   c. Draw diagrams like those in Figure 6-1 to represent the combination of sodium atoms with oxygen atoms.
   d. Write the balanced equation in terms of moles.
   e. How many moles of oxygen molecules will be required to completely react with 12 moles of sodium?
   f. How many moles of sodium oxide will be produced by the reaction in (e)?

1. a. $4Na(s) + O_2(g) \longrightarrow$
$2Na_2O(s) + energy$
   b. 4 molecules Na + 1 molecule
$O_2 \longrightarrow 2$ molecules Na_2O
   c.

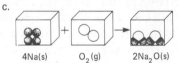

4Na(s) O_2(g) $2Na_2O$(s)

   d. 4 mol Na + 1 mol $O_2 \longrightarrow$
2 mol Na_2O
   e. 3 mol O_2
   f. 6 mol Na_2O
2. a. $2Na(s) + F_2(g) \longrightarrow 2NaF(s)$
   b. 2 atoms Na + 1 molecule $F_2 \longrightarrow$
2 molecules NaF
   c. 2 mol Na + 1 mol $F_2 \longrightarrow$
2 mol NaF
3. 2.50 mol Na_2O
4. 2.12 mol O_2
5. 1.20 mol $KClO_3$
6. 2.14 mol O_2
1.22 mol NO_2
1.83 mol H_2O

2. A mass of 46.0 grams of sodium reacts with 38.0 grams of fluorine to produce 84.0 grams of sodium fluoride, NaF.
   a. Write a balanced equation for this reaction.
   b. Express this equation in terms of atoms and molecules.
   c. Calculate the number of moles of each substance.

3. Calculate the number of moles of sodium oxide, Na_2O, that will be produced when 5.00 moles of sodium completely react with oxygen.

4. Calculate the number of moles of oxygen that will be required to completely burn 1.06 moles of methane, CH_4, to form carbon dioxide and water.

5. Calculate the number of moles of potassium chlorate, $KClO_3$, that must decompose to produce potassium chloride, KCl, and 1.80 moles of oxygen.

6. Calculate the number of moles of oxygen needed to burn 1.22 moles of ammonia, NH_3, and find the number of moles of nitrogen dioxide, NO_2, and water produced.

Stoichiometry and Predicting Yields

You are going to a party! There will be 12 people coming, and at the last minute, you have been asked to help out by making a vegetable dip. At home that afternoon, the only recipe you can find is the one below.

Vegetable Dip

sour cream
2 times as much mayonnaise as sour cream
$\frac{1}{4}$ as much fresh parsley as mayonnaise
$\frac{1}{4}$ as much onion as parsley
$\frac{1}{12}$ as much salt as onion

Could you translate this recipe into something that is edible?

What makes this recipe different from the one in Section 6-1? You should see that it only provides information about the proportions of ingredients to use and does not include the specific amounts of each one. A recipe that gives just proportions is not as easy to use as one that gives the actual measurements to use.

Like the recipe above, a balanced chemical equation also gives information on the proportions of reactants and products. It does not specify the amount of each substance that you would use if you were trying to perform the reaction in the lab. The next sections will show you how to translate the mole ratios of equations into more useful "recipes."

CONCEPTS

■ mole ratios can be used to predict the masses of reactants and products in a reaction
■ molarity can be used to predict the amounts of reactants and products involved in replacement reactions

6-5 Mole-Mass Calculations

"The mole is at the heart of chemistry."

When you were introduced to that idea in Chapter 4, it accompanied a diagram showing moles at the center of many mathematical relationships. For example, the diagram illustrates that moles can be used to convert from the number of molecules of a compound to the mass of a compound.

This section and Section 6-6 should convince you that moles also are at the heart of stoichiometry.

CHEM THEME

Calculations involving equations for reactions answer the question, "How much?" They use most of the concepts discussed earlier including unit factors, formula and equation writing and balancing, moles, atomic masses, and ratios and proportions.

Example 6-2

How many moles of sodium metal would be needed to react with chlorine gas and make 737 grams of sodium chloride?

Do You Know?

When table salt is prepared by evaporation of seawater, the salt contains magnesium chloride as an impurity. Magnesium chloride absorbs water very well and causes the table salt to appear moist on a humid day. Pure table salt will not absorb water as it has had the magnesium chloride removed.

▶ Suggested solution _____

The question is asking about a relationship between reactants and products in a chemical reaction, and that relationship is shown by a balanced equation. Write the balanced equation for the reaction of sodium and chlorine to produce sodium chloride.

$$2Na + Cl_2 \longrightarrow 2NaCl$$

The equation is written in terms of moles of reactants and products. However, in this example, you were asked about moles of reactants but are given grams of product. You might find that a kind of "road map" helps you organize your thinking. Begin your road map by showing what information you were given in the problem and what information you want to find.

$$737 \text{ g NaCl} \longrightarrow ? \text{ mol Na}$$

This road map reminds you that you know grams of sodium chloride from the problem and that you must find the moles of sodium required to produce it.

After diagraming the road map, look for relationships that allow you to calculate what you are after. Sometimes this can be done directly, as in Example 6-1. In that problem, you were asked about *moles* of oxygen and *moles* of magnesium oxide. The balanced equation gives the relationship between moles of oxygen and moles of magnesium oxide, so the question was answered by using one unit factor.

At times, there is not a single relationship that will allow the "trip" you want to take from the known to the unknown, so you must look for alternative routes. This must be done in the problem you are working now. Keeping in mind that "the mole is at the heart of chemistry," the road map can be changed as follows:

$$737 \text{ g NaCl} \longrightarrow ? \text{ mol Na}$$
$$\searrow \text{mol NaCl} \nearrow$$

There is no simple relationship that will allow calculation of moles of sodium from grams of sodium chloride. But there is a relationship that allows calculation of moles of sodium chloride from grams of sodium chloride and another relationship that allows calculation of moles of sodium from moles of sodium chloride. Those two relationships are

$$\frac{1 \text{ mol NaCl}}{58.5 \text{ g NaCl}} \quad \text{and} \quad \frac{2 \text{ mol Na}}{2 \text{ mol NaCl}}$$

Note that the first relationship requires information that is not in the problem. This common situation may cause difficulty at first. You must recall that you can add the atomic masses of elements in a compound to find the molar mass (Section 4-4) and you must recall that the molar mass in grams is the mass of one mole. This relationship between mass and moles is used so frequently in chemical calculations you need to know it solidly.

Now that you have the unit factors, show them on the road map. Each one is written over the appropriate arrow in the road map.

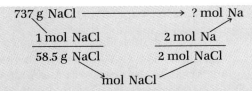

Use the relationships to answer the question that was asked.

$$737 \text{ g NaCl} \times \frac{1 \text{ mol NaCl}}{58.5 \text{ g NaCl}} \times \frac{2 \text{ mol Na}}{2 \text{ mol NaCl}} = 12.6 \text{ mol Na}$$

Many students will not follow the logic when several factors are used in a single expression like this. Encourage these students to solve the problem in steps until they can see that the result is the same and that each factor corresponds to one logical step in the solution of the problem.

Example 6-3

How many grams of potassium chlorate, $KClO_3$, must be decomposed to produce potassium chloride and 1.45 moles of oxygen?

▶ Suggested solution

Begin by showing the relationship between reactants and products in the balanced equation for the reaction.

$$2KClO_3 \longrightarrow 2KCl + 3O_2$$

Again, produce a road map to show where to start and where you want to end. In this problem, you were given moles of oxygen and are asked to find the number of grams of potassium chlorate required. This can be represented as

$$1.45 \text{ mol } O_2 \longrightarrow ? \text{ g KClO}_3$$

After diagraming the road map, look for relationships that can be used to obtain unit factors that will allow you to calculate what you want to find. As in the previous example, there is no direct way of calculating grams of one substance from moles of another. Devise an alternate route. There is a relationship between moles of potassium chlorate and moles of oxygen, and there is another relationship between moles of potassium chlorate and grams of potassium chlorate.

$$1.45 \text{ mol } O_2 \longrightarrow ? \text{ g KClO}_3$$
$$\searrow \qquad \nearrow$$
$$\text{mol KClO}_3$$

Every "trip" will involve the use of the mole ratios stated by the balanced equation and unit factors required to convert moles to grams or grams to moles. The first relationship for calculating moles of $KClO_3$ can be obtained by looking at the mole ratios in the balanced equation. The second relationship for converting moles of potassium chlorate to grams of potassium chlorate requires you to find the molar mass of potassium chlorate. This is done by adding the atomic masses of each element. Remember that the molar mass corresponds to the mass of one mole.

$$\frac{2 \text{ mol KClO}_3}{3 \text{ mol O}_2} \quad \text{and} \quad \frac{122.5 \text{ g KClO}_3}{1 \text{ mol KClO}_3}$$

Now you can construct a complete road map.

$$1.45 \text{ mol O}_2 \longrightarrow ? \text{ g KClO}_3$$

$$\frac{2 \text{ mol KClO}_3}{3 \text{ mol O}_2} \qquad \frac{122.5 \text{ g KClO}_3}{1 \text{ mol KClO}_3}$$

$$\text{mol KClO}_3$$

Putting all these relationships together, you get

$$1.45 \text{ mol O}_2 \times \frac{2 \text{ mol KClO}_3}{3 \text{ mol O}_2} \times \frac{122.5 \text{ g KClO}_3}{1 \text{ mol KClO}_3} = 118 \text{ g KClO}_3$$

6-6 Mass-Mass Calculations

Students need to observe chemical reactions directly. Some that can be done easily, safely, and inexpensively are zinc metal + 1.0*M* HCl, magnesium metal + 1.0*M* H$_2$SO$_4$, solid anhydrous sodium carbonate + 1.0*M* HCl. These reactions can be performed as demonstrations or in the lab.

The balanced equation gives the mole ratios of the reactants and products. When the amounts of each substance to be used in a reaction are measured, the mass is usually measured in grams. There may be times in the laboratory when you are asked to calculate the number of grams of product yielded by a certain mass of reactant. The mole ratios from the balanced equation can be used to construct the required unit factors for converting the moles of each substance to grams.

Example 6-4

How many grams of potassium chloride, KCl, are produced by decomposing 118 grams of potassium chlorate, KClO$_3$?

▶ Suggested solution _____

The first step in solving any stoichiometry problem is to write the balanced equation showing the mole ratios of the reactants and products. This is true even though all the measurements in the problem are given in grams.

$$2KClO_3 \longrightarrow 2KCl + 3O_2$$

The next step is to develop a road map to remind you of where you are beginning and where you will end up.

$$118 \text{ g KClO}_3 \longrightarrow ? \text{ g KCl}$$

As in the examples in Section 6-4, there is no obvious relationship that allows the direct calculation of *grams* of potassium chloride from *grams* of potassium chlorate. However, there is an indirect route for this kind of calculation that involves the mole relationships of the balanced equation.

$$118 \text{ g KClO}_3 \longrightarrow ? \text{ g KCl}$$
$$\text{mol KClO}_3 \longrightarrow \text{mol KCl}$$

You will notice that this road map looks similar to the ones used in Examples 6-2 and 6-3. The only difference is that an extra step has been added to convert the grams of reactant given in the problem to moles. In the earlier examples, the amount of one substance was given in moles. Next, find the relationships that allow calculation of each step. These relationships are

$$\frac{1 \text{ mol KClO}_3}{122.5 \text{ g KClO}_3} \quad \text{and} \quad \frac{2 \text{ mol KCl}}{2 \text{ mol KClO}_3} \quad \text{and} \quad \frac{74.5 \text{ g KCl}}{1 \text{ mol KCl}}$$

The first and last relationships are the unit factors showing that one mole of each substance corresponds to the molar mass of that substance. The middle relationship shows the mole ratio of potassium chloride and potassium chlorate given in the balanced equation. Fill these in on the road map.

$$118 \text{ g KClO}_3 \longrightarrow ? \text{ g KCl}$$
$$\frac{1 \text{ mol KClO}_3}{122.5 \text{ g KClO}_3} \qquad \frac{74.5 \text{ g KCl}}{1 \text{ mol KCl}}$$
$$\text{mol KClO}_3 \longrightarrow \frac{2 \text{ mol KCl}}{2 \text{ mol KClO}_3} \longrightarrow \text{mol KCl}$$

Showing all these relationships together gives

$$118 \text{ g KClO}_3 \times \frac{1 \text{ mol KClO}_3}{122.5 \text{ g KClO}_3} \times \frac{2 \text{ mol KCl}}{2 \text{ mol KClO}_3} \times \frac{74.5 \text{ g KCl}}{1 \text{ mol KCl}} = 71.8 \text{ g KCl}$$

Example 6-5

How many grams of oxygen will be required to react completely with 14.6 grams of sodium to form sodium oxide, Na_2O?

▶ Suggested solution _____

Begin by writing the balanced equation for the reaction.

$$4Na + O_2 \longrightarrow 2Na_2O$$

The question asks:

$$14.6 \text{ g Na} \longrightarrow ? \text{ g O}_2$$

Since there is no direct way of calculating grams of oxygen from grams of sodium, relationships are needed that provide a way of indirectly calculating the amount. As a reminder of what information is necessary for these calculations, show the alternative route to the answer.

CHEM THEME

Macroscopic observations indicate that a chemical reaction is taking place. The existence of new substances with different properties verifies the chemical change.

$$14.6 \text{ g Na} \longrightarrow ? \text{ g O}_2$$
$$\text{mol Na} \longrightarrow \text{mol O}_2$$

The conversions that are needed to work this problem are

$$\frac{1 \text{ mol Na}}{23 \text{ g Na}} \quad \text{and} \quad \frac{1 \text{ mol O}_2}{4 \text{ mol Na}} \quad \text{and} \quad \frac{32.0 \text{ g O}_2}{1 \text{ mol O}_2}$$

The first and last conversions contain information that is not in the problem but which you are expected to know how to obtain from the atomic masses of the elements given on the periodic table. The middle relationship shows the mole ratio of oxygen and sodium given by the balanced equation. Write each conversion over the appropriate arrow in the road map.

$$14.6 \text{ g Na} \longrightarrow ? \text{ g O}_2$$
$$\frac{1 \text{ mol Na}}{23 \text{ g Na}} \qquad \frac{32.0 \text{ g O}_2}{1 \text{ mol O}_2}$$
$$\text{mol Na} \longrightarrow \frac{1 \text{ mol O}_2}{4 \text{ mol Na}} \longrightarrow \text{mol O}_2$$

The calculations can be done in three separate steps or together as one problem. In three steps

$$14.6 \text{ g Na} \times \frac{1 \text{ mol Na}}{23.0 \text{ g Na}} = 0.635 \text{ mol Na}$$

$$0.635 \text{ mol Na} \times \frac{1 \text{ mol O}_2}{4 \text{ mol Na}} = 0.159 \text{ mol O}_2$$

$$0.159 \text{ mol O}_2 \times \frac{32.0 \text{ g O}_2}{1 \text{ mol O}_2} = 5.09 \text{ g O}_2$$

As one calculation:

$$14.6 \text{ g Na} \times \frac{1 \text{ mol Na}}{23.0 \text{ g Na}} \times \frac{1 \text{ mol O}_2}{4 \text{ mol Na}} \times \frac{32.0 \text{ g O}_2}{1 \text{ mol O}_2} = 5.08 \text{ g O}_2$$

Students can avoid such rounding errors by retaining one more digit than that to which they will round the final answer in all interim calculations.

It is interesting that the answers, 5.08 grams and 5.09 grams, differ in the last significant digit. Recall from Chapter 2 that the last digit in calculations like this is uncertain. In this case, the difference is due to rounding errors, but both answers are considered to be the same within experimental uncertainty.

6-7 Molarity and Replacement Reactions

Figure 6-2, on the left, shows a copper wire which has been placed in a solution of silver nitrate. Figure 6-2, on the right, shows the copper and silver nitrate reaction one day later. The blue color of the water is characteristic of the formation of copper(II) ions. The assumption is made that copper has replaced silver in the original solution. The metallic silver atoms are attached to the remaining copper wire.

Figure 6-2 On the left, the copper wire is just beginning to react. The solution is still colorless but some silver crystals have been produced on the copper wire. On the right, the reaction is complete. Silver crystals have covered the wire and the blue color shows that copper(II) nitrate is in solution.

How could you predict how much silver forms when you are dealing with a reactant that is a solution? To find the answer, go back to what you know about the concentration of a solution. If you know the molarity of the original silver nitrate solution, you can deal with the solution in terms of the number of moles of silver nitrate it contains. The following example shows how this problem could be solved.

Example 6-6

How many grams of copper will react to completely replace silver from 208 mL of 0.100M solution of silver nitrate, $AgNO_3$?

▶ Suggested solution _____

First, write the balanced equation for the reaction.

$$Cu + 2AgNO_3 \longrightarrow Cu(NO_3)_2 + 2Ag$$

Construct a road map to show the information given in the problem and what you need to find out. You are asked to find

$$208 \text{ mL of } 0.100M \text{ } AgNO_3 \longrightarrow ? \text{ g Cu}$$

With no direct way of calculating grams of copper from milliliters of silver nitrate, an alternative route is needed. Remember that moles are at the heart of chemistry!

208 mL 0.100M $AgNO_3$ $\longrightarrow$? g Cu

mol $AgNO_3$ $\longrightarrow$ mol Cu

Fill in the unit factors relating liters, molarity, moles, and grams on the revised road map. You will remember from Section 4-11 that molarity can be used to calculate the number of moles of silver nitrate in the 208 mL of solution. The balanced equation gives the unit factor that relates moles of silver nitrate to moles of copper.

208 mL 0.100M AgNO$_3$ $\longrightarrow$? g Cu

$$\frac{1 \text{ L AgNO}_3}{1000 \text{ mL AgNO}_3}$$

$$\frac{63.5 \text{ g Cu}}{1 \text{ mol Cu}}$$

L of 0.100M AgNO$_3$

$$\frac{0.100 \text{ mol AgNO}_3}{1 \text{ L AgNO}_3}$$

mol AgNO$_3$ $\longrightarrow$ $\dfrac{1 \text{ mol Cu}}{2 \text{ mol AgNO}_3}$ $\longrightarrow$ mol Cu

Molarity is a unitary rate that expresses the number of moles of a dissolved substance per liter of solution. The molarity of the silver nitrate solution is

$$0.100M \qquad \text{or} \qquad \frac{0.100 \text{ mol}}{1 \text{ L}}$$

The volumes of these solutions usually are measured in milliliters, so it is useful to note that the volume first must be converted to liters. The volume in liters is

$$208 \text{ mL AgNO}_3 \times \frac{1 \text{ L}}{1000 \text{ mL}} = 0.208 \text{ L AgNO}_3$$

To calculate the number of moles of silver nitrate in that solution, you multiply the volume in liters by the molarity, which is the number of moles per liter.

$$0.208 \text{ L AgNO}_3 \times \frac{0.100 \text{ mol}}{L} = 0.0208 \text{ mol AgNO}_3$$

The equation shows a 1-to-2 mole ratio between copper and silver nitrate. Moles of silver nitrate can be converted to moles of copper using that relationship.

$$0.0208 \text{ mol AgNO}_3 \times \frac{1 \text{ mol Cu}}{2 \text{ mol AgNO}_3} = 0.0104 \text{ mol Cu}$$

The last step is to convert moles of copper to grams of copper by multiplying by the mass of copper equivalent to 1 mole (the molar mass).

$$0.0104 \text{ mol Cu} \times \frac{63.5 \text{ g Cu}}{1 \text{ mol Cu}} = 0.660 \text{ g Cu}$$

The problem can be solved by combining these three separate steps into a single step.

$$208 \text{ mL AgNO}_3 \times \frac{0.100 \text{ mol AgNO}_3}{1000 \text{ mL AgNO}_3} \times \frac{1 \text{ mol Cu}}{2 \text{ mol AgNO}_3} \times \frac{63.5 \text{ g Cu}}{1 \text{ mol Cu}}$$
$$= 0.660 \text{ g Cu}$$

In a similar way, the amount of silver produced in the reaction of copper and silver nitrate can be calculated.

CHEM THEME

A study of many reactions allows you to see patterns that can be generalized to represent similar reactions. Once these patterns are known and can be expressed as balanced equations, they can be used to predict the amount of each substance involved.

Example 6-7

When copper replaces silver in 208 mL of 0.100M silver nitrate, $AgNO_3$, how many grams of silver will be produced?

▶ **Suggested solution** _____

The balanced equation for the reaction is

$$Cu + 2AgNO_3 \longrightarrow Cu(NO_3)_2 + 2Ag$$

You are asked to find:

$$208 \text{ mL of } 0.100M \text{ AgNO}_3 \longrightarrow \text{? g Ag}$$

To find the number of grams of silver produced, a revised road map is constructed.

$$208 \text{ mL } 0.100M \text{ AgNO}_3 \longrightarrow \text{? g Ag}$$
$$\searrow \qquad \nearrow$$
$$\text{mol AgNO}_3 \longrightarrow \text{mol Ag}$$

Show the relationships that are needed to obtain the answer.

208 mL 0.100M AgNO$_3$ —————————————→ ? g Ag

$$\dfrac{1 \text{ L AgNO}_3}{1000 \text{ mL AgNO}_3}$$

L of 0.100M AgNO$_3$

$$\dfrac{108.0 \text{ g Ag}}{1 \text{ mol Ag}}$$

$$\dfrac{0.100 \text{ mol AgNO}_3}{1 \text{ L AgNO}_3}$$

$$\text{mol AgNO}_3 \longrightarrow \dfrac{2 \text{ mol Ag}}{2 \text{ mol AgNO}_3} \longrightarrow \text{mol Ag}$$

The volume in liters of silver nitrate and the moles of silver nitrate present are calculated as in Example 6-6.

From the equation, the mole ratio of silver nitrate to silver is 2 to 2, so for moles of silver

$$0.0208 \text{ mol AgNO}_3 \times \dfrac{2 \text{ mol Ag}}{2 \text{ mol AgNO}_3} = 0.0208 \text{ mol Ag}$$

To convert moles of silver to grams of silver

$$0.0208 \text{ mol Ag} \times \dfrac{108 \text{ g Ag}}{1 \text{ mol Ag}} = 2.25 \text{ g Ag}$$

When the last three steps are combined you get the same answer.

$$0.208 \text{ L AgNO}_3 \times \dfrac{0.100 \text{ mol AgNO}_3}{1 \text{ L AgNO}_3} \times \dfrac{2 \text{ mol Ag}}{2 \text{ mol AgNO}_3} \times \dfrac{108 \text{ g Ag}}{1 \text{ mol Ag}}$$

$$= 2.25 \text{ g Ag}$$

Review and Practice

1. a. 32.0 g O_2
   b. 124 g Na_2O
   c. 3.20 g O_2
   d. 12.4 g Na_2O
2. a. $2 Fe + 3Cl_2 \longrightarrow 2FeCl_3$
   b. 2.000 mol Fe, 3.000 mol Cl_2
   c. 324.4 g $FeCl_3$
3. 155 g Na_2O
4. 67.8 g O_2
5. 147 g $KClO_3$
6. 56.1 g NO_2
7. 155 g Na_2O
8. 3.92 g KCl
9. a. 2.75 g Cu
   b. 0.0255 mol Cu
10. a. 0.3778 mol HCl
    b. 125.9 mL HCl
11. 0.1889 mol H_2
12. 55.8 g H_2O
13. 9.32 g $Pb(NO_3)_2$

1. a. How many grams of oxygen will be required to react completely with 92 grams of sodium to form sodium oxide?
   b. How many grams of sodium oxide will be produced by the reaction in (a)?
   c. How many grams of oxygen will be required to react completely with 9.2 grams of sodium?
   d. How many grams of sodium oxide will be produced by the reaction in (c)?

2. When 111.7 grams of iron and 212.7 grams of chlorine gas completely react, iron(III) chloride, $FeCl_3$, is formed.
   a. Write the balanced equation for the reaction.
   b. Calculate the number of moles of each reactant.
   c. How many grams of the product are formed?

3. Calculate the number of grams of sodium oxide, Na_2O, that will be produced when 5.00 moles of sodium react with oxygen.

4. Find the number of grams of oxygen that will be needed to burn 1.06 moles of methane, CH_4, to produce carbon dioxide and water.

5. Calculate the number of grams of potassium chlorate required to produce potassium chloride and 1.80 moles of oxygen.

6. Calculate the number of grams of nitrogen dioxide formed when 1.22 moles of ammonia, NH_3, react with 2.14 moles of oxygen to produce nitrogen dioxide and water.

7. Calculate the number of grams of sodium oxide that will be produced when 115 grams of sodium react with oxygen.

8. Calculate the number of grams of potassium chloride that will be formed by the decomposition of 6.45 grams of potassium chlorate.

9. For the reaction

$$Cu + 2AgNO_3 \longrightarrow Cu(NO_3)_2 + 2Ag$$

   a. How many grams of copper are required to replace 9.35 grams of silver in the solution of silver nitrate?
   b. If 5.50 grams of silver are produced in the above reaction, how many moles of copper were reacted?

10. For the reaction

$$Zn + 2HCl \longrightarrow ZnCl_2 + H_2$$

   a. How many moles of hydrochloric acid are needed to completely react with 12.35 grams of zinc?
   b. What volume of 3.00 molar hydrochloric acid is required to react with 12.35 grams of zinc?

11. How many moles of hydrogen are produced when 12.35 grams of zinc are reacted with the correct amount of hydrochloric acid?

12. A camping lantern uses the reaction of calcium carbide, CaC_2, and water to produce acetylene gas, C_2H_2, and calcium hydroxide, $Ca(OH)_2$. How many grams of water are required to produce 1.55 moles of acetylene gas?

13. When 7.52 grams of lead(II) carbonate, $PbCO_3$, are reacted with 27.5 milliliters of 3.00M nitric acid, HNO_3, what mass of lead(II) nitrate, $Pb(NO_3)_2$, will be formed?

Do You Know?

Nitrogen makes up about 80 percent of Earth's atmosphere. It is a vital part of all living organisms. Most of the nitrogen in commercial compounds is produced by the reaction of nitrogen with hydrogen to make ammonia, under special conditions not indicated by the equation.

Adjusting to Reality

It is important to remember that a balanced equation can be used to make many predictions. However, there are a number of factors associated with reactions that are not described by the equation. The equation describes what might happen. Balanced equations can be written for reactions that do not occur. For example, consider the following equation for the production of sugar from carbon and water.

$$12C + 11H_2O \longrightarrow C_{12}H_{22}O_{11}$$

At the present time, no one knows how to make sugar using only carbon and water.

An equation does not describe the exact conditions needed to make a reaction occur. If a specific temperature must be maintained, if constant mixing is required, or if a solvent like water or alcohol must be used, this information is determined through experimentation. Equations also do not describe the behavior of atoms during a reaction. Bonds will be broken and new bonds formed. Electrons may be exchanged, collisions occur, and intermediate products may be formed. All these things are important in understanding how a chemical reaction occurs, but the equation for the reaction does not provide this information.

Have you ever attempted a reaction that failed to occur in the lab? Missing or implied information may be responsible for your unsuccessful experiment. Have you ever tried a recipe that did not turn out? Often there is some hidden condition necessary that may not be mentioned in the recipe.

To represent the actual results obtained in a reaction accurately, specific types of calculations are used.

CONCEPTS

■ an excess reactant is one that is added to a reaction in larger than necessary amounts

■ the limiting reactant determines the amount of products

■ the experimental mass of products may be less than the theoretical mass predicted

■ the ratio of experimental mass produced to theoretical mass predicted gives the percent yield

6-8 Predicting Yields When One Reactant Is in Excess

The equation for the combustion of magnesium shows atoms of magnesium and oxygen combining in a 1-to-1 ratio in the product, magnesium oxide.

$$2Mg + O_2 \longrightarrow 2MgO$$

If magnesium atoms contact an oxygen molecule under the right conditions, the atoms will react to form magnesium oxide.

However, it is virtually impossible to have every atom or molecule of the reactants come together so that they will combine. Consequently, it is a common practice to add more than is necessary of one reactant, that is, an **excess** of one reactant. It is usually the cheaper one or the one that can be easily separated from the chemical system. In this way, one reactant will be completely used up and what is left of the other reactant can be recovered to be used again.

Figure 6-3 shows a reaction in which there is an excess of oxygen reacting with magnesium. How much magnesium oxide can be produced? By examining Figure 6-3, you can see that there are 6 atoms of

Figure 6-3 The drawing shows 6 atoms of magnesium and 7 molecules of oxygen reacting to produce 6 molecules of magnesium oxide. Four oxygen molecules remain unreacted or in excess.

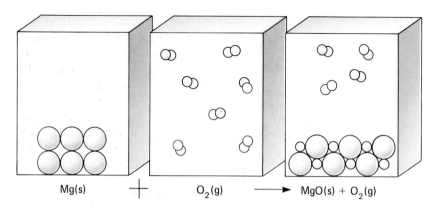

Mg(s) $+$ $O_2(g)$ $\longrightarrow$ MgO(s) + $O_2(g)$

magnesium and 14 atoms (7 molecules) of oxygen present. Since magnesium and oxygen are in a 1-to-1 ratio in magnesium oxide, there are 8 atoms of oxygen (4 molecules) in excess. The amount of magnesium oxide formed is determined by the number of magnesium atoms present. Once all the magnesium atoms are reacted, no more magnesium oxide is produced. Since magnesium is the reactant that limits the amount of product obtained, magnesium is called the **limiting reactant**.

Go back to the brownie recipe. Suppose you were intent on making brownies for dessert but discovered that you had only one square of chocolate instead of the two squares called for in the recipe. Could you still make the brownies? Of course, but you would need to cut the amount of each ingredient in the recipe by $\frac{1}{2}$. In this example, you could think of the chocolate as the limiting reactant.

When two reactants are not present in the exact ratio given in the equation, one reactant will be the limiting reactant. The reactant that is in excess, the amount of excess, and the amount of product can be determined.

Have students determine the mass of an empty test tube and a piece of zinc. Add 15.0 mL of 3.00*M* HCl. Have them make a drawing which represents what they will observe when the reaction has stopped. Zinc will be in excess if the piece you started with has a mass greater than 1.5 grams.

Example 6-8

If 1.21 moles of zinc are added to 2.65 moles of hydrochloric acid, HCl, then zinc chloride, $ZnCl_2$, and hydrogen gas, H_2, are formed. Determine which reactant is in excess and by what amount, and calculate the number of moles of each product.

▶ Suggested solution

As in all stoichiometry problems, begin with the balanced equation.

$$Zn + 2HCl \longrightarrow ZnCl_2 + H_2$$

You are asked four questions.

1.21 mol Zn + 2.65 mol HCl $\longrightarrow$? reactant in excess
$\longrightarrow$? amount of excess
$\longrightarrow$? mol $ZnCl_2$
$\longrightarrow$? mol H_2

To answer the first two questions, the relationship between the reactants is needed. Look at the balanced equation. The mole ratio is 1 mole of zinc to every 2 moles of hydrochloric acid. You could eliminate one step of the calculation if you compare the mole ratio and the information in the problem and make an accurate *prediction* about which reactant *seems* to be present in excess.

You might predict that hydrochloric acid is present in excess after looking at the numbers of moles given. The mole ratio given in the equation can be used to calculate the amount of hydrochloric acid, in moles, necessary to react with 1.21 moles of zinc.

$$1.21 \text{ mol Zn} \longrightarrow \text{? mol HCl}$$

$$1.21 \text{ mol Zn} \times \frac{2 \text{ mol HCl}}{1 \text{ mol Zn}} = 2.42 \text{ mol HCl required}$$

Since the problem shows that 2.65 moles of HCl are available to react, the hydrochloric acid is in excess and zinc is the limiting reactant. The amount of excess is found in subtracting the amount required from the amount actually present.

$$2.65 \text{ mol HCl} - 2.42 \text{ mol HCl} = 0.23 \text{ mol HCl in excess}$$

Suppose you predicted that zinc was present in excess. If so, you could check your prediction by calculating the moles of zinc needed to react with 2.65 moles of hydrochloric acid.

$$2.65 \text{ mol HCl} \longrightarrow \text{? mol Zn}$$

$$2.65 \text{ mol HCl} \times \frac{1 \text{ mol Zn}}{2 \text{ mol HCl}} = 1.33 \text{ mol Zn required} \quad \text{\textit{limiting}}$$

This shows that 1.33 moles of zinc would be needed to react with 2.65 moles of hydrochloric acid. However, there only are 1.21 moles of zinc available to react. Zinc is not present in excess and is found to be the limiting reactant. Hydrochloric acid is in excess. To determine how much excess hydrochloric acid there is, you would need to do the first two calculations shown above.

Once zinc is found to be the limiting reactant, you can answer the last two questions about how many moles of each product are made in the reaction. Calculate the moles of each product obtained when 1.21 moles of zinc are used. Use the mole ratios provided in the balanced equation.

Use limiting reagent!

$$1.21 \text{ mol Zn} \longrightarrow \text{? mol ZnCl}_2$$

$$1.21 \text{ mol Zn} \times \frac{1 \text{ mol ZnCl}_2}{1 \text{ mol Zn}} = 1.21 \text{ mol ZnCl}_2$$

$$1.21 \text{ mol Zn} \longrightarrow \text{? mol H}_2$$

$$1.21 \text{ mol Zn} \times \frac{1 \text{ mol H}_2}{1 \text{ mol Zn}} = 1.21 \text{ mol H}_2$$

When this experiment is done in the laboratory, the amount of zinc is measured in grams and the amount of hydrochloric acid is measured as a volume in liters or milliliters. Now look again at this reaction.

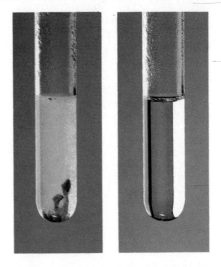

Figure 6-4. On the left, the zinc and hydrochloric acid are just beginning to react. The bubbles of hydrogen can be seen forming at the surface of the zinc and escaping into the air. When the reaction is complete, no zinc remains in the test tube. What substances will be in the test tube?

Example 6-9

When 79.1 grams of zinc are reacted with 1.05 liters of 2.00M hydrochloric acid, HCl, to produce zinc chloride, $ZnCl_2$, and hydrogen gas, H_2, which reactant will be in excess and by how much? Calculate the number of grams of each product.

▶ **Suggested solution** _____

First, write the balanced equation.

$$Zn + 2HCl \longrightarrow ZnCl_2 + H_2$$

You are asked

79.1 g Zn + 1.05 L of 2.00M HCl ⟶ ? reactant in excess
⟶ ? amount of excess
⟶ ? g ZnCl
⟶ ? g H_2

To answer any of the four questions, some relationship between the reactants and products must be established. That relationship is given in the mole ratio of the balanced equation. However, the information in the problem is given in grams and liters. There is no direct way of calculating grams of product from grams or liters of reactant. The amounts first must be converted to moles. Remember, moles are at the heart of chemistry! Draw a road map showing the alternative way of determining grams of each product.

79.1 g Zn + 1.05 L 2.00M HCl ⟶ ? $ZnCl_2$ + ? g H_2

mol HCl ⟶ mol $ZnCl_2$
mol Zn ⟶ mol H_2

Fill in the necessary relationships.

79.1 g Zn + 1.05 L 2.00M HCl ⟶ ? g $ZnCl_2$ + ? g H_2

$$\frac{1\,mol\ Zn}{65.4\,g\ Zn} \qquad \frac{2.00\,mol\ HCl}{1\,L\ HCl} \qquad \frac{136.4\,g\ ZnCl_2}{1\,mol\ ZnCl_2} \qquad \frac{2.02\,g\ H_2}{1\,mol\ H_2}$$

mol HCl ⟶ mol $ZnCl_2$
mol Zn ⟶ mol H_2

To answer the first two questions, the mole ratio from the balanced equation is necessary. To be able to compare the mole ratios in the equation to the amounts given in the problem, first convert grams of zinc and liters of hydrochloric acid to moles.

$$79.1\ \cancel{g\ Zn} \times \frac{1\,mol\ Zn}{65.4\ \cancel{g\ Zn}} = 1.21\ mol\ Zn$$

$$1.05\ \cancel{L\ HCl} \times \frac{2.00\,mol\ HCl}{1.00\ \cancel{L\ HCl}} = 2.10\ mol\ HCl$$

Look at the mole ratio of the equation and the amounts in moles of each reactant. Predict which reactant is in excess. You may predict that hydrochloric acid is in excess. Calculate how many moles of hydrochloric acid are necessary to react with 1.21 moles of zinc.

$$1.21 \text{ mol Zn} \longrightarrow ? \text{ mol HCl}$$

$$1.21 \text{ mol Zn} \times \frac{2 \text{ mol HCl}}{1 \text{ mol Zn}} = 2.42 \text{ mol HCl required}$$

The problem states that there are 2.10 moles of hydrochloric acid which is less than the amount required to react with the available zinc. This means zinc is in excess, and hydrochloric acid is the limiting reactant.

To find the amount of excess zinc, first calculate the number of moles needed to react with the 2.10 moles of hydrochloric acid present.

$$2.10 \text{ mol HCl} \longrightarrow ? \text{ mol Zn}$$

$$2.10 \text{ mol HCl} \times \frac{1 \text{ mol Zn}}{2 \text{ mol HCl}} = 1.05 \text{ mol Zn required}$$

Subtract to find the moles of Zn in excess.

$$1.21 \text{ mol Zn} - 1.05 \text{ mol Zn} = 0.16 \text{ mol in excess}$$

To answer the last two questions, use the number of moles of the limiting reactant, hydrochloric acid, to calculate the mass of each product that can be formed. The steps for calculating the number of moles of product and for converting moles into grams have been combined. Remember, the molar mass is obtained by adding the atomic masses of the elements in the compound. The molar mass corresponds to the mass of one mole of the substance.

$$1.05 \text{ mol HCl} \longrightarrow ? \text{ g ZnCl}_2$$

$$1.05 \text{ mol HCl} \times \frac{1 \text{ mol ZnCl}_2}{2 \text{ mol HCl}} \times \frac{136.4 \text{ g ZnCl}_2}{1 \text{ mol ZnCl}_2} = 71.6 \text{ g ZnCl}_2$$

$$1.05 \text{ mol HCl} \longrightarrow ? \text{ g H}_2$$

$$1.05 \text{ mol HCl} \times \frac{1 \text{ mol H}_2}{2 \text{ mol HCl}} \times \frac{2.02 \text{ g H}_2}{1 \text{ mol H}_2} = 1.06 \text{ g H}_2$$

In this example, you had to do two calculations to determine the amount of excess reactant. In the first attempt, hydrochloric acid was predicted to be in excess. When that prediction proved to be false, the calculation had to be repeated to show that zinc was in excess.

6-9 Extension: Using Theoretical Yield to Find Percent Yield

When a chemical reaction takes place, a given amount of reactant will produce a given amount of product. The assumption has been made that all the chemicals mentioned in the examples are 100 percent pure. It also has been assumed that the amounts of reactants and products can be contained and measured exactly. In the laboratory, of course, it is not likely that the chemicals used in a reaction are 100 percent pure. It also is difficult to recover all the products of a reaction and measure them accurately.

The amount of a product that it would be possible to obtain, assuming perfect conditions, is called the **theoretical yield**. The amount actually obtained in the experiment will be less than or equal to the theoretical yield. By dividing the **actual yield** by the theoretical yield and expressing that number as a percent, you can calculate the **percent yield**.

Example   6-10

When 45.8 grams of potassium carbonate, K_2CO_3, are reacted completely with an excess of hydrochloric acid, HCl, 46.3 grams of potassium chloride, KCl, are produced. Water and carbon dioxide also are formed. Calculate the theoretical yield and the percent yield of potassium chloride.

▶ **Suggested solution**

First, write the balanced equation for the reaction.

$$K_2CO_3 + 2HCl \longrightarrow 2KCl + H_2O + CO_2$$

You are asked to find the theoretical yield of potassium chloride and the percent yield. In other words, you first must answer what is the most potassium chloride that could be made from 45.8 grams of potassium carbonate under ideal conditions. Although the problem does not directly state that you are to find the yields in grams, you can see that grams of product would be a more useful measurement to calculate than moles of product.

$$45.8 \text{ g } K_2CO_3 \longrightarrow \text{ ? g KCl}$$

As in the previous problems you have solved, there is not likely to be a unit factor available in your memory or in a book that provides you with a one-step method of calculating the theoretical yield in grams of potassium chloride from grams of potassium carbonate. First, the moles of each substance can be calculated. The mole ratio from the equation allows calculation of moles of potassium chloride from moles of potassium carbonate. Show the revised road map.

45.8 g K_2CO_3 ⟶   ? g KCl

mol K_2CO_3 ⟶   mol KCl

Fill in the relationships that allow these calculations to be made.

45.8 g K_2CO_3 ⟶   ? g KCl

$$\frac{1 \text{ mol } K_2CO_3}{138 \text{ g } K_2CO_3} \qquad \frac{74.5 \text{ g KCl}}{1 \text{ mol KCl}}$$

mol K_2CO_3 ⟶   mol KCl

When all the unit factors are put together in one step, the theoretical yield of potassium chloride is

$$45.8 \text{ g } \cancel{K_2CO_3} \times \frac{1 \cancel{\text{ mol } K_2CO_3}}{138 \text{ g } \cancel{K_2CO_3}} \times \frac{2 \cancel{\text{ mol } KCl}}{1 \cancel{\text{ mol } K_2CO_3}} \times \frac{74.5 \text{ g KCl}}{1 \cancel{\text{ mol } KCl}}$$
$$= 49.5 \text{ g KCl}$$

To calculate percent yield, compare the actual yield given in the problem and the theoretical yield. The actual yield usually is smaller than the theoretical yield, so the ratio is set up with the actual yield on top to give a percent less than 100.

$$\text{The percent yield} = \frac{\text{actual yield}}{\text{theoretical yield}} \times 100$$

The actual yield of potassium chloride is given in the problem as 46.3 grams. The theoretical yield of potassium chloride when starting with 45.8 grams of potassium carbonate was calculated to be 49.5 grams.

$$\text{The percent yield of KCl} = \frac{46.3 \text{ g KCl}}{49.5 \text{ g KCl}} \times 100 = 93.5\%$$

Review and Practice

•**1.** Give three pieces of information that the following equation does *not* give about the reaction.

$$Zn(s) + MgCl_2(aq) \longrightarrow ZnCl_2(aq) + Mg(s)$$

2. **a.** If 20 molecules of hydrogen are added to 20 molecules of oxygen to form water, which reactant will be in excess?
b. How many molecules of water can be produced?

3. A mixture of zinc and hydrochloric acid is placed in a beaker. After several days, what substances would be present in the beaker?

4. If 10.45 grams of aluminum are 66.55 grams of copper(II) sulfate, $CuSO_4$, then aluminum sulfate, $Al_2(SO_4)_3$, and copper are formed.
a. Which reactant is in excess?
b. Calculate the mass of the excess.
c. Calculate the mass of each product.

5. If 15.50 grams of lead(II) nitrate, $Pb(NO_3)_2$, are reacted with 3.81 grams of sodium chloride, NaCl, then sodium nitrate, $NaNO_3$, and lead(II) chloride, $PbCl_2$, are formed.
a. Which reactant will be in excess?
b. Calculate the mass of the excess.
c. Calculate the mass of lead(II) chloride produced.

6. If 12.5 grams of copper are reacted with an excess of chlorine, then 25.4 grams of copper(II) chloride, $CuCl_2$, are obtained. Calculate the theoretical yield and the percent yield.

7. If 6.57 grams of iron are reacted with an excess of hydrochloric acid, HCl, then hydrogen gas and 14.63 grams of iron(II) chloride are obtained. Calculate the theoretical yield and the percent yield.

1. the behavior of the atoms, the temperature, the speed of the reaction
2. a. oxygen
   b. 20 molecules of water
3. $ZnCl_2$, H_2O and either excess Zn or excess HCl
4. a. Al
   b. 2.95 g
   c. 47.54 g $Al_2(SO_4)_3$
      26.48 g Cu
5. a. $Pb(NO_3)_2$
   b. 4.70 g
   c. 9.06 g $PbCl_2$
6. theoretical yield = 26.4 g
   percent yield = 96.2%
7. theoretical yield = 14.92 g
   percent yield = 98.1%

Careers in Chemistry

Ellen R. M. Druffel
Title: Associate Scientist
Job Description: Collects samples of coral from reefs in Hawaii, Australia, and the Florida Keys. Analyzes samples to determine present and past ocean chemistry and modes of circulation.
Educational Qualifications: Advanced degree in chemistry.
Future Employment Outlook: Good

Ellen Druffel works at the Woods Hole Oceanographic Institution, studying samples of coral to learn more about the geochemical cycle of carbon in the oceans. In the last 100 years, the amount of carbon dioxide in the atmosphere has risen by 20 percent, due to burning fossil fuels and the reduction of terrestrial plant life. This increase in carbon dioxide has caused Earth's atmosphere to warm up. Druffel and other scientists are trying to determine how effectively the oceans can absorb the excess carbon dioxide. She studies present and past ocean conditions by examining banded layers formed by corals. "By studying what happened in the past," says Druffel, "we can attempt to understand and predict future geochemical cycles."

Druffel first became interested in chemistry when she was in high school. Besides her ability in chemistry, she feels that many other skills contribute to her success in research. "One has to be a plumber, an electronics technician, a manager, and a machinist and have various other skills to build and operate a lab of this type. Most importantly, having competent, cooperative people working with you is essential for the success of our research."

Working conditions at Druffel's laboratory are quite good. "I work with a fine group of talented and crazy people, and this makes it all worthwhile," she explains. "I am usually quite busy due to proposal writing, publishing, and lab work, but the rewards of my work far outweigh the pressures involved."

Druffel encourages students to get direction from competent counselors and to set goals and try to attain them. She has a special message for young women: "You *can* have a family and a career in science at the same time. It takes organization and an understanding spouse, but it is attainable, and, I might add, very rewarding."

Numbers in red indicate the appropriate chapter sections to aid you in assigning these items. Answers to all questions appear in the Teacher's Guide at the front of this book.

Summary

■ The coefficients in a balanced equation give the ratio in which atoms, molecules, or formula units combine and form the products during a reaction. That combining ratio is the mole ratio. The mole ratio described by the equation is proportional. A fraction or multiple of these ratios also would be correct. Mole ratios can be used to predict the mass of one chemical needed to react with another.

■ Stoichiometry is a general term used to describe all the quantitative relationships given in the equation. Stoichiometric calculations can be used to predict the amount of reactant or product involved in a reaction.

■ Unitary rates and unit factors can be used to show relationships between mass and moles. They can be used to convert the mole ratios of the balanced equation into specific masses. Molarity can be used to measure the moles of a substance dissolved in a solution.

■ A balanced equation does not describe the specific conditions that may be necessary in order for a reaction to take place, and it does not describe how atoms and molecules behave during the reaction. The reaction described by an equation may not take place at all.

■ In many experiments, the reactants usually are not combined in the exact amounts given in the equation, and so one of the reactants is in excess. The reactant that is completely used up determines how much of the products will form and is called the limiting reactant.

■ The amount of product predicted under ideal reaction conditions is called the theoretical yield. The actual amount measured experimentally is usually less than or equal to the theoretical amount. The actual yield divided by the theoretical yield gives the percent yield.

Chemically Speaking

actual yield	percent yield
excess	stoichiometry
limiting reactant	theoretical yield
mole ratio	

Review

1. What is the ratio in which hydrogen molecules will react with nitrogen molecules to form ammonia, NH_3? (6-2)

2. Make a drawing of 3 molecules of nitrogen reacting with the correct number of molecules of hydrogen, and show the number of molecules of ammonia produced. (6-2)

3. If 2 molecules of oxygen react with one molecule of methane, CH_4, how many moles of oxygen will react with 1 mole of methane? (6-2)

4. Make drawings of 10 molecules of hydrogen added to 3 molecules of nitrogen to form ammonia, NH_3. (6-2)

5. Use the drawing in item 4 to determine which reactant is in excess and the number of molecules of excess. (6-8)

6. What is the reason for using an excess of one reactant during a chemical reaction? (6-8)

7. Explain why the percent yield of a product in a reaction is usually less than 100. (6-9)

Interpret and Apply

1. Write the balanced equation for hydrogen peroxide, H_2O_2, decomposing to form water and oxygen.
   a. What is the ratio of mole of hydrogen peroxide decomposed to mole of water produced? (6-2)
   b. What is the ratio of mole of hydrogen peroxide decomposed to mole of oxygen produced? (6-2)
   c. When one mole of hydrogen peroxide decomposes, how many moles of oxygen are produced? (6-2)

2. Which of these statements are not implied by this equation? (Adjusting to Reality introduction)

$$Fe(s) + CuSO_4(aq) \longrightarrow FeSO_4(aq) + Cu(s)$$

   a. Iron is more active than copper.
   b. Iron is losing electrons.
   c. One mole of iron will produce 63.5 grams of copper metal.
   d. This reaction takes place in water.
   e. The reaction will be 90 percent complete in 5 hours.
   f. Water participates in the reaction.
   g. Copper changes state in the reaction
   h. The color of the reaction mixture will change.

3. The equation for the reaction of nitrogen and hydrogen to produce ammonia, NH_3, is

$$N_2 + 3H_2 \longrightarrow 2NH_3$$

   a. Given 6 molecules of nitrogen and 12 molecules of hydrogen, make a drawing that represents the

reaction container before the reaction. (6-2)
b. Make a drawing that represents the same reaction container after the reaction. (6-2)
c. How many molecules of ammonia are formed? (6-2)
d. Which reactant is in excess? (6-8)
e. How many molecules are in excess? (6-8)

• 4. The equation for the reaction between nitric oxide, NO, and oxygen, O_2, is:

$$2NO + O_2 \longrightarrow 2NO_2$$

a. Given 10 molecules of nitric oxide and 8 molecules of oxygen, how many molecules of nitrogen dioxide, NO_2, can be formed? (6-2)
b. Make drawings that illustrate your answer. (6-2)
c. Which reactant is in excess? (6-8)

5. Nitric oxide and oxygen are colorless while nitrogen dioxide is red-brown. What macroscopic observations could you make when the reaction in item 4 takes place? (6-8)

• 6. a. When an excess of copper is reacted with 208 milliliters of 0.100M silver nitrate solution, what is the theoretical yield of silver in grams? (6-9)
b. The silver crystals produced in above are shaken off the copper wire, washed, dried, and weighed. They are found to have a mass of 2.07 grams. Suggest two circumstances that could explain the loss of silver. (6-9)

Problems

• 1. Silver reacts with nitric acid, HNO_3, to form nitrogen monoxide, NO, silver nitrate, $AgNO_3$, and water.
a. How many grams of nitric acid are required to react with 5.00 grams of silver? (6-6)
b. How many grams of silver nitrate will be formed? (6-6)
c. How many moles of nitrogen monoxide will be produced? (6-2)

• 2. Hydrazine, N_2H_4, and hydrogen peroxide, H_2O_2, react exothermically to produce nitrogen and water. When these are used as rocket fuel, how many grams of hydrogen peroxide are needed to react with 100.0 grams of hydrazine? (6-6)

• 3. If 17.5 grams of zinc are reacted with phosphoric acid, H_3PO_4, then zinc phosphate, $Zn_3(PO_4)_2$, and hydrogen are produced.

a. How many moles of phosphoric acid are required? (6-3)
b. If the phosphoric acid is a 3.00 molar solution, how many liters of the solution are needed? (6-7)
c. What mass of zinc phosphate will be produced? (6-6, 6-
d. How many moles of hydrogen will be given off? (6-5)

• 4. If 5.45 grams of potassium chlorate, $KClO_3$, are decomposed to form potassium chloride, KCl, then 1.95 grams of oxygen also are given off.
a. Calculate the theoretical yield of oxygen. (6-9)
b. Calculate the percent yield of oxygen. (6-9)
c. Explain why the percent yield of oxygen is less than 100. (6-9)

5. If 15.5 grams of aluminum are reacted with 46.7 grams of chlorine gas, then aluminum chloride, $AlCl_3$, is formed.
a. Which reactant is in excess? (6-8)
b. Calculate the number of grams of excess. (6-8)
c. Calculate the mass of aluminum chloride produced. (6-6

• 6. Large amounts of uranium metal are produced by reacting uranium(IV) chloride, UCl_4, with magnesium metal to produce magnesium chloride, $MgCl_2$, and uranium metal.
a. Write the balanced equation for the reaction. (6-2)
b. How many grams of magnesium are required to completely react with 155 grams of uranium(IV) chloride? (6-6)
c. How many grams of uranium will be produced? (6-6)

7. In a car battery, lead, lead(IV) oxide, PbO_2, and sulfuric acid, H_2SO_4, are reacted to produce lead(II) sulfate, $PbSO_4$, and water.
a. Write the balanced equation for the reaction. (6-2)
b. When 10.45 grams of lead, 15.66 grams of lead(IV) oxide, and 25.55 grams of sulfuric acid are mixed, which is the limiting reactant? (6-8)
c. How many grams of lead(II) sulfate will be produced? (6-6, 6-8)

• 8. During photosynthesis, a sweet potato plant combines carbon dioxide and water to produce sucrose, $C_{12}H_{22}O_{11}$, and oxygen gas. How many moles of carbon dioxide are required to produce 455 grams of sucrose? (6-5)

• 9. The charcoal briquettes used to cook in an outdoor grill are composed of 99 percent carbon. How many grams of oxygen are required when a 100.0 gram briquette is burned to produce carbon dioxide and energy? (6-6)

• 10. The methyl alcohol, CH_3OH, used in alcohol burners combines with oxygen to form carbon dioxide and water. How many moles of oxygen are required to burn 34.2 grams of methyl alcohol? (6-5)

Challenge

1. When 14.97 grams of iron react with chlorine gas, 43.47 grams of product are formed. Write the correct balanced equation for this reaction.

2. When 10.68 grams of manganese(IV) oxide are decomposed, oxygen is given off and 9.37 grams of a solid compound remain. Write the balanced equation for the reaction.

3. If 75.0 mL of 0.100M mercury(II) nitrate, $Hg(NO_3)_2$, are reacted with 150.0 mL of 0.100M sodium iodide, NaI, then orange, solid mercury(II) iodide, HgI_2, and sodium nitrate, $NaNO_3$, are formed. Calculate the mass in grams of the mercury(II) iodide that is formed.

• 4. What volume of 0.55M nickel(II) nitrate, $Ni(NO_3)_2$, will react with 85 mL of 0.25M potassium carbonate, K_2CO_3, to form nickel(II) carbonate, $NiCO_3$, and potassium nitrate, KNO_3?

• 5. What volume of 0.60M copper(II) sulfate, $CuSO_4$, will react with 45 mL of 1.50M sodium hydroxide, NaOH, to form copper(II) hydroxide, $Cu(OH)_2$, and sodium sulfate, Na_2SO_4?

• 6. What volume of 0.45M sodium carbonate, Na_2CO_3 will react with 82 mL of 0.25M iron(III) chloride, $FeCl_3$, to form iron(III) carbonate, $Fe_2(CO_3)_3$, and sodium chloride, NaCl?

7. A strip of zinc with a mass of 19.43 grams is placed in a beaker containing 425 cm^3 of 0.25M chromium(III) nitrate.
   a. Which reactant is in excess?
   b. Make a labeled drawing of the beaker and contents when the reaction is complete.
   c. Calculate the mass of chromium metal produced.

Synthesis

1. Iron metal reacts with a copper(II) sulfate solution to produce copper metal and iron(II) sulfate.
   a. Which is more active, iron or copper?
   b. What will you observe when the reaction takes place?

c. How many moles of copper will be produced if 5.45 grams of iron are placed in 135 milliliters of a 0.65M solution of copper(II) sulfate?
   d. How many grams of copper will be produced?
   e. How many atoms of copper will be produced?

2. When 26.42 grams of molybdenum metal are reacted with an excess of oxygen gas, a black compound with a mass of 33.04 grams is formed.
   a. Write the balanced equation for the reaction.
   b. What is the name of the compound formed?

3. A strip of chromium with a mass of 21.55 grams is placed in a beaker containing 635 mL of a 0.25M cobalt(II) chloride solution. When the reaction is complete, the chromium strip is removed, washed, dried, and the mass is found to be 16.05 grams.
   a. Write the balanced equation for the reaction.
   b. Describe the macroscopic changes observed during this reaction.
   c. Make a labeled drawing of the beaker and contents just before the chromium strip is removed.

4. When 125 cm^3 of 0.55M silver nitrate is reacted with 85 cm^3 of 0.25M aluminum chloride, solid silver chloride is formed.
   a. Write a balanced equation for the reaction.
   b. Calculate the number of moles of each reactant.
   c. Which reactant is in excess?
   d. Calculate the mass of silver chloride produced.
   e. Make a drawing of the beaker and contents at the completion of the reaction.

Project

Given samples of magnesium, chromium, iron, aluminum, nickel, copper, and tin metals and solutions of magnesium chloride, chromium(III) chloride, iron(II) chloride, aluminum chloride, nickel(II) chloride, copper(II) chloride, and tin(IV) chloride, plan experiments that help you determine an activity series for these elements.

CHAPTER 7

We live at the bottom of an ocean of air. The atmosphere surrounding Earth is composed of a mixture of gases that sustain most life on our planet. Oxygen, carbon dioxide, and nitrogen each play vital roles in the existence of living things. This chapter concerns itself with the physical laws that describe the behavior of gases.

A gas is the least complex of the three states of matter. Understanding the properties of gases is an important part in probing the nature of molecular systems.

Gases

Properties of Gases

The composition of Earth's atmosphere varies slightly depending on the location and the time of year. Most people are familiar with the differences in the quality of the air in winter and in summer. Pollutants such as nitrogen dioxide, NO_2, and sulfur dioxide, SO_2, are greatly increased in the summer in some locations by seasonal changes in the atmosphere. Generally, the atmosphere contains a mixture of gases, subject to variation, as found in Table 7-1.

Table 7-1

COMPOSITION OF THE ATMOSPHERE (percent of molecules in dry air)			
Nitrogen	78.08	Carbon dioxide	0.03
Oxygen	20.95	Helium	5×10^{-4}
Argon	0.93	Hydrogen	5×10^{-5}

Like all gases, the atmosphere exerts pressure on its surroundings. You can feel this pressure change in your ears as you climb a mountain, fly in an airplane, or ride in an elevator. The pressure is less on top of Mount Evans in Colorado than it is on the beach at Asbury Park, New Jersey. Atmospheric pressure increases as you move closer to the surface of Earth. What factors produce the pressure differences at varying heights above the planet's surface? What causes gas pressure in the first place? These are some of the questions you will probe in the beginning of this chapter.

CONCEPTS

- differences between solids, liquids, and gases
- compression of gases
- describing pressure
- nature of gas pressure
- the meaning of atmospheric pressure
- partial pressure

7-1 Comparing Solids, Liquids, and Gases

You are familiar with three of the four states of matter that were introduced in Chapters 1 and 2. Solids, liquids, and gases are states that you have seen, handled, worn, eaten, and breathed. It is helpful to summarize some of the things you may already know about these three states. A **solid** is generally rigid. Solids have their own shapes, and their volumes change only slightly in response to changes in temperature or pressure. Examples of solids are wood, metal coins, crystals of table salt, and ice. A **liquid** is a substance that takes the shape of its container. Liquids like water, oil, alcohol, and mercury metal at room temperature have volumes that change very little in response to changes in temperature or pressure. Unlike a solid or a liquid, a **gas** completely fills its container, and its volume is drastically changed by

This section should represent a review of some material presented in Chapter 1.

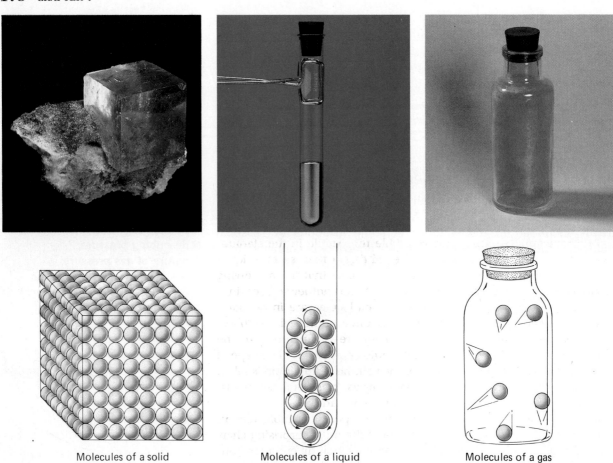

Molecules of a solid

Molecules of a liquid

Molecules of a gas

Figure 7-1 Compare the structured shape of the solid in this figure with the shapes of the liquid and the gas shown in the next two figures. Only the solid has a definite shape and volume.

Figure 7-2 Molecules in a liquid have greater freedom of movement than in a solid. They easily tumble around each other, yet they are essentially in contact with each other at all times.

Figure 7-3 Gas molecules move with greater freedom than those of solids and liquids. Spaces between gas molecules are large in comparison to the spaces between molecules of the other two states.

temperature and pressure changes. Examples of gases are those comprising Earth's atmosphere, as well as nitrogen dioxide and chlorine at room temperature. You can compare the appearance of solids, liquids, and gases in Figures 7-1, 7-2, and 7-3.

You know that a substance in one physical state can exist in another. You even make use of a substance differently depending on its state. Water makes a comfortable environment for swimming if it is at the right temperature. If it is cooled too much, it freezes to a solid and is better for skating. You can use the expansive character of gases to blow up a balloon. What is different about each of the three states of matter? How can the observations you make on the macroscopic level be viewed microscopically?

On the molecular scale, solids, liquids, and gases can be compared to the people in a baseball stadium. Imagine that the baseball stadium is filled to capacity. During the game, people are animated and energetic but remain in their specific seats. (Assume no one is getting up for refreshments at the moment!) At the end of the game, everyone streams out of the stadium. They are all moving to a greater extent now, confined by the shapes of the aisles but still very close together. When the people reach the exits, they disperse in many different directions, and the space between them increases enormously.

Figure 7-1 shows that the molecules of a solid are, for all practical purposes, arranged in an orderly manner like the people sitting in their stadium seats. Does this suggest why solids are rigid and have constant volume? A solid is composed of atoms or molecules packed closely together with little space between them. The particles that compose a solid do not move from place to place to any noticeable extent. As a result, a solid maintains a specific shape. The volume of the solid is composed almost entirely of the volume of the molecules themselves. Therefore, compressing the solid to any great extent is not possible.

On a molecular scale, liquids resemble the people in the stadium aisles, who move more freely than when sitting in their seats. A liquid is an example of a **fluid**, which is any substance that flows. Flowing occurs when molecules are free to slide past one another and continually change their relative positions. The molecules are in constant motion. They are free to flow under, over, and around their neighboring molecules, as shown in Figure 7-2. They are confined only by the borders of their container. As a result, liquids conform to the shape of their containers. Like a solid, most of the volume of the liquid is taken up by the volume of the molecules themselves. In general, compressing a liquid does not change its volume significantly.

Gases, like liquids, are fluids and are composed of molecules in constant, random motion. However, one of the most surprising characteristics of a gas is that its molecules have a much different environment than those of either a solid or a liquid. As illustrated in Figure 7-3, gas molecules are not neatly "packaged" or arranged, nor are they touching each other most of the time. The gas molecules resemble somewhat the crowd leaving the baseball stadium. Gas molecules will move throughout their container, but without a container, they disperse freely.

On the average, gaseous molecules are many times farther from each other than molecules of solids and liquids. To give you an idea of the distance between gas molecules, consider this example: One mole of solid carbon dioxide (dry ice) occupies 28 cm^3, a space a little bigger than a large ice cube. Almost all of the volume of the solid carbon dioxide is taken up by the molecules themselves. Yet, as a gas at room temperature, the same number of CO_2 molecules occupies 25 000 cm^3, a volume almost 1000 times greater! Only a small fraction of the total volume of the gas is occupied by the molecules themselves. The rest of the volume is empty space. Molecules in the gaseous state have a tremendous amount of room to move about. From another point of view, there also is lots of room for the molecules to move closer together, as when the gas is compressed. You probably have seen tanks of compressed helium gas that is used to fill balloons at amusement parks, parades, sporting events, fairs, etc. A typical 44-liter tank is able to fill over 6500 liters worth of balloons, roughly 500 balloons that are each 25 cm in diameter! The compressibility of gases is a unique characteristic that is applied to such diverse purposes as filling scuba tanks, inflating tires, and pressurizing airplanes.

It helps to spend some time on this concept so that students can better visualize the tremendous amount of empty space in a sample of gas.

Figure 7-4 A brick weighs the same amount no matter what position it has on a table. However, the pressure exerted by the brick when it stands on its smallest face is nearly four times the pressure exerted when the brick rests on its widest face.

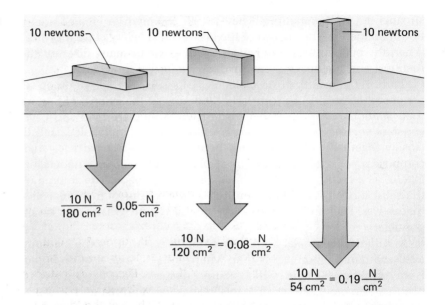

$$\frac{10 \text{ N}}{180 \text{ cm}^2} = 0.05 \frac{\text{N}}{\text{cm}^2}$$

$$\frac{10 \text{ N}}{120 \text{ cm}^2} = 0.08 \frac{\text{N}}{\text{cm}^2}$$

$$\frac{10 \text{ N}}{54 \text{ cm}^2} = 0.19 \frac{\text{N}}{\text{cm}^2}$$

7-2 Pressure

Much of what you will study in this chapter involves information about pressure, so it is useful for you to know exactly what pressure is. A simple example can be used to introduce the concept. Imagine a brick resting on a tabletop as shown in Figure 7-4. Due to the gravitational force exerted on the brick, the weight of the brick becomes a force exerted on the table. What exactly is the difference between force and pressure?

Weight, as just mentioned, is the force that gravity exerts on the brick. Weight is proportional to the mass of the brick. **Pressure** is a force exerted on some given area. A brick laying on its broad side, narrow side, or end has the same weight. The force exerted on the table is the same in all three cases. The pressure, however, will not be the same because the force will be distributed over different areas.

A common brick weighs about 10 newtons. (One newton exerts a force equal to about the weight of 30 pennies in your hand.) The brick's largest side has an area of about 180 cm², its smaller face has an area of about 120 cm², and its end has an area of about 54 cm². The pressure exerted by the brick is obtained by dividing the weight of the brick by the area over which the weight is distributed. Referring to the diagram, you will note that the pressure under the brick varies. When the brick is resting on its end, the pressure is the greatest. When the brick rests on its largest face, the pressure is the lowest.

The difference between force and pressure is very clear if someone wearing high-heeled shoes accidentally steps on your toes with the heel. You feel the pressure, and, unfortunately, the pain too! The weight of a woman is concentrated over a small area producing a huge amount of pressure. The same woman in a pair of sneakers would exert much less pressure (although your foot may still hurt). Remember that pressure is measured in force *per area*. The high heel transmits the force to a much smaller area than the area covered by the sneaker; the resulting pressure is larger.

Do You Know?

Pressure is a unitary rate. It describes the force exerted on each square centimeter of surface and can be used to find the total force exerted on any size area.

The concept of force per unit area is dramatically demonstrated when a person falls through thin ice and must be rescued from the cold water below. A rescuer can move more safely by lying flat on the ice, because the rescuer's weight (force) is distributed over a greater area than by moving on foot. The rescuer can reach the victim without falling to the same fate. The same principle allows a person to move over snow on skis, while someone in street shoes would simply sink.

7-3 Pressure of a Gas

How can the concept of pressure be applied to moving molecules of a gas? Perhaps an analogy can help you picture what is happening. Suppose you have a clear plastic box that contains some Ping-Pong balls and you begin to shake the box. What will happen? Of course, you know what you will see. The balls will begin to fly around, colliding with each other and with the walls of the box. Figure 7-5 illustrates what occurs. The balls seem to fill the entire box. If you were to place your hands on the outside of the box, you might feel the Ping-Pong balls bounce off the sides. Each time one rebounds from the wall, there is a push or force exerted on the box.

If this force is measured for a given unit of area on the wall of the box, the "pressure" of the moving Ping-Pong balls is obtained. In Figure 7-5, there are a small number of Ping-Pong balls in the box, so only a few collisions per second occur within the target area. The pressure exerted by the Ping-Pong balls is relatively low. On the other hand, the box in Figure 7-6 contains many Ping-Pong balls. More collisions with the target area occur per second because there are more balls that are moving. In this situation, the pressure of the Ping-Pong balls against the container walls at any one instant is greater than before. Can you think of a way of increasing the number of collisions with the target area in Figure 7-5 without adding more balls to the box?

Molecules of a gas behave much the same way as do the Ping-Pong balls in the box. When gas molecules strike the walls of their container, they exert a force on the walls. If this force is measured for a given unit of area, the gas pressure may be obtained. In SI units, pressure is measured in **pascals** (Pa). One pascal is equal to one newton (N) of force per square meter of area.

$$1 \, Pa = \frac{1 \, N}{m^2}$$

Gases have a tremendously wide range of pressures. The pressure of Earth's atmosphere on your body is 10^5 pascals. The pressure in a "filled" tank of compressed helium gas is 10^7 pascals. By contrast, the pressure of hydrogen gas in deep space is on the order of 10^{-23} pascals. This pressure is equivalent to one molecule of hydrogen gas per cubic meter. For all practical purposes, deep space is a **vacuum**, a system in which the pressure is zero pascals.

Why would it be useful to know the pressure of a gas? When working with gases, you need to know how much gas you have, but measuring the mass of a gas can be difficult. Measuring the volume is much easier. You have read that the volume of a gas is dramatically affected by pressure and temperature. As you will soon find out, the amount of

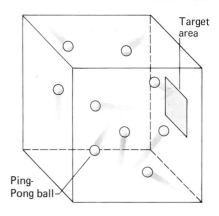

Figure 7-5 Each time the box is shaken, the Ping-Pong balls bounce in every direction. Collisions with the walls and other balls occur constantly. The force of the collisions on the marked target area is representative of the overall pressure the balls exert in the box.

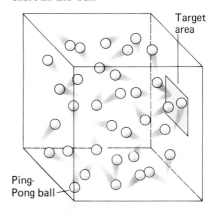

Figure 7-6 The pressure on the target area is greater here than in Figure 7-5. The force has increased because more Ping-Pong balls hit the target area at any instant.

Let students think of solutions to the questions on their own. Pressure may be increased in Fig. 7-5 by shaking the box more vigorously (increasing the energy). The greater mass of the balls would increase the force of collisions and, therefore, increase pressure.

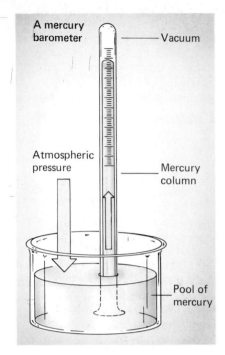

Figure 7-7 The atmospheric pressure supports the column of mercury at a specific height inside the tube. Atmospheric pressure balances the pressure exerted by the mercury column. If atmospheric pressure decreases, the column will fall. If atmospheric pressure increases, the column will rise.

Convince the class that the inside diameter of the tube will not affect the height of the mercury column. Pressure is force per unit area. In a wider tube, the force of the mercury will be greater but the area to which the force is applied is proportionally greater too.

Demonstrate the surprising force of air pressure. Rinse out an empty gallon metal can, e.g., a duplicator fluid can. Place about 200 cm³ of water in it. Heat (cap off) until the water is boiling vigorously. Remove the can from the heat. Immediately cover the can tightly and place it in a pan of cold water. The force of atmospheric pressure will crush the can. CAUTION: Do not reheat the sealed can!

If you have students who have had physics or who are good in math, you may want to ask them to show how mmHg are converted to pascals.

gas in a particular volume can be calculated easily if you know the pressure and temperature of the gas.

The question then becomes, how can the pressure of a gas be determined? The answer is to allow the gas to exert a pressure against something that exerts a pressure back—a pressure that can be measured. Often this something is a column of mercury.

Mercury is a convenient substance for measuring pressure for several reasons. It is dense so a small amount of mercury will balance rather large pressures. It does not easily react with other substances, and mercury is a liquid at normal temperatures, making it easy to construct an instrument to measure a wide range of pressures. Two such instruments are described here.

A **barometer** is any device that is used to measure Earth's atmospheric pressure. A simple experiment will illustrate how a barometer made with a column of mercury works. Starting with a meter-long glass tube that is sealed at one end, enough mercury is poured in to fill the tube. Then, while the open end of the tube is plugged, the tube is inverted into a pool of more mercury. When the plug is removed, what do you think will happen to the column of mercury in the tube? Remember that the upper end of the tube is sealed. Figure 7-7 depicts the result of the procedure. Some of the mercury will run out of the tube and into the mercury pool. Now no air has been able to get into the tube above the column of mercury, so a vacuum exists in the space at the sealed end of the tube. A vacuum cannot exert a force. Yet the mercury column in the tube must be supported by some force, otherwise all of the mercury in the tube would run out into the pool. The force is Earth's atmosphere, which is pushing down on the pool and is therefore supporting the column of mercury. When the force of the mercury is balanced by the force of the atmosphere, the mercury stops running out.

The apparatus shown in Figure 7-7 is calibrated to measure the height of the mercury column. The height can be converted to units of pressure.* As the atmospheric pressure decreases, the column of mercury drops. Similarly the column will rise if atmospheric pressure increases. The mercury-column barometer is only one of many kinds of instruments used to measure atmospheric pressure. Several different barometers are shown in Figure 7-8.

The average pressure at sea level is 101.325 kilopascals (101 325 pascals). Pressure also can be measured in **atmospheres**. One atmosphere is equivalent to 101.325 kilopascals. Both values are called **standard pressure**.

$$1 \text{ atm} = 101.325 \text{ kPa}$$

You should become proficient at interchanging pascals, kilopascals, and atmospheres because you will work with all of these units in your study of chemistry. When the temperature is 0°C and the pressure of a gas is 101.325 kilopascals (one atmosphere), standard conditions prevail. Standard temperature and pressure are often designated by the letters **STP**.

---

*A column of mercury 76 cm high exerts a pressure of 1 atmosphere (101.325 kPa). Knowing how to make the conversion is not necessary to understand how a barometer works.

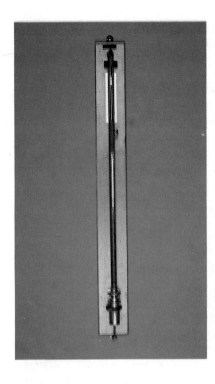

Figure 7-8 Weather forecasters, scientists, airplane pilots, and sailors are examples of people who need to know atmospheric pressure. As a result, barometers come in a variety of shapes and sizes. Do you recognize any of the instruments pictured here?

Do You Know?

At sea level on an average day, a cubic mile of air weighs 1.1 × 10^{10} lbs (equivalent to a mass of 5 × 10^9 kg). At an elevation of 220 miles, a cubic mile of air weighs 2 oz (a mass of 56.7 g).

Example 7-1

Change 99 500 pascals into kilopascals and then into atmospheres.

▶ **Suggested solution**

Since the prefix *kilo* means "one thousand of," the equivalence in pascals is

$$1 \text{ kPa} = 1000 \text{ Pa}$$

Therefore, to change 99 500 Pa to kPa

$$99\,500 \text{ Pa} \times \frac{1 \text{ kPa}}{1000 \text{ Pa}} = 99.5 \text{ kPa}$$

In order to solve the second part of the problem, you need to use the relationship between atmospheres and kilopascals:

$$101.325 \text{ kPa} = 1.00 \text{ atm}$$

The equation is written as

$$99.5 \text{ kPa} \times \frac{1.00 \text{ atm}}{101.325 \text{ kPa}} = 0.982 \text{ atm}$$

In each expression, check your answer by inspecting the arrangement of the units in the solution setup. Does the solution provide you with the units you want? Does the answer seem reasonable? The pressure in kilopascals should have a smaller numerical value than the pressure in pascals. The value in atmospheres should be smaller still.

Standard pressure in SI units is a rather large value for students to memorize. Regularly providing the value to them or having them round to 100 kPa are alternatives you may wish to consider.

STP can be a useful concept to know. However, after high school, students may see little of STP even in advanced chemistry courses. It is helpful for them to become accustomed to working with other conditions as well.

A device similar to a barometer can be used to measure the pressure of gases other than the atmosphere. Such an apparatus is called a **manometer**. Figure 7-9 illustrates how one kind of manometer operates. In this case, the closed mercury tube is connected by a U-bend to a flask containing the gas in question. Gas molecules moving within the flask rebound against the walls of the flask and the surface of the mercury. As the molecules push against the mercury, they exert a force which supports the column at a specific height. The difference in height of the mercury in the two sides of the U-tube can be used to find the gas pressure in pascals.

Figure 7-9 Gas molecules push against the mercury surface, exerting a pressure that can be measured by finding the difference in the height of the mercury columns. This value can be converted into SI units of pressure.

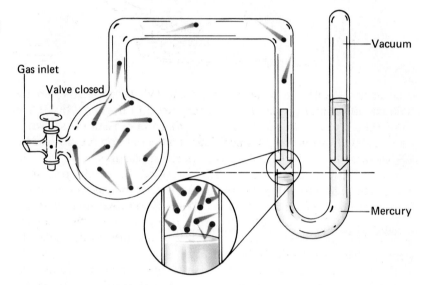

7-4 Dalton's Law of Partial Pressure

When you are working in the lab, collecting pure samples of gas is a problem. Basically the difficulty is getting an empty container to put the gas in. A glass bottle makes a fine container, but how do you empty the bottle to make room for the gas? As the introduction to this chapter mentions, Earth is surrounded by an ocean of air. Any "empty" bottle actually contains air. If you try to fill the bottle with another gas, the gas mixes with the air already present.

The common solution to this problem is to first fill the collecting bottle with water to get rid of the air, invert the bottle in a tank of water, and then bubble the gas into the bottle to displace the water. The setup is shown in Figure 7-10. In cases where the gas is not very soluble in water, this arrangement works very well, except for one thing. A little of the water evaporates and mixes with the gas being collected. Fortunately an easy adjustment can be made to account for the effect of the water. To see how this is done, you must understand what John Dalton discovered in 1801. The experiment shown in Figure 7-11 should help you understand Dalton's findings.

In Figure 7-11, there are three one-liter bulbs at 25°C. The first bulb contains 0.0050 mole of air. The pressure indicated by the manometer is equivalent to 12.4 kPa. There is 0.0011 mole of water vapor in the

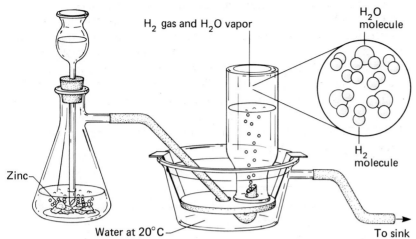

H₂ gas and H₂O vapor

H_2O molecule

H_2 molecule

Zinc

Water at 20°C

To sink

second bulb, and the pressure is 2.7 kPa. What happens when the volumes of air and water vapor are combined in the same bulb? The third bulb shows this situation. Notice that the pressure is shown to be 15.1 kPa. Provided no chemical reaction occurs, the pressure of the gas mixture is the sum of the two separate pressures.

The individual pressure of each gas is called the **partial pressure**. The partial pressure of a gas is the pressure that the gas would exert if it were the only gas in the container. There is so much space between molecules in a gas that molecules of another gas can readily share the space. Each gas behaves independently of the other and makes its own contribution to the total pressure.

The same addition technique successfully predicts the total pressure when several gases are mixed together. **Dalton's law of partial pressure** can be expressed in the form of an equation.

$$P_t = P_a + P_b + P_c \ldots P_z$$

P_t is the total pressure and P_a, P_b, P_c, etc., symbolize the partial pressures of the individual gases in the mixture.

In the laboratory, you will have to correct for the partial pressure of water vapor when you collect a gas over water. The problem that follows is typical of how to make such a correction.

Figure 7-11 The total pressure of a mixture of gases equals the sum of the partial pressures of all the gases in the container.

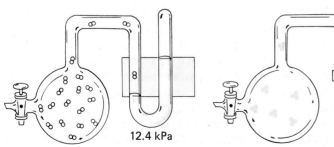

12.4 kPa

0.0050 mole air in one liter at 25°C

2.7 kPa

0.0011 mole water vapor in one liter at 25°C

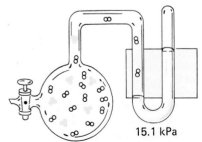

15.1 kPa

0.0050 mole air + 0.0011 mole water vapor in one liter at 25°C

Example 7-2

What is the pressure of hydrogen gas collected over water at 24.0°C? The water levels inside and outside the collecting bottle are equal at the end of the experiment. The atmospheric pressure is 94.4 kPa.

▶ **Suggested solution** _____

When a gas such as hydrogen is collected over water, the pressure inside the vessel is due to two gases. One gas is the H_2 gas. The other gas is the water vapor, whose pressure is dependent on temperature. (In Chapter 15, you will learn why this dependency occurs.) Dalton's law of partial pressure is expressed for this problem as

$$P_t = P_{H_2} + P_{H_2O}$$

The total pressure inside the container is the same as the pressure outside (atmospheric pressure) because the water levels inside and outside the tube are equal.

$$P_t = P_{atm}$$

By substitution, the equation for total pressure may be rewritten

$$P_{atm} = P_{H_2} + P_{H_2O}$$

The pressure of the water vapor at 24.0°C is 3.0 kPa. This value is constant for the temperature given in the problem and may be obtained by referring to a table of water vapor pressures. Substituting this value and the stated atmospheric pressure into the equation gives

$$94.4 \text{ kPa} = P_{H_2} + 3.0 \text{ kPa}$$

Rearranging and solving for P_{H_2} yields the pressure of the hydrogen gas:

$$P_{H_2} = 91.4 \text{ kPa}$$

Review and Practice _____

1. Briefly explain how a gas differs on the molecular level from a liquid or a solid.

2. How is pressure different from weight?

3. Explain why the atmospheric pressure is lower on top of a mountain than at sea level.

4. What causes the pressure of a gas? How would increasing the speed at which the molecules move affect the pressure of the gas?

5. A sample of nitrogen is collected over water at 21.5°C. The vapor pressure of water at 21.5°C is 2.6 kPa. What is the partial pressure of the nitrogen if the total pressure is 99.4 kPa?

1. Gas molecules are far apart from each other in comparison to the molecules in a liquid or a solid.
2. Weight is a force exerted by gravity on a mass. Pressure is a force per unit area.
3. On a mountaintop there is less atmosphere "overhead" than at sea level.
4. Gas molecules collide with the sides of the container. Gas pressure increases.
5. 96.8 kPa

Behavior of Gases

A part of science deals with studying relationships between the structure of matter and its behavior. For example, geologists investigate the relationship between rock structure and earthquakes. From the data gathered by such investigations, geologists hope to develop a theory which will successfully predict future earthquakes.

So far, you have explored some of the general characteristics of gases. Now you will study the relationships between different physical properties of a gas. Volume, temperature, pressure, and number of moles will be considered. You will investigate these properties to determine what regularities exist among them. Do gases behave in a predictable manner? Is there a theory which explains the observed relationships of a gas?

CONCEPTS

■ relationship between temperature and volume
■ relationship between pressure and volume
■ combined effects of pressure, temperature, and volume
■ nature of an ideal gas
■ ideal gas law
■ the meaning of molar volume

7-5 Charles's Law

At the beginning of the nineteenth century, science took to the air. Ballooning was the rage. Two French scientists who had a thirst for adventure made great strides in the art and science of ballooning. They were Jacques Charles and Joseph Gay-Lussac. Both men developed hydrogen-filled balloons. In 1804, Joseph Gay-Lussac used a hydrogen-filled balloon to ascend to 7000 meters. This record stood unmatched for nearly 40 years.

A hot-air balloon, like the one shown in Figure 7-12, is open at the bottom in order for a heat source to increase the temperature of the air inside. As the air in the balloon gets warmer, it becomes less dense than the surrounding atmosphere. The balloon rises for the same reason that a piece of wood floats on water—the balloon, like the wood, is less dense than its surroundings. The question in the case of the

Do You Know?

Around the World in Eighty Days *is a famous book written by Jules Verne in 1873. It tells the rollicking, adventurous tale of an eccentric gentleman who bet his fortune that he could travel around the world by hot-air balloon in 80 days.*

Figure 7-12 The adventurous souls who today launch themselves aloft in hot-air balloons for fun and sport, can find their roots with the balloonists of the early 1800's. Note the open-ended bottom of the balloon that exposes the air to the heat source.

hot-air balloon is, why does the air become less dense? What is happening when the air is heated that causes a change in the gas?

Jacques Charles is given credit for correctly describing a fundamental relationship between the volume of a gas and its temperature. **Charles's law**, as the relationship is now called, can be demonstrated easily, using apparatus such as that shown in Figure 7-13.

Figure 7-13 The ammonia gas is at room temperature in the syringe on the left. On the right, the ammonia is being heated as the water is heated. The plunger moves up as the volume of the gas increases.

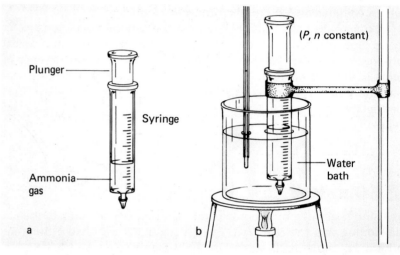

Do You Know?

As is often the case with scientific discoveries, Charles was not alone in his study of gas behavior. Both Dalton and Gay-Lussac observed regularities involving gas temperature and volume. However, Dalton's data were not very accurate, and Gay-Lussac deferred the privilege of scientific discovery to his fellow countryman.

A calibrated syringe is partially filled with ammonia gas at room temperature, and the volume is read. The syringe is then placed in a beaker of water, as shown in Figure 7-13b. The plunger of the syringe can move up and down to allow a constant pressure to be maintained. As the water is heated, the gas expands. The new volume is measured at increments of 5°C. The amount of gas in the syringe does not change. Table 7-2 shows the data that have been collected from such an experiment.

Table 7-2

VOLUME OF A SAMPLE OF NH_3 AT A CONSTANT PRESSURE OF 1 ATM		
TEMPERATURE (°C)	**VOLUME (cm³)**	**v/t (cm³/°C)**
20.0	64.8	3.24
25.0	64.9	2.60
30.0	66.4	2.21
35.0	68.1	1.95
40.0	68.6	1.72
45.0	70.0	1.56
50.0	71.3	1.43

If you look closely at the data, you will notice a regularity. As the temperature of the trapped ammonia gas increases, the volume of the trapped gas also increases. This relationship holds true for most gases at moderate temperatures when the pressure and the number of moles remain constant.

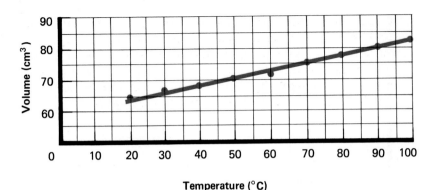

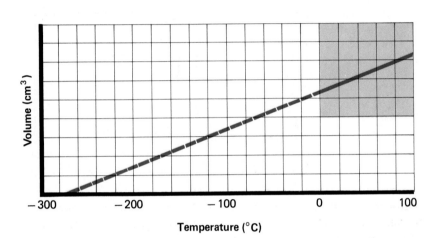

Figure 7-14 A graph (top) of the data in Table 7-2 indicates that the volume of the ammonia gas increases with increasing temperature. However, if the line is extended (bottom), it does not pass through the origin as it would for a direct proportion.

A graph of the data in Table 7-2 appears in Figure 7-14 (top). Recall from your study of graphs in Chapter 3 that the slope of this graph looks like one for a direct proportionality. If that is the case, then V/t should equal a constant. However, as you can see from the third column of the data table, the values calculated do not show a constant relationship. Furthermore, if the line of the graph is extended to meet the x-axis (Figure 7-14, bottom), the point of intercept is at $-273°C$ rather than at the origin like the graphs shown in Chapter 3. If you could set up a different temperature scale so that the graph passes through the origin, then V/t would equal a constant.

Figure 7-15 (next page) shows a graph of the same data using such a temperature scale. The scale is arrived at by redefining the zero temperature as the point at which the graph crosses the x-axis. This new temperature scale varies from the Celsius scale by 273 units. The scale is called the **kelvin scale** and is divided into increments called **kelvins**. If T represents temperature in kelvins and t represents temperature in Celsius degrees, then the relationship between the two temperature scales may be expressed as

$$T = t + 273$$

For example, 0°C is the same as 273 kelvins.

$$T = 0°C + 273$$
$$T = 273 \text{ K}$$

Figure 7-15 When the data from Table 7-2 are graphed using the Kelvin scale, the extended line passes through the origin. In actuality, the gas would liquify before reaching 0 kelvins.

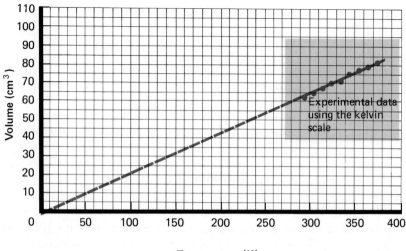

It may be helpful to point out that the kelvin scale also includes another constant relationship. The volume of a gas changes by 1/273 for every temperature change of 1 K. This could be expressed as a unitary rate $\left(\dfrac{\frac{1}{273}\text{V}}{1\text{ K}}\right)$ if it is desirable to reinforce that idea.

Note that kelvin temperature is read as "273 kelvins" and not "273 degrees kelvin." The degree sign is not used when writing kelvin temperatures. Table 7-3 shows the data plotted in Figure 7-15 using the new temperature scale.

Table 7-3

VOLUME OF A SAMPLE OF NH_3 AT A CONSTANT PRESSURE OF 1 ATM		
TEMPERATURE (K)	VOLUME (cm³)	V/T (cm³/K)
293	64.8	0.221
303	65.8	0.217
313	68.6	0.219
323	71.4	0.221
333	72.9	0.219
343	75.5	0.220
353	78.0	0.221
363	80.2	0.221
373	82.4	0.220

Why go to all this trouble of changing temperature scales? Remember that you are looking for a way to express a proportionality between the volume of a gas and its temperature. In Chapter 3, you learned that if the line on a graph of a proportionality passes through the origin, you can express the proportionality as

$$\frac{A}{B} = k \text{ (where } k \text{ is a constant)}$$

In this case, A is volume and B is temperature, so

$$\frac{V}{T} = k \text{ (when T is in kelvins; moles and pressure are constant)}$$

The constant k is a unitary rate telling you how much the volume changes for every change in temperature of 1 K. (Be careful not to confuse k, the constant, with K, the symbol for kelvin temperature.) Note the value of k for these data is about 0.220 cm³/K. Therefore the volume of the gas changes by 0.220 cm³ per kelvin.

Recall from Chapter 3 that the unitary rate also can be determined by finding $\Delta y/\Delta x$. Prove to yourself that 0.220 cm³/K is correct by calculating $\Delta y/\Delta x$ for the data graphed in Figure 7-15. (Remember, when using experimental data to find $\Delta y/\Delta x$, it is best to choose points from the best-fit line rather than adjacent data points on the table. The actual data points have some amount of uncertainty. The uncertainty is reduced by drawing the best-fit line when making the graph.)

The analysis of the data for this example is typical of any gas at moderate temperatures. Charles's law recognizes that *when the pressure and number of moles of a gas are held constant, the volume of the gas is directly proportional to its kelvin temperature.* For any sample of gas, a constant may be calculated for the V/T ratio. This constant is a unitary rate that can be used to find the volume of that particular sample at any other temperature (pressure and number of moles being constant).

Example   7-3

If a sample of gas occupies 100 cm³ at 200 K, what volume would it occupy at a temperature of 150 K? Pressure is held constant.

▶ **Suggested solution**

Since Charles's law states that $V/T = k$ when the pressure and amount of gas do not change, you can use the information given in the problem to find the value of the constant k.

$$\frac{V}{T} = \frac{100 \text{ cm}^3}{200 \text{ K}} = k$$

Dividing yields

$$\frac{0.50 \text{ cm}^3}{\text{K}} = k$$

This constant k is a unitary rate indicating that the volume changes 0.50 cm³ each time the temperature changes 1 K. The problem asks for the volume at 150 K. The problem is solved in the same way you have solved previous problems involving unitary rates:

$$? \text{ cm}^3 = 150 \cancel{K} \times \frac{0.50 \text{ cm}^3}{\cancel{K}}$$

$$? \text{ cm}^3 = 75.0 \text{ cm}^3$$

Check to see if the answer is reasonable. The temperature has decreased, so you would expect the volume to decrease. The new volume is indeed smaller than the old. The units work out correctly too, which should increase your confidence in the answer.

An alternate way of using the relationship $V/T = k$ is to set the new conditions equal to the constant you calculated.

$$\frac{V}{150 \text{ K}} = \frac{0.50 \text{ cm}^3}{1 \text{ K}}$$

Solve the equation by multiplying both sides by 150 K:

$$V = 150 \cancel{\text{K}} \times \frac{0.50 \text{ cm}^3}{\cancel{\text{K}}} = 75.0 \text{ cm}^3$$

Notice that this is exactly the same mathematical operation you did when k was treated as a unitary rate.

Example 7-4

A gas occupies a volume of 473 cm³ at 36.0°C. What will be the volume of the gas when the temperature is raised to 94.0°C? Assume that the pressure and number of moles are held constant.

▶ **Suggested solution**

This problem is very similar to example 7-3. The temperature of a gas is changed, and you are asked to find the new volume. In this case, the temperature has increased. Based on Charles's law, you should expect the volume to increase also. As in the previous problem, you can use the information given to find the unitary rate k. However, notice that the temperature in this case is measured in degrees Celsius. You must convert to kelvins in order to solve the problem. Recall that if T is kelvin temperature and t is Celsius temperature, then

$$T = t + 273$$

The initial temperature is 36.0°C. The kelvin temperature is calculated as

$$T = 36.0 + 273$$
$$T = 309 \text{ K}$$

To find the unitary rate, you can again use the equation for Charles's law $V/T = k$. Substituting the values for V and T given in the problem, you get

$$\frac{473 \text{ cm}^3}{309 \text{ K}} = 1.53 \text{ cm}^3/\text{K}$$

The volume of the gas changes 1.53 cm³ for every temperature change of 1 K. The problem asks for the volume at 94.0°C. Again, remember to change to kelvin temperature before working out the rest of the problem.

$$T = t + 273$$
$$T = 94.0 + 273 = 367 \text{ K}$$

Using the kelvin temperature and the unitary rate you calculated, you can find the new volume:

$$? \text{ cm}^3 = 367 \text{ K} \times \frac{1.53 \text{ cm}^3}{\text{K}}$$

Multiplying and adjusting for significant figures, the new volume is

$$? \text{ cm}^3 = 562 \text{ cm}^3$$

The new volume is larger than the old volume, which is what you expected.

In chemistry, as in many other subjects, there is more than one way to solve a problem. Another approach to using Charles's law is presented below. The following problem is the same as Example 7-4, but the solution is different. As you read the suggested solution, note the similarities to what you have already seen.

Example 7-5

A gas sample occupies a volume of 473 cm³ at 36.0°C (309 K). What will be the volume of the gas when the temperature is raised to 94°C (367 K)? Assume that the pressure is held constant.

Suggested solution

The problem states that the temperature of the gas has increased. Based on Charles's law, you should expect the volume to increase also. In fact, the volume will change in the same ratio as the change in temperature because volume and temperature are directly proportional. The new volume will be equal to the original volume multiplied by the temperature ratio. If V_1 represents the original volume and V_2 is the new volume, then

$$V_2 = V_1 \times \text{temperature ratio}$$

There are two ways of setting up the temperature ratio. It may be written as

$$\frac{309 \text{ K}}{367 \text{ K}} \quad \text{or} \quad \frac{367 \text{ K}}{309 \text{ K}}$$

How do you know which one to use? If you expect the volume to increase, then you would want to multiply by a ratio that is greater than 1. The correct ratio then must be 367 K/309 K. Substituting into the equation gives

$$V_2 = 473 \text{ cm}^3 \times \frac{367 \text{ K}}{309 \text{ K}}$$

Remind students that their answers may vary slightly depending on the method they choose to solve a problem. Rounding to significant figures during calculations will affect answers slightly. When several calculations are done in succession to arrive at a final result, it is good practice to carry at least one digit more than will show when the final answer is rounded to the proper number of digits. If this is done, no rounding error is introduced.

Solving for V_2 and adjusting for significant figures, you get

$$V_2 = 562 \text{ cm}^3$$

The new volume is larger than the original volume, and the units produced are correct for volume so the answer seems reasonable.

Returning to the question about the hot-air balloon, perhaps you now can see why it rises. As the air inside the balloon is heated, it expands and some of it is forced out of the open bottom. The warm air is now less dense than the surrounding atmosphere, and the balloon rises.

Do You Know?

Scuba divers need to be aware of Boyle's law. For every 10 meters in water depth, the pressure is increased by one atmosphere. A four-liter balloon would be reduced to one liter at a depth of only 30 meters.

7-6 Boyle's Law

When you use a bicycle pump to inflate a tire, you are using another regularity of gas behavior. As you push down on the bicycle pump, you reduce the volume of the gas inside the pump. At the same time, the pressure of the gas inside the pump increases. Air then flows into the tire because the pressure in the pump is greater than the pressure in the tire.

The principle involved can be demonstrated using the Ping-Pong ball model described earlier. Figure 7-16 shows a box with a movable piston on the top. Inside, the balls are moving, colliding with each other and the container walls. If the piston is lowered so that the available space is decreased, the number of collisions will increase. As you have already learned, if the number of collisions with the container walls increases, the pressure will increase. Provided the

Figure 7-16 As the piston is lowered, the space available to the moving balls decreases. The number of collisions with the walls of the container increases, and pressure goes up.

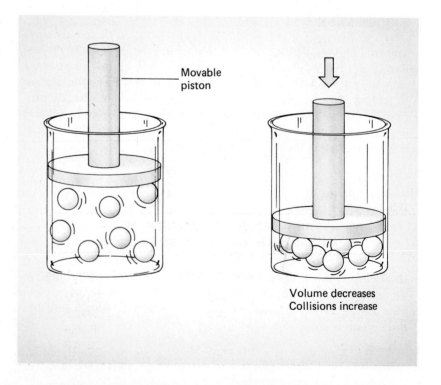

Movable piston

Volume decreases
Collisions increase

temperature and the number of moles are constant, as volume goes down, pressure goes up and vice versa.

In 1660, Robert Boyle, a British scientist, performed an experiment that measured the volume of a gas as the pressure on the gas changed. **Boyle's law** describes the observed relationship between pressure and volume when temperature and moles are held constant. The principle of Boyle's law may be demonstrated in a simple experiment such as the one illustrated in Figure 7-17.

A calibrated syringe is filled with air, and the volume is read. A weight placed on top of the plunger increases the pressure on the air in the syringe. While temperature and number of moles of gas are held constant, more weight is added above the plunger. The volume is measured for several different pressures. Table 7-4 shows the data that have been collected from such an experiment. A glance at the data in Table 7-4 shows you that as the pressure is increased from 66.7 kPa to 222.3 kPa, the volume of the gas decreases from 50 cm³ to 15 cm³. If the

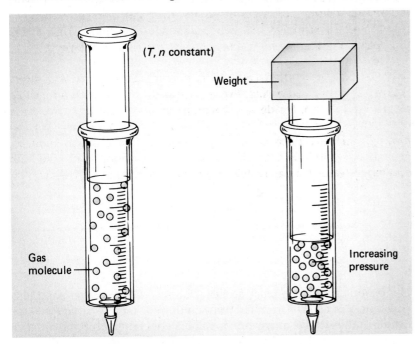

Figure 7-17 As the pressure on a gas increases, the volume of the gas decreases.

Table 7-4

COMPRESSION OF A SAMPLE OF AIR (temperature and moles are constant)		
VOLUME (cm³)	PRESSURE (kPa)	V × P (cm³·kPa)
50.0	66.7	3335.0
31.0	107.5	3332.5
21.0	158.6	3330.6
15.0	222.3	3334.5

data are graphed, as in Figure 7-18, the result is a curved line. Note that this graph is different from the graph illustrating Charles's law. Charles's law describes a direct proportionality in which volume increases with increasing temperature. In the case of pressure and volume, the relationship is an *inverse* (or indirect) proportionality. Volume decreases with increasing pressure. Mathematically, an inverse proportionality may be expressed as

$$AB = k \text{ (where } k \text{ is a constant)}$$

In this case, A is pressure and B is volume so that the equation may be rewritten as

$$PV = k \text{ (when temperature and moles are constant)}$$

Figure 7-18 A plot of the data in Table 7-4 indicates a relationship different from Charles's law. The pressure and volume of a gas are inversely related. When one increases, the other decreases.

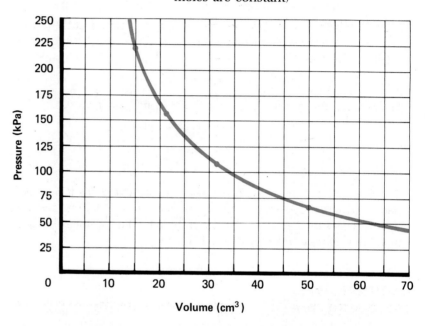

In the case of Boyle's law, the ratio of the pressure and the reciprocal of the volume ($1/V$) are a direct proportion. Thus $P/\dfrac{1}{V}$ describes a unitary rate.

However, reciprocal units of volume are difficult to visualize, and the unitary rate lacks the kind of meaning normally associated with such rates.

The product of the pressure and the volume of any one gas sample is constant. You can see from the third column of Table 7-4 that $P \times V$ for the air in the syringe at any one instant is constant within the range of experimental uncertainty. Boyle's law recognizes that *when the temperature and the number of moles are held constant, the volume of a gas is inversely proportional to the pressure applied on the gas.*

Although it is possible to calculate unitary rates for the relationship described by Boyle's law, it is not particularly useful to do so. It is easier to solve problems involving inverse proportions by simply keeping in mind what is constant. The following example illustrates how this can be done.

Example 7-6

If a sample of gas has a volume of 100 mL when the pressure is 150 kPa, what is its volume when the pressure is increased to 200 kPa? Temperature and amount of gas are constant.

▶ Suggested solution _____

The problem states that the pressure is increasing. Since the temperature and the number of moles remain constant, Boyle's law seems to apply to this situation. Therefore, you can expect the volume of the gas to decrease and the product of pressure and volume to be constant.

$$PV = k$$

The value of the constant can be found from the information in the problem.

$$150 \text{ kPa} \times 100 \text{ mL} = k$$
$$15\,000 \text{ kPa·mL} = k$$

Knowing the constant for this sample of gas, you can find the volume of the gas at any pressure. The problem asks for the volume at 200 kPa. Therefore,

$$200 \text{ kPa} \times ? \text{ mL} = 15\,000 \text{ kPa·mL}$$

Rearranging the equation to solve for volume, you get

$$? \text{ mL} = \frac{15\,000 \text{ kPa·mL}}{200 \text{ kPa}}$$

$$? \text{ mL} = 75.0 \text{ mL}$$

Check your answer to see if it makes sense. The volume was expected to decrease. According to the solution, the volume has indeed decreased. The arrangement of the equation also leaves units of volume for the answer, which is correct.

An interesting demonstration of Boyle's law is to place a marshmallow into a thick-walled 250-cm³ flask. Aspirate the air in the flask. The marshmallow will expand due to the reduced pressure.

Example 7-7

What pressure would be required to increase the volume of the gas described in the previous example to 150 mL?

▶ Suggested solution _____

You already know the PV constant for this gas sample is 15 000 kPa·mL. Substitution of the constant and the new volume into Boyle's equation should yield the new pressure.

$$PV = 15\,000 \text{ kPa·mL}$$
$$? \text{ kPa} \times 150 \text{ mL} = 15\,000 \text{ kPa·mL}$$

Rearranging the equation to solve for pressure gives

$$? \text{ kPa} = \frac{15\,000 \text{ kPa·mL}}{150 \text{ mL}}$$

$$? \text{ kPa} = 100 \text{ kPa}$$

Increasing the volume will lead to a reduction in pressure and that is what the answer suggests. The units are appropriate, so the answer looks reasonable.

As when working with Charles's law, in the case of Boyle's law there is more than one way to solve a problem. The following is another solution to problem 7-6. Note the similarities to the solutions you have already seen.

Example 7-8

If a sample of gas has a volume of 100 mL when the pressure is 150 kPa, what will the volume be of the gas if its pressure is increased to 200 kPa? Temperature and amount of gas are constant.

▶ **Suggested solution** _____

The problem states that the pressure on the gas has increased from 150 kPa to 200 kPa. From Boyle's law, you can predict that the volume will decrease. The original volume changes in a ratio related to the pressure change because pressure and volume are inversely proportional.

$$V_2 = V_1 \times \text{pressure ratio}$$

What pressure ratio is the correct one to use? Again, there are two possibilities:

$$\frac{150 \text{ kPa}}{200 \text{ kPa}} \quad \text{or} \quad \frac{200 \text{ kPa}}{150 \text{ kPa}}$$

If you expect the volume to decrease, then you would want to multiply by a ratio that is less than 1. The correct ratio then must be 150 kPa/200 kPa. Substituting into the equation gives:

$$V_2 = 100 \text{ mL} \times \frac{150 \text{ kPa}}{200 \text{ kPa}}$$

Solving the equation for V_2 gives

$$V_2 = 75.0 \text{ mL}$$

As predicted, the new volume is smaller than the original volume. The answer is the same as the one found in Problem 7-6.

Review and Practice _____

1. State Charles's law in words.

2. If a sample of gas has a volume of 3.00 dm³ (or liters) at a temperature of 52.3°C, what will the volume be if the temperature of the gas system is lowered to −27.0°C? Assume that the pressure is held constant. Remember to use the Kelvin temperature scale.

3. A bicycle tire pump has a volume of 1.0 dm³ inside the cylinder at a pressure of 1.1 kPa. What will the volume be inside the cylinder when the pressure is increased to 315 kPa? Assume that the temperature remains constant.

1. The volume of a gas is directly proportional to its Kelvin temperature when pressure and moles are held constant.
2. 2.27 dm³
3. 3.5 × 10⁻³ dm³

4. The pressure of a 5.71-L sample of neon gas is 23.4 kPa. Calculate the new pressure when the volume becomes 3.40 L. Assume that the temperature and number of moles of gas remain unchanged.

5. Use the solution method outlined in Example 7-5 to solve Example 7-3.

6. Use the solution method outlined in Example 7-8 to solve Example 7-7.

4. 39.3 kPa

5. $V_2 = 100 \text{ cm}^3 \times \dfrac{150 \text{ K}}{200 \text{ K}}$

   $V_2 = 75 \text{ cm}^3$

6. $P_2 = 150 \text{ kPa} \times \dfrac{100 \text{ mL}}{150 \text{ mL}}$

   $P_2 = 100 \text{ kPa}$

7-7 Application: Why Popcorn Pops

Almost everyone loves popcorn. Americans eat 10 billion liters of it each year. How does something as unappetizing as a kernel of corn become so delectable? Part of the answer lies in the starchy composition of corn and part in the physical behavior of gases.

A kernel of corn is composed mostly of carbohydrates and water. The kernel has three parts. The outer shell is called the hull (pericarp). The inside of the kernel is composed of starch (endosperm) and an embryo plant (see Figure 7-19). The endosperm acts as food for the embryo as it germinates.

When the kernels are placed into heated oil or a hot-air popper, the water inside the kernel becomes superheated, which means it is hot enough to vaporize but does not have the room to do so. The rapidly moving water molecules cause a tremendous increase in pressure on the hull (the kernel's container). The superheated water penetrates the starch structure under a pressure of about 900 kPa. Because the pressure is so great, the hull ruptures. The pressure surrounding the starch is then greatly reduced (to about 100 kPa, or atmospheric pressure). As the pressure drops, the water vaporizes and expands as described by Boyle's law. This change causes the starch to expand to about 30 times its original size. The expansion is so rapid that the kernel explodes, or pops.

Figure 7-19 Under proper conditions, a corn kernel produces a snack that so many people love to eat. The white fluff of the popcorn kernel is the expanded endosperm.

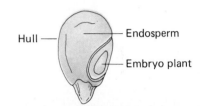

7-8 Combining the Gas Laws

For most gas systems like a bicycle pump, a change in pressure also will be accompanied by a change in temperature. The pump will get hotter as the air inside is compressed. Neither Charles's law nor Boyle's law alone can be used to solve such a problem. You have to account for the effects that both of these laws describe in order to solve a problem in which temperature and pressure change. The method used to solve Examples 7-5 and 7-8 are helpful in this situation.

Example 7-9

A sample of oxygen gas has a volume of 7.84 cm³ at a pressure of 71.8 kPa and a temperature of 25.0°C. What will be the volume of the gas if the pressure is changed to 101 kPa and the temperature is changed to 0°C?

▶ Suggested solution

In this problem, temperature is decreasing, so the volume of the gas should decrease. Pressure is increasing; so based on Boyle's law, you would expect the volume to decrease even more. The original volume changes in a ratio related to the temperature change and in a ratio related to the pressure change. An equation that takes both changes into account may be written as

$$V_2 = V_1 \times \text{temperature ratio} \times \text{pressure ratio}$$

Since the volume is expected to decrease, the temperature ratio must have a value less than 1. The correct temperature ratio is (remembering to change to kelvins)

$$\frac{273 \text{ K}}{298 \text{ K}}$$

The volume also decreases as a result of the pressure change, so the pressure ratio is

$$\frac{71.8 \text{ kPa}}{101 \text{ kPa}}$$

Substituting these ratios and the original volume into the equation gives

$$V_2 = 7.84 \text{ cm}^3 \times \frac{273 \cancel{K}}{298 \, K} \times \frac{71.8 \, \cancel{kPa}}{101 \, \cancel{kPa}}$$

Working out the calculations yields the new volume:

$$V_2 = 5.11 \text{ cm}^3$$

The solution to a problem such as this one also can be done as a stepwise process using methods you learned in Examples 7-3 and 7-6. First, imagine the pressure to be constant, and find a new volume using the unitary rate for changing temperature. Then, imagining the temperature to be constant and using the volume *just calculated,* find the *PV* constant for the original pressure. Use this constant and the final pressure to find the final volume. Prove to yourself that this method works by trying it yourself. Your answer may vary slightly from the one found in Example 7-9 due to variation in significant figures.

Review and Practice

1. When 3.5 dm^3 of carbon dioxide gas at 2.0 atm and 56°C is heated and compressed, the new pressure is 3.0 atm and the new temperature is 109°C. Find the new volume.

1. 2.71 dm^3
2. 238 mL
3. Temperature decreases with decreasing pressure and increasing volume just after the kernel pops.

2. The volume of a sample of gas is 200 mL at 275 K and 92.1 kPa. What will the new volume be of the gas at 350 K and 98.5 kPa?

3. Based on your knowledge of the behavior of gases, tell how the temperature within a popcorn kernel differs just before and after it pops.

7-9 An Ideal Gas and the Kinetic Molecular Theory

Look again at the graph in Figure 7-15. You may notice that when the kelvin temperature is zero, the volume of gas seems to be zero also. Since there is no such thing as negative volume, zero kelvins must be the lowest temperature possible. This temperature is known as **absolute zero**. (The kelvin scale is sometimes called the absolute scale.) Of course, a gas cannot really have zero volume because the gas molecules themselves occupy a certain amount of space. Zero volume at absolute zero is only a characteristic of an ideal gas.

What in the world is an ideal gas? The word *ideal* is often used to describe something that conforms perfectly to a given set of conditions. This definition is applied to the behavior of gases, although there is really no such thing as an ideal gas. At this point, you may be asking yourself, why study something that does not exist? Using ideal gases, scientists developed a theory to explain the gas behavior. You soon will see how this theory can be applied to real gases. At moderate temperatures and pressures, ordinary gases act as if they were ideal.

No single concept defines an ideal gas. Around 1860, two scientists, Ludwig Boltzmann from Germany and James Clerk Maxwell from Scotland, proposed the **kinetic molecular theory** to describe the behavior of an ideal gas. The theory includes four sets of conditions to which an ideal gas would conform.

1. *The molecules of an ideal gas are considered to be dimensionless points that have no volume.* Do you remember from your study of geometry that a point is considered to be dimensionless? It has no length, width, or height, and therefore no volume. In Section 7-1, you learned that the volume of gas molecules is very small in comparison to the total volume of the gas in its container. As a result, the molecular volume is ignored in most gas calculations.

2. *The molecules of a gas are in constant straight-line motion.* This motion is only interrupted by the collision of the molecules with the walls of the container and with each other. Recall the Ping-Pong ball analogy in Section 7-3. The balls travel in a straight line until they hit the inside walls of the container or each other. The average distance between two successive collisions is the **mean free path**. The mean free path is shorter when molecules are crowded together.

3. *The collisions that occur between molecules are perfectly elastic.* No energy is lost or gained when two molecules collide or when a molecule hits the wall of the container. In the case of the Ping-Pong balls, if energy were lost in collisions, the balls would eventually settle to the bottom of the container. The pressure would become zero.

4. *The molecules of an ideal gas do not exert any attractive forces on each other.* If there were attractions between the molecules, they would eventually bunch together and the pressure would decrease. Refer again to the Ping-Pong ball analogy. Think about how the pressure in the container would be affected if the balls were coated with molasses.

Do You Know?

In very efficient vacuums produced in the laboratory, the pressure is 10^{-11} kPa. At this low pressure, the mean free path of a gas molecule is 2500 kilometers.

A super ball dropped (not bounced) from shoulder height will not return to your hand because its elasticity is not perfect. Demonstrate elasticity using several kinds of balls.

Have students relate Boyle's law to the KMT. Use the Ping-Pong ball analogy. Increased temperature increases velocity and, therefore, the number of collisions. Momentum of each collision also increases. Pressure will increase unless volume increases proportionally to temperature.

The kinetic molecular theory (KMT) is one of the most enduring theories in science. It has been reformulated very little since the days when Boltzmann and Maxwell first proposed it. The theory explains the behavior of most gases you will study in chemistry. A real gas behaves as if it were ideal so long as conditions of pressure and temperature are moderate. When the pressure on a gas is large or when the temperature of the gas is low or both, the KMT no longer applies and gases behave differently. Absolute zero is an example of an extreme condition under which Charles's and Boyle's theories no longer work. Real gases liquify as they cool and cannot compress to zero volume because their molecules are not dimensionless points.

In this chapter, the KMT is used as a basis for explaining the relationships of gases on the molecular scale. Keep this theory in mind for the discussions that follow.

Do You Know?

Some metals become super conductors near absolute zero. Electrons experience almost no resistance in flowing through the metal.

7-10 The Ideal Gas Law

Charles's law and Boyle's law summarize some interesting information about gases, but they are of little value to chemists. As mentioned earlier, the chemist is interested in the behavior of gases because a convenient method is needed for finding the amount of gas in a sample. Chemists usually want to know how many moles of a substance are present. Neither of the laws described help them to do that. A relationship derived from the kinetic molecular theory does. It is called the **ideal gas law**, and it is stated mathematically as

$$PV = nRT$$

P, V, and T have their usual meanings; n is the number of moles of gas, and R is a constant known as the **ideal gas constant**. The derivation of this equation is beyond the scope of this book, but it is easy to show that the equation is equivalent to the two laws just discussed.

You have learned that Boyle's law applies when the number of moles and the temperature of the gas are constant:

$$PV = k \ (n, \ T \text{ constant})$$

In the ideal gas equation, R is a constant. If n and T also are constant, then the product nRT must be constant:

$$PV = nRT = k$$

Therefore,

$$PV = k$$

This equation is Boyle's law.

In the case of Charles's law, V/T equals k when the number of moles and the pressure are held constant. The ideal gas equation may be rearranged as follows to solve for V/T:

$$\frac{V}{T} = \frac{nR}{P}$$

If n and P are constant, then the term on the right side of the equation is constant:

$$\frac{V}{T} = \frac{nR}{P} = k$$

Therefore,

$$\frac{V}{T} = k$$

When P and n are held constant, the ideal gas equation is identical to Charles's law.

The advantage of the ideal gas equation is that it is a single equation that holds for any gas under moderate conditions of temperature and pressure. There is still a constant, R, in the expression, and it is evaluated in the same way that k was evaluated for Boyle's law or Charles's law. R can be calculated from experimental data as in the following example.

Example 7-10

Precisely 1 mole (32.0 g) of oxygen gas was collected in the laboratory at a temperature of 24.0°C and a pressure of exactly 100.0 kPa. The volume occupied by the mole of gas at this temperature and pressure was measured as 24.686 L. What is the value of R?

▶ **Suggested solution** _____

The ideal gas equation is

$$PV = nRT$$

Rearranging the equation to solve for R gives

$$R = \frac{VP}{nT}$$

The problem gives values (or information from which the value can be determined) for everything in the equation except R. Make a list of what values are known:

$$P = 100.0 \text{ kPa}$$
$$V = 24.686 \text{ L}$$
$$n = 1.00 \text{ mol (the molar mass of O}_2 \text{ is } 32.0 \text{ g/mol)}$$
$$T = 297 \text{ K}$$

The equation is valid only for kelvin temperature, so 24.0°C is converted to kelvins. These values now may be substituted into the rearranged ideal gas equation:

$$R = \frac{(24.686 \text{ L})(100.0 \text{ kPa})}{(1.00 \text{ mol})(297 \text{ K})}$$

Once the calculations are done, the value of R is

$$R = 8.31 \frac{\text{L·kPa}}{\text{mol·K}}$$

The numerical value of R is limited to three significant digits because the mass of oxygen was given to three digits. Notice the units for R. The ideal gas equation holds true only for kelvin temperatures. Also, the amount of substance used is always measured in moles, so neither of these units changes. However, pressure and volume may be expressed in other units. A value for R can be calculated for conditions when these other units are used.

Example 7-11

Convert the value of R just calculated to units of L·atm/mol·K.

▶ **Suggested solution**

Earlier in this chapter (Section 7-3) you were given the relationship between atmospheres and kilopascals:

$$1.00 \text{ atm} = 101.325 \text{ kPa}$$

Using this relationship, you can change units of pressure to atmospheres as

$$R = 8.31 \frac{\text{L·kPa}}{\text{mol·K}} \times \frac{1 \text{ atm}}{101.325 \text{ kPa}}$$

Dividing, you get

$$R = 0.0820 \frac{\text{L·atm}}{\text{mol·K}}$$

Note that the only units to change in the operation are those for pressure. When solving problems using the ideal gas equation, be sure to use a value for R that is consistent with the units in the problem. Do you need to memorize different values of R to match any possible combination of units given in a problem? The answer is no. In addition to finding R by the methods shown in Examples 7-10 and 7-11, you can convert the values and units in a problem to units matching a value for R that you already know.

When students work on gas law problems, you may find it worthwhile to provide values of R as an alternative to having them memorize several constants.

Example 7-12

Calculate the pressure of 1.65 g of helium gas at 16.0°C and occupying 3.25 L.

▶ **Suggested solution**

The problem provides data on the number of moles (calculated from the mass), pressure, volume and temperature, so the ideal gas law can be used to solve it.

$$PV = nRT$$

Rearranging the equation to solve for P gives

$$P = \frac{nRT}{V}$$

Volume is given in the problem. Temperature is provided in Celsius degrees, so it must be converted to kelvins:

$$T = 16.0°C + 273 = 289 \text{ K}$$

The molar mass of helium is 4.00 g/mol, so in this problem, the number of moles is

$$1.65 \text{ g} \times \frac{1 \text{ mol}}{4.00 \text{ g}} = 0.412 \text{ mol}$$

What value should you use for R? The problem uses liters as the volume measurement but does not specify what pressure units are used. Therefore, whatever value you choose for R will determine the units of your answer. Try using

$$R = 8.31 \frac{\text{L·kPa}}{\text{mol·K}}$$

Substitute all the known values into the ideal gas equation:

$$P = \frac{0.412 \text{ mol} \times 8.31 \dfrac{\text{L·kPa}}{\text{mol·K}} \times 289 \text{ K}}{3.25 \text{ L}}$$

Before working out the calculations, check the setup of your solution. Do the units seem to be correctly placed? Will you have units of pressure when you are finished? Multiplying and dividing the numbers in the expression yields

$$P = 304 \text{ kPa}$$

Suppose you had chosen a different value for R? You could use

$$R = 0.0820 \frac{\text{L·atm}}{\text{mol·K}}$$

The ideal gas equation then would be

$$P = \frac{0.412 \text{ mol} \times 0.0820 \dfrac{\text{L·atm}}{\text{mol·K}} \times 289 \text{ K}}{3.25 \text{ L}}$$

Working out the calculations gives

$$P = 3.00 \text{ atm}$$

Convince yourself that this answer is equal to 304 kPa by using the equivalence 1 atm = 101.325 kPa to change units.

Example 7-13

What will be the volume of a 0.70 g of nitrogen gas at standard temperature and 150 kPa pressure?

▶ Suggested solution _____

This problem is very similar to the previous one except that you are asked to find the volume instead of the pressure. Again, you can use the ideal gas equation to solve the problem.

$$PV = nRT$$

Rearranging the expression to solve for V gives

$$V = \frac{nRT}{P}$$

You are given information about the temperature, pressure, and number of moles:

$$P = 150 \text{ kPa}$$
$$T = 273 \text{ K}$$

The molar mass of nitrogen is 28.0 g/mol. Therefore the number of moles present is

$$n = 0.70 \text{ g} \times \frac{1.0 \text{ mol}}{28.0 \text{ g}}$$

$$n = 0.025 \text{ mol}$$

What value for R would be appropriate? Since the problem gives pressure in kPa, you can use

$$R = 8.31 \frac{\text{L·kPa}}{\text{mol·K}}$$

Do you see that using this value for R will give you an answer in liters? Substituting the values into the equation gives

$$V = \frac{0.025 \text{ mol} \times 8.31 \frac{\text{L·kPa}}{\text{mol·K}} \times 273 \text{ K}}{150 \text{ kPa}}$$

Check the setup of your solution to see if your answer will be in units of volume. Working out the calculations yields

$$V = 0.38 \text{ L}$$

You have seen several equations in this chapter. You may be wondering whether you have to remember them all. It really is not necessary to memorize all of the equations. If you remember $PV = nRT$ and think carefully about what you are being asked to do, you can solve any problem you are likely to face.

Review and Practice

1. What are the four assumptions made in the kinetic molecular theory?

2. Define absolute zero. What is its value in kelvins? In Celsius degrees?

3. Under what conditions is a gas most likely to behave as if it were ideal?

4. What are two possible values of the gas constant R? How are the values different?

5. Find the value for R in Example 7-10 if the volume is measured in milliliters instead of liters.

6. What is the volume of 6.73 g of oxygen gas at 0.0367 atm and 78°C? Remember to select the correct value for R.

7. How many moles of argon are there in a 27.3-L sample of gas at 85.9 kPa and 50.5°C?

8. Which of the four conditions of an ideal gas are most similar to the conditions of a real gas?

1. a) Molecules are dimensionless points; b) Molecules are in constant motion; c) Collisions are elastic; d) Molecules exert no attractive force.

2. the lowest temperature possible; $-273°C$, 0 K

3. low pressure and relatively high temperature.

4. $\dfrac{8.31 \text{ L} \cdot \text{kPa}}{\text{mol} \cdot \text{K}}$; $\dfrac{0.082 \text{ L} \cdot \text{atm}}{\text{mol} \cdot \text{K}}$; the R values differ in units of atmospheric pressure. $\left(\times \dfrac{1000 \text{ ml}}{1 \ell} \right)$

5. $\dfrac{8.31 \times 10^3 \text{ mL} \cdot \text{kPa}}{\text{mol} \cdot \text{K}}$

6. 165 L

7. 0.872 moles of argon.

8. Molecules of a real gas move in straight-line motion unless they collide with something. Collisions between molecules are elastic.

7-11 Molar Volume

One application of the ideal gas law is its use in calculating the volume of one mole of a gas at standard conditions (0°C and 1 atm). When the ideal gas law is solved for volume, the law is expressed as follows:

$$V = \frac{nRT}{P}$$

Since the problem presumes 1.00 mole of the gas, the values for the equation are

$$V = \frac{1.00 \text{ mol} \times 0.0821 \dfrac{\text{dm}^3 \cdot \text{atm}}{\text{mol} \cdot \text{K}} \times 273 \text{ K}}{1.00 \text{ atm}}$$

Solving for V yields

$$V = 22.4 \text{ dm}^3$$

One mole of a gas occupies 22.4 dm³ (or 22.4 L) at STP. This value is known as the **standard molar volume**. Recall in Chapter 4 that Avogadro's hypothesis states that equal volumes of gases, at equal temperature and pressure, contain the same number of particles. A molar volume of *any* gas contains one mole of the gas. Similarly a molar volume of a gas has a mass equal to its molar mass.

Molar volume sometimes serves as a convenience when working with gases because knowing the volume allows you to find the number of moles present and therefore the mass of the gas sample. You do not always have to use the longer calculations of the ideal gas law.

Figure 7-20 A mole of any gas will occupy a different molar volume at different temperatures. (Pressure is constant.) Is it reasonable to expect a molar volume of gas to occupy a greater volume at 25°C than at 0°C?

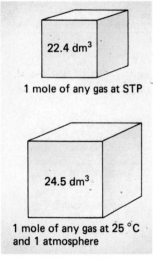

22.4 dm³

1 mole of any gas at STP

24.5 dm³

1 mole of any gas at 25 °C and 1 atmosphere

Since room temperature (25°C) is a more common condition that students will encounter in the lab, learning to work with molar volume at this temperature will be helpful. Drill students to become proficient in using a wide variety of temperature-pressure situations.

Example 7-14

A sample of oxygen gas occupies 5.6 dm³ at STP. How many moles of oxygen are present? What is the mass of the gas sample?

▶ **Suggested solution**

You have learned that 1 mole of oxygen at STP occupies 22.4 dm³. This relationship is a unitary rate that can be used to find the number of moles of another volume of the gas at STP. The number of moles is

$$5.6 \; \cancel{dm^3} \times \frac{1 \; mol}{22.4 \; \cancel{dm^3}} = 0.25 \; mol$$

Note the arrangement of the unitary rate. Check to see if it is written to yield the desired units. Also, note that the answer is 0.25 mole, which is less than 1 mole. This is the answer you would expect if the volume is less than the molar volume.

The molar mass of oxygen is 32 g/mol. Therefore, the mass of this gas sample is

$$0.25 \; \cancel{mol} \times \frac{32 \; g}{\cancel{mol}} = 8.0 \; g$$

Remember that the ideal gas law always can be used to solve a problem such as this one. In some instances, when molar volume can be used, the solution steps are simplified.

Molar volumes need not be calculated at STP. If the temperature of a gas is 25°C (room temperature), the molar volume at one atmosphere is obviously not 22.4 dm³. At 25°C and one atmosphere, a mole of any gas will have a volume of 24.5 dm³. Convince yourself that the molar volume is 24.5 dm³ at 25°C (298 K) by substituting 298 K for 273 K in the ideal gas equation. Figure 7-20 illustrates a comparison between the two molar volumes.

Review and Practice

1. What laws describing the behavior of ideal gases are combined to make the ideal gas law?

2. Solve the ideal gas law for R: that is, rearrange the ideal gas law so that R appears alone on one side of the equation. What are the units of R?

3. What is the numerical value for standard molar volume? Why is the concept of molar volume helpful?

4. What is the mass of 3.5 dm³ of NO_2 gas at STP?

5. What mass does a molar volume of carbon dioxide gas have at 25°C? (The answer is not 24.5 g.)

6. What properties of a gas must remain constant in Boyle's law?

1. Boyle's law
   Charles's law
2. $R = \dfrac{VP}{nT} = \dfrac{volume \cdot pressure}{mol \cdot K}$
3. 22.4 L; sometimes avoids need for more time-consuming calculations
4. 7.2 g
5. 44.0 g
6. temperature and number of moles

Applications of the Kinetic Molecular Theory

Gas molecules are in random motion. They collide with each other and with the walls of their container. Do all molecules in a sample of gas travel at the same velocity? Is there a way of thinking about the velocities of gas molecules that will make sense out of their chaotic motion? Several scientists have formulated explanations that address these questions.

In this chapter, you have seen some of the physical relationships of gases. Gases also react chemically with other substances. For example, hydrogen gas and oxygen gas will react under certain conditions to form water and release explosive amounts of energy. In the last part of this chapter, you will study some chemical reactions of gases. You will be using concepts you learned in Chapters 4, 5, and 6.

CONCEPTS

■ **the velocity of molecules**
■ **the distribution of velocities**
■ **Graham's law of gas diffusion**
■ **chemical reactions of gases**

7-12 Velocities of Molecules

Look again at the kinetic molecular theory in Section 7-9. The second part of the theory needs to be highlighted here. The molecules of a gas are in constant motion. These molecules are said to have kinetic energy. Recall from Chapter 1 that kinetic energy is the energy a particle has due to its motion. Mathematically kinetic energy (K.E.) is represented as

$$K.E. = \frac{1}{2}mv^2$$

Kinetic energy is equal to one half the mass of the particle (m) multiplied by the particle's velocity squared. (Be careful not to confuse the lowercase letter v used to represent velocity with the uppercase letter V for volume.)

Since a sample of gas, such as one mole of hydrogen, contains many billions of molecules, it seems safe to presume that at any given instant not all the molecules will be traveling at the same velocity. Figure 7-21

Describe how mass and velocity determine the kinetic energies of various objects such as golf balls, Ping-Pong balls, autos, etc.
Technically velocity and speed are two different concepts. *Speed* is a rate of motion, while *velocity* is speed in a specific direction. Students will need to know the difference between the two concepts if they study physics.

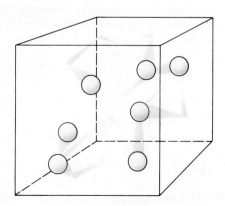

Figure 7-21 Gas molecules are in continuous random motion, changing direction and velocities when they collide with other molecules or with the container walls.

shows a typical sample of only a few molecules of a gas. Notice that in this "stop-action" diagram some molecules are colliding and others are in motion. Therefore, in a sample of gas composed of many billions of particles, there exists a tremendous range of velocities. It would seem an impossible task to take into account the almost infinite number of velocities in a gas sample at any instant.

From the kinetic molecular theory developed by Maxwell and Boltzmann, a scheme was formulated to describe the velocities of molecules in a gas. Although the actual mathematics of the work is beyond the scope of this text, the ideas are certainly understandable.

Suppose some small fraction of the gas molecules is moving at a relatively slow velocity, another fraction is moving very fast, and still a greater fraction of molecules moves at a moderate velocity. This idea is more clearly represented by a graph like the one shown in Figure 7-22.

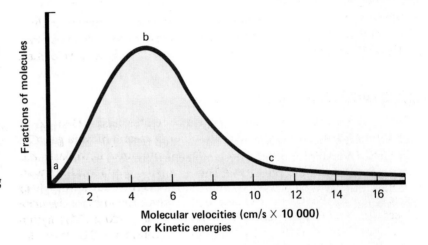

Figure 7-22 At any given instant, the molecules of a gas sample are moving at a variety of velocities. The curve represents the distribution of velocities of these molecules. At point *a*, the velocity is low, and the height of the curve is low on the *y*-axis, meaning that only a few molecules are moving at low velocity. At point *b*, many molecules are moving at a moderate velocity. At *c*, only a few molecules are moving at a high velocity.

Perhaps you could show the students a distribution curve from a recent chemistry test.

Make sure the students understand the relationship between kinetic energy and the temperature of a gas. Extend this concept to the temperature of liquids and solids. A liquid and a gas at the same temperature will have the same distribution of velocities of molecules.

The resulting curve shows the distribution of kinetic energies in the sample of gas. Distribution curves can be drawn for such data as the heights of students in your school or the letter grades a class of chemistry students received on a recent test. There is a tendency for the more moderate grades, for example, C, to occupy the highest points on the test-results graph, just as the moderate heights of most students will take up the highest points on the height graph. Looking at the curve in Figure 7-22, you can see that the largest fractions of molecules of the gas have moderate kinetic energies.

Maxwell and Boltzmann found that *the square of the average velocity of gas molecules is directly proportional to the temperature of the gas* (expressed in kelvins). If you increase the temperature of gas molecules, their average velocity increases by a constant rate. Volume, pressure, and the number of moles of the gas do not matter. Only the velocity of the molecules and the temperature of the gas are significant. When you heat a gas, you are making the molecules move faster. If you cool a gas, you are slowing them down.

If the sample of gas that yielded the graph in Figure 7-22 is heated, the average velocity of the molecules increases. The distribution is then shifted, or *skewed*, to the right to represent the increased average velocity of the molecules. In Figure 7-23, you can see a comparison of the original curve (shaded in yellow) and a distribution curve that

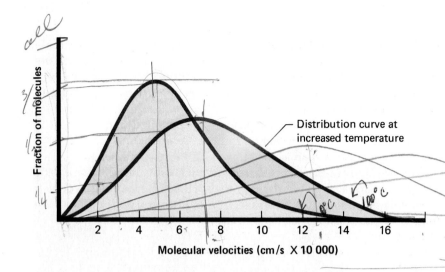

results when the temperature of the gas is increased (shaded in blue). Notice that the hotter the sample of gas, the faster the molecules move on the average and the wider the distribution of molecular velocities.

7-13 Graham's Law

Figure 7-24 shows two gases, each in one-liter flasks. One gas is brown nitrogen dioxide, NO_2, and the other is green chlorine gas, Cl_2. The gases are at the same temperature and pressure. Both flasks are connected to larger containers partially filled with air. When the valves of the flasks are opened simultaneously, the gas molecules move into the containers. The gases eventually will disperse throughout the containers, and the color in each tube will become uniform. The upper container will be light brown, and the lower container will be very pale green. Do you think both containers will become uniform in color at exactly the same time? The answer is no, the gases will not fill their respective containers simultaneously.

Since the temperatures of the two gases are equal, the average kinetic energies of the molecules of both gases also are equal. (See Section 7-11.)

Graham's law of effusion applies to a gas dispersing in a vacuum. The law of diffusion applies to a gas dispersing into another gas. Both laws actually describe the same phenomenon.

$$\frac{1}{2}mv^2 \text{ (of } NO_2) = \frac{1}{2}mv^2 \text{ (of } Cl_2)$$

Figure 7-24 The color of the gases in the tubes is an indicator of the location and concentration of the gas molecules. When they are evenly dispersed in the larger tubes, the colors will be uniform. The chlorine will be very pale.

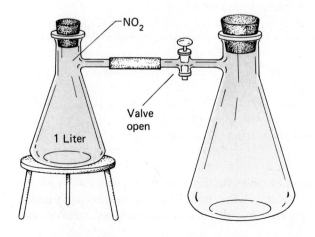

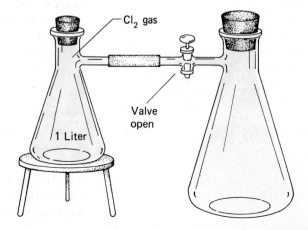

The molar mass of NO_2 (46.0 g/mol) is smaller than the molar mass of Cl_2 (71.0 g/mol). In order for the kinetic energies to remain equal, the average velocity of NO_2 must be greater than that of Cl_2. Therefore, NO_2 will diffuse faster than Cl_2. How can you determine how much faster the NO_2 diffuses?

Thomas Graham, a Scottish scientist, developed a mathematical explanation for this phenomenon in 1829. A comparison of the rates at which two gases diffuse is the subject of **Graham's law**.

Graham's law is based on the observation that the average kinetic energies of any two gas samples at the same temperature are equal. If a represents one gas and b represents the other gas, then this equality may be written as

$$\frac{1}{2}m_a v_a^2 = \frac{1}{2}m_b v_b^2$$

If both sides of the equation are multiplied by 2, you get

$$m_a v_a^2 = m_b v_b^2$$

The equation can be rearranged to show the relationship between the ratio of the masses and the velocities.

$$\frac{v_a^2}{v_b^2} = \frac{m_b}{m_a}$$

Finding the square root of both sides of the equation gives

$$\sqrt{\frac{v_a^2}{v_b^2}} = \sqrt{\frac{m_b}{m_a}}$$

Simplifying the equation yields

$$\frac{v_a}{v_b} = \sqrt{\frac{m_b}{m_a}}$$

The equation above is the usual way Graham's law is expressed. The ratio of the velocities of two gases equals the square root of the inverted ratio of the masses of the two gases. Recall the data in the original experiment involving NO_2 and Cl_2. In this case, gas a is NO_2 and gas b is Cl_2. Substitution of the molar masses of NO_2 and Cl_2 into Graham's law expression gives

$$\frac{v_a}{v_b} = \sqrt{\frac{71.0 \text{ g/mol}}{46.0 \text{ g/mol}}}$$

Dividing 71.0 g/mol by 46.0 g/mol yields

$$\frac{v_a}{v_b} = \sqrt{1.54}$$

Finding the square root of the quotient gives

$$\frac{v_a}{v_b} = 1.24$$

The NO_2 gas diffuses 1.24 times faster than the Cl_2 gas.

One application of Graham's law is the determination of the molar mass of an unknown gas. Suppose you have a colorless, odorless gas and do not know what it is. How might you find out? Knowing the molar mass might help. Let a gas of known molar mass diffuse through a hole and time it. Then let an equal amount of the unknown gas diffuse through the same hole and time it. You can calculate the molar mass of the unknown gas using Graham's law.

Example 7-15

The rate of diffusion of an unknown gas is four times faster than oxygen gas. Calculate the molar mass of the unknown gas.

▶ **Suggested solution**

Graham's law states

$$\frac{\text{rate a}}{\text{rate b}} = \sqrt{\frac{\text{molar mass b}}{\text{molar mass a}}}$$

If a is the unknown gas and b is oxygen gas, then the ratio of rates of diffusion is

$$\frac{\text{rate a}}{\text{rate b}} = \frac{4.00}{1}$$

The molar mass of oxygen is 32.0 g/mole. Substitution into Graham's equation gives

$$4.00 = \sqrt{\frac{32.0 \text{ g/mol}}{\text{molar mass a}}}$$

Squaring both sides of the equation yields

$$16.00 = \frac{32.0 \text{ g/mol}}{\text{molar mass a}}$$

Rearranging both sides of the equation to solve for molar mass gives

$$\text{molar mass a} = \frac{32.0 \text{ g/mol}}{16.00}$$

$$\text{molar mass a} = 2 \text{ g/mol}$$

The unknown has a molar mass of 2 g/mol and thus is hydrogen.

7-14 Chemical Reactions of Gases

Gases are involved in a wide range of chemical reactions from the production of carbon dioxide in a rising cake to the brewing of noxious gases in the atmosphere on a smoggy day. Not only are chemists interested in analyzing a reaction for the presence of a gas, but also they are concerned with the quantity of gases involved.

This section allows the students to integrate some concepts of the chapter with previous chapters.

In Chapter 6, you studied chemical reactions in a quantitative sense. You learned that the coefficients in a balanced chemical equation give you information about the number of molecules (particles) and moles that react. The coefficients also give you information about the number of dm³ (or liters) of gaseous substances in the equation.

In order to explain volume relationships of the gaseous substances in chemical reactions, recall Avogadro's hypothesis: Equal volumes of gases contain equal numbers of particles (when temperature and pressure are constant). This relationship is useful in solving volume calculations in chemical reactions of gases. It can be used as a model to predict the outcome of experiments when you are not in the lab. Table 7-5 lists the relationships that can be derived from the coefficients of a balanced chemical reaction involving the reaction of hydrogen and nitrogen to produce ammonia.

Table 7-5

GAS-VOLUME RELATIONSHIPS IN THE PRODUCTION OF AMMONIA (T, P CONSTANT)				
balanced equation	$3H_2(g)$	$+$ $N_2(g)$	$\rightarrow$	$2NH_3(g)$
reacting molecules	3	1		2
reacting moles	3	1		2
reacting volume ratios	3	1		2
predicted reacting	$6\ dm^3$	$2\ dm^3$		$4\ dm^3$
volumes	$12\ dm^3$	$4\ dm^3$		$8\ dm^3$
	$15\ dm^3$	$5\ dm^3$		$10\ dm^3$

The table shows you that the coefficients in a balanced chemical equation not only describe the ratio of reacting molecules and moles but also the ratio of reacting volumes. In this case, the ratio is $3:1:2$ for hydrogen to nitrogen to ammonia.

Example 7-16

How many dm^3 of water vapor are produced when $4\ dm^3$ of hydrogen gas react completely with $2\ dm^3$ of oxygen gas? All gases are measured at the same temperature and pressure.

▶ **Suggested solution**

Your first step is to write a balanced chemical equation:

$$2H_2(g) + O_2(g) \rightarrow 2H_2O(g)$$

You know that the coefficients in the equation represent the ratio of reacting volumes of the gases. Therefore, the ratio of hydrogen reacting with oxygen to produce water is

$$2:1:2$$

Note that the problem states $2\ dm^3$ of O_2 are used. According to the ratio of coefficients in the equation, the volume of water vapor formed must be twice the volume of the oxygen. Therefore,

$$?dm^3\ H_2O(g) = 2 \times vol\ O_2$$
$$?dm^3\ H_2O(g) = 2 \times 2\ dm^3$$
$$?dm^3\ H_2O(g) = 4\ dm^3$$

In this example and the ones following, the conventional symbolism to express ratio is used. If some students have difficulty following it, try expressing the ratios as unitary rates like those introduced earlier:

$$\frac{2\ vol\ H_2}{1\ vol\ O_2};\ \frac{2\ vol\ H_2O}{1\ vol\ O_2};\ \frac{1\ vol\ H_2}{1\ vol\ H_2O};$$

etc.

Example 7-17

How many liters of carbon dioxide gas are formed when 4.0 liters of methane gas (CH_4) react with 0.50 liter of oxygen gas to form carbon dioxide and water vapor? All gases are measured at the same temperature and pressure.

▶ **Suggested solution**

First, write a balanced equation.

$$CH_4(g) + 2O_2(g) \rightarrow CO_2(g) + 2H_2O(g)$$

The coefficients show you that the ratio of combining volumes of CH_4 to O_2 is $1:2$. For the gases to completely react, this ratio must always be maintained. Only half as much volume of CH_4 is needed as O_2. There is clearly an excess of CH_4 because 4.0 L is much more than half of 0.50 L. Therefore, O_2 is the limiting reagent here. The balanced equation indicates that the ratio of O_2 present to CO_2 formed is

Some students may follow the logic better using the unitary rate derived from the equation:

$$? \text{ L } CO_2 = 0.50 \text{ L } O_2 \times \frac{1 \text{ L } CO_2}{2 \text{ L } O_2}$$

$$2:1$$

The problem states that 0.50 L of O_2 is present. According to the ratio, this volume must be twice the volume of the CO_2 produced. Expressed mathematically,

$$? \text{ L } CO_2 \times 2 = 0.50 \text{ L}$$

Solving for the volume of CO_2 yields

$$? \text{ L } CO_2 = \frac{0.50 \text{ L}}{2}$$

$$? \text{ L } CO_2 = 0.25 \text{ L}$$

Some chemical reactions involve substances in the solid phase and others in the gaseous phase. For example, solid $CaCO_3$ (limestone) decomposes to form CO_2 gas and solid CaO. If you are to solve problems that ask for the quantity of the substances involved in the reaction, you need to combine some of the ideas you learned in Chapter 6 and in this chapter.

Example 7-18

When $CaCO_3$ decomposes, how many grams of $CaCO_3$ are needed to produce 9.0 liters of CO_2 measured at STP?

▶ **Suggested solution**

The first thing to do is to write a balanced chemical equation.

$$CaCO_3(s) \rightarrow CO_2(g) + CaO(s)$$

This can also be expressed as the unitary rate $\dfrac{1 \text{ mol CaCO}_3}{1 \text{ mol CO}_2}$.

Using unitary rates, the problem can be solved as

$$? \text{ g CaCO}_3 = 9.0 \text{ L CO}_2 \times \frac{1 \text{ mol CO}_2}{22.4 \text{ L CO}_2}$$

$$\times \frac{1 \text{ mol CaCO}_3}{1 \text{ mol CO}_2} \times \frac{100.1 \text{ g CaCO}_3}{1 \text{ mol CaCO}_3}$$

Better students can follow the solution when all three unitary rates are applied in one equation. Less able students will probably follow the logic better when the problem is solved in three independent steps.

The volume of CO_2 does not directly indicate how many grams of $CaCO_2$ will be needed. However, the balanced equation shows you that for every 1 mole of $CaCO_3$ reacted, 1 mole of CO_2 gas is generated. Thus

$$\text{moles CaCO}_3 = \text{moles CO}_2$$

The problem states that 9.0 L of CO_2 are formed. You have learned in this chapter that 1 mole of any gas at STP occupies 22.4 L. Using this relationship, you can determine the number of moles of CO_2 formed.

$$? \text{ mol CO}_2 = 9.0 \text{ L} \times \frac{1 \text{ mol}}{22.4 \text{ L}}$$

$$? \text{ mol CO}_2 = 0.40 \text{ mol}$$

You know

$$\text{moles CaCO}_3 = \text{moles CO}_2$$

Therefore,

$$\text{moles CaCO}_3 = 0.40 \text{ mol}$$

By using the molar mass of $CaCO_3$ you can convert to grams.

$$? \text{ g CaCO}_3 = 0.40 \text{ mol} \times \frac{100.1 \text{ g}}{1 \text{ mol}}$$

$$? \text{ g CaCO}_3 = 40 \text{ g}$$

Check the steps of the solution by following the units used in each step. Do they cancel out in each case to give the units desired? In any problem work that you do, you always should be able to use unit factors to check if your solution is reasonable.

Review and Practice

1. Using the kinetic molecular theory briefly explain what happens to the molecules of a gas when they are heated.

2. State Graham's law of gas diffusion in words.

3. What is the molar mass of an unknown gas if it diffuses 0.906 times as fast as argon gas?

4. How are the coefficients in a balanced chemical equation related to the number of liters of gaseous reactants and products?

5. How many liters of phosphorus pentachloride gas, PCl_5, are formed when 7.0 liters of phosphorus trichloride gas, PCl_3, react with 9.0 liters of chlorine gas, Cl_2? All gases are at equal temperature and pressure.

6. When $PCl_5(g)$ is heated, it decomposes into $PCl_3(g)$ and $Cl_2(g)$. If 23.2 g of PCl_5 decompose, what is the volume of Cl_2 produced at STP?

7. How many liters of hydrogen gas measured at 1 atm pressure and 25°C are formed from reacting 34.56 g of zinc in excess hydrochloric acid according to the following reaction?

$$2HCl(aq) + Zn(s) \rightarrow H_2(g) + ZnCl_2(aq)$$

1. Average velocity of molecules increases. They bounce off each other and the walls of the container more frequently.
2. The relative rates of diffusion of two gases is proportional to the square root of the inverse ratio of their molar masses.
3. 48.6 g/mol
4. They are the same.
5. 7 L of PCl_5
6. 2.49 L of Cl_2
7. 12.95 L of H_2

Numbers in red indicate the appropriate chapter sections to aid you in assigning these items. Answers to all questions appear in the Teacher's Guide at the front of this book.

Summary

■ A gas is a state of matter. A substance in the gaseous state takes the shape and volume of its container. Gas molecules are in constant motion colliding with each other and with the walls of their container.

■ Gases are compressible because the molecules are far apart and can be forced closer together by reducing the volume of the gas.

■ The pressure of a gas is caused by the collisions of the gas molecules with the sides of their container. In a mixture of gases, the total pressure is equal to the sum of the component pressures.

■ The volume of a gas is directly proportional to its kelvin temperature. The volume of a gas is inversely proportional to the pressure on the gas.

■ Gases at relatively high temperature and low pressure obey the ideal gas law, $PV = nRT$. Gases under high pressure and/or low temperature do not.

■ The kinetic molecular theory states that the molecules of a gas are in constant motion and the collisions are perfectly elastic (no loss of energy). Gas molecules travel at various velocities which can be plotted on a Maxwell-Boltzmann distribution graph.

■ The rate at which a gas diffuses through a small opening is dependent on its molar mass and temperature.

■ Gases react with each other, with liquids, or with solids under appropriate conditions. Balanced chemical equations give information about the volume relationships of the reactant and product gases.

Chemically Speaking

absolute zero	kinetic molecular theory
atmospheres	liquid
barometer	manometer
Boyle's law	mean free path
Charles's law	partial pressure
Dalton's law of	pascals
partial pressure	pressure
fluid	solid
gas	standard molar volume
Graham's law	standard pressure
ideal gas constant	standard temperature
ideal gas law	STP
kelvin scale	vacuum
kelvins	

Review

1. Compare and contrast the gas state with the condensed solid and liquid states. (7-1)

2. What property of gases enables them to be compressed? (7-1)

3. Give an example of units used to measure pressure. (7-2)

4. What is partial pressure? (7-4)

5. Charles's law relates what two properties of a gas? (7-5)

6. State Boyle's law in words. (7-6)

7. What is the function of R in the ideal gas law? (7-10)

8. State and explain the significance of 22.4 dm³. (7-11)

9. What are the four postulates of the kinetic molecular theory? (7-9)

10. What is kinetic energy? What property of a gas is proportional to the average kinetic energy of the molecules? (7-12)

Interpret and Apply

1. Explain the concept of gas pressure (force per unit area) in terms of the kinetic molecular theory. (7-3, 7-9)

2. Explain the concept of compressibility in terms of the kinetic molecular theory. (7-9)

3. At high pressures, such as 2×10^4 kPa, the volume occupied by gas molecules is significant in comparison to the total gas volume. Under this condition, would the real volume of the gas be larger or smaller than predicted by the ideal gas law? Briefly explain your answer. (7-9)

4. Use the Maxwell-Boltzmann distribution graph to explain why all the molecules in a sample of gas will not increase in speed when the gas is heated. (7-12)

5. Two gases, He and O_2, are at the same temperature. Which gas molecules are moving faster on the average? Why? (7-13)

6. Use the concepts of pressure and temperature to explain why it is dangerous to throw an aerosol can into a fire. (7-5, 7-6)

7. Two glass containers have the same volume. One is filled with hydrogen gas, the other with carbon diox-

ide gas. Both containers are at the same temperature and pressure. (7-11, 7-12)

a. Compare the number of moles of the two gases.
b. Compare the number of molecules of the two gases.
c. Compare the number of grams of the two gases.
d. The temperature of the container of hydrogen gas is increased. Now compare the pressure, volume, number of moles, and average kinetic energy of the hydrogen with the carbon dioxide.

8. A balloon can burst when too much air is added. Describe what happens to air pressure, volume, temperature, number of moles, and the balloon itself as it inflates and finally bursts. (7-9, 7-10)

Problems

1. Calculate the pressure of dry hydrogen gas in each of the following instances in which the hydrogen was collected over water. (7-4)

a. total pressure is 94 000 Pa, vapor pressure of water is 1200 Pa.
b. total pressure is 100.3 kPa, vapor pressure of water is 2600 Pa.

2. Change the following volumes of gases from the conditions given to the new conditions. Assume that pressure, temperature, or number of moles is constant if not given. (7-5, 7-6, 7-8)

a. 85 cm^3 at 61°C to 35°C
b. 7.3 dm^3 at 228°C to −48°C
c. 1.15 × 10^3 cm^3 at 75.2 kPa to 14.0 kPa
d. 94.7 dm^3 at 1.00 kPa to standard pressure
e. 4.03 × 10^3 dm^3 at STP to 200 K and 90.0 kPa
f. 139 cm^3 at 25°C and 78.4 kPa to STP

3. What is the pressure of 7.85 g of CO$_2$ at 27.0°C in a volume of 19.6 dm^3? (7-10)

4. What is the volume of 2.3 g of helium gas at 75.0°C and 101 000 Pa? (7-10)

5. In an experiment, it takes an unknown gas 1.5 times longer to diffuse than the same amount of oxygen gas, O$_2$. Find the molar mass of the unknown gas. (7-13)

6. What is the ratio of the rates of diffusion of hydrogen gas (H$_2$) to ethane gas (C$_2$H$_6$)? (7-13)

7. A weather balloon on the ground has a volume of 1.70 × 10^3 dm^3 at 23°C when filled with helium to a

pressure of 99.3 kPa. What is the volume of the balloon at 8500 meters where the temperature is −46°C and the pressure is 20.6 kPa? (7-8)

8. When coke (almost pure carbon) is burned in the presence of air, the product is carbon dioxide in the following equation:

$$C(s) + O_2(g) \rightarrow CO_2(g)$$

How many liters of CO$_2$ are produced from burning enough coke to react with 2.5 × 10^{10} dm^3 of oxygen gas? (7-14)

9. What will be the pressure in each container if (7-6)

a. just valve A is opened?
b. just valve B is opened?
c. just valve C is opened?
d. any two valves are opened?

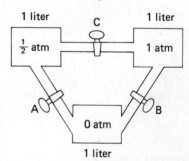

10. A 12.7-L sample of gas is under a pressure of 9.3 kPa. What will be the pressure of the gas when the volume increases to 20.1 L (temperature is held constant)? (7-6)

11. Determine the partial pressure of oxygen collected over water if the temperature is 30°C and the total pressure is 95.3 kPa. (7-4)

12. What is the volume occupied by 10.0 dm^3 of gas at 100.2 kPa after it has been compressed at constant temperature to 325.5 kPa? (7-6)

13. Find the volume of 7.2 g of argon gas at 0.53 atm pressure and 78.0°C. (7-10)

14. Iron(III) oxide, Fe$_2$O$_3$, reacts with carbon monoxide to produce iron metal and carbon dioxide. If 95.34 g of iron(III) oxide are present, how many liters of carbon monoxide, measured at 25°C and 1.00 atm pressure, are needed to completely react with the Fe$_2$O$_3$? (7-14)

15. Ammonia, NH$_3$, burns in oxygen to produce nitric oxide, NO$_2$, and water. If 20.3 dm^3 of ammonia are burned, how many dm^3 of O_2 are required? All gases are measured at STP.

16. If 2.55 dm^3 of nitrogen gas are collected at a pressure of 67.4 kPa, what volume will the gas occupy when the pressure is changed to 145.1 kPa? Assume that the temperature is held constant.

Challenge

1. Calculate the density in g/dm^3 of oxygen gas at STP.

2. Calculate the density of nitric oxide gas, NO_2, in g/dm^3 at standard pressure and 25°C.

3. What is the molar mass of an unknown gas whose density is 1.25 g/dm^3 at STP?

4. How many liters of $H_2(g)$ measured at STP are formed from the complete reaction of 1.31 g of magnesium with hydrochloric acid? The reaction is

$$Mg(s) + 2HCl(aq) \rightarrow H_2(g) + MgCl_2(aq)$$

5. Kernels of corn contain about 15 percent water by mass. How many dm^3 of water vapor measured at 100°C and 101 kPa pressure are formed from popping 155 g of popcorn?

6. Given the following data

VOLUME OF NITROGEN GAS (L)	TEMPERATURE (K)
4.28 L	303
5.79 L	410

   a. Draw a graph of the relationship between volume and temperature.
   b. Determine the slope of the line (m).
   c. Find the slope of a line in relationship to Charles's law expressed as $V/T = k$.
   d. Calculate the expected volume of the gas when the temperature reaches 1200 K.

7. What is the temperature of 85.9 cm^3 of oxygen gas if 70.0 cm^3 of the gas were heated from 35°C? Presume that the pressure is held constant.

8. A sample of carbon dioxide has a mass of 3.929 g and occupies 2000.0 cm^3 at STP. What is the density (in g/dm^3) of the carbon dioxide?

9. One mole of liquid sulfur dioxide occupies 43.9 cm^3. Calculate the percentage of space taken up by the sulfur dioxide molecules themselves in comparison to the total volume of the gas, if one mole of gaseous sulfur dioxide occupies 22 400 cm^3.

Synthesis

1. The Solvay process for making sodium bicarbonate uses this reaction:

$$H_2 + NH_3 + CO_2 + NaCl \rightarrow NH_4Cl + NaHCO_3$$

   How many dm^3 of NH_3 and CO_2 both at 150.2 kPa pressure and 54°C are needed to make 1.00 kg of sodium bicarbonate?

2. NO_2 is a toxic gas that forms acid rain when mixed with water in the air. The major source of NO_2 is the combustion of $N_2(g)$ in the cylinders of automobile engines. How many metric tons (1 × 10^6 grams) of NO_2 are produced when 5.0×10^8 dm^3 of N_2 react with excess oxygen in the auto engines in a city? Assume 25°C and 1.00 atm pressure.

3. The average molar mass of the gases in dry air is 29 g/mol. The molar mass of water is 18 g/mol. Use these facts and Avogadro's hypothesis to explain why a "falling barometer" (a decrease in atmospheric pressure) is a good indication that rainy weather is likely.

Projects

1. Explore the universe! Research the nature and conditions under which gases exist on the planets of the solar system. What conditions of temperature and pressure make Earth unique in the system?

2. Gases do not behave ideally under great pressure. Research the causes of this deviation from ideal behavior. What are the equations that can be used to predict real volumes of gases at high pressure?

3. Cryogenics is the study of the super cold—at temperatures near absolute zero. Find out how the characteristics of common materials change at temperatures near absolute zero. What can be said about helium at these conditions?

4. Divers need to understand the gas laws. Research the effects of temperature and pressure on divers.

What does a blueberry muffin have in common with an atom? As outrageous as it may seem, the muffin is a good model of the atom, according to a concept popular at the turn of this century. The blueberry muffin model does not mean to suggest that atoms are little cakes. However, it does help to set the stage for what is to follow in this chapter. Though the structure of the atom does not really resemble a blueberry muffin, such models played an important role in helping scientists reach the understanding of the atom as it is known today. Models help make familiar that which is unfamiliar.

Composition of the Atom

Early Discoveries of Subatomic Particles

Since earliest times, people have wondered what matter is and what it is made of. The ancient Greeks postulated the existence of atoms but did not know how to look for them. Hundreds of years later, scientists began searching for knowledge about these tiny theoretical particles. Why was there such a need to know about atoms? Scientists believed that learning about atoms was the key to explaining the nature and behavior of matter in the everyday world.

The structure of the atom eluded analysis for so long because it is so small; the largest atom is approximately 10^{-8} cm in diameter. It takes many millions of atoms to compose just the period at the end of this sentence.

The discovery of the composition of atoms is a fascinating example of the application of the scientific method: observation, experimentation, and verification. The modern theory of the atom was developed through a series of discoveries that began in the late 1800's.
Emphasize that modeling is part of the scientific method.

8-1 Scientific Modeling

In previous chapters, you were introduced to the idea of using a scientific model to convey information. (Remember the Ping-Pong balls in Chapter 7?) The results of experiments are used to suggest models that can be modified as new experimental results become available. A model helps you picture the reality being investigated, and it facilitates the communication of that knowledge to others.

There is nothing strange or unusual about how scientists develop models to explain things that they cannot observe directly. People do it all the time. When you were younger, you probably devised models of your own to explain things you could not understand. For example, thunder is a natural phenomenon that often perplexes children, so they develop mental pictures, or models, to explain the source of the sound. Children may imagine two huge, rain-laden clouds crashing into each other with great force. Or they may have heard the story of Rip Van Winkle and the bowling balls and imagine giant bowling alleys in the sky. The only thing different about scientific models is the persistence with which they are refined to make them ever more useful. Scientists use models to make predictions, propose experiments, or otherwise try to understand nature.

The process of modeling for the atomic scientist is very much like the modeling that took place with the discovery of dinosaurs. Investigators of both atoms and dinosaurs could not see the subjects to develop the models.

CONCEPTS

- scientific modeling
- experiments using the cathode-ray tube
- the discovery of electrons
- the discovery of protons
- charge and mass of electrons and protons

Thunder results from the intense heating of air inside the channel of the lightning flash. The air expands as if exploding, which causes the sound wave.

Figure 8-1 Children often create simple models to explain something that is unfamiliar.

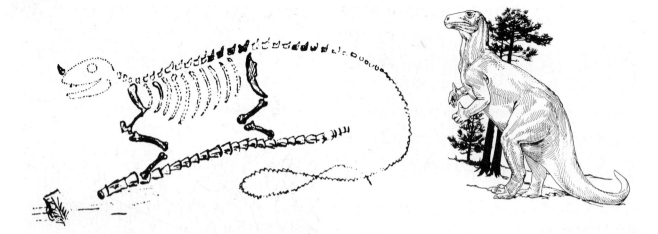

Figure 8-2 Gideon Mantell based his sketch (left) on the anatomy of a modern lizard. Notice that he placed the spiked thumb on the nose. The modern representation of an Iguanodon (right) has developed through many years of research. Note that the thumblike claw is positioned on the animal's forelimb.

It is helpful for students to see that models can be wrong and still be useful. When a model is tested and does not work, the new data are the basis for revising the model.

Dalton Model

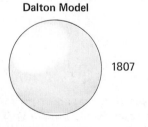

1807

Figure 8-3 John Dalton thought of the atom as an indivisible, uniformly dense, solid sphere that entered into chemical reactions but was left unchanged by the reaction.

About 150 years ago, scientists realized that dinosaurs once existed on this planet. An English doctor, Gideon Mantell, stumbled upon a few teeth and a horn-shaped bone in southern England. From these scant findings, he sketched his mental picture of the ancient animal. His illustration is shown in Figure 8-2 (left). Mantell's model was incorrect because what he thought was a horn on the animal's head was actually a claw. Figure 8-2 (right) is a modern model of the dinosaur based on evidence uncovered later. Although Mantell's initial model was inaccurate, his contribution was important for it began the study of an ancient world never before explored.

Present-day models of dinosaurs represent years of refinement based on new data. Scientists continue to study dinosaur fossils in order to develop even better models of what these puzzling creatures were like.

Like the models of dinosaurs, models of atoms are based on scientific discoveries. In order to understand modern atomic theory, it is helpful to study the early history of these discoveries. At the turn of the century, scientists devised experiments that probed the structure of the atom. When new data were derived through these experiments, scientists were able to construct a model of the atom that best fitted all the evidence gathered until that time.

In the early 1800's, an English schoolteacher, John Dalton, was the first person to develop a model of the atom. Experimental data at the time had shown that elements combined in the same percent by mass each time the same compound was formed. Dalton surmised a simple explanation for these data. He proposed that the atom was the smallest particle of matter, an indivisible sphere with a uniform density throughout. He further stated that all atoms of the same element had the same mass and the same chemical behavior, while atoms of different elements had different mass and differed in their chemical behavior. Dalton theorized that atoms of different elements are able to combine in fixed number ratios to produce specific compounds. Though Dalton based his model on chemical evidence, he knew nothing about the electrical nature of the atom. Yet his atomic model was an essential beginning for the refining process that was to follow.

Hold up a ball as a representation of Dalton's atom. Hold up another ball and bring the two together to form a "chemical bond." Ask the students how Dalton would explain bonding. (Remind them that he knew nothing of the electrical nature of the atom.) They may suggest that atoms are sticky or that atoms have hooks. Or the students may have still other ideas.

 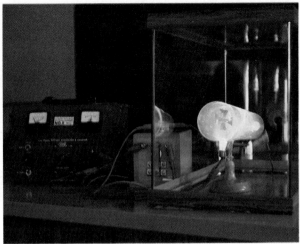

8-2 The Discovery of the Electron

From early experiments, nineteenth century scientists believed the atom was electrical in nature. English scientist, Sir William Crookes, investigated the nature of electrical discharges in gases. His work led to discoveries that provided evidence for new atomic models.

Crookes experimented with a partially evacuated tube like the one pictured in Figure 8-4. The tube contained a gas at low pressure. Implanted in the tube were two **electrodes**, which are conductors used to establish electric contact. Note that one of the electrodes was in the shape of a broad, flat cross. The other electrode was a disk. The electrodes were connected to an electric power source. When the power was turned on, a greenish glow was seen at the larger end of the tube.

The greenish glow fascinated Crookes. As he continued to observe the tube, he noticed a glowing beam coming from the disk electrode. Was this beam composed of atoms with an electric charge, or was it a beam of light? In order to answer this question, he placed a magnet close to the tube and noted that the beam was deflected from its straight-line course. Since light could not be influenced by a small magnet, Crookes concluded that the beam was composed of particles and that the particles carried some kind of charge.

Sir William also noticed that the beam produced a shadow of the cross on the large end of the tube. He assumed the location of the shadow meant that the beam was coming from the disk-shaped electrode. Because this electrode was the negative electrode, he concluded that the beam of particles must be negative.

In this case, the negative electrode is called a cathode, and the positive electrode is called an anode. Since the beam originated at the negative electrode, the beam was called a cathode ray, and the tube is now known as a **cathode-ray tube** (CRT).

Crookes repeated these experiments many times, using different gases in the tubes. Yet each time the greenish glow appeared along with the stream of particles. What was the greenish glow? Could the particles be coming from the atoms of the gas in the tube? Is it possible that atoms are not the smallest particles of matter as John Dalton assumed? These questions fascinated the first atomic scientists.

Figure 8-4 A Crookes tube (left) is a glass tube containing a gas at very low pressure. The electrodes are disks embedded in the glass. The cathode (negative electrode) is at the narrow end of the tube, and the anode (positive electrode) is at the wide end of the tube. When a cathode-ray tube containing a gas at very low pressure is subjected to about 10 000 volts of electricity, a greenish glow is seen around the edges of the glass at the wide end of the tube (right). The glow is called fluorescence. X rays also are given off by the anode upon bombardment with electrons. X rays can be dangerous, and precautions should be taken to prevent unnecessary exposure to them.

Do You Know?

Research for one purpose often gets applied to something else. Television picture tubes and computer monitors are cathode-ray tubes. A magnetic field is used to move the electron beam along one axis of the screen; and an electrostatic field is used to move it along the other.

Figure 8-5 In Thomson's cathode-ray tube, the ray passes through the hole in the anode and travels in a straight line to the fluorescent screen.

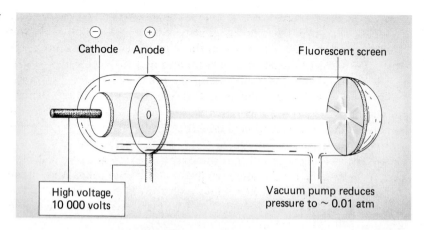

In 1897, another English scientist, J. J. Thomson, devised an experiment that would test the nature of cathode rays. He constructed a CRT much like the one shown in Figure 8-5. Thomson used a fluorescent screen at the far end of the tube. Such a screen is sensitive to the impact of a beam of charged particles, glowing wherever it is struck. The screen thus provided Thomson with a way of measuring the amount of deflection that occurred. The deflection of a charged particle depends on the mass of the particle and the size of its charge. By measuring the deflection of the beam, Thomson hoped to gain information about the particles that composed it.

The concept that two factors affect deflection of a charged particle is important for the students to understand here. It comes up several times later in the chapter. See the suggested demonstration in the Teacher's Guide at the front of this book.

Thomson duplicated Crookes' experiment by positioning a magnet close to the tube (Figure 8-6). As before, the cathode ray was bent. In the next stage of his experiment, Thomson added a second set of

Figure 8-6 The cathode rays are deflected in the same direction by the magnetic field as negatively charged particles would be.

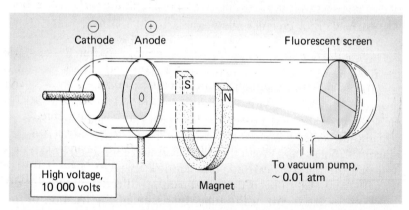

Figure 8-7 In a cathode-ray tube with a magnet and a second set of electrodes, the deflected ray is now bent again, this time upward, by the electric field located between the positive and negative plates. Recall that like charges repel and opposite charges attract.

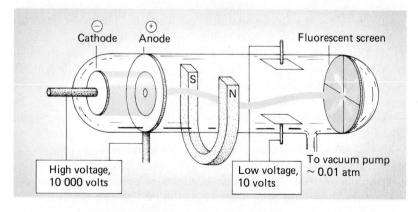

electrodes above and below the beam (Figure 8-7). The beam was first deflected downward by the magnetic field and was then deflected back upward as it passed between the second set of electrodes. Look at the positions of the second positive and negative electrodes in Figure 8-7. Why is the beam deflected upward as it passes between them?

Thomson used many different cathode-ray tubes in his experiments. He changed the metal of the electrodes and placed different gases in the tube. Regardless of how the conditions of the experiment were changed, the deflection of the particle beam was always the same. Thomson concluded that the particles composing the cathode rays were the same no matter what gas was in the tube. Were these particles some kind of basic component found in all atoms? Were there other particles that existed? Would these particles be the key to understanding the nature of matter?

The cathode rays that Thomson studied were made of the negatively charged particles that were later identified as **electrons**. As a result of Thomson's experiments, the first clue to the structure of the atom was uncovered, and a second clue was on the verge of being discovered.

Do You Know?

J. J. Thomson called the negative particles corpuscles. The name electron *had been coined in 1891 by English physicist G. J. Stoney, who studied electricity and proved that a fundamental unit of charge existed. The term was applied to cathode rays sometime later.*

8-3 The Proton

Once it became apparent that negative particles could be separated from atoms, researchers assumed there must also be positive particles because intact atoms have no apparent charge. In fact, a beam of such particles was discovered in 1885. It was not until the early 1900's, however, that they were investigated.

J. J. Thomson conducted further experiments with a cathode-ray tube containing hydrogen gas at very low pressure. When the tube was subjected to high-voltage electricity, the expected negative beam of electrons moved to the anode. But careful observations showed that yet another beam of particles was moving toward the cathode! The tube was modified to contain a cathode that was perforated with holes, similar to the one shown in Figure 8-8. Some of these mysterious particles passed through the holes of the cathode and beyond to the end of the tube. Thomson realized that this new beam had to be composed of positive particles. Why did he come to this conclusion?

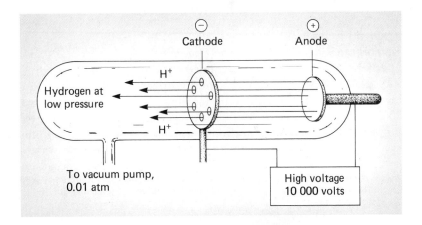

Figure 8-8 Thomson used a tube like the one illustrated here to analyze rays that were later identified as protons. He knew the particles were positive because they moved toward the cathode (negative electrode).

Do You Know?

We live in an information-based society that has grown up around the processing and transportation of information through electronic media. Telephones, television, radio, and computers depend on the flow of electrons. In recent decades, the trend for trafficking information has been away from electrons to light beams because of development of fiber optics. Soon most information will be transmitted optically, not electrically.

What exactly had taken place in the CRT? Thomson surmised that when sufficient energy is available in the tube, molecules are broken into atoms and the atoms are ionized. This process can be illustrated in the following equations:

$$H_2 + energy \longrightarrow 2H$$
$$H + energy \longrightarrow H^+ \quad + \quad e^-$$

anode rays   cathode rays
(electrons)

In the first equation, a molecule of hydrogen gas is separated into two hydrogen atoms. In the second equation, a hydrogen atom is broken down further into two oppositely charged particles: the negative electron and the mysterious positive ion. By 1920, this positive particle was known as a **proton**. Another piece of the atomic puzzle had been revealed. Thomson would soon suggest a model for the atom far different from Dalton's.

8-4 Extension: Physical Characteristics of Electrons and Protons

Once the existence of the electron and proton was established, scientists wanted to know more about the characteristics of the particles. What were their masses? How big were the charges they carried? Careful observations of the behavior of the particles in the cathode-ray tubes uncovered some of these data.

A real problem faced the early atomic scientists. How could they determine the masses of these particles? It was impossible for these investigators to put a few electrons on a balance and calculate a unitary rate as you did with rice grains in Chapter 3. An indirect method had to be used. In Section 8-2, you read that the amount of deflection of a particle beam depends on two factors: the mass of the particles and the size of their electric charge. A ratio of the size of the charge, e, to the mass, m, can be calculated based on the voltage needed to return the deflected beam to its original path. Thomson's experiments provided him with the data he needed to calculate such a charge-to-mass ratio for a single particle of the beam. He found the ratio for a single electron to be the following:

$$\frac{e}{m} = \frac{1.76 \times 10^8 \text{ coulombs}}{\text{gram}}$$

The coulomb is a unit of electric charge.*

Earlier experiments had determined the e/m ratio for the hydrogen ion. When Thomson compared the hydrogen ion's e/m ratio to the electron's, he found that the ratio for the electron was 2000 times the value obtained for H^+. How could these results be interpreted? On a comparative basis, if the e/m ratio of the electron is large, then one of two things must be true: Either the charge of the electron was very

The physics of e/m ratios is a little beyond most chemistry students. If you have students who have taken physics, perhaps you could have them prepare a brief analysis of the math needed to find the e/m ratio of the electron.

If the canal-ray tube, as it is called, is filled with a gas other than hydrogen, the amount of deflection of the ionized atoms would be less than that of hydrogen.

*The coulomb is equivalent to the amount of electric charge passing a point per second if the current is one ampere. A coulomb can also be thought of as the electric charge on 6.25×10^{18} electrons.

large, or the mass was very small. Thomson correctly presumed that the charges of the electron and proton (the H^+ ion) were equal but opposite. Therefore, the mass of the electron must be very small.

From this experiment Thomson was not able to determine separately either the charge or the mass of the electron. The charge-to-mass ratio was as far as his results went toward understanding the nature of the particle. However, Thomson's work was a major step in the right direction, and he was awarded a Nobel prize in 1906 for his accomplishments.

A few years later, an American physicist, Robert Millikan, added substantially to the growing knowledge about electrons. In 1909, Millikan successfully measured the charge on an electron using the apparatus pictured in Figure 8-9 and diagrammed in Figure 8-10. Oil drops were sprayed into the chamber and exposed to X rays, which ionized some of the drops positively or negatively. If an oil drop was negatively

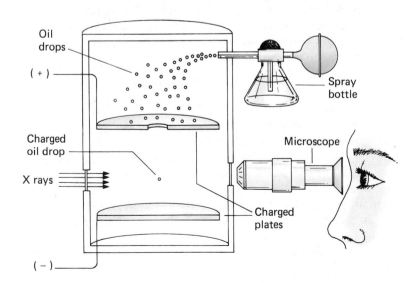

Figure 8-9 Millikan's oil drop experiment can be demonstrated safely and easily with just the equipment found in a physics lab.

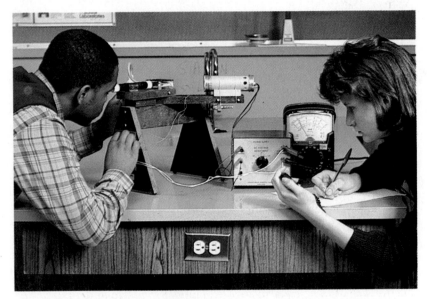

Figure 8-10 The upper plate has a positive charge and the bottom plate has a negative charge. A falling, negatively charged oil drop can be made to rise between the plates.

charged, Millikan adjusted the electric field so that the drop would move slowly upward in front of a grid in the telescope. Knowing the rate at which the drop was rising, the strength of the electric field, and the weight of the drop, he was able to calculate the charge on the drop.

Millikan was then able to combine this information with the results obtained by Thomson to calculate a value for the mass of a single electron. Although Millikan's work was an approximation, he paved the way for further investigations. Eventually, the mass of the electron was found to be 9.11×10^{-28} gram.

As Thomson had speculated, the mass of one electron was indeed very small. Even on the atomic scale, electrons are incredibly tiny. It would take 1837 electrons to equal the mass of one hydrogen atom!

Thomson performed experiments on the positive beam in much the same fashion as he did with the negative cathode ray. Studying the positive beam in magnetic and electric fields, he was able to determine the charge-to-mass ratio of a single particle. This ratio was found to be much smaller than the one for electrons. Because the e/m value was smaller, the mass of the particle had to be larger than that of the electron.

The e/m value for electrons was the same for all gases Thomson tested in the CRT, telling him that electrons from all atoms were the same. However, the e/m value for the positive particle varied with the gas in the tube. The e/m value for the hydrogen ion was the largest of any positive ion. Thomson interpreted this evidence to mean that the mass of the hydrogen ion was the smallest. He assumed that the positive ion formed from a hydrogen atom was a single particle with an equal but opposite charge to the charge on the electron. The symbol for this newly identified proton became H^+. Calculations later showed that its mass was 1.67×10^{-24} gram, which is very close to the mass of a hydrogen atom!

Investigations that occurred after Thomson's work showed that the varying e/m ratios he obtained were the result of different numbers of protons in the atoms of each of the different gases he tested. The more protons, the greater the mass. You will read more about this characteristic of atoms later in the chapter.

Do You Know?

J. J. Thomson was brilliant at devising experiments, but he was also notoriously clumsy. Many times other people actually carried out his plans to ensure against unexpected mishaps.

1. to explain phenomena and facilitate communication of knowledge; also, to picture reality
2. deflection of the beam by a magnet
3. The shadow cast by the anode showed the beam originated at the negative electrode.
4. the charge-to-mass ratio
5. e = charge in coulombs
   m = mass in grams
6. The proton has more mass and is positively charged; the electron is negatively charged.

Review and Practice_____

1. Name two ways in which a scientific model is useful.

2. What evidence led Sir William Crookes to conclude that the beam he saw in his experimental tube was actually a particle beam?

3. What evidence showed that the particles in the beam of the Crookes tube were negatively charged?

4. What properties of the electron were determined by J. J. Thomson in his experiments using cathode-ray tubes?

5. What do the symbols e and m represent in the e/m ratio Thomson found in his experiment?

6. How is a proton different from an electron?

Early Models of the Atom

At the turn of the century, some scientists suggested that all the natural laws of science had already been discovered. All that was left to do was to refine these laws. The truth, however, was that a new age of science was just beginning. J. J. Thomson was one of the heralds of the new age. His experiments revealed the tip of the atomic iceberg. His investigations into the nature of the atom spawned new activity in researching the atom.

A new branch of science dealing with radioactivity was born at this time. It gave scientists a powerful tool with which to investigate the atom. Ernest Rutherford made giant strides in understanding the nature of the atom with his experiments on radioactivity. He started with the atomic model suggested by Thomson.

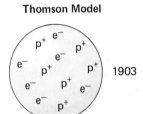

Thomson Model

1903

CONCEPTS

■ Thomson's model of the atom
■ the discovery of radioactivity
■ Rutherford's gold foil experiment
■ Rutherford's model of the atom
■ the discovery of the neutron

Figure 8-11 Thomson's plum pudding model of the atom was the first major modification since Dalton's.

Do You Know?

Dalton popularized the term atom, *but he did not invent it. The Greek philosopher Democritus was the first to envision matter as being composed of tiny particles. He named them atoms, from the Greek word* atomos, *which means "indivisible."*

8-5 Thomson's Model of the Atom

By 1903, the evidence was becoming quite clear that atoms were not solid spheres as John Dalton proposed 100 years earlier. Atoms had components. Instead of being the smallest particles of matter, atoms were made of even smaller *subatomic* particles. J. J. Thomson's model of the atom is pictured in Figure 8-11. His model incorporated the newly discovered evidence that atoms are composed of negatively charged particles (electrons) and positively charged particles (protons). Thomson thought the bulk of the atom was composed of positive charges, with electrons dispersed throughout in order to give the atom an overall neutral charge. The Thomson atom can be seen as a soft blueberry muffin like the one in the picture at the beginning of the chapter. The blueberries are the negatively charged electrons, and the muffin is a bulk of positive charge. The negative charges are embedded in the positive mass of the atom. (Thomson used an analogy that was a bit more common at his time in England. He called his model of the atom the plum pudding model.)

The Thomson model of the atom was an improvement over the Dalton model. Thomson's model is important because it accounted for the electrical nature of the atom, and for the first time the atom was seen as being composed of smaller particles.

CHEM THEME

The nature of models in science varies. While the ideal gas law and the kinetic molecular theory are mathematical and abstract, atomic models can be visual. Each, however, serves as a tool for understanding the unknown.

X rays are discussed in more detail in Chapter 10.

8-6 The Discovery of Radioactivity and the Gold Foil Experiment

The next chapter is on radioactivity. If you choose not to cover that chapter with the students, it would be appropriate to do a demonstration here on radioactivity. Perhaps a physics teacher could suggest a proper demonstration. A Geiger counter and a radioactive source are all that you need.

Radioactivity in naturally occurring isotopes involves either α- or β-decay accompanied by gamma radiation. A number of artificial radioisotopes emit only gamma radiation, and do not actually "decay" into another nuclide.

Do You Know?

Many problems in science are interrelated. Rutherford's work required a way to detect alpha radiation. Once radiation was recognized to be potentially hazardous, detecting it became important. Hans Geiger invented the Geiger counter, one of many machines that monitor intensity of radiation.

The nature of the atom was further revealed in an accidental discovery made by the French scientist Henri Becquerel. In 1896, he was investigating the nature of X rays, which he knew had the ability to fog a photographic plate even when the plate was covered by black paper. He coated some black paper with a uranium compound. When exposed to sunlight, the plate showed the outline of the compound from the paper. Becquerel thought that the sun's rays produced X rays in the compound that in turn exposed the photographic plate. He planned to repeat the experiment, but the next several days in Paris were cloudy, and he shut the plates away in a closet. On a whim, he developed the plates anyway, and to his surprise the outline of the uranium compound was there. Apparently the substance itself was capable of fogging the photographic plate. Becquerel concluded that the uranium must have given off some kind of rays on its own without the outside influence of sunlight. Later the rays were identified as the products of radioactivity. **Radioactivity** is a spontaneous process by which certain atoms emit particles and very penetrating rays (different from X rays). Becquerel was credited with the discovery of radioactivity, and the subject became the object of intense scientific curiosity. Scientists such as Marie Curie and Pierre Curie joined the investigations.

It is interesting that solving one problem in science often leads to the solution of another. As more scientists investigated radioactivity, the model of the atom would change again.

One scientist who made outstanding contributions to research into radioactivity and atomic structure was Ernest Rutherford of New Zealand. While at McGill University in Montreal, Canada, he discovered that radioactivity involved the emission of two kinds of particles, alpha (α) and beta (β), named for the first two letters of the Greek alphabet. An **alpha particle** is a high-speed, positively charged particle that has a mass equal to a helium nucleus. **Beta particles** are high-speed electrons. Later, a third kind of emission was discovered that had no mass but had very strong penetrating power. It was determined that this emission consisted of rays rather than a beam of particles, and they were named **gamma rays** (γ).

In 1909, Rutherford and his colleagues designed an experiment to investigate the deflection of alpha particles by atoms with large atomic masses. The experiment is pictured in Figure 8-12. Rutherford used thin sheets of foil made of gold, platinum, copper, or tin as targets for radioactive emissions. The source of alpha particles was the radioactive element polonium (discovered earlier by Marie Curie). The alpha particles are ejected from the polonium at incredible speeds, around 16 000 km/s. A fluorescent screen was placed around the source and target. This screen detected the alpha particles emitted because they produced a faint flash of light at the point of impact with the screen.

Rutherford fully expected the alpha particles to pass through the metal foil with little or no deflection from their original path. He based this expectation on Thomson's model of the atom. The experiment did indeed produce these results. Most of the high-energy alpha particles

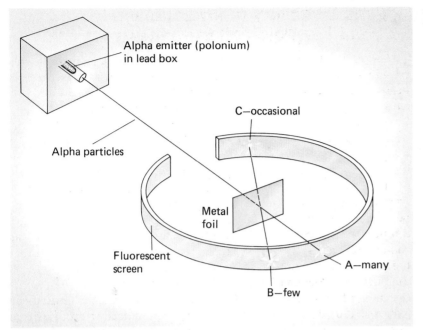

Figure 8-12 Rutherford fully expected alpha particles from the radioactive polonium to be deflected slightly to points *A* and *B*. The Thomson model did not predict the wild deflections to points such as *C*. As a result of these tests, the model of the atom had to be revised.

passed straight through the metal foil to hit the fluorescent screen at point *A*. Occasionally an alpha particle would be deflected through a small angle to strike the screen at point *B*. Then a co-worker named Hans Geiger suggested that perhaps some alpha particles could be repelled by the atoms in the foil and would strike the screen near the source of the particle beam. Rutherford was skeptical. But when Geiger performed the experiment, a few alpha particles *were* observed to be deflected through very large angles, some striking the screen at point *C*. Rutherford was truly amazed. He described his surprise in a lecture given some years later: "It was about as believable as if you had fired a 15-inch shell at a piece of tissue paper, and it came back and hit you."

8-7 The Rutherford Atom

The results of the Rutherford-Geiger experiment could not be explained using Thomson's model of the atom. Figure 8-13 illustrates what would have happened if Thomson's model had been correct. Rutherford believed the alpha particles would be uniformly repelled by the evenly distributed positive charges in the Thomson atom. Small angle deflections would not be a surprise. However, large-angle deflections could not be explained. Rutherford had to rethink the nature of the atom. The only reasonable explanation that would account for such a dramatic change in the path of some alpha particles was that the atom contained a very small, densely packed center of positive charge. Rutherford called this center of the atom the **nucleus**.

Figure 8-14 (left) represents the results of the experiment using Rutherford's idea. If the positive particles were clustered in a small center of the atom, few of the alpha particles would come close enough to be deflected. Some particles, passing near the nucleus, would be only slightly deflected. However, every once in a while an

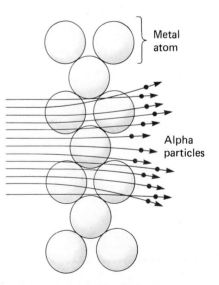

Figure 8-13 Using Thomson's model of the atom as a guide, Rutherford expected the alpha particles to pass through the atoms of the metal foil almost undisturbed.

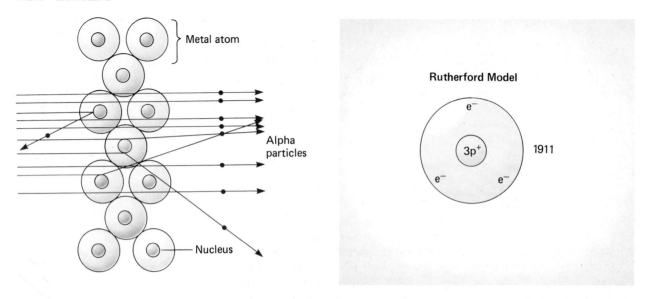

Figure 8-14 Rutherford's experimental results are explained by assuming that the atom contains a dense, positively charged center. As illustrated on the left, only the particles passing close to or colliding with a nucleus were greatly deflected. In the Rutherford model of the atom (right), the protons are located in the densely packed center, or nucleus. The electrons move around the nucleus at some distance.

alpha particle would collide almost head-on with a nucleus, and these particles would change their direction dramatically through large angles.

From the results of the experiment, Rutherford was able to develop a new model of the atom. He suggested that the positive charge in an atom is located in the densely packed nucleus. Most of the mass would also be in the nucleus. The electrons, he surmised, moved around the nucleus like bees around a hive. Figure 8-14 (right) depicts the Rutherford model of the atom. Most of an atom is empty space that separates the tiny nucleus from the distant electrons. In fact, the size of the nucleus is only 1/100 000 the diameter of the atom. To give you an idea of the enormity of distance within an atom, consider this analogy: If a hydrogen atom had a nucleus the size of a Ping-Pong ball, its electron would most likely be found 2 kilometers away! If you were an electron standing at the edge of this supersized atom, would you be able to see the nucleus in the center?

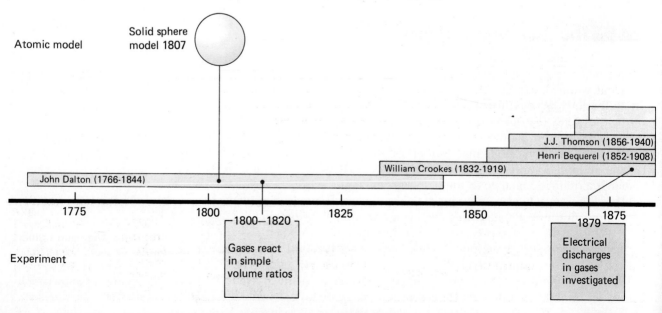

8-8 The Neutron

In 1932, another subatomic particle was discovered. Irene Joliot-Curie (daughter of Pierre and Marie Curie) bombarded beryllium with alpha particles. A beam with high penetrating power was formed. At first, this beam was believed to be gamma rays. However, British scientist James Chadwick demonstrated that the beam was made of particles that had approximately the same mass as protons. Furthermore the beam was *not* affected by electric or magnetic fields. Thus the particles were named **neutrons** because they did not possess a charge. Neutrons, like protons, are found in the nucleus of the atom. As you will see in Section 8-10, their presence is an important factor in maintaining the stability of the nucleus.

A summary of the early research concerning the nature of the atom will be helpful before you move on to the next part of the chapter. Figure 8-15 represents highlights of the history of the discovery of the atom from Dalton to Rutherford. It is important to realize that none of the scientists listed in the time line could claim that his findings stood alone. The discovery of the nature of the atom was a progression of ideas and experiments carried out by many people. Sir Isaac Newton said it well in the following quotation: "If I have seen farther than others, it is because I have stood on the shoulders of giants."

Neutrons were not discovered until 1932 because they are difficult to produce, and being neutral, they are difficult to detect.

Chadwick proved that the new beam from the beryllium reaction was made of particles with a mass similar to that of protons. He did this by bombarding hydrogen gas with the beam. The resulting collisions were seen to be elastic like the collision of two billiard balls of the same mass.

Review and Practice

1. Describe J. J. Thomson's model of the atom.

2. What are three kinds of radioactive emissions?

3. What unexpected results did Rutherford's gold foil experiment produce?

4. How did Rutherford explain the results of the gold foil experiment?

5. What is a neutron?

1. The bulk of the atom is protons, with electrons scattered inside.
2. alpha particles, beta particles, and gamma rays
3. large angle deflections
4. by suggesting the existence of a small, dense nucleus
5. a subatomic particle with approximately the same mass of a proton and no charge

Figure 8-15 The development of early models of the atom depended on the work of many people.

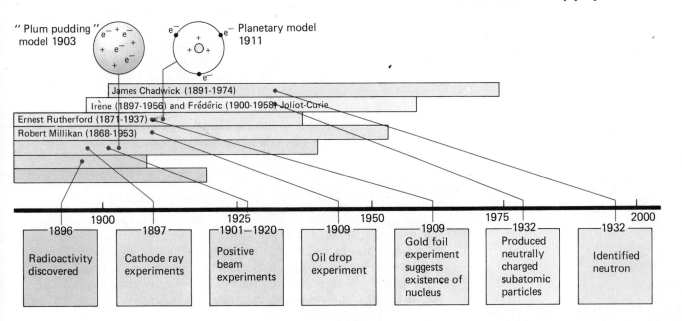

Atomic Number and Isotopes

CONCEPTS

■ protons in the nucleus
■ protons, neutrons, and mass number
■ the nature of isotopes
■ use of the mass spectrometer
■ chemical analysis of Mars

The instruments used by the founders of atomic physics were primitive by today's standards. Equipment in the laboratories of Thomson, Millikan, and Rutherford can be purchased today for several hundred dollars. Yet, the simplicity of the investigations conducted by these scientists reveal true genius. These early researchers were able to determine that an atom contains protons, neutrons, and electrons. They even were able to determine the masses of these invisible particles using rather simple equipment. Table 8-1 compares the relative charges and masses of the three particles.

Table 8-1

ELECTRONS, PROTONS, AND NEUTRONS			
	ELECTRON	PROTON	NEUTRON
Symbol	e^-	p	n
Charge	$1-$	$1+$	0
Mass	$\dfrac{1}{1837}$	1	1

If J. J. Thomson were alive today, he would be amazed at the progress made toward unraveling the complexity of the atom. Over 100 subatomic particles have been discovered. Yet, paradoxically, science has also revealed the utter simplicity of matter. Elements differ from one another only by the number of protons the atoms contain in their nuclei. This difference is the key to the arrangement of elements on the periodic table and to the chemical properties of the elements.

The masses of atoms are determined primarily by the number of protons and neutrons contained in the nucleus. Contrary to John Dalton's idea that atoms of the same element are exactly alike, not all atoms of the same element have the same mass.

Modern research has revealed much about the nature of atoms. As you might expect, the atomic model continues to change as more data are accumulated from research. One primary tool used in investigations is the mass spectrometer, which is capable of separating the atoms of an element that do not have the same mass. This instrument is used for chemical analysis in places as ordinary as a science lab or as exotic as outer space.

8-9 Atomic Number and Mass Number

Today's chemists need to know the number of protons, neutrons, and electrons in the atoms that they investigate. This information tells them most things necessary to understand the basic chemical and physical properties of the elements. Physicists have discovered that protons and neutrons are composed of even smaller particles, and you will read about these subatomic mysteries in Chapter 9. For now, it is sufficient to think of the atom in terms of three basic particles: protons, neutrons, and electrons.

Ernest Rutherford's gold foil experiment was important because it led him to postulate the existence of the nucleus within the atom. Furthermore, through painstaking calculations of the angle of deflection of the alpha particles, Rutherford determined the amount of positive charge in the nuclei of the atoms composing each foil. The larger the amount of positive charge in the foil, the more the alpha particles were deflected. For example, if tin was used in the experiment, the angle of particle deflection would be less than that for gold. Consequently, the amount of positive charge in a tin nucleus must be less than in gold. Rutherford also found that the positive nuclear charge is always equal to the charge of a whole number of protons.

What in an atom's nucleus gives it a positive charge? The answer, of course, is protons. In Sections 8-2 and 8-3, you studied J. J. Thomson's experiment with the cathode-ray tube. He determined that the hydrogen nucleus has the smallest positive charge and is therefore made of one proton. Other elements have larger positive charges corresponding to a larger number of protons. The **atomic number** of an element is the number of protons in the nucleus of its atoms. For hydrogen this number is one. The nucleus of a hydrogen atom carries a 1+ charge. For calcium the atomic number is 20. The calcium nucleus has a 20+ charge. The elements are listed in the periodic table according to increasing positive charge in the nucleus.

A careful look at the periodic table shows that no two elements have the same atomic number. Atomic number *defines* an element. For example, silver's atomic number is 47. All atoms of silver have 47 protons in their nuclei. If one more proton is added to a silver atom's nucleus, the atom is no longer silver; it becomes cadmium.

Protons and electrons have equal but opposite charges. In a neutral atom, then, the number of electrons must equal the number of protons. In the case of a neutral silver atom, there are 47 protons and 47 electrons.

Electrons can be removed from an atom if enough energy is applied to it. The loss of one or more electrons from an atom upsets the balance between the positive and negative charges. In the case of a silver atom, if energy is applied, one electron can be removed. What remains is an ion with a net charge of 1+. The symbol Ag^+ represents a silver ion with 47 protons and 46 electrons. Compare the silver atom with the silver ion, as illustrated in Figure 8-16.

The significance of Thomson's findings becomes clearer when taken in the context of atomic numbers. The *e/m* ratio for the gases varied with the identity of the gas because the atomic numbers varied. The *e/m* ratio for electrons did not vary because electrons are the same in all atoms.

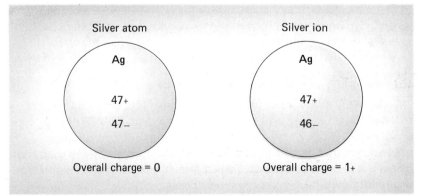

Figure 8-16 The silver atom has 47 protons and 47 electrons so its overall, or net, charge is zero. If one electron is lost, the remaining ion only has 46 negative charges and so has a net charge of 1+.

The increase in atomic mass from one element to the next is due to the increased number of protons and electrons. However, the atomic mass of each naturally occurring isotope is also dependent on the mass lost for binding energy. The average atomic masses on the periodic table reflect the relative abundances of these isotopes.

As you learned in Section 8-8, almost all atoms contain neutrons in their nuclei. A neutron has about the same mass as a proton but is neutral in charge. If you look at the periodic table you will notice that the atomic numbers increase by one whole number as you move from one element to the next (corresponding to one more proton in each case). However, the atomic masses of elements increase by amounts greater than one. This difference is largely due to the neutrons in the nucleus. Neutrons add mass to the atom but do not change its charge. The total number of protons and neutrons in an atom's nucleus is called the **mass number**. The mass number and the atomic number can be included in the symbol of an element. By convention, the atomic number is written as a subscript at the lower left of the symbol, and the mass number is written as a superscript above the symbol, as the following example indicates.

$$\text{mass number} \longrightarrow {}^{39}_{19}\text{K} \longleftarrow$$
$$\text{atomic number} \longrightarrow {}_{19}$$

Do You Know?

For brevity, atomic number is often symbolized by the letter Z and mass number by the letter A. For carbon, Z = 6 and A = 12; for fluorine, Z = 9 and A = 19.

A potassium atom with 19 protons and 20 neutrons in its nucleus has a mass number of 39. If the mass number of an atom and the number of protons in the nucleus are known, the number of neutrons present can be calculated easily.

Example 8-1

How many neutrons are present in an atom of chlorine that has a mass number of 37?

▶ Suggested solution _____

The example states that the chlorine atom has a mass number of 37. If p represents the number of protons and n is the number of neutrons, then the mass number can be written as

$$\text{mass number} = p + n$$

The expression may be rearranged as

$$n = \text{mass number} - p$$

According to the periodic table, chlorine atoms (and only chlorine atoms) contain 17 protons in their nuclei. Substituting this value and the mass number into the equation above gives:

$$n = 37 - 17$$
$$n = 20$$

Check your answer by adding the number of neutrons calculated to the number of protons given in the problem. The sum should be equal to 37.

8-10 Isotopes

Dalton's concept of the atom described all atoms of the same element as identical. However, not all atoms of the same element contain the same number of neutrons. (Remember that the number of protons in an atom determines its identity; the number of neutrons does not.) For example, some atoms of lithium contain three neutrons while others have four neutrons. These two forms of lithium are called isotopes. **Isotopes** are atoms of the same element that contain different numbers of neutrons. Many elements exist in isotopic form. Hydrogen has two naturally occurring isotopes; tin has ten. Table 8-2 lists some elements and their naturally occurring isotopes. The table is only a partial list.

The atomic number is always the same for any one element so it is often dropped when writing the symbol. Obviously, when dealing with

You may want to discuss in more detail the nature of radioactive isotopes if you are not going to teach the next chapter to your students.

You may want to mention that there are other radioactive decay modes besides alpha and beta. More exotic forms of decay are generally not covered in a high school chemistry class.

Table 8-2

				NUCLEUS			MASS OF ATOM (IN AMU)	NUMBER OF
ISOTOPE	**ABUNDANCE IN NATURE (%)**	**ATOMIC NUMBER**	**MASS NUMBER**	**NUMBER OF PROTONS**	**NUMBER OF NEUTRONS**	**ELECTRIC CHARGE**	**$^{12}C = 12.0000$**	**ELECTRONS IN NEUTRAL ATOM**
Hydrogen-1	99.985	1	1	1	0	1+	1.0078	1
Hydrogen-2	0.015	1	2	1	1	1+	2.0140	1
Helium-3	1.3×10^{-4}	2	3	2	1	2+	3.0160	2
Helium-4	~100	2	4	2	2	2+	4.0026	2
Lithium-6	7.5	3	6	3	3	3+	6.0151	3
Lithium-7	92.5	3	7	3	4	3+	7.0160	3
Beryllium-9	100	4	9	4	5	4+	9.0122	4
Boron-10	19.6	5	10	5	5	5+	10.0129	5
Boron-11	80.4	5	11	5	6	5+	11.0093	5
Carbon-12	98.89	6	12	6	6	6+	12.0000*	6
Carbon-13	1.11	6	13	6	7	6+	13.0034	6
Nitrogen-14	99.64	7	14	7	7	7+	14.0031	7
Nitrogen-15	0.36	7	15	7	8	7+	15.0001	7
Oxygen-16	99.76	8	16	8	8	8+	15.9949	8
Oxygen-17	0.04	8	17	8	9	8+	16.9991	8
Oxygen-18	0.20	8	18	8	10	8+	17.9992	8
Fluorine-19	100	9	19	9	10	9+	18.9984	9
Chlorine-35	75.77	17	35	17	18	17+	34.9689	17
Chlorine-37	24.23	17	37	17	20	17+	36.9659	17
Uranium-235	0.72	92	235	92	143	92+	235.0439	92
Uranium-238	99.28	92	238	92	146	92+	238.0508	92

*By definition, the mass of carbon-12 is exactly 12.0000—the standard for atomic masses, as discussed in Chapter 4.

Figure 8-17 Nuclear stability can be represented in a graph of the ratio of neutrons to protons in different atoms. Notice that as the atomic number increases, the number of neutrons needed to maintain stability increases faster.

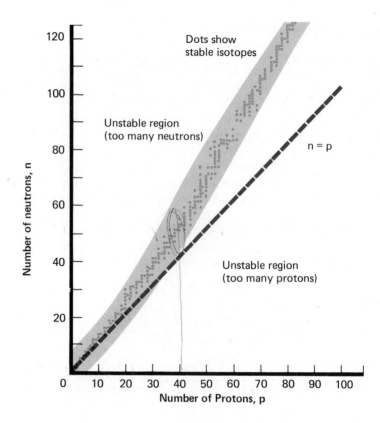

different isotopes, the mass number must be retained. Often you will simply see isotopes symbolized as ^{16}O and ^{18}O, or they may be written as they are in Table 8-2 as oxygen-16 and oxygen-18.

Not all isotopes are radioactive. The stability of an isotope depends on the number of neutrons present relative to the number of protons. Figure 8-17 shows a graph comparing the number of neutrons (y-axis) to the number of protons (x-axis) in the nuclei of atoms. The shaded area indicates the nuclei that are nonradioactive, or stable. You can see that there are many isotopes in the shaded area of the graph. Some isotopes, however, contain too many neutrons (to the left of the shaded area) or too few neutrons (to the right) and are unstable, or radioactive. Radioactive isotopes will decay, giving off alpha or beta particles to become stable.

Looking at the periodic table, you have probably noticed by now that the atomic masses are not whole numbers. These masses are averages based on the percent abundances of all the isotopes of that element. The atomic mass of any one isotope is not a whole number for reasons you will read about in Chapter 9.

Example 8-2

Using the atomic masses for the isotopes of chlorine listed in Table 8-2, find the average atomic mass of chlorine.

▶ Suggested solution _____

Two isotopes for chlorine are listed in Table 8-2: chlorine-35 and chlorine-37. Note that the abundance for chlorine-35 is considerably greater than that for chlorine-37. Logic suggests that the average atomic mass for chlorine should be closer to 35 than to 37.

In any sample of chlorine, you can assume that 75.77% of the atoms are chlorine-35 and 24.23% of the atoms are chlorine-37. The average mass will be equal to the sum of the masses of the fraction of Cl-35 atoms and the fraction of Cl-37 atoms.

average mass = (% Cl-35)(mass Cl-35) + (% Cl-37)(mass Cl-37)

Substituting the values from Table 8-2 and changing percents to decimals gives:

average mass = (0.7577 × 34.9689 amu) + (0.2423 × 36.9659 amu)

Multiplying the values for each isotope and adjusting for significant figures yields

average mass = 26.50 amu + 8.96 amu
average mass = 35.46 amu

By referring to the periodic table, you can verify that the answer is correct within experimental uncertainty.

The atomic mass of an element with one stable isotope can be measured very precisely, but this is not always the case for elements that exist as several stable isotopes. Depending on the origin and history of a particular sample, small differences in the percent abundances of the isotopes composing an element may vary. The study of these variations is part of current research in geology and astrophysics.

You have read a great deal about the atom in this chapter. You have learned something about the early experiments that led to the first understanding of atomic structure and how the scientists who made these inroads developed their models. It is helpful to pause here and summarize some of what you now know about atoms.

1. Atoms are very small.
2. Atoms are made of charged particles that may be separated from each other.
3. The identity of each element is determined by its atomic number (number of protons in each atom).
4. There are isotopes of the same element that differ only in the number of neutrons.
5. The mass of an electron is negligibly small compared to the mass of a whole atom.

Figure 8-18 The mass spectrometer helps make chemical analysis easier and more accurate. Instruments such as this one are found in analytical laboratories all over the world.

8-11 Extension: Mass Spectrometer

The principles first used by J. J. Thomson and Ernest Rutherford in their investigations of the atom are the basis for instruments used in modern research. The **mass spectrometer**, shown in Figure 8-18, is an extremely valuable instrument that is really a descendant of the cathode-ray tube.

The forerunner of the mass spectrometer was the mass spectrograph, which operates on a principle already familiar to you. The gas to be analyzed is injected into a vacuum chamber while being ionized by a stream of electrons. Once inside the chamber, the ions are accelerated through a magnetic field that causes their path to be deflected into a curve. Finally, they strike a photographic plate leaving a record of their impact. The amount of curvature is dependent on the mass of the ion and on its charge. Lighter ions are deflected more than heavier ones. Ions with a greater charge are deflected more than those with smaller charges. Measurement of the position of the lines of impact on the photographic plate allows the mass of an ion to be determined. Figure 8-19 illustrates an analysis of neon gas.

Students have a difficult time analyzing the mass spectrogram. They tend to get confused with the difference between the mass of the isotopes and charges on the ions.

Figure 8-19 In this example, a narrow beam of neon atoms is bombarded by electrons as it enters from the bottom of the chamber. Two kinds of neon ions result, Ne^+ and Ne^{2+}. The paths of the ions are bent in the magnetic field before they expose a photographic plate. Upon close analysis of the photograph, the Ne^+ line and the Ne^{2+} line are seen to be divided into two lines each, representing the two naturally occurring isotopes of neon gas—neon-20 and neon-22. The two lines are not of equal intensity on the photographic plate. Neon-20 produces a line about ten times more intense than neon-22 produces. The intensity of each line reveals the relative abundance of each naturally occurring isotope of neon. (A third isotope, neon-21, leaves a line too faint to be easily seen.)

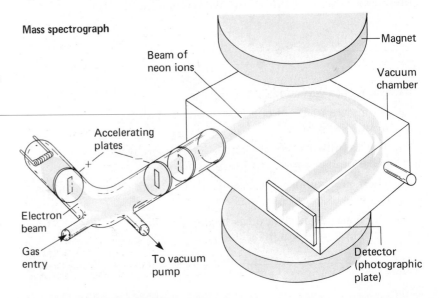

The mass spectrograph is such a versatile tool that it not only will separate an ion beam into its component isotopes, but also can determine the relative abundance of each isotope. The intensity of each line on the photographic plate is used to determine the abundance of that isotope. A heavy, distinct line indicates relatively high abundance. A thinner, less intense line indicates low abundance.

The mass spectrograph and the mass spectrometer operate on the same basis. They only differ in the way the information is reported, with the mass spectrometer projecting the results on an electronic recording device rather than in a photograph. In modern laboratories, the mass spectrometer is used more frequently.

While newer and more precise instruments than the mass spectrometer have been developed, its flexibility and usefulness make it valuable in a variety of research. Different mass spectrometers are used to identify chemical structures, to separate isotopes, and to make precise determinations of isotope mass. Applications of the mass spectrometer range from analyzing the components of air pollution to searching the Martian soil for traces of life.

You may conduct several experiments in the laboratory when you will be asked to analyze chemical unknowns. If you would like to have a mass spectrometer to help, you will have to convince your chemistry teacher to set aside about $300 000 out of the school budget! Mass spectrometers are as expensive as they are valuable.

8-12 Application: The Search for Life on Mars

The first of two fully automated *Viking* landers touched down on the surface of Mars on July 20, 1976. One of the purposes of the mission was to answer the question that astronomers have been asking for hundreds of years: Is there life on the red planet?

Figure 8-20 *Viking* reached Mars after traveling for a year and a half in space. When it landed, *Viking's* mechanical arm scooped up samples of Martian soil that were used in several analytical experiments.

Do You Know?

The first weather report from **Mars** was transmitted back to Earth by the computers aboard Viking: "Light winds from the east in the late afternoon, changing to light winds from the southeast after midnight. Maximum winds were 15 miles per hour . . . Temperature ranged from −122°F just after dawn to −22°F . . . Pressure steady at 7.7 millibars."

Years of research and engineering went into not only the *Viking* lander itself but also its sophisticated, state-of-the-art instrument package. The mission payload carried instrumentation for experiments in meteorology, geology, biology, and physics.

Because of weight considerations for lift-off and landing, the instruments were reduced in size. Some were 1/100 the size of comparable instruments used on Earth. Amazing as it may seem, there were two mass spectrometers on board the *Viking*!

As the lander was gently lowered through the thin Martian atmosphere, an upper atmosphere mass spectrometer (UAMS) sniffed out the rarified air to test for the presence of various gases. This spectrometer found that 95% of the atmosphere is carbon dioxide. Nitrogen gas was found to comprise 2.7 percent of the atmosphere, and 1.6 percent was argon. Less than 1 percent of the Martian atmosphere was a combination of oxygen gas, carbon monoxide, krypton, neon, xenon, and ozone, O_3.

When *Viking* touched down on the surface, a mechanical scoop was employed for soil analysis. The search for life in the Martian soil was accomplished by a very sophisticated set of experiments. One experiment used a gas chromatograph/mass spectrometer (GC/MS) capable of finding organic (carbon-based) molecules in the soil. The GC/MS searched for organic substances containing complex combinations of carbon and hydrogen. Such substances can be either precursors of life or remains of organisms.

After painstaking analysis of the data received from Mars, no evidence of life was found. However, unusual chemical reactions in the soil pointed to the possibility that life may have existed on the planet in the distant past, although some chemical reactions may have erased all traces of such life.

Even though evidence of life was not found on Mars, scientists were not disappointed in the mission. The information obtained from the *Viking* experiments helped advance knowledge in the physical and biological sciences and provided another clue to the mystery of the origin of the solar system.

Review and Practice

1. The number of which subatomic particle determines the identity of a specific element?

2. How is the mass number of an element determined?

3. An atom of zinc has an atomic number of 30 and a mass number of 65. How many protons, electrons, and neutrons are present in this zinc atom?

4. What are isotopes?

5. Boron exists in the form of two stable isotopes, boron-10 and boron-11, which occur in an abundance of 19.6 percent and 80.4 percent respectively. Using the atomic masses in Table 8-2, calculate the average mass of boron.

6. List two uses of the mass spectrometer (or mass spectrograph).

7. How was the mass spectrometer used to learn more about the planet Mars?

1. the proton
2. the number of protons plus the number of neutrons
3. p = 30; e⁻ = 30; n = 35
4. atoms of the same element with different numbers of neutrons
5. 10.8 amu
6. isotope identification and chemical analysis of the atmosphere or of other planets; also, identification of chemical structures, determination of isotope mass
7. analysis of the Martian atmosphere and soil for carbon-based compounds

Numbers in red indicate the appropriate chapter sections to aid you in assigning these items. Answers to all questions appear in the Teacher's Guide at the front of this book.

Summary

■ The discovery of the structure of the atom occurred at the turn of the century. The early models of the atom were made possible by a series of experiments which delved into the electrical nature of the atom.

■ The masses of the electron and proton were discovered by the combined work of J. J. Thomson and Robert Millikan. Their experiments also showed that the electron and proton are opposite in charge. By convention, electrons are negative and protons are positive. The nucleus was discovered by Ernest Rutherford. He recognized that the atom has a densely packed center composed of positively charged particles.

■ The atomic number of an element is the number of protons in the nucleus of its atoms. The mass number of an atom is the number of protons and neutrons in the nucleus. In a neutral atom, the numbers of protons and electrons are equal. If an atom loses or gains electrons, it becomes an ion.

■ Not all atoms of the same element have the same mass. Atoms of the same element with a different number of neutrons are called isotopes. Some isotopes contain unstable numbers of protons and neutrons in the nuclei of their atoms. These isotopes are radioactive, and they emit alpha or beta particles in the process of becoming more stable.

■ The mass spectrometer descended from experiments of Thomson and Rutherford to become a valuable tool used in modern chemical analysis. It can analyze elements for their isotopes and will detect trace substances with an accuracy of up to one part in ten million.

Chemically Speaking

alpha particles
atomic number
beta particles
cathode-ray tube
electrode
electron
gamma rays
isotopes
mass number
mass spectrometer
neutron
nucleus
proton

Review

1. Briefly describe Dalton's model of the atom. (8-1)

2. What is the name applied to the negative electrode in a cathode-ray tube? (8-2)

3. What happens to an electron beam when it is placed in a magnetic field? In an electric field? (8-2)

4. What is the mass of an electron? (8-4)

5. What part of the results of Rutherford's experiment were the most surprising to him? (8-6)

6. How was the neutron discovered? (8-8)

7. What particles are contained in the nucleus of the atom? (8-7, 8-8)

8. Where is most of the mass of an atom found? (8-7)

9. How do isotopes of the same element differ from each other? How are they the same? (8-10)

10. How is the mass spectrometer used as an analytical tool? (8-11)

11. What two properties of an ion influence its curvature in the magnetic field of a mass spectrometer? (8-11)

12. Using the periodic table, identify the elements whose atomic numbers are 2, 20, 33, and 84. (8-9)

13. What is the only element whose atoms do not always contain neutrons? (8-10)

Interpret and Apply

1. How are the methods used in studying dinosaurs similar to those used in the study of the atom? (8-1)

2. What was incorrect about J. J. Thomson's model of the atom? (8-7)

3. If the charge-to-mass ratio of the electron was accurately known, but the mass of the electron was found to be larger than it is now, would the charge on the electron be larger, smaller, or the same as it is now? (8-4)

4. Write a balanced equation for the ionization of a gaseous argon atom into a positive ion. (8-3)

5. In Rutherford's experiment, how would the use of foils made from different metallic elements affect the amount of deflection of the alpha-particle beam? (8-9)

6. Use symbol form to rewrite the information supplied below: (8-9)

ELEMENT	magnesium	iron	antimony
ATOMIC NUMBER	12	26	51
MASS NUMBER	26	57	123

7. Which would have a larger charge-to-mass ratio, a Na^+ or a K^+ ion? (8-4, 8-9)

8. What would be one neutron-to-proton ratio for a stable atom that has 40 protons? (8-10)

9. The atomic number for an element can be included in the symbol as a subscript at the left. Why would writing the subscript at the right of the symbol be a problem? (8-9)

10. Using Thomson's model of the atom as a guide, draw a sketch of what a boron atom would be. (8-5)

11. Why was it difficult to detect and identify the neutron even though most atoms contain neutrons? (8-8)

12. Which will have a greater e/m ratio: He^{2+} or Ar^{2+}? (8-4, 8-9)

Problems

1. How many electrons are needed to furnish a mass of one gram? What would be the mass of one mole of electrons? (8-4)

2. There are three naturally occurring stable isotopes of oxygen—oxygen-16, oxygen-17, oxygen-18—and two naturally occurring stable isotopes of hydrogen—hydrogen-1 and hydrogen-2. How many different water molecules can be made using the various combinations of all the isotopes of oxygen and hydrogen? (8-10)

3. Helium, as found in nature, consists of two isotopes. Most of the atoms have mass number 4, but a few have mass number 3. For each isotope indicate the following: (8-9)
   a. atomic number
   b. number of protons
   c. number of neutrons
   d. mass number
   e. nuclear charge

4. The average dimension for the radius of an atom is 1×10^{-8} cm, and the radius of a nucleus is 1×10^{-13} cm. Determine the ratio of atomic volume to nuclear volume, assuming both the atom and the nucleus are spheres. (Volume of a sphere = $4/3 \pi r^3$.) (8-7)

5. Precise atomic masses of each isotope of magnesium are given below along with the percent abundance of each isotope: (8-10)

magnesium-24	23.98504	78.70%
magnesium-25	24.98584	10.13%
magnesium-26	25.98259	11.17%

Calculate the atomic mass of magnesium.

6. Copy the incomplete table below. Presuming each symbol represents a neutral atom, fill in the missing information. (8-9)

SYMBOL	Kr	?	?	Dy
ATOMIC NUMBER	36	?	?	?
MASS NUMBER	?	?	235	?
NUMBER OF PROTONS	?	25	?	?
NUMBER OF NEUTRONS	48	55	?	98
NUMBER OF ELECTRONS	?	?	92	?

7. If a pure sample of uranium has 10 billion atoms, how many atoms of the uranium-235 isotope would probably be in the sample? (8-10)

8. Determine how many protons, neutrons, and electrons there are in an ion of iodine-131 that has a charge of 1−. (8-9, 8-10)

Challenge

1. The radius of a carbon atom in many compounds is 0.77×10^{-8} cm. If the radius of a Styrofoam ball used to represent the carbon atom in a molecular model is 1.5 cm, how much of an enlargement in volume is this? (Volume of a sphere = $4/3 \pi r^3$.)

2. Elements in the periodic table are ordered according to increasing atomic number, not increasing atomic mass. Identify two places where the atomic number increases but the atomic mass decreases and explain why these exceptions exist.

3. Examine the information in the table below.

ELEMENT	ATOMIC NUMBER	NATURALLY OCCURRING ISOTOPES	NATURAL RADIOACTIVE ISOTOPES
oxygen	8	3	0
nickel	28	5	0
samarium	62	7	3
uranium	92	3	3

Oxygen and nickel are elements with small masses in comparison to samarium and uranium. How do you account for the fact that the number of naturally occurring radioactive isotopes is greater for the more massive elements?

4. How many times more massive is an alpha particle than an electron?

5. Describe how the positions of the spectrographic lines would differ in the analysis of a mixture of silicon-28, silicon-29, and silicon-30. Assume that silicon ions with 1^+ and 2^+ charges are formed.

6. Find the mass of a proton using the following e/m ratio:

$$e/m = 9.58 \times 10^4 \text{ coulomb/gram}$$

Presume that the proton has the same charge as the electron, 1.60×10^{-19} coulomb.

7. The following words are terms that are related to the atom: *proton, neutron, nucleus, electron, charge, mass, atomic number, mass number, isotopes.* Using the method of concept mapping first described in Chapter 2, draw a concept map that logically relates these terms to the atom and/or to each other.

8. Using the molar masses of the electron and the hydrogen atom, calculate the molar mass of the proton.

Synthesis

1. The pressure in a cathode-ray tube is approximately 10^{-2} atmosphere when the glow appears. When the pressure is decreased to about 10^{-6} atmosphere, the gas stops glowing. How many molecules would there be in one cubic centimeter of gas when the glow first appears? How many molecules would there be in one cubic centimeter of gas when the glow disappears? The temperature is 0°C.

2. When helium atoms are energized in a cathode-ray tube, they can lose electrons and form a beam of positive ions. Write a charge-balanced equation for the case where a sample of helium atoms become helium ions, each with a charge of 2+.

3. Assume that the nucleus of the fluorine-19 atom is a sphere with a radius of 5×10^{-13} cm. Calculate the density of matter in the fluorine-19 nucleus.

4. Thomson thought that the charge of the proton was distributed evenly over the entire hydrogen atom, which has a radius of 10^{-8} cm. Rutherford realized that the proton is the hydrogen nucleus with a radius of only 10^{-13} cm. Given that the charge of a proton is 1.6×10^{-19} coulombs, compare the positive-charge density in coulombs/cm^3 of the Thomson hydrogen atom with the positive-charge density of the Rutherford hydrogen atom. (Volume of a sphere = $4/3\pi r^3$.)

Projects

1. Prepare a biographical report of one of the scientists mentioned in this chapter. Include background information not covered in the text about the scientist's experiments.

2. Until 1912, the elements were listed on the periodic table according to their atomic masses. Henry G. J. Moseley developed the concept of atomic number and rearranged the order of the elements as you see them today. Find out what experiment he performed in order to accomplish this task.

3. Find out how a television transforms wave impulses from a transmitting antenna into a picture. Emphasize how the picture tube focuses and directs the electron beam.

4. Nobel prize winners are a small part of the scientific community. There are many scientists around you. Contact and interview someone who works in your area and write a report about the person and the experience. Scientists can be found in such places as a water treatment plant, an agricultural extension station, a university, a county health department, or one of the local industries.

CHAPTER 9

M odern civilization depends on energy. Today's life styles are made possible by a heavy reliance on fuels that run machinery. Small wonder that the United States and other nations continue to search for replacements for the waning stores of fossil fuels that made widespread industrialization possible.

The atomic blasts that brought an end to World War II also brought hope for cheap, clean energy for peaceful uses. However, that hope has not been fully realized. In the United States, fears of disaster, concern about nuclear waste disposal, and escalating costs have fueled a continuing debate about using nuclear energy. Your future ability to make wise decisions concerning nuclear energy will depend partly on how well you understand the nature of nuclear chemistry.

Nuclear Chemistry
Exploring Radioactivity

Science introduced the world to the nuclear age more than 40 years ago. Today, the results of nuclear reactions have tremendous impact on your life. Nuclear reactions are used to fuel electric power plants, while radioactive isotopes are a tool used by chemists to find out how chemical reactions occur. In medicine, nuclear materials are used routinely in the diagnosis of some diseases and in the cure of some forms of cancer. Nuclear chemistry has even found its way into household smoke detectors. On a larger scale, nuclear weapons are a reality that the world struggles to deal with wisely.

How do nuclear reactions occur? What is happening to atoms that undergo nuclear change? Can such changes be measured? What effects do nuclear changes produce on you and the environment? If these questions are to be answered, it is necessary to take a much closer look at the tiny nucleus you first learned about in Chapter 8.

CONCEPTS
- how nuclear reactions differ from chemical reactions
- forms of radiation
- radiation in everyday life
- why ionizing radiation is significant
- alpha and beta emissions
- spontaneous vs. induced nuclear reactions

9-1 Nuclear Reactions vs. Chemical Reactions

You have already learned that chemical changes can be understood in terms of atoms. However, those changes involve only part of the atom—the electrons that surround the nucleus. In a nuclear reaction, this situation is vastly different. The changes that occur among the protons and the neutrons within the nucleus are responsible for an atom's nuclear properties.

In Chapter 8 you studied the composition of the nucleus. You learned that elements exist in different isotopic forms and that some isotopes are radioactive. Each isotope of an element is called a **nuclide**, and is identified by its number of protons and neutrons. Though the chemical nature of different nuclides of any one element is the same, such nuclides differ in their number of neutrons. The differences influence the kinds of nuclear reactions each can undergo.

9-2 Characteristics and Effects of Radiation

Radiation is a general term for energy or particles that are emitted from a source and travel through an intervening medium or space. Light, heat, and alpha particles, beta particles, and gamma rays (Chapter 8), are forms of radiation. Alpha particles are recognized as rapidly moving helium nuclei with two protons and two neutrons. As such, they can be symbolized as $_2^4He$ (or just α). Beta particles are electrons moving at tremendous speeds, thus having high kinetic energies. They have negligible mass and a charge that is opposite that of a proton. A beta particle is often represented as $_{-1}^{0}e$ or just β^-.

Radiation is classified as ionizing and nonionizing radiation. **Ionizing radiation** (such as X rays and gamma rays), produces ions out of the atoms and molecules in matter. In contrast, radio waves and light rays are called **nonionizing radiation** because they are not energetic enough to ionize matter.

When ionizing radiation encounters other matter, the results are of interest because they leave the world different from before. Essentially, the radiation collides with atoms causing electrons to be displaced. As a result, ions form. If the atoms are bonded to other atoms, the disruption of the electrons may break chemical bonds. Often, the ions lose their excess positive or negative charge by interacting with neighboring particles to form new bonds. In living cells, some ions that are formed from ionizing radiation come from water. When a gamma ray enters a cell, it can lose most of its energy in the cell by causing a water molecule to split apart. Two typical reactions are:

$$H_2O + gamma\ ray \longrightarrow H^+ + OH^-$$
$$H_2O + gamma\ ray \longrightarrow H_2O^+ + e^-$$

H^+ plus OH^-, and H_2O^+ plus e^- are examples of **ion pairs**, charged particles formed from irradiated matter. These ion pairs can cause changes in living cells. Many of these changes have no overall impact on the life of an organism; other changes do.

Immature cells and cells that are undergoing cellular division are the most sensitive to radiation. If affected, these cells (such as those in bone marrow, reproductive organs, and the lining of the intestine) may die or grow abnormally. If the radiation happens to strike a chromosome in a sperm or an ovum, the organism produced from that sperm or ovum may be affected in some way.

Remind students that although radiation may cause deleterious effects, "abnormal" is not synonymous with "bad." Many effects are negligible.

Figure 9-1 Buildings made of granite, clay, and some kinds of brick emit more radiation than those made of wood. If you lived in New York City's Grand Central Station (far right), you would receive 570 millirems of additional radiation per year. Building materials do not emit enough radiation to pose a threat to life.

As you read this chapter, whether you are in your home, a class-room, the library, or outdoors, you are being bombarded with atomic nuclei, subatomic particles and high-energy rays. Some radiation comes from outer space. Other sources of radiation are Earth itself and common building materials. **Background radiation** is this radiation that is constantly around you. One unit used to measure the amount of radiation absorbed by humans is the **rem**. The average citizen of the United States is exposed to 100 millirems of radiation per year.

Exposure to radiation may come from routine medical treatment and other ordinary sources. A chest X-ray will deliver 50 millirems, a dental X-ray 20 millirems. The human body is a source of radiation from carbon-14 and potassium-40, obtained in common foods. Some foods, such as beans and fruits, are low in radiation, whereas breads and potatoes are higher in radiation. Does this information mean that you must stop eating some of the foods you like? Not really. The effects of radiation exposure depend mostly on what kind of radiation and how much of it you receive.

Are there any ill effects from long-term exposure to low-level radiation? Most people are aware that massive doses of radiation are harmful. On the other hand, questions about the effects of repeated exposure to small doses are still being investigated.

A rad = dose of radiation that deposits 1×10^{-2} J/kg of tissue.

A rem = unit of dosage that causes the same amount of biological injury as 1 roentgen of X rays or gamma rays.

Do You Know?

A household smoke detector contains ionizing-type radioactive Americium. If held 20 cm from your body for one year, you would receive 2 millirems of radiation.

Bring to the classroom a Geiger counter and a few safe radioactive sources. These materials may be obtained from a physics teacher or through a scientific supply company. Monitor the radiation given off by the environment and compare it to the radiation given off by the safe sources.

9-3 Types of Nuclear Decay

The source of radioactivity is an unstable atomic nucleus. In Chapter 8, you learned that the stability of a nucleus depends on its ratio of neutrons to protons. If the nucleus contains too many or too few neutrons, it may eventually decay. There are many kinds of nuclear decay. In this section, you will examine two that are common.

Thorium-230 has an unstable nucleus, and undergoes radioactive decay through **alpha emission**. In this type of decay, an alpha particle is released and the mass of the resulting nucleus is less than the original. A "nuclear equation" can be written to describe the reaction:

$$_{90}^{230}\text{Th} \longrightarrow {}_2^4\text{He} + {}_{88}^{226}\text{Ra}$$

Because the alpha particle is composed of two protons and two neutrons, the resulting radium nucleus contains two fewer protons than thorium. Alpha emission also causes the mass number to be reduced by 4. Recall that the atomic number (number of protons) determines the identity of an element. The product in this reaction has an atomic number of 88, radium. In a decay reaction, the initial element (in this case thorium-230) is called the **parent nuclide** and the resulting element (radium-226) is called the **daughter nuclide**. Alpha decay always results in a daughter nuclide with an atomic number that is two less than the parent nuclide.

A nuclear equation is balanced when the sum of the atomic numbers on the right is equal to the sum of the atomic numbers on the left. The sums of the superscripts, or mass numbers, on the left and right also must be equal.

Alpha decay may be accompanied by the emission of a gamma ray, but this loss does not affect the atomic numbers or the mass numbers of the nuclides.

atomic numbers: $90 = 88 + 2$
mass numbers: $230 = 226 + 4$

Beta emission is the emission of an electron from a nucleus. You may be wondering how an electron can come from within a nucleus in the first place. Electrons were described in Chapter 8 as being *outside* the nucleus of the atom. A neutron can be thought of as being composed of a proton and an electron. Remember the combined charge of a proton and an electron is zero, and the masses of a neutron and proton are about the same. (The mass of the electron is negligible.) A neutron, $_0^1n$, can decay in an unstable nucleus producing a proton and an electron, as represented by the following equation. (Because a hydrogen nucleus is a proton, $_1^1H$ is customarily used to symbolize a proton when writing nuclear equations.)

$$_0^1n \longrightarrow _1^1H + _{-1}^0e$$

The decay of a neutron also produces a neutrino. It does not affect the nuclear equation. Neutrinos are mentioned in the pupil text in Section 9-14.

Protactinium-234 is a radioactive nuclide that undergoes beta emission. The following equation describes the decay:

$$_{91}^{234}Pa \longrightarrow _{92}^{234}U + _{-1}^0e$$

The disintegration of a neutron provides a new proton in the nucleus, which changes the atomic number to that of uranium. The loss of the electron does not affect the mass numbers. Beta decay results in a daughter nuclide containing one more proton than the parent nuclide. Gamma radiation often accompanies alpha and beta decay, but does not cause a change in either atomic number or mass number.

Other kinds of radioactive decay take place besides alpha, beta, and gamma decay. In orbital electron capture (K capture), an electron from an orbital surrounding the nucleus combines with a proton in the nucleus, forming a neutron. For example, $_4^7Be$ becomes $_3^7Li$. Some nuclides also decay by positron emission, such as $_6^{11}C$ becoming $_5^{11}B$.

There are other kinds of changes an unstable nucleus can undergo. Some of these changes involve subatomic particles that are not mentioned here. You may study these particles and related reactions in a physics course.

9-4 Instigating Nuclear Decay

The students may not be familiar with the concept of spontaneity. You need not be chemically technical, but a class discussion on the meaning of spontaneity may be in order.

The nuclides mentioned so far exhibit **spontaneous decay**, meaning their alpha and beta emissions occur naturally. Some nuclides undergo a series of spontaneous decays before a stable nuclide forms. The series is a natural phenomenon in which the daughter nuclide from one decay becomes the parent nuclide of the next decay. Figure 9-2 depicts the series of disintegrations of uranium-238 to lead-206. Two other naturally occurring series are known: the decay of uranium-235 into lead-207 and the decay of thorium-232 into lead-208. These isotopes of lead are stable.

Nuclear reactions can be induced when an otherwise stable nucleus is changed to an unstable one by bombardment with other particles. Experiments involving such nuclear reactions have produced some interesting results. The experiments led to the discovery of the neutron. They also resulted in the synthesis of new isotopes.

In 1919, Ernest Rutherford bombarded nitrogen-14 with alpha particles. The reaction produced an isotope of oxygen accompanied by the emission of a proton. The equation describing the reaction is:

$$_7^{14}N + _2^4He \longrightarrow _8^{17}O + _1^1H$$

Similar bombardment of other elements having low atomic numbers also produced free protons. These results supported Rutherford's theory that protons were contained in the nuclei of atoms.

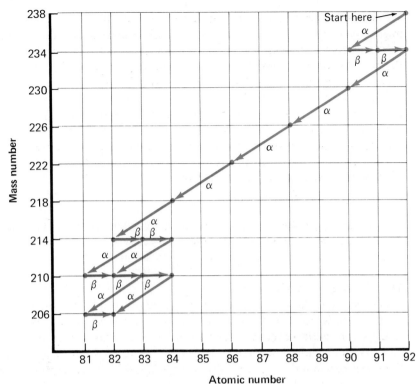

Figure 9-2 The radioactive decay series of uranium-238 to lead-206 involves several steps. Alpha emissions result in the loss of two protons and thus a decrease in atomic number. Beta emissions involve the gain of a proton and an increase in atomic number.

The discovery of the neutron was the result of experiments you first read about in Chapter 8 (Section 8-8). The nuclear equation for the reaction is written as:

$$\ce{^9_4Be} + \ce{^4_2He} \longrightarrow \ce{^{12}_6C} + \ce{^1_0n}$$

Neutrons are very effective for inducing nuclear reactions because their lack of charge allows them to approach a nucleus easily and disrupt its stability.

Since the first experiments, over a thousand radioactive isotopes have been manufactured in the laboratory. Many of them are used for biological and chemical research. The next part of the chapter will discuss some of these uses.

Review and Practice

1. How is a nuclear reaction different from a chemical reaction?

2. What is the difference between ionizing and nonionizing radiation? Give three examples of each kind of radiation.

3. What is background radiation?

4. Write a nuclear equation for the alpha decay of $^{218}_{84}\text{Po}$.

5. Write a nuclear equation for the beta decay of $^{214}_{82}\text{Pb}$.

6. How is it possible to cause a stable nucleus to undergo radioactive decay?

1. Nuclear reactions involve changes among protons and neutrons in the nucleus. Chemical reactions result from the interaction of electrons between atoms.
2. Ionizing radiation produces ion pairs in matter, non-ionizing radiation does not.

IONIZING	NON-IONIZING
X rays	visible light
cosmic rays	infrared light
gamma rays	ultraviolet light

3. naturally occurring radiation present in the environment
4. $^{218}_{84}\text{Po} \longrightarrow {}^{214}_{82}\text{Pb} + {}^4_2\text{He}$
5. $^{214}_{82}\text{Pb} \longrightarrow {}^{214}_{83}\text{Bi} + {}^{\ 0}_{-1}\text{e}$
6. Bombard the nucleus with a particle of sufficient energy to produce instability.

Using Nuclear Reactions for Research

CONCEPTS

■ synthesizing new elements
■ particle accelerators
■ half-life of radioactive nuclides
■ dating biological and geological samples using radioactive isotopes
■ radioactive isotopes as tracers

As soon as researchers were able to unravel some of the mysteries of the nucleus, they began looking for possible applications of the new science. Some applications were stumbled upon accidentally, like the synthesis of new isotopes. From this step, nuclear scientists moved quickly to synthesize entirely new elements. New and better instruments were developed to move subatomic particles at greater speeds. Studying the rates of decay of different radioactive isotopes has enabled scientists to use these materials in experiments related to geology, biology, medical research, and other fields. Radioactive tracers are used in determining the age of the rocks from the moon, how chemical reactions work, and when certain organs in the body malfunction.

9-5 Synthetic Elements

Do the reactions you have been reading about in Sections 9-3 and 9-4 seem somewhat unusual? The alpha and beta emissions described result in elements different from the ones that began each reaction. Known as **transmutation**, the process of changing one element into another has fascinated people since ancient times. One transmutation people frequently tried, hoping to become rich, was the chemical conversion of lead into gold. Much to everyone's dismay, no successful process was discovered. What no one knew is that regardless of the kind of chemical or physical reaction an element undergoes, it never changes its identity. Only in a nuclear reaction, involving a change in atomic number, can one element become another. With this knowledge, can people now devise nuclear reactions that will create enormous riches from a glob of lead? Unfortunately, the cost of such reactions far exceeds the value of the gold that could be obtained. Instead, transmutation reactions serve as a method for gaining information about nuclear structure.

Figure 9-3 Alchemists practiced a mixture of science, magic, and religion and flourished until the early 1700's. Among other things, they attempted to change less costly metals into gold. Their dabblings left a legacy of knowledge including the discovery of five elements—zinc, phosphorus, bismuth, arsenic, and antimony—and the establishment of some common laboratory procedures.

Figure 9-4 A tiny sample of curium (near right) is less than 0.4 mm in diameter. At the far right is plutonium trichloride in a test tube.

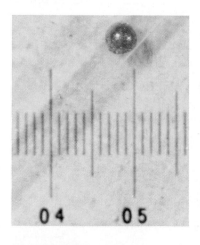

Since 1939, American teams led by Glenn Seaborg and more recently Albert Ghiorso, have successfully produced elements with larger atomic numbers than that of naturally occurring uranium. These elements, which do not exist in nature, are known as **transuranium elements**. (See Figure 9-4.) Most of the elements produced in the United States were made at the Lawrence Radiation Laboratory of the University of California, Berkeley. Scientists in both the United States and the Soviet Union claim to be the first to have synthesized elements 105 and 106. Recently, elements 108 and 109 have been synthesized in West Germany. Synthesis of element 107 has been reported in both the Soviet Union and West Germany. All of the transuranium elements that have so far been synthesized are radioactive, that is, they decay naturally, mostly by alpha emission.

9-6 Extension: Particle Accelerators

Tremendous amounts of energy are needed to produce the transuranium elements. Particles such as deuterium (the $_1^2H$ isotope of hydrogen), helium nuclei, and carbon nuclei have to be accelerated to great velocities to fuse with target nuclei. Particle accelerators are used to force such reactions. The **cyclotron**, diagrammed in Figure 9-5, was one of the earliest particle accelerators. It consists of two hollow, semicircular electrodes (dees) through which the particles travel in a spiral path. Charged particles introduced at the center line of the cyclotron are accelerated across the space between the electrodes under the influence of an electric field. The field alternates from plus to minus, giving the particles first a pull and then a push. The particles move in a gradually widening circular path as they gain the desired speed. They are then deflected out of the cyclotron toward a target.

The CERN **synchrotron** in Geneva, Switzerland, is a more powerful particle accelerator than the cyclotron. It is made of a ring of focusing electromagnets in an underground tunnel two kilometers in diameter.

Elements 107, 108, and 109 are very unstable. Their syntheses have not been completely verified.

Do You Know?

The first synthetic element was actually produced in Italy in 1937. Technetium, element 43, has no stable isotopes and does not exist in nature on Earth.

Elements beyond atomic number 104 are currently symbolized by combinations of three letters. Conflicting claims to being first to synthesize the new elements have caused disagreement about who should name them. The IUPAC has suggested a temporary system. Glenn Seaborg suggests dropping names in favor of using the atomic number alone.

SYNTHESIS OF SOME TRANSURANIUM ELEMENTS

$$_{92}^{238}U + _1^2H \longrightarrow _{93}^{238}Np + 2_0^1n$$

$$_{93}^{238}Np \longrightarrow _{94}^{238}Pu + _{-1}^0\beta$$

$$_{94}^{239}Pu + _2^4He \longrightarrow _{95}^{240}Am + _1^1p + 2_0^1n$$

$$_{92}^{238}U + _7^{14}N \longrightarrow _{99}^{246}Es + 6_0^1n$$

$$_{98}^{249}Cf + _6^{12}C \longrightarrow _{104}^{257}Unq + 4_0^1n$$

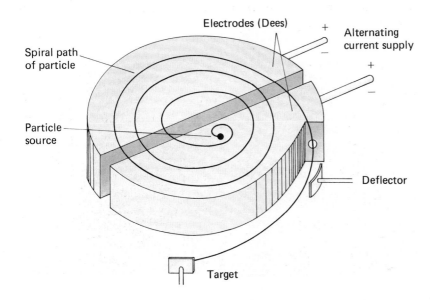

Figure 9-5 A cyclotron is a positive particle accelerator. The two electrodes are called Dee's because they are shaped like the letter D. The particles leaving the cylcotron hit the target at great speeds, resulting in high-energy impacts.

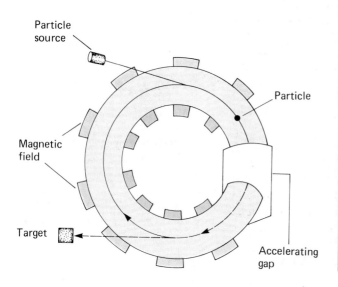

Particle
source

Particle

Magnetic
field

Target

Accelerating
gap

Figure 9-6 The underground tunnel of CERN's synchrotron particle accelerator (right) is located on the Swiss-French border. Particles are accelerated through the same track again and again instead of in a spiral like the cyclotron. (See diagram above.)

Do You Know?

CERN is an acronym for Conseil Européen Pour La Recherche Nucléaire (the European Laboratory for Nuclear Research).

Protons are accelerated a few times per revolution until they reach 500 000 revolutions in two seconds. The beam of high-velocity protons is then aimed at the target nuclei. The most powerful synchrotron in the world is at the Fermi National Accelerator Laboratory (Fermilab) near Chicago, Illinois. In late 1985, a particle collision yielding record-high reaction energies was successfully conducted there. When fully operational, the Fermilab facilities will produce collision energies over three times that of any other existing accelerator. Experiments such as those conducted at CERN and Fermilab go far beyond the synthesis of new elements. The work has led to several recent discoveries in the field of particle physics. Discoveries include information about the nature of the nucleus, subnuclear particles, and the forces that affect matter. CERN is constructing another kind of particle accelerator that is 27 kilometers in circumference. Larger particle accelerators, called *superconducting supercolliders*, or SSC's, are planned.

Figure 9-7 An aerial view of the track above the Fermilab synchrotron provides a sense of how big the particle accelerator really is. The underground tunnel of the synchrotron is 6.3 kilometers in circumference.

9-7 Half-Life and Geologic Dating

In a sample of radioactive nuclides, the decay of an individual nuclide is a random event. It is impossible to predict which nuclide will be the next one to undergo a nuclear change. How then, do you make sense out of things that cannot be predicted on an individual basis? One approach is to predict change for a given amount of substance—for example, one half. You can measure how much time is required for the level of radiation in a sample to decrease by half. You then have a rate that is useful for making predictions about collections of nuclides. Scientists have developed a standard for measuring the rate of radioactive decay of a nuclide. The time it takes for one half of the nuclei in a radioactive sample to decay is known as the **half-life**, $T_{\frac{1}{2}}$. For example, the half-life of uranium-238 is 4.5 billion years. If a sample of uranium-238 contains one million atoms, then 500 000 of the nuclei will decay in 4.5 billion years. After another 4.5 billion years, 250 000 nuclei (one half of those remaining) will decay, and so on. By contrast, a sample of 1 million nuclei of polonium-212 would decay much faster because the half-life of polonium-212 is 3.0×10^{-7} second. Half-lives of different radioactive nuclides vary tremendously.

One useful application of half-life measurements is in the determination of the ages of fossils and rocks. For example, carbon-14 is a radioactive nuclide constantly produced in the atmosphere. It has a half-life of 5730 years, and undergoes beta emission, decaying to nitrogen-14. During photosynthesis, green plants absorb carbon in the form of carbon dioxide. A percentage of this carbon dioxide is made from carbon-14. Once the plant dies, photosynthesis stops, and no more radioactive CO_2 is absorbed. However, decay of the carbon-14 continues. Careful measurements of the amount of radioactivity of carbon-14 in a once-living plant yields the approximate time in history when the plant died.

Ask the students what would be left after a radioactive nuclide completely decays. Some may say "nothing" is left. Clarify this misconception.

A blackline master listing half-lives of different nuclides is available in the Teacher's Resource Binder. Additional half-lives can be obtained from the *CRC Handbook* or *Lange's Handbook of Chemistry*.

The following graph may be used to illustrate to students the relationship between time and the fraction of C-14 not decayed in a sample.

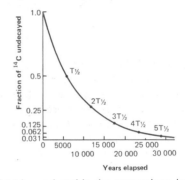

C-14 is produced in the atmosphere by the bombardment of stable N-14 with cosmic neutrons.

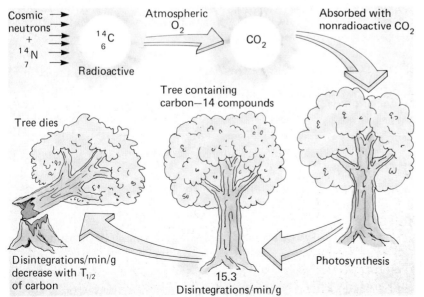

Cosmic neutrons + $^{14}_{7}N$

$^{14}_{6}C$ Radioactive

Atmospheric O_2

CO_2

Absorbed with nonradioactive CO_2

Tree containing carbon–14 compounds

Tree dies

Disintegrations/min/g decrease with $T_{1/2}$ of carbon

15.3 Disintegrations/min/g

Photosynthesis

Figure 9-8 Radioactive carbon is formed when neutrons from space collide with nitrogen-14 in the atmosphere. Plants absorb C-14 in the form of CO_2 and convert it to other carbon compounds during photosynthesis. When a plant dies, radioactivity from carbon-14 decreases because no more of the isotope is being absorbed. The rate of radioactivity is used to find the age of the fossil plant.

The radioactivity can be measured in the number of disintegrations that occur per minute per gram. This rate changes as the amount of radioactive material decreases with time. Measurements show that a living plant gives off 15.3 ± 0.1 disintegrations per minute per gram of material containing carbon-14. A plant that was living 5730 years ago (the half-life of carbon-14) today will have a rate of radioactivity one half as large.

$$\frac{1}{2} \times 15.3 \text{ disintegrations/min/g} = 7.6 \text{ disintegrations/min/g}$$

A plant that died 2×5730 years ago (two C-14 half-lives) will give off:

$$\frac{1}{2} \times \frac{1}{2} \times 15.3 \text{ disintegrations/min/g} = 3.8 \text{ disintegrations/min/g}$$

Knowing how many disintegrations previously living materials give off today, archaeologists can determine the geologic age of an organism.

Before discussing the solution to Example 9-1, you may want to have students work in small groups to decide how they would solve it. See the suggested problem-solving strategy in the Teacher's Guide at the front of this book.

Example   9-1

How old is a piece of ancient wood that is giving off β-emissions from carbon-14 at the rate of 1.9 ± 0.1 disintegrations/minute/gram?

▶ Suggested solution _____

The problem asks for an answer measuring time (the age of the wood). The radioactivity in the wood comes from the carbon-14 absorbed by the tree when it was alive. It is known that the half-life of carbon-14 is 5730 years. Can this information be related to what is asked for in the problem? Remember that when the wood was alive it was giving off 15.3 disintegrations per minute per gram. How many half-lives did the wood go through to reach its current radioactivity? After one half-life (5730 years) had elapsed, the radioactivity coming from the wood would be

$$\frac{1}{2} \times 15.3 \text{ disintegrations/min/g} = 7.65 \text{ disintegrations/min/g}$$

After another half-life, the radioactivity again would be reduced by one half, or

$$\frac{1}{2} \times 7.65 \text{ disintegrations/min/g} = 3.82 \text{ disintegrations/min/g}$$

After a third 5730-year time span, the radioactivity would be reduced by one half again, or

$$\frac{1}{2} \times 3.82 \text{ disintegrations/min/g} = 1.91 \text{ disintegrations/min/g}$$

Looking back at the problem, this calculated rate of radioactivity is close to what was measured coming from the wood. Therefore, the total time that must have elapsed since the death of the tree can be thought of as three half-lives.

$$5730 \text{ years} \times 3 = 17\,190 \text{ years}$$

Of course, the tree did not die exactly 17 190 years ago. Remember that the calculated value is an approximate age within a range of uncertainty. The wood is about 17 000 years old.

The half-life of carbon-14 is short in comparison to the age of most fossils. Radiocarbon dating is not effective for fossils older than 60 000 years. Determinations of the ages of older fossils, rocks, and minerals are made using radioactive isotopes with longer half lives, such as uranium, potassium, and argon.

Review and Practice

1. What is transmutation?

2. Describe how scientists are able to synthesize elements with atomic numbers larger than that of uranium.

3. Why are transuranium elements difficult to synthesize?

4. Why are particle accelerators important to the synthesis of transuranium elements?

5. What is half-life?

6. What is the age of a plant fossil that emits 7.65 beta particles per minute per gram?

1. changing one element into another by way of a nuclear reaction
2. Large nuclei are bombarded with smaller nuclei. The nuclei combine forming an element with an atomic number greater than either original nucleus.
3. Tremendous amounts of energy are needed to cause atomic nuclei to combine.
4. Particle accelerators give particles enough kinetic energy to penetrate the target nuclei and cause transmutation.
5. the time it takes for half of the nuclei in a radioactive sample to decay
6. 5730 years

9-8 Radioactive Isotopes as Tracers

Isotopes of the same element behave the same way in a chemical reaction. Those isotopes that are radioactive can be detected by instruments sensitive to radiation. Suppose a radioactive isotope of an element is substituted for a similar nonradioactive one in a chemical reaction. Instruments can follow the steps that take place in the chemical reaction because all compounds containing the radioisotope (a shorthand name for radioactive isotopes) will be radioactive. Such radioisotopes are called **tracers**. They are used to keep track of what happens to a chemical during a physical or chemical change.

If you studied biology, you probably learned about several chemical reactions that occur in living organisms. Most biology courses include the study of such topics as photosynthesis, the importance of enzymes for speeding up chemical reactions in living cells, and how vitamins and minerals are used in cell functions. How was the information that you studied obtained? Of course, thousands of different experiments have contributed to the accumulation of this knowledge. Radioisotopes are useful tools used by researchers in some of these experiments.

Radioactive tracers play a role in industry and environmental studies. Tracers can be used to detect the movement of groundwater through soil, the paths of certain industrial pollutants in the air and water, and the shifting of sand along coastlines. In industry, tracers help manufacturers test durability of mechanical components and identify structural weaknesses of equipment.

Nuclear medicine has been practiced almost since the days when nuclear reactions were first conducted in laboratories. Today, radioisotopes are used in medicine for both the diagnosis and the treatment of diseases.

A radioactive tracer can be substituted for the nonradioactive form of a chemical that is normally used by a specific organ. Doctors can track the tracer to help diagnose a possible malfunction. The isotope chosen for a procedure depends on the dosage, half-life, and chemical activity that are most suitable. As the organ absorbs the isotope, a scanner produces an image of the organ on a monitor. Areas of extremely high or extremely low radioactivity signal the existence of cells that may be functioning improperly.

One isotope commonly used for medical diagnosis is iodine-131. Iodine is a crucial chemical for the proper functioning of the thyroid gland. Iodine-131 accumulates in the gland as the nonradioactive form would, and can be detected easily. A half-life of 8.1 days is long enough to reveal thyroid malfunctions, but short enough to avoid endangering the patient. Table 9-1 lists some common radioisotopes used as tracers in medicine.

Table 9-1

COMMON RADIOISOTOPES USED IN MEDICINE	
RADIOISOTOPE	**TARGET ORGAN**
chromium-51	spleen
iodine-131	thyroid gland, lungs, kidneys
gallium-67	lymph glands
selenium-75	pancreas
technetium-99m*	brain, lungs, liver, spleen, bones

*Technetium-99m is a less stable form of Tc-99. Tc-99m emits gamma radiation becoming the more stable Tc-99.

Figure 9-9 A nuclear scan shows where radioactive iodine has been absorbed in a thyroid gland. Doctors use radioactive iodine-131 to identify areas of dysfunction and to treat tumors.

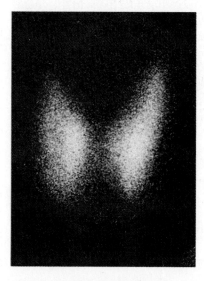

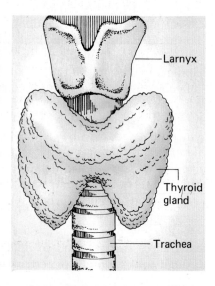

Larnyx

Thyroid gland

Trachea

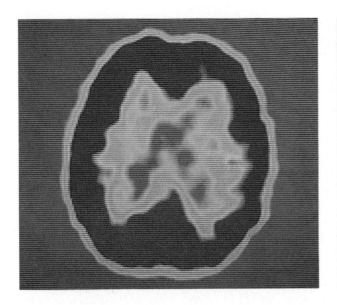 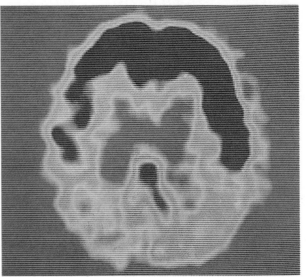

In recent years, several new instruments have been developed that help doctors diagnose medical problems. The procedures depend on the use of radioisotopes that emit measurable radiation. Two of these diagnostic tools are PET (positron emission tomography) and SPECT (single-photon emission computed tomography). In each case, the patient is given a low dose of a radionuclide that can be absorbed by the organ in question. Measurement of the radiation from within the patient enables doctors to diagnose malfunctions far more accurately than with simple X-ray procedures. PET is very expensive because it requires radionuclides that have short half-lives. A cyclotron must be maintained at the institution to produce the nuclides when needed. SPECT, which requires more conventional nuclear medicine components, is rapidly becoming commonplace.

Radioisotopes are also used in the treatment of malignant tumors. In the case of thyroid cancer, radioactive iodine is taken internally for a prescribed length of time. The nuclide accumulates in the gland (in amounts larger than those used for diagnosis) and radiation destroys both tumor cells and healthy tissue. In other situations, a malignant tumor is treated with radiation from an external source (such as cobalt-60). By constantly changing the angle of bombardment, healthy cells are subjected to small amounts of radiation while tumor cells receive a concentrated dose. In either case, radiation disrupts the fragile bonds of DNA molecules. Though healthy tissue is affected, the rapidly growing cancer cells are more susceptible to irreparable damage and are destroyed in greater numbers.

Figure 9-10 The photo on the left is a PET scan of a healthy human brain. The scan on the right is from a patient who has Alzheimer's disease.

Students are not expected to be familiar with the words that the acronyms PET and SPECT represent. The full names are provided here for reference.

Radioisotopic therapy is the name applied when a nuclide is administered internally (as with iodine-131). Treatment delivered from an external source is termed *radiation therapy*. In addition to radiation from a nuclide, patients may be treated with X rays or high-energy electrons.

1. a radioactive nuclide used to follow chemical or physical changes in a material or process; A tracer behaves chemically like a nonradioactive form of the same element and can be detected at each stage in a chemical reaction.
2. Cancer cells are more vulnerable to the effects of radiation than some healthy cells.
3. Radioactive iodine accumulates in the gland and subsequently destroys tumor cells.

Review and Practice

1. What is a radioactive tracer? Why can a tracer be used to study chemical reactions?

2. Why are radioisotopes helpful in the treatment of cancer?

3. Why is iodine-131 useful in the treatment of thyroid disease?

Nuclear Energy

When you look at the stars, you are seeing the results of some of the most energetic reactions in the universe. From what you have already studied, you know that chemical reactions absorb or release energy (Chapter 5). Nuclear reactions involve amounts of energy about one million times greater than the energy derived from chemical reactions. The light and heat that reach Earth are the results of a nuclear inferno raging within the closest star, the sun.

During World War II, nuclear science grew out of its infancy. Enrico Fermi, an Italian scientist, led the first team to conduct a self-sustaining nuclear reaction at the University of Chicago in 1942. Soon after, scientists in the United States worked feverishly to develop a bomb that would make use of the enormous energy released during nuclear fission. It was believed that Nazi Germany was close to finishing the development of the same bomb. The world had stepped into the atomic age.

The energy from fission has been harnessed in electric power plants to produce energy for consumption in many parts of the world. Scientists are investigating the possibility of using a fusion reaction like the sun's as a practical energy source on Earth. If efforts are successful, worries of meeting future energy needs would be eliminated.

Ironically, while nuclear reactions can be used to satisfy the growing need for energy, the waste materials from nuclear power plants pose a problem that is not easily solved. People are uncomfortable with the fact that they are exposed to more radiation today than 100 years ago. The effects of this exposure on populations and the environment has an impact on how society views the future uses of nuclear energy.

9-9 Energy and Mass in Nuclear Reactions

When the total mass of the starting materials in a nuclear reaction is compared with that of the products, a small portion of the mass seems to disappear. You have learned that it is not possible for matter to be destroyed. How is it then, that nuclear experiments conducted over and over again show that some matter disappears? What happens to the missing matter?

Albert Einstein postulated a relationship between matter and energy that explains how they can be interchanged. When a nuclear reaction takes place, the loss of mass is proportional to the energy given off in the reaction. Einstein's theory can be expressed in the famous mathematical equation:

$$E = mc^2$$

E is energy in joules, m is mass in kilograms, and c is the velocity of light (3.00×10^8 m/s). Einstein's equation can be applied to calculate the amount of energy released when a sample of material undergoes a nuclear change.

Do You Know?

Chemical reactions also involve the loss of mass. However, in chemical reactions the loss of mass is so small it cannot be measured.

Example 9-2

When the nuclei in one mole of radium-226 atoms decay by alpha emission to produce radon-222, how many kilojoules of energy are released? The mass of one mole of each of the nuclei involved is: radium-226 = 225.9771 g; radon-222 = 221.9703 g; helium-4 = 4.0015 g.

Nuclear masses rather than atomic masses are given here in order to eliminate the problem of accounting for the mass of the electrons when calculating mass-energy conversions.

▶ **Suggested solution**

The problem asks for energy. Information is given about mass. What do you know about the relationship of energy and mass? According to Einstein's equation, the amount of energy released in a nuclear reaction can be calculated from the loss of mass. Therefore, the first thing required is to find out how much mass was lost. The nuclear equation for the reaction is:

$$^{226}_{88}\text{Ra} \longrightarrow {}^{222}_{86}\text{Rn} + {}^{4}_{2}\text{He}$$

The loss of mass is:

$$\text{mass}_{\text{Ra}} - (\text{mass}_{\text{Rn}} + \text{mass}_{\text{He}}) = \text{``lost'' mass}$$
$$225.9771 \text{ g} - (221.9703 \text{ g} + 4.0015 \text{ g}) = 0.0053 \text{ g}$$

The products have lost 0.0053 gram. Using Einstein's equation you can calculate the energy generated from this loss of mass:

$$E = mc^2$$

The energy unit used in Einstein's equation is the joule. When energy is measured in joules, mass is expressed in kilograms. Yet the units in the problem are kilojoules and grams. You will have to adjust for this in the solution. The change to kilograms can be included when you set up the equation. Substitution yields the following expression:

$$E = \left(0.0053 \text{ g} \times \frac{1 \text{ kg}}{1000 \text{ g}}\right)\left(3.00 \times 10^8 \frac{\text{m}}{\text{s}}\right)^2$$

Dividing the mass expression by 1000 and squaring the velocity of light yields:

$$E = (5.3 \times 10^{-6} \text{ kg})\left(9.00 \times 10^{16} \frac{\text{m}^2}{\text{s}^2}\right)$$

Multiplying through the expression and adjusting to two significant digits yields:

$$E = 4.8 \times 10^{11} \frac{\text{kg} \cdot \text{m}^2}{\text{s}^2}$$

One joule is equivalent to 1 kg·m²/s² so substitution yields:

$$E = 4.8 \times 10^{11} \text{ J}$$

Recall that the problem asked for kilojoules. Converting from joules, you get:

$$E = 4.8 \times 10^{11} \text{ J} \times \frac{1 \text{ kJ}}{1000 \text{ J}}$$

$$E = 4.8 \times 10^8 \text{ kJ}$$

Modern theory states that particle exchanges between neutrons and protons account for binding energy.

Just how much energy is 4.8×10^8 kJ? The alpha decay of one mole of radium results in enough energy to melt the steel beams and girders in a 30-story office building! Compare this value to the energy produced during the combustion of octane, a chemical reaction you studied in Section 5-9.

The relationship between mass and energy as expressed by Einstein's equation helps to explain why a stable nucleus holds together. When a nucleus forms there is a loss of mass. Measurements show that the mass of a nucleus is slightly less than the sum of the individual masses of the protons and neutrons that comprise it. This discrepancy is known as the **mass defect**. Where does the lost mass go? As the nucleus forms, it is converted into energy. The lowered energy of the nucleus is the reason it is stable. In order to disrupt the nucleus, the same amount of energy would have to be added to the system. This energy is known as the **binding energy**, or the amount of energy needed to break a nucleus into its individual protons and neutrons.

Binding energy can be calculated from the mass defect. For example, the mass of one mole of nuclei of mercury-200, $^{200}_{80}\text{Hg}$, is observed to be 199.9244 grams. However, the mass of 80 moles of protons and 120 moles of neutrons is 201.6228 grams. The nucleus of mercury-200 is less than the mass of its component particles.

$$201.6228 \text{ g} - 199.9244 \text{ g} = 1.6984 \text{ g}$$

The mass defect for one mole of mercury-200 nuclei is 1.6984 grams. You can use Einstein's equation to find the binding energy of a mercury-200 nucleus. The procedure is similar to the previous sample problem.

Example 9-3

Calculate the binding energy in joules for one nucleus of mercury-200 if the mass defect of one mole of mercury-200 nuclides is 1.6984 grams.

▶ Suggested solution _____

As in Example 9-2, you are looking for the amount of energy related to a specific loss of mass. Einstein's equation can be used to solve the problem. If energy is to be expressed in joules, then mass must be expressed in kg and the speed of light must be expressed in m/s. So 1.6984 grams will have to be converted to kilograms in the solution to the problem.

Einstein's equation is:

$$E = mc^2$$

By substituting the mass and the speed of light (3.00×10^8 m/s) into the equation, you get

$$E = \left(1.6984 \text{ g} \times \frac{1 \text{ kg}}{1000 \text{ g}}\right)\left(3.00 \times 10^8 \frac{\text{m}}{\text{s}}\right)^2$$

Dividing the mass expression by 1000 and squaring the velocity of light gives

$$E = (1.6984 \times 10^{-3}\,\text{kg})\left(9.00 \times 10^{16}\,\frac{\text{m}^2}{\text{s}^2}\right)$$

Multiplying through the expression and adjusting to three significant digits yields:

$$E = 1.53 \times 10^{14}\,\frac{\text{kg}\cdot\text{m}^2}{\text{s}^2}$$

Recall that the joule is equal to 1 kg·m²/s² so you can substitute joules to get

$$E = 1.53 \times 10^{14}\,\text{J}$$

According to the problem, this value is for one mole of mercury-200 nuclei. To find the binding energy for one nucleus, divide by the number of nuclei in one mole (6.02×10^{23}).

$$\frac{1.53 \times 10^{14}\,\text{J}}{1\,\text{mol}} \times \frac{1\,\text{mol}}{6.02 \times 10^{23}\,\text{nuclei}} = \frac{2.54 \times 10^{-10}\,\text{J}}{\text{nucleus}}$$

Check your answer to see if it is reasonable. The problem asks for binding energy of one nucleus. Logic tells you that the magnitude of the answer in joules for one nucleus should be much smaller than for one mole. A check of the solution setup shows that the units also seem right.

Do You Know?

Albert Einstein (1879–1955) developed his famous theory of relativity by the time he was 26 years old. Many capable people work in science, but Einstein was considered unique. His ideas greatly advanced knowledge in nuclear chemistry, physics, and astronomy.

Binding energy often is calculated in joules per nuclear particle. What would you have to do to the answer in Example 9-3 to find this value for mercury? Looking back at the problem, note that the mass number for this isotope of mercury is 200 (meaning the sum of protons and neutrons is 200). If you divide the answer calculated in joules per nucleus by 200 particles per nucleus, you get the binding energy per nuclear particle:

$$\frac{2.54 \times 10^{-10}\,\text{J/nucleus}}{200\,\text{particles/nucleus}} = \frac{1.27 \times 10^{-12}\,\text{J}}{\text{particle}}$$

Binding energy varies from nuclide to nuclide. Figure 9-11 (on the next page) relates the binding energy in kJ/mol of nuclear particles to the mass number of the nuclide. From the graph, you can see that nuclides with a mass number of approximately 60 have the greatest binding energy, and are the most stable. The graph helps to explain why there are two ways to get energy out of atomic nuclei. The lighter nuclei give off energy and gain stability when they fuse together. The heavy nuclei (on the right side of the graph) give off energy when they break apart into smaller nuclei. In Sections 9-11 and 9-13 you will learn more about these processes.

Figure 9-11 Binding energy is greatest in the stable isotopes with mass numbers at the top of the curve. A tremendous amount of energy would be needed to break apart the nuclei of such isotopes.

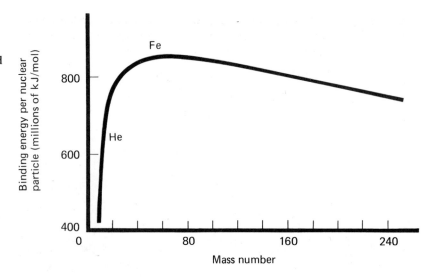

1. Matter is converted to energy.
2. mass and energy
3. the difference between the mass of the nucleus and the sum of the masses of its protons and neutrons
4. 7.29×10^8 kJ/mol
5. Binding energy is the amount of energy needed to break a nucleus into individual protons and neutrons.
6. 1.48×10^{-11} J/nucleus

Review and Practice

1. What happens to the matter that seems to disappear during a nuclear reaction?

2. Einstein's equation represents a relationship between what two entities?

3. What is the mass defect of a nucleus?

4. When the nuclei in one mole of polonium-211 decays by alpha emission to produce lead-207, how many kilojoules of energy are released? The mass in grams of one mole of each of the nuclei involved is: polonium-211 = 210.9405 g; lead-207 = 206.9309 g; helium-4 = 4.0015 g.

5. Briefly explain the concept of binding energy.

6. Calculate the binding energy in joules per nucleus of carbon-12 if the mass defect of one mole of carbon-12 is 0.0990 gram.

9-10 Splitting Atoms—Nuclear Fission

Fission is the process in which a nucleus is broken into smaller nuclei by bombardment with relatively low-energy neutrons. Scientists first became aware of the phenomenon during efforts to manufacture transuranium elements early in this century. It was discovered that the uranium-235 nucleus breaks apart after absorbing a neutron.

The addition of a neutron makes the uranium nucleus unstable. As shown in Figure 9-12, the nucleus splits apart, forming two unequal nuclei, emitting a number of neutrons, and releasing a large amount of energy. The energy is the result of the difference in binding energies of the uranium and the products of the reaction. (Recall from the graph in Figure 9-11 that unstable nuclides with high mass numbers lose energy when they split apart to form smaller, more stable nuclides.) The neutrons are likely to hit other uranium-235 nuclei and cause them to undergo fission, which in turn releases more neutrons and more energy. This self-propagating reaction is called a **chain reaction**. As the sample of uranium reacts, it can form many different nuclides. Several possible reactions are listed below:

$$^{235}_{92}U + ^1_0n \longrightarrow ^{139}_{56}Ba + ^{94}_{36}Kr + 3^1_0n$$

$$^{235}_{92}U + ^1_0n \longrightarrow ^{146}_{57}La + ^{87}_{35}Br + 3^1_0n$$

$$^{235}_{92}U + ^1_0n \longrightarrow ^{144}_{55}Cs + ^{90}_{37}Rb + 2^1_0n$$

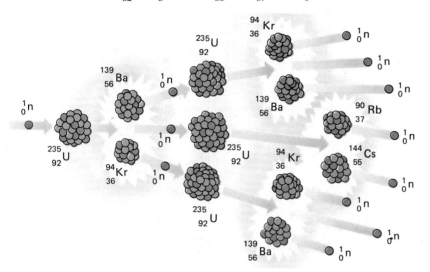

Figure 9-12 In a fission chain reaction, a uranium-235 nucleus absorbs a neutron, becomes unstable, and breaks into two smaller nuclei. The neutrons produced are absorbed by other uranium nuclei, and the process continues. Moderators slow down neutrons that are moving too quickly to be absorbed.

If the mass of U-235 is small, many of the neutrons produced by the reaction pass through the material without striking a nucleus to cause another disintegration. As the mass of U-235 is increased, the chance of a neutron hitting a nucleus also increases. There are more opportunities for additional disintegrations to occur to keep the reaction going. The minimum amount of material needed to sustain the chain reaction is called the **critical mass**. The critical mass depends on the number of neutrons produced by each disintegration and on the shape of the uranium metal.

Several heavy nuclei are capable of fission. In any fission reaction, tremendous amounts of energy are produced. Scientists became interested in how this energy could be controlled and used.

CHEM THEME

The idea of a chain reaction is not limited to nuclear reactions. Many chemical reactions occur in a series of steps. The products of the step reactions often react with the initial substances to allow the reactions to proceed.

9-11 Nuclear Power Plants

The explosion of a fission bomb is an uncontrolled fission reaction. Controlled fission reactions occur routinely in nuclear power plants and provide energy to generate electricity.

Uranium-235 is a nuclide used as fuel in many nuclear reactors. (Plutonium-239 is another common fuel.) Because uranium-235 makes up only 0.7 percent of naturally occurring uranium, the material used must be purified, or *enriched*, to increase the concentration of uranium-235.

Figure 9-13, on the next page, is a photograph of the core of a fission reactor. Fuel rods containing the enriched uranium are inserted into

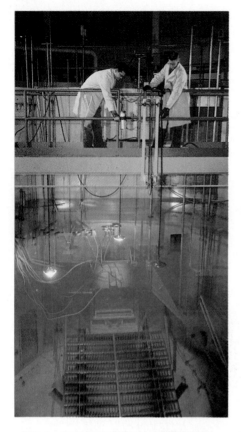

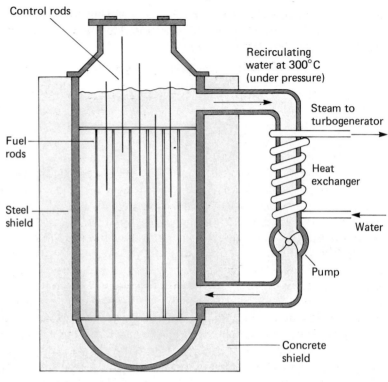

Control rods

Recirculating water at 300°C (under pressure)

Steam to turbogenerator

Heat exchanger

Water

Pump

Concrete shield

Fuel rods

Steel shield

Figure 9-13 The water-covered core of a nuclear reactor contains the fuel rods and mechanisms for controlling the reaction.

Figure 9-14 A typical commercial reactor contains 200 fuel rods, each 4 meters long. The control rods are moved in or out to control the rate of reaction. Pressurized water is pumped through the reaction chamber to absorb heat energy. At the heat exchanger, this water heats a separate water system making steam that runs a turbogenerator to make electricity.

The water used as a moderator is "heavy" water containing deuterium.

Do You Know?

A breeder reactor is a fission reactor that produces more fissionable material as it operates. Some nations discourage the construction of breeder reactors because the materials produced are the type that also are used in building nuclear weapons.

the reaction chamber within the core. Figure 9-14 is a diagram of the reaction chamber of a pressurized nuclear reactor. As the fission reaction proceeds, neutrons collide with other uranium atoms to continue the chain reaction. If the neutrons are moving too fast, they cannot be readily absorbed by the U-235 nuclei. Therefore, in reactors using U-235 as a fuel, the neutrons are slowed down by collisions with **moderators**. The most effective moderators are usually molecules of water, beryllium, or graphite.

The rate of reaction is regulated by **control rods**, usually made of cadmium, that absorb some of the neutrons. The further the control rods are lowered into the fuel-rod assembly, the slower the rate of reaction. If control rods were not present inside the reactor, the heat of the reaction would melt the reactor core. Water acts as a coolant and is used as the medium to transfer heat between the reactor and the turbines that produce electricity. **Shielding** of steel and high-density concrete protect personnel from radiation.

9-12 Building Atoms—Nuclear Fusion

The nuclear reaction that fuels the stars is far more powerful than fission. Enormous amounts of energy are released from stars when small nuclei join to form larger nuclei in a process known as **fusion**. It is this energy that sustains a star through its lifetime. In the sun and in most stars, hydrogen, the most abundant element in the universe, is converted into helium. The reaction, illustrated in Figure 9-15, is represented by the following equation.

$$\ce{^2_1H + ^3_1H -> ^4_2He + ^1_0n} + 1.7 \times 10^9 \text{ kJ/mol}$$

The symbols $\ce{^2_1H}$ and $\ce{^3_1H}$ represent two isotopes of hydrogen, also called *deuterium* and *tritium*, respectively. Note the enormous amounts of energy released in the reaction given above. In one day, the energy reaching Earth from the sun equals all the energy ever used by humans on this planet!

Fusion occurs for the same reason that fission does. The process leads to a more stable nucleus. Look again at the graph of binding energies in Figure 9-11. Lighter nuclei gain stability when they are fused into a larger nucleus. The energy output of a fusion reaction is a reflection of the greater binding energy of the new nucleus. The fission of 1 kilogram of uranium-235 produces energy equal to that obtained from 2 million kilograms of coal, but the fusion of 1 kilogram of hydrogen into helium produces energy 20 times greater than that!

Fusion would seem to be the answer to the energy problem. It is no wonder that scientists in the United States, the Soviet Union, and other countries are trying to duplicate the reaction of the stars. However, a huge input of energy is required to initiate the fusion process, which reduces the net energy gained from the reaction. Furthermore, in order to sustain the reaction, a temperature of 200 million kelvins is needed. At this temperature atoms and molecules are stripped of their electrons and revert to the fourth state of matter known as plasma. Although matter exists easily in stars as plasma, no conventional fuel container can sustain the high temperatures required for the reaction.

Several methods have been proposed to produce a sustained fusion reaction. One approach involves using a mixture of deuterium, tritium, and the lithium-6 nuclide, which requires less energy to get the reaction started. The tokamak reactor, first developed in the Soviet Union, uses a doughnut-shaped magnetic field to contain the nuclear fuel.

Have students compare this energy value to that for the octane combustion reaction discussed in Section 5-9.

Do You Know?

The sun converts 4 million tons of hydrogen into helium every second. In addition to helium, fusion reactions in stars also produce elements such as carbon, nitrogen, silicon, and oxygen. The heavier elements (above iron in the periodic table) form during explosions of supernovas.

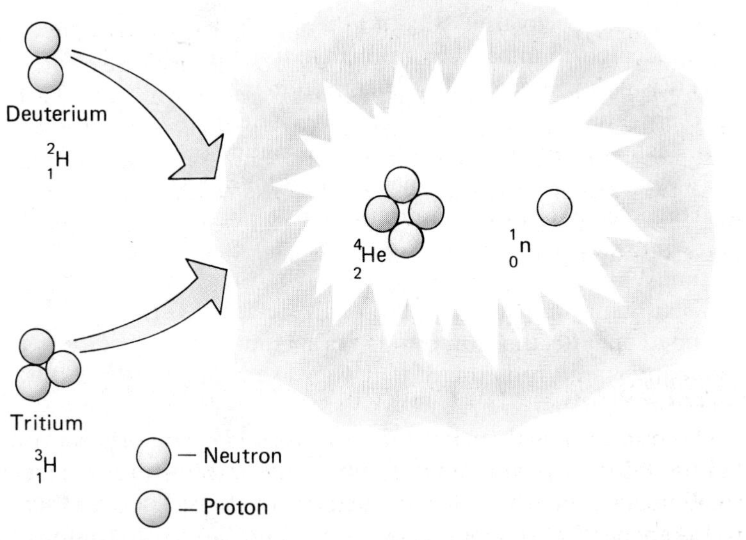

Figure 9-15 The combination of deuterium and tritium to produce helium is one example of fusion. Nuclei of other elements of low atomic mass also can undergo fusion reactions to produce heavier, more stable nuclei and release energy.

Figure 9-16 The Tokamak Fusion Test Reactor at Princeton University is one of the sites for fusion research in the United States. The reactor requires an electrical power supply equal to that of a city with a population of one half million.

Another approach, the laser fusion reactor, uses a laser pulse aimed at glass pellets of tritium and deuterium fuel to start the reaction. The process is expensive.

Do You Know?

Only 10^{-13}*% of the deuterium available in seawater would be sufficient to meet the world's total energy needs for one year.*

When nuclear plants were first built, the original plan was to reprocess the uranium. Wastes were to be stored on site for only a year or two. However, reprocessing has proved to be economically unfeasible, so the stores of radioactive leftovers are increasing.

Recently scientists using a tokamak reactor at the Massachusetts Institute of Technology, announced that they had produced a reaction at the "breakeven point." At the breakeven point, there is the potential to obtain as much energy from the reaction as was used to initiate it.

The search for an economically feasible fusion reaction is important for several reasons. First, the deuterium fuel is more readily available than uranium. In fact, deuterium can be extracted from ordinary seawater. Second, unlike the products of fission reactions, the helium produced in fusion is nonradioactive and does not pose a problem for waste disposal. Third, the reaction could be stopped at any time without the fear of a nuclear meltdown, which is a possibility with nuclear fission. Last, the reaction would be a source of virtually inexhaustible energy and could eliminate most of the world's dependence on other fuels. Scientists disagree about if and when a practical fusion reaction will be achieved. In the meantime, the research continues.

9-13 Extension: Nuclear Waste

No one likes to think about waste materials. Yet, waste products have become an unsettling problem in today's world. Industrial chemical wastes and waste from consumption habits of the general population are a hazard to the health and well-being of all plant and animal life. Disposing of nuclear wastes from commercial power plants, military weapons testing, and medical treatment also is of great concern.

The spent fuel rods from a nuclear power plant contain the radioactive products from fission as well as 90 percent of the original uranium. What happens to these rods when they are no longer useful? Presently they are stored at the plant facilities. As nuclear plants continue to operate, the stored rods are accumulating. Currently there are so many stored rods that if they were placed end to end, they would circle the planet.

Nuclear power plants are not the sole sources of radioactive wastes. In the United States, millions of gallons of wastes from military applications are being stored in federal repositories. Nuclear wastes from industry and from the use of radioactive materials in medicine are accumulating as well. The demand for nuclear materials is unlikely to decrease in the future. Consumers want and use the products that industry provides. Furthermore, few people would suggest that life-saving medical techniques be abandoned because of the radioactive wastes they produce.

The problem of what to do with these wastes needs a reliable solution. Many nuclear wastes have half-lives of thousands of years and must be placed where they will not endanger life. Consider some suggestions for waste disposal that have been rejected. Ocean dumping is not a viable solution because of the danger of canisters corroding and releasing their contents. Incineration produces gaseous radioactive products that would be released into the air. Sending the wastes into outer space to burn up near the sun is possible but expensive, and nobody can guarantee that the vehicle will successfully clear Earth's atmosphere without a mishap. The issue is not an easy one to resolve.

A promising solution to the growing nuclear waste disposal problem is to bury the material in canisters deep underground in beds of salt. Several problems need to be overcome in order for this disposal method to prove safe. Will the canisters be able to maintain their integrity over many hundreds or thousands of years? Is the site for storage geologically stable? Can the risk of accident in transporting dangerous materials to the storage sites be minimized? What about future generations that may uncover these sites accidentally?

A possible solution to the problem of disposal of spent nuclear fuel rods is reprocessing. The used rods are chemically dissolved in an aqueous solution, and the long-lived alpha emitters (nuclides of uranium, plutonium, and americium) are extracted and made into new fuel rods. The remaining radioactive wastes are then converted into a solid, usually glass, by a process known as *vitrification*. The glass is stored underground. France has had some success with vitrification, but the process has not fared well in the United States. At this time, the rate of waste production exceeds the rate at which the materials can be reprocessed, and the cost is high.

The potential for use of nuclear materials is enormous. Yet, developing the technology needed to dispose of the waste products safely and economically will affect the growth of nuclear industries and the expansion of nuclear research in the future. People will have to make responsible decisions about how nuclear materials should be used. Understanding the issues is the first step in making those decisions.

Figure 9-17 Radioactive wastes are mixed into molten glass that is then poured into graphite molds to harden. The wastes become trapped in the solid glass cylinders that result. The cylinders are placed in containers and can be stored underground.

Review and Practice

1. How do fission and fusion differ?

2. Describe the concept of a chain reaction.

3. How is a chain reaction controlled in a nuclear power plant?

4. What role does water play in a nuclear power plant?

5. Briefly describe the greatest problem scientists face in successfully causing a nuclear fusion reaction to be cost effective.

6. Why is the disposal of nuclear wastes such a great problem?

1. In fission, large nuclei are split into smaller ones; in fusion, small nuclei are fused together into larger ones.
2. Neutrons bombard atomic nuclei causing them to split apart. More neutrons are released and bombard other nuclei, and so on.
3. Control rods placed between fuel rods absorb neutrons and slow the reaction.
4. Water is a moderator. It slows the speed of the neutrons so that they can be absorbed by the nuclei. Water also transfers heat from the reactor to the steam turbine.
5. sustaining a sufficiently high temperature in a confined volume
6. There is so much; it is highly radioactive; it is dangerous for a long time.

The Atom Revisited

Over the years, new discoveries about the atom led to more questions. Researchers were not satisfied with the three-particle concept of the atom. Experiments suggested that the atom contained many more pieces. Was it possible that protons, neutrons, and electrons were not the basic building blocks of atoms? Could they be composed of even smaller particles? Questions like these stimulated further studies of the atom and the search for *elementary* particles, particles from which the others are made.

As scientists built more powerful accelerators to investigate the atom, they found a surprising number of subatomic particles. In the 1960's, these atomic prospectors, searching for the ultimate pieces of the atom, were surrounded by over 100 different subatomic particles. The more the atom was probed, the more complex it became. Some theorists wondered if the search for elementary particles would end in a hopelessly long parade of unclassifiable entities.

The picture is much brighter today. Recent research suggests a new order and a kind of simplicity to the particles that compose matter. This section provides you with a brief glimpse of the subnuclear world.

9-14 What Is a Quark Anyway?

When atoms or other particles are made to crash into each other in accelerators, numerous other particles come flying out from the point of contact. The tracks of these particles are recorded by detectors and studied. Figure 9-18 illustrates what the tracks from a collision might

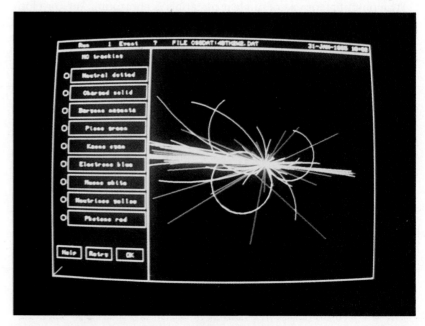

Figure 9-18 This photo is a computer simulation of a collision "event" from the Fermi National Laboratory (Fermilab). The red and green tracks show the bending of charged particles in the detector's magnetic field. The white lines represent the directions of the quarks. Quarks themselves are never observed.

look like. The results of such experiments left scientists wondering if further study would only compound the problem of trying to find elementary particles.

In the midst of this turmoil, Murray Gell-Mann at the California Institute of Technology, and George Zweig at CERN in Switzerland, announced a solution to the problem. They accounted for the bewildering number of subatomic particles by suggesting that they are combinations of smaller particles called quarks. A **quark** is a theoretical particle with a fractional charge of either 1/3 or 2/3 the charge of an electron. Although evidence for the quark is limited, its existence has been postulated to simplify the subatomic menagerie of particles that do exist.

Originally three quarks were postulated, but the number has grown to six. (Only two quarks are believed to account for the subatomic particles you are studying in this book.) Each quark is identified by its particular characteristic, or "flavor." Table 9-2 lists the six quarks and their fractional charges.

A subnuclear particle is termed a *hadron*. Those hadrons made of 3 quarks (protons and neutrons) are in turn called *baryons*. A *meson* is made of a quark-antiquark pair. All six quarks have antiquark counterparts.

Do You Know?

The sources of particle names are not always scientific. The word "quark" was selected by Murray Gell-Mann from an obscure line in the book Finnegans Wake, by James Joyce: "Three quarks for Muster Mark." He liked the whimsical sound of it. Sheldon Glashow, who predicted the fourth quark, called it charmed because he was so charmed with the solution it supplied to his theoretical subnuclear puzzle.

Table 9-2

THE SIX KINDS OF QUARKS	
PARTICLE NAME	ELECTRIC CHARGE
up	2/3
down	−1/3
charmed	2/3
strangeness	−1/3
top (truth)	2/3
bottom (beauty)	−1/3

Notice the unusual names in the table. With respect to quarks, scientists have developed new meanings for common words. Quarks are obviously not beautiful or charming in the everyday sense, nor do they come in flavors of vanilla and chocolate. "Flavor," "charmed," and "strangeness" are just part of the new language of quarks.

Two quarks, "up" and "down," are thought to compose most of the ordinary matter in the universe. Evidence for the other quarks has been found in collision experiments. Protons and neutrons are composed of different combinations of "up" and "down" quarks, as shown in Table 9-3. Can you see how the combination of quark charges yields a positive charge for a proton and no charge for a neutron?

Table 9-3

PROTON		NEUTRON	
up	2/3	down	−1/3
up	2/3	down	−1/3
down	−1/3	up	2/3

Do You Know?

Neutrino in Italian means "little neutron." Neutrinos can easily travel through Earth from New York to Australia without hitting a single electron or nucleus!

There are six identified leptons: electrons, electron neutrinos, muons, muon neutrinos, tau particles, and tau neutrinos. Each lepton has an antiparticle.

Do You Know?

Many science fiction authors make use of the matter/antimatter connection. The reaction between matter and antimatter was the scientific basis for the fictional warp-drive engines in the famous Star Trek *stories.*

1. a theoretical particle with a charge of either 1/3 or 2/3 that of an electron
2. proton = 2 up, 1 down
neutron = 2 down, 1 up
3. Leptons and quarks are both thought to be fundamental particles, that is, particles not composed of any smaller particles. An electron is a lepton.
4. antiparticles with the same characteristics as specific particles of matter but with opposite charge

Individual quarks have never been isolated. Scientists believe that in protons, neutrons, or other particles made of quarks, there is force binding the quarks together. Many researchers think it is impossible to overcome this force and separate the quarks. Nevertheless, some scientists continue to search for the particle.

Are electrons made of quarks? Scientists do not think so. Electrons are not thought to be composed of smaller particles. They are classified as elementary particles called **leptons**. Neutrinos are another example of leptons. A *neutrino* is an elementary particle with infinitely small mass and no charge. Some leptons have been observed to leave tracks in particle collisions.

9-15 The Other Side of Matter

If you have encountered science fiction in books or movies, you probably have heard of antimatter. What is antimatter? Scientists state that for every particle that exists, there is an antiparticle, or an opposite. Antiparticles have charges that are opposite those of their counterparts. For example, the antiparticle to the electron is the positron, a particle of the same mass but with a positive charge. The antiparticle to the proton is the antiproton. Likewise, there are antiquarks, antineutrinos, etc. **Antimatter** is the collective name given to these antiparticles. When a particle meets its antiparticle, the two annihilate each other and release energy. After their discoveries, antiparticles quickly became a tool for further subnuclear research. Scientists study matter-antimatter collisions in the hopes of finding answers to fundamental questions about the basic building blocks of all matter. Fermilab reported its first successful proton-antiproton collision in late 1985.

If you are somewhat bewildered by the modern mysteries of the atom, you are not alone. The secrets of matter and energy are not easily discovered. Recent work has produced an explosion of ideas and evidence that encourages researchers to keep going. Scientists are on the brink of understanding more than ever before about matter and the forces of the universe. American scientists propose building a mammoth-sized particle accelerator, about 100 kilometers in diameter, to provide the giant energies necessary for these studies. It is ironic that the largest machine ever built may one day lead to the discovery of the smallest particle in existence.

Review and Practice

1. What is a quark?

2. Describe the combinations of quarks that compose a proton and a neutron.

3. How is a lepton thought to be similar to a quark? Give an example of a lepton.

4. What is antimatter?

Numbers in red indicate the appropriate chapter sections to aid you in assigning these items. Answers to all questions appear in the Teacher's Guide at the front of this book.

Summary

■ A nuclear change differs from a chemical change because nuclear reactions involve protons and neutrons of an atom, while chemical reactions involve electrons.

■ Radiation is commonly found in the environment. Radioactivity can affect living organisms. Ionizing radiation causes ion pairs to form in cells. These ions can change the chemistry of the cell and affect cell function.

■ Nuclear decay can occur as a result of alpha or beta emission. Stable nuclides can be changed into unstable ones by bombardment with atomic and subatomic particles. Nuclear reactions are used to synthesize transuranium elements and radioactive isotopes not found in nature. The results of such reactions have led to greater knowledge about atoms.

■ The half-lives of radioisotopes can be measured. This information can be applied to research in biology, medicine, chemistry, geology, and physics.

■ Much larger amounts of energy are released during nuclear reactions than in chemical reactions. A detectable amount of mass is lost, and the resulting energy produced can be calculated from the equation $E = mc^2$.

■ The fission of U-235 and Pu-239 is harnessed in modern nuclear reactors. The products of a fission reaction are energy, smaller nuclei, and neutrons that can split other atoms in a chain reaction. Fusion energy is a topic of much research today. Small atomic nuclei are fused into larger ones as tremendous energy is released. If scientists could successfully "capture the sun" in the laboratory, Earth would have an almost unlimited supply of energy.

■ An enormous number of subatomic and subnuclear particles are known to exist. It is believed that many of these particles are composed of quarks. Experiments to learn more about the nature of matter and energy continue as more powerful particle accelerators are built.

Chemically Speaking

alpha emission	critical mass
antimatter	cyclotron
background radiation	daughter nuclide
beta emission	fission
binding energy	fusion
chain reaction	half-life
control rods	ionizing radiation

ion pairs	radiation
lepton	rem
mass defect	shielding
moderators	spontaneous decay
nonionizing radiation	synchrotron
nuclide	tracer
parent nuclide	transmutation
quark	transuranium elements

Review

1. Describe what happens during an alpha-emission reaction. (9-3)

2. What is the importance of critical mass to the continuation of a chain reaction? (9-10)

3. What is the half-life of carbon-14? (9-7)

4. Briefly describe why carbon-14 becomes incorporated into plants. (9-7)

5. Describe how binding energy is calculated. (9-9)

6. What particles of an atom are not involved in a chemical change? (9-1)

7. List five transuranium elements named after people or places you have heard of. (9-5)

8. Briefly explain how a cyclotron gives very high energies to subnuclear particles. (9-6)

9. What is the average background radiation in the United States? (9-2)

10. Where does the energy come from that is released during nuclear fission? (9-9, 9-10)

11. Describe two methods under consideration for the disposal of nuclear wastes. (9-13)

12. What is nuclear fusion? (9-12)

Interpret and Apply

1. What accounts for the energy released in a ' reaction? (9-9)

2. If the production of carbon-14 varied thro tory, would this variation have any effec' racy of radiocarbon dating? Explain. (⁵

3. Sodium-23, the only naturally occurring isotope of sodium, has a molar mass of 22.980 g/mol. If protons and neutrons each have a molar mass of 1 g/mol, explain why the molar mass of sodium-23 is not 23.000 g/mol. (9-9)

4. The half-life of uranium-238 is 4.5 billion years. Explain why there is so much of this isotope still undecayed on Earth. (9-7)

5. It is desired to use radioactive sulfur as a tracer in an experiment. Two beta-emitting isotopes are available: $^{35}_{16}S$ ($T_{\frac{1}{2}} = 87$ days) and $^{37}_{16}S$ ($T_{\frac{1}{2}} = 5$ minutes). Which would you choose and why? (9-7, 9-8)

6. Why is it difficult to study the physiological effects of low doses of radiation over a long period of time? (9-2)

7. Why is it not possible for a particle to be composed of three "up" quarks? (9-14)

8. What is one combination of quarks that would produce a particle with a 1− charge? (9-14)

9. The major component of bone is $Ca_3(PO_4)_2$. How are Ca-47 and P-32 useful for finding bone lesions? (9-8)

6. Calculate the energy evolved in kJ for the following reaction: (9-9)

$$^{222}_{86}Rn \longrightarrow {}^{218}_{84}Po + {}^{4}_{2}He$$

The mass of one mole of each nuclide is:

$$^{222}Rn = 221.9703 \text{ g}$$
$$^{218}Po = 217.9628 \text{ g}$$
$$^{4}He = 4.0015 \text{ g}$$

7. What is the binding energy in kJ of one nucleus of deuterium, $^{2}_{1}H$? The difference between the mass of one mole of deuterium and the sum of the masses of the protons and neutrons in one mole of deuterium is 0.003 40 g. (9-9)

8. The difference between the sum of the individual particles of which uranium-238 is composed and the actual molar mass of uranium-238 is 1.9353 g/mol. Calculate the binding energy in kJ of one mole of uranium-238. (9-9)

9. A typical fission process occurs after $^{235}_{92}U$ absorbs a neutron and becomes the unstable isotope $^{236}_{92}U$. This isotope can break apart, producing a $^{137}_{52}Te$ nucleus, a $^{97}_{40}Zr$ nucleus, and 2 neutrons. Write an equation to represent this nuclear reaction. (9-3, 9-10)

Problems

1. Fill in the blanks with the proper symbol: (9-3, 9-4)
   a. $^{238}_{92}U \longrightarrow$ _____ $+ {}^{4}_{2}He$
   b. $^{231}_{90}Th \longrightarrow$ _____ $+ {}^{0}_{-1}e$
   c. _____ $\longrightarrow {}^{211}_{83}Bi + {}^{0}_{-1}e$
   d. $^{226}_{88}Ra \longrightarrow {}^{222}_{86}Rn +$ _____
   e. $^{40}_{19}K \longrightarrow {}^{40}_{20}Ca +$ _____

2. After 10 half-lives, the radioactivity in a sample is considered to be negligible. How long should strontium-90 be stored if $T_{\frac{1}{2}} = 28$ years? How long should iodine-131 be stored if $T_{\frac{1}{2}} = 8.05$ days? (9-7)

3. The half-life of $^{125}_{53}I$ is 60 days. What percent of original radioactivity would be present after 360 days? (9-7)

4. What is the age of a piece of wood that gives off β-emissions from carbon-14 of 0.96 disintegration/minute/gram? (9-7)

5. Calculate the energy evolved in kJ for a reaction in which the mass defect is 1.68×10^{-4} g/mol. (9-9)

Challenge

1. A research institute requests 1.00 g of bismuth 214, which has a half-life of 20.0 minutes. How many grams of bismuth-214 would have to be prepared if the shipping time to the institute is one hour and 20 minutes?

2. The half-life of radium-226 is 1590 years. What fraction of a sample of radium-226 would remain after 9540 years?

3. Titanium-51 decays with a half-life of 6 minutes. What fraction of the radioactive material present at time zero would still be available after 1 hour?

4. Compare the experiments you would design to find the half-lives of two different elements if one element has a half-life between 1 and 8 minutes, and the other element has a half-life between 150 and 200 days.

5. Show mathematically why the radioactivity in a sample is considered to be negligible after 10 half-lives.

Synthesis

1. In a nuclear fusion reaction two nuclei come together to form a larger nucleus. For example, deuterium, ^2_1H, and tritium, ^3_1H, can fuse to form helium, ^4_2He, and a neutron in the reaction:

$$^2_1H + {}^3_1H \longrightarrow {}^4_2He + {}^1_0n + 1.69 \times 10^9 \text{ kJ}$$

If the chemical burning of one mole of hydrogen liberates 241.8 kJ, how much hydrogen in grams would have to be burned to liberate the same amount of heat as liberated by the fusion of one mole of deuterium, ^2_1H, nuclei?

2. One mole of radium-226 decays by alpha emission to yield 4.8×10^8 kJ of energy. How much energy is evolved when one gram of radium-226 decays? (The nuclear mass of radium-226 is 225.9771 g/mol.)

3. During World War II, the only practical technique developed for the separation of the two forms of uranium hexafluoride gas was one which took advantage of the different rates of diffusion of the gases. Even though it is a tedious and time-consuming process, the usable $^{235}UF_6$ was separated from $^{238}UF_6$ by diffusion of the gases through an opening. Calculate the ratio of rates of diffusion of the two gases.

Projects

1. Research and report on the events that took place at the Three-Mile Island nuclear power plant in March, 1979.

2. Research the use of radioisotopes in cancer therapy. Find out details not covered in the chapter about how the radioisotopes are used, when they are used, and how effective they are.

3. Research the scientific contributions of the Curie family: Marie and Pierre, their daughter Irene, and her husband Frederic Joliot. Describe the history of their work and the roles they played in the development of nuclear science.

4. Investigate the irradiation of food. Find out what foods are good candidates for irradiation, how the process works, and why it is beneficial. Describe the controversies that surround the use of irradiation to treat food.

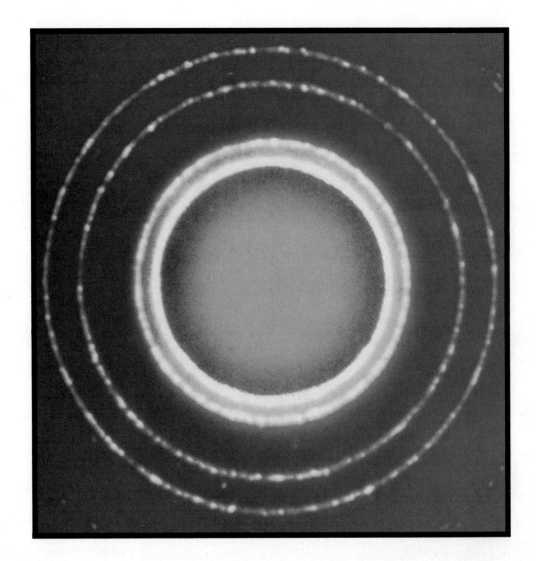

This chapter completes the model of the atom that was introduced in Chapter 8. In previous chapters, electrons were considered as particles. The behavior of electrons often can be understood if they are thought of this way. For example, a cathode-ray tube produces a beam of electrons that can be analyzed for the charge-to-mass ratio of a single particle. In other experiments, however, it is easier to interpret the behavior of electrons using the mathematics of waves. In the picture above, a beam of electrons produces a wave pattern as it passes through a thin metal foil. This chapter will introduce you to waves and how the properties of waves are used in explaining the current model of the atom. An understanding of the behavior of electrons in the atom is vital for predicting how atoms react with other atoms to form ions and molecules.

Electrons in Atoms

Waves and Energy

The modern picture of the atom is a radical departure from previous models. It is based on a mathematical description of the energy of an electron within an atom. The model does not tell where electrons are located within the atom. And, although the energies of the electrons are referred to as "energy levels," these "levels" do not correspond to the location of electrons in the atom as the "level" of water corresponds to its height above a given reference point.

The modern view of the atom can be understood more fully with a knowledge of waves. You are familiar with some kinds of waves—water waves that you can see and sound waves that you cannot. There are other wave phenomena, important to the model of the atom, that you will learn about in this chapter.

The atomic model used today views electrons as having only certain amounts of energy, and not others. The energy of an electron changes by certain fixed values, so it is said to be *quantized*. What kind of model is consistent with this view of the atom? What conclusions can be drawn about the reactivity of atoms when the energy of electrons is described in this manner? These issues will be addressed after you have had an introduction to waves.

CONCEPTS

- reviewing changes in the model of the atom
- familiar and unfamiliar waves
- frequency, wavelength, and amplitude—characteristics of waves
- light as a wave
- photons—packages of energy
- quantization
- electromagnetic waves

10-1 The Atomic Model Changes

In Chapter 8, you studied models of the atom proposed by John Dalton, J. J. Thomson, and Ernest Rutherford. Each of these scientists contributed to the understanding of atomic structure. John Dalton viewed the atom as a solid sphere. His model of the atom was based on evidence discovered from the combining ratios of elements and compounds in chemical reactions. Almost one hundred years later, J. J. Thomson proposed a new model that incorporated evidence of the component particles of the atom—protons and electrons. Further experimental evidence suggested to Ernest Rutherford that the atom contained a nucleus of positive particles and electrons traveling about the nucleus at great speeds. The model of the atom was later refined to include neutrons in the nucleus.

Why was it that Rutherford's model had to be changed? The nuclear atom he proposed had provided an explanation for the experiments on the scattering of alpha particles. However, it also set up a contradiction of facts. Rutherford suggested that the atom had negative charges surrounding a very small positive nucleus. Many experiments had shown that objects with unlike electric charges attract each other. If there is no force holding them back, the objects move toward each

Have the students review the concept of scientific modeling and the development of models of the atom as described in Chapters 2 and 8.

other. According to all experiments dealing with electric charge, the electrons should be pulled into the nucleus. Rutherford's model for the atom suggested an unstable structure, yet atoms are stable. They do not collapse. Evidently something was missing from the model.

10-2 Waves

You can study wave motion by trying this simple experiment. Half fill your kitchen sink with water. Float a cork or similar object in the center of the sink. Then generate some waves by dipping the back of a teaspoon into the water with a regular motion, and observe what happens. You should notice that the waves spread out from the spoon rather quickly in all directions on the surface of the water. Does the cork bounce up and down with the rhythm of the passing waves? Does the cork move along the surface of the water with the waves?

Scientists have formalized some ideas to describe waves that include those you can observe in the sink experiment. The **frequency** of a wave is the number of waves that pass a point per unit of time. In the experiment, the cork bobs up and down every time a wave passes. This motion can be used to measure the frequency of the waves in the sink.

Figure 10-1 illustrates the effect of waves on a boat in water. As waves pass, the boat moves up and down. If the boat rises 10 times in one minute, the frequency is 10 cycles per minute. Frequency is usually symbolized by the Greek letter *nu*, ν, so for the boat, $\nu = 10$ cycles per minute.

Another property of waves is **wavelength**, the distance between similar points in a set of waves, such as from crest to crest or trough to trough. The Greek letter *lambda*, λ, is often used to symbolize wavelength. Waves of differing wavelengths are illustrated in Figure 10-2.

A third measurable property of a wave is amplitude. Imagine a line, stretching in the direction of the wave, that is midway between the troughs and the crests. The **amplitude** of the wave is the distance from the crest to the imaginary line. The waves illustrated on the right in Figure 10-2 all have the same wavelengths but have different amplitudes.

If you did the floating cork experiment, you may have noticed that the cork did not move along the surface with the wave. Instead, it only

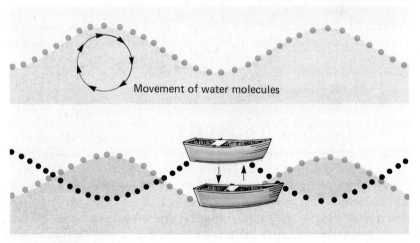

Movement of water molecules

Figure 10-1 The wave moves past the boat but does not carry the boat forward. The number of times the boat bobs up and down per unit of time is a measure of the frequency of the water waves.

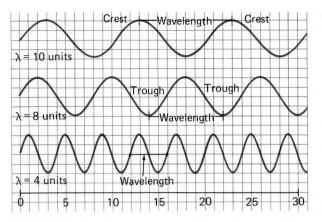

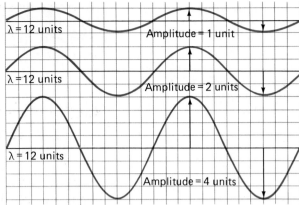

bobbed up and down as the wave passed. The wave does not *carry* the water; it transmits energy *through* the water. The energy you used to move the spoon up and down at the surface of the water was transmitted by the wave. The amount of energy used to move the cork depends on the mass of the cork, how far it moves up and down (amplitude), and how often a crest reaches the cork (frequency). It is the energy of waves that provides evidence for the current theory describing electrons in atoms.

Figure 10-2 Waves of different wavelengths but with the same amplitude are shown on the left. On the right, the waves shown have the same wavelengths but varying amplitudes.

10-3 Light

You are familiar with waves that you can see, for example, water waves. What about waves you cannot see? Sound waves travel through the air creating a ripple effect analogous to the spoon in the kitchen sink. You "observe" the sound if your ear is able to detect the wave. Light is made of waves you ordinarily do not notice, but these waves can be detected and measured under certain circumstances.

Like water waves, light waves can be characterized by wavelength and frequency. For water waves on an ocean, the wavelength may be 12 meters and the frequency may be one wave every 20 seconds, or 0.05 cycle per second. For light, wavelengths are much shorter and frequencies are much greater. Table 10-1 gives some examples of wave measurements for comparison.

Table 10-1

FREQUENCY AND WAVELENGTH OF DIFFERENT KINDS OF WAVES		
WAVE	FREQUENCY ν (number per second or hertz*)	WAVELENGTH λ (meters)
Water	5.0×10^{-2}	1.2×10^{1}
Red light	4.3×10^{14}	7.0×10^{-7}
Yellow light	5.2×10^{14}	5.8×10^{-7}
Blue light	6.4×10^{14}	4.7×10^{-7}
Violet light	7.5×10^{14}	4.0×10^{-7}

*Hertz is a measure of frequency. One hertz (Hz) is one cycle per second.

Figure 10-3 A beam of white light is refracted by a prism and separated into a rainbow of colors. One mnemonic device to remember the sequence of colors from longest to shortest wavelength is *ROY G BIV:* red, orange, yellow, green, blue, indigo, violet.

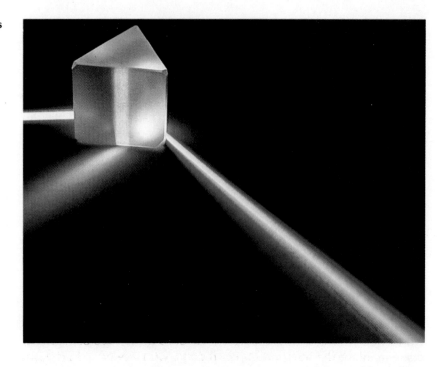

Make a spectrum of colors for the students with a slide projector. After a discussion of frequencies and color, hold up a transparent red filter in the beam of light. Notice that only the red end of the spectrum is visible. Hold up filters of other colors for comparison.

Planck developed the equation $E = nh\nu$ (where $n = 1, 2, 3, \ldots$) to explain blackbody radiation. Planck suggested atoms of a solid oscillate with definite frequencies, but he was uncomfortable with the idea of "quantization." However, Einstein saw it as an explanation for the photoelectric effect. He postulated the existence of packets of light with $E = h\nu$.

Note that red, yellow, blue, and violet lights have different wavelengths and frequencies. Each color of light can be described accurately by giving its frequency or its wavelength.

The nature of light can be examined more closely using a filmstrip projector. If the white light coming from the projector passes through a narrow slit and prism, a rainbow of colors, or a **continuous spectrum**, is produced, such as the one in Figure 10-3. The light beam passing through the prism is bent, or refracted. Different frequencies that compose white light are bent through different angles, causing the light waves to separate into colors. Each color corresponds to waves of light within a particular range of frequency and wavelength.

In Section 10-1, you learned that a water wave transmits energy. Light waves also transmit energy. This energy depends on several factors, including the length of time of exposure and the area over which the light is spread. The amount of energy transmitted by the light also depends on the frequency of the light wave. Knowledge about the relationship between energy and frequency was advanced by the work of Max Planck and Albert Einstein.

In 1900, Max Planck (a German scientist) studied the emissions of light from hot, glowing solids. He observed that the color of the solid varied in a definite way with temperature. Planck suggested a relationship between the energy of atoms in the solid and the wavelength of light being emitted. According to Planck, the energy values of the atoms varied by small whole numbers. Soon after, Albert Einstein expanded on Planck's work by dealing specifically with light energy. He suggested that light was emitted in "packets" of energy, and that the energy was proportional to the frequency of the light wave.

The idea that light energy comes in discrete packets may seem strange. You are more familiar with the concept that matter comes in packets called atoms; or that charge comes in packets, like the charge on an electron. The packets of light suggested by Einstein eventually became known as **photons**. The energy of a photon is described in an equation based on the work of Planck and Einstein.

$$E = h\nu$$

Energy, E, is measured in joules and is directly proportional to frequency, ν. The quantity h is known as Planck's constant, and has a value of 6.6262 $\times$ 10^{-34} joule-second. In the visible light spectrum, the frequency of light increases from red to violet. If the frequency increases, so does the energy per photon from red to violet. Each frequency of light has its own specific energy per photon *and no other*. Light is said to be quantized.

Einstein's suggestion—that light energy could be thought of as both particles and waves—was a tremendous leap in thinking for the time. Yet, this idea and the concept of specific amounts, or quantization, of energy are the basis for a theory that revolutionized atomic physics.

The concept of quantization is important for understanding the current model of the atom. The following experiment may help explain the idea more fully.

In some circumstances, you can make a wave appear to *stand still*. A rope about the length of a jump rope, a rubber hose, or a Slinky spring, can be used for this experiment. Secure the rope or spring at one end. Move the other end up and down with one hand in rhythm until you notice that the crest of the wave produced does not seem to move along the rope, as shown in Figure 10-4. You have produced a standing wave. If you produced a wave that looks like Figure 10-4, the wave has three nodes—one in the middle and one at either end. A

Strike several different tuning forks. Discuss the meaning of frequency and pitch.

Students may find it interesting to know that the ideas of Planck, Einstein, and later Bohr and de Broglie were sometimes lucky "shots in the dark," based on educated speculation.

Give students other examples of items that come in quantized packets, such as raisins, paperclips, or staples.

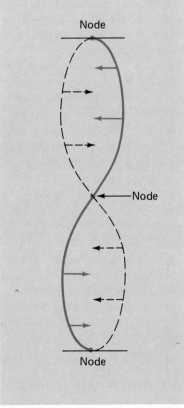

Wavelengths with a denominator greater than 2 do not occur in the standing wave system. If the wavelength is greater or less that $n(\lambda/2)$, where n = an integer, the forward and reverse waves would interfere with each other.

The rope analogy is not perfect. Energy used for the turning rope is not *only* a function of wavelength. As the wave passes, it does work on the rope, and work is a function of mass, amplitude, and frequency. The relationship between frequency, wavelength, and energy only holds if the other variables are constant. Nevertheless, even in its limited sense, the analogy may help students understand the concept of waves before they tackle quantum mechanics.

Figure 10-4 A Slinky anchored at one end will produce a standing wave only when the oscillation (up-and-down hand motion) at the other end is constant and at a particular rate. The standing wave pictured here has three nodes (as labeled in the diagram).

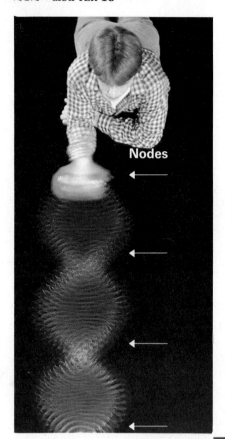

Nodes

Figure 10-5 A faster oscillation will produce a standing wave with a shorter wavelength. The standing wave pictured here contains four nodes.

node is the location on the wave that has an amplitude of zero. With three nodes, the distance across the spring is 1 wavelength. Now gradually increase the speed of your hand. For a short time the wave is out of sync and no standing wave is present. However, at just the right speed of your hand, a standing wave with four nodes can be produced (Figure 10-5). The distance across the rope now is $1\frac{1}{2}$ wavelengths. By further increasing the speed of your hand, it may be possible to generate a standing wave with five nodes or even more.

Several relationships are important here. First, only a standing wave of 1 wavelength, $1\frac{1}{2}$ wavelengths, 2 wavelengths, and so on can be produced. It is impossible to produce a standing wave, given the conditions of the experiment, with $1\frac{2}{3}$ or $1\frac{3}{4}$ wavelengths. Wavelengths will only change by factors of one half—nothing less and nothing more. The wavelength of the standing wave is said to be quantized. It will be helpful to keep the standing wave in mind for the discussion of the modern view of the atom.

Another relationship involves the frequency and the wavelength of the wave. All other things being equal, when you increase the up-and-down motion of your hand to produce three to four nodes, the frequency of the wave increases. At the same time, the wavelength decreases. Compare the wavelengths of the waves in Figures 10-4 and 10-5. The relationship between frequency and wavelength will be described in more detail in the next section.

10-4 Extension: The Electromagnetic Spectrum

When you walk outside on a sunny day, your face feels the warmth of the sun. The sun radiates energy that reaches Earth in the form of waves. The energy you feel as warmth and the light energy you see are part of a wide variety of energies that are categorized as electromagnetic radiation. **Electromagnetic waves** are produced by a combination of electrical and magnetic fields. You already have heard of many kinds of these waves. They include radio waves, microwaves, infrared radiation, ultraviolet radiation, X rays, gamma rays, and visible light. All electromagnetic waves travel at the speed of light, 3.00×10^8 m/s in a vacuum. (The speed of light in air is only negligibly slower.)

Electromagnetic waves differ from each other by wavelength and frequency. The waves can be ordered by increasing frequency on a continuum, or spectrum, like the one pictured in Figure 10-6. Radio waves have the smallest frequency on the continuum and gamma rays have the largest. Visible light is only a small part of the spectrum, with frequencies of about 10^{14} cycles per second.

When you studied Table 10-1 earlier, did you see a regularity in the data? Look at the table again. For each color of light, frequency increases as wavelength decreases. Each electromagnetic wave has its own wavelength and frequency. A mathematical relationship exists between a wavelength and frequency for any electromagnetic wave:

$$c = \lambda \times \nu$$

The quantity c is the velocity of light. In this equation, c is a proportionality constant that relates wavelength to frequency. Look at the

Do You Know?

In about one second, a beam of light (if it were to curve) could travel around Earth's equator seven times.

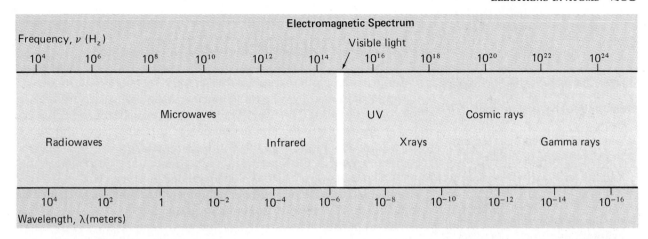

Figure 10-6 The named sections of the electromagnetic spectrum have no precise boundaries. They serve as a convenient way to classify radiation according to source.

electromagnetic spectrum in Figure 10-6. Note the same trend in the figure as in Table 10-1. For any electromagnetic wave, the product of wavelength and frequency must be equal to the velocity of light. Thus wavelength and frequency are inversely proportional. If you know the wavelength of the wave, you can calculate the frequency, and vice versa.

In the previous section, you read that the greater the frequency of a wave, the greater its energy. Electromagnetic waves with very large frequency, and therefore a large amount of energy, are X rays. They are produced by the bombardment of an anode (for example, tungsten) by high-energy electrons. X rays are important in a variety of fields in science. You are familiar with dental and chest X rays. X rays also are used to investigate the structure of matter (X-ray crystallography), and they are used in analytical chemistry to determine the composition of substances. X rays are even used to find metal fatigue in aircraft and machinery. The usefulness of these waves stems from their high energy and small wavelength. The wavelength of an X ray is about 10^{-8} cm, or about the diameter of an atom.

1. It suggested an unstable situation in which electrons would be pulled into the nucleus and the atom would collapse.

Do You Know?

In January, 1986, Voyager 2 detected a magnetic field around the planet Uranus. Characteristic radiowaves emitted by electrons that were escaping the planet through its magnetic field provided evidence for the field's presence.

Review and Practice

1. Describe a significant weakness in Rutherford's model of the atom. ✓

2. List five examples of waves.

3. Describe the difference between the frequency, wavelength, and amplitude of a wave. ✓

4. Describe evidence from the kitchen sink experiment to support the statement: "Water waves transmit energy, not matter."

5. What is a photon? What is the difference between a photon of yellow light and a photon of violet light?

6. List the seven colors of the visible light spectrum in *increasing* energy per photon.

7. How are frequency and wavelength of an electromagnetic wave related?

8. What does it mean to describe something as being quantized?

9. What wave characteristic determines the energy of a light wave?

8. existing only with or at specific amounts of energy
9. Energy is dependent on (proportional to) frequency.

2. ocean or water waves, sound waves, light waves, radio waves, microwaves (other possible answers: infrared radiation, ultraviolet radiation, X rays, gamma rays)
3. Frequency is the number of waves passing a point per unit of time. Wavelength and amplitude are not measured over time. Wavelength is the distance between corresponding points of a wave, and amplitude is the distance the wave deviates from an imaginary mid-line running in the direction of wave propagation.
4. The cork moves up and down in the same location but does not move in the direction of the wave.
5. A photon is a packet of energy of an electromagnetic wave. A photon of violet light carries more energy.
6. red, orange, yellow, green, blue, indigo, violet
7. Frequency times wavelength is equal to a constant, c (the speed of light). As frequency increases, wavelength decreases proportionally.

The Hydrogen Atom

CONCEPTS

■ the bright-line spectrum of hydrogen
■ energy of the hydrogen atom
■ the significance of the Bohr model of the atom
■ using quantum mechanics to explain the hydrogen atom
■ the meaning of orbitals
■ quantum numbers used to describe electrons in orbitals
■ the spectra of stars

Show the bright-line spectrum to the students. Have them hold up replica diffraction gratings to view the spectrum given off by the classroom fluorescent lights. If you do not have spectrum tubes and a power source, perhaps a physics teacher has some you can borrow.

Figure 10-7 The unrefracted lights from glowing gases (left) produce different bright-line spectra for each element (right). A continuous spectrum is provided for comparison.

In the opening photograph of this chapter, there is a pattern shown that was produced by a beam of electrons as they passed through a thin piece of metal foil. The diffraction (scattering) pattern produced by the electrons is similar to the pattern produced by X rays diffracted through a crystal. In 1927, when this phenomenon was first discovered, the idea that the behavior of particles could be described in the mathematics of wave motion seemed strange. However, it is a powerful idea that leads to a clearer understanding of the behavior of matter in general, and of electrons in atoms in particular.

The relationship between waves and atoms can be demonstrated in a rather striking, yet simple experiment. The atom itself can be made to give off waves of definite frequency, which serve as clues to the behavior of electrons in the atom. Hydrogen is the simplest of all atoms; it contains only one electron in the neutral state. Therefore, the hydrogen atom will be described first.

10-5 The Puzzle of the Bright-Line Spectrum

You are familiar with neon lights that glow a brilliant red on advertising signs. If you were to view the sign through a prism, you would see something unusual. Rather than a continuous spectrum, the light from the sign would be separated, or refracted, into several individual colors. This phenomenon can be reproduced in the laboratory. Figure 10-7 shows several gases, each in a different tube connected to a source of electricity. When viewed through a prism, each gas is seen to have its own spectrum of bright lines. The **bright-line spectrum** of an element consists of several distinct (separate) lines of color, each with

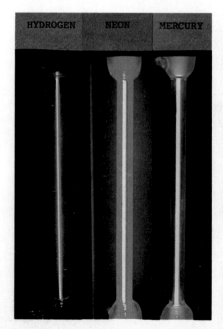

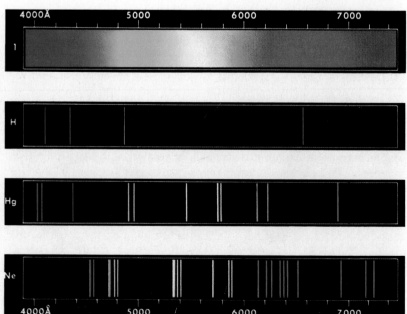

its own frequency. Compare the bright-line spectra in Figure 10-7, and you will notice that the spectrum for each element is unique.

Consider the bright-line spectrum of hydrogen gas. Table 10-2 lists the wavelengths and frequencies of visible light emitted by glowing hydrogen under low pressure.

Table 10-2

THE BRIGHT-LINE SPECTRUM OF HYDROGEN		
COLOR	FREQUENCY, ν (Hertz)	WAVELENGTH, (meters)
Red	4.57×10^{14}	6.565×10^{-7}
Blue-green	6.17×10^{14}	4.863×10^{-7}
Blue	6.91×10^{14}	4.342×10^{-7}
Violet	7.31×10^{14}	4.103×10^{-7}

Angstrom units are not part of standard SI measurements, but they remain in common use. The unit appears here in the photographs of spectra samples. You may wish to give your students the metric equivalent of the unit: $1 \text{ Å} = 10^{-8}$ cm.

From the table, you can see that each bright line in the spectrum of hydrogen has a specific wavelength and frequency. Scientists were baffled by this phenomenon and were not able to explain why hydrogen or any other glowing gas should form a discontinuous spectrum. They reasoned that since there were billions of atoms in the tube, there should be billions of different frequencies visible. Moreover, scientists wondered why each element exhibited a unique bright-line spectrum.

In 1885, a Swiss schoolteacher named Johann Balmer studied the bright-line spectrum of hydrogen. He discovered a mathematical progression in the sequence of lines. The progression can be expressed in the form of an equation:

Balmer's original equation concerned the mathematical relationship of the wavelengths of the spectral lines. The connection of wavelength to energy was made later.

$$E = \text{constant} \left(\frac{1}{2^2} - \frac{1}{n^2} \right) \qquad n = 3, 4, 5, \ldots$$

E is the energy of a spectral line, and n is an integer with a value greater than 2. The constant need not be specified here. The significance of Balmer's work was to become apparent as another scientist tackled the puzzle of the hydrogen atom.

10-6 Energy Levels and the Hydrogen Atom

In 1913, a Danish scientist, Niels Bohr, successfully explained the bright-line spectrum of hydrogen. His reasoning went something like this. Energy is being added to the hydrogen gas in the tube in the form of electricity. Energy is leaving the tube in the form of light. The energy of the light leaving the tube is quantized, that is, only certain frequencies of light are observed (the frequencies for hydrogen are listed in Table 10-2). Bohr reasoned that the atoms of hydrogen in the tube must be absorbing energy, then releasing it in the form of specific frequencies of light. He suggested that the hydrogen atoms themselves are quantized, that they exist only in certain, definite energy states, now called **energy levels**. The atoms absorb specific amounts of energy and then exist for a short time in higher energy levels. Such atoms

Avoid drawing Bohr orbits on the chalkboard. Students have seen enough of shells and two-dimensional atoms in junior high school and in biology. The task here is to set up the notion of energy levels not orbit energy.

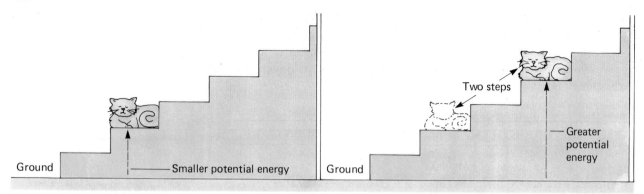

Ground —— Smaller potential energy

Two steps

Ground —— Greater potential energy

Figure 10-8 The cat may move either up or down on the staircase but must do so by whole numbers of steps. Since the cat's potential energy corresponds to the height at which it sits, only certain energy values are possible. Thus the cat's potential energy may be said to be quantized.

A difficult concept for students in the chapter is energy. Students usually do not have a good physics background when they take chemistry. Continue to give examples of different forms of energy.

Use pith balls to show attractions of oppositely charged particles. You may have done this demonstration earlier, but the visual reassociation here is helpful.

are described as "excited." Excited hydrogen atoms will emit energy as they return to lower energy levels. The light energy given off by an "excited" hydrogen atom is equal to the energy lost as the atom returns to a lower level.

A helpful analogy to Bohr's idea of energy levels in the hydrogen atom is a staircase. Consider a housecat sitting on the first step of a staircase, as pictured in Figure 10-8 (left). The cat's potential energy is considered to be small because it is close to the ground. (In this case, potential energy is due to the gravitational attraction between the cat and Earth.) If the cat climbs to a higher step, as shown in Figure 10-8 (right), its potential energy increases. If it climbs higher, its potential energy increases even more. The energy change can be measured in specific values that correspond to the heights of the steps. If the cat moves from one step to a lower step, its energy decreases by a fixed amount, and only that amount.

There are two points to be emphasized in this analogy. First, the cat must always change positions by an integral number of levels. It can move one, two, three, or more steps, but there is no way for it to move up or down by a half or a third of a step. Second, a change between two specific steps, for example, the second and the fourth, always involves the same amount of change in energy. These ideas can be applied to the hydrogen atom with a few modifications.

A hydrogen atom is made up of a single electron-proton system. The negatively charged electron is attracted to the positively charged proton. This attraction gives the electron-proton system potential energy. Bohr concluded that the greater the electron-proton distance, the greater the potential energy in the system, and therefore, the higher the energy level. (He wrongly concluded that the electron in a given energy state keeps the same distance from the proton while moving around it in a circular path.) He surmised that since excited hydrogen atoms emit light energy of only specific frequencies, hydrogen atoms have specific energy levels, and no others.

There are some particularly significant facts about Bohr's theory. First, he used Einstein's idea of photons of energy to explain the specific amounts of energy observed in hydrogen's bright-line spectrum. Second, when Bohr did his calculations to find the energy levels of hydrogen, his values for the spectral lines matched the mathematical relationship found by Johann Balmer. Furthermore, Bohr predicted the existence of other spectral lines for hydrogen in the infrared and ultraviolet regions of the electromagnetic spectrum. The subsequent discovery of these spectral lines was a triumph for his theory.

At this point, it may be helpful to summarize how Bohr's conclusions are applied to the bright-line spectrum of hydrogen:

1. Hydrogen atoms exist in only specified energy states.
2. Hydrogen atoms can absorb only certain amounts of energy, and no others.
3. When excited hydrogen atoms lose energy, they lose only certain amounts of energy, emitted as photons.
4. The different photons given off by hydrogen atoms produce the color lines seen in the bright-line spectrum of hydrogen. The greater the energy lost by the atom, the greater the energy of the photon.

The energy of the electron and the energy of the hydrogen atom are meant to convey the same concept; *energy of the hydrogen atom* refers to electronic energy.

Figure 10-9 is a diagram showing the energy levels of the hydrogen atom. Unlike the staircase in Figure 10-8, the "steps" of the hydrogen atom are not evenly spaced. (You will see later that the energy level "steps" are not actual distances between the electron and the nucleus of the atom.) Colored arrows indicate the amount of energy per photon that is given off when the atom changes from a higher level to level two. These energies correspond to the lines in the bright-line spectrum of hydrogen shown in Figure 10-7. The lines are known as the Balmer series. In a sense, analysis of the spectral lines of hydrogen provide a picture of the atom's "hidden staircase." Note in Figure 10-9, that the energies emitted per photon are much larger when an atom returns from a higher level to level one. These values are in the ultraviolet region and correspond to those predicted by Bohr. The difference in energy between level one and another level is considerably greater than between any other two levels.

The Lyman series (in the UV region) corresponds to electrons dropping to level one from other levels. Remind students that UV light has greater energy per photon than visible light. The energy changes for moving in and out of level one are the greatest. The Paschen series involves changes to the third level from even higher levels. Energy changes are smaller and appear in the infrared region. You may want to expand on the "hidden staircase" ideas, but emphasize to students that "steps" in the atom are not physical locations in the same sense as those of a staircase.

Energy level

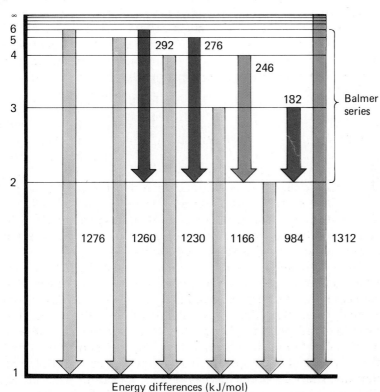

Energy differences (kJ/mol)

Figure 10-9 Energy changes in the hydrogen atom between higher levels and level two are shown in the colors that correspond to its bright-line spectrum (the Balmer series). Energy lost in the transition from higher levels to level one yield photons with energy in the ultraviolet region (changes shaded in gray).

Do You Know?

A red sweater appears red because the molecules of the dye in the sweater are absorbing energy in the blue-green region of the visible spectrum. You see the wavelength of light not absorbed, red.

An important value to note in the energy level diagram is 1312 kJ. If this amount of energy is absorbed by one mole of hydrogen atoms, all the electrons are removed, leaving one mole of hydrogen ions. The reaction can be expressed in an equation:

$$H(g) + 1312\,kJ \longrightarrow H^+(g) + e^-$$

The value 1312 kJ is the ionization energy of hydrogen. **Ionization energy** for any element is the amount of energy needed to remove an electron from the gaseous atom. Ionization energies are different for different elements.

Consider that if hydrogen ions and electrons recombine to form a mole of neutral hydrogen atoms, the energy emitted is 1312 kJ:

$$H^+ + e^- \longrightarrow H(g) + 1312\,kJ$$

The fact that this energy is greater than any of the other energies listed in Figure 10-9 makes sense and lends support to the idea of energy levels in the hydrogen atom. The other energies in the diagram reflect instances when the electron moves from one energy level to another *without being removed* from the atom.

Niels Bohr viewed the hydrogen atom in some sense as a planetary system. According to Bohr, the circling electron remains at a particular distance from the nucleus until the atom absorbs enough energy to "boost" the electron to a higher energy level, or larger circular orbit. The Bohr theory was successful in predicting the frequencies of the bright-line spectrum of hydrogen. Yet, when his equations were applied to atoms with more than one electron, they failed to account for the bright-line spectra observed. Bohr's theory had a fundamental flaw because it could not be applied to explain the experimental spectra of any element besides hydrogen. It seemed necessary to rethink the model of the atom once more.

10-7 The Modern Hydrogen Atom

A new model of the atom was needed that would account for the bright-line spectra of atoms with many electrons. The model also would have to be compatible with the periodic table. Elements are grouped together in the periodic table according to certain chemical and physical properties. Chemical properties of elements are directly related to the behavior of the electrons in atoms. The new theory would have to explain why groups of elements react similarly.

The new theory also needed to account for the energy of electrons in atoms described by Niels Bohr. Two new ideas in physics paved the way for the new model. In 1923, a French scientist named Louis de Broglie made a rather bold suggestion. Thinking about Einstein's theory that waves can have properties of particles (photons), de Broglie reasoned that perhaps particles could have properties of waves. If so, the behavior of electrons in atoms might better be understood using the mathematics of waves.

Soon after, Erwin Schrödinger, a German scientist, applied the idea of particle waves to electrons in atoms. He developed equations to describe the energy of electrons. Although the actual mathematics of

Louis de Broglie proposed his "particles with wavelike properties" idea in his doctoral thesis, but his committee thought it was rubbish. They would have rejected the thesis if not for Einstein's support of the idea.

Schrödinger's work is beyond the scope of this book, it is important to note that his conclusions provided further support to the concept of using the language of waves to describe electrons in atoms. (Schrödinger's equations also correctly described the behavior of waves in the electromagnetic spectrum.) The branch of physics that stems from the work of de Broglie and Schrödinger was first called wave mechanics. Since it describes the behavior of electrons in terms of quantized energy changes, it is more commonly called **quantum mechanics**.

There is no experimental evidence that the electron has a "path" around the nucleus, as Bohr had thought. Therefore, the notion of a path was discarded as meaningless. Instead, the behavior of the electron is described in terms of the probability of finding the electron in a given region of space. **Probability** is the likelihood of an occurrence. For example, the probability of finding a student in the hallway of your school during class time is small because the vast majority of students are in class at that time. However, the probability of finding a student in the hallway between class periods is much greater.

It is not possible to pinpoint both the speed and position of an electron in an atom. Any experiment designed to measure one of these characteristics will automatically change the other. On the other hand, the mathematics of quantum mechanics makes it possible to calculate the probability of finding an electron in a certain volume of space around the nucleus.

Quantum mechanics incorporates both particle-wave theory and probability into its description of the energy of an electron in the atom. The name **orbital** is given to the region of space around a nucleus in which an electron is most likely to be found. Descriptions of orbitals have been developed from the mathematics of wave mechanics. Orbitals of the simplest atom, hydrogen, will be discussed here.

When a hydrogen atom is in its lowest energy state, often called the **ground state**, the energy of the electron is defined by an orbital shown in the model in Figure 10-10. Notice that the sphere does not have the same density of color throughout. You can think of the density of color as corresponding to the probability of finding the electron in that region of space. The relatively high density of the region close

Demonstrate probability by having students (from their seats) throw Velcro covered Ping-Pong balls at a felt target. Over a period of time, the students see that the balls are more likely to land close to the target.

The Heisenberg uncertainty principle states that for a particle as small as an electron, it is not possible to accurately determine the position and momentum (mass × velocity) at the same time.

Remind students not to confuse the words *orbit* and *orbital*. They need to avoid thinking in terms of paths.

Some educators advocate abandoning the orbital in high school chemistry. They may have a point. You may choose to avoid overburdening the students with probability density and the Schrödinger equation. However, orbitals are used in this book to interpret the periodic table and will be used in the chapter on bonding.

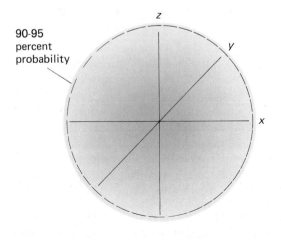

1s orbital

Figure 10-10 The 1s orbital of hydrogen is represented by a shaded sphere. Diagrams depicting orbitals normally enclose the region in which there is about a 95% probability of finding the electron (represented here by the dotted line). Areas of darker shading represent regions where the probability of finding the electron is greatest.

to the nucleus signifies that the electron is found closer to the nucleus more often than it is found farther away.

An orbital is sometimes referred to as a charge cloud. The negative charge of the electron can be thought of as "smeared" over the entire orbital. Charge density is greatest near the nucleus, where the probability of finding the electron is great.

The orbital description for the electron in an atom can be compared with the pattern of holes in a dart board. After it has been used for a long time, the dart board has many holes near the bull's-eye. The number of holes per unit area of the dart board decreases as you look farther from the center. The number of holes per unit area at any distance from the bull's-eye is a measure of the probability that the next dart will land there. The holes in the dartboard do not indicate anything about the order in which they were made or where the next dart will land. The situation in the atom is similar. The orbital describes the probability that an electron will be at a particular distance and direction from the nucleus, but it does not indicate where the electron was, is now, or will be next.

Figure 10-11 This *Landsat* photo shows the distribution of populated areas around Baltimore, Maryland. The blue areas are those of greatest population density. Although the photo does not enable you to predict the location of any single person, it does show you those regions in which a person is most likely to be found.

In the previous section, you learned that the electron of the hydrogen atom can be found only in certain energy levels. If the electron absorbs energy, it can be found in the second, third, or even higher levels. The energy levels of the electron of the hydrogen atom are each designated by a number, n. For the first energy level (the ground state), $n = 1$. For the second energy level, $n = 2$, and so on. The letter "n" is called the **principal quantum number**. Quantum numbers are used to describe an electron in an orbital. These labels serve much like an address that distinguishes one house from another. (The values of the quantum numbers are derived from Schrödinger's wave equations. These values define the energies and probability densities of the electrons. You will not need to work with these equations here.)

In any one energy level, there is a specific number of orbitals possible. The value n^2 determines the number of orbitals for each energy level. Table 10-3 lists the total number of orbitals for the first four energy levels of hydrogen.

Table 10-3

ORBITALS OF THE HYDROGEN ATOM	
ENERGY LEVEL	NUMBER OF ORBITALS
(n)	(n^2)
1	1
2	4
3	9
4	16

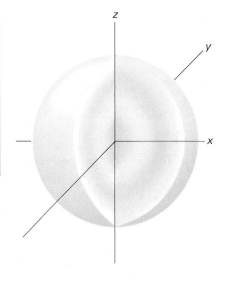

2s orbital

There is only one orbital for the first energy level. There are several orbitals that describe energy levels two, three, and four. The first orbital in an energy level is designated by the letter "s." For the first energy level ($n = 1$), the lone orbital is designated as the "1s" orbital. Figure 10-10, shown earlier, illustrates the 1s orbital of hydrogen.

If the electron absorbs enough energy to be in level two, four orbitals completely describe the energy of the electron. The 2s orbital (Figure 10-12) is shaped similarly to the 1s orbital, but is larger. The larger size seems reasonable because the 2s orbital describes a higher energy state for the electron than does the 1s orbital. The probability of finding the electron in the 2s orbital is spread out over a larger region of space.

Three other orbitals exist in the second energy level. These three orbitals are not spherically shaped like the 2s orbital, but rather they are each double lobes, as represented by the structures in Figure 10-13. The double-lobe structure is signified by the letter "p." Since the 2p orbitals are oriented around different axes in space, they are further differentiated by the labels $2p_x$, $2p_y$, and $2p_z$.

Perhaps you are wondering how an electron can move from one lobe in a 2p orbital to another if the lobes are not touching. It is a good question. If you think of an orbital as describing the path of an electron, there is no answer to your question. Strictly speaking, an orbital does not describe a "house" in which an electron roams about. The orbital is a description of an electron wave. Recall from Section 10-3 that a standing wave has places, or nodes, where the amplitude is zero. In much the same fashion, some orbitals have places where the probability of finding an electron is zero. These places are also called

Figure 10-12 The 2s orbital, like the 1s orbital, is spherically shaped. The 2s orbital for hydrogen has two regions where the probability of finding the electron is greatest.

Figure 10-13 The 2p orbitals of hydrogen are represented by three pairs of flattened lobes. Each 2p orbital has a region, known as the nodal plane, in which the probability of finding the electron is zero. When all three p orbitals are taken together, they describe a spherical shape.

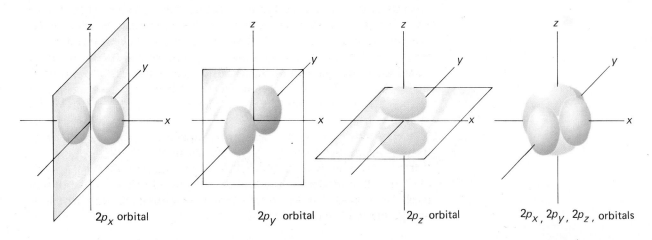

$2p_x$ orbital $2p_y$ orbital $2p_z$ orbital $2p_x$, $2p_y$, $2p_z$, orbitals

Figure 10-14 The energy levels of the hydrogen atom.

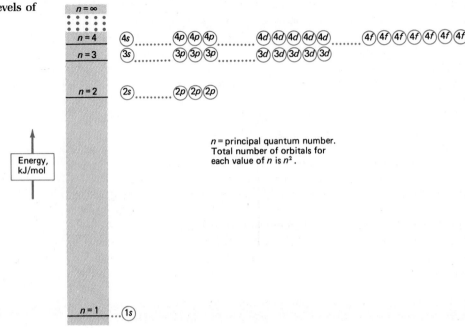

n = principal quantum number. Total number of orbitals for each value of n is n^2.

Energy, kJ/mol

The five $3d$ orbitals are $3d_{x^2-y^2}$, $3d_{xy}$, $3d_{z^2}$, $3d_{xz}$, and $3d_{yz}$. The shapes of the d orbitals are included on a transparency master in the Teacher's Resource Binder.

For an isolated hydrogen atom, the energies in any one level are degenerate; i.e., the energy of the $3s$, $3p$, and $3d$ orbitals are the same. If the atom is brought under the influence of some other electromagnetic force, the difference in the angular momentum of the electron in various orbitals would come into play, and the energies would no longer be degenerate.

nodes. Each $2p$ orbital has a node lying along the plane of the nucleus. The $2s$ orbital also has a node. Look carefully at Figure 10-13 and see if you can locate where the probability of finding the electron in the $2s$ orbital is zero.

In level three, there are nine orbitals ($n^2 = 9$). One is the $3s$ orbital and three more are the $3p$ orbitals ($3p_x$, $3p_y$, and $3p_z$). Each of these has either a spherical or a double-lobed shape, but larger than their second level counterparts. Level three has five more orbitals, designated as $3d$ orbitals. They are differentiated in space on the x, y, and z axes as well. The shapes of the $3d$ orbitals are more complex than the s and p orbitals and are not shown.

Figure 10-14 summarizes the energy levels of hydrogen and the orbitals present in each level. Notice that level four can have 16 orbitals ($n^2 = 16$). In level four, there is one $4s$ orbital, plus three $4p$ orbitals, five $4d$ orbitals, and seven additional orbitals, designated by the letter "f." The s, p, d, and f regions of an energy level are often called *sublevels*.

When the electron of the hydrogen atom absorbs enough energy to move to the fourth energy level, the electron can occupy any of the sixteen orbitals possible in that level. Each orbital has exactly the same energy. If the electron loses energy and moves from the fourth to the second energy level, it can occupy any of the four orbitals for that level. (The loss of energy is observed as light in the blue-green region of the visible spectrum.)

10-8 Quantum Numbers

You have been introduced to one of the four quantum numbers used to describe an electron in an orbital. These numbers specify an "address" for the electron in the atom. Because orbitals are regions of space oriented around a particular point (the nucleus), three aspects of the volume must be described in order to specify an orbital. You are already familiar with the principal quantum number, *n*, that describes the *size* of the orbital. The value for *n* is an integer that ranges in value from 1, 2, 3, . . . to infinity.

A second quantum number, "*l*," designates the *shape* of an orbital. The letters *s*, *p*, *d*, and *f* are used to specify this second quantum number. An *s* orbital is spherical, a *p* orbital is composed of double lobes. The *d* and *f* orbitals also are composed of lobes, but are more complex.

The third quantum number, "*m*," describes the orbital's *orientation* about the *x*, *y*, and *z* axes. For example, a $2p_x$ orbital is a double-lobed orbital of the second energy level, oriented around the *x*-axis. Three quantum numbers must be used to define a given orbital, and no two orbitals can have the same three quantum numbers.

A fourth quantum number designates the *spin* of the electron. Its meaning will be discussed when you read about atoms with many electrons (Section 10-10).

10-9 Application: Eyes on the Universe— The Spectra of Stars

Some late evening when the sky is clear, take a good look at the stars. Are they all white? Actually, they range in color from red to blue. If you look carefully you will see some differences. Temperatures of the stars' surfaces determine their color. Hotter stars (about 10 000 K) appear blue. Cooler stars (about 2000 K) appear red.

The naked eye is not able to distinguish the spectrum of light that comes from a star. Astronomers use a spectroscope attached to a powerful telescope for this task. The spectroscope uses a prism or other device to bend light into its component wavelengths. When starlight is analyzed in a spectroscope, an almost continuous spectrum is seen. However, dark lines also are visible in the spectrum, such as those shown in the first spectrum in Figure 10-15. The dark lines are caused by *absorption* of light, at specific wavelengths, coming from the star. What is absorbing the light at these wavelengths?

Do You Know?

The letters s, p, d, and f once stood for sharp, principal, diffuse, and fundamental, which were classifications of spectral lines in the emission spectra of elements.

No mathematical treatment of quantum numbers is intended here. Emphasize that three characteristics of a volume in space need to be known in order to describe it: size, shape, and orientation. If you wish to establish the connection between the word quantum *number* and the numbers used in the solutions to the Schrödinger equation, you might tell the students that the quantum number *l* is an integer from 0 to *n* − 1, and the quantum number *m* is an integer from −*l* to +*l*.

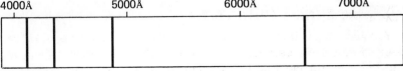

Figure 10-15 The first picture shown here is a photograph of a portion of the solar spectrum. Dark lines result from the absorption of energy by the relatively cooler gases at the surface of the star. The second picture is a diagram of the absorption spectrum of hydrogen. Compare it with the element's bright-line spectrum in Figure 10-7.

In an experiment, if white light is passed through a tube containing hot hydrogen gas at low pressure, a spectrum is produced that contains dark lines at the same wavelengths where the bright-line spectrum of hydrogen would show red, blue-green, blue, and violet (as shown earlier in Figure 10-7). The dark lines occur because the atoms of the hot hydrogen gas absorb light energy only at frequencies that they can later emit! (The bright-line spectrum of an element also is called the *emission* spectrum.)

The absorption spectrum for any element, like the bright-line spectrum, is unique. Starlight can be analyzed using its absorption spectrum to determine the elements in a star. Spectroscopy is the most powerful tool astronomers have for analyzing the chemical composition of the stars. It has been determined that stars are composed mostly of hydrogen (between 60 and 80 percent). Hydrogen and helium together make up about 96 percent of the composition of stars. The other 4 percent is composed of neon, oxygen, nitrogen, carbon, magnesium, argon, sulfur, iron, and chlorine. Molecules can be detected in the cooler stars.

Spectral analysis also reveals information about stellar temperature, pressure, rotation, turbulence, and magnetic fields. This wealth of data comes from the light of stars. The nature of the universe, in a very real sense, is placed at our fingertips by an understanding of electrons in atoms.

1. a series of colored lines, each with its own specific frequency

2. It is a way of describing the total energy of an electron in reference to the nucleus of the atom. Electrons can be found only in certain specific energy levels and no others.

3. When electrons of hydrogen drop from higher energy levels to lower ones, the energy lost is emitted as photons of specific energy and thus specific wavelength.

4. a region in space around the nucleus in which electrons are likely to be found

5. The bright-line spectrum of hydrogen. Bohr successfully predicted the line spectrum of hydrogen in the IR and UV regions.

6. Ionization energy is the energy needed to remove an electron from the influence of the nucleus. Spectral lines result from energy of electrons still associated with the nucleus.

7. The region of 0.99 probability of finding the electron in the 3s orbital is larger than that in the 2s.

8. Energy of a specific wavelength is absorbed by the atom.

9. An orbital does not describe the path of an electron.

10. It is derived from the mathematics of particle waves. It also describes the behavior of electrons in atoms in terms of probability rather than specific paths.

11. They would be the same orbital.

12. An absorption spectrum is continuous with dark lines at the frequencies corresponding to the energies absorbed by the atoms. The dark lines occur in the same location where the colored lines of an element's bright-line spectrum would be.

Review and Practice

1. What is a bright-line spectrum?

2. What is an energy level of an atom? Why are energy levels described as quantized?

3. Briefly explain why the bright-line spectrum of hydrogen is composed of discrete lines and is not a continuous spectrum.

4. What is an orbital?

5. What evidence provided support for Bohr's model of the atom?

6. Why is the ionization energy of hydrogen greater than the energy of any frequency in the bright-line spectrum for hydrogen?

7. How is a 3s orbital for the hydrogen atom different from a 2s orbital?

8. Explain what happens when the electron of a hydrogen atom changes from a 2s orbital to a 5s orbital.

9. Why is it incorrect to say that because a 1s orbital is spherical, the electron travels around the nucleus in a circle?

10. How is the quantum mechanical model of the atom different from previous atomic models?

11. Why is it not possible for two different orbitals to have the same first three quantum numbers?

12. How does an absorption spectrum of an element differ on the atomic level from a bright-line spectrum?

Electrons in Many-Electron Atoms

The hydrogen atom is the simplest of all atoms because it consists of only one electron and one proton. The electrons in more complex atoms (atoms with many electrons) can be understood in terms of the hydrogen atom if some modifications are made to the model. The material you study in this lesson provides the basis for what you will learn in the next two chapters. A great deal of chemistry is derived from knowledge about electrons in atoms. The modern theory of the atom successfully explains the relationships of elements in the periodic table. The theory also has been successful in explaining the behavior of atoms when they react with each other. Credibility in the model of the atom as described through quantum mechanics is strengthened because it is compatible with experimental data and it can be used to predict chemical behavior.

CONCEPTS

- organizing electrons in many-electron atoms
- electron spin and the Pauli exclusion principle
- electron configurations of neutral atoms
- electron configurations of ions

10-10 Many-Electron Atoms

Quantum mechanics can be applied to atoms that have more than one electron. Since most atoms have more than just a few electrons, it is reasonable to assume that many orbitals are occupied simultaneously by electrons in these atoms. This situation is different from hydrogen, where the single electron can occupy only one orbital of the atom depending on the energy of the electron.

Experimental data from the spectral analysis of elements also suggests that two electrons can occupy one orbital of a many-electron atom. In order to account for the small differences in the responses of atoms to magnetic fields, scientists have suggested that each electron "spins" on an axis.* If two electrons occupy the same orbital, the electrons will spin in opposite directions. By convention, one electron is assigned a spin of $+\frac{1}{2}$ and the other electron has a spin of $-\frac{1}{2}$. The spin characteristic of electrons is the fourth quantum number, and is designated by the letter "m_s." It is important to note here that if two electrons occupy the same orbital they must have opposite spins. The energy of electrons is defined by their quantum numbers. No two electrons can have the same four quantum numbers, or they would be the same electron. This concept is known as the **Pauli exclusion principle** (for Wolfgang Pauli, who devised it). Two electrons can have the same first three quantum numbers (if they are in the same orbital), but their spins will be opposite. No more than two electrons can be in any one orbital.

The procedure of organizing electrons in atoms from the orbital with the lowest energy to the orbital with the highest energy leads to the **electron configuration** of the atom. Organizing electrons in orbitals for the many-electron atoms is a relatively easy mechanical task.

The spin characteristic applies to the electron as a particle.

Pauli's exclusion principle is based on experimental evidence as related to the position of the elements on the periodic table. Even though the dynamics of the wave-mechanical model is better understood today, rigorous mathematics has been worked out for an atomic system with only a few electrons.

*The spinning of an electrically charged particle generates a small magnetic field around the particle.

Orbital notation seems overwhelming for many students. Encourage them to persevere. After the students practice writing some configurations, they usually understand and even like the subject.

You can do it. The sequence of filling many-electron atoms follows the scheme of the hydrogen atom to some extent. Table 10-4 shows the electron configurations of the first five elements in the periodic table, written in increasing order of atomic number. It is presumed that each atom is neutral in charge, that is, the number of protons in the nucleus and the number of electrons is equal.

Table 10-4

ELECTRON CONFIGURATIONS FOR ELEMENTS 1 TO 5				
ELEMENT	SYMBOL	$1s$	$2s$	$2p$
Hydrogen	$_1$H	⊘	○	○○○
Helium	$_2$He	⊗	○	○○○
Lithium	$_3$Li	⊗	⊘	○○○
Beryllium	$_4$Be	⊗	⊗	○○○
Boron	$_5$B	⊗	⊗	⊘○○

All electron configurations are listed in the atom's ground state, that is, in the lowest energy possible. The symbol ⊘ represents one electron in an orbital. Two electrons in the same orbital are represented by the symbol ⊗. The opposite directions of the slashes represent the opposite spins of the electrons. The order of "filling" up orbitals has a simple rule:

Start with the $1s$ orbital. Then begin filling the next highest energy orbitals, for example, $2s$, $2p$, until all the electrons are accounted for.

The electron configurations for the next three elements after boron are given in Table 10-5.

Table 10-5

ELECTRON CONFIGURATIONS FOR ELEMENTS 6 TO 8				
ELEMENT	SYMBOL	$1s$	$2s$	$2p$
Carbon	$_6$C	⊗	⊗	⊘⊘○
Nitrogen	$_7$N	⊗	⊗	⊘⊘⊘
Oxygen	$_8$O	⊗	⊗	⊗⊘⊘

Hund's rule states that when electrons fill orbitals within a sublevel, each orbital is occupied by a single electron before any orbital has two electrons, and all electrons in singly occupied orbitals have the same direction of spin.

There are some exceptions to the filling of electrons in electron configurations. Cu, for example, has a completed $3d$ sublevel. You may wish to limit discussion of the exceptions until students grasp the basics. Exceptions can be discussed when students study Chapter 11.

Notice that in carbon, two electrons are placed in separate $2p$ orbitals. The correct way of placing them is shown above. If both electrons are placed in one $2p$ orbital, there would be a slight repulsion between them (even if they had opposite spins) that would increase the energy of the system. The atom would then not be in the ground state. By placing the electrons in different $2p$ orbitals, this repulsion is lessened. When adding electrons to a sublevel (such as the $2p$ sublevel system, or the $3p$ or $3d$) the following rule is helpful to remember:

Put one electron in each orbital of the sublevel first, then go back and pair electrons in each orbital as necessary.

Figure 10-16 shows an energy level diagram for many-electron atoms. Study the chart carefully. The $1s$ orbital has the lowest energy and is filled first. Notice, however, that for the many-electron atoms, the 2s and 2p orbitals are not shown to have the same energy even though their energies are the same in the hydrogen atom. What may seem even more puzzling is that the 4s orbital appears to be at a lower energy than the 3d orbital system. What causes the differences in orbital sequencing in the many-electron atoms from the hydrogen atom?

Figure 10-16 The filling of orbitals in a many-electron atom occurs in the sequence that provides for the lowest potential energy of the system.

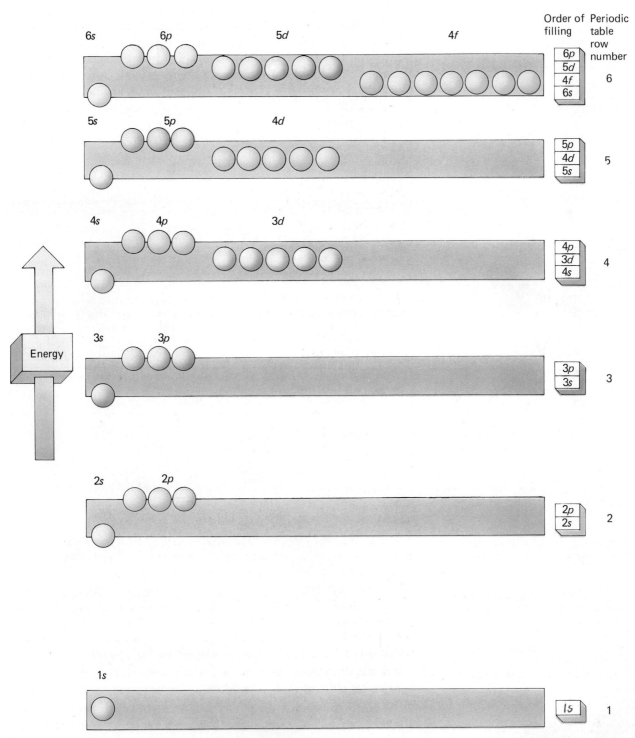

A wall chart of Figure 10-16 is extremely helpful to students.

Hydrogen has only one electron and one proton. All other atoms have many electrons, and these electrons repel one another because they have like charges. At times, placing electrons in the orbital associated with the lowest available energy for the hydrogen atom brings the electrons of many-electron atoms close together. The result is a great deal of repulsion and a higher energy overall. Consequently, the atom as a whole may have a lower potential energy if some of the electrons "skip" to higher energy levels. These "outer electrons" occupy a higher energy orbital, which puts them farther from other electrons in the atom. This explanation is supported by experimental evidence that shows that the 4s orbitals fill before the 3d orbital system for some elements. The 4s orbital is always a higher energy level than the 3d. But if there are other electrons present, placing electrons in the 4s orbital before filling the 3d orbital results in a lower energy for the *atom as a whole*.

Two factors affect the energy of electrons in orbitals. (1) The energy as predicted from the orbital notation of the hydrogen atom. (2) The increase in potential energy due to the repulsion of electrons.

Example 10-1

Write the electron configuration for the following elements: neon, calcium, and manganese.

▶ **Suggested solution** _____

(Cover the solution below and try to write the electron configurations without referring to the answers.) You need to know the total number of electrons present for any one element. Recall that in a neutral atom, the number of electrons is equal to the number of protons (atomic number). Once you know the number of electrons, fill in the electron configurations.

	1s	2s	2p	3s	3p	4s	3d
$_{10}$Ne	⊗	⊗	⊗⊗⊗	○	○○○	○	○○○○○
$_{20}$Ca	⊗	⊗	⊗⊗⊗	⊗	⊗⊗⊗	⊗	○○○○○
$_{25}$Mn	⊗	⊗	⊗⊗⊗	⊗	⊗⊗⊗	⊗	⊘⊘⊘⊘⊘

Notice that in manganese the 3d orbitals are filled after the 4s orbital because this arrangement of electrons gives the manganese atom the lowest energy as a whole.

The diagrams that you have drawn to represent the electron configurations in Example 10-1 are also known as **orbital diagrams**. As you probably have noticed, it is time consuming to draw little slashes and circles. (Orbital diagrams will be useful when you study the periodic table and bonding in Chapters 11 and 12.) For now, it is simpler to use a shorthand method chemists have devised for writing electron configurations. The electron configuration for nitrogen can be written more succinctly as

$$_{7}N \qquad 1s^2\, 2s^2\, 2p^3$$

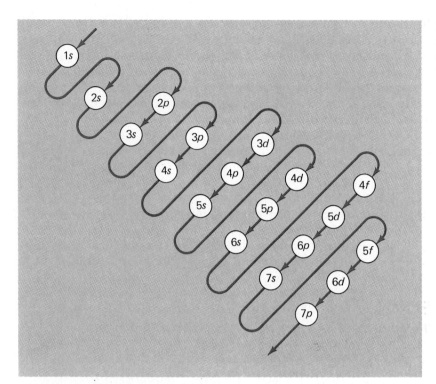

The configuration is read as: "1s two, 2s two, 2p three (not 1s squared, etc.) Keep in mind the one drawback of the shorthand method: It does not tell you explicitly that the three electrons in the 2p sublevel are arranged one electron each in the $2p_x$, $2p_y$, and $2p_z$ orbitals.

There is no real need to memorize the order of filling electrons for the many-electron atoms. However, Figure 10-17 is a mnemonic device that may help you work out configuration schemes.

10-11 Electron Configurations for Common Ions

Electron configurations can be written for simple ions. Chapter 11 will make clearer the relationship of electron configurations of ions and the periodic table. Here you need only understand that electron configurations of ions can be successfully predicted from what you already know. Several examples are given below in Table 10-6.

Once students become accustomed to writing configurations, you may want to show them how to use the more abbreviated notation of the noble-gas core plus the outer electrons. For example, the configuration of sodium may be written as [Ne]3s^1. This notation is described in Chapter 11.

Table 10-6

ELECTRON CONFIGURATIONS OF ATOMS AND IONS			
ELEMENT	ELECTRON CONFIGURATION (NEUTRAL ATOM)	ION	ELECTRON CONFIGURATION
$_{12}$Mg	$1s^2\,2s^2\,2p^6\,3s^2$	Mg^{2+}	$2s^2 2s^2 2p^6$
$_{17}$Cl	$1s^2\,2s^2\,2p^6\,3s^2\,3p^5$	Cl^-	$1s^2\,2s^2\,2p^6\,3s^2\,3p^6$
$_{26}$Fe	$1s^2\,2s^2\,2p^6\,3s^2\,3p^6\,4s^2\,3d^6$	Fe^{2+}	$1s^2\,2s^2\,2p^6\,3s^2\,3p^6\,3d^6$

In the case of the magnesium ion, the symbol Mg^{2+} means that two electrons have been removed from the neutral atom. The two electrons that are removed are the ones with the highest energy, $3s^2$. For Cl^-, one electron has been added to the neutral chlorine atom. Therefore, the chloride ion has a full $3p$ orbital system. Notice that in the case of Fe^{2+}, the $4s$ electrons were removed because the electrons occupying the $4s$ orbital have the highest energy.

Example 10-2

Write the electron configurations for the following ions:

$$S^{2-} \quad Na^+ \quad Zn^{2+}$$

▶ Suggested solution _____

The sulfur atom (atomic number = 16) gains two electrons to become S^{2-}. The electron configuration of S^{2-} is:

$$S^{2-} \quad 1s^2\, 2s^2\, 2p^6\, 3s^2\, 3p^6$$

Sodium (atomic number = 11) loses one electron to become Na^+. It has an electron configuration of:

$$Na^+ \quad 1s^2\, 2s^2\, 2p^6$$

Zinc (atomic number = 30) loses two electrons to become Zn^{2+}. The electrons with the highest energy for the zinc atom are in the $4s$ orbital. Thus they are the electrons lost, and the electron configuration of Zn^{2+} is

$$Zn^{2+} \quad 1s^2\, 2s^2\, 2p^6\, 3s^2\, 3p^6\, 3d^{10}$$

1. An electron configuration represents the ordered series of orbitals occupied by electrons in an atom.

2. If the $3d$ orbitals filled first, there would be more energy in the system due to electron-electron repulsion. The atom has lower energy as a whole if the $4s$ fills first.

3.

	1s	2s	2p	3s	3p
Na	⊗	⊗	⊗⊗⊗	⊘	○○○
Cl	⊗	⊗	⊗⊗⊗	⊗	⊗⊗⊘
S	⊗	⊗	⊗⊗⊗	⊗	⊗⊘⊘
Mg	⊗	⊗	⊗⊗⊗	⊗	○○○

Na $1s^2\, 2s^2\, 2p^6\, 3s^1$
Cl $1s^2\, 2s^2\, 2p^6\, 3s^2\, 3p^5$
S $1s^2\, 2s^2\, 2p^6\, 3s^2\, 3p^4$
Mg $1s^2\, 2s^2\, 2p^6\, 3s^2$

4. Ca^{2+} $1s^2\, 2s^2\, 2p^6\, 3s^2\, 3p^6$
Al^{3+} $1s^2\, 2s^2\, 2p^6$
P^{3-} $1s^2\, 2s^2\, 2p^6\, 3s^2\, 3p^6$
F^- $1s^2\, 2s^2\, 2p^6$
Co^{2+} $1s^2\, 2s^2\, 2p^6\, 3s^2\, 3p^6\, 3d^7$

5. They have opposite spins (+1/2 and −1/2).

6. No two electrons can have the same four quantum numbers.

7. There is slightly less repulsion between electrons (and thus lower potential energy for the atom) if each one occupies a separate orbital until no empty d orbitals remain.

Review and Practice_____

1. What is meant by the term *electron configuration?*

2. Briefly explain why the $4s$ orbital system "fills" before the $3d$.

3. Draw an orbital diagram and write the electron configuration for atoms of each of the following elements: Na, Cl, S, and Mg.

4. Write the electron configuration for each of the following ions: Ca^{2+} Al^{3+} P^{3-} F^- Co^{2+}.

5. How do two electrons that occupy the same orbital differ?

6. What does the Pauli exclusion principle state about electrons?

7. Explain why the electron configuration of manganese in Example 10-1 shows the five $3d$ electrons each in a separate orbital rather than grouped in pairs.

Numbers in red indicate the appropriate chapter sections to aid you in assigning these items. Answers to all questions appear in the Teacher's Guide at the front of this book.

Summary

■ The model of the atom used by scientists today is described by wave mechanics, or quantum mechanics. The modern model complements scientific knowledge in the areas of chemical periodicity, chemical bonding, and chemical reactivity.

■ All waves have certain characteristics such as frequency, wavelength, and amplitude. The wavelength of a standing wave is quantized.

■ Light is an electromagnetic wave. It occupies a small portion of the electromagnetic spectrum. White light contains all the colors of the rainbow. A prism refracts white light into its component colors, each having a wavelength and frequency different from the others.

■ The spectrum of hydrogen contains separate color lines. These lines are explained by the concept of quantized energy levels in hydrogen atoms. An excited hydrogen atom can lose energy and "fall" to a lower energy level. In the process, an exact amount of energy in the form of a specific wavelength of light is emitted.

■ The modern theory of the hydrogen atom is built upon wave mechanics and quantization of energy. The position and path of an electron in an atom cannot be determined. Instead, an electron is described in terms of an orbital—a volume in which an electron is likely to be found. Orbitals are identified by symbols, such as $1s$, $2s$, $2p$, etc., which distinguish the orbital's energy level, shape, and orientation in space.

■ The structure of electrons around the nucleus of a many-electron atom is denoted in an electron configuration or drawn as an orbital diagram. Electron configurations of elements correspond to the arrangement of elements in the periodic table.

Chemically Speaking

amplitude	node
bright-line spectrum	orbital
continuous spectrum	orbital diagrams
electromagnetic waves	Pauli exclusion principle
electron configuration	photon
energy levels	principal quantum number
frequency	probability
ground state	quantum mechanics
Hertz	wavelength
ionization energy	

Review

1. Describe a characteristic that is similar and a characteristic that is different between an ocean wave and a standing wave. (10-3)

2. List four regions of the electromagnetic spectrum. (10-4)

3. What is the significance of Planck's constant? (10-3).

4. Which colors appear in the bright-line spectrum of hydrogen? (10-5)

5. In what form does a hydrogen atom lose energy from level three to level two? (10-5, 10-6)

6. How do astronomers use spectroscopy to analyze stars? (10-9)

7. What is an orbital? (10-7)

8. List all the orbitals available for the hydrogen atom when $n = 3$. (10-7)

9. What orbital will the sixth electron of nitrogen occupy? (10-10)

10. What is ionization energy? What is the value of the ionization energy of one mole of hydrogen atoms? (10-6)

11. How was Rutherford's model of the atom an improvement over Thomson's model? (10-1)

12. Describe the horizontal and vertical motion of a boat floating on ocean waves. (10-2)

13. What are the three quantum numbers that completely describe an orbital? (10-8)

14. How are electron configurations for neutral atoms different from the electron configurations of the corresponding ions? (10-11)

15. Explain how the first three quantum numbers describe an orbital. (10-8)

Interpret and Apply

1. Explain the statement: "The wavelength of a standing wave is quantized." (10-3)

2. Give four examples of waves that carry energy. (10-2, 10-3)

3. What is the significance of the equation $E = h\nu$? (10-3)

4. Describe why glowing hydrogen atoms in a spectrum tube emit energy in the blue region of the visible spectrum. (10-5 to 10-8)

5. Briefly explain what a standing wave has in common with electrons in atoms. (10-3, 10-7)

6. What is the probability of finding an electron in the node of an orbital? Compare the probability of finding an electron close to the nucleus of an atom for the $1s$ orbital versus the $3s$ orbital. (10-7)

7. What is wrong with the following descriptions of an orbital? (10-7)
   a. An orbital is a ball of uniform density.
   b. An orbital is a solar system like a planet circling the sun at a fixed distance.

8. Which of the following statements about orbitals is false? (10-7)
   a. Orbitals are distributed in space around the nucleus.
   b. Orbitals are regions in which electrons are likely to be found.
   c. Orbitals show the path of the electron.
   d. Orbitals are part of one model for atomic structure.

9. What must be done to a hydrogen atom to change its $2s$ electron to a $3s$ electron? What happens when a hydrogen atom with a $3s$ electron becomes a hydrogen atom with a $2s$ electron? (10-6, 10-7)

10. Name the elements that correspond to each of the following electron configurations. (Assume all are neutral atoms.) (10-10)
    a. $1s^2 2s^2 2p^1$
    b. $1s^2 2s^2$
    c. $1s^2 2s^2 2p^6 3s^2 3p^2$

Problems

1. If you make a standing wave of wavelength 2.5 meters with a rope, what must be done to the rope in order to decrease the wavelength to 1.25 meters? (10-3)

2. How much energy must a mole of hydrogen atoms absorb if the atoms are to increase in energy from the first level to the fifth level? (10-6)

3. What is the frequency of electromagnetic radiation emitted when the electrons of one mole of hydrogen atoms change from the $4s$ orbital to the $1s$ orbital? (10-6)

4. Write the electron configurations for the following elements: arsenic, krypton, bromine, and phosphorus. (10-10)

5. Write the electron configurations for the following ions: K^+, O^{2-}, Br^-, and Ga^{3+}. (10-11)

6. Draw the orbital diagrams, and write the electron configurations for the following elements: potassium, silicon, fluorine, and argon. (10-10)

7. Write the electron configurations for Be, Mg, Ca, and Sr. What is the similarity in the configurations of the outermost electrons of these elements? (10-10)

8. Write the electron configurations for Sc, Ti, Ni, and Zn. Which sublevel is changing in these configurations? (10-10)

Challenge

1. What is the energy given off by one hydrogen atom as it loses energy from level 3 to level 2 in joules per atom?

2. For each of the following electron configurations of neutral atoms, determine the name of the element listed and determine if the configuration as written is the ground state or an excited state:
   $1s^2 2s^2 2p^6$
   $1s^2 2s^2 2p^5 3s^2$
   $1s^2 2s^2 2p^6 3s^2 3p^6 4s^2 3d^3$
   $1s^2 2s^2 2p^6 3s^2 5s^1$

3. For each of the following electron configurations of neutral atoms, determine if the configuration as written is the ground state, an excited state, or if it is an impossible configuration:
   N $1s^2 2s^2 2p^3$
   Na $1s^2 2s^2 2p^6 4s^1$
   Ne $1s^2 2s^3 2p^5$
   V $1s^2 2s^2 2p^6 3s^3 3p^6 4s^2 3d^2 4p^1$

4. Write the electron configurations for elements 37, 38, and 39 (rubidium, strontium, and yttrium).

5. A radio station broadcasts at a frequency of 105.4 MHz. What is the wavelength of this electromagnetic wave? (MHz is the symbol for megahertz.)

6. A very bright line in the bright-line spectrum of sodium has a wavelength of 590 nm. What is the frequency of this line?

7. Calculate the frequency of the electromagnetic waves produced when the electrons in a mole of hydrogen atoms change from the fifth to the second energy level. (Planck's constant has a value of 3.98×10^{-13} kJ·s/mol.)

Synthesis

1. Write the electron configurations for each of the following ions: O^{2-} Mg^{2+} Cl^- Sc^{2+} Cd^{2+} As^{3-}. Which elements do the electron configurations of these ions resemble?

2. The outermost s and p electrons in atoms determine the way many elements react. Using electron configurations explain why aluminum forms a stable 3+ ion. Predict the charge on a stable positive ion of gallium.

Projects

1. One of the principles of the quantum theory is that molecules absorb only definite and distinct amounts of energy. The principle helps to explain how the greenhouse effect can take place in Earth's atmosphere. Research and report on the greenhouse effect and include the quantum theory in the report.

2. The year 1905 has been described as an important time for Albert Einstein. Investigate the work of Einstein, and report on why 1905 was such a significant year.

3. Research and report on how a microwave oven cooks some foods so quickly.

4. Research and report on the information gained about the sun through analysis of its absorption spectrum.

The clear, harmonious sound of a symphony orchestra is produced using about twenty different instruments. The sounds you hear depend on the skill of the musicians and conductor, and on the grouping of the various instruments on the stage. A random seating arrangement might produce sounds more aptly labeled as noise rather than music.

In this chapter you will see how classifying the elements was once related to music, and how order in nature forms the basis of the periodic table.

The Periodic Table

The Need to Organize the Elements

By the 19th century, chemists had made many careful observations of the elements and their compounds. They then saw the need to organize their observations by regularities in the properties of the elements. Deciding how to classify the elements was difficult. Metallic elements had some properties in common, but others had properties that varied greatly. The idea of classifying the elements in a meaningful way became a perplexing problem.

One effort to develop an element classification system was based on sequencing the known elements by mass. Another early attempt at organizing the elements by properties included the arrangement of elements into chemical groups depending on how they reacted with oxygen. However, these early classification schemes also posed problems in that they did not provide an adequate model for predicting. A useful model was to be formulated in the late 1860's.

CONCEPTS

- the need for a systematic organization
- classifying by properties
- the structure and form of the table
- predicting the properties of new elements

Review with the students the information about the periodic table from Chapter 2.

11-1 Organizing by Properties

Dmitri Mendeleev (1834–1907), a Russian scientist, published his classification scheme for the elements in 1869. His first periodic table is seen in Figure 11-1 and listed the elements in order of increasing molar mass. Mendeleev categorized the elements by matching similar properties. You can think of what he was trying to do as like putting together a large puzzle. Some elements were placed with others because they had similar properties and seemed to fit together. But in Mendeleev's puzzle there seemed to be pieces missing.

In Figure 11-1, you notice some spaces (after Ca, Zn, Hg, etc.). Mendeleev explained his reasoning for including these gaps by presuming that some elements were yet to be discovered. He then used this table

Figure 11-1 Mendeleev's periodic table did not contain rows with an equal number of elements. The blanks indicated the place where he thought elements would be discovered at a later date.

1								H						Li			
2								Be	B	C	N	O	F	Na			
3								Mg	Al	Si	P	S	Cl	K	Ca	—	Er?	Y?	In?
4	Ti	V	Cr	Mn	Fe	Ni,Co	Cu	Zn	—	—	As	Se	Br	Rb	Sr	Ce	La	Di	Tb
5	Zr	Nb	Mo	Rh	Ru	Pd	Ag	Cd	U	Sn	Sb	Te	I	Cs	Ba		
6	—	Ta	W	Pt	Ir	Os	Hg	—	Au	—	Bi	—	—	Tl	Pb		

Do You Know?

John Newlands applied his knowledge of music to his observations of the behavior of the elements. European music is based on a repeating scale of octaves. Every eighth note in the scale repeats: A, B, C, D, E, F, G, A, B, etc. Newlands observed a similar repetition in the properties of the known elements. He was able to organize the elements into a table that was seven elements wide. The eighth element in the table had properties similar to the first element.

The scale of notes from *do* to the next *do* is an octave.

CHEM THEME

The periodic table is a model that plausibly accounts for past observations and accurately predicts future ones.

There is a wealth of information in Table 11-1. It may seem overwhelming to the students. Therefore, help them to see some of the relationships given in the table, especially formulas and ratios.

to predict the properties of the missing elements. He was later found to be surprisingly accurate in his predictions. Imagine, at the time Mendeleev was making these predictions, electrons, protons, and neutrons had not yet been discovered. Mendeleev's predictions of the properties of new elements were based on the organizational scheme of his table. He presumed that the missing elements had chemical and physical properties similar to other elements in each vertical column.

Mendeleev and others working on organizational schemes for the elements recognized that the repetition in properties of the elements was a fundamental pattern in nature. This fundamental pattern is known as the periodicity of the elements and forms the basis of the periodic law. The modern **periodic law** states: The properties of the elements recur periodically when the elements are arranged in increasing order by their atomic numbers.

The periodic table used today is similar to Mendeleev's table. Since new elements have been discovered or synthesized since Mendeleev's time, the table has been altered to accommodate no less than 49 new elements.

11-2 The Periodic Table Today

Table 11-1 represents the first 21 elements and some of their physical and chemical properties. You should use it to note regularities among these elements for the purpose of classifying them as Mendeleev did by their properties.

In looking at Table 11-1 note that helium, neon, and argon are the only elements shown that do not form hydrides and fluorides. You could put these elements together in a group. These three elements do not react under usual conditions and had not been discovered when Mendeleev first proposed his table. (In 1870, helium was discovered in the sun using spectroscopy.)

Next, there seems to be a regularity in the combining ratios of the elements in forming hydrides and fluorides (columns 6 and 8). The ratios are in groups with the following sequence 1, 2, 3, 4, 3, 2, 1. Note that sodium, potassium, fluorine, and chlorine react with hydrogen and fluorine in the same way lithium reacts. Moreover, magnesium, calcium, oxygen, and sulfur react the same way with hydrogen and fluorine as beryllium.

Now look at the ionization energies of lithium, sodium, potassium, fluorine, and chlorine. These elements could be further classified into groups by their ionization energies. Ionization energy will be described further at the end of this chapter.

Table 11-2 shows the same 21 elements in Table 11-1. These elements have been arranged by the regularities in forming fluorides and the similarities in ionization energies.

The complete modern version of the periodic table is shown in Table 11-3. Look at the arrangement of the elements in vertical columns. Elements are grouped in vertical columns by similar chemical properties. These vertical **groups** are sometimes called chemical families and may have special names. Groups 1, 2, 13, 14, 15, 16, 17, and 18 are known as the representative elements.

Table 11-1

SOME INFORMATION ABOUT THE FIRST 21 ELEMENTS

SYMBOL	MOLAR MASS (grams)	ATOMIC NUMBER	PHYSICAL STATE AT 25°C 1 atm	FORMULA FOR HYDRIDE	RATIO TO H	FORMULA FOR FLUORIDE	RATIO TO F	IONIZATION ENERGY (kJ/mol)
H	1.01	1	gas	H_2	1	HF	1	1312
He	4.00	2	gas	—		—		2372
Li	6.94	3	solid	LiH	1	LiF	1	519
Be	9.01	4	solid	BeH_2	2	BeF_2	2	900
B	10.8	5	solid	B_2H_6	3	BF_3	3	799
C	12.0	6	solid	CH_4	4	CF_4	4	1088
N	14.0	7	gas	NH_3	3	NF_3	3	1406
O	16.0	8	gas	H_2O	2	OF_2	2	1314
F	19.0	9	gas	HF	1	F_2	1	1682
Ne	20.2	10	gas	—		—		2080
Na	23.0	11	solid	NaH	1	NaF	1	498
Mg	24.3	12	solid	MgH_2	2	MgF_2	2	736
Al	27.0	13	solid	Al_2H_6	3	AlF_3	3	577
Si	28.1	14	solid	SiH_4	4	SiF_4	4	787
P	31.0	15	solid	PH_3	3	PF_3	3	1063
S	32.1	16	solid	H_2S	2	SF_2	2	1000
Cl	35.5	17	gas	HCl	1	ClF	1	1255
Ar	39.9	18	gas	—		—		1519
K	39.1	19	solid	KH	1	KF	1	418
Ca	40.1	20	solid	CaH_2	2	CaF_2	2	590
Sc	45.0	21	solid	not known		ScF_3	3	223

Group 1 is the **alkali metals** (hydrogen is usually not included in the alkali metal family). These metals react with water to form what is known as an alkali or basic solution. Sodium, for example, reacts with water in a spectacular way. This reaction can be represented by the following equation.

$$2Na(s) + 2H_2O(l) \longrightarrow 2Na^+(aq) + 2OH^-(aq) + H_2(g)$$

Table 11-2

PERIODIC GROUPS FOR THE ELEMENTS HELIUM TO SCANDIUM

							He 2
Li 3 LiF	Be 4 BeF_2	B 5 BF_3	C 6 CF_4	N 7 NF_3	O 8 OF_2	F 9 F_2	Ne 10 —
Na 11 NaF	Mg 12 MgF_2	Al 13 AlF_3	Si 14 SiF_4	P 15 PF_3	S 16 SF_2	Cl 17 ClF	Ar 18 —
K 19 KF	Ca 20 CaF_2	Sc 21 ScF_3					

The periodic table lists the groups from 1 to 18. This system avoids the confusion of naming some groups A and others B.

TABLE 11-3

PERIODIC TABLE OF THE ELEMENTS

(based on $^{12}_{6}C = 12.0000$)

1*								
1 **H** Hydrogen 1.007								

2
3 **Li** Lithium 6.941

4
4 **Be** Beryllium 9.012

☐ Solid	☐ Liquid	☐ Gas

14 _____ Atomic number **Si** _____ Symbol Silicon _____ Name 28.0855 _____ Atomic mass

11 **Na** Sodium 22.98977	**12** **Mg** Magnesium 24.305

TRANSITION METALS _____

3	4	5	6	7	8	9
19 **K** Potassium 39.098	**20** **Ca** Calcium 40.08	**21** **Sc** Scandium 44.955	**22** **Ti** Titanium 47.88	**23** **V** Vanadium 50.9415	**24** **Cr** Chromium 51.996	**25** **Mn** Manganese 54.938	**26** **Fe** Iron 55.847	**27** **Co** Cobalt 58.933

| **37**
Rb
Rubidium
85.467 | **38**
Sr
Strontium
87.62 | **39**
Y
Yttrium
88.905 | **40**
Zr
Zirconium
91.224 | **41**
Nb
Niobium
92.906 | **42**
Mo
Molybdenum
95.94 | **43**
Tc
Technetium
(98) | **44**
Ru
Ruthenium
101.07 | **45**
Rh
Rhodium
102.906 |

| **55**
Cs
Cesium
132.905 | **56**
Ba
Barium
137.3 | **57**
La
Lanthanum
138.906 | **72**
Hf
Hafnium
178.49 | **73**
Ta
Tantalum
180.948 | **74**
W
Tungsten
183.85 | **75**
Re
Rhenium
186.207 | **76**
Os
Osmium
190.2 | **77**
Ir
Iridium
192.22 |

| **87**
Fr
Francium
(223)† | **88**
Ra
Radium
(226.0) | **89**
Ac
Actinium
227.028 | **104**
(261) | **105**
(262) | **106**
(263) | **107**
(262) | **108** | **109**
(266) |

Lanthanide series

58 **Ce** Cerium 140.12	59 **Pr** Praseodymium 140.908	60 **Nd** Neodymium 144.24	61 **Pm** Promethium (145)	62 **Sm** Samarium 150.36
90 **Th** Thorium 232.038	91 **Pa** Protactinium 231.036	92 **U** Uranium 238.029	93 **Np** Neptunium (244)	94 **Pu** Plutonium (244)

Actinide series

INNER TRANSITION METALS

*The numbers heading each column represent group numbers recommended by the American Chemical Society Committee on nomenclature.
†Masses in parentheses are the mass numbers of the most stable isotope.

				13	14	15	16	17	18
									2 He Helium 4.0026
				5 B Boron 10.81	6 C Carbon 12.0111	7 N Nitrogen 14.0067	8 O Oxygen 15.9994	9 F Fluorine 18.998	10 Ne Neon 20.179
10	11	12		13 Al Aluminum 26.9815	14 Si Silicon 28.0855	15 P Phosphorus 30.973	16 S Sulfur 32.06	17 Cl Chlorine 35.453	18 Ar Argon 39.948
28 Ni Nickel 58.69	29 Cu Copper 63.546	30 Zn Zinc 65.39	31 Ga Gallium 69.72	32 Ge Germanium 72.59	33 As Arsenic 74.92	34 Se Selenium 78.96	35 Br Bromine 79.904	36 Kr Krypton 83.80
46 Pd Palladium 106.42	47 Ag Silver 107.868	48 Cd Cadmium 112.41	49 In Indium 114.82	50 Sn Tin 118.71	51 Sb Antimony 121.75	52 Te Tellurium 127.60	53 I Iodine 126.905	54 Xe Xenon 131.29
78 Pt Platinum 195.08	79 Au Gold 196.967	80 Hg Mercury 200.59	81 Tl Thallium 204.383	82 Pb Lead 207.2	83 Bi Bismuth 208.980	84 Po Polonium (209)	85 At Astatine (210)	86 Rn Radon (222)

63 Eu Europium 151.96	64 Gd Gadolinium 157.25	65 Tb Terbium 158.925	66 Dy Dysprosium 162.50	67 Ho Holmium 164.930	68 Er Erbium 167.26	69 Tm Thulium 168.934	70 Yb Ytterbium 173.04	71 Lu Lutetium 174.96
95 Am Americium (243)	96 Cm Curium (247)	97 Bk Berkelium (247)	98 Cf Californium (251)	99 Es Einsteinium (252)	100 Fm Fermium (257)	101 Md Mendelevium (258)	102 No Nobelium (259)	103 Lr Lawrencium (260)

Figure 11-2 Sodium and calcium have properties that are characteristic of metals. Sulfur and bromine are both classified as nonmetals. As you can see they look very different from the metals. To which group does each of these elements belong?

Do You Know?

Elements in you:
 Very common: H, O, C, N
 Scarce: Na, Mg, P, S, Cl, K, Ca
 Very Scarce: F, Si, V, Cr, Mn, Fe, Co, Cu, Zn, Se, Mo, Sn, I
The very scarce elements make up only 0.6% of your body mass, but they are very essential to your health.

If the periodic table is used as a model for predicting, you might expect that if sodium reacts with water to produce a base and water, it is likely that the other metals in the group will react similarly.

Group 2 is composed of the **alkaline earth metals**. Group 17 includes the halogens. The elements in the last group in the periodic table, group 18, are called the **noble gases**, because they are generally nonreactive under most conditions.

Groups 3 to 12 are called the **transition metals**. Chemists once believed that these elements behaved in a manner that is intermediate between the active metals and nonmetals. Transition metals are those elements in the middle of the periodic table. Table 11-4 shows the periodic table labeled with some family names.

Horizontal rows of elements are called **periods**. Period 1 contains two elements, hydrogen and helium. Periods 2 and 3 contain eight elements each. Periods 4 and 5 contain 18 elements each. How many elements are in periods 6 and 7?

Periods 4 and 5 each contain ten transition metals. Period 6 contains 24 transition metals. The elements with atomic numbers 57 to 71 are transition metals that were once known as the rare earth metals. This name is a misnomer as not all of these elements are rare. Today, rare earth metals are replaced by the terms *inner transition metals* or *lanthanide series*. The lanthanide series has similar properties. For convenience in displaying the table on a page, the lanthanides are placed in a separate row below the major part of the table. Chemically, they all fit in the space for element 57. You can think of the lanthanide series as playing cards stacked one on top of the other. This stack of elements is placed in position 57 in the periodic table. The periodic table would then be three-dimensional.

Period 7 also contain 24 transition metals. The elements with atomic numbers 89 to 103 are also referred to as part of the inner transition metals or *actinide series*.

Do You Know?

Yttrium and Europium are used to enhance the red in color TV.

Table 11-4

ELECTRON CONFIGURATIONS AND THE PERIODIC TABLE

1	2											13	14	15	16	17	18
3 Li	4 Be											5 B	6 C	7 N	8 O	9 F	10 Ne
11 Na	12 Mg	3	4	5	6	7	8	9	10	11	12	13 Al	14 Si	15 P	16 S	17 Cl	18 Ar
19 K	20 Ca	21 Sc	22 Ti	23 V	24 Cr	25 Mn	26 Fe	27 Co	28 Ni	29 Cu	30 Zn	31 Ga	32 Ge	33 As	34 Se	35 Br	36 Kr
37 Rb	38 Sr	39 Y	40 Zr	41 Nb	42 Mo	43 Tc	44 Ru	45 Rh	46 Pd	47 Ag	48 Cd	49 In	50 Sn	51 Sb	52 Te	53 I	54 Xe
55 Cs	56 Ba	57 La	72 Hf	73 Ta	74 W	75 Re	76 Os	77 Ir	78 Pt	79 Au	80 Hg	81 Ti	82 Pb	83 Bi	84 Po	85 At	86 Rn
87 Fr	88 Ra	89 Ac	104	105	106	107	108	109									

58 Ce	59 Pr	60 Nd	61 Pm	62 Sm	63 Eu	64 Gd	65 Tb	66 Dy	67 Ho	68 Er	69 Tm	70 Yb	71 Lu
90 Th	91 Pa	92 U	93 Np	94 Pu	95 Am	96 Cm	97 Bk	98 Cf	99 Es	100 Fm	101 Md	102 No	103 Lr

Alkali Metals

Alkaline Earth Metals

Transition Elements

Lanthanide Elements

Actinide Elements

Halogens

Noble Gases

Review and Practice

1. State the periodic law in your own words.

2. Name two groups containing metallic elements and two groups containing nonmetallic elements.

3. You have seen the equation for the reaction of sodium with water. Write a reaction that seems reasonable for cesium and water.

4. Using the periodic law and the information in Table 11-1, write reasonable formulas for the following substances:
a. strontium hydride
b. hydrogen selenide
c. hydrogen iodide
d. rubidium fluoride
e. gallium fluoride
f. tellurium fluoride

1. When elements are arranged according to their increasing atomic numbers, chemical and physical properties recur periodically.
2. metals: alkali and alkaline earths
nonmetals: halogens and noble gases
3. $Cs(s) + 2H_2O \longrightarrow Cs^+(aq)$
$+ 2OH^-(aq) + H_2(g)$
4. a. SrH_2 d. RbF
b. SeH_2 e. GaF_3
c. HI   f. TeF_2

Patterns in Electron Structure

CONCEPTS

■ **energy levels of electrons**
■ **the relationship between electron arrangements and chemical behavior**
■ **periodicity of outermost electrons**
■ **electronic forces acting within the atom**

Mendeleev's organization of the elements was based entirely on their macroscopic properties. The modern periodic table is based on the microscopic structure of atoms. The wave mechanical model of the atom became popular among scientists because it provided a reasonable explanation for the periodic nature of properties. The patterns in electron configurations and the arrangement of elements in the periodic table seem to go hand in hand.

Electrons and protons in the atom create forces of attraction and repulsion. These forces influence the internal structure and reactivity of the atom. Therefore, studying these interactions will help you understand the trends in the periodic table and the nature of bonding among atoms.

11-3 The Periodic Table and Electron Configurations

In Chapter 10, you studied the electron configurations of the elements. Electron configurations describe the orbital arrangement of electrons in an atom and help to explain why we notice similarities in chemical properties for some elements.

Table 11-5 will give you an opportunity to practice writing electron configurations. On a separate sheet of paper copy the table, then write in the missing electron configurations for the elements listed. You will probably need the periodic table to find the atomic number of each element. Presume that each element is neutral so that the atomic number or number of protons in the nucleus, equals the number of electrons in the orbitals. Some of the electron configurations are already given so that you can be sure you are on the right track.

Table 11-5

ELECTRON CONFIGURATIONS FOR SOME ELEMENTS			
ALKALI METALS	**ALKALINE EARTHS**	**HALOGENS**	**NOBLE GASES**
Li	Be $1s^2 2s^2$	F	Ne
Na $1s^2 2s^2 2p^6 3s^1$	Mg	Cl	Ar
K	Ca	Br $1s^2 2s^2 2p^6 3s^2$ $3p^6 4s^2 3d^{10} 4p^5$	Kr

Stress the importance of the outermost electrons. Make sure all the students know which electrons in an electron configuration are in the outermost s and p orbital system.

Do you notice a similarity in the electron configurations for the noble gases in the last column? All of them have eight electrons in the outermost s and p orbitals. The *outermost s and p orbitals* are the orbitals with the highest principal quantum number. For example, the outermost s and p orbitals for argon are $3s$ and $3p$.

The outermost orbitals of krypton are occupied by eight electrons, $4s^2$ and $4p^6$. The electrons in the outermost orbitals are most often involved in chemical bonding. For this reason, electrons in the outermost s and p orbitals are referred to as **valence electrons**. Most noble gases have outermost s and p orbitals that are completely full. (The s and p orbitals can be occupied by a maximum of eight electrons.) Helium is an exception as it has only two electrons which occupy the $1s$ orbital.

Now look at the configuration you have written for the alkali metals. Each alkali metal has one valence electron. Write the electron configuration of rubidium, and see if it follows the pattern. Does it? Yes, the electron with the highest energy occupies the $5s$ orbital.

For the alkaline earth metals the two electrons with the highest energy occupy the outermost s orbital. Write the electron configuration of strontium. What is the last orbital to be filled?

The electron configurations of the halogens, group 17, also show a pattern. Each atom has seven valence electrons.

A pattern exists for each group of representative elements. Using this information and other observations, chemists could now explain why elements of a particular group in the periodic table react similarly. Each group is characterized by a similar outermost energy level configuration.

One final note about electron configurations and the periodic table. Count the number of transition elements in period 4. There are ten from scandium to zinc. What is the maximum number of electrons that can occupy the $3d$ orbital system? The answer is ten. It is no coincidence that there are ten transition metals and ten electrons in a completely filled $3d$ orbital system. Scandium has one $3d$ electron while zinc has ten $3d$ electrons, the maximum number. In the fourth period transition elements, the $3d$ orbital system is being filled one electron at a time from scandium to zinc. Which orbital system is being filled in the fifth period transition elements?

Table 11-6 is a periodic table that shows the relationship between the organization of the table and electron configurations. The outermost s and p electrons are listed for the representative elements. The outermost d or f orbital system is listed for the transition elements and the inner transition elements.

The $4d$ orbital system is being filled in the fifth period transition elements.

CHEM THEME

The chemical and physical properties of elements depend on the number and arrangement of electrons in the atom.

11-4 Electron Configurations and Chemical Behavior

The noble gases generally are nonreactive, and they have completely full outermost orbital systems. A generalization can be made.

When an atom's outermost orbital system is full, the atom tends to be nonreactive.

Helium is one exception to this generalization. Helium is very nonreactive but it has only two electrons occupying the $1s^2$ orbital.

Table 11-6

ELECTRON CONFIGURATIONS AND THE PERIODIC TABLE

Sublevels being filled

s-block

	1	2														**p-block**	13	14	15	16	17	18

1s — 1 H, 2 He

2s — 1: 3 Li, 2: 4 Be

d-block: 3 4 5 6 7 8 9 10 11 12

3s — 11 Na, 12 Mg

2p — 5 B, 6 C, 7 N, 8 O, 9 F, 10 Ne

3p — 13 Al, 14 Si, 15 P, 16 S, 17 Cl, 18 Ar

4s — 19 K, 20 Ca

3d — 21 Sc, 22 Ti, 23 V, 24 Cr, 25 Mn, 26 Fe, 27 Co, 28 Ni, 29 Cu, 30 Zn

4p — 31 Ga, 32 Ge, 33 As, 34 Se, 35 Br, 36 Kr

5s — 37 Rb, 38 Sr

4d — 39 Y, 40 Zr, 41 Nb, 42 Mo, 43 Tc, 44 Ru, 45 Rh, 46 Pd, 47 Ag, 48 Cd

5p — 49 In, 50 Sn, 51 Sb, 52 Te, 53 I, 54 Xe

6s — 55 Cs, 56 Ba

5d — 57 La, 72 Hf, 73 Ta, 74 W, 75 Re, 76 Os, 77 Ir, 78 Pt, 79 Au, 80 Hg

6p — 81 Tl, 82 Pb, 83 Bi, 84 Po, 85 At, 86 Rn

7s — 87 Fr, 88 Ra

6d — 89 Ac, 104, 105, 106, 107, 108, 109

7p —

f-block

4f — Lanthanide series — 58 Ce, 59 Pr, 60 Nd, 61 Pm, 62 Sm, 63 Eu, 64 Gd, 65 Tb, 66 Dy, 67 Ho, 68 Er, 69 Tm, 70 Yb, 71 Lu

5f — Actinide series — 90 Th, 91 Pa, 92 U, 93 Np, 94 Pu, 95 Am, 96 Cm, 97 Bk, 98 Cf, 99 Es, 100 Fm, 101 Md, 102 No, 103 Lr

---

The alkali metals all contain one electron in the outermost s orbital. The alkali metals are very reactive. An alkali metal atom easily loses one electron. When a neutral atom loses an electron, the number of negative particles outside the nucleus is one less than the number of positive particles in the nucleus. Therefore, the overall charge on the ion is 1+. Sodium is a member of the alkali metal group. Its atomic number is 11, so its atoms have 11 electrons and 11 protons. Its electron configuration is:

$$\text{Na} \quad 1s^2 2s^2 2p^6 3s^1$$

The electron configuration of the sodium ion is:

$$\text{Na}^+ \quad 1s^2 2s^2 2p^6$$

The ion contains 11 protons and 10 electrons. If the Na^+ ion has 10 electrons, then it has the same electron configuration as neon, a noble gas. Na^+ and Ne are **isoelectronic** because they have the same electron configuration. Potassium forms a 1+ ion, with which noble gas is the K$^+$ ion isoelectronic? Observations show that the alkali metals form 1+ ions in chemical reactions. As neutral atoms, they have one electron in the outermost *s* orbital.

The K$^+$ ion is isoelectronic with Ar.

The alkaline earth elements (group 2) have two electrons in their outermost *s* orbital. Magnesium is an alkali earth metal. Its atomic number is 12 and its electron configuration is:

$$\text{Mg} \quad 1s^2 2s^2 2p^6 3s^2$$

The alkaline earth metals tend to form 2+ ions in chemical reactions. The electron configuration for the magnesium ion is:

$$\text{Mg}^{2+} \quad 1s^2 2s^2 2p^6$$

Note that this ion also is isoelectronic with neon. When an atom of the alkali earth group loses two electrons to become a 2+ ion, it has an electron configuration like a noble gas.

The halogens (group 17) contain seven electrons in the outermost *s* and *p* orbitals. Fluorine atoms have nine electrons and nine protons. The electron configuration for a fluorine atom is:

$$\text{F} \quad 1s^2 2s^2 2p^5$$

Fluorine is one electron short of a complete set of eight electrons in the second energy level. The fluoride ion has a charge of 1−. An ion with a charge of 1− has one extra electron. The fluoride ion has 10 electrons and nine protons. The fluoride ion configuration is:

$$\text{F}^- \quad 1s^2 2s^2 2p^6$$

Note that the fluoride ion also is isoelectronic with neon. Observations show that a halogen atom can gain an electron resulting in an ion with 1− charge. Any halogen element that gains an electron becomes a 1− ion and is isoelectronic with the nearest noble gas.

Figure 11-3 An atom gains or loses electrons in the presence of atoms from other substances. In this reaction of magnesium metal with hydrochloric acid, the magnesium metal loses electrons to form Mg^{2+} ions. These ions are in solution and cannot be seen.

Perhaps you have noticed in writing chemical formulas that most metals form positive ions. *Metals* may *lose* electrons in chemical reactions and become isoelectronic with the nearest noble gas. Nonmetals generally form negative ions. *Nonmetals* may *gain* electrons in some reactions and become isoelectronic with the nearest noble gas. (Nonmetal elements also can share electrons in chemical reactions.)

Example 11-1

Write the electron configuration for each of the elements Li, Ca, S, and Cl. Use this configuration to predict the most likely charge of an ion for each.

▶ Suggested solution _____

The atomic number of lithium is 3. The electron configuration for lithium is $1s^2 2s^1$. If the lithium atom loses the $2s^1$ electron, it becomes isoelectronic with helium. Li^+ has an electron configuration of $1s^2$.

The atomic number of calcium is 20. Calcium has an electron configuration of $1s^2 2s^2 2p^6 3s^2 3p^6 4s^2$. If calcium loses two electrons, it is isoelectronic with argon. The calcium ion would have a 2+ charge.

Sulfur is element number 16. It has an electron configuration which is $1s^2 2s^2 2p^6 3s^2 3p^4$. Sulfur is two electrons short of being isoelectronic with the nearest noble gas which is argon. If sulfur were to gain two electrons (it is a nonmetal so it may gain electrons), the S^{2-} ion has 18 electrons which would make it isoelectronic with argon. The electron configuration of a sulfide ion is $1s^2 2s^2 2p^6 3s^2 3p^6$.

Chlorine is element number 17. Its 17 electrons have a configuration of $1s^2 2s^2 2p^6 3s^2 3p^5$. If chlorine were to gain one electron, it would be isoelectronic with argon. The charge of the chloride ion is 1−. The electron configuration for the chloride ion is the same as the sulfide ion and argon.

Review and Practice _____

1. How are the electron configurations of the noble gases similar?

2. What does isoelectronic mean?

3. What is the electron configuration for the oxide ion? With which noble gas is this ion isoelectronic?

4. Name the chemical group or family of which each of the following elements is a member. Then determine the number of electrons in the outermost *s* and *p* orbitals for each group: Rb, Te, Ca, and Xe.

5. Without writing out the electron configuration for nickel, use its position in the periodic table to determine how many electrons it has in its 3*d* orbital system.

1. All have full outermost s and p orbital systems (except for He).
2. Two atoms or ions are isoelectronic if they have the same electron configurations.
3. $1s^2 2s^2 2p^6$ Ne
4. Rb is an alkali metal.   1
Te is in the oxygen group.   6
Ca is in the alkaline earth group.   2
Xe is a noble gas.   8
5. eight

Periodic Trends

So far you have seen that the periodic table can be of tremendous help in organizing information about the elements. Referring to an element's position in the periodic table allows you to assume that elements in the same family will react similarly. The table is organized in a manner that enables you to assume certain patterns in the electron configurations of the elements. From these patterns, you now have an idea why elements form certain ions.

In the final lesson of this chapter you will study some other properties of elements that show specific patterns related to the organization of the periodic table. Such patterns are often referred to as periodic *trends*. An awareness of certain periodic trends is necessary for an understanding of the next chapter which deals with chemical bonding.

CONCEPTS

■ **sizes of atoms versus their ions**
■ **ionization energy**

11-5 Atomic and Ionic Radii

How big is an atom? This question may seem rather simple. However, it is not an easy question to answer. If someone were to ask you to measure the circumference of a volleyball, you might wrap a string around the ball and then measure the distance from the end of the string to the point on the string where the ball was wrapped once. The atom is not as easily measured because it is incredibly small when compared to the size of a volleyball. Also, the atom does not have a definite circumference like a volleyball. Electrons are described by orbitals. Remember that orbitals are not physical objects with a definite size and shape. The pictures you have seen of orbitals are models to show a high probability volume for an electron. The probability density of an orbital fades to zero at an infinite distance from the nucleus.

However in spite of the problems, the size of an atom can be estimated in several ways. Atomic size is generally described by the radius of an atom rather than a circumference. One method for determining the **atomic radius** involves determining the distance between the nuclei of metal atoms in a crystal. Such measurements can be made using a technique called X-ray diffraction. For elements that exist in pure form as molecules, such as chlorine, measurements can be made

Show the students a styrofoam® ball. Call it an atom. Measure its circumference. Then ask students why this method cannot be applied to measuring the size of an atom.

Do You Know?

The size of an orbital is not constant from atom to atom. The size is dependent on the attractive force of the nucleus and the electrons, and the repulsion among electrons that occupy orbitals.

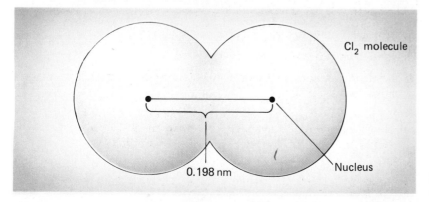

Figure 11-4 Covalent radius of the chlorine molecule.

Table 11-10

COVALENT AND IONIC RADII OF SOME ELEMENTS (in nm)

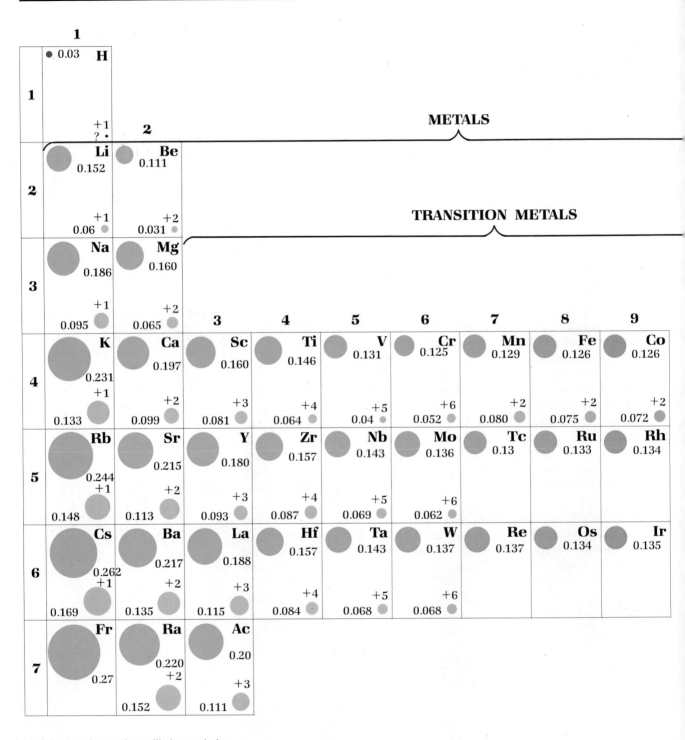

	1	**2**	**3**	**4**	**5**	**6**	**7**	**8**	**9**
1	• 0.03 **H**					METALS			
2	+1 ? • **Li** 0.152 +1 0.06 •	**Be** 0.111 +2 0.031 •			TRANSITION METALS				
3	**Na** 0.186 +1 0.095	**Mg** 0.160 +2 0.065							
4	**K** 0.231 +1 0.133	**Ca** 0.197 +2 0.099	**Sc** 0.160 +3 0.081	**Ti** 0.146 +4 0.064	**V** 0.131 +5 0.04 •	**Cr** 0.125 +6 0.052	**Mn** 0.129 +2 0.080	**Fe** 0.126 +2 0.075	**Co** 0.126 +2 0.072
5	**Rb** 0.244 +1 0.148	**Sr** 0.215 +2 0.113	**Y** 0.180 +3 0.093	**Zr** 0.157 +4 0.087	**Nb** 0.143 +5 0.069	**Mo** 0.136 +6 0.062	**Tc** 0.13	**Ru** 0.133	**Rh** 0.134
6	**Cs** 0.262 +1 0.169	**Ba** 0.217 +2 0.135	**La** 0.188 +3 0.115	**Hf** 0.157 +4 0.084	**Ta** 0.143 +5 0.068	**W** 0.137 +6 0.068	**Re** 0.137	**Os** 0.134	**Ir** 0.135
7	**Fr** 0.27	**Ra** 0.220 +2 0.152	**Ac** 0.20 +3 0.111						

Call attention to the unlikely regularity that as the number of electrons increases across a period, the size of the atoms decreases.

					17	18

	0.03 **H**	
	−1	
	0.208	

		13	14	15	16	17	18

0.088 **B**	0.077 **C**	0.070 **N**	0.066 **O**	0.064 **F**	**Ne**
+3	+4	−3	−2	−1	
0.020	0.015	0.171	0.140	0.136	

Al	0.117 **Si**	0.110 **P**	0.104 **S**	0.099 **Cl**	**Ar**
0.143					
+3	+4	−3	−2	−1	
0.050	0.041	0.212	0.184	0.181	

10	11	12

Ni	**Cu**	**Zn**	**Ga**	**Ge**	**As**	**Se**	**Br**	**Kr**
0.124	0.128	0.133	0.122	0.122	0.121	0.117	0.114	
+2	+1	+2	+3	+4	−3	−2	−1	
0.070	0.096	0.074	0.062	0.053	0.222	0.198	0.195	

Pd	**Ag**	**Cd**	**In**	**Sn**	**Sb**	**Te**	0.133 **I**	**Xe**
0.138	0.144	0.149	0.162	0.14	0.141	0.137		
+2	+1	+2	+3	+4	+5	−2	−1	
0.050	0.126	0.097	0.081	0.071	0.062	0.221	0.216	

Pt	**Au**	**Hg**	**Tl**	**Pb**	**Bi**	**Po**	0.140 **At**	**Rn**
0.138	0.144	0.155	0.171	0.175	0.146	0.14		
+2	+1	+2	+3	+4	+5			
0.052	0.137	0.110	0.095	0.084	0.074			

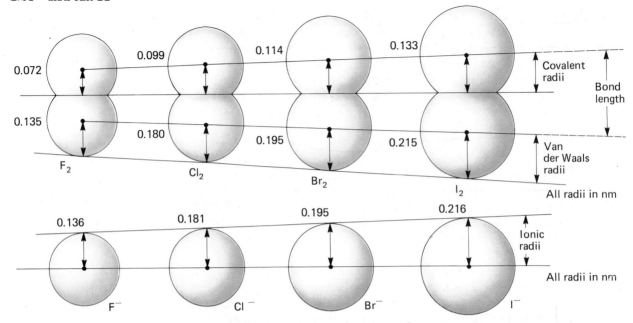

Figure 11-5 Differences in the covalent radii of the halogens are shown in the top drawing. Compare these values to the ionic radii values for the same elements in the bottom drawing.

of the distances between nuclei for two atoms bonded together. This measurement is referred to as the covalent radius and is shown in Figure 11-4.

The sizes of ions also can be estimated using X-ray diffraction. The **ionic radius** is a measure of the size of the electron probability volume for an ion. Table 11-7 shows the relative atomic and ionic radii for some elements. As you scan the table two trends should be apparent.

1. Atomic radii decrease from left to right across a period in the periodic table (until group 18).
2. Atomic and ionic radii increase from top to bottom in a group or family.

Let's now look at some explanations to account for these two trends in size. First, you know that the higher the atomic number, the greater the number of protons in the nucleus. The charge on the nucleus increases with increasing atomic number. The force of attraction between the nucleus and any electron around the nucleus also increases. With an increased force of attraction, the electrons will be closer to a nucleus of greater charge. Thus, within a period, as the nuclear charge increases, the atom gets smaller. Note from Table 11-7 that an oxygen atom is smaller than a boron atom. The electrons in an oxygen atom are drawn in closer to the nucleus by the larger nuclear charge of oxygen (8+ versus 5+ for boron).

The size of the atoms will increase as you move down a group for two reasons. First, the number of electrons increases. These electrons will occupy orbitals further and further from the nucleus. Second, repulsive forces among the inner electrons will shield the outer electrons from the attractive force of the nucleus. This weakening of the force between the nucleus and the valence electrons is called the **shielding effect** of the inner electrons. Shielding effect provides a reasonable explanation for why some atoms are larger than predicted.

How do you account for changes that occur between the atomic radius and ionic radius? As you learned in the previous section, metal

Once the students learn the term "shielding effect," they tend to overuse it. Make sure that the students understand where the shielding effect comes from; that is, from the force of repulsion of inner and outer electrons.

atoms lose electrons and nonmetals gain electrons to reach a more favorable energy state. Let's look at some of the changes that occur when sodium forms an ion. If the valence electron ($3s^1$) were removed, the 1+ ion is smaller than the atom. This is because only the first and second energy-level electrons would remain. Also, the total nuclear charge is greater than the negative charge on the electron cloud. In general, the positive ion formed when an atom loses its valence electrons is smaller than the neutral atom of the element.

If a neutral atom gains an electron, the ion is larger than the neutral atom. For example, if fluorine captures an electron, from an atom such as sodium the fluoride ion is larger than the neutral fluorine atom. This is because the increased negative charge results in greater mutual repulsion among the electrons. The overall greater repulsion will cause the ion to increase in size. Note also that the excess charge on the electrons cloud reduces the effect of the attractive force of the nucleus.

Why do chemists need to know about the size of atoms and ions in the first place. Technology has advanced to a stage that allows scientists to design and build molecules for specific purposes. For example, biochemists are interested in the size of atoms composing proteins and enzymes. In order for an enzyme or drug to function properly, sections of its surface must match parts of the surface of the substance with which it is reacting. If atoms on the reacting surface are too large or too small, they will interfere with the functioning of the enzyme or drug.

Review and Practice

1. What happens to the size of atoms as you move from left to right across the periodic table?

2. Explain why the size of atoms increases with an increase in molar mass for any group or family of elements.

3. Why does the size of an atom increase when another electron is added to the outer energy level of a neutral nonmetal atom?

4. Predict the charge on the most common ion for each of the following elements. Then predict which ions would be smaller than their neutral atoms and which would be larger.
   Mg   Cl   Al   S   Cs   I   O

1. The size of the atoms decreases.
2. The size of the atoms increases down a group because atoms with larger molar masses have electrons in higher energy levels.
3. The size increases because of the added repelling force among the electrons.
4. Mg^{2+} smaller
   Cl^- larger
   Al^{3+} smaller
   S^{2-} larger
   Cs^+ smaller
   I^- larger
   O^{2-} larger

11-6 Ionization Energy

Another important property that shows periodic trends is ionization energy. Recall from Chapter 10 that **ionization energy** is the energy needed to remove an electron from a neutral gaseous atom. The ionization process can be expressed in an equation:

$$\text{Element(g)} + \text{ionization energy} \longrightarrow \text{Ion}^+(g) + e^-$$

Ionization energies of many atoms were determined in the 1920's by bombarding gaseous samples of an element with high-energy electrons. The energy of the electrons is known precisely. When electrons

When the students graph the ionization energies they get involved in the concept. Have students note the pattern seen when the lines between the ionization energies are completed.

have enough kinetic energy, their collisions with atoms in the gas cause the atoms to ionize. First ionization energy is the energy that the bombarding electrons need to eject the most weakly held electron from the atoms to form positive ions.

Ionization energies were studied intently because they form a basis for predicting which elements may form the positive ions in ionic substances. An element with a low ionization energy forms a positive ion because only a small amount of energy is needed to remove an electron from the neutral atom. Elements with high ionization energies rarely form positive ions in chemical reactions. However, they may form negative ions or no ions at all.

Ionization energies vary among the elements. The ionization energies for the first 20 elements are listed in Table 11-8. IE_1 is the energy needed to remove the outermost electron from the atom; IE_2 is the energy needed to remove the next outermost electron from the atom; IE_3 and IE_4 are the energies needed to remove the third and fourth outermost electrons, respectively.

Table 11-8

THE FIRST FOUR IONIZATION ENERGIES FOR SOME ELEMENTS					
ATOMIC NUMBER	**ELEMENT**	**IONIZATION ENERGY (kJ/mol)**			
		IE_1	IE_2	IE_3	IE_4
1	H	1312	—	—	—
2	He	2372	5510	—	—
3	Li	519	7285	11770	—
4	Be	900	1756	14860	21010
5	B	799	2427	3660	25030
6	C	1088	2353	4621	6223
7	N	1406	2855	4592	7459
8	O	1314	3384	5315	7458
9	F	1682	3372	6040	8401
10	Ne	2080	3962	6122	8548
11	Na	498	4561	6913	9543
12	Mg	736	1450	7731	10540
13	Al	577	1817	2745	11580
14	Si	787	1577	3232	4355
15	P	1063	1903	2910	4955
16	S	1000	2251	3361	4562
17	Cl	1255	2297	3822	5160
18	Ar	1519	2689	3947	5909
19	K	418	3067	4504	6258
20	Ca	590	1145	4936	6752

A regular pattern becomes apparent when these values are graphed as shown in Figure 11-6. Do you notice a regularity? There seems to be a gradual increase in ionization energy as you move across a row in the periodic table. A noble gas has the highest ionization energy in any period. An alkali metal has the lowest ionization energy in the period.

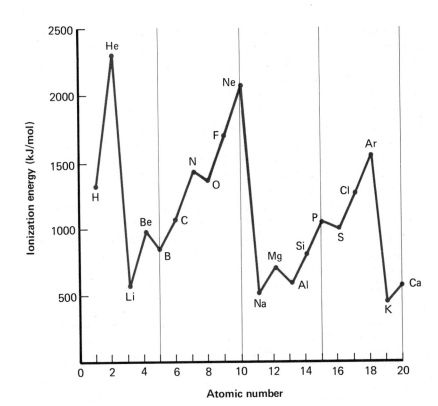

Figure 11-6 The ionization energies of the first 20 elements are graphed. Look at this graph and the periodic table, what similarities do you see in the arrangements?

A low ionization energy for an atom means that the electron that is removed is not held very tightly by the nucleus. The *alkali elements* (group 1) have low ionization energies. Sodium, like the other alkali metals, has one electron in the outermost energy level.

$$Na \quad 1s^2 2s^2 2p^6 3s^1$$

This electron is shielded from the effective attractive force of the nucleus by the inner electrons in the first and second energy levels. Thus, the outermost electron is easily removed from the sodium atom.

At the other end of the periodic table, the noble gases have complete outermost energy levels. Since all the outermost electrons are in the same energy level, they are all attracted almost equally to the nucleus. Compare the electron configuration of sodium to that of argon:

$$Ar \quad 1s^2 2s^2 2p^6 3s^2 3p^6$$

Note that the outermost electrons for sodium and argon are in the third energy level. But also note that the positive charge on the sodium nucleus is +11 while that for argon is +18. The attraction of the nucleus for the electrons in argon is far stronger than that for sodium. It will take more energy to overcome the pull of the nucleus for the outermost argon electron. The effect is that the noble gases have the highest ionization energy in a period of elements. Another approach to this trend is to look at the relative sizes of the atoms as you move from left to right across the periodic table. The size of the atoms decreases across a period. The nuclear attractive force is greater in the smaller atoms. Therefore, more energy is required to remove an electron from the noble gases than any other atom in a given period.

You may have noticed that ionization energies decrease as the molar masses increase in a group or family. This trend is explained in terms of the shielding effect of the inner electrons.

Look at the second ionization energies of helium, lithium, and beryllium in Table 11-8. The energy needed to remove a second electron from each of the atoms is listed in the IE_2 column. Notice that it is more difficult to remove the second electron from Li than from He. What explanation could account for these data? When lithium loses its first electron, the Li^+ ion is isoelectronic with helium. Yet, the second ionization energy of Li is greater than the first ionization energy of He even though both have the same electron configuration. The reason is that lithium has a nuclear charge of $+3$ and helium has a nuclear charge of $+2$. The attraction between the Li^+ nucleus and its electrons is greater than the attraction between the helium nucleus and its electrons.

Figure 11-7 Some periodic trends are summarized on the table shown.

SOME OF THE TRENDS IN THE PERIODIC TABLE

Atomic radii increase →
Ionization energies increase →
Metallic properties decrease →

Atomic radii increase ↓
Ionization energies increase ↓
Metallic properties increase ↓

(handwritten: decrease)
(handwritten: error ↓)

1 H 1																	2 He 4.003
3 Li 6.941	4 Be 9.012											5 B 10.81	6 C 12.01	7 N 14.01	8 O 16.00	9 F 19.00	10 Ne 20.18
11 Na 22.99	12 Mg 24.31											13 Al 26.98	14 Si 28.09	15 P 30.97	16 S 32.06	17 Cl 35.45	18 Ar 39.95
19 K 39.10	20 Ca 40.08	21 Sc 44.96	22 Ti 47.90	23 V 50.94	24 Cr 52.00	25 Mn 54.94	26 Fe 55.85	27 Co 58.933	28 Ni 58.69	29 Cu 63.55	30 Zn 65.38	31 Ga 69.72	32 Ge 72.59	33 As 74.92	34 Se 78.96	35 Br 79.90	36 Kr 83.80
37 Rb 85.47	38 Sr 87.62	39 Y 88.91	40 Zr 91.22	41 Nb 92.91	42 Mo 95.94	43 Tc (98)	44 Ru 101.1	45 Rh 102.906	46 Pd 106.42	47 Ag 107.9	48 Cd 112.4	49 In 114.8	50 Sn 118.7	51 Sb 121.8	52 Te 127.6	53 I 126.9	54 Xe 131.3
55 Cs 132.9	56 Ba 137.3	71 Lu 175.0	72 Hf 178.5	73 Ta 178.5	74 W 183.9	75 Re 186.2	76 Os 190.2	77 Ir 192.22	78 Pt 195.08	79 Au 197.0	80 Hg 200.6	81 Tl 204.4	82 Pb 207.2	83 Bi 209.0	84 Po (209)	85 At (210)	86 Rn (222)
87 Fe (223)	88 Ra (2260)	103 Lr (260)	104 (261)	105 (262)	106 (263)	107 (262)	108	109 (266)									

1. The energy needed to remove one electron from a neutral gaseous atom.
2. When barium loses $2e^-$, it becomes isoelectronic with Xe. It takes a great deal of energy then to remove another electron.
3. Each noble gas in a period has the largest positive nuclear charge. Electrons are attracted to a large nuclear charge, more so than the other elements in the period.

Review and Practice

1. What is ionization energy?

2. The second ionization energy of barium is relatively small. The third ionization energy is very large. Briefly explain the large difference between these two ionization energies.

3. Explain why the first ionization energies of the noble gases are the highest in any given period.

Numbers in red indicate the appropriate chapter sections to aid you in assigning these items. Answers to all questions appear in the Teacher's Guide at the front of this book.

Summary

■ Elements were initially classified by properties into metals and nonmetals. As more elements were discovered, it became apparent that metals and nonmetals was too broad a classification system. Elements were then grouped into chemical families according to physical and chemical properties by Dmitri Mendeleev.

■ The modern periodic table consists of elements listed according to increasing atomic number. Periods are listed horizontally in the periodic table and chemical families (groups) are listed vertically. Metals are on the left and nonmetals are on the right.

■ Electron configurations can explain the chemical and physical properties of the elements. The most stable (nonreactive) elements are the noble gases which have eight electrons in their outermost s and p orbitals. Other elements enter into reactions and become stable ions (metals become positive ions and nonmetals become negative ions) which are often isoelectronic with the nearest noble gas.

■ Two forces acting within the atom account for many of the chemical and physical properties of the elements: proton-electron attraction and electron-electron repulsion. It is these forces coupled with the differences in energy levels of electrons that help explain the nature of periodicity.

■ Trends in the periodic table (atomic radii and ionization energy) help to explain the chemical properties of elements.

Chemically Speaking

alkaline earth metals
alkali metals
atomic radius
chemical family
halogens
ionic radius
ionization energy

isoelectronic
noble gases
period
periodic law
shielding effect
transition metals
valence electrons

Review

1. List five metals and five nonmetals. (11-2)

2. List by symbol, and name the elements which compose the halogen family. Do the same for the noble gases. (11-2)

3. Define the terms "period" and "group" in reference to the periodic table. (11-2)

4. What is periodic in the periodic table? (11-2)

5. How did Dmitri Mendeleev know to make room in his periodic table for "missing" elements? (11-2)

6. Electrons occupying which orbitals most influence the chemical properties of an element? (11-3)

7. What happens to the size of atoms as one moves from left to right across the periodic table? (11-5)

8. Write a generalized equation for ionization. (11-6)

9. Aluminum's electron configuration is $1s^2 2s^2 2p^6 3s^2 3p^1$. Write the symbol for the most stable ion of aluminum. (11-6)

Interpret and Apply

1. If Rb_2O, MgO and Al_2O_3 are stable compounds, what would be the formulas for stable oxides of Sr, K, and Ga? (11-2)

2. Give two examples of periodic relationships in everyday life (other than the periodic table). (11-1, 11-2)

3. Name the group on the periodic table in which each of the following elements fall, and categorize each as a metal or nonmetal.
Rb, Nb, As, Xe, Eu, Ag, element 106 (11-2)

4. Name three elements that have an incomplete $4d$ orbital system. (11-3)

5. Determine the number of protons and electrons in a stable ion of iodine and in a stable ion of radium. (11-4)

6. Why are the noble gases considered stable? (11-4)

7. Which has a larger radius, Na^+ or Ne? (11-5)

8. Identify the three elements using the information listed below: (11-6)

Element X:

has a relatively high ionization energy

generally forms a 2− ion

has an outermost electron configuration of $3s^23p^4$

Element Y:

reacts with oxygen to form Y_2O

has a very low ionization energy

is in the fourth period of the periodic table

Element Z:

is a transition metal

is used in U.S. coinage

has eight electrons in the third orbital system

9. Briefly explain why Ba has a lower first ionization energy than Ca. (11-6)

10. Which of the following elements would have the largest third ionization energy: K, Ca, or Ga? (11-6)

Problems

1. Chlorine is commonly used to purify drinking water. When chlorine dissolves in water, it forms hypochlorous acid.

$$Cl_2(g) + H_2O(l) \longrightarrow HOCl(aq) + H^+(aq) + Cl^-(aq)$$

Predict what happens when iodine, I_2, dissolves in water. Write the chemical equation for this reaction. (11-2)

2. Here is a list of chemicals: NaF, NaOH, H_2Se, CS_2, $AlCl_3$, Na_3PO_4. Use them and the periodic table to suggest values of x and y in the following formulas: (11-2)

a. $Al_x(OH)_y$
b. $Tl_x(PO_4)_y$
c. Sn_xSe_y
d. $Ba_x(PO_4)_y$

3. Use the periodic table to suggest formulas for compounds of the following pairs of elements: (11-5)

a. strontium-sulfur
b. gallium-fluorine
c. beryllium-tellurium
d. chlorine-iodine
e. arsenic-bromine

4. In general, the molar mass of elements increases as the atomic number increases. Find pairs of elements in the periodic table that are exceptions to this generalization. (11-2)

5. The size of an atom can be expressed as the closest distance of approach by another atom. For the halogens, the estimates of atomic size are F, 0.064 nm; Cl, 0.099 nm; Br, 0.114 nm. Estimate the value for iodine in nanometers. (11-5)

6. The metals of group 2 of the periodic table combine with the halogens to form ionic solids. Write a general equation to represent these reactions, using M for the group 2 metal and X for the halogen. (11-2)

7. The first four ionization energies in kJ/mol for the element Al are as follows:

IE_1 = 577 kJ/mol IE_3 = 2745 kJ/mol
IE_2 = 1817 kJ/mol IE_4 = 11580 kJ/mol

What is the charge for the most common ion of aluminum? How many outermost electrons does aluminum have? (11-6)

Challenge

1. Which of the following electron configurations would you expect to have the lowest second ionization energy? Give reasons for your choice.
a. $1s^22s^22p^6$
b. $1s^22s^22p^63s^1$
c. $1s^22s^22p^63s^2$

2. Which of these elements—Na, Si, Cl, or CS—has:
a. the highest first ionization energy?
b. the smallest radius?
c. the most metallic character?

3. Use the periodic table to predict which substance in each of these pairs has the smaller radius:
a. K, Br
b. Ne, F^-
c. K^+, Ga^{3+}
d. S, Se

4. In order to predict whether the arrangement of anions in an ionic crystal will form a tetrahedral hole or octahedral hole, the radius ratio rule is often used. The radius ratio rule states that if the ratio of the radius of the cation to the anion has a value from 0.414 to 0.732, the hole will be octahedral; if the ratio of the radius of the cation to the anion has a value of 0.225 to 0.414, a tetrahedral hole is present. Predict the crystal arrangement (octahedral or tetrahedral) for the following ionic compounds:

a. NaCl radii $Na^+ = 0.095$ nm; $Cl^- = 0.181$ nm
b. LiBr radii $Li^+ = 0.060$ nm; $Br^- = 0.195$ nm
c. FeS radii $Fe^{2+} = 0.075$ nm; $S^{2-} = 0.184$ nm

Synthesis

1. When 25.0 grams of sodium metal reacts with excess oxygen gas, how many grams of sodium oxide will be formed?

✓2. The electron configuration for the following neutral atoms are given for use in answering questions a to e:

A $1s^2 2s^2 2p^6 3s^2$
B $1s^2 2s^2 2p^6 3s^1$
C $1s^2 2s^2 2p^6$
D $1s^2 2s^2 2p^5$
E $1s^2 2s^2 2p^3$

a. Which of the electron configurations given above would you expect to have the lowest ionization energy?
b. Which of the electron configurations given above would you expect for a noble gas?
c. List the five configurations in predicted order of increasing ionization energy (lowest to highest).
d. Predict the configuration that should have the highest second ionization energy (IE_2).
e. Predict the configuration that should have the lowest second ionization energy (IE_2).

3. Strontium-90 is a radioactive product of nuclear fission. From its position in the periodic table, determine why this isotope is likely to be absorbed by the body into the bone structure.

Projects

1. Research the discovery of the noble gases, such as helium, neon, and argon.

2. Draw a large periodic table. Find as many examples of elements as you can. Attach (if feasible) the samples of the elements to the periodic table. (This can be done as a class project.)

The hydrogen and oxygen atoms comprising water molecules, and even the water molecules themselves, are too small to be seen. But when bonded together, the atoms and molecules produce a beautiful, uniquely shaped snowflake. Every substance visible to the naked eye is much larger than an atom, so it is logical to conclude that atoms tend to group together. In previous chapters, you learned that these atoms are not just close to each other but are bonded, or "stuck together," some more strongly than others.

Experiments have demonstrated that the bonds in some molecules are hard to break. Do chemical bonds differ from each other, and if so, how? Why is it that some bonds are easily broken while others are not? What kinds and shapes of molecules result from chemical bonds? These are some of the questions that you will explore in this chapter.

Chemical Bonding

An Introduction to Bonding

Most people hate to study chemical theory. Tantalizing tastes, seductive aromas, and alluring colors are the kinds of chemistry more often appreciated, though not necessarily understood.

There comes a time when it is worth the effort to understand. Understanding the chemistry of photosynthesis may help farmers to increase food supplies. Analyzing the molecules that carry genetic codes may lead researchers to find a cure for disease. Determining the mechanism of reactions occurring in the environment may help everyone protect it for future generations. This kind of understanding will not come unless the nature of molecules is studied—how they come together, how they do not, and why. What is the glue that holds atoms of less than 100 kinds of elements together in millions of living, breathing forms?

In this lesson, you will begin to look at bonding and develop a simple model that can be used to talk about differences between molecules. You will still enjoy sights and smells, but you also will begin to understand something about them.

CONCEPTS

■ the nature of covalent bonds
■ potential energy changes in bonding
■ electronegativity of atoms in bonds
■ polar covalent bonds
■ formation of ionic bonds
■ ionic crystals

12-1 What Questions Need to be Answered?

In this chapter you will investigate the electrical nature of chemical bonding. You will see what theories of bonding have been devised to explain much of the diversity observed in matter. Several important questions should be considered as you examine the simple models presented.

How many atoms are connected? In the final analysis, the only way to know the formula of a compound is to determine the amount of each reactant in the compound and calculate the proportions, as discussed in Chapter 4. Still, it would be nice to be able to predict the *most likely* formula of a compound. A good theory of bonding helps you to explain the chemical fact that $LiCl$ is more likely to form than Li_2Cl or $LiCl_2$.

The formulas for *all* compounds cannot be predicted. From the law of multiple proportions, you know that more than one compound of the same two elements are possible. There is H_2O and H_2O_2, CO and CO_2, $CuCl$ and $CuCl_2$. However, a good theory of bonding allows intelligent guesses about the structure of many molecules.

Which atoms are connected? The structure of molecules implies two things: what atoms are connected and what shape they produce in the molecule. Information about the atoms that are connected can be important in predicting chemical properties. Conversely, information about chemical properties provides important clues to structure. The discussions that follow about bonding will address both of these ideas. ***What is the shape of the molecule?*** Shape is the other aspect of structure, and it helps determine chemical properties. If you have worked a jigsaw puzzle, you know that pieces of the wrong shape will not fit. The same is true of molecules. Many chemical reactions—especially those in living systems—depend on the ability of two molecules to get close to each other. If a molecule has the wrong shape, it cannot get close enough to another molecule to react. The action of drugs, the digestion of food, the production of food in green plants, and the reproduction of cells in your body involve chemical reactions in which molecular shape plays an important role. A good theory of bonding predicts the shape of molecules.

How strong is a bond? The energy involved in the processes of bond forming and bond breaking is known as the **bond energy**. It is a measure of bond strength and can be very useful. It can tell you how much energy you get from a gram of sugar and how much energy is stored when a tree produces cellulose. The strength of a bond can indicate which bonds in a molecule come apart easily and which stubbornly remain stuck. A good model of bonding provides a basis for understanding these differences in bond energy.

As you can see, a good theory of bonding offers better understanding of many chemical facts. The simple theories that will be introduced in this chapter can provide you with some of this understanding. You may find it helpful to focus on these questions as you read the chapter.

Whenever possible, use space-filling or ball-and-stick models when discussing concepts from the chapter to help students think in three dimensions.

12-2 Covalent Bonding

In Chapter 8, you learned that atoms are made of particles that have positive electrical charges and particles that have negative electrical charges. Unlike charges attract each other. How can this information be used to develop an explanation for bonding?

The first picture in Figure 12-1 represents two separate hydrogen atoms. The + in the center of each atom represents the single proton in the nucleus, and the shaded areas around the nucleus represent the single electron in the $1s$ orbital. If you imagine the two atoms moving very close to one another, you get something similar to the second picture shown in Figure 12-1. What has happened in terms of forces of attraction?

When the atoms are far apart, the only force of attraction is between the proton of one atom and the electron of the *same* atom. As long as the atoms are far apart, there are no other forces large enough to have any major effect. As the atoms approach each other, however, other forces become important. There are repulsive forces between the electrons of the two atoms and between the protons of the two atoms. There are also new attractive forces. The proton of one atom can attract the electron of the other atom, and vice versa. If these new attractive forces are greater than the new repulsive forces, the two atoms stick together to form a molecule. The result is a **chemical bond**—a situation in which a pair of electrons is simultaneously attracted to two atomic nuclei.

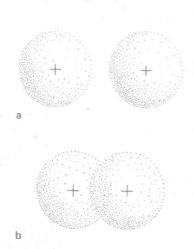

a

b

Figure 12-1 Two hydrogen atoms are represented before and after bonding.

Another way to look at the situation is in terms of energy changes. As two *oppositely* charged particles approach each other, the potential energy of the system *decreases*. Conversely, if two particles of the same charge approach each other, the potential energy of the system increases. Other effects being equal, natural systems tend to decrease potential energy. For many atoms, the formation of a chemical bond with another atom leads to a decrease in the energy of the system.

Consider the formation of a chemical bond in the hydrogen gas molecule. Figure 12-2 illustrates the energy changes taking place when two hydrogen atoms come together. The potential energy of the system is plotted on the *y*-axis. Follow the position of the circle from one drawing to the next to track how the energy changes as the chemical bond forms.

As the two atoms approach, the potential energy appears to decrease. This change can be explained by increased attractive forces as both electrons begin to interact with both protons. However, the

Figure 12-2 The potential energy for two hydrogen atoms changes as they approach and form a covalent bond. The same principle can be applied to the formation of other covalent bonds.

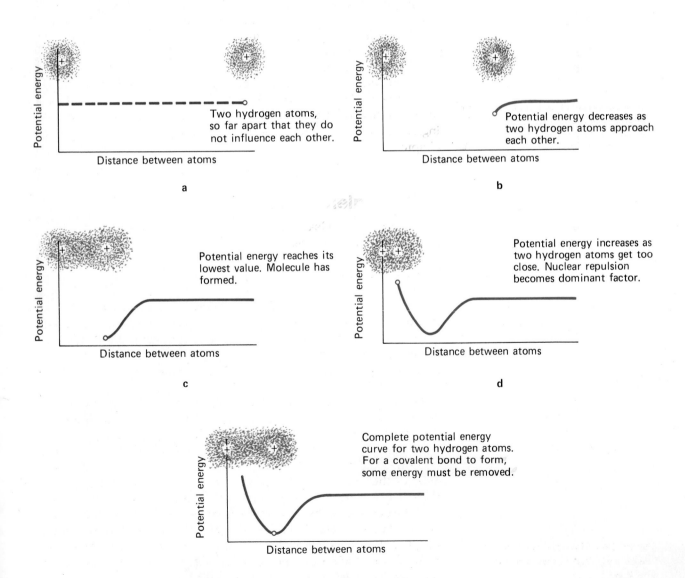

a — Two hydrogen atoms, so far apart that they do not influence each other.

b — Potential energy decreases as two hydrogen atoms approach each other.

c — Potential energy reaches its lowest value. Molecule has formed.

d — Potential energy increases as two hydrogen atoms get too close. Nuclear repulsion becomes dominant factor.

e — Complete potential energy curve for two hydrogen atoms. For a covalent bond to form, some energy must be removed.

Students may have some difficulty with the changes in potential energy. Allow sufficient time for them to study Figure 12-2 and to ask questions.

See further explanation in the teacher's chapter notes concerning the overlapping of orbitals in the formation of the H_2 molecule. You may want to explain to students why a molecule such as H_3 or He_2 is not possible.

orbitals cannot merge completely because forces of repulsion are also operating. The protons repel each other, as do the electrons, when the distance between the atoms continues to decrease. The minimum in the potential energy graph occurs when there is a balance between the electrical forces in the molecule. The fifth drawing represents the two atoms at a point where the system has the lowest possible potential energy.

Note in the diagram that the orbitals of the atoms are shown to overlap. The electrons are being shared, and the probability of finding them between the two protons increases. A chemical bond has formed. This type of chemical bond, which results from the sharing of a pair of electrons between atoms, is called a **covalent bond**.

How can the formation of a covalent bond be represented in terms of orbitals of the atoms? You know that each atom has one electron in a $1s$ orbital. The orbital diagram for the hydrogen atom is

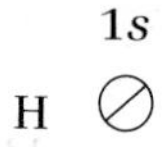

$$1s$$

$$\text{H} \quad \oslash$$

Recall from Chapter 10 that any orbital can hold a maximum of two electrons. As the two hydrogen atoms approach each other and the orbitals overlap, each atom appears to have two electrons in its $1s$ orbital. This can be represented by drawing the orbital diagram for each atom and shading the orbitals that overlap:

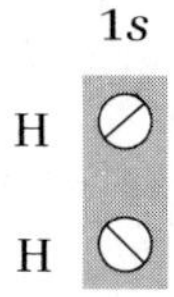

The fluorine molecule is another example of two atoms bonded by a covalent bond. Recall that the fluorine atom has the electron configuration $1s^2 2s^2 2p^5$. One $2p$ electron is unpaired. An orbital diagram may be used to represent the overlap of two $2p$ orbitals between the fluorine atoms:

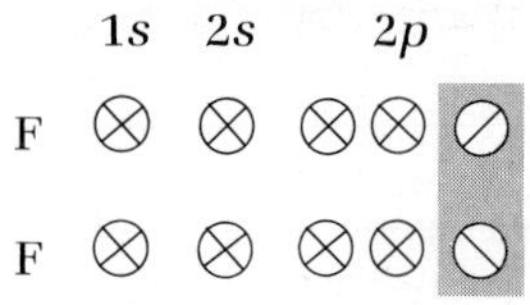

Covalent bonds occur between a variety of atoms. As you continue to study this chapter, you will become familiar with many molecules that contain covalent bonds. You also will see how covalent bonds may differ.

12-3 When Atoms Share Electrons Unequally

Suppose you had a bag of peanuts you wanted to share with a friend. You may share it fifty-fifty. On the other hand, you may not be in a munching mood, so you give most of the peanuts to your friend and have only a little yourself. Or, perhaps you are ravenous, so you give just a small amount to your friend. In all three cases, you are sharing but not necessarily fifty-fifty. The same can be said for atoms in a covalent bond. They share electrons, but not necessarily equally. Hydrogen and several other substances will be used here to illustrate a difference in covalent bonds of this nature.

The two atoms of hydrogen in a H₂ molecule are alike, so the electrons are shared equally between the two atoms. However, in the case of HCl where the two atoms are different, the sharing of electrons is not equal. The chlorine atom has a tendency to draw the shared pair of electrons closer to itself and the hydrogen atom is left somewhat electron deficient, as seen in Figure 12-3. The attraction an atom has for the shared pair of electrons in a covalent bond is called the atom's **electronegativity**. Chlorine has a greater attraction for the shared pair of electrons in a bond with hydrogen, so chlorine is said to have a larger electronegativity.

Linus Pauling, an American scientist, devised a method of calculating the electronegativities of atoms in chemical bonds. Figure 12-4 is a periodic table that lists Pauling's values for electronegativities of the elements. When the values are compared, the trends become obvious. Notice that the numbers increase from left to right across the periodic table and decrease down a group, or family. The most electronegative element is fluorine and the least is cesium.

The difference between the electronegativities of two atoms in a bond can be used as a guide to determine the degree of electron sharing in the bond. As the difference increases, the degree of sharing decreases. If the difference in electronegativities between the two bonding atoms is 0.2 or less, the pair of bonding electrons is shared equally for all practical purposes. On the other hand, if the difference between electronegativities is 1.7 or greater, the electron is transferred considerably from one atom to the other. In such a case, the element of greater electronegativity is said to exist as a negative ion, while the element of lesser electronegativity exists as a positive ion. The electro-

Emphasize to students that electronegativity differences apply to individual bonds and not to isolated atoms or many bonds within the same molecule.

Spatial model
of HCl

Figure 12-3 There is an unequal distribution of the shared electrons in the bond between the hydrogen and the chlorine.

Francium has no long-lived isotopes, and little of its chemistry is known.

Figure 12-4 There are several systems for indicating the electronegativities of the elements. Pauling calculated the one most commonly used today.

ELECTRONEGATIVITIES OF THE ELEMENTS

1	2	3	4	5	6	7	8	9	10	11	12	13	14	15	16	17
2.1 H																
1.0 Li	1.5 Be											2.0 B	2.5 C	3.0 N	3.5 O	4.0 F
0.9 Na	1.2 Mg											1.5 Al	1.8 Si	2.1 P	2.5 S	3.0 Cl
0.8 K	1.0 Ca	1.3 Sc	1.5 Ti	1.6 V	1.6 Cr	1.5 Mn	1.8 Fe	1.8 Co	1.8 Ni	1.9 Cu	1.6 Zn	1.6 Ga	1.8 Ge	2.0 As	2.4 Se	2.8 Br
0.8 Rb	1.0 Sr	1.2 Y	1.4 Zr	1.6 Nb	1.8 Mo	1.9 Tc	2.2 Ru	2.2 Rh	2.2 Pd	1.9 Ag	1.7 Cd	1.7 In	1.8 Sn	1.9 Sb	2.1 Te	2.5 I
0.7 Cs	0.9 Ba	1.1–1.2 La–Lu	1.3 Hf	1.5 Ta	1.7 W	1.9 Re	2.2 Os	2.2 Ir	2.2 Pt	2.4 Au	1.9 Hg	1.8 Tl	1.8 Pb	1.9 Bi	2.0 Po	2.2 At
0.7 Fr	0.9 Ra	1.1–1.7 Ac–No														

Pauling derived his values from measurements of bond energies. The values given are averages because electronegativity varies with different compounds. This method of determining bond character can serve as a guide, but it is not without exception.

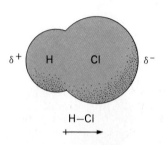

H–Cl

Figure 12-5 The arrow shown here is a common symbol for a dipole. By convention, the arrow points to the more electronegative end of the bond.

Alert students not to confuse the two symbols for delta, δ and Δ, which are used differently in chemistry.

static attraction between the two oppositely charged ions is called an **ionic bond**. In KCl, potassium has an electronegativity of 0.8, while the value for chlorine is 3.0. The difference is 2.2, and the bond is considered ionic.

Molecules that contain bonds with electronegativity differences between 0.3 and 1.6 are considered to be covalent but with unequal sharing of electrons. In the HCl molecule, for example, chlorine has the greater electronegativity (3.0 in comparison to 2.1 for hydrogen). The difference between the two electronegativities is 0.9. The electrons are shared unequally. In such a case, the atom with the greater electronegativity takes on a partial negative charge (between zero and 1−) as the shared electrons spend more time nearby. The other atom takes on a partial positive charge (less than 1+). Thus a bond of this nature has somewhat negative and somewhat positive poles that together are said to make up a **dipole**. Figure 12-5 shows two ways dipoles may be illustrated. The lower-case Greek letter *delta*, δ, is used to indicate the partial charges at each end of the bond. A bond with a dipole is called a **polar covalent bond** or simply a polar bond. (Covalent bonds without dipoles are termed nonpolar.) Molecules such as HCl, HF, NH_3, H_2O, SO_2, and CCl_4 contain polar bonds.

A continuum is shown in Figure 12-6 that describes the three kinds of bonds—ionic, polar covalent, and covalent. Remember that these three categories do not reflect different mechanisms for bonding. Rather, they represent a way to classify the degree of sharing of electrons that occurs in any bond.

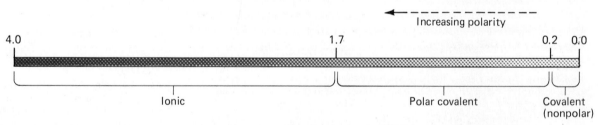

Electronegativity difference

Increasing polarity

4.0   1.7   0.2   0.0

Ionic   Polar covalent   Covalent (nonpolar)

Figure 12-6 The distinction between nonpolar, polar, and ionic is a matter of degree. Also some polar bonds are more polar than others, and some ionic bonds are more ionic than others.

Often ionic and covalent bonding is taught with an emphasis on the differences between the two. You may want to point out to the students that the lowering of potential energy is a key factor in the stability of both kinds of bonding.

All ionic bonds have some covalent character. Bonds with an electronegativity difference of 1.7 are generally classified as a little more than 50% ionic. Remind students that there are no sharp delineations.

Example   12-1

Classify the bond in each of the following as ionic, polar covalent, or covalent:

$$HBr \quad\quad Cl_2 \quad\quad NaF$$

Suggested solution

One way to classify by bond type is to find the differences in electronegativities for the atoms in each case. Refer to the table in Figure 12-4 for the values that you need.

For HBr, the difference in electronegativities is

$$2.8 - 2.1 = 0.7$$

$$Br \quad\quad H$$

From the chart in Figure 12-6, you can see that the bond would be classified as polar covalent.

In Cl_2, the atoms are the same and therefore have the same electronegativity (3.0). The difference is 0.0, so the bond is covalent.

For NaF, refer again to Figure 12-4 for the electronegativity value of each element. The difference in electronegativities is

$$4.0 - 0.9 = 3.1$$
$$\text{F} \qquad \text{Na}$$

The bond is considered to be ionic.

Polarity of chemical bonds can produce important chemical properties in molecules. The presence of polar bonds often (but not always) makes a molecule polar. If molecules have slightly positive and negative ends, there is more attraction than usual between the molecules. It should take more energy to separate the molecules, and you might expect to see higher boiling points for such substances.

Many of the properties of water are due to its polar character. As you will read later in the chapter, the polar bonds of the water molecule produce a partial positive charge on the side with the hydrogen atoms and a partial negative charge on the oxygen side. The polar nature of water partially accounts for the solubility of ionic compounds in water. Water molecules surround the positive and negative ions and are attracted to them. The attachment of the water molecules is normally not strong, but it is sufficient to supply some of the energy needed to separate the oppositely charged ions that are found in a crystal of an ionic solid.

Polarity helped explain why intermolecular distances were smaller and bond energies were larger than those that had been predicted previously for some molecules.

Do You Know?

The degree of sharing in a bond as determined by the differences in electronegativities of the atoms is sometimes called the character of the bond. NaCl has more ionic character than NaI.

Review and Practice

1. What is a chemical bond?

2. Does potential energy increase or decrease as two oppositely charged particles approach each other? Explain.

3. How can the differences between a covalent bond, a polar covalent bond, and an ionic bond be categorized using the concept of electronegativity?

4. Describe what happens to the orbitals in the bond between two hydrogen atoms. Where are the bonding electrons to be found most of the time?

5. What is electronegativity?

6. Identify the kind of bond contained in each of the following: HI, F_2, CsCl, MgO, O_2, KBr, AsH_3, PbI_3, PCl_3.

1. the coming together of two atoms when a pair of electrons is simultaneously attracted to both nuclei.
2. Potential energy decreases; attractive forces grow stronger as the particles approach each other, which lowers the energy of the system.
3. Bonds are categorized by the degree of sharing of a pair of electrons in the bond. The greater the difference in electronegativity of the two atoms, the more unequally the electrons are shared.
4. As two H atoms approach each other, their orbitals overlap. The electrons are drawn simultaneously to the two nuclei, and are found between the two nuclei most of the time.
5. a measure of an atom's attraction for the shared pairs of electrons in a bond
6. HI = polar covalent; F_2 = covalent; CsCl = ionic; MgO = ionic; O_2 = covalent; KBr = ionic; AsH_3 = covalent; PbI_3 = polar covalent; PCl_3 = polar covalent

12-4 The Nature of Ionic Bonding

When the difference in the electronegativities of two atoms in a bond is great, the bond is classified as ionic. The probability of finding the shared pair of electrons near the more electronegative atom is so high that the electrons are considered to be removed from one atom and transferred to the other atom in the bond. Ionic bonding occurs between metallic and nonmetallic elements. The metal loses one or

Refer students to the material they learned in Chapter 11 regarding the periodic table. Ask them to summarize what factors within a group or period contribute to a particular element's tendency to gain or lose electrons.

Figure 12-7 As the Na^+ and Cl^- ions approach each other, the potential energy in the system decreases and is released as heat.

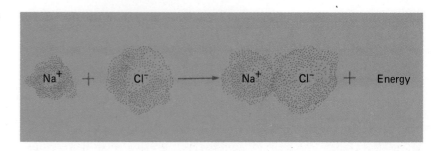

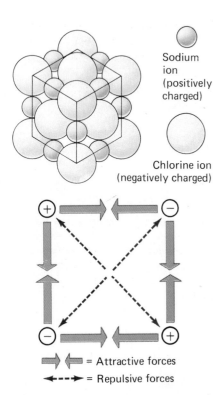

Sodium ion (positively charged)

Chlorine ion (negatively charged)

⟶◄═══ = Attractive forces
◄---► = Repulsive forces

Figure 12-8 The arrangement of Na^+ and Cl^- ions in a crystal lattice leads to a cubic crystal on a macroscopic level. Attractive forces occur between Na^+ and Cl^- ions, while similarly charged ions repel each other.

more electrons (becoming a positive ion, or cation) and the nonmetal gains one or more electrons (becoming a negative ion, or anion).

When metallic sodium and chlorine gas interact, the product is sodium chloride, as represented by the following equation:

$$2Na(s) + Cl_2(g) \longrightarrow 2NaCl(s)$$

Each sodium atom loses an electron to become a 1+ ion, and each chlorine atom gains an electron to become a 1− ion. As a sodium ion and a chloride ion approach each other, the attractive force between the particles increases, and the potential energy of the system decreases. (See Figure 12-7.) As more oppositely charged ions approach the first pair, the potential energy decreases even more, and a crystal begins to form. The crystal takes on a regular, three-dimensional arrangement of positive and negative ions that is called a **crystal lattice** (Figure 12-8). Forces acting in the crystal lattice of NaCl are represented in the figure also. For sodium and chloride ions, the attractive forces are stronger than the repelling forces. The crystal continues to grow as long as positive and negative ions are available to be added to the crystal, and as long as the heat given off can be effectively removed.

Ionic crystals are very stable substances as demonstrated by their high melting temperatures. Their stability comes from the strong attractive forces between oppositely charged ions composing the crystal lattice. Some examples of ionic substances are NaF, $BaBr_2$, KCl, and Li_2O. Such substances are not described as molecular because they do not exist as individual molecules in the solid state.

Ionic bonds occur most frequently between metals from groups 1 or 2 and the halogens, group 17. Some nonmetals such as oxygen, nitrogen, and sulfur also may form ionic bonds with some of the metals from groups 1 or 2. You can become accustomed to judging whether a specific bond is ionic or covalent by comparing the electronegativities of the elements in the bond. As the difference between the values becomes smaller, the bonds become less ionic and more covalent.

Review and Practice

1. Distinguish between covalent, polar covalent, and ionic bonds.

2. Using the three classifications of bonds discussed in this lesson predict which is likely to be present in compounds made from elements of groups 1 and 17 and from groups 16 and 17.

3. Explain how an ionic bond forms.

4. How are the ions in a crystal lattice arranged? Why are such arrangements not described as molecular?

1. Electrons are shared more or less equally in covalent bonds, and unequally in polar covalent bonds. In ionic bonds, the pair of electrons is said to be associated with the more electronegative atom most of the time.
2. Ionic bonding is likely to occur between elements in group 1 and 17; covalent or polar bonding between group 16 and 17.
3. Ionic bonds form when positive and negative ions attract each other.

4. Ions are arranged in characteristic patterns of tightly packed aggregates of alternating + and − ions. *Molecule* usually refers to a group of atoms bonded covalently.

Describing Molecular Structures

One of the goals of this chapter is to increase your understanding of why molecules behave the way they do. You have already learned something about how atoms come together to form chemical bonds. The next step is to *think* in terms of molecules. Being able to visualize molecules makes thinking about them easier.

A mental picture of an ionic crystal may be easy to conjure up in your mind. However, the shape and structure of molecules are more difficult. Chemists have devised methods of representing molecules that contain covalent bonds. There are millions of different kinds of molecules. Because of their overwhelming variety and because of the different sorts of bonding involved, no single system of representing molecules is the best for all cases.

Yet, one of the important things to know about a molecule is which atoms are bonded together. As you have seen, orbital diagrams can be used to represent the overlap of orbitals between two individual atoms, but they become more cumbersome when several atoms are involved. If it is not necessary to indicate which orbitals overlap, a simpler system of dot structures can be substituted. The system is not foolproof and serves more as a guide than an errorless, absolute principle. However, it does provide a simple representation of many molecules. Current descriptions of molecules are supported by and derived from experimental evidence. Measurements of bond energies and data from spectroscopic analyses provide clues to how atoms are bonded together, even in molecules that differ from the simple models described here.

CONCEPTS

- the octet rule
- drawing electron dot structures
- single, double, and triple bonds
- bond energy and the strength of bonds
- spectroscopic evidence for bonding
- resonance

12-5 Electron Dot Structures for Representing Atoms

In 1916, G. N. Lewis, an American chemist, developed a system of arranging dots to represent outermost electrons around the symbols of the elements. The symbol is used to denote the nucleus and all the electrons except the valence electrons. The number of dots placed around the symbol of the element is equal to the valence electrons. These **electron dot structures** (often called Lewis structures) and the electron configurations for the elements of the second period are listed in Table 12-1 on the next page.

The number of dots placed around each symbol matches the number of electrons in the element's outermost s and p orbitals. The arrangement of the dots around the symbols has no special significance and should not be interpreted as the actual location of the electrons around the nucleus. It is customary to write the dots in pairs on the four sides of the symbol as a reminder that electrons are paired in orbitals (Chapter 10). Each of the other elements in a family of ele-

Table 12-1

ELECTRON CONFIGURATIONS, ORBITAL DIAGRAMS, AND ELECTRON DOT STRUCTURES OF THE SECOND PERIOD ELEMENTS

ELEMENT	ELECTRON CONFIGURATION	ORBITAL DIAGRAM			ELECTRON DOT STRUCTURE
		$1s$	$2s$	$2p$	
Li	$1s^2\ 2s^1$	⊗	⊘	○ ○ ○	Li·
Be	$1s^2\ 2s^2$	⊗	⊗	○ ○ ○	Be·
B	$1s^2\ 2s^2\ 2p^1$	⊗	⊗	⊘ ○ ○	·B·
C	$1s^2\ 2s^2\ 2p^2$	⊗	⊗	⊘ ⊘ ○	·C·
N	$1s^2\ 2s^2\ 2p^3$	⊗	⊗	⊘ ⊘ ⊘	·N:
O	$1s^2\ 2s^2\ 2p^4$	⊗	⊗	⊗ ⊘ ⊘	·O:
F	$1s^2\ 2s^2\ 2p^5$	⊗	⊗	⊗ ⊗ ⊘	:F:
Ne	$1s^2\ 2s^2\ 2p^6$	⊗	⊗	⊗ ⊗ ⊗	:Ne:

The arrangement of the dots around the atoms is arbitrary. Electrons can be paired as their number increases from Li to Ne, but students may become confused at this point when working with carbon. Hybridization, which results in four unpaired electrons in carbon, is covered in the Appendix.

If you discussed the Pauli exclusion principle and Hund's rule when teaching Chapter 10, you may want to remind students how these principles deal with the pairing of electrons in orbitals.

ments will have the same number of dots around it. For example, the electron dot formula for chlorine is

$$:\overset{\cdot\cdot}{\underset{\cdot\cdot}{Cl}}:$$

The formula has the same number of dots as that for fluorine because both have the same number of outermost electrons, seven.

Example 12-2

Write the electron dot (Lewis) structure for each of the following elements: phosphorus, iodine, sulfur, sodium, hydrogen, and krypton.

▶ Suggested solution _____

Dots are used to represent only the outermost s and p electrons of an atom. By referring to the periodic table, you can determine the number of these electrons for most atoms. For example, phosphorus is in group 15, the same group as nitrogen, and has five outermost s and p electrons. The electron dot structure for phosphorus is:

$$·\overset{\cdot}{P}:$$

The electron dot structures for the other elements listed in the problem are determined similarly:

$$·\overset{\cdot\cdot}{\underset{\cdot\cdot}{I}}:\quad ·\overset{\cdot\cdot}{\underset{\cdot\cdot}{S}}:\quad Na·\quad H·\quad :\overset{\cdot\cdot}{\underset{\cdot\cdot}{Kr}}:$$

Electron dot structures can be used to construct simple representations of atoms bonded in molecules. The simplest example is hydrogen gas. As you read earlier, the orbital of each electron overlaps with the other. The two atomic nuclei share the pair of electrons. This can be written as:

$$H:H$$

The pair of dots between the two symbols represent the electrons involved in the covalent bond.

In the next section, you will have an opportunity to learn how to represent more complex molecules using electron dot structures. The structures will serve as a first glimpse of how atoms are organized in some molecules.

12-6 The Octet Rule—Predicting How Atoms are Arranged in Molecules

How is it possible to predict the bonding arrangements that occur between atoms? In Chapter 10, you learned that the noble gases do not form many compounds. It was suggested that the electron configuration of these gases, with filled s and p orbitals, is particularly stable. You also know that when elements from groups 1, 2, 17, and some elements from groups 3 and 16 form stable ions, they do so by losing or gaining electrons. The resulting configurations are like those of the noble gases. These observations have led chemists to formulate the "rule of eight," or the **octet rule**, a simple explanation that allows the prediction of bonding arrangements between certain atoms. The assumption of the octet rule is that *bonded nonmetallic atoms have eight electrons in their outermost energy levels*. The rule holds true for a large number of molecules, although there are many exceptions. One of the most common exceptions is hydrogen. The hydrogen atom in a covalent bond will only be associated with one shared pair of electrons, which is the configuration of stable helium.* Although the octet rule does not accurately predict the stable configuration of all molecules, it can serve as a convenient tool to picture how atoms are connected in certain compounds.

In order to use the octet rule to write electron dot structures for molecules, you need to know four things:

1. *How many of each kind of atom are in the molecule?* This can be determined from the formula of the molecule.
2. *Which atoms are connected?* Only a sophisticated chemical analysis of compounds can tell you that for certain, but the following generalizations usually lead to reasonable predictions:
   a. If the formula of the molecule contains only one atom of an element and several atoms of another element, the single atom is in the center, with the other atoms bonded to it.
   b. If it is not obvious which atom is the central atom, draw a structure that is most symmetrical.

Emphasize to students that the octet rule is a general guide that does not explain all molecular arrangements.

In most cases, the central atom is a matter of common sense, and will be obvious to students once they start to draw the structures.

*Like the other noble gases, the outermost orbital of helium is filled. In this case, only two electrons are needed because only the first energy level is involved.

3. *How many dots should appear in the structure?* The number of dots will be equal to the sum of the valence electrons available from all the atoms in the molecule.
4. *Where do the dots go in the structure?* Place the dots around the atoms so that each nonmetallic atom has eight electrons—an octet structure.

Electron dot structures are not difficult to construct. The easiest way to learn is by doing several simple examples. You can then adapt the method to more complex molecules. The following examples will give you some practice.

Example 12-3

Draw the electron dot structures for the following molecules: CCl_4, NH_3, and C_2H_6.

Problem-solving skills would be strengthened by asking students to propose structures for C_2H_6 before reading the text discussion.

Remind students that hydrogen can only bond to one atom, and that the arrangement shown for C_2H_6 is the most reasonable. It is also symmetrical.

Call students' attention to the fact that two electrons of nitrogen are not involved in bonding.

▶ **Suggested solution**

Following the suggestions for writing electron dot structures, construct the diagrams in a step-by-step process:

	CCl_4	NH_3	C_2H_6
Number of atoms	C = 1 Cl = 4	N = 1 H = 3	C = 2 H = 6
How atoms are arranged	Cl Cl C Cl Cl	H N H H	H H H C C H H H
Total number of dots per element	C = 4 Cl = 28 ——— 32	N = 5 H = 3 —— 8	C = 8 H = 6 —— 14
Arrangement of dots	:C̈l: :C̈l:C:C̈l: :C̈l:	H:N̈:H H	H H H:C:C:H H H

The pairs of dots that appear between atoms represent a shared pair of electrons or a covalent bond between the two atoms. The dots that are *not* between atoms represent electrons not involved in bonding.

For convenience, a bonded pair of electrons is frequently represented as a line instead of a pair of dots. The three structures above can be drawn as

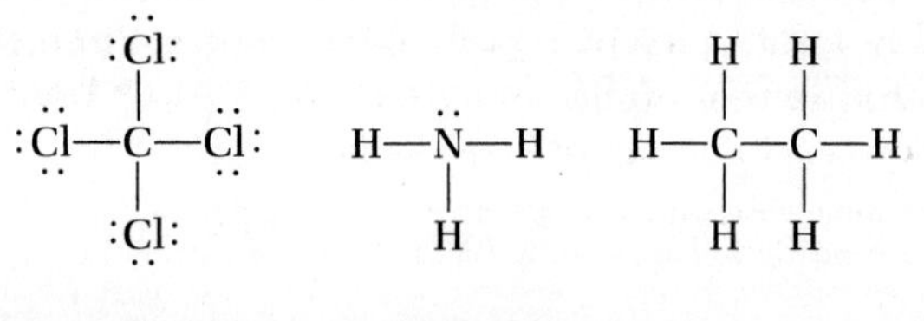

$$\begin{array}{ccc} :\ddot{C}l: & & \\ | & & H \quad H \\ :\ddot{C}l-C-\ddot{C}l: \qquad H-\ddot{N}-H \qquad H-C-C-H \\ | & | & | \quad | \\ :\ddot{C}l: & H & H \quad H \end{array}$$

Example 12-4

Draw the electron dot structure for CO_2.

▶ Suggested solution _____

Using the same procedure as you did in Example 12-3, you know that CO_2 has one carbon atom and two oxygen atoms. According to the second rule for drawing dot structures, the two oxygen atoms will each be attached to the carbon atom:

$$O \quad C \quad O$$

The total number of electrons to be represented is 16:

$$C = 4$$
$$O = \underline{12}$$
$$16$$

If you try to draw the structure using the procedures you have already learned, there is a small problem. There does not seem to be enough dots to give each atom an octet structure:

$$:\ddot{O}—C—\ddot{O}: \quad \text{or} \quad :\ddot{O}—\ddot{C}—\ddot{O}:$$

In either case, carbon or oxygen will not have eight electrons represented. For this situation, a modification of rule 4 will help:

1. Determine how many dots are needed to represent eight electrons around each atom *without any bonds*. For CO_2, the number is 24.
2. Subtract the actual number of electrons available from this total:

$$24 - 16 = 8$$

This difference is the number of bonding electrons. For CO_2, there should be eight electrons located in bonds between the carbon atom and the two oxygen atoms. The dot structure would now be drawn as:

$$:\ddot{O}::C::\ddot{O}: \quad \text{or} \quad :\ddot{O}=C=\ddot{O}:$$

Each oxygen atom shares two pairs of electrons with the carbon atom. A **double bond** is a covalent bond in which four electrons (two pairs) are shared by the bonding atoms. Double bonding is an arrangement that accounts for all electrons and is compatible with the octet rule. There is sound evidence for the existence of double bonds in molecules. Atoms believed to be bonded by double bonds are more difficult to break apart than similar atoms bonded by single bonds. Also, experimental measurements show that the bond distance of a suggested double bond between two atoms is shorter than when the atoms are believed to be bonded by only a single bond.

Evidence also has been found for the existence of triple bonds. A **triple bond** is a bond in which two atoms share three pairs of electrons. An example of a molecule containing a triple bond is nitrogen gas. The electron dot structure for N_2 is

$$:N:::N: \quad \text{or} \quad :N{\equiv}N:$$

Do You Know?

Double and triple bonds are most commonly found in compounds that contain combinations of carbon, oxygen, and nitrogen. A "quadruple" bond between two carbon atoms is unstable.

Example 12-5

Draw the electron dot structure for the polyatomic ion SO_3^{2-}.

▶ **Suggested solution**

Following the first two rules as before, you can conclude that the sulfur atom is at the center of the molecule, with each oxygen atom bonded to it:

$$O \quad O$$
$$S$$
$$O$$

For a polyatomic ion, the number of dots in the structure will be equal to the total number of valence electrons of the atoms plus or minus the electrons that account for the charge of the ion. In the case of SO_3^{2-}, there are two more electrons because the ion has a charge of $2-$:

$$
\begin{aligned}
S &= 6 \\
O &= \underline{18} \\
& 24 + 2 = 26
\end{aligned}
$$

First, draw a single bond between the sulfur atom and each oxygen atom. Then fill in enough dots to give each oxygen atom a total of 8.

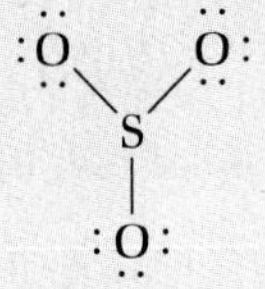

You have used 24 dots. At this point, sulfur only has 6 electrons. Place the two remaining electrons with the sulfur atom.

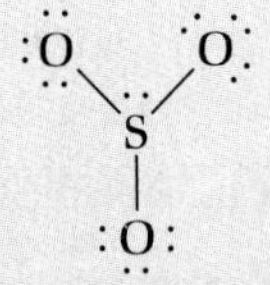

The structure seems reasonable because each atom has 8 electrons and all electrons are accounted for. The two added electrons next to sulfur are not necessarily the two electrons in the $2-$ charge of the ion. The two extra electrons are not associated with sulfur exclusively. One more addition to the structure and it will be complete. Square brackets are placed around it and the charge on the ion is indicated:

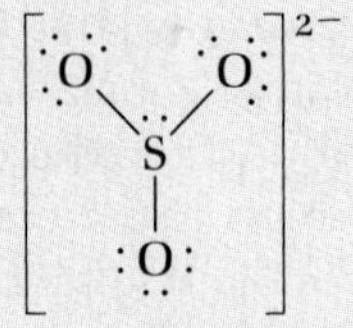

12-7 Exceptions to the Octet Rule

The octet rule is a device used to picture the structure of molecules as perceived through experimental evidence. Molecules, of course, do not "obey" the octet rule or any other rule. The concept serves only as a scientific tool or rule of thumb. Not surprisingly, the bonding in some molecular structures cannot be explained using the octet rule.

Molecules that contain an odd number of valence electrons cannot be diagrammed adequately using the octet rule as a guide. For example, nitric oxide, NO, has 11 valence electrons. Two electron dot structures can be drawn, though neither is an accurate representation of the molecule. Situations like this one are described in Section 12-10.

$$\cdot \ddot{N}\!=\!\ddot{O} \quad \text{or} \quad \ddot{N}\!=\!\ddot{O}$$

There are some stable gaseous molecules that possess a central atom with fewer than eight electrons. Beryllium fluoride, BeF_2, and boron trifluoride, BF_3, are examples. Beryllium has two valence electrons. In the gaseous state, it can be involved in a stable covalent bond with two fluoride atoms, as represented by the structure below:

$$:\!\ddot{F}\!:\!Be\!:\!\ddot{F}\!:$$

Boron has three valence electrons. It can bond covalently in the gaseous state with three fluoride atoms:

$$:\!\ddot{F}\!:\!B\!:\!\ddot{F}\!:$$
$$:\!\ddot{F}\!:$$

There are some central atoms, especially phosphorus and sulfur, that can be surrounded by more than eight electrons in a stable molecule. The phosphorus atom can share five covalent bonds (ten electrons). Sulfur can share as many as six bonds (12 electrons). The structures of the molecules that form can be explained by assuming that bonding occurs using *s*, *p*, and *d* orbitals. The theory that deals with this type of bonding is discussed in the appendix.

You may want to use this opportunity to introduce students to isomerism. If so, ask students to draw structures for C_4H_{10}, C_5H_{12}, and $C_2H_2Cl_2$. The students will arrive at more than one possibility for each.

The five bonds in phosphorus and the six bonds in sulfur are explained by hybridization (five sp^3d orbitals in phosphorus and six sp^3d^2 orbitals in sulfur).

The two structures for NO are resonance structures. Resonance is discussed in Section 12-10.

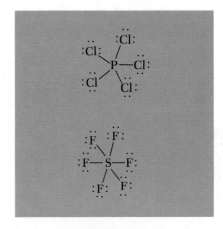

Figure 12-9 Phosphorus pentachloride and sulfur hexafluoride are examples of molecules in which the central atom is bonded to more than four atoms.

Review and Practice

1. simple structure using an atomic symbol and dots to represent an atom
1. What is an electron dot (Lewis) structure?

2. Briefly describe how the electron dot structure for an element is determined.

3. Write the electron dot structure for each of the following elements: potassium, arsenic, bromine, silicon, tellurium, aluminum.

4. Draw the electron dot structure for each of the following molecules: Cl_2, SiF_4, H_2O, C_3H_8.

5. Draw the electron dot structure for each of the following formulas: C_2H_4, SO_4^{2-}, HCN.

6. What is a double bond? Give an example of a molecule containing a double bond.

2. It is the symbol of the element surrounded by the same number of dots as there are outer *s* and *p* electrons in atoms of that element.

3. $K\cdot$; $\cdot\ddot{As}\!:$; $:\!\ddot{Br}\cdot$; $\cdot\ddot{Si}\cdot$; $\cdot\ddot{Te}\!:$; $\cdot Al\cdot$

4. $:\!\ddot{Cl}\!:\!\ddot{Cl}\!:$;   with F structures; $H\!:\!\ddot{O}\!:\!H$; $H\!:\!\ddot{C}\!:\!\ddot{C}\!:\!H$

5. structures for C_2H_4, SO_4^{2-}, and $H\!:\!C\!:::\!N$

(Students may substitute dashed lines for shared pairs of electrons.)

6. the sharing of two pairs of electrons between two atoms; CO_2

12-8 Bond Energy

In Section 12-2 you studied the energy effects of chemical bonding. When bonds form, the potential energy of the system decreases. It seems reasonable to think that the potential energy of a system increases when chemical bonds are broken.

Bond energy is the energy required to separate two bonded atoms. It takes a great deal of energy to break a strong bond, and only a little energy to break a weak bond. *It should be remembered that energy must be put into a system in order to break a bond, and energy is always released when a bond is formed.*

Because atoms and molecules are so small—much too small to see or measure individually—measurements are made on large aggregates of molecules. The bond energy normally is expressed as the amount of energy in kilojoules required to break one mole of bonds. For example, experiments have shown that 436 kJ of energy must be added to one mole of hydrogen molecules in order to break all the covalent bonds. This relationship can be expressed in the equation:

$$H_2(g) + 436\,kJ \longrightarrow H(g) + H(g)$$

The reaction is endothermic, that is, energy must be absorbed to produce individual gaseous atoms. Bond energies have been measured for a number of covalent bonds, and some of them are listed below.

> **CHEM THEME**
>
> The energy effects of all chemical reactions come from the breaking and forming of bonds.

Table 12-2

BOND ENERGIES (kJ/mol)*			
H—H	436	H—Cl	431
C—H	414	H—Br	364
C—C	347	H—I	299
C=C	619	N—N	159
C≡C	812	N≡N	941
O—O	138	N—H	389
O=O	494	Cl—Cl	243
O—H	463	Br—Br	193
C—O	335	I—I	151
C=O	707	C—Cl	326
H—F	569	C—Br	276

*The bond energies in the table are averages. Values vary depending on the structure of the molecule.

The bond energies in Table 12-2 can be used to calculate the amount of energy absorbed when bonds are broken, or the amount of energy released when bonds form. For example, the formation of one mole of HBr bonds releases 364 kJ of energy to the surroundings. The reaction is expressed in the following equation:

$$H(g) + Br(g) \longrightarrow HBr(g) + 364\,kJ$$

The energy usually is released as heat.

The table allows you to compare the strengths of some bonds and helps to explain patterns of reactivity in certain molecules. Compare the bond energies listed for a single, double, and triple carbon-carbon

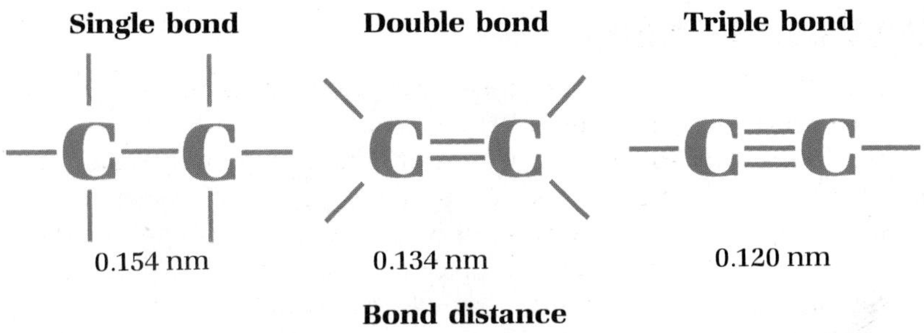

Single bond   **Double bond**   **Triple bond**

0.154 nm   0.134 nm   0.120 nm

Bond distance

Figure 12-10 A carbon atom is capable of a single, double, or triple bond with another carbon atom. Triple bonds are the strongest and shortest of the three carbon-carbon bonds.

bond. The strength of the double bond is greater than that of the single bond, and the triple bond is greatest of the three. Measurements of the bond distances for each bond in Figure 12-10, indicate that the atoms are bonded more closely as the bond strength increases.

Interestingly, the bond energy of the carbon-carbon double bond is not twice that of the single bond. The energy difference between the triple and double bonds is even less than between the double and single bond. How can these data be interpreted?

The data suggest that the two bonds of a double bond are not the same strength and that the second bond can be more easily broken. The same is true for the second and third bond in a triple bond. Not surprisingly, experiments show that a compound such as C_2H_4, containing a double carbon bond, is more reactive than C_2H_6, which has only a single bond between the carbon atoms. The triple-bonded C_2H_2 molecule is even more reactive under certain circumstances.

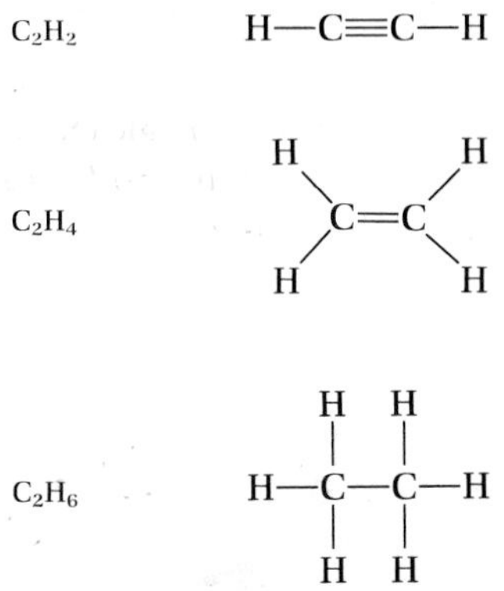

The reactivity of double and triple carbon-carbon bonds varies with the type of reaction. Section notes for Sections 23-7 and 23-8 provide more detail about this topic.

12-9 Extension: Experimental Evidence for Bonding Theory

A theory is a good one if it makes predictions that are consistent with experimental data and if it enables you to make predictions before experiments are done.

In Chapter 10, you learned that spectroscopic studies of elements helped provide information to build a theory about electrons in atoms. Spectroscopic data may also be used to develop the theory about the structure of molecules. Chemical bonds are not static. The atoms that are bonded vibrate toward and away from the molecule's center of mass. Molecules also tumble, or rotate, around their centers of mass. (See Figure 12-11 on the next page.) Molecules vibrate and rotate only in specific energy modes. Vibrations and rotations in between these specific states are not observed. Therefore, it may be said that the energy of molecules, as well as atoms, is quantized. Like atoms, molecules can absorb electromagnetic radiation of specific wavelengths, resulting in characteristic absorption spectra.

Scientists have taken advantage of the quantum characteristics of molecules in order to:

1. classify bonds between elements according to bond lengths and bond strengths, and
2. identify an unknown chemical by matching its energy absorption pattern against a known pattern.

Show the CHEM study film *Molecular Spectroscopy* if it is available.

Figure 12-11 Molecules rotate at different velocities because of their shape and mass. Additionally the bonded atoms in a molecule vibrate. The absorption of energy by molecules causes changes in vibrational and rotational motion at specific energy levels. Different kinds of bonds produce different spectral patterns.

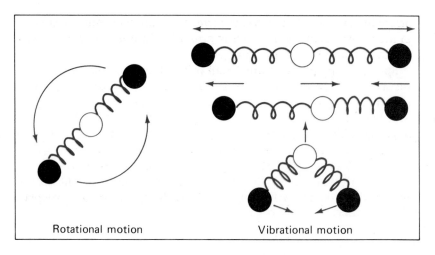

Rotational motion   Vibrational motion

Energy for the transition of electrons in orbitals can be detected in the visible or ultraviolet regions of the electromagnetic spectrum (Chapter 10). Energy for the transitions of molecular vibrations (vibrational energy) is detected in the infrared region, close to the visible spectrum. Energy for the transitions of molecular rotations (rotational energy) is detected in the infrared and microwave regions.

If a known sample of the compound cyclohexene, C_6H_{10}, is placed in an infrared spectrometer, a spectrum much like the one in Figure 12-12 is produced on a recording machine. The long peaks downward indicate the wavelengths at which the cyclohexene absorbed energy. The infrared spectrum of cyclohexene is unique. If an unknown substance produces the same spectrum as that shown in Figure 12-12, the data serve as evidence that the unknown is cyclohexene.

Information about bonding can be gathered through infrared and microwave spectroscopy. Bond angles, bond distances, and bond

Figure 12-12 The carbon-carbon double bond in cyclohexene produces a characteristic peak in the molecule's infrared spectrum. Analysis of spectral patterns provides information about the nature of bonds in molecules and the identities of the molecules themselves.

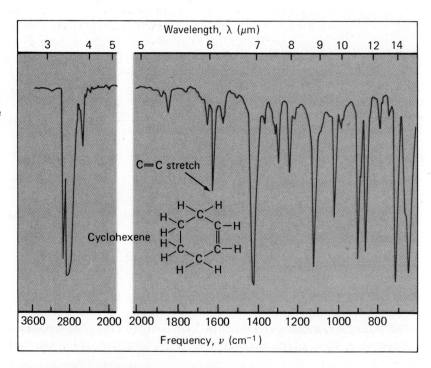

strengths are revealed by the changes in rotation of molecules. Much of the information about bonding presented in this chapter was accumulated using these powerful tools.

Spectroscopy plays an important role in forensic science (crime detection). Instruments such as the infrared or the ultraviolet-visible spectrometer provide evidence in criminal cases that was not obtainable before. Spectrometers are routinely used to test for the presence of drugs, glue, and many other organic and inorganic substances. In one case, an infrared spectrometer was used to positively match the paint on the body of a hit-and-run victim with the paint on the suspect's car. These instruments also can be used to match clothing materials found at the scene of a crime with the clothing of suspects.

12-10 Extension: Resonance

In Section 12-6, it was mentioned that there are limitations to using electron dot structures to represent molecules. Some molecules and polyatomic ions have properties that are not adequately explained by electron dot structures. The molecule sulfur trioxide, SO_3, is a good example. If you follow the method described previously, you could arrive at a structure with a double bond and two single bonds:

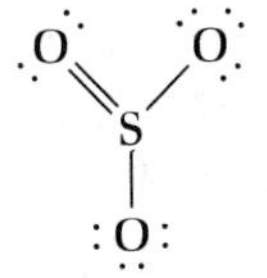

If that is the case, the molecular spectrum for the molecule should show two peaks—one corresponding to absorption for the double bond and one corresponding to absorption for the single bonds. However, the spectrum for SO_3 has only a single peak, suggesting that all the bonds are identical. Furthermore, experimental measurements of bond strengths indicate that the strength of each bond is greater than would be expected for a single bond but less than expected for a double bond. Obviously, an electron dot structure cannot show 1 1/3 bonds. There are limitations to the usefulness of the method.

The SO_3 molecule represents an example of situations when the simple theory of bonding presented here must be modified. As you continue your study of chemistry, you will learn more complex theories of bonding that enable chemists to explain the experimental evidence that exists.

There are situations when it is useful to be able to draw molecules like SO_3. To do so, try to imagine the double bond connecting any one of the oxygen atoms to sulfur. You would then be able to draw *three* structures:

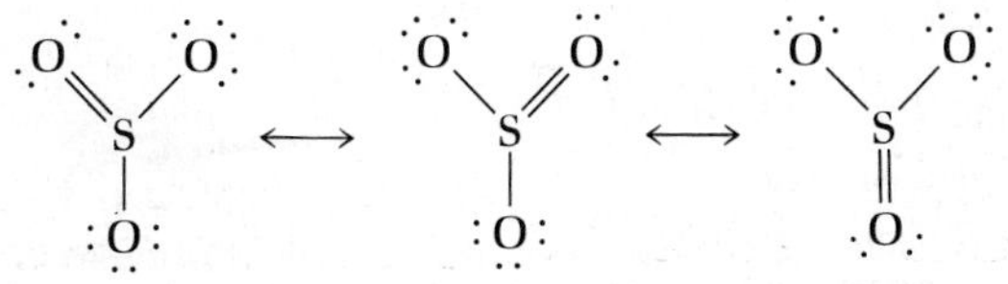

The actual SO_3 molecule is described as a composite of the three pictured here. The various structures used to represent a molecule that cannot be represented with one diagram are known as **resonance structures**. (The double-ended arrows indicate resonance structures.) Resonance structures are not meant to imply that three different molecular structures are present in SO_3. They merely indicate that *no single structure adequately represents the molecule*.

Example 12-6

Draw an electron dot structure for the nitrate ion, NO_3^-.

▶ **Suggested solution**

The ion contains 24 electrons, 23 from the valence electrons of the atoms and one additional electron from the negative charge. By following the steps used in previous examples, you may arrive at the following structure:

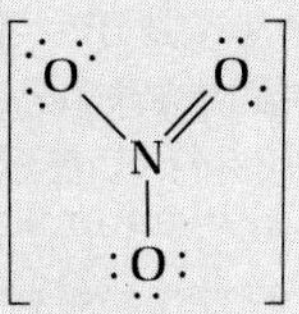

It is also equally possible to draw

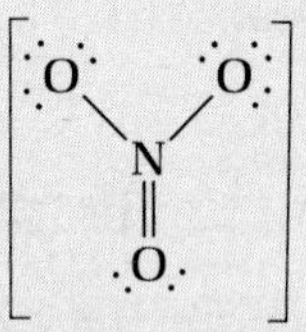

and

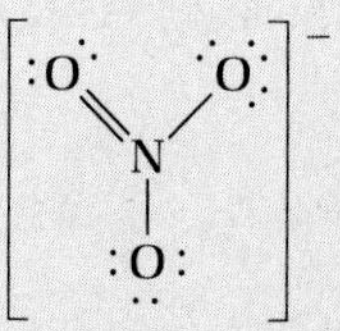

It should be remembered that none of these structures actually exists for the NO_3^- ion. The three structures together simply help to provide a better representation than any one of the diagrams alone.

1. when no single electron dot structure adequately represents the molecule
2.

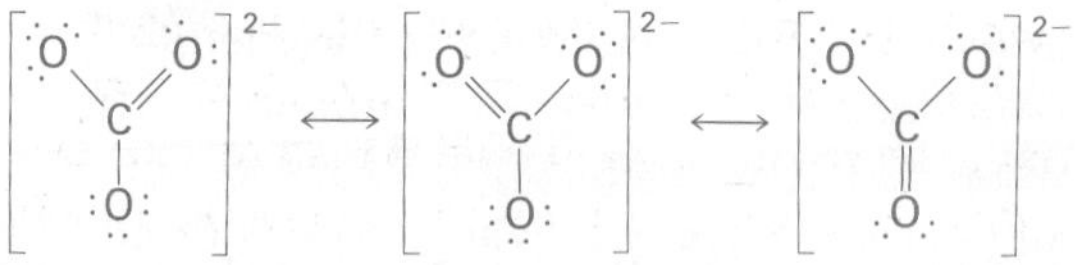

Review and Practice

1. Briefly explain when a resonance structure is helpful in representing a molecule.

2. The carbonate ion, CO_3^{2-}, can be represented by resonance structures. Draw the electron dot structures for the carbonate ion.

3. What is bond energy?

4. In Table 12-2, find the bond energies of a single and a double oxygen-oxygen bond. What conclusions can you draw from the data?

5. In what ways do the quantum characteristics of chemical bonds and infra-red spectrometry provide information about molecular structure?

3. the energy required to separate two bonded atoms
4. The bond energy of the double oxygen bond is much greater than that of the single bond. Two doubly bonded oxygen atoms would be more difficult to separate than two singly bonded ones.
5. Quantum characteristics of molecules can be used to 1) classify bonds between atoms, and 2) identify unknown chemicals by matching their energy absorption patterns against known patterns.

The Shape and Behavior of Molecules

Several activities in school involve only two dimensions. Teachers write on rectangular blackboards, students write on flat pieces of paper, and print is two dimensional. Even work with a computer is restricted to the two dimensions of a display screen. It is sometimes difficult to think in three dimensions. Electron dot structures may give you the impression that molecules are two dimensional, but they are not. This part of the chapter will introduce you to the three-dimensional world of molecules. It is important to see molecules in this fashion, because it is more realistic and because the shapes of molecules influence some of the properties of matter.

Liquids and solids are condensed states of matter whose behaviors are affected by molecular interactions. These interactions depend on molecular shapes. One liquid, water, has a relatively high boiling point when compared to analogous compounds in the oxygen family. How can this irregularity be explained? Understanding something about the shapes and interactions of molecules can help you answer this question. Later you will learn how some of these concepts are being applied to studies of biochemical processes.

CONCEPTS

- repulsion of electrons in molecules
- predicting the shapes of molecules from their number of bonds
- polar molecules vs. nonpolar molecules
- forces between molecules
- hydrogen bonds—a special intermolecular force
- computer-built molecules
- attractions affecting nonpolar molecules

12-11 Molecules and Geometric Shapes

What information do you already have that can help you modify two-dimensional Lewis models to more accurately picture molecules in three dimensions? You know that all electrons have the same charge, and that like charges repel. It is reasonable to assume that the electron pairs in bonds will be oriented in a molecule as far from each other as possible. The valence electrons should be expected to occupy regions of space so that the shared pairs are evenly spread out around the central atom.

Consider how these assumptions can be applied to a molecule containing two single covalent bonds like BeH_2. In Figure 12-13, balloons are used to help you visualize what the molecule would be like. Imagine that the beryllium atom is located at the spot where the two balloons are tied, and the two hydrogen atoms are located almost at the far end of each balloon. For the electron pairs to be as far apart as possible, the two hydrogen atoms would be on opposite sides of the molecule. The angle formed when lines are drawn between the beryllium and two hydrogen atoms is 180°, and the shape of the molecule is linear. This model is supported by experimental measurements of the bond angle of BeH_2.

Next, consider a molecule containing three covalent bonds such as boron trifluoride, BF_3. The dot structure for the molecule can be drawn as

$$:\!\ddot{F}\!-\!B\!-\!\ddot{F}\!:$$
$$|$$
$$:\!\ddot{F}\!:$$

The actual shapes of many molecules have been determined experimentally using X rays, microwaves, and NMR spectroscopy.

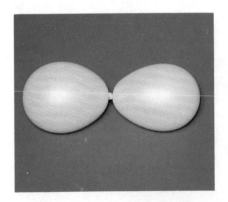

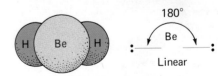

Figure 12-13 As illustrated by the balloon model, a linear arrangement of atoms for a molecule such as beryllium hydride allows for the greatest distance between the electron pairs.

Figure 12-14 Three balloons tied together are farthest apart when they are at angles of 120°. A molecule having this geometry, such as boron trifluoride, is called trigonal planar.

Figure 12-15 When four balloons are arranged so that they are as far apart as possible, a structure called tetrahedral results. The angles between the balloons are all 109.5°.

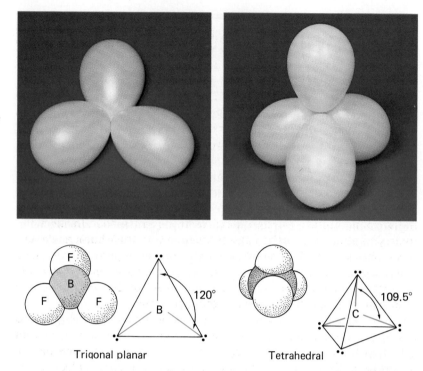

Trigonal planar   Tetrahedral

The term *valence shell electron pair repulsion* is introduced at the end of the section. At this point, it may be more important for the students to focus on the concept rather than trying to remember its name.

In all examples given here, you may want to explain to students that the angles are between the atomic kernels involved in bonding.

How might this molecule actually appear in space? Again, balloon models help you to picture the possibilities. Figure 12-14 shows three balloons representing the three bonding pairs of electrons. If the balloons are arranged to take positions that are as far apart as possible, they are in a plane around boron. The angle formed by any two fluorine atoms with the central boron atom is 120°. The molecular shape is described as trigonal planar. This model is also supported by measurements of the bond angles of BF_3.

Methane, CH_4, has four covalent bonds. The dot structure of methane is quite different from the actual three-dimensional geometry of the molecule.

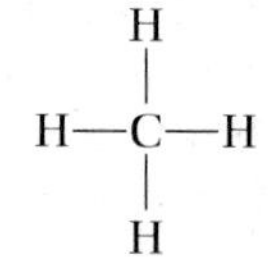

$$H-\underset{\underset{H}{|}}{\overset{\overset{H}{|}}{C}}-H$$

If all four bonds are in the same plane, then the maximum distance between the bonding pairs of electrons would make equal bond angles of 90° between hydrogen atoms. However, analysis shows that the H—C—H bond angles in methane are 109.5°. Are the assumptions made in the beginning of this section no longer valid? The balloon model in Figure 12-15 shows how you can orient the hydrogen atoms to give bond angles greater than 90°. A three-dimensional figure can form so that the carbon atom is at the center and each of the shared pairs of electrons is directed toward the corners of a tetrahedron. (A tetrahedron is a geometric shape with four sides, each side an equilateral triangle.) Now the hydrogen atoms are positioned around the central atom so that they are as far apart as possible. Each H—C—H bond is 109.5°, and the structure is described as tetrahedral. The initial assumptions are still supported by experimental evidence.

In the molecules you have seen so far, all of the valence electrons of the central atom are bonding electrons. How is the shape of a molecule affected if one or more pairs of these valence electrons are not involved in a bond? For example, in ammonia, NH_3, nitrogen has one lone pair of electrons.

Ammonia has a shape that is derived from a tetrahedral structure. Nitrogen is at the center of the molecule with each hydrogen atom directed towards a corner of a tetrahedron. The lone pair of electrons is directed towards the fourth corner. Experimental measurements indicate that the H—N—H bond angle is 107°, slightly less than the 109.5° of methane. One possible explanation for the difference is that due to electron repulsion, the lone pair of electrons occupies a slightly larger orbital than the electrons involved in bonding. In addition, the overlapping orbitals of the bonding electrons are smaller. In effect, the hydrogen atoms are pushed closer together and their bond angle narrows. The situation is similar to a four-balloon arrangement where one balloon is larger than the others. The shape of the ammonia molecule and other molecules in which there are three bonded atoms and one lone pair of electrons is called trigonal pyramidal.

The water molecule has two hydrogen atoms bonded to oxygen and two lone pairs of electrons. The shape of the molecule is derived from a tetrahedral structure. Each hydrogen atom and each lone pair of electrons is directed towards a corner. Like the NH_3 molecule, the H—O—H bond angle in water is less than 109.5°. Experiments show it to be 104.5°. In this case, there are two lone pairs of electrons that occupy larger orbitals than the electrons in bonds. The hydrogen atoms are pushed even closer together. The shape of the water molecule and other molecules with three bonded atoms and two lone pairs of electrons is described as angular, or bent.

Some molecules have more than four pairs of electrons in bonds around a central atom. In Section 12-7, you read that phosphorus can share five bonds, and sulfur can share six bonds. The balloon representation for PCl_5 is shown in Figure 12-18 on the next page. Five balloons are needed to represent the shape of PCl_5. The name of this shape is trigonal bipyramidal. Three pairs of shared electrons in a plane describe a triangle around the center of the structure. The other two pairs are arranged one above and one below the plane. Six balloons are needed to suggest the shape of the SF_6 molecule, which is

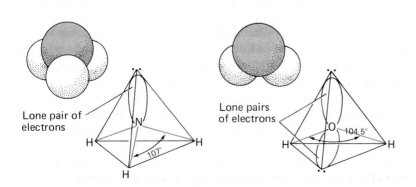

Figure 12-16 In the ammonia molecule, the hydrogen atoms are at the corners of an equilateral triangle, but the nitrogen atom does not lie in the plane of that triangle. Nitrogen is above the plane at the apex of a pyramid, with the three hydrogen atoms forming the base.

Figure 12-17 The atoms in a water molecule lie in a plane and form an angle of 104.5°. The two lone pairs of electrons do not lie in that plane.

Figure 12-18 Can you see why a molecule with five atoms bonded to a central atom is described as trigonal bipyramidal?

Figure 12-19 Six balloons illustrate the octahedral shape of molecules with six bonded atoms around the central atom.

Figure 12-20 Molecules can be classified by their geometry. When predicting the shape of a molecule, the number of bonds and of lone pairs of electrons need to be considered. You cannot always determine the shape of a molecule by looking at the formula.

octahedral (Figure 12-19). There are other combinations with five or six balloons, but these combinations are variations on the same theme. Figure 12-20 provides a summary of all the structural shapes you have been introduced to in this section.

EXAMPLES OF MOLECULAR SHAPES

NAME	STRUCTURAL SHAPE	ATOMS BONDED TO CENTRAL ATOM	LONE PAIRS OF ELECTRONS	BOND ANGLE	EXAMPLE	
					FORMULA	ELECTRON DOT DIAGRAM
Linear		2	0	180°	BeH_2	H:Be:H
Trigonal planar	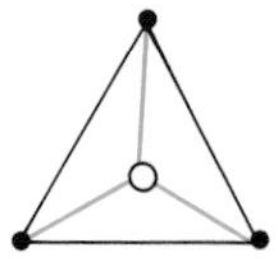	3	0	120°	BF_3	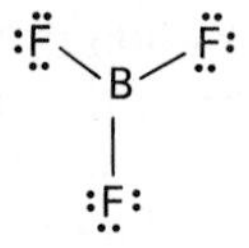
Tetrahedral	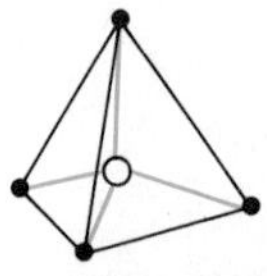	4	0	109.5°	CH_4	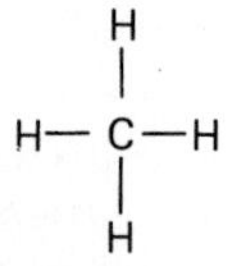
Trigonal pyramidal	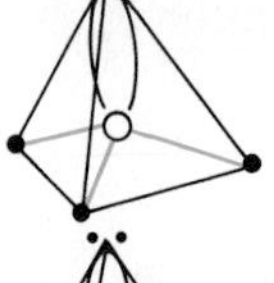	3	1	107°	NH_3	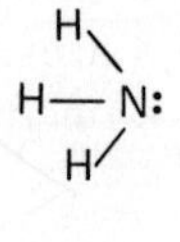
Angular	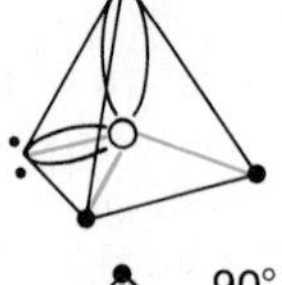	2	2	104.5°	H_2O	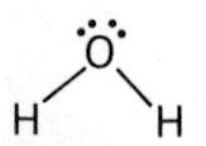
Trigonal bipyramidal	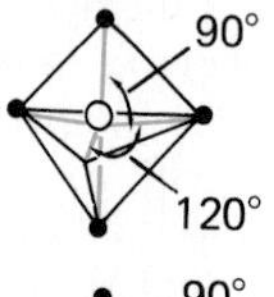	5	0	90° 120°	PCl_5	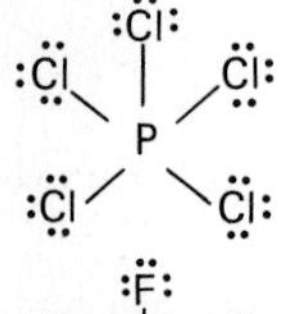
Octahedral	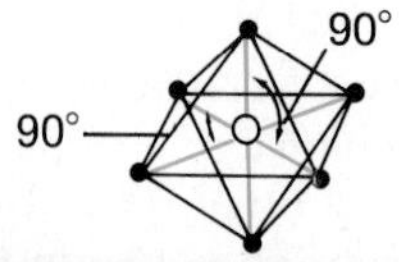	6	0	90°	SF_6	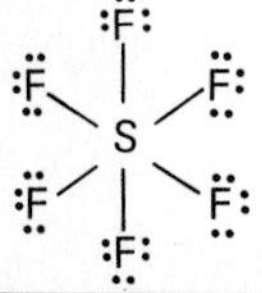

Example 12-7

Predict the shape of the following molecules:

$$H_2S, \quad SiH_4, \quad InCl_3, \quad AsH_3$$

▶ **Suggested solution**

Use what you know about elements and their patterns of bonding to predict the shapes of unknown molecules. You already have learned that elements in the same family in the periodic table behave in similar ways chemically. It is reasonable to assume that such elements form similar types and numbers of bonds. The simplest method for anticipating the shape of an unknown molecule is to compare it with a molecule whose central atom is in the same family as the central atom of the unknown. In the case of H_2S, sulfur and oxygen are in the same family and therefore have the same number of valence electrons. It is reasonable to assume that H_2S has a shape similar to H_2O, which is angular.

Silicon and carbon are in the same family, so you could predict that the shape of SiH_4 will be similar to CH_4, which is tetrahedral.

$InCl_3$ most likely has the same shape as BF_3, which is trigonal planar.

AsH_3 would have the same shape as NH_3, trigonal pyramidal.

Do You Know?

Although "octa" means eight, an octahedral structure only has six atoms attached to the central atom. Eight planar surfaces are needed to completely enclose the six bonded atoms in an octahedral molecule.

Notice that $InCl_3$ and AsH_3 would not have the same shape even though their formulas are similar (a central atom bonded to three atoms). AsH_3 contains an unshared pair of electrons belonging to the arsenic atom.

There are several theories in chemistry that suggest explanations for bonding in molecules. Each theory leads chemists to predict the same molecular shapes you have been introduced to in this section. The theory you have studied here, describing the equal distribution of electron pairs around a central atom, is known as the **valence shell electron pair repulsion theory** (VSEPR for short). It provides a good introduction to predicting the shapes of molecules because it is rather easy to understand and it is comprehensive, accounting for the shapes of many molecules. The theory does raise some questions, however. You learned that the s and p valence orbitals are oriented around the x, y, and z axes at angles of 90°. Yet the VSEPR theory indicates that the central atom may have pairs of electrons arranged in a tetrahedral shape, for example, with a bond angle of 109.5°. Chemists have devised a theory that describes how atomic orbitals become bonding orbitals with orientations predicted by VSEPR theory. The theory uses a concept called hybridization and is presented in the Appendix.

1. Bonded atoms and lone pairs of electrons will be arranged about a central atom as far apart as possible to minimize repulsion.
2. Their charges are the same, negative.
3. No. Electron dot structures are two-dimensional; molecules are three-dimensional. The dot structure of CH_4 implies that the C—H bonds are 90° apart. The actual bond angle is 109.5°.
4. They have different numbers of bonded electrons and lone pairs of electrons around their central atoms.
5. GaH_3 = trigonal planar; GeH_4 = tetrahedral; PCl_3 = trigonal pyramidal; SO_3 = trigonal planar
6. linear

Review and Practice

1. State the basic assumption of the VSEPR theory.

2. Why do sets of shared pairs of electrons in a molecule repel each other?

3. Do electron dot structures accurately describe the shapes of molecules? Explain with an example.

4. Why do bond angles in CH_4, NH_3, and H_2O differ?

5. Describe the shapes of the following molecules: GaH_3, GeH_4, PCl_3, and SO_3.

6. Predict the shape of the HCl molecule, which has only one covalent bond.

12-12 When Do Polar Bonds Make Polar Molecules?

If molecules were isolated units and did not interact with each other, the topic of molecular shapes would be of very little value. Molecules, however, do interact, and their shapes determine the extent and nature of the interactions. Recall that in Chapters 1 and 7, you learned that molecules are very close to each other in the solid and liquid states of matter. It is fairly easy to understand that positive and negative ions can form condensed states because of the intense electrostatic attractions present. How is it that neutral molecules can form liquids and solids? There must be some sort of forces that attract molecules to form solids and liquids. If such forces exist, their effects can be observed and the data may be used as evidence to confirm the theoretical shapes of molecules you studied in Section 12-11.

Recall the discussion in Section 12-3 about polar bonds. In the HCl molecule, the more electronegative atom, chlorine, attracts the shared pair of electrons more than hydrogen does. The result is a polar bond, or dipole, that is somewhat negative near the chlorine atom and somewhat positive near the hydrogen atom. Because HCl consists of only one polar bond, it also is a polar molecule. As molecules of HCl approach each other, there is an attraction between them (Figure 12-21). The slightly positive end of one molecule is attracted to the slightly negative end of the other molecule. The attraction is intermolecular because it takes place *between* molecules. It is logical to state that other molecules, consisting of two different atoms sharing a polar bond, are polar molecules. However, the situation is not as simple for more complex molecules. Polar bonds do not always produce a polar molecule.

Take, for example, a molecule such as BeH_2. Each Be—H bond is a dipole, and because hydrogen is more electronegative than beryllium, the partial negative charge of the bond is near the hydrogen atom. It was suggested that BeH_2 is a linear molecule with a bond angle of 180°. If so, the two dipoles in the molecule can be represented as shown in Figure 12-22. The individual dipoles would seem to be directed opposite to each other so that they cancel. An analogy to this situation is a tug-of-war. When two competing teams are pulling in opposite directions on a rope with equal force, the rope and the teams do not move. The resultant force is zero.

In a similar way, the BeH_2 molecule would have a resultant dipole of zero because the hydrogen atoms are attracting the shared electrons equally, but in opposite directions. BeH_2 contains two polar bonds, yet a linear shape suggests a nonpolar molecule. Is there experimental evidence that can give credibility to what has been described?

One reason polarity in molecules can be detected is because the molecules are affected by an electric field generated between two charged plates. In Figure 12-23 (left), molecules are oriented randomly, but when an electric field is generated (right), the molecules tend to align themselves so that the negative ends point toward the positive plate and vice versa. The presence of the dipoles has a measurable effect on the charge held between the plates. The physics and mathematics of what happens are beyond the scope of this book. What is important is the fact that molecules can be identified experimentally as polar or nonpolar. In the case of BeH_2, the molecule behaves as if it is nonpolar. A linear structure is a reasonable shape for a nonpolar molecule that contains two polar bonds.

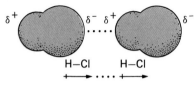

Intermolecular
attraction

Figure 12-21 Two HCl molecules will be attracted to each other at their oppositely charged ends.

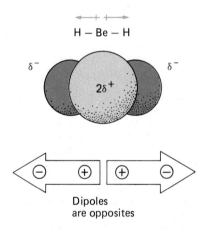

Dipoles
are opposites

Figure 12-22 BeH_2 has two polar bonds. However, the dipoles are oriented in opposite directions. The molecule is nonpolar overall.

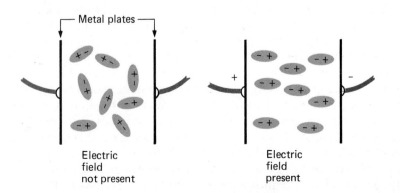

Electric
field
not present

Electric
field
present

Figure 12-23 The activation of an electric field causes polar molecules to orient themselves so that their charged ends are turned toward the oppositely charged electric plate. Nonpolar molecules would be unaffected.

Figure 12-24 The three dipoles in BF_3 result in a nonpolar molecule. In methane, there are four dipoles that cancel. The molecule is nonpolar. Note that in a C—H bond, the negative end of the dipole is at the carbon atom. The cancellation effect in the molecule is still the same.

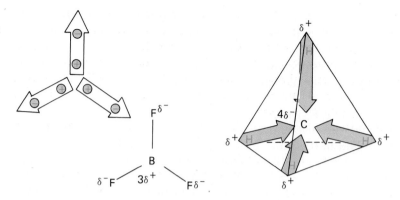

You may want to explain to students that the dipole is a displacement of charge rather than a "force." However, the analogy may help them understand what is happening.

Make sure the students understand the difference between bond polarity and molecular polarity.

Figure 12-25 In ammonia, the atoms are oriented in such a way that the dipoles do not cancel. Also the end of the molecule that contains the lone pair of electrons is more negative. The molecule is polar.

The experimental behavior of molecules can be used to lend support to the shapes of other molecules described in Section 12-11. BF_3 and CH_4 both have been shown in experiments to be nonpolar. A trigonal planar shape for BF_3 and a tetrahedral shape for CH_4 would account for the cancellation of dipoles within each molecule. The first diagram in Figure 12-24 illustrates the direction of each polar B—F bond. The three bonds are arranged such that they are equally spread around the central boron atom. A comparable situation would occur in the CH_4 molecule, where the four slightly polar C—H bonds are spaced equally around the central carbon. In each case, the shape leads you to conclude that the molecule is nonpolar.

The situation is different for the ammonia molecule, NH_3, which has been shown experimentally to be polar. Does the suggested trigonal pyramidal shape predict a polar molecule? Figure 12-25 shows the orientation of the dipoles in the NH_3 molecule. Note that each N—H bond is more negative toward the nitrogen atom, and the dipoles are not pointing directly toward or away from each other in an equilateral way. Recall that the unshared pair of electrons of nitrogen pushes the shared pairs of electrons toward one end of the molecule. The dipoles are not equally distributed and the resultant dipole will not be zero. In the ammonia molecule, there is a partial negative charge and a partial positive charge (indicated in the figure by the red dipole symbol). Although the dipole does not have as great a charge separation as that between a positive and negative ion, it becomes important when another ammonia molecule is nearby. There is an attraction between the

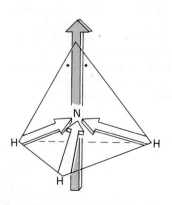

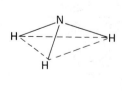

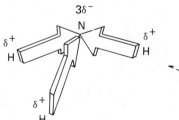

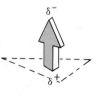

negative end of one molecule and the positive end of the other molecule. This situation is similar to the attraction between two HCl molecules described earlier. The attraction will not be as strong as the force of attraction between two oppositely charged ions, yet the attraction is enough to cause molecules that are polar to interact differently than molecules that are nonpolar. The force of attraction between polar molecules is called a **dipole-dipole force**. Dipole-dipole forces can be intermolecular, that is, occurring between molecules. The forces also can be intramolecular, occurring *within* the same molecule, as is the case of very long or large molecules like proteins.

Water, like ammonia, is a polar molecule because the two dipoles do not cancel. The bonds are not oriented in opposite directions and the molecule has a resultant dipole.

12-13 Hydrogen Bonds

In the introduction to this lesson, you read that water boils at an unusually high temperature. Because of its rather low molar mass, water should be a gas at room temperature and one atmosphere pressure. Even molecules with higher molar masses, such as SO_2 or CO_2, are gases under these conditions. There must be some additional force of attraction between water molecules in the liquid and solid state.

When the molecules of a compound contain hydrogen bonded to a very electronegative atom (as in water), an attraction occurs between the slightly positive hydrogen atom of one molecule and the electronegative atom of another molecule. This particular type of intermolecular attraction is known as a **hydrogen bond**. It usually occurs in compounds whose molecules contain hydrogen bonded to fluorine, oxygen, or nitrogen.

Figure 12-26 illustrates the interaction between several water molecules. Notice that a hydrogen atom of one water molecule is attracted to the oxygen atom of another water molecule. Hydrogen is very small and has no unbonded electrons to interfere with its approach to the more negative atom of another molecule. As a result, a hydrogen bond is a slightly stronger force than other dipole-dipole forces, having about one tenth the strength of covalent or ionic bonds.

Hydrogen bonding also will occur between hydrogen and the electrons in the π ring of aromatic compounds, such as benzene.

Hydrogen bonding accounts for the strength of the fibers in wood due to the orientation of the fibers. Tear a piece of newspaper vertically *across* the lines of print. The tear line is rather straight. If you try tearing the paper *along* the line of print, the tear is ragged and tends to return to the vertical. Hydrogen bonds hold the paper fiber together. You are breaking fewer hydrogen bonds by tearing down the paper than in the same direction as the printed lines.

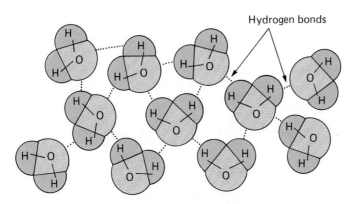

Hydrogen bonds

Figure 12-26 Molecules that contain hydrogen and either oxygen, fluorine, or nitrogen are most likely to form hydrogen bonds with each other. In the case of water, each oxygen atom can attract a hydrogen atom from two separate water molecules.

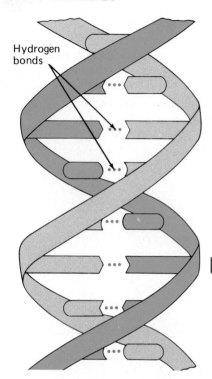

Hydrogen bonds

It is precisely because of hydrogen bonds that Earth is a planet with liquid water. Hydrogen bonds provide the attractive force that keeps water molecules together. The force overcomes the kinetic energy of the molecules that would otherwise cause water to be gaseous at room temperature. The fact that water is a liquid between 0°C and 100°C is one of the reasons life can exist on Earth. Hydrogen bonds also provide a force that helps maintain the shapes of complex organic molecules like proteins and DNA (Figure 12-27). Even the beautiful snowflake pictured at the beginning of this chapter owes its shape to hydrogen bonds. As the water molecules solidify, they become arranged in a hexagonal pattern produced by the effects of hydrogen bonds.

Figure 12-27 The DNA molecule resembles a twisted ladder whose sides are made of alternating sugars and phosphate groups. The "rungs" are composed of nitrogen-base molecules held together in pairs by hydrogen bonds. Although one hydrogen bond is weak in comparison to a covalent bond, the presence of many thousands of hydrogen bonds in DNA holds the two strands of the molecule together.

London forces (also known as dispersion forces) are named for Fritz London, who first suggested instantaneous dipoles.

12-14 Forces Between Nonpolar Molecules

Polar molecules have a relatively strong attraction for each other, resulting from the molecular dipoles present. However, there are attractive forces even for molecules having no electric dipole. For example, two nitrogen molecules must attract each other slightly to explain the fact that the gas liquefies and forms a crystalline solid under suitable conditions. However, the molecules interact through a somewhat different process than polar molecules.

The attractive forces between nonpolar molecules can be illustrated more simply by examining the attraction between two noble gas atoms. Quantum mechanics helps explain the origin of the forces involved. In Chapter 10, you learned that the electrons in an atom are best described in terms of probability. For a noble gas, the electrons can be thought of as evenly distributed *on the average*, so there is no permanent dipole. But at *any given instant*, the distribution may not be even, so the atom would then have a **momentary dipole**. Figure 12-28 shows two helium atoms with the electrons in "stop action." It is possible for this arrangement of electrons to exist at a fraction of an instant, allowing for the existence of a temporary attraction between the atoms. Over a time, these short-lived attractions will result in a slight general attraction of the two atoms for each other. These very weak interactions are called **London forces**. They exist between atoms or

Figure 12-28 When two helium atoms approach each other, it is possible that the electrons of one helium atom are attracted to the nucleus of the other atom for a fraction of an instant. The attractions resulting from momentary dipoles in helium are very weak, but they are enough to lead to the formation of helium liquid at a very low temperature, 4.2 K.

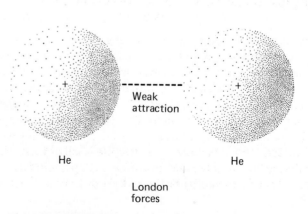

Weak attraction

He

He

London forces

molecules that are not ordinarily attracted to each other by dipole forces. As a result of London forces, nonpolar molecules may overcome their kinetic energy and condense to form a liquid at low temperatures.

London forces are about one tenth the force of most dipole-dipole interactions, and are the weakest of all the electrical forces that act between atoms or molecules. The forces increase with increasing numbers of electrons and increasing sizes of molecules. Substances composed of nonpolar molecules, such as methane and carbon dioxide, have rather low boiling points because the forces holding the molecules together as liquids are weak. Substances composed of polar molecules, such as water and ethanol (ethyl alcohol), have relatively high boiling points because they are held together as liquids by much stronger dipole-dipole forces. London forces and dipole-dipole forces are collectively known as **van der Waals forces**, for the Dutch scientist who first suggested their importance.

Intermolecular forces produce differences in the properties of nonpolar and polar compounds. For example, compare the boiling points of CO_2 ($-78.44°$), H_2O ($100°C$), and H_2S ($-60.33°C$). Of the three, water has the highest boiling point. Why? Each of the molecules are small, containing two or three atoms, and have fairly low molar masses, so you would not expect to see such differences. One explanation for the measurements involves the effects of intermolecular forces. CO_2 is a linear molecule in which the two C—O dipoles cancel. Water on the other hand is polar. The H—O dipoles do not cancel. The attractions between the water molecules mean that more energy is needed to separate the molecules into the gaseous state than is needed for carbon dioxide. How then does H_2O compare to H_2S, which has a similar geometry to water and also is a polar molecule? Water molecules form hydrogen bonds, which are even stronger than other dipole-dipole interactions. H_2S does not exhibit hydrogen bonding, and thus does not require as much energy as water to separate the molecules. H_2S boils at a lower temperature than H_2O, but not as low as the nonpolar CO_2. The behavior of nonpolar versus polar molecules has dramatic importance in the properties and uses of chemical compounds. As you continue your study of chemistry, you will learn more about these effects.

CHEM THEME

It has been said that all of chemistry can be reduced to the interaction of electrical forces. Ions form salts, atoms bond together to form molecules, and molecules are attracted to other molecules all because of forces that are electrical in nature.

12-15 Application: Building Molecules by Computer

You have seen the importance of thinking about molecules in three dimensions. The shapes of molecules influence their behavior. Molecular biologists, scientists who analyze the structure of molecules in living systems, are concerned with very large and complex molecules. For example, proteins are molecules that have molar masses in the many thousands of grams per mole. The molecules are composed of long chains of amino acids that are twisted, bent, and turned into a maze of bonded atoms resembling a crumpled ball of yarn. The maze is not randomly bent and twisted, however. Each protein has a definite architecture that in turn affects the nature of how it functions biochemically.

Do You Know?

A change in the concentration of CO_2 induces a resting mosquito to start flying. The mosquito then moves in a random fashion until it detects either infrared energy from a warm surface or water vapor. If you are nearby, the mosquito will head for your warm, moist skin. A useless bit of information? Chemists have discovered that molecules of good mosquito-repellents are spherical in shape rather than long and thin. These molecules do not actually repel a mosquito but instead seem to block the sensory nerves in the mosquito's antennae that respond to the presence of CO_2, warmth, and moisture. The mosquito continues its random flight instead of homing in on you.

With practice, it is not very difficult to visualize the three-dimensional nature of simple molecules like water or methane. Proteins present an entirely different reality. With large molecules, there are so many factors that influence their structure (atomic bonds, hydrogen bonds, van der Waals forces, and bond angles) that it is difficult to keep all the variables in mind. The computer has made it much easier to visualize the large molecules of living organisms.

Computer graphics can be used to design pharmaceutical chemicals for treating various medical conditions. For example, a body protein called *renin* is linked to high blood pressure. Computer graphics help scientists design drugs that will interact only with renin and no other protein in the body. The interaction occurs solely if the drug fits into the receptor sites of the protein (determined by the shape of the protein molecule). The arrangement is called lock-and-key fit. The manufactured drug interacts with the renin at specific sites on the molecule, making the renin unavailable for other reactions. Molecular geometry is an important factor in the success of the drug. The "docking" of molecules in computer simulations is illustrated in Figure 12-29. The docking must take place on the surface of the molecules, which is why researchers are keenly interested in their surface structures.

In 1985, a procedure was developed by Michael Rossmann at Purdue University for viewing rhinovirus 14 (one of about 100 common cold viruses) in three dimensions. In order to see inside the virus, Rossmann had to make a crystal of pure virus. An analytical method called X-ray crystallography was used to generate images of slices of the virus (Figure 12-30). X-ray crystallography produces a series of dots on film. Since the film is two-dimensional, and the virus is three-dimensional, the research team needed to take many shots of the virus

Figure 12-29 Computer simulations provide a medium in which researchers can design new drugs. The effectiveness of the drugs depends on their ability to fit together with a target molecule.

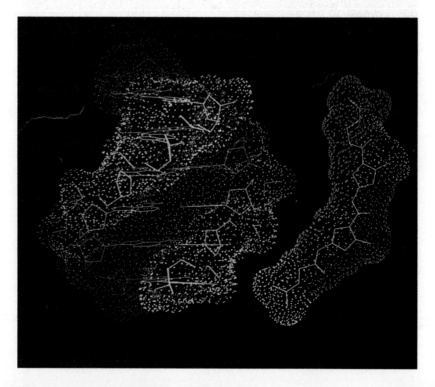

Figure 12-30 X-ray crystallography provided Rossmann's team with two-dimensional images of cold viruses, such as the one shown in the top photo. The images were combined to make the composite, color model shown below.

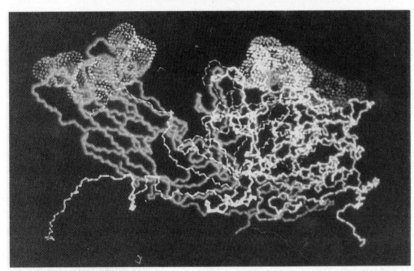

from different angles and put them together later in a three-dimensional picture. An ordinary computer would take about 10 years to analyze the six million pieces of information produced by the technique. However, by using the Cyber 205, a supercomputer at Purdue, the data was processed in only a month. The result was a model of the rhinovirus at the bottom of Figure 12-30.

The information from the procedure is used to learn how the virus attacks living cells. The surface of the virus, diagrammed on the next page, appears to consist of an undulating surface of ridges and valleys. It is believed that the key to the virus's ability to attach itself to a cell depends on the deep canyons in the virus surface. On the other hand, antibodies cannot fit into the canyons but can attach only to the tops of the ridges. Furthermore, every cold virus differs and also frequently changes by mutation. The result is that no single type of antibody can successfully fight all the viruses. The outlook for the development of a cold vaccine is not promising, but with the information now being learned about the geometry of cold viruses, the synthesis of drugs to help people fight the viral attacks becomes more likely.

Figure 12-31 The recently mapped surface of a common cold virus is a series of hills and valleys. Human antibodies do not have surface geometries that match the virus well enough to completely inhibit its action in the body.

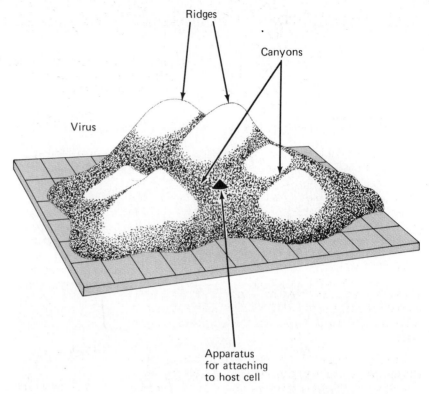

Ridges

Canyons

Virus

Apparatus
for attaching
to host cell

The use of computers to study molecular architecture is a new science. The extent to which computer technology can help in analyzing and synthesizing molecules is unknown. The possibility exists for making a program that will depict the exact structure of a protein merely by analyzing the specific sequence of its amino acids. The problem in composing a program of such scope is not so much a chemical one (scientists know the chemical composition of many proteins) but a geometrical one—how to mathematically account for all the variables that determine the shapes of proteins. Future research will determine the feasibility of such a program.

1. A hydrogen bond exists between the H atom in one polar molecule with an O, N, or F atom in it and another molecule of the same kind. Hydrogen bonds exist between molecules rather than within molecules.
2. covalent, hydrogen bonds, dipole-dipole forces, London forces
3. Each of these arrangements allows for maximum distance and therefore, minimum repulsion between all the bonded pairs of electrons.
4. CO_2 and BeH_2; NH_3 and H_2O
5. Temporary unequal distribution of electrons in the molecules leads to momentary dipoles. Weak attractions between molecules result.
6. Biochemicals often function by interacting with other chemicals. Interactions are limited by whether or not the different molecules have compatible shapes.

Review and Practice

1. What is a hydrogen bond? How is it different from other chemical bonds?

2. Rank the following bonds and forces from the strongest to the weakest: covalent, London forces, hydrogen bonds, dipole-dipole forces.

3. For molecules having two, three, or four covalent bonds respectively and no lone pairs of electrons, explain why the shapes linear, trigonal planar, or tetrahedral are predicted.

4. Give examples of two molecules, containing polar bonds, that are nonpolar. Give examples of two molecules, containing polar bonds, that are polar.

5. How is it possible for attractions to exist between nonpolar molecules?

6. How are the shapes of biochemicals important to their ability to function?

CHAPTER REVIEW

Numbers in red indicate the appropriate chapter sections to aid you in assigning these items. Answers to all questions appear in the Teacher's Guide at the front of this book.

Summary

■ It is convenient to classify chemical bonds as ionic or covalent. Covalent bonds occur when bonded atoms share a pair of electrons. In some cases the electrons are not shared equally, which results in a bond that is polar covalent. Ionic bonding occurs between metallic and nonmetallic elements when the positive metallic ion is attracted to the negative nonmetallic ion.

■ Covalent bonding can be represented by electron dot structures. The valence electrons are drawn as dots around the symbols of the elements. Eight dots represent a full outermost *s and p* orbital system. This arrangement has a lower potential energy and tends to make the atom stable.

■ There are exceptions to the stable octet rule. Hydrogen can have only two valence electrons. Molecules such as BF_3, PCl_5, and SF_6 have structures to which the octet rule does not apply.

■ When chemical bonds are formed, energy is released; when bonds are broken, energy is absorbed. The energy needed to break a mole of a particular kind of bond is called the bond energy.

■ Molecules are three dimensional. The shape of a molecule depends on the number of bonded atoms and the number of unshared pairs of electrons surrounding the central atom. According to the valence shell electron pair repulsion theory, pairs of bonding electrons are oriented in space as far apart as possible. This theory can account for the shape of many molecules and explain why some molecules containing polar bonds are nonpolar while others are polar.

■ Molecules interact with each other. Molecules with relatively high boiling points have rather strong interactions. Dipole-dipole forces and sometimes hydrogen bonds account for strong intermolecular interactions. London forces account for the weak intermolecular interaction among nonpolar molecules. London forces and dipole-dipole forces are collectively called van der Waals forces.

■ Knowledge about bonding and the shapes of molecules is applied to research in a variety of ways. Energy absorption in bonds can be used to "fingerprint" molecules for future identification. The shapes of complex biochemicals influence their reactivity in living organisms. Scientists are using computer models of molecular shapes to study biochemical reactions and synthesize new compounds.

Chemically Speaking

bond energy
chemical bond
covalent bond
crystal lattice
dipole
dipole-dipole force
double bond
electron dot structures
electronegativity
hydrogen bond

ionic bond
London forces
momentary dipole
octet rule
polar covalent bond
resonance structure
triple bond
valence shell electron pair repulsion
van der Waals forces

Review

1. Briefly explain what happens to the potential energy of a system when two oppositely charged particles approach each other. (12-2)

2. What is a crystal lattice? (12-4)

3. How does a covalent bond form between two non-metallic atoms? (12-2)

4. What is electronegativity? How is it sometimes used to classify chemical bonds? (12-3)

5. What characteristics about molecules should a good bonding theory help to explain? (12-1)

6. State the octet rule. (12-6)

7. What is a double bond? A triple bond? (12-6)

8. What is a resonance structure? Give an example of a resonance structure. (12-10)

9. Does the potential energy of a bonded system increase or decrease when the bond is broken? Explain. (12-8)

10. Briefly explain how the valence shell electron pair repulsion theory is used to predict molecular shapes. (12-11)

11. Describe what happens in hydrogen bonding. (12-13)

12. What is the difference between a polar bond and a nonpolar bond? (12-3)

13. How is it possible for a molecule to be nonpolar if it contains polar bonds? (12-12)

14. Name and describe two kinds of intermolecular forces. (12-12, 12-13, 12-14)

363

15. In general, what conditions cause two atoms to combine to form: (12-3)
   a. a bond that is mainly covalent?
   b. a bond that is mainly ionic?

Interpret and Apply

1. Which of the properties listed here are characteristic of an ionic solid? (12-4)
   a. low melting temperature
   b. conducts electricity as a solid
   c. dissolves in water to form a solution that contains mostly ions
   d. dissolves to form a solution that contains mostly molecules
   e. conducts electricity when melted
   f. forms solid crystals

2. Which are more stable under ordinary conditions, hydrogen atoms or hydrogen molecules? Give a reason for your answer. (12-2)

3. Where would you look on the periodic table for elements that frequently are involved in covalent bonding? For elements that are frequently involved in ionic bonding? (12-3)

4. Relate the octet rule to the electron configuration of an element's outermost s and p orbitals. (12-6)

5. Which molecule has the greatest bond energy, O_2, N_2, or Cl_2? (12-8)

6. Why would you expect London forces to be weaker than hydrogen bonds? (12-13, 12-14)

7. What type of intermolecular attraction (hydrogen bonds, London forces, dipole-dipole forces) would you expect for each of the following substances? (12-12, 12-13, 12-14)
   a. NH_3 d. HCl
   b. BF_3 e. C_2H_6

   c. H—C—Cl (with H above and below C)   f. H—C—C—O—H (with H's above and below the two C's)

8. Predict whether a molecule with the formula CH_3F is polar or nonpolar. Explain your reasoning. (12-12)

9. Use concepts of structure and intermolecular forces to explain why the freezing point of CO_2 is lower than that of water. (12-12, 12-13, 12-14)

10. DNA (deoxyribonucleic acid) is the main information-bearing molecule in a cell. Molecules of DNA contain the kinds of bonds you have been studying in this chapter. Below is a portion of a DNA molecule in which certain bonds have been labeled. Identify each bond as a covalent, polar covalent, or hydrogen bond. (12-2, 12-13)

Problems

1. Classify the bonds in each of the following substances as covalent, polar covalent, or ionic: (12-3)
   a. K_2O e. KCl
   b. BeO f. CBr_4
   c. KH g. N_2
   d. SiF_4

2. Draw an orbital diagram for each of the following molecules: (12-2)

   Br_2, N_2, HCl, H_2O_2

3. Draw the electron dot (Lewis) structures for: (12-6)

   CH_3Cl, CH_2Cl_2, $CHCl_3$, CCl_4 H_2O_2

4. Draw electron dot structures for: (12-6)

   NF_3, $FNNF$, $HCCH$, H_2NOH

5. What is the electron dot structure for each of the following isoelectronic substances? (12-6)

   N_2, CN^-, NO^+, CO

364

CHAPTER REVIEW

6. Draw resonance structures for the following substances: (12-10)

   a. NO_2^- **b.** O_3

7. Predict the shape of each of the following substances: (12-11)

   a. PH_3 **d.** PH_4^+
   b. H_2S **e.** CF_4
   c. SeF_6

8. Which of these molecules would you expect to be hydrogen-bonded in the liquid or solid state? (12-13)

   a. H—C(H)(H)—O—H   **d.** H—C(H)(H)—N(H)(H)

   b. CH_4

   c. Se—H (with H)   **e.** H—O—Cl

9. Using the bond energies listed in Table 12-2, determine which of the following molecules is the most stable: F_2, Cl_2, or I_2. (12-8)

10. Draw electron dot structures for the following: OH^-, H_3O^+, SCN^-. (12-6)

Challenge

1. Write the formula you would expect for a combination of each of the following. Identify each combination as ionic, covalent, polar covalent, or no bond.
   a. F and Na   **d.** Ne and Ne
   b. C and F   **e.** As and Cl
   c. F and F

In each case describe the forces involved between the molecules or ions in the solid state.

2. Predict the shapes of each of the complex ions below. Use the metal ion in each case as a central atom. CN^- and H_2O surround the central atom and are called ligands.

   $Fe(CN)_6^{3-}$ $Al(H_2O)_6^{3+}$ $Zn(NH_3)_4^{2+}$

3. Using the concepts of structure, electronegativity, and the shapes of molecules, determine which molecule in each pair below has the greater molecular dipole.
   a. NH_3, NF_3
   b. BF_3, NH_3
   c. CH_4, CCl_4
   d. HCl, HF
   e. SO_2, BeH_2

Synthesis

1. Why are the ammonium ion, NH_4^+, and the methane molecule, CH_4, considered to be isoelectronic? Explain why they do not have similar chemical properties.

2. Using the concepts of potential energy and ionic radii (Chapter 11), account for the trend in melting temperatures of the following ionic substances.

IONIC SUBSTANCE	MELTING TEMPERATURE
KF	858°C
KCl	770°C
KBr	734°C
KI	681°C

3. Both the first ionization energy and the electronegativity values of elements increase across the periodic table from left to right. Also, both decrease from top to bottom of the table. Using what you have learned about atoms, bonding, and other periodic characteristics of the elements, explain the two trends.

Projects

1. Find out how to make ionic crystals and then make several different kinds.

2. Construct three-dimensional molecular models of some of the molecules discussed in this chapter.

3. Research and report on the molecular orbital theory of bonding.

4. Research and report on the use of spectrometers to test for the presence of very small amounts of drugs. Describe how these techniques are applied, such as in testing athletes who compete in the Olympics.

The production of today's automobiles involves elements from almost every family in the periodic table. The use of metals may be obvious to you, since engines and body parts of many cars are made of metallic elements. Less obvious is the role of nonmetals such as oxygen and carbon, that are used in processing the metals that make up these parts. Nonreactive gases, such as argon and krypton, surround metals during welding. Silicon, a metalloid, is in some of the compounds in the windshield, tires, electronic circuitry, and electric wire insulation. Headlights may contain an element from the halogen family, while paint and other surface coatings contain transition metals or their compounds. In this chapter, you will find out about the properties of these and other elements in the periodic table.

Elements: A Closer Look

Representative Metals

Metals make up the majority of elements in the periodic table. Most metals are so reactive that they are not found in the elemental state in nature. It was not until the nineteenth century that scientists were able to isolate all of the alkali and alkaline-earth elements. This lesson highlights elements from the first two families of metals, and also highlights a metal often used in conjunction with one of the alkaline-earth metals.

CONCEPTS

■ the metallic bond
■ properties of metallic elements
■ reasons for the high reactivity of Group I and II metals
■ alloys as mixtures of metals

13-1 Properties of Metals

Most properties of metals are due to the way in which the metal atoms are bonded to each other. Each metal atom has more outer energy levels available than electrons to fill those levels. In a metallic solid, these levels shift slightly. Because of this shift, the distances between energy levels in metallic solids are very small. Only a tiny amount of energy is needed to raise an electron from one energy level to a higher level. The loss of a tiny amount of energy causes an electron to fall from a higher to a lower level. A picture of metallic solids that suggests this easy shifting of energy levels is the metallic bond. In the **metallic bond**, very closely spaced energy levels permit the electrons considerable freedom of movement within the arrangement of the metallic solid. Each metal atom is bonded to all the atoms around it by a surrounding "sea" of electrons.

The idea of a more or less uniform sea emphasizes an important difference between metallic and covalent bonding. In covalent bonds the electrons are localized in a way that rigidly fixes the positions of the atoms, even when stressed. In contrast, the electrons in the outermost energy levels of a metal are spread throughout the crystal. The atoms are arranged in a crystal structure, but can slide past each other when stressed and still remain bonded, as illustrated in Figure 13-1. Chemists talk about *nonlocalized electrons* in a metal. The sliding of atomic layers gives metals *malleability*, the ability to be shaped without breaking, and *ductility*, the ability to be drawn into wire. The mobility of the outermost electrons is also responsible for the good electrical and thermal conductivity of metals.

13-2 Alkali Metals—Sodium and Potassium

Recall from Chapter 11 that the periodic table is arranged so that elements with similar electron arrangements are placed in the same column. Each column makes up a family in which the elements have

The Chem Study film "Chemical Families" can now be used to show students samples of rare elements. Many chemical reactions which are too dangerous to be done in high school classrooms are also shown in the film.

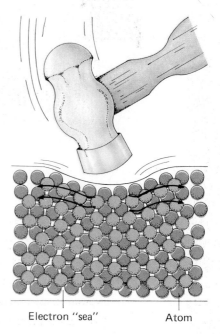

Electron "sea"   Atom

Figure 13-1 In a metal, atoms can slide past each other yet remain bonded. This is due to the metallic bond.

similar chemical properties. The family of alkali metals includes the elements listed in Table 13-1.

Figure 13-2 The alkali metals, sodium (top) and potassium (bottom).

Table 13-1

PROPERTIES OF THE ALKALI METALS				
ELEMENT	PHYSICAL DATA	BONDING IN COMPOUNDS/ CHARGE	ABUNDANCE	OUTER LEVEL CONFIGURATIONS
Lithium Li	silvery white	ionic 1+	rare	$2s^1$
Sodium Na	silvery white	ionic 1+	sixth most common element	$3s^1$
Potassium K	silvery white	ionic 1+	seventh most common element	$4s^1$
Rubidium Rb	silvery white	ionic 1+	very rare	$5s^1$
Cesium Cs	silvery white	ionic 1+	very rare	$6s^1$
Francium Fr	silvery white; radioactive	ionic 1+	extremely rare	$7s^1$

The alkali metals are not found in nature in the elemental state due to their high reactivity. This reactivity is related to the ease with which the outer electrons are transferred to other elements. All of the alkali metals are difficult to purify, handle, and store because they react with air, water, and many other substances. They are usually stored in kerosene or some similar "dry" (nonaqueous) liquid hydrocarbon.

Alkali metals have several metallic physical properties in common, which are provided by the metallic bond. These include a silvery-white luster, high conductivity of heat and electricity, and high malleability. Pure alkali metals are soft enough to be cut with a knife.

All ordinary compounds of the alkali metals are ionic. The compounds are quite soluble in water, and the solutions are colorless. Sodium and potassium are the most common of the alkali metals. Sodium is prepared from sodium chloride, which is obtained from the oceans or from underground deposits. Most sodium produced in the United States is used in sodium vapor lamps and during the manufacture of detergents and of tetraethyl lead, the "lead" still used in some gasolines.

Sodium reacts vigorously with water to produce sodium hydroxide, as shown in Figure 13-3. All alkali metals react similarly. The reaction for sodium is represented by the following equation:

$$2Na(s) + 2H_2O(l) \longrightarrow 2NaOH(aq) + H_2(g) + energy$$

This formation of hydroxides accounts for the name given to this family. One term applied to compounds containing hydroxide ions is alkali; another term is base. Sodium hydroxide is a very corrosive compound and is important in industry, especially in the manufacture of other chemicals.

Potassium and sodium have some similar physical properties. However, potassium metal is more reactive than sodium metal. This greater reactivity, along with the lesser abundance and higher cost of potassium, limits its uses in industrial processes where sodium is effective. Potassium and sodium are both essential for life in plants and animals. Fertilizers usually contain potassium chloride, and this compound also can be substituted for sodium chloride in the diets of people who must reduce their intake of sodium.

One example of the role of potassium and sodium in body functions is the transmission of nerve impulses, as illustrated in Figure 13-4. The membranes of nerve cells are able to maintain different concentrations of sodium and potassium ions inside and outside the cell. The transmission of a nerve impulse is accompanied by the inward flow of sodium ions and the outward flow of potassium ions across the cell membrane. The original ion concentrations are then restored by "pumps" in the cell membrane. The amount of time required to restore the initial ion concentrations determines how soon a second impulse can be transmitted, usually thousandths of a second.

In a normal human body there are complex systems of checks and balances on ion concentration. It is not easy to maintain the correct balance of ions in the body artificially. "Water pills," which reduce swelling due to water retention (edema), can have the side effect of removing too many potassium ions from the body. In such cases, other pills containing a potassium ion compound may be prescribed. However, administering the exact dosage required, and no more, can be difficult, and a high potassium ion concentration can create problems such as muscle weakness and reduction in heart rate. A low concentration of potassium and sodium ions can result from periods

Figure 13-3 Sodium reacts with water to produce hydrogen gas, sodium ions, and hydroxide ions.

An excess of potassium salts, rather than sodium salts, also can lead to difficulties in body functions. Potassium salts are generally a safe substitute for sodium salts because most foods contain far less potassium than sodium.

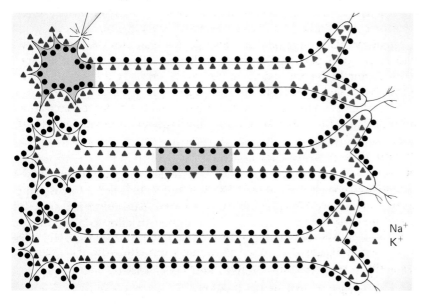

Na^+
K^+

Figure 13-4 Nerve cells maintain different concentrations of ions on either side of the cell membrane. (top) Stimulus of the nerve initiates an impulse. (center) The impulse moves as sodium ions flow into the cell and potassium ions flow out. (bottom) After the impulse has passed through, "pumps" in the membrane transport the ions back to their original positions.

Recent research suggests that the practice of taking salt pills and electrolyte sports beverages during training may cause the body more stress than relief. Instead, natural foods and juices are to be encouraged.

of intense physical training. To avoid the stresses of ion imbalance, athletes commonly consume foods high in potassium, such as bananas. This method provides the necessary elements in a form that does not interfere with the natural checks and balances in the human body.

13-3 Alkaline-Earth Metals—Magnesium and Calcium

The attraction between the metal nuclei and the free electrons of the metallic crystal is stronger in alkaline-earth metals than in alkali metals. This attraction accounts for the difference in physical properties.

The alkaline-earth metals are denser, harder, and have higher melting and boiling temperatures than the corresponding alkali metals. Alkaline-earth metals are less reactive than alkali metals, but are too reactive to be found in the elemental state in nature. Table 13-2 lists the alkaline-earth metals and some of their properties.

Table 13-2

PROPERTIES OF THE ALKALINE-EARTH METALS				
ELEMENT	PHYSICAL DATA	BONDING IN COMPOUNDS/ CHARGE	ABUNDANCE	OUTER LEVEL CONFIGURATION
Beryllium Be	gray-white	ionic 2+	rare	$2s^2$
Magnesium Mg	silvery white	ionic 2+	eighth most common element	$3s^2$
Calcium Ca	silvery white	ionic 2+	fifth most common element	$4s^2$
Strontium Sr	silvery white	ionic 2+	rare	$5s^2$
Barium Ba	yellowish white; lumpy	ionic 2+	rare	$6s^2$
Radium Ra	brilliant white; radioactive	ionic 2+	very rare	$7s^2$

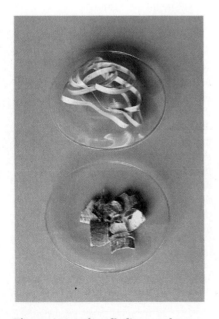

Figure 13-5 The alkaline-earth metals, magnesium on the top and calcium on the bottom.

Magnesium and calcium, shown in Figure 13-5, are the most common of the alkaline-earth metals. The metallic form of each is extracted from chloride salts, $MgCl_2$ or $CaCl_2$. Both of these compounds are common in seawater. Of these two metals, magnesium is the one most easily used in the metallic state. Magnesium has a low density, forms a hard protective oxide coating when exposed to air, and will produce very strong alloys with aluminum. An **alloy** is a mixture of two or more elements. An alloy has different properties than the pure elements of which it is made. Magnesium alloys are popular in sports equipment that requires strength and light weight, such as bicycles, backpack frames, and tennis racquets.

When heated in air, magnesium burns to produce a mixture of magnesium oxide and magnesium nitride, along with a great deal of energy. The brilliant white light from burning magnesium is often used in fireworks. The affinity of magnesium for oxygen is so great that at high temperatures it will even react with the oxygen bound in carbon dioxide and in steam. It is very difficult to put out a magnesium fire.

Magnesium oxide, also called magnesia, is produced when magnesium carbonate is heated:

$$MgCO_3(s) + energy \longrightarrow MgO(s) + CO_2(g)$$

Magnesia can be reacted with water to form the slightly soluble magnesium hydroxide. A mixture of magnesium hydroxide in water is known as milk of magnesia. The low solubility of magnesium hydroxide makes it an excellent antacid. Excess hydrochloric acid (stomach acid) can cause discomfort and stomach sores, known as ulcers. Magnesium hydroxide slowly dissolves as needed to neutralize excess acid. The remaining magnesium hydroxide stays in suspension and acts as a mild laxative.

The most common compound of calcium is calcium carbonate, $CaCO_3$, the main ingredient of limestone. The carbonates of calcium and magnesium are only slightly soluble in water, unless dissolved carbon dioxide is present. The reaction with carbon dioxide produces the soluble calcium hydrogen carbonate, as shown in this equation:

$$CaCO_3(s) + H_2O(l) + CO_2(g) \longrightarrow Ca(HCO_3)_2(aq)$$

When rainwater with dissolved carbon dioxide filters through the ground to limestone deposits, the rock is dissolved and carried away. In this way, any opening or crack in the rock formation will gradually enlarge. Many years of this process will result in a limestone cavern, like the one shown in Figure 13-8. The opposite reaction occurs when the calcium hydrogen carbonate solution is exposed to air in the cavern. Carbon dioxide is given off by the solution, and solid calcium carbonate is deposited as stalactites and stalagmites.

Figure 13-6 Sports equipment made using magnesium combines strength and light weight with the ability to "dampen" the vibrations.

Figure 13-7 When underground water containing dissolved $Ca(HCO_3)_2$ emerges in a cavern, CO_2 and H_2O are released to the air, leaving $CaCO_3$ in the form of stalactites. Some water may drip off the stalactites to the floor, building up stalagmites by the same process.

Do You Know?

Napoleon III served his most distinguished guests on aluminum plates. Less important guests were served on plates of gold or silver.

Groundwater often contains dissolved calcium hydrogen carbonate and magnesium hydrogen carbonate, as well as various metallic salts and carbonates. When this "hard water" evaporates or is heated, carbon dioxide is driven off and the insoluble carbonates remain. These compounds are deposited in devices used to heat hard water, and in soil that has been irrigated with hard water. Problems are developing in areas where hard groundwater has been used for irrigation for several years. The deposits of insoluble carbonates, and of metallic salts and other elements, make the soil difficult to farm and decrease crop yields. Removal of calcium and magnesium carbonates from a small device, such as a teakettle or steam iron, is easy using a mild acid such as vinegar. Large-scale removal poses greater challenges.

13-4 Aluminum

Aluminum is the third most abundant element in Earth's crust, where, due to its reactivity, it usually is combined with oxygen. The processes first used to produce the element from its compounds were so costly that pure aluminum was more expensive than gold or silver. Inexpensive production of aluminum from aluminum oxide began in the late nineteenth century. Since that time, the uses of aluminum have continually increased. Aluminum has low density, good electrical conductivity, and, although highly reactive, it forms a protective, corrosion-resistant oxide coating when exposed to air. Its alloys are strong, lightweight, workable, and can be adapted for construction, fuel storage tanks, car, truck, and airplane parts, window frames, and numerous other uses. About half the aluminum produced in the United States is converted to special-use alloys.

When powdered aluminum and iron(III) oxide are mixed and ignited by a magnesium fuse, a vigorous reaction occurs.

$$2Al(s) + Fe_2O_3(s) \longrightarrow 2Fe(l) + Al_2O_3(s) + energy$$

Note that the iron produced is molten. This *thermite reaction* produces a great deal of heat energy, making it useful in welding.

Alum compounds are used in papermaking, pickling, water purification, and in some baking powders. *Alum* is an aluminum sulfate compound. The most common is potassium alum, $KAl(SO_4)_2 \cdot 12H_2O$. Other alums have other alkali metals, silver, or ammonium in the place of potassium, and other ions with a 3+ charge may take the place of aluminum.

Do You Know?

The method for inexpensive production of aluminum was developed in the 1880's by Charles Hall, then a student at Oberlin College in Ohio. It was discovered almost simultaneously by Paul-Louis Heroult in France. The Hall-Heroult process is still in use today.

1. The outer electrons in metals are free to move into other nearby available energy levels because only a small amount of energy is required. These "free" electrons are bonded to all of the positive ions which remain. This "sea" of electrons bonds all the metal atoms and is also responsible for the strength, malleability, and conductivity that are characteristic of metals.
2. metal hydroxide, ~~water~~ and energy
3. Magnesium forms a hard stable coating of magnesium oxide which prevents further reaction.
4. A mixture of other metals or elements with a metal that alters the properties of the metal.
5. The calcium carbonate in limestone is dissolved in the presence of CO_2 in rainwater.

Review and Practice

1. Describe a metallic bond.

2. What are the products of a reaction between an alkali metal and water?

3. What properties make magnesium usable in elemental form while other alkaline-earth metals are not?

4. What is an alloy?

5. Summarize the chemical events involved in limestone cave formation.

Transition Metals

In chemistry, a metal is an element that tends to lose electrons easily in the presence of other substances. By this definition, the alkali and alkaline-earth elements are the most metallic. However, metals are commonly thought of as having properties such as strength, hardness, durability, malleability, and a high melting temperature. The transition metal family includes elements, such as copper, iron, and silver, that have those properties typically associated with metals. Transition metals are also good conductors of heat and electricity.

CONCEPTS

- metallic bonding in transition metals
- trends in properties of fourth-row transition elements
- properties and uses of sample transition metals
- multiple oxidation states and the formation of transition metal compounds
- energy levels in transition metal atoms and absorption of visible light

13-5 Some Properties and Regularities of Fourth-Row Transition Elements

Most of the "typical" properties of metals are determined by the number of outer electrons available for metallic bonding and by the relative positions of the electron orbitals in the atom. Transition elements have from two to six valence electrons available for bonding, whereas alkali and alkaline-earth metals have only one or two valence electrons, respectively. The number of valence electrons and closeness of energy levels in transition elements accounts for their ability to form ions with different charges. Figure 13-8 shows the position of the transition metals in the periodic table. Properties of transition metals in the fourth row of the periodic table are summarized in Table 13-3.

Note that, for several properties, regularities can be observed as you proceed across the row.

Show the students some solid colored compounds and solutions of the transition metals.

Electron Configurations The element scandium contains one more electron than calcium. The 21st electron enters the lowest-energy orbital that is not fully occupied. Figure 13-9 shows that this is a $3d$ orbital. There are five $3d$ orbitals. Putting a pair of electrons in each of these orbitals means that ten electrons can be accommodated before

CHEM THEME

The organization of the periodic table reveals the patterns of behavior that elements follow as atomic number increases. These patterns can be used to predict the behavior of undiscovered elements.

3	4	5	6	7	8	9	10	11	12
21 Sc	22 Ti	23 V	24 Cr	25 Mn	26 Fe	27 Co	28 Ni	29 Cu	30 Zn
39 Y	40 Zr	41 Nb	42 Mo	43 Tc	44 Ru	45 Rh	46 Pd	47 Ag	48 Cd
57 La	72 Hf	73 Ta	74 W	75 Re	76 Os	77 Ir	78 Pt	79 Au	80 Hg
89 Ac	104	105	106						

Figure 13-8 The position of the transition elements in the periodic table.

Table 13-3

SOME PROPERTIES OF THE FOURTH-ROW TRANSITION METALS										
PROPERTY	**Sc**	**Ti**	**V**	**Cr**	**Mn**	**Fe**	**Co**	**Ni**	**Cu**	**Zn**
Atomic Number	21	22	23	24	25	26	27	28	29	30
Molar mass (g)	45.0	47.9	50.9	52.0	54.9	55.9	58.9	58.7	63.5	65.4
Abundance* (% by mass)	0.005	0.44	0.015	0.020	0.10	5.0	0.0023	0.008	0.0007	0.01
Melting Temperature (°C)	1400	1812	1730	1900	1244	1535	1493	1455	1083	419
Boiling Temperature (°C)	3900	3130**	3530**	2480**	2087	2800	3520	2800	2582	907
Density (g/cm³)	2.4	5.4	6.0	7.1	7.2	7.9	8.9	8.9	8.9	7.1
First ionization energy (kJ/mol)	644	656	648	648	715	752	752	732	736	903

*In Earth's crust.   **Estimated.

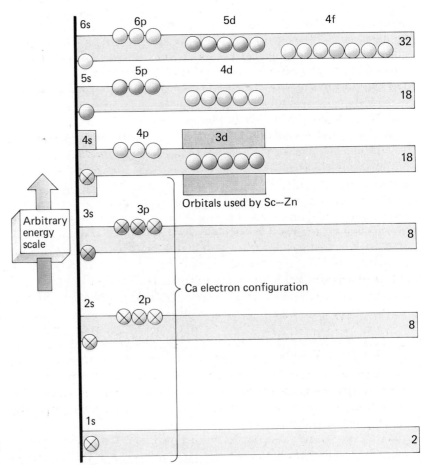

Figure 13-9 The fourth-row transition elements use 3d orbitals beyond the configuration for calcium.

Table 13-4

THE ELECTRON CONFIGURATIONS OF THE FOURTH-ROW TRANSITION ELEMENTS

ELEMENT	SYMBOL	ATOMIC NUMBER	ELECTRON CONFIGURATION
Scandium	Sc	21	[Ar] $3d^1$ $4s^2$
Titanium	Ti	22	[Ar] $3d^2$ $4s^2$
Vanadium	V	23	[Ar] $3d^3$ $4s^2$
Chromium	Cr	24	[Ar] $3d^5$ $4s^1$
Manganese	Mn	25	[Ar] $3d^5$ $4s^2$
Iron	Fe	26	[Ar] $3d^6$ $4s^2$
Cobalt	Co	27	[Ar] $3d^7$ $4s^2$
Nickel	Ni	28	[Ar] $3d^8$ $4s^2$
Copper	Cu	29	[Ar] $3d^{10}$ $4s^1$
Zinc	Zn	30	[Ar] $3d^{10}$ $4s^2$

the higher-energy $4p$ orbitals are needed. The fourth-row transition elements have the electron configurations shown in Table 13-4.

Valence electrons for the transition elements include d electrons. Notice that the configurations for chromium and copper interrupt the regular buildup of electrons. In chromium, an atom in the gaseous state has lower energy if one of the $4s$ electrons moves into a $3d$ orbital. This gives chromium a half-filled set of $3d$ orbitals and a half-filled $4s$ orbital. In copper, the atom has lower energy if the $3d$ set is completely filled with ten electrons and the $4s$ orbital is half-filled.

Molar Mass   The molar mass increases regularly across the row, with the exception of cobalt and nickel. The molar mass of nickel might be expected to be higher than that of cobalt because there are more protons (28) in the nickel nucleus than in the cobalt nucleus (27). The reason for the seemingly "backwards" placement lies in the abundance of the naturally occurring isotopes of these elements. Natural cobalt consists entirely of one isotope, cobalt-59. Natural nickel is made up primarily of two isotopes, nickel-58 and nickel-60. The isotope of nickel with mass number 58 is about three times as common as the isotope having mass number 60.

Melting Temperature   Except for zinc at the end of the group, melting temperatures for transition elements are quite high. This is reasonable, since these elements have a large number of outer electrons and also a large number of vacant outer orbitals which provide stronger metallic bonding. Toward the end of this group of elements, the $3d$ orbitals become filled and the melting temperature is relatively low.

Ionization Energy   The ionization energies for the transition elements are similar. The values are intermediate between low values for the alkali metals and high values for the noble gases. Increasing nuclear charge tends to increase the ionization energy. However, this usual trend seems to be nearly offset by the extra screening from the nucleus provided by the added electrons.

Color   Many compounds of the transition metals and their aqueous solutions absorb light in the visible region of the spectrum. The energy

Do You Know?

Pure "white gold" does not exist. White gold is an alloy of gold and white metals such as nickel or silver.

levels that account for this absorption are relatively close together and involve unoccupied d orbitals. The environment of the ion will influence the color by changing the spacing of these levels. An example of this is the Ni^{2+} ion, which changes from green when surrounded by water molecules to blue when ammonia is added in place of the water, as you can see in Figure 13-10.

Figure 13-10 The Ni^{2+} ion is characteristically green in water (left), blue in ammonia (right).

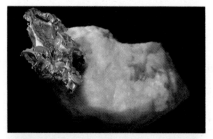

Figure 13-11 Some transition metals, such as copper, gold, and silver, occur in the element state in nature.

Abundance With the exception of iron and titanium, the transition elements are not very abundant in Earth's crust. There is evidence that the center of the earth is predominantly iron and nickel. One theory for the formation of the earth suggests that the temperature rose to several thousand degrees during the early stages of development. Metals such as iron would have been present as liquids of fairly high density. These liquids would have moved toward the center of the earth. The silicate rocks, which make up the outer levels of Earth's crust, would have floated on top of the liquid metals.

Activity Some transition elements, such as copper, are said to be less reactive than hydrogen because they do not react with hydrogen ions, H^+. A few of these metals, including silver, gold, and platinum, are called "noble metals" because of this low activity. Low reactivity is one property that has contributed to the widespread use of these metals.

13-6 Some Transition Metals—Uses and Compounds

Properties of transition metals have led to many applications of these metals and of their alloys and compounds. This section describes some specific transition metals. It also explores the variety of compounds that can be formed by one transition metal. Several transition metals are shown in Figure 13-11.

Copper Copper is a transition metal which is low in activity and abundant in supply. It often is found in the metallic, elemental state in nature. Even when combined with another element, usually sulfur, it is easily separated through a process called roasting, which involves heating the ore with access to air. You probably are familiar with copper because of its widespre d use. Its principal use is as an electrical conductor; the electrical conductivity per mass of copper is second only to silver, which is far less abundant. Copper is preferred over iron for applications involving transport of water. Water pipes made of copper resist both corrosion and formation of hard water deposits better than pipes made of iron or steel.

Copper will not replace hydrogen in dilute acids, but it will react with some concentrated acids such as nitric and sulfuric. Reaction with sulfuric acid produces copper(II) sulfate.

$$Cu(s) + 2H_2SO_4(l) \longrightarrow CuSO_4(s) + SO_2(g) + 2H_2O(l) + energy$$

Copper(II) sulfate in solution with water displays a striking blue color characteristic of the hydrated Cu^{2+} ion. Copper(II) sulfate is commonly used in electroplating, pigments, and extraction of ores.

Silver Pure silver will not react with oxygen in air, but it tarnishes in the presence of sulfur-containing compounds. Silver tarnish is a thin film of silver sulfide formed by a reaction of silver and hydrogen sulfide in the air.

$$4Ag(s) + 2H_2S(g) + O_2(g) \longrightarrow 2Ag_2S(s) + 2H_2O(l) + energy$$

Silver halides, the compounds formed when silver reacts with halogens such as bromine and chlorine, are sensitive to light. This property accounts for a major use of silver in industry—photography. Photographic film contains microscopic crystals of silver halides in a gelatinous coating. Color film contains color-producing dyes in addition to the silver compounds. The process for producing black-and-white images is outlined here and in Figure 13-12. When film is exposed to light, the halide crystals react to form tiny amounts of pure silver. The amount of silver formed depends on the amount of light falling on a particular crystal. An invisible pattern, or image, of silver is

Figure 13-12 The processing of black-and-white film to produce a photo. (*a*) Beginning at left, film contains crystals of light-sensitive silver halides. Exposure to light causes microscopic crystals of metallic silver to form on the surface of the halides. (*b*) Developing the film releases more silver in the exposed crystals, making the image visible. (*c*) Fixer solution stops the reaction of the developer and removes unexposed silver halides. (*d*) The developed negative is darker in areas where more light hit the film, lighter where less light hit. (*e*) During printing, light is blocked by the darker areas of the negative and transmitted by the lighter areas. The image that forms on the photographic paper is the reverse of the negative.

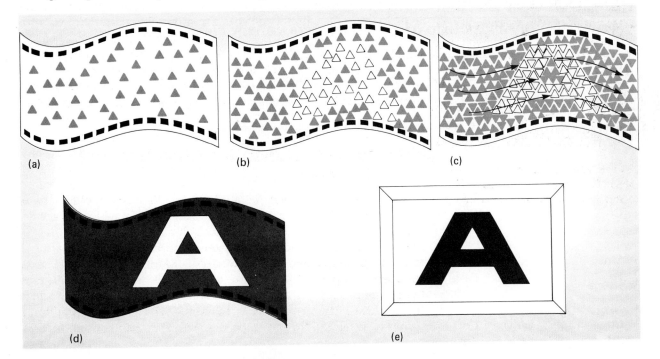

(a)　　　　(b)　　　　(c)

(d)　　　　(e)

Figure 13-13 Chromium has different oxidation states and forms compounds of different colors.

present in the exposed silver halide crystals before developing. During development, silver atoms serve as catalysts that allow most of the silver compound in the exposed crystals to become silver metal, which becomes visible as an image. The unexposed silver halide crystals do not react with the developer fast enough to form silver metal during development. After development, unexposed silver halides are removed from the negative by the fixer solution. The silver halides removed by the fixer solution can be recovered by reprocessing. If the unexposed silver halides were not removed, they, too, would blacken as the photograph or negative was viewed in light.

Chromium Pure chromium, although highly reactive, is resistant to corrosion because it forms a protective chromium oxide layer when exposed to air. Many objects, such as car bumpers, are chromium plated to protect against corrosion. Most "chrome plating" is actually 90% nickel. Since a nickel surface would eventually lose its gleam, a top layer of chromium is added.

Suppose you go into a chemical stockroom to see what kinds of chromium compounds can be found. First you find a bottle of green powder labelled Cr_2O_3, chromium(III) oxide. Next to it there is a bottle containing a red powder, CrO_3, chromium(VI) oxide, and some black powder marked CrO, chromium(II) oxide. Elsewhere in the stockroom there would be bright-yellow potassium chromate, K_2CrO_4, next to a bottle of orange potassium dichromate, $K_2Cr_2O_7$.

Based on the stockroom search, you would conclude that chromium forms a number of stable compounds, most of them colored solids. Chromium may have different charges, including 2+, 3+, and 6+, in its compounds. Another stockroom search of other transition metal compounds would lead to similar conclusions. Most transition metals form many different compounds due to the multiple oxidation states they achieve. Table 13-5 summarizes some of the information that chemists have found for the transition elements.

Table 13-5

								NUMBER OF VALENCE ELECTRONS IN NEUTRAL ATOM	
TYPICAL OXIDATION NUMBERS FOUND FOR FOURTH-ROW TRANSITION ELEMENTS									
	REPRESENTATIVE COMPOUNDS								
SYMBOL	OXIDATION NUMBER OF TRANSITION ELEMENT								
	1+	2+	3+	4+	5+	6+	7+	$3d$	$4s$
Sc			Sc_2O_3					1	2
Ti		TiO	Ti_2O_3	TiO_2				2	2
V		VO	V_2O_3	VO_2	V_2O_5			3	2
Cr		CrO	Cr_2O_3			CrO_3 K_2CrO_4 $K_2Cr_2O_7$		5	1
Mn		MnO	Mn_2O_3	MnO_2		K_2MnO_4	$KMnO_4$	5	2
Fe		FeO	Fe_2O_3					6	2
Co		CoO	Co_2O_3					7	2
Ni		NiO	Ni_2O_3					8	2
Cu	Cu_2O	CuO						10	1
Zn		ZnO						10	2

13-7 Iron: Properties, Production, and Uses

The most abundant of the transition metals is iron, ranking fourth among the elements found in Earth's crust. Of all transition metals, iron has the best combination of abundance, strength, cost, ease of purification, and adaptability to a variety of uses. This makes iron one of the most important metals in construction and manufacturing.

The processing of iron ore to produce iron is an excellent example of how chemical reactions are controlled on a large scale. Iron ore is typically a mixture of the minerals hematite, magnetite, and quartz. Ore is fed in at the top of a vertical reactor called a blast furnace. Limestone, $CaCO_3$, and coke, a form of solid carbon, are added to the furnace along with the iron ore, while air is blown in at the bottom as illustrated in Figure 13-14. When the rising oxygen gas encounters the carbon, carbon monoxide forms. The carbon monoxide then reacts to remove the oxygen in the iron ore, probably in three stages:

$$CO(g) + 3Fe_2O_3(s) \longrightarrow 2Fe_3O_4(s) + CO_2(g)$$
$$CO(g) + Fe_3O_4(s) \longrightarrow 3FeO(s) + CO_2(g)$$
$$CO(g) + FeO(s) \longrightarrow Fe(s) + CO_2(g)$$

The above reactions are greatly simplified. There are a number of side reactions occurring at the same time. For example, limestone must be decomposed to form lime and carbon dioxide. The lime then combines with the quartz, SiO_2, in the ore to form a byproduct called slag, which is mostly calcium silicate, $CaSiO_3$. The carbon dioxide that is produced when limestone decomposes forms more carbon monoxide and removes more oxygen from the iron ore. The molten products of these reactions come to rest at the bottom of the furnace, where the lower density slag floats to the top of the mixture and is separated

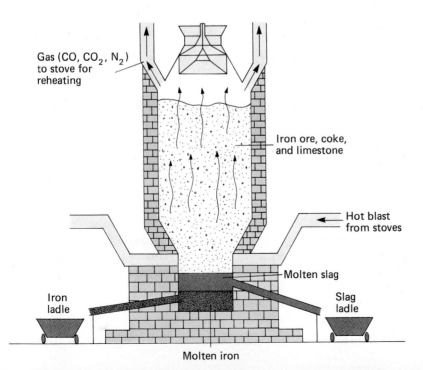

Gas (CO, CO_2, N_2) to stove for reheating

Iron ore, coke, and limestone

Hot blast from stoves

Molten slag

Iron ladle

Slag ladle

Molten iron

Figure 13-14 Cross-sectional view of a blast furnace used in the production of metallic iron from iron ore.

from the iron. The iron that is left has a high carbon content and several impurities, which are later removed in the production of steel.

Iron often is combined with other metals to make alloys for specific purposes. Stainless steel, an alloy of steel (an iron and carbon alloy), chromium, and nickel, is an example of how metals can be combined to form a substance with characteristics unlike those of the pure elements. Iron and nickel readily oxidize in air, and chromium is too reactive to occur in the elemental state in nature. Stainless steel, an alloy of the three, resists almost all corrosion and reaction. The inclusion of some carbon strengthens the alloy. Table 13-6 is a listing of some common alloys of iron and their uses.

Table 13-6

SOME ALLOYS OF IRON		
ALLOY	REPRESENTATIVE COMPOSITION (% BY MASS)	COMMON USES
Pig Iron	93 Fe 3–4 C, other elements	refined into steel or wrought iron
Cast Iron	90–95 Fe 2–4 C, other elements	construction, cookware, woodstoves
Steel	99 Fe 0–1 C, other metals	automobiles, construction
Stainless Steel	76 Fe 16 Cr 8 Ni	Surgical tools, food preparation, utensils, other uses where resistance to corrosion is important
Alnico	61 Fe 10 Al 17 Ni 12 Co	permanent magnets
Tempered Steel (Tool Steel)	steel heated to high temperature, then cooled quickly	hand and industrial tools

Iron is essential to animal life, because of its role in the heme complex of hemoglobin. The iron atom in the heme group binds with an oxygen molecule and carries it to the cells, where the oxygen plays a vital role in cellular respiration. However, other molecules also can be bound to the iron in hemoglobin. The carbon monoxide molecule, when available, is picked up by the hemoglobin in the red blood cells two hundred times more easily than oxygen, preventing enough oxygen from reaching the cells.

Do You Know?

Copper, rather than iron, has the role of oxygen transport in the blood of mollusks and crustaceans.

13-8 Application: Transition Metals and Their Compounds as Colors for Dyes and Gems

The colorful history of chromium is demonstrated in the discovery and synthesis of one of the first artificial dyes. In 1856 the German chemist, August Wilhelm von Hofmann, wondered if the antimalarial drug quinine might be synthesized from aniline. Aniline has a chemical composition somewhat similar to that of quinine. Hofmann's assistant, William Henry Perkin, a student, decided to try this synthesis. Neither Hofmann nor Perkin knew the complex structure of quinine or they would have realized that the task was impossible at that time. Perkin, unaware of this difficulty, treated aniline with potassium dichromate in the hope of obtaining quinine. He was about to throw away the resulting sticky mess when he noticed a purplish glint. When he added alcohol, he dissolved a purple substance from the mixture.

At the time of Perkin's experiments, the homespun clothing worn by most people came only in shades of gray. Colorful dyes from plant and animal sources were costly. Purple was so expensive that it was synonymous with royalty. Perkin suspected he had a dye. He left school, borrowed money from his family, and within six months his factory was producing "aniline purple." This inexpensive dye was so popular that this period in history became known as the "Mauve Decade" after the French word for purple. Perkin, having founded the synthetic dye industry, retired a very wealthy man at thirty-five.

Dyes like aniline purple, as well as natural dyes, are composed of large molecules whose characteristic color depends on the energy levels available in the molecular bonds. As more is learned about the bonding and energy levels of such compounds, it may become possible to design and synthesize a molecule of any desired color.

The color of gems is due to bonding patterns and the absorption of visible light. For example, a crystal of the ionic compound aluminum oxide, when pure, is white or colorless. When chromium atoms take the place of one in every hundred aluminum atoms, the resulting deep red crystal is the gemstone ruby. Titanium atoms and iron atoms in the place of aluminum produce a blue crystal, sapphire. Yet pure chromium is not red, nor is titanium blue.

How, then, is color produced in gems? There are several mechanisms by which impurities affect the colors of compounds. All of these

Do You Know?

Quinine was finally synthesized by Robert Woodward and William Doering in 1944, when Japan cut off natural supplies of the drug to western nations during World War II.

Figure 13-15 All of these gemstones owe their color, in some way, to transition metals.

mechanisms involve the absorption of light energy by electrons in the atoms of the impurity. The way that chromium affects the color of ruby is an example of one of these mechanisms. Recall from Chapter 10 that the electron configuration of chromium is

$$1s^2 2s^2 2p^6 3p^6 3d^5 4s^1$$

In ruby, chromium forms a 3+ ion by losing the $4s^1$ and two of the $3d$ electrons, with the resulting electronic structure of $3d^3$. The three unpaired electrons in the $3d$ orbitals can absorb energy and reach higher energy levels. The energies that the electrons can absorb correspond to yellow-green and violet wavelengths of light. When white light enters the gem, the yellow-green and violet wavelengths are absorbed by the unpaired electrons. The red wavelengths, and some of the blue wavelengths, are not absorbed but are transmitted to the eye of the viewer. The perceived result is a deep red color with blue overtones. Table 13-7 lists some gems that result from trace impurities of transition metals.

Differences in the exact shade and depth of color in gemstones depend on the amount of impurity, its exact placement, and the presence of more than one impurity.

"Garnet" is a term applied to a group of silicate minerals with the generalized formula $A_3B_2Si_3O_{12}$. A is filled with 2+ ions (Mg^{2+}, Fe^{2+}, or Mn^{2+}) and B with 3+ ions (Al^{3+}, Fe^{3+}, or Cr^{3+}).

Table 13-7

	IMPURITIES AND THE COLOR OF GEMS		
GEM	**COLOR**	**IMPURITY**	**SUBSTANCE**
amethyst	violet	iron	silicon dioxide (SiO_2)
emerald	green	chromium	aluminum beryllium silicate $Be_3Al_2(Si_6O_{18})$
garnet	red green	magnesium, iron chromium	various silicates
ruby	red	chromium	aluminum oxide (Al_2O_3)
sapphire	blue	titanium	aluminum oxide (Al_2O_3)

Do You Know?

The "starring" in star sapphires is caused by needles of rutile, TiO_2, in triangular patterns throughout the stone.

Review and Practice

1. strength, high melting point, malleability, durability, colorful compounds
2. There are more electrons available for metallic bonding.
3. The molar mass increases regularly except for nickel.
4. photography, jewelry
5. good conductor, malleable, corrosion resistant, colorful compounds
6. Stainless steel is shiny, hard, corrosion resistant and is used in jewelry, pipes, sinks, tanks and tools. Alnico is used to make strong, permanent magnets.
7. copper, silver, gold

1. Name three properties commonly associated with transition metals.

2. Explain why transition metals are harder and stronger than the alkali metals.

3. Describe how the molar masses of fourth-row elements change as you proceed across the row from left to right.

4. What is a primary use of silver compounds?

5. Name four properties of copper that have led to its widespread use.

6. Describe two common alloys of iron, and their uses.

7. Name two transition metals that will not replace hydrogen from dilute acids.

The Metalloids

In chapter 11 you studied those properties which characterize metals and those of nonmetals. There are some elements that are difficult to classify as one or the other. For example, carbon is difficult to classify. The diamond form of carbon is a poor conductor, yet the graphite form conducts fairly well. Neither form looks metallic, so carbon is classified as a nonmetal.

Silicon looks like a metal. However, its conductivity properties are closer to those of carbon. Since some elements are not distinctly metallic or nonmetallic, a third class, called the metalloids, is used.

CONCEPTS

- properties of metalloid elements
- definition and characteristics of allotropes
- some structures and properties of silicones

13-9 Properties of the Metalloids

The properties of **metalloids** are intermediate between metals and nonmetals, as outlined in Table 13-8. Whereas most metals form ionic compounds, metalloids as a group may form ionic or covalent bonds. Under certain conditions pure metalloids conduct electricity, but do so poorly, and are thus termed semiconductors. This property makes

Table 13-8

PROPERTIES OF SOME METALLOIDS				
ELEMENT	PHYSICAL DATA	OXIDATION STATES	ABUNDANCE	OUTER LEVEL CONFIGURATIONS
Boron B	crystals very hard	3+	fairly common	$2s^22p^1$
Silicon Si	gray-black; metallic luster	4+	second most abundant element	$3s^23p^2$
Germanium Ge	gray-white; metallic luster	4+, 2+	rare	$3d^{10}4s^24p^2$
Arsenic As	As: shiny gray, metallic As$_4$: yellow nonmetallic	3+, 5+	widely distributed	$3d^{10}4s^24p^3$
Antimony Sb	Sb: silvery; metallic Sb$_4$: yellow; nonmetallic	3+, 5+	rare	$5s^25p^3$
Tellurium Te	gray-white; metallic	2+, 4+, 6+	rare	$5s^25p^4$
Polonium Po	radioactive and unstable	4+	very rare	$6s^26p^4$

Figure 13-16 The location of metalloid elements on the periodic table.

	13	14	15	16	17
	5 **B**	6 **C**	7 **N**	8 **O**	9 **F**
12	13 **Al**	14 **Si**	15 **P**	16 **S**	17 **Cl**
30 **Zn**	31 **Ga**	32 **Ge**	33 **As**	34 **Se**	35 **Br**
48 **Cd**	49 **In**	50 **Sn**	51 **Sb**	52 **Te**	53 **I**
80 **Hg**	81 **Ti**	82 **Pb**	83 **Bi**	84 **Po**	85 **At**

Other elements that form allotropes include carbon, antimony, and phosphorus, sulfur, and oxygen, described in the next section.

metalloids important in microcircuitry, which will be discussed in greater depth in Chapter 14.

Notice that the boundary between metals and nonmetals in the periodic table runs between the metalloid elements, as shown in Figure 13-16. Although aluminum falls along this line, it is considered a metal due to its metallic properties.

13-10 Some Metalloids and Their Uses

Boron Boron is a typical metalloid even though its crystal structure and some of its compounds have unusual properties. Boron and boron carbide form crystals with structures similar to that of diamond, as illustrated in Figure 13-17. The hardness resulting from these very strong structures is put to use for grinding, cutting, and polishing a variety of materials.

Boric oxide, B_2O_3, forms boric acid with hot water.

$$3H_2O(l) + B_2O_3(s) \longrightarrow 2H_3BO_3(aq)$$

Boric acid is used as a mild antiseptic in eyewashes.

Arsenic An **allotrope** is one of two or more forms of an element, existing in the same physical state but having different molecular structures. Arsenic occurs in two allotropic forms: the gray, metallic form, and as As_4, a yellow, nonmetallic solid. The yellow form is easily vaporized. Arsenic, along with other "heavy metal poisons" such as lead, mercury, and cadmium, has an extraordinary affinity for sulfur. Sulfur is a part of nearly all enzymes. Enzymes are the essential catalysts for chemical reactions in cells. The reaction of heavy metals and sulfur in the enzyme dramatically alters the functioning of the body. After time, cellular reactions can no longer take place normally, and death of the affected organism results. The major uses of arsenic and its compounds depend on their poisonous nature. Arsenic compounds are used in weed killers and as insecticides on plants and livestock.

Figure 13-17 Boron and boron carbide form very hard crystals. Boron nitride can be treated with high pressure to form similar crystals.

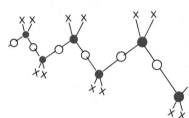

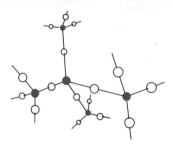

Silicon Silicon is perhaps the best known of the metalloids. As the second most abundant element in Earth's crust, it is also the most common metalloid. Clays, sands, and most rock-forming minerals are composed chiefly of *silicates*. Silicate minerals exist as a variety of silicon and oxygen frameworks, some of which are shown in Figure 13-18.

Figure 13-18 Silicate minerals exhibit a variety of crystal forms, suggesting the structures illustrated here.

A naturally abundant compound of silicon is silicon dioxide, SiO_2, also known as silica, and as the mineral quartz. Silicon dioxide is processed with carbon or magnesium to obtain elemental silicon:

$$SiO_2(s) + 2C(s) \longrightarrow Si(s) + 2CO(g)$$
$$SiO_2(s) + 2Mg(s) \longrightarrow Si(s) + 2MgO(s)$$

Silicon can be made to form very long chain molecules with oxygen. Silicon has four electrons available for bonding, but does not easily form double bonds. When a hydrocarbon, usually obtained from petroleum, is attached to silicon in a silicon-oxygen chain, the resulting compound is known as a **silicone**.

All silicones are synthetic. They do not occur in nature and are remarkably resistant to the actions of heat, chemicals, moisture, and staining agents. Silicones are widely used as coatings to waterproof paper and fabrics, and to protect metals from corrosion. Silicone compounds also are added to natural latex to increase durability.

Review and Practice

1. Which metalloid is most abundant in Earth's crust?

2. Which metalloid forms compounds almost as hard as diamond?

3. Describe the allotropes of arsenic. What conditions must be true for two substances to be called allotropes?

4. Name two uses of silicones.

1. silicon
2. boron
3. As—gray metal metallic bonding, high melting point, malleable, strong; As_4—yellow, molecular compound, low melting point
4. protect carpet, clothing, fabrics, furniture heat resistant paper and fabric water resistant fabric

Representative Nonmetals

CONCEPTS

■ the importance of several non-metals to life
■ properties of some nonmetals
■ characteristics of the halogens
■ properties and uses of the noble gases

In the lesson on transition metals, metals were defined chemically as elements that tend to lose electrons in the presence of other elements. Nonmetals are elements that have a strong tendency to accept electrons in the presence of other elements. When two nonmetals combine, the electrons are shared in a covalent bond, as you learned in Chapter 12. This lesson focuses on nonmetallic elements, their properties, uses, and importance to the functions of living things.

13-11 The Nitrogen Family

Nitrogen It has been known since the last century that nitrogen is an essential element in protein molecules, including enzymes.

Although nitrogen is an essential element for life, it is rare in the environment except in air. The nitrogen gas molecule is held together by a triple bond, which has the following structure:

$$N\equiv N$$

The nitrogen triple bond is strong and so difficult to break that nitrogen molecules are not reactive. Nitrogen gas is so unreactive that a common means of preserving food is to package it in nitrogen gas. Nitrogen molecules in the air cannot be used directly by most organisms. An exception is the nitrogen-fixing bacteria that live in the roots of some plants. These bacteria are able to convert atmospheric nitrogen to a nitrogen compound. Plants, and ultimately animals, rely on nitrogen compounds produced by nitrogen-fixing bacteria, or compounds that are in the soil or added to it through fertilization. Nitrogen compounds for fertilizer are produced commercially from atmospheric nitrogen using the Haber process, which will be described in Chapter 19.

Phosphorus One characteristic of nonmetals is the tendency to form allotropes. There are three common allotropes of phosphorus, each of which displays its own properties.

White phosphorus has individual molecules, P_4. It is sometimes called yellow phosphorus when the presence of impurities affects its color. The tetrahedral structure of the white phosphorus molecule is illustrated in Figure 13-19. Solid white phosphorus has a waxy texture and is highly reactive, igniting in moist air at 30°C. Phosphorus vapor is given off when P_4 is exposed to light.

Black phosphorus has a double layer structure in which each atom is attached to three neighboring atoms in both layers, as depicted in Figure 13-19. It is made under pressure from white phosphorus. Black phosphorus is stable in air at almost any temperature and has a texture similar to that of graphite.

Red phosphorus consists of tetrahedrons of phosphorus bonded into a random structure. It is stable in air, but will ignite at 260°C.

Phosphorus does not occur uncombined in nature, but it is found in the mineral apatite, $3Ca_3(PO_4)_2 \cdot CaF_2$, and in the impure form of apa-

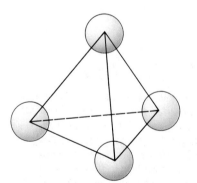

White phosphorus

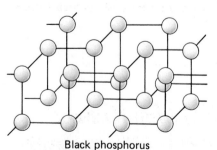

Black phosphorus

Figure 13-19 (top) The tetrahedral structure of white phosphorus, P_4. (bottom) The double layer structure of black phosphorus.

tite, phosphate rock. Phosphate rock occurs primarily in ancient deposits of sea bird droppings, and is commercially mined and processed for fertilizer and for the various forms of pure phosphorus.

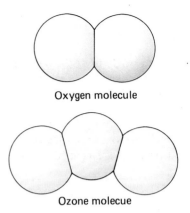

Oxygen molecule

Ozone molecue

Figure 13-20 The structures of oxygen (top) and ozone (bottom).

13-12 The Oxygen Family

Oxygen Oxygen forms compounds with almost every other element. This activity makes oxygen useful in a number of ways. Pure oxygen is obtained commercially from pressurized liquid air. Industrial applications of oxygen include the removal of impurities during the processing of iron and other metals. A medical use is in treatment of patients who cannot get enough oxygen into their systems from breathing air normally.

Oxygen is vital to life due to the roles it has in cellular functions, of which cellular respiration is the most important. During cellular respiration, cells release the energy stored in glucose molecules, $C_6H_{12}O_6$, by breaking the molecules down into water and carbon dioxide. This process is complex and involves many steps, but can be summarized by this net equation.

$$C_6H_{12}O_6(s) + 6O_2(g) \longrightarrow 6CO_2(g) + 6H_2O(l) + energy$$

In the final stage of cellular respiration, electrons must be removed from the system. This is done by oxygen molecules, which will readily accept electrons. Once electrons are accepted by oxygen, the oxygen combines with hydrogen to produce water, one of the waste products of cellular respiration. Other elements, such as nitrogen, form molecules that are too stable to accept electrons, and could not remove electrons from the system.

While oxygen molecules can perform this function during cellular respiration, an allotrope of oxygen, ozone, O_3, can not. Ozone is a blue gas produced from a reaction of oxygen with an electric spark, lightning, or another form of energy. You may have noticed the sharp odor of ozone in small quantities near electric motors, photocopy machines, or during an electric storm. Ozone also is a byproduct of gasoline combustion. Figure 13-20 compares the structures of oxygen and ozone. Ozone is more reactive than oxygen. The ability of ozone to break down a variety of compounds has applications in water and sewage treatment, where it is used to destroy bacteria.

Sulfur Sulfur, like phosphorus, exists in allotropic forms at different temperatures. The most stable allotrope is a ring structure of eight sulfur atoms. At standard temperatures, these molecules form the rhombohedral crystals shown in Figure 13-21. The molecules can be heated and slowly cooled to form elongated monoclinic crystals, also shown. Other allotropes of sulfur include a chainlike molecule believed to exist in melted sulfur at 165°C. A rubbery, plastic form of sulfur is produced by the quenching, or immediate cooling, of boiling sulfur. Most allotropes of sulfur will revert to the stable S_8 rhombic form when kept at room temperature.

Sulfur occurs in nature both as elemental sulfur and combined with other elements as sulfates or sulfides. Large deposits of sulfur occur in Texas and Louisiana. The most common process of extracting sulfur is the Frasch process, which uses boiling water to melt sulfur-containing

Do You Know?

Although ozone gas is poisonous, its disinfecting properties in water led people to believe that it also was part of clean, pure air. Neighborhoods were named "Ozone Heights" or "Ozone Park," and people spoke of "smelling the ozone" in country air. Ozone actually has a sharp, disagreeable odor.

Do You Know?

In the late 1970's, researchers discovered animal communities located near deep-sea heat vents. Hydrogen sulfide from the vents is the energy source for these communities, rather than sunlight.

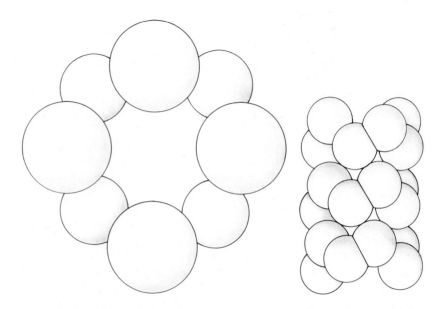

Figure 13-21 (left) The ring structure of sulfur, S_8, is stable at room temperature. (right) S_8 molecules form rhombohedral crystals at standard temperatures.

deposits underground. The sulfur, which is almost pure, is forced to the surface with compressed air.

13-13 The Halogens

The members of the halogen family are reactive, nonmetallic elements. Each has a distinctive color, and some have characteristic odors as well. They have relatively high electronegativities, and will gain one electron to complete their outer energy level, resulting in a 1− charge. Properties of the halogens are summarized in Table 13-9.

Table 13-9

PROPERTIES OF THE HALOGENS				
ELEMENT	PHYSICAL DATA	BONDING IN COMPOUNDS	ABUNDANCE	OUTER LEVEL CONFIGURATION
Fluorine F	pale yellow gas	ionic or covalent	common	$2s^2 2p^5$
Chlorine Cl	yellow-green gas	ionic or covalent	very common	$3s^2 3p^5$
Bromine Br	dark orange liquid; irritating odor	ionic or covalent	fairly rare	$4s^2 4p^5$
Iodine I	violet solid	ionic or covalent	widely distributed	$5s^2 5p^5$
Astatine At	radioactive and unstable	?	extremely rare	$6s^2 6p^5$

Some samples of halogen elements are shown in Figure 13-22. The halogens react readily with metals, resulting in ionic compounds that are sometimes called salts. Table salt, NaCl, is one such compound. The equations below represent reactions that form two other salts:

$$Fe(s) + Br_2(l) \longrightarrow FeBr_2(s)$$
$$2Cs(s) + F_2(g) \longrightarrow 2CsF(s)$$

The formation of many such salts by similar reactions between metals and elements of this group led to the name halogen, which means "salt former."

All halogens form diatomic molecules in which two atoms are joined by a nonpolar, covalent bond. Halogens also form covalent compounds with other nonmetals. The compound formed depends on the activity of the halogen, as illustrated by the formation of sulfur compounds with halogens:

$$S_8(s) + 24F_2(g) \longrightarrow 8SF_6(g)$$
$$S_8(s) + 8Cl_2(g) \longrightarrow 8SCl_2(g)$$
$$S_8(s) + 4Br_2(l) \longrightarrow 4S_2Br_2(l)$$
$$S_8(s) + I_2(s) \longrightarrow \text{no reaction}$$

The ratio of halogen to sulfur in the compound is related to the reactivity of the halogen. The greater the reactivity of the halogen, the greater the ratio of halogen to sulfur. As the atomic mass of the halogen increases, its reactivity decreases.

Recall from Chapter 5 that each halogen can replace any other halogen which is less active:

$$Cl_2(aq) + 2KBr(aq) \longrightarrow 2KCl(aq) + Br_2(aq)$$

Fluorine cannot be replaced because it is the most active member of the family. In fact, fluorine is the most active of all elements.

Figure 13-22 The halogen elements (left to right) fluorine, chlorine, bromine, iodine.

Chlorine bleach is a solution of sodium hypochlorite, NaClO, with water. When bleach is mixed with ammonia solution, highly poisonous gases are evolved. Caution students against using these two cleaners, or products containing them, simultaneously.

Do You Know?

A Belgian chemist named Louyet was killed by escaping hydrogen fluoride gas in his attempt to isolate fluorine. For many years hydrogen fluoride appeared to be the "universal solvent," capable of dissolving everything.

Do You Know?

Hydrofluoric acid, HF, is used to frost light bulbs to soften the light. HF also is used in the production of fluorocarbons such as Teflon®, a coating, and Freon®, which is a refrigerant.

Fluorine was originally discovered around 1780 in the poisonous gas, hydrogen fluoride. It was not isolated from other elements for a century, however, in spite of the attempts of many of the best chemists. The first attempts were unsuccessful because they involved attempts to replace fluorine with a more active element. Using electricity to separate fluorine from a compound was tried in the early 1800's. Fluorine could be isolated in this way, but it immediately reacted with the container, often with disastrous results. Finally, in 1886, Ferdinand Frederic Moissan succeeded in keeping fluorine pure. His method involved cooling a solution of potassium fluoride to $-50°C$ to reduce the activity of fluorine, and using platinum for the containers. The pale yellow gas that he isolated was the element fluorine.

CHEM THEME

The structure and bonding of a molecule determine whether it is generally active or inert.

13-14 Noble Gases

The first discovery of a noble gas, argon, is an example of serendipity, or finding something of greater value while searching for something else. In 1882 the British chemist John William Strutt (Lord Rayleigh) was trying to determine accurately the relative atomic masses of oxygen, hydrogen, and nitrogen. Avogadro's hypothesis predicted that equal volumes of each gas would contain the same number of molecules. Lord Rayleigh prepared and purified each element by several methods. He got the same measurements for the mass of hydrogen and oxygen regardless of the method of preparation. The mass values for nitrogen, however, were different, depending on whether the nitrogen gas was prepared from air or from ammonia. Ten years later a Scottish chemist, William Ramsay, became interested in this problem when he came to work for Rayleigh. Ramsay suspected an unknown gas. After removing all known gases except nitrogen from a sample of air, he reacted the remaining gas with red hot magnesium to remove the nitrogen. A small bubble of nonreactive gas remained. Ramsay and Rayleigh used the spectrum of the gas to confirm that this nonreactive gas was indeed a new element. Since the gas would not react with any other element, it was named argon, meaning "no work."

The atomic mass of argon, about 40, placed it in the region of chlorine, potassium, and calcium. The nonreactive nature of argon meant that it did not fit in with any previously known chemical family. Ramsay placed it in a new family and began to look for other nonreactive gases. In the next four years, Ramsay discovered helium, neon, krypton, and xenon, and identified them by the spectra they emitted.

Once noble gases were discovered, chemists tried to combine them with other elements. All early attempts to form compounds were unsuccessful, and it became a principle of chemistry that noble gases would not react. In 1962 Neil Bartlett was working with the extremely active compound, platinum(VI) fluoride, PtF_6, and tried using it to make a compound of a noble gas. When he found a compound containing xenon, $XePtF_6$, the announcement was greeted with skepticism, since the inertness of noble gases had become a major chemical "truth." However, the announcement did spur a renewed interest in making noble gas compounds. To date, a large number of compounds

Do You Know?

Before 1962, the only combinations of the noble gases with other elements was in the formation of "clathrates." In a clathrate, the molecules form cagelike structures in which atoms of an inert gas are trapped, but are not chemically combined.

of xenon and krypton have been prepared, but none of argon, neon, or helium.

Today, most noble gases are obtained by separation of liquid air. Helium is obtained from underground wells, usually as a byproduct of natural gas production. The major use of noble gases is in welding, where a nonreactive atmosphere is needed due to the high temperatures involved. Another use, especially of argon and krypton, is in light bulbs like the one shown in Figure 13-23. An inert gas in the bulb contributes to the life of the filament. Some fluorescent bulbs and advertising signs also contain noble gases. These gases glow when excited by a small amount of electricity.

Figure 13-23 Welding shown on the left is done in an atmosphere of noble gas. The high temperatures involved make the use of other atmospheres dangerous. An inert gas, such as argon, is used in incandescent bulbs shown on the right to minimize the evaporation of metal from the filament.

Review and Practice

1. What property makes nitrogen gas generally unreactive?

2. What is the role oxygen plays in cellular respiration?

3. Which group of nonmetals is the most reactive?

4. How does the activity of the halogens change as the atomic mass increases?

5. Name the primary property of the noble gases.

6. List two uses of noble gases.

1. the strong stable triple bond of shared electrons
2. Oxygen is an electron acceptor.
3. the halogens
4. reactivity decreases
5. nonreactive
6. welding, light bulbs

Careers in Chemistry

Carlos Aguilar
Title: Industrial Chemist
Job Description: Performs chemical analysis of metals with the use of complex equipment. Supervises the work of several technicians and checks their results. Is responsible for adjusting instruments so they operate correctly. Writes daily reports of analyses.
Educational Qualifications: College degree in chemistry, with additional courses in physics, electronics, and biochemistry.
Future Employment Outlook: Excellent

Carlos Aguilar became interested in chemistry at an early age. "Ever since I was a kid, I've had an interest in science. When I took my first chemistry course, I knew it was my field. It related to everything I could see in life—everything I observed in the world around me." Today Carlos Aguilar works as a chemist in a plant that produces zirconium and hafnium. These materials are used in the manufacture of fuel control rods in a nuclear reactor. Aguilar is responsible for measuring the concentration of impurities in the zirconium and hafnium.

Aguilar describes a typical day: "The first thing I do is check the night shift report and then prioritize the work I need to get done. After carefully checking to make sure the instrumentation is operating correctly and adjusting any errors, I give the work and samples needed to be checked to the technicians under me. When the sample checks are returned to me, I write up my daily reports. Any problems encountered must be turned over to my immediate supervisor. Finally, I review what should be left for the night shift to do."

Careful measurement and precise results are important in Aguilar's work. An error in detecting impurities could lead a reactor core to malfunction or even melt down.

The characteristics that have helped Aguilar in pursuing his career include being highly observant, taking initiative, and having an analytical mind. Says Aguilar, "A good chemist must pay close attention to detail and have a solid knowledge of the field." He also recommends taking courses in physics, electronics and biochemistry in high school if they are available.

Aguilar's working conditions are excellent. Safety is put first, and the equipment is the best available. The salary is excellent and job advancement good—Aguilar was promoted from assistant chemist to chemist during his first year on the job.

Aguilar concludes: "The future for a chemist in the nuclear industry is exciting. Progress in nuclear power is increasing. The work we do is vitally important for my children and their children, who will use nuclear power far into the future."

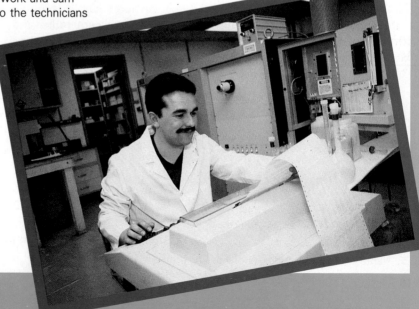

CHAPTER REVIEW

Summary

■ Most elements are metallic, that is, they tend to lose or give up electrons in the presence of other substances. Most properties of metals can be explained using the model for the metallic bond.

■ Alkali elements are the most reactive of the metals and all tend to give up one electron in reactions to form 1+ ions. Sodium and potassium are the most common alkali metals, and are necessary to life.

■ Alkaline-earth elements are slightly less reactive than alkali elements, and give up two electrons to form 2+ ions.

■ A mixture of two or more metals, or of a metal and a nonmetal, is called an alloy. Alloys can have properties that are different from those of pure elements, and can be formulated for many different purposes.

■ Transition metals exhibit the familiar properties commonly associated with metals, such as strength, durability, and a high melting temperature.

■ Metalloids are elements displaying properties intermediate between those of metals and nonmetals.

■ Some nonmetals and metalloids exist in different structural forms called allotropes. ✓

■ Nonmetals tend to accept or share electrons. Living matter contains many covalent bonds because of the stability of this type of bond.

■ The halogen family is the most reactive of the nonmetals. The activity of halogens decreases with increasing atomic mass. ✓

■ The noble gas family tends to be nonreactive, although compounds of heavier noble gases have been prepared. The nonreactive nature of noble gases has led to their use in a number of applications.

Chemically Speaking

allotrope
alloy
metallic bond
metalloid
silicone

Review

1. What are the three broad categories of elements? (13-1)

2. Describe the positions of positive metal ions and of outer level electrons in the metallic bond model. (13-1)

3. Why are pure alkali metals stored under kerosene? (13-2)

4. Which two elements play a key role in the transmission of nerve impulses? (13-2)

5. Give three properties of the alkali metals. (13-2)

6. Name three elements that are never found uncombined in nature. (13-2, 13-3, 13-13)

7. List three common properties of the alkaline-earth metals. (13-3)

8. What element is usually alloyed with magnesium? (13-3)

9. Explain why magnesium hydroxide is a better antacid than a strong, soluble base such as sodium hydroxide. (13-3)

10. What is the most common compound of calcium? (13-3)

11. What type of compound forms on the surface of pure aluminum, magnesium, and chromium, when these metals are exposed to air? What roles do these compounds play in determining the useful applications of these metals? (13-3, 13-4, 13-6)

12. List three properties of aluminum that make it useful. (13-4)

13. Give the chemical definition of a metal. (13-5)

14. Name two noble metals. (13-6)

15. What substances are used to remove impurities from iron to make steel? (13-7)

16. Which broad category of elements has properties that are intermediate between those of the other two groups? (13-9)

17. Which is the second most abundant element in the earth's crust? (13-10)

18. What property of the heavy metals arsenic, lead, and mercury makes them poisonous? (13-10)

19. What are the primary elements in silicate minerals? How common are silicate minerals? (13-10)

20. White, black, and red phosphorus are made of the same element, but the structures and properties of these substances are different. What is the name for this phenomenon? (13-10)

21. Which molecules in the body contain nitrogen? (13-11)

22. Which family of elements is known as the salt-formers? (13-13)

23. What is the most active element? (13-13)

24. What family of elements is used to create atmospheres for welding? Why? (13-14)

Interpret and Apply

1. Write the equations for the following reactions.
   a. sodium with aluminum chloride (13-1)
   b. lithium with water (13-2)
   c. potassium with bromine (13-2)
   d. silicon dioxide with calcium (13-3)
   e. calcium oxide with water (13-3)
   f. aluminum oxide with water (13-4)

2. Why were transition metals, especially the "noble metals," discovered centuries before the alkali or alkaline-earth metals? (13-2, 13-3, 13-6)

3. In a campfire the mineral covellite, or copper(II) sulfide, reacts with oxygen to form copper(II) oxide and sulfur dioxide. The copper(II) oxide then reacts with charcoal, or carbon, to produce copper metal and carbon dioxide. Write the balanced equation for both of these reactions. (13-6)

4. Which would be harder, iron containing 1% carbon or iron containing 0.5% carbon? Explain your answer. (13-7)

5. Phosphorus, sulfur, and oxygen are not the only elements that form allotropes. What kinds of conditions, or differences in conditions, might lead to the formation of allotropes from one element? (13-11)

6. Two separate one-mole samples of sulfur are reacted with six moles of fluorine and six moles of bromine, respectively. In which reaction will there be sulfur left over? Explain your answer. (13-13)

7. Draw the Lewis structure for a diatomic iodine molecule. (13-13)

Concepts needed to do these problems were covered in previous chapters. Sections in which the featured elements are discussed are noted for each problem.

Problems

1. When 17.5 grams of calcium are reacted with 15.3 grams of sulfur to form calcium sulfide, which reactant is in excess? (13-3)

2. Calculate the volume of the hydrogen produced at a pressure of 95.0 kilopascals and a temperature of 43°C when 13.5 grams of calcium react with an excess of water. (13-3)

3. Calculate the mass of magnesium oxide produced when 17.5 grams of magnesium are reacted with an excess of oxygen. (13-3)

4. How many grams of hydrogen chloride will react with a tablespoon of milk of magnesia which contains 6.75 grams of magnesium hydroxide? (13-3)

5. Calculate the volume of hydrogen produced at STP when 13.5 grams of calcium react with an excess of water. (13-3)

6. How many grams of potassium are required to replace the aluminum in aluminum chloride, $AlCl_3$, to produce 145 grams of aluminum? (13-4)

7. What mass of carbon, when combined with oxygen, is required to produce 585 moles of carbon monoxide? (13-7)

8. How many kilograms of iron can be produced by 585 moles of carbon monoxide reacting with an excess of iron(II) oxide to produce iron and carbon dioxide? (13-7)

9. When 25.8 liters of sulfur dioxide are combined with oxygen at STP, how many moles of sulfur trioxide will be produced? (13-12)

10. Oxygen and sodium hydroxide are produced when sodium peroxide, Na_2O_2, reacts with water. What volume of oxygen at a pressure of 98.3 kilopascals and a temperature of 28°C will be produced when 10.6 grams of sodium peroxide react with an excess of water? (13-12)

11. What volume of fluorine at a temperature of 22°C and a pressure of 105 kilopascals will just react with 35.6 grams of titanium to form titanium(IV) fluoride? (13-13)

12. Uranium(IV) fluoride and water are formed by reacting hydrogen fluoride with uranium(IV) oxide. What mass of hydrogen fluoride is required to produce 8.45 kilograms of uranium(IV) fluoride? (13-13)

Challenge

1. A teakettle contains 34.6 grams of boiler scale which is 60% calcium carbonate and 40% magnesium carbonate. What volume of vinegar, which is a 4.5% solution by mass of acetic acid, $HC_2H_3O_2$, will be required to react with the boiler scale to remove it? (13-3)

2. When 2.16 grams of magnesium are heated in air, it combines with oxygen and nitrogen and the mass is increased to 3.43 grams. When the resulting mixture of compounds is reacted with water, ammonia is formed and 3.58 grams of magnesium oxide remain. What percent of the magnesium initially combined with nitrogen? (13-3)

3. A copper ore is 6.0% copper(II) oxide. The copper(II) oxide is separated from the ore and reacted with carbon to form copper metal and carbon dioxide. What mass of copper can be obtained from 3500 kilograms of ore? (13-6)

4. When 15.0 mL of 20% hydrogen peroxide, with a density of 1.11 g/cm^3, are decomposed completely to form water and oxygen gas, what volume of oxygen gas will be produced at a pressure of 95.0 kilopascals and a temperature of 35°C? (13-12)

5. When 15.0 mL of 30% hydrogen peroxide, with a density of 1.16 g/cm^3, are decomposed completely, what volume of oxygen will be produced at a pressure of 95.5 kilopascals and a temperature of 27°C? (13-12)

6. There is very little helium in Earth's atmosphere, even though helium has been produced in enormous quantities for millions of years by radioactive decay processes. This is because, at 20°C, the average velocity of a helium molecule exceeds the escape velocity of Earth's gravitational field, which is about 1200 m/s. Using Graham's Law, and given the average velocity of nitrogen at 20°C as 480 m/s, show that the average velocity of a helium molecule is greater than the escape velocity. (13-14)

Synthesis

1. Both magnesium and calcium can be used to convert uranium(IV) fluoride to uranium. If both magnesium and calcium are the same price per kilogram, which would cost the least to produce a kilogram of uranium?

2. If magnesium costs $11.50 per kilogram, and calcium costs $15.00 per kilogram, which would cost the least to produce a kilogram of uranium?

3. Predict any other elements that could be used to produce uranium from UF_4.

4. Draw the Lewis structure for diatomic bromine. Is the molecule polar or nonpolar?

5. Which is more dense, dry air or air that is 50% saturated with water vapor?

Projects

1. Household ammonia is a solution of ammonia, NH_3, in water. Design an experiment to find out the concentration of ammonia in water.

2. The addition of calcium oxide, CaO, to water produces limewater, which is often used to detect the presence of carbon dioxide. Research the reactions involved and demonstrate the use of limewater to detect the presence of CO_2.

3. Research the problem of metallic salts, carbonates, and other deposits, in irrigated farmland. How widespread is the problem? What solutions are in use, or have been proposed? Contact your state Soil Scientist, who is often associated with the state university, to get started.

Each time a space shuttle returns from a mission in space, high technology has made that mission possible. Many of the shuttle's systems and components are examples of state-of-the-art technology.

By nature the high tech world is in a constant state of change, with new advancements and applications continually being introduced. This chapter is somewhat different than the others you have studied. It has three important goals—first, to introduce you to some applications of chemistry in the high-tech world; second, to give you an idea of the risks and benefits associated with technological advancements; and third, to show you how to explore the rapidly changing fields of high tech. The ability to seek out and evaluate information in different fields is an important skill, one that you will be required to use often after leaving this course.

High-Tech Chemistry

Silicon in the High-Tech World

Once in space, the shuttle begins a host of tasks that include improving global communications, repairing damaged satellites and launching new ones, and conducting scientific experiments in an environment very different from that on Earth. It would be impossible for the shuttle and its crew to do their complex tasks without the help of computers. Figure 14-1 shows the flight deck of the *Columbia*. No fewer than five on-board digital computers free the astronauts for other responsibilities. Three computers are assigned to guidance and navigation, one is in charge of equipment on board, and another is used for performance monitoring. Computers on the ground also monitor the systems for each space mission.

Computers have changed the way we live. Billions of pieces of information are being processed and exchanged every day by this electronic medium. The advantage of the computer is its ability to store, retrieve, and process information in a very short period of time. At the heart of every computer are microchips, or integrated circuits made mostly of silicon. Silicon and other semiconducting elements have played a critical role in the development of the computer, and are the basic elements of the electronics industry.

CONCEPTS

- **properties of silicon crystals**
- **doping semiconductors**
- **making integrated circuits**
- **photovoltaics and satellite communications**

The aluminum in the solid fuel used for the shuttle boosters is mixed into a synthetic rubber binder (polybutadiene acrylic acid acrylonitrile). The solid fuel is contained within a tube without access to atmospheric oxygen. Ammonium perchlorate is included in the fuel as an oxidizer. The iron oxide catalyst controls the rate of burning.

Figure 14-1 The flight deck of space shuttle *Columbia*. The flight computers are located in the center of the picture between the two seats. Data keyboards are located on the console between the seats, and the monitors are in the center of the main control panel.

This chapter is slightly different from the others. It encourages students to find out more about various aspects of high technology, and introduces them to strategies for doing research. See the teacher's chapter notes at the front of this book for suggestions on possible uses of this chapter.

14-1 What Makes a Computer?

Semiconductors   Recall from Chapter 13 that some elements, called metalloids, have chemical and physical properties intermediate between metals and nonmetals. Silicon, germanium, and arsenic are examples of metalloids. Pure silicon has some metallic properties but, unlike true metals, it forms a crystalline structure like that of a diamond. The crystalline structure of silicon suggests that each silicon atom is covalently bonded to four other silicon atoms, as illustrated in Figure 14-2. The four outermost electrons of one silicon atom ($3s^23p^2$) are able to bond with the outermost electrons of other silicon atoms, resulting in eight electrons around each silicon atom. Figure 14-2 also shows some examples of silicon crystals. Pure germanium and pure boron also form diamondlike crystal structures.

Figure 14-2 Pure silicon crystals on the left are composed of atoms of silicon covalently bonded to each other in a tetrahedral structure. Silicon crystals on the right need to be grown with great care. They are very expensive to manufacture because of the purity required.

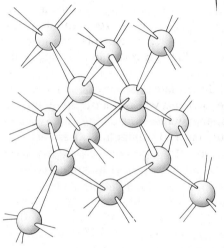

Review with the students the location of the semiconductors or metalloids on the periodic table.

The electrons in silicon are not free to roam about in the crystal as they would be in true metals and are said to be localized. A crystal of pure silicon is, therefore, a poor conductor of electricity. However, certain impurities can be added while the crystal is growing, to provide "extra" electrons. The extra electrons serve to increase electrical conductivity. Arsenic has five outermost electrons ($4s^24p^3$), but only four of these electrons are bonded to neighboring silicon atoms. There are eight electrons around each arsenic atom, plus one electron that is free to flow in the direction dictated by an electric field, as illustrated in Figure 14-3. Crystals of this nature, containing deliberately added impurities, are described as **doped** crystals. Semiconductors that have been doped with impurities to produce mobile electrons are called **n-type semiconductors**.

Pure silicon also can be doped with an impurity that results in a deficiency of electrons, such as gallium. Gallium has three outermost electrons ($4s^24p^1$). These three electrons form bonds with three neighboring silicon atoms, but there will be a deficiency at the site of the fourth neighboring silicon atom, as illustrated in Figure 14-3. The result is a positive "hole" in the crystal. Semiconductors that have been doped to produce holes, or electron deficiencies, are called **p-type**

Circuit components include transistors, resistors, capacitors, and diodes, each of which serves different functions.

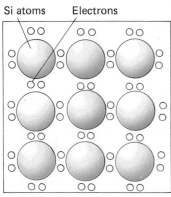

Perfect Si crystal

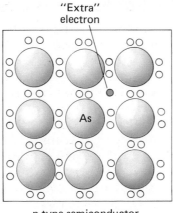

n-type semiconductor

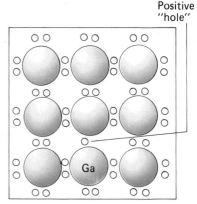

p-type semiconductor

semiconductors. When a voltage is applied to the crystal, electrons flow to fill the holes, leaving behind other holes. You can think of a p-type semiconductor as a crystal in which positive holes move across the crystal in response to an electric field, traveling in a direction opposite to the flow of electrons.

Why are n-type and p-type semiconductors important? They are the basic substances that determine how and when electrons will flow in electronic equipment. The proper combination of the two types of semiconductors, along with insulators and connecting wires, make up the tiny "electronic brains" that operate computers, calculators, stereos, and any piece of equipment you can think of that is called electronic.

Integrated Circuits Silicon can be doped to produce p-type and n-type semiconductors on different areas of the surface. When p-type and n-type semiconductors are in contact, they form a junction. The negative electrons in the n-type semiconductor will be attracted to the positive holes of the p-type semiconductor. Electrons will have a tendency to flow across the junction toward the p-type side, establishing a voltage across the junction. Figure 14-4 illustrates the flow of electrons across one such junction. With the proper combinations of junctions, various parts of electronic circuits can be fabricated.

Figure 14-3 (left) In a simplified illustration of a silicon crystal, each silicon atom is surrounded by eight electrons. Each atom has four covalent bonds. (center) In an n-type semiconductor, "extra" electrons come from the outermost electrons of added impurities, in this case arsenic. (right) The impurities in a p-type semiconductor can contribute only three electrons to four bonds. The result is a positive "hole" in the crystal at the site of the impurity.

Direction of Electric Field

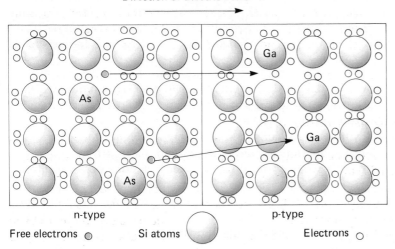

n-type

p-type

Free electrons ◉   Si atoms ◯   Electrons ◯

Figure 14-4 At a junction, electrons flow from the n-type semiconductor to fill the positive holes of the p-type semiconductor. Combinations of junctions make up the electron switching devices of a circuit.

Figure 14-5 The integrated circuit shown has been magnified 75 times. Its actual size also is shown. The complex structure includes many components and connections, all contained in one block of semiconducting material.

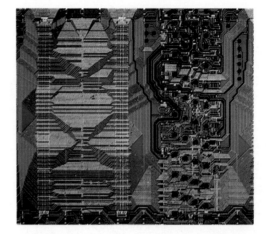

A transistor has one of two purposes in a circuit, amplifying or switching. In early computers, transistors acted primarily as switches.

Electronic circuits used for even simple applications are made up of many individual parts, called components. The proper arrangement of components on a single tiny piece of semiconductor material forms an **integrated circuit**, or "chip." Figure 14-5 shows an integrated circuit. In the 1960's, integrated circuits typically contained about 30 components. Today, some integrated circuits contain over one million electronic components in a volume of only 12.5 mm³.

Figure 14-6 The first electronic digital computer, ENIAC, transmitted signals by relays (electronic switches) and electron tubes, such as the one shown. Tubes required a lot of space and energy, and had to be artificially cooled to prevent breakage.

Figure 14-7 An early breakthrough in electronics was the transistor. Transistors are far superior to the tubes they replaced, but all the components in every circuit still had to be tested and assembled individually.

The revolution in computers would not be possible without the technology of integrated circuits. In 1946, the first computer, ENIAC, was composed of 18 000 electron tubes and 1500 relays. Part of ENIAC is shown in Figure 14-6. Note its size. Electron tubes were later replaced by transistors, such as those shown in Figure 14-7. Transistors provided a reduction in power usage and size, and an increase in reliability. The work involved in assembling circuits out of individual components was still prohibitive, however. With further reductions in size and power usage due to integrated circuits, the same functions performed by ENIAC can now be performed by a silicon chip four millimeters square, like the one shown in Figure 14-5.

Miniaturization The features of integrated circuits are so small (about 13 μm) that a circuit is first drawn and photographed at about 400 times larger than its final size, then the photograph is reduced. The circuit is layered onto the surface of the silicon through a process called *photoengraving*. In this process, a layer of semiconductor is deposited. Then the layer is doped to produce the components of the circuit. Another layer is deposited and doped, and another, until all the layers of the circuit are complete. Figure 14-8 is a diagram of a cross section of several layers of one integrated circuit. Very Large Scale Integration (VLSI) chips contain from 65 000 to 2 000 000 components with details that are about 1 μm in length.

Lasers are presently used for deposition of materials, engraving, direct wiring of conducting materials, and even for wet chemistry (aqueous solutions) on semiconductor surfaces.

Electrical conductor

Insulator

neutral

n-type

p-type

p-type

n-type

n-type

capacitor
p-type

transistor

neutral

resistor

0.005 cm

(layers of components)

silicon wafer base

0.007 cm

Figure 14-8 An integrated circuit is composed of carefully applied layers of materials that form components. In this illustration a capacitor (stores electrons) and resistor (controls current) are neighbors to a transistor (amplifies current).

In actual practice, many integrated circuits are layered onto the silicon at once. The surface of the silicon becomes a grid with the same circuit layered onto each segment. Figure 14-9 shows an engraved silicon wafer, containing hundreds of identical integrated circuits. These identical segments will be individual chips when the wafer is cut apart. After the individual chips are separated, leads are attached, as shown in Figure 14-10. The leads connect the chip to the device in which it functions.

Miniaturization has itself brought about a new technology. Integrated circuits are now being fabricated by adding a few atoms and molecules at a time. This technology involves very precise work, a new understanding of the surface of materials, and an understanding of how surfaces can be changed by chemical reaction.

Figure 14-9 An etched silicon wafer, containing hundreds of chips. Each chip will be electronically tested and defective ones marked. The wafer is then sawed apart into the individual chips.

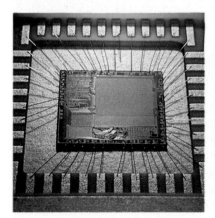 

Figure 14-10 (left) Very fine gold leads are used to connect the integrated circuit to larger leads. (right) The chip is packaged in plastic for protection and convenience.

The laser is now being used to deposit material, only tens of atoms thick, onto silicon chips. Conductors, semiconductors, and insulators can all be deposited. For example, when a conductor is needed to fill a channel between components only several micrometers wide, $Mo(CO)_6$ gas is energized with the monochromatic light from a laser. The compound dissociates and deposits a thin ribbon of molybdenum metal, a conductor, onto the semiconductor as represented in the following equation:

$$Mo(CO)_6(g) + energy \longrightarrow Mo(s) + 6CO(g)$$

The added molybdenum atoms conduct electrons from one component to another in an integrated circuit. The laser has brought a level of precision and reliability to microelectronics that was not possible without it.

Microelectronics has advantages and limitations. Miniaturization by way of integrated circuits has brought about a reduction in the size of electronic equipment such as computers. Integrated circuits are cheaper and more reliable than electron tubes and individual components. They use less electricity and they last longer because of cooler operating temperatures. One limitation of miniaturization is that the impurities needed to produce the n-type and p-type semiconductors tend to diffuse away from their locations. When this diffusion occurs, the chip becomes useless. (However, this takes from 40 to 100 years to occur.)

Natural radioactivity also can cause components to develop defects. Positive and negative radioactive particles can damage junctions by turning an n-type semiconductor into a p-type, or a p-type into an n-type. The more that components are miniaturized, the more important each individual atom becomes to the circuit. The effects of radiation are thus heightened as miniaturization increases. Another limit on miniaturization is the behavior of electrons. One model that explains electron flow pictures electrons as bumping each other along in a steady stream. As pathways become narrower, electrons tend to flow as individual particles rather than in a steady stream. This causes uncertainty in the path of electron flow and would cause ultraminiature circuits to be unreliable.

14-2 Extension: "Seeing" the Atom

As miniaturization becomes more important, the surface detail of semiconductors becomes critical. Scientists at the IBM Zurich Research Laboratory have developed a microscope that is capable of "seeing" details down to the atomic level. The scanning tunneling microscope, as it is called, does not use visible light energy to view the atoms. The wavelengths of visible light are 2000 times larger than the diameters of atoms, and are not suitable for detailed atomic work. This special microscope uses the energy already existing in the electrons of the surface atoms. The orbitals of surface atoms extend somewhat beyond the surface of the material under study. If a probe is brought close to the surface of the material and a small voltage is applied, electrons will flow between the orbital channels (tunnels) of the surface atoms and the probe, as illustrated in Figure 14-11. The change in

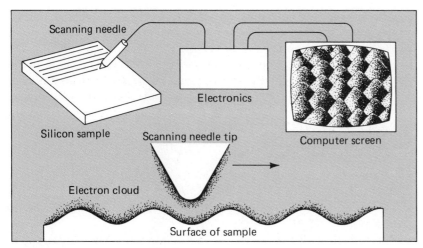

Figure 14-11 A flow of electrons from a scanning tunneling microscope passes through the cloud of outer orbitals of surface atoms. The tip of the needle is kept at a constant distance from the surface atoms at all times. The computer reads the up-and-down movement of the tip as it travels across the surface. The readout is viewed on a monitor.

current as the probe moves slowly across the surface is detected and translated into computer graphics. Figure 14-12 is an image of the surface of silicon, as "seen" by the scanning tunneling microscope. The bumps correspond to surfaces of individual atoms.

Applications of the scanning tunneling microscope are not limited to microelectronics. The microscope also has diagramed the zig-zag structure of the DNA helix, and has given biologists a glimpse at individual parts of viruses. The microscope has much potential in chemical research. The energy needed to "see" the surface of atoms also can cause chemical reactions. By tuning the electron beam of the microscope to specific energies, scientists can cause a designated reaction to occur and observe the reaction on the atomic level. The scanning tunneling microscope will give chemists the opportunity to "see" chemical reactions as they occur.

Electron microscopes have projected images of atomic rows. The need for the scanning tunneling microscope is in the details it provides of the surfaces of materials. Electrons from standard electron microscopes probe deep *under* the surface of the material studied.

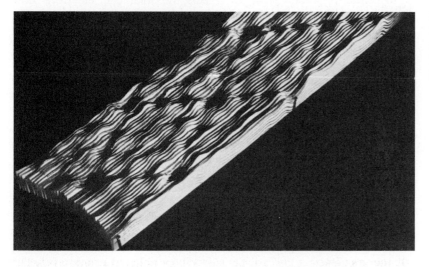

Figure 14-12 The surface of silicon is depicted in this illustration by the scanning tunneling microscope. The bumps correspond to surfaces of individual silicon atoms.

14-3 Solar Cells

The space shuttle has deployed a number of satellites that function as communication stations and scientific instruments. One of these satellites is pictured in Figure 14-13. Satellites must function for years

Figure 14-13 Communication satellites depend on large arrays of solar panels for electrical energy.

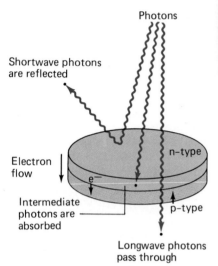

Figure 14-14 A solar, or photovoltaic, cell can produce electricity because photons of a certain wavelength excite electrons, which flow to create a current. Photons of shorter wavelength are reflected by the cell, and photons of longer wavelength pass through.

if they are to be cost-effective. Batteries do not last long enough for satellite use, and fuel cells are unreliable. Solar panels have been found to be the only economical power source for satellites.

The solar panels used on satellites are composed of many thousands of solar cells. When the intense sunlight in space hits the solar cells, electricity is generated. The solar cell, also known as a **photovoltaic cell**, changes radiation into electricity by taking advantage of the properties of semiconductors. A solar cell is a thin wafer of pure silicon doped with arsenic to render it an n-type semiconductor. A thin surface of p-type semiconductor, silicon doped with boron for example, is deposited over the wafer. A junction is established in much the same fashion as in an integrated circuit. One electric lead is attached to the wafer and another lead is attached to the surface of the p-type semiconductor. When sunlight strikes the solar cell, electrons flow from the surface layer to the body of the cell. The electrons moving through the leads, as illustrated in Figure 14-14, is an electric current.

Many kinds of materials are used to make solar cells. Some solar cells employ layers of silicon dioxide, SiO_2, and thin films of chromium, copper, silver, gold, or aluminum over the n-type or p-type silicon. Others use thin layers of compounds, such as cadmium sulfide or zinc sulfide, over the silicon base. Photovoltaic cells presently used by satellites have about a 20% efficiency rating, which means that 20% of the energy from the sunlight striking the cells is converted into electrical energy. The greatest disadvantage of photovoltaic cells is the high cost of manufacturing pure silicon.

1. arsenic
2. the circuit consists of electronic components on a tiny piece of semiconductor material
3. The energy comes from a small voltage applied near the surface being studied. The electrons in the orbitals of the surface atoms flow in orbital channels.
4. An n-type semiconductor is in contact with a p-type semiconductor. When light of the proper frequency strikes the semiconductor material, an electric current is produced.

Review and Practice

1. Name an element that may be used to dope a pure germanium crystal in order to make an n-type semiconductor.

2. Describe an integrated circuit.

3. What is the source of energy used to "see" surface atoms in the scanning tunneling microscope?

4. Briefly describe how a solar cell produces electricity.

Ceramics and Glass

At the conclusion of a mission, the space shuttle returns to Earth in a ball of fire. Passage through Earth's atmosphere causes friction that generates heat on the surface of the spacecraft. Temperatures on the nose and leading edges of the wings and tail reach 1650°C. The shuttle and its crew must be protected from this extreme heat. Several thermal protection systems are used to provide shielding. One system is made up of 31 000 heat-resistant silica tiles that protect the underside of the shuttle. They are specially manufactured ceramics, 90% air by volume, yet strong enough to withstand the vibrations and forces of takeoff and atmospheric impact. Ceramics, an ancient material, has found a place in the Space Age.

Glass is another material that has been known and used for thousands of years. Glass and ceramics may not seem to be good candidates for high-tech materials. However, the chemical and physical properties of these ancient substances are well matched for the needs of high technology. The transparency of glass and the extreme heat resistance of ceramics are among the characteristics that make these materials so desirable. In addition, glass and ceramics are abundant and inexpensive.

CONCEPTS

- the nature of ceramics
- high-tech uses of ceramics
- compositions of some forms of glass
- use of glass in fiber optics

14-4 Ceramics, Old and New

The word ceramics usually conjures up thoughts of a pot formed of clay and fired in a kiln. A ceramic is composed of small crystals of aluminosilicate in a matrix of glassy cement. Silicates are salts of the orthosilicate ion, SiO_4^{4-}. In this ion, the silicon atom is covalently bonded to four oxygen atoms, forming a tetrahedral structure with an atom of silicon, Si, in the center, as depicted in Figure 14-15. In an aluminosilicate, a number of these tetrahedrons are centered on an aluminum atom, rather than a silicon atom. The aluminosilicate tetrahedrons form thin platelike crystals, which are cemented in silica glass, SiO_2, to form a ceramic.

Porcelain, bricks, earthenware, and bone china (containing calcium phosphate) are all ceramic materials. The Romans made a combination of ceramics and glass called opal glass by adding antimony oxide,

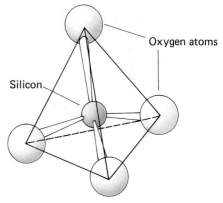

Figure 14-15 The SiO_4^{4-} ion is the basis of most silicate material. The silicon atom is covalently bonded to four oxygen atoms, forming a tetrahedron.

Figure 14-16 (left) Ancient Romans knew of the beauty in ceramic glass (called opal glass). (right) Other kinds of glass and ceramic materials also have been used to create works of art throughout history.

Sb$_2$O$_3$, to the glass slurries. Today, Li$_2$O, Al$_2$O$_3$, TiO$_2$, and TiO are added to silica to form modern ceramic glass cookware.

Ceramics offer many advantages over other materials. Ceramics are hard, porous, heat resistant, and generally unreactive. These characteristics have made ceramic materials important to high-tech industries. Figure 14-17 shows some of the items made out of modern ceramics. Magnets, rocket nose cones, nuclear fuel rods, spark plugs, engine parts, windows in integrated circuits, and even human bone replacements all contain ceramic materials.

Figure 14-17 All these items use high-tech ceramic materials.

14-5 Glass and Optical Fibers

The mineral quartz and pure silica glass have the same chemical composition, SiO$_2$. However, quartz has a crystal structure that is regular throughout the crystal, whereas glass is less ordered, as illustrated in Figure 14-18. Most commercial glasses contain sodium or calcium

Optical fibers can be used for interesting demonstrations. They can be obtained from some science supply houses. A kind of decorative lamp also uses optical fibers. It is best to use plastic fibers when students will be handling them.

Figure 14-18 The structures of quartz and glass differ drastically. Quartz has a regular pattern, as do all crystals. Glass, however, does not have such regularity and cannot be considered a true crystal.

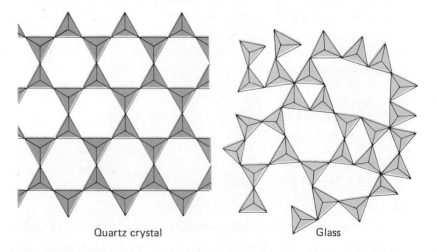

Quartz crystal   Glass

oxides or carbonates. These salts are added because pure melted silica is too difficult to process. The salts break up the long chains of covalent bonds by interspersing ionic bonds throughout the material. Lead(II) oxide is sometimes added to an alkaline glass melt to improve the clarity and density of the glass, forming what is commonly called "lead crystal."

Glass long has been recognized for its beauty and practicality. Today, however, it has taken on a new importance in the communications industry. Glass is an excellent medium for carrying light energy. **Optical fibers** are very fine fibers of glass that transmit light, even around corners. The light beam will follow the path of the fiber even at a bend because the light beam is reflected back and forth within the glass. This phenomenon, illustrated in Figure 14-20, is called *total internal reflection.*

Do You Know?

Phototropic glass is used in manufacturing self-darkening eye glasses. The ability of the glass to change color in sunlight is due to 0.27% silver oxide and traces of the halogen elements and copper. In light, microcrystalline silver forms in the solid glass to give the darkened color. In the absence of light, the silver recombines with the other elements to form a transparent compound.

Optical fibers are strong, but breaks do occur. A crack represents a barrier to light transmission. Splicing an optical fiber is not easy. Splicing is one problem that must be addressed as the use of fiber optics increases.

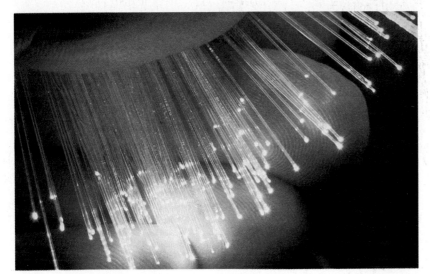

Figure 14-19 Light is transmitted through the length of an optical fiber with almost no loss of energy.

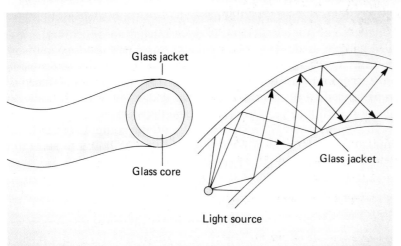

Figure 14-20 (left) The core of an optical fiber is made of one kind of glass, and the jacket of another kind. (right) Light passing laterally inside the fiber is reflected within the fiber because of a difference in physical properties between the core and jacket.

Highly transparent glass must be used in the manufacture of optical fibers. Typically, the SiO_2 used is formed from silicon tetrachloride, as in a reaction represented by this equation:

$$SiCl_4 + O_2 \longrightarrow SiO_2 + 2Cl_2$$

The fibers contain small amounts of P_2O_5, GeO_2, or B_2O_3, added to improve the fiber core's internal reflection. Figure 14-21 shows a step in the manufacture of optical fibers.

Figure 14-21 Optical fibers are extremely difficult to manufacture and are relatively expensive. Glass can be drawn into fibers 16 kilometers long. One fiber can carry 10^4 times more phone calls than 500 copper wires.

There are many advantages to using fiber optics as a means of transmitting voice or numerical data. Two advantages are speed and accuracy. A gallium arsenide (GaAs) solid state laser can produce a binary code from light pulses (a light pulse means "one," no pulse means "zero") that are transmitted at the rate of 420 million bits of information per second in a fiber 203 km long. It does so with virtually no errors and without the help of the signal boosters needed in copper wire. In one second, a GaAs laser could send the contents of this book 3 times through an optical fiber. It would take 21 hours to send the same information through copper wire.

Review and Practice

1. small crystals of aluminosilicate in a matrix of glassy cement
2. hard, porous, heat resistant, nonreactive, and inexpensive
3. SiO_2 with sodium or calcium oxides or carbonates
4. The light beam is continually reflected back and forth within the optical fiber.
5. light energy

1. What is the basic structure of ceramic materials?

2. Name three characteristics that make ceramic materials useful.

3. What is the chemical composition of most commercial glass?

4. What happens to a light beam in an optical fiber?

5. What form of energy is used to transmit information through optical fibers?

Chemical Research

You learned in Chapter 1 that the communication of accurate information is vitally important to science. Research is not just experiments done in a lab or in the field, but it involves utilizing the results of other scientists.

You now have the opportunity to become an expert on a particular topic of chemistry that is related to current technology. You will find many ideas for research in the Chapter Review section of this chapter. In researching a topic, you should focus your attention not only on the scientific principles, but also on the social, moral, political, and economic implications scientists face working in this field. Some technologies involve risks for the people who work at them, but hopefully the benefits outweigh the risks.

The space program has provided countless innovations that have helped to save lives. Heart pacemakers, fire retardant materials, medicines, and life monitoring devices are products that are a direct result of space program research. However, such life-saving advances have not come without the loss of life in space research.

Computer technology involves the risk of depending on artificial intelligence rather than human judgement. Chemical technology runs the risk of disrupting the environment by the production of necessary yet toxic materials. We do not live in a perfect world and since technology is a part of this world, it suffers from that same lack of perfection.

CONCEPTS

- finding information
- evaluating information
- presenting information

14-6 Conducting a Literature Search

There are different reasons for conducting research on a topic in science or technology. For the scientist, this is the first step in conducting experimental research. Scientists and nonscientists often want to know the social, economic, or political impact of a particular issue. An engineer may have an invention in mind, and needs to know whether it has been done before. Whatever the motivation, the methods of conducting a literature search are fairly constant.

There are several points to keep in mind as you become the class "expert" on a chosen scientific topic. Plan to spend some time finding information in your school or public library. (If you have access to a college or university library, all the better.) You might begin by gathering information from an encyclopedia or other general reference source. Then you should delve into more specific sources by finding your topic in *The Reader's Guide to Periodical Literature*. Besides searching in books and periodicals, some of which are listed in the next section, you may want to ask specific questions of technical experts at a university, industrial research center, or manufacturing plant.

In research, it is not as important to gather volumes of information as it is to evaluate the facts you gather. As you do your research, ask yourself some questions. Is the information accurate? How do I decide? What do I do when information from different sources is conflicting? Which source is more reliable? How do I determine which information represents "good science," and which misrepresents scientific principles? What are the social, economic, and moral issues that are raised by this new technology?

These are difficult questions. The answers to them will come from further research and from your own conclusions about what you have read. Information often is verified when it is found in several reliable and independent sources. The reliability of a source depends on who is supplying the information. *Independent* means that the sources you find are not simply referring to each other for the same information. Your teacher can help you evaluate the materials you find and suggest what to look for when doing research.

There are additional questions you should ask yourself as you prepare to present your material. How should I communicate this information to my classmates, given the fact that I am the expert and they are the "general public"? What background does the public already have to evaluate the information I am giving them? What background information do I need to give them? What is the best way to present my research so that the public will understand it? How should controversial issues be handled in a public forum?

When you are in the habit of asking questions such as these, your skills at gathering and evaluating material for usefulness and accuracy will be greatly enhanced. Most importantly, you will learn by doing.

14-7 Sources of Information

When doing research on any topic, it is helpful to know where to start. The sources listed here are divided into general references, in the form of encyclopedias, and more specific references, in the form of magazines. The amount of information you will be able to obtain from a particular source may depend on the audience for whom the material is written. Some of the magazines are intended for practicing scientists, while others provide a "popular" approach to topics. Some magazines are intended for readers who are not scientists, but who do have a good background in science. The same is true of books. Keep in mind that books provide an opportunity for in-depth treatment and analysis of their subjects, and should not be overlooked. Check the libraries available to you for the books they have on your specific topic.

Although this list is primarily concerned with science or science-related issues, you can use a similar approach to organize research in many areas besides science. The sources listed can be found at most public libraries. From these sources, you may learn of others that can help you further.

A. General References (sources that will give you an overview of the topic you are researching)

Encyclopaedia Britannica

How It Works, illustrated encyclopedia of science and technology.

The Micro Dictionary, a dictionary of computer technology.

McGraw-Hill Encyclopedia of Science and Technology

The New Book of Popular Science, 6 volumes; Volume 3 on physical science and Volume 6 on technology may be of particular interest.

Science Year, the World Book Science Annual. Contains reviews of

major science happenings of the previous year.
Van Nostrand's Scientific Encyclopedia
The World Book Encyclopedia

B. Periodicals (dealing with scientific and technical information)

Aviation Week and Space Technology, weekly news of space and shuttle events.

Byte, monthly periodical focusing on the computer industry.

Compute, monthly periodical dealing with computers in the home.

Creative Computing, monthly issues including a special section on technology.

Discover, monthly magazine dealing with science and technology.

Environment, monthly periodical dealing with the issues concerning the environment.

Omni, monthly periodical of science fiction, but it contains a technology section.

Popular Mechanics, monthly periodical that sometimes covers issues related to high technology.

Popular Science, monthly magazine covering science issues for the general public.

Science, weekly, published by the American Association for the Advancement of Science. Contains many articles and reports on high technology. Subjects are treated in greater depth than in the "popular" magazines.

ScienceXX (XX, year of publication, is part of title), published monthly by the American Association for the Advancement of Science "to bridge the distance between science and citizen." Contains articles of interest on a variety of subjects.

Science Digest, monthly periodical that contains technology, computer, and energy sections.

Science News, weekly publication that contains summary articles on science and science news briefs.

Scientific American, monthly periodical containing in-depth articles of interest and a "Science and the Citizen" department.

Space World, monthly publication on space sciences.

C. Periodicals (dealing with science issues in a context of economics or society)

Bulletin of the Atomic Scientists, monthly publication containing articles by famous scientists, focusing on science as it impacts on world and national affairs.

Business Week, weekly periodical dealing with economics. Contains a technology section.

Forbes, biweekly magazine with a technology section.

Fortune, monthly publication with a technology section.

High Technology, monthly business magazine devoted to recent developments, applications, and trends in high technology.

Natural History, monthly publication with articles about how technology affects the environment.

Newsweek, weekly magazine with technology and science sections.

Time, weekly news magazine with a technology section.

Numbers in red indicate the appropriate chapter sections to aid you in assigning these items. Answers to all questions appear in the Teacher's Guide at the front of this book.

Summary _____

■ The space shuttle is a good example of high technology. It features many scientific advancements that were not possible even a decade ago. Many of the technologies used in space science have applications in other areas.

■ The microelectronics industry is based on several metalloid elements. When a crystal of silicon is doped with certain other elements, the resulting crystal is able to conduct electricity. Layers of semiconducting materials are fabricated to produce microelectronic components.

■ The miniaturization of integrated circuits has brought about a revolution in technology, and has made possible the complex computer systems of today. Laser technology has taken microelectronics to the limit of size reduction.

■ The practicality of ceramics has expanded to the fields of electronics and medicine. Because of durability, chemical and heat resistance, and reasonable cost, ceramic technology is a popular subject of research.

■ Fiber optics, composed mostly of glass, seems to be the most efficient method of transmitting large quantities of information. Fiber optics is capable of transmitting information thousands of times faster than by copper wire.

■ The rapidly changing world of high technology provides opportunities to develop research and analytical skills. Reliable reference materials are invaluable for doing research to better understand a topic.

Chemically Speaking _____

doped
integrated circuit
n-type semiconductors
optical fibers
photovoltaic cells
p-type semiconductors

Review _____

1. Write the complete electron configuration for silicon and germanium.

2. Briefly describe why electrons in the bonds of pure silicon crystals are not free. (14-1)

3. Describe how electron deficiencies, called holes, can be thought of as moving in a direction opposite to the movement of electrons in a p-type semiconductor. (14-1)

4. What device serves as the "brains" of electronic equipment? (14-1)

5. Name the process by which an integrated circuit is layered onto a semiconductor. (14-1)

6. How has the laser revolutionized the manufacturing of integrated circuits? (14-1)

7. How do the structures of quartz and silica glass differ? (14-5)

8. Name two uses for photovoltaic cells. (14-5)

9. What substance is added to silica glass to render it a crystal with a high lead content? (14-5)

10. Name one advantage that fiber optics has over copper wire for the transmission of information. (14-5)

Interpret and Apply _____

1. Write the outermost electron configurations for Ga, As, P, B, and Se. Which of these elements would you expect could be used to make n-type semiconductors? Which would you expect could make p-type semiconductors? (14-1)

2. What are some of the limitations on miniaturization of integrated circuits? Do you think it is reasonable to manufacture components of integrated circuits only one atom thick (about 10^{-8} cm)? Explain the reasons for your answers. (14-1)

3. List three advantages of integrated circuits over electron tubes. (14-1)

4. In light of the recent development of the scanning tunneling microscope, is it correct to say that individual atoms can now be seen in detail? Explain. (14-2)

5. What advantage does the scanning tunneling microscope offer a chemist interested in surface chemistry and in the nature of chemical reactions? (14-2)

6. Automobile engines produce mechanical energy from chemical energy. Light bulbs produce heat and light energy from electrical energy. What energy conversion is involved in the use of solar cells? (14-3)

7. Digital electronic transmission of information depends on the binary system, in which *on* is "one" and *off* is "zero." How does fiber optics transmit the binary code? (14-5)

Problems

1. If a component of an integrated circuit is 3 μm in length, how many components can you expect to fit across the length of an integrated circuit which is 4 mm long? The components need to be separated by a distance of 5 μm so they do not interfere with each other's operation. (14-1)

2. Lasers can be used to deposit tungsten directly onto a silicon substrate in order to make an integrated circuit. The starting materials for laser deposition of tungsten are tungsten hexafluoride and hydrogen gas. Write a balanced equation for this laser-induced reaction. (14-1)

Research Projects

1. Research the transistor. How is the transistor fabricated? What role(s) do transistors play in circuits?

2. Bring an integrated circuit to class. Ask a biology teacher to loan your class a stereo microscope. Identify the various parts of the circuit on an enlarged drawing. Make the magnified chip and drawing available for viewing.

3. Research the kinds of raw materials that make up computer chips. What are the sources of these materials? Has the cost of precious metals, such as gold, been affected by their use in high-tech industries?

4. Find out what economic factors affect the cost of computers. How much can the cost come down? What are the consequences to the way people live when there are lots of inexpensive computers?

5. Read *Megatrends* by John Naisbitt. Report on Chapter 2, "High Tech—High Touch."

6. Research the different compositions of glass, for example, silica, soda glass, and Pyrex®. How are these different materials produced? What processes are involved in the manufacture of these different types of glass? Bring examples of these different types of glass into class, and discuss the structure, uses, and advantages of each kind.

7. Find out the many advantages of fiber optics over conductors such as copper. Are there any disadvantages of using fiber optics? To what uses, other than communications, has fiber optics been applied? What materials other than glass have been used in fiber optics?

8. Research the field of liquid crystals. What are they? How do they work? What are the various kinds? What practical uses have they been put to? Bring some examples of liquid crystals to class.

9. Find out more about modern ceramic materials. What items are presently being made out of the new ceramics? What materials may eventually be replaced by ceramics? What advantages do high-tech ceramics have over these materials?

10. What does the term "artificial intelligence" mean? What is the state of the art in artificial intelligence? Where do present trends in artificial intelligence seem to be leading?

11. With a small group of students, report to the class on the principles of technological development as described in *Megatrends* by John Naisbitt. Use the format of a debate, with this thesis: *Has high technology stifled the human spirit?*

12. How has the computer affected the way people learn, and what people learn? Sources of information on this topic may include educational journals and interviews with professional educators. Be sure that you are able to separate data from opinions when exploring this topic.

13. Find out about error and error rates in computers. Circuits can be damaged by cosmic radiation. If you are building a machine to process information, how do you handle this kind of problem? What is redundancy, and how does it apply to this problem?

14. Today, analytical chemists are able to find minute traces of dangerous compounds and elements in the environment that were not detectable before. How is this research accomplished? Find out what compounds and elements pose a threat, to what extent, and in what amount. What is the responsibility of the analytical chemist in making the public aware of such hazards? What are the responsibilities of business and government? Does the public have any responsibility to keep itself informed?

Liquids and solids are called the condensed states of matter. Water is a unique substance in that it simultaneously exists in three states under normal conditions on Earth. Water also is unusual in that ice floats on liquid water.

Most of the matter you encounter is solid, although some is liquid. With the exception of air and what you learned in Chapter 7, your experience with gases has been limited. You have learned that the kinetic molecular theory was used as the model to explain the behavior of gases.

In this chapter, you will expand your knowledge of the kinetic molecular theory by using the model to explain the characteristics and behavior of water and other liquids and solids.

The Condensed States of Matter

Extending the Kinetic Molecular Theory

From elementary school you probably remember the standard definition for a liquid: a liquid has no definite shape; it takes on or assumes the shape of its container. If you think for a moment, you can probably list a number of other characteristics that would make the definition more precise. You know that liquids flow, that is, they move under the influence of gravity. Some flow quickly, like water. Others move slowly, like molasses. You know that liquids change to gases by evaporation or through boiling. Detecting the odor of perfume gives you evidence that the liquid perfume evaporates. You can probably list other characteristics, but do you know why liquids flow at different rates, evaporate, boil, or freeze?

15-1 A Microscopic View of Condensation Processes

You can begin developing a model to explain liquid behavior by referring to what you already know about gases. Gas molecules are assumed to be in constant random motion. Within any sample of gas molecules, there is a distribution of kinetic energies. Faster-moving molecules have more kinetic energy than slower molecules. It is assumed that there are no forces of attraction between molecules in an ideal gas. Here is where the kinetic molecular theory models are altered when applied to liquids. In a liquid, the molecules have less average kinetic energy than gas molecules. The energy of attraction between molecules in a liquid exceeds their kinetic energy so the molecules are held together. You studied intermolecular forces in Chapter 12. Now you will look at solids, liquids, and gases on a molecular level to gain a better understanding as to the effects of these forces.

Pretend that you are 5 billion times smaller than you really are. You are standing in a large box filled with nitrogen gas molecules at room temperature. Most of the box is empty space, and you can see nitrogen molecules moving in all directions at a variety of speeds. Although there are plenty of collisions taking place, an individual molecule usually moves quite a long distance before it collides with another molecule. What happens if the temperature goes down? The picture would

CONCEPTS

- a molecular view of condensation and freezing
- attractive forces between particles account for the behavior of solids and liquids
- evaporation is a surface phenomenon; boiling is an internal phenomenon
- a solid or liquid in a closed system reaches equilibrium with its vapor
- vapor pressure is a characteristic property of a substance
- change of state processes involve conversions between kinetic and potential energy

Begin by asking students why copper is a solid, why bromine is a liquid, and why neon is a gas at room temperature. Students should suggest that the forces between particles are greatest in copper and weakest in neon.

boiling point N$_2$ = −196°C
melting point N$_2$ = −210°C
For N$_2$, the liquid range is 14 C°.

be much the same, but it would seem to be like a slow-motion movie. The average kinetic energy and, therefore, the speed of the molecules, decreases as the temperature goes down.

At −196°C something new and startling happens. Clusters of molecules are falling to the bottom of the box, crowding in close. As you struggle to swim up, you can see that the molecules have almost no space between them. Although they touch each other, you can easily push them past one another. They still move around rapidly in all directions. But now they are so close to one another that they travel only a short distance before colliding. And, of course, you are getting bounced back and forth, too. When at last you swim high enough, you find yourself floating at the top of this churning crowd of molecules. Most of the box above you is empty now, and only an occasional nitrogen molecule can be seen darting by.

Every once in a while you can observe that a molecule seems to plunge from above and mix with the crowded scene around you. Just as often, one leaves the sea of molecules to move upward into the space near the top of the box.

As the temperature continues to drop, below −196°C, the molecular motion decreases. It is like the ocean calming down after a violent storm. At −210°C, you notice that another change is taking place. The molecules become firmly attached to each other in a regular pattern. You can jump around on top of the molecules because they are held in a rigid structure. They are about as far apart now as they were at −196°C. They still move around nearly as rapidly as they did before, but only occasionally does a molecule slide past another to fill an empty hole and leave a new hole in its wake. It is as if the molecules are trapped into specific places with respect to one another; they are no longer fluid.

There are still a few molecules high above you in the open space of the box. You can see a molecule move down from above to take a place in the regular pattern. Occasionally a molecule escapes from the pattern to move off toward the top of the box.

Did you recognize the processes being described in the story? Nitrogen gas is being condensed to form a liquid and then a solid by reducing the temperature. The evaporation process for liquid nitrogen also is being described. Note that when the liquid evaporates, gas molecules exist in the box with the liquid. The story was formulated based on experimental results. Now you can look at some of the experiments.

15-2 Phase Equilibria and Vapor Pressure

If a liquid is placed in a closed evacuated container, a gas pressure can be noted by using a manometer. If water is placed in an evacuated flask, the manometer shows that the gas pressure rises rapidly from 0 kPa to 3.17 kPa. Then no further pressure change occurs as long as the temperature is kept at 25°C. How can you interpret these observations?

Water is the only substance in the flask. Some molecules, near the surface of the liquid, would have enough energy to overcome attractions of nearby molecules and move out of the liquid and become

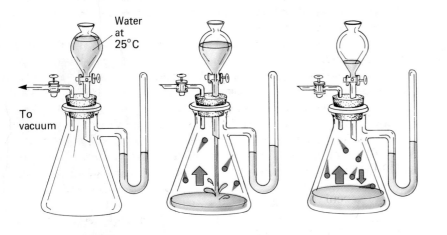

Evacuated flask   Water being added   Approaching equilibrium

Equilibrium attained
Constant pressure
Evaporation rate = Condensation rate

vapor. As the number of molecules in the vapor phase increases, the gas pressure rises. After a while the gas pressure reaches a constant value, 3.17 kPa. This pressure is called the **vapor pressure** of water. The pressure remains at 3.17 kPa as long as the temperature is kept at 25°C. Figure 15-1 shows a model for what we believe is happening to cause the vapor pressure to become constant. At first there are many more molecules leaving the liquid than are returning. The gas pressure goes up as the rate of evaporation is greater than the condensation rate. The number of molecules in the vapor phase increases. After time, the chance that a molecule will move back into the liquid phase also increases. Soon the rate of evaporation equals the rate of condensation and the two processes balance each other and a state of **equilibrium** is reached between the liquid and vapor phases. At equilibrium, the pressure measured by the manometer is constant.

If the temperature of the water is reduced to −10°C, you have ice and the pressure is 0.260 kPa. Some water molecules leave the crystal surface and enter the gas phase. The vapor pressure for ice at −10°C is 0.260 kPa. Now if the temperature is increased to −5°C, you should expect that more water molecules would have enough energy to leave the solid. The equilibrium vapor pressure also should go up. The vapor pressure for ice at −5°C has been measured at 0.402 kPa.

Ethanol (ethyl alcohol) also is a liquid at room temperature. Its vapor pressure at 25°C is 18 kPa. This value is higher than the vapor pressure of water at this temperature. Ethanol has a greater tendency to evaporate than water. At 40°C, ethanol has a vapor pressure of 30 kPa. At 60°C the vapor pressure is 47 kPa. The vapor pressure of ethanol also increases rapidly with increasing temperature. The vapor pressure of every liquid increases as the temperature is raised.

These results can be readily explained in terms of the kinetic theory. As the temperature increases, the average molecular velocity increases. You know from Chapter 7 that the fraction of molecules in a

Figure 15-1 As water molecules enter the vapor phase in the evacuated flask, a pressure is noted by the manometer. Once the rate of evaporation equals the rate of condensation, equilibrium is reached and the pressure remains constant.

CHEM THEME

The concept of equilibrium is a model used to explain why the pressure in a closed system remains constant.

CHEM THEME

The kinetic molecular theory provides reasonable explanations for the behavior of matter subjected to changes in temperature.

Figure 15-2 At 40°C, water and ethanol have different vapor pressures as shown by the manometers on the left. Ethanol is more volatile as shown by its higher vapor pressure. The graph on the right compares the vapor pressures of water, ethanol, and acetone. Which substance is most volatile?

liquid with sufficient energy to enter the vapor phase would increase at higher temperatures. Figure 15-2 shows a graph of vapor pressure curves for three different substances.

A substance whose molecules evaporate readily is said to be **volatile**. Look again at Figure 15-2. Notice the different vapor pressures for all three liquids at 20°C. From the graph you can see that acetone has a higher vapor pressure than ethanol, hence acetone is more volatile than ethanol. Volatile substances which are combustible must be handled with care as their vapors can ignite easily.

The evaporation of water serves to regulate your body temperature. Whenever your body gets overheated, you perspire. Your body is cooled by perspiration because heat is absorbed from the skin to evaporate the water on your skin. As the sweat evaporates, you lose the excess heat. Perspiring is one mechanism that helps keep your body temperature at equilibrium. Body temperature is closely regulated to remain at 37.0°C. The state of your health is often judged by determining if your body temperature is normal.

When the humidity is high, the air is said to be nearly saturated with water vapor. The gaseous water molecules in the air condense on your skin just about as fast as the water molecules escape in your perspiration. As a result, the rate of evaporation is reduced. The opposite situation occurs in some arid desert regions. High temperatures are not uncomfortable unless you are in the direct sun, because the humidity is so low that a high rate of evaporation keeps you cool. In those areas dehydration is a potentially serious problem.

Molecules can escape from the surface of a liquid at any temperature to enter the gas phase as vapor. When the vapor pressure of a

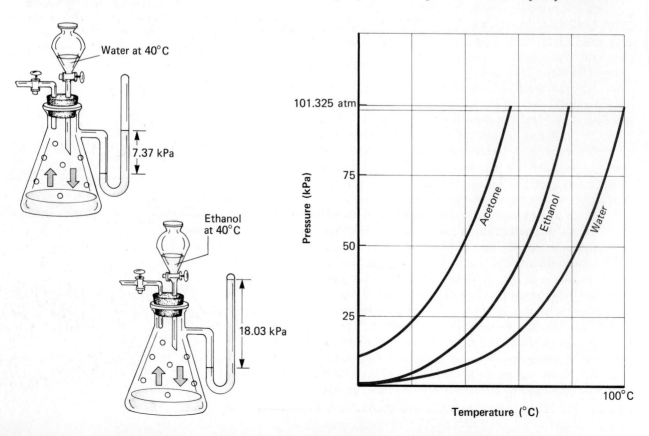

liquid is equal to the atmospheric pressure, a new phenomenon occurs. At this temperature, bubbles of vapor can form in the liquid and the liquid boils. Boiling is fixed by the external pressure. For example, if the pressure is 101.3 kPa or 1 atm, water boils at 100°C. At this temperature, the vapor pressure of water equals 101.3 kPa or 1 atm. Figure 15-4 shows a bubble forming in water at 100°C. The vapor pressure inside the bubble is equal to the external atmospheric pressure.

The normal boiling temperature of a liquid is the temperature at which the vapor pressure of that liquid is exactly one standard atmosphere. Suppose that the atmospheric pressure is 84.5 kPa. Then bubbles of vapor could form in liquid water at 95°C. The vapor pressure of water equals 84.5 kPa at 95°C. Water boils at this temperature when the atmospheric pressure is 84.5 kPa.

Vapor pressure also is a property of solids. If ice is placed in an evacuated flask at −10°C, you would note that no liquid water forms. Yet a manometer would detect a vapor pressure.

Solid carbon dioxide often is called dry ice. At normal pressure dry ice changes to gaseous CO_2 without melting. The process of a solid going directly to the vapor without first forming a liquid is called **sublimation**. Another example of sublimation can be seen if you live in an area where the winters are very cold. Some snow disappears without the temperature ever rising to 0°C. The solid phase of water sublimes under certain conditions.

Table 15-1 gives the vapor pressure for water at various temperatures. The vapor pressure of a liquid is the same whether or not other gases are present. Vapor pressure is a property of a liquid.

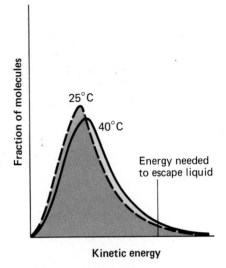

Figure 15-3 At higher temperatures more water molecules have enough kinetic energy to leave the surface of the liquid.

Figure 15-4 Boiling and evaporation are two different phenomena. Boiling occurs within the body of the liquid. Evaporation occurs at the surface of the liquid.

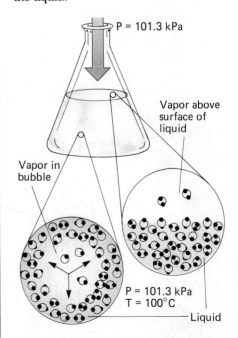

Table 15-1

VAPOR PRESSURE OF WATER AT VARIOUS TEMPERATURES	
TEMPERATURE (°C)	PRESSURE (kPa)
0.0	0.610
10	1.23
15	1.70
20	2.34
25	3.17
30	4.24
35	5.62
40	7.38
45	9.58
50	12.4
55	15.7
60	20.0
65	25.0
70	31.2
75	38.6
80	47.3
85	57.4
90	70.1
95	84.5
100	101.3

15-3 Change of State: Liquid to Gas

As liquid water is heated, its temperature rises. But once the liquid begins to boil, the temperature stops rising. This is unexpected. When scientists observe something unexpected, such as this failure of the temperature to rise when heat is added, they ask, "Why?"

A graph of the temperature of a liquid varying with time is a useful way to see how liquids behave during a change of state. At 30-second intervals the temperature is recorded. A graph of temperature vs. time is plotted in Figure 15-5. You can see that initially the temperature increases in a straight line. At point A the temperature ceases to rise with additional heat. The horizontal portion of the graph represents the change of state. Figure 15-6 shows changes of state for ethanol and acetone. Notice the similarities and differences among the three graphs. When the liquid is initially heated, the temperature increases, so the best fit among the data points is a straight line. At the boiling point for each compound, the temperature stops increasing as the compound changes from a liquid to a gas.

How can these observations be explained? The answer is based on the law of conservation of energy. The energy added to the liquid either increases the kinetic energy of the molecules (and the temperature goes higher) or it overcomes the forces of attraction between them. Before the boiling point of the liquid is reached, most of the added heat energy increases the average kinetic energy of the molecules in the liquid, so the temperature continues to rise. However, once the atmospheric pressure equals the vapor pressure, many of the molecules in the liquid have sufficient kinetic energy to escape the liquid as a gas. As a result, the average kinetic energy of the liquid no longer increases and the temperature remains constant. However, energy still must be supplied to overcome the attractive forces between the molecules. As that energy is supplied, more and more rapidly moving molecules escape the liquid as gaseous molecules. The temperature of the liquid stays constant until all of the liquid has been changed to a gas. At that point, if one continues to heat the gas, the

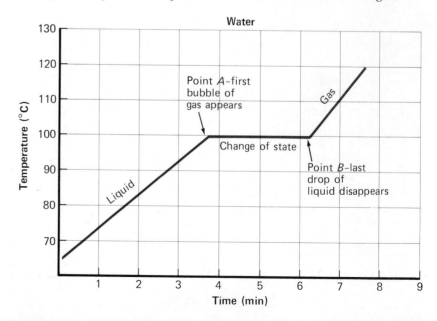

Figure 15-5 Change of state graph for liquid water.

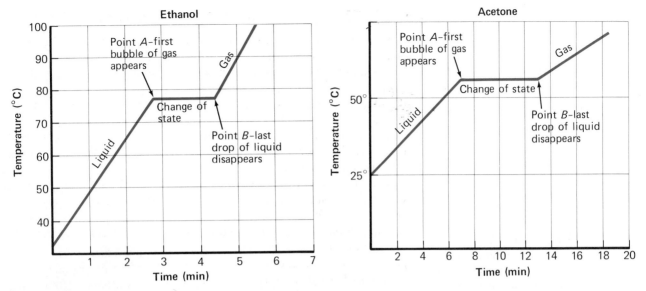

speed (and temperature) of the gaseous molecules again increases (point *B* on the graphs).

Figure 15-6 The graph on the left shows the change of state for liquid ethanol. The graph on the right shows the change of state for liquid acetone.

15-4 Change of State: Solid to Liquid

If the molecules in a solid are arranged in an orderly repeating pattern, the word **crystalline** is used to describe the solid. What happens when energy is added to a solid? At any temperature above 0 K there is vibratory motion in a crystal. As energy is added to a solid, the temperature and the motion increase. When the temperature gets high enough, the molecular motion disturbs the regular crystal pattern and the solid melts.

The melting curve for water is shown in Figure 15-7. You can see that something unusual is happening during the melting of the solid. Even

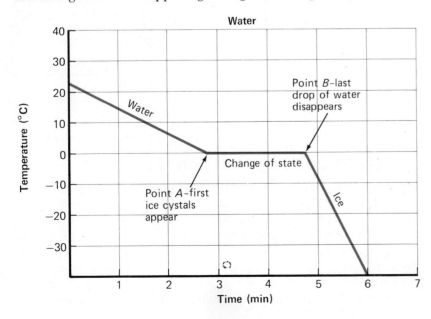

Figure 15-7 This change of state graph shows the conversion of ice to liquid water.

though energy is added continuously to the system, the temperature stays constant from the time the liquid appears until all the solid has melted. This temperature is the melting temperature of the solid.

Where does the energy go? During the melting process the temperature of the liquid is the same as the temperature of the solid. The energy supplied is not increasing the temperature of either state. This energy destroys the crystal lattice and is stored in the liquid as potential energy.

The amount of energy required to melt one mole of ice at 0°C is known from experiments. The measured value will be presented in describing energy changes in detail in Chapter 17.

Figure 15-8 shows the solid-liquid change of state for ethanol. Note the similarities between this graph and Figure 15-7. Also note that the plateaus occur at different temperatures. Ethanol freezes at −114°C, while water freezes at 0°C at 1 atm.

Figure 15-8 also shows a complete change-of-state diagram for nitrogen as it cools from −170°C to −240°C. Notice the same regularities in the graph for nitrogen as there are for water and ethanol. Notice the small temperature range over which nitrogen is a liquid. How does this compare to the range for water?

Figure 15-8 The graph on the left shows the conversion of solid ethanol to its liquid. The graph on the right shows the conversion of nitrogen gas to a liquid and solid. This process was described in Section 15-1.

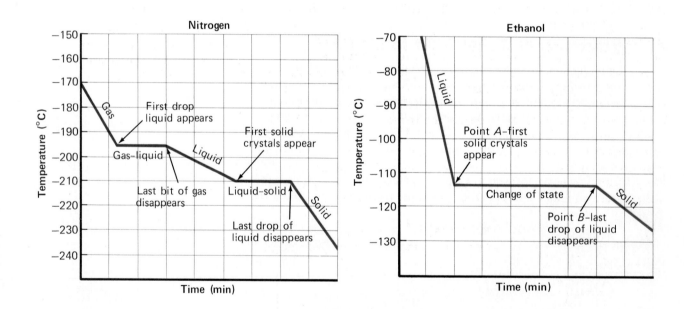

15-5 Water: The Uncommon Liquid

An uncommon but important property of water is the irregularity of its density with changes in temperature. Changes in the density of liquids can be explained in terms of the kinetic molecular theory. The highest density of a liquid occurs when the molecules are packed tightly. In most liquids, the colder the liquid is, the closer packed the molecules are. Water is different. Why does this happen? If you could look into the structure of ice, you might see a rigid network of water molecules. The crystal arrangements of ice has a lot of space incorporated in it. There are fewer molecules per unit volume in ice than there are in water. The structure of ice keeps molecules further from each other than does liquid water.

Since the density of ice is less than the density of water, you would expect ice to float on water, which it does. The actual changes in density values are not very large. However, these values are large enough to account for the fact that freezing water can break concrete walls of a swimming pool if too much water is left in the pool during freezing weather.

You saw in the photograph at the beginning of this chapter another interesting property of water. It can exist in all three states simultaneously. However, is this a unique property of water?

You know from the previous sections of the chapter that vapor pressure is dependent on temperature. For boiling to occur, the vapor pressure of water must exceed the atmospheric pressure. If the atmospheric pressure is reduced, boiling occurs at lower temperatures than normal. The relationships among pressure, temperature, and the phase of a substance can be summarized by a **phase diagram**. A portion of a phase diagram for water is shown in Figure 15-10. The data used to plot this type of graph were obtained from laboratory experiments used to determine how water behaves in a *closed system* under changing temperature and pressure. Phase diagrams can be made for a variety of substances.

Now study the diagram in detail looking at the information it provides. First, you can see the pressure and temperature ranges over which water exists as a solid, liquid, or vapor. Note the coordinates represented by point C on the diagram ($P = 1$ atm, $T = 100°C$). Point C represents the normal boiling point of water. The arc from points B to C represents changes in the boiling point as the pressure is reduced.

Point B on the diagram has a special significance. It is called the triple point for water. The **triple point** of a substance shows the temperature and pressure under which a substance exists simultaneously in all three phases at equilibrium (in a closed system). The triple point for water is special in that it is used as the SI standard for thermody-

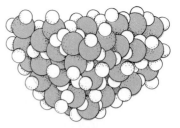

Liquid water

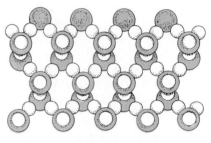

Ice structure

Figure 15-9 Ice is less dense than water. The rigid structure of ice keeps molecules further apart than they are in liquid water.

Phase diagram for water

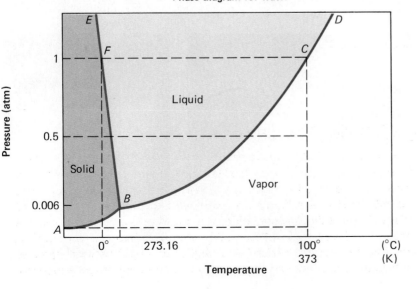

Figure 15-10 A phase diagram for water.

namic temperature. The SI temperature unit, kelvin, is defined as 1/273.16 of the thermodynamic temperature of the triple point of water.

The arc from points A to B represents those pressures and temperatures under which water will sublime in a closed system. What point on the diagram represents the normal melting point of water?

15-6 Surface Tension

Have you ever seen bugs walking on water? What holds them up? Water striders have specialized feet that enable them to stay on the surface of the water. But the water also has a feature which helps the bugs stay afloat. Water has high surface tension. The surface tension of water is due to the same attractive forces acting between water molecules in the liquid that were discussed earlier in this chapter. Beneath the surface, all the water molecules are attracted equally by all the surrounding molecules, as shown in Figure 15-11. However, the molecules on the surface are not acted upon equally by forces in all directions. These unbalanced forces cause surface tension. You can demonstrate surface tension fairly easily with a bowl of water, some forceps, and a needle. Carefully wipe the needle to remove any oil or dirt. With the forceps, lay the needle carefully on the water. It will float. The needle is supported by the surface tension of the water.

Surface tension prevents water from being a good wetting agent. Have you ever noticed how water tends to bead up on the surface of a freshly waxed car, but it tends to spread out on the surface of a dirty car. This spreading of a liquid which adheres to the surface of solids is called "wetting" the solid. It is due to greater attractive forces between the liquid and solid molecules than the attractive forces among the liquid molecules. How well a liquid wets a solid depends on the size of the forces between a solid and liquid compared to the surface tension of the liquid.

To make water a better wetting agent, ingredients in commercial detergents reduce the surface tension of water. If a drop of liquid soap is added to the pin floating on the surface of the water, the pin will quickly sink to the bottom. The soap reduces the surface tension.

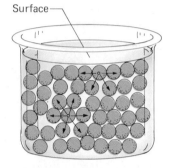

Figure 15-11 Unbalanced forces acting on a surface molecule produce a liquid "skin" that is due to surface tension.

Figure 15-12 The needle is supported by the surface tension of the water in the photo on the left. Surface tension causes the formation of spherical droplets in the photo on the right.

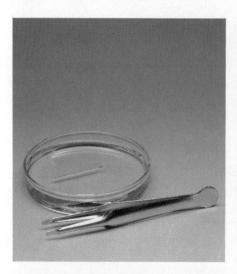

Review and Practice

1. Draw a complete change-of-state diagram for H_2O starting at $-20°C$ and adding energy until the final temperature is $120°C$.

2. Identify the regularities apparent in all change-of-state diagrams?

3. a. What happens to the energy being added to a liquid as the liquid is being warmed?
   b. What happens to the energy being added as a liquid changes to a gas?

4. If the atmospheric pressure over ethanol is increased, what happens to the boiling point of ethanol?

5. Acetone has a greater vapor pressure than ethanol at $25°C$. What does this difference indicate about the strength of the attractive forces between acetone molecules and ethanol molecules.

6. Figure 15-13 is a phase-change diagram for 50 grams of copper at 1 atmosphere.
   a. What is the boiling point of copper?
   b. What is the melting point of copper?
   c. What is the temperature range for liquid copper?

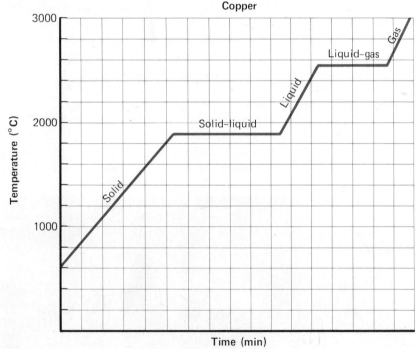

Copper

7. a. What happens to the average kinetic energy of ethanol as it freezes?
   b. What happens to the potential energy of ethanol as it freezes?

8. Draw a complete change-of-state diagram for ethanol as it is heated from $20°C$ to $100°C$. Use Figures 15-6 and 15-8 for the melting and boiling temperatures.

9. When more energy is added to a closed system of water and water vapor at $25°C$, the temperature of the system will increase, but not as much as would be expected. Explain why.

10. Explain what happens in terms of kinetic energy changes as ethanol is cooled from $90°C$ to $60°C$.

1. See the Teacher's Guide at the front of this book

2. At both the melting point and the boiling point the temperature remains steady until all the molecules have been changed.

3. a. as kinetic energy increases, molecules rotate more violently.
   b. potential energy increases as the molecules move farther apart.

4. increases

5. They are weaker.

6. a. $2582°C$
   b. $1083°C$
   c. $1499°C$

7. a. remains constant
   b. decreases

8. See the Teacher's Guide at the front of this book

9. Some of the added heat is used to increase the kinetic energy of the molecules but some of it is used to change liquid to gas phase.

10. As it cools from $90°$ to $80°C$, the kinetic energy decreases. As it condenses the kinetic energy remains constant while the potential energy decreases. Then below $80°C$ the kinetic energy decreases.

Figure 15-13 Change of state diagram for copper.

Classifying Solids

- solids have strong intermolecular and intramolecular bonds
- solids can be classified by packing arrangement
- network solids are very hard
- ionic crystals show various packing arrangements that depend on the number and size of the ions in the formula unit
- hydration
- alloying to achieve desired properties

Figure 15-14 shows four familiar solids—an aluminum pan which contains a block of wax, and a glass dish which contains a lump of salt. Now consider the behavior of these solids under different conditions. If all four are put in the oven at 200°C, only the wax melts. If water is added to the pan and dish only the salt dissolves. If struck by a hammer, the salt and glass shatter, but the wax and pan may just change shape.

All these solids are made of atoms. The aluminum pan is composed of only one kind of atom, while the other three solids are compounds. The differences in characteristics, such as melting point, solubility, density, and malleability, are caused by the different types of bonding in each solid. These characteristics may be used to classify the solids.

There are many substances that are rigid and which maintain their own shape, but they are not true solids because they are not crystalline. Examination of their structures on an atomic level reveals no regular arrangement of the atoms. These substances are called **amorphous solids** which means "without order." Ordinary bottle glass and window glass are among the most common examples of amorphous solids. When the glass is heated it does not show a distinct melting temperature. Rather, it gets softer and softer, melting gradually over a wide temperature range. As melted glass cools it does not solidify at a particular temperature. Glass flows more and more slowly as it is cooled so it is often called a super-cooled liquid.

In this lesson you will learn about the arrangement of the atoms that cause such different characteristics. You will begin by studying the simplest solids, metals. Recall what you learned in Chapter 13 about the nature of the metallic bond. The delocalized electrons in the bond

Figure 15-14 The photo on the left shows four solids with very different physical and chemical properties. Glass in an amorphous solid that softens rather than melts at a definite temperature.

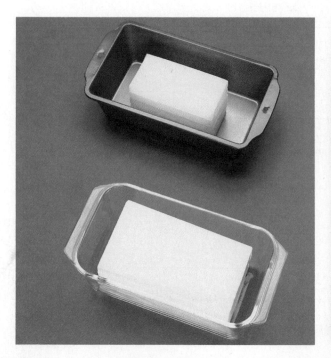

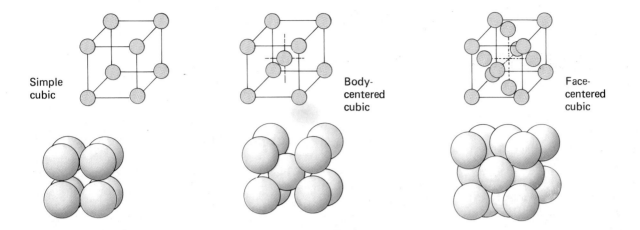

Simple cubic

Body-centered cubic

Face-centered cubic

account for the properties of metals. Now you will look more closely to see how the atoms are arranged.

Figure 15-15 Three types of unit cells.

15-7 Packing Arrangements

Metals are easiest to study since they consist of one type of atom. The crystal structure of a metal can be described in terms of the unit cell. The **unit cell** is the smallest unit which, when repeated in three dimensions, produces the crystal. Three types of unit cells are shown in Figure 15-15.

In the body-centered cubic cell, the spherical atoms occupy about 68% of the total volume of the crystal. In this arrangement, an atom in the center of a cube is surrounded by one atom on each of the eight corners of the cube. Each atom then has eight nearest neighbors.

There are two packing arrangements in which each atom has twelve nearest neighbors. These two closest packing arrangements are more efficient. Each has 74% of the total volume occupied by the atoms, but they are different in the way that the layers of atoms are stacked. A layer of spheres is shown in Figure 15-16. A second layer can be placed directly on top of the first. The most compact arrangement has the spheres of the second layer above the holes or spaces of the first layer. A third layer can be added in one of two ways. If the spheres in the

Plastic foam spheres can be used to make unit cells by gluing them together or by joining the spheres with wooden splints.

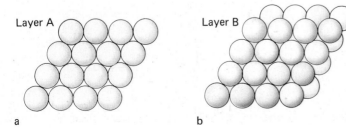

Layer A

Layer B

a

b

Figure 15-16 The two packing arrangements for hexagonal closest packed are the same in the way the second layer (layer B) is stacked.

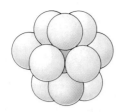

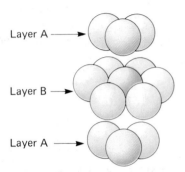

Layer A →

Layer B →

Layer A →

Hexagonal closest packing

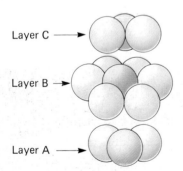

Layer C →

Layer B →

Layer A →

Cubic closest packing gives face-centered cubic unit cell

Figure 15-17 Multiple layer comparisons of the two types of hexagonal closest packing.

third layer are over the holes in the second layer, they are directly above the spheres in the first layer. The structure is called hexagonal closest packed and is shown in Figure 15-17. If the third layer of spheres differs from the first and second layers, the structure is called face-centered cubic. The unit cell for face-centered cubic consists of spheres at the eight corners of the cube and a sphere at the center of each face.

There is no simple relationship among the packing arrangement and properties such as density, melting point, heat of fusion, hardness, and conductivity. The common crystalline forms of the elements are shown in Table 15-2. In the body-centered cubic arrangement, the atoms occupy 68% of the space so you would expect those metals to be less dense. The alkali metals are body-centered cubic arrangements and low in density. However, molybdenum, tungsten, and platinum also are body-centered cubic and very dense. In addition, the properties of the elements given in Table 15-2 are those of the elements in their usual state. The presence of small amounts of impurities or imperfections in the crystal may affect these properties.

Table 15-2

PROPERTIES OF METALS WITH DIFFERENT PACKING ARRANGEMENTS		
HEXAGONAL CLOSEST PACKING		
ELEMENT	MELTING POINT (°C)	DENSITY (g/cm³)
beryllium	1276	1.85
magnesium	649	1.74
cobalt	1495	8.92
cadmium	321	8.65
osmium	3045	22.6
FACE-CENTERED CUBIC		
aluminum	660	2.70
calcium	839	1.55
strontium	769	2.45
nickel	1453	8.9
copper	1083	8.96
lead	327	11.4
BODY-CENTERED CUBIC		
lithium	180	0.53
sodium	98	0.97
potassium	64	0.86
rubidium	39	1.53
barium	725	3.5
chromium	1857	7.18
vanadium	1890	6.11
iron	1535	7.85
molybdenum	2617	10.2
tungsten	3410	19.3
platinum	1772	21.4

15-8 Application: Alloys and Cavities

The earliest metals discovered were silver, gold, mercury, copper, lead, and tin. In pure form all are soft and flexible. These metals were useful in making coins and jewelry because they do not corrode easily. They are not hard, strong, or durable enough to be used as tools.

The first known alloy, bronze, was made about 5000 years ago. An alloy is not a chemical compound because the two or more elements in it are not combined in a fixed percentage by mass. The elements can be combined in nearly any proportion, so an alloy is actually a mixture. Bronze is a mixture of about 75% copper and 25% tin which make bronze much harder and more durable than either copper or tin. Bronze can be used to make tools and weapons. The discovery of bronze led to an era called the Bronze Age. The Bronze Age followed the Stone Age and was replaced in turn by the Iron Age since iron became more widely used.

Most metals are solid at room temperature so they must be melted to form an alloy. The characteristics of the alloy depend on the proportions of each element, the radius, charge, chemical affinity of the elements used, and the type of packing arrangement of each element. In general the pure metal is softer, more flexible, and a better conductor. The alloy is harder and less flexible because when atoms of different sizes are present the planes of atoms do not easily slide past each other.

If the alloying atom is much smaller than the atoms to which it is being alloyed, it may fit into the spaces that exist between the atoms. These alloys are called interstitial alloys and tend to be very hard and brittle. The alloying atom must be quite small with a radius generally about one half of the alloyed atom. Atoms like hydrogen, carbon, nitrogen, and boron form this type of alloy. The alloy of iron and carbon to form steel is an interstitial alloy.

Mercury is the only metal that exists as a liquid at room temperature. Alloys made from dissolving metals in mercury are called amalgams. The amalgam you are most familiar with is the mercury amalgam used in dental fillings. The repair and preservation of natural teeth is a fairly recent medical advance. Tooth decay is a common human disease and source of pain which once had to be endured unless one was willing to face the greater pain of extraction. However, in 1844 a dentist discovered that breathing nitrous oxide gas before having a tooth pulled could alleviate the pain. When it was discovered that infected teeth caused infections in other parts of the body, that discovery, combined with painless extraction resulted in an era where most adults had false teeth by the age of fifty.

One of the most important advances in dentistry was the development of materials to repair damaged teeth. The diseased part of the tooth is drilled away, a chemical is applied to retard the spread of the infection, and the cavity is filled with a material that is durable, strong, hard, and nontoxic. The filling expands to just fit tightly, seal the opening in the tooth, and not react with the pulp of the tooth.

The amalgam is made by mixing mercury with an alloy of several metals just before it is used. The mercury mixes with the other elements in the alloy and hardens in a few minutes. The mercury continues to mix and harden for several days.

Do You Know?

Silver and gold are both too soft to be used for rings and other jewelry without alloying. Platinum, palladium, or silver are alloyed with gold to make white gold. Pure gold is 24 karat. Fourteen-karat gold is composed of 14 parts of gold and 10 parts of other metal.

CHEM THEME

Pure substances have a well-defined set of characteristic properties such as sharp melting point, density, and hardness. The presence of impurities changes the characteristic properties of a substance. The properties of a desired substance can be controlled by using the correct amounts of each component.

Figure 15-18 Dental technicians mix the mercury amalgams used in dental fillings. Liquid mercury is no longer used when making the amalgam.

Do You Know?

It is thought that fluorides prevent cavities in two ways. First, the fluoride makes the enamel of the tooth stronger and more resistant to attack. Recent studies suggest that fluorides also change the surface of the enamel and make it slick. Substances which cause tooth decay are then less likely to adhere to the tooth.

The components of a typical amalgam are shown in Table 15-3.

Table 15-3

A TYPICAL DENTAL ALLOY		
ELEMENT	% BY MASS	CONTRIBUTION TO PROPERTIES OF FILLING
silver	69.4	strength
tin	26.2	additional strength and hardness and adjusts expansion to match that of tooth
copper	3.6	adds hardness and strength
zinc	0.8	reduces brittleness, removes oxygen and other impurities

After mixing metals in the proportions shown in Table 15-3, the mixture is heated to melt the solids and form a homogeneous mixture. The alloy is then cooled, and thin layers are sliced off and ground until the particle size is about 35 micrometers. Small particles have greater surface area and harden faster with greater initial strength when mercury is added to form the amalgam.

The amalgam is formed by mixing the alloy and the mercury in approximately a one-to-one mass ratio. Very pure mercury is used in dentistry and research shows the toxicity to be very low. The small amount of mercury that does not combine with the alloy is eliminated from the body. The mercury does not form toxic mercury ions. Because of recent concern over levels of mercury in the human body, there has been some speculation that mercury is responsible for certain undiagnosed illnesses. The danger has been evaluated in numerous studies and there are no definitive data at present to indicate that mercury in dental fillings is a health hazard.

15-9 Network Solids

Network solids form an extensive web of covalent bonds in a giant three-dimensional crystal. A molecule can be considered to be composed of all the atoms present in a given crystal. The amount of energy required to break covalent bonds is large, and a large number of bonds

Do You Know?

An amorphous form of carbon is made by burning natural gas in a limited supply of air and collecting the soot on cold metal plates. The soot is called carbon black and is blended with natural rubber to produce shoe soles and auto tires with increased wear resistance.

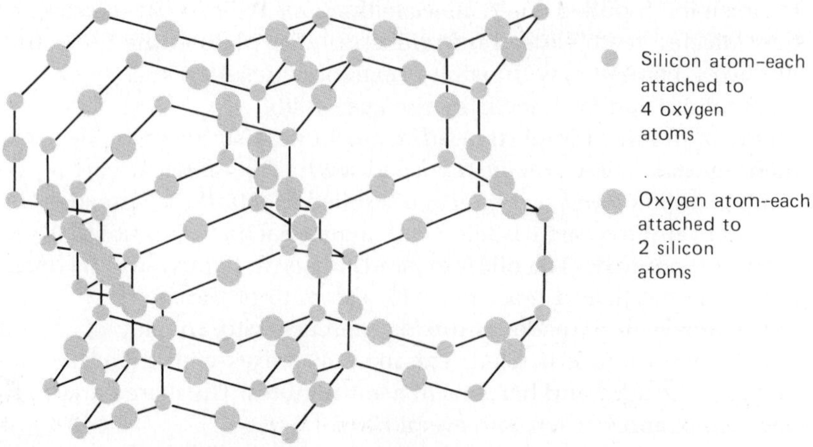

Silicon atom–each attached to 4 oxygen atoms

Oxygen atom–each attached to 2 silicon atoms

Figure 15-19 Silicates (silicon-oxygen compounds) are network solids.

must be broken before melting can occur. Therefore, network solids typically have high melting points. The orientation of the covalent bonds in network solids makes them extremely hard and brittle.

Carbon and silicon tend to form covalent bonds with tetrahedral geometry. The electrons in the covalent bonds are not free to move, so network solids at room temperature tend to be insulators rather than conductors. In the network solid boron nitride boron combines with nitrogen to form the network solid boron nitride, BN. Network solids have high melting points, are brittle, and are among the hardest substances known.

15-10 Ionic Solids

Ionic crystals are composed of alternating positive and negative ions in a three-dimensional arrangement as shown in Figure 15-20. The crystals are hard and brittle and most have high melting points. Crystal planes exist and the crystal can be cleaved or split along one of these planes as shown in Figure 15-21. The ions can continue indefinitely in any direction in a perfect crystal.

In a perfect crystal, every atom or ion would be in just the right place. If the water in a solution of sodium chloride is evaporated rapidly, the ions do not have time to get in the correct position and a white powder in formed that is composed of many tiny crystals bunched together. If the crystal is grown slowly, a large clear cubic crystal results.

Ionic substances can crystallize in different packing arrangements like metals. The packing of ions into a crystal structure is more complex because the ions are sometimes quite different in size and shape.

Two factors contribute to the crystal structure of an ionic compound: the relative numbers of positive and negative ions, and the relative sizes of these ions. For example, in sodium chloride there is one sodium ion for every chloride ion. But the chloride ion, with a radius of 0.181 nm, is nearly twice as large as the sodium ion whose radius is 0.095 nm. Aluminum oxide has two aluminum ions (0.050 nanometer in radius) for each three oxide ions (0.140 nanometer in radius). So the crystal structure of Al_2O_3 is quite different from that of sodium chloride.

Ionic crystals do not conduct electricity so you would assume that they contain no mobile electrons. However, you know that if an ionic solid like salt is melted, the ions are mobile in the liquid state and will conduct electricity.

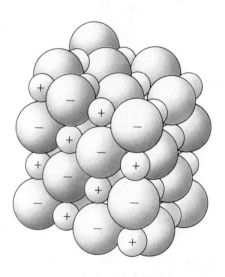

Figure 15-20 The arrangement of ions in a face-centered cubic crystal.

Figure 15-21 Cleaving a crystal involves placing the wedge or knife parallel to the cleavage planes.

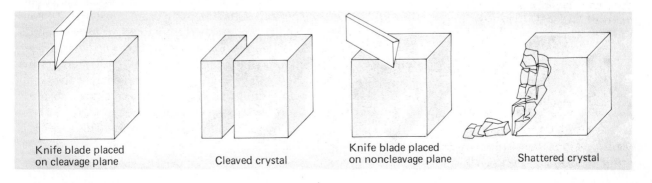

Knife blade placed on cleavage plane

Cleaved crystal

Knife blade placed on noncleavage plane

Shattered crystal

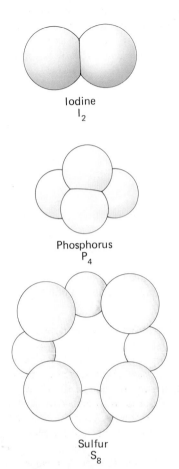

Iodine
I_2

Phosphorus
P_4

Sulfur
S_8

Figure 15-22 Iodine, phosphorus, and sulfur consist of discrete molecules held together by intermolecular forces.

15-11 Molecular Solids

Several elements, such as iodine, sulfur, and phosphorus, are solids at room temperature. These elements are composed of molecules that contain a few atoms joined by covalent bonds. The arrangement of the atoms in each molecule (iodine, phosphorus, and sulfur) is shown in Figure 15-22.

The intramolecular bonds are strong in iodine, phosphorus, and sulfur ranging from 150 kJ/mol to 215 kJ/mol, respectively. When these substances are heated, they melt and vaporize, but remain as molecules.

$$P_4(s) + energy \longrightarrow P_4(l)$$
$$P_4(l) + energy \longrightarrow P_4(g)$$

When sulfur is heated it changes molecular form. However each form consists of S_8 molecules.

The forces between molecules are very weak as evidenced by the fact that molecular solids generally have low melting points. You learned about these attractions called London forces in Chapter 12.

London forces between molecules are very weak; if these are the only intermolecular forces between molecules, then those substances with large molar masses will be solids at room temperature. These molecules are most subject to London forces due to their large numbers of electrons.

15-12 Hydrated Crystals

Many ionic solids are soluble in water. When an ionic crystal is placed in water, its polar molecules bond with the ions. The ions dissociate and are surrounded by water molecules. Each ion tends to have a certain number of molecules attached to it and these ions are called **hydrated ions**.

Many common ionic solids are prepared from reactions that take place in solution like those described in Chapter 5. When the water is evaporated from the solution, a fixed number of the water molecules is retained in the apparently dry solids. This water is called water of hydration and the resulting solid crystal is a **hydrate**. When copper(II) sulfate is crystallized from a water solution, five molecules of water are attached to each formula unit.

$$CuSO_4(aq) + energy \longrightarrow CuSO_4 \cdot 5H_2O(s) + H_2O(g)$$

In writing formulas for hydrates, a dot is used to separate the ionic salt from the water of hydration as in copper(II) sulfate pentahydrate, $CuSO_4 \cdot 5H_2O$, or magnesium sulfate heptahydrate, $MgSO_4 \cdot 7H_2O$. Four of the five water molecules are attached to the copper ion by coordinate covalent bonds. The remaining water molecule is hydrogen bonded to the sulfate ion. The blue color of $CuSO_4 \cdot 5H_2O$, and of hydrated copper ions in solution, is due to the presence of the water molecules. The copper ions together with the water molecules provide energy levels that absorb red and reflect blue light.

Water of hydration often can be removed by heating. This process is called dehydration.

$$CuSO_4 \cdot 5H_2O(s) + energy \longrightarrow CuSO_4(s) + 5H_2O(g)$$

Figure 15-23 Hydrated copper(II) sulfate crystals are shown in the test tube on the left. Anhydrous copper(II) sulfate is shown on the right.

The dehydration of copper(II) sulfate pentahydrate is accompanied by a color change as shown in Figure 15-23. The white solid is called anhydrous copper(II) sulfate. **Anhydrous** is from the Greek "an" which means without and "hydrous" which means water. The process can be reversed by adding water to the anhydrous copper(II) sulfate so the blue color reappears.

 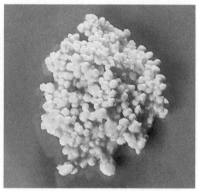

Figure 15-24 Anhydrous $CuSO_4$ can be hydrated by adding water as shown in the left photo. Anhydrous solids can absorb moisture from the air. Calcium chloride in the right photo absorbs so much water from the air that it may dissolve, such solids are described as deliquescent.

Water of hydration must be considered in measuring amounts of reagents. For example, if you want to prepare 1.00 liter of a $1.00M$ solution of copper(II) sulfate, you must consider the water present in the hydrated form in the reagent bottle. You need 1.00 mole of copper(II) sulfate which has a molar mass of 159.5 grams. If you measure 159.5 grams of the hydrate from the reagent bottle you will not have 159.5 grams of copper(II) sulfate since the hydrate is 36% water. To determine the molar mass for a mole of the hydrate, the mass of 5 moles of water must be added to the mass of $CuSO_4$. This addition gives $159.5 + 90$ for a molar mass of 249.5 grams. The 1.00 liter of a $1.00M$ solution of copper(II) sulfate is prepared by adding 249.5 grams of the hydrate, $CuSO_4 \cdot 5H_2O$, to enough water to make a final volume of 1.00 liter.

Display bottles of several hydrates so students can see that the water of hydration is in the formula on the label and is included in the formula mass. Examples of common hydrates are $MgSO_4 \cdot 7H_2O$, $FeSO_3 \cdot 7H_2O$, $CuSO_4 \cdot 5H_2O$, $Na_2CO_3 \cdot 10H_2O$, and $Na_2CO_3 \cdot H_2O$. If anhydrous forms of any of these are available display them as well. You can show that one cannot usually tell if a substance is a hydrate by just observing the crystals. Heat a small amount of each hydrate in a 20 × 200-mm Pyrex test tube gently so students can observe the liquid that condenses on the inside of the test tube.

Review and Practice

1. What kind of bonds exist between the atoms in molecular substances like carbon dioxide, carbon tetrachloride, and chlorine?

2. Write the formula for iron(II) sulfate heptahydrate.

1. covalent
2. $FeSO_4 \cdot 7H_2O$

Careers in Chemistry

JoAnne Morin
Title: Analytical/Environmental Chemist
Job Description: Investigates hazardous waste sites. Collects soil and water samples for laboratory analysis. Prepares reports containing results of analyses and recommendations for action.
Educational Qualifications: Degree in chemistry, with courses in both organic and inorganic chemistry, and additional courses in mathematics and physics.
Future Employment Outlook: Excellent

On a typical day, JoAnne Morin visits a lagoon where a manufacturing facility has been dumping wastes. The purpose of her visit is to sample for toxic materials that may be present in the wastes. She has researched the site thoroughly, contacting the state and the Environmental Protection Agency for records of soil or water contamination. She puts on protective clothing and breathing apparatus and checks for gases that might affect the lungs. She then collects soil or water samples and takes them to the laboratory for analysis. Finally Morin returns to her office to write a report containing the site history, its hydrologic and environmental setting, results of the analyses, conclusions, and her recommendations to the EPA.

Morin is employed by an environmental consulting firm that works under contract for the EPA. She sometimes has to travel in order to assist other EPA offices. Morin describes her work in this way: "Because of the contractual nature of our office, we do not have to hustle to get work. We are fully dedicated to whatever tasks EPA assigns throughout the length of our contract. Job security is not as certain for us after the contract ends as it might be in other types of companies, but the task is worth the sacrifice. The subject is interesting, working conditions are excellent, and being a member of a team is satisfying. This is an interdisciplinary endeavor. Chemical theory is applied in the field and lab, but we also deal with engineering problems too."

JoAnne Morin became interested in environmental chemistry when she was looking for a way to combine her interest in chemistry with service to the public. Writing and math skills, particularly calculus, are important in her job. A background in physics and organic chemistry is indispensable.

The future employment outlook for environmental consulting is excellent. There are many contracts available through the EPA and from nongovernmental sources as well. It is a growing field with rapidly developing technology. Says JoAnne Morin: "Environmental consulting will be around for a while."

Numbers in red indicate the appropriate chapter sections to aid you in assigning these items. Answers to all questions appear in the Teacher's Guide at the front of this book.

Summary

■ The kinetic theory can be applied in explaining the behavior of the condensed status of matter.

■ The magnitude of the forces between particles determines whether a substance exists as a solid, liquid, or gas at room temperature. When the molecules in a liquid have enough energy to overcome the attractive forces, the molecules escape from the liquid to become vapor. Vapor molecules thus create a pressure above the liquid known as the vapor pressure. Humidity is water vapor in the air.

■ Time vs. temperature graphs enable you to see the effect of heat energy on the temperature of a compound. The plateaus on the graph show the boiling and melting temperatures of the compound.

■ As liquid molecules lose kinetic energy, they slow down until finally becoming firmly attached to each other. The molecules are then in the solid state and thus are capable of vibratory motion only.

■ Sublimation involves molecules in the solid state that gain enough energy to overcome the forces holding them in the crystal structure so they can escape as a gas. These gaseous molecules exert pressure which is the vapor pressure of the solid.

■ Water is the most common liquid, yet it exhibits uncommon properties. It expands as it cools below 4°C so that ice at 0°C is less dense than water at 0°C, and the ice floats. Surface tension is high in water requiring wetting agents like soaps and detergents to make water a better cleaning agent.

■ A true solid is a crystalline material in which the atoms occur in a regular pattern throughout the crystal.

■ Atoms, molecules, or irons may be packed in several different ways. Face-centered cubic, body-centered cubic, and hexagonal closest packed are three common arrangements.

■ A substance which maintains its own shape but has no regular internal structure is amorphous and not a true crystalline solid.

■ Network solids consist of atoms joined by covalent bonds to form a giant crystal. Network solids such as diamond, are very hard, brittle, and have high melting points.

■ Ionic solids have strong, directional bonds. Ionic compounds are hard, brittle, and have high melting points.

■ The electrons in the ionic crystal are fixed in place, thus ionic solids do not conduct electricity. When an ionic solid is melted, the ions are mobile and can conduct electricity.

■ There are several different ionic crystal structures. The structure formed by a given substance is determined by the ratio of the ions and their relative sizes.

■ The forces between molecules in a molecular solid are very weak. Most molecular solids have low melting points.

■ When the ions break away from the crystal and become surrounded by water molecules, the ions are hydrated. A hydrate contains a specific number of water molecules in the crystal. If the hydrate is heated to remove the water, the anhydrous form of the crystal is produced.

■ The desired properties of an alloy, such as hardness, conductivity, and durability, can be obtained by selecting the appropriate mixture of metals.

Chemically Speaking

amorphous solid
anhydrous form
crystalline solid
equilibrium
hydrate
network solid

phase diagram
sublimation
triple point
unit cell
vapor pressure
volatile

Review

1. What is a phase? (15-2)

2. How does evaporation differ from boiling? (15-2)

3. How can you increase the vapor pressure of a liquid? A solid? (15-2)

4. What is the relationship between volatility and vapor pressure? (15-2)

5. What is the relationship between kinetic energy and temperature? (15-3)

6. Why is the temperature of a substance constant during melting? (15-4)

7. What is described by the triple point of water? (15-5)

8. What types of materials are used to reduce surface tension? (15-6)

9. How do the numbers of nearest neighbors differ for body-centered cubic and hexagonal closest packed? (15-7)

10. Is iron isomorphous or polymorphous? (15-7)

11. What metals were in the first alloy? (15-8)

12. What are amalgams? (15-8)

13. What is the usual spatial arrangement of atoms in a network solid? (15-9)

14. What are two factors that determine the shape of an ionic crystal? (15-10)

Interpret and Apply

1. At 30°C which would have a greater vapor pressure? (15-2)
   a. motor oil or gasoline ·
   b. water or acetone ·
   c. perfume·or salad oil
   d. water·or mercury

2. In order to kill bacterial spores, the water in an autoclave must reach a temperature of 121°C. How is this temperature possible when water normally boils at 100°C and then converts to steam before increasing in temperature? (15-2, 15-3, 15-5)

3. A beaker contains ice cubes and water at 0°C. An ice cube at −5°C is added to the beaker. After equilibrium is established the temperature is still 0°C. (15-4)
   a. What happened to the temperature of the new ice cube?
   b. What happened to the water?
   c. A small amount of warm water is added to the beaker. At equilibrium, the temperature is still 0°C. What happened to the warm water?
   d. What happened to some of the ice?

4. Containers A, B, and C contain H_2O at 120°C, 25°C, and −10°C, respectively, all at the same pressure. (15-4, 15-5)
   a. Which sample has molecules with the greatest kinetic energy?
   b. Which sample is the least compressible?
   c. Which sample has the greatest density?
   d. Which sample has the most regular arrangement molecules?
   e. Which sample has a definite volume?
   f. Which sample has a definite shape?

5. How can water be made to boil at room temperature? (15-2, 15-5)

6. Why are good wetting agents valued by soap and detergent makers? (15-6)

7. Why is the boiling temperature of water lower in Denver, Colorado (altitude 1609 m) than in Boston, Massachusetts (as sea level)?

8. In an experiment, a round-bottom flask is evacuated to near zero pressure. Some liquid alcohol is injected into the flask. Some liquid alcohol remains after the pressure becomes constant. A pressure gauge attached to the flask indicates a pressure of 0.0724 atmosphere. (15-2)
   a. What is the vapor pressure of the alcohol?
   b. What will happen to the vapor pressure of the alcohol if enough air is injected into the system to bring the total pressure to 1 atmosphere. Explain.

9. Explain why leftover food placed in a refrigerator tends to dry out if left uncovered. (15-2)

10. Both carbon tetrachloride, CCl_4, and mercury, Hg, are liquids whose vapors are poisonous to breathe. If CCl_4 is spilled, the danger can be removed by airing the room overnight. If Hg is spilled, it is necessary to pick up the liquid droplets with a "vacuum cleaner" device. Explain why the precautions for Hg and CCl_4 differ. (15-2)

11. A one-gallon can containing a small amount of water was heated. After the water boiled for a few minutes and the can was filled with water vapor, a stopper was used to seal the can. Explain why the can collapsed upon cooling. (15-2)

12. Imagine traveling across the Sonoran Desert in Arizona when the temperature is 49°C (120°F). Your car's air conditioner is broken but you have lots of water aboard. Suggest a way to keep cool as you continue traveling. (15-2)

13. Which would cause a more severe burn, 1 gram of $H_2O(g)$ at 100°C or 1 gram of $H_2O(l)$ at 100°C? Explain. (15-4)

14. Which has the strongest bonds between atoms—calcium or cobalt? (15-7)

15. If an iron wire is heated from room temperature to 900°C, it changes in packing arrangement from body-centered cubic to face-centered cubic. Will the length of the wire increase or decrease? (15-7)

16. When sulfur is added to iron, is the resulting steel harder or more flexible? (15-7, 15-8)

17. What element could be added to nickel to increase the hardness? (15-8)

Problems

1. The vapor pressure of acetone is about 54 kPa at 40°C. Estimate an approximate boiling temperature relative to water and ethanol. (15-2)

2. Calculate the theoretical percent of water in magnesium sulfate heptahydrate. (15-12)

3. How many moles of water are required to hydrate 0.283 mole of copper(II) sulfate? (15-12)

4. There are 23.6 grams of magnesium sulfate heptahydrate that are dehydrated. (15-12)
   a. How many grams of water will be given off?
   b. What will be the mass of the anhydrous magnesium sulfate?
   c. How many moles of water are given off?
   d. How many moles of anhydrous magnesium sulfate remain?
   e. What is the ratio of the number of moles of water given off to the number of moles of anhydrous magnesium sulfate remaining?

5. Describe how you would prepare 2.50 liters of 1.25M solution of magnesium sulfate using magnesium sulfate heptahydrate as the reagent. (15-12)

6. Calculate the volume of one mole of: (15-7)
   a. magnesium
   b. copper
   c. lead
   d. sodium
   e. iron
   f. tungsten
   g. molybdenum

7. Calculate the percent of water in barium chloride dihydrate. (15-12)

Challenge

1. Make a drawing of spheres with a radius of 1.00 centimeter in a simple cubic cell that is composed of eight atoms, one on each corner of the cube. Calculate the volume of one of the spheres. Calculate the volume of the enclosing cube. Show that 52% of the total volume is occupied by the spheres.

2. Pure iron is easy to magnetize but loses its magnetism rapidly. Explain why steel that contains carbon is more difficult to magnetize but remains magnetic longer.

3. If a crystal of NaCl with a mass of 0.0585 gram is formed in three days, how many Na^+ ions and Cl^- ions are deposited on the crystal each second?

Synthesis

1. Using Figure 15-2, calculate the total pressure of a mixture of acetone and ethanol at 30°C. Assume one mole of each substance is present in the solution.

2. Would pure iron or steel be preferable for use in an electromagnet? Explain your choice.

3. When 10.45 grams of hydrated iron(II) sulfate are dehydrated, the mass of the anhydrous iron(II) sulfate is 5.71 grams. Use this data to find the formula for the hydrate.

Projects

1. Most modern refrigerators use Freon as the coolant in the freezer compartment. Find out how this substance removes heat from the freezer. Where does the heat go?

2. Make a drawing of a face-centered cubic structure and calculate the percentage of the total volume occupied by an atom.

3. Use the technique described by B. W. van de Waal on page 293 of "The Journal of Chemical Education," Volume 62, Number 4, April 1985, to construct models of closest-packing arrangements. Use water displacement to determine the percentage of total volume occupied by the atoms.

CHAPTER 16

In the late 1800's, John Roebling designed and supervised the building of the Brooklyn Bridge. To insure a firm submarine foundation for the span, large underwater chambers known as caissons were constructed. A pressurized air supply was introduced into these chambers via surface ventilation fans. Upon returning to the surface, workers often developed a debilitating and sometimes fatal condition known as caissons disease. Roebling's son was one of those who succumbed to this puzzling illness.

Today this condition is more commonly referred to as the bends, or decompression sickness. Resulting from a change in blood gas solubility, nitrogen leaves solution and collects as tiny bubbles in various body tissues and joints. Its painful symptoms cause the sufferer to attempt relief of this stress, often by bending over. Scuba divers and other individuals breathing compressed air must remain aware of this process or suffer its potentially fatal consequences.

Solutions

Characteristics of Solutions

Many intriguing and important chemical reactions occur in solutions. The oceans, atmosphere, and precious gems within Earth's crust are examples of the diversity and magnitude among the types of solutions. Most life processes occur in solution and are dependent upon the physical and chemical properties of this state.

For centuries, people have investigated the properties of solutions. They have used their findings in a wide range of technological advances that have included the creation of metal alloys, refining of petroleum, desalination of seawater, and survival in environments lacking available oxygen.

You also have studied the properties of solutions though you may not be aware of it. Have you ever wondered why an oil and vinegar salad dressing never remains mixed? No matter how much you shake the dressing, these liquid components will never form an actual solution. As soon as the dressing is allowed to stand, oil and vinegar separate, forming two distinct layers or phases.

Have you ever tried to clean paint brushes with water? Some paints, such as water colors and certain latex paints, rinse from the brush. Oil base paints and most house paints require a special solvent for cleansing.

These common encounters illustrate the importance of understanding the solution process. Within this lesson you will be introduced to the microscopic models used to explain your macroscopic observations of the behavior of solutions.

CONCEPTS

■ characteristics of homogeneous and heterogeneous mixtures
■ components of solutions
■ types of solutions
■ solubility
■ miscibility
■ solvation
■ concentration standards

16-1 Homogeneous Mixtures

Seawater, air, carbonated water, vinegar, and window glass are materials composed of two or more substances. If a beaker of seawater is allowed to evaporate, crystals of various salts remain. Earth's atmosphere is composed of many gases including nitrogen, oxygen and carbon dioxide. Soda pop contains sugar, flavoring and even carbon dioxide gas. In each of these mixtures, the component substances are uniformly distributed throughout the solution. Materials demonstrating this type of uniform mixing of components are called **homogeneous mixtures** or solutions. Several common solutions are shown in Figure 16-1. Note that a solution does not have to be a liquid.

If sand particles are placed in a beaker of water, the particles will accumulate on the bottom of the beaker. Even if the mixture is stirred continually, the sand eventually settles out when the stirring is stopped. Mixtures characterized by such an observable segregation of component substances are called **heterogeneous mixtures**.

Figure 16-1 Solutions are not only found as liquids. Familiar solids like brass and window glass are solutions too.

Within a homogeneous mixture or solution, the substance present in the greater quantity is known as the **solvent**. The **solute** is the component of less abundance. These terms depend on relative quantities of solution components and may be reassigned as the composition of the solution changes. For example, in a 5% alcohol solution, alcohol is the solute and water is the solvent. If additional alcohol is added to the solution so that a 70% alcohol solution is produced, the roles of solvent and solute is reversed.

Solutions may be gaseous, solid, or liquid in nature. Dry air is a familiar example of a gaseous solution. Under similar environmental conditions, the components of air are usually found in fixed percentages. Table 16-1 illustrates the composition of smog-free air at sea level.

Table 16-1

COMPOSITION OF SMOG-FREE DRY AIR		
SUBSTANCE	**FORMULA**	**PERCENT OF MOLECULES**
Nitrogen	N_2	78.08
Oxygen	O_2	20.95
Argon	Ar	0.93
Carbon dioxide	CO_2	0.03

Trace quantities of neon, helium, krypton, hydrogen, and xenon are present in air. Their total contribution is less than 0.003%.

Sterling silver is an example of a solid solution. When copper atoms are dissolved in molten silver, they mix uniformly. Upon cooling, the solution solidifies and this fashionable alloy is formed. Other solid solutions include alloys and amalgams of which you are familiar; brass (copper and zinc), dental fillings (mercury and silver), and steel (iron and carbon).

Liquid solutions may contain solid, liquid, or gaseous solutes. Salt-water is a familiar example of a solid dissolved within a liquid. Vinegar is a solution containing two liquids, acetic acid and water. Carbonated water contains carbon dioxide gas molecules existing between molecules of water. Solutions having water as the solvent are referred to as aqueous solutions. Many reactions, including those vital for life processes, occur in aqueous solutions. Blood, spinal fluid, lymph and saliva are some of the more familiar solutions of biological importance.

16-2 Solubility and Miscibility

If a small amount of solid, such as NaCl, is added to a beaker of water, the solution process begins. The solute particles diffuse among the solvent molecules. If the entire mass of solute has entered into solution, you would say the solute dissolved. A solution that is able to dissolve more of a solute is **unsaturated**.

Eventually, added solute will not dissolve, the solution is **saturated**. At a specific temperature, the amount of solid solute capable of dissolving within a given volume of solvent is a fixed quantity. **Solubility** refers to the amount of substance needed to make a saturated solution at a specified temperature. Excess solid solute will accumulate at the bottom of the solution container.

The solubilities of solids in liquids vary widely. For example, sodium chloride dissolves in water at 25°C until the concentration is about six moles per liter. The solubility of NaCl is $6M$ at 25°C. In contrast, only a small amount of sodium chloride dissolves in ethyl alcohol at 25°C. The solubility is $0.0009M$. Even in the same liquid, solubilities differ over wide limits. The solids calcium chloride, $CaCl_2$, and silver nitrate, $AgNO_3$, have solubilities in water exceeding one mole per liter. Silver chloride, AgCl, has a solubility in water of only 10^{-5} mol/L.

Solubility data usually is expressed in handbooks in terms of the number of grams of solute that dissolve per 100 g of water at a specified temperature.

Temperature influences how much solid will dissolve and how fast. In general, the amount and rate of dissolving will go up as the temperature goes up. This relationship is shown in the solubility data in Table 16-2.

Table 16-2

SOLUBILITY DATA FOR VARIOUS COMPOUNDS (in g/100 g H₂O)		
	TEMPERATURE	
SUBSTANCE	0°C	100°C
NH_4Cl	29.4	77.3
NH_4I	155.0	250
$Ca(C_2H_3O_2)_2$	37.4	29.7
$CuSO_4 \cdot 5H_2O$	23.1	114
$FeCl_3$	74.4	535.7
PbI_2	0.044	0.42
$MgSO_4$	22.0	50.4
KI	128.0	206
NaCl	35.7	39.2
NaOH	42.0	347

Because of this range of solubilities, the word soluble does not have a precise meaning. There is an upper limit to the solubility of the most soluble solid. On the other hand, even the least soluble solid furnishes a few dissolved particles per liter of solution. Glass containers are used for much of the work done in the lab because glass has such low solubility in water. Yet in some experiments this solubility must be taken into consideration.

In this book, the word *soluble* will be used to mean that more than 0.1 mole of a substance dissolves per liter.

Saturated solutions in closed systems are characterized by a state of dynamic equilibrium between dissolved and undissolved solute particles. Though it appears that nothing is occurring within the system, there is activity on the microscopic level. As dissolved particles collide with undissolved solute particles, energy needed to maintain their solution state is lost. Eventually, such low velocity particles drop out of

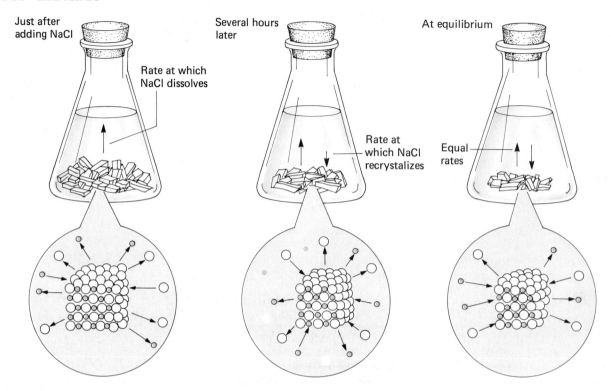

Just after
adding NaCl

Rate at which
NaCl dissolves

Several hours
later

Rate at
which NaCl
recrystalizes

At equilibrium

Equal
rates

Figure 16-2 The formation of a saturated solution in a closed container eventually becomes an equilibrium system. At equilibrium, the saturated solution shows no observable changes.

solution as undissolved solute. Other solute particles, initially kept from dissolving, may now enter the solution phase. A saturated solution shows a constant interchange between dissolved and undissolved solute which is the dynamic equilibrium. The rate at which dissolved solute crystallizes equals the rate at which solid solute dissolves as shown in the model in Figure 16-2. As a result, there is no observable change in the amount of undissolved solute.

A double arrow is used to describe equilibrium processes. The double arrow shows that two processes are occurring simultaneously. In the case of a saturated sodium chloride solution in a closed container, the equilibrium could be depicted as

$$NaCl(s) \rightleftharpoons Na^+(aq) + Cl^-(aq)$$

The rate at which most solid solutes dissolve can be increased by raising the temperature of the solvent, stirring the solution to enhance mixing, or increasing the surface area of the solute by crushing the crystals.

Certain liquid solute-solvent combinations form solutions in any proportion. Water and ethanol may mix in any desired proportion, always resulting in a solution. Components of gaseous solutions demonstrate a similar tendency. When there is no apparent limit to the solubility of one substance in another, the components are said to be completely **miscible**.

Solutions demonstrating complete miscibility are characterized by components of similar properties. When mixed within a container, water and ethanol molecules demonstrate mutual attraction and repulsion. The charged ends of these polar molecules tend to align next to the oppositely charged regions of their molecular neighbors as

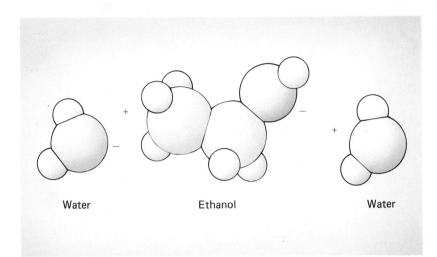

shown in Figure 16-3. Since both solution components demonstrate similar size and polarity, either molecule may occupy neighboring locations. This nonselective behavior allows for mixing of solution components in any proportions.

Miscibility in all proportions also can occur if both solution components are nonpolar and of similar size. Since the molecules have a uniform charge distribution, they mix freely, forming a solution.

When solution components have dissimilar polarities, they often demonstrate a limit to their mutual solubility. These components are referred to as partially miscible. Water and gasoline form such a solution. As water and gasoline are mixed, polar attraction will account for some of the observed solubility. Once the solubility limit is obtained, however, additional liquid will not enter into solution. Any component restricted from the solution state will form a distinct layer.

When mixed, cooking oil and water do not dissolve in each other. The nonpolar cooking oil molecules do not disperse among the polar water particles, forming two distinct layers as shown in Figure 16-4. Liquids that do not dissolve in each other are called **immiscible**.

Figure 16-4 Vegetable oil and water are immiscible in all proportions. Two distinct phases form as the polar water molecules have little attraction for the nonpolar oil molecules.

16-3 Solvation of Ionic Solids

The interaction between solute and solvent particles is called **solvation**. You can study solvation as applied to solutions by looking at what happens when an ionic solid such as sodium chloride is placed in water.

Crystalline sodium chloride consists of positive and negative ions arranged in a fixed geometric pattern. As discussed in Section 16-2, the water molecule is polar. This bond polarity accounts for water's active role in the solvation of ionic solids. When sodium chloride is placed in an aqueous environment, the polar water molecules associate with the surface layer of sodium and chloride ions. Figure 16-5 illustrates the solvation of a sodium chloride crystal.

Figure 16-5 Hydration of sodium chloride ions causes the ions in the crystal to dissociate. Note how the orientation of the water molecules differs depending on the charge on the ion.

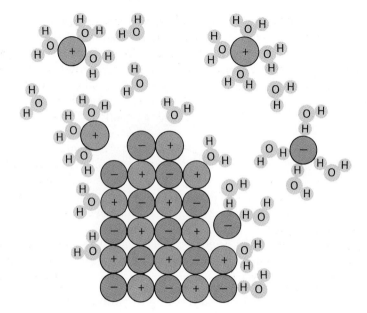

Positive sodium ions attract the negatively charged ends of the water molecules. Simultaneously, the negative chloride ions associate with water's positively charged hydrogen ends. Several water molecules may associate with each ion, isolating the particle from its neighboring ions. When the charged ions are surrounded by water molecules they are said to be hydrated.

Eventually the distances between ions is increased allowing the hydrated ions from the solute surface to diffuse into the solvent. Since solvation occurs at the surface of a solid solute, you can now see why increasing the surface area of a solid solute hastens solvation. This process of crystal decomposition into component ions is called **dissociation**. The dissociation of NaCl can be expressed by the following equation.

$$NaCl(s) \xrightarrow{\text{H}_2\text{O}} Na^+(aq) + Cl^-(aq)$$

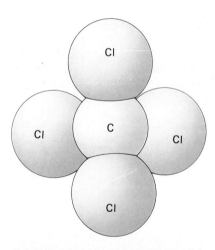

Most compounds you work with in the laboratory contain ionic or polar covalent bonds. Since these particles exhibit a localized charge, they are attracted to the polar water molecule. This attraction, and subsequent solvation, accounts for water's reputation as a near universal solvent.

How might this mechanism differ if an ionic solid is mixed with a nonpolar solvent, such as carbon tetrachloride? Since the solvent would have little attraction for the solute ions, very little solvation would occur. The components would remain segregated.

16-4 Solvation of Molecular Substances

Figure 16-6 Carbon tetrachloride is a nonpolar molecule. When mixed with water CCl₄ is immiscible and forms a two phase system.

There are two intermolecular forces that account for the attraction between molecular substances. The weaker of these are the van der Waals forces, and they occur in all interactions. The stronger, polar forces are found in molecules exhibiting a significant charge distribution.

When polar substances such as water and ethanol are mixed, they form a solution. The polar ends of the solute associate with the oppositely charged polar regions of the solvent. The solute molecules disperse throughout the solution. The polar attraction may draw the solution components into a smaller volume. This may be seen when 50 mL of ethanol are added to 50 mL of water. The resultant volume, as shown in Figure 16-7, is significantly less than the additive sum of 100 mL.

If polar and nonpolar substances are mixed, they demonstrate minimal attraction for each other. The particles remain segregated and solvation does not occur. An oil and vinegar salad dressing illustrates this relation. When agitated the components mix, but when left standing the polar vinegar separates from the nonpolar oil, generating a liquid with two distinct phases.

Solvation is determined by the compatibility of the solute and solvent charges. If both solution constituents are nonpolar or if both are significantly polar, solvation will occur. In summary, solvation compatibility may be simply stated as like dissolves like.

Do You Know?

If water gets into an automobile engine's fuel lines, it may interfere with the combustion process. To prevent this undesired effect, a solution known as dry gas is added to the fuel. Dry gas consists of a methanol solution that will solvate water found within the fuel system. Since the resultant methanol solution is flammable, it combusts within the engine. The water contaminant may now exit the engine along with combustion products.

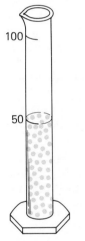
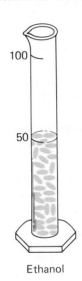
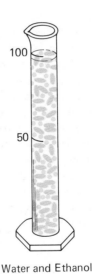

Water

Ethanol

Water and Ethanol

Figure 16-7 When equal volumes of water and ethanol are mixed the resultant volume is less than the sum of the individual volumes since water molecules take positions between ethanol molecules.

16-5 Extension: Molarity vs. Molality

Solution properties are dependent upon the relative amounts of the component substances. The terms "dilute" and "concentrated" are often used in describing solute concentration. Although these words may be useful when precise concentrations are not needed, they are inadequate in defining absolute solution concentrations. Scientists utilize more accurate methods for expressing the concentration of solutions which can be used as standards for comparison.

In Chapter 4, the concept of molarity was introduced as one concentration standard. Derived by dividing the number of solute moles by the number of liters of solution, molarity offers a simple method of expressing solution concentration.

$$molarity = \frac{number\ of\ solute\ moles}{liters\ of\ solution}$$

CHEM THEME

The mole concept is used in expressing concentrations in molarity and molality.

Certain properties of solutions depend upon the number of solute particles present within a given mass of solvent. Properties of a solvent such as boiling point and freezing point that can be altered due to the presence of solute particles are called colligative properties. For example, the temperature at which saltwater freezes is directly related to the moles of salt found within a particular mass of water. When studying the colligative properties of a solute, it is more convenient to express concentration directly in moles of solute per mass of solvent. Solution **molality** (m) offers such an option. Molality is obtained by dividing the moles of solution solute by the kilograms of solvent.

$$molality = \frac{moles\ of\ solute}{kilogram\ of\ solvent}$$

Example 16-1

How would you prepare a $0.5m$ solution of NaCl using 500 grams of water?

▶ **Suggested solution** _____

By definition $0.5m$ can be interpreted as a salt solution made from 0.5 mol NaCl mixed with 1 kg of H_2O. It should make sense to you that if you are preparing a solution with the same concentration but using half the amount of solvent (500 g vs. 1 kg), you should need half the amount of solute.

$$0.5\ mol\ NaCl \times 1/2 = 0.25\ mol\ NaCl$$

Converting moles to grams using the molar mass of NaCl gives:

$$0.25\ mol\ NaCl \times \frac{58.5\ g\ NaCl}{mol} = 15\ g\ NaCl$$

The solution is prepared by mixing 15 g NaCl with 500 g (500 mL) of water.

Review and Practice _____

1. Calculate the molarity of the following solution components:
  a. 33 g of NaCl in 1000.0 mL of solution
  b. 5.2 g of NH_3 in 500.0 mL of solution
  c. 0.10 g of $C_6H_{12}O_6$ in 10.0 mL of solution
  d. 8.6 g of HCl in 50.0 mL of solution
  e. 94 g of H_2O_2 in 450.0 mL of solution

2. Calculate the molality of the following solution components:
  a. 0.6 g of CCl_4 in 420 g of benzene
  b. 0.45 g of HNO_3 in 905 g of water
  c. 7.8 g of $MgCl_2$ in 5.24 kg of water
  d. 52 g of C_2H_5OH in 160 g of water
  e. 15.7 g of C_6H_6 in 400 g of CCl_4

1. a. 0.56 M
   b. 0.62 M
   c. $5.6 \times 10^{-2}\ M$
   d. 4.8 M
   e. 6.1 M
2. a. $9 \times 10^{-3}\ m$
   b. $7.8 \times 10^{-3}\ m$
   c. $1.6 \times 10^{-2}\ m$
   d. 7.1 m
   e. 0.50 m

The Parameters of Solubility

Have you ever placed a teaspoon of sugar in a cup of iced tea? If so, did all the sugar crystals dissolve or did a small amount remain at the bottom of the glass? You are probably aware that you can get more sugar to dissolve in hot tea than in iced tea. What is happening on the microscopic level to cause this difference in solubility?

There are many factors that affect the solubility of substances. Although you may not be aware of the chemistry involved, you often influence the solution process by controlling environmental conditions. In this lesson, you will review several of the parameters that have direct effects on solubility.

CONCEPTS

■ thermal effects on solubility
■ supersaturated solutions
■ behavior of gaseous solutes
■ Henry's Law
■ heat of solution

16-6 Effect of Temperature on Solubility

Although the solubilities of solids in liquids vary widely, increases in temperature *generally* are associated with increased solubility. This relationship between temperature and solubility for some solutes is illustrated in the solubility graph in Figure 16-8. Note that the solubility curves for some solutes do not have a constant slope. In addition, there are solutes whose solubilities decrease with increases in temperature.

If the temperature of a saturated solution is lowered, solubility is affected and an excess of solute is now found in the solution. The displaced solute particles often will precipitate out of solution. Certain solutions, however, are capable of supporting this additional solute mass. Solutions containing more solute than can normally be dissolved at a given temperature are known as **supersaturated**.

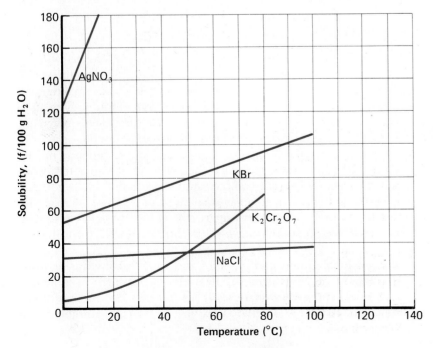

Figure 16-8 The solubility curves for various solids shows that some solids like $K_2Cr_2O_7$ do not have data points along a straight line.

Figure 16-9 A supersaturated solution of salicylic looks like an unsaturated solution. It is only after a tiny seed crystal is added that the excess solute recrystallizes.

A supersaturated solution represents an unstable situation. Such solutions are characterized by a tendency to precipitate excess solute. If a particle of solute, known as seed crystal, is added to a supersaturated solution, the excess solute precipitates.

This reaction can easily be observed using a salicylic acid solution. When water is heated to boiling, it can dissolve a large mass of salicylic acid. Once cooled, the solute may remain dissolved. Since the solubility of salicylic acid at the lower temperature is exceeded, the system is unstable. If a small amount of salicylic acid is added to the beaker, the entire solution appears to have solidified as shown in Figure 16-9.

Figure 16-10 The solubility of CO_2 in a carbonated beverage decreases with increases in temperature. Opening a bottle of warm soda shows how quickly the gas comes out of solution once the pressure on the solution is reduced.

16-7 Pressure and Gas Solubility

When a bottle containing a carbonated beverage is opened, bubbles are often observed rising to the liquid's surface. This sometimes violent release of carbon dioxide gas can be stopped if the container cap is retightened. If the beverage is stored in a warm room, an even more energetic explosion of gas occurs as shown in Figure 16-10. How might these observations be explained on a microscopic level?

Observations indicate that the solubility of a gas depends on the pressure acting upon the system. If the pressure of the system is reduced, the dissolved gaseous solute rapidly leaves the solution phase. The liberated molecules collect as small bubbles. As these bubbles rise within the solvent, they increase in size under reduced pressure. Upon reaching the liquid's surface, the bubbles burst, releasing the gas.

This relationship between pressure and gas solubility was first discovered by the English chemist, William Henry. **Henry's Law**, named in his honor, states this relationship as follows: The mass of a gas solute dissolved within a liquid is proportional to the pressure upon the system.

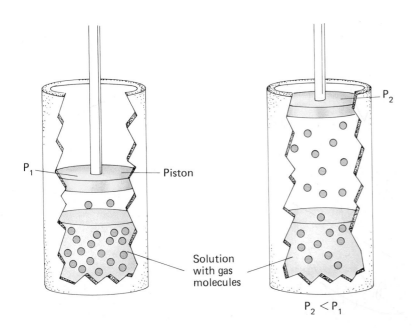

Figure 16-11 As the pressure above a liquid solution with a gaseous solute is reduced, the solubility of the gas is reduced.

Although it is not important that beginning students know the mathematical statement of Henry's Law, it is important that they be able to translate English statements such as this into mathematical statements and decide how experimental evidence could be gathered to test the validity of the statement. This would be an excellent place to focus on such intellectual skills.

This pressure-dependent effect can be seen by referring back to the behavior of a sealed container containing a carbonated beverage. Most carbonated drinks are aqueous solutions of sugar, additives, and carbon dioxide. As carbon dioxide molecules escape from the solution phase, they rise to the liquid's surface. Eventually, this CO_2 above the remaining beverage comes to equilibrium with dissolved CO_2.

If the container is opened, the pressurized gas escapes. This reduces the pressure on the liquid's surface, allowing additional gas molecules to leave the solution phase and rise to the liquid's surface. This behavior is frequently observed when a bottle of carbonated liquid is uncapped.

If the container is recapped, however, the pressure of the gas phase will once again increase, inhibiting the further release of gaseous solute. This is why carbonated beverages must be recapped in order to retain their "fizz."

When a gas/liquid solution is heated, the solution components collide with increased frequency. Energy transferred through intermolecular collisions may increase the velocity of individual particles. Gaseous solute particles may attain a velocity high enough to allow their escape from the solution phase. This increased tendency to exit the solution phase is seen as a corresponding decrease in gas solubility. It is therefore not surprising to observe the explosionlike escape of gas associated with the opening of a warm carbonated beverage.

Do You Know?

When breathing pressurized air, a potentially fatal quantity of nitrogen may dissolve within the blood. To rid the body of this undesired gaseous solute, a gradual decrease in pressure is required. Scuba divers accomplish this by ascending to predetermined decompression stops. At regular depth intervals, they wait a specified time period, allowing the nitrogen to slowly diffuse out of the blood.

A more difficult question is why shaking the capped beverage before opening greatly increases foaming. This question could be a challenging research project for interested students. It is one that the high school student could investigate through experiments.

16-8 Heat of Solution

There exists in nature the one tendency that determines physical and chemical reactability: the tendency toward maximum disorder in the universe. For a reaction to occur spontaneously, it must satisfy this requirement.

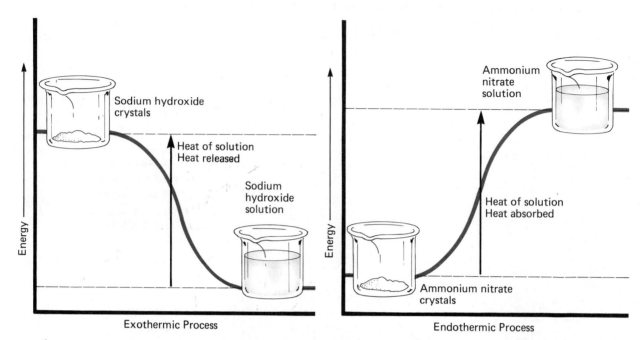

Figure 16-12 The difference in energy states between a solid solute and its resulting aqueous solution can be described by the term heat of solution. Note the differences in energy states for endothermic vs. exothermic solvation reactions.

When most solids are placed in water, the attraction between solvent and solute particles is large enough to overcome the energy maintaining the solute's crystalline structure. Additional energy is released by the solution components when hydration occurs. This change in energy associated with solvation is called the **heat of solution**.

If ammonium nitrate is added to a beaker of water, the resultant solution demonstrates a noticeable decrease in temperature. Energy from the immediate environment is absorbed in order for solvation to occur. The attractions between solute particles is greater than the attractions between the solute and solvent. Why does the solute dissolve spontaneously? The solution exhibits increased disorder. It makes sense that randomly distributed solute particles have less order than the ions remaining within the crystal lattice.

1. they contain more solute than a solution would normally hold at a particular temperature
2. dissolve as much NaCl as possible in boiling water, then cool the solution
3. atmospheric pressure is less than the pressure on the capped solution, gas molecules have enough kinetic energy to escape the liquid
4. pressure on the liquid, temperature of the liquid
5. energy must be absorbed by the surroundings in order to overcome the energy maintaining the ammonium nitrate crystal structure

Review and Practice

1. Why are supersaturated solutions unstable?

2. How would you prepare a supersaturated solution of sodium chloride?

3. Explain why uncapped carbonated beverages lose their "fizz."

4. What factors affect the solubility of a gas within a liquid?

5. Why is the solvation of ammonium nitrate accompanied by a decrease in temperature?

Colligative Properties of Solution

After a snowfall, people often spread a layer of salt onto sidewalks and driveways. As the salt mixes with the snow, it dissolves, forming a saltwater solution. The freezing point of a saltwater solution is lower than that of pure water; therefore, the snow may now melt.

If a different solute were added to the fallen snow, a similar effect would be observed. This change in freezing temperature is independent of the chemical identity of the solute. It depends solely upon the concentration of solute particles found within the solution.

CONCEPTS

- volatile and nonvolatile substances
- effect of solute upon vapor pressure
- boiling point elevation
- freezing point depression
- fractional distillation

16-9 Colligative Properties

Solute particles may physically alter certain properties of a solution. When these properties are independent of solute identity, they are known as **colligative properties**. Colligative properties of solutions include vapor pressure, osmotic pressure, and boiling and freezing temperatures.

As discussed in the previous chapter, all solids and liquids exhibit a vapor pressure. Those substances with a high vapor pressure are referred to as volatile. Those having low vapor pressure are nonvolatile.

If a nonvolatile solute is dissolved in a volatile solvent, the solution components compete for space at the liquid's surface. When a solute particle displaces a surface layer solvent molecule, it reduces the opportunity for the solvent to escape into its gaseous phase. Since fewer volatile molecules occupy the solution/vapor interface, a decrease in solution vapor pressure occurs. The observed drop in vapor pressure is directly proportional to the amount of solute particles found in solution.

16-10 Boiling Point Elevation

When the vapor pressure of a liquid equals the atmospheric pressure, boiling occurs. How might this process be affected by the addition of nonvolatile solute? If salt were added to a beaker of water, the sodium and chloride ions would compete with water molecules for space at the solution's surface. Nonvolatile ions would eventually displace some of the more volatile water molecules, and a drop in the solution vapor pressure would occur.

To raise the lowered solution vapor pressure to the atmospheric pressure, additional thermal energy is required. As this excess energy is absorbed by the solution components, rise in temperature occurs. The vapor temperature will also rise as solvent molecules at the interface have higher kinetic energy and more are able to leave the solution. Eventually a new temperature, characterized by equal atmospheric and vapor pressures, will be obtained. The solution may now boil. This solute-dependent effect is known as the **boiling point elevation**.

Figure 16-13 A distillation set-up.

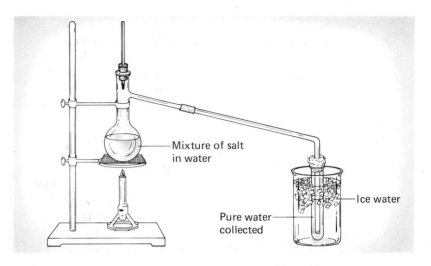

Recall the graph shown in Figure 2-6 on page 29. You could see how the boiling behavior of pure methanol differed from a mixture of methanol and water. Pure methanol boils at 65°C, the mixture boils at 86°C.

Do You Know?

To obtain large volumes of noble gases, air must be liquified. The extremely cold liquid solution is then introduced into a series of distillation columns. Once the oxygen and nitrogen components are removed, the remaining constituents may be separated by difference in their boiling temperatures.

16-11 Distillation

Although most of Earth's surface is covered with water, only a small amount is available as a freshwater resource. Even natural supplies of fresh water may contain a wide range of chemical contaminants. Distillation is a technique that utilizes the difference in component vapor pressures to produce a solute-free liquid.

A simple distillation apparatus is shown in Figure 16-13. As the solution is heated, the volatile water molecules obtain sufficient energy to escape from the solution phase. The nonvolatile solute cannot absorb the energy required for its vaporization and so remains behind. As the generated steam passes through a cold-water jacket, it is cooled. Eventually the cooled steam condenses into pure water.

Although simple distillation may be sufficient to separate solute components of dissimilar vapor pressures, how might solution constituents of similar vapor pressure be separated? Chemists use a process known as **fractional distillation** to separate solution fractions of similar vapor pressures. This technique is dependent upon the repeated vaporization and condensation of progressively purer solution fractions. The petroleum industry uses fractional distillation as a method to separate the fractions of crude oil.

16-12 Freezing Point Depression

Chemists have utilized the colligative properties of specific solutes for many practical uses. Ethylene glycol is a nonvolatile liquid that is commonly used as commercial antifreeze. It protects water-cooled automobile engines from freezing by lowering the freezing point of the coolant. It also elevates the boiling point of the solution so that it may more effectively remove heat from the engine.

How might the presence of solute particles affect the freezing process? Recall from the beginning of this lesson that saltwater causes snow to melt. If salt is added to a beaker of water, the dissociated ions diffuse throughout the solution. These particles interfere with the crystallization (freezing) of water. The effects of solute interference are overcome by further reducing the temperature.

As freezing occurs, salt is displaced from the crystalline matrix of the solvent water. Since the remaining solution will contain increasing amounts of solute, its freezing point temperature will steadily decrease. This colligative effect, caused by the presence of solute particles, is known as **freezing point depression**.

Looking at some actual numbers may help you better understand the colligative effects on freezing point. It is possible to calculate the reduction in the freezing point of water when rock salt is used to melt snow. At 0°C the solubility of salt is about 280 g/1000 g H_2O. Expressing this solubility as molality would require converting grams NaCl to the number of moles of NaCl.

$$280 \text{ g NaCl} \times \frac{1 \text{ mol}}{58.5 \text{ g NaCl}} = 4.8 \text{ mol}$$

$$4.8 \text{ mol NaCl/1 kg } H_2O = 4.8m$$

Since NaCl dissociates to form two ions per formula unit, there are actually 9.6 moles of ions in the solution.

Experiments show that the freezing temperature of 1000 grams of water drops 1.86 C° with the addition of 1 mole of solute ions. In the case of rock salt, the total temperature depression for this solution would be $(1.86°)(9.6) = 18°$.

This value, 18°, approximates how much the freezing point of water might be reduced. Theoretically the addition of rock salt to snow reduces the freezing point of water from 0°C to −18°C. However actual freezing point measurements for real solutions differ from the approximations. The deviations depend on the nature of the solute and the amount of interaction among particles.

Would a nonelectrolyte solute, such as sugar, have an identical effect on the freezing point? Although the chemical identity of the solute does not directly affect this property, it may affect the number of actual solute particles present. When NaCl is placed in solution, each salt particle dissociates into equal numbers of Na^+ and Cl^- ions. Therefore one mole of sodium chloride contains 2 moles of ions in solution. Since sugar is a molecular solid, it does not dissociate. If you compare the effect upon the freezing point of water with equal numbers of moles of NaCl and glucose, it should not be surprising to observe the salt exhibiting twice the effect as sugar.

Review and Practice

1. Compare the effect of one mole of $MgCl_2$ to one mole of KCl on the freezing point depression of an aqueous solution.

2. What is the difference between a volatile and nonvolatile substance?

3. Why is ethylene glycol added to the cooling system of automobile engines?

4. What technique is used to separate the components of crude oil?

1. One mole $MgCl_2$ dissociates to form 3 moles of ions as compared to one mole KCl which dissociates to form 2 moles ions. Theoretically, $MgCl_2$ would have 1½ times the effect on the freezing point of the solvent.
2. Volatile substances have high vapor pressures, nonvolatile substances have low vapor pressures.
3. Ethylene glycol lowers the freezing point of water and thus prevents water in the cooling system of a car from freezing at low temperatures.
4. fractional distillation

Reactions Within Solution

Many chemical reactions occur quickly when the reactants are solvated. For example, when crystals of silver nitrate are mixed with crystals of sodium chloride, product forms slowly. If, however, aqueous solutions of these salts are mixed, a white product is formed. The product, silver chloride, reacts with light and, therefore, can be used in the production of photographic film and paper. This lesson covers some properties of reactions in solution with particular emphasis upon solvated ion behavior.

16-13 Precipitation and Net Ionic Equations

Lead(II) nitrate and potassium iodide are ionic solids that are soluble in water. If a solution of $Pb(NO_3)_2$ and a solution of KI are combined, bright yellow crystals form within the mixture as shown in Figure 16-14. These colorful reaction products eventually "fall out" of the mixture, collecting on the vessel's bottom. How can you explain this observation in terms of solubility?

If you consider your experience in writing equations for reactions, it is obvious that a double replacement reaction has occurred. The identity of the precipitate is probably not obvious.

$$Pb(NO_3)_2(aq) + 2KI(aq) \longrightarrow PbI_2(s) + 2KNO_3(aq)$$

Water is not a reactant, but a medium for solvating individual ions, thus it does not appear in this chemical equation.

Examining the above equation, you see that products of different solubilities are formed. A chemistry handbook could show that KNO_3 is very soluble while PbI_2 has a very low solubility. Using this informa-

Figure 16-14 The formation of a PbI_2 precipitate from the reaction of $Pb(NO_3)_2(aq)$ with KI(aq).

tion you see how the state symbols in the equation were determined. Potassium nitrate, KNO_3, is a water-soluble salt. Upon its formation, it readily dissociates into its component ions. Lead(II) iodide, however, has a very low solubility in an aqueous solution. Since it does not readily dissociate into hydrated ions, it precipitates out of the solution phase. Table 16-3 shows some regularities noted on the solubilities of various compounds. This information is based on experimental results.

Table 16-3

SOLUBILITY OF SOME IONIC COMPOUNDS IN WATER				
NEGATIVE ION (ANION)	PLUS	POSITIVE ION (CATION)	FORM A COMPOUND WHICH IS	
Any anion	+	Alkali metal ions (Li^+, Na^+, K^+, Rb^+, or Cs^+)	"	Soluble, i.e., >0.1 mol/L
Any anion	+	Ammonium ion, NH_4^+	"	Soluble
Nitrate, NO_3^-	+	Any cation	"	Soluble
Acetate, CH_3COO^-	+	Any cation except Ag^+	"	Soluble
Chloride, Cl^-, or Bromide, Br^-, or Iodide, I^-	+ +	Ag^+, Pb^{2+}, Hg_2^{2+}, or Cu^+ Any other cation	" "	Not soluble Soluble
Sulfate, SO_4^{2-}	+ +	Ca^{2+}, Sr^{2+}, Ba^{2+}, Ra^{2+}, Ag^+, or Pb^{2+} Any other cation	" "	Not soluble Soluble
Sulfide, S^{2-}	+ + +	Alkali ions or NH_4^+, Be^{2+}, Mg^{2+}, Ca^{2+}, Sr^{2+}, Ba^{2+}, or Ra^{2+} Any other cation	" " "	Soluble Soluble Not soluble
Hydroxide, OH^-	+ + +	Alkali ions or NH_4^+ Sr^{2+}, Ba^{2+}, or Ra^{2+} Any other cation	" " "	Soluble Slightly soluble Not soluble
Phosphate, PO_4^{3-}, or Carbonate, CO_3^{2-}, or Sulfite, SO_3^{2-}	+ +	Alkali ions or NH_4^+ Any other cation	" "	Soluble Not soluble

In an aqueous environment, most soluble ionic substances dissociate into component ions. The dissociation of the three soluble components of the reaction to form lead(II) iodide are as follows:

$$Pb(NO_3)_2(s) \longrightarrow Pb^{2+}(aq) + 2NO_3^-(aq)$$
$$KI(s) \longrightarrow K^+(aq) + I^-(aq)$$
$$KNO_3(s) \longrightarrow K^+(aq) + NO_3^-(aq)$$

Showing the actual nature of the reactant and product ions involved in a reaction would be a more accurate representation. The equation presented for the formation of PbI_2 can be rewritten as follows:

$$Pb^{2+}(aq) + 2NO_3^-(aq) + 2K^+(aq) + 2I^-(aq) \longrightarrow$$
$$PbI_2(s) + 2K^+(aq) + 2NO_3^-(aq)$$

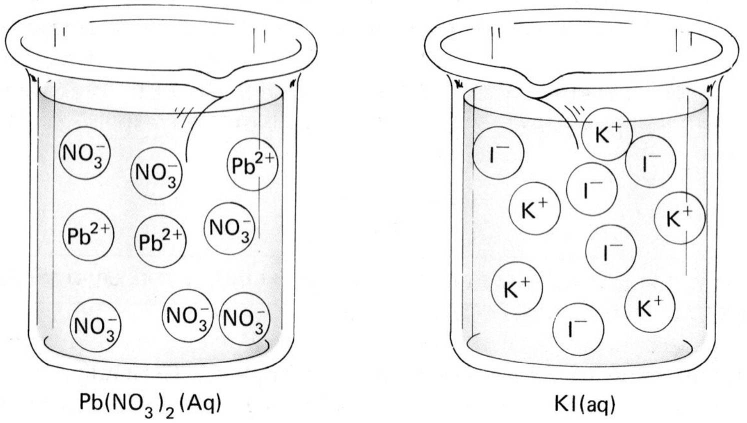

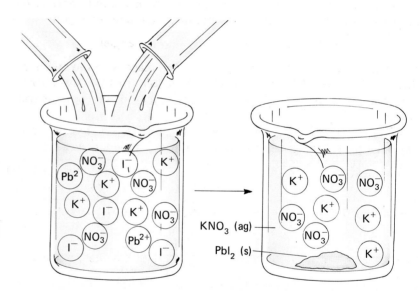

Figure 16-15 Soluble salts such as $Pb(NO_3)_2$ and KI dissociate in water.

Figure 16-16 The concentrations of Pb^{2+} and I^- ions in solution generally exceeds the solubility of PbI_2 and a precipitate forms. The aqueous solution contains the spectator ions of the PbI_2 reaction.

Known as an ionic equation, this representation can give an insight into the behavior of the reaction components.

The ions $K^+(aq)$ and $NO_3^-(aq)$ remained unchanged during the reaction. Since they do not participate in the reaction they are called **spectator ions**. To simplify the reaction representation, spectator ions may be deleted from the ionic equation. The resultant representation, called the **net ionic equation**, shows only the reacting species and may be illustrated as follows:

$$Pb^{2+}(aq) + 2I^-(aq) \longrightarrow PbI_2(s)$$

In the PbI_2 reaction a precipitate is formed. Precipitates do not always form. When a solution of sodium hydroxide and a solution of hydrogen chloride (hydrochloric acid) are mixed, the following reaction occurs:

$$Na^+(aq) + OH^-(aq) + H^+(aq) + Cl^-(aq) \longrightarrow$$
$$Na^+(aq) + Cl^-(aq) + HOH(l)$$

The net ionic equation for this reaction is:

$$OH^-(aq) + H^+(aq) \longrightarrow HOH(l)$$

You will recognize the product better when the formula is written in its ordinary form, H_2O. This reaction between solutions containing H^+ and OH^- to form the nonionic compound water is a very common reaction. It will be discussed at length in Chapter 20 on acids and bases.

In another reaction involving the H^+ ion, metals react and hydrogen gas forms. One such reaction is described by the following molecular equation and net ionic equation.

$$Zn(s) + 2HCl(aq) \longrightarrow H_2(g) + ZnCl_2(aq)$$
$$Zn(s) + 2H^+(aq) \longrightarrow H_2(g) + Zn^{2+}(aq)$$

From the second equation, you can see that any source of H^+ ions might cause the reaction that changes Zn to Zn^{2+}.

Perhaps you recognize these equations as examples of a replacement reaction like those discussed in Chapter 5. Generally, in a replacement reaction, a metal is reacting with a metal ion to form a new metal and a new metal ion. The reaction between copper and silver nitrate could be described by the following net ionic equation:

$$Cu(s) + 2Ag^+(aq) \longrightarrow Cu^{2+}(aq) + 2Ag(s)$$

You might ask yourself what could be taking place with the atoms in this reaction. How does the copper atom acquire that positive charge and become an ion? How does the positive silver ion become a neutral atom?

Still another ionic reaction is illustrated by the reaction between baking soda and vinegar. The reaction goes to completion because one of the products decomposes.

$$NaHCO_3(aq) + HC_2H_3O_2(aq) \longrightarrow NaC_2H_3O_2(aq) + H_2CO_3(aq)$$

The H_2CO_3 that forms decomposes readily to form H_2O and CO_2.

$$H_2CO_3(aq) \longrightarrow H_2O(l) + CO_2(g)$$

The overall reaction can be described by the following net ionic equation.

$$HCO_3^-(aq) + H^+(aq) \longrightarrow H_2O(l) + CO_2(g)$$

These are but a few examples of reactions that can occur when solutions containing ions are mixed.

16-14 Limitations of Net Ionic Equations

Net ionic equations provide a simplified description of many reactions. They can often clarify what is taking place in a chemical reaction. However, like most good things, they have limitations.

It is impossible to write an equation if you do not know what the reactants and products are. To write an ionic equation, you need even more information. You must know which compounds dissociate into ions when they are in solution and which compounds do not dissociate into ions. Until you can learn several chemical facts, this is not easily determined. Consequently, students in a beginning course may not be able to write ionic equations for many ionic reactions. This should not disturb you. As you learn more chemical facts, your ability to describe them with ionic equations will increase.

Another problem arises with net ionic equations. They provide a simplified description of a reaction, but it is possible to oversimplify. The net ionic equation implies that, regardless of the additional ions in solution, those ions mentioned in the net ionic equation combine as indicated. It just is not so. Any time other chemical species are present, other chemical reactions are possible.

The reaction between zinc metal and a compound containing the H^+ ion was described in the last section. If nitrate ions are present in solution (as from HNO_3), a different reaction may occur as represented by the following equation.

$$Zn(s) + 4H^+(aq) + 2NO_3^-(aq) \longrightarrow 2NO_2(g) + 2H_2O(l) + Zn^{2+}(aq)$$

As you can see, net ionic equations that ignore other ions in solution can be misleading.

16-15 Solubility Equilibrium

Using what you have learned about solubility, precipitates, and net ionic equations, it is possible to make predictions about the formation of precipitates when salt solutions are mixed. Consider the following situation. Will a precipitate form when solutions of NaOH and $FeCl_2$ are mixed? The answer is yes if the amount of product formed exceeds its solubility. You can predict the formulas of the products of the reaction. What new combinations of ions are possible if a reaction takes place?

$$Fe^{2+} \text{ with } OH^- \quad \text{and} \quad Na^+ \text{ with } Cl^-$$

In replacement reactions, the charges of the positive and negative ions do not change during the reaction. Write the correct formula for the products, and complete the equation for this reaction.

$$FeCl_2(aq) + 2NaOH(aq) \longrightarrow Fe(OH)_2 + 2NaCl$$

What you have written is the balanced equation for the displacement reaction if it occurs. It will occur if one of the products is insoluble in water. According to Table 16-3, is either product insoluble? $Fe(OH)_2$ is insoluble. The net ionic equation for the reaction is:

$$Fe^{3+} + 2OH^- \rightleftharpoons Fe(OH)_2$$

Note that the double arrow is used to show that the precipitate is in equilibrium with its ions in solution.

The solubility of $Fe(OH)_2$ is 0.000 15 g in 100 g of water. This value is represented as follows:

$$\frac{0.000\,15 \text{ parts solute}}{100 \text{ parts solvent}} \quad \text{or} \quad \frac{1.5 \text{ parts solute}}{1\,000\,000 \text{ parts solvent}}$$

One part per million would be a very low solubility. When solutions normally encountered in the laboratory are mixed, a solid precipitate appears when the product has a solubility of less than 1 g/100 g of water. The solubility of $Fe(OH)_2$ is well below this value. The solubility of NaCl is about 36 g/100 g of water. Therefore, you would not expect to see a precipitate of NaCl.

Predicting the results of this reaction was easy because there is a distinct difference in the solubilities of the two possible products.

Another method of determining whether a particular product will precipitate involves using the concentrations of the ions used in forming that product.

In any saturated solution of a substance you have excess solute, the solvent, and in the case of an ionic solid there are ions present in the solution. Recall from Section 16-2 that an ionic solid is in equilibrium with its ions in a saturated solution. The rate at which ions are solvated equals the rate at which solvated ions reattach to the solid crystals. In studying saturated solutions, chemists have found that the product of the ion concentrations (in moles per liter) equals a constant value, dependent on the identity of the saturated solute and the temperature of the solution. This constant is called the **solubility product constant**. The abbreviation K_{sp} is used to denote the solubility product constant. The K_{sp} values for several substances are shown in Table 16-4.

Table 16-4

SOME SOLUBILITY PRODUCT CONSTANTS AT 25°C			
COMPOUND	K_{sp}	COMPOUND	K_{sp}
AgCl	1.7×10^{-10}	$SrCrO_4$	3.6×10^{-5}
AgBr	5.0×10^{-13}	$BaCrO_4$	8.5×10^{-11}
AgI	8.5×10^{-17}	$PbCrO_4$	2×10^{-16}
$AgBrO_3$	5.4×10^{-5}	$CaSO_4$	2.4×10^{-5}
$AgIO_3$	3.1×10^{-5}	$SrSO_4$	7.6×10^{-7}
		$PbSO_4$	1.3×10^{-8}
		$BaSO_4$	~~1.5×10^{-5}~~ 1.6×10^{-9}
		$RaSO_4$	4×10^{-11}

Now consider how these data can be used in predicting precipitation formation. The K_{sp} of a substance can be used to calculate the solubility of a substance. Suppose you wish to know if calcium sulfate will precipitate when a solution has a $0.05M$ concentration of calcium ions and a $0.05M$ concentration of sulfate ions.

From Table 16-4, you see that the K_{sp} for $CaSO_4$ is 2.4×10^{-5}. By definition, K_{sp} is the product of the concentration of the calcium and sulfate ions in a saturated solution.

$$K_{sp} = [Ca^{2+}][SO_4^{2-}]$$

$$2.4 \times 10^{-5} = [Ca^{2+}][SO_4^{2-}]$$

The brackets are used to denote concentrations in moles per liter. For a precipitate to form, the product of the concentrations must exceed 2.4×10^{-5}. Now see if this is the case in your solution.

$$Ca^{2+} = [0.05]$$
$$SO_4^{2-} = [0.05]$$

Substitute these values into the ion-product expression.

$$K_{sp} = [Ca^{2+}][SO_4^{2-}]$$
$$K_{sp} = [0.05][0.05]$$
$$K_{sp} = 2.5 \times 10^{-3}$$
$$2.5 \times 10^{-3} > 2.4 \times 10^{-5}$$

The K_{sp} problems presented here have 1 to 1 ion ratios. When working with K_{sp}'s for compounds that dissociate giving mole ratios other than 1 to 1, the concentrations of the ions must be raised to powers equal to their mole ratios. For example the K_{sp} expression for $Ca(OH)_2$ is

$$K_{sp} = [Ca^{2+}][OH^-]^2$$

The ion product in the mixture is greater than the K_{sp}. Therefore the calcium sulfate will precipitate. You can see, however, that very dilute solutions of calcium ions and sulfate ions might not provide a high enough concentration of ions to exceed the K_{sp}. If such is the case, no precipitate forms.

Experimental data also can be used to calculate a K_{sp} as the following example shows.

Example 16-2

Calculate the solubility product constant of AgBr. When AgBr is dissolved in warm water, the [Br⁻] is found to be $8.77 \times 10^{-7}M$.

▶ **Suggested solution**

Write the dissociation equation for AgBr.

$$AgBr(s) \rightleftharpoons Ag^+(aq) + Br^-(aq)$$

The K_{sp} expression for that reaction is the product of the ion concentrations.

$$K_{sp} = [Ag^+][Br^-]$$

The silver ion concentration equals the bromide ion concentration since AgBr decomposition shows a 1 to 1 mole ratio between the ions.

$$[Ag^+] = [Br^-]$$

Thus the K_{sp} expression may be written in terms of the bromide ion or silver ion concentration.

$$K_{sp} = [Br^-]^2 = [Ag^+]^2$$

From the problem, the concentration of the bromide ion is given as $8.77 \times 10^{-6}M$ and substituted into the equation.

$$K_{sp} = [8.77 \times 10^{-6}]^2$$
$$K_{sp} = 7.7 \times 10^{-13}$$

Do You Know?

The basic photographic process involves the formation of insoluble silver. When exposed to light, photochemicals impregnated in film react to form silver which precipitates in regions of the film exposed to light. Since the precipitate is dark in color, a reversal or negative photographic image is formed.

Example 16-3

Calculate the solubility of $CaCO_3$ in water at 25°C. The K_{sp} of $CaCO_3$ is 4.8×10^{-9}.

▶ **Suggested solution**

When $CaCO_3$ dissolves, it produces calcium ions and the same number of carbonate ions.

$$CaCO_3(s) \longrightarrow Ca^{2+}(aq) + CO_3^{2-}(aq)$$
$$[Ca^{2+}] = [CO_3^{2-}]$$

Therefore the K_{sp} can be written in terms of either ion. You can use the calcium ions.

$$K_{sp} = [Ca^{2+}]^2 = 4.8 \times 10^{-9}$$

By taking the square root of both sides of the equation you get the value for the $[Ca^{2+}]$.

$$[Ca^{2+}] = \sqrt{4.8 \times 10^{-9}}$$
$$[Ca^{2+}] = 6.9 \times 10^{-5}$$
$$[Ca^{2+}] = 6.9 \times 10^{-5} \text{ and } [CO_3^{2-}] = 6.9 \times 10^{-5}$$

From this problem you should see that when one mole of $CaCO_3$ dissolves, only 6.9×10^{-5} mole of Ca^{3+} and 6.9×10^{-5} mole of CO_3^{2-} ions are produced. This value (6.9×10^{-5}) represents the number of moles of $CaCO_3$ dissolved to produce the ions in 1 liter of solution. So $6.9 \times 10^{-5}M$ is the solubility of $CaCO_3$.

16-16 Dissolving Precipitates

Suppose you want to remove paint from a surface. One way to remove the paint is to scrape or sand off the old paint. Another easier, quicker way to remove old paint is to use a chemical which will react with the paint to dissolve it. For some paints, turpentine is a good dissolving agent. Water is the best solvent known but some compounds are insoluble in water, just like paint. Chemists have learned from many experiments about some compounds which are good solvents for certain solids.

For instance, you may need to do an experiment using aluminum ions. The chemist has aluminum hydroxide available but $Al(OH)_3$ is insoluble in water as shown by its small K_{sp}.

$$Al(OH)_3 \rightleftharpoons Al^{3+}(aq) + 3OH^-(aq)$$
$$K_{sp} = [Al^{3+}][OH^-]^3$$
$$K_{sp} = 1.9 \times 10^{-33}$$

The chemist must add something to the aluminum hydroxide solution that will react with the hydroxide ion. The hydrogen ion of an acid reacts almost completely with the hydroxide ion.

$$H^+(aq) + OH^-(aq) \longrightarrow H_2O$$

If hydrogen ions from a strong acid form water by reacting with the few OH^- ions dissolved from the solid $Al(OH)_3$, then more $Al(OH)_3$ dissolves releasing more $OH^-(aq)$ ions. More water is formed from the reaction of the OH^- and hydrogen ions. This dissociation of the $Al(OH)_3$ continues until all of the solid is dissolved, producing aluminum ions in solution as well.

From many experiments with insoluble hydroxides and strong acids chemists have noted regularity: Insoluble hydroxides are soluble in excess strong acids. Iron(II) hydroxide, $Fe(OH)_2$, dissolves in hydrochloric acid using the same mechanism as the aluminum hydroxide example. The H^+ ions from HCl combine with OH^- ions in solution. $Fe(OH)_2$ dissolves to replace the lost OH^- ions that were removed by adding H^+.

Insoluble metal carbonates, such as $CaCO_3$, also can be dissolved by strong acids. In this case the acid reacts with the carbonate ion, making carbonic acid, which decomposes to produce CO_2 and H_2O.

$$CaCO_3(s) \rightleftharpoons Ca^{2+}(aq) + CO_3^{2-}(aq)$$
$$2H^+ + CO_3^{2-} \rightleftharpoons H_2CO_3$$
$$H_2CO_3 \rightleftharpoons CO_2 + H_2O$$
$$CaCO_3(s) + 2H^+ \rightleftharpoons Ca^{2+}(aq) + CO_2(g) + H_2O(l)$$

Review and Practice

1. What is the identity of the precipitate formed during the reaction between lead(II) nitrate and potassium iodide?

2. Write a balanced equation for the reaction between solutions of silver nitrate and sodium chloride.

3. Write the net ionic equation for the following reactions:
a. silver nitrate and sodium chloride
b. lead(II) nitrate and potassium bromide
c. sodium sulfate and barium chloride

4. Use Table 16-3 to determine which of the following compounds are insoluble in water.
a. sodium hydroxide
b. ammonia acetate
c. calcium sulfate
d. lead(II) chloride
e. potassium chloride
f. calcium bromide

5. Why are spectator ions deleted to form net ionic equations?

6. Write net ionic equations for the following reactions. Use Table 16-3 to help you in determining solubilities.
a. $NaOH(aq) + HCl(aq) \longrightarrow$
b. $Bi(NO_3)_2(aq) + NaOH(aq) \longrightarrow$
c. $Pb(C_2H_3O_2)_2(aq) + K_2SO_4(aq) \longrightarrow$
d. $CuSO_4(aq) + FeCl_3(aq) \longrightarrow$
e. $FeSO_4(aq) + (NH_4)_2S(aq) \longrightarrow$
f. $K_2CO_3(aq) + Sr(NO_3)_2(aq) \longrightarrow$
g. $NaCl(aq) + KOH \longrightarrow$
h. $NaI(aq) + AgNO_3(aq) \longrightarrow$
i. $Al_2(SO_4)_3(aq) + CaCl_2(aq) \longrightarrow$
j. $Na_2SO_4(aq) + CaCl_2(aq) \longrightarrow$
k. $K_2SO_4(aq) + Ba(C_2H_3O_2)_2(aq) \longrightarrow$
l. $MgSO_4(aq) + CaBr_2(aq) \longrightarrow$

7. What is the solubility product expression for AgCl?

8. Write the solubility product expression for the following solids:
a. AgBr
b. Ag_2SO_4
c. $Pb(OH)_2$
d. Hg_2SO_4
e. $Al_2(SO_4)_3$

9. Will a $CaCO_3$ precipitate form if the Ca^{2+} concentration in a solution is $0.0002 M$ and the CO_3^{2-} concentration is the same, $0.0002 M$? Use the K_{sp} given in Example 16-3.

10. Calculate the solubility of barium chromate in water using the data in Table 16-4.

11. Compare the K_{sp} values for AgCl, AgBr, and AgI. Which compound is most soluble in water? Which is least soluble?

1. PbI_2
2. $AgNO_3(aq) + NaCl(aq) \rightleftharpoons$
   $AgCl(s) + NaNO_3(aq)$
3. a. $Ag^+(aq) + Cl^-(aq) \rightleftharpoons AgCl(s)$
   b. $Pb^{2+}(aq) + 2Br^-(aq) \rightleftharpoons PbBr_2(s)$
   c. $Ba^{2+}(aq) + SO_4^{2-}(aq) \rightleftharpoons BaSO_4(s)$
4. c and d are insoluble
5. Net ionic equations show only those species reacting to form a precipitate, water, or a gas.
6. a. $H^+(aq) + OH^-(aq) \rightleftharpoons H_2O(l)$
   b. $Bi^{3+}(aq) + 3OH^-(aq) \rightleftharpoons Bi(OH)_3(s)$
   c. $Pb^{2+}(aq) + SO_4^{2-}(aq) \rightleftharpoons PbSO_4(s)$
   d. no reaction
   e. $Fe^{2+}(aq) + S^{2-}(aq) \rightleftharpoons FeS(s)$
   f. $Sr^{2+}(aq) + CO_3^{2-}(aq) \rightleftharpoons SrCO_3(s)$
   g. no reaction
   h. $Ag^+(aq) + I^-(aq) \rightleftharpoons AgI(s)$
   i. $Ca^{2+}(aq) + SO_4^{2-}(aq) \rightleftharpoons CaSO_4(s)$
   j. $Ca^{2+}(aq) + SO_4^{2-}(aq) \rightleftharpoons CaSO_4(s)$
   k. $Ba^{2+}(aq) + SO_4^{2-}(aq) \rightleftharpoons BaSO_4(s)$
   l. $Ca^{2+}(aq) + SO_4^{2-}(aq) \rightleftharpoons CaSO_4(s)$
7. $K_{sp} = [Ag^+][Cl^-]$
8. a. $K_{sp} = [Ag^+][Br^-]$
   b. $K_{sp} = [Ag^+]^2[SO_4^{2-}]$
   c. $K_{sp} = [Pb^{2+}][OH^-]^2$
   d. $K_{sp} = [Hg^+]^2[SO_4^{2-}]$
   e. $K_{sp} = [Al^{3+}]^2[SO_4^{2-}]^3$
9. $4 \times 10^{-8} > 4.8 \times 10^{-9}$, a ppt. forms
10. 9.2×10^{-6} moles per liter
11. AgCl – most soluble, AgI – least soluble

CHAPTER REVIEW

Numbers in red indicate the appropriate chapter sections to aid you in assigning these items. Answers to all questions appear in the Teacher's Guide at the front of this book.

Summary

■ Solutions are homogeneous mixtures. Although there may be solid, liquid, or gaseous solutions, all the constituents must exist within the same phase.

■ The process by which substances enter the solution phase is called solvation. Solvation is governed by the similarity of the solute and solvent charges. The statement "like dissolves like" summarizes this relationship. Most often, the solvent is water and the solute is an ionic solid.

■ Solution concentration may be expressed by several methods. Molarity and molality offer unique advantages for their specific usage as concentration standards.

■ Solutions demonstrate properties that are both dependent and independent of their component identity. Properties independent of the chemical identity of the solute are known as colligative properties. When a nonvolatile solute is added to a solution it will decrease the solution vapor pressure. A corresponding decrease in freezing temperature and elevation of boiling temperature also will occur.

■ Solution components of similar vapor pressure may be separated by fractional distillation. The petroleum industry utilizes this process to refine crude oil into its various components.

■ Hydrated ionic solids dissociate into component ions. Reactions occurring between these ions may produce an insoluble product. Known as a precipitate, this substance settles out of the mixture.

■ Reacting ionic species may be represented by a net ionic equation. This representation does not include nonreacting spectator ions from the full equation.

■ A form of the equilibrium expression for the dissociation of a salt in water is called the solubility product expression, K_{sp}. Knowing the K_{sp} value enables one to predict the solubility of a compound. Knowing the concentration of one of the ions and the K_{sp} value enables one to predict the limit of concentration of the other ion at which precipitation will occur.

■ Precipitates can be dissolved by adding reagents which react with one of the dissolved species producing a species that is removed from the reaction and thus upset the equilibrium. Strong acids dissolve both insoluble hydroxides and insoluble carbonates.

Chemically Speaking

boiling point elevation
colligative properties
dissociation
fractional distillation
freezing point depression
heat of solution
Henry's law
heterogeneous mixture
homogeneous mixture
immiscible
miscible

molality
net ionic equation
saturated
solubility product constant
solute
solvation
solvent
spectator ion
supersaturated
unsaturated

Review

1. How do heterogeneous materials differ from homogeneous materials? (16-1)

2. Give an example of how water may be either a solvent or a solute. (16-1)

3. Do all solute particles initially restricted from the solution remain outside the solution state? (16-2)

4. How does temperature relate to solubility? (16-2)

5. Describe the process in which sodium chloride dissociates within an aqueous solution. (16-3)

6. What two forces account for the attraction between molecular substances? (16-4)

7. Explain the statement, "like dissolves like." (16-4)

8. Why is it necessary to have a concentration standard other than molarity? (16-5)

9. What is molality? (16-5)

10. How will an increase in solution temperature affect the solubility of a gas dissolved within a liquid? (16-7)

11. What is meant by the term "heat of solution"? (16-8)

12. How is fractional distillation used by the petroleum industry? (16-11)

13. What effects will the addition of a solute have on the boiling and freezing points of a solution? (16-10, 16-12)

14. How does the solubility of the products of a reaction affect the net ionic equation? (16-13)

15. What is the advantage of representing a reaction by a net ionic equation? (16-13)

16. What is a spectator ion? (16-13)

17. What are the disadvantages of using net ionic equations? (16-14)

18. What is equal when a saturated solution is at equilibrium? (16-2, 16-15)

19. Why does recapping an opened soda pop bottle help keep the remaining liquid carbonated? (16-7)

Interpret and Apply

1. Compare and contrast a saturated and unsaturated solution. (16-2, 16-6)

2. Predict the miscibility of the following pairs of substances. (16-2)
  a. water and methanol
  b. methanol and carbon tetrachloride
  c. carbon tetrachloride and benzene
  d. benzene and ethanol
  e. ethanol and methanol
  f. ammonia and carbon tetrachloride
  g. ammonia and water

3. Will carbon tetrachloride solvate an ionic solid? Explain. (16-2)

4. How many phases would you encounter in a mixture containing oil, water, alcohol, and carbon tetrachloride? (16-2)

5. List three colligative properties that vary as the number of solute particles in the solution varies. (16-9)

6. Would ocean water in which ice has been formed and then removed be a better source of salt than untreated ocean water? Explain. (16-6)

7. Why is it safer to ice skate on a frozen lake than on a frozen oceanic bay exposed to the same conditions? (16-12)

8. How would a chemist prepare a supersaturated solution of sodium thiosulfate? (16-6)

9. How would an increase in pressure affect the solubility of a gaseous solute dissolved within a liquid solvent? An ionic solid dissolved within a liquid solvent? A liquid solute dissolved in a liquid solvent? (16-7)

10. Although submarines may dive to depths of tremendous water pressure, the air within their hulls remains at atmospheric pressure. Knowing this, explain why people on submarines need not worry about decompression sickness. (16-7)

11. When ammonium nitrate is placed in water, it dissociates into component ions. A measurable decrease in solution temperature accompanies this solvation process. What happens to the energy absorbed from the environment? If the solution contains more energy than the separate reactants, why did solvation occur? (16-8)

12. How would an increase in the solute surface area affect the rate at which a solute dissolves? (16-6)

13. Assume the following compounds dissolve in water to form separate, mobile ions in solution. Write the formulas and the names of the ions that can be expected. (16-2)
  a. HI
  b. Na_2CO_3
  c. $Ba(OH)_2$
  d. KNO_3
  e. NH_4Cl
  f. $Ca(C_2H_3O_2)_2$
  g. $(NH_4)_3PO_4$
  h. $(NH_4)_2SO_4$

14. Consider the reaction:

$$A^+(aq) + B^-(aq) \rightleftharpoons AB(s)$$
$$K_{sp} = 1.5 \times 10^{-6}$$

What happens if the product of the concentrations of A^+ ion and B^- ion is greater than the K_{sp} value? (16-15)

15. Using the reaction in item 14, if the concentration of A^+ is decreased, what happens to the concentration of B^-? (16-15)

16. When $12M$ HCl is added to a saturated NaCl solution, precipitation occurs. Explain why. (16-16)

17. When $10M$ NaOH is added to a saturated NaCl solution, precipitation occurs. Explain why. (16-16)

Problems

1. Compute the mass of solute needed to make 500 mL of solution at the indicated molarity. (16-5)
  a. $0.5M$ H_2SO_4 e. $0.1M$ $FeCl_3$
  b. $0.01M$ HNO_3 f. $3M$ NH_3
  c. $6.0M$ HCl   g. $5.0M$ KOH
  d. $0.50M$ NaOH

2. Calculate the molality of the following solutions. (16-5)
  a. 0.05 g CO_2 in 652 g of water (H_2O)
  b. 56 g NH_3 in 50 g of water (H_2O)

c. 3.21 g C_6H_{12} in 231 g of benzene (C_6H_6)
d. 320 g CCl_4 in 3.5 kg of benzene (C_6H_6)
e. 157 g H_2O in 1.25 kg of ethanol (C_2H_5OH)
f. 50.5 g C_7H_8 in 742 g of hexane (C_6H_{14})
g. 3.0 g I_2 in 286 g of carbon tetrachloride (CCl_4)

3. What mass of solute is needed to dissolve in the given amount of solvent to obtain the indicated solution molality? (16-5)
a. $FeCl_3$ to 1000 g H_2O for a 0.238m solution
b. Br_2 to 500 g CCl_4 for a 0.356m solution
c. C_6H_6 to 100 g C_7H_8 for a 0.550m solution
d. CCl_4 to 30.0 g C_6H_6 for a 2.25m solution
e. C_2H_3OH to 750 g H_2O for a 1.50m solution

4. How many grams of methanol must be added to 2.00 moles of water to make a solution containing equal numbers of H_2O and CH_3OH molecules? How many molecules (of all kinds) does the resulting solution contain? (16-5)

5. How many grams of ammonium chloride, NH_4Cl, are present in 0.30 liter of a 0.40M NH_4Cl solution? (16-5)

6. A chemist evaporates 25.0 mL of NaCl solution to dryness and finds 0.585 gram of NaCl. What was the molarity of the original solution? (16-5)

7. Which of the following ionic compounds are soluble in water? (16-13)
a. cesium chloride
b. lithium sulfate
c. strontium carbonate
d. potassium phosphate
e. magnesium sulfide
f. silver iodide

8. Seawater is saturated with AgCl. The $[Cl^-]$ in seawater is 0.53M, and the K_{sp} for AgCl is 1.8×10^{-10}. Calculate the $[Ag^+]$ in seawater. (16-15)

9. Write the K_{sp} expressions for the following compounds dissolved in water. (16-15)
a. $Al(OH)_3$
b. $Ca_3(PO_4)_2$
c. $CoSO_4$

Challenge

1. At the onset of arctic winter, large regions of the sea's surface freeze. Explain what happens to the freezing point of the ocean water found beneath the ice masses as winter progresses.

2. The temperature at which a pure liquid boils remains constant until all the liquid has changed to a gas. If a solution is heated to boiling, however, the temperature required to maintain the boiling state steadily increases. Explain.

3. Would spreading crystals of barium chloride on a layer of snow affect the melting process? Explain.

4. In the process of fermentation, yeast can produce a solution that is about 15% alcohol (30 proof). At higher concentrations of alcohol, the yeast cannot survive. How, then, are the more concentrated alcoholic beverages obtained?

5. Although honey is a thick liquid, it often contains noticeable crystals. What are these crystals and how did they get there?

6. The solubility product constant for PbF_2 is 3.7×10^{-8}. Calculate the grams of each ion dissolved in one liter of a saturated solution.

7. The concentration of Mg^{2+} in seawater is 0.0540M. How many liters of seawater must be processed to produce 1 g of Mg?

Synthesis

1. How might a compound's molecular geometry relate to its solubility in water?

2. How can a gaseous solution of oxygen, nitrogen and carbon dioxide be separated?

Projects

1. Prepare a report on the different methods of desalinating seawater.

2. Design an experiment to separate the components of a glucose-water solution.

3. Investigate the various gas solutions that deep sea divers breathe to avoid decompression sickness.

4. Prepare a report on the cooling system of an automobile engine.

Getting a midnight snack from the refrigerator is such a common event that you probably think little about it, but there is a lot of energy-related science in this everyday occurrence. That is the subject of this chapter.

Take the snack itself. Food is the source of energy for your body. The chemicals that make up food react within your body to provide energy for the many activities that fill your life. Similar reactions take place outside your body as well. Foods combine with oxygen in the air and "spoil." Foods are kept under refrigeration to slow down those reactions, preserving the food for use when you are gripped by hunger pangs.

The refrigerator itself is a wonderful machine that uses electrical energy to move heat from the molecules inside the refrigerator to the outside air. Building such machines was made possible by learning what energy is, the various forms that it takes, and the laws that govern the transformation of energy from one form to another.

Thermodynamics

Heat, a Form of Energy

Thermodynamics (from the Greek *thermos* meaning "heat" and *dynamics*, which implies motion) is the study of energy transformations. In this chapter, you will learn some of the language used to describe energy changes in chemical systems, and you will see how that language can be used to describe things as dissimilar as losing weight or building a battery. You also will learn how to predict which chemical changes can take place by themselves and which chemical changes can never occur without energy from an outside source.

CONCEPTS

■ the difference between heat and temperature
■ specific heat
■ heats of fusion and vaporization
■ heat and the kinetic energy of molecules

17-1 Heat and Temperature

In Chapter 1, you learned that energy can cause matter to change in some way. Heat, light, electricity, and energy of motion (kinetic energy) are common forms of energy.

Mention heat, and most people think of how hot something is—the temperature. When the room is too cold, you "turn up the heat." What you do, of course, is raise the temperature by adding heat to the room. How much heat is required to raise the temperature depends on what matter you are heating, how much matter you are heating, and whether the matter changes to some other form in the process.

Before giving names to the various measures of heat, the distinction between heat and temperature needs to be clarified. Heat is a form of energy; temperature is not. Temperature may be thought of as the intensity of heat in matter. It is similar to the concept of density which describes the intensity or concentration of matter, that is, the amount of matter (mass) in a given volume. Temperature describes the average kinetic energy of molecules or the amount of heat per molecule of substance. Strange as it may seem, there is far more heat in a large iceberg than in a cup of boiling water. The temperature or average energy of the water molecules in the iceberg is less than the temperature of water molecules in the boiling water, but the total energy in the large mass of ice is much larger than the total energy in the cup of boiling water.

Unlike density, temperature is not expressed as a unitary rate. It does not provide a numerical value of the heat in each cubic centimeter of matter as density provides a numerical value of the mass in each cubic centimeter of matter. Temperature measurements are defined by the instrument used to make them.

The common thermometer is a glass tube containing mercury or some other fluid. When heat is added to a thermometer, the molecules in the glass and the mercury move faster, collide more often and with

Most people think of "hotness" and "coldness" as separate entities. Convince students that all heat experiences can be explained by assuming that there is only one form of energy called heat. Cold is experienced when heat moves from a person's warm body to the surroundings that are at a lower temperature. Cold is not experienced because "coldness" moves from the cold air to one's body.

Figure 17-1 Zero on the Celsius scale is the normal freezing point of water. Zero on the Fahrenheit scale is the temperature of a salt and ice mixture.

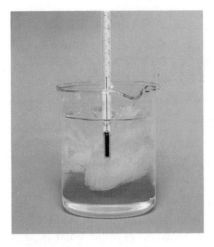 

Demonstrate the expansion of a liquid by making a water thermometer. Fill a flask with colored water and stopper with a rubber stopper into which a long glass tube has been inserted. Now warm the flask and show the water rising in the tube.

more energy, and move slightly farther apart, allowing room for more energetic motion. In other words, the materials expand. The liquid mercury expands more than the solid glass, and the column of mercury rises inside the thermometer. By making scratches on the glass tube, you can create a scale from which to read the temperature indicated by the thermometer. When the liquid in the thermometer quits moving, you can assume that the temperature of the thermometer is the same as the temperature of whatever it is touching. In this way, you can use the thermometer to measure the temperature of air, your body, or other matter.

You probably have never thought about the assumption made when you use a thermometer, but it is important. The assumption is made that two objects placed in contact will gain or lose heat until their temperatures are the same. You can make that assumption because you know from everyday experience that heat always flows from hotter objects to colder objects. Nobody would place a bottle filled with ice water in their bed, expecting the bed to get warm and the ice water to get colder! Neither would you place a hot water bottle on a sprained ankle, expecting it to cool the ankle while the hot water absorbs heat from the skin. These expectations sound ridiculous. You know that nature does not work that way. Heat always flows spontaneously from hotter objects to colder objects, and the flow continues until everything in contact is at the same temperature.

17-2 Heat Required to Change Temperature

Although it is obvious from experience that heat flows from hotter objects to colder ones, it is not obvious that different amounts of heat are required to raise the temperature of different kinds of matter. Specific heat is the name given to the amount of heat required to raise the temperature of matter. Since specific heat depends on how much matter is heated as well as how much the temperature rises, it is expressed as a *double* unitary rate. In SI units, **specific heat** is the energy (in joules) required to raise one kilogram of matter one kelvin. One kelvin is the same as one Celsius degree, so you also could say that the specific heat is the energy in joules required to raise the temperature of one kilogram of a material one degree Celsius. Table 17-1 gives the specific heats of several common materials.

Two things should be evident from Table 17-1. First, the amount of energy required to raise the temperature of one kilogram of substance one kelvin varies from one substance to another. It also is evident that the heat required to raise the temperature of a substance varies when the substance is solid, liquid, or gas. Compare the specific heat of water as a gas at 100°C and as a liquid at 100°C. The temperature at which the specific heat is measured is specified because the specific heat varies slightly from one temperature to another. This can be seen by comparing the three entries for liquid water.

A practical problem with energy is that it is not always where you want it when you want it. A house with windows facing south may collect more heat from the sun than is needed to heat the house during the day, but there will be no heat collected at night. Solar homes must store heat during the day for use at night.

Various materials are used for heat storage. The material used should be inexpensive, easy to handle, take up little space in the house, and have a high specific heat. A substance with a high specific heat makes a good "heat sponge." It can absorb a large amount of energy before its temperature rises much. The substance would heat

If a student lives in a solar home, have him/her explain how heat is stored.

Table 17-1

SPECIFIC HEAT OF COMMON MATERIALS		
MATERIAL	SPECIFIC HEAT J/kg · K	TEMPERATURE (°C)
air (g)	1000	0
aluminum (s)	910	17–100
ammonia (g)	2190	15
ammonia (l)	4710	20
asbestos (s)	820	20–98
carbon dioxide (g)	830	15
carbon (graphite) (s)	670	11
carbon (diamond) (s)	470	11
copper (s)	380	20
gold (s)	130	18
granite (s)	800	12–100
iron (s)	470	18–100
lead (s)	130	20–100
mercury (l)	140	20
paraffin (s)	2900	0–20
silver (s)	240	15–100
sodium chloride (s)	850	0
sucrose (sugar)	1250	20
water (s)	2060	0
water (l)	4220	0
water (l)	4180	25
water (l)	4210	100
water (g)	2070	100
wood (s)	1760	20

Students can make solar collectors by placing glass over a plastic foam box painted black inside. By placing equal masses of various materials inside identical collectors, exposing the collectors to the sun, and monitoring the temperature of the air inside the box, they can see that some materials store more heat than others.

up slowly during the day while absorbing energy and, at night, would cool down slowly as it released heat. Considering these factors, granite and water are used most commonly for heat storage of the materials listed in Table 17-1. Granite, a common rock, is inexpensive and easy to handle. It has a relatively high specific heat. Water is inexpensive and has a very high specific heat. One drawback of water is that it must be stored in leak-proof containers.

Both materials can be used to store heat by allowing them to warm in the sun during the day and then cool by releasing heat to the house during the night. Which material would be the best? The concept of specific heat may be useful in solving the problem.

Example 17-1

Which material, granite or water, should be used to store solar heat in a home? How much will be needed to store 100 000 kilojoules of heat?

▶ **Suggested solution**

The amount of heat stored will depend on both the *mass* of the material used and the *temperature change* allowed. Temperature changes of about 10 degrees will cause little discomfort inside the house. You can calculate the mass of granite and water that can store 100 000 kilojoules of heat when its temperature changes by 10 K.

From Table 17-1, the specific heat of granite is 800 joules per kilogram-kelvin. That means that each kilogram of granite will store 0.800 kilojoule of heat for each degree its temperature changes. Then if the temperature changes 10 K

$$\frac{0.800 \text{ kJ}}{\text{kg} \cdot \text{K}} \times 10 \text{ K} = \frac{8.00 \text{ kJ}}{\text{kg}} \text{ heat stored}$$

You now have a unitary rate indicating the heat that can be stored by each kilogram of granite. You can use it to find the mass of granite needed to store 100 000 kilojoules of heat:

$$100\,000 \text{ kJ} \times \frac{1 \text{ kg}}{8.00 \text{ kJ}} = 1.25 \times 10^4 \text{ kg granite}$$

This seems like a lot of granite. How much space will it take up? The density of granite is about 2.75 grams per cubic centimeter. You can use that value to calculate the number of cubic meters of granite like this:

$$1.25 \times 10^4 \text{ kg} \times \frac{\text{cm}^3}{2.75 \text{ g}} \times \frac{\text{m}^3}{10^6 \text{ cm}^3} \times \frac{1000 \text{ g}}{\text{kg}} = 4.5 \text{ m}^3 \text{ granite}$$

That seems possible, but it is a little larger than perhaps you would like. How much water would be needed to store the same amount of heat? The specific heat of water shown in Table 17-1 (at 25°C) is 4180 joules per kilogram-kelvin. Liquid water can store 4.180 kilojoules for each degree its temperature changes. When the temperature changes 10 K

$$\frac{4.180 \text{ kJ}}{\text{kg·K}} \times 10\text{ K} = \frac{41.80 \text{ kJ heat stored}}{\text{kg}}$$

The mass of water needed to store 100 000 kilojoules of heat would be

$$100\,000 \text{ kJ} \times \frac{\text{kg}}{41.80 \text{ kJ}} = 2.39 \times 10^3 \text{ kg water}$$

How much space would that much water take up? The density of water is 1.00 gram per cubic centimeter. The number of cubic meters required would be

$$2.39 \times 10^3 \text{ kg} \times \frac{\text{cm}^3}{1.00\,\text{g}} \times \frac{\text{m}^3}{1 \times 10^6 \text{ cm}^3} \times \frac{1000\,\text{g}}{\text{kg}} = 2.39 \text{ m}^3$$

Water takes up about half the space required by the granite. Which material would you use?

17-3 Heat Required to Change State

The specific heat of a substance describes the amount of energy that must be added to increase the temperature of a substance or the amount of energy that must be removed to lower its temperature. However, you know that when you change the temperature of a substance, it may change states. When gases cool, they may change to liquids; when liquids cool, they may change to solids. Relatively large amounts of energy are involved in such changes in state. Table 17-2 compares the specific heat of liquid water with its **heat of fusion**, the energy required to change one kilogram of water from solid to liquid, and its **heat of vaporization**, the energy required to change one kilogram from liquid to gas. How much heat is required depends on how much water is changed. Table 17-2 shows the kilojoules of energy required for a change in state of one kilogram of water or of one mole of water.

Table 17-2

SPECIFIC HEAT, HEAT OF FUSION, AND HEAT OF VAPORIZATION OF WATER		
	(J/kg · K)	(J/mol · K)
Specific heat at 0°C	4220	76.0
Specific heat at 100°C	4210	75.8
	(J/kg)	(J/mol)
Heat of fusion at 0°C	334 000	6010
Heat of vaporization at 100°C	2 260 000	40 700

It is amazing that so much energy is required to melt ice or boil water without changing its temperature, but it is true. There are many ways to capitalize on this characteristic of nature. Your own body's cooling system takes advantage of the large heat of vaporization of water. When you perspire, huge amounts of heat are used to convert water in your body to gaseous water that escapes into the air. Air conditioning systems in large buildings remove heat by evaporating water. The refrigeration system in air conditioners and your refrigerator absorb heat by vaporizing liquid Freon® or ammonia. The gaseous refrigerant is then compressed until it changes to a liquid. The heat that was absorbed to vaporize the liquid then raises the temperature of the liquid, and the heat is dumped into the outside air by circulating the hot liquid through cooling coils in contact with the cooler outside air.

Farmers often protect their crops from severe frost damage by spraying water into the air around their crops. The water freezes, releasing heat in the process. The heat released when the water freezes raises the temperature of the surrounding air, and the ice coating the crops insulates the plants from the air to further reduce the damage from subzero temperatures.

Do You Know?

Ammonia is frequently used as the working fluid in refrigeration systems because it has a high heat of vaporization.

![17-4] **Heat at the Microscopic Level**

The microscopic model of matter used to explain chemical changes also explains the heat changes just described. In Chapter 7 you learned that the behavior of gases can be explained by the kinetic molecular theory. This theory suggests that molecules are in constant random motion. The average speed of the molecules is proportional to the temperature of the gas. Keep in mind that temperature is proportional to the *average* speed of the molecules. At any temperature, some molecules move very slowly while others move fast. This explains how liquids evaporate at temperatures below their normal boiling points. Even though the average speed of the molecules is below the speed required to overcome the attractive forces between molecules in the liquid, some molecules have enough energy to overcome those forces of attraction. If those fast molecules are at the surface of the liquid, they escape to form a gas. The liquid evaporates.

If you have not previously done so, use an apparatus that simulates the motion of gas molecules, with metal spheres to simulate the motion of gases at different temperatures.

When the rapidly moving molecules in a liquid escape, the average speed of molecules left behind is lower. This is seen as a decrease in temperature as shown in Figure 17-2. If the liquid is in contact with a body at a higher temperature, for example, your skin, heat is transferred from the body to the liquid, raising the temperature. Once again, there are molecules moving rapidly enough to escape from the liquid, continuing the process of evaporation. The net effect of these changes is that the temperature of the liquid remains constant as heat is transferred from the body to the liquid, and molecules of the liquid evaporate. In effect, the heat supplied from the body is used to overcome attractive forces between molecules. This is the heat described as heat of vaporization.

Heat of fusion is explained in a similar manner. In any solid, some molecules vibrate slowly while others vibrate more violently. Energetic

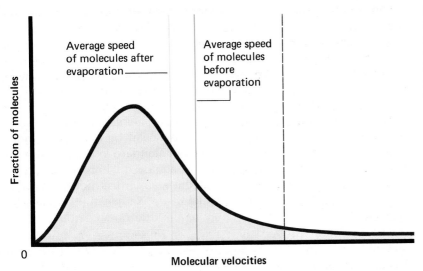

Figure 17-2 At any temperature, some molecules move fast and some move slowly. The blue line shows the average speed of the molecules before the faster molecules (blue region) escape the liquid as a gas. The yellow line shows the average speed of the molecules that remain (yellow region) after evaporation.

molecules at the surface of the solid have sufficient energy to overcome attractive forces holding the molecules in place. When this happens, the temperature of the solid decreases because the more energetic molecules are no longer part of the solid. If the solid is in contact with something at a higher temperature, heat is transferred to the molecules in the solid, and more molecules acquire the energy required to "melt." This process occurs rapidly, and no change in temperature occurs, even though energy is being acquired from the surroundings. The energy acquired is used to overcome the forces of attraction that hold molecules together. That energy is the heat of fusion.

Review and Practice

1. What does it mean to say that the specific heat of iron is greater than the specific heat of gold?

2. Solar homes often use rock or water to collect heat from the sun by day to warm the home at night. If the temperature of a metric ton (1000 kg) of granite is raised 10 degrees in one house and a metric ton of water is raised 10 degrees in another house, which house has collected more heat?

3. A kg of water and a kg of liquid ammonia are at the same temperature. Do the two liquids contain the same amount of heat energy? Explain.

4. The same amount of heat is added to a metric ton of iron and to a small iron nail.
   a. Which would reach a higher temperature?
   b. Why is there a difference? Give your explanation in terms of the concentration or intensity of heat in matter.
   c. What is the name given to the amount of heat per unit of substance?

5. At the microscopic level, what form does the energy take when we add heat to raise the temperature of a solid, a liquid, or a gas?

1. More energy is required to raise the temperature of 1 kg of iron one degree than to raise the temperature of 1 kg of gold one degree.
2. the one using water
3. No, the specific heats vary.
4. a. the iron nail
   b. The heat is more concentrated in the nail.
   c. temperature
5. It takes the form of kinetic energy as the particles move faster.

Measuring Energy

CONCEPTS

■ **comparing energy standards**
■ **calorimetry**

The energies shown in Tables 17-1 and 17-2 are measured values, and you will understand energy changes better if you have some idea of how those measurements are made.

You know from Chapter 1 that all measurements involve comparison with a standard. You measure your height by comparing your length to the length of a standard meter or standard foot. You measure your mass by comparing it to that of a standard kilogram. Energy is measured in a similar fashion, but the standard is not something stored in a vault like the standard kilogram. Rather, the standard is described as the amount of energy required to do something.

17-5 An Energy Standard

It takes energy to raise the temperature of water and make the molecules move faster. In early work, the energy required to raise the temperature of one gram of water one Celsius degree was defined as one **calorie** (from the Latin word for heat, *calor*). The energy needed to raise the temperature of one kilogram of water one Celsius degree is 1000 times larger, so it is called a **kilocalorie**. The kilocalorie is the unit used to describe the energy available in foods. It commonly is referred to as one **Calorie** (with a capital C).

The calorie or Calorie is seldom used in scientific work because the joule has been selected as the standard unit for any form of energy. A joule (J) is the energy involved when a force of one newton acts through a distance of one meter. One calorie is equal to 4.184 joules, and one Calorie (kilocalorie) is equal to 4184 joules. Joules are used to express energy in this text.

A joule is a small unit. One match, burned completely, liberates about 1050 joules or 250 calories.

17-6 Measuring Specific Heat

You will now see how this energy standard is used to determine the specific heat of some metal. You can heat or cool the metal with water and use the temperature change of this water to calculate the heat required to change the temperature of the metal.

Before seeing how this is done, you need to consider some practical problems. The procedure will not work unless all of the energy in the metal is transferred to the water so it can be measured. This never happens, but it is possible to come close.

An ordinary thermos bottle or a block of plastic foam is a very good thermal insulator. Heat is transferred very slowly through the thermos bottle or the plastic foam to the air surrounding it. This is why plastic foam cups are excellent containers for hot drinks and why thermos bottles keep soup or coffee hot for several hours. If heat changes take place inside a thermos bottle or a thick plastic foam container, very

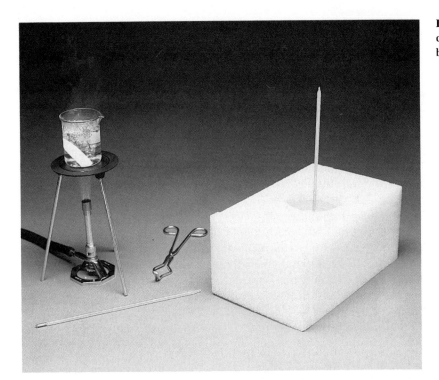

Figure 17-3 A simple calorimeter can be made by cutting a hole in a block of plastic foam.

little heat is lost to the surroundings, and a good estimate can be made of the heat transferred within the container. A well-insulated container in which heat measurements are made is called a **calorimeter**. Since *calor* means heat and *meter* means measure, the word literally means "heat measurer."

Figure 17-3 shows a block of plastic foam with a hole in it. It makes a good calorimeter. If there is a kilogram of water in the plastic foam container, each degree change in the temperature of the water will correspond to an increase or decrease of one kilocalorie or 4180 joules of energy. If the water in the plastic foam calorimeter is at room temperature, then there is no tendency for the water to exchange heat with the surroundings, and measurements will be even more exact. (The smaller the difference in temperature between the two objects that are in contact, the less energy is transferred from one to the other.)

Weigh a piece of aluminum, heat it, and place it into the calorimeter filled with water. The aluminum will lose heat to the water until the aluminum and water are at the same temperature. The temperature change of the water can be used to calculate the heat gained by the water. That also will be the heat lost by the aluminum, and you can calculate the specific heat of it.

There is one additional practical problem that must be overcome. It is impossible to stick a glass thermometer into a piece of aluminum to measure its temperature. How can you measure the original temperature of the aluminum? One solution is to heat the aluminum in boiling water. Measure the temperature of the water, and if the aluminum has been in the boiling water for several minutes, then you can assume that its temperature is the same as the temperature of the surrounding water.

By now you should see how you could calculate the specific heat of aluminum. Example 17-2 describes the experiment and illustrates the calculations required.

Example 17-2

Given the measurements provided below, calculate the specific heat of aluminum.

▶ **Suggested solution**

One kilogram of water is placed in the calorimeter shown in Figure 17-3 along with a thermometer used to record the temperature. A piece of aluminum is weighed and heated in a beaker of boiling water and a second thermometer is used to measure the temperature of the water.

After waiting a few minutes to be sure that the temperature of the water in the two containers does not change any more, the thermometers are read and the data recorded. Here are data from one experiment:

Mass of water in the calorimeter	1000 g
Initial temperature of water in calorimeter	22.4°C
Mass of aluminum sample	125.25 g
Temperature of boiling water bath and aluminum	99.3°C

After taking these readings, the aluminum is quickly taken out of the boiling water and placed into the water in the calorimeter. When the liquid in the thermometer stops rising, the final temperature of the water and aluminum has been reached:

Final temperature of water and aluminum in calorimeter	24.4°C

You have all the data needed to do the calculations. First, see how much heat was gained by the water in the calorimeter. The specific heat of water is 4180 joules per kilogram-kelvin. This indicates that 4180 joules of heat are added to each kilogram of water to increase the temperature one Celsius degree. (Remember, a kelvin is the same size as a Celsius degree.) Since the water was originally at a temperature of 22.4°C and the final temperature was 24.4°C, the change was 2.0 K. The mass of the water was 1.000 kilogram. The energy gained by the water is

$$\frac{4180\ \text{J}}{\text{kg} \cdot \text{K}} \times 1.000\ \text{kg} \times 2.0\ \text{K} = 8.4 \times 10^3\ \text{J}$$

If you do not have students do so in lab, demonstrate the measurement of the specific heat of a metal as a class experiment. Actually seeing the experiment will make Example 17-2 much more meaningful.

Assume that the calorimeter is insulated well enough so that all of this energy must have come from the aluminum placed in it. This is the energy lost by the aluminum.

With this assumption, you can say that the energy lost by the aluminum was 8.4×10^3 joules. The temperature change of the aluminum was 74.9 K (99.3 − 24.4). The mass of the aluminum was 0.125 25 kilogram. Specific heat is given in terms of joules per kilogram-kelvin. The specific heat of aluminum is

$$\frac{8.4 \times 10^3 \text{ J}}{0.125\ 25 \text{ kg} \times 74.9 \text{ K}} = 8.9 \times 10^2 \text{ J/kg·K}$$

This experimental value compares very favorably with the specific heat of aluminum recorded in Table 17-1. Since the values recorded in Table 17-1 are given to three significant digits, you know that they were the result of more precise measurements than the ones you can make in your laboratory. The thermometers used in those measurements can be read to the nearest hundredth of a degree, and corrections are made for the heat lost to the calorimeter and to the surrounding air. However, the experiments done to obtain the data in Table 17-1 are exactly like the experiment outlined in Example 17-2. The experiments represent measurements that you can easily make in your laboratory. Perhaps your teacher will have you do some.

Review and Practice

1. You were told that a calorie is equal to 4.184 joules. Write a unitary rate that would express the energy of 25 calories in units of joules. What unitary rate would you use to express 25 000 joules in units of calories?

2. One large apple supplies about 100 Calories (kilocalories) of energy. How many joules of energy would this be?

3. One cup (8 oz.) of yogurt contains about 250 Calories of energy. How many joules is this?

4. Standards of measurement normally are selected for convenience and accuracy. Why do you think the calorie was defined as the energy required to raise the temperature of water rather than the energy required to raise the temperature of some other material?

5. One cup of milk (244 g) provides 150 Calories, and 30 percent of the total calories comes from carbohydrates. When carbohydrates are metabolized in the body to carbon dioxide and water, they release about 4 Calories per gram. Determine the amount of energy derived from carbohydrates in one cup of milk.

6. Use this balanced equation to identify the false statement(s):

$$H_2(g) + 431 \text{ kJ} \longrightarrow 2H(g)$$

   a. The reaction is endothermic.
   b. Two grams of $H_2(g)$ contain more energy than two grams of $H(g)$.
   c. Burning one gram of $H_2(g)$ will produce more heat than burning one gram of $H(g)$.
   d. Energy is released when two hydrogen atoms form a hydrogen molecule.

1. a. 4.184 J/cal
   b. 0.2390 cal/J (1 cal/4.184 J is not a unitary rate.)
2. 4.184×10^5 J
3. 1.05×10^3 kJ
4. Water is a very common liquid, and we know how to get it in a very pure form. Also, it is inexpensive.
5. 45 calories (If 30% of the 150 calories comes from carbohydrates, then the answer is simply 0.3 × 150 cal.)
6. a. true
   b. false
   c. false
   d. true
(Students may not be able to answer this question, given the presentation in *this* chapter.)

Heats of Reaction

CONCEPTS

■ the energy of chemical bonds
■ measuring heats of reaction
■ determining available calories in food
■ useful chemical work
■ entropy and energy changes

Now you know how energy amounts are calculated and what happens at the microscopic level when heat is transferred from one object to another. You can study energy changes that take place during chemical reactions and use those energy changes to predict whether a particular reaction will occur, how much energy one can get from a particular reaction, and how much of that energy is available to do useful work.

17-7 Making and Breaking Bonds

In Chapter 12 you learned that energy is released to the surroundings when any chemical bond forms. Such energy-releasing processes are exothermic. In order to break those chemical bonds, the same amount of energy must be supplied from the surroundings. Such energy-absorbing processes are endothermic. The energy transferred between molecules undergoing a chemical change and the matter surrounding them may take many forms. When wood burns in a fireplace, you see light and feel heat as a result of the increased kinetic energy of the surrounding air molecules and the infrared radiation. You may even hear pleasant popping and crackling sounds as the fire burns. All of these forms of energy result from the exothermic reaction taking place as wood combines with oxygen in the air to form water, carbon dioxide, and other simple molecules.

The cellulose molecules in wood were constructed in living cells from carbon dioxide and water. Photosynthesis, the complex chemical process by which plants build large molecules like cellulose, sugar, and starch, is an endothermic reaction that uses sunlight as the energy source.

The sun is the primary source of energy on Earth. Hydroelectric power is possible because energy from the sun evaporates water which later falls as rain at high altitudes. As the water runs downhill, some of the sun's energy is recaptured by forcing the water to turn a turbine or waterwheel. Scientists commonly say that the sun's energy is stored in the water. In a similar way they often say that the sun's energy is stored in complex molecules put together in plant cells. Energy from the sun is used in the endothermic reaction of photosynthesis, and it is retrieved in the form of heat and light when wood is burned or foods are digested and metabolized in the body. Coal and petroleum are the fossil remains of plant and animal matter, and these substances can be burned to obtain energy that came from the sun millions of years ago.

Biologists and biochemists frequently talk about energy being stored in chemical bonds, and they may even say that energy is released when these "high energy bonds" are broken. It is easy to see how the language developed, but it can be confusing to people who know that energy is always absorbed when bonds are broken. It is only when bonds form that energy is released to the surroundings.

The apparent contradiction between the way biologists and chemists talk about energy changes in chemical reactions is due to the fact

Do You Know?

At our present rate of consumption, it is believed that all petroleum on Earth will be consumed by the middle of the 21st century. Coal reserves are much larger, but also limited.

that biologists frequently focus attention on only half of what is taking place. Changes that do not involve both the breaking and formation of bonds are rare. Virtually all chemical changes involve the breaking of bonds that already exist and the formation of new bonds to form new substances. It is possible to calculate the energy associated with the overall chemical change, but it is not possible to calculate the energy associated with only one part of the change.

17-8 Standard Heats of Formation

Because overall changes in energy can be calculated but the energy of particular bonds cannot, it is necessary to select a reference point and describe bond energies and the like as being greater than or less than that reference point. The situation is similar to that of measuring temperature.

To create a temperature scale, a particular temperature is arbitrarily assigned a value of zero. For the Celsius scale, the zero point is the temperature at which water freezes or melts. A positive temperature on the scale simply indicates that the average kinetic energy of molecules is greater than the average kinetic energy of water molecules at 0°C. A negative temperature indicates that the average kinetic energy of the molecules is less than the average kinetic energy of water molecules at 0°C.

A similar tactic is used to describe energy changes. Scientists have agreed to assign a zero value to the energy of all elements in their standard state. The precise meaning of standard state is subtle and you need not worry about it in this course. You may think of the standard state of an element as the conditions under which you would ordinarily find the pure element. For example, oxygen ordinarily exists as a diatomic gas at a pressure of one atmosphere and a temperature of about 25°C. That is its standard state. Aluminum ordinarily exists as a crystalline solid, and that is its standard state. It is important to keep in mind that energy is involved in any temperature change or pressure change, so the standard state of an element always applies to a particular temperature and pressure as well as a particular physical state (solid, liquid, or gas).

Once a starting point for measuring energies is accepted, the energy associated with any chemical change can be described. For example, you might ask how much energy is involved when hydrogen gas burns in oxygen to form water.

Both hydrogen and oxygen are gases, so you cannot burn them in a calorimeter like the one shown in Figure 17-3. They would not stay in the plastic foam block. Besides, the plastic foam would ignite, ruining the calorimeter as well as the experiment. However, hydrogen and oxygen gases can be forced into a very strong steel container and maintained at a constant pressure of one atmosphere. If two wires are built into the container, an electric spark can be made to jump between the wires to ignite the mixture inside the metal container. The hydrogen and oxygen will react rapidly, and the energy released from the exothermic reaction will warm the container and its contents. If this is done inside a plastic foam block filled with water, the tempera-

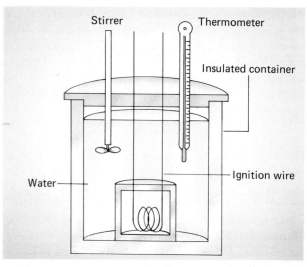

Figure 17-4 The photograph shows an actual bomb calorimeter and the equipment used with it to make careful measurements of heat. The diagram shows the essential features of the calorimeter.

ture change can be measured, and the heat generated by the reaction can be calculated in a manner similar to that shown in Example 17-2.

Figure 17-4 shows a schematic drawing of such a calorimeter and a photograph of the equipment. As you can see from the photograph, the actual equipment may be quite complex. The complex equipment is needed to ensure that all of the energy produced is captured and measured. However, the principle of the measurement is quite simple. Much of what one must learn in order to work as a chemist or other scientist is how to solve the practical problems that stand in the way of making precise measurements needed to understand natural events.

The equation for the reaction just described is

$$2H_2(g) + O_2(g) \longrightarrow 2H_2O(l)$$

Notice that the equation specifies the physical state of both reactants and products. It is important to have this information when using equations to describe energy changes, because the amount of energy that is measured depends on the physical state of the reactants and products. For example, if this reaction were carried out at a higher temperature and the water formed in the reaction remained in the gas phase, the heat released by the reaction would be less. It would be less by exactly the amount of energy released when gaseous water condenses to form a liquid, the heat of vaporization.

As you know, the amount of heat released during an exothermic reaction, such as the burning of hydrogen gas, depends on the amount of hydrogen that burns. Consequently, heat generated by reactions is always described in unitary rates that indicate the amount of energy released or absorbed for some given amount of matter. The heat of reaction may be described in terms of mass, as specific heat, heat of fusion, and heat of vaporization were described in Tables 17-1 and 17-2. However, chemists normally describe chemical changes with equations which are expressed in terms of *moles* of reactants and products. The equation for the combustion of hydrogen gas indicates that two moles of hydrogen gas react with one mole of oxygen gas to form two moles of water. If this reaction takes place at 25°C, the temperature of the water in the calorimeter would rise, and calculations would show that 572 kilojoules of energy were given off during the reaction. This can be shown in the equation for the reaction as follows:

$$2H_2(g) + O_2(g) \longrightarrow 2H_2O(l) + 572 \text{ kJ}$$

The equation indicates that 572 kilojoules of energy are transferred to the surroundings when *two* moles of water are formed from its elements in their standard state. Divide this value by two, and you have a unitary rate indicating how much energy is released when the bonds in one mole of hydrogen gas and one-half mole of oxygen gas are broken and new bonds are formed to make *one* mole of liquid water.

There are two terms used to describe this energy. It is energy observed during a chemical reaction, so it is called the **heat of reaction**. However, the reaction discussed is one in which a compound is formed from its elements when those elements are in their standard state. This particular heat of reaction is called the standard **heat of formation** of the compound. Standard heats of formation have been measured or calculated for many compounds, and they are recorded in reference books. The values are like this one for liquid water:

$$\Delta H_f^\circ = -286 \text{ kJ}$$

This symbolism is packed with information, and you need to know what it means. First, the delta, Δ, refers to a *change* in energy. The particular change described here is the change in energy when the bonds that exist between hydrogen atoms and oxygen atoms in their standard states as diatomic molecules of a gas are broken, and these atoms form new bonds as water molecules and weak bonds that hold the molecules together as a liquid.

The H in the symbolism stands for **enthalpy**, the term used to describe the heat measured when changes take place at constant pressure. The superscript, $^\circ$, is used to indicate a standard value, and the subscript, $_f$, is used to indicate a "formation" reaction. In other words, the superscript and subscript indicate that the energy recorded is for the reaction in which a compound is formed from its elements in their standard states.

Even though the unit of mole normally is not recorded in tables, it is understood. In other words, the energy recorded is the energy that would be measured when *one mole* of liquid water is formed. The complete unitary rate is -286 kilojoules per mole.

Finally, the negative sign in front of the energy value indicates that the reaction is exothermic. Energy is transferred from the metal container in which the reaction took place to the water surrounding it. If energy were transferred from the surroundings to the reaction vessel, a positive sign would be used to indicate an endothermic reaction had taken place.

If students do not do so in laboratory, measure the heat of a reaction as a demonstration. The reaction between hydrochloric acid and aqueous ammonia to form aqueous ammonium chloride works well.

Teachers and textbook authors often forget how much of our technical language is new and incomprehensible to students. Make a habit of checking students' understanding by asking them to explain the meaning of representations like this one.

CHEM THEME

The amount of energy released in an exothermic reaction is proportional to the amounts of the reactants.

17-9 Using Known Energies to Calculate Other Values

Not all heats of reaction can be calculated directly. For example, scientists cannot measure the energy change that takes place during photosynthesis. They do not know how to make that reaction take place in a beaker or calorimeter, and it takes place in plants too slowly to measure accurately. Fortunately, energies that cannot be determined experimentally can be calculated from energies that can be determined experimentally.

This can be illustrated using the water example. As indicated above, water can be formed as a liquid or as a gas. Simply change the temperature at which the reaction takes place. The standard heat of formation for liquid water is -286 kilojoules per mole as indicated above. The standard heat of formation for gaseous water is -242 kilojoules per mole. The difference between these two values is the same as the heat of vaporization of water at $25°C$, 44 kilojoules per mole. The $+44$ kilojoules per mole value is the energy required to change water from a liquid to a gas. Energy must be added to the system. If water were changed from a gas to a liquid, the value would be -44.0 kilojoules per mole, indicating that energy would be transferred from the water to its surroundings.

Imagine the formation of liquid water taking place in two steps. In the first step, imagine the gaseous elements reacting to form gaseous water, as indicated by the following equation:

$$H_2(g) + 1/2\,O_2(g) \longrightarrow H_2O(g) \qquad \Delta H = -242\,kJ/mol$$

After the gaseous water forms, imagine cooling the gas until it condenses to form a liquid, as indicated by this equation:

$$H_2O(g) \longrightarrow H_2O(l) \qquad \Delta H = -44\,kJ/mol$$

These two equations and the energy change associated with them can be added to give the equation and energy associated with the formation of liquid water from its elements like this:

$$H_2(g) + 1/2\,O_2(g) \longrightarrow H_2O(g) \qquad \Delta H = -242\,kJ/mol$$
$$H_2O(g) \longrightarrow H_2O(l) \qquad \Delta H = -44\,kJ/mol$$
$$\overline{H_2(g) + 1/2\,O_2(g) \longrightarrow H_2O(l) \qquad \Delta H = -286\,kJ/mol}$$

Notice that the mole of gaseous water appears on both the right and left of the equations being added. When the two are added, this common term is subtracted from both sides in the same way that ions in ionic equations were subtracted to produce a net ionic equation. If you know the enthalpy change for a series of reactions that can be added together to describe a reaction of interest, you can calculate the enthalpy change you want by adding the equations and enthalpies you know.

17-10 Counting Calories

The tables used to "count Calories" when on a diet represent one practical application of the principles just discussed. The chemical reactions that take place when food is digested and metabolized are complex, and they are not totally understood. However, the starting materials and products are known. Sucrose, ordinary table sugar, is $C_{12}H_{22}O_{11}$ and when it is metabolized, the final products are carbon dioxide, water, and energy. The overall reaction is the same as sugar burning in oxygen:

$$C_{12}H_{22}O_{11}(s) + 12O_2(g) \longrightarrow 12CO_2(g) + 11H_2O(l) + energy$$

The heat of vaporization for water at $25°C$ is 44 kJ/mol. Table 17-2 gives the heat of vaporization for water at $100°C$ as 40.7 kJ/mol.

When sugar is burned in a calorimeter like the one shown in Figure 17-4, the heat of reaction is 5647 kilojoules per mole or 1350 kilocalories per mole. Dividing these values by the molar mass of sucrose gives a value of 16.51 kilojoules or 3.947 kilocalories of energy released for each gram of sugar burned. The kilocalorie, you will recall, is the Calorie commonly used to describe the energy obtained from foods.

Sucrose is one of many compounds called sugars, and all sugars belong to the larger class, carbohydrates. Most foods contain one or more simple sugars with the formula, $C_6H_{12}O_6$. All of these sugars burn to produce carbon dioxide and water, releasing about 4 Calories of energy per gram in the process. Starch, another carbohydrate, also produces about 4 Calories of energy for each gram that the body burns.

Although the body can digest sugars and starch, it cannot digest cellulose, another common carbohydrate. This "fiber" in food will burn in oxygen or air, but it does not "burn" in the body. It passes through the digestive tract essentially unchanged. Consequently, burning a piece of celery or a peanut in a calorimeter will not give an accurate measure of the energy provided when the food is digested and metabolized in the body.

In practice, foods are dried to remove the water in them. They are then analyzed to determine the mass of digestible carbohydrate, fat, protein, and fiber. The mass of each digestible component is then multiplied by the heat of combustion for that component, and the separate heats are added to calculate the total energy available from the food. **Heat of combustion** is the energy released when a substance is burned or combusted. The number of Calories that are available from various foods are estimated using the procedure just described.

Notice that the calculations are for the energy available. Individuals vary in their body chemistry, and an individual's chemistry varies from time to time. Illness or emotional stress affects how much food is actually digested and absorbed by the body. Clearly, counting the Calories available in foods only provides an approximation of the energy the body actually uses.

The enormous amount of energy available in foods can be illustrated by burning a single peanut or a sugar cube in a crude calorimeter. Place water to be heated in a small juice can suspended inside a large juice can which acts as a heat shield. Ignite the food and place it under the small can of water. Measure the change in temperature and calculate the heat gained. (A sugar cube can be lit if cigarette ash is rubbed onto the face before lighting.)

17-11 Energy Available to Do Work

As you can see from the discussion of food Calories, there may be a difference between the energy available in foods and the energy actually utilized by the body. Similarly, when wood is burned in a fireplace or natural gas is burned in a furnace, only part of that energy heats the home. The rest goes up the chimney, does work on the air to expand its volume, or is wasted in some other way. Just how much energy is wasted depends on how clever people are in designing stoves, boilers, electric generators, and other machines used to convert energy into useful forms.

Most machines are designed to do work—to apply a force through a distance. People want to move things. Automobiles are machines that move themselves and their occupants from one place to another. Refrigerators are machines that move Freon® or some other substance

CHEM THEME

When a chemical reaction occurs, atoms, mass, and energy are conserved. The atoms are rearranged as old bonds are broken and new bonds are formed.

Figure 17-5 The beaker on the left shows a solution of copper(II) sulfate when a zinc strip is first placed in the beaker. The beaker on the right shows the solution when the reaction is complete. Notice the change in color of the solution and the copper "mud" formed.

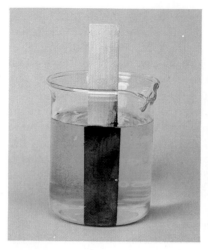

through the coils inside and outside of the box being cooled. Electric generators are machines that move electrons through wires. All of these machines use a source of energy to do work. None is capable of converting all of the energy available from the fuel into the useful work they are designed to do. Some of the energy is lost as heat to the surroundings.

Two conditions are necessary to get useful work from a chemical reaction. First, the reaction must be spontaneous. If a continuous input of energy must be supplied to make the reaction take place, it can never be used to do useful work.

If a piece of zinc metal is placed in a solution of copper(II) sulfate, a reaction takes place, as shown in Figure 17-5. The zinc metal dissolves, and copper metal forms as a brown "mud" that falls to the bottom of the container. The net ionic equation for the reaction is

$$Zn(s) + Cu^{2+}(aq) \longrightarrow Zn^{2+}(aq) + Cu(s)$$

In effect, two electrons are transferred from each atom of zinc metal to each copper(II) ion in solution. This change was spontaneous. No outside influence was required to make it happen, so the reaction has the potential to do useful work.

The reverse of this reaction would be the reaction of copper metal with a solution containing zinc ions. If you place a piece of copper metal in a solution of zinc sulfate, nothing happens. You could force the reaction to take place by placing a strip of copper in the solution along with a strip of zinc and attaching the two metals to a battery, as shown in Figure 17-6. The battery will provide the energy to move electrons from the copper strip to the zinc, and the following reaction will take place:

$$Cu(s) + Zn^{2+}(aq) \longrightarrow Cu^{2+}(aq) + Zn(s)$$

However, it does not take place spontaneously, so it could not be used to do work.

The reaction between zinc metal and copper(II) ions is spontaneous so it has the potential to do work. In fact, work is done in this reaction. Matter (electrons) moves from one place to another, and moving matter is a form of work. However, as this reaction took place, the movement of the electrons did nothing you could call useful work. It is

Figure 17-6 The reverse of the reaction shown in Figure 17-5 can be forced to take place by connecting copper and zinc to a battery as shown.

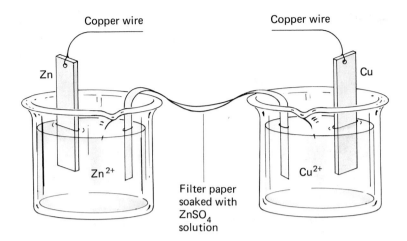

Figure 17-7 The zinc and copper(II) sulfate reaction is arranged to do work.

capable of doing work, but it does no useful work under the conditions shown in Figure 17-5. All of the energy was produced in the form of heat.

A second condition required to get useful work from a chemical reaction is to figure out a way that the work done can be useful. This is rather easy to do in the case of zinc reacting with copper(II) ions.

Instead of placing the zinc metal directly in the solution of copper(II) ions, place it in a solution containing zinc ions. Place a copper strip in a separate beaker containing the solution of copper(II) ions. Now join the two solutions with a piece of paper soaked in one of the solutions. This arrangement is shown in Figure 17-7.

Since the zinc metal is no longer in contact with the copper(II) ions, there is no way for the electrons to move directly from one to the other, but you can make a path for the electrons by connecting the zinc and copper strips with a copper wire. If the wire is connected to an electric motor so the electrons flow through the motor, you can use the energy to do work. The complete arrangement is shown in Figure 17-8. Batteries operate in a similar manner to produce useful work as you will discover in Chapter 21.

If you measure the voltage produced from a chemical cell, the voltage obtained will be lower than the $E°$ value in a table because energy is used to operate the meter used to make the measurement. The lower the current required by the meter, the less the discrepancy.

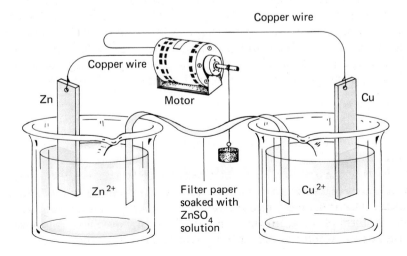

Figure 17-8 This copper-zinc battery is doing work. As electrons pass through the motor, the motor turns to lift the weight.

Even the arrangement in Figure 17-8 will not convert all of the energy of the reaction into useful work. There will be heat losses due to the electrical resistance in the wire, and frictional losses in the motor. The slower the current flows through the wire and the more efficient the motor, the more work that is done.

17-12 Electrical Work and Free Energy

The zinc-copper reaction was used to illustrate how energy can be converted into useful work. It is relatively easy to use such a system to measure the maximum possible work that can be obtained from the reaction. This maximum possible work is called the **Gibbs Free Energy** and is normally symbolized by G. It is named for J. Willard Gibbs, an American scientist.

Like all energies that have been discussed, there is no way to measure the free energy of a system. You can only calculate the *change* in free energy associated with a reaction. Such changes in free energy are described by the symbol, ΔG.

Like changes in enthalpy, changes in free energy can be calculated for reactions that cannot be done easily in the laboratory by using tables reporting experimental values for reactions that can be done. Equations that describe the possible reactions are added to obtain the equation for the reaction that is not possible. Such calculations are often done to predict whether a particular reaction can occur spontaneously.

Remember that ΔG represents the maximum amount of work that can be done by a chemical reaction, and no work will be accomplished unless the reaction occurs spontaneously. Reactions that occur spontaneously will always transfer energy to the surroundings and it can be used to do useful work. According to the sign convention discussed earlier, when energy is transferred from a chemical system to the surroundings, the energy change is described as negative. The ΔG for any reaction that occurs spontaneously will be negative. If the ΔG calculated for a chemical reaction is positive, the reaction described cannot occur spontaneously. The reverse reaction can occur spontaneously.

What if the value of ΔG is zero? In that case, neither the forward nor the reverse reaction described by an equation can take place spontaneously. In other words, the chemical system must be in equilibrium. Knowing this condition, it is possible to calculate the concentration of reactants and products that will produce an equilibrium. As you can see, the Gibbs Free Energy is a powerful tool if you know how to use it. You will be able to master the chemistry concepts in this text without it, so the actual calculations will be left for a later course.

Gibbs Free Energy is the maximum possible work that can be done at constant pressure. It excludes work due to expansion. The Helmholtz function describes the maximum possible work. It pertains to systems at constant volumes rather than at constant pressures. The Gibbs function is discussed here because students are usually interested in chemical systems at constant pressure and in the conditions of equilibrium that follow from them.

17-13 Why Changes Occur Spontaneously

You may wonder why some changes occur spontaneously while others do not. Why, for example, does heat flow from warm bodies to cool bodies and not the other way around? The answer to this question is randomness.

Nature behaves in a random manner. Molecules of a gas move at random, bumping into one another and transferring energy from one to another in the process. The air filling a bottle occupies all of the space in the bottle because the motion of the molecules is random.

Just as the location of molecules is randomly determined, so is the distribution of energies. As it has been said several times, at any temperature some molecules are moving slowly and some are moving rapidly. The speed and direction of each molecule changes as a result of the random collisions among molecules. The lowest possible speed is always zero. As the temperature of a gas increases, so does the range of speeds of the molecules.

Figure 17-9 shows the number of molecules plotted against the speed of the molecule. The yellow area of the graph shows the distribution at a low temperature, and the blue area shows the distribution at a higher temperature. The higher the temperature, the larger the range in energy of the molecules. There are more high energy molecules at higher temperatures.

The transfer of heat from hot objects to cold objects can be explained in two ways. The first explanation requires that you keep in mind that temperature is a measure of the average kinetic energy of matter, $1/2\,mv^2$, a relationship introduced earlier in this book. If matters are simplified by limiting the discussion to molecules of the same mass, hotter molecules are simply speedier molecules. What happens when a speedy object and a slower one collide? The slower one speeds up and the faster one slows down. Their average kinetic energies or temperatures come closer together. There is no way that objects could collide so that the slow one goes even slower and the fast one goes even faster. Energy is transferred from the hotter object to the colder one.

Another way to think about this change is to recognize that spontaneous processes always maximize randomness. Just as the molecules of a gas are randomly distributed in a bottle rather than being organized at one end, energy tends to become randomly distributed in nature. When a hot object first comes in contact with a cold one, the energy is not randomly distributed. It is concentrated in the hot object. As that heat energy becomes randomly distributed, some of it is

This point will require some discussion or experimentation. Using billiard balls or other spheres of the same mass, have students imagine or demonstrate collisions between fast and slow balls. What happens to their relative speeds after collision? The simplification made earlier must be kept in mind. What is said would be true only for particles of equal mass. If particles of different mass collide, it is possible for a slow-moving massive particle to transfer momentum to a fast-moving particle of lower mass and increase its velocity.

Figure 17-9 This graph shows the distribution of molecular speeds at two different temperatures.

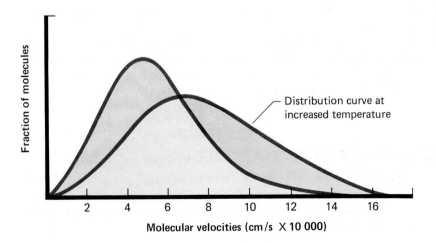

Fraction of molecules

Distribution curve at increased temperature

2   4   6   8   10   12   14   16

Molecular velocities (cm/s X 10 000)

Figure 17-10 In a system composed of a single atom, no distribution of energy is possible.

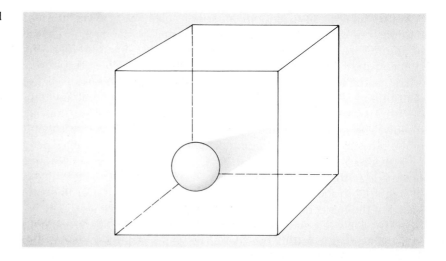

There are numerous ways to distribute energy between two atoms, A and B. One way would be to have 100% of the available energy on atom A and 0% on atom B. Another way the energy could be distributed would be to have 99% on atom A and 1% on atom B, etc.

transferred to the molecules in the colder object. The temperature of the colder object rises, and the temperature of the hotter object falls.

This randomness or disorder in matter is directly related to the number of ways that energy can be distributed within matter. Consider the simplest possible system, one atom of matter, as shown in Figure 17-10. There is no way to distribute the energy. The energy of the system is the energy of that one atom. Now consider the system shown in Figure 17-11 that has two atoms. Some of the energy is due to the motion of one atom and some is due to the motion of the other. The energy is distributed, and there are several ways to do it. By increasing the number of particles, the number of ways the energy of the system can be distributed is increased. The more atoms you have, the more ways there will be to distribute the energy among the atoms.

The number of ways energy can be distributed also is increased by increasing temperature. Figure 17-9 shows that the range of kinetic energies of particles increases when the temperature increases. Thus, increasing temperature increases the number of possible ways that energy can be distributed.

When atoms bond together to form molecules, some energy is in the form of vibrations and rotations within the molecule. There are more

Figure 17-11 In a system composed of two atoms, the energy is distributed between the two atoms.

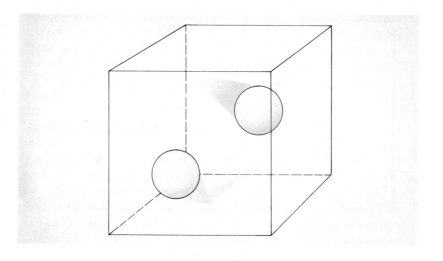

ways to distribute energy in a collection of molecules than in a collection of the same number of atoms.

When molecules are bound together in a solid, the molecules can vibrate, but they cannot move in space. The number of ways that energy can be distributed is lower than when the molecules are in a liquid where they can move past one another. Similarly, when molecules are in a gas, they have even more freedom of motion, and the number of ways that energy can be distributed within a gas is greater than in a liquid.

The term that scientists use to describe the way energy can be distributed in a system is **entropy**, and it is symbolized by S. Spontaneous changes always result in greater entropy—there is an increase in the number of ways the energy can be distributed. The greater entropy can occur because the change produces more particles, or because the change is from solid to liquid or from liquid to gas, or because the change raises the temperature.

As with all other energy terms, only *changes* in entropy can be measured. Entropy has units similar to those used for specific heat: joules per mole-kelvin. It describes the dispersal of energy in terms of joules per mole of particles and per kelvin. This differs from enthalpy and Gibbs free energy, the other energy terms that have been discussed. Both ΔH and ΔG are expressed in joules per mole. Temperature is not involved. In order to combine terms, the energy associated with a change in entropy must be multiplied by the absolute temperature at which the measurements are taken in order to have quantities that can be added. The relationship between ΔG, ΔH and ΔS is

$$\Delta G = \Delta H - T\Delta S$$

Some discussion of this equation may help you see what it is describing. Recall that ΔG is the free energy associated with any change. If the change is spontaneous, ΔG is less than zero (negative). It also has been said that all spontaneous changes result in an increase in entropy. Energy becomes more widely distributed throughout the universe.

Both terms on the right are involved in this overall increase in entropy. The first term, ΔH, states whether heat is transferred from the system being discussed to the surroundings or vice versa. A negative ΔH value indicates that energy is transferred to the surroundings where it can be distributed over all matter in the universe. A negative ΔH value must correspond to an increase in entropy of the universe.

Whereas a negative ΔH value is related to an increase in entropy of the surroundings, ΔS describes the entropy change of the system. It has a positive value if the entropy of the system increases. $T\Delta S$ also will be positive, and an increase in entropy results in a negative contribution to ΔG.

Spontaneous changes are more likely to be exothermic (negative ΔH) and are more likely to result in an increase in entropy of the system (positive ΔS). However, neither condition is essential. Exothermic reactions that result in a lower entropy for the system can take place spontaneously if the absolute value of ΔH is greater than the absolute value of $T\Delta S$. Similarly, endothermic reactions may take place spontaneously if the absolute value of $T\Delta S$ is greater than the absolute value of ΔH.

Entropy is often described as the amount of disorder. Strictly speaking, it is the possibility for disorder. It is related to the number of ways that energy can be distributed rather than the number of ways the energy is actually distributed at any point in time.

It may be useful to compare entropy to specific heat. In order to calculate an energy term, both specific heat and entropy must be multiplied by a temperature term. However, there is one important difference. In the case of specific heat, the temperature term is a change in temperature. In the case of entropy, the temperature term is the absolute temperature of the system under consideration.

1. endothermic
2. boiling away the water
3. enthalpy
4. $C(s) + 2H_2(g) \longrightarrow CH_4(g)$
$\Delta H = -74.9$ kJ
$\Delta H_f^{\circ} CH_4 = -74.9$ kJ/mol
5. Parts of the foods that will burn in a calorimeter do not get digested and metabolized in the body.
6. Some energy is always lost to the surroundings as heat. One hundred percent efficiency cannot be attained.
7. The reaction will not occur spontaneously.
8. zero
9. a. It is spontaneous.
   b. It increases.
10. negative
11. a. 783.0 kJ
    b. 3.28 kJ
    c. 4.11 kJ
12. a. unburned match head
    b. ear of corn
    c. unused flashbulb
13. a. endothermic
    b. positive

Review and Practice

1. Is the melting of ice an endothermic change or an exothermic change?

2. Which requires more energy, melting a ton of ice or boiling away a ton of water?

3. What is the term used to describe the heat produced or absorbed when a chemical reaction takes place at constant pressure?

4. Methane gas, CH_4, cannot be formed by the direct reaction of graphite (the standard state of carbon) and hydrogen gas. However, the reactions described by the following equations can be done in a calorimeter and the heats of reactions can be measured. Use these heats of reaction to calculate the heat of formation of methane.

$$C(s) + O_2(g) \longrightarrow CO_2(g) \qquad \Delta H = -393.5 \text{ kJ/mol}$$
$$2H_2(g) + O_2(g) \longrightarrow 2H_2O(l) \qquad \Delta H = -571.7 \text{ kJ/mol}$$
$$CH_4(g) + 2O_2(g) \longrightarrow CO_2(g) + 2H_2O(l) \qquad \Delta H = -890.3 \text{ kJ/mol}$$

5. Why is it inaccurate to burn foods in a calorimeter and measure the heat to determine the energy that your body gets from the food when it is digested and metabolized?

6. The heat produced when a fuel burns is not the energy available to do useful work. Why?

7. ΔG for a reaction has a positive value. What does that allow you to say about the reaction?

8. What is the value of ΔG when the reaction between hydrochloric acid and sodium hydroxide reaches equilibrium?

9. The value of ΔG for burning any fuel is negative.
   a. What does this tell you about the spontaneity of the reaction?
   b. What happens to the entropy of the universe when this reaction takes place?

10. When iron is exposed to air and water, the iron rusts. Is the value of ΔG for this reaction positive or negative?

11. Copper metal can be obtained from the reaction of copper(II) oxide and hydrogen gas.

$$CuO(s) + H_2(g) \longrightarrow Cu(s) + H_2O(l) \qquad \Delta H = +130.5 \text{ kJ}$$

Calculate ΔH for each of the following:
   a. 6 moles of CuO reacting
   b. 2 grams of CuO reacting
   c. 2 grams of Cu that is formed

12. Which contains more stored (potential) energy:
    a. a burned or unburned match head?
    b. an ear of corn or the chemicals a plant needs to make the ear of corn?
    c. a used or unused camera flashbulb?

13. In photosynthesis, plants convert the sun's energy into sugars in the reaction

$$6CO_2(g) + 6H_2O(l) \longrightarrow C_6H_{12}O_6(s) + 6O_2(g)$$

   a. Is this reaction exothermic or endothermic?
   b. What is the sign of the ΔH for this reaction?

Numbers in red indicate the appropriate chapter sections to aid you in assigning these items. Answers to all questions appear in the Teacher's Guide at the front of this book.

Summary

■ The study of heat and its transformations is called thermodynamics. Heat refers to the total amount of energy in a substance; temperature is a measure of the average kinetic energy of molecules within the substance. They are related, but they are not the same thing.

■ Specific heat is the amount of heat required to raise the temperature of a substance. Specific heats are normally recorded in joules per kilogram-kelvin or joules per mole-kelvin. The specific heat of a substance changes slightly with temperature, and physical state.

■ The heat required to change a substance from solid to liquid without changing the temperature is called the heat of fusion. The heat required to change a substance from liquid to gas without changing the temperature is called the heat of vaporization.

■ Energy can be stored by raising the temperature of a substance, by melting it, or by changing it to a gas. The energy can be retrieved by cooling the substance, freezing it, or condensing it.

■ At the molecular level, energy takes the form of kinetic energy of the molecules or the energy of bonding that holds particles together.

■ The heat content of a substance cannot be determined experimentally. Only changes in heat content, ΔH, can be calculated. Heat is measured in reference to a standard. The calorie is the heat required to raise one gram of water one degree Celsius. One calorie is 4.184 joules.

■ In order to calculate heats of reaction, the heat content of all elements in their standard states are assigned values of zero. Heat measurements are done in well-insulated containers called calorimeters. Heats of reaction that cannot be determined experimentally can be calculated from heat of reactions that can be so determined.

■ The energy available to do useful work is called the Gibbs Free Energy, ΔG, and it is not the same as the enthalpy, ΔH, of the change. All spontaneous changes have a negative ΔG. An increase in entropy is associated with an increase in disorder of the universe, made possible because of an increase in the ways that the total energy can be distributed.

■ The most common uses of thermodynamics are to calculate the amount of energy that can be obtained in a chemical reaction and to predict whether a reaction can occur spontaneously.

Chemically Speaking

calorie
Calorie
calorimeter
enthalpy
entropy
Gibbs Free Energy
heat of combustion

heat of formation
heat of fusion
heat of reaction
heat of vaporization
kilocalorie
specific heat
thermodynamics

Review

1. At the microscopic level, what happens to the energy added to change a solid to a liquid or to change a liquid to a gas? (17-4)

2. Why is it true that a cup of boiling water contains less heat than a large iceberg? (17-1)

3. How are calories and Calories different? (17-5)

4. Dry ice is solid carbon dioxide. It does not "melt" but instead turns from a solid into a gas in a process called sublimation.
   a. Is this change exothermic or endothermic?
   b. What sign would the ΔH value have? (17-8)

5. A one ounce bar of chocolate contains about 147 Calories. Is this the amount of energy your body gets every time you eat one? Explain. (17-10)

6. The ΔG value for a reaction has a negative value. What does that tell you about the reaction? (17-12)

7. What would the ΔG value be for a reaction at equilibrium? (17-12)

8. What two conditions are necessary for a system to produce useful chemical work? (17-11)

9. Do endothermic reactions have to result in a decrease in entropy of the system? Explain. (17-13)

10. Energy is conserved in an exothermic reaction. Where does the energy come from and where does it go? (17-7)

Interpret and Apply

1. If steam at 100°C accidentally passes over your hand, you are likely to be burned more severely than by water at that same temperature. Use the concept of heat of vaporization to explain why. (17-3)

2. Many solar homes store heat by letting the sun melt a solid material like Glauber's salt, sodium sulfate decahydrate, $NaSO_4 \cdot 10H_2O$, which has a melting point near normal room temperature. Even small amounts of such solids can store large quantities of heat. Use the concept of heat of fusion to explain how. (17-3)

3. Why would the heat produced by the reaction described below not be the standard heat of formation of water? (17-8)

$$H_2(l) + 1/2\ O_2(l) \longrightarrow H_2O(l)$$

4. Under standard conditions, which would have a higher entropy, a mole of carbon dioxide molecules or a mole of carbon atoms? Why? (17-13)

5. When ammonium nitrate, NH_4NO_3, dissolves in water, the beaker containing the solution gets cold. Explain why and indicate the sign of the ΔH for the reaction. (17-8)

6. The following reaction is spontaneous:

$$CaO(s) + SO_3(g) \longrightarrow CaSO_4(s)$$

Indicate the sign of each value: (17-13)
a. ΔG c. ΔH
b. ΔS

7. Consider the change in state of water to ice: (17-3)

$$H_2O(l) \longrightarrow H_2O(s)$$

a. Is the reaction spontaneous at 298 K?
b. What is the sign of the ΔH value?
c. How could the reverse reaction occur?

8. The following reaction is exothermic: (17-8)

$$2Na(s) + Cl_2(g) \longrightarrow 2NaCl(s)$$

a. What is the sign of ΔH for this reaction?
b. What is the sign of ΔH for the reverse reaction?
c. Is the heat content of the product greater or less than that of the reactants?

9. For the following reaction: (17-8, 17-13)

$$S(s) + O_2(g) \longrightarrow SO_2(g) \qquad \Delta H = -297\ kJ$$

a. Is this reaction exothermic or endothermic?
b. What is the value of ΔH for the reverse reaction?
c. How does the heat content of 1 mole of SO_2 compare to that of 1 mole of S plus 1 mole of O_2?

10. Write chemical equations for the following changes in state: (17-3)
a. the freezing of water ($\Delta H = -6.01\ kJ/mol$)
b. the condensation of water vapor ($\Delta H = -40.7\ kJ/mol$)

11. Translate the following statements into chemical equations. (17-8, 17-10)
a. The heat of formation of magnesium oxide, MgO, is −607.1 kJ/mol.
b. The heat of formation of ammonia, NH_3, is −46.0 kJ/mol.
c. The heat of combustion of octane, C_8H_{18}, is −5470.6 kJ/mol.

Problems

1. A solar heating specialist is considering paraffin as a possible solar heat collector. How many kilograms of paraffin will be necessary to collect as much energy as 4.78×10^3 kg of water? (17-2)

2. The heat of combustion for ethane is −1559.8 kJ/mol in the following reaction: (17-2, 17-10)

$$C_2H_6(g) + 7/2O_2(g) \longrightarrow 2CO_2(g) + 3H_2O(g)$$

How many kilograms of water could be heated from 10°C to 50°C in a calorimeter by the above reaction? (Use 4180 J/kg·K for the specific heat of water.)

3. When wood is burned, it releases 18.8 kilojoules of energy per gram. Heating oil used to heat some houses releases 47.3 kilojoules of energy per gram. How many kilograms of wood would you need to burn to supply the same amount of heat provided by 400 kilograms of oil? (17-10)

4. The heat of reaction of carbon dioxide is different in each of the following reactions: (17-9)

$$C(diamond) + O_2(g) \longrightarrow CO_2(g) \quad \Delta H = -395.4\ kJ$$
$$C(graphite) + O_2(g) \longrightarrow CO_2(g) \quad \Delta H = -393.5\ kJ$$

a. Find the heat of reaction for the manufacture of diamond from graphite.
b. Is the heat absorbed or given off as graphite is converted into diamond?

5. The heats of reaction for the following reactions can be measured directly: (17-9)

$$1/2N_2(g) + 1/2\ O_2(g) \longrightarrow NO(g)$$
$$\Delta H = +90.4\ kJ/mol$$
$$1/2N_2(g) + O_2(g) \longrightarrow NO_2(g)$$
$$\Delta H = +33.8\ kJ/mol$$

a. Find the heat of reaction for the combustion of nitric oxide, NO:

$$NO(g) + 1/2O_2(g) \longrightarrow NO_2(g)$$

b. Is the reaction exothermic or endothermic?

6. The heat of formation of sodium hydroxide, NaOH, is −427 kJ/mol. Calculate the heat of formation (ΔH_f) for each of the following: (17-8)
   a. 1.37 moles of NaOH
   b. 47.0 grams of NaOH

7. The heat of formation for phosphorous acid, H_3PO_3, is −972 kJ/mol. Calculate the heat of formation for each of the following: (17-8)
   a. 16.7 moles of H_3PO_3
   b. 691 grams of H_3PO_3

8. The energy from the combustion of hydrazine, N_2H_4, is used to power rockets into space in the reaction

   $$N_2H_4(g) + O_2(g) \longrightarrow N_2(g) + 2H_2O(l)$$
   $$\Delta H = -627.6 \text{ kJ/mol}$$

   How many kilograms of hydrazine would be necessary to produce 1.0×10^8 kJ of energy? (17-10)

9. In a calorimeter, 1.23 kg of water is heated from 21.0°C to 26.8°C. What is the ΔH for this reaction? (17-2)

10. The thermite reaction is spectacular and exothermic. Iron(III) oxide, Fe_2O_3, and metallic aluminum produce molten iron and aluminum oxide in a few seconds. (17-9)

    $$Fe_2O_3(s) + 2Al(s) \longrightarrow Al_2O_3(s) + 2Fe(s)$$

    a. Find ΔH for this reaction given the following:

    $$2Al(s) + 3/2\, O_2(g) \longrightarrow Al_2O_3(s)$$
    $$\Delta H = -1670 \text{ kJ/mol}$$

    $$2Fe(s) + 3/2O_2(g) \longrightarrow Fe_2O_3(s)$$
    $$\Delta H = -822 \text{ kJ/mol}$$

    b. Calculate the ΔH for the reaction of 4 moles of Fe_2O_3 with Al.
    c. How much heat energy is released when 1 kilogram of iron is produced in the thermite reaction?

Challenge

1. A simple fat molecule has the formula $C_3H_5(OH)_2O\text{-}CO(CH_2)_2CH_2$. The heat of reaction when it is combusted to CO_2 and H_2O is 6405 kJ/mol or 1531 kcal/mol. Find the amount of energy released per gram of fat. Compare it to the amount of energy released when a carbohydrate is burned (4 kcal/gram). Which provides more energy per gram?

2. The amount of solar radiation received in Arizona annually is about 8.4×10^6 kJ/m². How much coke (C)

must be burned to carbon dioxide in the following reaction to produce the same amount of energy?

$$C(s) + O_2(g) \longrightarrow CO_2(g) \qquad \Delta H = -393.7 \text{ kJ/mol}$$

Synthesis

1. Consider the following heat content diagram:

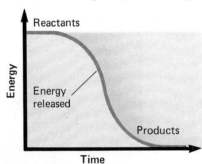

   a. Is this reaction endothermic or exothermic?
   b. What is the sign of ΔH?
   c. Write an equation for a reaction that would be represented by this diagram.

2. The heat of formation for sulfuric acid, H_2SO_4 is −908 kJ/mole. Find ΔH for each of the following:
   a. 0.12 mol H_2SO_4
   b. 373 g H_2SO_4
   c. amount of H_2SO_4 needed to make 500 mL of a 3M solution if the concentrated acid is 98% pure.

Projects

1. See what you can learn about the procedure used by nutritionists to determine the Calories in food. Some of the questions that you might learn about are:
   a. How are the carbohydrates, fats, proteins, and fiber in foods separated so the Calories of each component can be measured?
   b. How much variation does one find from one sample to another? When it is said that an average apple supplies 100 Calories, how accurate is that? Does it matter whether it is a Jonathan or a Golden Delicious? What mass is "average"?
   c. Will some people gain more weight than others even if they eat the same foods and exercise the same amount? Why?
   d. Does it matter whether you get your energy from sugar, starch, fats, or proteins?

2. After learning how food Calories are measured, try making some measurements yourself and see how closely your values compare to those of the experts.

When a candle burns, the carbon and hydrogen in the wax combine with oxygen from the air in an exothermic reaction that produces heat and light. Once a candle is lit the burning continues until the candle is consumed. What energy changes get this reaction started?

If bright, shiny strips of lithium, iron, and copper metals are placed where they can come in contact with air, the lithium changes immediately; the iron reacts noticeably after a few days, but changes in the copper take weeks to be noticeable. Why do similar reactions take place at different rates?

Reaction Rates

Collision Theory

What causes some reactions to occur quickly, when others occur slowly, and some others do not occur at all? You might expect reactions to occur as the result of collisions between reacting particles. So the rate of a reaction might have something to do with collisions. Air contains both nitrogen and oxygen molecules that are colliding constantly. At standard pressure, each molecule collides an average of more than four billion times per second. Nitrogen can react with oxygen to produce nitrogen dioxide, NO_2, a red-brown gas. Even though each molecule of nitrogen or oxygen collides billions of times every second, the amount of nitrogen dioxide present after a year is too small to be detected. Why is this reaction so slow?

It appears that all collisions do not result in the formation of new products. In this lesson, you will study the collision theory model that can be used to explain how and why reaction rates vary.

CONCEPTS

■ effective collisions cause reactions
■ concentration affects reaction rate
■ surface area affects reaction rate
■ activation energy
■ the activated complex is a transitional structure
■ effects of catalysts and inhibitors on reaction rate

18-1 Nature of Reactants

Lithium will react with water at 25°C to produce hydrogen gas and a solution that conducts electricity. In Figure 18-1 the bubbles of hydrogen gas are visible. As the reaction continues, enough positive and negative ions are produced to complete the circuit and light the bulb. Careful measurement of the temperature before and after the reaction shows that energy is given off. The conclusion that a reaction has

Figure 18-1 The photograph on the left shows the reaction of lithium with water. On the right, the presence of lithium ions in solution is shown by the lit bulb of the conductivity apparatus.

495

Figure 18-2 The photo on the left shows lithium hydroxide produced when the solution in Figure 18-1 evaporates. Compare the reaction of lithium in Figure 18-1 to the reaction of potassium shown on the right.

occurred can be tested by evaporating the water. A white ionic solid with properties identical to those of lithium hydroxide, not lithium metal, remains after evaporation.

If sodium is reacted with water, similar products are obtained that include hydrogen gas, a release of energy, and an electrolytic solution containing sodium ions and hydroxide ions. Potassium reacts in a similar way. The equations for the reactions are as follows:

$$2Li(s) + 2H_2O(l) \longrightarrow 2Li^+(aq) + 2OH^-(aq) + H_2(g) + energy$$
$$2Na(s) + 2H_2O(l) \longrightarrow 2Na^+(aq) + 2OH^-(aq) + H_2(g) + energy$$
$$2K(s) + 2H_2O(l) \longrightarrow 2K^+(aq) + 2OH^-(aq) + H_2(g) + energy$$

When pieces of each element that are of equal size, shape, and purity are reacted with water, potassium reacts faster than lithium. The difference in the rate of these similar reactions is due to differences in the nature of the reactants. The rate of a reaction can be expressed either in terms of the amount of reactant consumed per unit of time or as the amount of product formed per unit of time. The rate of the reaction of lithium with water can be expressed quantitatively several ways. Two possible expressions are:

$$reaction\ rate = \frac{mass\ of\ Li\ reacted}{time}$$

$$reaction\ rate = \frac{volume\ of\ H_2\ formed}{time}$$

Since most reactions occur in solution, it is common practice to describe reaction rate in terms of the change in concentration (decrease) of a reactant over a given period of time.

CHEM THEME

Studying reaction rates provides another important measurement in chemistry. The industrial chemist is keenly concerned with how much can I get and how fast?

18-2 Collision Theory and Concentration

You can use the reaction of nitric oxide, NO, with oxygen to produce nitrogen dioxide to expand the collision theory model to explain the effects of changing concentration. The equation for the reaction is:

$$2NO(g) + O_2(g) \longrightarrow 2NO_2(g) + 114\ kJ$$

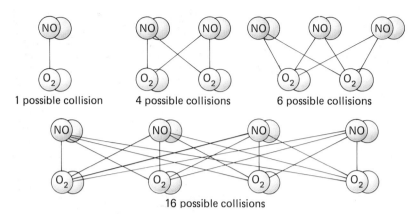

1 possible collision   4 possible collisions   6 possible collisions

16 possible collisions

Figure 18-3 Increasing the number of reacting particles (increased concentration) increases the chances for an effective collision.

When two moles of NO are mixed with one mole of O_2 in a reaction vessel at 25°C, the reaction to form NO_2 is essentially complete within one minute. If two moles of NO are mixed with two moles of O_2 in the same reaction vessel at 25°C, the reaction takes place faster. It makes sense to assume that when there are more molecules in the reaction vessel there will be more collisions. If the number of collisions per second between reacting particles increases, the reaction rate will increase. Using the collision theory model to explain observed reactions, it appears that the rate of a chemical reaction is increased as the concentration of the reactants is increased. Figure 18-3 shows how increasing the number of particles increases the number of collisions.

Figure 18-4 Hydrogen gas is produced on the surface of zinc metal when zinc reacts with hydrochloric acid.

18-3 The Effect of Surface Area

Figure 18-4 shows a reaction taking place on the surface of zinc metal. The atoms of zinc in the interior of the pieces cannot react until they come in contact with the acid. If a piece is cut in half, some additional zinc atoms will now be on the surface and can react. Increasing the amount of zinc available to react will make the reaction take place faster. Cutting the pieces into smaller pieces will further increase the rate of reaction. It makes sense that if a reaction takes place on the surface of a substance, increasing the surface area should increase the number of collisions between reacting molecules.

Figure 18-5 Cutting a cubic object increases its surface area. In the case of a solid reactant, like zinc, more surface area provides more reaction sites.

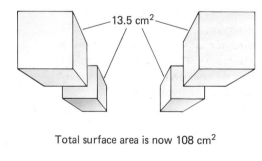

Total surface area is now 108 cm^2

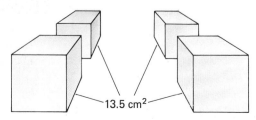

13.5 cm^2

Total surface area 54 cm^2 Total surface area 72 cm^2

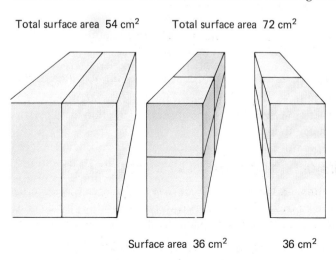

Surface area 36 cm^2 36 cm^2

18-4 Effect of Temperature

You have seen examples throughout this course where increases in temperature change rates. From Chapter 7, you know that an increase in temperature increases the average velocities of gas molecules. In making solutions, increasing the temperature of the solvent increases the rate at which most solutes dissolve. Therefore, it is not surprising that an increase in temperature would increase the rates of reactions.

Chemists have learned that chemical reactions occur only when collisions have sufficient energy to cause a rearrangement of atoms. You can understand this if you think of collisions between cars. If a stalled car is being pushed from behind by another car, there are gentle bumps from the car behind. No damage is done to either car. This situation is very different from a high-speed collision. High-energy collisions cause extensive car damage. High-energy molecular collisions cause the "molecular damage" called a chemical reaction.

In Chapter 7, you saw that gas molecules at a particular temperature show a wide range of molecular energies. A few molecules travel at very high speeds. They have high kinetic energy. If one of these molecules collide with another molecule, you would expect "molecular damage" to take place. Most molecules in the sample have much lower speed and energy and do not react.

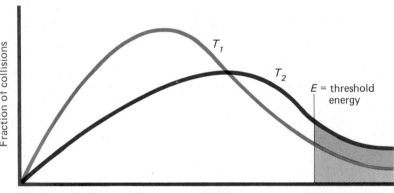

Figure 18-6 Collision energy distributions are shown for reacting particles at two different temperatures. The number of collisions with energy greater than the activation energy, E, is larger at the higher temperature, T_2.

The distribution curve at temperature T_1 showing the fraction of collisions having a particular energy is shown in Figure 18-6. This curve will be useful in discussing the effect temperature has on the rate of a chemical reaction. Let us propose that chemical reactions take place only if the two colliding molecules bring enough energy to the collision so that rearrangement of atoms occurs to form new molecules. The minimum amount of energy is called the "threshold energy," E. A vertical line is drawn in Figure 18-6 to indicate this energy. The shaded area to the right of this line shows the fraction of collisions with energy greater than E.

At a low temperature T_1, not many molecules have high energies. Not many collisions involve energies greater than the threshold energy. Very few of the collisions lead to a chemical reaction. What happens when the temperature is increased to T_2? The distribution curve changes shape. It flattens and spreads out as shown in Figure 18-6. The average speed of the molecules is greater at the higher temperature. There are more molecules now that have high kinetic energy.

Thus, more of the collisions involve energy greater than E. Therefore, the reaction rate is greater at a higher temperature.

Almost all reaction rates are increased when the temperature of a reaction is increased, although for very fast reactions the effect of temperature is small. A generalization that does not always apply is that for reactions which do not appear to be instantaneous, an increase in temperature of 10°C will double the rate of reaction. Increasing the temperature increases the velocity of the molecules and the number of collisions, but it does not double the number of collisions. An increase in temperature of 10°C increases the average energy of collision, but it does not double the energy of collision. It is the combination of these two factors—more collisions and higher energy of collision—that doubles the rate of reaction. It is difficult to separate these factors because changing the temperature changes both the collision energy and the number of collisions.

This rule of doubling the reaction rate for an increase in temperature of 10°C only applies to certain reactions over a moderate temperature range. However, the rule is useful in predicting the effect of temperature on reaction rates.

18-5 Activation Energy

Why is there a threshold energy for a chemical reaction to take place? Why do all collisions not result in the formation of new products? To answer these questions, consider the following analogy. Imagine someone trying to roll a bowling ball up a very steep hill. On most tries the bowling ball slows down and stops before it gets to the top of the hill. The kinetic energy of the bowling ball is converted to potential energy as the ball slows down. Then it rolls back down on the same side of the hill. The hill acts as a barrier. Look at Figure 18-7 to see what happens. Only occasionally does the bowler give the ball enough kinetic energy so that it gets to the top of the hill and rolls down the other side. A successful try is shown on the right of Figure 18-7.

Figure 18-7 In (a), the bowler did not give the ball enough energy to get over the energy barrier. In (b) the ball has enough energy to go over the energy barrier.

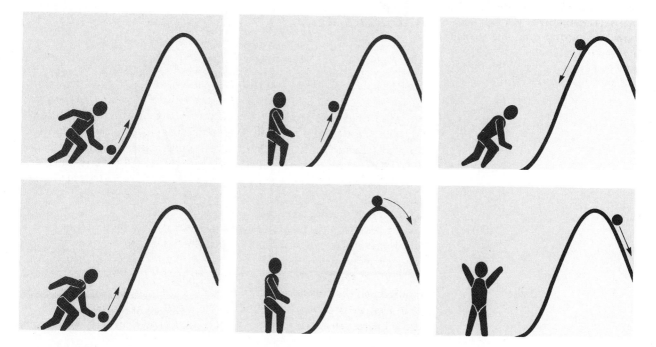

Do You Know?

Air balloons used for observation during the Civil War were filled with hydrogen gas.

Picture a similar situation for molecules in a chemical reaction. During molecular collisions, atoms can take up new bonding arrangements that have more potential energy than either the reactants or the products. These atomic arrangements have high potential energy like the bowling ball at the top of the hill. There is a minimum potential energy that must be achieved by colliding reactants before they can convert to some other form. Figure 18-8 shows the distribution curve for the kinetic energy of molecular collisions placed next to the potential energy hill for a chemical reaction. (Remember, however, there is no actual "hill" between reactants and products. It is only an energy barrier.) To react, molecules must collide with enough energy to assume the high-energy configuration of atoms represented by the top of the hill. With less energetic collisions the molecules do not react. Several values are marked on the energy distribution curve with circles. If molecular collisions having those energies occurred, the system could reach the corresponding circles on the potential energy diagram. Quite obviously these collisions are not successful ones.

The energy barrier shows the minimum threshold energy, E. Chemists have given E the name **activation energy**. The configuration at the top of the energy barrier is a "molecule" called the **activated complex**. At first, this molecule may seem unusual to you. It has a very short lifetime and gets rid of its high potential energy by breaking apart, forming either the reactants or the products. Both the reactants and products have less potential energy than the activated complex.

Figure 18-9 will help you visualize what happens when two hydrogen iodide molecules react to form hydrogen and iodine. Potential energy is plotted vertically, and the horizontal direction shows the

Figure 18-8 These graphs show a comparison between the kinetic energy distribution curve and the activation energy diagram. The graph on the left is turned on its side to relate the activation energy value, E on both curves. Note that only a small number of collisions have sufficient kinetic energy to reach the top of the potential energy hill.

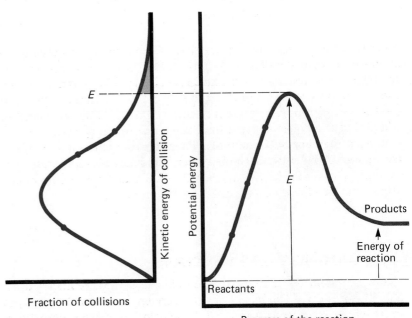

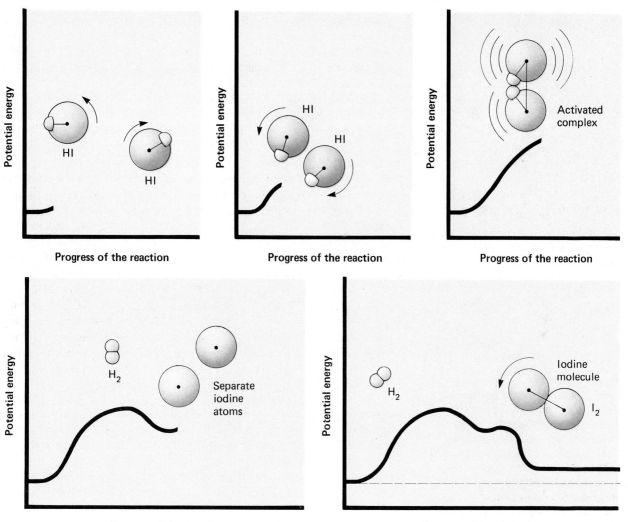

progress of the reaction. The position of the circle in each of these figures represents the potential energy for the entire system. The most favorable geometry for these collisions is shown. This geometry corresponds to the smallest possible value for E in the reaction. Thus you see that it is not only important for molecules of sufficient energy to collide but the collision must have the appropriate geometry.

In each diagram only one molecule is represented, but it is important to recall that any observable system contains *many* molecules. For example, a one-liter flask of H_2 and O_2 at 25°C and a pressure of one atmosphere will contain about 2.5×10^{22} molecules.

Figure 18-9 You can see the change in energy as two reactants approach each other (a and b). The formation of the activated complex (c) is the highest point on the potential energy curve. As the reaction proceeds and the H_2 molecule forms (d), the potential energy decreases. The unbonded I atoms keep the potential energy high (e). As the I_2 molecules form, the potential energy drops. The potential energy difference between the reactants and products is the amount of energy absorbed by this endothermic system.

18-6 Catalysts and Inhibitors

Aqueous solutions of hydrogen peroxide, H_2O_2, are commonly available in drugstores. The label gives the concentration as 3.0% H_2O_2 by mass which is equivalent to a $0.90M$ solution. Hydrogen peroxide is unstable and decomposes to produce water and oxygen.

Figure 18-10 The decomposition of hydrogen takes place very slowly. There is not enough oxygen produced to ignite the splint in the beaker on the left. The catalyzed decomposition of hydrogen peroxide proceeds quickly. The presence of O_2 product ignites the glowing splint in the beaker on the right.

$$2H_2O_2(aq) \longrightarrow 2H_2O(l) + O_2(g) + 108 \text{ kJ}$$

The rate of this reaction at 25°C is very slow as shown in Figure 18-10. When manganese(IV) oxide is added, the reaction rate increases dramatically. Enough oxygen is produced to cause the glowing splint to burst into flame as shown in Figure 18-10. When the reaction is complete, the manganese(IV) oxide can be recovered unchanged. The manganese(IV) oxide is a catalyst for this reaction. A **catalyst** is a substance that increases the rate of a chemical reaction without being used up itself.

A catalyst can speed up a reaction by providing a low-energy pathway from reactants to products. The catalytic path has a lower activation energy. The catalytic path involves a different reaction mechanism and the formation of a different activated complex. The use of a contact catalyst orients the colliding particles in such a way that more collisions are effective. Many enzymes catalyze reactions by improving the collision geometry.

Before other important chemical reactions are discussed, look at the arrangement of the energy diagrams shown in Figure 18-11. The curve showing distribution of collision energy and the potential energy diagram are placed next to each other. The two horizontal lines represent the activation energy for the catalyzed and noncatalyzed reactions. The shaded areas indicate the number of collisions having kinetic energy greater than E_{cat} and E. There are many collisions having energy greater than E_{cat}. Only a few collisions have energy greater than E. The catalyzed reaction has a higher reaction rate than the noncatalyzed reaction, since more collisions have the required energy.

An **inhibitor** reduces a reaction rate by preventing the reaction from occurring. The inhibitor may combine with a reactant or a catalyst to form a complex that is stable at that temperature so that a reaction will

Do You Know?

Catalysts used for automobile emission control use platinum or palladium that is attached to a honeycomb ceramic. The metals do not always adhere well to the ceramic. Tin(IV) oxide adheres well to both metals and to the ceramic surface and has made the catalytic converter ten times more effective.

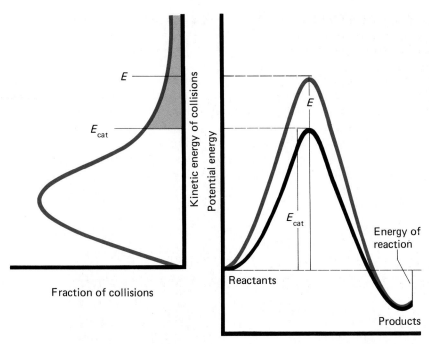

Figure 18-11 These graphs show another comparison of the curves in Figure 18-8 with the relationships for a catalyzed versus an uncatalyzed reaction. There are more collisions with sufficient kinetic energy to produce products for the catalyzed reaction. Note that the energy of the reaction does not change.

not occur. Preservatives used in foods and medical preparations are two examples of substances that contain inhibitors.

Review and Practice

1. Write and balance the equation for the reaction of sodium and water.

2. Which reacts faster with water, lithium or potassium?

3. Write the balanced equation for the reaction of magnesium with hydrochloric acid.

4. One piece of magnesium is reacted with 3.00M hydrochloric acid at 25°C. Another piece of magnesium of equal size and shape is reacted with 1.00M hydrochloric acid at 25°C. Predict which reaction occurs at a faster rate.

5. A chunk of zinc is reacted with 3.00M hydrochloric acid at 25°C. An equal mass of powdered zinc is reacted with 3.00M hydrochloric acid at 25°C. Compare the reaction rates.

6. State two ways that the number of effective collisions between reactants can be increased.

7. How can the energy of collision be increased?

8. What is the name of the transitional structure formed during the process of breaking bonds and reforming new bonds?

9. How does a catalyst increase collision efficiency?

10. How does increasing the concentration of a reactant affect the rate of a reaction?

1. $2Na(s) + 2H_2O(l) \longrightarrow$
$2Na^+(aq) + 2OH^-(aq) +$
$H_2(g) + energy$
2. potassium
3. $Mg(s) + 2HCl(aq) \longrightarrow$
$MgCl_2(aq) + H_2(g) + energy$
4. the reaction in 3.00M HCl(aq)
5. Powdered zinc reacts faster.
6. increase temperature, use a catalyst
7. increase the temperature
8. activated complex
9. A catalyst provides a reaction pathway with a lower activation energy.
10. The rate is directly proportional to the concentration of the reactant.

Rate Relationships

CONCEPTS

■ **rate constants are determined experimentally**

■ **reactions may occur in steps**

■ **rate-determining steps determine overall reaction rates**

Controlling reaction rates is economically important to industry. In synthesizing a substance, an industrial chemist looks for ways to produce as much product as quickly and cheaply as possible. Conversely, if a reaction is undesirable, such as the formation of rust, the chemist wants to keep this reaction rate at a minimum.

You can now apply what you have learned about the factors that affect reaction rates to see how chemists describe rate relationships mathematically.

18-7 Rate Equations and Rate Constants

At a fixed temperature, the rate of a given reaction depends on the concentrations of the reactants. To develop a mathematical expression for the rate, recall the decomposition reaction for hydrogen peroxide to form water and oxygen. The reaction rate for the decomposition of hydrogen peroxide can be measured by measuring the volume of oxygen gas produced as a function of time.

$$H_2O_2 \longrightarrow H_2O + \tfrac{1}{2}O_2$$

The rate of decomposition of hydrogen peroxide decreases with decreasing concentration. The rate at any instant is directly proportional to the concentration of hydrogen peroxide at that time. This proportionality can be expressed mathematically as:

$$rate = k[H_2O_2]$$

The proportionality constant, k, is called the **rate constant**. The brackets signify the molar concentration of the reactant. The rate constant varies with temperature and defines the fraction of H_2O_2 molecules that react per given time. The value of the rate constant also is dependent on the nature of the reacting substances.

For the general reaction

$$A + B \longrightarrow C + D$$

the rate equation has the form

$$rate = k[A]^x[B]^y$$

in which [A] and [B] represent molar concentrations of the reactants, and k is the rate constant for the particular reaction. The exponents x and y and the constant k must be determined experimentally. For a particular reaction, experiments would be carried out where one reactant would be held constant and the other reactant concentration would vary. Rate data would be collected to determine how the rate is affected for each concentration change. Once these data are evaluated, the values of x and y can be determined. Actual calculations of rate constants would require the use of calculus. Such calculations are beyond the scope of this book.

18-8 Reaction Mechanisms

Consider the reaction between oxygen and hydrogen bromide at 400°C. The equation for the reaction is:

$$4HBr(g) + O_2(g) \longrightarrow 2H_2O(g) + 2Br_2(g)$$

Four molecules of HBr react with one molecule of O_2. Does this mean that four molecules of HBr must simultaneously collide with one molecule of oxygen? The probability that five gaseous molecules will collide simultaneously is very small. Instead, experiments show that this reaction takes place in a series of steps.

$$HBr(g) + O_2(g) \longrightarrow HOOBr(g) \qquad \text{slow}$$
$$HOOBr(g) + HBr(g) \longrightarrow 2HOBr(g) \qquad \text{fast}$$
$$HOBr(g) + HBr(g) \longrightarrow H_2O(g) + Br_2(g) \quad \text{fast}$$

The first reaction takes place slowly while the second and third reactions are fast. The sequence of reactions that produce bromine and water are called the **reaction mechanism** for the overall reaction. Because the first reaction is the slowest reaction in the mechanism, it determines the overall rate. The second and third reactions cannot occur until HOOBr is formed. The slowest reaction in a reaction mechanism is called the **rate-determining step**.

An analogy might help you better understand the rate-determining step. Imagine five people working together to wash dishes. The first two clear the table and hand the dishes to a third person who washes them and places them on a draining rack. The last two people dry and stack them. Which step in this process is likely to be the rate-determining step?

Figure 18-12 The reaction mechanism for the decomposition of formic acid, HCOOH is shown on top. The bottom reaction shows the mechanism for the decomposition using an acid catalyst. Note the differences in the pathways shown and the molecular arrangements.

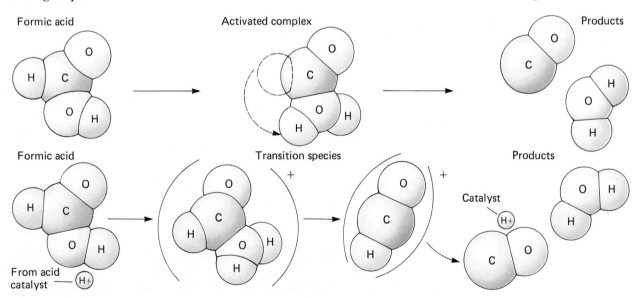

Review and Practice

1. For the dishwashing analogy in Section 18-8, describe the effects on the reaction rate if a sixth person is added to
   a. clear the tables.   **b.** wash dishes.   **c.** dry dishes.

1. a. no change   b. The rate is increased.   c. no change

Jesse C. H. Hwa
Title: Technical Consultant
Job Description: Offers advice regarding research, management, and technology for chemical companies. Assists in teaching newly developed technology to companies in other countries.
Educational Qualifications: Advanced degrees in chemistry with additional courses in mathematics and business.
Future Employment Outlook: Good

Jesse C. H. Hwa travels as a technical consultant from the United States to Europe, the Far East, and everywhere in between. Through his experience in chemical research and his background in strategy planning, business assessment, and investment, he assists chemical companies in developing sound business management.

"I first became interested in chemistry when I was attending the Senior Middle School in China," says Hwa. "I took a course in chemistry my second year. It was love at first sight. I had no question in my mind. The very nature of science was exciting to me."

Hwa spent many years doing chemical research regarding polymers (long-chained molecules) before becoming a consultant. Now he is able to use his love of chemistry in ways other than research. Hwa explains that his work as a consultant has three different phases: "The first phase is research. This involves solving problems and using one's head to decipher chemical puzzles. The second is management. Management of activities related to chemistry leads to the understanding of the relationships between chemistry

and our daily lives and industry. The third phase includes using the broad scope of my contacts and knowledge to assist other people who have problems." One of Hwa's major achievements has been using his Chinese background and his research in polymers to help China develop its own chemical industry. "This is a unique opportunity to assist the chemical industrial growth of another country."

The characteristics that make Hwa a good chemist have also helped him become a good consultant. He believes that being a successful student of chemistry demands both curiosity and intelligence. The world of chemistry is full of excitement as the frontiers of chemistry continue to expand into new areas. A good grasp of chemistry, its related sciences, and its applications in our daily lives all enable Hwa to be useful to others and to society.

A career in chemical consulting can provide many opportunities for advancement. At first, advancement comes through moving up from performing research to managing other researchers. Eventually, with experience in business management, comes the opportunity to achieve a goal important to Hwa—broadening and strengthening the relationship between chemistry and society.

Numbers in red indicate the appropriate chapter sections to aid you in assigning these items. Answers to all questions appear in the Teacher's Guide at the front of this book.

Summary

■ The rate of a chemical reaction depends on several factors that include the number and energy of the collisions, the nature of the reactants, and the energy of the pathway of the reaction mechanism available.

■ Collision theory provides explanations as to why reaction rates vary. Increasing the number of effective collisions between reactants increases reaction rate. The fraction of collisions that result in reactions is indicated by the collision efficiency. When the velocity of the particles is increased, both the number of collisions and the energy of the collisions are increased.

■ Only collisions that result in enough energy to form an activated complex will result in the formation of products. The amount of energy that must be added to form the activated complex is the activation energy. The activation energy is at a minimum when the particles collide with the most favorable geometry.

■ A catalyst can increase the rate of a reaction by providing an alternate pathway with a lower activation energy. The use of a contact catalyst improves the collision geometry of the molecules. An inhibitor slows a reaction by blocking a reaction pathway.

■ The rate of a reaction is proportional to the product of the concentrations of the reactants. For the general reaction

$$A + B \longrightarrow C + D$$

the rate equation has the form

$$rate = k[A]^x[B]^y$$

The rate constant k, and the exponents x and y must be determined experimentally.

■ Many reactions occur in a series of steps, called the reaction mechanism. For these reactions, the overall rate of the reaction is dependent on the speed of the slowest step, which is the rate-determining step.

Chemically Speaking

activated complex	inhibitor
activation energy	rate-determining step
catalyst	reaction mechanism
collision theory	

Review

1. If samples of sodium and lithium of equal size, shape, and purity are placed in a bottle of chlorine gas at 25°C, which would you expect to react fastest? Explain why. (18-1)

2. Use the collision theory to explain why increasing the concentration of hydrochloric acid would cause an increase in the rate of reaction with zinc. (18-2)

3. Hydrogen and iodine are reacting at 400°C according to the equation:

$$H_2(g) + I_2(g) \longrightarrow 2HI(g)$$

How would the rate of reaction be affected by:
a. increasing the temperature.
b. increasing the concentration of hydrogen.
c. increasing the concentration of both the hydrogen and the iodine.
d. adding an inhibitor. (18-2–18-4)

4. Explain why milk left on the kitchen table spoils faster than milk left in the refrigerator. (18-4)

5. Explain how an inhibitor can extend the shelf-life of a package of food. (18-6)

6. The equation for the formation of ammonia is:

$$N_2(g) + 3H_2(g) \longrightarrow 2NH_3(g)$$

Explain why this equation is not likely to represent the reaction mechanism. (18-8)

7. A group of students is assembling a ten-page document for mailing. There are 50 copies of each typed page in separate stacks. The pages must be (1) assembled in order, (2) straightened, (3) stapled, and (4) inserted into envelopes for mailing.
a. If four students work together, each performing a different operation, which of the above might be the rate-determining step?
b. What would be the effect on the overall rate if five more people assembled the pages (step 1)?
c. What would be the effect on the rate if the five helpers worked on the second step? The third step? The fourth step?
d. What would be the effect if the envelopes had to be addressed, stamped and mailed in the fourth step? (18-8)

Interpret and Apply

1. The speed of the space shuttle can be expressed in metric units, kilometers per hour. What metric units would you use to discuss the rate of:
   a. consumption of gasoline in a car?
   b. production of coal in a mine?
   c. formation of hydrogen gas when zinc reacts with hydrochloric acid? (18-1)

2. Which will react faster, zinc with $3M$ hydrochloric acid or zinc with $1M$ hydrochloric acid? (18-2)

3. Which will burn faster, a solid log, a split log, or wood shavings? (18-3)

4. Sketch a potential energy curve for an endothermic reaction. Label the parts representing the activated complex, activation energy, and change in enthalpy. (18-5)

5. White phosphorus reacts immediately and rapidly with oxygen when exposed to air. What can you say about the activation energy for this reaction? (18-5)

6. If fine copper wool is heated in a crucible it reacts with oxygen to form copper(II) oxide. At room temperature, however, no change is noticed in the copper wool. Compare the activation energy of copper and phosphorus (see item 5) with oxygen. (18-5)

7. A group of educators wish to have Jane Goodall give a lecture to a group of teachers on the behavior of chimpanzees. One of the educators knows Jane personally and goes to an adjacent office to telephone Jane's agent in New York. The agent calls to send a telegram via satellite to Africa. The telegram is typed and put into an envelope. The envelope is given to a messenger who must travel a few kilometers by boat and a few hundred meters on foot before handing the message to Jane Goodall. The messenger returns to the telegraph office with the reply and the process is reversed. Which is the rate-determining step in this process? (18-8)

8. Hydrogen peroxide reacts with hydrogen ions and iodide ions according to the following equation:

   (a) $H^+ + I^- + H_2O_2 \longrightarrow H_2O + HOI$

   The mechanism often suggested for this reaction is:

   (b) $H^+ + H_2O_2 \longrightarrow H_3O_2^+$ fast
   (c) $H_3O_2^+ + I^- \longrightarrow H_2O + HOI$ slow

   a. Show that adding equations (b) and (c) gives equation (a).

   b. How would you expect the rate to be affected if the concentration of I^- is doubled?
   c. How would you expect the rate to be affected if the concentration of H^+ is doubled? (18-7–18-8)

9. A liter of a $1.0M$ solution of hydrogen peroxide slowly decomposes into water and oxygen. If 0.50 mole of hydrogen peroxide decomposes during the first six hours of the reaction, explain why only 0.25 mole decomposes during the next six-hour period. (18-2–18-8)

Problems

1. At 20°C, a small strip of magnesium reacts with $3.0M$ hydrochloric acid to produce 12 mL of hydrogen gas in 20 seconds. (18-1)
   a. What is the rate of this reaction?
   b. What volume of hydrogen might be produced in 20 seconds at 30°C?

2. Calculate the surface area on a cube of zinc 1000 cm on each edge. If the cube is cut into smaller cubes that are 10 cm on each edge, find the area of each cube, the total number of cubes, and the total area of all the cubes. The cubes are then cut into cubes that are 1 cm on each edge. Find the area of one of these cubes, the number of these cubes, and the total area of all the 1 cm cubes. (18-3)

3. The catalyzed decomposition of 50 mL of 3% hydrogen peroxide is 90% completed in 60 seconds at 20°C. How long might it take for the reaction to be complete at 40°C? (18-4)

4. At 20°C, a 3% solution of hydrogen peroxide produces 15 mL of oxygen gas in 120 seconds.
   a. What is the rate of this reaction?
   b. How long might it take to produce 15 mL of oxygen at a temperature of 40°C? (18-1, 18-4)

5. In an experiment, a sample of $NaClO_3$ is 90% decomposed in 60 minutes at 20°C. How long would this decomposition take at a temperature of 40°C? (18-4)

6. Carbon dioxide reacts with water in animal cells via the following reaction:

   $$CO_2 + H_2O \longrightarrow H^+ + HCO_3^-$$

   Without a catalyst, two molecules of CO_2 react with two molecules of H_2O per minute when the temperature is 37°C. How many molecules of carbon dioxide will react in one day? A single molecule of the enzyme carbonic anhydrase can catalyze 3.6×10^7 molecules of carbon dioxide in one minute. How many molecules of carbon dioxide will react in one day using one molecule of the enzyme?

7. One reaction involved in the formation of smog is:

$$O_3(g) + NO(g) \longrightarrow O_2(g) + NO_2(g)$$

The rate of the reaction can be calculated using the following equation:

$$\text{rate} = k[O_3]^x[NO]^y \text{ where } x \text{ and } y$$

both equal 1.
a. What would be the effect on the reaction rate if the concentration of ozone is doubled?
b. What would be the effect on the reaction rate if the concentration of nitric oxide is tripled?
c. What would be the effect on the reaction rate if the concentration of ozone is doubled and the concentration of nitric oxide is tripled? (18-7)

8. In an important industrial process for producing ammonia, the overall reaction is:

$$N_2(g) + 3H_2(g) \longrightarrow 2NH_3(g) + 100.3 \text{ kJ}$$

A yield of about 98% can be obtained at 200°C and 1000 atm. The process makes use of a catalyst of finely divided iron oxides containing small amounts of potassium oxide and aluminum oxide.
a. Is this reaction endothermic or exothermic?
b. How many grams of hydrogen must react to form 25 grams of ammonia?
c. Does the equation for the overall reaction represent the reaction mechanism? (18-5–18-8)

Challenge

1. Using the reaction of hydrogen plus iodine to form hydrogen iodide, design and describe an experiment in which one could determine separately the effect on reaction rate of increasing:
a. the number of collisions
b. the average energy of collision

Synthesis

1. Explain why it takes longer to hard-boil an egg in a pan of boiling water at Pikes Peak than in Boston.

2. The catalyzed decomposition of 55 mL of 3.0% hydrogen peroxide, H_2O_2, is 90% complete in 60 seconds at 20°C. What is the volume of oxygen gas produced at 95.5 kPa in 60 seconds?

3. A strip of magnesium with a mass of 0.22 grams reacts with 65 mL of 3.0M HCl at a temperature of 30°C and a pressure of 92 kPa in 25 seconds. Calculate the rate at which hydrogen gas is produced in mL per second.

Projects

1. Considering that very little energy (about 1.9 kJ/mol) is required to convert graphite to diamond, find out why this process is so difficult.

2. The following equations represent the proposed mechanism by which chlorine-containing fluorocarbons could destroy ozone, O_3, in the atmosphere.

$$CFCl_3 \longrightarrow CFCl_2 + Cl$$
$$Cl + O_3 \longrightarrow ClO + O_2$$
$$ClO + O \longrightarrow Cl + O_2$$

a. Explain why chlorine atoms are said to catalyze ozone decomposition.
b. If the chlorine atoms catalyze ozone decomposition, will the amount of Cl presently in the atmosphere ever be used up?
c. Look for a current article on the effect of fluorocarbons on ozone.

The photograph above illustrates a system in balance. Children on a seesaw also are a system in balance, but their system differs from the juggler's, and both systems differ from the balance that is found in chemical equilibrium. How do the differences of these systems teach us something about chemical equilibrium?

The seesaw is an example of balance because once the children's weights have been distributed evenly, the seesaw stops moving. In equilibrium there is a balance, but the system is not at a standstill. The juggler illustrates a dynamic balance because the balls are constantly in motion, while the number of balls in the system remains the same. However, the juggler has to continuously add energy to maintain the balance. In chemical equilibrium, neither energy nor matter is added or subtracted from the system in order to maintain the balance. The seesaw and the juggler introduce characteristics of balance, but as you will see in this chapter, additional aspects are involved in chemical equilibrium.

Reaction Equilibrium

Conditions for Reaction Equilibrium

Recall from Chapter 16 that in a saturated solution two opposing processes are taking place. Think of a simple example, such as preparing a saturated solution of copper(II) sulfate in the lab. You know that as the solid dissolves, the blue color of the solution darkens until the saturation point is reached. Even though the saturated solution appears unchanged, you know that a saturated solution can reach equilibrium. Solid particles are dissolving while other dissolved solute particles are recrystallizing. These two processes are occurring at the same rate and can be written in equation form as:

$$\text{rate}_{\text{dissolving}} = \text{rate}_{\text{recrystallizing}}$$

A saturated solution in contact with undissolved solid is an example of phase equilibrium. There is no net change in the concentration of the solution nor in the mass of undissolved solid. Does a similar process occur in chemical reactions? If you drop a piece of copper in nitric acid, you see things change, as shown in Figure 19-1. The solution turns blue, the copper disappears, and a brown gas is formed. Eventually, everything seems to stop changing. What you will do in this chapter is look more carefully at situations such as this when it appears that nothing is happening.

CONCEPTS

■ characteristics of opposing reactions in a closed system
■ characteristics of reaction equilibrium
■ observable properties at equilibrium
■ properties of dynamic reaction systems
■ effects of changing pressure and temperature on gaseous equilibrium systems

Review solubility and vapor equilibrium studied in Chapters 15 and 16, and compare the characteristics of these systems to reaction equilibrium.

19-1 Reversible Reactions

If the brown gas just described is collected in a flask, it looks like Figure 19-2. The brown gas, nitrogen dioxide, is toxic. It is one of the gases that is produced in an automobile during the combustion of gasoline. It appears rather uninteresting, but if you do the same kinds of experiments that were done in Chapter 7 on ideal gases, strange results are obtained. Recall that the volume of a gas changes in a predictable way as you change the temperature. You would expect the volume of the nitrogen dioxide gas to be directly proportional to the temperature. However, if you change the temperature of the nitrogen dioxide gas you will make two interesting observations. First, you will find that the volume of the gas is *not* directly proportional to the temperature, as you would predict from your study of other gases. Second, there is the unmistakable observation that the gas becomes darker or lighter depending on whether the gas is heated or cooled, as shown in Figure 19-2.

Figure 19-1 Chemical changes take place when copper and nitric acid are in contact.

Figure 19-2 The flasks contain nitrogen dioxide gas, NO_2. The flask on the left shows the gas at 25°C; the flask in the center shows the gas at 0°C; and the flask on the right shows the gas at 100°C.

Another experiment can be done to see how the volume changes as the pressure on the gas is increased. If you place some of the gas in a glass syringe and close off the end, you can push hard on the plunger to increase the pressure. You would observe that the compressed gas looks much darker, but in a fraction of a second it lightens slightly (although the final color is still darker than the initial color.) The original darkening might be explained by the fact that you have pushed more molecules into a smaller space. The higher concentration of molecules produces the more intense color. But why does the color then fade? Do some of the molecules disappear?

If you record the volume of the gas at several different pressures, you will find that the product of the pressure and the volume of the red-brown gas does not produce a constant. We cannot explain these observations using the model for ideal gases discussed in Chapter 7. This is one of the natural and inevitable processes of science. When a model that has been useful in explaining observations is found to be inadequate, changes are made to solve the problem. The gas laws in Chapter 7 were valid because the number of molecules of gas was constant; the laws are not valid in this system because the total number of molecules is changing.

Analysis shows that the brown gas is not pure. Most of the molecules in the flask in Figure 19-2 have the formula N_2O_4, a colorless gas. Analysis also shows that when the gas is heated, the system turns darker brown and most of the molecules have the formula NO_2. When the system is cooled, the color fades and there are again more N_2O_4 molecules in the flask. It appears that a reaction involving both N_2O_4 and NO_2 must be occurring.

$$N_2O_4(g) + energy \longrightarrow 2NO_2(g)$$
colorless   brown

One colorless N_2O_4 molecule absorbs energy and decomposes to form two brown NO_2 molecules. It can be inferred that when the color becomes lighter the opposite reaction is occurring.

$$2NO_2(g) \longrightarrow N_2O_4(g) + energy$$
brown   colorless

Two brown NO_2 molecules combine with the loss of energy to form one colorless N_2O_4 molecule. It is possible to describe what is happening in both situations by using one equation.

$$2NO_2(g) \rightleftharpoons N_2O_4(g) + energy$$
brown   colorless

In this reaction, the product has reacted to reform the reactant. This type of reaction is called a **reversible reaction**. The double arrow $\rightleftharpoons$ is used to represent a reversible reaction. The formation of products and the reformation of reactants are two opposing reactions that are occurring simultaneously for a reversible reaction in a closed system. Such equations can be read in both the forward and reverse directions. When the two reactions occur at the same rate, and no observable changes are taking place, **chemical equilibrium** has been reached.

Do you see why the behavior of NO_2 at various temperatures and pressures cannot be explained by the ideal gas model? As the temperature or pressure changes, a reversible chemical reaction is occurring that changes the composition of the gas. This reaction will reach a different state of equilibrium for different conditions. The focus of this chapter will be on using the equilibrium model to explain what you observe happening in any reversible reaction system.

Another example of a reversible reaction equilibrium system is the reaction between carbon monoxide and hydrogen to produce methanol, as shown by the following equation.

$$CO(g) + 2H_2(g) \rightleftharpoons CH_3OH(g)$$

Initially only carbon monoxide and hydrogen are present in the reaction container. As time goes on the concentration of methanol increases. After a long time there is still some CO and H_2 in the reaction container, because the reverse reaction is occurring and CH_3OH molecules decompose. The forward reaction produces CH_3OH, and the reverse reaction produces CO and H_2.

$$Synthesis \longrightarrow$$
$$CO(g) + 2H_2(g) \rightleftharpoons CH_3OH(g)$$
$$\longleftarrow Decomposition$$

Most of the reactions you have studied at this point have not taken place in a closed system. For example, the reaction between copper and nitric acid that was used to produce the brown NO_2 gas is actually a two-step reaction in an open system.

$$3Cu(s) + 8HNO_3(aq) \longrightarrow 3Cu(NO_3)_2(aq) + 2NO(g) + 4H_2O(l)$$

$$2NO(g) + O_2(g) \longrightarrow 2NO_2(g)$$
colorless   brown

If the reactants are placed in a test tube, the brown NO_2 gas that is produced in the second reaction diffuses throughout the room. (Since this gas is toxic, the reaction should only be carried out under a fume hood.) If an attempt is made to close the system by sealing the test tube in some way, the production of NO gas in the closed system could create a pressure large enough to shatter the test tube.

CHEM THEME

Chemical equilibrium is a model. Like other chemistry models, it proposes a microscopic explanation for macroscopic observations. All models are subject to change to accommodate new observations.

The copper reaction differs from the methanol reaction. For the copper reaction, the reverse reaction does not occur to any significant extent because the NO_2 product given off in an open system cannot combine with the other products to reform the reactants. The forward reaction continues until either all of the copper is gone or all of the nitric acid is reacted.

19-2 Recognizing Reaction Equilibrium

The reversible reaction involving NO_2 and N_2O_4 described in the previous section reached a state of equilibrium and show the following characteristics:

1. The system was closed.

2. Opposite reactions occur at the same rate.

3. Equilibrium was reached by starting with either reactants or products.

4. The temperature was constant.

When the conditions above are met, all properties of a system in equilibrium will be constant even though the forward and reverse reactions continue to occur.

Chemical equilibrium is a dynamic process. Dynamic comes from the Greek word *dynamis*, meaning power, but in the context of chemical reactions it means in continuous motion. At equilibrium the forward and reverse reactions continue to occur. Although reactants are changing into products and products are changing into reactants, there is no net change because the rates of the forward and reverse reactions are equal.

$$rate_{forward} = rate_{reverse}$$

How do we know that the reactions are still occurring at equilibrium? In certain chemical systems it is possible to use radioactive tracers to "tag" atoms of an element in a reaction. In Chapter 9, you read about how these tracers work. Charting the change in radioactivity for reactants and products has provided the data to support the conclusion that even at equilibrium, reactions continue to occur. Since the individual molecules cannot readily be observed, the macroscopic properties of color and temperature usually are used to monitor the progress of the forward and reverse reactions until equilibrium is reached.

Figure 19-3 shows molecular representations of the NO_2-N_2O_4 equilibrium at different temperatures. These models can be used to describe what is happening on a microscopic level. The color of the system at the listed temperature is suggested by the ratio of molecules in each drawing. The relative concentration for each gas is shown by the pie graphs in Figure 19-4. The labeled parts of each graph are proportional to the number of molecules in Figure 19-3.

You know from Chapter 18 that raising the temperature increases the rate of a chemical reaction. The fact that the color of a sample of the NO_2 gas becomes darker in the flask on the right of Figure 19-2

CHEM THEME

Dynamic equilibrium exists whenever two opposing reactions are occurring at the same rate. Equilibrium also occurs in vapor pressure systems, saturated solutions, and in the reaction of weak acids and bases with water.

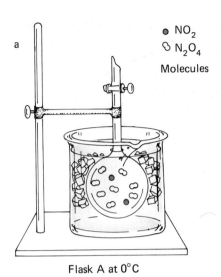
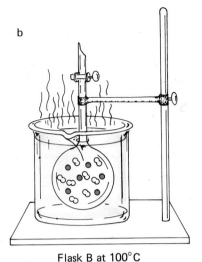

● NO₂
◑ N₂O₄
Molecules

Flask A at 0°C

Flask B at 100°C

Figure 19-3 When NO₂ gas on the left is cooled, the molecules react to form molecules of the colorless gas, N₂O₄. Heating the gas on the right results in decomposition of N₂O₄ to produce more molecules of the brown gas, NO₂.

indicates that initially, the rate of the reverse reaction is increased more by the temperature change than is the rate of the forward reaction, as represented in the right side of Figure 19-3.

The reasons for the initial favoring of one reaction over the other will be discussed in Section 19-3.

$$2NO_2(g) \rightleftharpoons N_2O_4(g) + energy$$
brown   colorless

The net concentration of NO₂ increases, which means more collisions between NO₂ molecules. This increase in collisions causes the rate of the forward reaction to increase. Soon forward and reverse reaction rates are equal once more, but the rates are greater than they were when the flask was at 0°C in Figure 19-2. Equilibrium is established again.

On the other hand, lowering the temperature of the system shows the opposite effect. At a constant temperature of 0°C, the rates of the forward and reverse reactions again equalize, but the lighter color of the flask in Figure 19-2 indicates that the concentration of N₂O₄ is now much greater than it was at 100°C, as shown in Figure 19-4. This must mean that the rate of the reverse reaction is decreased more by the temperature change than is the rate of the forward reaction. The net concentration of NO₂ therefore decreases. As before, the reaction rates soon become equal and a new state of equilibrium is reached.

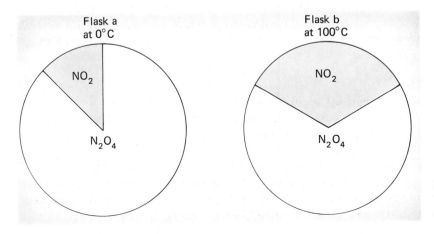

Flask a at 0°C

NO₂

N₂O₄

Flask b at 100°C

NO₂

N₂O₄

Figure 19-4 The relative concentrations of NO₂ and N₂O₄ at equilibrium at 0°C and 100°C.

$$2NO_2 \rightleftharpoons N_2O_4$$

Add energy, T ↑

Remove energy, T ↓

$$2NO_2 \rightleftharpoons N_2O_4$$

Figure 19-5 Adding energy to the NO_2—N_2O_4 system causes an increase in NO_2. Removing energy causes an increase in N_2O_4.

Note that equilibrium does *not* mean that the concentrations of the product and reactants are equal; rather, the rate at which NO_2 reacts to form N_2O_4 is the same as the rate at which N_2O_4 forms NO_2. You should recognize that at the temperatures specified in the pie graphs, the concentration of each gas is constant while reactions continue to occur at equilibrium.

Another way of looking at the concentrations of substances at equilibrium is shown in Figure 19-6. Here are two graphical representations of the NO_2-N_2O_4 equilibrium system at 100°C. In Figure 19-6a the reaction begins with 1.00 mole of N_2O_4 and no NO_2. In Figure 19-6b the reaction begins with 2.00 moles of NO_2 and no N_2O_4. What is the concentration of N_2O_4 at equilibrium in both graphs? Notice that even though the initial concentrations of N_2O_4 at 100°C were very different, the equilibrium concentrations of N_2O_4 in both reaction flasks are the same. How do the NO_2 concentrations compare?

Figure 19-6 Changing concentrations of NO_2 and N_2O_4 in a sealed flask at 100°C, as equilibrium is reached.

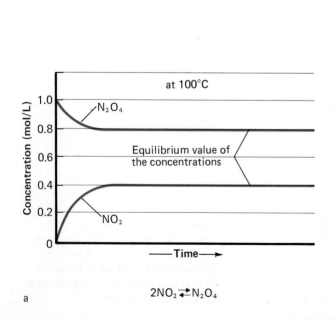

a

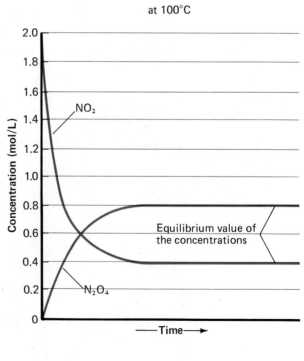

b

19-3 Effects of Altering Equilibrium Pressure and Temperature

Recall the experiment with the brown gas in the glass syringe described in Section 19-1. When pressure was applied to the gas the result was that the color of the gas became lighter. We wondered initially if some of the molecules disappeared. But now that you know that a chemical change is occurring in the syringe, you can use the equilibrium model to explain the changing color of the gas.

In increasing the pressure on the gas, the available volume of the syringe was reduced, compacting the molecules of NO_2 and allowing for more frequent collisions between molecules. The lightening of the gas color in the syringe indicates that NO_2 molecules are reacting to form N_2O_4 molecules, as shown in Figure 19-7. The increase in pressure initially favors the reaction that produces fewer moles per unit volume to counter the increase in pressure. This change makes sense when you consider the equation for the reaction.

$$2NO_2(g) \rightleftharpoons N_2O_4(g)$$

Two molecules of NO_2 are being converted to one molecule of N_2O_4. This forward reaction continues at a faster rate than the reverse reaction until a new equilibrium is reached. At this point there are fewer total particles in the smaller volume.

Conversely, a decrease in pressure results in an increase in the available volume for the system and favors the reaction that produces the greater number of moles per unit volume. For the NO_2-N_2O_4 equilibrium the reverse reaction will be favored by a decrease in pressure.

A change in pressure does not affect every equilibrium system. Consider the reaction for the decomposition of hydrogen iodide.

$$2HI(g) \rightleftharpoons H_2(g) + I_2(g)$$

In this system, the number of moles of reactant equals the number of moles of products, two moles on each side. An increase or decrease in

When the NO_2 gas is compressed the temperature also increases, which can cause a higher concentration of NO_2. The color shift is actually a result of the effects of both pressure and temperature. In this case, we are only dealing with the effect of changing pressure.

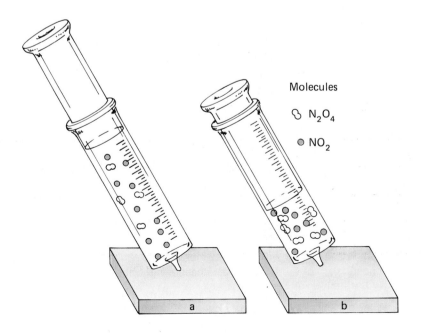

Molecules

$\text{\Large ʂ}$ N_2O_4

$\bullet$ NO_2

Figure 19-7 As the plunger is depressed on the sealed syringe, pressure is increased on the gas inside. Increasing pressure on the NO_2—N_2O_4 system favors the formation of N_2O_4 molecules.

Note that the formation of HI from solid I_2 is an endothermic reaction. This fact accounts for the positive ΔH listed for HI in handbooks.

Do You Know?

In 1955, the process to convert graphite to industrial diamonds was developed. High pressure favors the formation of diamond since its density is greater than that of graphite. The reaction is:

$$C_{graphite} + 188\ kJ \rightleftharpoons C_{diamond}$$

The conversion is carried out at temperatures near 2000°C and pressures between 50 000 and 100 000 atmospheres. Catalysts are used to obtain a suitable reaction rate.

pressure changes the available volume and increases the concentration of all components, but since there are the same number of reactant and product molecules the relative concentrations are not changed. As a result, the relative amount of reactants and products is unaffected by the change in pressure.

You have already observed how a change in temperature affects the NO_2-N_2O_4 equilibrium system. How does a change in temperature affect the HI equilibrium system?

The decomposition of hydrogen iodide is another reversible reaction in which color can be used to infer how changes affect this system at equilibrium.

$$2HI(g) + energy \rightleftharpoons H_2(g) + I_2(g)$$
$$\text{colorless} \qquad\qquad \text{colorless} \quad \text{deep purple}$$

The production of I_2 is signified by a deepening of the purple color in the reaction flask. If the color lightens you can assume that the rate of the reverse reaction momentarily increases to produce more colorless HI and restore equilibrium.

Reducing the temperature in a reaction flask containing HI, H_2, and I_2 at equilibrium causes the gas in the flask to lighten in color. Increasing the temperature in the flask causes the gas to darken. How does the equilibrium model explain these observations?

Increasing the temperature in the flask is, in effect, increasing the energy of the surroundings. You know from other discussions that substances react to achieve a lower energy state. However, it is difficult for molecules to release energy if the surrounding molecules already have a lot of energy. A way of overcoming this problem is for the endothermic reaction to occur. At higher temperatures, the endothermic reaction is favored because it essentially acts as a heat sink.

It then makes sense to conclude that the exothermic reaction is favored by a decrease in temperature. From the equation for the reaction, you can see that the forward reaction, the decomposition of HI, is endothermic. The reverse reaction, the formation of HI, is exothermic. The equilibrium model thus explains why the color of the gas darkens with an increase in temperature.

Review and Practice

1. pressure, color, concentration
2. No. All equilibrium macroscopic properties (like color) are constant.
3. As the color gets darker, the concentration of NO_2 is increasing and the concentration of N_2O_4 is decreasing.
4. the rates of opposing reactions

1. Name three macroscopic properties that can be observed to determine when a chemical equilibrium is attained.

2. In a flask initially containing colorless molecules of N_2O_4 the brown color characteristic of NO_2 appears, and the color steadily becomes darker. Is this system in equilibrium? Explain.

3. In item 2, if the brown color becomes darker, what is happening to the concentration of the NO_2? What is happening to the concentration of the N_2O_4 as the color deepens?

4. What specifically is equal in a chemical reaction that is in equilibrium?

Other Factors Affecting Equilibrium

Many of the reactions used in the commercial production of chemicals are reversible reactions that attain equilibrium. In the production of chemicals it is important to obtain the maximum yield of a certain product. Suppose you are in the business of making ammonia, NH_3. Ammonia is used extensively as a raw material in the production of fertilizers for agricultural. The production of ammonia from nitrogen and hydrogen is a reversible reaction that attains equilibrium.

$$N_2(g) + 3H_2(g) \rightleftharpoons 2NH_3(g)$$

Therefore, it is impossible to obtain a 100% yield of ammonia by using the appropriate mole quantities of nitrogen and hydrogen. The concentration of NH_3 at equilibrium can, however, be changed by altering the conditions under which the reaction occurs. Earlier in this chapter you looked at how changing the temperature and the pressure affected the production of both NO_2 and HI. What other factors could be used to increase the yield of NH_3 at equilibrium?

CONCEPTS

- **effects of changing concentration**
- **effects of catalysts**
- **using Le Chatelier's principle to predict the results of changing conditions on equilibrium systems**
- **applying the equilibrium model to the commercial production of ammonia**

Conditions for the commercial production of ammonia are covered in Section 19-7.

19-4 Effects of Changing Equilibrium Concentrations

Consider a reaction flask containing HI, H_2, and I_2 at equilibrium that has HI added to it. How will the reaction rates and concentrations be affected?

$$2HI(g) \rightleftharpoons H_2(g) + I_2(g)$$

If there are more HI molecules in the container, the chances of them colliding and decomposing are increased. Therefore, more H_2 and I_2 molecules are initially produced. The forward reaction is favored and the concentrations of H_2 and I_2 increase. Figure 19-8a shows how the system responds to the addition of HI, according to the model. The

Figure 19-8 The equilibrium model predicts an increase in molecules of iodine when HI is added to a flask containing H_2, I_2, and HI at equilibrium. As HI is added on the left; increased collisions between molecules produce more H_2 and I_2 molecules. The pie graph shows the relative concentrations of reactants and products at equilibrium.

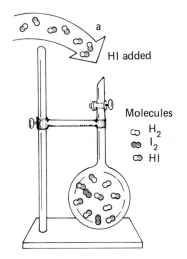

HI added

Molecules
- H_2
- I_2
- HI

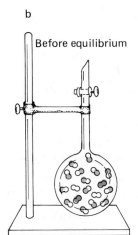

b   Before equilibrium

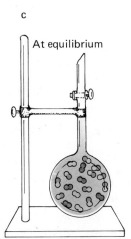

c   At equilibrium

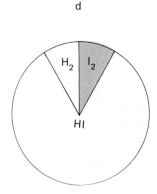

d

H_2 I_2

HI

Relative concentrations at equilibrium

Figure 19-9 The removal of HI from a flask containing H_2, I_2, and HI. The greater relative concentration of I_2 results in a darker color of the gas mixture. The model suggests decomposition of HI to H_2 and I_2 causing the color change. The pie graph shows the relative concentrations of substances at equilibrium.

actual reaction flask shows a darker violet color. Thus macroscopic observation supports the microscopic model explaining why the I_2 concentration increases.

Consider what will happen if you could remove H_2 from the flask. The gas in the flask darkens. Can you explain this observation using the equilibrium model? Figure 19-9 illustrates the removal of H_2 on a molecular level. Initially, some of the HI molecules decompose to replace the H_2 that was removed. For each H_2 molecule produced a molecule of I_2 also is formed. The forward reaction is favored for a time, but eventually the rates of the opposing reactions are equal again and the color is constant, but darker than before removal of H_2.

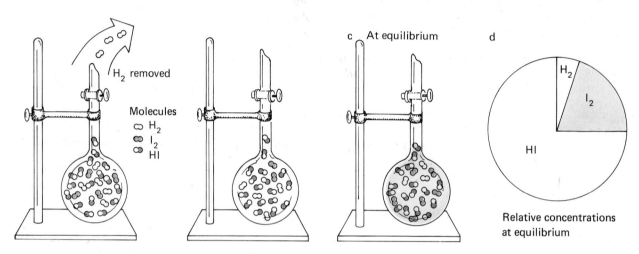

c At equilibrium

d

H_2 removed

Molecules
○ H_2
◉ I_2
∞ HI

H_2

I_2

HI

Relative concentrations at equilibrium

19-5 Effects of Catalysts on Equilibrium

An inhibitor has the effect of decreasing the rate of reaction. For example, an inhibitor may interfere with the functioning of a catalyst such that catalytic effects are cancelled and the original reaction rate restored.

In Chapter 18, it was noted that a catalyst provides an alternate lower energy pathway for a reaction. In the presence of a catalyst, both the forward and the reverse reactions will have increased rates and the system will reach equilibrium faster. The addition of a catalyst does not change the concentrations of the reactants and products at equilibrium. Therefore, a catalyst imposes no net change on an equilibrium system. Only the time required to reach equilibrium is affected.

19-6 Le Chatelier's Principle

CHEM THEME

Nature tends to counteract an imposed change, whether ecological, as in animal populations, physical, such as the law of action and reaction, and chemical, as described in this chapter.

Like yourself, the French chemist, Henry Louis Le Chatelier (1850–1936), studied the effects of changing conditions on a large number of chemical systems at equilibrium.

Let us review the characteristics of the equilibrium model that you have studied for gaseous reactions in the previous sections.

Temperature effects: Increasing the temperature of an equilibrium system favors the endothermic reaction and a new equilibrium state is reached.

Pressure effects: Increasing the pressure on an equilibrium system favors the reaction that produces the fewest number of particles per unit volume and a new equilibrium state is reached.

Concentration effects: Increasing the concentration of a reactant favors the forward reaction, whereas increasing the concentration of a product favors the reverse reaction, and a new equilibrium state is reached.

Catalytic effects: The addition of a catalyst has no net effect on an equilibrium system.

You can see that with the exception of adding a catalyst, one reaction tends to be favored until equilibrium is reached when the conditions are altered. Le Chatelier made the following generalization based on this same information that is now called **Le Chatelier's principle**.

If a system in equilibrium is subjected to a change, processes occur that tend to counteract the imposed change and the system reaches a new state of equilibrium.

How can Le Chatelier's principle be used to explain what is happening to the NO_2-N_2O_4 equilibrium when the temperature is increased? The change imposed on the system is a change in temperature. The deepening of color in the reaction flask suggests that N_2O_4 decomposes to increase the concentration of NO_2. This process is referred to as a *shift* in the original equilibrium. Since the equation is written as:

$$2NO_2(g) \rightleftharpoons N_2O_4(g) + \text{energy}$$

and the increase in temperature favors the reverse reaction, the increase is said to shift the equilibrium to the left. If the forward reaction had initially been favored, the equilibrium system is said to have shifted to the right. When the color remains constant, equilibrium has once more been attained and the rates of the opposing reactions are again equal. Figure 19-10 is a visual representation of the effects of shifting equilibria.

Be sure students understand that the so-called "shift in equilibrium" refers to the direction of the reaction that is favored until a new equilibrium is reached.

Do You Know?

Physicians determine the effects of some drugs by monitoring the concentrations of various chemicals in the blood. These concentrations depend on equilibrium reactions that shift with varying conditions. For instance, in the treatment of sickle-cell anemia, doctors observe the effects of drugs on the concentration of bound oxygen in the blood.

19-7 Application: The Haber Process

In 1910, the German chemist, Fritz Haber (1868–1934), developed an efficient process that combines nitrogen and hydrogen to make ammonia. The ammonia product could then be used as a raw material to make other nitrogen compounds. The production of ammonia using this process involves an equilibrium system.

$$N_2(g) + 3H_2(g) \rightleftharpoons 2NH_3(g) + 92.4 \text{ kJ}$$

The formation of ammonia represents a decrease in the total moles of gas from four moles to two moles. You know from Le Chatelier's principle that high pressure would increase the relative yield of NH_3 at equilibrium.

The reaction to form NH_3 is exothermic. From Le Chatelier's principle you can readily see that a high temperature favors the decomposition of NH_3. Low temperature favors the formation of NH_3. Yet at 25°C both the forward and reverse reactions take place slowly.

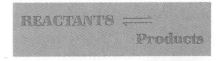

Figure 19-10 An increase in concentration of products indicates an equilibrium shift to the left. An increase in concentration of reactants indicates a shift to the right.

Do You Know?

Before the Haber process was developed, the only other commercial method of producing bound nitrogen was the distillation of coal to produce coke. Coal contains about 1% ammonia.

Do You Know?

Le Chatelier had tried an experiment in 1901 that was similar to Haber's, but a violent explosion took place and Le Chatelier stopped work on the process. Late in life Le Chatelier wrote, "I let the discovery of ammonia synthesis slip through my hands. It was the greatest blunder of my scientific career."

Thus it appears that high pressure and low temperature provide optimal conditions for producing ammonia. But are these conditions practical? It is expensive to build high-pressure equipment. In addition, a high temperature is necessary for a satisfactory reaction rate as time is money. Figure 19-11 shows yields of ammonia under different reaction conditions. The compromise between temperature and pressure used for this process involves an intermediate temperature of close to 500°C and a pressure of 350 atmospheres. Even then the success of the process depends on the use of a catalyst to achieve a suitable reaction rate. Under these conditions, only about 30 percent of the reactants are converted to NH_3. To increase the yield even further, ammonia is removed from the reaction container by liquefying it under conditions at which N_2 and H_2 will remain as gases. The unreacted N_2 and H_2 are recycled until the conversion yield is much higher. Pressures of up to 1000 atmospheres and temperatures close to 700°C are used in conjunction with a catalyst for this recycling process.

Almost all of the nitrogen compounds used today are derived from ammonia made by using the Haber-Bosch process. Haber's work was turned over to Carl Bosch for industrial development. Haber received the 1918 Nobel prize in chemistry for his work.

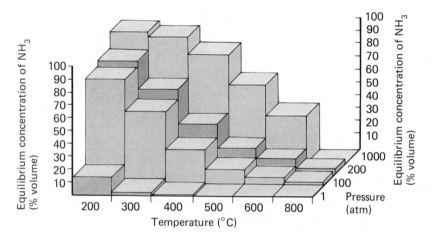

Figure 19-11 The equilibrium concentration of ammonia varies with differences in temperature and pressure on the system.

1. [I_2] is decreased.
2. [NO_2] is decreased.
3. a. right
   b. no shift
   c. left

Review and Practice_____

1. What is the effect on the concentration of I_2 when a small amount of H_2 is added to this system in equilibrium?

$$2HI(g) \rightleftharpoons H_2(g) + I_2(g)$$

2. What is the effect on the concentration of NO_2 when the volume of this system at equilibrium is increased?

$$N_2(g) + 2O_2(g) \rightleftharpoons 2NO_2(g)$$

3. State the direction of the equilibrium shift in these reactions when the pressure is increased.
   a. $PCl_3(g) + Cl_2(g) \rightleftharpoons PCl_5(g)$
   b. $2NO(g) \rightleftharpoons N_2(g) + O_2(g)$
   c. $4NH_3(g) + 5O_2(g) \rightleftharpoons 4NO(g) + 6H_2O(g)$

A Quantitative Look at Reaction Equilibrium

You have studied how changing conditions can affect the concentrations of reactants and products in a system by shifting the equilibrium. You are now ready for a mathematical description of equilibrium systems. A quantitative look at equilibrium will better help you see the effects of changing conditions to give the maximum yield of a desired product.

CONCEPTS

- equilibrium constants
- calculating equilibrium concentrations
- temperature effects on K_{eq}
- the roles of pure solids and liquids in the equilibrium expression
- complex ion equilibrium systems in solution

19-8 Equilibrium Expression and the Equilibrium Constant, K_{eq}

Consider the chemical system for the decomposition of hydrogen iodide at equilibrium at 423°C.

$$2HI(g) \rightleftharpoons H_2(g) + I_2(g)$$

Table 19-1 lists equilibrium concentrations for this reaction at 423°C based on laboratory measurements. Notice that the concentrations of the products, H_2 and I_2, are not necessarily equal. The equilibrium concentrations of H_2 and I_2 will be equal only if the initial concentrations of H_2 and I_2 are equal or if the experiment is begun with HI. Study the equation, and the table, and think about what is occurring on a microscopic level to explain the different equilibrium values.

Table 19-1

EXPERIMENT NUMBER	[HI]	[H₂]	[I₂]
	EQUILIBRIUM CONCENTRATIONS FOR HI, H_2, AND I_2 AT 423°C EXPRESSED IN MOLES PER LITER		
1	17.7×10^{-3}	1.83×10^{-3}	3.13×10^{-3}
2	16.5×10^{-3}	2.91×10^{-3}	1.71×10^{-3}
3	13.5×10^{-3}	4.56×10^{-3}	0.74×10^{-3}
4	3.53×10^{-3}	0.48×10^{-3}	0.48×10^{-3}
5	8.41×10^{-3}	1.14×10^{-3}	1.14×10^{-3}

The equilibrium concentrations of H_2 and I_2 will be equal if the reaction begins with equal concentrations of H_2 and I_2. Can you explain why these concentrations are not equal in experiments 1, 2, and 3? The data in Table 19-1 can be expressed in various ways mathematically. It would be useful to have an expression that provides a constant ratio for the equilibrium of reactants and products in this system at

The various relationships in Table 19-2 are included to illustrate a way in which mathematical relationships are worked out. Be sure the students understand that only the center expression and resulting values give a constant expression.

423°C. Recall from Chapter 18 that brackets are used to represent the concentrations of reactants and products as molarities or moles per liter. Thus $[I_2]$ is shorthand for saying "the concentration of iodine in moles per liter."

Table 19-2

VALUES FOR VARIOUS CONCENTRATION RATIOS FOR TABLE 19-1		
EXPERIMENT NUMBER	$\dfrac{[H_2][I_2]}{[HI]}$	$\dfrac{[H_2][I_2]}{[HI]^2}$	$\dfrac{[H_2][I_2]}{2[HI]}$
1	3.24×10^{-4}	1.83×10^{-2}	1.62×10^{-4}
2	3.02×10^{-4}	1.83×10^{-2}	1.51×10^{-4}
3	2.50×10^{-4}	1.85×10^{-2}	1.25×10^{-4}
4	0.65×10^{-4}	1.85×10^{-2}	0.32×10^{-4}
5	1.55×10^{-4}	1.84×10^{-2}	0.77×10^{-4}
	ratio not constant	ratio constant	ratio not constant

Thus it appears that the following expression provides us with a constant ratio regardless of the changing concentrations.

$$\frac{[H_2][I_2]}{[HI]^2} = \text{constant} = 1.84 \times 10^{-2} \text{ (at 423°C)}$$

This ratio is the product of the equilibrium concentrations of hydrogen and iodine, the products of the reaction, divided by the square of the concentration of the reactant, HI. This constant value is called the **equilibrium constant** and shows the ratio of product concentration terms to reactant concentration terms. The equilibrium constant often is represented by the symbol K_{eq}.

You may have wondered why the reciprocal of the ratio was not used. It too gives a constant. The decision was a matter of choice. Chemists have agreed to use the ratio with concentrations of products in the numerator and with the concentrations of reactants in the denominator. In this ratio, the power to which the concentration of each substance is raised is equal to its coefficient in the balanced equation. This use of exponents may be clearer if the chemical equation is written in a slightly different way.

$$HI + HI \rightleftharpoons H_2 + I_2$$

Then the constant ratio has the form:

$$K_{eq} = \frac{[H_2][I_2]}{[HI][HI]}$$

This expression can be rewritten as:

$$K_{eq} = \frac{[H_2][I_2]}{[HI]^2}$$

An equilibrium constant expression can be written for any system that attains equilibrium. For a reaction with this general form:

$$aA + bB \rightleftharpoons eE + fF$$

the expression for the equilibrium constant is:

$$K_{eq} = \frac{[E]^e[F]^f}{[A]^a[B]^b}$$

It is important to note that these equations provide a mathematical ratio that closely approximates what actually occurs in the lab. K_{eq} represents a kind of ideal equilibrium law just as the ideal gas law closely approximates the behavior of gases under ideal conditions. Using the general equation for K_{eq}, it is possible to calculate the K_{eq} for a reaction from laboratory data.

Equilibrium constants can be calculated using partial pressures for gas reactions or ion activities for reactions in aqueous solution. The units associated with K_{eq} can vary. The HI reaction has no units since moles per liter squared in the numerator are cancelled by moles per liter squared in the denominator.

Example 19-1

The equilibrium equation for the formation of ammonia is:

$$N_2(g) + 3H_2(g) \rightleftharpoons 2NH_3(g)$$

At 200°C, the concentrations of nitrogen, hydrogen, and ammonia at equilibrium are measured and found to be $[N_2] = 2.12$, $[H_2] = 1.75$, and $[NH_3] = 84.3$. Calculate the equilibrium constant at the temperature given.

▶ **Suggested solution**

First, write the equilibrium expression.

$$K_{eq} = \frac{[NH_3]^2}{[N_2][H_2]^3}$$

Substitute into this equilibrium expression the concentrations given in the problem.

$$K_{eq} = \frac{[84.3]^2}{[2.12][1.75]^3}$$

Simplifying the expression gives

$$K_{eq} = \frac{(7.11 \times 10^3)}{(2.12)(5.36)}$$

$$K_{eq} = 626$$

Table 19-3 lists various equilibrium systems and the corresponding K_{eq} at specific temperatures.

What does the numerical value for K_{eq} tell you? If K_{eq} is large, either the numerator in the equilibrium expression must be large or the denominator must be small. Either way, at equilibrium, there is a high concentration of products relative to reactants. A small value for K_{eq} means the opposite. At equilibrium, there is a high concentration of

Table 19-3

SOME EQUILIBRIUM SYSTEMS AND CONSTANTS		
REACTION	**EQUILIBRIUM EXPRESSION**	**K_{eq}**
$2NO_2(g) \rightleftharpoons N_2O_4(g)$	$K_{eq} = \dfrac{[N_2O_4]}{[NO_2]^2}$	1.20 at 55°C
$N_2(g) + 3H_2(g) \rightleftharpoons 2NH_3(g)$	$K_{eq} = \dfrac{[NH_3]^2}{[N_2][H_2]^3}$	626 at 200°C
$2HI(g) \rightleftharpoons H_2(g) + I_2(g)$	$K_{eq} = \dfrac{[H_2][I_2]}{[HI]^2}$	1.85×10^{-2} at 425°C 85 at 25°C
$2SO_2(g) + O_2(g) \rightleftharpoons 2SO_3(g)$	$K_{eq} = \dfrac{[SO_3]^2}{[SO_2]^2[O_2]}$	261 at 727°C
$PCl_5(g) \rightleftharpoons PCl_3(g) + Cl_2(g)$	$K_{eq} = \dfrac{[PCl_3][Cl_2]}{[PCl_5]}$	2.24 at 227°C 33.3 at 487°C
$COCl_2(g) \rightleftharpoons CO(g) + Cl_2(g)$	$K_{eq} = \dfrac{[CO][Cl_2]}{[COCl_2]}$	8.2×10^{-2} at 627°C
$2NO(g) + O_2(g) \rightleftharpoons 2NO_2(g)$	$K_{eq} = \dfrac{[NO_2]^2}{[NO]^2[O_2]}$	6.45×10^5 at 227°C
$C(s) + 2H_2(g) \rightleftharpoons CH_4(g)$	$K_{eq} = \dfrac{[CH_4]}{[H_2]^2}$	8.1×10^8 at 25°C
$CO(g) + H_2O(g) \rightleftharpoons CO_2(g) + H_2(g)$	$K_{eq} = \dfrac{[CO_2][H_2]}{[CO][H_2O]}$	1.02×10^5 at 25°C 10.0 at 690°C 3.59 at 800°C
$H_2(g) + Cl_2(g) \rightleftharpoons 2HCl(g)$	$K_{eq} = \dfrac{[HCl]^2}{[H_2][Cl_2]}$	1.8×10^{33} at 25°C
$C(s) + H_2O(g) \rightleftharpoons CO(g) + H_2(g)$	$K_{eq} = \dfrac{[CO][H_2]}{[H_2O]}$	1.96 at 1000°C

reactants relative to products. Whenever you see a numerical value for an equilibrium constant, think of it in this way:

A large K_{eq} means the products are favored at equilibrium.

A small K_{eq} means the reactants are favored at equilibrium. These ideas are illustrated in Figure 19-12.

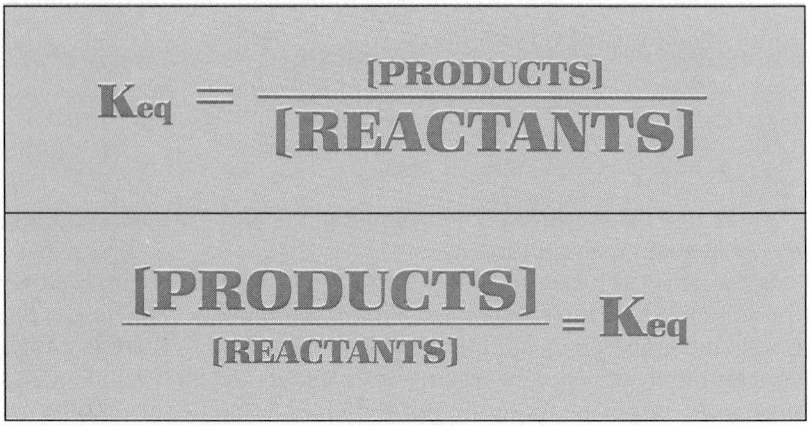

Figure 19-12 A small K_{eq} indicates a low concentration of products relative to reactants. A large K_{eq} indicates a large concentration of products relative to reactants.

$$K_{eq} = \frac{[PRODUCTS]}{[REACTANTS]}$$

$$\frac{[PRODUCTS]}{[REACTANTS]} = K_{eq}$$

Example   19-2

Suppose there are 2.00 moles of HI in a one-liter flask at 425°C, that react to produce H_2 and I_2. When equilibrium is reached, the concentrations of H_2 and I_2 are determined to each be 0.214 mole per liter. How can the equilibrium constant for this reaction be calculated?

▶ **Suggested solution**

From these data you assume the reaction does not go to completion because the balanced equation shows that the two moles of HI should react to produce one mole each of H_2 and I_2. Since the experimental data show there are 0.214 mole each of H_2 and I_2, you can assume that twice this amount (2×0.214) or 0.428 mole of HI has reacted. At equilibrium, the HI concentration equals the initial concentration minus that amount that reacted to form the products. The relationship between the initial and equilibrium concentrations are shown in Table 19-4.

Table 19-4

CONCENTRATIONS IN A 1 L REACTION FLASK AT 425°C			
GAS	**INITIAL**	**EQUILIBRIUM**	**METHOD USED TO FIND THE CONCENTRATION**
HI	2.00M	2.00M − 2(0.214) = 1.57M	amount of H_2 and I_2 produced, subtracted from original HI concentration
H_2	0	0.214M	determined experimentally
I_2	0	0.214M	determined experimentally

Substituting the equilibrium concentrations into the equilibrium expression gives:

$$K_{eq} = \frac{[H_2][I_2]}{[HI]^2} = \frac{[0.214][0.214]}{[1.57]^2} = 1.86 \times 10^{-2}$$

The value 1.86×10^{-2} is the K_{eq} for the HI reaction at 425°C and is affected only by a change in temperature. The value 1.86×10^{-2} holds within experimental uncertainty for this reaction at 425°C no matter how much the concentrations of each substance vary, and agrees with the value given in Table 19-3.

Example   19-3

Calculate the equilibrium concentration of HI at 425°C for the reaction

$$2HI(g) \rightleftharpoons H_2(g) + I_2(g)$$

if the equilibrium concentrations are $[H_2] = 2.90 \times 10^{-3}$ and $[I_2] = 1.70 \times 10^{-3}$.

▶ Suggested solution

The equilibrium expression for this reaction is:

$$K_{eq} = \frac{[H_2][I_2]}{[HI]^2}$$

Table 19-3 shows that at 425°C the value of K_{eq} is 1.85×10^{-2}. Since the concentrations of H_2 and I_2 are known, the concentration of HI can be calculated algebraically as x.

$$1.85 \times 10^{-2} = \frac{[2.90 \times 10^{-3}][1.70 \times 10^{-3}]}{[x]^2}$$

Simplifying the expression gives:

$$1.85 \times 10^{-2}x^2 = (2.90 \times 10^{-3})(1.70 \times 10^{-3})$$

To isolate the x^2 term on the left side of the equation, divide both sides by 1.85×10^{-2}. Solving for x^2 you get:

$$x^2 = 2.66 \times 10^{-4}$$

The value of x is determined by taking the square root of both sides of the equation.

$$x = \sqrt{2.66 \times 10^{-4}}$$
$$x = 1.63 \times 10^{-2}$$

The equilibrium concentration of HI is 1.63×10^{-2} mole per liter.

Example 19-4

Calculate the equilibrium concentration at 400°C of NH_3 for the reaction:

$$N_2(g) + 3H_2(g) \rightleftharpoons 2NH_3(g)$$

The equilibrium concentrations for the reactants at 400°C are $[N_2] = 0.45$ and $[H_2] = 1.10$. The K_{eq} at this temperature is 1.7×10^{-2}.

▶ Suggested solution

Begin by writing the equilibrium expression for the ammonia reaction.

$$K_{eq} = \frac{[NH_3]^2}{[N_2][H_2]^3}$$

Let x represent the value of the NH_3 concentration. Substituting the known values given in the problem, you have

$$1.7 \times 10^{-2} = \frac{[x]^2}{[0.45][1.10]^3}$$

$$x^2 = 1.0 \times 10^{-2}$$

Solving for x requires taking the square root of both sides of the equation.

$$x = \sqrt{1.0 \times 10^{-2}}$$

$$x = 1.0 \times 10^{-1}$$

Thus $[NH_3] = 1.0 \times 10^{-1}$.

Example 19-5

The equilibrium concentrations of SO_2 and O_2 are each $0.0500M$ and $K_{eq} = 85.0$ at 25°C for the reaction:

$$2SO_2(g) + O_2(g) \rightleftharpoons 2SO_3(g)$$

Calculate the equilibrium concentration for SO_3 at this temperature.

▶ ## Suggested solution

The equilibrium expression is:

$$K_{eq} = \frac{[SO_3]^2}{[SO_2]^2[O_2]}$$

Let x equal the concentration of SO_3. Substituting the value of the equilibrium constant given and the concentrations of SO_2 and O_2 you have:

$$85.0 = \frac{(x)^2}{[0.0500]^2[0.0500]}$$

$$x^2 = 0.0106$$

Taking the square root of both sides of the equation gives:

$$x = 0.103$$

Thus the concentration of SO_3 is $0.103M$.

Review and Practice

1. In an equilibrium expression for a reaction, do the concentrations of the reactants appear in the numerator or the denominator?

2. Write the equilibrium expression for the following reactions:
   a. $H_2(g) + S(g) \rightleftharpoons H_2S(g)$
   b. $PCl_5(g) \rightleftharpoons PCl_3(g) + Cl_2(g)$
   c. $4NH_3(g) + 5O_2(g) \rightleftharpoons 4NO(g) + 6H_2O(g)$
   d. $CH_4(g) + 2O_2(g) \rightleftharpoons CO_2(g) + 2H_2O(g)$

1. denominator

2. a. $K_{eq} = \dfrac{[H_2S]}{[H_2][S]}$

   b. $K_{eq} = \dfrac{[PCl_3][Cl_2]}{[PCl_5]}$

   c. $K_{eq} = \dfrac{[NO]^4[H_2O]^6}{[NH_3]^4[O_2]^5}$

   d. $K_{eq} = \dfrac{[CO_2][H_2O]^2}{[CH_4][O_2]^2}$

3. $K_{eq} = 0.50$
4. $[PCl_5] = 0.40$
5. $[NO_2] = 0.72$

3. The following reaction represents an equilibrium system.

$$PCl_5(g) \rightleftharpoons PCl_3(g) + Cl_2(g)$$

Calculate the equilibrium constant for this reaction if the equilibrium concentrations are $[PCl_5] = 0.32$, $[PCl_3] = 0.40$, and $[Cl_2] = 0.40$.

4. The following reaction reaches equilibrium at 25°C and has an equilibrium constant of 1.78.

$$PCl_5(g) \rightleftharpoons PCl_3(g) + Cl_2(g)$$

If the equilibrium concentrations of PCl_3 and Cl_2 are both 0.85 mole per liter, calculate the concentration of PCl_5 at equilibrium.

5. Two moles of $NO_2(g)$ are injected into an evacuated one-liter flask. At 55°C the equilibrium concentration of N_2O_4 is $1.7 \times 10^{-2}M$. What is the equilibrium concentration of $NO_2(g)$?

19-9 Effect of Temperature on K_{eq}

All chemical changes involve either the release or the absorption of energy. In the exothermic reaction for the formation of hydrogen iodide, heat is evolved.

$$H_2(g) + I_2(g) \rightleftharpoons 2HI(g) + 9.4 \, kJ$$

Lowering the temperature of the system favors the formation of hydrogen iodide. Raising the temperature favors the decomposition of hydrogen iodide. If the forward reaction is favored, more product forms and the value of K_{eq} increases. If the reverse reaction is favored, the amount of reactants increases. Hence the values of the denominator of K_{eq} also increase and the final value of K_{eq} decreases. The value of the equilibrium constant for the HI reaction decreases from 67.5 at 357°C to 50.0 at 400°C.

Refer to Section 19-3, if needed.

19-10 Equilibrium Systems Involving Liquids and Solids

Notice that the reactions studied in the previous sections involve reactants and products that exist as gases or are in solution. How are reactions involving solids and liquids treated?

Experiments involving solid and gaseous substances in a closed system show that once the equilibrium is established the equilibrium concentrations of a gaseous product are unaffected by changing the amount of solid in the reaction container. In reactions taking place in aqueous solution, the concentration of the water can usually be ignored since only a tiny fraction of water is reacting and the concentration hardly changes. (Some specific equilibrium systems involving water will be covered in the next chapter.)

In determining the value of K_{eq} for a reaction involving solids or liquids, the concentrations of these substances are omitted from the equilibrium expression because their concentrations are constant and as such are part of the K_{eq}.

At 900°C, solid calcium carbonate exists in equilibrium with its decomposition products solid calcium oxide and carbon dioxide gas. The reaction is:

$$CaCO_3(s) \rightleftharpoons CaO(s) + CO_2(g)$$

You may question why changing the initial mass of solid $CaCO_3$ used does not affect the equilibrium. The amount of CO_2 produced in this reaction is limited by the size of the reaction vessel and the temperature of the reaction. Figure 19-13 compares the reaction of different amount of $CaCO_3$ in containers of the same volume, 1000 cm³. Even if you started with 100 grams of $CaCO_3$ instead of 40 grams, the amount of CO_2 produced would still be close to 1.0×10^{-2} mole, as long as the temperature and the available volume of the container remained the same.

Using this reasoning, the equilibrium expression for the $CaCO_3$ reaction would be written as:

$$K_{eq} = [CO_2]$$

Do You Know?

Quicklime, CaO, is used in agriculture and in the production of mortar and plaster. In the commercial production of CaO from limestone, $CaCO_3$, the decomposition reaction is prevented from reaching equilibrium by removal of the product CO_2.

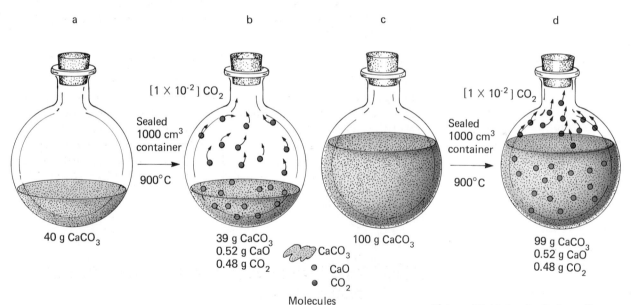

Figure 19-13 A mixed-phase (heterogeneous) equilibrium system, $CaCO_3 \rightleftharpoons CaO + CO_2$, at 900°C. Decomposition of $CaCO_3$ is favored until the limit on CO_2 concentration, set by the volume of the container, is reached. Changing the amount of solid $CaCO_3$ in the system does not affect the equilibrium concentration if the volume of the container is constant.

19-11 Equilibrium Reactions in Solution

Many important equilibrium systems occur in solution. One that can be investigated in the laboratory is the equilibrium between iron(III) ions and thiocyanate ions. This equilibrium system is prepared by mixing 5 drops of 0.20M iron(III) nitrate, $Fe(NO_3)_3$, with 5 drops of 0.20M potassium thiocyanate, KSCN, which can then be diluted to make any color changes visible. The solution of $Fe(NO_3)_3$ is an electrolyte. Recall from Chapter 16 that an electrolyte is a substance capable of conducting electricity. Figure 19-14 shows photos of solutions of $Fe(NO_3)_3$ and KSCN, and of the equilibrium between iron(III) ions and thiocyanate ions when the two solutions are mixed. For each

It is suggested that you demonstrate the $FeSCN^{2+}$ equilibrium system to students. The visual demonstration will really enhance their understanding.

Figure 19-14 Test tube *a* contains 0.20*M* Fe(NO₃)₃, test tube *b*, 0.20*M* KSCN. Test tube *c* is the solution equilibrium resulting when *a* and *b* are mixed. Diluting the system (test tube *d*) allows changes to be observed readily.

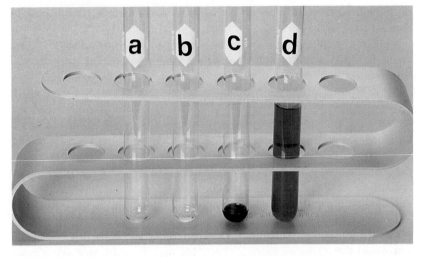

Fe^{3+} ion in solution there are three NO_3^- ions so the solution is electrically neutral. The KSCN solution is also electrolytic due to the K^+ ions and SCN^- ions. When Fe^{3+} ions come in contact with SCN^- ions, they combine to form a complex ion, $FeSCN^{2+}$, which is soluble in water. The K^+ ions and NO_3^- ions remain in solution and unchanged throughout this process. Recall that these ions are called spectator ions. Figure 19-15 illustrates how the ions in each solution appear before and after mixing.

Figure 19-15 Ions in test tubes *a*, *b*, and *c*, according to the model. Test tube *a* contains Fe^{3+} and NO_3^- ions. Test tube *b* contains K^+ and SCN^- ions. Mixing the solutions results in formation of the complex ion, $FeSCN^{2+}$. In test tube *c*, K^+ and NO_3^- remain in solution as spectator ions.

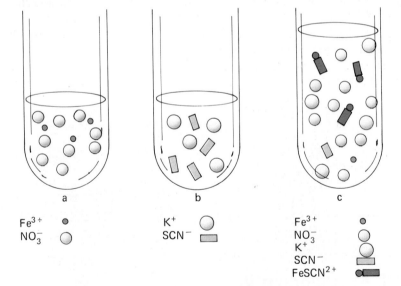

The equilibrium equation can be written as:

$$Fe(NO_3)_3(aq) + KSCN(aq) \rightleftharpoons FeSCN^{2+}(aq) + KNO_3(aq)$$

In ionic form this equation would be written as:

$$Fe^{3+}(aq) + 3NO_3^-(aq) + K^+(aq) + SCN^-(aq) \rightleftharpoons$$
$$FeSCN^{2+}(aq) + K^+(aq) + 3NO_3^-(aq)$$

Since K^+ and $3NO_3^-$ appear on both sides of the equation, they are cancelled to give the net ionic equation.

$$Fe^{3+}(aq) + SCN^-(aq) \rightleftharpoons FeSCN^{2+}(aq)$$

In the lab, the effect of adding various ions to this equilibrium system can be studied. Using Le Chatelier's principle, which you studied in Section 19-6, it is possible to explain the changes observed. When ions are added the color becomes darker, stays the same, or becomes lighter. The forward reaction makes the color darker and the reverse reaction makes the color lighter. If the rate of the forward reaction is increased (formation of $FeSCN^{2+}$), the color of the solution becomes darker as the equilibrium shifts to the right. If the reverse reaction is increased (decomposition of $FeSCN^{2+}$), the color becomes lighter as the equilibrium shifts to the left. When the color is again constant, a new equilibrium has been established with a higher or lower concentration of $FeSCN^{2+}$.

Consider what happens in the following experiment. First, a drop of solution containing K^+ and NO_3^- ions is added to a test tube containing $FeSCN^{2+}$. The color of the $FeSCN^{2+}$ solution does not change as shown in Figure 19-16. Why? Recall that both K^+ and NO_3^- are spectator ions. The addition of spectator ions does not affect the equilibrium.

To a separate sample of the $FeSCN^{2+}$ solution, a drop of solution containing Fe^{3+} ions and Cl^- ions is added. The color becomes darker, then remains constant as a new equilibrium is established. How could this observation be interpreted? Look at the equation for the reaction again:

$$Fe^{3+} + SCN^- \rightleftharpoons FeSCN^{2+}$$

The system reacts to remove the added Fe^{3+} ions by combining them with SCN^- ions already in solution. This shifts the equilibrium to the right. The addition is counteracted as predicted by Le Chatelier's principle as a new equilibrium is reached.

To another sample of the $FeSCN^{2+}$ at equilibrium, a drop containing Ag^+ ions and NO_3^- ions is added. The color becomes lighter and a

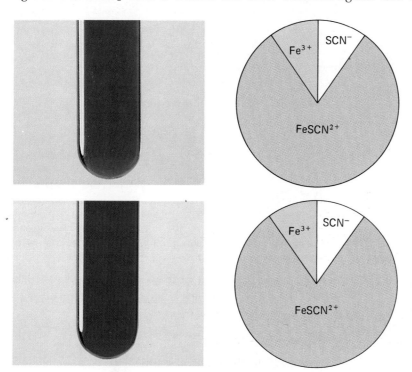

Figure 19-16 Macroscopic and microscopic illustrations of ion addition to the $FeSCN^{2+}$ equilibrium. The upper diagram is a control system for comparison. Addition of K^+ and NO_3^- has no observable effect on the lower test tube.

precipitate forms. Which way has equilibrium been shifted? The observable changes can be interpreted by assuming that the Ag^+ ions produced the change since the NO_3^- ions are known to be spectator ions. But what is the precipitate? The solution contains the following ions: Fe^{3+}, SCN^-, $FeSCN^{2+}$, Ag^+, and NO_3^-. At this point, checking a handbook of physical properties will reveal that of the ionic compounds indicated, AgSCN is the only possible identity for the precipitate. All other cation-anion combinations form soluble substances. So the Ag^+ ions combine with SCN^- ions (opposite charges) to form insoluble AgSCN(s), thus removing SCN^- from the system. The system responds to the decrease in available SCN^- by shifting the equilibrium to the left as some of the $FeSCN^{2+}$ complex ions decompose to replace the SCN^- ions. These changes are represented in Figure 19-17.

Figure 19-17 The addition of Fe^{3+} to the top test tube shifts the equilibrium to the right producing more $FeSCN^{2+}$. The addition of Ag^+ to the bottom test tube produces a precipitate and a shift to the left.

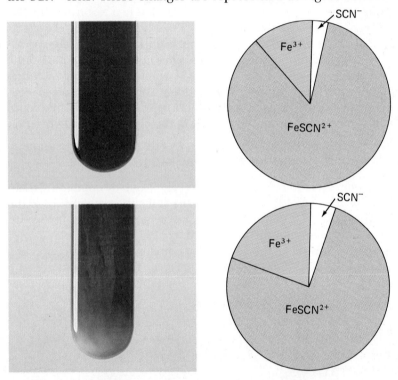

Review and Practice

6.45×10⁵ @ 227°C

1. products
2. increased

3. a. $K_{eq} = \dfrac{[H_2S]}{[H_2]}$

   b. $K_{eq} = [CO_2]$

   c. $K_{eq} = \dfrac{[NO_2]^4[H_2O]^6}{[NH_3]^4[O_2]^7}$

1. Does the equilibrium constant given for the formation of nitrogen dioxide from nitrogen oxide and oxygen shown in Table 19-3 indicate that reactants or products are favored?

2. If the temperature of an exothermic reaction at equilibrium is lowered, is the value of K_{eq} increased or decreased?

3. Write equilibrium expressions for the following reactions.
   a. $H_2(g) + S(l) \rightleftharpoons H_2S(g)$
   b. $CaCO_3(s) \rightleftharpoons CaO(s) + CO_2(g)$
   c. $4NH_3(g) + 7O_2(g) \rightleftharpoons 4NO_2(g) + 6H_2O(g) + energy$

Summary

■ Many important chemical reactions are reversible. When the forward and reverse reaction rates are equal in a closed system, a state of equilibrium occurs.

■ When a system in equilibrium undergoes a change, processes occur that tend to counteract the imposed change and establish a new equilibrium. Le Chatelier's principle can be applied to predict how changes in the temperature, pressure, or concentrations of the reactants and products will affect an equilibrium system.

■ The addition of a catalyst does not affect equilibrium but a catalyst can reduce the time required for a chemical system to reach equilibrium.

■ The equilibrium expression gives the quantitative relationships between the concentrations of the reactants and the products at a given temperature. The equilibrium constant, K_{eq}, can be calculated from experimental data. If the equilibrium constant is known, it can be used to calculate the concentrations of the unknown reactants or products for a specific temperature.

■ The quantitative aspects of chemical equilibrium can be shown using the expression for the equilibrium constant. For the reaction

$$aA + bB \rightleftharpoons eE + fF$$

K_{eq} has the form

$$K_{eq} = \frac{[E]^e[F]^f}{[A]^a[B]^b}$$

where $[A]$, $[B]$, $[E]$, and $[F]$ stand for the concentrations of these substances in moles per liter at equilibrium. The exponents a, b, e, and f are taken from the coefficients in the balanced equation.

■ The value of the equilibrium constant, K_{eq}, depends on the temperature of the system. An increase in temperature favors the endothermic reaction; a decrease in temperature favors the exothermic reaction.

■ The numerical value for the equilibrium constant indicates if the equilibrium favors the formation of reactants or products. When K_{eq} is small, the reactants are favored. When K_{eq} is large, the products are favored.

■ For equilibrium reactions involving solids or liquids, the concentrations of these substances are part of K_{eq}. This is because the concentrations of solids and liquids remain constant throughout the reactions.

■ Many equilibrium reactions occur in an aqueous solution. Changes in the concentration of ions can be predicted in accordance with Le Chatelier's principle.

Chemically Speaking

equilibrium constant
Le Chatelier's principle
reversible reaction

Review

1. Write the reverse of this synthesis reaction. (19-1)

$$2Hg(g) + O_2(g) \longrightarrow 2HgO(s)$$

2. The following reaction takes place in an open container.

$$MgCO_3(s) + 2HCl(aq) \longrightarrow$$
$$MgCl_2(aq) + H_2O(l) + CO_2(g)$$

This reaction will not reach equilibrium. Why? (19-1)

3. Name two properties that can be measured to determine whether a chemical system is in equilibrium. (19-2)

4. Describe the microscopic events taking place at equilibrium for this reaction. (19-2)

$$2Hg(g) + O_2(g) \rightleftharpoons 2HgO(s)$$

5. Name the four conditions necessary for a chemical system to reach equilibrium. (19-2)

6. State Le Chatelier's principle. (19-6)

7. When an equilibrium is shifted to the right, does the concentration of the reactants increase or decrease? (19-6)

8. How might the addition of a catalyst affect the equilibrium concentrations for this reaction? (19-5)

$$2NO(g) + O_2(g) \rightleftharpoons 2NO(g)$$

9. The color in the following equilibrium system darkens when another reagent is added. Is the equilibrium shifted to the right or to the left? (19-6)

$$Fe^{3+}(aq) + SCN^-(aq) \rightleftharpoons FeSCN^{2+}(aq)$$

10. Write the equilibrium expression for the following reactions. (19-8)
    a. $wW + xX \rightleftharpoons uU + vV$
    b. $H_2O(g) + CO(g) \rightleftharpoons H_2(g) + CO_2(g)$
    c. $COCl_2(g) \rightleftharpoons CO(g) + Cl_2(g)$
    d. $H_2(g) + Cl_2(g) \rightleftharpoons 2HCl(g) + energy$
    e. $CO(g) + NO_2(g) \rightleftharpoons CO_2(g) + NO(g)$
    f. $Zn(s) + 2Ag^+(g) \rightleftharpoons Zn^{2+}(aq) + 2Ag(s)$
    g. $C_2H_6(g) \rightleftharpoons H_2(g) + C_2H_4(g)$

Interpret and Apply

1. Use Le Chatelier's principle to predict how the changes listed will affect the following equilibrium reaction. **(19-3 to 19-8)**

$$2HI(g) \rightleftharpoons H_2(g) + I_2(g) + 25.9 \text{ kJ}$$

↑ a. What is the effect on the concentration of HI if a small amount of H_2 is added?
— b. What is the effect on the concentration of HI if the pressure of the system is increased?
↑ c. What is the effect on the concentration of HI if the temperature of the system is increased?
— d. What is the effect on the concentration of HI if a catalyst is added?
  e. Write the equilibrium expression for this reaction.
  f. At 425°C the equilibrium constant is 1.88×10^{-2}. Does equilibrium favor the formation of the reactant or the products?

2. Methanol (methyl alcohol) can be manufactured using the following equilibrium reaction. **(19-3 to 19-5)**

$$CO(g) + 2H_2(g) \rightleftharpoons CH_3OH(g) + energy$$

Predict the effect of the following changes on the equilibrium concentration of $CH_3OH(g)$.
↑ a. a decrease in temperature
↑ b. an increase in pressure
↑ c. addition of $H_2(g)$
— d. addition of a catalyst

3. In the equilibrium reaction:

$$4HCl(g) + O_2(g) \rightleftharpoons 2H_2O(g) + 2Cl_2(g) + 114.4 \text{ kJ}$$

Predict the direction of equilibrium shift if the following changes occur. **(19-3 to 19-5)**
R a. the pressure is increased
L b. energy is added
R c. oxygen is added
L d. HCl is removed
— e. a catalyst is added

4. In the equilibrium reaction:

$$2NO(g) + O_2(g) \rightleftharpoons 2NO_2(g) + 114.6 \text{ kJ}$$

What will be the change in the equilibrium concentration of NO_2 under each of the following conditions? **(19-3 to 19-5)**

↑ a. O_2 is added   ↓ c. energy is added
↓ b. NO is removed   — d. a catalyst is added

5. For the following reaction:

$$N_2O_4(g) + 58.9 \text{ kJ} \rightleftharpoons 2NO_2(g)$$

how will the equilibrium concentration of NO_2 be affected by the following conditions? **(19-3 to 19-5)**

↓ a. an increase in pressure
↑ b. an increase in temperature
— c. the addition of a catalyst

6. Suggest four ways to increase the concentration of SO_3 in the following equilibrium reaction. **(19-6)**

↓T add   add
↑P $2SO_2(g) + O_2(g) \rightleftharpoons 2SO_3(g) + 192.3 \text{ kJ}$

7. Nitric oxide, NO, releases 57.3 kJ/mol when it reacts with oxygen to give nitrogen dioxide. **(19-9)**
  a. Write the equation for this reaction.
  b. Predict the effect that increasing the temperature will have on:
    1. the equilibrium concentrations $NO\uparrow$ $O_2\uparrow$ $NO_2\downarrow$
    2. the numerical value of the equilibrium constant ↓
    3. the speed of formation of $NO_2\uparrow$
  c. Also, predict how increasing the NO concentration will affect 1, 2, and 3 above.

8. Consider the following solution equilibrium reaction.

$$Fe^{3+}(aq) + SCN^-(aq) \rightleftharpoons FeSCN^{2+}(aq)$$

Data was collected on adding various reagents to this system. **(19-10)**
$KNO_3(aq)$ = no change
$Fe_2(SO_4)_3(aq)$ = solution darkens
$KOH(aq)$ = solution lightens
$Na_2SO_4(aq)$ = no change
For each reagent determine:
  a. which ions are spectator ions
  b. which ions caused the color change

9. Predict the color change if each of the following reagents is added to the equilibrium system described in item 8. **(19-10)**
  a. $NaOH(aq)$ L   e. $NaCl(aq)$ O
  b. $NaSCN(aq)$ D   f. $LiSCN(aq)$ D
  c. $NaNO_3(aq)$ O   g. $LiNO_3(aq)$ O
  d. $LiOH(aq)$ L   h. $KSCN(aq)$ D

Problems

1. Use Table 19-3 to find the equilibrium constants for the following reactions. **(19-8)**
  a. $N_2O_4(g) \rightleftharpoons 2NO_2(g)$ at 55°C
  b. $PCl_3(g) + Cl_2(g) \rightleftharpoons PCl_5(g)$ at 227°C
  c. $CO(g) + Cl_2(g) \rightleftharpoons COCl_2(g)$ at 627°C

2. Consider the following equilibrium reaction.

$$H_2(g) + Br_2(g) \rightleftharpoons 2HBr(g) + energy$$

The K_{eq} for this reaction at 25°C is 1.02. At equilibrium the concentration of HBr is 0.50 mole per liter. Assuming H_2 and Br_2 are present in equal amounts, calculate the concentration of H_2 at equilibrium. **(19-8)**

CHAPTER REVIEW

3. For the reaction,

$$N_2(g) + 3H_2(g) \rightleftharpoons 2NH_3(g) + 92 \text{ kJ}$$

at an equilibrium temperature of 1000°C, a 1.00-liter flask contains 0.102 mole of ammonia, 1.03 moles of nitrogen, and 1.62 moles of hydrogen. Calculate the equilibrium constant at this temperature. (19-8)

4. For the following equilibrium reaction,

$$N_2O_4(g) + 58.9 \text{ kJ} \rightleftharpoons 2NO_2(g)$$

a liter flask at 55°C is found to contain 3.6 moles of $N_2O_4(g)$ in equilibrium with 1.75 moles of $NO_2(g)$. Calculate the value of K_{eq} for this reaction. (19-8)

5. Analysis of the following equilibrium reaction at 900°C provides the concentrations listed below.

$$CO(g) + H_2O(g) \rightleftharpoons CO_2(g) + H_2(g)$$

Experiment	[CO]	[H₂O]	[CO₂]	[H₂]
1	0.352	0.352	0.648	0.148
2	0.266	0.266	0.234	0.234
3	0.186	0.686	0.314	0.314

Write the equilibrium expression for the reaction and calculate the value of the equilibrium constant for each experiment. (19-8)

6. In the following reaction at 448°C, the equilibrium concentrations are $HI = 0.0040M$, $H_2 = 0.0075M$, $I_2 = 0.000043M$.

$$2HI(g) \rightleftharpoons H_2(g) + I_2(g)$$

Calculate the equilibrium constant at this temperature. (19-8)

7. Reactant W and X, each at a concentration of 0.80 mole per liter, react slowly to produce U and V according to the following equation.

$$W + X \rightleftharpoons U + V$$

At equilibrium, the molarity of U is 0.60. Calculate the value of the equilibrium constant. (19-8)

8. For the following reaction at equilibrium at 2000°C, the concentrations of N_2 and O_2 are both 5.2 moles per liter.

$$N_2(g) + O_2(g) \rightleftharpoons 2NO(g) \quad K_{eq} = 6.2 \times 10^{-4}$$

Calculate the concentration of NO. (19-8)

9. In the following reaction at equilibrium at 2000°C, $K_{eq} = 1.6 \times 10^3$.

$$2NO(g) \rightleftharpoons N_2(g) + O_2(g)$$

The concentration of NO is 0.13 mole per liter. If the concentrations of N_2 and O_2 are equal, calculate the concentration of N_2. (19-8)

Challenge

1. At equilibrium at 1120°C the concentration of the reactants and products are measured and found to be $CO = 0.010M$, $H_2O = 0.020M$, $CO_2 = 0.010M$, and $H_2 = 0.010M$. For the following reaction, does the equilibrium favor the formation of the products or the reactants? (19-8)

$$CO(g) + H_2O(g) \rightleftharpoons CO_2(g) + H_2(g)$$

2. A 1.00-liter reaction container initially contains 9.28×10^{-3} moles of H_2S. At equilibrium the concentration of H_2S is 7.06×10^{-3} moles. Calculate the equilibrium constant for this reaction.

$$2H_2S(g) \rightleftharpoons 2H_2(g) + S_2(g)$$

3. The equilibrium constant for the decomposition of phosphorus pentachloride gas to form phosphorus trichloride gas and chlorine gas is 0.0211 at a certain temperature. If the initial concentration of PCl_5 is $1.00M$, what is the equilibrium concentration of chlorine? (19-8)

4. At 800°C the equilibrium constant for the following reaction is 0.279.

$$CO_2(g) + H_2(g) \rightleftharpoons CO(g) + H_2O(g) \quad \Delta H = 42.6 \text{ kJ}$$

At a different temperature the equilibrium constant is 0.100. Is this different temperature higher or lower than 800°C? Give your reasoning. (19-9)

Synthesis

1. A 1.00-liter vessel initially contains 7.00 grams of carbon dioxide and 4.50 grams of water at equilibrium with the products carbon dioxide and hydrogen at 800°C. If the equilibrium constant is 3.59, how many grams of carbon dioxide are formed (assume that the concentrations of CO_2 and H_2 are the same)?

Projects

1. According to his Nobel lecture, Fritz Haber wanted to find an economical process for fixing nitrogen to make fertilizer. Compare the amount of ammonia being used today for fertilizer to the other uses of ammonia. List the five chief uses of ammonia other than fertilizer.

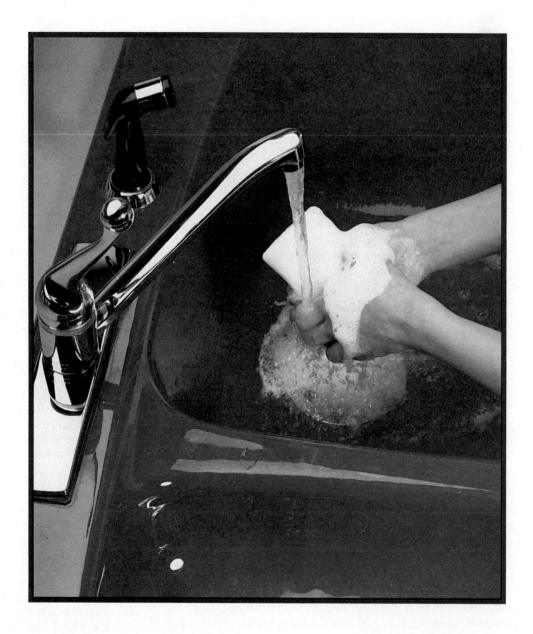

The English word *acid* comes from *acidus,* the Latin word for sour. Sour taste is a prominent characteristic of the compounds called acids. Many fruits taste sour. You can be quite confident that foods such as cherries, lemons, grapefruit, yogurt, and tomatoes all contain compounds that are classified as acids.

Bases also have a property that is familiar to you. A substance that has a soapy or slippery feel to it generally contains a compound classified as a base. In this chapter, you will study other characteristics used to describe acids and bases and why these compounds are so important to the study of chemistry.

Acid-Base Reactions

Identifying Properties of Acids and Bases

Understanding chemical behavior helps you understand your world. However, there are too many reactions to try to remember them all individually. It would be somewhat like trying to remember the names and occupations of all the people in New York City. You have seen many examples of how classifying compounds into groups with similar characteristics reduces the amount of information one must remember. Acids and bases comprise two groups of substances that have been categorized by their chemical behavior.

CONCEPTS

■ classifying compounds as acids and bases
■ Brönsted-Lowry definitions of acids and bases
■ conjugate acid-base pairs
■ characteristics of amphiprotic substances
■ preparation and properties of acids
■ preparation and properties of bases

20-1 General Characteristics of Acids and Bases

If you have ever tasted a lemon you are aware of its very sour taste. Lemon juice contains an acid. Grapefruit and spoiled milk also are sour due to the presence of acids. Sour taste is a property common to compounds classified as acids. If you have ever got soap in your mouth, you have noted a bitter taste. Bitterness is common to compounds classified as bases. Do not try the taste test because some acids and bases are harmful.

One property of acids is how they behave when they contact some metals. When hydrochloric acid and magnesium metal are mixed together, bubbles are produced and the magnesium metal reacts. When tested, the bubbles are found to be hydrogen gas. A similar reaction occurs when other acids are combined with magnesium. In Figure 20-1 you can see the long glass tube that is used to measure the amount of hydrogen generated by the acid and metal. When most metals are combined with a base, no noticeable change occurs.

An electric current is not conducted through pure water, but it is conducted through an acid solution. Basic solutions also conduct electricity. The apparatus shown in Figure 20-2 is used to test conductivity. The presence of ions allows current to flow between the two electrodes, and the light bulb glows. Solutions which conduct an electric current are called electrolytes. Table 20-1 lists the characteristic properties of acids and bases which are both electrolytes.

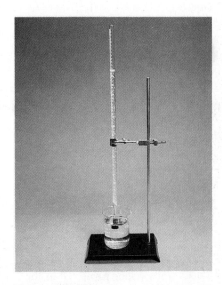

Figure 20-1 The reaction of magnesium metal with hydrochloric acid produces hydrogen gas.

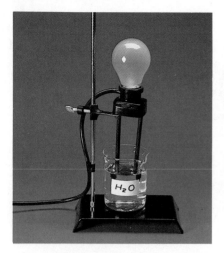

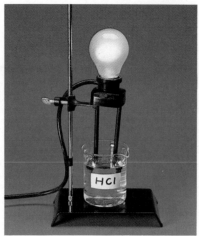

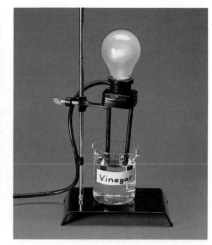

Figure 20-2 Conductivity testing of pure water on the left, hydrochloric acid in the middle, and acetic acid on the right. Note that pure water does not conduct and the conductivity of the two acid solutions varies.

Table 20-1

CHARACTERISTIC PROPERTIES OF ACIDS AND BASES

ACIDS

Dissolve in water to give a solution that:
1. tastes sour
2. conducts an electric current
3. causes certain dyes to change color
4. liberates hydrogen when it reacts with certain metals
5. loses the above properties when it is reacted with a base, though the resulting solution will conduct a current

BASES

Dissolve in water to give a solution that:
1. tastes bitter
2. conducts an electric current
3. causes certain dyes to change color
4. feels slippery
5. loses the above properties when it is reacted with an acid, though the resulting solution will conduct a current

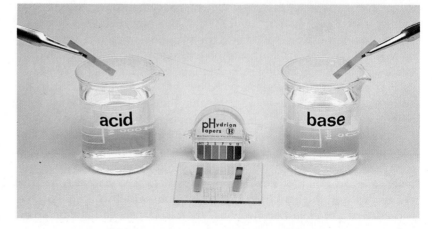

Figure 20-3 Litmus is an indicator. Blue litmus paper turns red in the presence of an acid. Red litmus turns blue in the presence of a base.

Figure 20-4 Bromthymol blue indicator turns yellow in the presence of an acid as shown in the photograph on the left. Silver nitrate is used to detect the presence of chloride ions in the beaker on the right.

20-2 Definition of Acids and Bases

Hydrogen chloride, HCl, is a choking gas at room temperature. If it is cooled below $-84.9°C$, it liquefies. The liquid state does not conduct an electric current so it must not contain ions. HCl readily dissolves in water, and the solution does conduct an electric current. What are the ions and from where did they come?

Certain compounds, called **indicators**, can be used to detect the presence of acids and bases. The indicator will appear as one color in the presence of high concentrations of H^+ (acid solution) and another color when the H^+ concentration is low (base solution). How indicators respond to H^+ concentration is discussed in Section 20-11. When an indicator such as bromthymol blue is added to the HCl solution, the color of the indicator changes to clear yellow. Bromthymol blue is always yellow in the presence of high concentrations of H^+. Thus you conclude that HCl in water contains H^+ ions.

If the HCl in water produced H^+, a positive ion, then the solution also must contain a negative ion. A possible candidate is the chloride ion, Cl^-. Adding silver nitrate is the qualitative test used to confirm the presence of chloride ions in a solution. If a drop of $AgNO_3(aq)$ is added to the HCl solution, a cloudy white precipitate forms as in Figure 20-4, indicating the presence of chloride ions. The precipitate is silver chloride.

$$Cl^-(aq) + Ag^+(aq) \longrightarrow AgCl(s).$$

When HCl reacts with water, it produces the H^+, which combines with the water molecule to produce a hydronium ion, H_3O^+

$$HCl + H_2O \longrightarrow H_3O^+ + Cl^-$$

The **hydronium ion** is a hydrated proton. A model for the formation of the hydronium ion is shown in Figure 20-5. Though more than one water molecule is involved, the hydrated proton is traditionally shown with a single water molecule. A number of different experiments support the idea that the hydrogen ion is hydrated when water is the

Do You Know?

One of the earliest definitions of an acid was proposed by Svente Arrhenius, a Swedish chemist. According to Arrhenius, an acid is any compound which produces hydrogen ions in solution, and a base is any compound which produces hydroxide ions in solution.

The hydronium ion represents a number of hydrated proton aggregates that are known to exist. Particles such as $H_5O_2^+$ and $H_7O_3^+$ are represented by the general formula H_3O^+.

Figure 20-5 The formation of the hydronium ion involves a proton transfer from the acid to the water molecule. This reaction is a Brönsted-Lowry acid-base reaction.

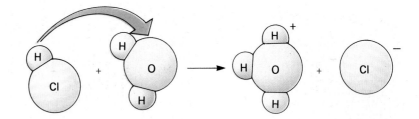

solvent. Notice how the water behaved in the reaction with HCl. Water accepted the H^+ in forming the hydronium ion.

Ammonia, NH_3, also is a gas at room temperature. When it is cooled below $-33.35°C$, it liquefies. This liquid does not conduct electricity, so it does not contain ions. However, when ammonia is dissolved in water, the solution conducts electricity.

Figure 20-6 The formation of the ammonium ion involves the transfer of a proton from the water molecule to the ammonia molecule. This reaction is a Brönsted-Lowry acid-base reaction.

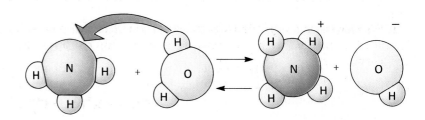

Do You Know?

Gilbert Lewis, an American chemist, expanded the definition of an acid and a base to include many compounds which do not contain hydrogen. A Lewis acid is an electron pair acceptor; a Lewis base is an electron pair donor.

Johannes N. Bronsted of Denmark and Thomas M. Lowry of England, but working in the United States, proposed the same definition of an acid and a base, at about the same time on two different continents. So both are credited with the discovery.

When ammonia dissolves in water, it accepts a H^+ from the water, producing the NH_4^+ and the OH^-.

$$NH_3 + H_2O \longrightarrow NH_4^+ + OH^-$$

The water donated a proton (H^+) which was used by NH_3 to form NH_4^+. The importance of the solvent in acid-base reactions led chemists to expand the definition of acid and base. Johannes Brönsted in Denmark, and Thomas Lowry in England, independently proposed that an **acid** is any substance that can produce (or donate) a proton, while a **base** is any substance that can accept a proton in a reaction.

Water can act as a proton acceptor or a proton donor. Compounds that exhibit this dual behavior are called **amphiprotic**. How water behaves depends on what other ions are present. In the case of HCl and H_2O, water acts as a base in accepting the proton from the HCl. Water is a Brönsted-Lowry base. In the case of NH_3, water donates a proton to produce the ammonium ion and the hydroxide ion, in this case, acts as a Brönsted-Lowry acid. All acid–base reactions can be described as competing for protons.

When an acid such as HCl and a base such as KOH combine, they react with each other to counteract their acidic and basic properties. When moles of protons donated by the acid and moles of protons accepted by the base are equal, **neutralization** occurs. A salt is produced only if the base is OH^- along with water.

$$HCl + KOH \longrightarrow KCl(aq) + H_2O$$

Figure 20-7 Bromthymol blue remains blue in the presence of a base such as aqueous ammonia. An aqueous ammonia solution causes red litmus paper to turn blue.

If an acid is accidentally spilled, a weak base may be applied to neutralize the acid and quickly nullify its effects. Likewise, if a base is accidentally spilled, a weak acid is applied to nullify the effects of the base.

If the neutralization reaction of hydrochloric acid and potassium hydroxide is written in terms of ions, you will see what happens when the salt is formed.

$$H_3O^+(aq) + Cl^-(aq) + K^+(aq) + OH^-(aq) \longrightarrow$$
$$2H_2O(l) + K^+(aq) + Cl^-(aq)$$

Notice that in this expression, there are terms common to both sides. To write the net ionic equation, the spectator ions are not included in the equation.

$$H_3O^+(aq) + OH^-(aq) \longrightarrow 2H_2O$$

H^+ from HCl   from NaOH   neutral water

20-3 Conjugate Acid–Base Pairs

When NH_3 is combined with water, the ammonia acts as the base, accepting a proton from the water, which acts as an acid.

$$H:\overset{..}{N}:H + H:\overset{..}{\underset{..}{O}}:H \longrightarrow \left[H:\overset{H}{\underset{H}{N}}:H \right]^+ + :\overset{..}{\underset{..}{O}}:H^-$$

Once the ammonia gets the proton from the water, the NH_3 becomes the ammonium ion, NH_4^+. This ammonium ion is capable of donating a proton to some other ion if the occasion arises. Molecules and ions that differ only by a proton (hydrogen ion) are sometimes referred to as **conjugate acid–base pairs**. NH_4^+ and NH_3 are a conjugate acid–base pair. Conjugate means paired together. H_2O and OH^- are another conjugate acid–base pair.

The acid form of the pair contains a proton that is missing in the base form. The presence of a proton is the only difference between members of a conjugate acid–base pair.

$$\text{NH}_4^+ \quad \text{and} \quad \text{NH}_3$$
$$\text{acid} \qquad\qquad \text{base}$$

$$\text{H}_2\text{O} \quad \text{and} \quad \text{OH}^-$$
$$\text{acid} \qquad\qquad \text{base}$$

You might ask why NH_3 does not act like an acid, produce a proton, H^+, and in turn become NH_2^-.

$$\text{NH}_3 \longrightarrow \text{H}^+ + \text{NH}_2^-$$
$$\text{acid} \qquad\qquad \text{base}$$

There are few molecules that attract hydrogen ions strongly enough to pull one off the ammonia molecule.

Acids and bases that donate or accept one proton are called monoprotic. If a substance is capable of accepting more than one proton, it is a polyprotic base. A polyprotic acid is capable of donating more than one proton. Examples of polyprotic acids are sulfuric acid, H_2SO_4, and phosphoric acid, H_3PO_4.

Polyprotic acids ionize in steps. In each step the acid donates a proton.

1st ionization step: $\text{H}_2\text{SO}_4 + \text{H}_2\text{O} \longrightarrow \text{HSO}_4^- + \text{H}_3\text{O}^+$
$$\qquad\qquad\qquad\qquad \text{acid} \qquad\qquad\qquad \text{base}$$

2nd ionization step: $\text{HSO}_4^- + \text{H}_2\text{O} \longrightarrow \text{SO}_4^{2-} + \text{H}_3\text{O}^+$
$$\qquad\qquad\qquad\qquad \text{acid} \qquad\qquad\qquad \text{base}$$

Phosphoric acid ionizes in three steps since it is capable of donating three protons.

1st ionization step: $\text{H}_3\text{PO}_4 + \text{H}_2\text{O} \longrightarrow \text{H}_2\text{PO}_4^- + \text{H}_3\text{O}^+$
$$\qquad\qquad\qquad\qquad \text{acid} \qquad\qquad\qquad \text{base}$$

2nd ionization step: $\text{H}_2\text{PO}_4^- + \text{H}_2\text{O} \longrightarrow \text{HPO}_4^{2-} + \text{H}_3\text{O}^+$
$$\qquad\qquad\qquad\qquad \text{acid} \qquad\qquad\qquad \text{base}$$

3rd ionization step: $\text{HPO}_4^{2-} + \text{H}_2\text{O} \longrightarrow \text{PO}_4^{3-} + \text{H}_3\text{O}^+$
$$\qquad\qquad\qquad\qquad \text{acid} \qquad\qquad\qquad \text{base}$$

Both HSO_4^- and H_2PO_4^- may act as either an acid or base. These ions are amphiprotic like water.

20-4 Preparation and Properties of Acids

Sulfuric acid   More sulfuric acid is produced yearly by industry than any other chemical in the world. Why is sulfuric acid used so widely? Primarily, it is cheap to make. Sulfur, oxygen, and water are the raw materials needed to make the acid. There are large, natural deposits of sulfur available, plus more sulfur is produced as a by-product of many metal-refining processes. Another raw material needed is oxygen, which is very plentiful in the air—another inexpensive resource. Water is necessary and it also is inexpensive. The commercial reaction to prepare the acid is simple, which also helps to keep the cost down. There are no by-products to pollute the environment, another cost-saving feature.

Do You Know?

Oleum is the Latin word for oil. Sulfuric acid was once called oil of vitriol.

If you strike a match to sulfur, it burns to produce sulfur dioxide, SO_2, that dissolves in water to produce an acid. H_2SO_4 is made from SO_3 dissolved in water. It is somewhat inconvenient to make SO_3 from SO_2, but it is worth the expense. Through the Contact process SO_2 is converted to H_2SO_4 in a series of steps. A catalyst is used to increase the number of effective collisions between the SO_2 molecules and O_2 molecules. Then, the SO_3 is bubbled through H_2SO_4 making $H_2S_2O_7$, which is diluted with H_2O to make H_2SO_4. A summary of the reactions in the commercial preparation of sulfuric acid follows:

$$S(l) + O_2(g) \longrightarrow SO_2(g)$$

$$2SO_2(g) + O_2(g) \xrightarrow{\text{catalyst}} 2SO_3(g)$$

$$SO_3(g) + H_2SO_4(l) \longrightarrow H_2S_2O_7(l)$$

$$H_2S_2O_7(l) + H_2O(l) \longrightarrow 2H_2SO_4(l)$$

Sulfuric acid has four properties which make it so useful in industry. It is a good neutralizer of bases. It is a good oxidizing agent. (Oxidizing agents play an important role in electron transfer reactions in electrochemistry.) It has a great affinity for water so it can be used as a dehydrating agent. Finally, it is a good electrolyte so it is used in lead storage batteries in cars. Reagent bottles of concentrated sulfuric acid contain about 96% H_2SO_4 in solution.

Hydrochloric acid Sulfuric acid also is used in the preparation of hydrochloric acid. In this reaction, the sulfuric acid reacts with NaCl to produce gaseous HCl (hydrogen chloride), which bubbles through water to make hydrochloric acid. The complete reactions are as follows:

$$2NaCl(s) + H_2SO_4(l) \longrightarrow Na_2SO_4(aq) + 2HCl(g)$$
$$H_2O(l) + HCl(g) \longrightarrow H_3O^+(aq) + Cl^-(aq)$$

Note from the equations that hydrochloric acid is an aqueous solution of hydrogen chloride gas.

Since sulfuric acid is so cheap to make, why bother to make any other acid if they have similar properties? So far, you have only learned about properties common to all acids. But each acid has unique properties as well. When sulfuric acid is combined with some metal ions, it produces insoluble sulfates. Such sulfates may not be desirable. Hydrochloric acid does not have a similar property. Concentrated nitric acid cannot be used in the presence of copper because NO_2, a very toxic gas, is produced. However, nitric acid is a good choice when high solubility is desired. Recall that all nitrate compounds are soluble in water. So other, more expensive acids are needed for their other properties.

Another widely used acid is HNO_3, nitric acid. It is tenth in the total volume of acids produced in the United States. Nitric acid is used to make ammonium nitrate and phosphate fertilizers (important in agriculture), explosives, plastics, dyes, and lacquers.

Nitric acid Nitric acid is made by the Ostwald process. In this process, ammonia, NH_3, reacts with oxygen in the presence of a catalyst to become nitrogen dioxide, NO_2. This reddish-brown gas dissolves in warm water to produce 60% (by mass) nitric acid. To get more concentrated nitric acid, another slower (and, therefore, more expensive) method must be used. The reactions for the formation of HNO_3 are as follows:

$$4NH_3(g) + 5O_2(g) \xrightarrow{\text{catalyst}} 4NO(g) + 6H_2O(g)$$

$$2NO(g) + O_2(g) \longrightarrow 2NO_2(g)$$

$$3NO_2(g) + H_2O(l) \longrightarrow 2HNO_3(aq) + NO(g)$$

The yellow stain from HNO_3 on skin is due to the action of HNO_3 on protein. HNO_3 is a nonspecific test for any type of protein.

Some of the distinguishing characteristics of concentrated nitric acid are its suffocating odor and the yellow stain it leaves on skin. The pale yellow color of nitric acid is due to the release of NO_2 as the acid is exposed to light.

Other acids Binary acids (containing two elements) can be made by the direct combination of the element with hydrogen gas, such as this reaction for the formation of hydrobromic acid.

$$H_2(g) + Br_2(l) \longrightarrow 2HBr(g)$$
$$HBr(g) + H_2O(l) \longrightarrow H_3O^+(aq) + Br^-(aq)$$

Other acids like phosphoric acid are formed when a nonmetallic oxide and water react.

$$P_4O_{10}(s) + 6H_2O(l) \longrightarrow 4H_3PO_4(aq)$$

Carbonic acid is formed in the same way, using carbon dioxide gas that is bubbled through water.

$$CO_2(g) + H_2O(l) \longrightarrow H_2CO_3(aq)$$

It is the presence of CO_2 in the air that causes natural rainwater to be slightly acidic. Carbonated water for soft drinks is made by this same reaction. The CO_2 is kept in solution by increasing the pressure on the solution.

Acids which contain three elements are ternary acids. Table 20-2 shows the names and formulas of some common binary and ternary acids.

Table 20-2

FORMULAS AND NAMES OF SOME COMMON ACIDS

BINARY ACIDS		TERNARY ACIDS	
HCl	hydrochloric	H_2SO_3	sulfurous
HBr	hydrobromic	H_2SO_4	sulfuric
HI	hydroiodic	HClO	hypochlorous
HF	hydrofluoric	$HClO_2$	chlorous
H_2S	hydrosulfuric	$HClO_3$	chloric
		$HClO_4$	perchloric
		HNO_2	nitrous
		HNO_3	nitric

20-5 Preparation and Properties of Bases

Sodium hydroxide is contained in some relatively common household products. Drano® is the trade name for a mixture of sodium hydroxide and flakes of aluminum. The active ingredient in most oven cleaners is sodium hydroxide. Large quantities of sodium hydroxide are used in the manufacture of soap from fats and oils.

The reactions that take place when sodium hydroxide opens your clogged drain or cleans your oven suggest important properties of sodium hydroxide and other bases. They also suggest why this compound is called caustic soda and must be used with extreme care.

Caustic means "capable of burning, corroding, or destroying living tissue," and is an accurate description of the properties of sodium hydroxide. Soda, as you probably have guessed, refers to the element sodium. It is just these caustic properties that make sodium hydroxide useful as a drain cleaner. The most common substances that clog drains are fats in the kitchen, and hair in the shower or bath. Sodium hydroxide reacts with fats to form soap, which is soluble in water. The reaction of sodium hydroxide and fats is shown in the following equation.

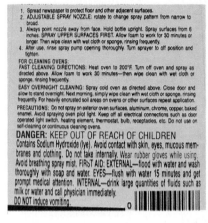

Figure 20-8 Oven cleaners are caustic substances that may contain bases. They should be handled with care.

$$3NaOH + \text{fat} \longrightarrow \text{glycerol} + 3CH_3(CH_2)_{18}C\begin{smallmatrix}O\\ONa\end{smallmatrix}$$

fat   glycerol   soap

This same reaction takes place when ordinary soap is made commercially and when oven cleaners are used to remove grease in the oven. The reaction is rather slow, which explains why the instructions with oven cleaners suggest that the cleaner be applied and left for a time before you attempt to remove the grime. This reaction proceeds faster at higher temperatures which accounts for the suggestion that the cleaner will work best if applied to a warm oven.

Aluminum flakes are added to sodium hydroxide in Drano® to speed the reaction with fats. Aluminum reacts rapidly with sodium hydroxide as described by the following equation:

$$2Al + 2NaOH + 2H_2O \longrightarrow 2NaAlO_2 + 3H_2 + \text{energy}$$

The energy produced by this reaction is transferred to the water in the drain and speeds up the reaction between sodium hydroxide and fats. The hydrogen gas produced agitates the liquid, helping to break up the clog. All the material in the drain does not have to dissolve; it just needs to be loosened enough to pass through the pipe as water is flushed down the drain. Does this reaction represent an acid–base reaction according to the Brönsted–Lowry theory? If not, how does it differ from those described as acid–base reactions?

Hair is a protein rather than a fat, but sodium hydroxide reacts with proteins, too. Again, the action of sodium hydroxide on protein

Figure 20-9 Drano® crystals contain small bits of aluminum metal that react once the crystals are dissolved.

Repeat application if necessary and let work for one hour. Avoid all splashing—pa and clothing. Keep children away from basins containing Liquid-plumr. Do not use

FOR CLOGGED AND SLUGGISH DRAINS—NO NEED TO REMOVE STANDING WA TO HELP PREVENT DRAIN PROBLEMS—USE ONE-EIGHTH BOTTLE WEEKLY. FOR GARBAGE DISPOSER DRAINS—FOLLOW SAME INSTRUCTIONS AS FOR OTHER DRAINS. FLUSH WITH DISPOSER RUNNING. WILL NOT HARM PIPES OR DISPOSERS.

DANGER: Keep out of Reach of Children. Contains sodium hypochlorite and sodium hydroxide. Harmful if swallowed. Injures eyes and mucous membranes on contact. Do not use Liquid-plumr with ammonia, toilet bowl cleaners or other drain openers; splashing or the release of hazardous gases may occur. Clean up any spillage immediately. Do not re-use empty container. Rinse container and replace cap before discarding.

EMERGENCY TREATMENT: EYES—Flush immediately with water. If wearing contact lenses, remove first. EXTERNAL—Wash immediately with water. INTERNAL—Drink large quantities of water. Do not induce vomiting. THEN, CALL PHYSICIAN.

Contents: Water, sodium hypochlorite, sodium hydroxide, sodium chloride and corrosion inhibitor. Questions or Comments? Call (800) 292-2200. In California call (800) 292-2212. Patent Pending

Figure 20-10 Many household products contain acids or bases. The caustic properties of the ingredients are a factor in their handling and use.

loosens the clog in the drain and allows you to flush the loosened material through the drain.

As you know from your study of biology, skin and muscle also are made of protein. In addition, oils produced by special cells in your skin provide natural lubrication that keeps the outer skin soft and free of cracks. What do you suppose will happen if sodium hydroxide comes in contact with your skin? At home, solutions of sodium hydroxide are used in cleaning products and in paint and varnish removers. Read the labels on some of your home cleaning products. If it says the product contains lye or caustic soda, it contains sodium hydroxide.

Sodium hydroxide is marketed commercially in pellet form. It is deliquescent (has the ability to absorb water vapor from the air). Deliquescence makes sodium hydroxide a good drying agent. Thus this chemical must be stored with the lid closed tightly to prevent the pellets from deliquescing. It is difficult to remove hydrated NaOH pellets from a jar.

You now have some idea why sodium hydroxide is an important commercial product; it has properties that are useful in many everyday applications. But you also know that in order for something to be a commercial success, it must be possible to produce it and deliver it to customers at a competitive price. From the previous discussion of industrial processes, what factors affect the cost of commercial products? Besides the ones mentioned earlier, the cost of energy supplied to run the reaction, plant costs and the need to dispose of any by-products also are important factors to be considered.

The raw materials used to make sodium hydroxide are inexpensive. One raw material, sodium chloride, is found in large salt mines and it can be mined fairly readily, which keeps the price down. The only other product is water—another inexpensive material.

The location of the salt and the large supply of fresh water must be considered when building a plant to make sodium hydroxide. Transportation costs must be kept as low as possible. Even if the raw materials are cheap, if the plant is isolated from the potential markets, transportation of the raw materials and the end product will increase the cost of production considerably.

You have already studied the electrolysis of water. An electric current is passed through the water and it decomposes into hydrogen and oxygen gases. Hydrogen is collected at the cathode and oxygen at the anode.

Sodium hydroxide is produced by the electrolysis of sodium chloride dissolved in water.

$$2NaCl(aq) + 2H_2O(l) \longrightarrow 2NaOH(aq) + H_2(g) + Cl_2(g)$$

The hydrogen gas collects at the cathode and chlorine gas collects at the anode. Care must be taken to keep the gases separate for economic and chemical reasons. Hydrogen gas ranks fifth in commercial volume and chlorine ranks eighth. Each gas sold separately is more valuable than hydrogen chloride gas, which forms explosively if the two gases mix. Also the HCl will react with the NaOH—which is the sought-after product. If chlorine gas reacts with the aqueous sodium hydroxide, then sodium hypochlorite, sodium chloride, and water are produced.

Figure 20-11 Chemicals are produced industrially in facilities like the plant shown here.

$$2NaOH(aq) + Cl_2(g) \longrightarrow NaOCl(aq) + NaCl(aq) + H_2O(l)$$

In the past, chlorine gas was not separated from the reaction mixture and NaOCl was one of the by-products. The production of this by-product led to its marketing as a household bleach. Here is an example of how a "need" was created by marketing specialists in order to use something produced for a totally different reason. Since people are convinced that white shirts are nicer than grey shirts, a market is born. The active ingredient in Clorox® is sodium hypochlorite, NaOCl.

Review and Practice_____

1. List four properties of an acid.

2. Why should you not taste an acid to see if it is sour?

3. What are the Brönsted–Lowry definitions of an acid and a base?

4. Why is water considered amphiprotic?

5. Why is Ca(OH)$_2$ considered a base? Write an equation which represents the reaction of Ca(OH)$_2$ in HCl.

6. What is neutralization?

7. Write a balanced equation for the reaction of potassium hydroxide and nitric acid.

8. Write the equation for the reaction of HI in water.

1. tastes sour; conducts an electric current; causes indicators to change color; produces H_2 gas when acid reacts with certain metals; neutralizes a base
2. Some acids are harmful.
3. An acid is a proton donor. A base is a proton acceptor.
4. Water can act as either a proton donor or proton acceptor.
5. Ca(OH)$_2$ is basic because it accepts protons from an acid.
6. when moles of protons from the acid equal moles of protons accepted by the base
7. KOH + HNO$_3$ $\longrightarrow$ H$_2$O + KNO$_3$
8. HI + H$_2$O $\longrightarrow$ H$_3$O$^+$ + I$^-$

Strengths of Acids and Bases

You may have heard the terms strong and weak used in describing an acid or base. These adjectives represent an important distinction. In everyday speech, strong and weak generally refer to concentration. However, in this next lesson you will see how strong and weak are defined in relation to acids and bases.

20-6 Strong Acids and Weak Acids

Remember that the H^+ ion, being just a tiny proton, does not remain alone in a solution, but immediately unites with a water molecule to produce the hydronium ion, H_3O^+. It is the presence of the hydronium ion that imparts to the solution the properties characteristic of an acid. The strength of an acid depends on the number of hydronium ions produced per mole of acid. Acids of similar concentration can differ in the amount of hydronium ion produced. Look at Figure 20-2. The photographs show the use of a conductivity apparatus to test the conducting properties of two acid solutions—acetic acid and hydrochloric acid. The molarity of each acid solution is the same, yet the photographs differ. The dimmer bulb indicates there is less electricity conducted, which leads to the conclusion that there are fewer ions present in the solution. In the discussion that follows, look at how the extent of the reaction differs for these two acids.

Recall how the equation for the reaction of HCl with water was written.

$$HCl + H_2O \longrightarrow H_3O^+ + Cl^-$$

Conductivity experiments indicate that many ions are produced in the solution which indicates that the forward reaction is favored. When

Figure 20-12 These diagrams present microscopic models to explain the conductivity differences shown for HCl and CH_3COOH in Figure 20-2. Note the presence of fewer ions for CH_3COOH affects the conductivity.

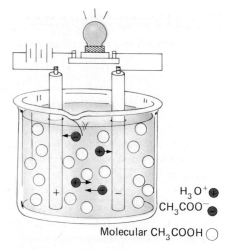

Cl^- ⊖ H_3O^+ ⊕

H_3O^+ ⊕
CH_3COO^- ⊖

Molecular CH_3COOH ◯

HCl reacts, no HCl molecules are left in solution. HCl ionizes completely. A **strong acid** is considered to react completely with water in a dilute aqueous solution. Another strong acid is nitric acid, HNO_3. Its reaction with water is the following:

$$HNO_3 + H_2O \longrightarrow H_3O^+ + NO_3^-$$

As in the case of HCl, all of the HNO_3 molecules react to form hydronium ions. Six common strong acids are shown in Table 20-3.

Table 20-3

COMMON STRONG ACIDS	
NAME	**FORMULA**
hydrobromic	HBr
hydrochloric	HCl
hydroiodic	HI
nitric	HNO_3
perchloric	$HClO_4$
sulfuric	H_2SO_4

Some substances, such as acetic acid, do not react completely when dissolved in water. Note in Figure 20-2 and the model in Figure 20-12 that the bulb is much dimmer. This reaction is represented by the following equation:

$$CH_3COOH + H_2O \rightleftharpoons CH_3COO^- + H_3O^+$$

Notice that this reaction is written as an equilibrium system. At equilibrium, a considerable amount of CH_3COOH molecules remain. There are only a small proportion of hydronium ions and acetate ions, so the solution is a weak conductor of electricity. An acid that does not react completely in an aqueous solution is called a **weak acid**.

20-7 Strong Bases and Weak Bases

A **strong base** is one that reacts completely when it dissolves.

$$NaOH(s) \longrightarrow Na^+(aq) + OH^-(aq)$$
$$KOH(s) \longrightarrow K^+(aq) + OH^-(aq)$$

It is the presence of the OH^- ion that gives the solution its basic character.

There are few strong bases. Bases formed from group 1 metals and the heavier metals of group 2 are strong, even though the group 2 bases are not completely soluble in water.

Ammonia is the most common **weak base**. You have seen in Section 20-2 that when ammonia reacts with water, it produces the ammonium ion, NH_4^+, and the hydroxide ion, OH^-. At equilibrium, the reverse reaction is favored.

20-8 Comparing Strengths of Acids

At the beginning of the chapter, acids and bases were described as classes of compounds. Rather than talking about the properties of hydrochloric acid, sulfuric acid, nitric acid, oxalic acid, etc., you talk about properties of acids. You have seen the limitations of the strategy—although some of the properties of these compounds are the same, other properties make an acid more useful for a particular need.

Table 20-4 can be used to make reasonable predictions about chemical reactions that you have never observed. For example, when sodium hydrogen carbonate and hydrofluoric acid are combined in an aqueous solution, what reaction is likely to occur? Although other factors may come to play that make some of these predictions wrong (just as you may incorrectly predict which of two students will score higher on an exam), the predictions will be correct often enough to make the process worthwhile.

To use this table, you must remember that a base is a proton acceptor and an acid is a proton donor. You cannot have one without the

Table 20-4

RELATIVE STRENGTH OF COMMON ACIDS AND BASES

ACID NAME	ACID FORMULA	BASE FORMULA	BASE NAME
hydrochloric	HCl	Cl^-	chloride ion
nitric	HNO_3	NO_3^-	nitrate ion
sulfuric	H_2SO_4	HSO_4^-	hydrogen sulfate ion
hydronium ion	H_3O^+	H_2O	water
oxalic	$H_2C_2O_4$	$HC_2O_4^-$	hydrogen oxalate ion
hydrogen sulfate ion	HSO_4^-	SO_4^{2-}	sulfate ion
phosphoric	H_3PO_4	$H_2PO_4^-$	dihydrogen phosphate ion
hydrofluoric	HF	F^-	fluoride ion
formic	$HCOOH$	$HCOO^-$	formate ion
hydrogen oxalate ion	$HC_2O_4^-$	$C_2O_4^{2-}$	oxalate ion
acetic	CH_3COOH	CH_3COO^-	acetate ion
carbonic	H_2CO_3	HCO_3^-	hydrogen carbonate ion
hydrogen sulfide	H_2S	HS^-	hydrogen sulfide ion
hypochlorous	$HClO$	ClO^-	hypochlorite ion
ammonium ion	NH_4^+	NH_3	ammonia
hydrocyanic	HCN	CN^-	cyanide ion
hydrogen carbonate ion	HCO_3^-	CO_3^{2-}	carbonate ion
monohydrogen phosphate ion	HPO_4^{2-}	PO_4^{3-}	phosphate ion
water	H_2O	OH^-	hydroxide ion

Strength of acid increases (left, pointing up)

Strength of base increases (right, pointing down)

other. A strong base readily accepts protons. A weak base does not accept protons as readily.

Now use the table to answer the question about sodium hydrogen carbonate and hydrofluoric acid. To answer this question, you must realize that the reacting species will be the hydrogen carbonate ion, HCO_3^-, from the sodium hydrogen carbonate and the HF, though not much HF reacts in solution. Now look at Table 20-4. You see that the HCO_3^- is a stronger base since it appears below HF and F^-. The reaction is:

$$HCO_3^- + HF \rightleftharpoons H_2CO_3 + F^-$$
$$\text{base} \quad \text{acid} \qquad \text{acid} \quad \text{base}$$

In other words, hydrofluoric acid is a stronger acid than the hydrogen carbonate so the reaction has a greater tendency to proceed to the right. Similarly carbonic acid is a stronger acid than the fluoride ion, which again indicates that the products are favored.

20-9 Weak Acid Equilibrium and K_a

Equilibrium expressions can be written for acids and bases and the equilibrium constants can be measured experimentally.

$$HF + H_2O \rightleftharpoons H_3O^+ + F^-$$
$$\text{acid} \quad \text{base} \qquad \text{acid} \quad \text{base}$$

As you have learned, acid strength is measured by the amount of hydronium ion produced per mole of acid in a solution. The conductivity apparatus that was shown in Figure 20-2 can be used to qualitatively determine if an acid solution is weak or strong. To quantitatively measure the relative strengths of acids, it is necessary to measure their hydronium ion concentrations.

How can the hydronium ion concentration be determined for solutions of weak acids? Since the ionization processes for weak acids are not complete, it is not possible to use the initial concentration of the acid in the solution. One possible way of determining these concentrations is shown in Figure 20-13.

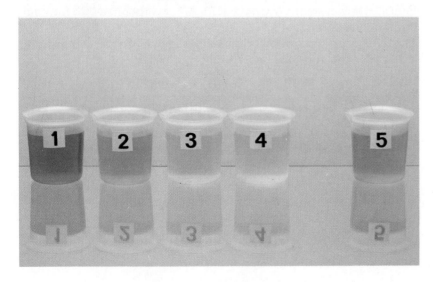

Figure 20-13 The five beakers contain four HCl solutions at different concentrations. The beaker on the right contains HF solution with an unknown H_3O^+ concentration. The H_3O^+ concentration in HF can be determined by comparison to the known solutions.

A set of standard solutions is prepared by diluting a HCl solution whose concentration is known. The volume of each solution is 1 liter. The molarity of each solution have been calculated as dilution problems. Since HCl reacts completely in water, the concentration of the H_3O^+ ion is the same as the calculated molarity of HCl. Samples of each solution are placed in beakers. The same amount of an indicator is added to each of the HCl solutions. The color varies from one beaker to another because the hydronium ion concentration differs for each. A weak acid now can be tested by comparing it to the standard solutions.

Hydrofluoric acid, HF, is a weak acid whose hydronium ion concentration is unknown. The equation for the ionization of HF is:

$$HF + H_2O \rightleftharpoons H_3O^+ + F^-$$

The solution on the right in Figure 20-13 is 0.100M HF. The same amount of indicator has been added to this beaker as was added to the HCl solutions. Notice that the color of the HF solution seems to match the second HCl beaker, where the hydronium ion concentration was calculated as:

$$[H_3O^+] = 8 \times 10^{-3} \text{ mole per liter}$$

Then, it can be assumed that the $[H_3O^+]$ for the HF solution also is $8 \times 10^{-3}M$.

The concept of this section, using molar concentrations to calculate K_a or K_b is limited to dilute solutions—0.100M or less. Equilibrium values for more concentrated solutions are calculated using activities. You may want to point this out to your students.

The ionization of hydrofluoric acid is an equilibrium reaction. For such reactions, the relative strength of that acid may be expressed quantitatively in terms of an equilibrium expression. Applying what you learned in Chapter 19, you would expect the equilibrium expression for the hydrofluoric acid ionization to be:

$$K_{eq} = \frac{[H_3O^+][F^-]}{[H_2O][HF]}$$

Using the 0.100M HF solution described above, you can look at the concentrations of each species in the solution. Table 20-5 organizes the data for comparison. Note that the value for the water concentration remains relatively constant at equilibrium. Recall from Chapter 19 that the concentrations of pure liquids are incorporated into the equilibrium constant.

Looking back at the equilibrium expression, it can be rearranged to account for the constant concentration of water.

$$K_{eq}[H_2O] = \frac{[H_3O^+][F^-]}{[HF]}$$

CHEM THEME

The concept of equilibrium as applied to acid-base reactions is a model used to explain our macroscopic observations and to make predictions for other acid-base reactions.

The product of two constants, $K_{eq}[H_2O]$, can be used to define another constant, the acid dissociation constant of a weak acid, K_a.

$$K_{eq}[H_2O] = K_a$$

For the HF reaction, the expression for the acid dissociation constant, K_a, is:

$$K_a = \frac{[H_3O^+][F^-]}{[HF]}$$

Table 20-5

CONCENTRATIONS IN A 0.100M HF SOLUTION			
SPECIES	**INITIAL CONCENTRATION**	**EQUILIBRIUM CONCENTRATION**	**METHOD USED TO FIND CONCENTRATION**
HF	0.100M	0.100M − 0.008M = 0.092M	$[HF] - [H_3O^+]$ The concentration of HF that ionizes must be subtracted from the initial HF concentration.
H_2O	55.6M	55.6M − 0.008M = 55.6M	The molarity of the water is calculated by*: $$\frac{1000 \text{ g}}{1 \text{ L}} \times \frac{1 \text{ mol}}{18.0 \text{ g}} = 55.6M$$
H_3O^+	—	$8 \times 10^{-3}M$	The color is matched with a HCl solution of known concentration.
F^-	—	$8 \times 10^{-3}M$	Using the mole relationship from the equation: $[H_3O^+] = [F^-]$

*Recall that 1 liter of water essentially is equivalent to 1000 grams of water.

The data from Table 20-5 can be used to calculate K_a for HF.

$$[H_3O^+] = 8 \times 10^{-3}$$
$$[HF] = 9.2 \times 10^{-2}$$
$$K_a = \frac{[8 \times 10^{-3}][8 \times 10^{-3}]}{[9.2 \times 10^{-2}]}$$
$$K_a = 7 \times 10^{-4}$$

Acid dissociation constants for various acids are listed in Table 20-6. The numerical value of K_a indicates the ability of an acid to donate a proton. The larger the value, the greater the ability. Since strong acids react 100% in dilute aqueous solution, their K_a values are so large that they are not used in problem work.

Table 20-6

DISSOCIATION CONSTANTS FOR SOME WEAK ACIDS		
ACID	**REACTION**	**K_a(AT 25°C)**
hydrofluoric	$HF + H_2O \longrightarrow H_3O^+ + F^-$	6.6×10^{-4}
formic	$HCOOH + H_2O \longrightarrow H_3O^+ + HCOO^-$	1.8×10^{-4}
acetic	$CH_3COOH + H_2O \longrightarrow H_3O^+ + CH_3COO^-$	1.8×10^{-5}
carbonic	$H_2CO_3 + H_2O \longrightarrow H_3O^+ + HCO_3^-$	4.4×10^{-7}
hydrogen sulfite ion	$HSO_3^- + H_2O \longrightarrow H_3O^+ + SO_3^{2-}$	6.2×10^{-8}
hypochlorous	$HClO + H_2O \longrightarrow H_3O^+ + ClO^-$	2.9×10^{-8}
hydrocyanic	$HCN + H_2O \longrightarrow H_3O^+ + CN^-$	6.2×10^{-10}
hydrogen carbonate ion	$HCO_3^- + H_2O \longrightarrow H_3O^+ + CO_3^{2-}$	4.7×10^{-11}

In using K_a values, it is important to specify a temperature for which that K_a value is valid. Most frequently, 25°C is used as the standard temperature.

Look again at the equation for the reaction of HF. As was mentioned earlier, the equilibrium expression for this reaction is:

$$HF + H_2O \rightleftharpoons H_3O^+ + F^-$$

$$K_a = \frac{[H_3O^+][F^-]}{[HF]}$$

Suppose for some reason, such as an increase in temperature, that more molecules dissociate into ions. How would that change affect the numerical value of the equilibrium constant? Would the value be larger or smaller? Suppose in another circumstance some of the ions associated to produce more molecules, and thus few ions. How would that change affect the numerical value of the equilibrium constant? Would it make it larger or smaller?

Suppose you have two acid solutions each at a concentration of $1M$. One acid has a value of $K_a = 1 \times 10^{-5}$ and the other acid has a value of $K_a = 10$. In which solution would you find that more of the acid molecules had ionized? How is the value of K_a related to acid strength?

You may want to introduce the use of the quadratic equation to solve other equilibrium problems. We did not use it in this text in an effort to reduce the information students are expected to deal with in this chapter.

Example   20-1

What is the $[H_3O^+]$ when 0.25 mole of acetic acid, CH_3COOH, is dissolved in enough water to make one liter of solution?

▶ **Suggested solution**

According to the equation for the reaction of CH_3COOH in water, the number of H_3O^+ ions produced equals the number of CH_3COO^- ions produced.

$$CH_3COOH + H_2O \longrightarrow H_3O^+ + CH_3COO^-$$

Look up the ionization constant for CH_3COOH in Table 20-6 and write the equilibrium expression for the reaction.

$$K_a = \frac{[H_3O^+][CH_3COO^-]}{[CH_3COOH]} = 1.8 \times 10^{-5}$$

From the problem you know that the concentration of the solution is $0.25M$. When acetic acid ionizes, a small amount of the molecules are consumed to produce the ions. For the acetic acid concentration in this problem the amount reacted is small enough to be insignificant. Thus in the calculation the concentration can be assumed to remain at $0.25M$.

$$K_a = \frac{[H_3O^+][CH_3COO^-]}{[0.25]} = 1.8 \times 10^{-5}$$

In the strictest sense equilibrium constants calculated from activities would be unitless. We have chosen not to include units for the equilibrium constants in this book.

From the balanced equation you know the mole ratio of H_3O^+ and CH_3COO^- are 1 to 1.

$$[H_3O^+] = [CH_3COO^-]$$

$$K_a = \frac{[H_3O^+]^2}{[0.25]}$$

Clear the denominator from the equation by multiplying both sides by 0.25M.

$$K_a(0.25) = [H_3O^+]^2$$

$$(1.8 \times 10^{-5})(0.25) = [H_3O^+]^2$$

To find the $[H_3O^+]$, you must take the square root of both sides.

$$[H_3O^+] = 6.7 \times 10^{-3}\,M$$

In most ways the equilibrium characteristics of weak base solutions are similar to weak acid solutions. As you know, the basic properties of a solution depend upon the concentration of hydroxide ion present in the solution.

The reaction of a weak base is an equilibrium reaction just like the dissociation of a weak acid. K_b represents the dissociation constant of a weak base. The calculations are similar.

Review and Practice

1. What ion species is characteristic of acidic solutions?

2. How do strong and weak acids differ?

3. Write the reaction of nitric acid with water.

4. Why is a weak acid reaction with water written as an equilibrium reaction?

5. What is the weakest conjugate base in Table 20-4?

6. From Table 20-4, which conjugate base is stronger—hydroxide ion, OH^-, or ammonia, NH_3?

7. What ion accounts for the basic character of a solution?

8. Write the equation for the complete dissociation of $Ba(OH)_2$ in water.

not 100% ionization

1. The presence of hydronium ion, H_3O^+, makes it acidic.
2. They differ in the completion of the reaction with water
3. $HNO_3 + H_2O \longrightarrow NO_3^- + H_3O^+$
4. It is equilibrium because all of the molecules do not react.
5. Cl^-
6. OH^- is stronger.
7. a proton acceptor
8. $Ba(OH)_2(s) \longrightarrow Ba^{2+}(aq) + 2OH^-(aq)$

20-10 Dissociation of Water and K_w

Water is considered a nonelectrolyte. However a few water molecules do react to form ions as represented by the following equation.

$$2H_2O \rightleftharpoons H_3O^+ + OH^-$$

This reaction is an equilibrium system. Therefore, an equilibrium constant for water can be determined. The equilibrium expression is written as:

$$K_{eq} = \frac{[H_3O^+][OH^-]}{[H_2O]^2}$$

CHEM THEME

The dissociation of water is explained by the equilibrium model.

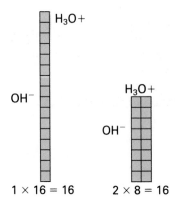

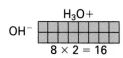

For the figure $H_3O^+ \times OH^- = 16$ a constant in this illustration

In real terms $[H_3O^+] [OH^-] = 1 \times 10^{-14}$

Figure 20-14 The product constant for these figures is 16. There are various multiples that will give this product. K_w for water is 10^{-14}. $[H_3O^+] \times [OH^-]$ always equals the product constant 10^{-14}.

equimolar = equal molarity

complete longhand

1. $[H_3O^+] = 1.00 \times 10^{-1}M$
2. $[OH^-] = 1 \times 10^{-13}M$
3. The $[H_3O^+]$ is higher in the HCl solution since it reacts completely with water to form H_3O^+.
4. $[H_3O^+] = 1 \times 10^{-7}$; $[OH^-] = 1 \times 10^{-7}$
5. K_w describes the proportionality constant (ion product constant), 1×10^{-14}
6. $[H_3O^+]$ decreases; $[H_3O^+]$ and $[OH^-]$ are inversely proportional

Note that the coefficient 2 for H_2O in the balanced equation now appears as the exponent 2 for the water concentration term. As you have seen in the previous sections, the concentration of water is constant within the uncertainty associated with changes of weak electrolytes. The water K_{eq} and $[H_2O]$ can be combined to define a new constant, K_w.

$$K_{eq}[H_2O]^2 = [H_3O^+][OH^-]$$
$$K_w = [H_3O^+][OH^-]$$

The value for K_w at 25° has been determined to be 1.00×10^{-14}. Thus,

$$[H_3O^+][OH^-] = 1.00 \times 10^{-14}$$

Since water is a neutral solution, it has as many hydronium ions as hydroxide ions. (Looking back at the equation, you will see that H_3O^+ and OH^- are in a 1-to-1 mole ratio.) Substituting in the equation gives:

$$[H_3O^+][H_3O^+] = 1.00 \times 10^{-14}$$

By finding the square root, the hydronium ion concentration can be determined.

$$[H_3O^+] = 1.00 \times 10^{-7}$$

In pure water the concentrations of H_3O^+ and OH^- are equal. But what happens if NaOH or gaseous HCl is added to the water? HCl acts as a strong acid when it dissolves in water, forming H_3O^+ and Cl^- ions. All acids increase the $[H_3O^+]$ and decrease the $[OH^-]$ in the system. On the other hand, NaOH acts as a strong base, forming OH^- and Na^+ ions. Bases increase the $[OH^-]$ and decrease the $[H_3O^+]$. In either case, the $[H_3O^+]$ and $[OH^-]$ are no longer equal. However, experiments show that the equilibrium relationship is still valid for every aqueous solution containing H_3O^+ and OH^- at 25°C.

$$K_w = [H_3O^+][OH^-] = 1 \times 10^{-14}$$

As you have already learned when studying Boyle's law concerning gases (Chapter 7), this kind of equation represents an inverse proportion. If $[H_3O^+]$ increases, then $[OH^-]$ must decrease and vice versa. The relation is represented schematically in Figure 20-15.

Review and Practice

1. What is the $[H_3O^+]$ in a 0.100 M solution of HCl?

2. What is the $[OH^-]$ in the solution described in item 1?

3. Would the $[H_3O^+]$ be higher in a 0.100M solution of HCl or CH_3COOH. Explain your reasoning.

4. What is the concentration of H_3O^+ in pure water? What is the concentration of OH^- in pure water?

5. What does K_w describe and what is its numerical value?

6. What happens to the $[H_3O^+]$ in an equilibrium solution if the $[OH^-]$ is increased? Why?

Typical Reactions of Acids and Bases

In the beginning of this chapter you learned that the reaction between an acid and a base can produce a salt and water. This reaction is called a neutralization reaction because the protons donated by the acid are accepted by a base.

In this lesson you will take the idea of neutralization further in seeing how this reaction is used in the lab analysis of different materials.

CONCEPTS

- characteristics of neutralization reactions
- the pH scale
- titration of an acid and base

20-11 Indicators

Acid-base indicators are large molecules that can act as weak acids or weak bases like the compounds and ions listed in Table 20-4. Some of them contain hydrogen and can donate protons in much the same way that HCl reacts in water to donate protons. In the following equation, HInd represents such an indicator.

$$\underset{\text{acid form}}{\text{HInd}} \rightleftharpoons H^+ + \underset{\text{base form}}{\text{Ind}^-}$$

Ind is not the symbol for an element. It is used here to represent any indicator anion. The H in the formula represents the hydrogen atom that becomes hydronium ion when the indicator acts as an acid.

Some indicators are weak bases, and the neutral molecule acts as the proton acceptor rather than the proton donor. This reaction is shown by the following equation.

$$\underset{\text{acid form}}{\text{HInd}^+} \rightleftharpoons H_3O^+ + \underset{\text{base form}}{\text{Ind}}$$

Note that this equation is very similar to the equation showing the NH_4^+ and NH_3 acid-base pair in Table 20-4.

Whether an indicator is a weak acid (as shown by the first equation) or a weak base (as shown by the second equation), it can exist in two forms that have a different color. For example, the acid form of methyl orange ($HInd^+$) is red, and the base form (Ind) is yellow. The acid form of phenolphthalein (HInd) is colorless, and the base form (Ind^-) is red.

When an indicator is in a solution with a high concentration of hydronium ions, most of the molecules exist in the acid form and the solution takes on the color of the acid form. When the indicator is in a solution containing a stronger base, most of the indicator molecules give up protons and exist in the base form. The solution then takes on the color of the base form of the indicator. Thus, the indicator provides information about the hydronium ion concentration.

Figure 20-15 The acid and base forms of bromthymol blue and phenolphthalein. Note that the base forms of both molecules are charged.

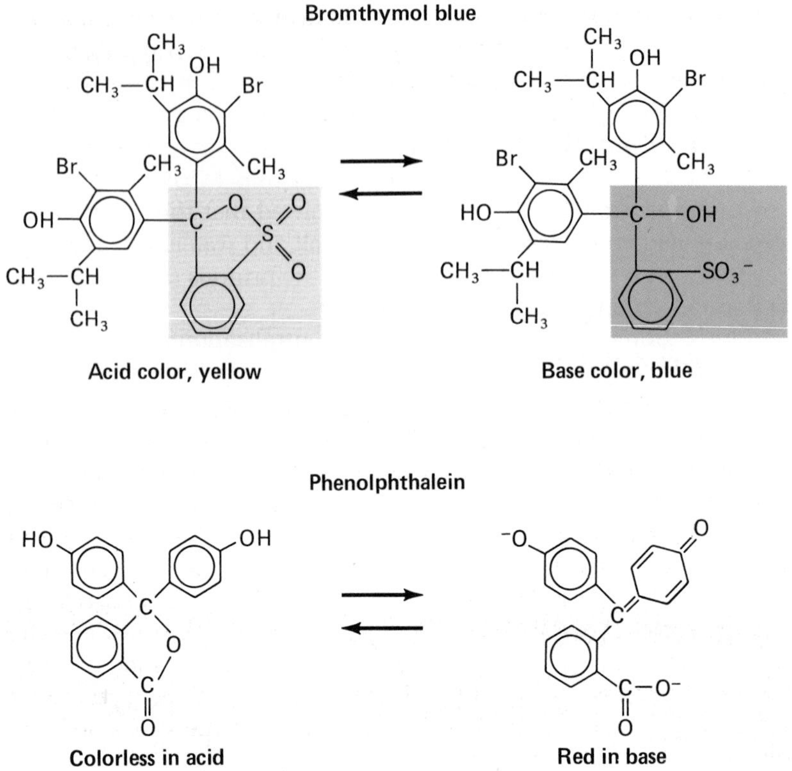

Bromthymol blue

Acid color, yellow

Base color, blue

Phenolphthalein

Colorless in acid

Red in base

20-12 Measuring Hydronium Ion Concentration

Many reactions in living systems take place in water solutions, and the hydronium ion concentration influences the reactions that can take place. For this reason, chemists have found that it is important to measure hydronium ion concentrations, even when they are very low.

As suggested by the $[H_3O^+]$ in pure water $(0.000\ 000\ 1M)$, hydronium ion concentrations do seem small. They can be much smaller. In a saturated solution of NaOH, the hydronium ion concentration may be as low as $0.000\ 000\ 000\ 000\ 001M\ (10^{-15}M)$. In a saturated solution of HCl, $[H_3O^+]$ may be as high as $10M$. As you can see, the concentration of hydrogen ions in water solutions can vary a great deal. In fact, it varies so much that it is difficult to imagine.

Early in this course you were asked to count a million of something. Now try to imagine a millionth of something. A millionth of a sheet of typing paper is about the size of the period at the end of this sentence. A millionth is 10^{-6}. The hydronium ion concentration in water is 10^{-7}, only one tenth as much. Can you imagine one tenth of the size of a period? Well, that area compared to the area covered by a whole sheet of paper is similar to the comparison between the hydrogen ion concentration in water and that in a $1M$ solution of HCl. It is a big difference, but it is only half the story.

Remember that in a saturated solution of NaOH, the hydrogen ion concentration can be as low as $10^{-15}M$. That is 100 million times less than $[H_3O^+]$ in pure water. To get some idea of the range of concentration, try to imagine spreading out 100 million sheets of typing paper. (They would cover 2.33 square miles or, said another way, a square measuring about 1.53 miles on each side.) Now compare the size of one tenth of a period with that huge square of typing paper. The difference in size is comparable to the difference in hydrogen ion concentration that can occur in water solutions.

20-13 A Scale for Large Changes: pH and pOH

There are difficulties in trying to represent something that varies over a wide range of values. To illustrate the problem, look at some data representing the $[H_3O^+]$ during an acid-base reaction. The reaction to be discussed is the neutralization of a $1M$ solution of NaOH with a $1M$ solution of HCl. Focus on the change in the $[H_3O^+]$ during the reaction. You will be starting with 100 mL of the $1M$ NaOH solution. The $1M$ HCl is added a little at a time until all of the NaOH has reacted and there is an excess of HCl. Each time some of the acid is added the $[H_3O^+]$ of the solution is measured with the instrument shown in Figure 20-16 to see how it changes. Figure 20-17 is an attempt to plot a graph of these data. The $[H_3O^+]$ is shown along the y-axis and the volume of HCl added is shown along the x-axis.

The problem is selecting a scale for the $[H_3O^+]$ that will allow you to place all of the data on a single graph and still give an accurate picture of what is taking place. Figure 20-17 shows a graph that includes all of

Figure 20-16 This equipment is used to measure changes in $[H_3O^+]$ concentration. It is called a pH meter.

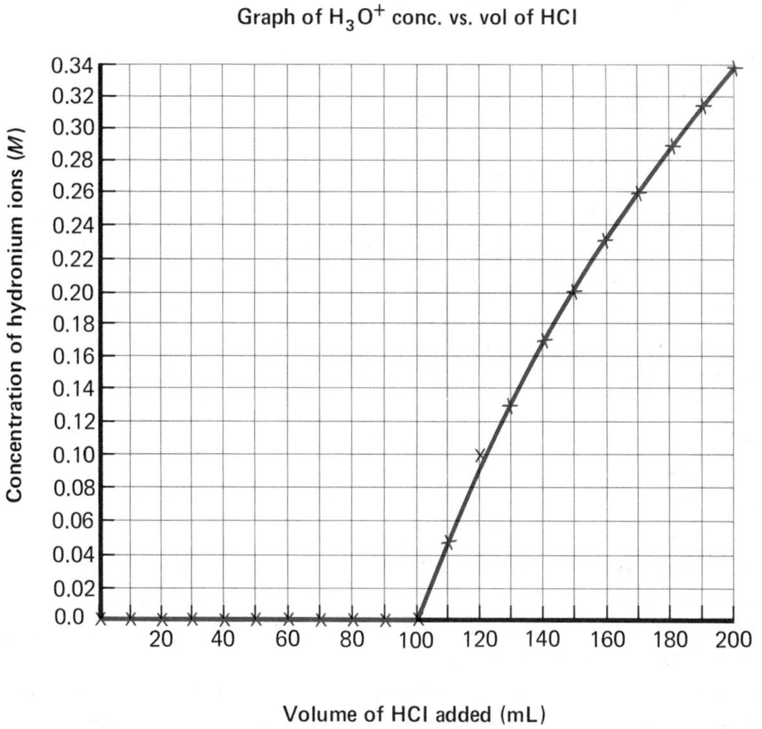

Figure 20-17 The data from Table 20-7 are graphed to show the change in hydronium ion concentration as $1M$ HCl is added to $1M$ NaOH. On this scale it appears that there is no change in $[H_3O^+]$ until 100 mL of the acid have been added.

Graph of H_3O^+ conc. vs. vol of HCl

[Graph: y-axis labeled "Concentration of hydronium ions (M)" ranging from 0.0 to 0.34 in increments of 0.02; x-axis labeled "Volume of HCl added (mL)" ranging from 20 to 200 in increments of 20]

the data points, but the scale is so small that there appears to be no change in the $[H_3O^+]$ until 100 mL of HCl have been added. Looking at Table 20-7, you know that this is not true. The $[H_3O^+]$ had doubled by the time 30 mL of acid had been added, and it had increased tenfold by the time 90 mL of acid had been added. This change does not show on the graph in Figure 20-17. The only way to make it show is to use a different scale.

Figure 20-18 shows the data from Table 20-7 plotted with a new scale. Now the change in $[H_3O^+]$ that occurs during the addition of the first 99 mL of acid can be seen. The problem is that the rest of the data cannot be shown. Is it possible to select a scale that shows the changes taking place and still get all the data on the graph?

One solution to this problem is to transform the wide range of values to a logarithmic scale. The **logarithm** of a number is the power to which 10 must be raised to equal that number. An increase of 1 in the logarithm of a number represents a tenfold increase in the number.

Figure 20-18 A portion of the graph in Figure 20-17 is plotted using a different scale for the hydronium ion concentration. Note that the hydronium ion concentration changes rapidly when 90–100 mL of acid are added. Using this scale makes it difficult to get all the data from Table 20-7 on a conveniently sized graph.

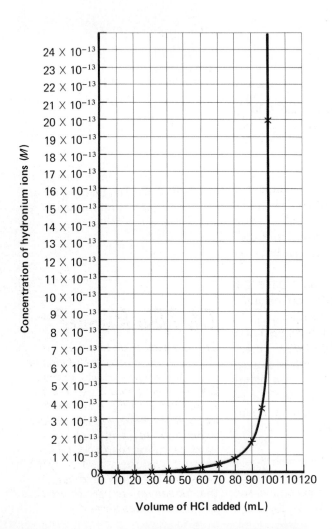

Volume of HCl added (mL)

Table 20-7

$[H_3O^+]$ AS $1M$ HCl IS ADDED TO 100 mL OF $1M$ NaOH	
VOLUME OF HCl ADDED (mL)	**HYDRONIUM ION CONCENTRATION**
0	1.0×10^{-14}
10	1.2×10^{-14}
20	1.4×10^{-14}
30	1.8×10^{-14}
40	2.3×10^{-14}
50	3.0×10^{-14}
60	4.0×10^{-14}
70	5.6×10^{-14}
80	9.1×10^{-14}
90	2.0×10^{-13}
95	3.8×10^{-13}
99	2.0×10^{-12}
99.9	2.0×10^{-11}
99.99	2.0×10^{-10}
100	1.0×10^{-7}
100.01	5.0×10^{-5}
100.1	5.0×10^{-4}
101	5.0×10^{-3}
105	2.4×10^{-2}
110	4.8×10^{-2}
120	9.1×10^{-2}
130	1.3×10^{-1}
140	1.7×10^{-1}
150	2.0×10^{-1}
200	3.3×10^{-1}

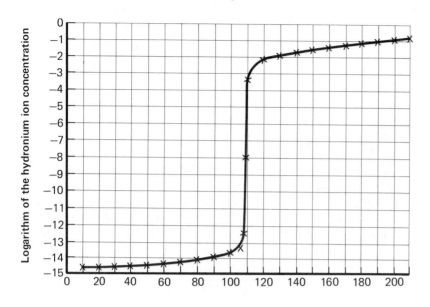

Graph of log H₃O⁺ vs vol of HCl.

Logarithm of the hydronium ion concentration

Volume of HCl added (mL)

Figure 20-19 In this graph, the hydronium ion concentrations now appear as logarithms. All the data from Table 20-7 now fit on a conveniently sized graph. Note that the logarithms are all negative giving a graph that does not have its origin in the lower left corner.

The $[H_3O^+]$ after 90 mL of acid were added is shown as 2.0×10^{-13}, and the logarithm of that number is -12.7. When 99 mL of acid had been added, the $[H_3O^+]$ had increased 10 times to a value of 2.0×10^{-12}. However, the logarithm increased only 1 to a value of -11.7.

When the $[H_3O^+]$ recorded in Table 20-7 varies from 0.000 000 000 000 01 to 0.33 (a change of over 10 million million), the $\log[H_3O^+]$ changes only from -14 to nearly 0. Although the $[H_3O^+]$ cannot be represented conveniently on a graph, the $\log[H_3O^+]$ can be. The data are plotted in Figure 20-19. Note that the changes in $[H_3O^+]$ during the early stages of the addition of acid are shown and that all of the data are plotted.

There is really only one thing that is inconvenient about the graph shown in Figure 20-19. All of the values for the $\log[H_3O^+]$ are negative. You are more accustomed to working with positive numbers.

Any time the log of the hydronium ion concentration is represented, you are likely to get a negative number. In most analytical experiments involving acids the hydronium ion concentration is seldom greater than $1M$, so the $\log[H_3O^+]$ is usually negative. (The logarithm of any number between 1 and 0 is negative.) Not willing to bother with negative numbers all the time, chemists defined a term to describe the log of the hydronium ion concentration. The term is **pH**.

$$pH = -\log[H_3O^+]$$

Since the logarithm of the hydronium ion concentration is almost always negative, minus the logarithm of the hydronium ion concentration is almost always positive. The $\log[H_3O^+]$ and pH are the same

Do You Know?

Enzyme activity depends on pH. Pepsin, an enzyme which helps in the digestion of proteins, is metabolically active in a pH of 1.5. It works in the stomach where HCl is secreted and where the pH is low enough to enable the pepsin to work. Once the food and pepsin pass into the duodenum, the pH goes way up as the secretions in the duodenum are very basic. Thus the pepsin is deactivated.

Figure 20-20 Using the pH scale the data from Table 20-7 can be plotted showing a graph with its origin in the lower left corner.

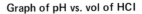

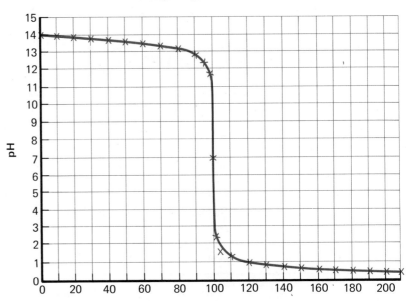

Volume of HCl added (mL)

Do You Know?

Flowers are sensitive to the pH of the soil. Hydrangeas produce blue flowers when grown in an acid soil and pink blossoms when grown in an alkaline soil.

number but with opposite sign. As a result, as $[H_3O^+]$ increases, the pH value decreases (-5 is larger than -6, but $+5$ is less than $+6$). With pH, you can use positive numbers and plot graphs in the upper right quadrant, where you are more accustomed to seeing them.

Figure 20-20 shows the data from Table 20-7 expressed as pH. It is just like Figure 20-19 but rotated 180° on the x-axis. (It is what you would see in a mirror placed along the x-axis of Figure 20-19.)

Since water is neutral, the midpoint or neutral point of the pH scale is labeled as 7. The $[H_3O^+]$ of water is $1.0 \times 10^{-7} M$. Thus it makes sense that the neutral point on the pH scale is 7. Relationships among $[H_3O^+]$, $[OH^-]$, and pH are shown in Figure 20-21.

Figure 20-21 From the number line for pH note that as pH increases $[H_3O^+]$ decreases. What is the relationship between pH and $[OH^-]$?

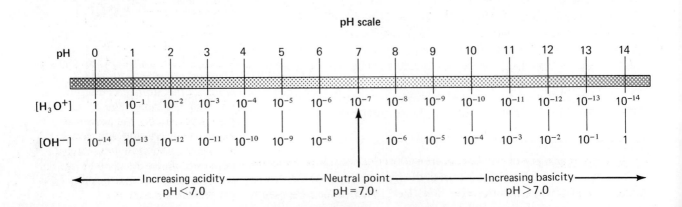

Table 20-8 lists pH values for some common substances. The Examples that follow will give you a better idea of how pH is calculated.

Table 20-8

pH VALUES FOR SOME COMMON SUBSTANCES*	
pH	**SUBSTANCE**
0	1*M* HCl
1	gastric (stomach) juice
2	lemon juice
3	grapefruit, soft drinks
4	tomato juice
5	black coffee
6	human saliva, cow's milk
7	pure water
7.4	human blood
8	seawater
9	
10	soap and detergent solutions
11	
12	saturated $Ca(OH)_2$ solution
13	
14	1*M* NaOH

*Note that the values listed represent an average point in a pH range for each material. For example, tomato juice has a pH that ranges from 4 to 4.5. Detergent solutions can vary from 9.5 to 10.5.

Example 20-2

What is the pH of a 0.01*M* HNO_3 solution?

▶ **Suggested solution**

HNO_3 is a strong acid, so it ionizes 100% in solution. Therefore,

$$[H_3O^+] = 0.01M = 1 \times 10^{-2}M$$

By definition $\quad pH = -\log[H_3O^+]$

$$\log \text{ of } 1 \times 10^{-2} = \log 10^0 + \log 10^{-2}$$
$$= 0 + -2.0$$

$$pH = -(-2.0) = 2.0$$

Using a calculator, enter 1×10^{-2}, press the log key, change the sign, and the display should read 2.

Example 20-3

What is the $[H_3O^+]$ concentration of a solution that has a pH of 3?

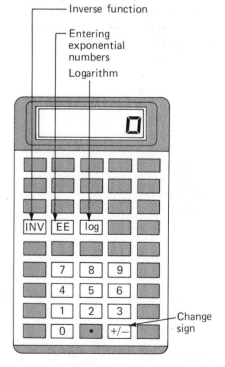

Figure 20-22 Calculators can be used to easily convert values to logs and back again. Though calculator keyboards may vary some of the function keys you may need to solve pH problems are shown.

CHAPTER 20

Suggested solution

When asked to calculate the $[H_3O^+]$ from the pH you must use antilogs. The antilog of number is the number equal to 10 raised to the power of a given number.

$$\text{antilog } x = 10^x$$

In this problem, you are being asked to calculate the antilog of -3.

$$pH = -\log[H_3O^+] = 3$$
$$\log[H_3O^+] = -3$$
$$[H_3O^+] = \text{antilog of } -3$$
$$[H_3O^+] = 10^{-3}$$

Using a calculator that has log and antilog functions, enter -3, press the INV and log keys, and the display will read 0.001.

Basic solutions can be ranked using the pOH scale. Such a scale appears exactly opposite of the pH scale shown in Figure 20-21. pOH is defined as:

$$pOH = -\log[OH^-]$$

Therefore, the following relationships exist for the pOH scale.

$$pOH < 7 \text{ for basic solutions}$$
$$pOH = 7 \text{ for neutral solutions}$$
$$pOH > 7 \text{ for acidic solutions}$$

Since there is a relationship between the $[H_3O^+]$ and $[OH^-]$ for any solution, a relationship between pH and pOH can be derived.

$$[H_3O^+][OH^-] = 1 \times 10^{-14}$$
$$pH + pOH = 14$$

Example 20-4

Calculate the pOH of a solution made by mixing 0.20 gram of NaOH with water to make a 500-mL solution.

▶ Suggested solution

First, calculate the molarity of the solution. This is done by calculating the moles of NaOH in 0.20 gram.

$$0.20 \text{ g NaOH} \times \frac{1 \text{ mol}}{40 \text{ g NaOH}} = 0.005 \text{ mol NaOH}$$

Thus the molarity of the solution is:

$$\frac{0.005 \text{ mol NaOH}}{0.500 \text{ L soln}} = 0.01M$$

The dissociation equation for sodium hydroxide shows that one mole of NaOH produces one mole of OH^- ions.

$$NaOH(s) \longrightarrow Na^+(aq) + OH^-(aq)$$

Therefore, for a $0.01M$ solution of NaOH:

$$[OH^-] = 1 \times 10^{-2}M$$

By definition $\quad\quad pOH = -\log[OH^-]$

$$\log 1 \times 10^{-2} = \log 1 + \log 10^{-2}$$
$$= 0 + -2 = -2.0$$
$$pOH = -(-2.0) = 2.0$$

Using a calculator, enter 1×10^{-2}, press the log key, and change the sign. The display should read 2.

Review and Practice

1. Define pH.

2. If $[H_3O^+] = 0.0001M$, what is the pH of the solution?

3. If pH = 9, what is the hydronium ion concentration?

4. If pH goes from 4 to 3, what has happened to the $[H_3O^+]$? What has happened to the $[OH^-]$?

1. pH is minus the log of the hydronium ion concentration.

2. 4

3. 1×10^{-9}

4. ~~decreased~~ 10 times; OH^- increases 10 times

20-14 Titration

The most dramatic thing that Figure 20-20 reveals is the very rapid change in hydronium ion concentration when 100 mL of the HCl solution have been added. Something seems to be taking place at that point.

Since the HCl and NaOH are of the same concentration, equal volumes contain equal numbers of solute molecules. Furthermore, the equation indicates that one mole of NaOH combines with one mole of HCl.

$$HCl + NaOH \longrightarrow NaCl + H_2O$$

When 100 mL of $1M$ HCl are added to 100 mL of $1M$ NaOH, the resulting solution should consist of only the products of the reaction. The proportion of reactants in the mixture is exactly the same as the proportion described by the balanced equation. This point is called the **equivalence point**. It is the point at which the proportions of reactants in the beaker are equivalent to the proportions described by the equation.

In a reaction between a strong acid and a strong base in a water solution, there is a dramatic change in the pH of the solution at the equivalence point. This fact provides a way to find the equivalence point and is the basis for one of chemistry's most important analytical tools, **titration**.

You have already seen that indicators are weak acids or weak bases that change color as $[H_3O^+]$ of a solution changes. Figure 20-23 shows the color of some common indicators at various pH values. As the shaded areas of the graph suggest, indicators do not change color instantaneously at a given pH. The color change is due to the changing proportion of the indicator molecules in the acid or base form.

CHEM THEME

One question commonly asked in chemistry is: "How much do I have?" Titration is one analytical technique that can be used to answer this question.

Figure 20-23 The pH range over which six indicators change color. Note that all of the indicators change color in the region of rapid change in pH.

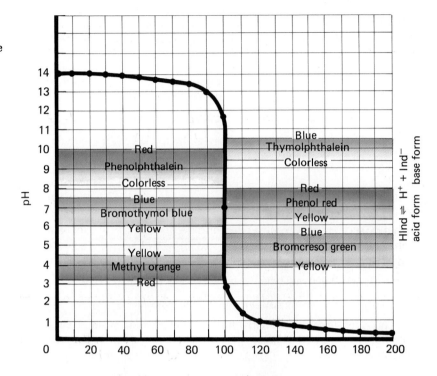

Below a pH of 3.2, virtually all of the methyl orange molecules have hydrogen ions attached and are in the acid form, which is red. Above that pH, the methyl orange molecules begin to donate protons to other bases in the solution, leaving the methyl orange in the base form, which is yellow. By the time the hydrogen ion concentration has decreased to a pH of 4.4, virtually all of the methyl orange molecules have lost protons and exist solely in the base form of the molecule. In similar fashion, phenol red is almost entirely in the acid form when the pH of the solution is below 6.6. As the hydrogen ion concentration falls (pH rises), molecules in the yellow, acid form dissociate to form the red, base form of the molecule. This transition is essentially complete at a pH of 8.0. Other color changes are shown in Figure 20-23.

The major thing to notice about Figure 20-23 is that all six of the indicators change color between the point where 99.9 mL of HCl have been added to the 100 mL of NaOH and the point where 100.1 mL of HCl have been added. In other words, all of these indicators change color at or very near the equivalence point for the reaction. Therefore, you can use this color change to indicate the equivalence point. (That is why they are called indicators.) Table 20-9 shows some common indicators and the pH range over which they change color.

Titration has many applications. It can be used to find the concentration of solutions that cannot be made accurately from dry solids. Another application might be quality control in the production of household vinegar. Ordinary vinegar is a 5% solution of acetic acid. This percentage is equivalent to about 0.83M. Titration can be used to be sure that the vinegar produced has this concentration.

Table 20-9

COMMON INDICATORS AND THEIR COLOR CHANGES				
INDICATOR	ACID	COLOR TRANSITION	BASE	pH OF TRANSITION INTERVAL
methyl violet	yellow	aqua	blue	0.0–1.6
methyl yellow	red	orange	yellow	2.9–4.0
bromphenol blue	yellow	green	blue	3.0–4.6
methyl orange	red	orange	yellow	3.2–4.4
methyl red	red	buff	yellow	4.8–6.0
litmus	red	pink	blue	5.5–8.0
bromthymol blue	yellow	green	blue	6.0–7.6
phenol red	yellow	orange	red	6.6–8.0
phenolphthalein	colorless	pink	red	8.2–10.6
thymolphthalein	colorless	pale blue	blue	9.4–10.6
alizarin yellow	yellow	orange	red	10.0–12.0

In order to explain the procedure used in titration, consider the neutralization reaction between NaOH and acetic acid.

$$NaOH + HC_2H_3O_2 \longrightarrow NaC_2H_3O_2 + H_2O$$

Note that, in this reaction, 1 mol of a NaOH reacts with 1 mol of $HC_2H_3O_2$. You could calculate the acetic acid concentration in vinegar if you knew how many moles of NaOH are needed to react with the acetic acid in a known volume of vinegar. According to the equation, the number of moles of NaOH that react is equal to the number of moles of acetic acid.

The key to the operation is determining when enough NaOH is added to the solution of vinegar to just react with the acetic acid present. In other words, the key is adding enough base to reach the equivalence point. This can be done with indicators.

Phenolphthalein commonly is used for this purpose, because it is easy to see the change from colorless in acid solution to red (faint pink when only a drop or two are used) in a base solution.

Add a drop or two of phenolphthalein to a few milliliters of any acid solution. Then, using a medicine dropper, add base to the solution. You will find that the solution remains colorless until you add enough base to neutralize all of the acid. Then the solution turns red, as shown in Figure 20-24. The amazing thing is that near the equivalence point, a single drop of base causes the color change. Adding a drop or two of acid changes the solution to colorless again. You can shift the color from colorless to pink and back again as often as you like by simply adding a drop or two of base and then a drop or two of acid.

Now you will see how to use this procedure to check the concentration of a vinegar solution.

First, 50 mL of vinegar are placed in a flask. (Any amount could be used but 50 mL is a convenient sample.) Next, one or two drops of phenolphthalein are added. The solution is colorless at this point.

NaOH of known concentration is added until the solution just turns pink. The NaOH usually is dispensed from a graduated piece of glass-

Figure 20-24 The equivalence point of an acid-base reaction is signified by a color change of the indicator used. In this photograph, the addition of one or two drops of base at the equivalence point changes the color of the phenophthalein in the solution.

ware called a buret so that the volume can be measured accurately. The apparatus is shown in Figure 20-25.

When the solution changes color, the equivalence point has been reached. The equation and the known concentration of the base can be used to calculate the concentration of the acid. The procedure will become clear as you work through the following Example.

Example   20-5

It is found that 42.5 mL of 1.02M NaOH have been added to 50.0 mL of vinegar when the phenolphthalein in the solution just turns pink. What is the concentration of the vinegar?

▶ **Suggested solution**

By now you know that quantities involved in a reaction are always calculated in terms of moles. First, calculate the number of moles of NaOH that were involved in the reaction.

$$? \text{ mol NaOH} = 42.5 \text{ mL NaOH soln} \times \frac{1.02 \text{ mol NaOH}}{1000 \text{ mL NaOH soln}}$$
$$= 0.0434 \text{ mol NaOH}$$

The color change occurs at the equivalence point—when the moles of base added are just enough to react with the moles of acetic acid in the vinegar. Therefore, you know that it takes 0.0434 mol of NaOH to neutralize the moles of acetic acid present in 50.0 mL of vinegar. The equation shows the proportions of acid and base that react.

Write a balanced equation for this reaction.

$$\text{NaOH} + \text{HC}_2\text{H}_3\text{O}_2 \longrightarrow \text{H}_2\text{O} + \text{NaC}_2\text{H}_3\text{O}_2$$

It is clear from the equation that the number of moles of acid that react is the same as the number of moles of base. However, this is not always true.

The equation indicates that 1 mol of sodium hydroxide reacts for each mole of acetic acid present. Since the ratio is one-to-one in this case, the calculation is trivial. You know that there must have been 0.0434 mol of acetic acid in the vinegar, because 0.0434 mol of NaOH was required to neutralize the acid.

$$? \text{ mol HC}_2\text{H}_3\text{O}_2 = 0.0434 \text{ mol NaOH} \times \frac{1 \text{ mol HC}_2\text{H}_3\text{O}_2}{1 \text{ mol NaOH}}$$
$$= 0.0434 \text{ mol HC}_2\text{H}_3\text{O}_2$$

The sample of vinegar used in this titration was 50.0 mL. Then you know that the vinegar contains 0.0434 mol of acetic acid in 50.0 mL of solution. What is the molarity of the vinegar?

$$\frac{0.0434 \text{ mol HC}_2\text{H}_3\text{O}_2}{50.0 \text{ mL soln}} \times \frac{1000 \text{ mL}}{\text{L}} = \frac{0.868 \text{ mol}}{\text{L soln}} = 0.868M$$

This sample of vinegar is a little more concentrated than the normal 5% solution, which is 0.83M.

Example 20-6

What is the concentration of acid in rainwater when 100 mL are titrated against 25.12 mL of 0.001 05M NaOH. Since acid rain contains several acids, HA will be used to represent the acids present.

▶ **Suggested solution**

Write the balanced equation for the neutralization reaction.

$$HA + NaOH \longrightarrow NaA + H_2O$$

In this balanced equation, one mole of base can neutralize one mole of acid. From the lab data, it is found that 25.12 mL of 0.001 05M NaOH was required to just reach the equivalence point. Moles of base used are calculated next.

$$25.12 \; \cancel{mL \; base} \times \frac{0.00105 \; mol \; base}{1000 \; \cancel{mL \; soln}} = 2.64 \times 10^{-5} \; mol \; base$$

2.64×10^{-5} moles of base are required to neutralize the moles of acid in 100 mL of rainwater. Thus, the moles of acid in the solution is 2.64×10^{-5}. Since the acid is present in 100 mL of rainwater, the molarity of HA is:

$$\frac{2.64 \times 10^{-5} \; mol}{0.100 \; L} = 2.64 \times 10^{-4} M$$

The concentration of acid in rainwater without any polluting acids normally is about $1 \times 10^{-6} \; M$. Acid rain is about 100 times more acidic than natural rainwater. In some areas of the northeast U.S. acid rain is even more acidic.

Figure 20-25 Most simple titrations involve the use of burets.

Review and Practice

1. What is the concentration of a NaOH solution when 30 mL of 0.5M HCl are needed to neutralize 50 mL of the base?

2. What is the concentration of acetic acid in vinegar when 32.5 mL of 0.56M NaOH are required to neutralize 15 mL of the vinegar?

3. What is the concentration of NH_3 in household ammonia when 48.25 mL of 0.5246M HCl are needed to neutralize 22.00 mL of the ammonia solution?

4. Oxalic acid, $H_2C_2O_4$, is a solid that is only slightly soluble in water. However, as it reacts with KOH or any other strong base, it continues to dissolve. When 6.25 g of oxalic acid are placed in water containing phenol red, the suspension is yellow. A solution of KOH is added until the color changes to red. What is the molarity of the KOH solution if 32.2 mL are required to change the phenol red to red?

5. A 5.0-g tablet of $Mg(OH)_2$ neutralizes 450 mL of stomach acid. What is the molarity of the HCl in the solution?

1. 0.3M
2. 1.2M
3. 1.151M NH_3
4. 4.25M
5. 0.38M HCl

Salts

In the last lesson the reaction between hydronium ions and hydroxide ions was explored in some detail. The spectator ions in those reactions were ignored. However, the reaction of the acid anion and the base cation forms a general group of compounds called salts. When the salt formed during an acid-base reaction is soluble in water, the salt remains as aqueous ions dissolved in the water. When the resulting salt of an acid-base reaction is insoluble, a precipitate forms.

Salts are ionic compounds. Many salts such as NaCl consist of a metal ion and a simple nonmetallic ion, or a metal ion and a polyatomic anion, such as Na_2SO_4.

CONCEPTS

■ salts are produced by acid-base reactions
■ hydrolysis reactions
■ behavior of anhydrides

20-15 Formation of Salts

Salts can be formed several different ways. As has already been mentioned, neutralization reactions between an acid and a base produce a **salt**.

$$HCl(aq) + NaOH(aq) \longrightarrow H_2O(l) + NaCl(aq)$$

When the water is evaporated the sodium ions and chloride ions unite to form crystalline salt. In Figure 20-26 you see large deposits of sodium chloride which have been deposited over many years as the water from a large inland sea has gradually evaporated. This salt deposit is very hard and level, making the Bonneville Salt Flats one of the best drag race courses in the world.

The salts deposited by receding ocean water are not produced by neutralization reactions, but rather by the solvation of ions from soil

Do You Know?

Sodium chloride is used to preserve food. Foods soaked in sodium chloride solutions, called brines, have large salt concentrations. When a microorganism lands on the food, it absorbs the salt. The osmotic pressure within the cells increases and the cell walls burst, killing the microorganism. The salt keeps microorganisms from multiplying on the surface of the food and thus prevents spoiling.

Figure 20-26 The Bonneville Salt Flats were once under water. The evaporation of the water left huge salt deposits.

and rocks during rain storms. These ions are eventually washed into the oceans.

Salts are also produced by direct union of two elements.

$$2Na(s) + Cl_2(g) \longrightarrow 2NaCl(s)$$

Another method of salt production is one you have seen in the lab when you reacted magnesium metal with hydrochloric acid.

$$Mg(s) + 2HCl(aq) \longrightarrow MgCl_2(aq) + H_2(g)$$

In general an acid reacts with some metals to produce hydrogen gas and a salt.

20-16 Hydrolysis

Salts are formed during the neutralization of an acid and a base. However, analysis shows that all salt solutions are not at pH7. When a salt is formed from a strong acid and a strong base, the resulting salt produces a solution at pH7 if soluble in water.

However, if the salt is the product of a strong acid and a weak base, analysis shows that the salt solution is slightly acidic. How can this result be explained? Look at the reaction between HCl and NH_3. HCl is a strong acid. NH_3 is a weak base. Recall that the ammonium ion will donate a proton in the presence of water to form the molecular base.

$$NH_4^+ + H_2O \rightleftharpoons NH_3 + H_3O^+$$

This reaction increases the hydronium ions is present in the solution, rendering the solution slightly acidic. This phenomena is called **hydrolysis.**

Likewise, if the salt is the product of a weak acid and a strong base, analysis shows that the salt solution is slightly basic. A hydrolysis reaction occurs when the conjugate base of the weak acid reacts with hydronium ions, causing an excess of hydroxide ions in the solution. Consider the neutralization reaction of acetic acid and sodium hydroxide. The acetate ion, CH_3COO^- readily forms molecular acetic

Figure 20-27 Some salts react with water (hydrolyze) to form acidic or basic solutions. Sodium acetate on the right is a basic salt. Ammonium chloride on the left is an acidic salt.

acid in the presence of water. This reaction increases the supply of hydroxide ions, OH^-.

$$CH_3COO^- + H_2O \longrightarrow CH_3COOH + OH^-$$

Thus only salts of a strong acid and a strong base produce a neutral solution (pH = 7). Salts of a strong acid and weak base produce solutions which are slightly acidic (pH < 7) whereas salts of a weak acid and a strong base produce solutions which are slightly basic (pH > 7).

20-17 Acidic and Basic Anhydrides

An oxygen-containing compound that reacts with water producing an acidic solution is called an **acidic anhydride**.

$$SO_3 + H_2O \longrightarrow H_2SO_4$$

An oxygen-containing compound that reacts with water producing a basic solution is called a **basic anhydride**.

$$Na_2O + H_2O \longrightarrow 2NaOH$$

When a basic anhydride and an acidic anhydride combine, they form a salt.

$$N_2O_5 + Na_2O \longrightarrow 2NaNO_3$$

The reaction is a neutralization without the formation of water. N_2O_5 is the acidic anhydride of HNO_3.

$$2HNO_3 \longrightarrow N_2O_5 + H_2O$$

acid   anhydride   water

Anhydride means without water so a basic anhydride is a base without water.

$$Ca(OH)_2 \longrightarrow CaO + H_2O$$

base   anhydride   water

What is the acidic anhydride of sulfurous acid?

Do You Know?

Grass thrives in an alkaline soil. Adding lime, CaO, increases the alkalinity or basicity of the soil by the following reaction:

$$CaO + H_2O \longrightarrow Ca(OH)_2$$

Table 20-10

Periodic Trends in Anhydrides						
Li_2O basic	BeO amphoteric	B_2O_3 acid	CO_2 acid	N_2O_3 acid		F_2O acid
Na_2O basic	MgO basic	Al_2O_3 amphoteric	SiO_2 acid	P_4O_6 acid	SO_3 acid	Cl_2O acid
K_2O basic	CaO basic	Ga_2O_3 amphoteric	GeO_2 amphoteric	As_4O_6 amphoteric	SeO_3 acid	Br_2O acid
Rb_2O basic	SrO basic	In_2O_3 basic	SnO_2 amphoteric	Sb_4O_6 amphoteric	TeO_3 acid	I_2O_5 acid
Cs_2O basic	BaO basic	Tl_2O_3 basic	PbO_2 amphoteric	Bi_2O_3 basic		

20-18 Application: Acid Rain

Whenever coal or petroleum products are burned, usually for fuel, some of the side products of that combustion are represented as SO_x and NO_x. SO_x and NO_x can represent SO, SO_2, NO, and NO_2.

All are toxic gases that escape through the smokestacks where the fuels are being burned and released into the atmosphere each day. When sun shines on those gases in the atmosphere, some gas molecules undergo a chemical reaction with the atmospheric oxygen.

$$2NO + O_2 \longrightarrow 2NO_2$$
$$2SO_2 + O_2 \longrightarrow 2SO_3$$

If you have ever seen smog, the reddish-brown color of it is due to the presence of the NO_2 in the atmosphere.

When it rains, these gases behave as acidic anhydrides producing acids in the rainwater.

$$NO_2 + H_2O \longrightarrow HNO_3$$
$$SO_2 + H_2O \longrightarrow H_2SO_3$$
$$SO_3 + H_2O \longrightarrow H_2SO_4$$

All these reactions produce acids. One is a weak acid and the other two are strong acids. The presence of these acids in the rain is having serious effects on the environment. The pH of rain has been lowered from the natural pH of about 6.4 to a pH range of 3 to 4 in some areas. As this acid rain soaks into soils and drains across rocks and fields into streams and rivers, it carries with it increased concentrations of some poisonous ions, such as aluminum and copper, which occur naturally in rocks and soil. These metallic ions in minute concentrations are not harmful to aquatic life, but with increased concentrations they cause suffocation and death of aquatic animals.

The acid rain falling on foliage is believed to damage the leaves of trees, and as the rain is absorbed through the roots of trees it is believed to weaken the plants. In a weakened condition, the plant is more prone to disease from other sources such as insect infestation and fungal infection.

Figure 20-28 Acid rain is causing foliage damage in many areas.

Smokestacks in the industrialized sections of the United States intentionally were built very high so that the pollutants would not fall back down onto the surrounding countryside. The intent was to let the air currents carry the smoke and other pollutants away in the wind. However, it is those pollutants that eventually come back down to Earth in the form of acid rain. Since the problem has been recognized, some effort has been made to reduce the amount of pollution by chemical means. One pollutant in particular, SO_3, can be reduced substantially by combination with CaO (lime from limestone) to produce $CaSO_4(s)$.

20-19 Buffers

You have seen how the addition of small amounts of acid or base can upset the $[H_3O^+] - [OH^-]$ equilibrium. However, in certain situations it is crucial to have a solution that can withstand the addition of an acid or base without substantially changing pH. Such solutions are said to be **buffers**.

Blood is a solution whose pH must remain constant in order for it to function properly in the body. Blood has the characteristic of maintaining pH with the addition of small amounts of acid or base. The presence of H_2CO_3 and HCO_3^- in blood maintains the pH. Carbonic acid and the hydrogen carbonate ion are a conjugate acid-base pair.

When excess hydronium ions enter the blood, the following reaction occurs to reduce their concentration.

$$H_3O^+ + HCO_3^- \longrightarrow H_2CO_3 + H_2O$$

When excess hydroxide ions enter the blood, their concentration is reduced by the following reaction

$$H_2CO_3 + OH^- \longrightarrow HCO_3^- + H_2O$$

Normal blood has a pH of 7.35. These reactions help to control the pH of blood within very narrow limits.

In looking at the blood example, you should note that the buffer pair is derived from a weak acid and the negative ion from the salt of a weak acid.

1. a. $(NH_4)_2SO_4$
   b. evaporation
   c. acidic
2. a. Na^+ increases, CH_3COO^- increases, CH_3COOH decreases, OH^- increases, H_3O^+ decreases
   b. CH_3COO^- increases, H_3O^+ decreases, CH_3COOH increases
   c. Na^+ increases, no effect on other ions
3. The addition of hydronium ions will shift a reaction in the direction that causes the formation of molecular acid. The addition of a base causes the acid molecules to react with water (shift in the opposite direction), thereby producing H_3O^+ to neutralize the OH^- from the base.

Review and Practice

1. When sulfuric acid and ammonia are combined in water,
   a. what salt is produced?
   b. how can that salt be extracted?
   c. is the salt acidic, neutral or basic?

2. Describe what happens to the concentrations of each species present in sodium acetate-acetic acid buffer solution when each of the following are added:
   a. sodium hydroxide solution
   b. potassium acetate
   c. sodium chloride solution

3. Explain how the buffer pair in item 1 behaves with the addition of acid or base using Le Chatelier's principle.

Numbers in red indicate the appropriate chapter sections to aid you in assigning these items. Answers to all questions appear in the Teacher's Guide at the front of this book.

Summary

■ Acids in solution conduct an electric current, produce hydrogen gas in the presence of active metals, taste sour, cause certain dyes to change color, and lose some of these properties in the presence of a base. Bases in solution conduct an electric current, taste bitter, cause certain dyes to change color, do not react with most metals at normal temperatures, and lose some of these properties in the presence of an acid.

■ The Brönsted-Lowry theory describes acids as proton donors and bases as proton acceptors. The hydronium ion is a hydrated proton that forms when an acid donates a proton to water.

■ If an acid or base is labeled as weak, the substance does not react completely in water. Strong acids and bases react completely.

■ The reaction of a weak acid or base with water is an equilibrium system. The relationships among the concentrations of the products and reactants of the reaction are defined by the K_a and K_b.

■ The dissociation of water is another equilibrium system. The relationship between the $[H_3O^+]$ and $[OH^-]$ of pure water is described by K_w.

■ A salt is the product of the combination of a base cation and acid anion. Salts also are formed from direct combination of the elements, or by acidic hydrogen displacement by a metal.

■ A salt is either acidic, basic, or neutral depending on the substances from which it was formed. Acidic salts are formed from strong acids and weak bases. Basic salts are formed from strong bases and weak acids. Neutral salts are formed from the reaction of a strong acid and strong base or the reaction of a weak acid with a weak base. The hydrolysis reaction causes the pH of some salt solutions to be above or below 7.

■ Anhydrides are oxides that dissolve in water to produce acidic or basic solutions. The behavior of anhydrides is a significant factor in the formation of acid rain.

■ Buffers can withstand small additions of acid or base without appreciably changing the H_3O^+ and OH^- balance of a solution. Buffers can be made from conjugate acid-base pairs.

Chemically Speaking

acid	K_b
acidic anhydride	K_w
amphiprotic	logarithm
base	neutralization
basic anhydride	pH
buffer	salt
conjugate acid-base pair	strong acid
equivalence point	strong base
hydrolysis	titration
hydronium ion	weak acid
indicator	weak base
K_a	

Review

1. What properties do acids and bases have in common? What properties differ?

2. Write the equation for the reaction of nitric acid reacting with water.

3. Write the equation for the reaction of cesium hydroxide dissolving in water. √

4. In the reaction of acetic acid in water, what is the acid, the base, the conjugate acid, and conjugate base?

√ 5. Write the Brönsted-Lowry acid-base reaction for water.

6. What is the K_a expression for acetic acid?

7. What indicators would be suitable for titrations with expected equivalence point pH's of 5.0 and 8.4?

√8. Write a reaction for the formation of KCl by:
   a. neutralization
   b. hydrogen displacement by a metal
   c. direct combination of the elements

Interpret and Apply

X 1. How does K_w relate to K_{eq}?

X 2. When a strong acid is added to water, why does the $[H_3O^+]$ become that of the acid added?

3. If the pH of a solution changes from 10 to 8, what has happened to the $[OH^-]$?

4. Why are the K_a values for weak acids less than one?

5. According to Table 20-4, which is the stronger acid?
a. $H_2PO_4^-$ or HPO_4^{2-} c. NH_4^+ or HN_3
b. H_2SO_4 or HSO_4^- d. HS^- or H_2S

6. Write a balanced equation for the reaction of magnesium and hydrochloric acid.

7. Write a balanced equation for the reaction of sulfur dioxide and water.

8. Write equations to show the step-wise reaction of the following polyprotic acids with water: H_2S, H_2CO_3, and H_3PO_4.

9. How can the choice of indicator affect the results of a titration?

10. What could you do besides start over, if you add too much standardized solution to a titration and overshoot the equivalence point?

11. An acid, HA, produces hydronium ions, H_3O^+, in the equilibrium.

$$HA + H_2O \longrightarrow H_3O^+ + A^-$$

a. Does equilibrium favor reactants or products for a strong acid?
b. Does equilibrium favor reactants or products for a very weak acid?
c. If acid HA_1 is a stronger acid than HA_2, is K_1 a larger or smaller number than K_2?

$$K_1 = \frac{[H_3O]^+[A_1^-]}{[HA_1]} \quad K_2 = \frac{[H_3O]^+[A_2^-]}{[HA_2]}$$

12. List the following compounds in order of increasing acid strength:

$$NH_4^+, HSO_4^-, H_2S$$

13. If $0.1M$ solutions are made of NH_4Cl, $KHSO_4$, and H_2S, in which will the $[H_3O^+]$ be highest and in which will it be lowest?

14. When sodium acetate, $NaCH_3COO$, is added to an aqueous solution of hydrogen fluoride, HF, a reaction occurs in which the weak acid HF loses hydrogen ions.
a. Write the equation for the reaction.
b. What base is competing with F^- for H^+?

15. a. Write the equation for the reaction that allows the acid-base reaction between hydrogen sulfide, H_2S, and the carbonate ion, CO_3^{2-}.
b. What are the two bases competing for the H_3O^+?

16. What salt is formed when zinc and sulfuric acid are combined? How is this salt extracted?

17. Describe two methods for producing potassium iodide.

18. Why is limestone (calcium carbonate or calcium hydroxide) added to acid soil? Why would gypsum (calcium sulfate) or aluminum sulfate be added to basic soil?

Problems

1. Calculate the concentration of a KOH solution that reaches the equivalence point when 30.0 mL are titrated against 42.7 mL of $0.498M$ HNO_3 used as the standardized solution.

2. How many milliliters of $0.28M$ NaOH would be required to neutralize 28.73 mL of $0.15M$ HCl?

3. What is the pH of a $0.01M$ HNO_3 solution?

4. Calculate the $[H_3O^+]$ of a solution which is $0.001M$ KOH.

5. For pure water, $K_w = 1.0 \times 10^{-14}$. Calculate the K_a for pure water.

6. A student dissolves 200 grams of CH_3COOH in enough water to make one liter of solution.
a. What is the molarity of this acetic acid solution?
b. What is the $[H_3O^+]$? Assume negligible change in CH_3COOH at equilibrium because of ionization.

7. Calculate $[H_3O^+]$ and $[OH^-]$ in a solution by mixing 50.0 mL of $0.200M$ HCl and 49.0 mL of $0.200M$ NaOH.

8. Calculate $[H_3O^+]$ and $[OH^-]$ in a solution made by mixing 50.0 mL of $0.200M$ HCl and 49.9 mL of $0.200M$ NaOH.

9. How much more $0.200M$ NaOH solution must be added to the solution in problem 7 to change $[H_3O^+]$ to 1×10^{-7}?

10. What is the $[H_3O^+]$ in a solution whose pH is 8? Is the solution acidic or basic? What is the $[OH^-]$ in the same solution?

11. A $0.25M$ solution of benzoic acid is found to have a $[H_3O^+]$ equal to 4×10^{-3}. Assuming the following reaction, calculate the K_a for benzoic acid. Use HA to represent benzoic acid.

$$HA(aq) + H_2O(l) \longrightarrow H_3O^+(aq) + A^-(aq)$$

12. Calculate the $[H_3O^+]$ when 0.056 grams of KOH are dissolved in one liter of solution.

13. How many grams of LiBr would be present after evaporation when 500 mL of $1M$ HBr and 500 mL of LiOH are titrated?

Challenge

1. If a certain acidic solution is 1.34% ionized, is that acid considered strong or weak?

2. What volume of $0.600M$ H_2SO_4 would be required to titrate a solution containing 2.50 g of sodium hydrogen carbonate?

$$2NaHCO_3 + H_2SO_4 \longrightarrow Na_2SO_4 + 2CO_2 + 2H_2O$$

3. The HF in a $0.100M$ aqueous solution is 8% dissociated. What is the value of its K_a?

4. Consider the following two equations:

$$Al(OH)_3(s) + OH^-(aq) \longrightarrow [Al(OH)_4]^-(aq)$$
$$Al(OH)_3(s) + 3H_3O^+(aq) \longrightarrow Al^{3+}(aq) + 6H_2O(l)$$

What term describes the behavior of $Al(OH)_3$? In which equation is the $Al(OH)_3$ acting as an acid? As a base?

Synthesis

1. Now that you have finished this chapter, you are well aware of many household items that are caustic in nature. Poison prevention is an important issue in storing caustic materials in the home. For example, if liquid Drano® is accidentally swallowed, the base causes severe damage to the esophagus. However, the base usually does not harm the stomach. Try to explain why. Directions on the Drano® container advise you NOT to induce vomiting in case of ingestion! Explain.

2. Commercial vinegar is usually 5% by mass acetic acid. Calculate the H_3O^+ concentration of acetic acid using Table 20-6 for the K_a value of acetic acid.

Project

1. Find out what chemicals are used in swimming pool maintenance. How is the concentration of these chemicals controlled? What role does pH play in the analysis of the water? What happens if the pH is not correct?

2. What is acid rain? What sources of toxic gases make acid rain different in the eastern and western United States? What is acid fog? What are the effects of acid rain and acid fog on living and nonliving materials?

A broad class of chemical reactions is being put to practical use by chemists. The plating of metal is only one example of this kind of reaction. Millions of kilograms of metals are electrically plated onto other metals each year. Plating beautifies, protects, and is less costly than making objects entirely from expensive metals.

Electrochemistry

Oxidation-Reduction

Many familiar chemical processes belong to a class of reactions called oxidation-reduction, or redox reactions. Every minute, billions of redox reactions are taking place inside your body. Reactions in batteries, wood burning in a campfire, the corrosion of metals, the ripening of fruit, and gasoline burning in an automobile engine are examples of the wide range of reactions included in this category.

The term *oxidation* first referred to the combining of oxygen with other substances, as in the processes of iron rusting and carbon burning. After more knowledge had been accumulated, chemists realized that other nonmetal elements besides oxygen would combine with metals in many similar reactions, and they began to use the word *oxidation* to describe a much larger class of reactions. *Reduction* first referred to the process of reducing metal ore, usually an oxide or a sulfide of the metal, to the pure metal. Chemists now use the term *reduction* in a much broader sense.

CONCEPTS

- oxidation and reduction reactions
- oxidation numbers
- balancing redox equations

21-1 The Nature of Redox Reactions

A very simple reaction can be used to demonstrate oxidation and reduction. When green, solid copper(II) chloride, $CuCl_2$, is dissolved in water, it produces an acid solution containing copper(II) ions, Cu^{2+}(aq), and chloride ions, Cl^-(aq), as seen in Figure 21-1. When a piece of aluminum foil is added to the solution, there is an immediate reaction. Bubbles, gas, and heat are produced. Changes are observed both in the solution containing the copper(II) ions and in the aluminum foil. The aluminum seems to change from its normal silvery color to a reddish brown color as seen in Figure 21-2. The solution loses its turquoise color and becomes colorless. The aluminum actually is

Perform this demonstration. Do not say too much about what is happening. The students should have enough chemical sense to describe the events. They may even remember that they did this reaction in the beginning of the course.

Figure 21-1 (left) Solutions of copper(II) ions are characteristically green or blue. (right) A reaction between aluminum and copper(II) ions in an acid solution. The solution is fizzing, and the aluminum is growing a brown, spongy material.

581

going through two reactions. In one of them, the aluminum reacts with the hydrogen ions present in the acid solution to produce bubbles of hydrogen gas. In the other reaction, the aluminum reacts with the copper(II) ions in solution. The unbalanced equation for the reaction between the two metals is

$$Cu^{2+}(aq) + Al(s) \longrightarrow Cu(s) + Al^{3+}(aq)$$

The chemical equation tells you that the copper ions in solution are forming copper metal atoms at the same time the aluminum foil is changing to aluminum ions. The reddish brown substance forming on the aluminum foil is the copper coming out of solution. The solution becomes colorless because the aluminum ions which are replacing the copper(II) ions in solution are colorless.

It is instructive to consider what is happening to the two metals independently. Each aluminum atom loses three electrons to form an aluminum ion as described by the following equation:

$$Al(s) \longrightarrow Al^{3+}(aq) + 3e^-$$

Each neutral aluminum atom acquires a 3+ charge by losing three electrons. This process is called *oxidation*.

The lost electrons will not exist alone. In this reaction, they are transferred to the copper(II) ions in solution as described by the following equation:

$$Cu^{2+}(aq) + 2e^- \longrightarrow Cu(s)$$

When a copper(II) ion acquires two electrons, it loses its 2+ charge and becomes neutral. This process is called *reduction*.

It may seem strange that an electron gain is called a reduction. Remember that the term reduction first was used to describe the process of reducing metal ore to elemental metal. Centuries later, after the electron was discovered (1897) and the existence of ions was established, chemists realized that the word reduction had been applied to a process in which electrons were being added to the positive metallic ions in ores.

Each equation written above describes only half of what takes place when aluminum reacts with copper(II) ions. Such equations are, logically enough, called descriptions of **half-reactions**. Two half-reactions, one that describes an oxidation and another that describes a reduction, are required to completely describe any **redox reaction**. Furthermore, the overall reaction will be described accurately only when the equations for the half-reactions are adjusted to show the same number of electrons lost in the oxidation half as are gained in the reduction half. In the example used above, each aluminum atom lost three electrons, but each copper(II) ion accepted only two.

Students often can be confused by this half-reaction. For some, the equation does not clearly show that three electrons are leaving an atom of aluminum.

It is very important that the students be able to visualize the relationship between the symbol Cu^{2+} and the number of protons and electrons in the ion.

Example 21-1

Balance the following redox reaction:

$$Cu^{2+}(aq) + Al(s) \longrightarrow Cu(s) + Al^{3+}$$

▶ Suggested solution

To balance this equation, begin by writing the equations for the half-reactions:

$$Cu^{2+}(aq) + 2e^- \longrightarrow Cu(s)$$
$$Al(s) \longrightarrow Al^{3+}(aq) + 3e^-$$

In order to balance the gain and loss of electrons, the equations can be multiplied by 3 and 2 respectively:

$$3Cu^{2+}(aq) + 6e^- \longrightarrow 3Cu(s)$$
$$2Al(s) \longrightarrow 2Al^{3+}(aq) + 6e^-$$

The number of electrons in both equations is now equal. The equations for the two half-reactions can be added together to yield a balanced redox equation:

$$3Cu^{2+} + 2Al(s) \longrightarrow 3Cu(s) + 2Al^{3+}(aq)$$

You should notice that a balanced oxidation-reduction equation does not show any electrons. The equation should be checked for correct balancing. By inspecting it, you can see that the number of copper and aluminum atoms and ions are equal on each side of the equation. Do the charges on each side balance? If you look at the left side of the equation, you see that three copper ions, each with a 2+ charge, would have a total charge of 6+. Since each of the two aluminum atoms is neutral, the total charge on the left is 6+. On the right side of the equation, the three copper atoms are neutral and the two aluminum ions, each with a charge of 3+, have a total charge of 6+:

$$3(2+) + 2(0) = 3(0) + 2(3+)$$
$$6+ = 6+$$

The equation is balanced in terms of charge with 6+ on each side.

In the copper(II) ion-aluminum metal reaction, the copper(II) ions were reduced, and simultaneously the aluminum atoms were oxidized. The electrons flow from the aluminum to the copper(II) ions. If the aluminum metal were not present, the copper(II) ions would not be reduced. Oxidation and reduction occur together. The aluminum brings about the reduction of copper(II) ions and is called the **reducing agent**. If the copper(II) ions were not present, the aluminum would not oxidize. The copper(II) ions are the **oxidizing agent**. Notice that the substance which is reduced is the oxidizing agent and the substance which is oxidized is the reducing agent.

As an exercise in problem solving, ask students to devise a way to balance the equation without writing half-reactions (balance charges). The two solutions should be compared to show that the results are the same.

Review and Practice

1. Determine which of the following processes are oxidations and which are reductions:
   a. Co^{2+} becomes Co
   b. $2I^-$ becomes I_2
   c. Fe^{3+} becomes Fe^{2+}
   d. Sn^{2+} becomes Sn^{4+}

1. a. reduction   b. oxidation
   c. reduction   d. oxidation

2. An oxidizing agent is the reactant which is reduced in a redox reaction.

3. A redox reaction is balanced when the number of atoms or ions of each element are equal on both sides of the equation. Electric charges must also balance.

4. a. $2Na(s) + Cl_2(g) \longrightarrow$
$$2Na^+(s) + 2Cl^-(s)$$
$Cl_2(g)$ is reduced. $Na(s)$ is the reducing agent.

b. $Cu^{2+}(aq) + Mg(s) \longrightarrow$
$$Cu(s) + Mg^{2+}(aq)$$
Cu^{2+} is reduced. Mg is the reducing agent.

c. $3Fe^{3+}(aq) + Al(s) \longrightarrow$
$$3Fe^{2+}(aq) + Al^{3+}(aq)$$
Fe^{3+} is reduced. Al is the reducing agent.

d. $2Au^{3+}(aq) + 3Cd(s) \longrightarrow$
$$2Au(s) + 3Cd^{2+}(aq)$$
Au^{3+} is reduced. Cd is the reducing agent.

You should explain to students that the system of oxidation numbers is an arbitrary but useful one set up by chemists so that they are better able to describe what happens in redox reactions.

2. What is an oxidizing agent?

3. How do you know when a redox reaction is balanced?

4. Balance the following redox reactions. Identify what has been reduced and what is the reducing agent:

a. $Na(s) + Cl_2(g) \longrightarrow Na^+(s) + Cl^-(s)$
b. $Cu^{2+}(aq) + Mg(s) \longrightarrow Cu(s) + Mg^{2+}(aq)$
c. $Fe^{3+}(aq) + Al(s) \longrightarrow Fe^{2+}(aq) + Al^{3+}(aq)$
d. $Au^{3+}(aq) + Cd(s) \longrightarrow Au(s) + Cd^{2+}(aq)$

21-2 Oxidation Numbers

In order to understand the process of electron transfer in redox reactions more clearly, chemists have assigned oxidation numbers to all atoms and ions. An **oxidation number** is the real or apparent charge an atom or ion has when assigned a certain number of electrons. The oxidation number of an atom is determined by pretending! The assumption is made that *all* bonds are ionic. This pretend-that-all-bonds-are-ionic game is played as though all bonding electrons belong to the atom with the higher electronegativity. The electronegativity of each atom determines whether it will more readily gain electrons or lose them to another atom. (To review electronegativity, see Section 12-2 and Figure 12-4.)

To see how oxidation numbers are assigned, consider the electron-dot structure for elemental hydrogen gas:

$$H:H$$

Since the atoms are identical, there is no difference in electronegativity. There is no reason to assign the bonding electrons to one atom in preference to the other, so the electrons are distributed equally:

$$H\frac{.}{.}H$$

Neither atom has gained nor lost an electron. Therefore, both atoms are still neutral, with a charge of 0. The oxidation number of each hydrogen atom also is 0. Since all pure elements are composed of identical atoms, all pure elements will have oxidation numbers of 0. All of the following elements, for example, would have oxidation numbers of zero: $Mg(s)$, $Na(s)$, $Hg(l)$, $He(g)$, $O_2(g)$, $S_8(s)$. The physical state of the element does not matter.

The oxidation number for simple ions such as those shown in Table 21-1 is merely the charge of the ion.

Table 21-1

OXIDATION NUMBERS OF SIMPLE IONS	
ION	OXIDATION NUMBER
Mg^{2+}	2+
Al^{3+}	3+
Cl^-	1−
S^{2-}	2−

Assigning oxidation numbers for compounds and complex ions is a little more difficult than assigning oxidation numbers for elements and simple ions. Water is a compound. In Chapter 12, you learned that the hydrogen atoms and oxygen atom in water are held together by covalent bonds. The atoms share pairs of electrons. However, since chemists try to understand what is occurring to electrons in oxidation-reduction reactions, they *pretend* that all bonds are ionic and assign charges as if the shared electrons were completely on one atom.

The electron-dot structure for water is shown below:

$$\text{H} : \overset{..}{\underset{..}{\text{O}}} : \text{H}$$

If the bonds between oxygen and hydrogen were ionic rather than covalent, which element would get the bonding electrons? In other words, which element is more electronegative? Figure 12-4 shows that oxygen is more electronegative than hydrogen. The bonding electrons are assigned to oxygen:

$$\text{H} \left[: \overset{..}{\underset{..}{\text{O}}} : \right] \text{H}$$

$$1+ \quad 2- \quad 1+$$

A neutral oxygen atom has 6 electrons in its outermost energy level. With the addition of the 2 extra electrons from hydrogen, oxygen now appears to have 8 electrons and, therefore, a $2-$ charge. The oxidation number of oxygen in water is $2-$.

A neutral hydrogen atom has 1 electron. The hydrogen atoms in the water molecule have no electrons assigned to them, so the charge is now $1+$. The oxidation number of hydrogen in water is $1+$. Since the overall charge on a water molecule is zero, the addition of the oxidation numbers for all the atoms in the molecule must be zero:

Atoms	Total oxidation number
2H	$2(1+) = 2+$
O	$1(2-) = 2-$
H_2O	$= 0$

In most cases, the oxidation number of oxygen will be $2-$ and that of hydrogen will be $1+$. However, there are some exceptions. Look at the electron-dot structure of hydrogen peroxide, H_2O_2:

$$\text{H} : \overset{..}{\underset{..}{\text{O}}} : \overset{..}{\underset{..}{\text{O}}} : \text{H}$$

The electrons bonding hydrogen to oxygen are assigned to oxygen as usual, but what about the electrons bonding the two oxygen atoms? There is no difference in electronegativity between the two atoms, so the bonding electrons are split evenly:

$$\text{H} \left[: \overset{..}{\underset{..}{\text{O}}} \middle| \overset{..}{\underset{..}{\text{O}}} : \right] \text{H}$$

$$1+ \quad 1- \quad 1- \quad 1+$$

The result is that each oxygen atom gains only one electron and its charge is $1-$. All peroxides contain an oxygen-oxygen bond, and the oxidation number of oxygen in these compounds is $1-$. The overall charge on the molecule is zero, and the sum of the oxidation numbers for all the atoms also is zero.

It is inconvenient to always write electron-dot structures to determine oxidation numbers. Several easy-to-follow rules usually are memorized to assign oxidation numbers to atoms and ions.

Rule 1   The oxidation number for any atom in its elementary state is 0.

Rule 2   The oxidation number for any simple ion is the charge on the ion.

   a. The oxidation number of alkali metals in compounds is 1+, as in Li^+, Na^+, and K^+.

   b. The oxidation number of alkaline-earth metals in compounds is 2+, as in Mg^{2+}, Ca^{2+}, and Ba^{2+}.

Rule 3   The oxidation number for oxygen usually is 2−. In peroxides, it is 1−.

Rule 4   The oxidation number for hydrogen is 1+ in all its compounds except in metallic hydrides like NaH or BaH_2, where it is 1−.

Rule 5   All other oxidation numbers are assigned so that the sum of oxidation numbers equals the net charge on the molecule or polyatomic ion.

These rules will allow you to determine the oxidation number of the atoms in most compounds or polyatomic ions. If you cannot find a rule that describes the compound or ion with which you are working, draw the electron-dot structure to predict their oxidation numbers.

Do You Know?

Only one element, fluorine, has an electronegativity greater than that of oxygen. When oxygen is bonded to fluorine, oxygen assumes a 1+ oxidation number. Such compounds are rare.

Example   21-2

What is the oxidation number of each atom in the sulfite ion, SO_3^{2-}?

▶ Suggested solution _____

The net charge on the ion is 2−. Therefore, the sum of the oxidation numbers of sulfur and the three oxygen atoms must be equal to 2−. According to Rule 3, each oxygen atom has an oxidation number of 2−. The expression below will help to predict the oxidation number of sulfur:

Atoms	Total oxidation number
S	1(?) = ?
3O	3(2−) = 6−
SO_3	= 2−

If you recall one of the basic principles of prealgebra, the number line, the oxidation number of sulfur becomes obvious. What number plus 6− equals 2−? The answer is 4+. Therefore, the solution to the example is shown below:

Atoms	Total oxidation number
S	1(4+) = **4+**
3O	3(2−) = 6−
SO_3	= 2−

The oxidation number of sulfur is 4+.

Oxidation numbers are useful in determining if an oxidation or a reduction has taken place in a reaction. Any change in the oxidation number of an atom during a chemical change shows that a redox reaction has taken place. An increase in the oxidation number of an element indicates **oxidation**. A decrease (or reduction) in the oxidation number of an element indicates **reduction**.

Example 21-3

Elemental iron can be made from iron ore, iron(III) oxide, Fe_2O_3. Is iron being oxidized or reduced?

▶ Suggested solution _____

To determine whether iron is being oxidized or reduced, you must find what happens to the oxidation number of iron during its conversion from ore to pure metal. First, find the oxidation number of iron in Fe_2O_3:

Atoms	Total oxidation number
2Fe	2(?) = **6+**
3O	3(2−) = 6−
Fe_2O_3	= 0

The oxidation number of each atom of iron in iron ore is 3+.

The product of the reaction is elemental iron and it has an oxidation number of 0 according to Rule 1 of Section 21-2. The oxidation number of iron is *reduced* from 3+ to 0:

$$Fe^{3+} \longrightarrow Fe^0$$

A decrease in the oxidation number of iron indicates that it was reduced.

Example 21-4

Is nitrogen oxidized or reduced when ammonia, NH_3, is decomposed to form elemental nitrogen?

▶ Suggested solution _____

In ammonia, the oxidation number of nitrogen is 3−:

Atoms	Total oxidation number
N	1(?) = 3−
3H	3(1+) = 3+
NH_3	= 0

Each atom of elemental nitrogen has an oxidation number of 0. When the oxidation number of the nitrogen atom increases from 3− to 0, it is considered to be oxidized.

1. Oxidation number is the real or apparent charge an atom or ion has when assigned a certain number of electrons.
2. Oxidation is an increase in the value of the oxidation number of an element. Reduction is a decrease in the value of the oxidation number.
3. a. C = 4−, H = 1+
   b. S = 4+, O = 2−
   c. Mn = 4+, O = 2−
   d. Cr = 3+
   e. P = 3−
   f. N = 5+, O = 2−
   g. H = 1+, O = 2−
   h. Na = 1+, O = 1−
   i. Mg = 2+, H = 1−
   j. Cr = 6+, O = 2−
   k. K = 1+, Mn = 7+, O = 2−
4. a. oxidation
   b. reduction
   c. neither (the oxidation number of Cr in each species is 6+)
   d. oxidation
   e. reduction

Review and Practice

1. Define *oxidation number*.

2. Define *oxidation* and *reduction*.

3. Determine the oxidation number of one atom of each element in the following substances:
   a. CH_4
   b. SO_2
   c. MnO_2
   d. Cr^{3+}
   e. P^{3-}
   f. NO_3^-
   g. H_3O^+
   h. Na_2O_2 (a peroxide)
   i. MgH_2
   j. $Cr_2O_7^{2-}$
   k. $KMnO_4$

4. Determine if each of the following changes is an oxidation, a reduction, or neither:
   a. $SO_3^{2-} \longrightarrow SO_4^{2-}$
   b. $CaO \longrightarrow Ca$
   c. $CrO_4^{2-} \longrightarrow Cr_2O_7^{2-}$
   d. $2I^- \longrightarrow I_2$
   e. $IO_3^{1-} \longrightarrow I_2$

21-3 Balancing Redox Reactions

A balanced chemical equation gives a chemist valuable information about a reaction. It lists all reactants and products and also indicates quantitative relationships between them. You learned in Chapter 5 that a balanced equation must show the same number of atoms or ions on each side in order to satisfy the law of conservation of mass. Since redox equations are concerned with the transfer of electrons, balancing them requires consideration of the number of electrons on each side of the equation. A redox equation is balanced when the number of atoms or ions of each substance and the number of electrons, or the electric charge on each side of the equation, is the same.

Some redox reactions are rather easy to balance by inspection, like the ones in Section 21-1. Other redox reactions are more complex, and require a step-by-step system for balancing. The following examples show the steps necessary to balance a complex redox reaction.

Example 21-5

There are many ways to balance redox reactions. An algebraic method using simultaneous equations is becoming popular. However, the half-reaction method will be understood by most students.

Balance the equation for the redox reaction between zinc metal and the vanadate ion, VO_3^-, in an acid solution: (The spectator ions are not included in the equation.)

$$H^+ + Zn + VO_3^- \longrightarrow VO^{2+} + Zn^{2+} + H_2O$$

▶ Suggested solution _____

Step 1. *Write the equation for the half-reactions of the reactants being oxidized and reduced without including electrons.*

Zinc metal is oxidized, since it increases in oxidation number from 0 to 2+:

$$Zn^0 \longrightarrow Zn^{2+}$$

At first glance, it may be difficult to see what is being reduced. Closer inspection of oxidation numbers shows that vanadium decreases from 5+ to 4+:

Atoms	Total oxidation number
V	1(?) = **5+**
3O	3(2−) = 6−
VO$_3$	= 1−

Atoms	Total oxidation number
V	1(?) = **4+**
O	1(2−) = 2−
VO	= 2+

The reduction half-reaction of vanadate is

$$VO_3^- \longrightarrow VO^{2+}$$

Step 2. *Make the number of all atoms or ions equal on both sides of each equation, with the exception of oxygen and hydrogen.*

There is only one zinc atom on each side of the first half-reaction. There is only one vanadium atom on each side of the second half-reaction.

Step 3. *Make the number of oxygen atoms equal on each side by adding water molecules to one side of the equation.*

It is important to remember that this reaction takes place in a water solution. Water molecules are present and can be involved in the reaction. There is no oxygen to balance in the zinc half-reaction:

$$Zn \longrightarrow Zn^{2+}$$

Two water molecules are added to the right side of the vanadate half-reaction to yield 3 oxygen atoms on each side:

$$VO_3^- \longrightarrow VO^{2+} + 2H_2O$$

Step 4. *Make the number of hydrogen atoms equal on each side by adding hydrogen ions, H+, where needed.*

Hydrogen ions are present, since the reaction takes place in an acid solution. There are no hydrogens to balance in the equation for the zinc half-reaction:

$$Zn \longrightarrow Zn^{2+}$$

Four hydrogen ions are added to the left side of the vanadate half-reaction to balance the four hydrogen atoms in two molecules of water on the right side:

$$4H^+ + VO_3^- \longrightarrow VO^{2+} + 2H_2O$$

Step 5. *Balance the electric charge on both sides of each half-reaction by adding electrons.*

To strengthen problem solving skills, encourage students to think of other strategies that might be used to balance such equations. Compare results from several procedures. Comparable results should strengthen faith that the answer derived is correct.

Two electrons are added to the right of the zinc half-reaction to yield zero electric charge on both sides:

$$Zn \longrightarrow Zn^{2+} + 2e^-$$

One electron is added to the left of the vanadate half-reaction to yield an electric charge of 2+ on both sides:

$$1e^- + 4H^+ + VO_3^- \longrightarrow VO^{2+} + 2H_2O$$

Note that it is not necessary, and is often impossible, to have a zero charge on each side of a half-reaction.

Step 6. *Equalize electrons in both half-reactions. The entire vanadate half-reaction is multiplied by 2 so that the electrons in both half-reactions will be equal:*

$$Zn \longrightarrow Zn^{2+} + 2e^-$$
$$2e^- + 8H^+ + 2VO_3^- \longrightarrow 2VO^{2+} + 4H_2O$$

Step 7. *Combine the two half-reactions.*

$$8H^+ + Zn + 2VO_3^- \longrightarrow 2VO^{2+} + Zn^{2+} + 4H_2O$$

Notice that the electrons cancel.

Step 8. *Check your work by making sure that the numbers of each kind of atom and ion and the electric charge balance.*

The charge balances for the combined reactions:

$$8(1+) + 0 + 2(1-) = 2(2+) + 2+ + 4(0)$$
$$6+ = 6+$$

Example 21-6

Balance the equation for the redox reaction between phosphorus and the iodate ion, IO_3^-, in an acid solution:

$$P_4 + IO_3^- \longrightarrow H_2PO_4^- + I^-$$

▶ Suggested solution _____

The steps for balancing this redox reaction are the same as in Example 21-5. See if you can solve this problem without referring to that example. Then refer to Example 21-5 to check your work before reading through the suggested solution here.

Notice before you begin that hydrogen ions and water molecules are not written in the equation now. They will be added later.

Step 1. *Write the equations for the half-reactions.*

Phosphorus is oxidized, since its oxidation number increases from 0 to 5+:

$$P_4 \longrightarrow H_2PO_4^-$$

Iodine is reduced, since its oxidation number decreases from 5+ to 1−:

$$IO_3^- \longrightarrow I^-$$

Step 2. *Make the number of all particles equal on both sides of each equation except for hydrogen and oxygen.*

$$P_4 \longrightarrow 4H_2PO_4^-$$
$$IO_3^- \longrightarrow I^-$$

Step 3. *To make the oxygen atoms equal, add water molecules.*

$$16H_2O + P_4 \longrightarrow 4H_2PO_4^-$$
$$IO_3^- \longrightarrow I^- + 3H_2O$$

Step 4. *To make the hydrogens equal, add hydrogen ions.*

$$16H_2O + P_4 \longrightarrow 4H_2PO_4^- + 24H^+$$
$$6H^+ + IO_3^- \longrightarrow I^- + 3H_2O$$

Step 5. *Balance the electric charge on both sides of each half-reaction by adding electrons.*

$$16H_2O + P_4 \longrightarrow 4H_2PO_4^- + 24H^+ + 20e^-$$
$$6e^- + 6H^+ + IO_3^- \longrightarrow I^- + 3H_2O$$

Step 6. *Equalize electrons in both half-reactions.*

$$48H_2O + 3P_4 \longrightarrow 12H_2PO_4^- + 72H^+ + 60e^-$$
$$60e^- + 60H^+ + 10IO_3^- \longrightarrow 10I^- + 30H_2O$$

Step 7. *Combine the two half-reactions.*

$$\overset{18}{\cancel{48}}H_2O + 3P_4 \longrightarrow 12H_2PO_4^- + \overset{12}{\cancel{72}}H^+ + \cancel{60e^-}$$
$$\cancel{60e^-} + \cancel{60}H^+ + 10IO_3^- \longrightarrow 10I^- + \cancel{30}H_2O$$
$$\overline{18H_2O + 3P_4 + 10IO_3^- \longrightarrow 10I^- + 12H_2PO_4^- + 12H^+}$$

The electrons cancel.

Step 8. *Check your work by making sure that the number of each kind of particle and the charge balance.*

The equation is charge balanced:

$$18(0) + 3(0) + 10(1-) = 10(1-) + 12(1-) + 12(1+)$$
$$10- = 10-$$

Review and Practice

1. Balance the redox equations for these reactions by inspection:
   a. $I^- + Cl_2 \longrightarrow Cl^- + I_2$
   b. $Co + Fe^{3+} \longrightarrow Co^{2+} + Fe^{2+}$

2. Balance the redox equations for these reactions:
   a. $Cr_2O_7^{2-} + Hg \longrightarrow Cr^{3+} + Hg^{2+}$
   b. $MnO_4^- + H_2O_2 \longrightarrow Mn^{2+} + O_2$
   c. $Zn + NO_3^- \longrightarrow NH_4^+ + Zn^{2+}$
   d. $Cl_2 \longrightarrow Cl^- + HClO$

1. a. $2I^- + Cl_2 \longrightarrow 2Cl^- + I_2$
   b. $Co + 2Fe^{3+} \longrightarrow Co^{2+} + 2Fe^{2+}$
2. a., b., c., d., See Chapter 21 notes for answers.

Electrochemical Cells

CONCEPTS

■ redox reactions and electrochemical cells
■ measuring the amount of electricity, the rate, and the tendency of electrons to flow
■ half-cell potentials
■ the voltage of spontaneous
redox reactions

Stop for a moment to think how strange and amazing it is that a chemical reaction can light a flashlight or make a toy car scoot across the floor. The reactions that cause battery-operated items to work are oxidation-reduction reactions. When electrons flow from the reducing agent to the oxidizing agent, an electric current is established. If the electrons flow in one direction in the cell, they can do work. An **electrochemical cell** is a chemical system in which a spontaneous oxidation-reduction reaction can produce useful electrical work. The current may run an electric motor which converts electrical energy into mechanical energy. In a flashlight, the current passes through a resistant material, such as tungsten, which transforms the energy of the electrons into heat and light energies.

Our society depends on electrochemical cells. A motor vehicle uses a storage battery to start the engine. Storage batteries are examples of electrochemical cells. A common battery, or dry cell, is also an electrochemical cell. Flashlights, toys, games, watches, and portable computers operate on power supplied by electrochemical cells.

21-4 The Nature of Electrochemical Cells

Chemistry has invaded the home in a very quiet and dependable way in the form of electrochemical cells. Such a cell (sometimes called a voltaic cell) is easily constructed so that it can do useful work as you saw in Chapter 17. Figure 21-2 shows the same electrochemical cell featured in Section 17-11. The beakers contain $1M$ solutions of copper(II) sulfate, $CuSO_4$, and zinc sulfate, $ZnSO_4$. Both copper(II) sulfate and zinc sulfate are good electrolytes, which, as you learned in Chapter 16, are substances whose water solutions conduct an electric current. A zinc strip is placed into the zinc sulfate solution, and a copper strip is placed into the copper(II) sulfate solution. These metal strips are electrodes, electrically conducting solids which are placed in contact with the electrolyte solutions. A salt bridge is positioned so that it is in both beakers. The **salt bridge** is composed of a concentrated solution of a strong electrolyte, usually potassium nitrate, in a U-tube container. The purpose of the salt bridge will become apparent soon. In order to complete the circuit, a wire is connected to each electrode, and those wires are attached to a light bulb. As soon as the circuit is complete, the bulb lights up. The electrochemical cell is doing useful electrical work. However, in a matter of minutes, the glow in the bulb begins to fade and will eventually go out.

A very simple, yet interesting experiment can be performed to explain how the cell works. When the mass of each electrode is measured before and after the reaction, it is found to be different. The copper electrode will have a greater mass after the experiment, and the zinc electrode will have less mass. The explanation for the change in mass is rather simple. As the cell was working, the zinc electrode was being oxidized, forming zinc ions that were released into solution:

$$Zn(s) \longrightarrow Zn^{2+}(aq) + 2e^-$$

Do You Know?

Battery *actually means a series of electrochemical cells such as in a 12-volt car battery. Flashlight batteries are more accurately called dry cells.*

Do not bring up half-cell potential yet. Let the qualitative concepts of electrochemical cells sink in first.

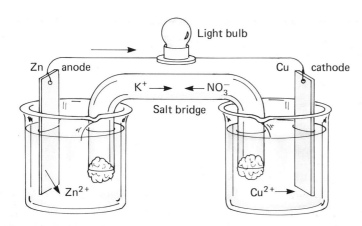

Figure 21-2 This electrochemical cell is a copper-zinc cell. Two $1M$ solutions, each containing one of the ions, are in contact with electrodes. The electrodes are attached to each other by a conducting wire. Electrons will not flow until a salt bridge connects the two solutions. The current will flow from the zinc electrode to the copper electrode.

The copper ions from the copper(II) sulfate solution were being reduced at the copper electrode:

$$Cu^{2+}(aq) + 2e^- \longrightarrow Cu(s)$$

The zinc electrode was losing electrons while the copper electrode was gaining electrons. The electrons were flowing in one direction from the zinc to the copper as seen in Figure 21-2. Sufficient current flows to cause the bulb to light up. The salt bridge is necessary to prevent the build-up in each beaker of excess positive or negative charges that would interfere with the flow of electrons. As positively charged zinc ions, Zn^{2+}, are formed on the left, negatively charged nitrate ions, NO_3^-, from the salt bridge enter the beaker. At the same time, positively charged copper(II) ions, Cu^{2+}, are removed from the solution on the right, and potassium ions, K^+, enter the beaker to take their place. The contents of both beakers remain electrically neutral even as the concentrations of reactants and products change during the reaction. The current, as depicted in Figure 21-2, will continue to flow in the cell until the cell reaches equilibrium.

Zinc metal is more easily oxidized than copper metal. Copper ions are more easily reduced than zinc ions. The spontaneous reaction in the electrochemical cell is expressed as the combination of the two half-reactions mentioned before:

$$Zn(s) + Cu^{2+}(aq) \longrightarrow Zn^{2+}(aq) + Cu(s)$$

In this example, the zinc electrode is called the anode, and the copper electrode is called the cathode. The **anode** is the site of oxidation, and the **cathode** is the site of reduction. Figure 21-3 is meant to help you remember which is which.

Metal elements vary widely in their relative ease of oxidation. Table 21-2 lists some common metals according to this property.

Table 21-2

OXIDATION PROPERTIES FOR SOME COMMON METALS
EASILY OXIDIZED ⟶ NOT EASILY OXIDIZED
Li K Cs Ca Na Mg Al Mn Zn Cr Fe Cd Co Cu Ag Au
STRONG REDUCING AGENTS ⟶ WEAK REDUCING AGENTS

Do You Know?

Biting into a piece of aluminum foil can graphically demonstrate the current an electrochemical cell is capable of generating. Aluminum acts as the anode, silver fillings in your mouth act as the cathode, and saliva acts as the electrolyte. The nerve of the tooth detects the current.

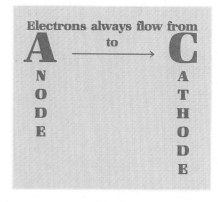

Figure 21-3 At the anode, oxidation takes place. The cathode is the site of reduction. Electrons flow from the anode to the cathode.

Point out to students that Table 21-2 shows the same property as Table 5-2 (Activity Series).

The table shows that lithium metal is the most easily oxidized of the group and is the strongest reducing agent. It is reasonable that the metals to the left in the table are not found free in nature but always in their oxidized states, that is, as positively charged ions in ionic compounds. Gold and silver, on the other hand, are found in nature in their unoxidized states as free metals.

21-5 Comparing Electrochemical Cells

There are so many different combinations of oxidation and reduction half-reactions that the number of possible electrochemical cells is enormous. How could you determine which cells would be the best for commercial use or which would be the most effective in producing electricity?

Three concepts about the nature of electricity will be helpful in comparing electrochemical cells. Perhaps these concepts will be more meaningful if the flow of electric current through a wire is compared to the flow of water through a pipe. Suppose you were a fire fighter trying to put out a raging fire. You hook up your hose to a fire hydrant and hope that the water flows through the hose as usual. As a fire fighter, you would be interested in three things:

Is there enough water available to put out the fire? It would not do you much good if you were able to put out only half the fire. The total amount of water available is important.

How fast can water flow from the fire hydrant through the hose? You could not effectively fight the fire, even with the Pacific Ocean as a reservoir, if the opening from the hydrant was small and allowed only a thin stream of water to flow out of it.

How high will you be able to reach with the water coming out of the hose? Even if you could deliver thousands of gallons of water per second to the fire, you would not be able to reach the fire on the third floor if the water pressure was low.

The *amount* of water, its *flow rate*, and its *pressure* are important in fighting a fire. The same kinds of things are important to consider when working with electrochemical cells. Cells need enough electricity, moving at an adequate rate, with sufficiently high pressure to do the job at hand. Table 21-3 and Figure 21-4 compare the names of important quantities in the water system with those in an electrochemical cell.

Table 21-3

A COMPARISON OF THE FLOW OF WATER TO ELECTRICITY		
QUANTITY MEASURED	UNITS USED WITH WATER	UNITS USED WITH ELECTRICITY
amount of flow	gallons	coulombs
rate of flow	gallons/minute	coulombs/second (amperes)
tendency to flow (pressure)	pounds/inch²	volts

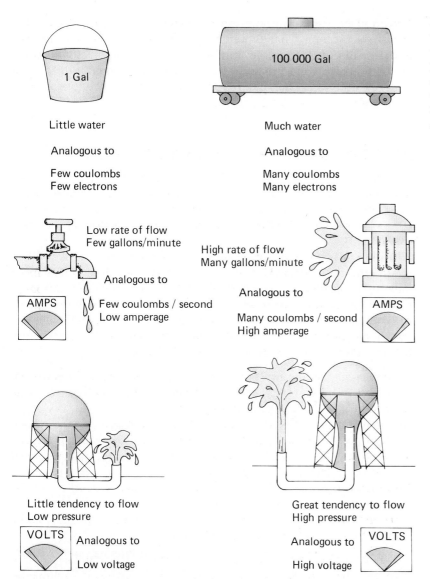

Figure 21-4 The flow of water through a hose and the flow of electrons through a circuit have similar properties. The terms used in describing an electric circuit are more easily understood when the more familiar terms describing water are applied.

A unit used to measure the amount of electricity flowing through a wire is the coulomb. A **coulomb** is 1.04×10^{-5} mole of electrons.

The rate of flow is the amount divided by time. Gallons per minute for water and coulombs per second for electricity are used to indicate rate of flow. One coulomb per second is also called an **ampere**. A circuit through which one coulomb is passing per second is said to be carrying a current of one ampere.

Electrical pressure is expressed in volts. **Voltage** is a measure of the tendency of electrons to flow. The electrical voltage corresponds to water pressure. The higher the voltage, the greater the tendency of the electrons to flow. Voltage is just one of the many ways to express potential energy. When a switch is closed in an electric circuit, electrons move through the circuit. Like water moving downhill, electrons spontaneously move from a position of high potential energy to one of lower potential energy. The voltage of a cell measures its ability to do electrical work.

Demonstrate the rather surprising reaction of a strip of Mg in 1M Cu^{2+} solution. The Mg is the anode and a Cu strip is the cathode. No salt bridge is needed but do not short out the cell by touching the two electrodes. See if you can play a small portable radio from the energy of this cell.

Do You Know?

Have you ever been shocked when you walked on a plush carpet and then touched a doorknob? The voltage in the shock is about 10 000 volts. However, the amperage is very low.

The size of the cell is also important. A tiny cell can do little work no matter what the voltage. The electrodes will be used up first.

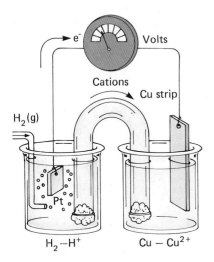

Figure 21-5 A half-cell containing hydrogen gas (at 1 atm) and hydrogen ions (1M) is called a standard half-cell. This half-cell is assigned a half-cell potential of 0.00 volts.

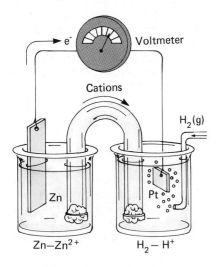

Figure 21-6 In a zinc-hydrogen cell, the hydrogen ions are being reduced to hydrogen gas. Even though the standard half-cell is reversed in comparison to Figure 21-10, the half-cell potential remains 0.00 volts. The total voltage of the cell is +0.76 volt which comes from the oxidation of zinc metal at the anode.

Many electrochemical cells can be tested for their ability to produce an electric current. What should be compared from cell to cell is the tendency of the electrons to flow, or the voltage. An electrochemical cell with a high voltage is one that can do a large amount of useful work per unit time. Each half-reaction contributes to the total voltage of the cell.

A copper-hydrogen cell is set up as shown in Figure 21-5. In the right beaker, the copper is being reduced:

$$Cu^{2+}(aq) + 2e^- \longrightarrow Cu(s)$$

In the left beaker, hydrogen gas is being oxidized:

$$H_2(g) \longrightarrow 2H^+(aq) + 2e^-$$

The voltmeter, a device for measuring voltage, reads 0.34 volt as the electric current flows.

In Chapter 20, you learned that Brönsted-Lowry acid-base systems can be thought of in terms of the competition for protons. In much the same way, oxidation-reduction reactions can be thought of as a competition for electrons. In a copper-hydrogen cell, copper has a greater tendency to gain electrons than do hydrogen ions. The electrons flow away from hydrogen toward the copper ions, not the other way around. The spontaneous reaction between copper ions and hydrogen gas can be expressed as

$$Cu^{2+}(aq) + H_2(g) \longrightarrow Cu(s) + 2H^+(aq)$$

In the competition for electrons, the copper(II) ions win over the hydrogen ions. The products predominate in this reaction. Apparently, copper metal does not easily give up electrons to hydrogen ions.

In Figure 21-6, a zinc-hydrogen cell is depicted. In this cell, the zinc is being oxidized and the hydrogen is being reduced. The cell voltage is 0.76 volt. The two half-reactions are

$$Zn(s) \longrightarrow Zn^{2+}(aq) + 2e^-$$
$$2H^+(aq) + 2e^- \longrightarrow H_2(g)$$

In the competition for electrons, the hydrogen ions win out over the zinc ions. Electrons flow from the zinc to the hydrogen ions. The overall spontaneous reaction is

$$Zn(s) + 2H^+(aq) \longrightarrow Zn^{2+}(aq) + H_2(g)$$

The copper-hydrogen and zinc-hydrogen cells provide information about the relative abilities of copper, zinc, and hydrogen ions to compete for electrons. Since both cells have a common reactant, hydrogen, the ability of the three ions to gain electrons can be compared. In the first cell, copper(II) ions are more easily reduced than hydrogen ions. However, in the second cell, hydrogen ions are more easily reduced than zinc ions. You can go back to the cell reactions and see that the tendency of an atom or ion to gain electrons can be determined by looking at the reactions in the electrochemical cell. These ions, listed in order of decreasing tendency to be reduced, are Cu^{2+}, H^+, and Zn^{2+}.

The voltage of a particular electrochemical cell is a measure of the tendency of electrons to flow from the substance which is oxidized to the substance which is reduced. The voltage for a cell is composed of two parts: the contribution made by each half-reaction. Chemists call the contributions to the cell voltage made by each half-reaction the **half-cell potential**, a term which emphasizes the relationship between voltage and potential energy. Half-cell potentials are symbolized by the letter E. The voltage of a cell depends on the concentration of the reactants and products present in the solution; therefore, a standard cell potential needs to be defined. The symbol $E°$ refers to the standard half-cell potential for 1-molar solutions, at 1 atmosphere of pressure for gases, and 25°C.

Chemists have decided to use the hydrogen half-cell as the standard to which all other half-cell potentials will be compared. Choosing a particular half-cell as the standard is something like designating the freezing point of water as zero on a Celsius thermometer. The decision to use the freezing point of water as the "standard" temperature was an arbitrary but reasonable one, considering the importance of water in most chemical reactions. All temperatures above the freezing point of water are assigned positive numbers. The temperatures below the freezing point are given negative numbers.

The hydrogen half-cell potential is written as

$$E° = 0.00 \text{ volts}$$

The copper-hydrogen cell potential is 0.34 volt and since the hydrogen half-cell potential is set equal to 0.00 volts, the half-cell potential for the reduction of copper is 0.34 volt:

$$Cu^{2+} + 2e^- \longrightarrow Cu \qquad E° = +0.34 \text{ volt}$$

In the zinc-hydrogen cell, the half-cell potential for the oxidation of zinc is 0.76 volt, since the hydrogen half-cell potential is again 0.00 volts:

$$Zn(s) \longrightarrow Zn^{2+} + 2e^- \qquad E° = +0.76 \text{ volt}$$

The above equation is for the oxidation of zinc. However, chemists have agreed to compare *reduction potentials*. Half-cell potentials are given as reduction potentials to demonstrate the relative ability of each type of ion to win the competition for electrons. Therefore, the zinc half-reaction is reversed and written as a reduction reaction for comparison with the others. The sign of the half-cell potential also is reversed:

$$Zn^{2+} + 2e^- \longrightarrow Zn \qquad E° = -0.76 \text{ volt}$$

The negative sign indicates that zinc ions have less tendency to be reduced than do hydrogen ions. The three ions you have been studying and their half-cell potentials are listed in order below:

$$Cu^{2+} + 2e^- \longrightarrow Cu \qquad E° = +0.34 \text{ volt}$$
$$2H^+ + 2e^- \longrightarrow H_2 \qquad E° = 0.00 \text{ volts}$$
$$Zn^{2+} + 2e^- \longrightarrow Zn \qquad E° = -0.76 \text{ volt}$$

You can see that if the sign of a half-cell potential is positive, the atom or ion will more easily accept electrons than will hydrogen ions. If the sign of a half-cell potential is negative, the atom or ion will not accept electrons as easily as hydrogen ions. Table 21-4 lists some substances in order of their ability to gain electrons. Notice that hydrogen is set in the middle of the table as a reference point with an $E°$ value of 0.00 volts.

Strictly speaking, it is activity rather than molarity that must be equal to 1. However, molarity is a reasonable approximation of activity in dilute solutions.

In most thermodynamic work, 25°C is the standard temperature, not the 0°C of STP.

Do You Know?

The brain acts in many ways like an electrochemical cell. Neurons conduct electrical activity by releasing neurotransmitters. These compounds interact with other neurons, and the interaction produces a flow of K^+ and Na^+ ions.

An electrochemical cell can be made from a lemon, a copper penny, and a strip of Zn. Stick the Zn and penny in a lemon about 1 cm apart. Attach a voltmeter. It should register about 0.15 volt.

Table 21-4

STANDARD REDUCTION POTENTIALS

	HALF-REACTION*		$E°$ (volts)	
EASILY REDUCED	$F_2(g)$	$+2\,e^- \longrightarrow 2F^-$	$+2.87$	NOT EASILY OXIDIZED
	$H_2O_2 + 2H^+$	$+2\,e^- \longrightarrow 2H_2O$	$+1.77$	
	$4H^+ + SO_4^{2-} + PbO_2$	$+2\,e^- \longrightarrow PbSO_4 + 2H_2O$	$+1.68$	
	$MnO_4^- + 8H^+$	$+5\,e^- \longrightarrow Mn^{2+} + 4H_2O$	$+1.52$	
	Au^{3+}	$+3\,e^- \longrightarrow Au$	$+1.50$	
	$Cl_2(g)$	$+2\,e^- \longrightarrow 2Cl^-$	$+1.36$	
	$Cr_2O_7^{2-} + 14H^+$	$+6\,e^- \longrightarrow 2Cr^{3+} + 7H_2O$	$+1.33$	
	$MnO_2 + 4H^+$	$+2\,e^- \longrightarrow Mn^{2+} + 2H_2O$	$+1.28$	
	$\frac{1}{2}O_2(g) + 2H^+$	$+2\,e^- \longrightarrow H_2O$	$+1.23$	
	$Br_2(l)$	$+2\,e^- \longrightarrow 2Br^-$	$+1.06$	
	$NO_3^- + 4H^+$	$+3\,e^- \longrightarrow NO(g) + 2H_2O$	$+0.96$	
	Ag^+	$+\,e^- \longrightarrow Ag$	$+0.80$	
	$NO_3^- + 2H^+$	$+\,e^- \longrightarrow NO_2(g) + H_2O$	$+0.78$	
	Fe^{3+}	$+\,e^- \longrightarrow Fe^{2+}$	$+0.77$	
	$O_2(g) + 2H^+$	$+2\,e^- \longrightarrow H_2O_2$	$+0.68$	
	I_2	$+2\,e^- \longrightarrow 2I^-$	$+0.53$	
	Cu^+	$+\,e^- \longrightarrow Cu$	$+0.52$	
	Cu^{2+}	$+2\,e^- \longrightarrow Cu$	$+0.34$	
	$SO_4^{2-} + 4H^+$	$+2\,e^- \longrightarrow SO_2(g) + 2H_2O$	$+0.17$	
	Cu^{2+}	$+\,e^- \longrightarrow Cu^+$	$+0.15$	
	Sn^{4+}	$+2\,e^- \longrightarrow Sn^{2+}$	$+0.15$	
	$\frac{1}{8}S_8 + 2H^+$	$+2\,e^- \longrightarrow H_2S(g)$	$+0.14$	
	$2H^+$	$+2\,e^- \longrightarrow H_2(g)$	0.00	
	Pb^{2+}	$+2\,e^- \longrightarrow Pb$	-0.13	
	Sn^{2+}	$+2\,e^- \longrightarrow Sn$	-0.14	
	Ni^{2+}	$+2\,e^- \longrightarrow Ni$	-0.25	
	Co^{2+}	$+2\,e^- \longrightarrow Co$	-0.28	
	$PbSO_4$	$+2\,e^- \longrightarrow Pb + SO_4^{2-}$	-0.36	
	Cr^{3+}	$+\,e^- \longrightarrow Cr^{2+}$	-0.41	
	Fe^{2+}	$+2\,e^- \longrightarrow Fe$	-0.44	
	Ag_2S	$+2\,e^- \longrightarrow 2Ag + S^{2-}$	-0.69	
	Cr^{3+}	$+3\,e^- \longrightarrow Cr$	-0.74	
	Zn^{2+}	$+2\,e^- \longrightarrow Zn$	-0.76	
	$2H_2O$	$+2\,e^- \longrightarrow H_2(g) + 2OH^-$	-0.83	
	Mn^{2+}	$+2\,e^- \longrightarrow Mn$	-1.18	
	Al^{3+}	$+3\,e^- \longrightarrow Al$	-1.66	
	Ti^{2+}	$+2\,e^- \longrightarrow Ti$	-1.75	
	Mg^{2+}	$+2\,e^- \longrightarrow Mg$	-2.37	
	Na^+	$+\,e^- \longrightarrow Na$	-2.71	
	Ca^{2+}	$+2\,e^- \longrightarrow Ca$	-2.87	
	Sr^{2+}	$+2\,e^- \longrightarrow Sr$	-2.89	
	Ba^{2+}	$+2\,e^- \longrightarrow Ba$	-2.90	
	Cs^+	$+\,e^- \longrightarrow Cs$	-2.92	EASILY OXIDIZED
NOT EASILY REDUCED	K^+	$+\,e^- \longrightarrow K$	-2.92	
	Li^+	$+\,e^- \longrightarrow Li$	-3.00	

Reducing strength increases (left margin) · Oxidizing strength increases (right margin)

*Ionic concentrations, $1\,M$ in water at 25°C.

21-6 Predicting Redox Reactions

Chemists use half-cell potentials to predict whether an oxidation-reduction reaction will occur. If the cell voltage for an overall reaction is positive, the reaction will proceed as written. If, however, the cell voltage is negative, the reaction will not proceed spontaneously. Some examples will demonstrate how potentials can be used to predict the outcome of an oxidation-reduction reaction.

Silver tarnish is a film of silver sulfide, Ag_2S, formed when silver is oxidized by hydrogen sulfide, H_2S, in the air. One way to restore the luster to silverware is to immerse it in an electrolyte solution and bring it into contact with a metal that is more easily oxidized than silver. The silver sulfide forms silver and sulfide ions. The more active metal is oxidized and transfers electrons to the silver, causing silver atoms to plate out on the silverware. The silver is reduced.

Example 21-7

Can aluminum foil immersed in an electrolyte solution be used to restore the luster to silverware? Show the balanced equation and the cell potential for the overall reaction. (Assume $1M$ solutions for all reactants.)

▶ Suggested solution

The first part of the question is asking whether aluminum will be more easily oxidized than silver. The answer comes from looking at Table 21-4, which shows reduction half-reactions and their potentials. To compare *oxidation* reactions of aluminum and silver, the half-reactions are reversed and the signs of the half-cell potentials also are reversed:

$$Ag(s) \longrightarrow Ag^+(aq) + e^- \qquad E° = -0.80 \text{ volt}$$
$$Al(s) \longrightarrow Al^{3+}(aq) + 3e^- \qquad E° = +1.66 \text{ volts}$$

The half-cell potential given for the oxidation of aluminum is greater than the half-cell potential for the oxidation of silver. In other words, aluminum will be more easily oxidized than silver and can be used to force silver ions to be reduced to silver atoms and plate out on the silverware. Remember that if one substance is oxidized, another substance must be reduced in any redox reaction. With aluminum present, silver ions will be reduced:

$$Ag^+(aq) + e^- \longrightarrow Ag(s) \qquad E° = +0.80 \text{ volt}$$

Aluminum will be oxidized:

$$Al(s) \longrightarrow Al^{3+}(aq) + 3e^- \qquad E° = +1.66 \text{ volts}$$

The balanced equation for the reaction and the cell potential for the overall reaction are obtained by adding together the two half-reactions and their voltages. To balance the electric charge, the silver half-reaction is multiplied by 3:

Students easily confuse this procedure with Hess's Law procedures used to calculate $H°$ for a series of reactions. (See Chapter 17.) Enthalpy is an *extensive* property; its value depends on the amount of substance that reacts, so ΔH values always have units of J/mol or J/g. The amount of reaction must be taken into account. $E°$ values describe an *intensive* property and do not depend on the extent of reaction. There are cases in which $E°$ values must be multiplied to calculate a cell potential (*J. Chem. Ed. 53* (1976): p. 453), but it is probably best to ignore these cases in this introductory course.

$$3Ag^+ + 3e^- \longrightarrow 3Ag \qquad E° = +0.80 \text{ volt}$$
$$Al \longrightarrow Al^{3+} + 3e^- \qquad E° = +1.66 \text{ volts}$$
$$\overline{3Ag^+(aq) + Al(s) \longrightarrow 3Ag(s) + Al^{3+}(aq) \qquad E° = +2.46 \text{ volts}}$$

The cell potential for the overall reaction is $+2.46$ volts.

Notice that in the silver half-reaction, the moles of silver were 3 times that listed in Table 21-4 so that the overall equation would be charge-balanced. The half-cell potential was *not* multiplied by 3. The presence of 3 times as many electrons does not mean that they will have 3 times the tendency to flow. The voltage remains unchanged. In a similar way, the pressure of water flowing from a larger reservoir would not necessarily be greater than the pressure of water flowing from a smaller reservoir.

As seen in Example 21-7, a positive cell potential means that the reaction proceeds spontaneously and that the products of the reaction predominate. Figure 21-7 is a helpful way to remember the relationship of cell potential and the direction of the reaction.

When E° is

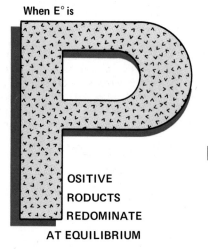

OSITIVE

RODUCTS

REDOMINATE

AT EQUILIBRIUM

Figure 21-7 When the cell potential is positive, the products predominate. The net redox equation proceeds spontaneously in a forward direction, producing the products.

Example 21-8

When silver metal is placed in $1M$ hydrochloric acid, HCl, under standard conditions, there is no observable reaction. However, when magnesium metal is placed in the same hydrochloric acid solution, the metal oxidizes and hydrogen gas is produced. Explain these two phenomena.

▶ **Suggested solution**

Perhaps it is easiest to look for an answer by first considering the reaction that does occur. When magnesium metal was put into a solution of HCl, it was oxidized:

$$Mg(s) \longrightarrow Mg^{2+}(aq) + 2e^-$$

For that reaction to occur, another substance must be reduced. In this case, hydrogen ions from the dissociation of HCl become hydrogen gas:

$$2H^+(aq) + 2e^- \longrightarrow H_2(g)$$

From Table 21-4, the reduction potential of magnesium is -2.37 volts. The potential for the *oxidation* of magnesium would then be $+2.37$ volts. The reduction potential for hydrogen is 0.00 volts. Adding the half-reactions and their potentials together gives

$$Mg(s) \longrightarrow Mg^{2+}(aq) + 2e^- \qquad E° = +2.37 \text{ volts}$$
$$\underline{2H^+(aq) + 2e^- \longrightarrow H_2(g) \qquad E° = 0.00 \text{ volts}}$$
$$2H^+(aq) + Mg(s) \longrightarrow Mg^{2+}(aq) + H_2(g) \qquad E° = +2.37 \text{ volts}$$

Since the cell potential for the overall reaction is positive, the reaction will proceed in the example.

If silver was to react with the same HCl solution, it would be oxidized in a similar manner. From Table 21-4, the half-cell reaction and its potential for the *oxidation* of silver is

$$Ag(s) \longrightarrow Ag^+(aq) + e^- \qquad E° = -0.80 \text{ volt}$$

The substance being reduced would again be hydrogen ions:

$$2H^+(aq) + 2e^- \longrightarrow H_2(g) \qquad E° = 0.00 \text{ volts}$$

Adding the half-reactions and their potentials together gives

$$2Ag(s) \longrightarrow 2Ag^+(aq) + 2e^- \qquad E° = -0.80 \text{ volt}$$
$$\underline{2H^+(aq) + 2e^- \longrightarrow H_2(g) \qquad\quad E° = 0.00 \text{ volts}}$$
$$2Ag(s) + 2H^+(aq) \longrightarrow 2Ag^+(aq) + H_2(g) \qquad E° = -0.80 \text{ volt}$$

The cell potential is negative. The reaction will not occur spontaneously under standard conditions.

The cell potential or voltage for the overall reaction is a measure of how easily an oxidation-reduction reaction will occur. Once equilibrium is reached, the tendency for the reaction to proceed in one direction equals the tendency to proceed in the reverse direction. The net tendency is zero. This is due to the changing concentrations of the reacting solutions. At equilibrium, the solutions are no longer $1M$, and they will not produce the same amount of voltage. The voltage for a reaction at equilibrium is zero.

Review and Practice

1. What is the difference between an anode and a cathode?

2. Explain the difference between the rate of electrical flow and the tendency to flow.

3. Calculate the cell potentials expected for the following redox reactions. Assume all ion concentrations are $1M$:
a. $Co + Fe^{3+} \longrightarrow Co^{2+} + Fe^{2+}$
b. $H_2 + Cl_2 \longrightarrow 2H^+ + 2Cl^-$
c. $I^- + MnO_4^- \longrightarrow I_2 + Mn^{2+}$ (in an acid solution)

4. Will silver metal in a solution of chloride ions produce silver ions and chlorine gas?

5. Ampere is defined as coulombs/second:
a. What is a general name for this type of expression?
b. For a current of constant amperage, what is the relationship between the number of electrons passing through the wire and the amount of time the current flows?
c. Draw a graph to show number of electrons plotted against time.

6. What will happen if an aluminum spoon is used to stir a solution of iron(II) nitrate, $Fe(NO_3)_2$?

7. What will happen if an iron spoon is used to stir a solution of aluminum chloride, $AlCl_3$?

8. Can a $1M$ solution of iron(III) sulfate, $Fe_2(SO_4)_3$, be stored in a container made of nickel metal? Explain your answer.

9. Most of the bromine produced in the United States is made by oxidizing bromine ions, Br^-, to bromine gas, Br_2, using chlorine gas, Cl_2. What is $E°$ for this reaction?

1. An anode is the site of oxidation. A cathode is the site of reduction.
2. The rate of electrical flow is a measure of the electrical charge (coulombs) flowing through a circuit per unit time (seconds). One coulomb per second is an ampere. The tendency to flow is a measure of electrical pressure. It is the potential difference between the anode and cathode in an electrochemical cell. This tendency is usually measured in volts.

3. a.
$$Co \longrightarrow Co^{2+} + 2e^-$$
$$\underline{2Fe^{3+} + 2e^- \longrightarrow 2Fe^{2+}}$$

$$E° = +0.28 \text{ v}$$
$$\underline{E° = +0.77 \text{ v}}$$
$$E° \text{ cell} = +1.05 \text{ v}$$

b.
$$H_2 \longrightarrow 2H^+ + 2e^-$$
$$\underline{Cl_2 + 2e^- \longrightarrow 2Cl^-}$$

$$E° = +0.00 \text{ v}$$
$$\underline{E° = +1.36 \text{ v}}$$
$$E° \text{ cell} = +1.36 \text{ v}$$

c.
$$10I \longrightarrow 5I_2 + 10e^-$$
$$\underline{2MnO_4^- + 16H^+ + 10e^- \longrightarrow 2Mn^{2+} + 8H_2O}$$

$$E° = -0.53 \cdot v$$
$$\underline{E° = +1.52 \text{ v}}$$
$$E° \text{ cell} = +0.99 \text{ v}$$

4. No. Both half-reactions are oxidations.
5. a. unitary rate or proportionality constant
b. The number of electrons are directly proportional to time.
c. It would be a straight line.
6. The aluminum spoon will be oxidized and the Fe^{2+} ions will be reduced to iron metal.
7. There will be no spontaneous reaction. $E° = -1.22$ v for the reaction.
8. No, the container will dissolve. $E° = +1.02$ v
9. $E° = +0.30$ v

Putting Redox Reactions to Work

CONCEPTS

- **electrochemical cells that can do useful work**
- **electrolysis and nonspontaneous redox reactions**
- **corrosion**

Both spontaneous and non-spontaneous oxidation-reduction reactions are useful in different ways. Spontaneous redox reactions power batteries in toys, watches, and automobiles. The way batteries work is rather easy to understand because it follows the principles that you have already studied in this chapter.

You are now familiar with redox reactions that occur spontaneously. It is possible to reverse spontaneous reactions. Millions of kilograms of materials are made every day by this valuable process. The plating of metals, the reduction of metals from their ores, and the separation of valuable elements from compounds all depend on reversing spontaneous redox reactions.

21-7 Electrochemical Cells and Storage Batteries

Write to Duracell or Eveready for very informative battery guides.

One of the most practical advantages of the common electrochemical cell, or battery, is that it can supply electrical energy at a desired time. Potential electrical energy stored in a battery is released when the circuit is completed and chemical reactions take place. Two familiar examples are the cells used in flashlights, radios, and calculators and the lead storage battery used for starting an automobile.

Not all electrochemical cells contain water solutions of their electrolytes like those you have been studying. A common battery, or **dry cell**, is an electrochemical cell in which the electrolytes are present as solids or as a paste. For example, one common type of dry cell, the Leclanché cell, is filled with a paste of manganese dioxide, MnO_2, water, and ammonium chloride, NH_4Cl. A zinc can acts as the anode, and the cathode is a carbon-manganese dioxide rod located in the center of the cell. Zinc is oxidized, and manganese dioxide is reduced to various compounds of Mn^{3+} ions.

The role of carbon in the alkaline battery is to carry electrons and increase the conductivity of the cathode.

Another common dry cell is shown in Figure 21-8. It is an alkaline manganese cell, commonly referred to as an alkaline battery. This cell consists of a powdered zinc anode and a cathode made of manganese dioxide and carbon. The cell uses potassium hydroxide, KOH, an alkaline substance, as the electrolyte which maintains the charge balance in the cell. The chemistry of the alkaline battery is expressed in the following redox equation:

$$Zn + 2MnO_2 \longrightarrow ZnO + Mn_2O_3$$

Zinc is oxidized, and the manganese in manganese dioxide is reduced. The alkaline battery is more efficient than the Leclanché cell because electrical resistance to the flow of electrons does not increase with use and because it yields two to ten times more current.

Review the connection between density and concentration so students understand why measuring the density of H_2SO_4 in a car battery can be useful.

The lead storage battery used in an automobile is not a dry cell. It contains several compartments, or cells, one of which is shown in

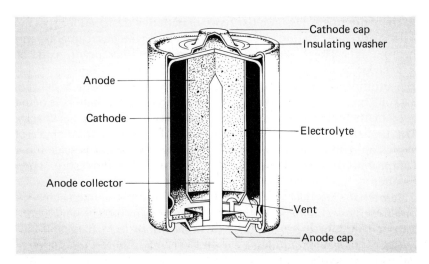

Cathode cap
Insulating washer
Anode
Cathode
Electrolyte
Anode collector
Vent
Anode cap

Figure 21-9. Lead is oxidized to form lead(II) sulfate, $PbSO_4$, at the anode, and lead(IV) oxide, PbO_2, is reduced to lead(II) sulfate at the cathode. Sulfuric acid, H_2SO_4, is the electrolyte used to maintain the charge balance of the cell. The reactions which occur during discharge (use) of this electrochemical cell can be represented in this manner:

$$
\begin{array}{ll}
Pb + SO_4^{2-} \longrightarrow PbSO_4 + 2e^- & E^\circ = +0.36 \text{ volt} \\
PbO_2 + SO_4^{2-} + 4H^+ + 2e^- \longrightarrow PbSO_4 + 2H_2O & E^\circ = +1.68 \text{ volts} \\
\hline
PbO_2 + Pb + 4H^+ + 2SO_4^{2-} \longrightarrow 2PbSO_4 + 2H_2O & E^\circ = +2.04 \text{ volts}
\end{array}
$$

The net equation shows that the sulfuric acid is absent from the product side of this equation. It is consumed when the battery is discharged. In addition, water is formed. This means that as the battery is used, the electrolyte, sulfuric acid, becomes more dilute. Measuring the density of the sulfuric acid can provide a simple way of knowing when a battery should be recharged. High density for the electrolyte indicates the sulfuric acid concentration is high, which in turn means the battery is charged. As the battery acid becomes more dilute, its freezing point rises and it is more likely to freeze at higher temperatures. In areas where the winter temperatures are low (below 0°C), it is important to keep a car battery charged.

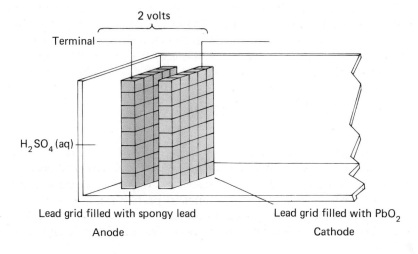

2 volts
Terminal
H_2SO_4 (aq)
Lead grid filled with spongy lead
Anode
Lead grid filled with PbO_2
Cathode

Figure 21-9 A 12-volt battery has six of these lead storage cells. More plates are added to each cell to increase the capacity to store energy.

There are many, less familiar types of batteries designed for specific uses. Miniature "button" batteries are used in watches, cameras, and hearing aids. These batteries often contain zinc-mercury(II) oxide or zinc-silver oxide combinations for the anode and cathode.

A lithium-sulfur dioxide battery is a powerful cell which delivers 3.0 volts per cell. It is lightweight and dependable. This battery is being used by the military as the power source for infantry radio backpacks. The battery is not available to the public because of the danger of explosion when exposed to heat. The sulfur dioxide gas is sealed in a compartment and, like all gases, will increase in pressure when heated.

Some batteries, like the lead storage battery, are rechargeable. Another rechargeable battery is the nickel-cadmium battery which contains a cadmium anode and nickel oxide cathode. A rechargeable battery seems ideal for running a car. However, cadmium is too scarce to be used on a wide scale.

21-8 Quantitative Aspects of Electrochemical Cells

Figure 21-10 shows an electrochemical cell made with a copper anode in a copper solution and a silver cathode in a silver solution. The half-reactions and half-cell potentials yield the following redox reaction and cell potential:

$$Cu \longrightarrow Cu^{2+} + 2e^- \qquad E° = -0.34 \text{ volt}$$
$$2Ag^+ + 2e^- \longrightarrow 2Ag \qquad E° = +0.80 \text{ volt}$$
$$2Ag^+ + Cu \longrightarrow 2Ag + Cu^{2+} \qquad E° = +0.46 \text{ volt}$$

As the cell discharges, the silver electrode increases in mass and the copper electrode decreases in mass. Is the change in mass the same at both electrodes? At first, it may appear so. However, the data from

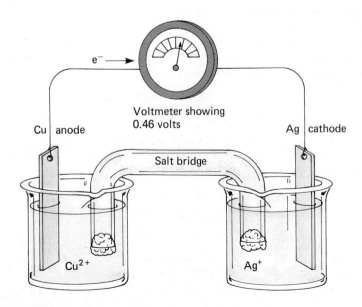

Figure 21-10 As this electrochemical cell drains, silver metal is produced at the cathode, and copper metal is oxidized at the anode. The copper electrode will have decreased in mass at the same time that the silver electrode increased in mass.

such an experiment indicate that the electrodes do not change equally in mass. In one experiment, the silver electrode gained 2.15 grams, and the copper electrode lost 0.635 gram. Why? As you recall from Chapter 5, chemical equations express molar relationships between reacting substances. When the experimental masses are changed to moles, an interesting relationship becomes evident:

$$2.15 \text{ g Ag} \times \frac{1 \text{ mol Ag}}{108 \text{ g Ag}} = 0.0199 \text{ mol Ag reacted}$$

$$0.635 \text{ g Cu} \times \frac{1 \text{ mol Cu}}{63.5 \text{ g Cu}} = 0.0100 \text{ mol Cu reacted}$$

The ratio of moles of silver reacted to moles of copper is

$$\frac{0.0199 \text{ mol Ag}}{0.0100 \text{ mol Cu}} = \frac{1.99}{1.00}$$

The ratio 1.99 to 1.00 can be changed to 2.00 to 1.00 without violating the limits of uncertainty in this problem. The ratio shows that for every two moles of silver ions reacted, only one mole of copper reacted. The net redox equation for the reaction indicates the same thing. The reason for this relationship is that each silver ion requires only one electron for reduction. However, each copper atom gives up two electrons during oxidation to become a copper(II) ion. When one copper atom loses two electrons, two silver ions are reduced. Therefore, one mole of copper atoms will produce two moles of silver atoms.

Example 21-9

By what mass in grams will a chromium cathode increase when it is coupled to a magnesium half-cell in which the magnesium anode loses 1.53 grams? (Assume the chromium ions in solution are Cr^{3+}.)

▶ Suggested solution _____

The first step is to write the half-reactions for the electrochemical cell. Oxidation of magnesium occurs at the anode, and reduction of chromium occurs at the cathode:

$$Mg \longrightarrow Mg^{2+} + 2e^-$$
$$Cr^{3+} + 3e^- \longrightarrow Cr$$

Multiply the equations by 3 and 2 respectively to balance the number of electrons. Find the balanced redox reaction by adding the half-reactions together:

$$3Mg \longrightarrow 3Mg^{2+} + \cancel{6e^-}$$
$$\underline{2Cr^{3+} + \cancel{6e^-} \longrightarrow 2Cr}$$
$$2Cr^{3+} + 3Mg \longrightarrow 2Cr + 3Mg^{2+}$$

The balanced equation shows the molar ratio of chromium and magnesium. For every 2 moles of chromium produced, 3 moles of magnesium must be oxidized. However, the problem gives the

CHEM THEME

Quantitative electrochemical problems emphasize again the need for moles.

There are many mental steps in a problem such as this. Allow the students to see the relationship between one step and the next.

To develop problem solving skills, have students attempt this problem before looking at the suggested solution. The concepts required to solve this problem should be familiar by now. If students attempt the problem in small groups, some groups should be able to solve it. With luck, more than one solution will be found, and class discussion can be used to focus on the various strategies used by students when "they don't know what to do." After the discussion, the example can be examined as *one* way to solve the problem. The example will be easier to understand after students have tried to solve the problem independently.

amount of magnesium lost in grams and asks for grams of chromium produced. You must translate

$$\text{g Mg} \longrightarrow \text{mol Mg} \longrightarrow \text{mol Cr} \longrightarrow \text{g Cr}$$

The calculations can be set up as one step:

$$1.53 \text{ g Mg} \times \frac{1 \text{ mol Mg}}{24.3 \text{ g Mg}} \times \frac{2 \text{ mol Cr}}{3 \text{ mol Mg}} \times \frac{52.0 \text{ g Cr}}{1 \text{ mol Cr}} = 2.18 \text{ g Cr}$$

The chromium cathode will increase in mass by 2.18 grams.

21-9 The Electrolytic Process

Each year millions of metric tons of aluminum and other valuable elements are reduced from ores or salts by a process called electrolysis. In **electrolysis**, an external source of electricity causes a nonspontaneous redox reaction to occur. Electrolysis also is used to separate commercially important substances from aqueous solutions.

Aluminum metal is obtained from its ore by electrolysis. Most aluminum ore is composed of the mineral bauxite, $Al_2O_3 \cdot 2H_2O$, and impurities of iron and silicon oxides. The bauxite is first refined to aluminum oxide which is then dissolved at high temperatures in the mineral cryolite. When enough electricity is applied to the molten solution, aluminum metal is produced at the cathode. Simplified versions of the half-reactions are

$$4Al^{3+} + 12e^- \longrightarrow 4Al$$
$$\frac{6O^{2-} \longrightarrow 3O_2 + 12e^-}{4Al^{3+} + 6O^{2-} \longrightarrow 4Al + 3O_2}$$

This process for manufacturing aluminum is effective yet expensive. The process uses about 60 kilojoules of electrical energy for each gram of aluminum produced. Approximately ten million metric tons of aluminum are produced each year worldwide. The amount of energy needed to produce all this aluminum is enormous. As a step toward energy conservation, aluminum is recycled. Only about one percent of the energy used to extract aluminum from its ore is needed to recycle aluminum.

The electrolysis of water is easily performed in the lab. When an electric current is passed through water acidified with sulfuric acid to provide enough conducting ions, the two component elements of water bubble off as gases. The half-reactions which occur at each electrode are

$$4H^+ + 4e^- \longrightarrow 2H_2 \qquad E° = \ \ 0.00 \text{ volts}$$
$$\frac{2H_2O \longrightarrow O_2 + 4H^+ + 4e^-}{2H_2O \longrightarrow 2H_2 + O_2} \qquad \begin{matrix} E° = -1.23 \text{ volts} \\ E° = -1.23 \text{ volts} \end{matrix}$$

The negative voltage indicates that the overall reaction is not spontaneous. Yet, when more than 1.23 volts are applied from an external source, the reaction proceeds. This reaction is used industrially to produce oxygen of very high purity. The hydrogen gas is sold as a by-product.

Do You Know?

The process of obtaining aluminum from its ore by electrolysis was discovered by Charles Martin Hall and, almost simultaneously, by a 21-year-old student at Oberlin College, Ohio, in 1886.

One kilogram of good quality bituminous coal provides 3.3×10^4 kJ of energy, enough to produce about 551 g of Al from its ore.

21-10 Extension: Oxidation-Reduction Gone Wild—Corrosion

When it comes to corrosion, iron is the champion. In the United States alone, the rusting (corrosion) of iron costs 50 billion dollars per year. Much of the iron produced is used to replace and repair items which are being destroyed by corrosion. What is the chemical nature of rusting and how can it be controlled?

There are many factors that influence the process of rusting. Water vapor and oxygen are necessary for rusting to occur. The presence of hydrogen ions speeds up the reaction. Corrosion occurs most rapidly at points where iron is strained by applied pressure or by bending, and where iron comes into direct contact with another metal.

A sad and troublesome example of iron corrosion is the Statue of Liberty. In 1985, teams of workers began the long process of strengthening a very weak Lady. Over many years, the Statue of Liberty has been corroding, as shown in Figure 21-11. The iron ribs which support the copper skin of the statue were subjected to the effects of wind, rain, salt air, and pollution. Iron corrodes in several steps. The iron oxidizes:

$$Fe \longrightarrow Fe^{2+} + 2e^-$$

Do You Know?

It takes about 100 years for a tin-coated steel can (a common tin can) to degrade completely, about 400 years for an aluminum can, and about 100 000 years for a glass bottle.

Figure 21-11 The Statue of Liberty needed renovation because of the extensive damage that corrosion caused to the statue's supporting framework.

Oxygen is reduced in the presence of water:

$$H_2O + 1/2O_2 + 2e^- \longrightarrow 2OH^-$$

The iron(II) and hydroxide ions combine to form iron(II) hydroxide:

$$Fe^{2+} + 2OH^- \longrightarrow Fe(OH)_2$$

This ionic substance is unstable in the presence of water and is quickly oxidized to iron(III) hydroxide:

$$2Fe(OH)_2 + H_2O + 1/2O_2 \longrightarrow 2Fe(OH)_3$$

Both iron(III) hydroxide and iron(III) oxide, Fe_2O_3, (formed when iron(III) hydroxide is heated by the sun) are components of rust.

There was another condition that speeded the corrosion process in the Statue. Because the iron ribs of the Statue of Liberty were in contact with the copper skin, the iron rusted about 1,000 times faster than if the copper had not been present. When two dissimilar metals are in contact with each other and surrounded by an electrolyte solution, one metal acts as an anode and is oxidized. The other metal serves as a conductor of electrons to the oxygen which is reduced. Since iron is more easily oxidized than copper, it acts as the anode, giving up electrons when in the presence of copper. The repair teams are replacing the iron ribs with a stainless steel alloy of iron, chromium, nickel, and molybdenum. Stainless steel resists corrosion.

One way of preventing the corrosion of iron is to prevent oxygen and water from coming in contact with the iron. Paint and grease work well. However, once a painted surface is scratched or chipped, the corrosion begins and is hard to stop as evidenced in rusting cars. A better way to protect iron from corrosion is to bolt pieces of zinc or magnesium metal to the surface of the iron. Both zinc and magnesium are more easily oxidized than iron. The other metal corrodes leaving the iron intact. This method of preserving iron objects is called cathodic protection. It is used to protect the hulls of large ships or the propeller shafts of speedboats.

1. The Alkaline battery is an electrochemical "dry" cell. It generates electricity from a redox reaction within the can whenever a complete circuit is available. Zn is the anode and MnO_2 is the cathode. The electrolyte is a solution of KOH in a gelling agent. The reaction is:

$$Zn + 2MNO_2 \longrightarrow ZnO + Mn_2O_3$$

2. Sulfuric acid is the electrolyte in the lead storage battery. The acid solution also provides hydrogen ions for the reduction of lead(IV) oxide to lead(II) sulfate.

3. 1.3×10^4 kg Na

4. The reaction is

$$Mg + 2H^+ \longrightarrow Mg^{2+} + H_2$$

The answer is 0.416 g H_2.

5. Iron rusts as Fe is oxidized to Fe^{2+}. Oxygen gas is reduced in the presence of H_2O to form OH^-. The Fe^{2+} and OH^- produce unstable $Fe(OH)_2$. This substance reacts with water and oxygen to form $Fe(OH)_3$.

Review and Practice

1. Briefly describe how an alkaline battery generates an electric current.

2. For what reason is acid used in the lead storage battery in a car?

3. Sodium metal is widely used in the manufacture of some sodium compounds and synthetic detergents. The electrolysis of molten sodium chloride produces liquid sodium metal, $Na(l)$, and chlorine gas, Cl_2. How many kilograms of liquid sodium metal are produced at the cathode when 2.0×10^4 kilograms of chlorine gas form at the anode?

4. How many grams of hydrogen gas would be produced from the oxidation of 5.00 grams of magnesium metal?

5. Briefly explain how iron rusts.

Numbers in red indicate the appropriate chapter sections to aid you in assigning these items. Answers to all questions appear in the Teacher's Guide at the front of this book.

Summary

■ Many of the reactions which chemists study are oxidation-reduction reactions. Oxidation is an increase in the oxidation number of an element, and reduction is a decrease in the oxidation number of an element. Oxidation numbers help chemists describe redox reactions more clearly. The ions and atoms in compounds are assigned oxidation numbers according to a set of rules.

■ An oxidizing agent is the substance in a reaction that is reduced, and the reducing agent is the substance that is oxidized. An oxidation reaction must occur together with a reduction reaction.

■ A balanced redox equation must be mass- and charge-balanced and should not contain electrons in the net equation.

■ An electrochemical cell is a chemical system capable of spontaneously producing an electric current. A cell is composed of electrodes and electrolytes. Electrical energy is produced as electrons flow from the substance that is oxidized to the substance that is reduced.

■ Cell voltage is a measure of the tendency of electrons to flow. If two substances have a great difference in their tendency to lose electrons, the voltage of the cell will be high. Cell voltage can be used to predict whether a redox reaction will occur spontaneously.

■ Batteries are electrochemical cells. They are examples of spontaneous redox reactions. Batteries can store electrical energy until it is needed.

■ An electric current can be used to drive redox reactions which do not occur spontaneously. The process is called electrolysis. Sodium, copper, and aluminum metals, as well as chlorine and oxygen gases, are a few of the substances which can be formed by electrolysis.

■ Rust is a product of the corrosion of iron. When iron is oxidized, it forms iron(III) oxide and iron(III) hydroxide. This material flakes off easily and exposes the surface of the iron to more moisture and corrosion.

Chemically Speaking

ampere	half-reaction
anode	oxidation
cathode	oxidation number
coulomb	oxidizing agent

dry cell	redox reaction
electrochemical cell	reducing agent
electrode	reduction
electrolysis	salt bridge
half-cell potential	voltage

Review

1. List three oxidation half-reactions. (21-1)

2. What are the steps for determining the oxidation number of each atom in NO_2? (21-2)

3. What two quantities must be equal in a balanced redox equation? (21-3)

4. Draw a complete diagram representing a copper-silver electrochemical cell. (21-4)
   a. Label all parts of the cell including the solutions.
   b. Show the half-reactions that are occurring.
   c. Give the cell voltage that would be generated for 1 molar solutions at standard temperature and pressure.
   d. Show the direction of electron flow through a voltmeter connected to the electrodes.

5. What is the function of a salt bridge in an electrochemical cell? (21-4)

6. Briefly explain voltage. What other words come to mind when you think of voltage? (21-5)

7. What is a half-cell potential? (21-5)

8. Describe a chemical reaction that occurs in an alkaline manganese cell. (21-7)

9. Define *electrolysis*. (21-9)

10. Describe the anode and cathode reactions in the corrosion of iron. (21-10)

Interpret and Apply

1. a. If a neutral atom becomes an ion with a 1+ charge, has it been oxidized or reduced? Write a general equation using M for the neutral atom. b. If an ion, X^-, acquires a 2− charge, has it been oxidized or reduced? Write a general equation. (21-2)

2. Determine which of the following elements are in their common oxidized state and which are in their common reduced state. Use Table 21-4 as a reference if necessary: (21-2)
   a. Mg   d. Na^+ g. Cr^{3+}
   b. I^- e. Al^{3+} h. Mn
   c. O_2 f. S^{2-}

3. Determine the oxidation number of each atom in the following substances: (21-2)
   a. H_2SO_4 d. U_2O_5 g. ClO_2^-
   b. P_4 e. VO_3^- h. HSO_3^-
   c. UO_3 f. $S_4O_6^{2-}$

4. State whether the change represents an oxidation, a reduction or neither: (21-2)
   a. $MnO_2 \longrightarrow Mn_2O_3$
   b. $NH_3 \longrightarrow NO_2$
   c. $HClO_4 \longrightarrow HCl + H_2O$
   d. $O_2 \longrightarrow O^{2-}$
   e. $P_2O_5 \longrightarrow P_4O_{10}$

5. One method of obtaining copper metal is to let a solution containing Cu^{2+} ions trickle over scrap iron. Write the equations for the two half-reactions. Assume Fe^{2+} forms. Indicate in which half-reaction oxidation is taking place. (21-2)

6. If you wished to replate a silver spoon, would you make it the anode or the cathode in a cell? What would you use as the other electrode? Use half-reactions in your explanation. (21-4)

7. Consider this equation: (21-2)

$$Mg + Co^{2+} \longrightarrow Co + Mg^{2+}$$

   a. Which substance is oxidized?
   b. Which substance is reduced?
   c. Which substance acts as the reducing agent?
   d. Which substance acts as the oxidizing agent?

8. Which of these elements might be used as an oxidizing agent and which might be used as a reducing agent? Use Table 21-4 as a reference if necessary: (21-2)
   a. Ca^{2+} c. Fe^{3+} e. F^-
   b. Au   d. Na   f. Mn

9. Explain how the process of electrolysis is different from what occurs in an electrochemical cell. (21-9)

10. Would copper metal make a good cathodic protector to prevent iron from rusting? Explain. (21-10)

11. You learned in Section 21-7 that the reactions that occur in a lead storage battery produce +2.04 volts of electricity. How can a lead storage battery be made to have 12 volts? (21-7)

Problems

1. Aluminum metal will react with the hydrogen ions in acidic solutions to liberate hydrogen gas. Write equations for the two half-reactions and the balanced redox reaction. (21-3)

2. Write a balanced equation for the reaction between tin(II) ions, Sn^{2+}, and permanganate ions, MnO_4^-, in an acid solution to produce tin(IV) ions, Sn^{4+}, and manganese(II) ions, Mn^{2+}. (21-3)

3. Write balanced equations for each of the following reactions: (21-3)
   a. $H_2O_2 + I^- + H^+ \longrightarrow H_2O + I_2$
   b. $Cr_2O_7^{2-} + Fe^{2+} + H^+ \longrightarrow Cr^{3+} + Fe^{3+} + H_2O$
   c. $Cu + NO_3^- + H^+ \longrightarrow Cu^{2+} + NO + H_2O$

4. Write balanced equations for each of the following reactions: (21-3)
   a. $HBr + H_2SO_4 \longrightarrow SO_2 + Br_2 + H_2O$
   b. $NO_3^- + Cl^- + H^+ \longrightarrow NO + Cl_2 + H_2O$
   c. $Zn + NO_3^- + H^+ \longrightarrow Zn^{2+} + NH_4^+ + H_2O$
   d. $BrO^- \longrightarrow Br^- + BrO_3^-$

5. Use the half-cell potentials in Table 21-4 to decide if the reactions in item 3 will occur spontaneously at standard conditions. (21-6)

6. Write balanced equations for each of the following reactions. Identify what has been oxidized and what is the oxidizing agent: (21-3)
   a. $ClO_2^- + MnO_2 \longrightarrow ClO^- + MnO_4^- + OH^-$
   b. $In + BiO^+ \longrightarrow In^{3+} + Bi$
   c. $V^{2+} + H_2SO_3 \longrightarrow V^{3+} + S_2O_3^{2-}$
   d. $Mn + IrCl_6^{3-} \longrightarrow Mn^{2+} + Ir + Cl^-$

7. Titanium metal will react with iron(II) oxide, FeO, to form titanium oxide, TiO_2, and iron metal:
   a. Write a balanced equation for the reaction.
   b. How many grams of iron will be produced from the reaction of 57.3 grams of titanium with excess iron(II) oxide? (21-8)

8. Equations for the two half-reactions taking place in an electrochemical cell are:

$$Cr \longrightarrow Cr^{3+} + 3e^-$$
$$Pb^{2+} + 2e^- \longrightarrow Pb$$

   How many grams of lead will deposit on the cathode when 1.56 grams of chromium dissolve from the anode? (21-8)

9. When copper is placed in a concentrated nitric acid solution, vigorous bubbling takes place as a brown gas is produced. The copper disappears, and the solution changes from colorless to greenish-blue. The

CHAPTER REVIEW

brown gas is nitrogen dioxide, NO_2, and the greenish-blue color of the solution is due to the formation of copper(II) ions, Cu^{2+}. Write the balanced redox equation for this reaction. (21-3)

10. Steel screws and bolts can be plated with cadmium to minimize rusting. From this balanced redox equation, find the half-cell potential for the reduction of cadmium. (21-6)

$$Cd^{2+} + Ni \longrightarrow Cd + Ni^{2+} \qquad E° = +0.15 \text{ volt}$$

11. Complete and balance each of the following equations. Using the half-cell potentials from Table 21-4, predict whether each reaction will proceed spontaneously under standard conditions. (21-6)
    a. $Mg(s) + Sn^{2+}(aq) \longrightarrow$
    b. $Mn(s) + Cs^+(aq) \longrightarrow$
    c. $Cu(s) + Cl_2(g) \longrightarrow$
    d. $Fe(s) + Fe^{3+}(aq) \longrightarrow$

12. When zinc metal is placed in a concentrated sulfuric acid solution, the zinc dissolves and hydrogen sulfide, H_2S, is produced. Write the balanced redox equation for this reaction. (21-3)

Challenge

1. If a piece of copper metal is dipped into a $1M$ solution of Cr^{3+} ions, what will happen? Explain, using $E°$ values.

2. Which of the following reducing agents will spontaneously reduce $Cu^{2+}(aq)$ to $Cu(s)$ under standard conditions?
    a. $Sn^{2+}(aq) \longrightarrow Sn^{4+}(aq)$
    b. $Cl^-(aq) \longrightarrow Cl_2(g)$
    c. $Au(s) \longrightarrow Au^{3+}(aq)$
    d. $MnO_4^-(aq) \longrightarrow Mn^{2+}(aq)$
    e. $Ag(s) \longrightarrow Ag^+(aq)$

3. Which of the following oxidizing agents will spontaneously oxidize $Fe^{2+}(aq)$ to $Fe^{3+}(aq)$?
    a. $Ag^+(aq) \longrightarrow Ag(s)$
    b. $Pb^{2+}(aq) \longrightarrow Pb(s)$
    c. $Al^{3+}(aq) \longrightarrow Al(s)$
    d. $MnO_4^-(aq) \longrightarrow Mn^+(aq)$
    e. $Li(s) \longrightarrow Li^+(aq)$

4. Iodine is recovered from iodates in Chilean saltpeter by the reaction described by this unbalanced equation:

$$HSO_3^-(aq) + IO_3^-(aq) \longrightarrow I_2(g) + SO_4^{2-}(aq) \\ + H^+(aq) + H_2O(l)$$

a. How many grams of sodium iodate, $NaIO_3$, will react with 1.00 mole of potassium bisulfite, $KHSO_3$?
b. How many grams of iodine are produced?

Synthesis

1. A half-cell consisting of a palladium electrode in a 1-molar palladium(II) nitrate solution, $Pd(NO_3)_2$, is connected with a standard hydrogen half-cell. The overall cell potential is 0.99 volt. The platinum electrode in the hydrogen half-cell is the anode. Determine the $E°$ value for the reaction:

$$Pd(s) \longrightarrow Pd^{2+}(aq) + 2e^-$$

2. A 12-volt lead storage battery contains 700.0 grams of pure hydrogen sulfate, H_2SO_4, dissolved in water.
    a. If this acid solution was spilled, how many grams of solid sodium carbonate, Na_2CO_3, would be needed to neutralize it (producing CO_2 gas and H_2O)?
    b. How many liters of $2.0M$ Na_2CO_3 solution would be needed?

3. Concentrated nitric acid, HNO_3, is $15.9M$ and contains 68 percent HNO_3 by weight in water. How many liters of concentrated acid are needed to react with 0.100 kg of copper metal in the following reaction?

$$Cu(s) + H^+(aq) + NO_3^-(aq) \longrightarrow Cu^{2+}(aq) \\ + NO_2(g) + H_2O(l)$$

Projects

1. Research the contribution Sir Humphry Davy made to the field of electrochemistry.

2. Research the chemical reactions that take place in several different kinds of batteries. Written information usually is available from the company that makes the batteries.

3. The voltage that an electrochemical cell can deliver is dependent upon the concentration of the reacting materials. Find out how concentration affects the voltage of a cell.

4. It is recommended that oxidizing agents and reducing agents be separated from each other on the shelves in chemical stockrooms. Of those reagents commonly found in a high school stockroom, find out which are the strongest oxidizing agents and which are the strongest reducing agents. How could they be stored safely?

611

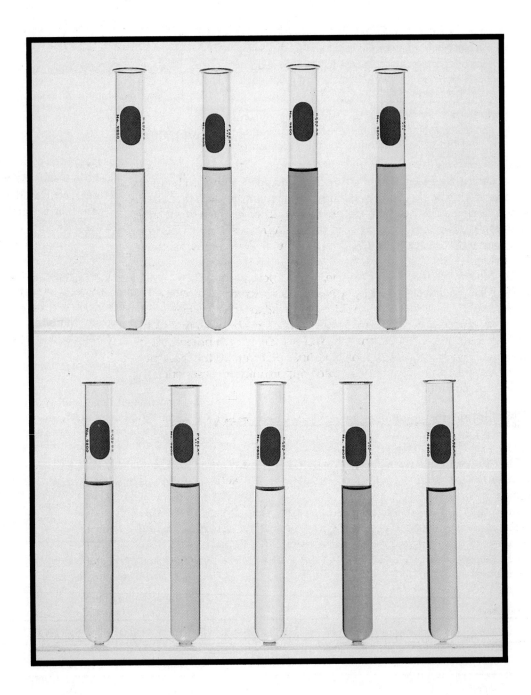

The nine test tubes in the photograph each contain some compound dissolved in water. Some of the solutions look very much alike, yet they are all different. Qualitative analysis involves the experimental determination of unknown substances. In this chapter, you will develop the problem-solving skills that will enable you to determine the identity of each solution in the photograph.

Analytical Problem Solving

Reasoning from Experimental Data

Most principles of chemistry were discovered after decades of full-time pursuit by many scientists. When you read about a chemistry concept in this book, you may get the mistaken impression that chemistry is a closed science with nothing left to discover. One of the most interesting aspects of chemistry is the on-going research to discover new materials or build new theories to simplify the study of matter. A good research project not only answers questions, it also raises additional questions that lead to further experimentation.

Qualitative analysis provides you with an opportunity to apply a research approach to solve a problem. You will use the chemistry concepts you learned in the previous chapters of this book as well as the skills you have acquired in the lab. You will design lab procedures to help you in identifying solutions like those pictured on the left. Qualitative analysis also offers an excellent opportunity to experience the thrill of discovery.

CONCEPTS

■ macroscopic observations indicate that a reaction occurs
■ the data from a series of reactions can be used to identify precipitates.
■ experimental data can be used to make predictions
■ unknown mixtures are analyzed by separation techniques
■ usefulness of flame tests

22-1 Looking for Regularities in Known Reactions

The first step in trying to determine the identity of an unknown compound in solution might be to observe the reaction behavior of known compounds in solution.

Table 22-1 shows the observations made when five different solutions are mixed together as shown in the table. If no noticeable change occurs, N.R. is used to represent no reaction. The abbreviation ppt. stands for precipitate. An X is used to denote when the same solutions are mixed.

The two most important aspects of this chapter are the higher level thought processes that are required, and the opportunity to review most of the important concepts in chemistry.

Table 22-1

REACTIONS OF SOME COMPOUNDS IN AQUEOUS SOLUTION					
	$NiCl_2(aq)$	$NH_4I(aq)$	$BaCl_2(aq)$	$CoSO_4(aq)$	$K_2CO_3(aq)$
$NiCl_2(aq)$	X	N.R.	N.R.	N.R.	green ppt.
$NH_4I(aq)$	N.R.	X	N.R.	N.R.	N.R.
$BaCl_2(aq)$	N.R.	N.R.	X	white ppt.	white ppt.
$CoSO_4(aq)$	N.R.	N.R.	white ppt.	X	red ppt.
$K_2CO_3(aq)$	green ppt.	N.R.	white ppt.	red ppt.	X

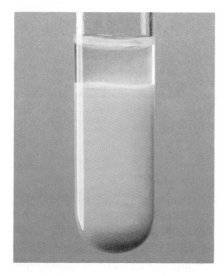

Figure 22-1 This precipitate forms when solutions of NiCl$_2$ and K$_2$CO$_3$ react.

The analysis of an unknown should not be taught as a formula. First add HCl, then add K$_2$CO$_3$, etc. Students should be encouraged to plan experiments of their own to analyze and verify their theories. This is the reason that a standard flow chart is not given.

Using this information, you will see how it can be applied to the analysis of an unknown. Now the detective work begins. You are given a solution that contains a compound from Table 22-1. You would make your analysis by mixing portions of your unknown with the other reagents listed in Table 22-1. A **reagent** is a substance used to detect, measure, or produce other substances. In this case, the different reagents are used to detect the presence of certain substances in the unknown.

How could you determine the identity of your unknown with as few tests as possible? Which reagents would you add and in what order? Which reagents would provide little or no information as to the identity of the unknown? Write out a procedure you would follow to find the identity of your unknown.

Once you have formulated your plan of attack, do some further detective work in trying to identify the precipitates in Table 22-1. The reactions that yield precipitates are double replacement reactions having the general equation covered in Chapter 5. You can probably write the equations easily enough but when you add the appropriate state symbols to the products, you have some decisions to make.

The first precipitate in the top right corner of the table comes from the reaction of nickel(II) chloride and potassium carbonate. Treating this reaction as a double replacement yields the following equation:

$$NiCl_2(aq) + K_2CO_3(aq) \longrightarrow NiCO_3 + 2KCl$$

Which of the two products is the precipitate? Are they both precipitates? Do you have enough information to decide? One way to know for sure might be to write equations for all of the reactions in Table 22-1 that produce precipitates. Starting from the top:

1. $NiCl_2(aq)$ + $K_2CO_3(aq)$ $\longrightarrow$ $NiCO_3(?)$ + $2KCl(?)$
2. $BaCl_2(aq)$ + $CoSO_4(aq)$ $\longrightarrow$ $BaSO_4(?)$ + $CoCl_2(?)$
3. $BaCl_2(aq)$ + $K_2CO_3(aq)$ $\longrightarrow$ $BaCO_3(?)$ + $2KCl(?)$
4. $CoSO_4(aq)$ + $BaCl_2(aq)$ $\longrightarrow$ $BaSO_4(?)$ + $CoCl_2(?)$
5. $CoSO_4(aq)$ + $K_2CO_3(aq)$ $\longrightarrow$ $CoCO_3(?)$ + $K_2SO_4(?)$
6. $K_2CO_3(aq)$ + $NiCl_2(aq)$ $\longrightarrow$ $NiCO_3(?)$ + $2KCl(?)$
7. $K_2CO_3(aq)$ + $BaCl_2(aq)$ $\longrightarrow$ $BaCO_3(?)$ + $2KCl(?)$
8. $K_2CO_3(aq)$ + $CoSO_4(aq)$ $\longrightarrow$ $CoCO_3(?)$ + $K_2SO_4(?)$

Look at equations 1 through 8, and you should notice some patterns. First, equations 1 and 6 are duplicates. What other equations are duplicates of each other? The next thing you should notice is that KCl appears as a product four times (equations 1, 3, 6, 7).

Now go back to the original problem of trying to determine the identity of the green precipitate.

$$NiCl(aq) + K_2CO_3(aq) \longrightarrow NiCO_3(?) + 2KCl(?)$$

From Table 22-1, you see that this reaction produces a green precipitate. If the precipitate is KCl, you should note a green precipitate everytime it is a product in a reaction. The results of equation 3 show that when BaCl$_2$(aq) and K$_2$CO$_3$(aq) are mixed, you get a white precipitate. Therefore, based on the evidence you have, it seems reasonable to conclude that the precipitate in equation 1 is $NiCO_3$. The complete equation is:

$$NiCl_2(aq) + K_2CO_3(aq) \longrightarrow NiCO_3(s) + 2KCl(aq)$$

Thus far you have focused your attention on reactions that occur but often times it is just as useful to look at what does not happen in analyzing your data. For example, look at equation 2.

$$BaCl_2(aq) + CoSO_4(aq) \longrightarrow BaSO_4(?) + CoCl_2(?)$$

This reaction produces a white precipitate that could be either $BaSO_4$ or $CoCl_2$. The only other reaction producing these products is the duplicate of equation 2. Now what can you do? Look at Table 22-1 again but this time focus your attention on the reactions that did not occur? If $CoCl_2$ is the white precipitate, should you see that same result when $CoSO_4(aq)$ and $NiCl_2(aq)$ are mixed together? Note there is no reaction, so it is reasonable to assume that $CoCl_2$ is soluble. The finished equation for reaction 2 is:

$$BaCl_2(aq) + CoSO_4(aq) \longrightarrow BaSO_4(s) + CoCl_2(aq)$$

Based on the information you have thus far, you should be able to complete the following exercises.

CHEM THEME

A substance is identified by its characteristic properties. Characteristic properties include solubility, color, odor, and the chemical reactions that take place when it is mixed with other reagents.

Review and Practice

1. Complete equation 3 from Section 22-1 by adding the appropriate state symbols. Give reasons for your answer.

2. Make a table to classify all of the compounds in equations 1–8 as soluble or insoluble. Which substances can't be classified based on the information you have?

3. Make a list of any patterns you note from the solubility table you make in item 2.

4. You are given an unknown solution containing one of the compounds in Table 22-1. Outline the procedure you would use to determine the identity of the unknown.

1. $BaCl_2(aq) + K_2CO_3(aq) \longrightarrow BaCO_3(s) + 2KCl(aq)$
If KCl is a precipitate it should be the same color each time. In equation 1 the ppt. is green, in 3 the ppt. is white. This can be explained by assuming the green ppt. is $NiCO_3$ in equation 1, and the white ppt. in equation 3 is $BaCO_3$.

2.
insoluble	soluble	?
$NiCO_3$	KCl	$CoCO_3$
$BaSO_4$	$CoCl_2$	K_2SO_4
$BaCO_3$	$NiCl_2$	
	$BaCl_2$	
	K_2CO_3	

3. chlorides are usually soluble
carbonates are often insoluble
4. Step 1. Add K_2CO_3 if a green ppt. forms $x = NiCl_2$
if a white ppt. forms $x = BaCl_2$
if a red ppt. forms $x = CoSO_4$
if N.R. occurs $x = NH_4I$ or K_2CO_3
Step 2. Add $CoSO_4$ if N.R. occurs $x = NH_4I$
if a red ppt. forms $x = K_2CO_3$

22-2 Expanding Your Data Base

By now you might think that being a chemical detective is not as difficult as you imagined. However, you must realize that in the previous section, you were dealing with five possibilities in relation to the vast number of compounds. Now you will see how you can build on your knowledge from the previous section to tackle a bigger problem.

Table 22-2 consists of ten different compounds, some of which were in Table 22-1. As you scan the table, note some other differences. Now you are confronted with signs other than precipitation that indicate a reaction has occurred.

Table 22-2

REACTIONS OF TEN COMPOUNDS IN AQUEOUS SOLUTION								
	$NiCl_2$	NH_4I	$BaCl_2$	$CoSO_4$	K_2CO_3	NaOH	HCl	$Pb(NO_3)_2$
$NiCl_2$	X	N.R.	N.R.	N.R.	green ppt.	green ppt.	N.R.	white ppt.
NH_4I	N.R.	X	N.R.	N.R.	N.R.	NH_3 odor	N.R.	yellow ppt.
$BaCl_2$	N.R.	N.R.	X	white ppt.	white ppt.	white ppt.	N.R.	white ppt.
$CoSO_4$	N.R.	N.R.	white ppt.	X	red ppt.	red ppt.	N.R.	white ppt.
K_2CO_3	green ppt.	N.R.	white ppt.	red ppt.	X	N.R.	bubbles of a gas	white ppt.
NaOH	green ppt.	NH_3 odor	white ppt.	red ppt.	N.R.	X	hot reaction	white ppt.
HCl	N.R.	N.R.		N.R.	bubbles of a gas	hot reaction	X	white ppt.
$CuSO_4$	N.R.	N.R.	white ppt.	N.R.	blue ppt.	blue ppt.	N.R.	white ppt.
KNO_3	N.R.	N.R.	N.R.	N.R.	N.R.	N.R.	N.R.	N.R.

When the information you have is not enough to identify the unknown, more information must be acquired. The lab is the best place to obtain information, but the use of reference books should also be encouraged.

There are two instances in Table 22-2 where the results indicate a noticeably exothermic reaction. If you look at the data you see that these reactions involve mixing an acid and a base. You know from Chapter 20 that neutralization reactions can produce a soluble salt and water. Your indication that a reaction occurs is by the noticeable increase in temperature of the contents of the test tube.

Assume you are given an unknown solution in a test tube that contains one of the nine compounds listed in Table 22-2. What procedure would you go through to identify your unknown using the fewest number of tests? Answering this question will take time and some trial-and-error analysis. What you are looking for are those tests that allow you to narrow the possibilities as quickly as possible.

Review and Practice

See the Teacher's Guide at the front of this book for answers to these questions.

1. Write equations for the reactions in Table 22-2 that yield precipitates.

2. Identify the precipitates for the tests in Table 22-2.

3. Expand the solubility table you made in answering item 2 on page 615 to include the new substances in Table 22-2.

4. Expand the list of solubility patterns you made for item 3 on page 615.

5. You are given an unknown solution containing one of the compounds in Table 22-2. Outline the procedure you would use to determine the identity of the unknown.

6. What would you predict as the identity of the gas produced when HCl(aq) is mixed with K_2CO_3(aq)?

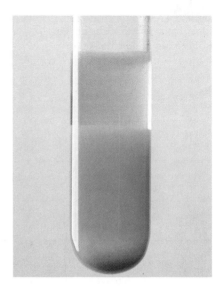

Figure 22-2 The precipitate in this test tube is actually a mixture of $NiCO_3$(s) and $CoCO_3$(s).

22-3 Dealing with Mixtures

Suppose you are given an unknown solution that contains one or more of the compounds listed in Table 22-2. Would you approach the analysis procedure in a different way?

Let's look at some of the other angles you must consider. First, the results of some tests are very similar. For example, you could add K_2CO_3 to your unknown and get a green precipitate. Based on this result you might conclude that your unknown is a nickel(II) chloride solution. The experienced detective would look for further evidence. You could also have a white precipitate in the test tube that might be masked by the color of the green precipitate. You could have a precipitate whose color does not seem to match any of those in Table 22-2. You must also consider that K_2CO_3 will not confirm the presence of NH_4I or NaOH or itself, yet they could be present.

A test that is specific for a certain compound or ion would be of great help in isolating the components of a mixture. For example, the presence of an ammonia odor when an unknown is tested with a strong base indicates that an ammonium compound is present in the unknown. No other ion would give this result. Therefore, the addition of NaOH is a specific test for the presence of ammonium compounds.

Another useful method in dealing with mixtures of compounds is to use tests that will enable you to separate substances. What if you were given a solution containing nickel(II) chloride and copper(II) sulfate. How could you precipitate these compounds in two steps using a single sample of the solution mixture? In looking at Table 22-2, you see that adding $BaCl_2$(aq) will cause the SO_4^{2-} ions to precipitate. The other ions remain in solution. If the $BaSO_4$ precipitate is allowed to settle, the liquid can be poured off and tested further with other reagents.

Figure 22-3 The flame color for lithium ions.

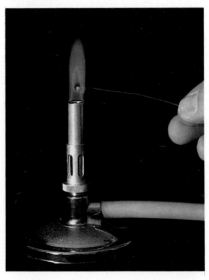

22-4 Other Analytical Techniques: Flame Tests

Potassium, sodium, and lithium salts are chemically similar, and they are members of the alkali metals (group 1). Most alkali metal salts are soluble, so precipitation cannot be used to confirm the presence of Li^+, K^+, or Na^+ in an unknown solution. Another technique called a **flame test** can be used to identify these and some other metal ions.

It has been known for centuries that certain substances would color a flame when burned. These observations were applied by the Chinese in making fireworks. Figure 22-3 shows a platinum wire that has been dipped in a solution of lithium chloride. The scarlet color of the flame is a characteristic of lithium. The same result can be obtained if the wire was dipped into a solution of any other soluble lithium salt.

Figure 22-4 Flame tests for various ions, from left to right on top— potassium, sodium, calcium; bottom—strontium, barium, copper.

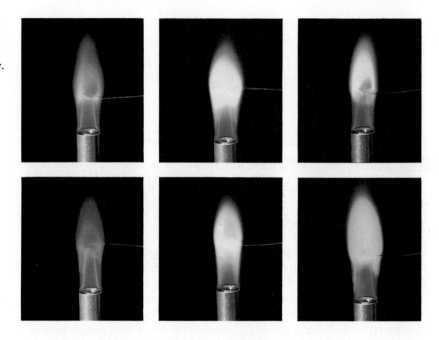

Figure 22-4 shows the flame tests for potassium, sodium, calcium, strontium, barium, and copper. Notice that the flame tests for lithium and strontium are very similar. With some experience, you should be able to tell the crimson of strontium from the scarlet of lithium.

In using flame tests as an analytical tool it is important to avoid contamination. The test wire must be clean and the solutions or solid salts tested must be free of contamination.

The unknowns can range from simple to analyze, to very complex, with separation techniques required. The difficulty of the unknown can be matched with the interest, ability, and persistence of the student.

22-5 Extension: Applying Your Analytical Skills to a Large-Scale Problem

You can apply the reasoning skills you have acquired in the previous sections to test unknown solutions containing many more possibilities than those listed in Table 22-2. Tests that separate ions in a mix-

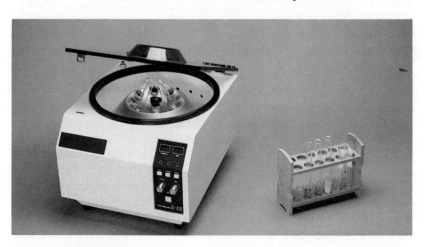

Figure 22-5 A centrifuge is used in separating precipitates from solutions. As the centrifuge turns, the force of gravity pulls the precipitate to the bottom of the test tube.

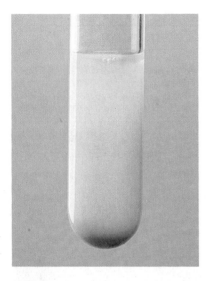 

Figure 22-6 A barium sulfate precipitate appears pink due to Co^{2+} ions in solution. After centrifuging and washing, the precipitate is distinctly white.

ture become much more important in large scale analysis. Precipitates can be separated more efficiently from soluble salts when a centrifuge is used. A **centrifuge** rapidly spins a test tube causing a dense precipitate to settle to the bottom of the test tube. The liquid in the test tube can be poured off easily.

You should recognize that laboratory analysis such as this does not always provide exact results. Students often see white precipitates as grey. Sometimes it is difficult to determine the exact color of a precipitate as colored soluble salts may be present with the precipitate. For example, if $CoSO_4(aq)$ is mixed with $Ba(NO_3)_2(aq)$, there appears to be a pink precipitate in the test tube. In this case, the pink color is due to the Co^{2+} ion which is in solution. If the precipitate is washed with a few milliliters of water, centrifuged, and the wash solution decanted, the white precipitate of barium sulfate remains.

Recall from Chapter 13 that transition metals form highly colored hydrated complex ions. Some solutions can be characterized by color, but you should not be fooled by using color alone to make an identification. You can be led astray very easily.

Table 22-3 on the next page shows the results of experiments using 30 different compounds. Use this information to expand your knowledge of specific tests and solubilities of compounds.

Descriptive analytical chemistry is most effectively learned in the laboratory. If the students share the data they obtain, the observations in Table 23-3 can be obtained in three or four 45-minute periods. Each student will have had an opportunity to observe some of the chemical reactions. You may find that the reagents suggested in the experiments that are in the Lab Book are slightly different. Substances that are especially toxic have not been included in the laboratory exercises, but some of them have been discussed in the chapter.

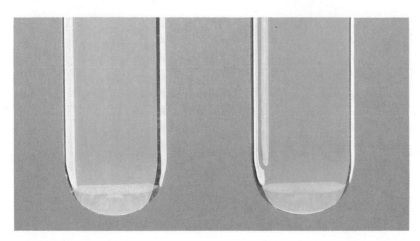

Figure 22-7 Do not be fooled by trying to identify a metal ion by color alone. Both of these solutions are yellow. Yet the test tube on the right contains $FeCl_3$ while the test tube on the left contains Na_2CrO_4.

Table 22-3

	NiCl$_2$	NH$_4$OH	Li$_2$CO$_3$	NH$_4$Cl	BaCl$_2$	KI	ZnCl$_2$	(NH$_4$)$_2$CO$_3$
FURTHER STUDIES OF COMPOUNDS IN AQUEOUS SOLUTION								
NH$_4$I	N.R.	N.R.	N.R.	N.R.	N.R.	N.R.	N.R.	N.R.
(NH$_4$)$_2$CO$_3$	green ppt.	N.R.	N.R.	N.R.	white ppt.	N.R.	white ppt.	X
CaCl$_2$	N.R.	faint white ppt.	white ppt.	N.R.	N.R.	N.R.	N.R.	white ppt.
CoSO$_4$	N.R.	red ppt.	red ppt.	N.R.	white ppt.	N.R.	N.R.	red ppt.
HCl	N.R.	hot reaction	gas bubbles no odor	N.R.	N.R.	N.R.	N.R.	gas bubbles no odor
Ba(NO$_3$)$_2$	N.R.	faint white ppt.	white ppt.	N.R.	N.R.	N.R.	N.R.	white ppt.
Pb(NO$_3$)$_2$	white ppt.	white ppt.	white ppt.	white ppt.	white ppt.	yellow ppt.	white ppt.	white ppt.
Hg$_2$(NO$_3$)$_2$	white ppt.	black ppt.	pale yellow ppt.	white ppt.	white ppt.	yellow ppt.	white ppt.	pale yellow ppt.
NaI	N.R.	N.R.	N.R.	N.R.	N.R.	N.R.	N.R.	N.R.
HNO$_3$	N.R.	hot reaction	gas bubbles no odor	N.R.	N.R.	N.R.	N.R.	gas bubbles no odor
NaCl	N.R.	N.R.	N.R.	N.R.	N.R.	N.R.	N.R.	N.R.
KBr	N.R.	N.R.	N.R.	N.R.	N.R.	N.R.	N.R.	N.R.
MgSO$_4$	N.R.	faint white ppt.	white ppt.	N.R.	white ppt.	N.R.	N.R.	white ppt.
K$_2$SO$_4$	N.R.	N.R.	N.R.	N.R.	white ppt.	N.R.	N.R.	N.R.
H$_2$SO$_4$	N.R.	hot reaction	gas bubbles no odor	N.R.	white ppt.	N.R.	N.R.	gas bubbles no odor
(NH$_4$)$_2$S	black ppt.	N.R.	N.R.	N.R.	N.R.	N.R.	N.R.	N.R.

SrCl$_2$	Pb(NO$_3$)$_2$	K$_2$CO$_3$	MgSO$_4$	NaOH	Na$_2$S	Na$_2$CO$_3$	NiSO$_4$	Zn(NO$_3$)$_2$
N.R.	yellow ppt.	N.R.	N.R.	NH$_3$ odor	N.R.	N.R.	N.R.	N.R.
white ppt.	white ppt.	N.R.	white ppt.	NH$_3$ odor	N.R.	N.R.	green ppt.	white ppt.
N.R.	white ppt.	white ppt.	N.R.	faint white ppt.	N.R.	white ppt.	N.R.	N.R.
white ppt.	white ppt.	red ppt.	N.R.	red ppt.	black ppt.	red ppt.	N.R.	N.R.
N.R.	white ppt.	gas bubbles no odor	N.R.	hot reaction	rotten eggs odor	gas bubbles no odor	N.R.	
N.R.	N.R.	white ppt.	white ppt.	white ppt.	N.R.	white ppt.	white ppt.	N.R.
white ppt.	X	white ppt.	white ppt.	white ppt.	black ppt.	white ppt.	white ppt.	N.R.
white ppt.	N.R.	pale yellow ppt.	white ppt.	black ppt.	black ppt.	pale yellow ppt.	white ppt.	N.R.
N.R.	yellow ppt.	N.R.	N.R.	N.R.	N.R.	N.R.	N.R.	N.R.
N.R.	N.R.	gas bubbles no odor	N.R.	hot reaction	rotten eggs odor	gas bubbles no odor	N.R.	N.R.
N.R.	white ppt.	N.R.	N.R.	N.R.	N.R.	N.R.	N.R.	N.R.
N.R.	white ppt.	N.R.	N.R.	N.R.	N.R.	N.R.	N.R.	N.R.
white ppt.	white ppt.	white ppt.	X	faint white ppt.	N.R.	white ppt.	N.R.	N.R.
white ppt.	white ppt.	N.R.	N.R.	N.R.	N.R.	N.R.	N.R.	N.R.
white ppt.	white ppt.	gas bubbles no odor	N.R.	hot reaction	rotten eggs odor	gas bubbles no odor	N.R.	N.R.
N.R.	black ppt.	N.R.	N.R.	NH$_3$ odor	N.R.	N.R.	black ppt.	white ppt.

Summary

■ When ions react in solution the following macroscopic observations can be made:
  1. heat is given off
  2. bubbles or an odor form
  3. an insoluble precipitate forms

■ If no macroscopic changes occur, assume the ions do not react.

■ To determine the identity of an unknown solution it is necessary to utilize experimental data and then reason correctly from the data.

■ The color of an ion in solution or a flame test can be used to help identify an unknown.

■ An experiment that positively identifies an ion provides a unique test. Some ions are most easily identified after a series of tests.

■ If an unknown is a mixture of ions, it is necessary to consider the possible effects of several ions when a reaction is done.

■ Some ions interfere with the tests for other ions and it may be necessary to separate the ions by precipitating, complexing, washing, and centrifuging.

■ The analysis of an unknown is an opportunity to apply many previously learned concepts which include formula writing, solubility, equilibrium, acids, and bases.

Chemically Speaking

centrifuge
flame test
reagent

Review

1. Write the formulas for these aqueous solutions: (22-1)
   a. sulfuric acid
   b. potassium hydroxide
   c. sodium carbonate
   d. strontium nitrate
   e. ammonium sulfate
   f. potassium sulfate

2. What are the cations in solutions a, b, and c of item 1? (22-1)

3. What are the anions in solutions d, e, and f of item 1? (22-1)

4. What ion shows a violet flame test? (22-4)

5. Which element has a crimson flame test? (22-4)

6. Which ions are green in solution? (22-5)

7. Write ionic equations and identify the pale-yellow precipitate in Table 22-3. (22-5)

8. Which of the following solutions could your unknown be if the solution is blue? (22-5)
   a. $NiCl_2(aq)$
   b. $KCl(aq)$
   c. $CuSO_4(aq)$
   d. $Sr(NO_3)_2(aq)$
   e. $Cu(NO_3)_2(aq)$
   f. $CoSO_4(aq)$

Interpret and Apply

1. List the different ions contained in an unknown that is a mixture of $KCl(aq)$ and $MgSO_4(aq)$. (22-1)

2. List the different ions contained in an unknown that is a mixture of $Na_2CO_3(aq)$ and $(NH_4)_2CO_3(aq)$. (22-1)

3. You have an unknown that might contain both $Na_2CO_3(aq)$ and $Na_2SO_4(aq)$. If you add $Sr(NO_3)_2(aq)$ to test for $Na_2SO_4(aq)$, the $Na_2CO_3(aq)$ will interfere. How could the $Na_2CO_3(aq)$ interfere? How could the $Na_2CO_3(aq)$ be removed before testing with $Sr(NO_3)_2$? (22-3)

4. An unknown is a solution of sodium hydroxide, sodium carbonate, or sodium chloride. When a solution of sulfuric acid is added, no reaction occurs. What is the unknown? (22-5)

5. An unknown is a solution of sodium hydroxide, sodium carbonate, or sodium chloride. Name a reagent which, when added to the unknown, will give results that indicate the identity of the unknown and explain why. (22-5)

6. An unknown is a solution of barium nitrate, barium chloride, or sodium sulfide. When a solution of silver nitrate is added, no reaction occurs. What is the identity of the unknown? (22-5)

CHAPTER REVIEW

7. An unknown is a solution of barium nitrate, barium chloride, or sodium sulfide. Name a reagent which, when added to the unknown, will identify it. Explain your answer. (22-5)

8. An unknown is a solution of zinc sulfate, cobalt (II) nitrate, or lead (II) nitrate. When a solution of barium chloride is added no reaction occurs. What is the identity of the unknown? (22-5)

9. An unknown is a solution of zinc sulfate, cobalt (II) nitrate, or sulfuric acid. Name one reagent which, when added to the unknown, will identify it. Explain your answer. (22-5)

10. An unknown is a solution of sodium iodide, sodium hydroxide, or sodium carbonate. When a solution of lead(II) nitrate is added a yellow precipitate is formed. What is the identity of the unknown? (22-5)

11. An unknown contains a solution of sodium iodide, sodium hydroxide, or sodium carbonate. Name a reagent which, when added to the unknown, will identify it. Explain your answer. (22-5)

12. An unknown is a solution of barium chloride, sodium carbonate, lead(II) nitrate, or potassium hydroxide. When a solution of sodium iodide is added a yellow precipitate forms. What is the identity of the unknown? (22-5)

13. An unknown is a solution of barium chloride, sodium carbonate, lead(II) nitrate, or potassium hydroxide. Name a reagent which, when added to the unknown, will identify it. Explain your answer. (22-5)

Challenge

1. When 20.0 milliliters of a $0.50M$ solution of silver nitrate are added to 20.0 milliliters of a $0.50M$ solution of calcium chloride, which reactant is in excess? What mass of silver chloride will be precipitated in the reaction? When the water in the filtrate is evaporated, what is the mass of the remaining solid?

2. When 12.4 grams of sodium hydroxide are dissolved in enough water to make 500 milliliters of solution, what will be the pH of the solution?

3. Calculate the pH of a solution when 40.0 milliliters of $1.00M$ hydrochloric acid are mixed with 60.0 milliliters of $1.00M$ sodium hydroxide.

Synthesis

1. How many milliliters of a $0.50M$ solution of sulfuric acid are required to completely react with 10.0 milliliters of a $0.50M$ solution of potassium hydroxide?

2. How many milliliters of a $0.25M$ solution of nitric acid are required to completely react with 20.0 milliliters of $0.05M$ solution of barium hydroxide?

3. When 40.0 milliliters of a $0.50M$ solution of sodium carbonate are reacted with an excess of hydrochloric acid, what volume of carbon dioxide gas will be formed at 25°C and a pressure of 95 kilopascals.

4. How many grams of copper(II) sulfate pentahydrate are required to prepare 2.00 liters of $0.25M$ $CuSO_4(aq)$?

5. Calculate the pH of a solution which contains 4.9 grams of sulfuric acid, H_2SO_4, per liter.

6. Describe how you would prepare 1.00 liter of a $0.50M$ solution of iron(II) sulfate heptahydrate, $FeSO_4 \cdot 7H_2O$.

7. When 20.0 milliliters of $0.50M$ lead(II) nitrate are reacted with 20.0 milliliters of $1.00M$ potassium iodide, what mass of lead(II) iodide will be formed?

8. What is the molarity of a solution in which 10.8 grams of sodium nitrate are added to enough water to make 175 milliliters of solution?

9. Describe how you would prepare 1.50 liters of a $0.50M$ solution of magnesium sulfate heptahydrate.

10. When 20.0 milliliters of $0.05M$ silver nitrate are added to 30.0 milliliters of a $0.020M$ solution of sodium sulfate, will a precipitate form? $K_{sp}Ag_2SO_4 = 1.18 \times 10^{-5}$

Projects

1. Do enough experiments or consult a handbook so you can determine a solubility pattern for fluorides.

2. Find a specific test for phosphate in water. Analyze the water in a local stream or lake for phosphates.

Look around you. Most of what you see is made from organic chemicals. Several kinds of organic compounds have literally changed life on Earth. Polymers are used for everything from plastic dishes to panty hose, from automobile tires to plumbing. Organic compounds are the ingredients of many life-saving medicines, and they are used in some of the fertilizers that increase crop production worldwide. There are more organic compounds known to exist than all other compounds combined. Besides the fact that your body and the foods you eat are made of organic chemicals, it is likely that you encounter materials synthesized from organic compounds dozens of times every day. This chapter provides you with an introduction to the identities and properties of some of these chemicals.

Organic Chemistry

Introduction to Carbon Compounds

In 1828, German chemist Friedrich Wöhler synthesized the compound urea. It is a substance normally found in cells as a waste product of protein metabolism and is composed of carbon, hydrogen, oxygen, and nitrogen. Using reactants of a mineral nature, Wöhler was the first person to produce an organic compound in the laboratory. Although at that time the discovery had little economic importance, its scientific significance changed the field of carbon chemistry forever. Compounds found in living organisms and thought to contain a nonreproducible "vital force" could now be prepared in a chemist's workshop.

The existence and behavior of carbon compounds are two of the most intriguing aspects of contemporary chemistry. Although most carbon compounds were originally isolated from coal and petroleum, the majority of carbon compounds are now produced in laboratories. In the first part of this chapter, you will be introduced to Earth's natural bounty of carbon compounds. You also will learn how carbon's chemical behavior accounts for over 90 percent of all known compounds.

CONCEPTS

- general properties of carbon compounds
- coal and petroleum—natural sources of organic compounds
- a variety of organic compounds
- shapes of carbon chains
- saturated vs unsaturated bonds

23-1 What Are Carbon Compounds?

It is obvious that carbon compounds contain carbon. However, unlike compounds formed from other elements, carbon compounds are far more numerous and varied. There are several million different carbon compounds known, and over 300,000 new ones are synthesized each year! Most of these compounds are combinations of carbon with a handful of other elements. The most common elements found with carbon are hydrogen, oxygen, nitrogen, sulfur, phosphorus, and the halogens.

Because there are so many carbon compounds, an entire branch of chemistry has developed around them. As a result of many experimental observations, some general chemical and physical properties of carbon compounds are known. For example, carbon compounds are either nonelectrolytes or very weak electrolytes, and as a group they tend to have low melting points. Carbon compounds made solely from carbon and hydrogen are generally nonpolar and insoluble in water, whereas other classes of carbon compounds include some substances that are quite soluble. In this chapter, you will take a closer look at the properties of a few groups of carbon compounds. You also will find out why they are significant in today's world.

23-2 Where Do Carbon Compounds Come From?

Millions of years ago, climatic conditions on Earth resulted in the rapid growth of vegetation. As plants and trees later died, they fell to the floors of forests, swamps, and marshes. The continuing cycle of growth and decay resulted in vast deposits of plant material. As the material accumulated, pressure and heat began altering the chemical structure of the buried compounds. In time, the layers were pressed into hard beds of **coal**, composed chiefly of carbon atoms and incorporating appreciable amounts of oxygen, hydrogen, nitrogen, and sulfur compounds into its structure.

Until the onset of the Industrial Revolution, the wealth of chemicals stored in coal deposits was seldom utilized. Through a process called **destructive distillation**, in which coal is heated in the absence of air, a wide array of carbon compounds was liberated. Coke, a residue of the process, was substituted for charcoal in iron making. Subsequently, people began to investigate how to use the volatile by-products of distillation, such as coal gas and coal tar. Coal tar was separated into over 200 different carbon compounds. These substances could then either be used directly or treated in the laboratory to produce additional compounds.

At about the same time that Earth's prehistoric environment was dominated by land plants, its shallow seas supported heavy growth of algae, bacteria, and plankton. As these microscopic organisms died, their bodies settled into the ooze of the seafloor. Microbes, living within the ooze, began decomposing the accumulated matter, producing methane gas, CH_4. As the ooze grew deeper, high temperatures and pressures altered the molecular structures of the compounds that were present (sugars and other carbon-based substances). Huge deposits of crude oil were produced during these reactions. Under tremendous pressure, these mixtures of compounds diffused into the porous space of nearby sandstone. There they waited, until an industrialized society recognized the value of petroleum and natural gas.

Plants and animals are highly efficient chemical factories. In order to maintain their living condition, organisms must synthesize numer-

Do You Know?

It has been estimated that Earth's atmosphere contains about 40 times as much carbon as found in all the fossil fuels and forests, despite the fact that only 0.03 percent of the atmosphere is carbon dioxide. The largest store of carbon, however, is found in carbonates and other minerals. These rocks contain about 400,000 times as much carbon as the coal, oil, and wood reserves.

Figure 23-1 The coal deposits of North America formed from swampy forests that existed during the Carboniferous Period, which began about 345 000 000 million years ago.

ous carbon-based molecules. These compounds include proteins, sugars, cellulose, starches, vitamins, plant oils, waxes, fats, gelatins, dyes, drugs, and fibers. Since all the sources of carbon compounds originally came solely from living organisms, the chemistry of carbon was called **organic chemistry**. This name comes from the belief, now discarded, that living materials were organized in a unique way, containing a special ingredient absent in nonliving matter. Although organic chemicals are now routinely synthesized in laboratories, the name continues to be applied to the chemistry of carbon compounds.

23-3 So Many Organic Compounds

Why are there so many different carbon compounds in comparison to compounds made from other elements? One reason is the bonding behavior of carbon. In Chapter 12, you learned that carbon atoms have four electrons available for bonding and that these electrons can be shared in four covalent bonds with other atoms. Some of these bonds may be double or triple bonds. Additionally, carbon atoms have an unusual ability to link together to form chains of varying length. This bonding behavior can produce carbon chains of astounding proportions. Occasionally, the chain may bond in such a fashion as to produce one or several rings made of carbon atoms.

Atoms in organic compounds often may be arranged in several different ways, much like the pieces of a set of building blocks can be assembled and reassembled to build different objects. Since the properties of any substance depend upon both the composition and the structure of a molecule, different arrangements of the same atoms result in different substances, each with a unique chemical and physical identity. For example, the 92 atoms of the molecule $C_{30}H_{62}$ may be arranged to form about four billion possible compounds! You will have an opportunity to look at some simpler molecules and the several structural arrangements the atoms of these molecules can have.

23-4 Sorting Out Organic Compounds

Carbon compounds containing only carbon and hydrogen atoms are called **hydrocarbons**. Organic compounds in which some or all of the hydrogen atoms have been replaced by other atoms are considered to be **derivatives** of hydrocarbons. Although simple in structure, hydrocarbons constitute the building blocks for more complex organic molecules. Coal, petroleum, natural gas, and certain species of trees (sources of turpentine and rubber) are the most common natural sources of these compounds.

The longest carbon chain in a hydrocarbon molecule is frequently referred to as the **carbon backbone**, or skeleton. When structural diagrams of hydrocarbons are made, often only the carbon skeleton and associated bonds are drawn. It is assumed that the identity of the atoms not shown is hydrogen. For example, the structural formula for ethane, C_2H_6, can be drawn in either of the following ways:

This description of spatial arrangement of electron pairs is based on the valence shell electron pair repulsion theory explained in Chapter 12. Hybridization is discussed in the Appendix.

The ability of carbon atoms to link in chains is known as *catenation*.

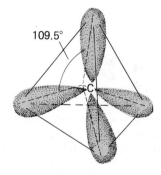

Figure 23-2 A carbon atom can form covalent bonds with up to four other atoms.

In the second case, only the carbon skeleton of two carbons and their bonds are represented. Hydrocarbons may contain a single linear chain of carbon atoms:

These simple molecules are called **straight-chain** or **unbranched** hydrocarbons. More complex hydrocarbons may be composed of several carbon chains that cross, such as those on the left. These molecules are called **branched** hydrocarbons. The carbon backbone of each molecule is shown in blue.

Hydrocarbons may contain all single bonds or combinations of single, double, and triple bonds. The reactivity of a hydrocarbon is dependent upon the number and type of multiple bonds found within the molecule. Molecules made entirely of single carbon-carbon bonds are relatively stable. They cannot incorporate additional atoms into their structure, so they are said to be **saturated**. When double or triple bonds are found between carbon atoms, the possibility exists for opening some of the bonds and introducing additional atoms to the structure. Hydrocarbons containing at least one double or triple carbon-carbon bond are referred to as **unsaturated**.

saturated   unsaturated

Carbon-carbon bonds themselves also may be described as saturated or unsaturated, depending on whether they are single bonds, or double or triple bonds.

1. It was the first time an organic compound was synthesized in the laboratory.
2. They are derived from decayed matter of both plants and animals that lived millions of years ago.
3. Bacteria decomposed the accumulated matter into new organic compounds including methane.
4. high temperatures and pressures
5. a carbon compound containing only carbon and hydrogen atoms
6. Saturated compounds contain only single bonds. Unsaturated compounds contain at least one double or triple carbon-carbon bond.
7.

straight-chained

8. They tend to be either nonelectrolytes or weak electrolytes and have low melting points.

Review and Practice

1. Why was Wöhler's synthesis of urea significant to the field of carbon chemistry?

2. Why are coal, natural gas, and petroleum often called fossil fuels?

3. What role did bacteria play in creating fossil fuel deposits?

4. What physical conditions are thought to have transformed organic compounds from organisms into petroleum?

5. What is a hydrocarbon?

6. What is the difference between a saturated organic compound and an unsaturated compound?

7. Draw both the complete structural diagram and the carbon skeleton that represent the formula C_3H_8. Is this molecule straight-chained or branched?

8. List two general characteristics associated with organic compounds.

A Closer Look at Hydrocarbons

How is it possible to make sense out of the tremendous numbers of organic chemicals and the vast body of information known about them? What makes some carbon compounds different from others? You have read that the simplest of organic compounds are hydrocarbons and that they vary in size, shape, and type of bonds. There clearly is a need to logically organize these chemicals into groups that can be studied. Furthermore, some hydrocarbons share the same chemical formula and differ only by how their atoms are arranged. It would be helpful to have a foolproof way of identifying each compound by a specific name. In this lesson, you will study the different groups of hydrocarbons, and have an opportunity to learn something about the system used to identify organic compounds.

CONCEPTS

■ classifying hydrocarbons
■ characteristics of the alkane series
■ a systematic way of naming hydrocarbons
■ structural isomers
■ characteristics of alkenes
■ alkynes—hydrocarbons containing triple bonds
■ cyclic aliphatic compounds
■ aromatic hydrocarbons—the nature of the benzene ring
■ processes used in refining petroleum

23-5 Alkanes

The more volatile portion of petroleum can be subjected to fractional distillation, in which substances with different condensation temperatures separate as they are cooled. Some of the products that result from this process are a group of closely related hydrocarbons called alkanes. **Alkanes** are straight- or branched-chain, saturated compounds containing only single bonds. The first eight members of the alkane series are listed in Table 23-1 on the next page.

Notice the relatively low boiling points of the compounds listed in the table. This observation should not come as a complete surprise. Boiling points of liquids depend, in part, on the degree of intermolecular attractions. In Section 23-1, you read that hydrocarbons are generally nonpolar. As a result, there is very limited opportunity for intermolecular attractions to keep these molecules in a liquid state.

On the other hand, the boiling points of compounds listed in the table increase with increasing numbers of atoms. How can this observation be explained? Recall from Chapter 12 (Section 12-13) that even nonpolar molecules are affected by London forces—weak intermolecular attractions resulting from momentary dipoles. The more atoms there are in the molecule, the more opportunity there is for these attractions to have an overall effect.

The simplest of the alkanes is methane gas, CH_4. It is produced during the anerobic decomposition of organic substances (decomposition without oxygen), and is the major constituent of natural gas. The presence of methane in coal mines and swamps supports its association with decaying organic materials. Some biologists believe methane was produced in Earth's primitive atmosphere from reactions with hydrogen and carbon gases. Further reactions involving methane, water, nitrogen, carbon dioxide, carbon monoxide, and hydrogen may have played a vital role in the evolution of life.

The next member of the alkane series is ethane, C_2H_6. The ethane molecule contains one carbon-carbon single bond and six carbon-

Table 23-1

REPRESENTATIVE ALKANES			
CHEMICAL FORMULA	**STRUCTURAL FORMULA**	**NAME**	**BOILING POINT ($^\circ$C)**

CHEMICAL FORMULA	STRUCTURAL FORMULA	NAME	BOILING POINT ($^\circ$C)
CH_4	H—C—H with H above and H below	methane	−162
C_2H_6	H—C—C—H	ethane	−89
C_3H_8	H—C—C—C—H	propane	−42
C_4H_{10}	H—C—C—C—C—H	butane	0
C_5H_{12}	H—C—C—C—C—C—H	pentane	36
C_6H_{14}	H—C—C—C—C—C—C—H	hexane	69
C_7H_{16}	H—C—C—C—C—C—C—C—H	heptane	98
C_8H_{18}	H—C—C—C—C—C—C—C—C—H	octane	126

The term *paraffin* is another name for the alkanes. The term stems from the fact that the alkanes have a low affinity for reaction with many reagents.

The prefix *n-*, as in *n*-pentane, sometimes is used to indicate the structure is straight-chained, or *normal*.

hydrogen bonds. Along with methane, ethane is a constituent of natural gas.

Propane, C_3H_8, also is found in natural gas, and often is used as a fuel in portable stoves, lanterns and low-temperature torches. Propane's molecular structure differs from ethane by the presence of a unit made of one carbon atom and two hydrogen atoms, represented as —CH_2—. Notice in Table 23-1 that as alkane molecules increase in length, each successive compound has an additional —CH_2— group in its structure. A series of compounds, whose members differ by the addition of the same structural unit, is called a **homologous series**.

The alkanes are an example of one homologous series. Study the chemical formulas in the table. Do you see a pattern? The general formula for an alkane is C_nH_{2n+2}, where n is the number of carbon atoms in the molecule. Can you predict the formulas for the nine-carbon and ten-carbon compounds in this series?

23-6 Naming Hydrocarbons

In the early history of organic chemistry, the number of compounds known was limited. Their names came from a variety of sources and have become well established. However, as the number of known compounds increased, there was a need for a more systematic method of naming new ones. The current procedure for naming organic compounds is based on a set of rules formulated by the International Union of Pure and Applied Chemistry (IUPAC).

Unbranched alkanes, like those in Table 23-1, are the simplest to name. Hydrocarbons above butane derive their names from the longest straight-chain of carbon atoms found in the molecule. For these compounds, the name is based on the Greek word for the number of carbons. The prefix *pent*, *hex*, *hept*, or *oct* is followed by the family ending "*ane*." The names for the next two compounds in the alkane series are nonane (nine carbons) and decane (ten carbons).

There are many saturated hydrocarbon compounds that have branched chains—for instance, molecule *A* to the right. Clearly the black portion of this structure is almost the same as heptane, and the colored portion is very much like methane. You could call the molecule methane-heptane, but that is a bit awkward to say. More importantly, it does not distinguish the first molecule from the others in the figure, which also are forms of methane-heptane. According to the IUPAC rules, the name of the smaller portion, or branch, is modified so that *yl* replaces the *ane* ending. (The branch is known as an alkyl group.) The name for molecule *A* then becomes methylheptane. This name is less awkward to say, but there is still a small problem to solve: how to distinguish between molecules *B* and *C*, which are different from molecule *A* and from each other. In molecule *A*, the methyl group is on the third carbon atom of the long chain, so the name of the compound is 3-methylheptane. If you count from the other end of the molecule, the methyl group is on the fifth carbon. However, by convention, the lowest "address" is always used to locate a branch. Molecule *B* is 2-methylheptane and molecule *C* is 4-methylheptane.

The three molecules shown here are examples of **structural isomers**, compounds with the same chemical formula but different arrangements of atoms. In the case of the methylheptanes, all three molecules have the formula C_8H_{18}, but the atoms are arranged differently. Many organic compounds exist as structural isomers. Such isomers differ somewhat in physical properties such as melting points, boiling points, and solubilities in different solvents. Differences in the properties of structural isomers can be observed experimentally. The numbering system described here is a way to distinguish between the compounds by name.

Now consider a slightly more complicated example in which two methyl groups are present. In the molecule shown on the next page, there are seven carbon atoms in the longest unbranched chain, so

Figure 23-3 Natural gas is a commonly used fuel. It consists mostly of methane mixed with other hydrocarbons.

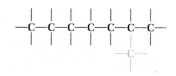

Molecule *A*

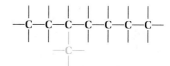

Molecule *B*

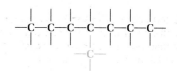

Molecule *C*

Carbon atoms bonded by single covalent bonds are free to rotate on the axis of the bond. The "straight chains" may twist and turn, but they are still the same molecule. The prefix of an alkane's name is based on the longest unbranched chain, which may not be straight. Flexible models will help get this point across.

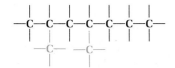

heptane is used as the last part of the name. One methyl group is on the second carbon atom and the other methyl group is on the fourth carbon atom. The name for the molecule could be 2-methyl-4-methylheptane. It is descriptive and accurate, but there is a shorter way of showing the presence of two methyl groups. The prefix *di*, meaning two, is used. The name for the compound is then 2,4-dimethylheptane.

Example 23-1

One compound with the chemical formula C_5H_{12} is pentane. Pentane has structural isomers. For each isomer, draw the structural diagram and determine its precise name.

▶ **Suggested solution**

You already know one possible combination for a carbon skeleton of five carbon atoms—a straight chain. Other possible ways to combine the carbon atoms and have sites available for twelve hydrogens are

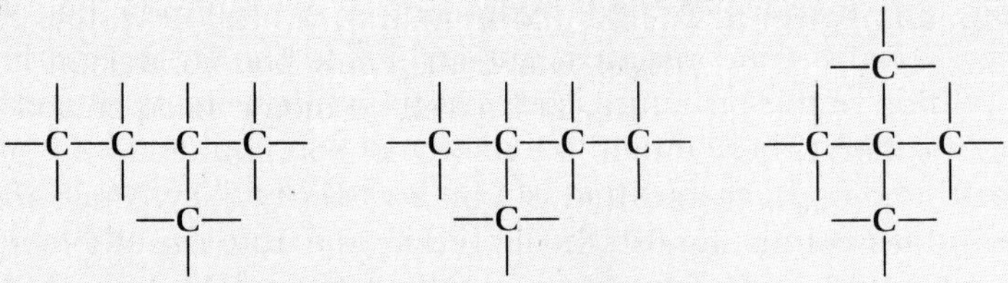

Filling in the missing hydrogens, you get

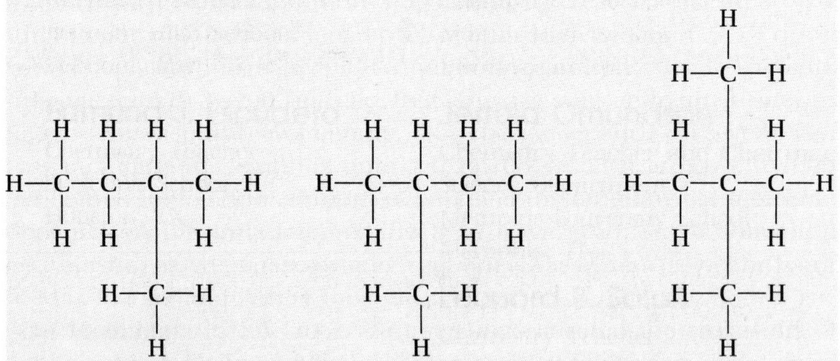

Use three-dimensional models to convince students that 3-methylheptane and 5-methylheptane are the same molecule.

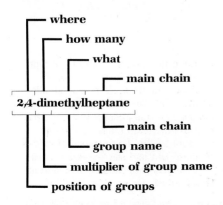

Examine the first two structures. The lowest "address" for the methyl group on both structures is the second carbon (counting from the right on the first molecule and from the left on the second molecule). Since the "address" of the methyl group should always be the lowest possible, the names of these two structures are the same, 2-methylbutane. (The answer is appropriate because the two structures are, in fact, the same molecule.) The third molecule has two methyl groups on the second carbon of a three-carbon skeleton. This molecule is 2,2-dimethylpropane. The diagram at the left summarizes the parts of an IUPAC name for a typical organic compound.

Review and Practice

1. Butane exists in the form of two structural isomers. Draw the structural formula and write the IUPAC name of each.

2. Draw the straight-chain structures for nonane and decane.

3. Match each name in a–d with the correct structure in e–h.

   a. 3-ethyl-2-methylhexane   c. 2,2,4-trimethylhexane
   b. 3-ethyl-4-methylhexane   d. 3-ethylhexane

e.

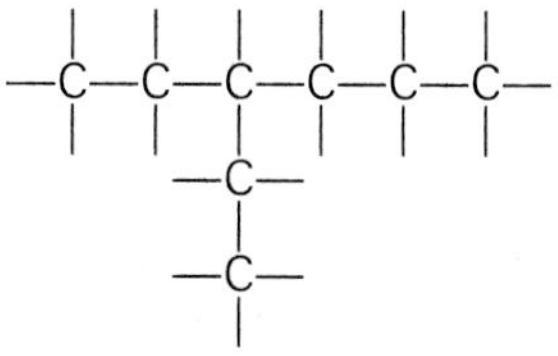

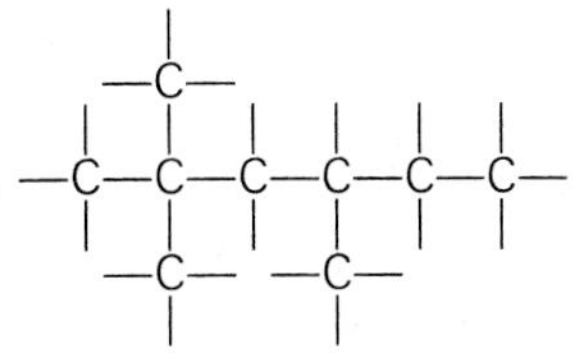

f.

g.

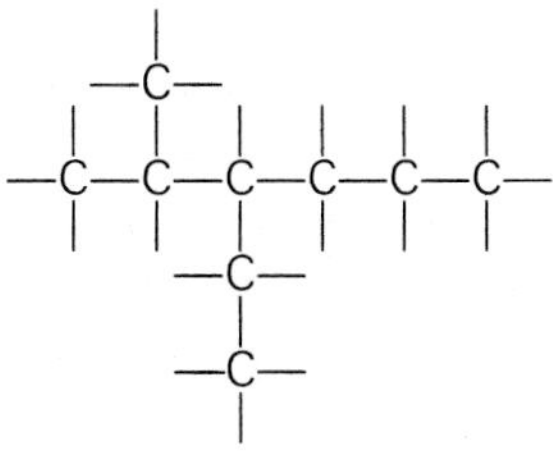

h.

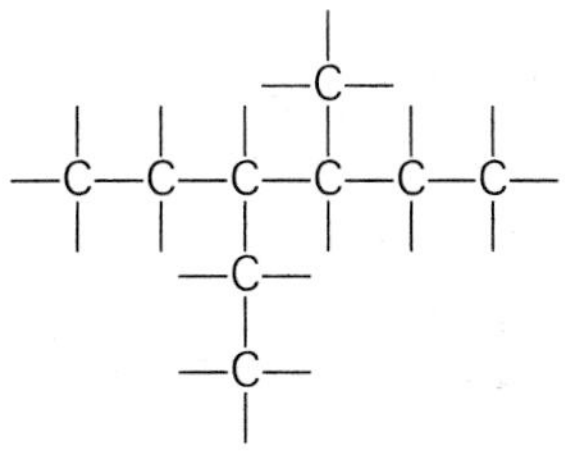

4. Write the names of the following alkanes:

a.

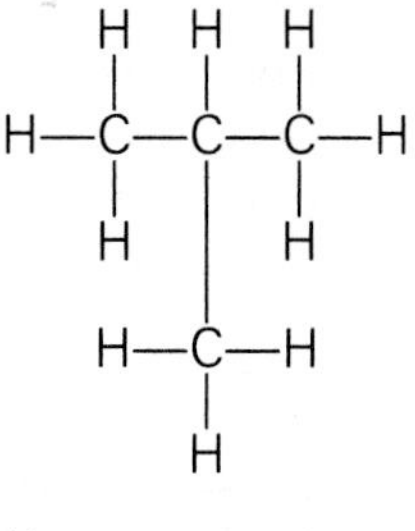

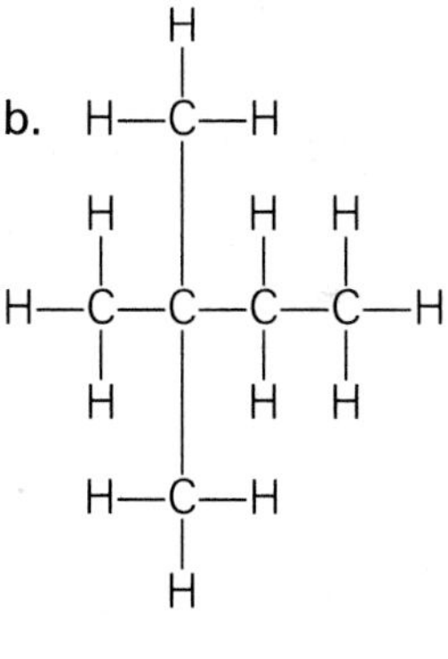

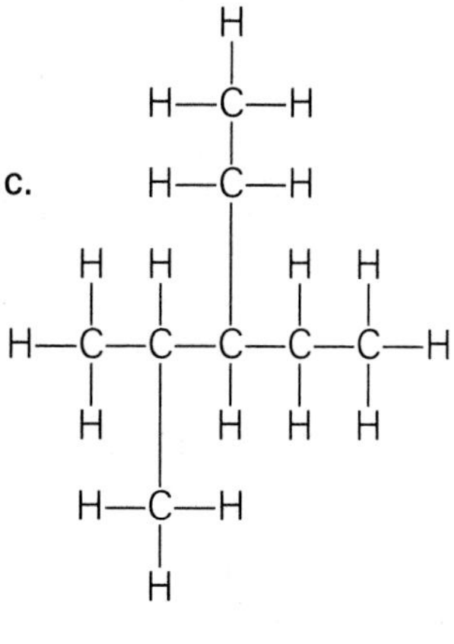

b.

c.

5. How many hydrogen atoms would be in a molecule of an alkane containing 15 carbon atoms? 50 carbon atoms?

Answers (right margin):

1. butane

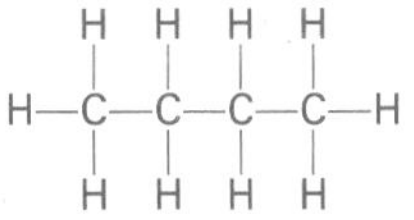

2-methyl propane

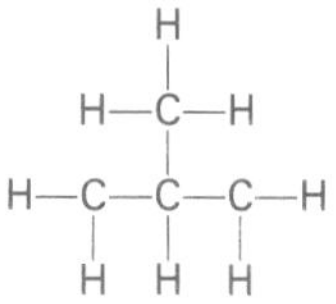

2.

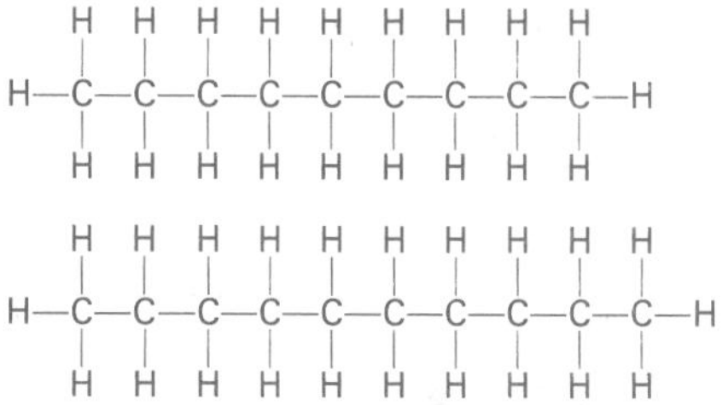

3. a = f; b = h; c = g; d = e
4. a = 2-methylpropane; b = 2,2-dimethylbutane; c = 3-ethyl-2-methylpentane
5. 32 hydrogen atoms; 102 hydrogen atoms

23-7 Alkenes

When petroleum is heated to high temperatures in the presence of certain catalysts, the larger alkanes composing it may be broken into smaller alkanes or they may form new types of compounds. This process, known as **cracking**, is used to produce organic molecules containing double-bonded carbon atoms. Straight- or branch-chained hydrocarbons containing at least one double carbon-carbon bond belong to the **alkene** series. Some examples of alkenes are listed in Table 23-2 on the following page.

Cracking that involves the use of high temperatures is known as *thermal* cracking and produces mainly smaller alkanes and alkenes (especially ethene). *Catalytic* cracking occurs when the hydrocarbon vapors are heated under pressure and in the presence of oxides of aluminum and silicon. This process yields highly branched alkanes and alkenes, which are used in gasolines.

The term *olefin* is applied to the alkenes. It means "oil-forming."

Table 23-2

MEMBERS OF THE ALKENE SERIES OF HYDROCARBONS		
CHEMICAL FORMULA	**STRUCTURAL FORMULA**	**NAME**
C_2H_4		ethene
C_3H_6		propene
C_4H_8		1-butene
C_4H_8		2-butene

Study the chemical formulas of the four compounds listed in the table. There is a pattern similar to that of the alkanes. In this case, the general formula for an alkene is C_nH_{2n}. Note also that the name of each alkene is similar to the alkane having the same number of carbon atoms. However, the ending of the name is changed from *-ane* to *-ene*.

The last two compounds in the table are structural isomers, differing only in the location of the double bond. Since the location of the double bond will influence the chemical behavior of the alkene, its position needs to be identified. Like the naming of a branched (alkyl) group in the alkanes, the double bond is identified by the lowest number of the carbon atom involved in the bond. For example, the following molecule is called 2-pentene. Note the double bond is between the second and third carbon atom.

The carbon atoms in a double bond are not free to rotate, which results in the existence of geometric isomers. For example, the 2-butene in Table 23-2 exists as two geometric isomers— *cis*-2-butene and *trans*-2-butene, where the methyl groups attached to the double-bonded carbons are on the same side and opposite side of the molecule, respectively:

cis-2-butene   *trans*-2-butene

In general, alkenes are more reactive than the corresponding alkanes. Recall from Chapter 12 (Section 12-9) that one bond of a carbon-carbon double bond is more easily broken than the other. As a result, the unsaturated alkenes are more susceptible to chemical reagents than similar alkanes.

Ethene, C_2H_4, is the simplest alkene. This compound (commonly known as *ethylene*) is a gas with a slightly sweet odor. Certain plants naturally produce ethene. It is chemically produced during the refining of hydrocarbons and is found in petroleum and natural gas. Ethene is one of the most important organic compounds in the chemical industry. It is used in the production of such varied products as ethyl alcohol, solvents, plastics, gasoline additives, antifreeze, detergents, and synthetics. It is even used to hasten the ripening of fruit.

23-8 Alkynes

The third series of hydrocarbon compounds contain a triple carbon-carbon bond and are known as **alkynes**. All members of this unsaturated series also are reactive. (Recall from Chapter 12 that very little energy is needed to break one bond of a triple carbon-carbon bond.) Alkynes have the general formula C_nH_{2n-2}. They are named by replacing the *-ane* ending of the parent alkane with the ending *-yne*. Table 23-3 lists a few typical alkynes.

Figure 23-4 An oxyacetylene torch uses a mixture of oxygen and ethyne (acetylene).

Table 23-3

MEMBERS OF THE ALKYNE SERIES OF HYDROCARBONS		
CHEMICAL FORMULA	STRUCTURAL FORMULA	NAME
C_2H_2	H—C≡C—H	ethyne
C_3H_4	H—C≡C—C(H)(H)—H	propyne
C_4H_6	H—C≡C—C(H)—C(H)—H	1-butyne
C_4H_6	H—C(H)(H)—C≡C—C(H)—H	2-butyne

As with the alkenes, the location of the triple bond affects the properties of the compound, and so it is identified by assigning it the lowest number of the carbon involved in the bond. Procedures for naming the branches on alkyne molecules are the same for alkenes and alkanes. Note that the last two compounds in the table are structural isomers.

The simplest and most common alkyne is ethyne, C_2H_2. More commonly known as *acetylene*, this explosive gas is an important industrial hydrocarbon. Its frequent use in oxyacetylene torches produces temperatures sufficient to cut and weld steel.

Caution students not to confuse the *-ene* ending of acetylene with the alkenes. Acetylene is not an IUPAC name, but it is more commonly used.

23-9 Cycloalkanes

In Section 23-7, you learned that the general formula for an alkene is C_nH_{2n}. Consider the formula C_5H_{10}. This compound could be pentene. But suppose you did not know anything about alkenes. Is there a

In cyclic molecules, numbers are assigned to the ring carbon atoms in a clockwise manner so that the lowest "address" is used for branch positions:

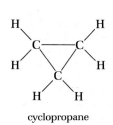

1,3-dichlorohexane

structural isomer for this formula that does not contain a double bond? If you try a few arrangements, you may discover that the following one works:

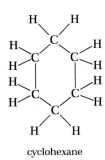

This substance is an example of a **cycloalkane**, a saturated hydrocarbon in the form of a ring. Cycloalkanes include rings made of three, four, five, or more carbon atoms, and they are named by using the prefix *cyclo-* in front of the appropriate alkane name. The structure above is cyclopentane. Others include

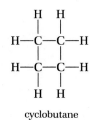

cyclopropane

cyclobutane

cyclohexane

cycloheptane

The structural formulas of cycloalkanes tend to be cumbersome when all the hydrogen atoms are included, so these compounds are often represented by the geometric shape of the basic ring only. Each point on the structure is understood to represent a carbon atom bonded to two hydrogen atoms:

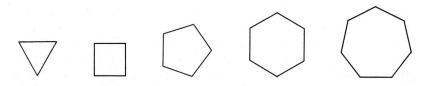

Do You Know?

Gasoline fuels, obtained by the simple distillation of petroleum, contain various volatile hydrocarbon compounds. Occasionally these fuels may explode prematurely, producing engine "knock." To reduce this problem, branched-chain alkanes, alkenes, and aromatic hydrocarbons are added to the fuel.

Three- and four-carbon ring structures tend to be less stable than their straight-chain counterparts because there is a strain put on the bond angles in the molecule. Recall from Chapter 12 that the bond angles for any one carbon sharing four covalent bonds is 109.5°. In order to form a ring, this angle is altered to a smaller size between carbon atoms (60° in cyclopropane and 90° in cyclobutane). As a result, there is a tendency for these compounds to undergo chemical reactions that will relieve the strain by opening the ring and forming straight-chain compounds.

Cyclic structures also exist among the alkenes and to a lesser degree among the alkynes. Two common cycloalkenes are shown at the left. Unsaturated cyclic hydrocarbons are frequently more reactive than their saturated or straight-chain counterparts and some are unstable.

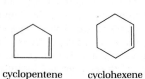

cyclopentene   cyclohexene

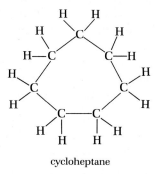

23-10 Aromatic Hydrocarbons

Another important class of organic compounds is the aromatic hydrocarbons. The name "aromatic" was originally used to describe these compounds because most of them possess rather distinctive fragrances or aromas. Aromatic hydrocarbons were previously obtained from the distillation of coal tar, which itself is a byproduct of the preparation of coke from coal. Some aromatic compounds are still derived this way, but the most important one, benzene, is obtained chiefly during the refining of petroleum.

Benzene, C_6H_6, is the simplest of the aromatics. All of these compounds have a basic structure of one or more rings made of six carbon atoms.

Experimental evidence shows that the atoms of the benzene molecule lie in a plane, with 120° angles between pairs of bonds formed by a given carbon atom. The carbon-hydrogen bond distance is consistent with the carbon-hydrogen bond distance found for hydrocarbons, so it is reasonable to conclude that each carbon atom and its neighboring hydrogen are bonded by a single bond.

The carbon-carbon bond is unusual, however. Experimental data show that the distance between two carbon atoms in a benzene ring is less than a carbon-carbon single bond in ethane, but greater than a carbon-carbon double bond in ethene:

C—C	in ethane	1.54×10^{-9} nm	known single bond
C ? C	in benzene	1.40×10^{-9} nm	
C═C	in ethene	1.34×10^{-9} nm	known double bond

A comparison of the bond distances suggests that the carbon-carbon bonds in benzene are somewhere between single and double bonds. The chemical behavior of ethane, benzene, and ethene leads to the same conclusion. Benzene is more reactive than ethane, but it reacts to a lesser degree and differently than ethene. In addition, all six carbon-carbon bonds are found to be the same in length and in chemical reactivity. These observations lead to the conclusion that the electrons of benzene are somehow shared equally around the ring. The electrons are said to be **delocalized** because they are not associated with any one carbon atom. To reflect this idea, the structure of the benzene ring can be represented as shown in Figure 23-5. The molecule is described as having a region of electron density above and below the plane of the ring. For convenience, chemists frequently draw the benzene ring as

The circle represents the delocalized electrons of the double bonds. The symbol serves as a reminder that the carbon-carbon bonds in benzene are different from either single or double bonds.

Benzene is the starting material for thousands of compounds. Combinations of other atoms besides carbon and hydrogen can be attached to the carbons of a benzene ring to create organic chemicals of incredibly varied characteristics. Benzene rings even may fuse together to form aromatic compounds of increasing complexity that in turn are used for raw materials in other processes. You come in contact with the products of aromatic compounds every day. Benzene

Molecular orbital theory, which is the theory of bonding that leads to the illustration in Figure 23-5, is not discussed in this book. If you introduced the theory yourself when teaching Chapter 12, you can return to the idea here. An alternative is to show the students the resonance structures of benzene:

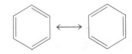

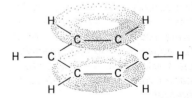

Figure 23-5 Bonding theories suggest the delocalized electrons of the benzene ring are shared equally by all six carbon atoms.

Students should be reminded that the alternating single and double bonds in these diagrams are misleading and that in the actual molecule, all bonds in the ring are alike. In spite of this oversimplification, these structural formulas lead to correct predictions for much of the chemistry of benzene and other aromatic compounds.

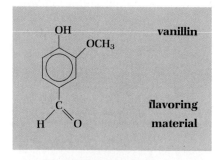

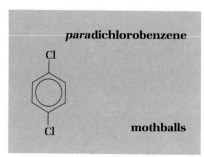

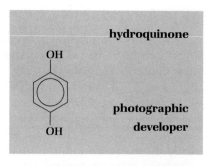

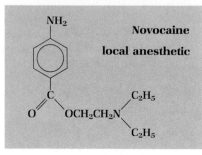

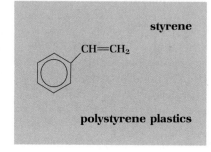

Figure 23-6 Here are just a few examples of familiar substances that are benzene derivatives.

and its derivatives are used to produce plastics, synthetic fibers, dyes, medicines, anesthetics, synthetic rubber, food additives, paints, and explosives—to name just a few.

23-11 Extension: Refining Petroleum

Before the invention of the automobile, gasoline was a useless by-product of the refining of petroleum. Refining occurred mostly to obtain kerosene, or lamp oil, and some fuel oils. With the advent of electric power and the introduction of automobiles, the demand for petroleum products shifted. New methods were developed to refine petroleum in order to extract more and better quality gasoline. In addition, the science of synthetic chemistry created a need for new kinds of organic raw materials. Petroleum continues to be the primary source for organic compounds used in the chemical industry.

Petroleum (in the form of crude oil) is a mixture of hydrocarbons that must be separated before they can be put to use. The main components of petroleum are alkanes and cycloalkanes. Other substances include oxygen, sulfur, nitrogen, sodium chloride, and some metallic elements. Separation of the hydrocarbons is the first step in the refining process. Impurities are removed at later stages.

Figure 23-8 is a diagram that outlines the fractional distillation of petroleum. Crude oil is heated in a furnace to a temperature between 320° and 370°C, until a large portion vaporizes. The material is then moved into the fractionating tower, where the vapors rise, cool, and subsequently condense at different levels. The lower the condensation (boiling) points of the compounds, the higher they rise in the tower. Materials are drawn off the tower at different heights corresponding to different condensation temperatures. This process does not separate the petroleum into individual substances, but rather into groups of

Figure 23-7 Oil refineries operate around the clock. Depending on their size, refineries may distill tens of millions of liters of oil per day.

Metallic elements found in crude oil include vanadium, nickel, and most of the metallic elements found in seawater.

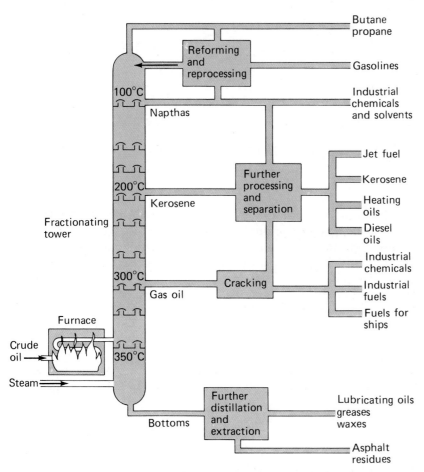

Figure 23-8 The demand for petro-leum-based products increases stead-ily. Almost every portion of crude oil has some usefulness.

While discussing the process of oil re-fining, you also can bring up the prob-lem of limited reserves of petroleum and the implications of the rate at which it is being used. This topic may be used for a class discussion or be assigned to students to research.

1. Major groups of hydrocarbons have differ-ent ranges of condensation temperatures.
2. C_nH_{2n}
3.

(The structures above are the *cis* isomers. Students may draw the *trans* isomers also.)

4. Cyclohexane has 12 atoms and cyclohex-ene has 10.

5.

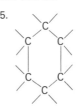

Some possible answers include

Some isomers the students draw may exist only on paper.
6. The bond distances and bond energies suggest that the carbon-carbon bonds in ben-zene are intermediate between the single and double bonds, whereas alkanes have all single bonds and alkenes contain bonds that are ei-ther single or double.

products whose condensation temperatures fall into a specific range. Some typical condensation ranges are shown in the diagram. The groups are then further processed. Long-chain alkanes are subjected to cracking in order to produce better quality gasolines and important alkenes like ethene. Other groups of materials are subjected to further refining, separation, and purification.

Review and Practice

1. Why is it possible to use fractional distillation to separate hydrocarbons in petroleum?

2. Write the general formula for an alkene.

3. Draw the structural formulas for the following alkenes: 2-pentene, 3-hexene, 4-dectene, cyclobutene.

4. How many hydrogen atoms are found in cyclohexane? In cyclohexene?

5. The formula C_6H_{12} can exist as several straight-chain and branched-chain structural isomers and as a cyclic structural isomer. Draw the structural formula for the cyclic compound and one each of the straight-chained and branch-chained isomers.

6. What evidence suggests the carbon-carbon bonds in benzene are different from bonds in alkanes and alkenes?

Reactions of Organic Compounds

''Organic chemistry nowadays almost drives me mad. To me it appears like a primeval tropical forest full of the most remarkable things, a dreadful endless jungle into which one does not dare enter for there seems to be no way out.''

—Friedrich Wöhler, 1835

When Wöhler made this statement, he had no notion of the vast number of carbon compounds yet to be synthesized. As chemists learned how to manipulate the reactions of compounds, the science of organic chemistry boomed. In addition to carbon and hydrogen, organic compounds can contain other elements whose atoms may be arranged in a variety of bonding combinations. Even the slightest difference in composition or structural arrangement may result in two compounds of very different chemical and physical properties. Through a process called polymerization, small molecules are linked together to form chains. The result is a class of giant molecules used to manufacture everything from videocassettes to parachutes to nonstick cookware. In this part of the chapter, you will have an opportunity to learn about several groups of organic compounds besides the hydrocarbons and find out what kinds of reactions these compounds can undergo.

CONCEPTS

- identifying functional groups in organic molecules
- esterification
- saponification
- substituting atoms in hydrocarbons
- addition of atoms across an unsaturated bond
- oxidation—converting one organic compound to another
- polymerization—building giant molecules

23-12 Functional Groups

The characteristics of organic molecules depend on their composition and their arrangement of atoms. For example, the properties of the alkenes depend to a large extent on the presence of the double bond. The position of the double bond also is responsible for certain characteristic properties. Any atom, group of atoms, or organization of bonds that determines specific properties of a molecule is known as a **functional group**. The functional group generally is the most reactive portion of a molecule, and its presence signifies certain predictable characteristics. It is possible for a molecule to have more than one functional group.

The double bond in the alkenes and the triple bond in the alkynes are functional groups. A functional group also may be an atom or group of atoms that is attached to a carbon atom in place of a hydrogen. The functional groups that are most commonly encountered often contain oxygen, nitrogen, or both. A functional group can even be a single halogen atom. (Sulfur and phosphorus also are components of functional groups.) Table 23-4 lists examples of some of the functional groups found in organic compounds. In each structural formula, **R—** represents the rest of the molecule to which the structural group is attached. R— may be a single carbon or a larger hydrocarbon portion of the molecule.

Table 23-4

COMMON FUNCTIONAL GROUPS FOUND IN ORGANIC COMPOUNDS

GENERAL STRUCTURE	GROUP NAME	EXAMPLE	
R—O—H	alcohol	$H-\overset{\displaystyle H}{\underset{\displaystyle H}{C}}-\overset{\displaystyle H}{\underset{\displaystyle H}{C}}-O-H$	ethanol
R—O—R′*	ether	$H-\overset{\displaystyle H}{\underset{\displaystyle H}{C}}-\overset{\displaystyle H}{\underset{\displaystyle H}{C}}-O-\overset{\displaystyle H}{\underset{\displaystyle H}{C}}-\overset{\displaystyle H}{\underset{\displaystyle H}{C}}-H$	ethoxyethane
$R-C{\overset{\displaystyle O}{\underset{\displaystyle H}{}}}$ (aldehyde, C=O and C—H)	aldehyde	$H-C{\overset{\displaystyle O}{\underset{\displaystyle H}{}}}$	methanal
$R-\overset{\displaystyle O}{\overset{\|}{C}}-R'$	ketone	$H-\overset{\displaystyle H}{\underset{\displaystyle H}{C}}-\overset{\displaystyle O}{\overset{\|}{C}}-\overset{\displaystyle H}{\underset{\displaystyle H}{C}}-H$	propanone
$R-C{\overset{\displaystyle O}{\underset{\displaystyle OH}{}}}$	acid	$H-\overset{\displaystyle H}{\underset{\displaystyle H}{C}}-C{\overset{\displaystyle O}{\underset{\displaystyle OH}{}}}$	ethanoic acid
$R-O-\overset{\displaystyle O}{\overset{\|}{C}}-R'$	ester	$H-\overset{\displaystyle H}{\underset{\displaystyle H}{C}}-\overset{\displaystyle H}{\underset{\displaystyle H}{C}}-O-\overset{\displaystyle O}{\overset{\|}{C}}-\overset{\displaystyle H}{\underset{\displaystyle H}{C}}-H$	ethyl acetate
$R-N{\overset{\displaystyle H}{\underset{\displaystyle H}{}}}$	amine	$H-\overset{\displaystyle H}{\underset{\displaystyle H}{C}}-N{\overset{\displaystyle H}{\underset{\displaystyle H}{}}}$	methylamine

*R and R′ symbolize the general hydrocarbon group of the molecule. They may be the same group or different groups.

Organic compounds that have the same functional group will behave similarly in chemical reactions. Some general characteristics and uses of the groups are summarized here.

Alcohols contain the functional group —OH. This group, known as the **hydroxyl** group, may be bonded to a carbon atom of a chain or cyclic hydrocarbon. Alcohols are named by replacing the *-e* ending of the parent alkane with *-ol*. Because of the significant difference in the electronegativities of oxygen and hydrogen in the O—H bond, the hydroxyl group is polar. From your studies in Chapters 12 and 16, you know that a polar structure such as —OH can be involved in hydrogen bonding. Predictably, alcohols of low molecular mass are more soluble in water than their corresponding alkane.

Methanol, CH_3OH, is the simplest alcohol. It is commonly known as methyl alcohol or wood alcohol, and is used in the synthesis of plas-

As the molar mass of the molecule increases, the effect of the —OH group is counterbalanced by the increasing portion that is nonpolar. Alcohols of larger molecular mass do not dissolve well in water.

Use three-dimensional models to convince students that there is only one structural form for methanol and for ethanol, whereas propanol exists as two isomers.

Students may find it helpful if you spend time giving them other examples of condensed formulas. Doing so may reinforce the convenience of using the formulas to represent the structure of organic molecules.

Denatured ethanol, for nonbeverage uses, has been treated with toxic additives (such as methanol or benzene) to make it undrinkable.

Alcohols in which the hydroxyl is attached to a carbon atom bonded only to one other carbon atom are known as *primary* alcohols. If the carbon atom has two other or three other carbon atoms attached, the molecule is a *secondary* or *tertiary* alcohol, respectively.

Do You Know?

Methanol is the main constituent of "dry gas," an additive for gasolines that helps to prevent frozen fuel lines in automobiles during very cold weather. Mixtures of methanol and ethanol make up gasohol, a fuel that is added to gasoline to extend the mileage per gallon.

Students may find it helpful to compare the structure of a simple ketone (2-propanone) with an ether (methoxyethane). They should recognize that the oxygen is double-bonded to one carbon in a ketone, but single-bonded between two carbons in an ether.

tics and fibers. Ethanol, CH_3CH_2OH* (known also as ethyl alcohol) has been produced by human societies since ancient times from the fermentation of fruits and grains. (Another name for ethanol is grain alcohol.) In addition to its presence in alcoholic beverages, ethanol is used in industrial processes as the starting material for other compounds and in the preparation of medicines. To meet the incredible demand for ethanol, it is commercially produced by a reaction between ethene gas and water (Section 23-14).

Alcohol molecules can have more than one hydroxyl group attached. The most important example of this type of compound is 1,2-ethanediol (more commonly known as ethylene glycol).

$$\begin{array}{ccc} & H & H \\ & | & | \\ H - & C - & C - H \\ & | & | \\ & OH & OH \end{array}$$

Ethylene glycol is used as antifreeze in automobiles and is one of the starting materials for the synthesis of polyester fabrics.

Ethers have an oxygen atom bonded between two hydrocarbon portions of the molecule. There is much less opportunity for a significant amount of polarity in the molecule, and as a result, ethers demonstrate little hydrogen bonding and have relatively low boiling points. The substance commonly called "ether" is ethoxyethane (also known as diethyl ether). Its structure is shown in Table 23-4. Diethyl ether is a volatile, highly flammable liquid that was once used extensively as an anesthetic. Its present role as an anesthetic is limited, but it continues to be used as a solvent for oils and fats.

Aldehydes and ketones both contain the **carbonyl** functional group:

$$\begin{array}{c} O \\ \| \\ -C- \end{array}$$

In aldehydes, the carbonyl group is attached to a carbon atom that has at least one hydrogen atom attached to it. The chemical formula for an aldehyde usually is written as RCHO. In ketones, the carbonyl group is attached to a carbon atom that is attached to two other carbon atoms. The chemical formula for a ketone usually is written as RCOR'.

Aldehydes are named by replacing the *-e* ending of the parent alkane with *-al*. The simplest aldehyde is methanal, HCHO, which commonly is known as formaldehyde. It has a sharp, penetrating odor and was once frequently used to preserve biological specimens. Its primary uses now are in the industrial preparation of plastics and resins. Aldehydes of larger molar mass have more pleasant odors. Some common flavorings, such as vanillin, oil of cinnamon, and oil of almonds, fall into this category.

Ketones are named by replacing the *-e* ending of the parent alkane by the ending *-one*. The simplest ketone is propanone, CH_3COCH_3, known better by its common name, acetone. It is a volatile liquid, used

---

*This type of chemical formula is a condensed version of the structural formula. It is a clearer way of distinguishing the isomers of the chemical formula C_2H_6O, without having to draw the entire structure.

extensively as a solvent for paints and lacquers, but most popularly used to dissolve nail polish.

Organic acids (or carboxylic acids) contain the **carboxyl** functional group, —COOH, which makes the molecules polar. These compounds also are weak acids due to the dissociation of the hydrogen atom from the carboxyl group. Organic acids are named by replacing the *-e* ending of the parent alkane with *-oic acid*. Many organic acids have common names, because they have been well known since before the establishment of a naming system. You probably have come in contact with a variety of organic acids, because some of them are components of common foods. Ethanoic acid, usually called acetic acid, is particularly well known. In diluted form (five percent solution with water), it is called vinegar and is produced by bacterial action on ethanol. Other organic acids are found in sour milk, yogurt, and a variety of fruits.

Have students compare the carbonyl group from aldehydes and ketones with the carboxyl group of an acid. They should be able to recognize the differences.

The extent of polarity of a carboxylic acid, like an alcohol, depends on the size of the nonfunctional part of the molecule.

Figure 23-9 Citrus fruits get their name from the citric acid they contain. Other foods contain different carboxylic acids.

Esters are produced from a reaction between organic acids and alcohols. The process, known as **esterification**, is a reversible dehydration reaction in which the alcohol loses an —OH group and the acid loses the hydrogen atom from its carboxyl group. Water is generated as a by-product.

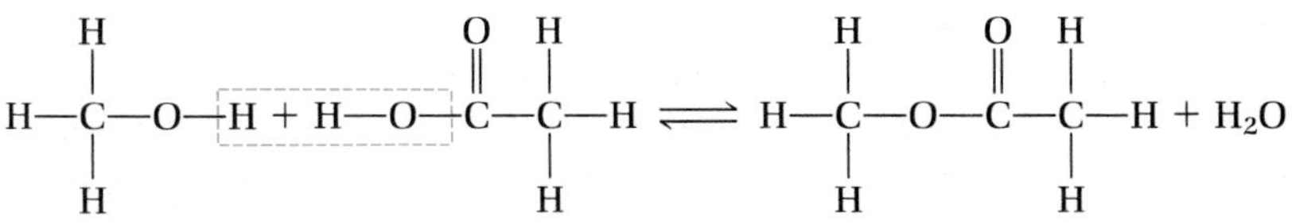

Esterification is an acid-catalyzed, equilibrium process. Saponification is the complete hydrolysis of an ester.

Most esters are better known by their common names. The IUPAC system is not discussed here. You may want to explain to students that the name of the ester is derived from changing the name of the acid from *-oic* to *-oate*, preceded by the name of the alcohol group. For example, methyl acetate is methyl ethanoate.

$$\begin{array}{c}\\[-0.3em]\end{array}$$

Most esters possess distinctive aromas and flavors. Fragrant foods, such as fruits, are a natural source of these compounds. Synthetic esters of low molar mass are often used as perfume additives and in

Table 23-5

	REPRESENTATIVE ESTERS	
AROMA	**FORMULA**	**ESTER**
Rum	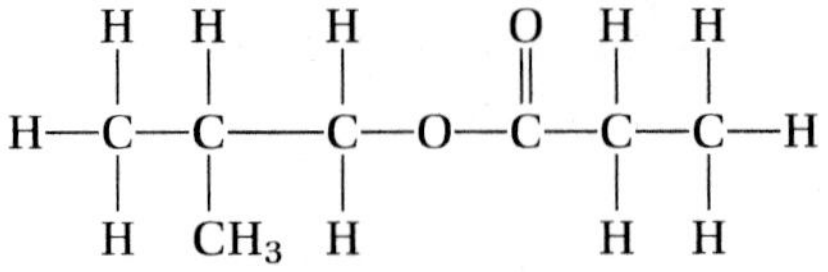	isobutyl propionate
Delicious apple	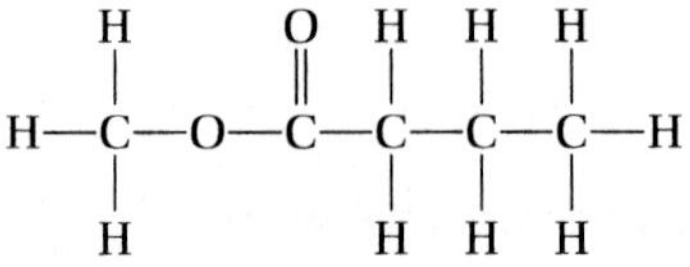	methyl butyrate
Pineapple	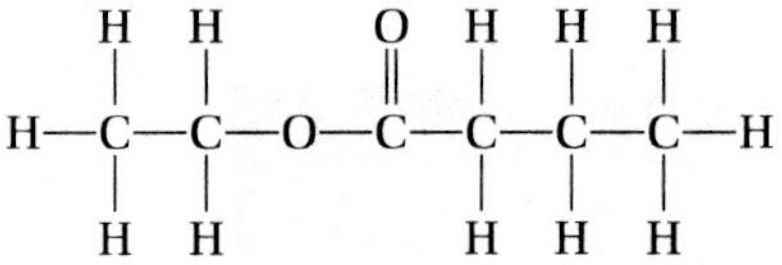	ethyl butyrate
Banana	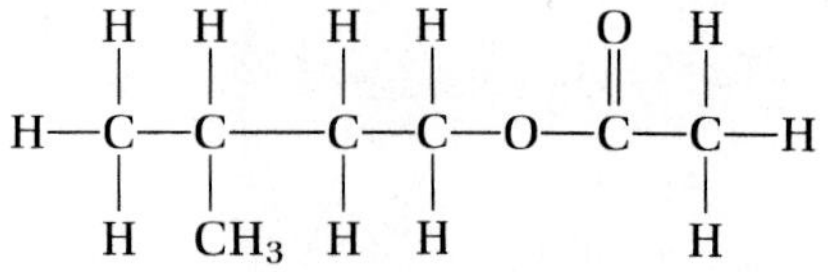	3-methyl butylacetate
Orange	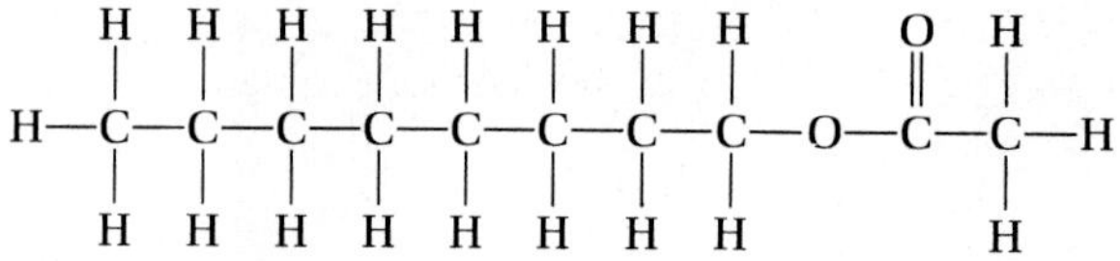	octyl acetate
Apricot	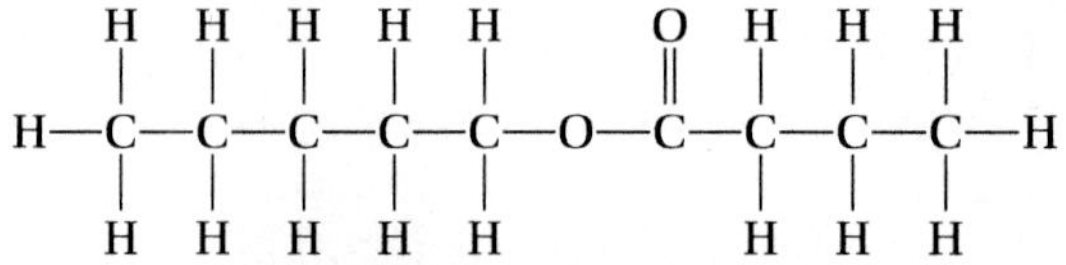	pentyl butyrate

artificial flavorings. Table 23-5 is a list of some common esters and their characteristic aromas.

Amines are organic compounds closely related to ammonia. It is possible for organic R— groups to replace one, two, or three hydrogens in the ammonia molecule to produce molecules like those illustrated here:

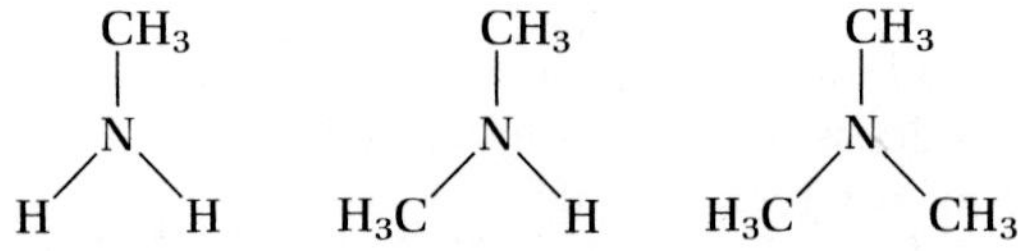

methylamine   dimethylamine   trimethylamine

Primary, secondary, and tertiary amines are distinguished by how many carbon atoms are attached to the nitrogen of the amino group.

If an amine or ammonia is treated with an organic acid, an important group of biological compounds called **amides** is produced. Amides contain a carbonyl group bonded to the nitrogen atom of an amine:

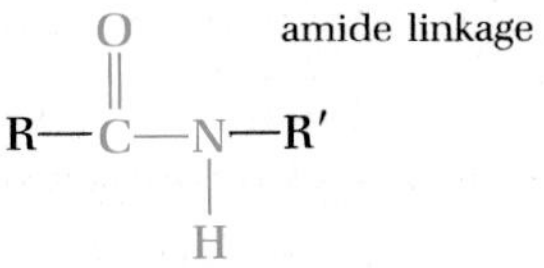

amide linkage

Do You Know?

If your eyes tear when you peel onions, this molecule is the culprit:

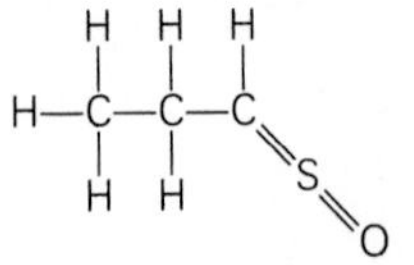

It is water soluble, so cut and peel onions under water. No tears.

Perhaps the most important property of the amides is their ability to form unusually strong intermolecular bonds. These bonds are responsible for the naturally occurring linkages between amino* acids in protein molecules. (You will read more about these bonds later in this chapter.) Amides also are important because they provide links between other molecular units (besides amino acids) to form larger molecules. Nylon is a product of this kind of linkage and will be discussed in Section 23-16.

Review and Practice

1. What features of an organic molecule influence its chemical behavior?

2. What is a functional group?

3. Write the name of the functional group contained in each of the four following molecules:

 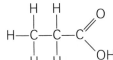 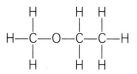

4. Why do ethers have low boiling temperatures?

5. From the names of the compounds listed, identify the functional group each molecule contains:

  a. propanol   **d.** methoxymethane
  b. ethylamine   **e.** ethanal
  c. butanone

6. Draw the structural formula for the ester made from ethanol and acetic (ethanoic) acid.

1. its number of atoms, structural arrangement, and number and type of functional groups

2. atoms, groups of atoms, or organization of bonds that determine a specific property

3. aldehyde; organic or carboxylic acid (carboxyl group); ketone (carbonyl group); ether

4. The central position of the oxygen atom leads to less of an opportunity for significant polarity. Intermolecular attractions are correspondingly low.

5. a. hydroxyl   d. oxygen
   b. amine   e. carbonyl
   c. carbonyl

6.
```
    H  H     O  H
    |  |     ‖  |
H—C—C—O—C—C—H
    |  |        |
    H  H        H
```

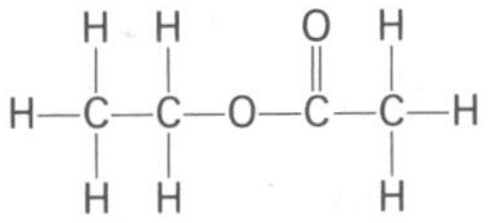

23-13 Substitution and Addition

Methane, the simplest hydrocarbon, can be treated with chlorine gas in the presence of heat or high-energy light waves. The reaction may be represented in the equation

$$CH_4 + Cl_2 \longrightarrow CH_3Cl + HCl$$

or

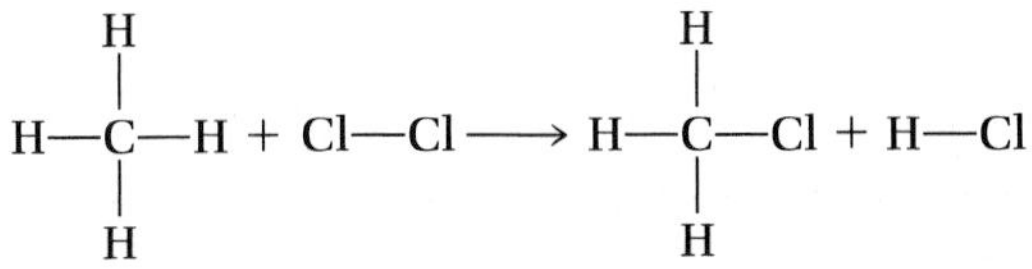

```
    H                    H
    |                    |
H—C—H + Cl—Cl ⟶ H—C—Cl + H—Cl
    |                    |
    H                    H
```

A chlorine atom has taken the place of one hydrogen atom in the methane molecule. The new organic compound that results from this reaction is an alkyl halide called chloromethane. This reaction, where a hydrogen atom is replaced by another atom or group of atoms, is

---

*The —NH$_2$ group on a carbon is referred to as an amino group; the molecule is called an amine.

The chlorination of methane leads to the production of all of these compounds simultaneously.

called **substitution**. Substitution of chlorine for additional hydrogen atoms in the methane molecule can result in three other compounds:

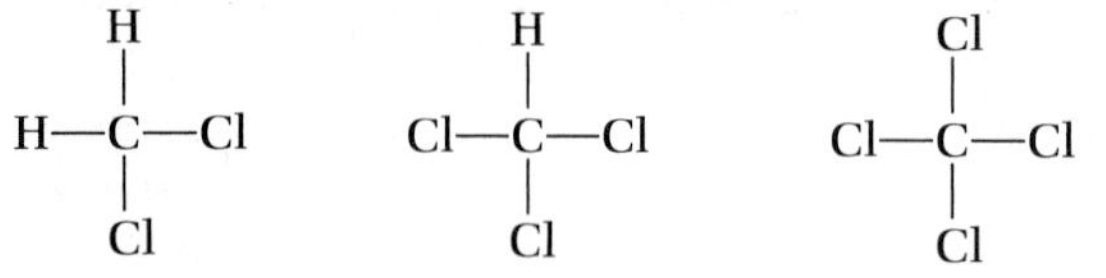

H—C—Cl with H above and Cl below (dichloromethane); Cl—C—Cl with H above and Cl below (trichloromethane); Cl—C—Cl with Cl above and Cl below (tetrachloromethane)

dichloromethane trichloromethane tetrachloromethane

Each of these compounds has unique chemical and physical properties. Chloromethane, CH_3Cl, is used as a refrigerant. Dichloromethane, CH_2Cl_2, is used as a solvent, a spray-can propellant, and most recently in a controversial method (due to possible health hazards) for extracting caffeine from coffee. Trichloromethane, $CHCl_3$, is better known as chloroform. Although it is seldom used today as an anesthetic, it remains an important organic solvent. The last product of chlorine substitution, CCl_4, is a grease solvent more commonly known as carbon tetrachloride.

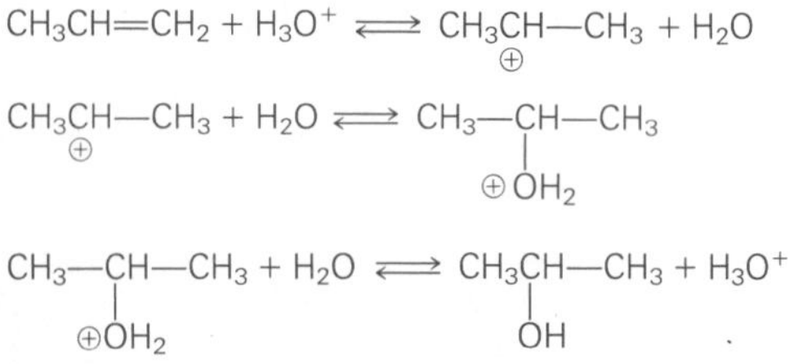

Some of the functional groups you learned about in the previous section can be synthesized by a reaction similar to substitution, called displacement. For example, an alkyl halide, such as chloromethane, can be treated with ammonia to produce methylamine.

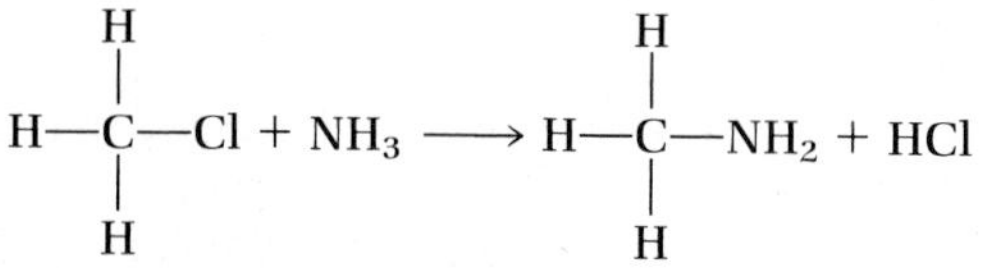

$$H\text{—}C\text{—}Cl + NH_3 \longrightarrow H\text{—}C\text{—}NH_2 + HCl$$

Displacement reactions are limited by the reactivity of the functional group and the group to be added.

Another reaction of organic molecules is addition. In an **addition** reaction, a carbon compound containing one or more double (or triple) bonds is reacted with another substance in order to open the double bond and add new constituents to the carbon atoms involved. For example, the demand for ethanol is much greater than what could be supplied through natural fermentation. To meet the demand, ethanol is synthesized in an addition reaction between ethene and water:

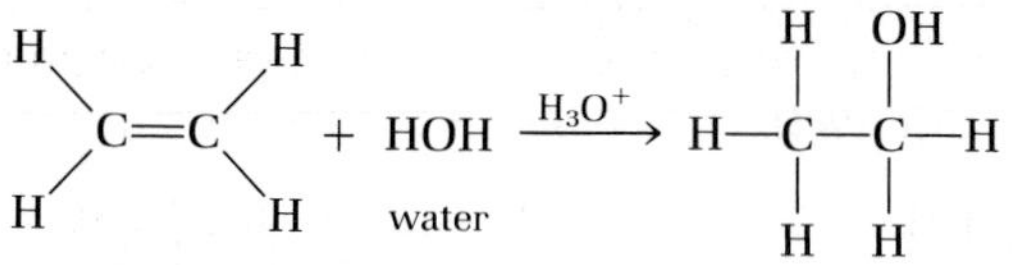

The addition of water to the double bond occurs in steps involving an electrophilic attack on carbon. In a slightly acidic solution, hydronium ions, H_3O^+, form. A proton from the hydronium ion is transferred to one of the carbon atoms in the double bond. When the carbon-hydrogen bond forms, the second carbon atom takes on a positive charge. The neutral OH_2 attaches to this carbon. Subsequently, a proton leaves this complex and attaches to an available water molecule forming another hydronium ion.

$$CH_3CH{=}CH_2 + H_3O^+ \rightleftharpoons CH_3CH\overset{\oplus}{\text{—}}CH_3 + H_2O$$

$$CH_3CH\underset{\oplus}{\text{—}}CH_3 + H_2O \rightleftharpoons CH_3\text{—}CH\text{—}CH_3 \;\; \underset{\oplus OH_2}{}$$

$$CH_3\text{—}CH\text{—}CH_3 + H_2O \rightleftharpoons CH_3CH\text{—}CH_3 + H_3O^+ \atop \oplus OH_2 \qquad\qquad OH$$

$$\underset{\text{ethene}}{H_2C{=}CH_2} + HOH \xrightarrow{H_3O^+} H\text{—}\underset{\text{ethanol}}{C\text{—}C}\text{—}H$$

ethene   water   ethanol

In the overall reaction (which is acid-catalyzed), the hydrogen of water is added to one of the carbon atoms and the —OH portion is added to the other carbon. A single carbon-carbon bond remains.

An addition reaction also may be used to synthesize an organic halide. For example, hydrochloric acid can be added across the double bond of ethene to produce chloroethane:

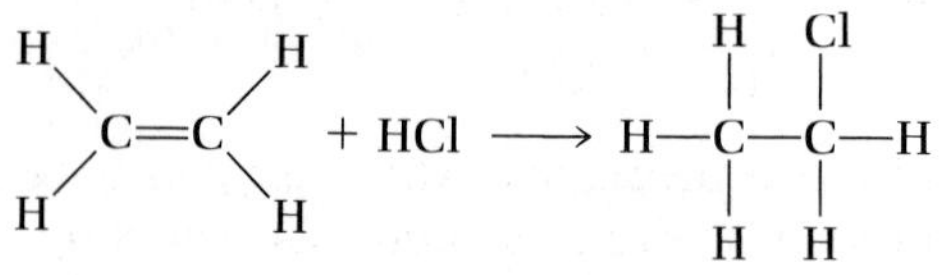

$$H_2C{=}CH_2 + HCl \longrightarrow H\text{—}C\text{—}C\text{—}H$$

Addition reactions do not easily occur across the double bonds of aromatic rings because their electronic structure is very stable. However, substitution of hydrogen atoms attached to an aromatic ring is common.

23-14 Oxidation and Reduction

Many organic compounds can be converted to other compounds through an oxidation reaction. The reaction you are probably most familiar with is the combustion of alkanes to produce carbon dioxide and water. (See Section 5-7.) For example, the combustion of methane, the main constituent of natural gas, may be burned as described by the equation

$$CH_4 + 2O_2 \longrightarrow CO_2 + 2H_2O$$

In organic chemistry, *oxidation* refers to the addition of oxygen to a molecule or the removal of hydrogen from a molecule. This meaning is consistent with what you learned in Chapter 21, where oxidation was defined as an increase in oxidation number. For example, an alcohol may be oxidized to an aldehyde:

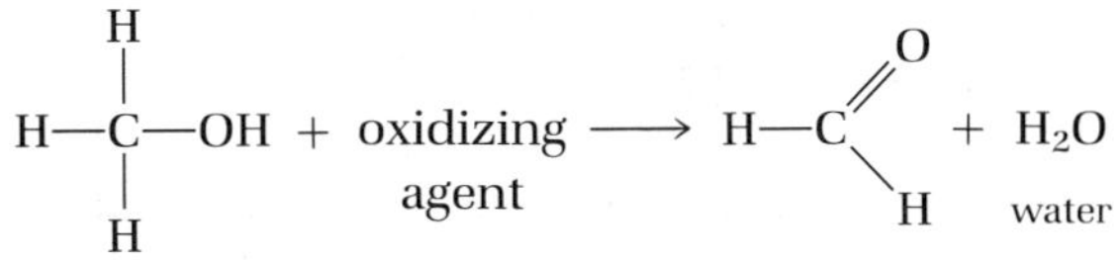

methanol → methanal + water

The oxidizing agent may be a substance such as copper at high temperatures, copper oxide, or potassium dichromate.

In the alcohol, the oxidation number of carbon is 2−. In the aldehyde, carbon has an oxidation number of zero. (Recall the rules you learned in Chapter 21 for finding oxidation numbers. Oxygen is usually 2− and hydrogen is 1+. The total for the molecule should be zero.) The oxidation number of carbon has increased, so carbon is said to be oxidized. (Note that hydrogen has been removed.)

Depending on the location of the —OH group, a ketone may form instead of an aldehyde:

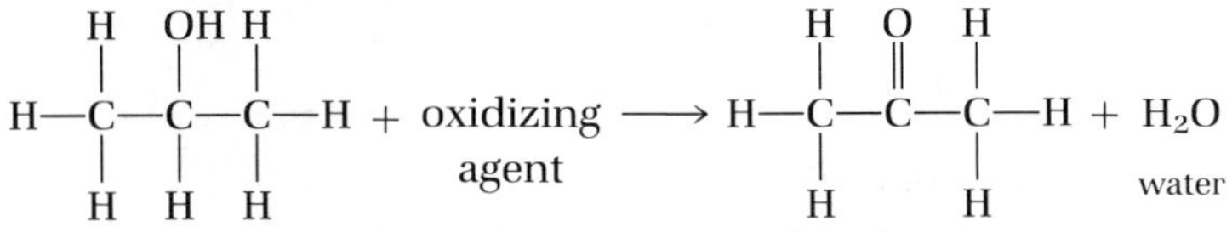

2-propanol → 2-propanone + water

It also is possible to oxidize an aldehyde to produce an organic acid:

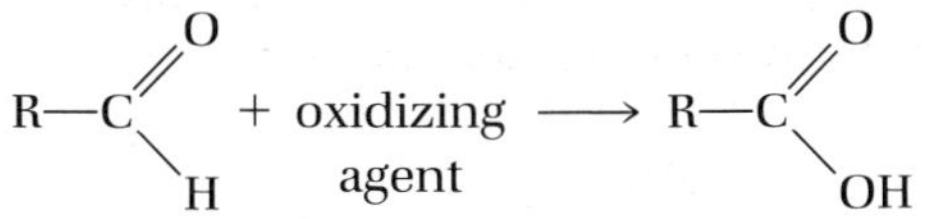

The reverse of these processes, where oxygen is removed from a molecule or hydrogen is added, is known as *reduction*. Again, the ter-

ɡy is consistent with what you learned in Chapter 21. If an
e is reduced to yield an alcohol, the oxidation number of car-
eases from zero to 2−. The oxidation number of carbon also
when an acid is reduced to an aldehyde.

chemists make use of oxidation and reduction to change
onal group into another. These processes, in combination
ution and addition reactions, serve as steps in the synthe-
d end products. This flexibility is one of the reasons the
lkenes derived from petroleum are so important as raw
example, ethene can be converted first to ethanol, then
taldehyde), then to ethanoic acid (acetic acid). Each of
es plays a role in the preparation of other useful sub-

⌐ymerization

As organic molecules of increasing complexity are formed, trends in
the physical and chemical properties of the resulting compounds may
be observed. For example, the boiling temperatures of the straight-
chain alkanes tend to increase as the number of carbon atoms in the
chain is increased. Ethane, C_2H_6, is a gas under standard conditions;
octane, C_8H_{18}, is a liquid; octadecane, $C_{18}H_{38}$, is a solid. Chemists have
learned to synthesize an alkane compound with a desired melting
point by controlling the length of the product chain.

Chemical reactivity may be similarly influenced by the presence of
functional groups that can be arranged in chains along the molecule.
The key to this chemical treasure chest is the process by which extended chain structures are formed. Known as polymerization, this
chemical behavior pattern may produce molecular chains of astounding lengths. The basic repeating units found within these complex
molecules are called **monomers**. Molecules composed of a repeating
sequence of monomers are known as **polymers**.

There are two types of polymerization reactions. The first is **addition polymerization**, which involves the bonding of monomers without the elimination of atoms. Bonding is accomplished by opening
unsaturated bonds between carbon atoms in the molecules. One frequently used monomer is ethylene (ethene), C_2H_4. In the presence of a
suitable catalyst, ethylene undergoes an addition reaction:

$$\underset{H}{\overset{H}{}}C=C\underset{H}{\overset{H}{}} + \underset{H}{\overset{H}{}}C=C\underset{H}{\overset{H}{}} \longrightarrow H-\underset{H}{\overset{H}{C}}-\underset{H}{\overset{H}{C}}-\underset{}{\overset{H}{C}}=C\underset{H}{\overset{C}{}}$$

As polymerization continues, more ethylene units are incorporated
into the structure to form the polymer polyethylene. The final product
may contain thousands of monomers, defined by the number n.

$$H-\underset{H}{\overset{H}{C}}\left(\underset{H}{\overset{H}{C}}\right)_n\underset{}{\overset{H}{C}}=C\underset{H}{\overset{H}{}}$$

polyethylene

Figure 23-10 At one time, parachutes were made from silk. The
less expensive, synthetically produced polymer Dacron is now the
more frequently used material.

One or more of the hydrogen atoms in ethylene may be replaced by groups such as —F, —Cl, —CH$_3$, and —COOCH$_3$. Synthetic polymers with trade names such as Teflon, Saran, and Lucite, or Plexiglas result. By varying additional components of the molecule, it is possible to create compounds with customized properties.

Condensation polymerization results when the formation of a polymer is accompanied by the elimination of atoms. For example, monomeric units known as amino acids may combine to form chains of polypeptides. If the chains are very long, they are called proteins. **Amino acids** contain both an amine group and an organic acid group. The simplest amino acid is glycine.

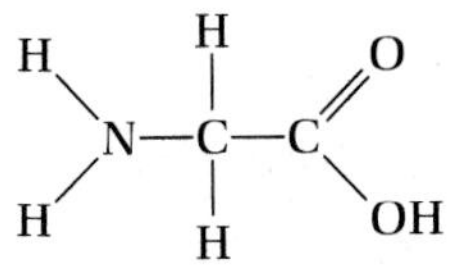

Do You Know?

One of the largest organic molecules is found in the chromosome of the fruit fly Drosophila melanogaster. The chromosome is about 2 centimeters long but only 0.000 002 cm in diameter. Its formula is approximately $C_{614\,000\,000}H_{759\,000\,000}N_{217\,000\,000}P_{62\,000\,000}O_{496\,000\,000}$ or a little over 2 billion atoms! Its molar mass is about 23 100 tons!

When the amine end of the molecule is joined with the acid end of another molecule, a molecule of water is eliminated.

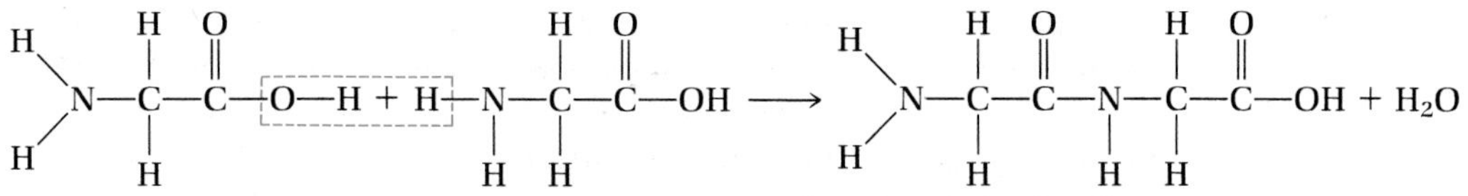

The result is a longer molecule made from the two amino acids linked by an amide bond. This particular kind of bond is called a **peptide bond**. You will read more about amino acids in Chapter 24.

Another example of a condensation polymer is the synthetic material Dacron. It results from a reaction between ethylene glycol, $HOCH_2CH_2OH$, and an aromatic organic acid called terephthalic acid.

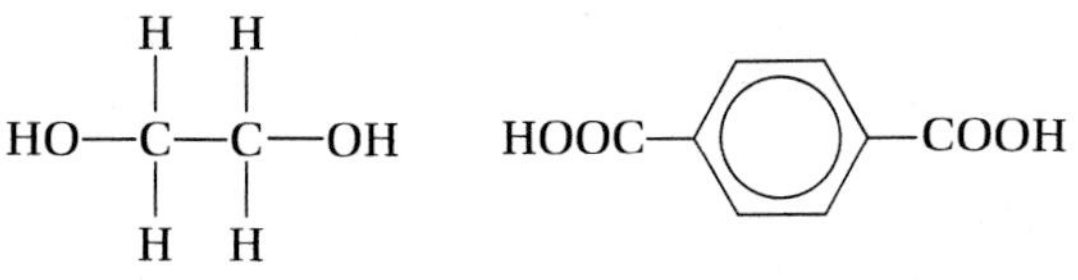

ethylene glycol   terephthalic acid

The equation below shows the first stages of this process:

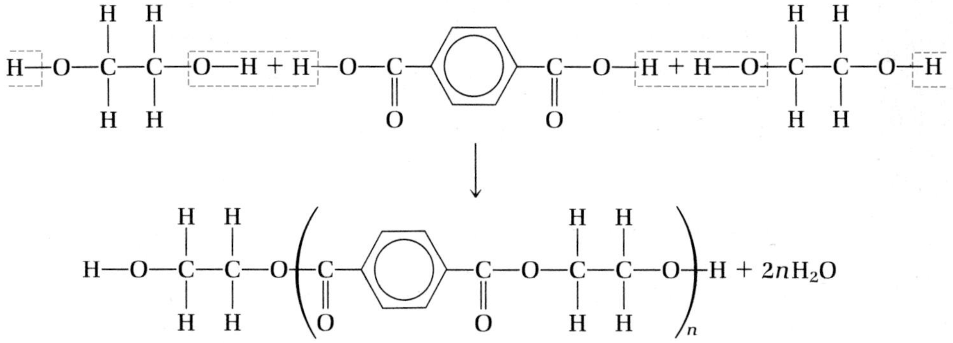

Dacron is an example of a polyester. When one —OH group of ethylene glycol reacts with one —COOH group of the terephthalic, an ester is formed and water is released. This reaction continues, producing a polymer with alternating molecules and a water molecule for every bond formed.

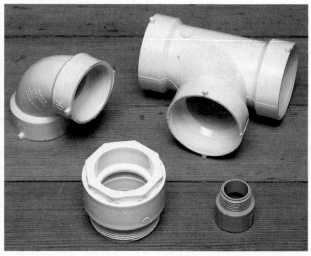

Figure 23-11 Plastic housewares and polyvinyl chloride (PVC) plumbing materials are just two examples of polymers.

The results of polymerization can be found in a variety of places. Vinyl siding for buildings, new plumbing materials, clothing, automobile tires, cassettes, records, and Styrofoam are all products of polymerization.

23-16 Application: Nylon—Fabric from a Beaker

Figure 23-12 Nylon fibers

The discovery of nylon is a good example of planned research coupled with unexpected observations. Researchers were in search of a synthetic fabric that could take the place of silk, a cloth widely used, but one that is very expensive. Silk is a natural product made from the fibers of the silkworm cocoon. The fabric is soft, durable, versatile, and warm. However, production of silk required so much manual labor that the cost was beyond the reach of most people. In Roman times, silk was literally worth its weight in gold.

Silk is a protein. Most protein molecules are polymers made of hundreds or even thousands of amino acids. Silkworms combine these acids into long chains to form silk fibers. (Two other naturally occurring polymers are cellulose and rubber.) In 1927, the Du Pont company decided to finance a research project to learn more about the characteristics of polymers and to learn how to make them. If polymers could be produced on an industrial scale, the fibers would be less expensive and therefore available to more people. A young American chemist, Wallace Carothers, was employed to organize the research program.

Since other chemists previously had difficulty making molecules join each other in long chains, Carothers decided to use longer starting molecules. Two that led to success were:

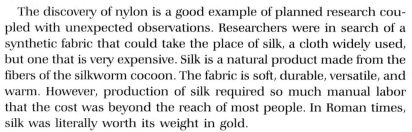

H_2N—CH_2—CH_2—CH_2—CH_2—CH_2—CH_2—NH_2

or

1,6-diaminohexane

$HOOC$—CH_2—CH_2—CH_2—CH_2—$COOH$

or

adipic acid

The combination of these two compounds into a polymer chain produces a substance known as nylon-66.

When heated, the monomer units bond by the formation of an amide between the two different functional groups. For each new bond formed, one water molecule is eliminated.

The 66 in nylon's name comes from the fact that both starting compounds contain six carbon atoms. It was originally called nylon 6,6.

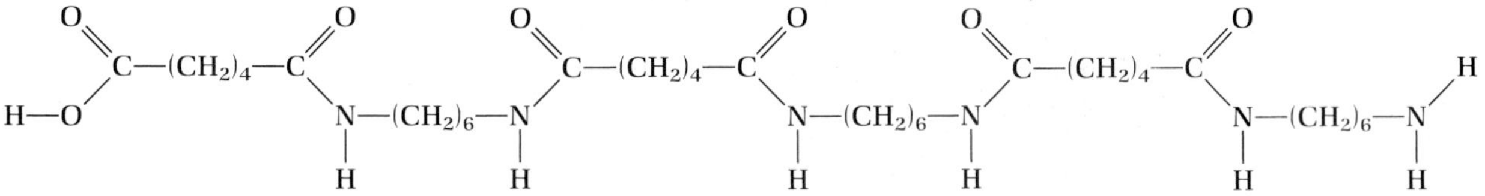

Ironically, the successful production of nylon was partially an accident. Carothers and his associates worked with various combinations of starting monomers. At one point, the two substances similar to the ones described here were in a beaker, separated into two layers. An associate happened to pull a stirring rod out of the solution. Long threads that formed where the two liquids interfaced were attached to the stirring rod. The fibers, which did not break, solidified on cooling. When pulled, the threads stretched like rubber but did not snap back. The first synthetic fiber had been made.

Carothers and his team were using the acid chloride and the amine.

Review and Practice

1. How do substitution and addition reactions differ?

2. Write the equation for the addition of hydrochloric acid to ethene. What is the name of the product formed?

3. What is polymerization?

4. What is the monomeric unit of a chain of molecules joined by peptide bonds?

5. Is the synthesis of nylon an example of a condensation or an addition polymerization?

6. Identify the type of organic reaction represented by each of the following:

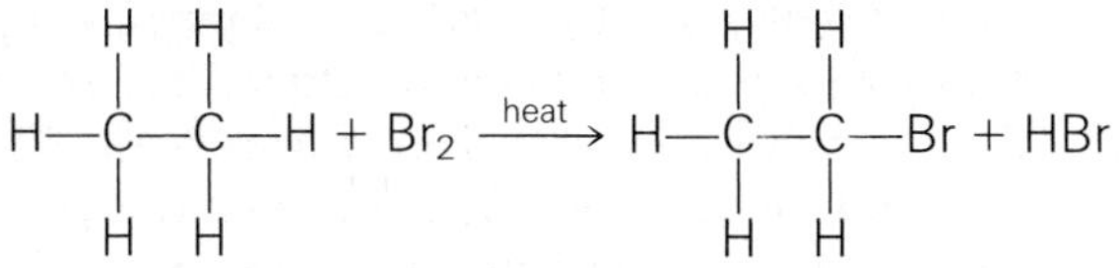

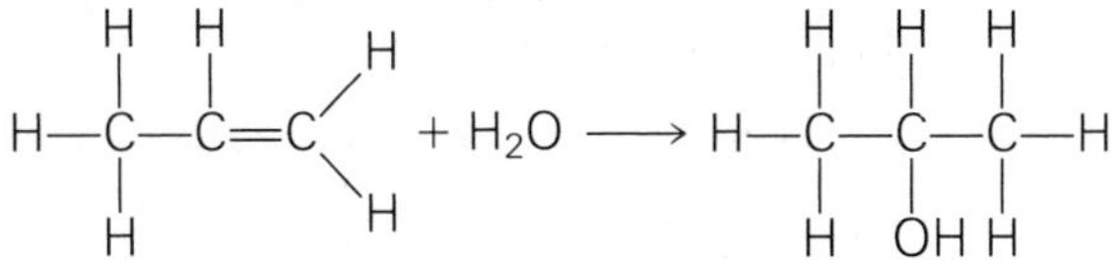

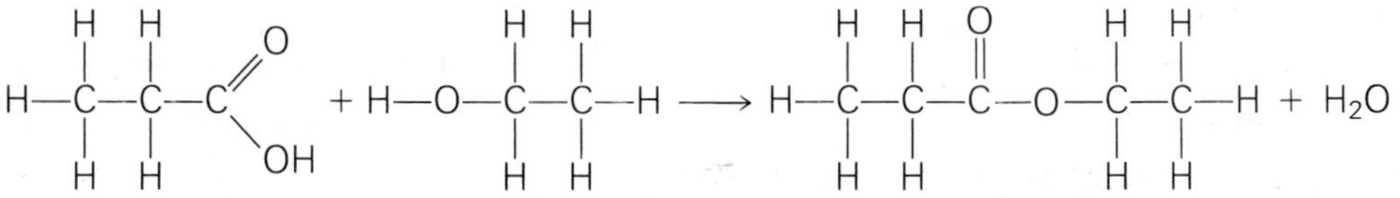

1. In substitution, an atom or group of atoms is replaced by another atom or group. In addition, new constituents are added across a double or triple bond.

2. $C_2H_4 + HCl \rightarrow CH_3CH_2Cl$ chloroethane

3. a process by which extended chain structures are formed from smaller units of molecules

4. an amino acid

5. condensation

6. a. substitution
   b. addition
   c. esterification or condensation

Careers in Chemistry

Todd Blumenkopf
Title: Organic Chemist
Job Description: Synthesizes chemical substances derived from plant sources.
Educational Qualifications: Advanced degree in chemistry, with courses in mathematics, physics, and biology; some experience in working with computers.
Future Employment Outlook: Excellent

Todd Blumenkopf has been successfully performing research in organic chemistry since his junior year in college. His research involves studying plants used for medicinal purposes and trying to synthesize the specific medicinal chemical. Currently Blumenkopf is working to synthesize a substance from a plant used in Mexico to induce labor leading to birth.

Organic chemists are employed in large numbers in the pharmaceutical industry, where career options are excellent and jobs are very secure. New areas of research, such as gene splicing, continue to develop. There are many opportunities to travel and interact with other scientists.

When asked what makes for a successful organic chemist, Blumenkopf says that the ingredients are a combination of curiosity, motivation, and patience. Since research projects are long-term, commitment is important. Besides a thorough background in chemistry, a chemist needs to take some courses in math, physics, and biology. Blumenkopf stresses that a knowledge of computer science may be valuable. Although most calculations can be done with a pocket calculator, computers are now being used frequently for information retrieval, project planning, and modeling studies. Many modern laboratory instruments are computer-controlled.

One minor problem Blumenkopf has faced is misconceptions regarding his physical condition. Blumenkopf uses a wheelchair because of a congenital spinal cord injury. He has found that educators unfamiliar with physically impaired persons often focus on perceived limitations faced by a physically impaired student. In time, however, these misconceptions fade away.

For a physically impaired chemist, proper planning is necessary. Lowering counters when possible, stacking materials from the floor, and extending tubing are some of the ways in which Blumenkopf has adapted his environment.

Blumenkopf's impairment has not inhibited his success. "Others perceive a physically impaired person as being limited," he says. "Have confidence in yourself!" He points out that his physical condition does not have a negative effect on his skills in thinking, writing, library research, and operating instruments and computers. With his degree of skill and motivation, Blumenkopf should serve as a model to any aspiring chemist.

CHAPTER REVIEW

Numbers in red indicate the appropriate chapter sections to aid you in assigning these items. Answers to all questions appear in the Teacher's Guide at the front of this book.

Summary

■ Originally obtained by processing fossil fuels, carbon compounds are now routinely synthesized within the laboratory. The properties of these compounds have played a major role in the evolution of industrialized society. Carbon compounds have had global applications because of their use as building materials, fabrics, dyes, explosives, drugs, flavorings, solvents, and fuels.

■ Hydrocarbons contain only two elements, carbon and hydrogen, yet these simple molecules are the building blocks for complex organic compounds. Hydrocarbons are classified according to structure and composition into alkanes, alkenes, alkynes, or aromatics. The first three groups are composed of simple carbon chains arranged in open- or closed-ring structures. The aromatics have a more complex structure due to the delocalization of electrons involved in the double bonds.

■ Hydrocarbon molecules may incorporate atoms other than hydrogen and carbon into their structure. Known as functional groups, these components often alter the chemical behavior of the resulting molecule. Many of the functional groups contain the element oxygen. Included in this category are such significant compounds as alcohols, ethers, carboxylic acids, aldehydes, ketones, and esters. Amines and amides are functional groups that contain the element nitrogen.

■ Organic compounds can undergo a variety of chemical reactions, which allows for the production of endless numbers of compounds. Depending on the functional group or groups in a molecule, compounds may react through substitution, addition, esterification, oxidation, polymerization, or combinations of any of these. One compound can be converted to another or used to synthesize something new.

■ Chemists can synthesize compounds with desired characteristics by controlling chain length and functional groups. Long molecules, composed of repeating molecular units (monomers), are produced during polymerization reactions. Resulting polymers include familiar synthetic materials such as nylon and polyethylene.

Chemically Speaking

addition	alkane
addition polymerization	alkene
alcohol	alkyne
aldehyde	amide

amine	ether
amino acid	functional group
aromatic hydrocarbon	homologous series
branched chain	hydrocarbons
carbon backbone	hydroxyl group
carbonyl group	ketone
carboxyl group	monomer
coal	organic acid
condensation	organic chemistry
polymerization	peptide bond
cracking	polymer
cycloalkane	saturated
delocalized	straight chain
derivatives	structural isomer
destructive distillation	substitution
ester	unbranched chain
esterification	unsaturated

Review

1. How is coal tar produced? (23-2)

2. Why was the chemistry of carbon originally called organic chemistry? (23-2)

3. Describe two factors that account for the multitude of carbon compounds. (23-3)

4. Compare the chemical stability of alkanes to alkenes. (23-5, 23-7)

5. What is a saturated organic compound? (23-4)

6. What is the general formula for members of the alkane series? (23-5)

7. Name four gases that may have played a vital role in the evolution of life. (23-5)

8. Why are methane, ethane, and propane considered members of a homologous series? (23-5)

9. What elements are most commonly found in organic compounds? (23-1)

10. What are the chief sources of organic compounds that are used in industry? (23-2)

11. Describe how the location of a branch in an alkane molecule is identified. (23-6)

12. How many carbons are in each branch of the molecule 2,2-diethyl-3-methyl-4-propyldecane? (23-6)

13. What is a structural isomer? (23-6)

14. What is the simplest alkyne? How is it used? (23-8)

15. What symbol is often used to represent cyclohexane? (23-9)

16. How are the carbon-carbon bonds in the benzene molecule different from those in alkanes and alkenes? (23-10)

17. How does the presence of a hydroxyl group influence an alcohol's solubility in water? (23-12)

18. What functional group is present in vinegar? (23-12)

(23-12) **19.** How do aldehyde and ketone structures differ?

20. What is a common source of esters? (23-12)

(23-15, **21.** Name two substances that contain amide bonds?
23-16)
22. What characteristics of a carbon chain may be altered to produce "tailor-made" compounds? (23-15)

23. What is a polymer? (23-15)

24. What two classes of organic compounds contain the carbonyl functional group? (23-12)

Interpret and Apply

1. Draw an example of a branched, unsaturated hydrocarbon. (23-4)

2. Why is it unnecessary to identify the double bond location in propene? (23-7)

3. How may carbon tetrachloride be produced? (23-13)

4. How might the presence of air affect the process of destructive distillation? (23-2, 23-14)

5. Can a cyclic hydrocarbon that is not aromatic contain double bonds? Explain. (23-9)

(23-5) **6.** Suggest a reason for the chemical stability of alkanes.

7. How many carbon atoms are found in the simplest ketone? Explain why there can be no less. (23-12)

(23-12) **8.** Can an ether contain three R— groups. Explain.

9. How do oxidation and reduction of organic compounds differ from each other? (23-14)

10. What products would result from the hydrolysis of an ester? (23-12)

11. Explain why the electrons in a benzene ring are said to be delocalized. (23-10)

12. Explain why most hydrocarbons are insoluble in polar solvents. (23-1, 23-5)

13. Is cyclopentane an isomer of the straight-chain hydrocarbon pentane? Explain. (23-6, 23-9)

14. Why is substitution on an aromatic ring an easier reaction to accomplish than addition to the ring? (23-13)

Problems

1. Draw three structural isomers of pentane. (23-6)

2. Draw the four structural isomers of C_4H_9Cl. (23-6)

3. Draw possible structures for compounds with the molecular formula C_2H_7N. (23-6)

4. Draw the structural formula of each of the following as derived from a parent ethane molecule: (23-6, 23-12)
   a. an alcohol
   b. an aldehyde
   c. a carboxylic acid

5. Draw the structural formula for each of the following: (23-6, 23-12)
   a. methoxymethane
   b. 1-pentanol
   c. 2-aminobutane
   d. propanal

6. Draw the structural formulas for the first three carboxylic acids. (23-12)

7. Draw the structural formulas for the following compounds: (23-6, 23-12)
   a. 2-methylbutane
   b. 3,3-dichlorohexane
   c. 4-ethyl-3,3,4-trimethyldecane
   d. 1,3-dimethylcyclohexane

8. Draw the structural formulas for the following compounds: (23-6, 23-12)
   a. 2-propanol
   b. 2-methyl-4-octanol
   c. 2-methoxypropane
   d. propyl acetate

9. Write the correct name for each of the following structures: (23-6, 23-12)

a.
```
      H   Br  Br  H   H   H
      |   |   |   |   |   |
   H—C — C — C — C — C — C—H
      |   |   |   |   |   |
      H   Br      H   H   H

              H—C—H
                |
                H
```

b.

```
     H  H      H  H  H  H  H
     |  |      |  |  |  |  |
H——C——C——————C——C——C——C——C——H
     |  |      |  |  |  |  |
     H  H      H  H  H  H  H

        H——C——H   H——C——H
           |         |
           H      H——C——H
                     |
                     H
```

c.

```
     H  Cl H  H  H
     |  |  |  |  |
H——C——C——C——C——C——H
     |  |  |  |  |
     H  H  Cl H  H
```

d.

```
              H            H
              |            |
           H——C——H      H——C——H
              |            |
     H  H     |   H        H   H
     |  |     |   |        |   |
H——C——C——C——C————C——C——C——H
     |  |     |            |   |
     H  H     |   H        H   H
              H——C——H
                 |
                 H
```

10. Write the correct name for each of the following structures: (23-6, 23-12)

a.

```
     H      H  H
     |      |  |
H——C——C==C——C——H
     |      |  |
     H      H  H
  H——C——H
     |
     H
```

b.

```
     H  H     H
     |  |    /
H——C——C==C
     |       \
     H        H
```

c.

```
     H  H        H
     |  |        |
H——C——C——C≡C——C——H
     |  |        |
     H  H        H
```

d. H——C≡C——H

e.

f.

g.

11. Write the correct name for each of the following structures: (23-12)

a.

```
     H    O
     |   //
H——C——C
     |   \
     H    OH
```

b.

```
     H  O  H  H
     |  ||  |  |
H——C——C——C——C——H
     |     |  |
     H     H  H
```

c.

```
        O
       //
H——C
       \
        H
```

d.

```
     H  OH H
     |  |  |
H——C——C——C——H
     |  |  |
     H  H  H
```

e.

```
     H        H  H  H
     |        |  |  |
H——C——O——C——C——C——H
     |        |  |  |
     H        H  H  H
```

f.

```
     H  H     H
     |  |    /
H——C——C——N
     |  |    \
     H  H     H
```

g.

```
     H  H  O  H  H
     |  |  ||  |  |
H——C——C——C——C——C——H
     |  |     |  |
     H  H     H  H
```

h.

```
     H  H  H  H
     |  |  |  |
H——C——C——C——C——OH
     |  |  |  |
     H  H  H  H
```

i.

```
  H     H  H  H     H
   \     |  |  |    /
    N——C——C——C——N
   /     |  |  |    \
  H     H  H  H     H
```

j.

```
     H  H  H     O
     |  |  |    //
H——C——C——C——C
     |  |  |    \
     H  H  H     OH
```

12. Identify the type of organic reaction represented by each of the following equations: (23-14 to 23-18)

a.

b.

c.

d.

e. $R-OH + HO-\overset{\displaystyle O}{\underset{\displaystyle R'}{C}} \xrightarrow{H_3O^+} R-O-\overset{\displaystyle O}{\underset{\displaystyle R'}{C}} + H_2O$

f.

Challenge

1. What is the maximum number of triple bonds that can be present in a three-carbon hydrocarbon? A four-carbon hydrocarbon? A five-carbon hydrocarbon? (Consider both straight-chain and cyclic forms.)

2. Ethane reacts with chlorine to substitute first one chlorine for hydrogen, then two, and so on until C_2Cl_6 is formed. Draw all the derivatives of ethane from this series of reactions.

3. Draw the structural formula for each of the following compounds:
a. 1,3-cyclopentadiene
b. 2-methyl-3-heptene
c. bromobenzene
d. 1-ethyl-2-methyl-3-propylbenzene
e. 3-methylphenol
f. 1-ethyl-3-methylcyclohexene
g. methylacetylene

4. Using structural formulas, write the reaction of the esterification of propanoic acid and ethanol.

5. Lucite is an addition polymer of methylmethacrylate:

Draw a portion of the Lucite structure.

6. Draw the structural formula for each of the following compounds:
a. 2-phenylphenol
b. methyl formate
c. diethyl amine
d. ethyl-methyl-propylamine
e. 1,4-diaminobutane
f. 1,2 dichloro-1,2-dibromoethene
g. 1,3-dibromopropene

7. Write the correct name for each of the following structures:

a.
$$H-\overset{\overset{\displaystyle H}{|}}{\underset{\underset{\displaystyle H}{|}}{C}}-O-\overset{\overset{\displaystyle O}{\|}}{C}-\overset{\overset{\displaystyle H}{|}}{\underset{\underset{\displaystyle H}{|}}{C}}-\overset{\overset{\displaystyle H}{|}}{\underset{\underset{\displaystyle H}{|}}{C}}-H$$

b.

c.

8. The chemical formula C_2H_6O exists as two structural isomers. Draw each isomer and identify the functional group in each case. Predict how the boiling points and solubilities in water of the two compounds differ. Explain the basis of your predictions.

Synthesis

1. Use the kinetic molecular theory to explain why the boiling points of the alkanes in Table 23-1 increase with increasing molecular mass.

2. Science-fiction writers sometimes describe alien worlds whose life forms are based upon the element silicon. What characteristics of the silicon atom might justify such a premise?

3. When low-grade coal is burned, toxic sulfur dioxide gas is produced. In the atmosphere, this gas is oxidized to sulfur trioxide. Sulfur trioxide associates with water to form sulfuric acid.

a. Write a set of three balanced equations showing the reactions responsible for the formation of this acid rain component.
b. Describe how acidic precipitation may be reduced.

4. When hydrocarbons are heated to high temperatures in the absence of air, they often decompose into related compounds of similar properties. Describe a method that might be used to separate these mixtures into pure substances.

5. In the preparation of methyl acetate, the yield of the ester is rather low at equilibrium. Study the equation below and apply Le Chatelier's principle to explain what can be done to increase the yield.

$$CH_3OH + CH_3COOH \overset{H_3O^+}{\rightleftharpoons} CH_3-O-\overset{\overset{\displaystyle O}{\|}}{C}-CH_3 + H_2O$$

methyl alcohol   acetic acid   methyl acetate   water

Projects

1. Using toothpicks to represent bonds and Styrofoam spheres to represent atomic nuclei, construct molecular models of the first five members of the alkane series.

2. Prepare a report on the techniques used by geologists to locate and extract petroleum.

3. Find out what type of fuel is burned at the nearest power plant. Investigate what problems may be associated with released air pollutants.

4. Investigate the difference between saturated and unsaturated fats in human diets. How is the body's use of these materials affected by the differences?

5. Investigate how the cracking process affects the efficiency of gasoline fuels.

Biochemistry is the study of complex organic molecules called biomolecules. These molecules were once thought to be produced only by living organisms. Today, the creation and large-scale manufacture of biomolecules is a new and rapidly growing industry whose importance has only begun to be imagined. Many of the new techniques and the discoveries they made possible have occurred in the last 10 years. Computer-generated images, like this one of DNA, deoxyribonucleic acid, looking down the axis of the double helix, help chemists to understand molecular structure and function.

In this chapter you will be introduced to the basic structure of the biomolecules and will be able to glimpse the infinite and exciting possibilities that lie ahead in this branch of chemistry.

Biochemistry

Introduction to Biomolecules

The molecules that make up individual cells and multicellular organisms come in an astonishing array of shapes, sizes, complexities and reactivities. They take part in billions of reactions per minute. All of them can be classed into four different groups according to similarities of structure and function. These groups are the carbohydrates, lipids, proteins, and nucleic acids.

CONCEPTS

■ structure and function of carbohydrates, lipids, proteins, and nucleic acids
■ carbohydrates and lipids as energy sources
■ selective binding of proteins
■ transcription of RNA from DNA
■ translation of RNA into protein

24-1 Carbohydrates

Anyone who has tried to diet has heard about carbohydrates. Many diet plans suggest cutting back on carbohydrates, and some suggest eating carbohydrate-free meals.

Carbohydrates are organic molecules that have an empirical formula of CH_2O, or an atom of carbon (*carbo-*) seemingly attached to a molecule of water (*hydrate*). Carbohydrates include sugars, starches, and cellulose.

The simplest carbohydrates are the simple sugars or **monosaccharides** (*mono-* meaning one, *saccharide* for sugar) that have the formula, $C_6H_{12}O_6$. Glucose is one of the most important monosaccharides. It is the simple sugar into which all carbohydrates are metabolized and is the major constituent of honey. Another naturally occurring monosaccharide is fructose, also $C_6H_{12}O_6$. It is an isomer of glucose (having the same molecular formula but a different structural formula). Fructose is the sugar responsible for the sweet taste of fresh fruit and some vegetables. Both glucose and fructose exist as straight-chain molecules and as cyclic structures. In solution, the cyclic form predominates.

Carbohydrates made of two monosaccharide units are called **disaccharides**. Table sugar, or sucrose, is a disaccharide consisting of one molecule of glucose and one of fructose joined together with the elimination of water. The formula is $C_{12}H_{22}O_{11}$.

Chitin is a polysaccharide found in the exoskeleton of insects and crustaceans.

Do You Know?

On the average, every American eats 100 pounds of sugar per year.

glucose + fructose → sucrose + H_2O

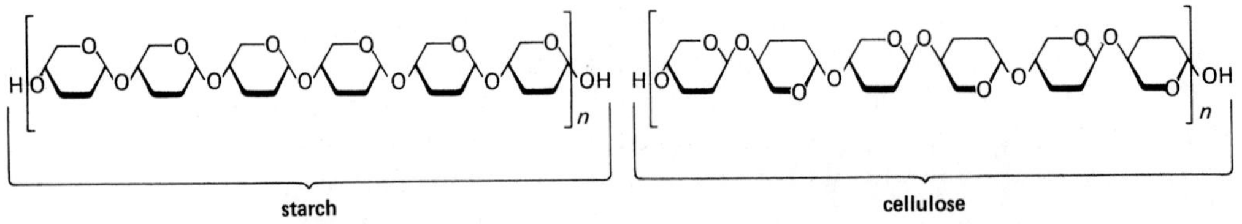

starch   cellulose

Figure 24-1 The differences in the glucose rings of cellulose and starch causes them to have different chemical properties. Only the ring atoms of glucose and the two —OH groups are shown for simplicity. See text for the complete molecular structure of glucose.

Polysaccharides are polymers made of glucose molecules, joined together into large, complex chains. Two important polysaccharides are cellulose and starch. Cellulose occurs in plant cell walls, and makes up the structural material in woody plants. Starch is the form in which plants store the glucose made during photosynthesis. Both are molecules made of long chains of glucose. However, the glucose in cellulose is a different form than that in starch. These two forms have different chemical properties. Your digestive enzymes are capable of digesting the glucose in starch but not the glucose in cellulose.

Carbohydrates provide the fuel for all life processes. Disaccharides and polysaccharides are broken down into glucose molecules that are oxidized in a complex series of chemical reactions to produce carbon dioxide and water. The complete breakdown of one mole of glucose makes 2870 kilojoules (686 kilocalories) of energy available for the vast number of energy-requiring processes in the body.

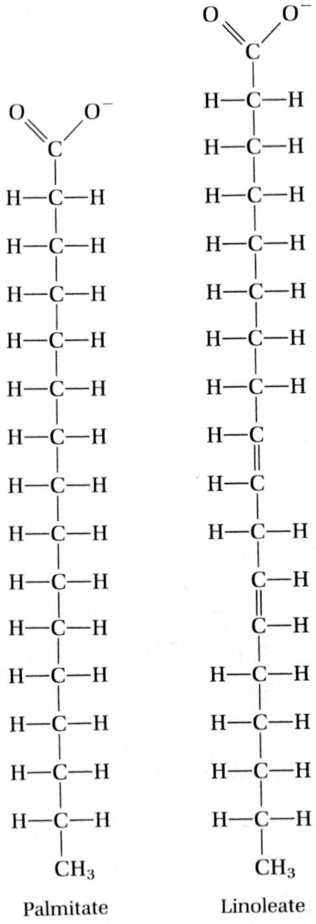

24-2 Lipids

Fats and oils belong to the class of biomolecules known as **lipids**. Most of the fats and oils you eat are triple esters of glycerol, an alcohol. Typically, three carboxylic or fatty acids are attached to one molecule of glycerol to form a **triglyceride**. R_1, R_2, and R_3 represent long chains of carbon and hydrogen atoms.

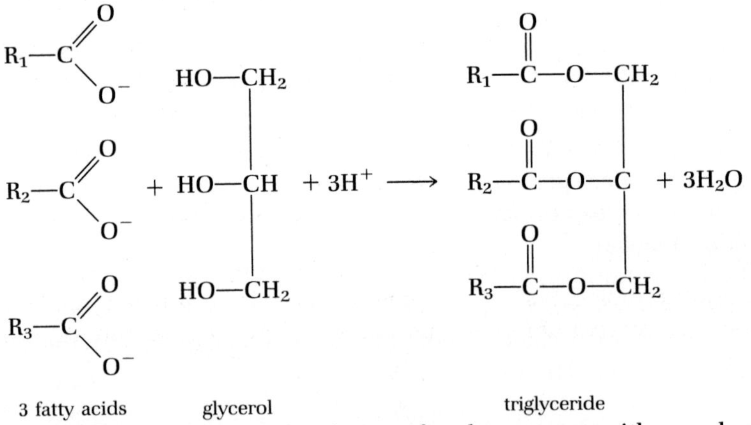

3 fatty acids   glycerol   triglyceride

Palmitate   Linoleate

Figure 24-2 Palmitate is a saturated fatty acid with no carbon-carbon double bonds. Linoleate is a polyunsaturated fatty acid with two carbon-carbon double bonds.

Fatty acids are long straight chains of carbon atoms with a carboxyl group, —COOH, at one end. Each carbon in the chain can be attached to another carbon atom by either a double bond or a single bond. If the carbons of a fatty acid each have as many hydrogen atoms as possible and no double bonds, the fatty acid is saturated with hydrogen. If the fatty acid has one carbon-carbon double bond, it is unsaturated. If it has two or more double bonds, it is **polyunsaturated**, a term that is probably familiar to you from television advertising. Most animal fats are saturated, while vegetable oils are unsaturated or polyunsaturated. The degree of saturation affects the melting point of the substance. Polyunsaturated fatty acids have a much lower melting point than saturated fatty acids which explains why vegetable oils are liquids at room temperature, while butter (an animal fat) is a solid.

Only one or two percent of the total calories in a diet need to come from *essential* fatty acids, fatty acids that the body cannot manufacture. These and other fatty acids consumed are used to construct and repair cell membranes. Lipids stored in fat tissue provide insulation for the body and cushion certain organs.

Lipids also are used as an energy source in the body. The amount of energy produced from the breakdown of fats is almost twice that produced from the oxidation of carbohydrates. One gram of fat produces 39.6 kilojoules of energy (9 kilocalories), whereas one gram of carbohydrate produces 17 kilojoules of energy (4 kilocalories). Stored fats from the body tissues act as an energy reserve. About 10 percent of the average person's body weight comes from fat. It can provide a month's worth of energy for the body. In contrast, the body's supply of glucose is gone in 24 hours.

Steroids are another class of lipids having a four-ring carbon skeleton. Steroids are varied in function, and include some of the vitamins, the sex hormones, and cortisone. Cholesterol is a steroid found naturally in almost every tissue of the body, especially the brain and spinal cord. The cell membranes of most cells contain cholesterol.

It has been shown that there is a correlation between increased cholesterol levels in the blood and the incidence of heart disease. Evidence suggests that by increasing the use of polyunsaturated fats in the diet, the level of blood cholesterol is lowered.

24-3 Proteins

Protein molecules are polymers made from smaller molecules known as **amino acids**. Amino acids all have a common structure.

Each has a central carbon atom to which is attached a hydrogen atom, a carboxyl group, an amine group and a side chain (R). It is only in the nature of the side chain that amino acids differ from one another. There are about twenty amino acids that occur in the proteins of living organisms, although there are more than 2000 natural and artificially-made amino acids. There are ten essential amino acids that cannot be synthesized by the body and must be included in the diet.

A protein molecule is built by linking the amino group of one amino acid to the carboxyl group of another. The bonding is accomplished by

Do You Know?

Americans have a love/hate relationship to fats. Fats make up about 40 percent of the average American diet, about twice what is necessary. Over half of the calories in such foods as eggs, hard cheese, peanut butter, and cold cuts comes from fat.

Amino acids

glycine

alanine

tyrosine

Figure 24-3 The three-dimensional shape of an enzyme is determined by the sequence of its amino acids. The substrate fits exactly into a specific area of the enzyme's surface called the active site.

Nucleotide

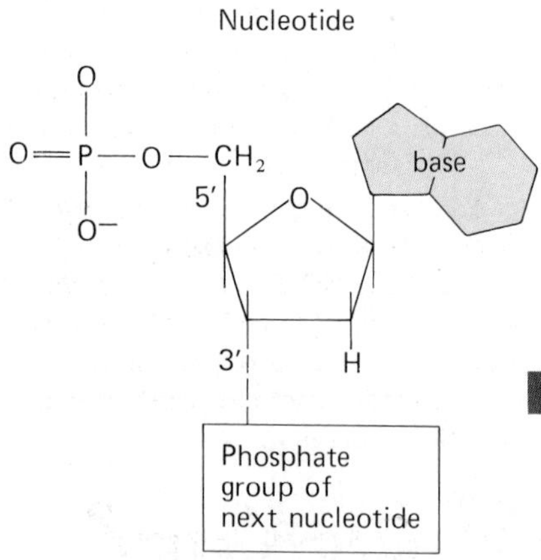

Figure 24-4 The backbone of a DNA molecule is made of repeating units called nucleotides. Each nucleotide consists of a phosphate group linked to either a ribose sugar (in RNA) or a 2,deoxyribose sugar molecule (in DNA) which is, in turn, bonded to one of five nitrogen bases. The placement of the 3′ and 5′ carbon atoms of the sugar molecule define the direction of the DNA molecule.

removing a molecule of water, or condensation polymerization. The resulting carbon-nitrogen linkage created in this way is an amide linkage known as a peptide bond (see Section 23-15), and the protein chain is referred to as a **polypeptide**.

How can twenty amino acids make the hundreds of millions of different proteins found in living organisms? The physical and chemical properties of a protein molecule are determined by the way the chain folds up in three-dimensional space. The folding pattern is, in turn, a result of the amino acid sequence of the particular protein. The twenty common amino acids can make millions of different proteins because each protein contains hundreds to thousands of amino acids and there are no limits to how they can combine. For example, in a protein of 100 amino acids, there are 20^{100} possible sequences of amino acids.

Hydrogen bonding between atoms of a protein helps to stabilize the molecule. Hydrogen bonds are weak, however, making proteins susceptible to heat. If the hydrogen bonds break, the three-dimensional structure of the protein falls apart and the protein is said to be **denatured**. Cooking an egg is a graphic example of denaturation.

Cysteine is the only common amino acid having a —SH side chain. **Cross-linking** between sulfur atoms in two cysteine molecules can form covalent, disulfide (S—S) bonds. This type of bonding in proteins is much stronger than hydrogen bonding. The protein in human hair contains a large amount of cysteine. The cross-linking provided by the large number of disulfide bonds gives hair its shape and strength. When a permanent is applied to hair, it first breaks the disulfide bonds, then as the hair is set in a new style and dried, new cross-links form to hold the hair in its new shape.

The folding of the protein chain produces complex three-dimensional shapes. Some proteins form long fibers. Keratin is a fibrous protein found in hair, skin, fingernails, and wool. Other proteins are globular. Globular proteins usually include an area of specific shape, like a fold or crevice, called the **active site**. Another molecule can fit into the active site and be acted upon by the protein. Enzymes are a class of globular proteins that catalyzes reactions in this way.

A chain of less than 50 amino acids is referred to as a polypeptide. A chain of more than 50 amino acids is called a protein.

24-4 Nucleic Acids

If proteins are the building materials of life, then nucleic acids are the blueprint. Included in this class of molecules is the famous double helix of the **DNA** molecule and its less celebrated partner, **RNA**. These biomolecules direct the functioning of cells and the synthesis of proteins. They also transfer genetic information across generations.

Nucleic acids are long-chain polymers made up of repeating units called **nucleotides**. Each nucleotide is made of a phosphate group linked to a five-carbon sugar molecule. The sugar is either a ribose sugar molecule (in ribonucleic acid, RNA) or a 2,deoxyribose sugar molecule (in deoxyribonucleic acid, DNA). Each nucleic acid chain has a backbone in which phosphate groups alternate with sugar molecules to form a covalently linked polymer. Attached to the sugar ring of each nucleotide by loss of a water molecule is one of four nitrogen bases: adenine (A), guanine (G), thymine (T), or cytosine (C). Adenine

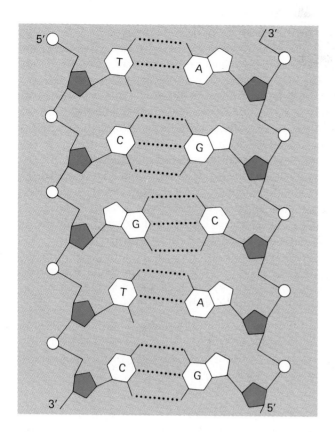

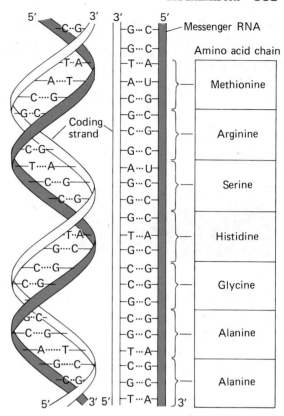

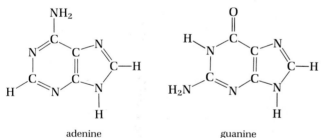

adenine   guanine

and guanine are called purines. Thymine and cytosine are called pyrimidines. In RNA, thymine is replaced with another pyrimidine called uracil (U).

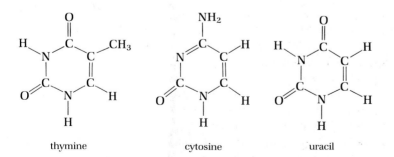

thymine   cytosine   uracil

The bases stick out in a plane that is perpendicular to the plane of the sugar-phosphate backbone. The whole chain has a right-handed twist.

Figure 24-5 In a double strand of DNA each nitrogen base is hydrogen bonded (dotted lines) to a complementary base on the opposite strand. Due to their shapes, guanine only will bond to cytosine and adenine only will bond to thymine (or to uracil in RNA).

Figure 24-6 Shown here in simplified form is the process by which DNA is transcribed into messenger RNA and the translation of the messenger RNA into a sequence of amino acids. The end product is a functional protein.

Do You Know?

Rosalin Franklin was responsible for most of the X-ray diffraction work that helped to determine the structure of DNA. She worked with Maurice Wilkins at Kings College in Cambridge, England, though she did not share the acknowledgement he received when Wilkins won the 1962 Nobel Prize with James Watson and Francis Crick for their research on DNA.

The natural form of DNA in cells is supercoiled. To demonstrate this kind of structure, imagine a rubber band that was cut, and one end twisted. If the ends were then rejoined, the rubber band would coil about itself into a supercoiled helix.

In the case of DNA, each base is linked by hydrogen bonds to a base that lies parallel to it on another strand of DNA. Given the structure and size of each base, adenine (A) can only pair with thymine (T) on the opposite strand, and guanine (G) can only pair with cytosine (C). This is the double helix structure first proposed by James Watson and Francis Crick in 1953. The nucleotide sequence of one molecule of DNA thus determines the sequence on the other. For example, if the sequence of bases on one DNA strand is ATTGCACC, the base sequence on the complementary strand would be TAACGTGG.

RNA consists of a single sugar-phosphate chain with unpaired bases. Uracil (U) is substituted for thymine (T), and ribose sugar takes the place of 2,deoxyribose sugar in the backbone. RNA molecules are capable of bonding to single strands of DNA so that uracil (U) bases in the RNA pair with the adenine (A) bases of the DNA.

Adenosine triphosphate, or **ATP**, is a single nucleotide consisting of the nitrogen base, adenine, and a ribose sugar molecule with three phosphate groups attached to it. ATP is found in all living organisms where it acts as an energy storage molecule.

DNA acts as a set of operating instructions for the cell by interacting with enzymes that copy the DNA into a strand of RNA (called messenger RNA), and with other enzymes that translate the RNA base sequence into the amino acid sequence of a protein. A section of DNA that codes for a protein is called a **gene**.

1. Monosaccharides—glucose, fructose; disaccharides—sucrose; polysaccharides—cellulose, chitin, starch
2. A saturated fat molecule has no C—C double bonds and is fully hydrogenated. An unsaturated fat has one or more C—C double bonds.
3. Saturated fats have higher melting points than unsaturated fats.
4. High blood cholesterol levels have been implicated in heart disease.
5.

6. the amino acid sequence
7. They are both nucleic acids. DNA is translated into RNA by enzyme action.

Review and Practice

1. Name three types of carbohydrates, and give an example of each type.

2. What is the difference between a saturated and an unsaturated fat?

3. What is the relation between the melting point and the degree of unsaturation in a fat?

4. How can high levels of cholesterol in the diet affect you?

5. Draw the general structure of an amino acid.

6. Explain how it is possible for twenty amino acids to make the millions of proteins that exist in living organisms.

7. How are DNA and RNA related?

Chemical Controls

What controls the billions of chemical reactions happening each minute inside your body? How does one part of the body communicate with another? How does your left hand know what your right hand is doing? The answers to these questions involve communication between cells. The processes are complex and have only recently been described in detail. Many of them involve chemicals that are manufactured, transmit a signal, and are destroyed—all within a tenth of a second. Studying these chemical messengers is necessarily difficult.

In a one-celled organism, all parts of the cell are in direct contact with all other parts. In a multicellular organism there are many kinds of cells with distinct tasks to perform. They may be far apart from each other. To coordinate all the various functions, there must be communication between these cell populations.

■ **communication through hormones and neurotransmitters**
■ **enzyme action**
■ **role of vitamins and minerals**

24-5 Hormones and Neurotransmitters

In most multicellular organisms there are two primary methods of communication between cells, hormones and nerve cells. In both systems, cells "talk" to one another by means of chemical signals. The main difference between the two systems is how directly the message is sent. A nerve cell sends a direct message to a particular set of target cells: muscle cells, glands cells, or other nerve cells. To send a message, the nerve cell releases a chemical called a **neurotransmitter**. Cell-to-cell communication takes place where two cells come in contact at a site called the synapse. Molecules of neurotransmitter become attached to **receptor** proteins on the surface of the target cell and produce chemical changes in the cell.

Hormone action is not so direct. The glands of the endocrine system release chemicals called **hormones** that act on cells or organs scattered throughout the body. The hormones are secreted into the bloodstream which then carries them to their destination within minutes or hours. Each target cell has receptor proteins that recognize only the

Figure 24-7 The communication provided by most hormones is not as direct as that provided by the nervous system. (a) Hormones from the endocrine glands enter the bloodstream and are carried throughout the body. The receptor proteins on target cells recognize a specific hormone and combine with it, removing it from the bloodstream. (b) Nerve cells communicate by releasing neurotransmitters close to the target cells.

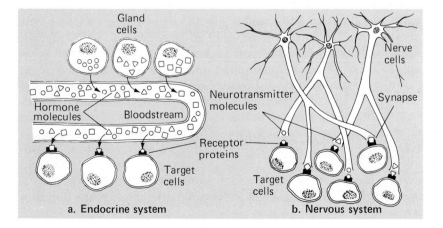

a. Endocrine system   b. Nervous system

hormone molecules meant to act on that cell. The receptors combine with the hormone molecules and bring them into the cell.

Recently it has been discovered that many of the messenger molecules employed by nerve cells also work as hormones. For example, norepinephrine, as a hormone, is released by the adrenal gland to stimulate contractions of the heart, to dilate the bronchial tubes, and to increase the contractile strength of arm and leg muscles. Norepinephrine also is a neurotransmitter. It works to constrict blood vessels and therefore increase blood pressure. The same molecule can carry a very different message as a hormone and as a neurotransmitter.

Molecules that act as hormones fall into two categories: peptide hormones and steroids. The peptide hormones are strings of amino acids. The steroid hormones are large molecules that are derived from cholesterol (a steroid lipid) and have the same basic structure. Small differences in the side chains attached to the four carbon rings of steroids produce hormones that differ widely in function.

Among the major steroids in humans are the glucocorticoids. They regulate the metabolism of glucose and of other substances. The mineral steroids are hormones that affect the body's salt balance. The sex steroids are involved in the maturation and functioning of the reproductive system.

Many of the peptide hormones act as chemical messengers in both the endocrine and nervous systems. The enkephalins, two peptides that differ slightly from each other, act as opiates in the brain to block pain. In the intestine, they are hormones that regulate the movement of food through the digestive system. Until the mid-1970's, it was not known that the peptides could act as neurotransmitters, although many of the peptide hormones were well known. The enkephalins were not discovered to act as neurotransmitters until 1975.

24-6 Enzymes

The enormous number of biochemical reactions occurring within cells is regulated by another class of substances called enzymes. **Enzymes** are biological catalysts that speed up chemical reactions, as well as control the rate at which reactions occur. They are globular protein molecules manufactured by each cell. More than 2000 enzymes have been identified on the basis of the chemical reactions they catalyze. All of them are structurally different from one another. The chemical and physical properties of an enzyme molecule, a protein, depend on how its chain of amino acids folds up in three-dimensional space. This, in turn, is determined by the sequence of amino acids.

An enzyme recognizes a specific molecule called a **substrate** and binds to it. Some enzymes are so specific they only act on one substrate, while others can act on a class of substrates. What makes an enzyme different from other proteins is that it can bring about some chemical change in the molecule to which it binds. The change usually involves the forming or breaking of a covalent chemical bond. Enzyme action may split the substrate into two pieces, may add a chemical side group to the molecule, or may simply rearrange the bonds in the substrate.

Enzyme action has three stages. First, the enzyme binds to the substrate, then the chemical reaction occurs and, finally, the altered substrate is released from the enzyme. All three stages are reversible. If an enzyme binds to molecule A and converts it into molecule B, the same enzyme also can bind to B and change it back into A.

The enzyme itself does not determine the direction of the reaction. The direction is determined by the relative concentrations of substrate molecules and product molecules at equilibrium. The enzyme speeds up the rate at which the reaction happens. In the absence of an enzyme, most biochemical reactions are extremely slow. The appropriate enzyme can speed them up by a factor of a million or more.

An enzyme speeds up a reaction by lowering the amount of energy required to start the reaction. Some enzymes do this by providing a medium that is more favorable than the surrounding one, or by bringing the reactants into close contact. Other enzymes take a more active role. They might add or remove a proton from the substrate, strain the substrate molecule's bonds, or even form temporary covalent bonds between the substrate and some part of the enzyme itself. Some enzymes require accessory molecules called coenzymes in order to accomplish their job (see Section 24-7).

Enzymes catalyze the important biochemical reactions that are responsible for energy, repair and growth in an organism. They also are becoming more and more important in industry. The diversity and efficiency of enzymes make them attractive options for use in chemical processes that have until now only been accomplished with difficulty or great expense. Some examples are given in Section 24-9.

Do You Know?

Areas on the surface of a protein molecule that tightly bond to another molecule are rougher in texture than areas that form loose bonds. The active site of an enzyme molecule, an area that forms a transient complex with the substrate molecule, is relatively smooth.

24-7 Vitamins and Minerals

Many enzymes require the presence of a nonprotein substance to function properly. Without these components, the catalyzed reaction does not occur fast enough to be useful to the cell. There are two kinds of substances that act with enzymes. **Minerals** are inorganic substances, usually metal ions. **Vitamins** are organic compounds.

Metal ions are important for the action of most enzymes, though their function is not completely understood. They can turn enzymes on and off or modify the rate at which the enzyme works. Iron, molybdenum, and copper are involved with the enzymes that carry out oxidation-reduction reactions within cells. Magnesium is important in all reactions where a phosphate group is transferred, such as in the reversible reaction of ATP, and in the functioning of the plant pigment chlorophyll a which is involved in photosynthesis.

In the above reactions, the metal ions are only in loose association with the enzyme. In other cases, the metal ion is firmly bonded to the protein molecule. Hemoglobin is the protein found in red blood cells that is responsible for carrying oxygen throughout the body. The capacity of hemoglobin to bind to oxygen depends on the presence of a nonprotein unit called the heme group. At the center of each heme molecule is an atom of iron. The iron atom is an integral part of the heme molecule and another atom cannot be substituted for it. Iron also is necessary for the functioning of myoglobin in muscles.

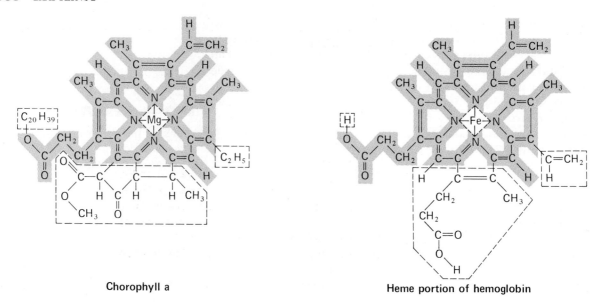

Chorophyll a

Heme portion of hemoglobin

Figure 24-8 Chlorophyll *a* has a central magnesium atom. Its structure resembles that of heme. The heme portion of hemoglobin is shown here with the central iron atom.

Vitamins are a diverse group of nonprotein, organic molecules. They often are referred to as coenzymes. A **coenzyme** binds to a specific site on a protein molecule and provides chemical functions that an enzyme alone cannot provide. Vitamins are classified by their solubility. Some are soluble in water, while others are soluble in fat.

Your body can make the enzymes necessary for its proper functioning, but it cannot make all the minerals or vitamins it needs. They must be included in your diet.

Table 24-1

MINERALS AND THEIR FUNCTIONS	
MINERAL	**FUNCTION**
Ca	bone growth and maintenance
Fe	combines with oxygen in hemoglobin of red blood cells; participates in electron transfer during carbohydrate metabolism
Zn	allows rapid absorption and release of CO_2 in red blood cells
Co	essential for functioning of vitamin B_{12} and for maturation of red blood cells
Mn	necessary for proper male sexual functioning; participates in conversion of NH_4^+ to urea
Cu	increases absorption of iron through small intestine
Mg	functions in carbohydrate metabolism
Na	necessary for nerve impulse transmission, acid-base buffering in blood, and absorption of glucose into cells; a major extracellular fluid cation
K	important in buffering, regulation of nerve and muscle functioning, and protein synthesis; a major intracellular fluid cation
P	aids in bone formation; aids in absorption of glucose, glycerol and fatty acids; important in synthesis of ATP
Cl	important in buffering and water balance; a major extracellular fluid anion
S	functions in protein synthesis; activates enzymes; important in detoxification

Table 24-2

WATER SOLUBLE VITAMINS		
VITAMIN	**FUNCTION**	**DISEASE OR PROBLEM IF DEFICIENT FROM DIET**
thiamine (B_1)	carbohydrate metabolism	beriberi, loss of appetite, depression
riboflavin (B_2)	protein and carbo-hydrate metabolism	inflammation of skin, swollen tongue
niacin	carbohydrate, fat, protein metabolism	pellagra, irritability, skin eruptions
pyridoxine (B_6)	protein metabolism	anemia, irritability,
pantothenic acid	carbohydrate, fat, protein metabolism	lowered resistance to infection
cobalamine (B_{12})	formation of proteins and nucleic acids	pernicious anemia
biotin	carbohydrate metabolism, protein synthesis	skin irritation
folacin	blood formation, synthesis of nucleic acids	anemia
ascorbic acid (C)	formation of bone, teeth, cartilage; absorption of Fe and Ca	scurvy, slow wound healing

Table 24-3

FAT SOLUBLE VITAMINS		
VITAMIN	**FUNCTION**	**DISEASE OR PROBLEM IF DEFICIENT FROM DIET**
A	visual pigments in eye	night blindness, dry skin and mucous membranes
D	deposition of Ca, P in bone	rickets in children
E	antioxidant	breakdown of red blood cells, anemia
K	aids in blood clotting	hemorrhage

Review and Practice

1. How does a hormone differ from a neurotransmitter?

2. a. Name two types of hormones.
  b. Which type can act as both a hormone and a neurotransmitter?
  c. Give an example of a hormone that also can be a neurotransmitter.

3. Why is only a small amount of enzyme necessary to catalyze a reaction?

4. What is the difference between a vitamin and a mineral?

1. Neurotransmitters are quick-acting chemicals produced by nerve cells. Hormones are produced by the endocrine glands and act over longer periods of time. Some chemicals can be both.
2. a. peptides and steroids
  b. peptides
  c. norepinephrine, enkephalins
3. It is not used up in the reaction.
4. Minerals are inorganic, usually metal ions; vitamins are organic compounds.

Biomolecular Engineering

CONCEPTS

■ **methods of gene cloning**
■ **manufacture of specialized proteins**
■ **applications of gene cloning**

Biochemistry is a radically different field than it was just ten years ago. The change is primarily due to new investigative techniques that allow not only observation of biochemical reactions, but also the ability to change the molecules involved in specific ways. These techniques have given rise to a whole new industry—biomolecular engineering. More importantly, it has changed the way that people think about living things. Among the many kinds of biomolecules available for reactions, three have attracted the most attention: DNA, RNA, and proteins.

24-8 DNA Cloning

In the last ten years biochemists have learned ways to cut DNA apart, modify it, reassemble it, and make unlimited copies. Perhaps most importantly, with DNA as the starting material, scientists can generate RNA and then protein molecules of a particular size and kind. The key to all these manipulations is gene cloning. **Gene cloning** is the process by which many identical copies of a particular DNA sequence are made. It is cloning, more than any other single technique, that has changed the nature of biochemistry.

The genetic makeup of complex organisms resisted analysis until recently. Even the DNA content of a bacterial cell is very large. In mam-

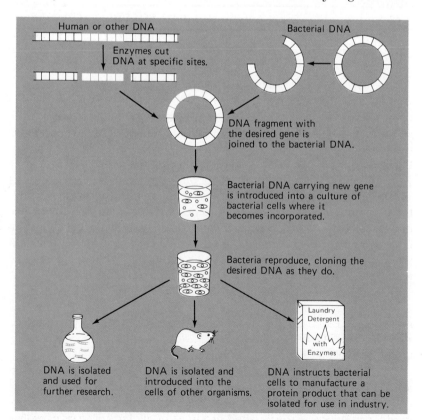

Figure 24-9 The diagram shows how foreign genes can be inserted into the DNA of a bacterial cell where it will be copied exactly (cloned) as the cell multiplies.

mals, each cell carries some 2.5 billion bits of information in the base pairs arranged along the chromosomal DNA. The base pairs are arranged in a specific order and divided into discrete working units— genes. There are between 50 000 and 100 000 genes in every mammal cell. Each one codes for the structure of a particular gene product, usually a protein. Studying a single gene was impossible until recently because single genes could not be isolated from the rest of the DNA, and because there was no way to make more copies of it.

A partial solution to these problems came from studying viruses. The DNA content of viruses is very small compared with that of a mammalian cell and yet their genes are similar. The DNA of one much studied virus, the SV40 virus of monkeys, has only 5243 base pairs arranged into five genes. An individual gene could be more easily isolated from a virus. Also, the viral DNA content will be copied exactly several hundred thousand times within the cell that it infects. It is not difficult to separate the viral DNA from the cell's DNA.

In the mid-1970's, procedures were developed for isolating genes and for analyzing the base sequences of DNA. DNA is structurally similar in all living things. As a result, biochemists can determine the base sequence of the DNA from a desired gene and can successfully blend it with the DNA of bacteria or viruses. In this way, a specific gene can be introduced into a virus or bacterium where it will be copied hundreds of thousands of times. This is the process of gene cloning.

The technology now exists to clone any of the genes coding for the hundreds of enzymes studied by biochemists and to insert those genes into plants and animals. Genes for important proteins such as insulin, interferon, and a number of growth hormones have been isolated.

Cloned genes cannot only be inserted into microorganisms like bacteria, but also can be incorporated into whole animals. The introduction of a cloned gene into some cells of a plant or animal can alter the genetic makeup of only those few cells that acquire the gene. The rest of the organism will remain unchanged. Obviously it would be more interesting and advantageous to be able to make the change in all the cells of the organism. This only can be done by inserting the cloned genes into the reproductive cells, eggs and sperm of the organism. In this way, mice have been developed that carry genes for extra growth hormone. Giant mice (one and a half times normal size) result. Cattle with altered growth potential will soon be available.

Another new advance in biomolecular engineering allows genes and the proteins for which they code to be redesigned so that they have new functions. These "designer genes" are produced by changing DNA sequences, or by adding or deleting a part of a gene. Genes altered by these techniques then can be reintroduced into the cells in which they normally occur. An enzyme that does not easily bind to its substrate can be made to associate readily with it, or to bind with a totally different substrate molecule. A gene that is usually activated by a particular metal ion can be made to shut down in the presence of that same ion. In short, there is less and less that is impossible for the biomolecular engineer to accomplish.

As gene cloning techniques are improved and as additional genes are cloned, the possibilities for altering an organism's characteristics will expand enormously. The biochemical engineer will no longer regard living things as the finished products of millions of years of evolution, but will be an active participant in the process of change.

Figure 24-10 The mouse on the left results when rat growth hormone genes are added to a fertilized egg. Because the gene was incorporated into all the cells of this mouse, it will be passed on to its offspring.

24-9 Proteins Made to Order

Industrial chemists have long known that for efficiency and economy, it is hard to beat enzymes. Enzymes are used in a wide variety of chemical processes, including the production of food, medicines, and detergents. However, there are problems. The protein structure of enzymes make them susceptible to destruction by heat, high pressure, mechanical stress, and organic solvents—all conditions routinely found in industry. Because most naturally occurring enzymes are so easily inactivated, very few new applications have arisen until recently.

There are several commercially important enzymes that have been in use for years. Rennin coagulates milk in cheesemaking. Three naturally occurring enzymes are used together to convert corn starch into high-fructose corn syrup. Proteases (enzymes that break down proteins) are added to laundry detergents to help remove stains.

Biochemists are now using computer graphics, DNA synthesis techniques, and gene cloning to tailor-make new enzymes that are built for a specific purpose or made to work in a certain set of conditions. This "protein engineering" promises to open the doors to a host of new applications and new products. Artificial enzymes could result in new drugs and agricultural chemicals. They could catalyze reactions that no living organism can. They could be modeled to replace naturally occurring materials, such as silk or wool.

Gene cloning is involved in the creation of new proteins. As explained in Section 24-8, the DNA of genes can be isolated and rearranged, then copied hundreds of thousands of times by bacterial cells. Finally, the cloned gene is translated into protein by the bacterial cells and a large number of protein molecules are produced.

A protein that is synthesized only in limited amounts in its normal environment can be produced in large amounts when its gene is redesigned for large scale production by a bacterial or yeast cell. It is on this fact that the biomolecular engineer depends. Microorganisms bearing cloned genes can be grown inexpensively in large volume.

One way to make proteins more stable at high temperatures is to increase the number of disulfide bonds (S—S) between nearby cysteine amino acids. The sulfur-sulfur bonds reduce the tendency of the protein molecule to unfold under the influence of high heat. Computer graphics have been used to find the sites where these bridges could be inserted without changing the overall nature of the protein.

The greatest challenge facing protein engineers is to create enzymes and other proteins that do not exist in nature. These synthesized proteins could be used to catalyze reactions that are not catalyzed by natural enzymes. Gene modification and cloning are used to make these new proteins. Another approach is to make proteins in test tubes through more conventional organic chemistry. One method involves growing chains of amino acids on microscopic polystyrene beads and then linking them together in a predetermined sequence.

Pollution control could offer one potential new market for enzymes. In the paper industry, for example, the waste water that results from the use of chlorine in the production of paper pulp creates a serious disposal problem. Such pollution would be substantially reduced if the chlorine were replaced by an enzyme similar to a naturally occurring one that breaks down the structural components of wood.

Figure 24-11 Genetically altered yeast cells are grown in large vats like this one where they multiply and produce large quantities of desired DNA and/or protein products.

24-10 A Glimpse Ahead

Many of the applications of gene cloning have not been invented. Others are just now becoming available. Biomolecular engineering is an exciting and controversial field of study.

One of the most promising areas in biomolecular engineering is in gene cloning for gene-deficient conditions. A condition that is being treated this way is growth hormone deficiency in children. The gene is cloned and translated into growth hormone protein. This protein is then extracted from the bacteria, purified, concentrated, and finally, used to treat the condition in children. After treatments with growth hormone, the child can resume a normal growth pattern.

The recognition of genetic diseases in humans has stimulated research on ways of repairing defective genes, both in the affected individual and in any offspring. Techniques are now available for introducing cloned genes into certain tissues such as bone marrow and skin cells. The cloned genes can be complete, healthy versions of defective ones in the tissue cells. Such a transfer of genes could help to reverse the effects of genetic defects. Insertion of cloned genes into human reproductive cells, where they would be passed on to all the cells of a resulting embryo, might be possible within the next ten years.

Harvesting food from the sea may become less luck and more science in the near future. Scientists have isolated a chemical from a marine bacterium that, when added to water containing the larvae of oysters, causes them to settle out of the water and onto the bottom. This makes them less vulnerable to being eaten and can greatly increase the number of oysters that make it to maturity. The bacterial chemical soon may be used in commercial oyster hatcheries.

In other research on food sources from the sea, scientists have isolated the gene that permits the winter flounder, a fish that inhabits the cold ocean bottom of the North Atlantic, to remain alive and functioning in water that is barely above freezing. It will soon be possible to clone the gene and transfer it into other fish that are not able to survive at freezing temperatures.

There are daily accounts in newspapers and magazines of new discoveries and techniques. As genetic manipulation becomes more of a reality, difficult questions arise about how it will be used. When human intelligence, body build, and temperament can be "improved" through gene cloning and transfer, should it be done? How will the environment be affected by the release of organisms with altered genes? What lies ahead in the future is only limited by the knowledge and understanding of the students of today.

Review and Practice

1. What is gene cloning?

2. Name two advantages that an artificially altered protein might have when compared to a naturally occurring protein.

3. Why have proteins been used only rarely to catalyze reactions in industry?

4. Give two ways gene cloning has been used to change organisms.

1. the process by which many identical copies of a gene are made
2. It may be less susceptible to heat. It could catalyze novel reactions or work more quickly.
3. Proteins are easily destroyed by common industrial conditions of high heat, pressure, mechanical stress, and organic solvents.
4. It has been used to change growth patterns, life cycles, tolerance to cold, and to correct defective genes.

Summary

■ Biochemistry is the study of complex biomolecules that function in a variety of ways in living organisms. Some of these molecules are being manufactured for use in industry.

■ Carbohydrates are molecules made of carbon, hydrogen, and oxygen with the empirical formula CH_2O. The breakdown of carbohydrates into monosaccharides provides much of the energy needed by the body.

■ Lipids also are used as an energy source by the body. Most lipids in the diet are triglycerides, composed of fatty acids and glycerol. Cholesterol is a steroid lipid that has been implicated in heart disease.

■ Proteins give cells their three-dimensional structure, direct the function of the cell, and are important as catalysts for biochemical reactions. They are made of amino acids linked by peptide bonds. The sequence of amino acids and the characteristics of the amino acid side chains give the protein molecule its specific shape and function.

■ Nucleic acids like DNA and RNA are responsible for the manufacture of specific proteins within a cell and also for the transmission of genetic information from generation to generation.

■ Hormones and neurotransmitters provide a means of communication within a cell. Most hormones are steroid derivatives and most neurotransmitters are peptides. Recent research has shown that some peptide hormones also can act as neurotransmitters.

■ Enzymes are protein catalysts that reduce the energy requirements of a reaction and speed up the reaction rate. Enzymes act on specific substrates or on classes of substrates.

■ Some enzymes require a mineral or vitamin for their activity. Minerals usually are metal ions that bind loosely or tightly to the protein. Vitamins are nonprotein coenzymes that work with enzymes. Vitamins and minerals must be included in the diet.

■ Techniques now exist to isolate DNA, modify it, and clone multiple copies of it using microorganisms like bacteria. Gene cloning is used to produce proteins for study and industry. It also can be used to treat gene-deficient conditions.

Chemically Speaking

active site	mineral
amino acids	monosaccharide
ATP	neurotransmitter
carbohydrates	nucleic acid
coenzyme	nucleotide
cross-linking	polypeptide
denatured	polysaccharide
disaccharide	polyunsaturated
DNA	protein
enzyme	receptor
fatty acid	RNA
gene	steroid
gene cloning	substrate
hormone	triglyceride
lipid	vitamin

Review

1. Name the four classes of biomolecules. (24-1, 2, 3, 4)

2. Give the primary function(s) of each class of biomolecules. (24-1, 2, 3, 4)

3. How does heat affect a protein? (24-3)

4. Name the units that make up the following large molecules:
   a. proteins (24-3)
   b. carbohydrates (24-1)
   c. nucleic acids (24-4)

5. a. What are the four nitrogen bases found in DNA?
   b. What are the four nitrogen bases found in RNA? (24-4)

6. Show the base pairs that can form in a DNA molecule. (24-4)

7. Show the base pairs that can form in a RNA molecule. (24-4)

8. Why is the order of amino acids important in a protein? (24-3)

9. What is the difference between the lipids in butter and the lipids in vegetable oil? (24-2)

10. What is an enzyme? (24-6)

11. How does the order of bases on one DNA strand determine the order on the other? (24-4)

CHAPTER REVIEW

Numbers in red indicate the appropriate chapter sections to aid you in assigning these items. Answers to all questions appear in the Teacher's Guide at the front of this book.

Interpret and Apply

1. Draw the possible isomers of C_3H_8O. (24-1)

2. Show how the disaccharide sucrose can be formed by splitting out a water molecule from one molecule of glucose and one of fructose. Represent both glucose and fructose in their cyclic form. (24-1)

3. Why is an enzyme usually specific for one process? (24-6)

4. Explain what happens to the protein in an egg when it is cooked. (24-3)

5. Palmitic acid is a saturated fatty acid with a melting point of 63°C. Linolenic acid is a polyunsaturated fatty acid that has a melting point of −11°C. Predict whether the following are saturated or polyunsaturated fatty acids:
   a. lauric acid mp = 44°C
   b. oleic acid mp = 4°C
   c. stearic acid mp = 70°C (24-2)

6. Draw the molecular structures of fructose and glucose. How do they differ? (24-1)

7. Which one of the following foods would provide the most energy per gram when it was metabolized completely: (24-2)
   a. steak
   b. butter
   c. potato
   d. lettuce

8. Name two conditions that could change the reactivity of an enzyme. (24-6)

9. What is the connection between nucleic acids and protein synthesis? (24-4)

Problems

1. A fragment of DNA has the base sequence TAGAGCCTCAGG: (24-4)
   a. If another strand of DNA paired to this fragment, what sequence of bases would the new DNA have?
   b. If a strand of RNA paired to this fragment, what sequence of bases would the RNA have?

2. Draw the structures for a DNA nucleotide and an ATP molecule. (24-4)

   a. How are they different?
   b. How are they similar?

3. When organic molecules in the cell are synthesized, energy is required. Where does it come from, and how is it stored? (24-1, 24-4)

4. If you consume 2500 Calories (kcals) of carbohydrates, how many grams of fat would you have to eat to supply the same number of Calories? (24-2)

Challenge

1. Cellulose is a polysaccharide with a molar mass of approximately 600 000 grams. Starch has a molar mass near 4000 grams. The monosaccharide unit in cellulose and in starch both have the empirical formula $C_6H_{10}O_5$. These units are about 0.50 nm long.
   a. Approximately how many units occur in cellulose and in starch?
   b. How long are the molecules of cellulose and starch?

Synthesis

1. A 100-milligram sample of a compound containing only C, H, and O was found by analysis to give 149 milligrams of CO_2, and 45.5 milligrams of H_2O when burned completely. Calculate the empirical formula.

2. Given the reaction:
   $$C_6H_{12}O_6(s) + 6O_2(g) \longrightarrow 6CO_2(g) + 6H_2O(l)$$
   a. What is the ΔH for the reaction?
   b. Is the reaction exothermic or endothermic?

Projects

1. Research the controversy surrounding the release of genetically altered organisms into the environment. What are the dangers? How can the laboratories that create these organisms be regulated?

2. Look for stories about other applications of biomolecular engineering in magazines and newspapers. How might they affect your life? How might they affect the next generation?

Appendix A

Description of a Burning Match

My match was a pale yellow stick about 6 cm long. The stick was square and about 2 mm across. One end of the stick had a dark blue solid covering about 0.5 cm of the stick, tipped by a light blue solid about 1 mm wide. I could bend the stick a little, but it broke after bending about 1 mm. The break was not straight across, but seemed to follow fibers that made up the stick.

When I rubbed the head of the match (the blue end) across a brick, I heard a scratching sound and then a distinct hiss. A flame came from the head and flared out in all directions. The flame extended from 0.5 to 1.5 cm from the head. As the hiss stopped, the flame steadied. Now it was shaped like a flickering triangle with a base of 0.5 to 1.5 cm and a height of 1 to 2 cm. The match itself went through the flame about 2 mm. It didn't seem to quite touch the flame. Near the match, the color of the flame was blue, but farther out it was yellow. Heat came from the flame as it moved along the match. As the flame moved, the match turned black and curled slightly.

When the flame reached my fingers, I blew it out. A red glow remained near the head of the match. Near the red glow, the black stick almost totally disappeared, leaving a small amount of white solid in its place.

When I first struck the match, black smoke rose from the match for a second or two. When I blew it out, white smoke rose for about the same length of time. There was also a smell that I have often smelled around fires, and that lasted for 5 to 10 seconds after I blew out the match.

Appendix B

Hybrid Orbitals

Hybrid orbitals is a model used by chemists to explain the shapes of some molecules. This model takes into account that atomic orbitals are changed when atoms bond.

Use atomic orbitals to determine the shape of a molecule such as methane, CH_4. The electron configuration of a carbon atom is $1s^2 2s^2 2p^2$. Experimental evidence shows that CH_4 has a tetrahedral shape. How can this shape be explained? It is certainly not what you might expect based on the available bonding sites for carbon.

Chemists have devised a model that ties together electron configurations with the geometry of molecules. As the carbon and hydrogen atoms bond, some energy is used to rearrange the $2s$ and $2p$ orbitals of the carbon atoms. The rearrangement results in four orbitals of equal energy and shape that form a tetrahedral structure. The $1s$ orbitals of the hydrogen atoms bond with each of the four new orbitals of the carbon atom. A methane molecule is formed. The potential energy of the molecule thus formed would be less than the energy of a molecule formed directly from the original atomic orbitals of carbon.

A closer look at the process may be of more help. In order to arrive at the finished product, several steps are taken along the way.

First, a $2s$ electron is promoted to a $2p$ orbital in the carbon atom:

C	1s	2s	2p
	⊗	⊗	⦸ ⦸ ○
C	⊗	⊘	⦸ ⦸ ⦸ promoted

Second, the newly formed $2s$ and $2p$ system hybridize so that all four orbitals will have the same energy, and the system will be arranged in a tetrahedron.

C	1s	New orbital system
	⊗	⦸ ⦸ ⦸ ⦸

Third, the hydrogen atoms bond to the new orbital system forming methane with a stable octet.

C   1s   New orbital system

The new orbital system is a hybrid of the original $2s$ and $2p$ orbitals. There is no energy difference between hybrid orbitals as there was between the original $2s$ and $2p$ orbitals. This new system is called a sp^3 (sp three) hybrid orbital system. The name refers to the origin of the new orbitals.

Hybridized orbitals also explains the linear structure of BeH_2. In order to form BeH_2 the hybrid model can be broken down into the following steps.

1. Promotion:

Be   1s   2s   2p

Be   promoted

2. Hybridization:

Be   1s   sp

3. Bonding:

Be   1s   sp

H H

The sp hybrid orbital system describes a molecule that is linear. The orbitals will repel each other and the bond angle will be 180°.

A hybrid system can be drawn for many molecules. Water, for example, has a sp^3 hybrid system. This system also describes the geometry of expanded octets as PCl_5 and SF_6. Table C-1 summarizes the relationship between hybridized orbital systems and the geometry of molecules.

Table C-1

HYBRID ORBITAL SYSTEMS					
NO. OF BONDED ATOMS	NO. OF UNSHARED PAIRS OF e-	TYPE OF HYBRID	BOND ANGLE	GEOMETRY OF MOLECULE	EXAMPLE
2	0	sp	180°	linear	BeF_2
3	0	sp^2	120°	trigonal planar	BF_3
4	0	sp^3	109.5°	tetrahedral	CH_4
3	1	sp^3	>109°	pyramidal	NH_3
2	2	sp^3	>105°	angular	H_2O
5	0	sp^3d	varies	trigonal bipyramidal	PCl_5
6	0	sp^3d^2	90°	octahedral	SF_6

GLOSSARY

Pronunciations by permission from *Webster's Ninth New Collegiate Dictionary* © 1983 by Merriam-Webster Inc., publishers of the Merriam-Webster® Dictionaries.

A

absolute zero   zero kelvin temperature (p. 201)

accuracy   closeness to the accepted value or amount of something (p. 75)

acid   a substance that can donate a proton in a reaction (p. 542)

acidic anhydride   oxygen-containing compound that produces an acidic solution when reacted with water (p. 574)

activated complex   the peak of the energy barrier being reached by the successful collision of reactant particles (p. 500)

activation energy   the minimum amount of energy required to overcome the energy barrier in a reaction (p. 500)

active site   the crevicelike portion of a globular protein that can encompass another molecule (p. 662)

activity   the ability of an element to replace another element in a compound (p. 136)

actual yield   the amount of product obtained in a chemical reaction (p. 170)

addition   a process in which atoms are added to an unsaturated organic molecule in a reaction (p. 646)

addition polymerization   a reaction in which monomers are bonded without the elimination of atoms (p. 648)

aldehydes   molecules in which the carbonyl group is attached to a carbon atom having at least one hydrogen atom (p. 642)

alkali metals   the group 1 elements in the periodic table, characterized by their ability to give up one electron to reach the electron configuration of a noble gas (p. 307)

alkaline earth metals   group 2 elements in the periodic table, characterized by their ability to give up two electrons to reach the electron configuration of a noble gas (p. 310)

alkanes   single- or branched-chain hydrocarbon compounds in which the carbon atoms are connected by only single covalent bonds (p. 629)

alkene series   straight- or branched-chain hydrocarbons that have at least one double carbon-carbon bond (p. 633)

alkynes (al'-kīnz)   hydrocarbon compounds that have a triple carbon–carbon bond (p. 635)

allotropes   forms of an element that may have different molecular structures (p. 384)

alloy   a mixture of two or more metals or a metal with a nonmetal (p. 370)

alpha emission   the release of an alpha particle in radioactive decay (p. 249)

alpha particle   a high-speed, positively charged particle emitted by a radioactive substance (p. 230)

amides   products of the reaction between an amine or ammonia with an organic acid (p. 644)

amines (ə-mēnz')   organic compounds derived from ammonia (p. 644)

amino acids   organic acids that are the chief components of proteins (p. 661)

amorphous solids   solids with random particle arrangement (p. 426)

ampere   a measurement of the rate of flow of electric current (p. 595)

amphiprotic (am(p)-fē-prō'-tik)   describing a compound that can act as a proton donor or a proton acceptor (p. 542)

amplitude   a measure of the displacement of a wave from its midpoint (p. 278)

anhydrous   describing a substance that is without water (p. 433)

anode   a positively charged or electron-poor electrode; the site of oxidation (p. 593)

antimatter   collectively all antiparticles, that is, the opposites of electrons, protons, etc. (p. 272)

area   the surface of an object (p. 17)

ATP   adenosine triphosphate, the energy-storing molecule of all living organisms (p. 664)

atmosphere   a unit of measurement of pressure equal to 101.325 kilopascals (p. 182)

atom   the smallest unit of an element (p. 36)

atomic mass   quantity of matter in an atom (p. 96)

atomic mass unit   a unit for expressing masses of atoms, equal to $\frac{1}{12}$ of the mass of carbon 12 (p. 96)

atomic number   the number of protons found in the nucleus of an atom of an element (p. 235)

atomic radius   half the distance between the nuclei of two unbonded atoms (p. 317)

Avogadro's hypothesis   principle stating that equal volumes of different gases at the same temperature and pressure contain equal numbers of particles (p. 97)

Avogadro's number (6.02×10^{23}) a convenient number to express the relative mass of a mole of atoms (p. 98)

B

background radiation   radiation from the natural environment (p. 249)

barometer   a device for measuring atmospheric pressure (p. 182)

base   1: the number in scientific notation that is to be multiplied (p. 64) 2: a substance that can accept a proton in a reaction (p. 542)

basic anhydride   an oxygen-containing compound that produces a basic solution when reacted with water (p. 574)

beta emission   the release of an electron from a nucleus in a radioactive decay (p. 250)

beta particle   a high-speed electron emitted from the nucleus of an atom during radioactive decay (p. 230)

binary compounds   compounds that contain only two elements (p. 45)

binding energy   the amount of energy needed to separate a nucleus into its individual protons and neutrons (p. 262)

boiling point   temperature at which a liquid vaporizes or changes to a gas (p. 26)

boiling point elevation   the raising of the boiling temperature of a pure liquid due to the presence of a solute (p. 451)

bond energy   the energy associated with the formation or breaking of a bond (p. 330)

Boyle's law   the principle describing the relationship between the pressure and volume of a gas when temperature and moles are constant (p. 195)

branched chain   describing hydrocarbon molecules composed of several carbon chains that cross (p. 628)

bright-line spectrum   distinctive line of colored light seen when viewed through a prism; produced by excited gaseous atoms or molecules (p. 284)

buffers   solutions that maintain a fairly constant pH even when more acid or base is added (p. 576)

C

calorie   the energy required to raise the temperature of one gram of water one degree Celsius (p. 474)

Calorie   unit of energy equal to one kilocalorie; used to describe the energy in foods (p. 474)

calorimeter   an instrument for measuring the heat content of a substance (p. 475)

carbohydrates   organic molecules, such as sugars and starches, that have the empirical formula CH_2O (p. 659)

carbon backbone   the longest carbon chain in a hydrocarbon molecule (p. 627)

carbonyl (kär'-bə-nil) **group**   the functional group —C— found in aldehydes and ketones (p. 642)

$$\overset{\displaystyle O}{\overset{\displaystyle \|}{}}$$

carboxyl (kär-bäk'-səl) **group**   the functional group, —COOH found in organic acids (p. 643)

catalyst   a substance that increases the rate of a chemical reaction and may be recovered unchanged at the end of the reaction (p. 502)

cathode   the negatively charged or electron-rich electode; the site of reduction (p. 593)

cathode-ray tube (CRT)   a gas discharge tube in which a beam of negative, glowing particles is produced (p. 223)

centrifuge   a machine whose spinning action separates substances of different densities (p. 619)

Charles's law   a principle describing the relationship between the volume of a gas and its temperature when pressure and moles are constant (p. 188)

chemical bond   the result of a mutual attraction of two atomic nuclei for a pair of electrons (p. 330)

chemical change   a change in which a new substance is formed having different properties (p. 31)

chemical equilibrium   condition when no observable changes are taking place in a reversible reaction (p. 513)

chemical family   relationships and similarities among elements as grouped in the periodic table (p. 311)

chemical formula   expression of a compound in symbols (p. 44)

chemical symbol   abbreviation of an element (p. 40)

classification   a grouping together of like things because of their similarity (p. 2)

coefficient (kō-a-fish'-ənt)   a number that precedes and multiplies the atoms in a chemical equation (p. 129)

coenzyme   nonprotein, organic compound that binds to a protein molecule, providing specific chemical functions (p. 668)

colligative properties   properties that depend on the numbers of particles and not on the nature of the particles (p. 451)

collision theory   principle that states that chemical reactions depend on collisions between reacting particles (p. 495)

combustion   chemical reaction of a substance with oxygen which produces energy in the form of heat and light (p. 134)

compound   a resultant substance formed from the chemical combination of simpler substances (p. 33)

concentration   the number of moles of solute in a given amount of solution (p. 117)

condensation polymerization   a reaction in which monomers are bonded and some atoms are eliminated (p. 649)

conjugate acid–base pairs   molecules or ions that differ only by a proton (p. 543)

continuous spectrum   the rainbow of colors caused by separating white light into different frequencies and wavelengths when it is projected through a prism (p. 280)

control rods   devices usually made of cadmium and used to regulate a nuclear reaction by absorbing some neutrons (p. 266)

coulomb (kül'-läm)   a unit used to measure the amount of electricity flowing through a wire (p. 595)

covalent bond   a chemical bond in which two atoms share a pair of electrons (p. 332)

cracking   the process of breaking down complex organic molecules by the use of heat and certain catalysts (p. 633)

critical mass   the smallest amount of material needed to continue the chain of disintegrations in nuclear fission (p. 265)

cross-linking   a form of bonding between sulfur atoms (p. 662)

crystal lattice   a pattern created by the arrangement of positive and negative ions in a crystal structure (p. 336)

crystalline   describing a solid structure in which the molecules or

ions are arranged in a definite, geometric, and repeating pattern (p. 421)

cycloalkane   saturated hydrocarbon composed of a ring of three or more carbon atoms (p. 636)

cyclotron   a particle accelerator in which particles move in a widening spiral path (p. 253)

D

Dalton's law of partial pressure   principle stating that the total pressure of a gas is equal to the sum of the individual pressures in a mixture (p. 185)

daughter nuclide   the resulting element in radioactive decay (p. 249)

decomposition (dē-käm-pə-zish'-ən)   a chemical reaction in which a substance breaks down to form two or more simpler substances (p. 31)

delocalized   refers to the distribution of electrons not associated with any one carbon atom in a hydrocarbon (p. 637)

denatured   describes proteins whose three-dimensional structure is destroyed due to the breakdown of hydrogen bonds (p. 662)

density   mass per unit volume of a material (p. 89)

derivatives   organic compounds having other atoms or groups of atoms in place of hydrogen atoms (p. 627)

diatomic molecule   a molecule consisting of two atoms (p. 107)

dipole   a polar bond or molecule with a partial positive charge and a partial negative charge (p. 334)

dipole–dipole force   the attraction between polar molecules (p. 357)

disaccharide (dī-sak'-ə-rīd)   a carbohydrate made of two monosaccharides (p. 659)

dissociation   separation of a crystal into component ions through the breakdown of the crystal lattice (p. 444)

distillation   evaporation followed by condensation of vapors; separates components of a mixture (p. 28)

DNA   biomolecules containing deoxyribose that store the information for cell activity, synthesis of proteins, and genetic transfer (p. 662)

doped   condition of pure silicon crystals after impurities have been added to increase the conductivity (p. 398)

double bond   a covalent bond in which two pairs of electrons are shared by two atoms (p. 341)

double replacement   a chemical reaction in which elements replace each other in a compound (p. 136)

dry cell   an electrochemical cell in which the electrolytes are present as solids or as a paste (p. 602)

E

electrochemical cell   a system of electrodes and electrolytes in which a spontaneous oxidation-reduction reaction can be used as a source of electric current (p. 592)

electrode   an electrically conducting solid that is used to make electrical contact (p. 223)

electrolysis   a breakup of the composition of matter by electric current (p. 32)

electromagnetic waves   energy waves produced by a combination of magnetic and electrical fields (p. 282)

electron   a negatively charged particle found in atoms and that make up cathode rays (p. 225)

electron configuration   organization of electrons in an atom from the lowest energy orbital to the highest energy orbital (p. 295)

electron dot structure   structure with dots placed about an element's symbol indicating the number of valence electrons (p. 337)

electronegativity   the attraction of an atom for a shared pair of electrons in a covalent bond (p. 333)

element   a pure substance that cannot be decomposed to simpler substances (p. 34)

empirical formula   the smallest whole number ratio of atoms found in a compound; determined through experimentation (p. 110)

endothermic (en-də-thər'-mik)   type of reaction in which more energy is absorbed to break bonds than is released to form new bonds (p. 137)

energy anything that can change the condition of matter (p. 7)

energy levels quantized levels of energy associated with electrons in atoms (p. 285)

enthalpy the heat content of a system (p. 481)

entropy a measure of the degree of disorder in a substance or system (p. 489)

enzymes biological catalysts that accelerate and control the rate of chemical reactions in cells (p. 666)

equilibrium the condition in any reversible process in which the forward and reverse processes occur at the same rate (p. 417)

equilibrium constant a ratio of the product of the concentrations of the substances to the product of the concentrations of reactants (p. 524)

equivalence point the point in acid–base titrations when the proportions of acid and base are equivalent to the proportions described by the equation (p. 567)

esterification a reaction between an organic acid and an alcohol (p. 643)

esters class of products of the reaction between an organic acid and an alcohol (p. 643)

ethers organic molecules having an oxygen atom bonded between two hydrocarbon portions (p. 642)

excess intentional inclusion of a larger quantity of one reactant in the combining process (p. 165)

exothermic (ek-sō-thər'-mik) describing a chemical reaction in which energy is released to the surroundings (p. 140)

exponent a number that indicates how many times another number is multiplied by itself (p. 64)

extensive property a characteristic property that depends on the amount of matter being measured (p. 90)

F

fatty acids chains of carbon atoms linked to a carboxyl group (p. 660)

fission a process in which an atomic nucleus is broken down into smaller nuclei by being bombarded with low-energy neutrons (p. 264)

flame test a procedure used to identify the presence of a metal in a compound by the color of the flame (p. 617)

fluid any substance that flows (p. 179)

formula unit the smallest whole-number ratio of ions of different elements in a compound (p. 105)

fractional distillation a process that separates a mixture into fractions by differences in their boiling points (p. 452)

freezing point the temperature at which a liquid freezes or becomes a solid (p. 30)

freezing point depression a condition when the freezing point of a solution is decreased due to the presence of solute particles (p. 453)

frequency the number of waves that pass a point in a unit of time (p. 278)

functional group the presence of an atom, group of atoms, or organization of bonds that determines specific properties of a molecule (p. 640)

fusion the combining of small atomic nuclei to form larger nuclei (p. 266)

G

gamma rays high-energy rays emitted in radioactivity (p. 230)

gas one of the four states of matter; characterized by molecules that are relatively far apart and that completely fill their container (p. 177)

gene a portion of a DNA molecule that codes for a protein (p. 664)

gene cloning a process used to produce many copies of a DNA sequence (p. 670)

Gibbs Free Energy the maximum possible work that results from a chemical reaction (p. 486)

Graham's law a mathematical formula for the comparison of the rates at which two gases diffuse (p. 212)

ground state the lowest energy state of an atom (p. 289)

groups vertical arrangements in the periodic table of elements with similar chemical properties (p. 306)

H

half-cell potential the contribution of cell voltage made by each half-reaction in an electrochemical cell (p. 597)

half-life ($T_{1/2}$) the time it takes for one half of the nuclei of a radioactive sample to decay (p. 255)

half-reactions two equations that describe oxidation or reduction reactions in a redox reaction (p. 582)

halogens a family of elements in the periodic table that has seven valence electrons (p. 312)

heat of combustion the energy released when a substance is burned (p. 483)

heat of formation the energy observed in the formation of a compound (p. 481)

heat of fusion the heat required to change one kilogram of matter from a solid to a liquid (p. 471)

heat of reaction the energy observed during a chemical reaction (p. 481)

heat of solution the change in energy required for solvation (p. 450)

heat of vaporization the heat required to change one kilogram of matter from a liquid to a gas (p. 471)

Henry's law principle stating that the mass of a gas solute dissolved within a liquid is proportional to the pressure upon the system (p. 448)

heterogeneous mixtures mixtures in which the components are segregated when at rest (p. 439)

homogeneous mixtures mixtures in which the components are uniformly distributed (p. 439)

homologous series a series of similar compounds that differ in number of the same structural unit (p. 630)

hormones chemicals released by the glands of the endocrine system (p. 665)

hydrated crystal a solid compound that has water molecules as an integral part of the crystal (p. 432)

hydrated ions ions surrounded by water molecules when in solution (p. 432)

hydrocarbons compounds made only from hydrogen and carbon (p. 135)

hydrogen bond the weak attraction of a positive hydrogen atom in one molecule to an electronegative atom of another similar molecule (p. 357)

hydrolysis (hī-dräl'-ə-səs) the reaction of a salt with water to produce an acidic or basic solution (p. 573)

hydronium (hī-drō'-nē-əm) **ion** a hydrated proton; H_3O^+ (p. 541)

hydroxyl (hī-dräk'-səl) **group** the functional group —OH found in alcohols (p. 641)

hypothesis a tentative explanation for observed facts or events (p. 2)

I

ideal gas constant a constant factor in the ideal gas equation relating pressure, temperature, volume, and number of moles, and represented by R (p. 202)

ideal gas law a mathematical equation based on the kinetic molecular theory that relates the amount of gas in a sample to pressure, volume, and temperature (p. 202)

immiscible describing substances that are insoluble in each other (p. 443)

indicator compounds used to detect the presence of acids and bases (p. 541)

inertia the property of matter that causes it to resist change in motion (p. 5)

inference a conclusion drawn from observed information (p. 1)

inhibitor a substance that reduces a reaction rate (p. 502)

integrated circuit an arrangement of electronic circuit components on a piece of semiconductor material (p. 400)

intensive property a characteristic or property not dependent on the mass being measured (p. 90)

ion pairs ions created by exposure to radiation (p. 248)

ionic bond the attraction between two oppositely charged ions (p. 334)

ionic radius the size of the electron probability volume for an ion (p. 320)

ionization energy energy needed to remove an electron from an atom (p. 288)

ionizing radiation radiation that produces ions from atoms or molecules (p. 248)

ions atoms or groups of atoms that are positively or negatively charged (p. 39)

isoelectronic having the same number of valence electrons (p. 314)

isotopes atoms of the same element that contain different numbers of neutrons (p. 237)

J

joule (J) SI unit for the amount of energy produced when a force of one newton acts over a distance of one meter (p. 140)

K

K_a the acid dissociation constant of a weak acid (p. 554)

K_b the base dissociation constant of a weak base (p. 557)

kelvin a unit of temperature on the kelvin scale (p. 189)

kelvin scale a scale for measuring temperature; based on a zero value equal to $-273.15°C$ (p. 189)

ketones organic molecules in which a carbonyl group is attached to a carbon atom that is bonded to two other carbon atoms (p. 642)

kilocalorie the energy required to raise the temperature of one kilogram of water one degree Celsius (p. 474)

kinetic energy energy of motion (p. 8)

kinetic molecular theory a theory that describes the sets of conditions for and behavior of ideal gases (p. 201)

K_w the ion product constant for water, which equals 1.0×10^{-14}

L

law of conservation of mass the principle stating that the total quantity of reactants is constant despite a change in form (p. 128)

law of definite composition the principle stating that the proportion of elements in a specific compound is a fixed quantity (p. 34)

law of multiple proportions the principle stating that different compounds can be made having different proportions of the same elements (p. 34)

Le Chatelier's principle a generalization stating that when conditions are changed, a system in equilibrium will adjust to produce a new equilibrium (p. 521)

leptons elementary particles of matter (p. 272)

limiting reactant a chemical that limits the amount of product that can be obtained in the combining process (p. 166)

lipids a group of biomolecules that includes fats and oils (p. 660)

liquid one of the four states of matter; characterized by molecules that are close together but that move freely (p. 177)

logarithm the power to which 10 must be raised to equal a specific number (p. 562)

London forces short-lived attractions between nonpolar molecules due to momentary dipoles (p. 358)

M

macroscopic observations the more obvious observations made by seeing, feeling, and smelling (p. 36)

macroscopic properties obvious properties such as boiling point or mass (p. 36)

manometer a device used to measure pressure of gases other than the atmosphere (p. 184)

mass a property of matter that causes it to have weight (p. 5)

mass defect the difference between the mass of a nucleus and the sum of the masses of protons and neutrons that comprise it (p. 262)

mass number the total number of protons and neutrons in the nucleus of an atom (p. 236)

mass spectrometer an instrument used for a variety of chemical analyses; makes use of differences in deflection of particles of different charges and different masses (p. 240)

matter anything that has mass and occupies space (p. 4)

mean free path the average distance between two successive collisions of molecules moving in straight-line motion (p. 201)

melting point the temperature at which a solid melts or becomes a liquid (p. 30)

metallic bond a bond between atoms with closely spaced energy levels that permit electrons to shift easily in metallic solids (p. 367)

metalloids a class of elements with intermediate properties between metals and nonmetals (p. 383)

metals category of elements that have particular properties, such as being able to conduct heat and electricity (p. 43)

microscopic model a small, rather than macro-, representation of matter (p. 36)

minerals inorganic substances (p. 667)

miscible unlimited ability to mix in solution (p. 442)

mixture a combination of different kinds of matter that retain their own properties (p. 26)

model a physical or mental representation that gives information about another object (p. 3)

moderators molecules of matter, such as water, used to slow down the neutrons in nuclear reactors (p. 266)

molality the concentration of solute per mass unit of solvent (p. 446)

molar mass the mass in grams of a mole of atoms or molecules (p. 105)

molarity the concentration of moles of solute per one liter of solution (p. 118)

mole an amount of matter that contains 6.02×10^{23} particles (p. 98)

molecular formula constituent elements and number of atoms of each element in a molecule; some multiple of the empirical formula (p. 113)

molecules particles that consist of more than one atom (p. 38)

momentary dipole the temporary, uneven distribution of electronic charge in an atom or molecule at a given instant (p. 358)

monomers single, repeating units of a polymer (p. 648)

monosaccharides (män-ə-sak'-ə-rīdz) the simplest carbohydrates, simple sugars (p. 659)

N

net ionic equation the ionic equation after spectator ions are deleted (p. 456)

network solids crystalline solids in which the atoms are joined by a network of covalent bonds (p. 430)

neutron an uncharged subatomic particle found in the nucleus of the atom (p. 233)

neurotransmitter a chemical released by a nerve cell to communicate with other cells (p. 665)

neutralization to destroy the distinctive or active properties of an acid or base (p. 542)

noble gas(es) the last group (group 18) on the periodic table, unreactive under most conditions (p. 310)

node a point on a wave with an amplitude of zero (p. 282)

nonionizing radiation radiation that does not produce ions from atoms or molecules (p. 248)

nonmetal an element generally characterized as a nonconductor and lacking the properties of a metal (p. 44)

nonvolatile solute substance that has a low vapor pressure (p. 451)

n-type semiconductors semiconductors that contain doped crystals producing mobile electrons (p. 398)

nucleic acids long-chained polymers that contain nucleotides of RNA or DNA molecules that direct the functioning of cells (p. 662)

nucleotides smaller units that comprise DNA and RNA molecules (p. 662)

nucleus the positively charged center of an atom (p. 231)

nuclide one of the isotopes of an element identified by its number of protons and neutrons (p. 247)

O

octet rule a generalization stating that bonded nonmetallic atoms have eight electrons in their outermost energy levels (p. 339)

optical fibers glass fibers used for the transmission of light (p. 407)

orbital a limited region of space surrounding a nucleus in which an electron most probably will be found (p. 289)

orbital diagrams drawings that represent electron configurations (p. 298)

organic acids compounds containing —COOH (the carboxyl group) (p. 643)

organic chemistry the study of carbon compounds (p. 627)

oxidation a chemical reaction in which the oxidation number of an atom, a group of atoms, or an ion increases (p. 587)

oxidation number the number of electrons assumed to be gained or lost in compound formation (p. 584)

oxidizing agent a substance that is reduced in a redox reaction (p. 583)

P

parent nuclide the initial element in radioactive decay (p. 249)

partial pressure the pressure exerted by only one gas in a mixture of gases (p. 185)

pascal (pas'-kəl) a unit for measuring pressure; equal to one newton of force per square meter of area (p. 181)

Pauli exclusion principle a concept stating that no two electrons can have the same four quantum numbers (p. 295)

peptide bond the specific amide bond that links two amino acid molecules (p. 649)

percent composition the mass of an element in a compound compared to the mass of the compound (p. 114)

percent yield a measure of the yield of a reaction; expressed as a percent by dividing the actual yield by the theoretical yield (p. 170)

periodic law an organizational scheme for elements that creates an identifiable pattern (p. 306)

periodic table the chemical elements arranged in a table according to their atomic structure (p. 43)

periods the horizontal rows of elements in the periodic table (p. 310)

pH a measure of hydronium ion concentration of a solution; $-\log[H_3O^+]$

phase diagram a pictorial representation of the relationship among pressure, temperature, and the phase of a substance (p. 423)

photons packets of light energy (p. 281)

photovoltaic cell an n-type semiconductor that changes solar radiation into electricity (p. 404)

physical change matter changing in appearance without forming new substances (p. 31)

physical states the common conditions, or states, in which matter exists (p. 6)

plasma the physical state of matter similar to a gas but composed of charged particles (p. 7)

polar covalent bond a bond in which there is a dipole, or unequal attraction for shared electrons (p. 334)

polyatomic ions charged groups of atoms (p. 48)

polymers compounds formed by two or more simpler molecules (p. 648)

polypeptide a protein chain consisting of amino acid molecules linked together (p. 662)

polysaccharides (päl-i-sak'-a-rīdz) carbohydrate polymers consisting of many monosaccharide units (p. 660)

polyunsaturated describing fatty acids that have more than one carbon–carbon double bond; characterized by a low melting point (p. 660)

potential energy the capacity for changes inherent in a body of matter due to its position or configuration (p. 7)

precision the amount of uncertainty in a measurement (p. 75)

pressure a force per unit area (p. 180)

principal quantum number a number that designates the energy level of an electron in an atom (p. 290)

probability the likelihood of occurrence where certainty is impossible (p. 289)

products the end results, or changed forms, after a chemical reaction (p. 128)

proportional describes a corresponding relationship of two quantities (p. 62)

protein polymers composed of large numbers of amino acids (p. 661)

proton a positively charged particle found in atoms (p. 226)

p-type semiconductors semiconductors that have been doped to produce electron deficiencies (pp. 398–399)

Q

quantity the amount of something that can be measured and described (p. 9)

quantum mechanics a branch of physics that deals with quantized energy changes of electrons in atoms (p. 289)

quark a theoretical particle believed to compose a vast array of subatomic particles (p. 271)

R

radiation energy in waves or particles that is emitted from a source and travels through space (p. 247)

radioactivity the spontaneous activity of certain atoms during which particles and penetrating rays are emitted (p. 230)

rate constant a proportionality constant used in quantitatively describing the rate of a specific reaction (p. 504)

rate-determining step the slowest step in a chemical reaction sequence (p. 505)

reactants (re-ak'-tənts) elements or compounds used in a chemical reaction (p. 128)

reaction mechanism a series of steps in a chemical reaction sequence (p. 505)

reagent a substance used to identify, quantify, or produce other substances (p. 614)

receptor a type of protein in cells that reacts to neurotransmitters and produces changes in the cells (p. 665)

redox reaction a chemical reaction consisting of an oxidation half-reaction and a reduction half-reaction (p. 582)

reducing agent a substance that is oxidized in a redox reaction (p. 583)

reduction a chemical reaction in which the oxidation number of an atom, a group of atoms, or an ion decreases (p. 587)

relative mass a mathematical comparison of the mass of one object to the mass of another (p. 95)

relative precision See relative uncertainty.

relative uncertainty the amount of uncertainty in a measurement comparted to all that is measured (p. 76)

rem a measure of the amount of ionizing radiation absorbed by humans (p. 249)

reversible reaction a chemical reaction in which products can react to form the reactants (p. 513)

RNA biomolecules, containing ribose, that act as messengers in the activities of cells; works in connection with DNA (p. 662)

S

salt a compound that may be produced from the reaction of the cation of a base and the anion of an acid (p. 572)

salt bridge a concentrated solution of an electrolyte which prevents the buildup of excess positive or negative charges in an electrochemical cell (p. 592)

saturated 1: describing a solution in which a maximum quantity of solute has dissolved (p. 440) 2: describing organic molecules containing no double or triple bonds (p. 628)

scientific notation method of expressing numbers in exponential form to simplify calculations (p. 64)

shielding materials of high density used to block the radiation between a nuclear reactor and operating personnel (p. 266)

shielding effect a weakening of the force between the nucleus and the outermost valence electrons by electrons in inner levels (p. 320)

SI (Le Système International d'Unités)   an international measuring system based on the number 10 (p. 10)

SI base units   the seven fundamental standards used to derive measurements in SI (p. 11)

significant digit   all the certain digits plus one uncertain digit in a measurement (p. 76)

silicone   a compound formed by attaching a hydrocarbon to silicon in a silicon–oxygen chain (p. 385)

single replacement   a chemical reaction in which one element replaces another in a compound (p. 135)

slope   refers to how fast a line ascends or descends in graphing (p. 88)

solid   one of the four states of matter; characterized by a definite shape and volume (p. 177)

solubility   a measure of the amount of substance that will dissolve in a given amount of solvent at a given temperature (p. 440)

solubility product constant   the ion-product constant for a compound, dependent on the concentrations of ions for a saturated solute and the temperature of the solution (p. 459)

solute   substance of lesser quantity in a homogeneous mixture (p. 440)

solution   a combination of two or more substances that exists as a homogeneous mixture (p. 27)

solvation   a chemical reaction between the solute and solvent resulting in a solution (p. 443)

solvent   the substance of greater quantity in a homogeneous mixture (p. 440)

specific heat   the amount of heat required to raise the temperature of one kilogram of matter one kelvin (or one degree Celsius) (p. 468)

spectator ions   ions present that do not participate in a chemical reaction (p. 456)

spontaneous decay   a natural process of nuclear decay of unstable nuclides (p. 250)

standard molar volume   the volume occupied by one mole of gas at STP (p. 207)

standard pressure   the average atmospheric pressure at sea level; 101.325 kilopascals or 1 atmosphere (p. 182)

steroids   lipids with a four-ring carbon skeleton (p. 661)

stoichiometry   (stoi-kē-äm'-ə-trē)   measuring quantitatively the amounts of elements or compounds involved in a chemical change (p. 153)

STP   an acronym for *standard temperature and pressure* (p. 182)

straight-chain   describing simple hydrocarbon molecules that contain a single, linear chain of carbon atoms (p. 628)

strong acid   a substance that reacts completely with water to donate protons to a base (p. 551)

strong base   a substance that dissociates completely with water and accepts protons from an acid (p. 551)

structural isomers   compounds that have the same chemical formula but whose atoms are arranged differently (p. 631)

sublimation   the process in which a solid becomes a vapor without first forming a liquid (p. 419)

subscript   a number used in chemical formulas to indicate the number of atoms of an element (p. 44)

substitution   a reaction in which a hydrogen atom in an organic compound is replaced by another atom or group of atoms (p. 645)

substrate   a specific molecule that undergoes a chemical change when affixed by an enzyme (p. 666)

supersaturated   describing a solution containing more solute than can normally be dissolved at a given temperature (p. 447)

synchrotron   a powerful particle accelerator that accelerates charged particles along a circular path (p. 253)

synthesis   a reaction in which two or more substances combine to form a compound (p. 134)

T

theoretical yield   the amount of product that would be obtainable under ideal conditions (p. 170)

theory   a proposal that predicts future events based on the regularity or similarity of past observations (p. 3)

thermodynamics   the study of energy transformations (p. 467)

titration   an analytical technique used to determine the concentration and composition of a solution using known quantities of standardized reagents (p. 567)

tracers   radioactive isotopes used to monitor a chemical in a physical or chemical change (p. 257)

transition metals   elements in groups 3 to 12 in the periodic table that behave in a manner that is intermediate between active metals and nonmetals; elements that are *d*-orbital fillers (p. 310)

transmutation   the process of changing one element into another in a nuclear reaction (p. 252)

transuranium elements   synthetic elements with atomic numbers larger than that of uranium (p. 253)

triglycerides   compounds formed by three acids and the alcohol glycerol (p. 660)

triple bond   a covalent bond in which three pairs of electrons are shared by two atoms (p. 341)

triple point   the temperature and pressure points at which a substance exists simultaneously as a gas, a liquid, and a solid (p. 423)

U

unbranched chain   *See* straight-chain.

unit cell   the smallest portion of a crystal structure (p. 427)

unitary rates   ratios with a denominator of one that are used in mathematical operations and conversions (p. 14)

unsaturated   1: describing a solution in which additional solute can be dissolved (p. 440) 2: describing hydrocarbons that contain at least one double or trible carbon–carbon bond (p. 628)

V

vacuum   space devoid of matter and where the pressure is zero pascals (p. 181)

valence electrons   electrons in the outermost orbitals of an element, which are most often involved in chemical bonding (p. 313)

valence shell electron pair repulsion theory (VESPR)   a theory that describes the equal distribution of electron pairs around a central atom in a molecule (p. 354)

van der Waals (van'-dər-wölz) **forces**   collective name for intermolecular forces; includes London forces and dipole-dipole forces (p. 359)

vapor pressure   the pressure of a vapor above its liquid or solid when the two states are at equilibrium (p. 417)

vitamins   nonprotein, organic compounds known as coenzymes (p. 667)

volatile   describing a substance whose molecules evaporate readily (p. 418)

volatile solute   a substance with a high vapor pressure (p. 451)

voltage   a unit for measuring the tendency of electrons to flow (p. 595)

W

water-forming   describing a chemical reaction in which water is formed (p. 136)

wavelength   a measurable distance between corresponding points of successive waves (p. 278)

weak acid   a substance that does not react completely with water to donate protons to a base (p. 551)

weak base   a substance that does not dissociate completely in water to accept protons (p. 551)

INDEX

ACKNOWLEDGEMENTS

Design Assistants: Julie Fair, George McLean, Nancy Smith-Evers

Illustration: Horvath and Cuthbertson, George Nichols, George Ulrich

Photo Credits: *Photo Research:* Fay Torresyap
Photo Styling: Elizabeth Willis

Teacher's Edition Design: Ann Curtis

All photography by Ken O'Donoghue, Boston, Massachusetts, unless otherwise noted below. D.C. Heath wishes to thank the following people for their assistance in technical photo sessions: Professor James F. Hall, Chemistry Department, Northeastern University, Boston, Massachusetts and Mr. Ted Hall, Chemistry Instructor, Wayland High School, Wayland, Massachusetts.

Cover: Richard Megna (Fundamental Photos)

Chapter 1: 1: Standard Scholarship. **8:** John Cunningham (Visuals Unlimited). **9:** *l,r* National Bureau of Standards. **18:** *l* John Bird.

Chapter 2: 24: Ted Cordingley. **25:** Three Lions. **56:** Candace Cochrane.

Chapter 3: 60: Science Source/Photo Researchers; *inset* Science Photo Library (Science Source/Photo Researchers). **72:** *l,r* Courtesy of D.C. Heath Collamore Division. **73:** Alex MacLean (Landslides). **75:** E.R. Degginger. **82:** Historical Division, Cleveland Health Sciences Library.

Chapter 4: 94: Nancy Sheehan.

Chapter 5: 124: Larry Lefever (Grant Heilman Photography). **143:** Joel Landau (Phototake). **144:** Bohdan Hrynewych.

Chapter 6: 172: Courtesy of Ellen Druffel.

Chapter 7: 176: Story Litchfield (Stock, Boston). **178:** E.R. Degginger. **187:** Ruth Dixon.

Chapter 8: 222: *l,r* Edwin Colbert, *The Great Dinosaur Hunters and Their Discoveries*, Dover Publications, Inc., N.Y., 1985. **223:** *l,r* David Kukla. **227:** Bruce Iverson. **240:** Dan McCoy (Rainbow). **241:** NASA.

Chapter 9: 246: George Schwartz (FPG); *inset* Robert Eckert (Stock, Boston). **248:** *l* Virginia Carleton (Photo Researchers); *r* Pastner (FPG). **252:** *l* Derby Art Gallery; *m,r* Courtesy of Lawrence Berkeley Laboratory. **254:** *t* C.E.R.N.; *b* Courtesy of Fermi National Laboratory. **258:** Phototake. **259:** *l,r* Dan McCoy (Rainbow). **266:** *l* FPG. **268:** Princeton University, Plasma Physics Laboratory. **269:** Battelle, Pacific Northwest Laboratories.

Chapter 10: 276: Courtesy of Education Development Center, Inc., Newton, MA. **280:** David Parker (Photo Researchers). **284:** *l* Tom Pantages; *r* adapted by permission from C.W. Keenan, D.C. Kleinfelder, and J.H. Wood, *General College Chemistry*, Sixth Edition, Harper and Row Publishers, Inc., 1980, plate 2, figure 4.6. **290:** NASA (Peter Arnold). **293:** *both* Courtesy of Mount Wilson and Las Campanas Observatories, Carnegie Institution of Washington.

Chapter 11: 304: Don Hunstein (Photo Trends).

Chapter 12: 328: Richard Walters. **360:** Robert Langridge and Dan McCoy (Rainbow). **361:** *t,b* Courtesy of Purdue University, Department of Biological Sciences.

Chapter 13: 366: Peter Menzel (Stock, Boston). **368:** *t* E.R. Degginger. **369:** Russ Kinne (Photo Researchers). **371:** *b* Sutton (FPG). **376:** *t,b* E.R. Degginger; *m* Tom McHugh (Photo Researchers). **381:** Malcolm S. Kirk (Peter Arnold). **384:** Lester V. Bergman, N.Y. **385:** *l* John Cancalosi (Tom Stack & Associates); *m* John Buitenkant (Photo Researchers); *b* John Pearson (FPG). **391:** *l* Terry Qing (FPG). **392:** Norman Ross Productions.

Chapter 14: 396: Photri. **397:** NASA (Photo Researchers). **398:** Frank Wing (Stock, Boston). **400:** *tl* IBM (E.R. Degginger); *tr* Science Photo Library (Science Source/Photo Researchers); *bl* Tom Pantages; *bm* Courtesy of Sperry Corporation; *br* Dan McCoy (Rainbow). **401:** *t* Charlie Falco (Photo Researchers); *bl* Manfred Kage (Peter Arnold); *br* A. Hart-Davis (Photo Researchers). **403:** IBM. **404:** Malcolm S. Kirk (Peter Arnold). **405:** *l* Michael Holford (collection of the British Museum); *r* Vic Lipka (FPG). **406:** Lou Jones. **407:** John Walsh (Photo Researchers). **408:** Courtesy of Corning Glass Works.

Chapter 15: 414: Grant Heilman Photography. **424:** *r* Tom Branch (Photo Researchers). **426:** *r* Martin Rogers (FPG).

Chapter 16: 438: Jeff Rotman (Peter Arnold).

Chapter 17: 480: Phil Degginger.

Chapter 18: 496: E.R. Degginger. **506:** Victoria Beller-Smith.

Chapter 20: 549: John Zoiner (Peter Arnold). **561:** Corning Glass Works. **572:** E.R. Degginger. **575:** Tom McHugh (Photo Researchers).

Chapter 21: 580: Bill Gallery (Stock, Boston). **607:** Judy Walker (Peter Arnold).

Chapter 23: 626: Courtesy of Department of Library Services, American Museum of Natural History. **631:** Adrian Atwater (Shostal Associates). **635:** David Halpern (Photo Researchers). **638:** Ken Biggs (Photo Researchers). **648:** Phil Degginger. **650:** *tl* John Marmaras (Woodfin Camp); *b* Donald Clegg. **652:** Bruce Iverson.

Chapter 24: 658: Copyright Regents, University of California, Computer Graphics Laboratory, University of California at San Francisco. **662:** Richard Feldmann, National Institutes of Health. **671:** R.L. Brinster, University of Pennsylvania. **672:** Hank Morgan.

Field Test Participants

The authors and editors of the **Heath Chemistry** program gratefully acknowledge the contributions made by the field test teachers. Their comments and suggestions as well as those of their students have enabled us to build a stronger chemistry program. Among those who participated are:

Mary Ann Lippin, Marlboro High School, NJ; **Virginia Kerr,** Middletown High School North, NJ; **Marianne Lisowski,** Middletown High School North, NJ; **Jerry Reed,** Rich Central High School, IL; **Jerry Deany,** Rich Central High School, IL; **Dan Vandercar,** School District #218, Oak Lawn, IL; **Robert L. Hinkle,** Maine East High School, IL; **Florence K. Williams,** Avon Grove High School, PA; **Thomas Brown,** Walpole High School, MA; **Michael Levine,** H. Lehman High School, NY; **Richard Smith,** Downtown Business Magnet High School, CA; **Richard A. Brown,** Minnechaug Regional High School, MA; **Diane W. Burnett,** Warren Central High School, IN; **Mary C. Christian,** North Providence High School, RI; **Marvin T. Curry,** Albemarle High School, VA; **John Davidson,** Oak Park and River Forest High School, IL; **Sheryl B. Dominic,** Deering High School, ME; **Diane L. Gerlach,** Temper High School, WI; **Thomas J. Russo,** Bayonne, NJ; **George H. Stevens,** Lansing, NY; **Natalie F. Tiernan,** Warren Township High School, IL; **Carol Cummings,** Warren Township High School, IL; **Jane H. Tiers,** St. Paul Academy and Summit School, MN; **Patricia Corcoran,** St. Paul Academy and Summit School, MN; **Edward L. Waterman,** Rocky Mountain High School, CO; **Jean Hamilton,** St. Thomas Aquinas High School, CT; **Barry Rosen,** Winthrop High School, MA; **Joyce M. Cooksey,** Weymouth North High School, MA; **Pauline Powers,** Walpole High School, MA; **William F. McKinney,** Grimsley Senior High School, NC; **Agnes Whitesell,** Independence High School, NC; **Annabelle L. Lerch,** Pocatello High School, ID; **Douglas K. Mandt,** Sumner High School, WA; **Philip H. Ogata,** Boulder High School, CO; **Jerrold W. Omundson,** Memphis University School, TN; **Sister Mary Ethel Parrott,** Notre Dame Academy, KY; **Sister Agnes Joseph,** Marian High School, MI; **Donald Russell,** Dublin High School, OH; **Edward Duzak,** St. Paul Catholic High School, CT; **Joseph W. MacQuade,** Marblehead High School, MA.